Lecture Notes in Electrical Engineering

Volume 505

** Indexing: The books of this series are submitted to ISI Proceedings, EI-Compendex, SCOPUS, MetaPress, Springerlink **

Lecture Notes in Electrical Engineering (LNEE) is a book series which reports the latest research and developments in Electrical Engineering, namely:

- Communication, Networks, and Information Theory
- Computer Engineering
- Signal, Image, Speech and Information Processing
- Circuits and Systems
- Bioengineering
- Engineering

The audience for the books in LNEE consists of advanced level students, researchers, and industry professionals working at the forefront of their fields. Much like Springer's other Lecture Notes series, LNEE will be distributed through Springer's print and electronic publishing channels.

For general information about this series, comments or suggestions please use the contact address under "service for this series".

To submit a proposal or request further information, please contact the appropriate Springer Publishing Editors:

Asia:

China, *Jessie Guo, Assistant Editor* (jessie.guo@springer.com) (Engineering)

India, *Swati Meherishi, Senior Editor* (swati.meherishi@springer.com) (Engineering)

Japan, *Takeyuki Yonezawa, Editorial Director* (takeyuki.yonezawa@springer.com) (Physical Sciences & Engineering)

South Korea, *Smith (Ahram) Chae, Associate Editor* (smith.chae@springer.com) (Physical Sciences & Engineering)

Southeast Asia, *Ramesh Premnath, Editor* (ramesh.premnath@springer.com) (Electrical Engineering)

South Asia, *Aninda Bose, Editor* (aninda.bose@springer.com) (Electrical Engineering)

Europe:

Leontina Di Cecco, Editor (Leontina.dicecco@springer.com)
(Applied Sciences and Engineering; Bio-Inspired Robotics, Medical Robotics, Bioengineering; Computational Methods & Models in Science, Medicine and Technology; Soft Computing; Philosophy of Modern Science and Technologies; Mechanical Engineering; Ocean and Naval Engineering; Water Management & Technology)

(christoph.baumann@springer.com)
(Heat and Mass Transfer, Signal Processing and Telecommunications, and Solid and Fluid Mechanics, and Engineering Materials)

North America:

Michael Luby, Editor (michael.luby@springer.com) (Mechanics; Materials)

More information about this series at http://www.springer.com/series/7818

José Machado · Filomena Soares
Germano Veiga
Editors

Innovation, Engineering and Entrepreneurship

 Springer

Editors
José Machado
School of Engineering
University of Minho
Guimarães
Portugal

Germano Veiga
Faculty of Engineering
University of Porto
Porto
Portugal

Filomena Soares
School of Engineering
University of Minho
Guimarães
Portugal

ISSN 1876-1100   ISSN 1876-1119   (electronic)
Lecture Notes in Electrical Engineering
ISBN 978-3-319-91333-9   ISBN 978-3-319-91334-6   (eBook)
https://doi.org/10.1007/978-3-319-91334-6

Library of Congress Control Number: 2018942336

Printed on acid-free paper

This Springer imprint is published by the registered company Springer International Publishing AG part of Springer Nature
The registered company address is: Gewerbestrasse 11, 6330 Cham, Switzerland

Preface

The emergence of new business models based on innovation, cooperation networks and the enhancement of endogenous resources are assumed to be a strong contribution to the development of competitive economies.

Regional Helix 2018 was an event that, according to the Smart Entrepreneurial Ecosystems logic, was based on not only purely academic but also technological and entrepreneurial dimensions, across the most diverse sectors of activity.

The objective of this conference was to show the set of entrepreneurial and infrastructural capacities as well as to transmit to all participants an overview about potentiating the foundations for the creation of networks of knowledge and entrepreneurial cooperation, involving Engineering, Innovation and Entrepreneurship stakeholders.

This edition of the Regional Helix 2018 conference, the 3rd conference in the successful series of conferences on Innovation, Engineering and Entrepreneurship, was organized at School of Engineering of University of Minho, by MEtRICs and Algoritmi Research Centres.

This edition was especially focused on Knowledge and Technology Transfer from Academia to Industry and Society, highlighting and proposing solutions for some main problems related to industrial and societal challenges, having as main target the creation of added value for real economy.

A wide range of topics were covered by the papers published in this volume. They were distributed among the topics: Control, Automation and Robotics; Mechatronics Design, Medical Devices and Well-being; Cyber-Physical Systems, IoT and Industry 4.0; Innovations in Industrial Context and Advanced Manufacturing; New Trends in Mechanical Systems Development; Advanced Materials and Innovative Applications; Waste to Energy and Sustainable Environment; Operational Research and Industrial Mathematics; Innovation and Collaborative Arrangements; Entrepreneurship and Internationalization; and Oriented Education for Innovation, Engineering and/or Entrepreneurship.

We would like to thank the support of the Organizing Committee and the invaluable contributions of the Scientific Committee members, external reviewers, invited speakers and session chairs. A special appreciation is given to Springer's team who support this edition. Last but not least, we want to thank the authors, for whom and by whom this event was made to happen.

The 224 papers submitted to the conference were peer-reviewed by the Scientific Committee. Based on the reviewers' ratings, 160 submissions were accepted for publication after following the reviewers' recommendations.

We believe that the papers in this volume show expressively how Innovation, Engineering and Entrepreneurship can be used to support the scientific community.

José Machado
Filomena Soares
Germano Veiga

Organization

Steering Committee

Luís Farinha	Polytechnic Institute of Castelo Branco/University of Beira Interior
Domingos Santos (CEDER Coordinator)	Polytechnic Institute of Castelo Branco
João Ferreira (NECE Coordinator)	University of Beira Interior
José Machado	University of Minho

Organizing Committee

General Conference Chair

José Machado	University of Minho/MEtRICs Research Centre

Conference Vice-chair

Filomena Soares	University of Minho/Algoritmi Research Centre

Local Organizing Committee

Cândida Vilarinho	University of Minho/MEtRICs Research Centre
Celina P. Leão	University of Minho/Algoritmi Research Centre
Demétrio Matos	University of Minho/MEtRICs Research Centre
Filomena Soares	University of Minho/Algoritmi Research Centre
João Pedro Mendonça	University of Minho/MEtRICs Research Centre
Jorge Cunha	University of Minho/Algoritmi Research Centre

José Machado	University of Minho/MEtRICs Research Centre
José Meireles	University of Minho/MEtRICs Research Centre
Leonilde Varela	University of Minho/Algoritmi Research Centre
Vítor Carvalho	University of Minho/Algoritmi Research Centre

International Scientific Committee

Adam Hamrol	Politechnika Poznańska, Poland
Adina Astilean	Universitatea Tehnica din Cluj-Napoca, Romania
Alain Fayolle	EMLYON Business School, France
Aldina Correia	Polytechnic Institute of Porto, Portugal
Alexandra Braga	Polytechnic Institute of Porto, Portugal
Aminul Islam	Technical University of Denmark, Denmark
André Catarino	University of Minho, Portugal
Arminda do Paço	University of Beira Interior, Portugal
Artur Rosa Pires	University of Aveiro, Portugal
Aurora Teixeira	University of Porto, Portugal
Bernardo Providência	University of Minho, Portugal
Björn Asheim	Lund University, Sweden
Bożena Skołud	Politechnika Śląska, Poland
Camelia Avram	Universitatea Tehnica din Cluj-Napoca, Romania
Carina Guimarães	University of Beira Interior, Portugal
Carla Marques	University of Trás-os-Montes e Alto Douro, Portugal
Carlos Rodrigues	University of Aveiro, Portugal
Caroline Loss	University of Beira Interior, Portugal
Catarina Moura	University of Beira Interior, Portugal
Catarina Sales	University of Beira Interior, Portugal
Celina P. Leão	University of Minho, Portugal
Christian Brackmann	Institute of Education, Science and Technology Farroupilha (IFFAR), Brazil
Clara Cruz Santos	University of Coimbra, Portugal
Cristina Broega	University of Minho, Portugal
Cristina Estevão	Polytechnic Institute of Castelo Branco, Portugal
Cristina Fernandes	Polytechnic Institute of Castelo Branco, Portugal
Cristina Figueiredo	University of Lisbon, Portugal
Damian Krenczyk	Politechnika Śląska, Poland
Dariuz Sedziak	Politechnika Poznańska, Poland
David Rodeiro-Pazos	University of Santiago de Compostela, Spain
David Urbano	Autonomous University of Barcelona, Spain
Dinis Leitão	University of Minho, Portugal
Domingo Ribeiro Soriano	University of Valência, Spain
Domingos Santos	Polytechnic Institute of Castelo Branco, Portugal

Dumitru Olaru	University of Iasi, Romania
Eduardo Perondi	Universidade Federal do Rio Grande do Sul, Brazil
Eliana Silva	Polytechnic Institute of Porto, Portugal
Elias C. Carayannis	The George Washington University, USA
Elsa Montenegro	Higher School of Social Work, Porto, Portugal
Erika Ottaviano	Università degli Studi di Cassino e del Lazio Meridionale, Italy
Eurico Seabra	University of Minho, Portugal
Fernando Ferreira	ISCTE Business School, Portugal
Filomena Soares	University of Minho, Portugal
Flávio Almeida	IADE–Universidade Europeia | University of Beira Interior, Portugal
Francisco Duarte	Bosch Car Multimedia, Portugal
František Gazdoš	Tomas Bata University in Zlín, Czech Republic
Gábor Páy	University of Nyíregyháza, Hungary
Géza Husi	Debreceni Egyetem, Hungary
Gianni Montagna	University of Lisbon, Portugal
Gintaras Dervinis	Kaunas University of Technology, Lithuania
Gheorghe Gheorghe	National Institute of Research and Development in Mechatronics and Measurement Technique, Romania
Gheorghe Popan	National Institute of Research and Development in Mechatronics and Measurement Technique, Romania
Gheorghe Prisacaru	University of Iasi, Romania
Gilberto Santos	Polytechnic Institute of Cávado and Ave, Portugal
Guilherme Kunz	Unioeste do Paraná, Brazil
Hans Georg Gemünden	Norwegian Business School, Norway
Hans Landström	Lund University, Sweden
Helder Carvalho	University of Minho, Portugal
Helena Alves	University of Beira Interior, Portugal
Helen Lawton Smith	Birkbeck, University of London, UK
Isabel de Sousa	Lusíada University, Portugal
Isabel Gouveia	University of Beira Interior, Portugal
Jan Hosek	Czech Technical University in Prague, Czech Republic
Jiří Vojtěšek	Tomas Bata University in Zlín, Czech Republic
Joana Cunha	University of Minho, Portugal
João J. Ferreira	University of Beira Interior, Portugal
João Matias	University of Aveiro, Portugal
John Parm Ulhøi	Aarhus University, Denmark
João Paulo Coelho Marques	Polytechnic Institute of Coimbra, Portugal
João Pedro Mendonça	University of Minho, Portugal

Joaquim Gabriel   University of Porto, Portugal
Jorge Cunha   University of Minho, Portugal
Jorge Hernandez   Rio Cuarto University, Argentina
José Carlos Teixeira   University of Minho, Portugal
José Lucas   University of Beira Interior, Portugal
José Machado   University of Minho, Portugal
José Manoel Carvalho   Federal Fluminense University, Brazil
  de Mello
Judith Terstriep   Research Unit «Innovation, Space & Culture»
                                    at IAT, Germany
Justyna Trojanowska   Politechnika Poznańska, Poland
Julie Hermans   Namur University, Belgium
Kim Klyver   University of Southern Denmark, Denmark
Leonilde Varela   University of Minho, Portugal
Loet Leydesdorff   University of Amsterdam, The Netherlands
Luis Farinha   Polytechnic Institute of Castelo Branco, Portugal
Luís F. Silva   University of Minho, Portugal
Luísa Branco   University of Beira Interior, Portugal
Magdalena Diering   Politechnika Poznańska, Poland
Manuel Nogueira   Higher School of Social Work, Porto, Portugal
Marcus Dejardin   Université de Namur, Belgium
Maribel Guerrero   Newcastle Business School, UK
Marina Ranga   European Commission, Joint Research Centre,
                                    Seville, Spain
Mario Lino Raposo   University of Beira Interior, Portugal
Marta Peris-Ortiz   Polytechnic University of Valencia, Spain
Maximiliano Cristiá   National University of Rosario, Argentina
Mehmet Bozca   Yildiz Technical University, Turkey
Miguel Duarte   University of Minho, Portugal
Monia Najjar   University of Monastir, Tunisia
Mustafa Kemal Apalak   Erciyes University, Turkey
Natacha Antão Moutinho   University of Minho, Portugal
Nuno Belino   University of Beira Interior, Portugal
Nuno Fernandes   Polytechnic Institute of Castelo Branco, Portugal
Nuno Peixinho   University of Minho, Portugal
Orlando Lima Rua   Polytechnic of Porto, Portugal
Óscar Afonso   University of Porto, Portugal
Paul Benneworth   University of Twente, The Netherlands
Paula Odete Fernandes   Polytechnic Institute of Bragança, Portugal
Paula Vieira   Higher School of Social Work, Porto, Portugal
Paulo Almeida   University of Minho, Portugal
Paulo Eigi Miyagi   University of São Paulo, Brazil
Paulo Mendonça   University of Minho, Portugal
Paulo Torres   Polytechnic Institute of Castelo Branco, Portugal
Pedro Neves   University of Beira Interior, Portugal

Pedro Rosário	University of Minho, Portugal
Pedro Souto	University of Minho, Portugal
Pedro Torres	Polytechnic Institute of Castelo Branco, Portugal
Peter-Jan Engelen	Utrecht University, The Netherlands
Petr Lepšík	Technical University of Liberec, Czech Republic
Petr Tůma	Technical University of Liberec, Czech Republic
Petr Valášek	Czech University of Life Sciences Prague, Czech Republic
Philip Cooke	University of Wales, UK
Pierluigi Rea	Università degli Studi di Cassino e del Lazio Meridionale, Italy
Reggie Davidrajuh	University of Stavanger, Norway
Renaldas Urniezius	Kaunas University of Technology, Lithuania
Ricardo Gonçalves	Mew University of Lisbon, Portugal
Ricardo Gouveia	University of Beira Interior, Portugal
Ricardo Hernández Mogollón	UEX, Spain
Rita Salvado	University of Beira Interior, Portugal
Rochdi El Abdi	University of Rennes 1, France
Roel Rutten	Tilburg University, The Netherlands
Rogério Dionísio	Polytechnic Institute of Castelo Branco, Portugal
Rui Gama Fernandes	University of Coimbra, Portugal
Rui Miguel	University of Beira Interior, Portugal
Sahin Yildirim	Erciyes Üniversitesi, Turkey
Salvador Roig	University of Valencia, Spain
Sam Tavassoli	CIRCLE, Lund University, Sweden & School of Management, RMIT University, Melbourne, Australia
Sara Nunes	Polytechnic Institute of Castelo Branco, Portugal
Sara Velez	University of Beira Interior, Portugal
Sascha Kraus	University of Liechtenstein, Liechtenstein
Saulo Dubard Barbosa	EMLYON Business School, France
Sharmistha Bachi-Sen	University at Buffalo, the State University of New York, USA
Soumodip Sarkar	University of Évora, Portugal
Teresa Morgado	New University of Lisbon, Portugal
Vijaya Kumar Manupati	VIT Vellore, India
Vitalii Ivanov	Sumy State University, Ukraine
Vitor Braga	Polytechnic Institute of Porto, Portugal
Vitor Carvalho	Polytechnic Institute of Cávado and Ave, Portugal
Wael Dghais	Université de Sousse, Tunisia

Contents

Control, Automation and Robotics

TExtractor: An OSINT Tool to Extract and Analyse Audio/Video Content

António Magalhães and João Paulo Magalhães$^{(\boxtimes)}$

CIICESI, ESTG, Politécnico do Porto, 4610-156 Felgueiras, Portugal
{8110244,jpm}@estg.ipp.pt

Abstract. Hacking, data breaches, and information loss are a growing concern for organizations. Aware of the escalation of cyber threats, organizations are looking for ways to detect and mitigate cyberattack scenarios. Cyber intelligence, that is, knowledge produced through data and information on cyber threats and its actors is one of the means explored for this purpose. OSINT (Open Source INTelligence) is one of the areas of data collection for the production of cyber intelligence.

In this paper we propose an OSINT tool (TExtractor) to facilitate the process of obtaining information about cyber threats. The TExtractor tool consists of extracting text from video/audio in open sources and searching for keywords linked to the activities of malicious actors. To support the development of TExtractor, we conducted a study to measure the effectiveness of different text extraction tools in audio/video sources. The results are presented in the paper and show that a tool like TExtractor can detect references to cyberattacks on audio/video sources in real time, with an accuracy between 60% and 70%.

Keywords: OSINT · Cyber security · Cyber intelligence
Text extraction

1 Introduction

In the cyber security field is commonly said that "you have been hacked, but you just do not know it yet". System failures data breaches, and information loss are currently occurring quite frequently and affecting governments, businesses, and individuals. The EU Cybersecurity Agency (ENISA) publishes on a yearly basis a report summarizing cyber threats accesses by collecting publicly information. In the 2017 report [1], it is highlighted that the: complexity of attacks and sophistication of malicious actions in cyberspace continue to increase; the monetization of cybercrime is becoming the main motive of threat agents; the need for related training programs and educational curricula remains almost unanswered. The conclusions to be drawn from the reports year after year are that the traditional layers of defense are not enough to protect the organizations. In this context, obtaining information on the threats and mode of operation of malicious actors is seen as one more layer of security to take into account. Cyber-intelligence

© Springer International Publishing AG, part of Springer Nature 2019
J. Machado et al. (Eds.): HELIX 2018, LNEE 505, pp. 3–9, 2019.
https://doi.org/10.1007/978-3-319-91334-6_1

thus gains an important role in collecting, analyzing and providing information related to cybercriminal, hacktivist and cyber espionage activities. It aims to provide context and relevance to a tremendous amount of data, improving the decision-making during and following the detection of a cyberattack and to drive momentum toward a predictive cybersecurity posture.

The benefits as well as the challenges imposed by the cyber-intelligence process are recognized. In 2001, NATO published, in [2], information on the subject of Open Source Intelligence (OSINT). At that time the focus was on the collection of sources, not information. More than 15 years have passed and regardless of the cyber-intelligence initiatives and technologies developed to detect and mitigate cyber threats, the ENISA 2017 report [1] stills refer that the automation of cyber-intelligence needs to further advance to include new levels of intelligence.

In this paper we propose a tool, named TExtractor, for the collection and analysis of audio and video data in OSINT sources. The malicious actors talk about their targets before the attack, they publish videos, use forums, social networks, and other channels. All of these channels should be monitored in order to detect the planning of attacks and leaks of information at an early stage. Considering the frequency with which audio/video files are created and shared, the development of a tool that automatically collects and analyzes this type of data poses several challenges. The first challenge has to do with how data is treated for analysis. Audio/video content involves a lot of data e to allow its analysis we explore the automatic extraction of text from the input data. This first challenge unleashes another that needs to be addressed, and that deals with the accuracy of text extraction tools from audio/video sources. Accuracy is a relevant aspect of the project and therefore in this work we conduct a study to verify the: accuracy of the extraction of text from audio/video sources considering different speech speeds and various comparison tools; accuracy of the extraction of text from audio/video sources considering different languages and several comparison tools; accuracy rate considering the automatic translation of the text to a default base language (e.g., english). Tools that contribute to the anticipation of cyberattack scenarios or situations where unknown attacks are made public are very important. The continuous monitoring audio/video channels, along with other tools, facilitate the cyber security analysts tasks, improve the cyber threat visibility by analyzing automatically large volumes of data and allowing the adoption of best mitigation/preventive controls.

This paper is organized as follows: Sect. 2 presents the state-of-the-art. The third section, presents the proposal for the implementation of TExtractor. Section 4 includes the accuracy analysis study. Section 5 concludes the paper.

2   Related Work

The volume and complexity of cyber attacks requires the development of advanced solutions able to detect and stop these attacks in time. Both academia and industry are focused and actively proposing solutions aimed at addressing the problem in the best way.

The Open Source Intelligence (OSINT) is still considered an emerging area. It consists in searching and retrieving relevant information from open sources and then filtering and processing the vast amounts of data in order for an analyst to derive useful intelligence [3]. In [4] authors discuss several topics related with OSINT. What techniques can be used to filter data and select information flooding in from the Internet? How to apply automated information extraction techniques to automate the data processing? How to select, sort, tag and index information automatically? How can an analyst quickly access and make sense of their domain of interest using a OSINT tool? These are some of the question addressed in the paper. Authors conclude the paper refering that the main challenge today is not the lack of information but rather identifying the relevant information enhancing the importance of developing core knowledge systems that allow for the efficient indexing and archiving so that patterns and cross-references are highlighted. In [5], authors present a division of OSINT according to the type of data sources. Newspapers, radio, television, Internet, Online publications, Social websites, public government reports, budgets, telephone directories, press conferences, academic publications, speeches, patents are some of the data sources referred. Considering the variety, volume and speed at which data is generated in these data sources, the creation of tools to assist the collection and analysis process is vital. Aliprandi et al. in [6] presents a platform (CAPER) for the prevention of organized crime, created in cooperation with european law enforcement agencies. Pfeiffer et al. in [7], presents the Sail Labs Media Mining System, a system which is capable of collecting vast amounts of data gathered from open sources in unstructured form and processing the data through a series of steps, which include speaker-identification, language-identification, automatic speech-recognition, face-and-object recognition as well as named-entity and topic-detection. Khelif et al. in [8], present a framework that allows to collect data from satellite, telephone network (PSTN), cellular, telecom VOIP and Internet VOIP applications and analyzes the data in order to identify the speaker, its gender and age, the language and accent and also to spot keywords contained in the input. The above work describes the tools but does not present studies on their accuracy. Saidon et al. in [9] present an invention that relies on a set of systems and methods for receiving live speech, converting the speech to text, and transferring the text to a user. Authors refer that the invention provides accurate transcription of live events to 95–98%.

3   TExtractor: Proposal and Methodology

Open source intelligence (OSINT) refers to intelligence that has been derived from publicly available sources. There are numerous OSINT tools, mainly of them used in the reconnaissance phase to gather as much information about the target as possible. Maltego [10], Shodan [11] and Censys [12] are three of the most referred tools. A complete list of OSINT free tools, categorized by type and updated, is maintained by the authors of the OSINT Framework [13]. On the page maintained by the authors there are several links to data sources and also some pointers to OSINT tools. From the list there is no specific tool for analyzing the audio/video contents.

The TExtractor tool aims to extract text from audio/video content for analysis. The text extracted is analyzed in order to detect the occurrence of keywords/patterns that might be related with cyber-attacks (e.g. DDoS, deface-ment, IP addresses, domains, emails addresses, names of the board members). The tool is composed by several build blocks, as illustrated in Fig. 1.

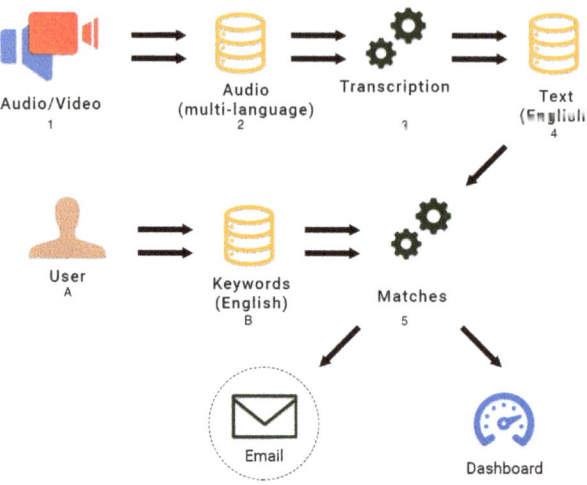

Fig. 1. TExtractor: building blocks

As illustrated in Fig. 1 there are two main inputs of data. The user/operator (A) provides a list of keywords that are stored in a database. The keywords should be defined in English. The other input identified with the number 1, corresponds to different audio/video data sources that can be configured (e.g. Youtube channel, video streaming channel, radio channel) for data consumption. The extractors extract text from the input, translate the text to english and store the result in a database (step 2–4). As the text is captured and translated the search for keywords/patterns are applied. Every match found send an alert to the user. The results of the analysis can also be viewed through a dashboard included in the Web application.

The development of the TExtractor involves several tools namely: extraction tools to transcribe text from audio/video sources; translators to translate the original text to english; comparison tools to match the keywords in the text.

The Speech to Text; Web speech API; Speech Logger; Speech notes are some of the most well-known tools in the field. The IBM Watson Speech to Text service [14] uses speech recognition capabilities to convert different language speech to text. The Web Speech API [15] is a Google project that applies neural network models to convert audio to text. It provides an easy to use API that recognizes over 110 languages. The Speechlogger [16] is another speech to text tool that makes use of the Google speech-to-text technologies for the best results and includes auto-punctuation and language translation features. Speechnotes [17] also takes advantage of the speech-to-text engines provided by Google, and according to the documentation it achieves accuracy rates above 90%.

After extracting the text from audio/video sources, we need to translate it (e.g. using the Google translator) and then search for the occurrence of keywords using existing tools. The lack of studies on the effectiveness of text extraction tools poses a challenge for the implementation of TExtractor. To overcome this challenge, we conducted an accuracy study that is presented in the next section.

4 Extracting Text from Audio/video: Accuracy Analysis

In this section we present an analysis study conducted to assess the accuracy of the text transcribers. For the assessment we considered three different types of data sources: Audiobook, Music and Speech. Each of the sources are in different languages: English, Portuguese and Spanish. The transcription was done by using the Speech to Text, Web Speech API, Speechlogger and Speechnotes transcribers. For the analysis in question we translated the extracted text to English. The Google translator was used for the translation. For each of the inputs we have the complete list of words in English. The analysis consists on measuring the similarity index between the original text and the text obtained from the transcription. The similarity index is given by Eq. 1.

$$Similarity = \frac{Copied\_words * 100}{Total\_words} \tag{1}$$

To allow the comparison between the extracted text and the original text we used to text comparison tools: Copyleaks and wDiff. Copyleaks [18] makes use of artificial intelligence algorithms to detect plagiarism and paraphrased content. wDiff [19] is a GNU program used to compare files on a word per word basis.

The environment used to execute this study was composed by a laptop installed with the MacOS Sierra Operating System version 10.12.5 installed with the Musical Instrument Digital Interface (MIDI) Interface version 3.1. The MIDI interface was used to feed the extractors with the audio data sources.

4.1 Experimentation Results

Each experiment was repeated three times and the results (average) are presented in Table 1.

The results reveal a big difference in terms of similarity between the Copyleaks and the wDiff comparison tools. The results achieved with Copyleaks are worse when compared with wDiff and this is explained by the way the tool compares the text. It only counts words as copied words if they were in the same sentence and position. From the table we also notice the difference between the type of input data. The extraction of text from music-type audio sources have

Table 1. Experimentation Results: 4 extractors, 2 comparison tools, 3 languages and 3 types of input

Extractors/Similarity (%)	Copyleaks			wDiff		
	English	Portuguese	Spanish	English	Portuguese	Spanish
Music						
Speech to Text	0.00%	0.00%	0.00%	5.64%	15.13%	23.9%
Web speech API	0.00%	6.90%	0.00%	4.97%	13.16%	1.06%
Speechlogger	0.00%	0.00%	0.00%	5.37%	6.58%	3.01%
Speechnotes	0.00%	0.00%	0.00%	2.28%	3.95%	0.35%
Speech						
Speech to Text	74.60%	0.00%	0.00%	72.30%	23.54%	11.18%
Web speech API	22.40%	9.60%	8.70%	67.47%	45.48%	34.76%
Speechlogger	17.60%	55.60%	15.80%	00.00%	40.89%	58.20%
Speechnotes	22.60%	5.30%	16.90%	64.55%	45.85%	49.59%
AudioBook						
Speech to Text	31.10%	0.00%	11.80%	68.46%	19.54%	52.57%
Web speech API	18.90%	6.20%	14.80%	63.42%	43.45%	56.19%
Speechlogger	100.00%	64.00%	14.40%	62.56%	34.93%	51.51%
Speechnotes	17.80%	4.80%	15.10%	63.08%	40.96%	56.19%

a low similarity rate, revealing the limitations of the extraction tools for this type of input. The amount of "noise" is commonly pointed as the main issue regarding the transcription of music. The results achieved when the input data is in English have always provided better similarity results.

5   Conclusion

Cyber threats have been around for many years, but in recent years we have witnessed further development and greater involvement of organizations in their management. Unfortunately, it is a very difficult area, where "bad guys" tend to always be one step ahead of the "good ones". To cope with the problem, coordinated efforts must be made to foster the adoption of the best technologies and processes as well as to educate and sensitize people to the problem.

In this paper we presented the basis for the creation of an OSINT tool, named TExtractor. TExtractor allows to extract text from audio/video, index information in databases, search for pre-defined keywords and to trigger alerts whenever keywords are detected. The accuracy of such a tool is vital and in this sense we presented a study that measured the efficiency of different transcribers, considering different types of input and base languages. From the results we observed that the values grouped by type and language are very similar, being the amplitude of their difference of about 3%. This reveals that the text extraction tools from the audio/video are consistent. The similarity observed between the transcription and the original text ranges from 60% to 70% for audio/video without noise and at a moderate frequency (speech). Although the results may not be excellent, it is important to note that, considering the volume of information generated and consumed in the area of cyber threats, having an automatism that makes 60% to 70% of manual work is very significant. It is also important to mention that an application such as the one proposed is not exclusive to the collection and analysis of data in the scope of cyber security. TExtractor can also be used for monitoring a brand or automating the clipping process, i.e. identifying references to brands or products in audio or video channels.

References

1. ENISA - Threat landscape report 2017. TR, EU Cybersecurity Agency
2. NATO. NATO open source intelligence handbook. NATO 2001
3. Glassman, M., Kang, M.J.: Intelligence in the internet age: the emergence and evolution of open source intelligence (OSINT). Comput. Hum. Behav. **28**(2), 673–682 (2012)
4. Best, C.: Challenges in open source intelligence. In: 2011 European Intelligence and Security Informatics Conference, pp. 58–62, September 2011
5. Richelson, J.T.: The U.S. Intelligence Community. Avalon Publishing (2015)
6. Aliprandi, C., Irujo, J.A., Cuadros, M., Maier, S., Melero, F., Raffaelli, M.: Caper: collaborative information, acquisition, processing, exploitation and reporting for the prevention of organised crime. In: Stephanidis, C. (ed.) HCI International 2014 - Posters' Extended Abstracts, pp. 147–152. Springer International Publishing, Cham (2014)
7. Pfeiffer, M., Avila, M., Backfried, G., Pfannerer, N., Riedler, J.: Next generation data fusion open source intelligence (OSINT) system based on MPEG7. In: 2008 IEEE Conference on Technologies for Homeland Security, pp. 41–46, May 2008
8. Khelif, K., Mombrun, Y., Motlicek, P., Backfried, G., Kelly, D., Sahito, F., Hazzani, G., Scarpatto, L., Chatzigavriil, E..: Towards a breakthrough speaker identification approach for law enforcement agencies (2017)
9. Saindon, R.J., Estrin, L.II., Brand, D.A., Brand, S.: Systems and methods for automated audio transcription, translation, and transfer with text display software for manipulating the text, November 2004. US Patent 6,820,055
10. Maltego - data mining tool, February 2018. https://www.paterva.com/
11. Shodan - Search engine internet-connected devices, February 2018. https://www.shodan.io
12. Censys - security driven by data, February 2018. https://censys.io
13. Nordine, J.: Osint-framework, February 2018. http://osintframework.com/
14. IBM Watson Speech to Text, February 2018. https://speech-to-text-demo.ng.bluemix.net/
15. Google Web speech api - text conversion powered by machine learning, February 2018. https://cloud.google.com/speech/
16. Speech recognition & instant translation, February 2018. https://speechlogger.appspot.com
17. Speech to text online notepad, February 2018. https://speechnotes.co
18. Copyleaks - plagiarism checker, February 2018. https://copyleaks.com/
19. Gnu wdiff, February 2018. https://www.gnu.org/software/wdiff/

Planning, Managing and Monitoring Technological Security Infrastructures

Jorge Ribeiro[1] , Victor Alves[1] , Henrique Vicente[2] ,
and José Neves[3(✉)]

[1] ARC4DigiT – Applied Research Center for Digital Transformation,
School of Technology and Management,
Polytechnic Institute of Viana do Castelo, Viana do Castelo, Portugal
jribeiro@estg.ipvc.pt, victoralves.at.vnc@gmail.com
[2] Department of Chemistry, School of Sciences and Technology,
Évora Chemistry Centre, University of Évora, Évora, Portugal
hvicente@uevora.pt
[3] Department of Informatics, School of Engineering, ALGORITMI Centre,
University of Minho, Braga, Portugal
jneves@di.uminho.pt

Abstract. Over the past few decades many different Information Technologies (IT) policies have been introduced, including COSO, ITIL, PMBook, CMM, ISO 2700x, Six Sigma, being COBIT IT (Control Objectives for IT) the framework that encompasses all IT and Information Systems (IS) governance activities at the organization's level. As part of the applicability of quality services certification (ISO 9001) in all IT services of a public institution, it is presented a case study aimed at planning, managing and monitoring technological security infrastructures. It followed the guidelines for the ISO 2700x family, COBIT, ITIL and other standards and conducted a survey to complement the IT process's objectives. With regard to an action-research methodology for problem-solving (i.e., a kind of attempt to improve or investigate practice) and according to the issue under analyze, the question is put into the terms, viz. "How can the ISO 2700x, COBIT, ITIL and other guidelines help with the planning, management and monitoring of technological security infrastructures and minimize the risk management of IT and IS?". Indeed, it may be resolved that it is possible to achieve the goals of planning, managing and monitoring a technological security infrastructure. In the future, we will use Artificial Intelligence based approaches to problem solving such as Artificial Neural Networks and Cased Based Reasoning, to evaluate this issue.

Keywords: Artificial intelligence · Artificial neural networks
Case based reasoning · Security infrastructures · COBIT · ITIL
ISO 27001 · ISO 27005

© Springer International Publishing AG, part of Springer Nature 2019
J. Machado et al. (Eds.): HELIX 2018, LNEE 505, pp. 10–16, 2019.
https://doi.org/10.1007/978-3-319-91334-6_2

1 Introduction

Many standards and frameworks have been developed in recent years to manage IT models and tools to assess the maturity of IT governance at the organization's level [1–7]. Indeed, in terms of policies that are aimed at managing and control specific IT areas one may attend to ISO 9001 [1], ITIL (Information Technology Infrastructure Library), ISO 27001, CMM (Capability Maturity Model), COSO (Committee of Sponsoring Organizations) and, in particular, to COBIT (Control Goals for Information and Technology), which was stated by ISACA (a nonprofit, independent association that advocates for professionals involved in information security, assurance, risk management and governance) and is structured as follows [2], viz.

- Criteria for information;
- IT resources; and
- IT procedures.

The IT processes are structured into 4 domains with 34 IT processes, which correspond to tasks and activities to be carried out at the organization's level. COBIT domains are divided into 4 areas, namely Planning and Organization, Procurement and Implementation, Provision and Support, Monitoring and Evaluation. In addition, COBIT includes a set of indicators that need to be effectively monitored to ensure that the control and monitoring of IS and IT is effective. Looking at another dimension related to IT governance, risk management can be defined as follows, viz.

"The level of uncertainty that the business needs to understand and effectively manage to achieve its goals and create value" (James W Deloach – Enterprise-wide Risk Management: Strategies for Linking Risk and Opportunity) [6].

Several studies on the importance and applicability of risk management have been presented, which illustrate the importance of this topic [3–7]. The process structure of ISO 27005 standard is divided into the following areas, viz.

- Context building process;
- Risk assessment;
- Risk analysis (risk identification, risk estimation, and risk evaluation);
- Risk treatment;
- Risk acceptance;
- Risk communication; and
- Risk monitoring and review.

The ISO 2700x family standards are acknowledged for information security management in relation to the definition of the Code of Conduct, as it is set in ISO 17799, and the guidelines for its applicability, presented in ISO 27002. From ISO 27003 to ISO 27050 the focus was on the definition of security techniques, viz.

- Critical aspects of ISO 27001;
- Guidelines for the implementation of ISO 27001; and
- Guidelines for monitoring, measurement, analysis and evaluation of security practices in information security management systems.

In addition to these standards, to define requirements for auditing and certification of the management of the security of information systems, guidelines for conducting audits, definition of the security controls that auditors should follow and in the requirements of the safety techniques associated with the application of the ISO 27001 standard. On the definition of communication security techniques in the internal departments/divisions of companies and the external communication between them. A focus on the presentation of a code of practice for security controls centered in defining the requirements for the management of services in information security, security techniques for the governance of information systems, security techniques for the management of information systems associated with financial services. A emphasis in the area of cloud computing, privacy in cloud computing services, definition of the competency requirements that professionals must have in order to manage the security of information systems, proposals or standards of areas not directly associated with the security of information systems and the standard for service continuity. An attention on the importance of cybersecurity, security in computer networks, management of security incidents in information systems, guidelines for the identification, collection, acquisition of digital forensic evidence, requirements to be analyzed in the operation and administration of systems, in intrusion detection systems, in the forensic area that regulates investigative methods to acquire digital evidence as well principles and processes for investigating information security incidents. Notice on the definition of guidelines for the management of information security events, proposals in digital forensic standards in order to contribute to the capture of digital evidence [1]. Indeed, the main contribution of this work is focused on defining a methodological approach to using the COBIT, ITIL, ISO 2700x guidelines (namely ISO 27001 and 27005) for planning, managing and monitoring security infrastructures. In our research, we conducted action research planning that analyzes the relationship between the introduction of groupware into an organization and the resulting changes in work habits and on the organizational structure.

The article is organized as follows, viz. Following the introduction, a case study, applied research methodology and the use of COBIT, ITIL, ISO 27005 and other guidelines will be presented. Finally, a conclusion is placed and suggestions for future work will be made.

2 Case Study

2.1 Research Method and Context Case Study

We followed the action research conducted by Olesen and Myers [8], which construes their research perspectives and allows them to focus on how individuals attempted to understand a specific situation. Therefore, we conduct a seven-step action research cycle that includes the following steps, viz.

- Identifying the Research Question, i.e., in this case the question "How the ISO 2700x, COBIT, ITIL and other guidelines can help with planning, management and monitoring of Technological Security Infrastructures and how to mitigate the Risk Management of IT and IS";

- Action planning, i.e., the definition of measures to solve the research question;
- Measures, i.e., the implementation and monitoring of the planned actions;
- Assessment, i.e., to decide if the measures addressed the research question. In the practical implementation, we follow a diagram with a set of frameworks, guidelines, and standards that should be related to the PDCA - plan, do, check approach.
- Action Planning, i.e., to determine the actions to be undertaken to address the research question;
- Action Taking, i.e., to conduct and monitor the planned actions;
- Evaluation, i.e., to determine if the actions have addressed the research question. In the practical implementation we follow a diagram of a set of frameworks, guidelines and standards related with the PDCA – plan, do, check approach.

This case study introduces the implementation and use of COBIT for IT governance at the Polytechnic Institute of Viana do Castelo (IPVC) (http://www.ipvc.pt) and integrates six organic units or schools, namely Education High School, Agrarian High School, Technology and High School of Administration, Management Sciences High School, Nursing High School, Central Services and Social Services. The Quality Management System (QMS) implemented at the IPVC and the processes that represent it, strive to implement the quality policy ISO 9001 [1, 9]. The unique IS and IT processes were set as the Information System Management (ISM), which was developed to manage and control COBIT, ITIL, and other Quality-of-Service policies on the IT/IS infrastructure. As a result, all these processes have been improved, which has led to a re-qualification of the ISO 27005 standard.

2.2   The Application of COBIT, ITIL, ISO 27005 and Other Guidelines

Based on the structure of ISO 9001 [1, 9, 11] it was defined and implemented a number of sub-processes based on COBIT domains, namely Planning and Organizing Information Systems IT goals, Acquiring and Implementing Information Systems, Delivering and Supporting Information, and Monitoring and Evaluation of Information Systems. In order to implement the planning, management and monitoring of technological security infrastructures and to minimize the risk management of IT and IS, the definition of risk management plans (i.e., IS and IT Infrastructure Inventory, Probability of Occurrence, Levels of Impact, Risk Levels, Security Threats and Concerns, Definition of Safeguards, Information Security Risk Management Policies, Security Policies (Backup Policy, Security Plans)) was accomplished. Data protection, access to the server room, access to the software applications of the services, registration and storage of the configurations of the components of the technological infrastructure, control of the backups, maintenance of the components of the technological infrastructure, monitoring and evaluation of the components of the technological infrastructure), policy disaster recovery is responsible for computer resources, data storage and retention, security for information systems, removal of institutional information, passwords, corporate e-mail, internet usage, workstation use, network infrastructure status, IT/IS social rules, and confidentiality and confidentiality strategies. In addition, a Best Practices Guide on storage and retention of institutional information should be defined and communicated. On the other hand, the institution

Results for each Topic

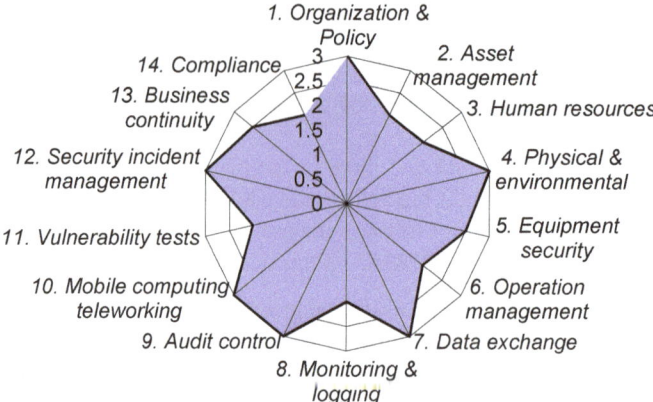

Fig. 1. Example of the result of the state of the scenario of the IT governance in a specific moment.

must define and characterize procedures related to violations of policies, warnings and penalties. To plan and manage the technology security infrastructure, we follow the guidelines of COBIT, ITIL, and ISO 2700x, but extended the methodological approach with a survey based on five general domains, namely security, software, backups, hardware, and infrastructure conditions. Survey is divided into 156 questions, which are structured in fourteen topics. By limiting the number of pages, we present the subdomains of the questions and not all 156 questions by themselves, structured as follows, viz. 1 - Organization & Policy; 2 - Asset Management; 3 - Human Resources; 4 - Physical and Environmental; 5 - Equipment Safety; 6 - Operation; 7 - Exchange of Data; 8 - Monitoring and Logging; 9 - Audit; 10 - Mobile Computing and Teleworking; 11 - Vulnerability Tests; 12 - Security Incident Management; 13 - Business Continuity; and 14 - Compliance.

In order to quantify the security's state of the infrastructure a survey was carried out three months after its implementation, and according to a probabilistic measure outside such period. For each posted question it was recorded its applicability in the form *Applicable/Not Applicable* or as *1/0*. Each question was also weighted in terms of integer values taken from the interval *0...3*, where *0* stands for "bad" and *3* for "good", using the formula *(AP/NAP) * Weight * Score*. Figure 1 shows an example of the outcome of the *IT Governance Scenario* at a given moment in time.

2.3   Results

Based on the case study research methodology in use, we define a set of policy procedures and tools that conclude that it is possible to achieve the goals being set and minimizing the risk management of IT and IS. Based on the research method, the results of the case study were announced as "Research Question's Identification". In this case, the question becomes, viz. 1 - Action Planning. After reviewing the various standards and frameworks that focus on managing and controlling the IT space, COBIT is a well-known framework for IT governance and security. Therefore, it was decided to implement the guidelines of COBIT, ITIL, ISO 2700x and other standards in the IPVC, especially in the IS and IT (Information Systems Management Process of the QMS) area. In order to undertake this goal, we proceed with the actions, viz. Analysis and specification of the technological needs of the institution; Study of existing standards and frameworks for monitoring and control of IT/IS; Elaborate procedures for the various activities based on COBIT and ITIL; Select indicators to assess IT performance and improve service quality; Set priorities for the implementation of COBIT, ITIL and other standards; Monitor the implementation of the procedures on each system in each organic entity; Consistent support to employees in each organic unit to promote the use of COBIT and ITIL documentation and practices. 2 - Actions. In this case study, and in particular for this institution, required the analysis and specification of all diagnosis needs and difficulties in existing ISS information services. According to the COBIT guidelines, this process, SMP, has been divided into four sub-processes that correspond to the four COBIT domains, viz. IS plan and organize; acquire and implement IS; deploy and support IS; and monitor and evaluate IS. 3 - Evaluation. To assess the use and implementation of the methodological approach, two indicators were defined, viz. One based on the COBIT specification and the other grounded on the score of the four topics described in COBIT Standards ISO 27001 and 27005. 4 - Learning's Specification. The learning obtained was as follows, viz. Improving the quality of care provided by the administrative services; manage the IS more efficiently by defining processes and indicators; reducing the tasks execution time; help to define specific indicators to assess the performance of IT services. It was also set policies and plans for IT management.

3 Conclusion and Future Work

This work reflects the ongoing development of a practical application of the ISO 9001, COBIT, ITIL and ISO 2700x standards to answer the question "How the ISO 2700x, COBIT, ITIL and other guidelines can assist in the planning, management and monitoring of Technical Security infrastructures as well as for risk minimization of IT and IS?". Based on the case study research methodology used, it was defined a set of policy procedures and tools that conclude that it is possible to achieve the defined goals and minimize the risk management of IT and IS. Over time, and in order to evaluate the best conditions in the IT/IS environment domains, we intend to apply Artificial Intelligence approaches to problem solving such as Artificial Neural Networks or Cased Based Reasoning [10–12].

References

1. ISO – International Organization for Standardization. https://www.iso.org/standards.html. Accessed 15 Feb 2018
2. ISACA – The Official Site of Information Systems Audit and Control Association. https://www.isaca.org/pages/default.aspx. Accessed 18 Feb 2018
3. Agrawal, V.: A framework for the information classification in ISO 27005 Standard. In: Qiu, M. (ed.) Proceedings of the 4th IEEE International Conference on Cyber Security and Cloud Computing (CSCloud 2017), pp. 264–269. IEEE Edition (2017)
4. Gonzalez-Granadillo, G., Dubus, S., Motzekc, A., Garcia-Alfaro, J., Alvarez, E., Merialdo, M., Papillon, S., Debar, H.: Dynamic risk management response system to handle cyber threats. Future Gener. Comput. Syst. **83**, 535–552 (2017)
5. Wangen, G.: Information security risk assessment: a method comparison. Computer **50**(4), 52–61 (2017)
6. Deloach, J., Temple, N.: Enterprise-Wide Risk Management: Strategies for Linking Risk and Opportunity. Prentice Hall, London (2000)
7. Cayirci, E., Garaga, A., Oliveira, A.S., Roudier, Y.: A risk assessment model for selecting cloud service providers. J. Cloud Comput. Adv. Syst. Appl. **5**(14), 12 (2016)
8. Olesen, K., Myers, D.: Trying to improve communication and collaboration with information technology: an action research project which failed. Inf. Technol. People **12**(4), 317–332 (1999)
9. Ribeiro, J., Gomes, R.: Information system to support quality management systems: a case study in a Portuguese high educational institution. In: Proceedings of the 4th Mediterranean Conference on Information Systems (MCIS 2009), paper 2 (2009)
10. Ramos, J., Oliveira, T., Satoh, K., Neves, J., Novais, P.: An orientation method with prediction and anticipation features. Artif. Intell. **20**(59), 82–95 (2017)
11. Neves, J., Fernandes, A., Gomes, G., Neves, M., Abelha, A., Vicente, H.: International Standard ISO 9001 – a soft computing view. In: Hammoudi, S., Maciaszek, L., Teniente, E., Camp, O., Cordeiro, J. (eds.) Enterprise Information Systems. Lecture Notes in Business Information Processing, vol. 241, pp. 153–167. Springer, Cham (2015)
12. Fernandes, B., Freitas, M., Analide, C., Vicente, H., Neves, J.: Handling default data under a case-based reasoning approach. In: Proceedings of the 7th International Conference on Agents and Artificial Intelligence (ICAART 2015), vol. II, pp. 294–304. Scitepress, Lisbon (2015)

Integrating MIT App-Inventor in PLC Programming Teaching

P. B. de Moura Oliveira[1,2]($\boxtimes$), J. Boaventura Cunha[1,2],
and Filomena Soares[3]

[1] ECT, UTAD-University of Trás-os-Montes and Alto Douro,
5000-801 Vila Real, Portugal
oliveira@utad.pt
[2] INESC TEC Technology and Science,
Campus da FEUP, 4200-465 Porto, Portugal
[3] Centro Algoritmi, School of Engineering,
Universidade do Minho, 4800-058 Guimarães, Portugal

Abstract. The potentialities of using mobile devices such as smartphones for teaching/learning purposes are huge. However, in some teaching areas its use is still residual. The use of mobile applications in the context of teaching PLC programming techniques is addressed in this work. The MIT App-Inventor II is deployed to develop mobile applications for learning purposes. An android based application entitled Time-Counts is proposed here, developed to support the teaching/learning process of timers. Preliminary results regarding its use by students are presented.

Keywords: MIT App Inventor · PLC programming
Control engineering education

1 Introduction

Currently, mobile devices are part of university student's everyday life. Thus, it can be advantageous to use such mobile devices to promote teaching and learning activities. Indeed, studies about the use of mobile devices in students teaching and learning activities have been recently reported by [1]. Among these mobile devices, smart-phones are exhaustively used by students, inside and outside the classroom. These devices can provide an efficient and flexible mean to access information, anywhere and anytime, enabling a more personalized learning, which has been termed M-learning [2–4]. The use of smart-phones for teaching/learning activities requires the development of mobile applications. A highly desirable feature is the possibility of teachers and students to develop tailored made mobile applications to suit their particular interests. However, the development procedure of mobile applications can be an obstacle, particularly for teachers/students with no specific programming skills for developing such applications. For Android operating system based devices, the MIT App Inventor 2 (MITApp2) [5] is a simple software tool which can provide solutions to empower both teachers and students in this domain [6, 7]. MITApp2 uses a drag and drop methodology based on building blocks, avoiding the tedious process involved with syntax

© Springer International Publishing AG, part of Springer Nature 2019
J. Machado et al. (Eds.): HELIX 2018, LNEE 505, pp. 17–24, 2019.
https://doi.org/10.1007/978-3-319-91334-6_3

based programming languages. Indeed, users with elementary logic and programming concepts can easily accomplish to make their own applications. MITApp2 has been previously used to provide students with applications addressing feedback control systems and industrial automation topics, as reported in [8] as well as to allow students to develop their own applications [9].

Within industrial automation courses, some major topics concern programmable logic controllers (PLC) and respective programming languages. Some of the main programming languages are: ladder diagrams (LD), (also commonly known as relay diagrams), function block diagrams (FBD), instruction list (IL), structured text (ST) and sequential function charts (SFC). These languages are standardized by the norm IEC-61131-3 [10, 11]. PLC programming languages concepts can be provided to students through specific target made mobile applications. The idea, is to use MITApp2 to develop simple Apps addressing specific topics. An example of such an application entitled Time-Counts (TC) addressing Timers concepts is proposed in this paper. The relevance of timers within the programming languages for logic and sequential controllers is high. As it will be described, TC allows to perform simple simulations for the three most used timers: on-delay timer, TON, off-delay timer, TOF, and pulse timer, TP. The general TC aim is to provide students with a simple App they can use to help perceiving some timers concepts, by running some simulations.

The rest of the paper is organized as follows: Sect. 2 presents a brief description of the MITApp2 potential for developing apps for control engineering teaching/learning; Sect. 3 presents the TC application description. Section 4 presents simulation results and finally Sect. 5 concludes the paper and outlines further work.

2   Introduction MIT APP-Inventor in Control Engineering Teaching

Based on this paper authors experience by using the MITApp2 to develop simple Apps for teaching/learning purposes, some of this tool remarkable features are the following:

- It is easy to use by users with elementary programming skills;
- The time that goes from the App development stage to a ready to try application is very short;
- It is a freeware tool, with plenty of supporting information provided in the tool internet page, such as tutorials [12].

The application design process is organized in two parts; (i) the graphical user interface and (ii) the block behavior programming. Both are based on a drag and drop approach. Figure 1(a) presents an example of the design used for an application within a curricular unit of modeling and control systems (MCS_2015) in the first degree course of Biomedical Engineering in UTAD University. In this case several buttons are used in the main App screen to open other screens regarding the topics of Theory, Problems, Quizzes, etc. Figure 1(b) presents an example of the block programming corresponding to the three buttons shown in Fig. 1, that when clicked opens the respective screen.

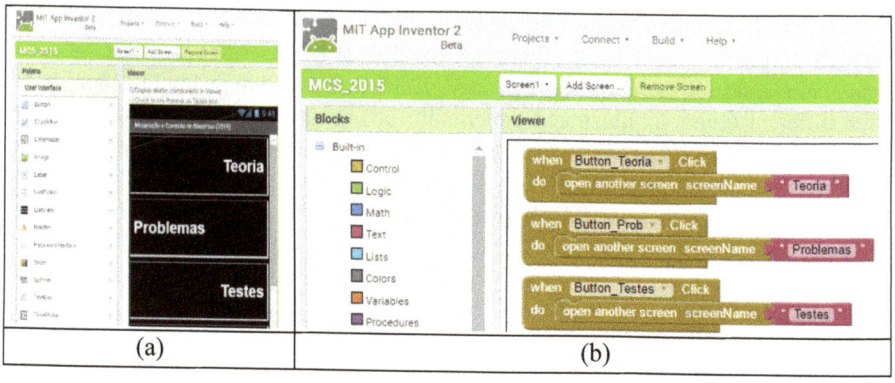

Fig. 1 (a) MITApp2 user interface design menu; (b) block programming menu. *Example extracted from MCS_2015 App (in Portuguese)*

MITApp2 enables to develop applications which can provide several useful features in the teaching/learning process, such as the following:

- information regarding specific theoretical topics (e.g. using a sliding show approach);
- problems to be solved, with or without solution;
- quizzes so they can test their learning regarding specific aspects, etc.

Currently it is a common procedure for a lecturer to prepare a slide show presentation for classes, made available for students via a learning management system. The general idea is that very soon the same procedure can also, as commonly, be used for mobile applications. Moreover, MITApp2 can also be used by students to perform projects and even to report assignment results, as published in [9].

3 TimeCounts Application

The application TimeCounts (TC) main objective is to provide students with simple animations which may allow them to better perceive some details regarding timers operation. The simulations are based on ladder diagrams, and the timers in question are the following:

- On delay timer, TON,
- Off delay timer, TOF and
- Pulse delay timer, TP.

All these timers can be represented with a generic b! is to enable students
Fig. 2(a) [11], where *IN* represents the timer input, Q then perceive what is
time and *ET* the elapsed time. One of the TC learn; input and time variables.
to simulate different possibilities used to con se TC to understand timers
supposed to occur to the timer output rel
After the topic introduction in classe

…m presented in
input, *PT* the preset

operation and in practical classes, consolidate the learning process by replicating the simulation in the physical PLC. In this case, the SIEMENS S7-1200 [13] PLC is used in practical classes for the purpose. A pedagogical LD illustrative example is deployed for the different timers as illustrated generically in Fig. 2(b). It is an elementary LD with a single rung, using as the timer input a normally open switch, S1, and connecting to the timer output an output coil. The output coil can represent the state of a lamp L. As it will be explained, the TC App user will define the timer preset time among several possibilities.

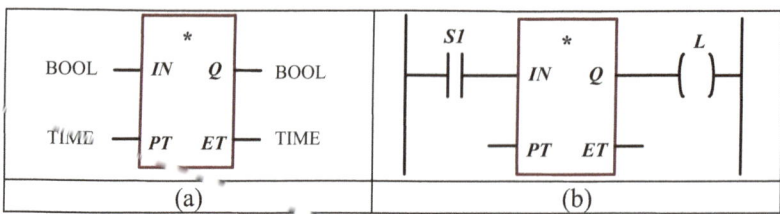

Fig. 2 (a) General block graphical representation for timers where * can be replaced by TON, TOF or TP; (b) General LD example used

The TC App first screen is illustrated in Fig. 3(a) with three interface buttons for the respective timer's types: TON, TOF and TP. Figure 3(b) presents the TON delay timer example. As it can be observed the normally open switch S1 represents the logical state of the normally open contact S1. The default preset time is PT = 2 s. However it can be selected among several other options as illustrated in Fig. 3(c). Figure 3(d) illustrates the TON delay timer with the preset time selected as PT = 6 s, before S1 is switched to the on state. The screenshot presented in Fig. 4(a) shows the result of enabling the timer with S1 on, while the PT was not reached, and finally Fig. 4(b), corresponds to an instant after the PT has been reached and S1 was not switch off.

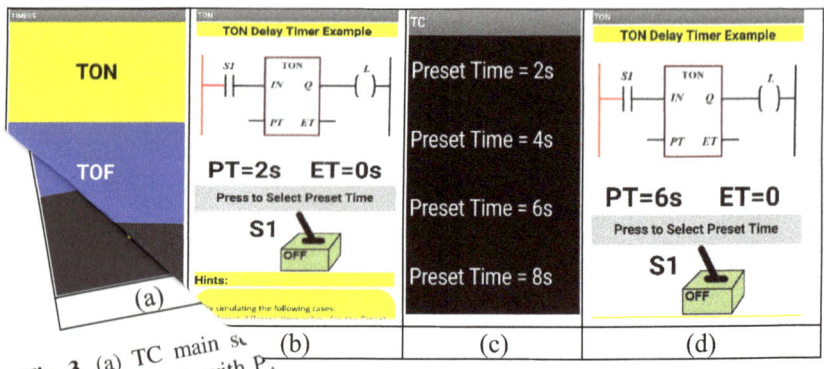

Fig. 3 (a) TC main screen (b) TON delay timer example; (c) Preset time selection menu. (d) TON example with PT in the off state

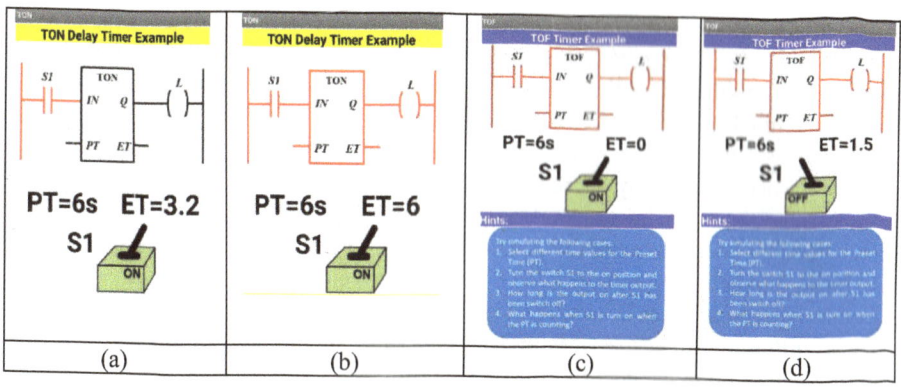

Fig. 4 (a) TON running with switch on; (b) TON when the preset time has been reached and S1 is on; (c) TOF example: timer output is on as soon as S1 is switched on (d) TOF example: the PT time is being counted since S1 was switched off

It can be helpful to better perceive the different timers working principles to try to replicate chronograms such as the one presented in Fig. 5, for the TON and TOF timer cases. Firstly by using the TC simulation and later on the real PLC. Thus, some exploratory hints are provided in the TC, for each timer. In the case of the TON timer some specific learning objectives are stated by the following questions:

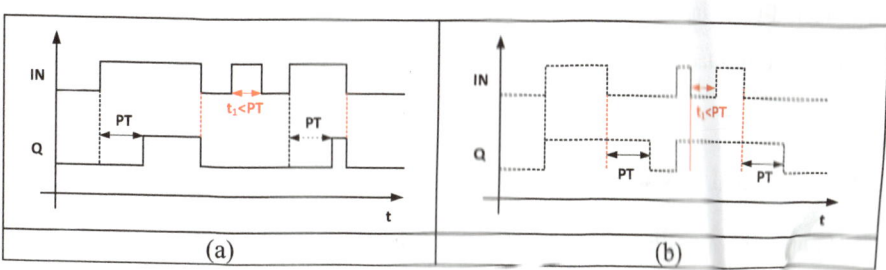

Fig. 5 Chronogram example for (a) TON Timer; (b) TOF Timer

- What happens to the timer output, Q, when the input changes to the on state?
- How long must the input IN remain in the on state in order to change the timer output Q to the on state?
- After the output reaches the on state, how is the time period in w it remains in the on state, controlled?
- What happens to the timer output, when the time period that ir N stays on is less than PT?
- What is the relation between the state of the output Q, and tivation of elements connected to Q in the LD?

A very interesting and relevant practical issue reg the TON timer is the adequate selection of the contact to be used as the IN Frequently, in practical classes, student use as TON input sensors logic state (with positive logic), which are in the on state, for brief instants (e.g. the detection of components passing in a conveyor belt). This causes the TON timer to start counting the preset time, but as the input is switched off much before the PT is reached (see case illustrated by t_1 in Fig. 5(a), the timer output is never activated. One way to avoid this problem is by memorising the event occurrence detected by the sensor, using a PLC binary memory register, and then use the corresponding contact as TON input. In this way the timer input is controlled by the PLC programmer, ensuring the activation of its output, as well as the period that it stays in the on state.

Figure 4(c) presents a TC screenshot, regarding a TOF timer simulation, where it can be seen that the timer output is on since the timer input is on. In this case ET will begin to count as soon as the switch S1 is turn off (see example in Fig. 4(d)). Based on the chronogram example provided in the App and presented in Fig. 5(b), some TOF learning objectives are stated by the following questions:

- What happen to the timer output, Q, when the IN input changes to the on state?
- When does the preset time PT starts to be counted?
- How long must the input IN remain on in order to change the timer output Q to the on state?
- What happens to the timer output, when the input IN passes from the on to the off state, and changes again to the on state before PT (example represented as t_1 in Fig. 5(b))?
- What is the relation between the output Q state, and the activation of elements connected to Q in the LD?

Due space limitations illustrations regarding the TP timer simulations are omitted form this paper.

4   Results and Discussion

The TC application has been proposed and it currently being used by students of the following curricular units: industrial automation to the 4th year of Electrical and Computers Engineering of UTAD University; 3rd year of Engineering and Industrial Management Master Course of UMinho University, in the first semester of the academic years 2016–2017 and 2017–2018. In the period that is being used, the oral feedback received from students in classes is positive. Suggestions about how to improve the tool has been received as well as to make other applications addressing topics such as counters, shift registers, etc.

5   Conclusion and Further Work

In this paper the MIT App-Inventor 2, was reported as a simple to use and effective tool to develop teaching mobile applications for Android based devices. The mobile devices in question are smartphones, and the topics addressed are related to feedback control system and industrial automation. An application entitled

Time-Counts (TC) was proposed, addressing the specific topic of Timers. Timers is a subject of the upmost relevance within programmable logic controllers programming languages. The general aim of the proposed application, is to allow students to prior (or after) the practical use of a PLC in classes, use simple simulations of a ladder diagram with the main normalized timers: on-delay timer, off-delay timer and pulse delay timer. The simulations enable students to perceive specific details regarding the timers operation, namely concerning the logic state of its input and output, as well as the relation with the preset time and elapsed time variables. Indeed, these details are difficult to be apprehended by students before simulation and practical PLC testing. So far, the feedback received from students is quite positive, as well as suggestions to develop more application addressing other PLC programming topics. It is important to state that as the MIT App-Inventor II is (still) an Android operating system tool, this may constitute a disadvantage, as there are students with smart-phones using other operating systems. When possible, part of the information included in this paper reported applications is also provided to students in other digital formats such as PDF files.

As future work, a formal pedagogical enquire is going to be proposed, in order to receive more detailed feedback from students. This data is going to be analyzed and reported as soon as possible. The potentialities of using MIT-App Inventor II for other applications regarding other timers (e.g. on-delay retentive timers), counters and other PLC programming function is going to be explored in the nearby future.

Acknowledgements. The authors would like to express their acknowledgments to COMPETE: POCI-01-0145-FEDER-007043 and FCT – Fundação para a Ciência e Tecnologia within the Project Scope: UID/CEC/00319/2013.

References

1. Sung, Y.-T., Chang, K.-E., Liu, T.-C.: The effects of integrating mobile devices with teaching and learning on students' learning performance: a meta-analysis and research synthesis. Comput. Educ. **94**, 252–275 (2016)
2. Roschelle, J.: Keynote paper: unlocking the learning value of wireless mobile devices. J. Comput. Assist. Learn. **19**(3), 260–272 (2003)
3. Georgiev, T., Georgieva, E., Trajkovski, G.: Transitioning from e-Learning to m-Learning: present issues and future challenges. In: 7th ACIS International Conference on Software Engineering, Artificial Intelligence, Networking, and Parallel/Distributed Computing, pp. 349–353 (2006)
4. So, S.: Mobile instant messaging support for teaching and learning in higher education. Internet High. Educ. **31**, 32–42 (2016)
5. MITApp2: MIT App Inventor (2018). http://www.appinventor.org/about. Accessed 1 Feb 2018
6. Pokress, S.C., Veiga, J.J.D.: MIT App Inventor, enabling personal mobile computing. In: PROMOTO 2013. ACM, Indianapolis, USA, October 26 (2013)
7. Roy, K.: Position statement: App Inventor instructional resources for creating tangible apps. In: IEEE Blocks and Beyond Workshop, pp. 119–120 (2015)

8. Moura Oliveira, P.B.: Teaching automation and control with App Inventor applications. In: Global Engineering Education Conference EDUCON, Tallinn, pp. 879–884. IEEE (2015)
9. Soares, F., Oliveira, P.M., Leão, C.P.: Control engineering learning by integrating App-Inventor based experiments. In: 12th Portuguese Conference on Automatic Control, vol. 402, pp. 845–855 (2016)
10. IEC: International Electrotechnical Commission (2018). http://www.iec.ch/. Accessed 1 Feb 2018
11. Heinz, K.J., Tiegelkamp, M.: IEC 61131–3: Programming Industrial Automation Systems. Springer, Heidelberg (2010)
12. MITApp2: MIT App Inventor. Tutorials (2018). http://appinventor.mit.edu/explore/ai2/tutorials. Accessed 1 Feb 2018
13. Siemens S7-1200 (2018): http://w3.siemens.com/mcms/programmable-logic-controller/en/basic-controller/s7-1200/pages/default.aspx. Accessed 1 Feb 2018

Error Correction Repetition Codes
with Arduino and Raspberry PI

Paulo Torres$^{(\boxtimes)}$ and Sergio Malhão

Instituto Politecnico de Castelo Branco (IPCB), Castelo Branco, Portugal
`paulo.torres@ipcb.pt`, `smalhao@ipcbcampus.pt`

Abstract. This work consists in the development of software with the main objective of increasing the quality of the communication system with Repetition Codes and GFSK modulation. To evaluate the probability of error we used simulation in software Python and implemented with hardware (Raspberry Pi, Arduino and transceiver $NRF24L01$).

Keywords: Arduino · Raspberry Pi · Python · NRF24L01 · GFSK

1 Introduction

Digital code theory began in the late 1940s with the works of Golay, Hamming and Shannon. Since then, a great effort has been made to obtain efficient codes, and the use of digital codes goes through the coding and decoding processes, which is usually the most difficult of the two.

The most studied codes are the cyclic codes, as these are the easiest to encode, supporting an algebraic structure, allowing to find various methods of decoding. In addition, they include an important family of codes, which are the Bose, Chaudhuri, Hocquenghem (BCH) codes, which are the generalization of the Hamming codes for multiple error correction. The importance of BCH codes is as much the smaller the "number of errors expected" vs "code block length". However, there is also a simple procedure to decode these codes.

In coding theory, the repetition code is one of the most basic error-correction codes. To transmit a message through a channel with noise that may corrupt the transmission in some places, the idea of the repetition codes is just repeat the message several times. The hope is that the channel corrupts only a minority of these repetitions, and in this way, the receiver will notice that a transmission error occurred, since the data flow received is not the repeat of a single message and in addition, the receiver can recover the original message by observing the message received in the data stream that occurs more often.

The aim of this study was to study and evaluate the probability of error in the real context of a radio-frequency transmission through the Gaussian Frequency Shift Keying (GFSK), with the application of repetition codes to correct errors to compare the communication quality with the situation without correction of errors, both in the theoretical context and in the real context. We used Arduino and Raspberry PI with transceiver $NRF24L01+$, in the practical experiments, with transceiver $NRF24L01+$ [1].

© Springer International Publishing AG, part of Springer Nature 2019
J. Machado et al. (Eds.): HELIX 2018, LNEE 505, pp. 25–31, 2019.
https://doi.org/10.1007/978-3-319-91334-6_4

2 Channel Coding

A designer of a digital communication system, faces the task of providing a system, both in economic terms, as well as, in terms of a rate, level of reliability and quality in the communication system. Two essential parameters, are the signal power and the bandwidth, which in conjunction with the spectral density of the channel noise, determine the spectral density of power ($\frac{E_b}{N_0}$). In practice, it is usually used to modify the quality of the data, as well as to obtain an acceptable quality, to do so, the coding of error control is used. Another incentive to use coding is to reduce the $\frac{E_b}{N_0}$, being this reduction exploited by the reduction of transmitted power and reduction of hardware costs.

Error control, it is used in order to obtain the integrity of the data and can be executed through direct error correction Forward Error Correction (FEC), i.e. the discrete font, generates information in the form of binary symbols, in which the channel encoder on the transmitter accepts the message bits with redundancy, according to a predefined rule, thus producing data encoded at a higher bit rate, already the decoder in receiver, will exploit the redundancy in order to decide which bits of the message were transmitted. The goal of the encoder and decoder is to minimize the noise effect on the channel [2].

2.1 Repetition Code

It can be said that the repetition code is one of the most basic codes of error correction with linear block codes, and the idea of this code is to repeat several times the message transmitted to minimize the errors in the transmission of the message, retrieving the original message observing the message received in the data stream more frequently [3–5]. The probability of error for this code, with $t = \frac{n-1}{2}$, $d_{min} = 2t + 1$ and $C_k^n = \frac{n!}{k!(n-k)!}$, is obtained through the following relationship:

$$P_{repet} = \sum_{i=t+1}^{n} C_i^n p(1-p)^{n-i} \cong C_{t+1}^n p^{t+1}. \tag{1}$$

2.2 Building Code

The construction of the code vector, is given by, $c = [b \mid m]$ in which b is the parity bits and the message bits, replacing in the equation and factoring the vector m of the message we have $c = [P|I_k]$, where I_k is the identity matrix k by n:

$$P = \begin{bmatrix} 1 & p_{00} & p_{0,n-k-1} \\ 1 & p_{11} & p_{1,n-k-1} \\ \vdots & \vdots & \vdots \\ 1 & p_{k-1,1} & p_{k-1,n-k-1} \end{bmatrix} \tag{2}$$

Defining the generating array (G) its canonical form through the relationship $G = [P \mid I_k]$, already your parity matrix (H) is given by the following relationship $H = [I_{(n-k)} \mid P^T]$. Both the generating matrix and the parity matrix are fundamental in describing a linear block code [2].

2.3   Examples of Repetition Codes

Now, we will present examples of repetition codes $(3, 1)$, $(5, 1)$ used in this work. For the above codes, through its generator polynomials we will be able to build its generating matrix and parity matrix. Polynomial generator for repeat code $(n = 3, k = 1)$ is $g(X) = 1 + x + x^2$, being its generating matrix $G = [1\ 1\ 1]$ and its parity matrix (H):

$$H = \begin{bmatrix} 1 & 0 & 1 \\ 0 & 1 & 0 \end{bmatrix} \tag{3}$$

In the Fig. 1 we can observe the value of the minimum distance of Hamming, that is, the minimum distance value of Hamming for all pairs of code words (Table 1).

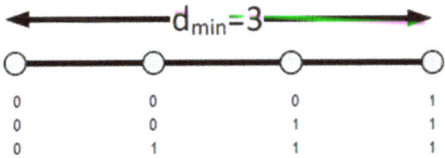

Fig. 1. Minimum distance for repetition code 3.

Table 1. Comparison table between the repetition codes with $k = 1$ and $\frac{E_b}{N_0} = 9.54\,\text{dB}$.

Code characteristics	Code 1	Code 2
Repetition code	$n = 3$	$n = 5$
Coding rate $(R = \frac{k}{n})$	0.333	0.2
Minimum distance (d_{min})	3	5
Probability of error (P_b)	6.75×10^{-6}	3.38×10^{-8}

Polynomial generator for repetition code $(n = 5, k = 1)$ is $g(X) = 1 + x + x^2 + x^3 + x^4$, being its generating matrix $G = [1\ 1\ 1\ 1\ 1]$, and its parity matrix:

$$H = \begin{bmatrix} 1 & 0 & 0 & 0 & 1 \\ 0 & 1 & 0 & 0 & 1 \\ 0 & 0 & 1 & 0 & 1 \\ 0 & 0 & 0 & 1 & 1 \end{bmatrix} \tag{4}$$

3   Hardware and Software

3.1   Transceiver $NRF24L01+$

The module used as the basis for the development of this work was the integrated circuit $NRF24L01+$, being this produced by Nordic Semiconductor [6]. Developed for wireless applications that uses the protocol Enhanced ShockBurstTM

Fig. 2. Pinout Transceiver $NRF24L01+$.

Table 2. Table of the main characteristics of the $NRF24L01+$.

Parameters	Typical values
Feeding voltage	1.9 V − 3.6 V (3.3 V)
Power current	1 μA
Frequency of operation	2.4 GHz
Transmission rates	256 kbps, 21 Mbps, 2 Mbps
Modulation type	$GFSK$
Communication	126 RF channel
Current (TX) to 0 dBm	11.3 mA
Current (RX) to 2 Mbps	13.5 mA
Current at rest	900 nA
Programmable transmission power	0, −6, −12, −18 dBm
Reception sensitivity with a BER of 0.1%, 2 Mbps	−82 dBm
Buffer	1 to 32 bytes of data at a time
Working temperature	−40° to 85°
Transmission range	100 m in free space 1 km with outdoor antenna
SPI	4 pins, 10 Mbps, 32 bytes TX and RX FIFO

in your data link layer. It is set up through the Serial Peripheral Interface (SPI) using up to 128 different channels of bandwidth 1 MHz, is designed to operate in the band of 2.4 GHz of Industrial, Scientific and Medical (ISM). This module consists of an integrated antenna, however, there is the same module with exterior antenna, connectors and other passive components. We will be able to observe some of the characteristics of the module $NRF24L01+$ in Fig. 2 and Table 2 [7].

3.2   Arduino

Arduino is an open source platform and is used for the construction of electronic projects. This platform consists of a physical programmable hardware board (micro-controller) and an Integrated Development Environment (IDE) software, which is the software used to develop and load the code for Arduino. The Arduino IDE uses a simplified version of the C++ software language, and we can observe some of the characteristics of Arduino [7].

3.3   Raspberry PI

The Raspberry Pi is essentially a small computer, developed in the UK, by the Raspberry Pi Foundation. This is made up of a hardware board, developed with the aim of promoting the teaching of basic programming sciences at school level. This is a relatively easy-to-use platform with a multitude of uses in the most diverse areas. Similarly, to Arduino, this is made up of two parts, hardware and software. The software to be used with Raspberry Pi, is the Raspbian operating system, which is based on Debian (Linux) optimized for the Raspberry PI hardware [7].

4   Practical Results

In the following figures we will be able to observe the way the hardware was interconnected (see Fig. 3). To develop this work, we used Python scripts with and without repetition code.

To compare the theoretical with practical simulations we evaluate the probability of error with GFSK modulation ($NRF24L01+$) (see Fig. 4).

In Arduino we used a script developed in $C/C++$, with data to transmit through $NRF24L01+$. It is intended that when the data were received in the receiver, the errors were corrected and at the same time it could obtain the number of errors in the receiver that were obtained with the transmission of

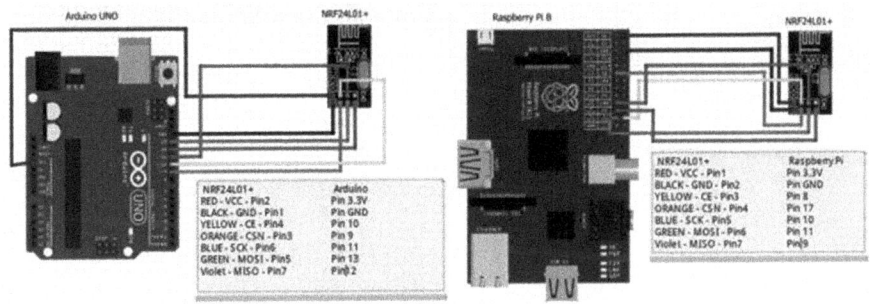

Fig. 3. (a) Link scheme between $NRF24L01+$ and Arduino. (b) Link of $NRF24L01+$ with Raspberry Pi.

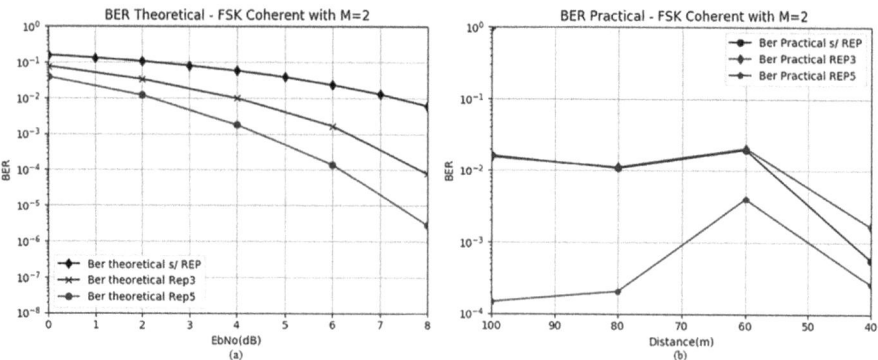

Fig. 4. Error probability with and without repetition codes for theoretical (a) and practical (b) values

1000 words. We will be able to observe through the Fig. 5 how the hardware interconnection was performed through the diagram of blocks presented, as well as its interconnection in a real context.

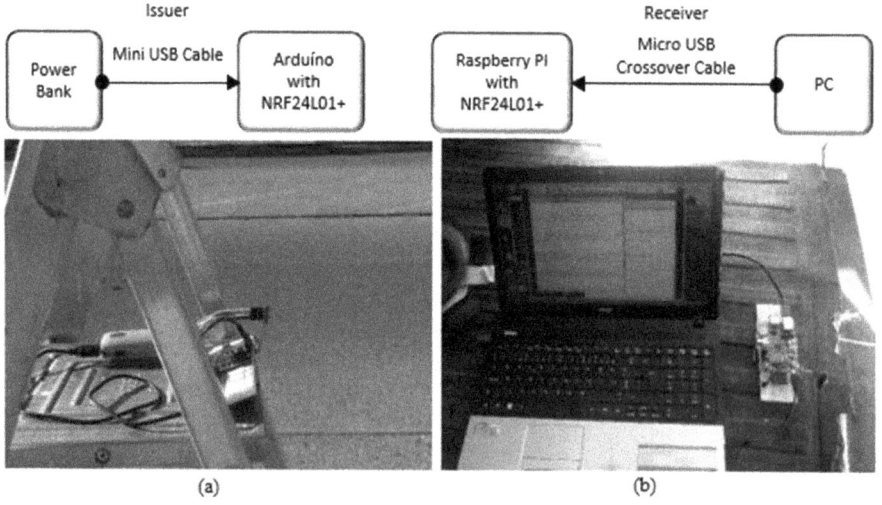

Fig. 5. (a) Issuer in a real environment. (b) Receiver in a real environment.

The block diagram of Fig. 6 shows all the steps of this work to implement the process receive a communication with high quality.

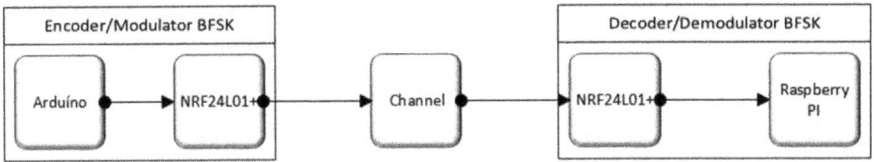

Fig. 6. Block diagram of communication system with the description of hardware.

5   Conclusions

More than ever the coding and correction of errors in transmissions, it is fundamental in a telecommunications system, notably the need for encryption, the combination of various sources of different debts, for example. It is necessary and indispensable to know the various types of error correction codes as well as their development at the software level.

With the realization of this work we can understand the development of error correction codes, since they have some complexity in their development, hence the importance to know them as we can see in Fig. 4. We can improve the quality of the communication system, with repetition codes with a higher number of repetitions (for example 7 repetitions) but the transmission rate will decrease.

However, when it came to the realization the practical tests were observed some influences at the level of transmission between the transmitter and the receiver, namely the temperature, the wind, and the relative humidity of the air, since, these parameters in some way influence.

References

1. Sestematizacao da Codificação e Discodificação de Codigos BCH. DEEC, FEUP (1995)
2. Haykin, S.: Communication Systems, 4th edn. Wiley, New York (2001)
3. Moreira, J.C., Farrell, P.G.: Essentials of Error Control Coding. Wiley, New York (2006)
4. Guimarães, P.D.D.A.: Sistemas de Comunicação II (2012). http://www.inatel.br/docentes/dayan/easyfolder/EE210/NotasDeAulaEE210.pdf. Accessed 22 Sep 2017
5. Doniak, M.: Comunicação Digital Demodulação em Banda Passante, p. 144. http://www.sj.ifsc.edu.br/~mdoniak/CDI_20705/CDI_Cap3.pdf. Accessed 20 July 2017
6. Nordic, nRF24L01 Single Chip 2.4GHz Transceiver Product Specification (2007). Accessed 03 Oct 2017
7. Linhares, F., Pires, B.: Controlo de um carro eletrico atraves de Arduino (2016)

Omnidirectional Mobile Robot Platform with Four *Mecanum* Wheels Featuring Remote Motion Control Through Either a Graphical Application or an Inertial Measurement Unit

Carlos Arantes[1] and João Sena Esteves[2]

[1] Department of Industrial Electronics, University of Minho,
Campus of Azurém, 4800-058 Guimarães, Portugal
a58765@alunos.uminho.pt
[2] Centre Algoritmi, Department of Industrial Electronics,
University of Minho, Campus of Azurém, 4000-058 Guimarães, Portugal
sena@dei.uminho.pt

Abstract. Omnidirectional mobile platforms allow simultaneous translation and rotation, which leads to the optimization of their trajectories, a possible reduction of the distance traveled and, consequently, a reduction of the energetic consumption. This paper presents an omnidirectional mobile robot platform based on four *Mecanum* wheels, a graphical application which allows remote motion control and monitoring of several parameters of the platform, and an inertial remote control that uses an IMU (Inertial Measurement Unit) and an AHRS (Attitude and Heading Reference System) to set the movements of the platform.

Keywords: *Mecanum* wheels · Omnidirectional mobile platform
Motion control · AHRS · IMU

1 Introduction

In many applications, the use of non-omnidirectional platforms constitutes an important limitation. An example is a classic electric wheelchair such as the one presented in [1]. Its main purpose is to facilitate the transportation of disabled people but it cannot perform some movements in an *x0y* plane because it uses the Ackermann steering system, which has 2 DOF (degrees of freedom). Usually, non-omnidirectional steering systems require complex algorithms for the dynamic management of trajectories and they may imply large distances traveled, a waste of energy and high response times.

Omnidirectional mobile platforms allow simultaneous translation and rotation, which leads to the following benefits:

(1) **Optimization of the trajectories of the mobile platform** – The distance traveled is minimized due to the combination of simultaneous rotation and translation;
(2) **Possible reduction of the energetic consumption** – Since the distance traveled is minimized, the required energy may also be decreased.

© Springer International Publishing AG, part of Springer Nature 2019
J. Machado et al. (Eds.): HELIX 2018, LNEE 505, pp. 32–38, 2019.
https://doi.org/10.1007/978-3-319-91334-6_5

Mobile platforms with four *Mecanum* wheels [2, 3] are omnidirectional. Each wheel has its own motor and platform motion control requires controlling each motor individually. The great variety of movements achievable through simultaneous translation and rotation cannot be properly controlled by traditional steering wheels used in non-omnidirectional platforms. More adequate solutions have been developed, usually based in joysticks [4, 5]. A motion control system based in hand gestures recognition is presented in [6].

This paper presents an omnidirectional mobile robot platform based on four *Mecanum* wheels (which is inherently more stable than those with only three wheels), a graphical application which allows remote motion control and monitoring of several parameters of the platform, and an inertial remote control that uses an IMU (Inertial Measurement Unit) and an AHRS (Attitude and Heading Reference System) to set the movements of the platform. The complete system is suitable, for example, for educational purposes.

2   System Architecture Overview

The developed system (Fig. 1a) has three main components, which will be detailed in this section: (1) Omnidirectional mobile robot platform, (2) Graphical application, and (3) Remote control.

2.1   The Omnidirectional Mobile Robot Platform (OMRP)

The omnidirectional mobile robot platform (OMRP) exchanges data either with the graphical application or the remote control. As shown in Fig. 1a, it has several subsystems. The power supply subsystem, which has both ultracapacitors (they allow ultra-fast charging) and a conventional lead-acid battery (for increased autonomy) as energy-storing devices, provides the required electrical power for each part of the OMRP. It is detailed on [7].

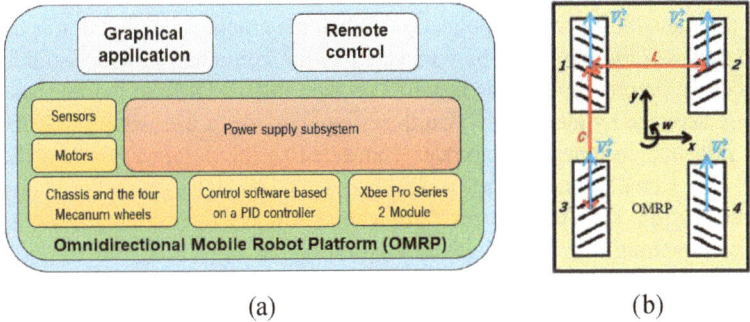

(a)   (b)

Fig. 1. (a) Block diagram of the developed system; (b) Schematic top view of the OMRP, clarifying the positions of the four *Mecanum* wheels and the referential used.

The OMRP has four *Mecanum* wheels placed as suggested in Fig. 1b. For a two-dimensional *xOy* plane, its steering system provides three independent possible movements: a translation along the x-axis, a translation along the y-axis and a rotation about the z-axis.

Controlling the movements of the OMRP requires defining its instantaneous velocity, which may be given as a three-dimensional vector: the linear velocity in the x-axis v_x, the linear velocity in the y-axis v_y and the angular velocity about the z-axis w_z. Taking this into consideration, it is necessary to compute the instantaneous velocity of each *Mecanum* wheel. Such computation requires the following conversion model that specifies the relationship between the velocities v_x, v_y *and* w_z, and the velocity of each *Mecanum* wheel, v_1, v_2, v_3 and v_4 (*C* and *L* are the distances highlighted in Fig. 1b; International System units are considered) [8–10]:

$$\begin{bmatrix} v_1 \\ v_2 \\ v_3 \\ v_4 \end{bmatrix} = \begin{bmatrix} 1 & 1 & \frac{-C+L}{2} \\ -1 & 1 & \frac{C+L}{2} \\ -1 & 1 & \frac{-C+L}{2} \\ 1 & 1 & \frac{C+L}{2} \end{bmatrix} \cdot \begin{bmatrix} v_w \\ v_y \\ w_z \end{bmatrix} \tag{1}$$

A PID controller system was used in order to ensure that the measured velocity is as close as possible to the computed reference velocity. The PID controller has four instances, each one matching one *Mecanum* wheel. An anti-reset windup block was also implemented into the PID controller in order to minimize the risk of high over-shooting on its output. Finally, a wireless module was integrated on the OMRP, enabling it to communicate with either the graphical application or the remote control. It was set a peer-to-peer network based on *XBee Pro Series 2* modules [11].

2.2 The Graphical Application

The graphical application is a program, written in C# programming language (running on a .NET virtual machine), used to wirelessly control the movements of the OMRP using a personal computer. It also provides remote monitoring of some parameters of the OMRP. It was designed in order to have two main windows. The main goal of the first one is to establish the wireless communication between the OMRP and the computer that runs the graphical application. When the graphical application is used, an *XBee Pro Series 2* module must be connected to the computer so as to establish the communication. Another module is installed on the OMRP. The two modules establish the wireless network by themselves. On the computer side, a user simply has to select the serial port that matches the modules connected to the computer. When the right serial port is selected, some frames are exchanged in order to verify the state of the wireless connection. If no error occurs, the second window opens. Otherwise, an error message is shown and the second window is not opened.

The second window aims to provide a way for both OMRP movements definition and monitoring of the OMRP. It was designed to have four separators in order to keep the data and settings organized. The first separator provides real-time information about the status of the wireless connection, some electrical parameters of the power supply subsystem of the OMRP, the environment temperature and humidity. The second separator provides feedback about the performance of the PID controller that manages each of the four *Mecanum* wheels. The separator shows in real-time the reference velocity, the measured velocity and the error (given by the difference between the reference velocity and the measured velocity). This feedback is provided using real-time graphs. The third separator offers the possibility of controlling the movements of the OMRP. The design of the separator was done having into account that it should have an intuitive interface and, at the same time, it should allow the execution of any trajectory. To achieve that, a set of buttons was created, each one to set a standard movement of the OMRP. Furthermore, a set of three numerical input boxes was designed to ensure the possibility of defining the velocities v_x, v_y and w_z independently. The fourth separator provides some independence to the OMRP movements. In other words, the graphical application should be able to command the OMRP in order to ensure that it performs some autonomous and useful movements. Two operation modes were considered in the implementation. The first one consists in following an object autonomously – using its infrared sensors, the OMRP follows an object trying to keep a constant distance to it. In the second operation mode, the OMRP autonomously tries to maintain a constant distance to the wall while moving forward or backward at a certain speed.

2.3   The Remote Control

The remote control provides an intuitive way of wirelessly controlling the movements of the OMRP, using an IMU.

An IMU has, at least, an accelerometer and a gyroscope. The remote control uses a 9-DOF IMU *Pololu MinIMU-9 V2*. It contains a 3 DOF accelerometer, a 3 DOF gyroscope and a 3 DOF compass. Using the data given by the three sensors and the DCM (Direction Cosine Matrix) algorithm [12], it is possible to set up an AHRS, a system that collects the data from an IMU and converts it to a new referential whose origin is the centre of mass of the remote control. Its rotation axes are *roll, pitch* and *yaw*.

The computed values of *roll, pitch* and *yaw* must be converted into the three velocity components of the OMRP (v_x, v_y and w_z), which is accomplished through the model

$$\begin{bmatrix} v_x \\ v_y \\ w_z \end{bmatrix} = \begin{bmatrix} 0.8 & 0.8 & 1 \end{bmatrix} \cdot \begin{bmatrix} roll \\ pitch \\ yaw - yaw_0 \end{bmatrix} \tag{2}$$

where the angles of the AHRS *(roll, pitch* and *yaw)* are given in degrees, linear velocities are given in cm/s and angular velocity is given in rad/s. Angle yaw_0 is a reference heading and yaw_1 is the actual heading. The gains 0.8, 0.8 and 1 were obtained empirically and they set the sensitivity of the movements of OMRP with respect to the AHRS.

Besides the IMU, the remote control incorporates an *XBee Pro Series 2* module and an *Atmel Atmega 328P* microcontroller for handling the collected data through the DCM algorithm. The system is powered by a 9 V rechargeable Ni-MH battery. A block diagram of the developed remote control is shown in Fig. 2.

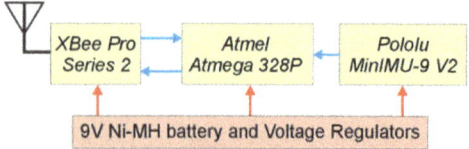

Fig. 2. Block diagram of the developed remote control.

3 Results

This section presents some results obtained with the previously described system.

3.1   The Omnidirectional Mobile Robot Platform (OMRP)

Figure 3 shows the final aspect of the developed OMRP and the prototype of the remote control. Coupled to each wheel of the OMRP there is a gearbox, an optical encoder and a brushed DC motor. In order to measure the distance to objects and obstacles, the OMRP also has four analog infrared sensors Sharp GP2Y0A21YK0F [13] (one at each side of the platform).

(a)   (b)

Fig. 3. (a) Final aspect of the developed OMRP; (b) Prototype of the remote control.

The OMRP was tested. It is capable of exchanging data with either the graphical application or the remote control. Each *Mecanum* wheel can be controlled independently, enabling the platform to perform any movement on an *xOy* plane. A video demonstrating the OMRP operating may be found here:

https://www.youtube.com/watch?v=Nm4K14rrIr8&feature=youtube.

It was validated that the OMRP is capable of following a moving body, keeping a predefined constant distance to it. It was also tested that it is able to maintain a predefined constant distance to a wall while moving forward or backward at a specific speed.

3.2   The Graphical Application

The first window of the developed graphical application enables a user to select the right serial port of the computer in order to start the connection with the OMRP. Then, the second window comes up evidencing its four separators. The first separator shows the values of several parameters of the OMRP, such as electric output power, ultra-capacitors voltage and environment temperature. The second separator presents graphs related to the speed of each *Mecanum* wheel (Fig. 4a) – the green line is the reference speed, the blue line is the real speed and the red line is the error (difference between reference speed and real speed). The third separator enables a user to command the OMRP through a set of intuitive buttons (Fig. 4b). Finally, the fourth separator is focused on enabling the OMRP to follow an object autonomously (keeping a specific distance to it) or to maintain a constant distance to the wall while moving forward or backward at a certain speed.

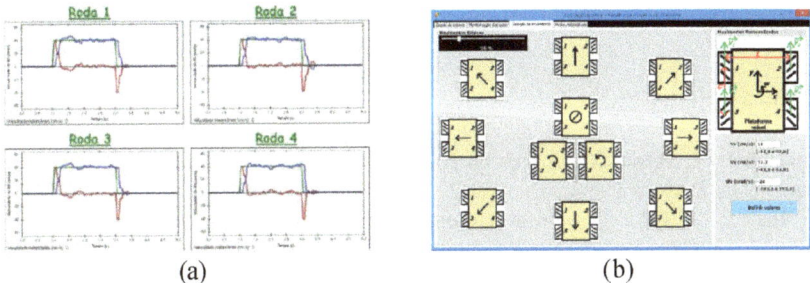

(a)   (b)

Fig. 4. (a) Second separator of the second window of the graphical application; (b) Third separator of the second window of the graphical application.

3.3 The Remote Control

The prototype of the remote control was developed and tested. It is capable of computing its values of *roll, pitch* and *yaw*. Two buttons were added to ensure safety – data is only sent to the OMRP while both buttons are pressed simultaneously. It was found that the computed values of the velocities v_x, v_y and w_z, sent to the OMRP, may be updated at a frequency of 50 Hz, considering a system clock of 16 MHz. The energy consumption of the remote control was measured. It needs 585 mW of electric power and its 9 V battery lasts for, approximately, 1 h and 13 min.

4 Conclusions and Future Developments

This paper presented an omnidirectional mobile robot platform (OMRP) with four *Mecanum* wheels. It is capable of (1) following a moving body, keeping a constant distance to it, and (2) maintaining a constant distance to a wall while moving forward or backward at a specific speed. The motion control of the platform is also possible through the use of either a graphical application or a remote control.

Besides controlling the movements of the OMRP, the graphical application monitors some parameters of both the environment and the platform, including showing the performance of the PID controller that tries to keep the measured velocity of the *Mecanum* wheels as close as possible to a reference velocity. This application is useful in order to study how the error (difference between the reference velocity and measured velocity) affects the trajectory described by the mobile platform.

The developed system may be used for educational purposes in the scope of omnidirectional platforms based on *Mecanum* wheels and AHRS. Furthermore, since the OMRP has the advantages inherent to omnidirectional platforms in terms of freedom of movements, it may be used in places in which the space is reduced.

As future developments, it is suggested to add a LIDAR sensor in order to enable the OMRP to have a very precise mapping of the space around it. Adding sensors for detecting harmful gases such as carbon monoxide would also be useful in some applications.

Acknowledgements. This work has been supported by COMPETE: POCI-01-0145-FEDER-007043 and FCT – Fundação para a Ciência e Tecnologia within the Project Scope: UID/CEC/00319/2013.

References

1. Tsalicoglou, C., Perrin, X.: Survey on Navigation Assistants for People with Disabilities, Autonomous Systems Laboratory, ETHZ, Zurich, Switzerland (2010)
2. van Haendel, R.P.A.: Design of an omnidirectional universal mobile platform, DCT 2005.117, DCT traineeship report, Eindhoven, September 2005
3. Nasir, O., Yousuf, M.: Introducing: The Mecanum Wheel. IEEE PNEC Perspective, vol. 4, Autumn 2012
4. Sarmento, L., Nunes, F., Martins, R.S., Sepúlveda, J., Esteves, J.S.: Remote control system for a mobile platform with four Mecanum wheels. Int. J. Mechatron. Appl. Mech. **2017**(1), 274–281 (2017). ISSN 25596497, EID: 2-s2.0-85029820801
5. Maldeniya, M.K.S.H., Madurawe, R.C., Thilakasiri, L.B.H.T., Thennakoon, T.M.S., Rajakaruna, R.M.T.P.: Remote Controlled 4WD Omni Directional Robot Using Mecanum Wheels. Department of Mechatronics, Faculty of Engineering, South Asian Institute of Technology and Medicine (SAITM), Sri Lanka (2015)
6. Vilaça, R., Ramos, J., Sepúlveda, J., Esteves, J.S.: Mobile platform motion control system based on human gestures. Int. J. Mechatron. Appl. Mech. **2017**(1), 267–273 (2017). ISSN 25596497, EID: 2-s2.0-85029814404
7. Arantes, C., Esteves, J.S., Sepúlveda, J.: A new energetically optimized power supply system for a mobile robot platform, using batteries and ultracapacitors to ensure both ultrafast charging and autonomy. In: ICINCO 2015, 12th International Conference on Informatics in Control, Automation and Robotics (2015)
8. Thomas, R.: Omni-Directional Mobile Platform for the Transportation of Heavy Objects. Massey University, Palmerston North, New Zealand (2011)
9. Rohrig, C., Heb, D., Kirsch, C., Kunemund, F.: Localization of an omnidirectional transport robot using IEEE 802.15.4a ranging and laser range finder. In: The 2010 IEEE/RSJ International Conference on Intelligent Robots and Systems, Taipei, Taiwan, 18–22 October 2010
10. Kim, J., Park, J., Kim, S.: Inertial navigation system for omni-directional AGV with Mecanum Wheel. Adv. Mech. Eng. **2**(1) (2012). ISSN 2160-0619
11. Digi International: XBee®/XBee-PRO® ZB RF Modules, 11001 Bren Road East, Minnetonka, MN 55343, November 2010
12. Macias, E., Torres, D., Ravindran, S.: Nine-axis sensor fusion using the direction cosine matrix algorithm on the MSP430F5xx Family. Texas Instruments, Application Report, SLAA518A - February 2012
13. Sharp Corporation: "GP2Y0A21YK0F", Datasheet, December 2006

Building a Behaviour Architecture: An Approach for Promoting Human-Robot Interaction

Bruno Amaro, Vinicius Silva$^{(\boxtimes)}$ ⓘD, Filomena Soares ⓘD,
and João Sena Esteves ⓘD

Department of Industrial Electronics, Algoritmi Research Centre,
University of Minho, Campus of Azurém, 4800-058 Guimarães, Portugal
{a70785, a65312}@alunos.uminho.pt,
{fsoares, sena}@dei.uminho.pt

Abstract. Human distraction behaviour is a paramount subject to take into account. Several woks in the literature try to tackle this topic mostly in the automotive industry. Following this trend, the present work proposes a system to detect the patterns of distraction/attention during an interaction activity between a human and a robot. The goal is to analyse selected patterns of distraction, such as eye gaze, head pose, blinking rate, among others, and adapting the robot behaviour, consequently promoting a more fluid interaction. A behavioural state machine that takes into account the engagement and the performance of the user in the activity is proposed.

Keywords: Human–computer interaction · Human–robot interaction
ZECA robot · Patterns of attention · Machine learning

1 Introduction

To do a task carefully is an important issue, since a little distraction may cause moderate to severe damages. With that in mind, new systems under development are able to detect patterns of distraction in a person. One of the industries where many studies have been conducted is the automobile industry. Generally, in these studies, computer vision systems with machine learning algorithms (e.g. SVM - Support Vector Machines and HMM - Hidden Markov Models, among others) are used. In one of these studies where SVM is used, a method of real-time detection of cognitive distraction and degraded driving performance has been developed. For the development of this project the data were collected in a simulator experiment in which the participants interacted with an in-vehicle information system (IVIS) during the driving [1]. The results obtained showed that the SVM models were able to detect the distraction of the driver with an average accuracy of 81.1%. The conclusion obtained in this study is that SVMs provide a viable mean of detecting cognitive distraction in real-time, overcoming the traditional approach of logistic regression. Additionally, this study demonstrated that the eye movements and simple measurements of the performance of drivers can be used to detect, in real time, the driver distraction.

© Springer International Publishing AG, part of Springer Nature 2019
J. Machado et al. (Eds.): HELIX 2018, LNEE 505, pp. 39–45, 2019.
https://doi.org/10.1007/978-3-319-91334-6_6

Another study presenting some early assessment results also uses SVM method in a facility that monitors the distraction of a driver. The module is able to detect the driver's visual and cognitive workload by merging stereo vision and tracking data by running SVM-based machine classification methods. The results showed more than 80% of success in the detection of visual distraction and a success of 68–86% in the detection of cognitive distraction. In the authors' opinion the results were satisfactory. In the course of the study, a comparison is made with the HMM method. It is concluded that the advantage of HMM is that it takes into account the transitions from one state to another. However, SVM can adapt better to momentary changes. In view of this, it is concluded that HMM may be better at detecting drowsiness, as this is a process that passes through some states. SVM could obtain better results in detecting a distraction caused by the phone ringing, because it is a momentary distraction. In conclusion, the classification algorithms obtained satisfactory results, and it is still necessary to reduce the cost of the system [2].

Although most studies are focused on driving, it is also important to detect distraction patterns in other situations, such as in a classroom setting, to find out if students are inattentive or not getting the most out of the lesson. In the present work, the goal is to detect patterns of distraction through the eye gaze, blink and head pose, in an activity where ZECA (Zeno Engaging Children with Autism), a *RoboKind Zeno R50* robotic platform from *Hanson Robotics*, is used to interact with children with Autism Spectrum Disorder (ASD). Ultimately, the goal is to detect whether or not the child is concentrated in the activity, thus getting the best out of it. In the scope of ASD research, although there are already several systems that interact with children, there is still no system capable of knowing if a child is really attentive to the robot. This is very important because, by detecting these patterns, the robot may capture the child's attention again, making the communication between the robot and the child more fluid. The intention throughout this project is to use the present framework in three activities with ZECA: (1) recognition of emotions, (2) imitation of emotions, and (3) storytelling [3, 4].

The recognition-imitation activities consist in the robot simulating human emotions through facial expressions, teaching a child to recognize and differentiate the different emotions, validating the response as soon as the child imitates the emotion in the correct way. In the last activity, storytelling, ZECA tells a story to a child and, in the end, it questions the child about what the character of the story was feeling. Taking into account these three activities, the present work consists in detecting patterns of distraction of a child and, whenever one of those patterns is detected, making the robot perform a suitable action to capture the child's attention again, in order to continue the activity.

This paper is divided into four sections. Following the introduction made in Sects. 1 and 2 details the proposed framework. In Sect. 3, the experimental outcomes are presented. Finally, in Sect. 4, the final remarks and the future work are addressed.

2 Proposed Framework

2.1 Experimental Setup

The experimental setup uses an RGB camera, a computer, and ZECA [5], a humanoid robot with 34 degrees of freedom (4 in each arm, 6 in each leg, 11 in the head, and 1 in the waist). This robot is capable of expressing facial cues thanks to the servo motors mounted on its face and a special material, *Frubber,* which looks and feels like human skin.

2.2 Experimental Methodology

An initial study was conducted to determine which patterns should be extracted in order to classify a user as attentive or not attentive. The following patterns were identified: the eye gaze, the head orientation, and the eyes blinking frequency. These patterns are detected during one of the activities described in the previous section.

The activities are conducted according to the following procedure: (1) ZECA greets the experimenter and the user; (2) ZECA asks which activity shall be played; (3) The selected activity starts and continues until the experimenter decides to end it [6].

In each activity it is verified if the user is distracted or not, by analysing the selected patterns. Accordingly to the classifier output, four conditions may occur:

- the user is attentive and responds the question,
- the user is attentive but does not respond the question,
- the user is distracted and does not respond the question,
- the user is distracted but responds the question.

A State Machine is a simple model to track the events triggered by external inputs. This is done by assigning intermediate states to decide what happens when a specific input comes in and which event is triggered. In this case, the events are being attentive or not being attentive and answering or not answering the prompted question.

The working of the State Machine (Fig. 1) to be used in the present work is described in the following points:

- The time between a stage and an inter-stage is 5 s and the time between stages is 10 s;
- The time without a response until the robot takes an action is 10 s;
- The user unanswered timeout until the robot moves to the next stage of the activity is 30 s.

The explanation for the above diagram is given below:

- The inputs of the State Machine are whether or not the user is attentive and whether or not the user answers the question. The output is accordingly to the correctness of the user's response and the detected distraction behaviour. The bubbles represent the states.
- The initial state is 1. This state corresponds to the moment when the robot asks the question and goes directly to stage 4, if the user is attentive and responds to the

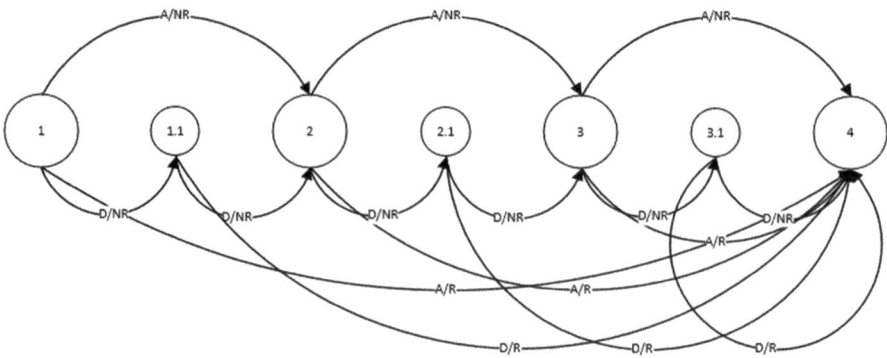

Fig. 1. State Machine with the workflow for the behaviour architecture of the robot (in this State Machine, A = Attentive, R = Responds, D = Distracted, and NR = does Not Respond)

question, or goes through all the different stages, if the user does not respond, or if he/she responds in one of the following stages, he then moves on to stage 4 as well.

- In the stages 2 and 3, if the user is attentive but does not respond, the robot repeats the question, motivating the user to answer.
- The sub-stages between each of these steps are considered for the cases when the user is inattentive and does not respond, causing the robot to try to regain his/her attention. These also serve to detect false positives that could happen if the user is classified as inattentive, but answers the question, which causes the robot to go directly to stage 4.
- In a stage 4, the robot verifies if the user response was correct, taking action accordingly. Then, it moves on to the next phase/question of the activity.
- If the user does not respond in 30 s and he/she is considered attentive, it means that the user does not know the answer, despite having been attentive to the robot.

According to the correctness of the user's answer, ZECA prompts a positive or negative reinforcement through sounds, gestures, and verbal communication [3].

2.3 Patterns Detection

For the detection of patterns of distraction an algorithm based on the DLIB library [7, 8], and OPENCV functions was developed. Thus, the system is able to track the face and eyes of the user, and in the future by using machine learning methods such as Support Vector Machine (SVM), the system will be able to recognize these patterns [9].

The DLIB is used as it makes available a collection of facial landmarks, which are used for the tracking of the face and the eyes, as well as other useful functions. OPENCV is used due to its suitability and applicability in computer vision solutions. For the calculation of the head orientation angles (yaw, pitch and roll), the points provided by the DLIB were used, as well as some OPENCV functions. For the blink detection it was used a state machine [8] with OPENCV functions. SVM will be used because, considering the state of the art, this is one of the most used methods for detecting these type of patterns, and tests will be conducted in a final stage, in order to verify if this method is adequate to this type of activities.

2.4 Interface Architecture

In order to control the system, a Graphical User Interface (GUI) is under development. This interface has three tabs: the "Main" tab, the "Configuration" tab, and the "Info" tab (Fig. 2).

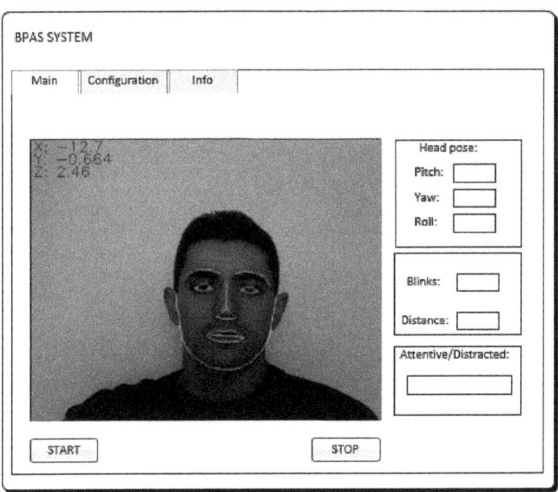

Fig. 2. "Main" tab of the interface

In the "Main" tab, it is possible to see the user's face with the rendering of selected facial landmarks of the DLIB. This tab also presents others metrics such as the number of blinks, the head orientation values (pitch, yaw, and roll), the direction of the pupil, and the distance that the user is from the robot. This last metric, distance between the user and the robot, is also an important parameter, because it can be used to detect if the user is close or far from the robot and, according to that, maintain an appropriate and safe position during the interaction. The experimenter can initiate or pause the system by pressing the buttons, 'Start', and 'Pause'. The experimenter will be able to select and control, by starting, pausing, and ending one of the three activities.

In the "Configuration" tab, the experimenter will have access to extra configuration steps for the chosen activity (e.g. setting the activity duration time) and he/she will also be able to train and configure the classifier model.

The 'Info' tab will display the user's information. The operator will have access to the database for the users, which will be organized by 'code', 'name', 'age', and 'gender'.

At the end, the goal is to register the user's performance in a session: the session time, the number of correct and wrong answers, the user's progress per session, the moments when the user was distracted, and the time between each prompted question.

3   Experimental Outcomes

The results obtained in a laboratorial environment allow computing the angles of orientation of the head – yaw, pitch and roll. It is also possible to calculate the number of blinks as well as to obtain the eye gaze, as shown in Fig. 3. These results are displayed in the "Main" tab of the interface, the on/off buttons are also functioning, as well as the displays where the data of the obtained patterns are placed.

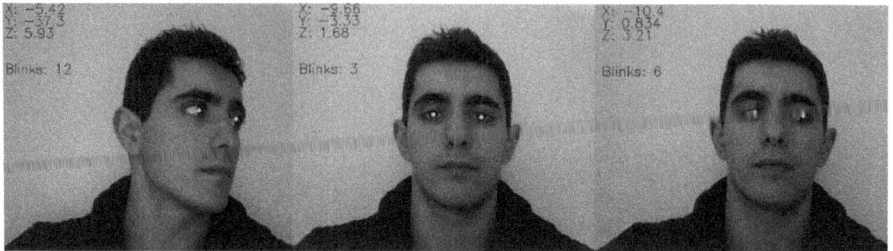

Fig. 3. Results obtained in a laboratorial environment

The system is not yet able to classify whether or not the user is attentive, leaving this to future work, where machine learning methods will be used for this classification.

4   Final Remarks

The work described in this paper is part of a research project in which the robot ZECA is used to promote the interaction with children with ASD, helping them to recognize emotions and to acquire new knowledge in order to promote social interaction and communication with the others.

Although there are several systems for detecting patterns of distraction, most of them are applied to driving. One of the situations in which the detection of these patterns is important is in robot activities that interact with children with ASD. But, in general, the robotic systems used in those activities cannot know if a child is attentive or not.

The main goal of the present work is the development of an algorithm capable of recognizing attention/distraction patterns of a user/child during an activity (e.g., detecting if he/she is engaged/interested in the activity by tracking his/her eye gaze and head motion). The algorithm already developed detects and tracks the face. It also computes the head pose orientation (pitch, roll, and yaw). In addition, it detects 70 facial landmarks, including the pupil landmark, making it possible to track the user's eye gaze and blinking rate. In order to determine the course of the activity, a state machine taking into account both the selected patterns and the user performance is proposed. This state machine influences the robot behaviour accordingly to the statement conditions. The developed system has a graphical user interface displaying the results, as well as the parameters and information of the user.

As future work, the SVM method will be applied to the training of a classification model based on the detected patterns, in order to detect whether the user is attentive or not. A database with a large amount of data related to characteristics of attentive patterns for adults and children will be developed. The overall system will be evaluated in a controlled laboratorial environment to see if it is able to detect distraction/attention patterns and allow the robot to respond properly. The final goal is to test the complete system with children with ASD in order to promote a more dynamic interaction.

Acknowledgements. The authors would like to express their acknowledgments to COMPETE: POCI-01-0145-FEDER-007043 and FCT – Fundação para a Ciência e Tecnologia within the Project Scope: UID/CEC/00319/2013. Vinicius Silva also thanks FCT for the PhD scholarship SFRH/BD/SFRH/BD/133314/2017.

References

1. Liang, Y., Reyes, M.L., Lee, J.D.: Real-time detection of driver cognitive distraction using support vector machines. IEEE Trans. Intell. Transp. Syst. **8**(2), 340–350 (2007)
2. Kutila, M., Jokela, M., Markkula, G., Rue, M.R.: Driver distraction detection with a camera vision system. In: 2007 IEEE International Conference on Image Processing, pp. VI-201–VI-204 (2007)
3. Costa, S.C.C.: Affective robotics for socio-emotional development in children with autism spectrum disorders, October 2014
4. Silva, V., Soares, F., Esteves, J.S.: Mirroring and recognizing emotions through facial expressions for a RoboKind platform. In: 2017 IEEE 5th Portuguese Meeting on Bioengineering (ENBENG), pp. 1–4 (2017)
5. "RoboKind | Advanced Social Robotics." http://robokind.com/. Accessed 09 Feb 2018
6. Costa, S., Soares, F., Pereira, A.P., Santos, C., Hiolle, A.: Building a game scenario to encourage children with autism to recognize and label emotions using a humanoid robot. In: The 23rd IEEE International Symposium on Robot and Human Interactive Communication, pp. 820–825 (2014)
7. "dlib C++ Library." http://dlib.net/. Accessed 13 Dec 2017
8. "Computer Vision for Faces | School of AI." https://courses.learnopencv.com/p/computer-vision-for-faces. Accessed 09 Feb 2018
9. Silva, V., Soares, F., Esteves, J.S., Figueiredo, J., Leão, C.P., Santos, C., Pereira, A.P.: Real-time emotions recognition system. In: 8th International Congress on Ultra Modern Telecommunications and Control Systems and Workshops (ICUMT), Lisboa, Portugal, 18–20 October 2016 (2016)

Virtual Reality Training of Practical Skills in Industry on Example of Forklift Operation

Przemysław Zawadzki, Paweł Buń[(⊠)], and Filip Górski

Chair of Management and Production Engineering,
Poznan University of Technology, Piotrowo 3 Street, 60-965 Poznan, Poland
{Przemyslaw.Zawadzki,Pawel.Bun,
Filip.Gorski}@put.poznan.pl

Abstract. The paper presents development and test procedure of a Virtual Reality application for training of practical skills. The presented case pertains to operation of forklifts. Need for the application is explained, as well as basic ideas, concepts of knowledge forms and organization of the virtual course. The test procedure is presented, along with obtained results. It has been found that the virtual forklift course is more time efficient than the traditional one and it is enthusiastically perceived by users from the target group.

Keywords: Virtual reality systems · Virtual training · Education

1 Introduction

Virtual Reality (VR) systems are solutions known for several decades. Recently, thanks to emerging of new, consumer devices, the VR branch is gaining popularity, also in industrial applications. The virtual environments allows users to freely move and interact with three-dimensional objects in real time [1]. VR has many professional applications. One of the most prevalent is education and trainings – the VR systems allow conducting engineering education [1–3], traditional education [1, 4] or medical training [5]. In industry, VR is quite often used for support of design processes [6], simulation of machines and devices [7] or for decision-making [8].

The VR systems are currently readily used for industrial education and training [9, 10]. There is a very broad scope of various equipment, allowing not only to obtain near-photorealism in visualization (VR goggles – Head Mounted Devices, HMDs), but also allowing user directly interact with elements inside a visualization (haptic devices, systems for tracking and gesture recognition), intensifying feelings of immersion. By integrating specific hardware with a properly prepared VR application, a specialized simulation environment can be obtained, in which the training is performed according to conditions present in reality. Here, it must be mentioned that VR is different to other industrial simulations, such as numerical simulations [11], in that it employs a human-centric approach. The VR training allows increasing effectiveness of knowledge transfer. It is far easier to repeat activities in VR, as well as train undesired or dangerous situations. VR applications are also easily scalable and they can be created in a way to

© Springer International Publishing AG, part of Springer Nature 2019
J. Machado et al. (Eds.): HELIX 2018, LNEE 505, pp. 46–52, 2019.
https://doi.org/10.1007/978-3-319-91334-6_7

allow adjustment of its content for a given user. Blümel and coauthors pointed out advantages of using VR training in their review work [12].

Many authors indicate applications for learning and training in industrial conditions [13–16]. It is therefore clear, that VR is beginning to be present everywhere, where there is a need for a flexible training solution, reducing costs and time of training in real conditions. It makes it suitable for implementation in industrial conditions.

This paper presents building and evaluation of a virtual training system for teaching technical skills – how to operate a forklift. The skill was selected for studies due to the target group of trainees – they usually are not high educated nor qualified in using advanced computer systems. Therefore, their feedback is more valuable – in authors' opinion – than studies on qualified engineers or engineering students, often found in existing literature. The authors created and tested the first prototype.

2 Aim of Work

The paper presents a case study of a virtual reality training solution, aimed at developing practical, technical skills in inexperienced workers, on example of forklift truck operation. The application comprises both theoretical lessons and practical exercises of forklift operation. This example was selected, as it easily allows to bring out VR training advantages – safety, repeatability and availability of virtual equipment at almost no costs. The proposed solution should become an alternative to traditional courses for industrial company workers. Aim of the pilot study, presented in this paper, was to evaluate if the proposed approach is correct and effective.

3 Training Program Scenario

Scenario of the course was prepared in cooperation with a professional training company. The course consists of 9 theoretical lessons and 4 practical exercises. The lessons and exercises are further divided into blocks. The whole knowledge necessary for a trainee to obtain is contained in form of interactive and non-interactive 2D and 3D content, placed in virtual space. The content has the following form:

- text – descriptions of problems, especially regarding information difficult to present in a visual form (names of standards, formulas etc.), as well as instructions,
- audio – complementary to the text content, read by a lector and launched by interaction with a character of virtual teacher, always present in the virtual space,
- 2D graphics – pictures, schemes, diagrams, infographics illustrating blocks,
- 3D content – interactive models of objects (machines and devices, elements of work environment etc.), along with animations, illustrating a current block.

The most important constant elements of the lesson scene are:

- virtual teacher – an animated character helping the user, replacing a living teacher,
- flipchart – space in the center of the scene, where 2D content is displayed,
- environment – e.g. a hall – 3D background space, that the user can navigate,

When realizing a theoretical lesson, the user can move freely between subsequent blocks of content. When performing an exercise, the user has a certain task to do in a single block and he cannot move forward until the task is done successfully. The time of each task is measured. After all the lessons and exercises are finished, the user can test his knowledge in an exam (a test with a set of closed questions).

4 Virtual Application Development

4.1 3D Models Preparation

All the necessary 3D models were prepared in 3DS Max software (Fig. 1), or downloaded from open digital libraries. Their geometry was optimized for better framerate in the VR application. Their structure was also prepared for later animation.

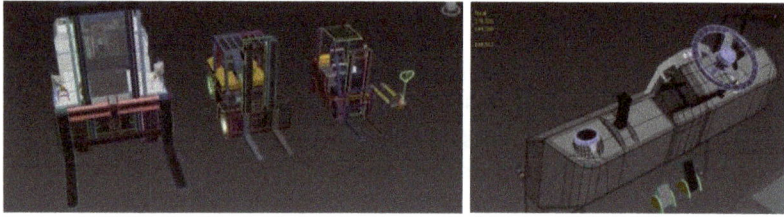

Fig. 1 Different forklift models in 3DS Max software (left), model of driver's cockpit, animation requirements considered (right)

4.2 Application Development

The virtual training application was prepared in the Unity 3D environment. The application was programmed using a flexible approach [17], so most of the course contents can be edited externally. This will facilitate expanding the knowledge library for teaching various skills. The authors have prepared a library of predefined interaction mechanisms (e.g. "click to animate"), so creating interactive blocks is easy and does not require writing code.

The user can move between the menu and scenes of lesson, exercise and exam. The menu allows selection of lessons/exercises (Fig. 2), as well as exam mode. The lesson and exercise scenes are presented in Fig. 3, where constant elements (teacher, flipchart) are visible, as well as 3D objects (forklift and its environment).

Basic interaction is performed via the "gaze and click" mechanism – the user looks at an interactive object first, the object is then highlighted (often a label is displayed), user can then click a button on the joystick (remote) to activate the object. This mechanism is simple enough to not require long training inside the virtual environment.

Fig. 2 Main interface for choosing a lesson or an exercise

Fig. 3 Left – lesson scene, right – exercise scene

4.3   System Integration – First Prototype of the System

The first prototype of the virtual forklift training, presented in this paper, contains all the functionalities of the planned full version, but it contains only 3 lessons and 1 exercise (out of 9 and 4, respectively). The hardware of the system consisted of a PC computer (Intel i7-4770 3.4 GHz, GeForce GTX 980, 8 GB RAM), a full HD TV for preview and a consumer VR solution - Oculus Rift CV1 HMD, with a tracking sensor and a remote, used for movement and interaction. The application installed on the system was tested internally and then a group of target users were invited for tests.

5   Tests of the System

5.1   Test Procedure

The tests were performed on a group of 17 people. The group consisted of target users – young persons without higher education, as well as already working persons, wanting to improve their qualifications. Both men and women were tested. The group also contained two instructors and three experienced forklift operators. Aim of the tests was to generally evaluate the application and gather constructive feedback, as well as initially assess its teaching effectiveness. In the beginning, a short introduction was made, presenting the application and available interaction methods, along with a short training (Fig. 4), to enable participants work later on their own. It must be emphasized, that for the most participants it was the first opportunity to use a VR solution.

The tests were not disturbed, individual times were not measured precisely. The participants could stop the experience at any time if they were bored, felt sick or simply

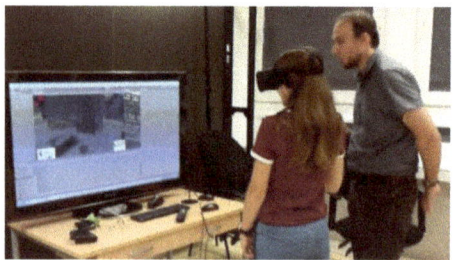

Fig. 4 Tests of the virtual training

got through all the available content. After the testing was done, two surveys were performed – a Simulation Sickness Questionnaire [18] for symptoms of cyber sickness and an evaluation survey. The participants were also separately asked about their most important observations. The evaluation survey consisted in awarding a score between 1 and 5 to each of the following criteria:

1. The HMD comfort of use and adjustment
2. Quality of 3D content
3. Quality and comfort of reading of 2D content (infographics, text).
4. Quality and usefulness of audio content (ambient, objects, lector)
5. Fluency of looking around the virtual scene
6. Fluency and intuitiveness of object interaction
7. Comfort of movement using the pilot (joystick)
8. Comfort of interaction with 3D objects by gaze and click mechanism
9. Animations
10. Realism of presented situations and objects, feeling of immersion

5.2 Test Results

All the participants except one finished all the available content, spending between 40 to 70 min inside the virtual environment. One person had to shorten the experience due to minor symptoms of cyber sickness, which occurred while driving the forklift in the practical exercise. All the other participants had practically no symptoms of cyber sickness, so their detailed results are omitted here.

The application was rated positively (see Fig. 5 and Table 1 for detailed results). The most important observations from the participants were as following:

– the VR training is far more attractive and compelling than a traditional one, all the test participants preferred the VR form if given a choice,
– interactive 3D models and animations were rated very high, although participants already skilled in forklifts had small doubts regards forklift physics (1-2 cases),
– possibility of safe, no-risk practice was also praised, although experienced forklift operators marked a need for more complex practical tasks,
– there is a clearly visible learning effect, as each consecutive task in the practical exercise was performed more and more effectively by the participants,
– for improving realism, forklift control should employ a physical driving wheel.

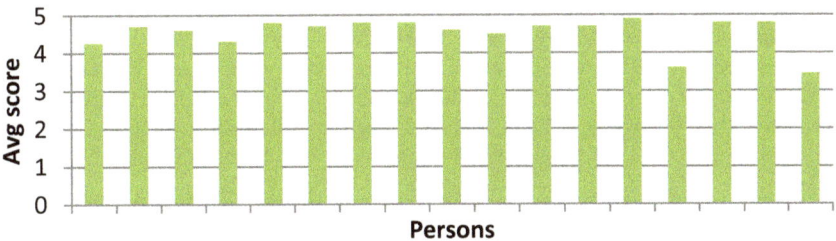

Fig. 5 Average results by persons

Table 1 Average results by survey questions

Question	1	2	3	4	5	6	7	8	9	10
Avg score	4.76	3.76	4.60	4.12	4.71	4.71	4.50	4.82	4.82	4.60

6 Conclusions

The training solution presented in the paper is currently evaluated for possibility of industrial implementation. The tests have confirmed its high effectiveness, as well as attractive form. Therefore, it should be emphasized, that advantage of virtual training over a traditional one has been proved, for a specific technical skill presented in the paper. From the viewpoint of an industrial company management, implementation is worthy of consideration, also in economical aspect (low-cost devices used). The training may be performed on one's own, without continuous participation of an instructor and it is shorter than a traditional course – most participants spent less than one hour learning a material usually presented in one working day, so the yield is approx. 80%, depending on learning capabilities of a given person.

The authors will further develop and test the solution, as the need has been recognized for more types of technical skills. In further work, problems of building of this type of applications in a maximally effective way will be considered, to ensure their continuous, efficient use in industrial conditions.

References

1. Martín-Gutiérrez, J., Mora, E., Añorbe-Díaz, B., González-Marrero, A.: Virtual technologies trends in education. EURASIA J. Math. Sci. Technol. Educ. **13**(2), 469–486 (2017)
2. Grajewski, D., Górski, F., Zawadzki, P., Hamrol, A.: Application of virtual reality techniques in design of ergonomic manufacturing workplaces. Procedia Comput. Sci. **25**, 289–301 (2013)
3. Rodriguez, J., Gutiérrez, T., Sánchez, E.J., Casado S., Aguinaga, I.: Training of procedural tasks through the use of virtual reality and direct aids, virtual reality and environments (2012)
4. Sigitov, A., Hinkenjann, A., Roth, T.: Towards VR-based Systems for school experiments. Procedia Comput. Sci. **25**, 201–210 (2013)

5. Escobar-Castillejos, D., Noguez, J., Neri, L., Magana, A., Benes, B.: A review of simulators with haptic devices for medical training. J. Med. Syst. **40**(4), 104 (2016)
6. Ye, J., Badiyani, S., Raja, V., Schlegel, T.: Applications of virtual reality in product design evaluation. In: Proceedings of the 12th International Conference on Human-Computer Interaction (HCI International 2007), Beijing, China, 22–27 July 2007. Published in: HCI'07 Proceedings of the 12th International Conference on Human-Computer Interaction: Applications and Services, vol. 4553 (Part 4), pp. 1190–1199 (2007)
7. Pandilov, Z., Milecki, A., Nowak, A., Górski, F., Grajewski, D., Ciglar, D., Klaić, M., Mulc, T.: Virtual modelling and simulation of a CNC machine feed drive system. Trans. FAMENA **39**(4), 37–54 (2016)
8. Colombo, S., Nazir, S., Manca, D.: Immersive virtual reality for training and decision making: preliminary results of experiments performed with a plant simulator. SPE Econ. Manag. **6**(4), 165–172 (2014). https://doi.org/10.2118/164993-pa
9. Wickens, C.D.: Virtual reality and education. In: IEEE International Conference on Systems, Man and Cybernetics 1992, pp. 842–847. IEEE (1992)
10. Seidel, R.J., Chatelier, P.R.: Virtual Reality, Training's Future?. Plenum Press, New York (1997)
11. Ivanov, V., Mital, D., Karpus, V., Dehtiarov, I., Zajac, J., Pavlenko, I., Hatala, M.: Numerical simulation of the system "fixture – workpiece" for lever machining. Int. J. Adv. Manufact. Technol. **91**(1), 79–90 (2017). https://doi.org/10.1007/s00170-016-9701-2
12. Blümel, E., Termath, W., Haase, T.: Virtual reality platforms for education and training in industry. Int. J. Adv. Corp. Learn. **2**(2) (2009)
13. Dostatni, E., Grajewski, D., Diakun, J., Wichniarek, R., Buń, P., Górski, F., Karwasz, A.: Improving the skills and knowledge of future designers in the field of ecodesign using virtual reality technologies, international conference virtual and augmented reality in education. Procedia Comput. Sci. **75**, 348–358 (2015). ISSN 1877-0509
14. Grajewski, D., Górski, F., Zawadzki, P., Hamrol, A.: Application of virtual reality techniques in design of ergonomic manufacturing workplaces. In: 2013 International Conference on Virtual and Augmented Reality in Education, vol. 25, pp. 289–301 (2013)
15. Langley, A., Lawson, G., Hermawati, S., D'Cruz, M., Apold, J., Arlt, F., Mura, K.: Establishing the usability of a virtual training system for assembly operations within the automotive industry. Hum. Factors Ergonomics Manufact. Serv. Ind. **26**(6), 667–679 (2016)
16. Gorecky, D., Khamis, M., Mura, K.: Introduction and establishment of virtual training in the factory of the future. Int. J. Comput. Integr. Manufact. **30**(1), 182–190 (2017)
17. Górski, F.: Building virtual reality applications for engineering with knowledge-based approach. Manag. Prod. Eng. Rev. **8**(4), 64–73 (2017)
18. Kennedy, R.S., Lane, N.E., Berbaum, K.S., Lilienthal, M.G.: Simulator sickness questionnaire: an enhanced method for quantifying simulator sickness. Int. J. Aviat. Psychol. **3**(3), 203–220 (1993)

The Typographic Object, Matter and Programming: Facilitated Interactions in an Ephemeral and Digital World

Karine Itao Palos[1(✉)] and Gisela Belluzzo de Campos[2]

[1] Design, Anhembi Morumbi University, São Paulo, Brazil
karineipa@hotmail.com
[2] Communications and Semiotics,
Pontifical Catholic University of São Paulo, São Paulo, Brazil
giselabelluzzo@uol.com.br

Abstract. This article analyzes the possibilities of user interaction with typography in the digital environment. These interactions are feasible due to the fact that the mathematical code allows for the predefinition of user response parameters based on their own actions. Thus, approaches with the ephemeral qualities that typography acquires in the digital context are carried out, caused by the polyvalence of the algorithmic code that, from generative programming, allows the user to interact with the typographic object. The codes allows to simulate behaviors and materials, allowing for the typographic character to transit through the screen or deform its characteristics according to the stipulated commands This ephemerality of the digital environment allows for the sensation that it is not composed of matter. A parallel was drawn between the concept of "matter" in the computational scenario, proposed by the philosopher of design Vilém Flusser [1], and the quality of "fluidity" observed in images created by digital generative programs, applying as study cases the Beowolf (1989) Anitype (2013) and Lettree (2004).

Keywords: Typography · Interaction · Generative system · Design

1 Introduction

This article proposes to discuss how the fluidity and ephemerality present in digital typography can promote interactivity through the generative algorithm, as well as understand typography potential to allow messages to appear through a character or a phrase, dispensing with the entire text.

The theoretical and methodological procedures adopted herein were the concept of "matter" present in the computational environment as developed by Flusser [1] and the conceptual discussion of what a program is. Subsequently, the concept of a program was related to generative systems in the computational environment, in order to clarify the relationship between code and typographic interactions. Finally, three design projects available on the internet, both for computers and for mobile devices, were chosen for this analysis, namely Beowolf (1989) by Just van Rossum and Erik van Blokland, Anitype (2013) by Jono Brandel and Lettree (2004) by Ricard Marxer Piñón.

© Springer International Publishing AG, part of Springer Nature 2019
J. Machado et al. (Eds.): HELIX 2018, LNEE 505, pp. 53–59, 2019.
https://doi.org/10.1007/978-3-319-91334-6_8

2   The Relations Between Matter and the Typographic Object

In the printed medium, typography acquires the physical characteristics of the support and materializes from the moment the pigment is impregnated on the paper sheet and fills its shape. It remains motionless and unchanged until the degradation of the material to which it has been confined. In digital systems, types have left their physical properties to become objects consisting of digitized binary code sequences, displayed on the computer screen [2].

Flusser [1] explains that, according to ancient Greek philosophers, what humans experience through the senses is called the "world of phenomena." The physical medium is an "amorphous jelly" that covers the outer forms, which are perceptible thanks to the suprasensibility of the theory. According to the author, the matter that presents this amorphous characteristic has its origin in the Greek word "hylé", which means wood used as raw material. When transporting this conception of matter to the construction of a typographic cliché, this cliché is similar to a stamp with the design of a letter carved in wood or metal. The typographer, when sculpturing the character, imposes the idea of the letter to this material, although it is subject to be destroyed physically and lose its form, through time, wear or even by a fateful accident. However, the theoretical concept of the shape of that character will always exist, because it can be retrieved through imagination.

Still according to Flusser [1], when a designer establishes the form of an object conferring characteristics and concepts pertinent to him/her, the designer also distorts the idea of the object. Thus, in the case of typographic cliché, it would be impossible to create an ideal character, since, although two clichés of the same character resemble each other, they will never be totally equal due to the qualities of the material used and the manual processes of creating it, which according to Bringhurst [3] makes exact repetition impossible. Designers can also not fully adhere to a theoretical elaboration of the previous form. This is more evident in the printing of two typographic characters impregnated on paper by the same cliché, as their ink deposits vary, as well as the pressure imposed on the sheet [2].

Rocha [2] (p. 26) points out that there have been changes in the forms of the letters of the alphabet from the first typographic presses to the present: "[…] The trajectory of typography in Europe covered the characteristic features of the culture of each country and each historical milieu in which they were produced". These changes occurred both because of the addition of new ideas to letter characteristics, as well as due to the limitations of the materials and the techniques used to make the cliché.

The typography is a living organism that adapts to the needs of the language. "This radically modifies the concept of 'theory': it no longer means a passive, religious contemplation of eternal forms, nor an empty skirmish of words… it means an advanced modeling of ever improving types." [4] (pp. 83–84).

The materialization of typography is usually associated with the physical environment, as argued above. According to Flusser [1] (p. 24) typography occurs in the virtual environment with even more propriety, since "[…] under the impact of information technology, it begins to return to the original concept of 'matter' as a transitory fulfillment of temporal forms." In this environment, the adaptive character inserted in the typography is intensified, which in the past was attributed to culture, pertinent to the phonetic adaptations of the language. In the digital environment, the theoretical form of the character has the possibility of much greater adaptation, since the abstraction associated with the theoretical sphere is attributed to the mathematical code that manipulates and conceives the form, allowing this form to be filled with simulations of physical materials. Thus the form of an object is increasingly volatile, passing very rapidly from solid to intangible, and 'matter' loses its direct relation to physical space.

Flusser [1] takes as a major issue in this discussion of 'matter' in the digital environment the fact that, since the ancient Greek philosophers, what mattered was the configuration of matter, leaving the form visible in the midst of the amorphous world, but what matters today is to introject "matter" into forms that emerge from highly intricate mathematical codes, which creates shapes in an uncontrollable manner.

This polyvalent feature of the pixel, in which it can simulate any kind of molecule in the digital world, allows for the replication of the properties of environmental phenomena in the physical world, such as wind friction or gravity affecting object movement [5]. However these computer-designed worlds require a mediation interface, for they cannot be reached by the senses of man. In turn, this mediator is an object that has its "matter" formalized in the world of phenomena and, for this "matter" to be created and formalized in the virtual environment, a physical support is necessary.

3   The Imperfections Executed in Theory

By designing the theoretical perspectives used to create and fill shapes, the mathematical code has greater control over graphic objects, allowing for the existence of an ideal object. The theoretical perspective is formally calculated before being filled with matter. Flusser [1] explains that, although it seems a contradiction, what is seen on the computer screen are forms that come from eternal, immutable formulas. These formulas, however immutable, can, however, deform, rotate, shrink, and magnify, in order to generate another equally immutable formula in this process.

The Beowolf typeface experiment (see Fig. 1), developed in 1989 by Just van Rossum and Erik van Blokland, which together form the LettError group, is a font initially created in a PostScript programming language. Later, with the advent of OpenType, it was adapted to this language and became more accessible from the point of view of the democratization of use. However, from a design point of view, it has become more restricted, since it allows modification only of the predefined griffins in relation to the other letters that surround it [6, 7]. This source was one of the first to allow for the simulation of the inaccuracies of typographic impressions on a computer by means of previously established rules that allow for the deformation, without the characterization of the letter, so that the software generates random contours in the characters each time the type is used, from fixed coordinates, although it is located in the same text [8].

On the computer, the "failures" are not simply obtained by chance, since they are precalculated by the machine. Through these algorithms, they render the shape of the character on the screen and fill it with matter, which makes it visible and accessible to the human senses.

Fig. 1. Beowolf font – character variations. Source: https://www.moma.org/collection/works/ 139326, last accessed 2017/08/14

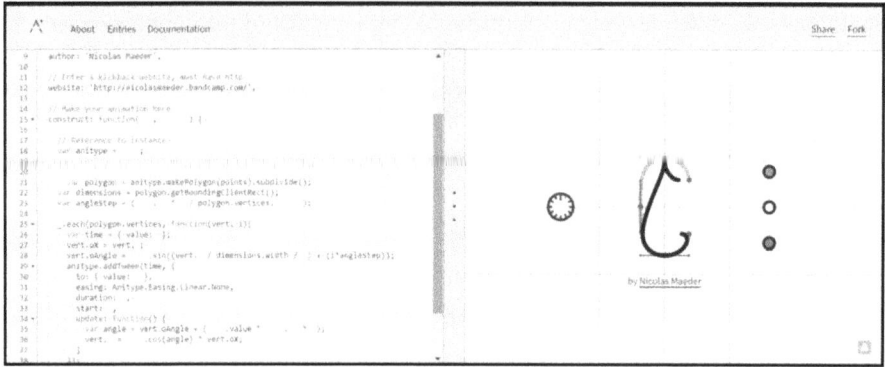

Fig. 2. Anitype – character animated by code. Source: http://www.anitype.com/, last accessed 2017/08/28

Another example is the Anitype experiment (see Fig. 2), created by Jono Brandel in 2013, generated by JavaScript code and made available for the web, which invites the user to interact with the code that moves and materializes each type.

By choosing one of the forty-two characters on the home screen and displayed side by side in rows and columns, the Anitype platform displays a retractable panel in the left-hand area of the window with the part of the source code that can be changed by the user and is responsible for the animation and formation of the chosen letter. These modifications are applied with a click of a button that resembles the video player located between the panels. On the right side, the user can see the character and visually track how the code changes imply in movement. In the same panel on the right side of the letter there are always three buttons: the first is blue and shows the bezier curves; the second is orange and displays the original form of the chosen character and the third, in green, shows the guiding lines. On the left side of the character there is a clock icon with a running pointer that represents the time of the character animation. It is possible to pause the movement with a click of the mouse on this icon. By creating a user on the platform, the user can save and share the created code by clicking on the 'Save' button in the upper right corner of the screen [9].

An experiment created by Ricard Marxer Piñón in 2004 named Lettree proposes to discuss calligraphy and share content. The project was programmed with Processing - an open programming language, ideal for images, animations and audio, with the intention of exploring the new dimension of textual representation that computers allow for. The author considers calligraphy a form of manifestation and expression of human subjectivity [10]. By typing, Lettree (see Fig. 3) allows for the insertion of white characters on a black background. Each typographic character is linked to the previous one by means of the tips of the serifs present in stems, ascending, descending, legs, tails and finials. Depending on the characteristics of the letter, the number of connections varies. The result of this interaction is a typographic composition that can generate an infinity of drawings, depending on the letters that the user chooses and from the point where the program makes the connection.

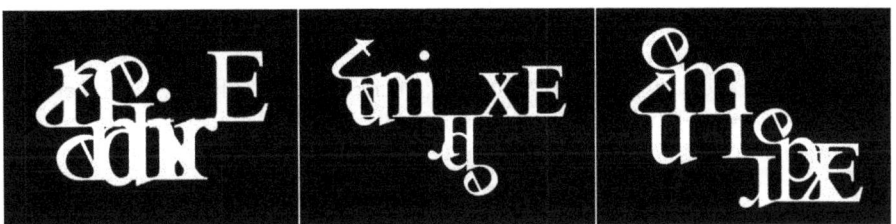

Fig. 3. Lettree – compositions generated with the word 'experiment'. Source: http://www.caligraft.com/, last accessed 2018/01/12

4 The Typographic Object and the Possibilities of Programming

The Beowolf, Anitype and Lettree projects use programming to enhance the typography experience. Programming is not exclusively associated with the computer, since it is a sequence of pre-established instructions. It does not limit creation, but rather enriches it, because what is rationalized are the processes, which frees the user from the need to make definitive decisions [11].

Computational thinking makes it possible to retrieve the concept of a program as a sequence of instructions to be followed in a more intense manner, since, on a computer, programming occurs through instructions and parameters pre-established by the human being. The programming language allows for manipulation of the binary code and causes the machine to be able to interpret it for the purpose of forming images that are produced by the actions of adding and subtracting pixels [4].

These instructions allow the computer to create shapes and fill them with "matter" or even mold them as many times as necessary, simulating both the movement and the behavior and texture corresponding to a given body. Programming, besides allowing for shaping the pixels, influences how they are perceived, as it allows for movement simulation, making it more organic or mechanical. This relationship approximates the design to the user who manipulates it, since "We classify and categorize elements in the world by how they move or not move" [7] (p. 291).

However, not all computational programming is generative. Even if one uses external input variables or random generators, to be considered generative it is necessary that the algorithm produce non-repetitive variations, in other words, it must have a certain autonomy from its creator [8]. Nevertheless, the idea of loss of control on the part of the designer, the programmer or the artist associated with this type of work is not true. It is just distinct from the usual way of working, where the professional accurately plans within the layout elements such as position, shape and animation. With programming, the project author begins to determine the rules that the code will operate by for the creation of these elements.

According to Omine [8] (p. 86), this type of project opens up several possibilities, since "[…] through generative practice, computational design values the dynamic relationships of the elements, systems, and processes that involve that which is designed and not just the result of the project as a static entity." With the reality of computational algorithm, which allows typography not to be tied to a functional form or a static support, a discussion on whether this character mutant can be recognized by the reader is created. According to Maçãs [6], character recognition is not something entirely associated with the repetition of form, but it is a mental process, and generative programs enable their internal and external typographic variations.

5   Discussion

Considering the questions raised regarding programming in the digital context, it is perceived that the typographic object, when subjected to parameters pre-established by the designer, artist or programmer, can be thought of as an object that adapts to user action through dynamic behavior.

In Lettree, what is in evidence is the external distortion of the letter that encompasses the relation between the characters, which overlap and have their spacing modified, whereas in Anitype and Beowolf the changes that occur in the type are internal, and interfere directly in the form of the character. However in Beowolf, the one that dictates the parameters that interfere in the formation of the character is the designer him/herself, since in Anitype the one responsible for creating these parameters is the user, who becomes the programmer, which can lead to a total decharacterization of the character

The Beowolf and Lettree experiments are experiences in which one does not have control over the final result, since in the former, the programmer has no control over the character format and, in the latter, it is not possible to control which character or the order in which the user will choose it, and also the point at which the characters will connect, as displayed in Fig. 3, where we see the word "experiment" as different compositions.

Due to generative programming, typography can appear in a project in an autonomous manner, whose result depends on user inputs, which is the opposite of what occurs in the printed medium, where the planning of the support, the grid, the hierarchy, the type of letter and the spacing is all in the designer's hand.

6   Conclusions

Generative algorithms are driven by pre-established rules that allow designers to determine parameters that influence the formatting and appearance of the typographic character. Therefore, this research seeks to understand the potential of the generative code by allowing the user or reader to participate in the interaction with the typographic character itself. This possibility of participation acts as a way to offer to this user feedbacks executed as a function of data inputs captured from the physical environment generated by that interacting person, like a kind of game. These ephemeral characteristics attributed to the character, such as the distortion of the letter form, are imaginary and stand out to the detriment of their verbal qualities. In future studies, we intend to investigate the potential of programming software specialized in image production, such as Processing, in order to integrate the user with the algebraic code using typography as an interactive object.

References

1. Flusser, V.: O Mundo codificado: por uma filosofia da comunicação, 5th edn. Cosac Naify, São Paulo (2015)
2. Rocha, C.: Projeto tipográfico: análise e produção de fonts digitais, 3rd edn. Rosari, São Paulo (2005)
3. Bringhurst, R.: Elementos do Estilo Tipográfico, 3rd edn. Cosac Naify, São Paulo (2015)
4. Flusser, V.: Escrita, há um futuro para a escrita?. Anna Brume, São Paulo (2011)
5. Richardson, A.: Data-Driven Graphic Design: Creative Coding for Visual Comunication. Bloomsbury Publishing, London (2016)
6. Maçãs, C.: Comportamentos da Tipografia Generativa: Uma Proposta para um Tipo Generativo [master's thesis], 129 p. Universidade de Coimbra, Coimbra (2013)
7. Reas, C., Fry, B.: Processing: A Programming Handbook for Visual Designers and Artists. Mit Press, London (2014)
8. Omine, E.: Design gráfico computacional: comparação aplicada no projeto [master's thesis], 147 p. Universidade de São Paulo, São Paulo (SP) (2014)
9. Brandel, J.: Anitype. Homepage. http://www.anitype.com/. Accessed 28 Aug 2017
10. Piñón, R.M.: Caligraft. http://www.caligraft.com/. Accessed 12 Jan 2018
11. Gerstner, K.: Disenãr Programas. Gustavo Gili, Barcelona (1979)

Designing the Mechanical Parts of a Low-Cost Hand Rehabilitation CPM Device for Stroke Patients

Husam Almusawi[1(✉)], Syeda Adila Afghan[1], and Husi Géza[2]

[1] Faculty of Informatics, University of Debrecen, Debrecen, Hungary
{husam, adila}@eng.unideb.hu
[2] Faculty of Engineering, Department of Electrical Engineering
and Mechatronics, University of Debrecen, Debrecen, Hungary
husi@eng.unideb.hu

Abstract. The aim of this paper is to introduce hand rehabilitation end-effector-based mechanical design that can be used by the post stroke patients for improving their hand recovery process after the hand injury caused by stroke. The reason behind of creating this mechanical design is to help the therapists in treatment of the hand after the stroke and follow the natural movement of the fingers during the rehabilitation therapy. Stroke victims often lose proper function of at least one hand and fingers, experiencing delays in gripping and releasing ability. This paper is focused on designing the mechanical assemblies of the Continues Passive Motion (CPM) device. The mechanical design has been divided to two assemblies, one discussing the fingers rehabilitation CPM device after the hand injury or stroke, secondly, it presents the mechanism movement and the Degree of Freedom (DOF) of the hand rehabilitation machine. Strict clinical requirements are involved in designing this system like the Mechanical parts must be sterilisable, movable, safe, fingers attachment setting duration, and effective system having low cost profile.

Keywords: Hand rehabilitation system · Stroke
End-effector based mechanical design · CPM · Hand recovery

1 Introduction

The common deficits after stroke includes the disturbance in Activity of Daily Living (ADL) hemiparesis, spasticity and incoordination [1], the main cause behind the impairment of hand motor from severe to mild is the weakness [2], specifically in Finger extensors and wrist muscles [3], it is being surveyed that 75% of stroke survivors encounters the motor deficit of the upper limb [4], The paralysis of the upper limb is the most frequent consequence of brain injury, and very often the rehabilitation procedures deal with repetitive passive movements, with the aim to restore if possible the damaged functions, or alternatively to teach how to handle differently those functions [5], It became a fundamental research to recover the hand after stroke, robotic advancement has been developed to help the purpose of hand recovery, one of the main

© Springer International Publishing AG, part of Springer Nature 2019
J. Machado et al. (Eds.): HELIX 2018, LNEE 505, pp. 60–66, 2019.
https://doi.org/10.1007/978-3-319-91334-6_9

challenging for the researchers of designing a device that could be used for medical purposes, specifically for hand rehabilitation is how the device will be attached to the fingers and do the movement that could exercise them, and does all the movement and the functions.

2 Literature Review

Due to the complex structure of human hand the mechanical parts designing became the hardest part. One of the popular mechanical design is End-effector based device design, contacts a subject's limit only at its most distal part, it simplifies the structure of the device. However, it may complicate the control of the limb position in cases with multiple possible degrees of freedom [6], NEREBOT; is a cable driven system for sustaining and moving the forearm of the stroke patients during the rehabilitation therapy. This system provides 3-dimensional wide assistance to the motion of patient's arm which is pronation-supination, abduction-abduction and elbow-flexion based exercises. The robot is composed of wheel and overhead structure which is manually adjustable [7], MIME (Mirror Image Movement Enabler) device consists of 6 DOFs with unilateral and bilateral modes along with four therapy levels (active-assisted, passive, constrained, bimanual) [8], Assisted Rehabilitation and Measurement (ARM)-Guide was developed by W. Zev Rymer is used to guide the reaching point and measures the range of motion of the arm, targets the abnormal conditions and spastic reflexes. The device is a linear constraint that is instrumented with a six-axis load cell and an optical encoder [9], Currently, the cost of complex robotic devices makes them unsuitable for use in a domestic or local clinical setting. Developing lower cost devices could help to address this and the development of a prototype low cost robotic device for post-stroke rehabilitation in a local clinic or domestic setting is the focus of this research.

3 Methodology

The mechanical parts of CPM machine design are to mimic the natural hand movement. Some parts are designed with Autodesk Inventor software and some of them are drawn by hand. Other parts are printed with 3D printer and few others are milled, whereas, some other parts are hand made with mechanical equipment. The mechanical design is divided into two assemblies, the first assembly for the device's home design and the other assembly for the fingertips attachments. Figure 1 shown the general overview of the our CPM machine.

3.1 Assembly 1: CPM Machine

The first step of designing of the assembly for the basic movements of the CPM machine. as it shown in Fig. 2. The design meant to convert the rotational mechanical movement to linear mechanical movement, by taking the measurement for hand length and the distance between each finger.

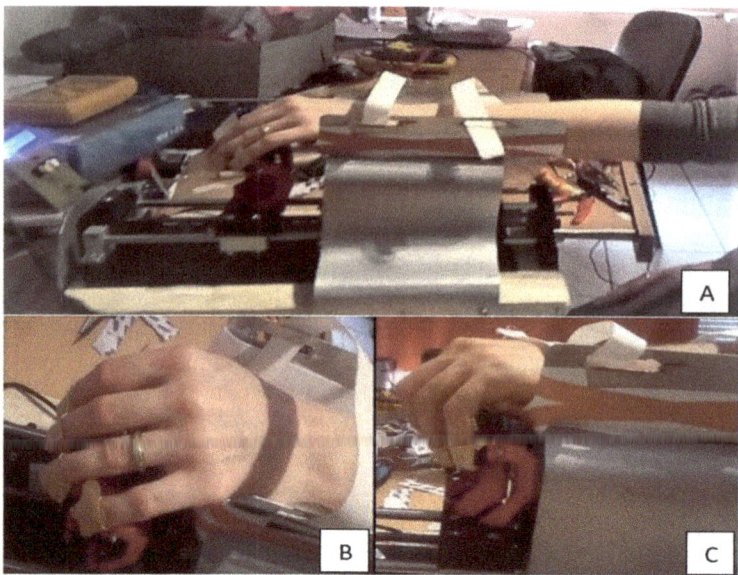

Fig. 1. General overview of Hand CPM machine from different views

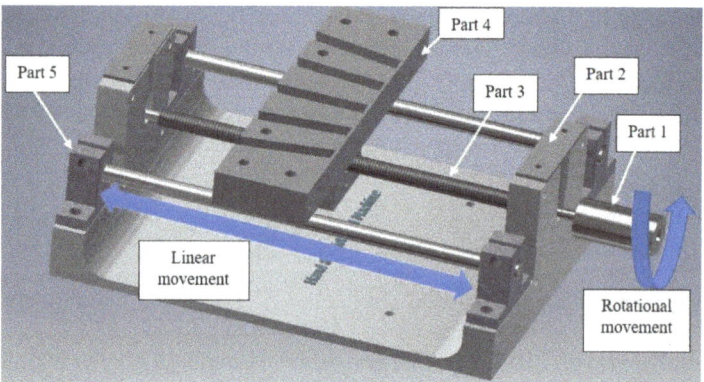

Fig. 2. Labelled parts of the assembly 1 and movement mechanism

Figure 2 demonstrating some the parts of the assembly. Part 1 is the coupler which is responsible to connect the motor shaft with the screw of Part 3, it is the coupler which is designed and milled at the milling lab of the Faculty of Electrical engineering and Mechatronics. Part 2: is the bearing holder which also will force the bearing to stay at the bearing home, this part can be replaced and positioned easily since it is removable. Part 3: is the screw of M8 size that is milled for making it suitable for different sizes. The next important fragment is the Part 4: is the fingertips holder which is designed to set the fingertips on it. This part has five movable slides inside which can use for left hand fingers and right-hand fingers, the reason behind these slides is to make the fingertips holder easy to move inside of it depending on the hand size. The mechanical design parts are easy to reassemble and removable in case some damage will happen to any part or for sterilization issues the therapist can sterilize each part individually if it needed.

3.2   Assembly 2: Fingertips Attachment

The second assembly design contains the fingertips and the attachments. They aimed to follow the natural movement of the fingers. Most of the parts are designed by free hand drawing, after that the parts are milled and few parts are just hand made by making them at milling lab. The main challenge was to design the mechanical joint as shown in Fig. 3 which demonstrates the step by step procedure for making the mechanical joints. The first step is to make the design for the joints by considering the kinematic movement of human fingers joint and the range of motion. The next step is to select the material, in this design we have used aluminium material due to its strong feature, availability and easy to sterilize. Then it is being cut by the cutting tool according to the design. The mechanical joint should have fingertip on the top of it and moveable with finger movement.

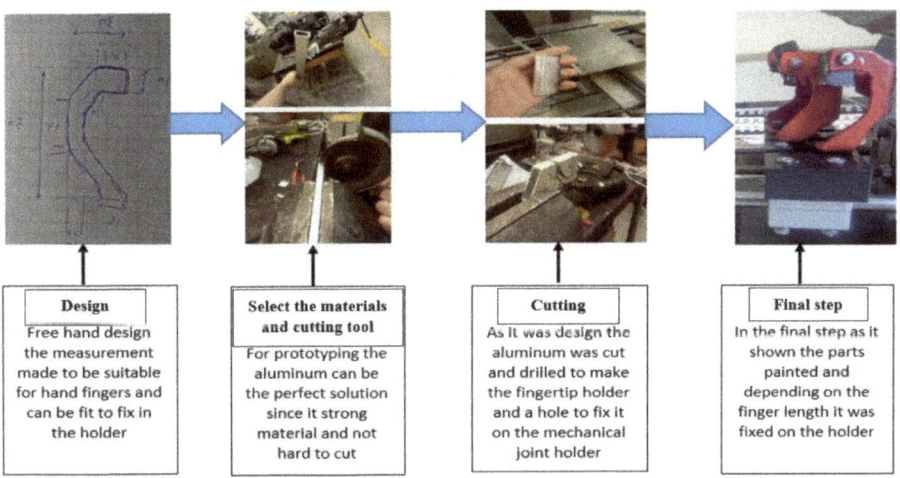

Design	Select the materials and cutting tool	Cutting	Final step
Free hand design the measurement made to be suitable for hand fingers and can be fit to fix in the holder	For prototyping the aluminum can be the perfect solution since it strong material and not hard to cut	As it was design the aluminum was cut and drilled to make the fingertip holder and a hole to fix it on the mechanical joint holder	In the final step as it shown the parts painted and depending on the finger length it was fixed on the holder

Fig. 3. Steps for making the mechanical joint for the finger tips

The hand forearm support holder which shown in Fig. 4 is also designed by hand sketching, the forearm support is attached on the top of the machine, it is important that in any kind of therapy, the hand must be fixed first to start the therapy. The hand forearm support is made to be perfect for different hand sizes and shapes therefore, the hand will sit comfortably on the forearm support. It is covered by medical bandages which could be replaced for sterilization purposes.

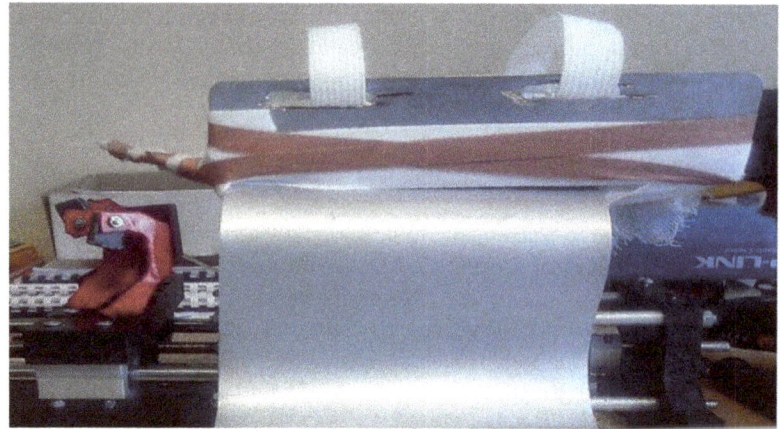

Fig. 4. Hand and Forearm Support

4 Mechanism Movement and DOF

To understand mechanism movement of the machine as illustrated in Fig. 5 which shows the DOF and ROM. In part A the system has one DOF which can be moved, when A will move, it will take all other parts with it like (B, C and D). In this movement after fixing the fingers on the fingertip which is placed on (E), the fingers will force to move in the same distance of (A), part C with same axis of DOF of A, but the reason of making to fix the finger holder on specific distance is the dependence on the fingers lengths and it is able to move within fixed distance inside the slides. D has other kind of DOF which can rotate around the centre, which make it to be more comfortable to the flow of natural movement of the hand. The mechanical joint (B) has

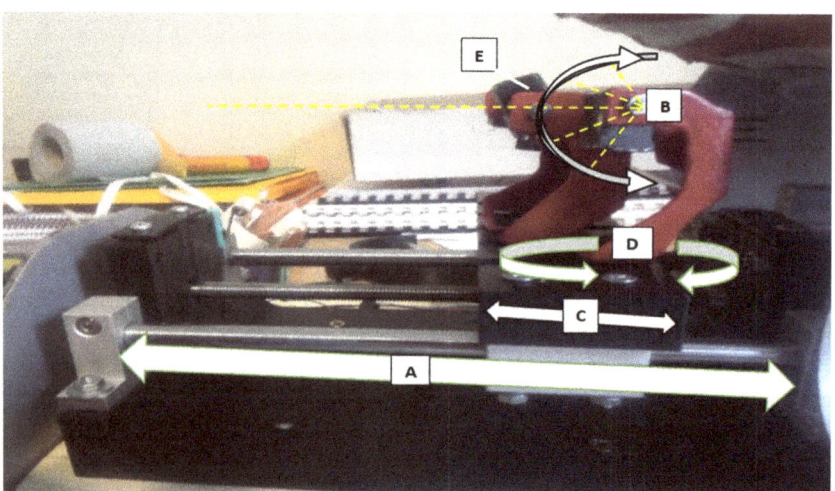

Fig. 5. The Mechanism Movement and DOF

different kind of DOF which can move up and down following the fingers movement in different degrees and different ROM depending on the distance of that part (A).

5   Actuation and Control

In actuation system the stepper motor is used, after setting fingers attachments and securing them, the first user (therapist) will selects the therapy type and, then the microcontroller receives the signal and process it. The processed data is sent to the motor driver which transforms it to the actuator in which connected coupler in Fig. 2 The software design is based on the use of two microcontrollers. These microcontrollers are connected to each other by serial communication. We have used Arduino Nano as first microcontroller and Arduino Uno as the second one. Figure 6 demonstrating the state diagram of system control flow.

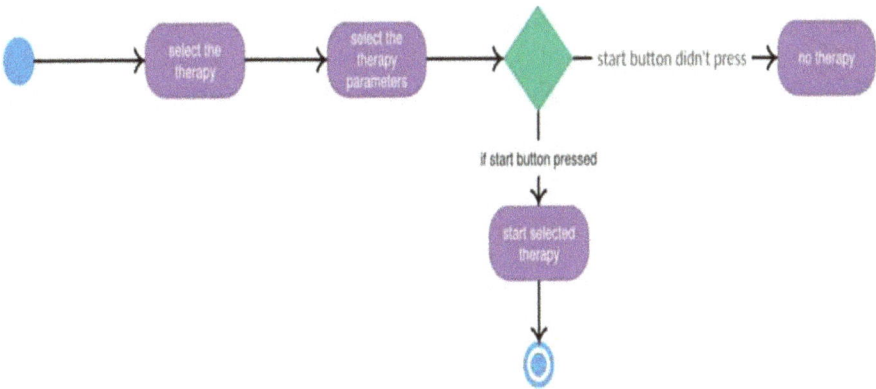

Fig. 6. State diagram of system control flow

6   Conclusion and Future Work

In general, the mechanical design made to be rigged in some parts and flexible and movable in other parts and always removable to sterilize it individually or to replace some parts in case of damage to avoid replacing the whole system. The hand structure could fit in this design and the fingers DOF would follow the movement performed by this CPM machine. The design is transportable. However, fingers could be set within few steps, the device are designed to be movable for sterilization issues therefore it can be set and reset easily. Opening and closing the fingers can be easily performed with good and smooth movement of the fingers. Device operates very well and has been successful during the experiments, still there are many possible open areas for improvement and future work. Some additional mechanical parts needed to add in future work for hand support.

References

1. Schaechter, J.D.: Motor rehabilitation and brain plasticity after hemiparetic stroke. Prog. Neurobiol. **73**(1), 61–72 (2004)
2. Bourbonnais, D., Vanden Noven, S.: Weakness in patients with hemiparesis. Am. J. Occup. Ther. **43**, 313–319 (1989)
3. Cauraugh, J., Light, K., Kim, S., et al.: Chronic motor dysfunction after stroke recovering wrist and finger extension by electromyography-triggered neuromuscular stimulation. Stroke **31**, 1360–1364 (2000)
4. Lawrence, E.S., Coshall, C., Dundas, R., Stewart, J., Rudd, A.G., Howard, R., Wolfe, C.D.: Estimates of the prevalence of acute stroke impairments and disability in a multiethnic population. Stroke **32**(6), 1279–1284 (2001)
5. Tappeiner, L., Ottaviano, E., Husty, M.L.: Cable-driven robot for upper limb rehabilitation inspired by the mirror therapy
6. Paweł, M., Jörg, F., Kurt, G.-H, Arne, J. T., Steffen, L.: A survey on robotic devices for upper limb rehabilitation. J. Neuroengineering Rehabil. **11**(1), 3 (2014)
7. Timmermans, A.A., Seelen, H.A.M., Willmann, R.D., Kingma, H.: Technology-assisted training of arm-hand skills in stroke: concepts on reacquisition of motor control and therapist guidelines for rehabilitation technology design. J. NeuroEngineering Rehabil. **6**(1), 1 (2009)
8. Stefan, H., Henning, S., Cordula, W., Anita, B.: Upper and lower extremity robotic devices for rehabilitation and for studying motor control. Current Opin. Neurol. **16**, 705–710
9. Reinkensmeyer, D.J., Schmit, B.D., Rymer, W.Z.: Assessment of active and passive restraint during guided reaching after chronic brain injury. Ann. Biomed. Eng. **27**, 805–814 (1999)

Simulating the Electrical Characteristics of Solar Panels Based on Practical Single-Diode Equivalent Circuit Model: Analyzing the Influence of Environmental Parameters on Output Power

Syeda Adila Afghan[1]([⊠]), Husam Almusawi[1], and Husi Géza[2]

[1] Faculty of Informatics, University of Debrecen, Debrecen, Hungary
{adila,husam}@eng.unideb.hu
[2] Faculty of Engineering, Department of Electrical Engineering
and Mechatronics, University of Debrecen, Debrecen, Hungary
husigeza@unideb.hu

Abstract. This research paper presents the simulation based study of two commercial solar modules to predict the real-time characteristics, modelled by the single mechanism five parameters (1M5P) solar equivalent circuit. The mathematical equations have been used to calculate the non-linear behavior of the system which is mandatory to characterize the I – V and P – V curves of the panels. The Photovoltaic systems of 300 W; Mono-crystalline and Poly-crystalline solar panels have been used to evaluate the model. Comprehensive simulations have been carried out to effectively asses the solar irradiation and temperature dependence in the performance of output power of the panels.

Keywords: Photovoltaic cell (PV) · 1M5P · I – V curve · P – V curve
LTSpice · MATLAB · Lambert - W function

1 Introduction

Global demand for energy resources continues to rise, specifically by the developing countries that embodies an expansion of industrialization, global economy and revised energy access. Nevertheless, the limited fossil fuels are utmost challenge to balance the supply and demand with the implementation of renewable energy sources. After hydro and wind power, Solar energy is the third most important renewable energy resource available in terms of global installed capacity [1]. According to the statistics report by Asian Photovoltaic Industry Association (APVIA) demonstrates that Asia now accounts for more than 65% of the global solar market, with 34.5 GW installed in China last year, followed by 8.6 GW in Japan and 4.1 GW in India [2]. Notwithstanding, European market has firm declaration regarding their policies for effective usage of Photovoltaic systems, even though added 4.6 GW to the grid in the year 2016. However, 4 European countries still belong to the global top 10 in 2017- Germany, UK, France and Spain [3]. When it comes to the selection of solar panels by the end

© Springer International Publishing AG, part of Springer Nature 2019
J. Machado et al. (Eds.): HELIX 2018, LNEE 505, pp. 67–74, 2019.
https://doi.org/10.1007/978-3-319-91334-6_10

users, still there is lack of understanding the PV system performance and power generation at its optimal level. For this purpose, modelling is the necessary tool to predict real time behavior of any solar cell. Many mathematical models have been developed to represent the highly nonlinear behavior of the PV cell [4]. In this case, we consider single diode model also known as Single mechanism five parameters (1M5P) equivalent circuit [5]. In this paper two commercial solar cells are being investigated to analyze the I – V and P – V characteristics of the Photovoltaic system. Moreover, to understand the effectiveness of the optimal power generation of the PV system, the panels are being tested under various temperature and solar irradiation levels.

2   Modelling of Solar Cell

To assess the electrical characteristics of any solar system, an equivalent circuit model is being tested at given set of operating conditions. Three models are used for modelling of the PV cell, array or module and amongst all the most preferable modelling system is Single Mechanism Five Parameters (1M5P) as shown in Fig. 1. The circuit consists of a current source, a single diode connected in parallel and two resistors; one in series and another in parallel [6]. It is mandatory to identify the five unknown parameters as the basic step towards the modelling. The datasheet includes the experimented data like short circuit current (Isc), circuit voltage (Voc) and maximum power points (MPP) as shown in Fig. 2 [7].

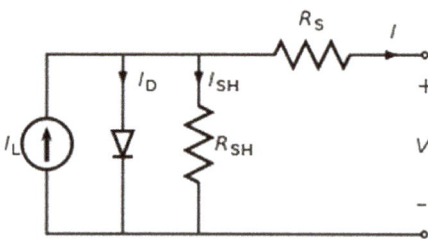

Fig. 1. Solar Equivalent Circuit

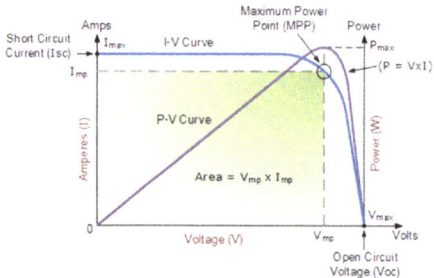

Fig. 2. Typical I – V characteristics of a Solar Cell

3 Mathematical Modelling of 1M5P Equivalent Circuit

It is important to derive the characteristics equations for a solar cell for calculating the five-parameters of an equivalent solar circuit, but due to limited data and extracted four boundary parameters are not sufficient to determine five-parameters, for this purpose the ideality factor a is suggested by many authors for reducing the number of equations [8], by applying Kirchhoff's law the equation is given as:

$$I = I_L - I_O - I_{sh} \tag{1}$$

The relationship between PV cell output current and terminal voltage is given by:

$$I = I_L - I_o \left[\exp\left(\frac{V + IR_s}{aV_T} \right) - 1 \right] - \frac{V + IR_s}{R_{sh}} \tag{2}$$

Along with Thermal Voltage;

$$V_T = n \frac{KT}{q} [V] \tag{3}$$

Where I_L denotes the photocurrent; I_O is the reverse saturation current; losses are represented by the R_S; leakage current across the p-n junction demonstrated by the R_{Sh}; and a as the ideality factor considering the Shockley diffusion theory. V_T represents the thermal voltage of the diode, k as Boltzmann constant; electron charge, q; Temperature, T; and number of cells in series, n.

The series resistor value is calculated by transforming the implicit expression as an explicit expression by an analytical technique. For making such calculation, Lambert W function is applied [9].

For calculating Rs

$$Rs = A \left(W_{-1}(Bexp(C)) - (D + C) \right) \tag{4}$$

For calculating Rsh:

$$R_{sh} = \frac{\left(V_{mp} - I_{mp}R_s \right)\left(V_{mp} - R_s\left(I_{sc} - I_{mp} \right) - aV_T \right)}{\left(V_{mp} - I_{mp}R_s \right)\left(I_{sc} - I_{mp} \right) - aV_T I_{mp}} \tag{5}$$

For calculating I_0:

$$I_0 = \frac{(R_{sh} + R_s)I_{sc} - V_{oc}}{R_{sh}\exp\left(\frac{V_{oc}}{aV_T}\right)} \tag{6}$$

For calculating I_L:

$$I_L = \frac{R_{sh} + R_s}{R_{sh}} I_{sc} \tag{7}$$

4 Methodology

For examining the electrical characteristics of solar panels, two commercial solar modules by Jinko Solar; JKM300P [10] and JKM300M [11] are tested. The electrical characteristics are given in the Table 1 which demonstrates the required data for insertion into the mathematical equations. The input parameters from the datasheet at (STC) Standard Test Conditions are 1000 W/m² incident normal radiance, Cell temperature is 25 °C and Air Mass AM 1.5 g.

Table 1. Electrical Specifications of JKM300P and JKM300M

JKM300P (Poly-crystalline)			JKM300M (Mono-crystalline)		
I − V Curve Ch:	*Parameters*	*Units*	*I − V Curve Ch:*	*Parameters*	*Units*
N	72	–	N	72	–
Voc	45.3	V	Voc	45.5	V
(Isc)	8.84	A	Isc	8.64	A
Vmpp	36.6	V	Vmpp	37.0	V
Pmax	300	W	Pmax	300	W
α	0.06	%/°C	alpha	0.05	%/°C
β	−0.31	%/°C	beta	−0.29	%/°C

First step is to calculate the unknown parameters (I_L, I_O, Rs, Rsh) extracted by the four boundary conditions, by the given set of equations. Furthermore, the ideality factor (a) should be considered as generic value as $a = 1.10$ [12]. The equations have been calculated on MATLAB due to the availability of Lambert W function in it. The parameters are calculated as follows:

- Approximating the ideality factor, a as 1.10
- Calculating the Series Resistance Rs with Eq. (4)
- Calculating the Shunt Resistance Rsh with Eq. (5)
- Calculating the Saturation Current Io with Eq. (6)
- Calculating the Photodiode Current I_L with Eq. (7)

The evaluated parameters are listed in Table 2 for both modules, where the unknown parameters have been calculated by the defined equations and pushing the ideality factor as constant and independent.

Table 2 Calculated parameters of JKM300P and JKM300M

JKM300P (Poly-Crystalline)		JKM300M (Multi-Crystalline)	
a	1.10	a	1.10
Ipv (A)	8.8494	Ipv (A)	8.6416
Io (A)	7.5642e−09	Io (A)	6.8337e−09
Rs (Ω)	0.3130	Rs (Ω)	0.2896
Rsh (Ω)	292.9495	Rsh (Ω)	1.5906e+03

5   Simulation and Results

The computed data are being modelled as an equivalent circuit for simulation purpose in the Linear Technology's LTSpice (Linear Technology Simulation Program with Integrated Circuit Emphasis) software. Numerous simulations have been performed based on two variations to substantiate the real-time performance of the panels (i) Solar irradiation on I – V and P - V curves (ii) The Temperature dependence P – V curves has been modelled to examine the PV cell response and the characteristics points to adequate the output performance of the cell.

5.1   Influence of Solar Irradiation

The plots of I – V and P – V characteristics curve have been produced by varying certain parameters one at a time keeping other parameters constant at STC. Figures 3 and 4 describes the I – V and P – V curve for varying solar irradiance value G from 200 W/m^2 to 1000 W/m^2 by keeping temperature Temp constant (at STC 25 °C) for both JKM300P and JKM300M panels. On observing the curve, it is indicated that by increasing the solar irradiation it also increases the cell current proportionally, it means that generated current depends on the incident light, whereas the slight increment on cell voltage is also observed. The maximum power produced by the JKM300P at 25 °C temperature is 280.56 W, whereas for JKM300M is 281.13 W.

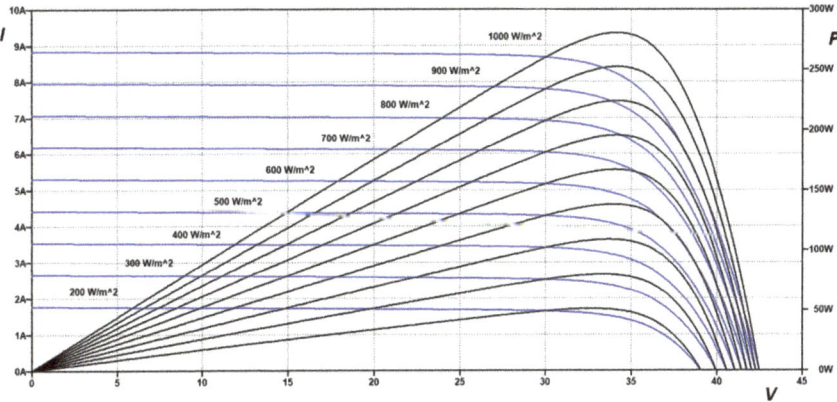

Fig. 3. I – V and P – V Curves of JKM300P simulated with LTSpice for Solar Irradiation levels from 200 W/m^2 to 1000 W/m^2 at 25 °C

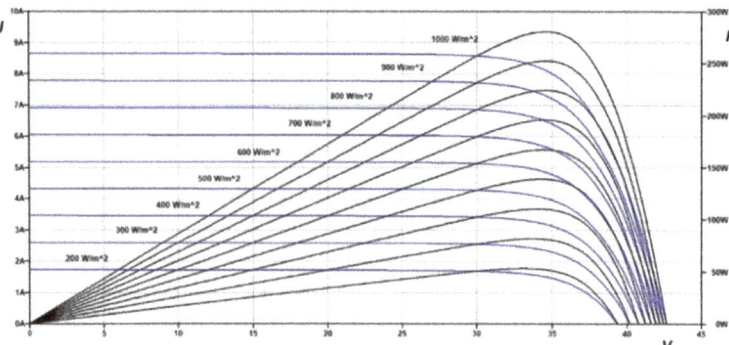

Fig. 4. I – V and P – V curves of JKM300 M simulated with LTSpice for Solar Irradiation levels from 200 W/m² to 1000 W/m² at 25 °C

5.2   Influence of Temperature Dependence

In this simulation, the temperature is being varied from 15 °C till 85 °C while keeping solar irradiation G constant (at STC 1000 W/m²). Figures 5 and 6 describes P – V characteristics curves for both modules for varying temperature value Temp. It is analyzed that if temperature increases the cell voltage decreases, hence effecting the output power of the solar panels.

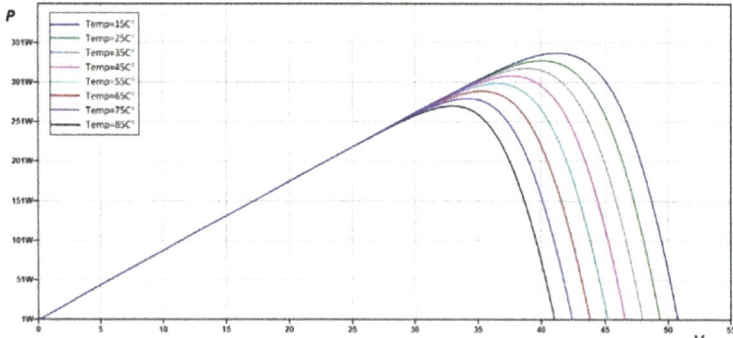

Fig. 5. P – V curves of JKM300P simulated with LTSpice for Temperature variation between 15 °C to 85 °C at 1000 W/m²

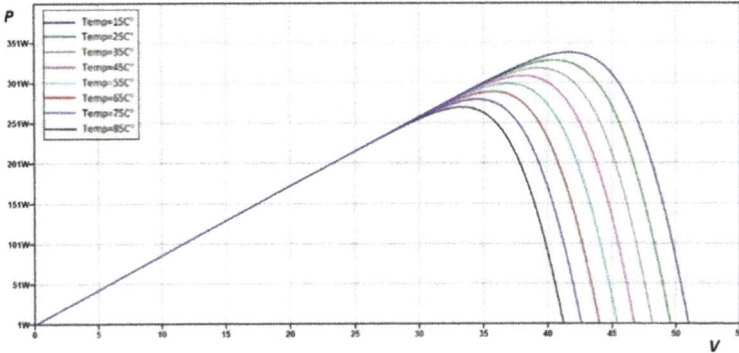

Fig. 6. P - V curves of JKM300M simulated with LTSpice for Temperature variation between 15 °C to 85 °C at 1000 W/m^2

6   Conclusion

In this research work, two commercial photovoltaic systems have been simulated by using the simple and accurate method under different working conditions for verifying the realistic modelling. The single mechanism five parameters diode equivalent circuit has been chosen to investigate the I – V and P – V curves for both modules. The manufacturer's datasheet is the only available source to obtain the unknown parameters which were calculated on MATLAB software. The obtained data are being simulated on LTSpice simulator by varying environmental conditions like solar irradiation and temperature dependence to examine the maximum output power of both panels. As a result, this study provides an insight to the end users for constructing realistic models and to be able to estimate the required performance when planning to produce their own solar panel system.

References

1. Blaabjerg, F., Lonel, D.M.: Renewable energy devices and systems – State-of-the-Art technology, research and development, challenges and future trends. Electr. Power Compon. Syst. **43**(12), 1319–1328 (2015)
2. PV Magazine, June 2017. https://www.pv-magazine.com/2017/05/26/apvia-sees-steady-q1-growth-in-asian-pv/
3. REN21. Renewables Global Status Report 2017: REN21 Secretariat: REN21 Publications, Paris, May 2017. http://www.ren21.net/gsr-2017/
4. Belhaouas, N., Cheikh, M.S.A., Malek, A., Larbes, C.: Matlab-Simulink of photovoltaic system based on a two-diode model simulator with shaded solar cells. Revue des Energies Renouvelables **16**(1), 65–73 (2013)
5. Yetayew, T.T., Jyothsna, T.R.: Improved single-diode modeling approach for photovoltaic modules using datasheet. In: Annual IEEE India Conference (INDICON), Bombay, pp. 1–6 (2013)

6. Azzouzi, M., Popescu, D., Bouchahdane, M.: Modeling of electrical characteristics of photovoltaic cell. J. Clean Energy Technol. **4**(6), 414–419 (2016)
7. Tobnaghi, D.M., Madatov, R., Naderi, D.: The effect of temperature on electrical parameters of solar cells. Int. J. Adv. Res. Electr. Electron. Instrum. Eng. (IJAREEIE) **2**(12), 6404–6407 (2013)
8. Villalva, G.M., Gazoli, R.J., Filho, R.E.: Comprehensive approach to modeling and simulation of photovoltaic arrays. IEEE Trans. Power Electron. **24**(5), 1198–1208 (2009)
9. Javier, C., Santiago, P., Carlos, M.D.: Explicit expressions for solar panel equivalent circuit parameters based on Analytical Formulation and the Lambert W- Function. Energies **7**(7), 4098–4115 (2014)
10. Jinko Solar: JKM300P - Poly-Crystalline, April 2017. https://www.jinkosolar.com/ftp/US-MKT-300P_v1.0_rev2013%B4+%A2%B4+%A21%B4+%A2%B4+%A2.pdf
11. Jinko Solar: JKM300M – Mono- Crystalline, April 2017. https://www.jinkosolar.com/ftp/EN%20JKM300M-72.pdf
12. Cuce, P.M., Cuce, E.: A novel model of photovoltaic modules for parameter estimation and thermodynamic assessment. Int. J. Low-Carbon Technol. **7**, 159–165 (2011)

Modified Hildreth's Method Applied in Multivariable Model Predictive Control

Marek Kubalcik[1(✉)], Vladimir Bobal[1], and Tomas Barot[2]

[1] Department of Process Control, Faculty of Applied Informatics,
Tomas Bata University in Zlín, Nad Stráněmi 4511, 760 05 Zlín, Czech Republic
{kubalcik, bobal}@utb.cz
[2] Department of Mathematics with Didactics, Faculty of Education,
University of Ostrava, Mlynska 5, 701 03 Ostrava, Czech Republic
Tomas.Barot@osu.cz

Abstract. A significantly important part of model predictive control (MPC) with constraints are algorithms of numerical optimization. Reduction of the computational complexity of the optimization methods has been widely researched. The reason is that in certain cases of predictive control of fast dynamics processes an optimization algorithm may not be feasible within the sampling period time. This situation occurs particularly when requirements on control are more complex, e.g. in the multivariable control. Hildreth's method based on the dual-problem-optimization-principles has been widely applied and implemented in model predictive control. However, modifications of this method are not widely described in context of model predictive control. This paper proposes a modification of Hildreth's method, which reduces the computational complexity of the algorithm, and its application in the multivariable predictive control.

Keywords: Model predictive control · Multivariable control
Quadratic programming · Hildreth's method · Algorithm-complexity

1 Introduction

Model predictive control (MPC) [1, 2] has been widely applied in controlling of industrial processes thanks to its ability to deal with control difficulties such as constrained variables, time-delay and nonlinearity. Predictive control is also one of the most effective approaches for control of multivariable systems [3]. An advantage of model predictive control is that the multivariable systems can be handled in a straightforward manner.

A structure of a predictive controller can be divided into two parts: a predictor [3] and an optimizer [4]. The basic idea of MPC is to use a model of a controlled process to predict future outputs of the process. A trajectory of future manipulated variables is given by solving an optimization problem [5] incorporating a suitable cost function [5] and constraints [5]. Only the first element of the obtained control sequence is applied. The whole procedure is repeated in following sampling period. This principle is known as the receding horizon strategy [1].

© Springer International Publishing AG, part of Springer Nature 2019
J. Machado et al. (Eds.): HELIX 2018, LNEE 505, pp. 75–81, 2019.
https://doi.org/10.1007/978-3-319-91334-6_11

A significantly important part of the model predictive control with constraints [2] are algorithms of numerical optimization [6]. Reduction of the computational complexity [7] of the optimization methods has been widely researched [8, 9]. Hildreth's method [10, p. 67–68] based on the dual-problem-optimization-principles [6] has been widely applied and implemented in the model predictive control. However, modifications of this method are not widely described in context of the model predictive control. The main aim of this paper is to propose a modification of Hildreth's method, which reduces the computational complexity of the algorithm, and its application in the multivariable predictive control. Simulation results of control with implemented modification in the optimization part of the predictive controller are shown in this paper. Also a comparison of computational demands on the predictive controller both with classical and modified optimization is presented. This comparison proves improvement of the computational efficiency when using the proposed modification.

2 Implementation of Multivariable Predictive Control

Further a system with two inputs and two outputs will be considered. The two – input/two – output (TITO) processes [3] are the most often encountered multivariable processes in practice and many processes with inputs/outputs beyond two can be treated as several TITO subsystems [3].

A mathematical model of TITO process can be considered in the form of matrix fraction

$$G(z^{-1}) = A^{-1}(z^{-1})B(z^{-1}) \tag{1}$$

This model can be transcribed to difference equations [3] which can be used for computation of the systems output predictions [3] over minimum N_1 and maximum N_2 horizons. Further the polynomials in the matrices A and B will be considered to be of the second order.

Further we will consider constraints of the manipulated variables. In a constrained case, a vector of future increments of the manipulated variables is given by solving an optimization problem given by (2), where y, u, and w are vectors of controlled, manipulated and reference signals. The matrices I, P, G leads from the prediction equations [3].

$$\Delta\mathbf{u} = \arg\min\left\{\frac{1}{2}\Delta\mathbf{u}^T\mathbf{H}\Delta\mathbf{u} + \mathbf{b}^T\Delta\mathbf{u}|\mathbf{M}\Delta\mathbf{u} \le \gamma\right\};$$

$$\mathbf{H} = \mathbf{G}^T\mathbf{G} + \mathbf{I}; \mathbf{b} = \mathbf{G}^T\left(P\begin{bmatrix} \mathbf{y}(k) \\ \mathbf{y}(k-1) \\ \mathbf{y}(k-2) \\ \Delta\mathbf{u}(k-1) \end{bmatrix} - \mathbf{w}\right) \tag{2}$$

Only the first element of the obtained control sequence Δu is applied. The vector Δu has a length that corresponds to the control horizon N_u.

3   Quadratic Programming Optimization in MPC

Optimization problem (2) [4] is then solved numerically by quadratic programming in each sampling period. In practical applications of the multivariable MPC, Hildreth's method is frequently applied for the quadratic programming optimization solution, as it was recommended in [10]. The Hildreth's method is based on numerical iterations, in which particular subresults are gradually improving. The method can be categorized as a dual optimization method [5].

The first setting of dual vector variable (3) can be considered in form of a multi-dimensional extreme [6] without any constraints. The dual solution (3, 4), which is further directly applied into the current step of the discrete MPC, is transformed into the form of the primary solution [6] (5).

$$d = \begin{bmatrix} d_1 & \cdots & d_{N_u} \end{bmatrix}^T = \arg\min \left\{ \frac{1}{2} d^T N d + o^T d \,|\, d \geq 0 \right\}; d \in \mathcal{R}^{N_u,1} \qquad (3)$$

$$N = MH^{-1}M^T; N_{ij} \in N \in \mathcal{R}^{N_u,N_u}; o = MH^{-1}b^T + \gamma; o_{ij} \in o \in \mathcal{R}^{N_u,1} \qquad (4)$$

$$\Delta u(k) = -H^{-1}(M^T d + b^T) \qquad (5)$$

In Hildreth's method, the first computation of dual variable (3) is further improved in the iteration based cycle [10, p. 67–68]. An index of iteration is denoted as ω and the improved solution of the dual variable is defined by (6).

$$d(\omega) = \begin{bmatrix} d_1(\omega) & \cdots & d_{N_u}(\omega) \end{bmatrix}^T \qquad (6)$$

As MPC has specific forms of matrices in the definition of the optimization problem, the multidimensional extreme does not occur frequently. For the constrained manipulated variable, the matrices M and γ have a form (7), where T is a lower triangular matrix.

$$M = \begin{bmatrix} -T \\ T \end{bmatrix}; \gamma = \begin{bmatrix} [u(k-1) - u_{\min} & \cdots & u(k-1) - u_{\min}]^T \\ [u_{\max} - u(k-1) & \cdots & u_{\max} - u(k-1)]^T \end{bmatrix} \qquad (7)$$

4   Proposal of Modified Hildreth's Numerical Method

The computational complexity of the optimization problem of the quadratic programming (2) is significantly increased by constraints in the MPC and by higher values of the horizons. In case of the multivariable control, the descreasing of the algorithm-complexity is particularly important. In [11], an explicit approach is presented which is based on off-line pre-computed optimal solutions. In other researches, the numerically based optimization has not been so widely proposed as in this explicit type of the optimization. Some other researches have tried to modify the numerical method itself, e.g. in the algorithms of the Sequency Linear Programming [9] or Interior Point Methods [8].

In this paper, a modification of the dual Hildreth's method, which is recommended for using in multivariable predictive control in [10], is proposed. The proposed modification is based on an addition of a new condition (8) in the iterative algorithm. The proposal assumes, that the new condition (8) can occur in the computational cycle of the algorithm more often than the standard exit condition based on equality of the last subresults.

$$M\left[-H^{-1}(M^T d(\omega) + b^T)\right] \leq \gamma \tag{8}$$

However, the new condition for aborting of the numerical cycle would spent a lot of operations with the evaluation of all included matrix operations. Therefore, condition (8) can be rewritten into (9) by means of the dual theory [5]. This form contains rapidly lower number of relation operations than condition (8). The same results will be achieved with both conditions applied in the algorithm

$$d(\omega) \geq \begin{bmatrix} 0 & \cdots & 0 \end{bmatrix}^T; d(\omega) \in \mathcal{R}^{N_u, 1} \tag{9}$$

5   Simulation Results

For the purpose of comparison, MPC of TITO process (1) with polynomial matrices (10, 11) both with and without the proposed modification was simulated in MATLAB.

$$A(z^{-1}) = \begin{bmatrix} 1 - 1.3264z^{-1} + 0.3271z^{-2} & 0.024z^{-1} - 0.0029z^{-2} \\ -0.0711z^{-1} + 0.0759z^{-2} & 1 - 1.0911z^{-1} + 0.134z^{-2} \end{bmatrix} \tag{10}$$

$$B(z^{-1}) = \begin{bmatrix} 0.2983z^{-1} - 0.097z^{-2} & 0.093z^{-1} + 0.0682z^{-2} \\ 0.1755z^{-1} + 0.0688z^{-2} & 0.1779z^{-1} + 0.1065z^{-2} \end{bmatrix} \tag{11}$$

The effect of the proposed modification is apparent only in a constrained case. The constraints are as follows: $u_{min} = -2$, $u_{max} = 2$. The quality of the control can be evaluated by means of criteria (12, 13).

$$J_1 = \sum_k [\Delta u_1(k)]^2 + \sum_k [\Delta u_2(k)]^2 \tag{12}$$

$$J_2 = \sum_k [w_1(k) - y_1(k)]^2 + \sum_k [w_2(k) - y_2(k)]^2 \tag{13}$$

The control quality criterions J_1(12) and J_2(13) were measured as well as the total number of all floating point operations performed during run of both MPC algorithm without modifications (F) and the modified algorithm (F^*). The numbers of floating-point-operations (flops) were determined for each program-command in the MATLAB [7].

$$N_1 = 1; \; N_u = N_2 = 5 + 5\varepsilon; \; \varepsilon \in \langle 0; 8 \rangle \qquad (14)$$

The complexity of the realized predictive control (Fig. 1) can be then determined using the polynomial regression [12] and can be expressed using the complexity function $O = O(\mu)$ described in detail in [7]. A problem-dimension-parameter μ is equal to both horizons N_2 and N_u.

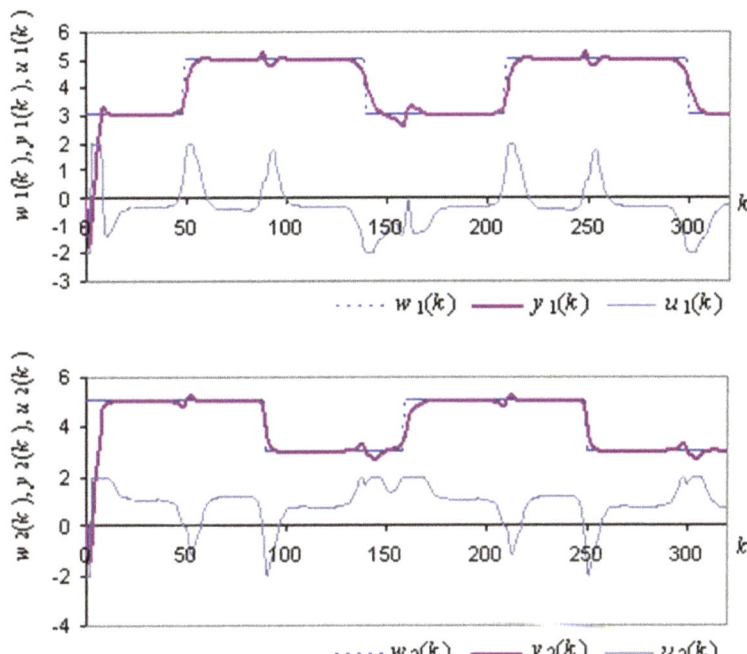

Fig. 1. Constrained MPC of two-input and two-output process

In Table 1, the numbers of floating-point-operations were analyzed for case of the MPC algorithm (F) and for the modified MPC ($F*$) by means of the complexity functions $O = O(\mu)$ expressed for all regression-polynomial-aproximated values of F and $F*$ for variable settings of parameters (14). As can be seen in Table 1, both algorithms have approximately similar control quality; however, the complexity of the proposed modified approach $O^*(\mu) = 276494\mu^3$ is significantly lower than complexity of the MPC without the modification $O(\mu) = 854310\mu^3$.

Table 1. Control quality and complexity using MPC and modified MPC

	MPC			MPC with modification		
μ	J_1	J_2	F [flops]	J_1	J_2	F^* [flops]
5	92,3328	268,4459	2,E+07	92,3334	268,4449	1,E+07
10	88,4763	262,1586	2,E+08	88,4768	262,1578	1,E+08
15	87,6638	261,5773	8,E+08	87,6642	261,5767	4,E+08
20	87,5846	261,5629	2,E+09	87,585	261,5623	1,E+09
25	87,5877	261,542	5,E+09	87,588	261,5415	2,E+09
30	87,6066	261,4577	1,E+10	87,6068	261,4572	4,E+09
35	87,6489	261,3348	2,E+10	87,649	261,3342	7,E+09
40	87,7746	261,1032	3,E+10	87,7747	261,1026	1,E+10
45	88,102	260,6048	4,E+10	88,1021	260,6042	2,E+10

6   Conclusion

The proposal of the modification of the Hildreth's dual optimization method was presented and efficiently implemented in the MPC of the TITO process. By the simulation in MATLAB, the decreased number of the floating-point-operations of the control algorithm was proved when using the proposed modification. Constrained multivariable MPC causes a higher computational complexity of the MPC's algorithm. Hildreth's method is an appropriate numerical method for the optimization part of MPC which can be further modified, as it was presented in this paper. The proposed modification presents possibilities of improving of the optimization methods using the extended rules of the dual optimization theory. Then these modifications can provide decreasing of number of operations in the optimization part of MPC.

References

1. Kwon, W.H.: Receding Horizon Control: Model Predictive Control for State Models. Springer, Heidelberg (2005). ISBN 1-84628-024-9
2. Corriou, J.P.: Process Control: Theory and Applications. Springer, London (2004). ISBN 1-85233-776-1
3. Kubalcik, M., Bobal, V., Barot, T.: Predictive control of two-input two-output system with non-minimum phase. In: 31st European Conference on Modelling and Simulation, pp. 342–347. European Council for Modelling and Simulation (2017). ISBN 978-0-9932440-4-9
4. Lee, G.M., Tam, N.N., Yen, N.D.: Quadratic Programming and Affine Variational Inequalities: A Qualitative Study. Springer, Boston (2005). ISBN 0-387-24277-5
5. Dostal, Z.: Optimal Quadratic Programming Algorithms: With Applications to Variational Inequalities. Springer, Boston (2009). ISBN 978-0387848051
6. Luenberger, D.G., Ye, Y.: Linear and Nonlinear Programming. Springer, Boston (2008). ISBN 978-0-387-74502-2
7. Hunger, R.: Floating point operations in matrix-vector calculus (Version 1.3). Technical report. Technische Universität München, Associate Institute for Signal Processing (2007). https://mediatum.ub.tum.de/doc/625604/625604.pdf

8. Hertog, D.: Interior Point Approach to Linear, Quadratic, and Convex Programming: Algorithms and Complexity. Kluwer Academic Publishers, Dordrecht (1994)
9. Rao, S.S.: Engineering Optimization: Theory and Practice. Kluwer Academic Publishers, Hoboken (2009). ISBN 978-0-470-18352-6
10. Wang, L.: Model Predictive Control System Design and Implementation Using MATLAB. Springer, London (2009). ISBN 978-1-84882-330-3
11. Ingole, D., Holaza, J., Takacs, B., Kvasnica, M.: FPGA-based explicit model predictive control for closed loop control of intravenous anesthesia. In: 20th International Conference on Process Control (PC), pp. 42–47. IEEE (2015). ISBN 978-1-4673-6627-4
12. Krivy, I., Tvrdik, J., Krpec, R.: Stochastic algorithms in nonlinear regression. Comput. Stat. Data Anal. **33**(12), 277–290 (2000)

Analytic Model Predictive Controller in Simple Symbolic Form

Daniel Honc$^{(\boxtimes)}$ and Milan Jičínský

University of Pardubice, nám. Čs. legií 565, 530 02 Pardubice, Czech Republic
`daniel.honc@upce.cz`

Abstract. Paper deals with an analytic solution of Model Predictive Controller in simple symbolic form. Process is approximated with a first order dynamical model. Special choice of prediction and control horizons is considered, so the symbolic solution is still applicable, and the controller has interesting "predictive" feature in case of known future set-point course. Such a controller can be used in simple devices like PLCs or microcontrollers without need of matrix operations. Its advantage is that the controller reacts to the process model parameters and penalty parameter change so the control can be very fast and efficient even in adaptive manner.

Keywords: Model Predictive Controller · MPC · First order process model
Analytical solution · Symbolic

1 Introduction

Model Predictive Control (MPC) is very spread and popular time-domain optimization-based controller design methodology. Plenty different process models, cost functions (performance indexes), analytical and numerical solutions with lot of choices and parameters give arise huge family of methods studied from theoretical point of view but also being applied in industry in different versions for decades. Strong potential of MPC methods lies in natural and graspable formulation (MPC origins can be found in industry), ability to control large systems with constraints and transport delays and work with known future disturbances or set-points – see [1, 2].

Connections between MPC and existing analytic control methods has been published in [3]. Standard analytic control methods can be considered as a special case of MPC. Both methods are identical in unconstrained case but MPC does not exhibit poor performance of analytical control methods when a constraint is present. In [4] class of nonlinear and linear plants for which MPC admits an analytical solution was characterized. Optimal control sequence takes significantly less time to calculate in case of analytical solution. On the other hand, quadratic programming can handle different types of constraints.

Presented approach is slightly different – we want to get analytical MPC which is parametrized with model parameters and penalty parameter and does not require any matrix operations or numerical methods. Then such a controller can be used in simple systems with less memory and computational power. Because of its parametric feature the controller reacts to process changes (changes in model parameters) and penalty

J. Machado et al. (Eds.): HELIX 2018, LNEE 505, pp. 82–88, 2019.
https://doi.org/10.1007/978-3-319-91334-6_12

parameter so the user can tune the controller online iteratively or use it in adaptive manner. The complexity of analytical controller depends on choice of prediction and control horizons. The challenge was to find such horizons that the formula is still simple and the controller has predictive behavior – it will react to the set-point change in advance.

2 Controller Derivation

2.1 Performance Index

Control aim is set-point following and disturbance rejection together with performance index minimization - to minimize sum of the squares of the future control error and future control changes (control moves)

$$J = \sum_{j=N_1}^{N_2} (\hat{y}(k+j) - w(k+j))^2 + q \sum_{j=1}^{N_u} \Delta u(k+j-1)^2. \tag{1}$$

Controlled variable predictions based on actual state and future control actions are needed. We will use dynamical process model for derivation of such a predictor.

2.2 Process Model and Predictor

We consider first order process model with time constant T and gain Z

$$T\frac{dy}{dt} + y = Zu . \tag{2}$$

After discretization with sample time T_s we get discrete-time process model

$$y(k) + a_1 y(k-1) = b_1 u(k-1), \quad a_1 = -e^{\frac{-T_s}{T}}, \quad b_1 = Z(a_1 + 1) . \tag{3}$$

We suppose disturbance model as a random walk process - summation of correlated prediction error e by polynomials C/A. The result is that the controller has integrating character (Fig. 1).

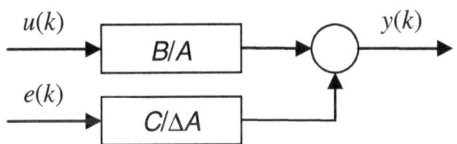

Fig. 1. Discrete-time process and disturbance model

The process model equation is

$$y(k) = \frac{B}{A}u(k) + \frac{C}{\Delta A}e(k) , \qquad (4)$$

where prediction error $e(k) = y(k) - \hat{y}(k)$ and process polynomials $A = 1 + a_1 z^{-1}$ and $B = b_1 z^{-1}$, second order filtering polynomial $C = 1 + c_1 z^{-1} + c_2 z^{-2}$ and $\Delta A = (1 + a_1 z^{-1})(1 - z^{-1}) = 1 + (a_1 - 1)z^{-1} - a_1 z^{-2}$. The Δ is backward difference operator $\Delta = 1 - z^{-1}$. By multiplying (4) with ΔA we get prediction equations in matrix form as

$$\underbrace{\begin{bmatrix} 1 & 0 & 0 & 0 \\ a_1 - 1 & 1 & 0 & 0 \\ -a_1 & a_1 - 1 & 1 & 0 \\ 0 & -a_1 & a_1 - 1 & 1 \end{bmatrix}}_{\mathbf{A}_p} \begin{bmatrix} \hat{y}(k+1) \\ \hat{y}(k+2) \\ \hat{y}(k+3) \\ \hat{y}(k+4) \end{bmatrix} = \underbrace{\begin{bmatrix} b_1 & 0 & 0 & 0 \\ 0 & b_1 & 0 & 0 \\ 0 & 0 & b_1 & 0 \\ 0 & 0 & 0 & b_1 \end{bmatrix}}_{\mathbf{B}_p} \begin{bmatrix} \Delta u(k) \\ \Delta u(k+1) \\ \Delta u(k+2) \\ \Delta u(k+3) \end{bmatrix}$$

$$+ \underbrace{\begin{bmatrix} -a_1 + 1 & a_1 \\ a_1 & 0 \\ 0 & 0 \\ 0 & 0 \end{bmatrix}}_{\mathbf{A}_m} \begin{bmatrix} y(k) \\ y(k-1) \end{bmatrix} + \underbrace{\begin{bmatrix} c_1 & c_2 \\ c_2 & 0 \\ 0 & 0 \\ 0 & 0 \end{bmatrix}}_{\mathbf{C}_m} \begin{bmatrix} e(k) \\ e(k-1) \end{bmatrix}$$

$$(5)$$

We suppose the prediction and control horizons $N_1 = 1$, $N_2 = N_u = 4$ and optimal predictions - zero future prediction error. By multiplying (5) with $\mathbf{A}_p^{-1}$ we get predictor in matrix form as

$$\underbrace{\begin{bmatrix} \hat{y}(k+1) \\ \hat{y}(k+2) \\ \hat{y}(k+3) \\ \hat{y}(k+4) \end{bmatrix}}_{\mathbf{Y}} = \underbrace{\begin{bmatrix} b_1 & 0 & 0 & 0 \\ A_1 b_1 & b_1 & 0 & 0 \\ A_2 b_1 & A_1 b_1 & b_1 & 0 \\ A_3 b_1 & A_2 b_1 & A_1 b_1 & b_1 \end{bmatrix}}_{\mathbf{G} = \mathbf{A}_p^{-1}\mathbf{B}_p} \underbrace{\begin{bmatrix} \Delta u(k) \\ \Delta u(k+1) \\ \Delta u(k+2) \\ \Delta u(k+3) \end{bmatrix}}_{\mathbf{U}} + \underbrace{\begin{bmatrix} A_1 & a_1 \\ A_2 & A_1 a_1 \\ A_3 & A_2 a_1 \\ A_4 & A_3 a_1 \end{bmatrix}}_{\mathbf{F}_y = \mathbf{A}_p^{-1}\mathbf{A}_m} \underbrace{\begin{bmatrix} y(k) \\ y(k-1) \end{bmatrix}}_{\mathbf{Y}_m}$$

$$+ \underbrace{\begin{bmatrix} c_1 & c_2 \\ A_1 c_1 + c_2 & A_1 c_2 \\ A_2 c_1 + A_1 c_2 & A_2 c_2 \\ A_3 c_1 + A_2 c_2 & A_3 c_2 \end{bmatrix}}_{\mathbf{F}_e = \mathbf{A}_p^{-1}\mathbf{C}_m} \underbrace{\begin{bmatrix} e(k) \\ e(k-1) \end{bmatrix}}_{\mathbf{E}_m} \qquad , \quad (6)$$

$$A_1 = -a_1 + 1, \quad A_2 = a_1^2 - a_1 + 1, \quad A_3 = -a_1^3 + a_1^2 - a_1 + 1, \quad A_4 = a_1^4 - a_1^3 + a_1^2 - a_1 + 1$$

$$\mathbf{Y} = \mathbf{G}\mathbf{U} + \mathbf{F}_y\mathbf{Y}_m + \mathbf{F}_e\mathbf{E}_m = \underbrace{\mathbf{G}\mathbf{U}}_{\text{forced response}} + \underbrace{\underbrace{\begin{bmatrix} \mathbf{F}_y & \mathbf{F}_e \end{bmatrix}}_{\mathbf{F}_p} \cdot \underbrace{\begin{bmatrix} \mathbf{Y}_m \\ \mathbf{E}_m \end{bmatrix}}_{\mathbf{x}_p}}_{\text{free response } \mathbf{f}} . \qquad (7)$$

2.3   Performance Index in Matrix Form and Analytic Solution

Cost function (1) in matrix form can be written as

$$J = (\mathbf{Y} - \mathbf{W})^T (\mathbf{Y} - \mathbf{W}) + \mathbf{U}^T \mathbf{Q} \mathbf{U}, \tag{8}$$

Where $\mathbf{W} = \begin{bmatrix} w(k+1) \\ w(k+2) \\ w(k+3) \\ w(k+4) \end{bmatrix}$, $\quad \mathbf{Q} = \begin{bmatrix} q & 0 & 0 & 0 \\ 0 & q & 0 & 0 \\ 0 & 0 & q & 0 \\ 0 & 0 & 0 & q \end{bmatrix}$.

By substitution (7) into (8) we get following quadratic form

$$J = \mathbf{U}^T \underbrace{\left(\mathbf{G}^T\mathbf{G} + \mathbf{Q}\right)}_{\mathbf{H}} \mathbf{U} + \mathbf{U}^T \underbrace{\mathbf{G}^T(\mathbf{f} - \mathbf{W})}_{\mathbf{g}} + \underbrace{(\mathbf{f} - \mathbf{W})^T \mathbf{G} \mathbf{U}}_{\mathbf{g}^T} + \underbrace{(\mathbf{f} - \mathbf{W})^T(\mathbf{f} - \mathbf{W})}_{k} \tag{9}$$

and the unconstrained solution can be written as

$$\mathbf{U} = -\mathbf{H}^{-1}\mathbf{g} = \left(\mathbf{G}^T\mathbf{G} + \mathbf{Q}\right)^{-1}\mathbf{G}^T\left(\mathbf{W} - \mathbf{F}_p\mathbf{x}_p(k)\right) = \mathbf{L}\left(\mathbf{W} - \mathbf{F}_p\mathbf{x}_p(k)\right). \tag{10}$$

The control law (actual control change) is

$$\Delta u(k) = \mathbf{K}\left(\mathbf{W} - \mathbf{F}_p\mathbf{x}_p(k)\right), \tag{11}$$

where $\mathbf{K}$ is the first row of matrix $\mathbf{L}$. This is analytic form of predictive controller but matrix operations must be used to calculate $\mathbf{K}$ and $\mathbf{F}_p$.

2.4   Simple Symbolic Forms of Predictive Controller

Simple MPC symbolic forms can be obtained for special choices of prediction and control horizons. If $N_1 = N_2 = 3$ and $N_u = 1$ we get control law as

$$\Delta u(k) = \frac{pb_1}{q + p^2 b_1^2} \left[w(k+3) + \left(a_1^3 - a_1^2 + a_1 - 1\right)y(k) - a_1 p y(k-1) \right. \\ \left. + \left(-c_1 p + c_2(a_1 - 1)\right)e(k) - c_2 p e(k-1) \right] \tag{12}$$

$$p = a_1^2 - a_1 + 1. \tag{13}$$

There is only one point in time $k + 3$ considered for the set-point following.

Only one control change is considered – control action is supposed to be constant for whole control horizon.

For horizons $N_1 = N_2 = 4$ and $N_u = 1$ we get

$$\Delta u(k) = \frac{rb_1}{q + r^2 b_1^2} \left[-w(k+4) + \left(a_1^4 - a_1^3 + a_1^2 - a_1 + 1\right)y(k) \right. \\ \left. - a_1 r y(k-1) - \left(c_1 r - c_2(a_1^2 - a_1 + 1)\right)e(k) - c_2 r e(k-1) \right] \tag{14}$$

$$r = a_1^3 - a_1^2 + a_1 - 1 \tag{15}$$

Controlled variable prediction used to calculate prediction error $e(k) = y(k) - \hat{y}(k)$ is the same for both versions

$$\hat{y}(k) = b_1 \Delta u(k-1) - (a_1 - 1)y(k-1) + a_1 y(k-2) + c_1 e(k-1) + c_2 e(k-2). \tag{16}$$

3 Control Experiments

Firstly we simulated control with model of the laboratory system GUNT RT 050 - speed control (see Fig. 2) – motor with mass flywheel and generator

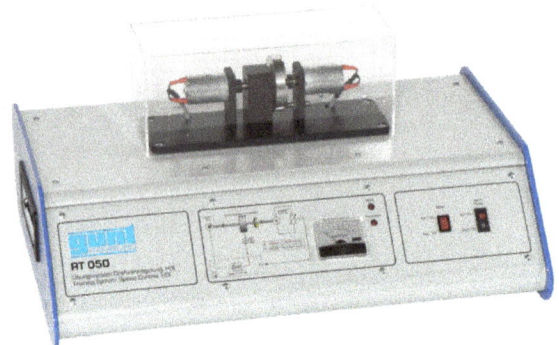

Fig. 2. GUNT RT 050 – speed control laboratory system

We have identified first order continuous-time transfer function model from measured dynamical responses as

$$F_1 = \frac{1.96}{3.16s + 1} \quad . \tag{17}$$

With sample time $T_s = 0.5$ s and penalty parameter $q = 1$ we got following control responses for both versions of the controller – by using Eqs. (12–16) (Fig. 3).

We have also tested how the model mismatch and measurement noise will influence the control responses. We have identified second order model of the same system - this is the right order of the controlled process

$$F_2 = \frac{1.95}{(2.36s + 1)(0.71s + 1)} \quad . \tag{18}$$

Second version of the controller ($N_1 = N_2 = 4$) uses model F_1 (17) like first order approximation of the real process. For simulation purpose we have added normally

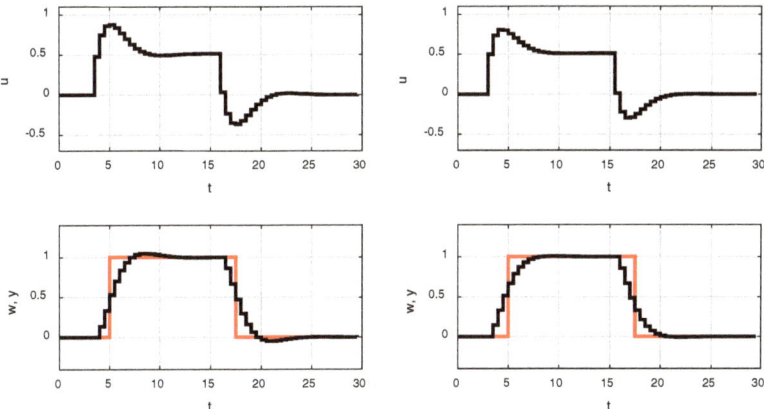

Fig. 3. Simulated control responses with horizons 3 (left) and 4 (right)

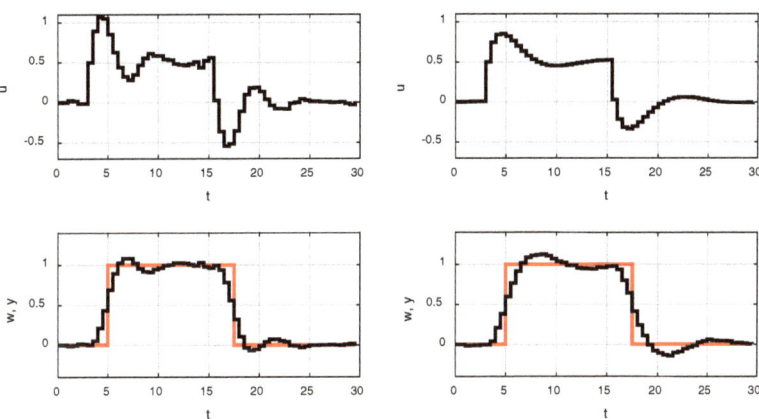

Fig. 4. Simulated control responses without (left) and with filtering polynomial (right)

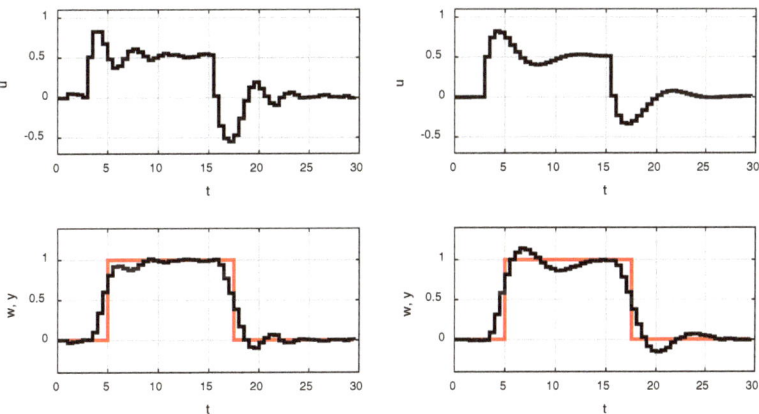

Fig. 5. Real control responses without (left) and with filtering polynomial (right)

distributed pseudorandom noise with standard deviation 0.01 to the controlled output of the process F_2 (18) to emulate measurement noise.

Two versions according to the filtering polynomial are applied - without filtering polynomial ($C = 1$) and with second order filtering polynomial $C = (1 - 0.8z^{-1})^2$. Simulated control responses are in Fig. 4 and real control responses in Fig. 5.

4   Conclusion

Two fast and easy to use symbolic forms of MPC are presented and applied in laboratory scale. Controllers are parametrized with first order process model parameters a_1 and b_1, one penalty parameter q and two filtering polynomial parameters c_1 and c_2. Penalty parameter allows to tune the control quality. Filtering parameters can be seen also as tunable parameters. The controller is quite sensitive to measurement noise without filtering ($c_1 = c_2 = 0$). First order polynomial (e.g. $c_1 = -0.8$, $c_2 = 0$) will give smoother control actions. Second order polynomial (e.g. $c_1 = -1.6$, $c_2 = 0.64$) filter improves control actions even more but also increases risk of oscillations. Some trade-off in filtering tuning is necessary similarly to state observing problem.

Acknowledgment. This research was supported by SGS project Modern technology for large-volume data processing and optimal control of technological processes (in Czech) at FEI, University of Pardubice. This support is very gratefully acknowledged.

References

1. Camacho, E.F., Bordons, C.: Model Predictive Control. Springer, New York (2004)
2. Rossiter, J.A.: Model-Based Predictive Control: A Practical Approach. CRC Press, Boca Raton (2003)
3. Soroush, M., Muske K.R.: Analytical model predictive control. In: Nonlinear Model Predictive Control. Progress in Systems and Control Theory, vol. 26. Birkhäuser, Basel (2000)
4. Soroush, M.: Plants for which model predictive control admits an analytical solution. In: Proceedings of the American Control Conference, pp. 3745–3750 (2007)

Overview of Ball & Plate Application for Collaborative Robot YuMi

Lubos Spacek[✉] and Jiri Vojtesek

Department of Process Control, Faculty of Applied Informatics, Tomas Bata University in Zlin, Nad Stranemi 4511, 760 05 Zlin, Czech Republic
lspacek@utb.cz

Abstract. Most industrial robotic manipulators are used for specific type of operation with predefined movements. This is obvious because industrial robots are constructed to perform repetitive tasks with certain precision and cycle time. This paper introduces another way to exploit advantages of robots by changing their movement dynamically with reaction to external forces and environment. This is partially solved by integrated force control sensors for manipulator grippers, but this article deals with the extension to control of the unstable system with fast dynamics. The best representative of such a system is a Ball & Plate model. This paper deals with the overview of the designed structure of the project and testing the feasibility of solutions.

Keywords: YuMi · Ball & Plate · LQ controller
Dynamic Control Unit · Collaborative robotics

1 Introduction

Using an industrial robot as a motion system for the Ball & Plate model is not the traditional way of controlling a ball on a plate. Extra degrees of freedom in this setup can be redundant, but it also offers additional opportunities in extending the model further. It has the same advantages as the Stewart platform [1], but with a more flexible usage, as the robot also offers movement in 3D space. There are many hobby projects for Ball & Plate model control and this paper tries to expand solutions in an industrial environment. The 2DoF LQ controller for polynomial approach design procedure [2,3] is presented based on the results of previous work of authors [4]. It is flexible and reliable enough for easy implementation to the programming environment of the robot.

2 Collaborative Robots

Collaborative robots are needed in applications where human access is expected in a workspace during the operation [5]. Although the Ball & Plate application does not necessarily require this type of access, it is surely helpful during testing. As the best candidate is collaborative industrial robot ABB IRB

© Springer International Publishing AG, part of Springer Nature 2019
J. Machado et al. (Eds.): HELIX 2018, LNEE 505, pp. 89–95, 2019.
https://doi.org/10.1007/978-3-319-91334-6_13

14000 YuMi (Fig. 1). Its advantages are two independent 7DoF manipulators with 0.02 mm repeatability and precise control [6]. YuMi can hold one platform (plate) in each hand, thus extending capabilities of the system. The Ball & Plate system can be obviously controlled with a 2DoF system and more degrees of freedom may be redundant, but they offer a flexibility to move in another dimension (similar to 6DoF Stewart platform [1] or classic robotic 6DoF manipulator).

Fig. 1. ABB IRB 14000 YuMi

3   Ball & Plate

The Ball & Plate model is well known unstable system in control engineering, which is suitable for testing the quality of designed controllers and control laws. Its relatively fast and unstable behavior makes it the perfect experimental plant for this task. The general structure is based on the idea of a ball moving on a plate with controlled inclination as seen in Fig. 2.

This general structure can be described for one dimension in linearized and simplified form shown in Eq. 1. That applies for the ball on the plate without friction and dynamics of motion structure.

$$\ddot{x} = K\alpha \quad \rightarrow \quad G(s) = \frac{K}{s^2} \tag{1}$$

where x is ball coordinate from the center of the plate ($\ddot{x}$ is its 2nd time derivation), α is angle of the plate, K is constant dependent on the gravitational acceleration g and the momentum of the ball and $G(s)$ is continuous-time transfer function with complex variable s.

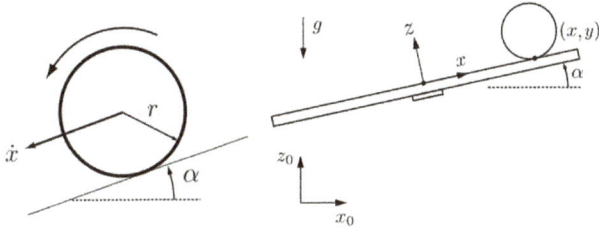

Fig. 2. Ball & Plate setup [7]

4   Control Unit

Although the software of the robot is quite advanced with state-of-the-art motion control and I/O system [8], it doesn't have the needed flexibility for implementation of various controllers, mainly the complex ones. It also does not have built-in real-time capabilities for even cycle period and sampling. Thus a control unit is needed to fill this gap. Since the application is using the industrial robot, it is also appropriate to use industrial-type control unit (instead of devices used broadly for hobby projects, such as Arduino or Raspberry Pi - although they would be also sufficient).

The DCU56IO Dynamic Control Unit from PROSYSTEMY Ltd. [9] will be used in this application (shown in Fig. 3). It fulfills industry-type requirements for precision, safety, connectivity, speed and reliability with excellent support from company's side, while it maintains reasonable cost compared to other products on the market. It is also possible to program the unit from Simulink or Scilab's Xcos environment, which can be considered as a great advantage for controller implementation together with its vast range of I/O ports.

Fig. 3. PROSYSTEMY DCU56IO Dynamic Control Unit [9]

5   Realization

5.1   Preparation

The position of the ball will be determined by resistive touchscreen foil placed on the plate. The output from the foil will be fed to the control unit, processed and desired angle of the plate will be sent to the robot. Because this paper deals with the feasibility of this solution and applicability to the real system, only simulation results are presented. These simulations are however executed in RobotStudio environment, which provides "semi-real" results. The ball is also simulated in RobotStudio because it offers rigid body dynamics simulation and tools for measuring the position of rigid bodies by default. RobotStudio is still just an ideal environment, but it is relatively precise in virtualization of the robot with all dynamics and delays. It serves as a good background for the feasibility evaluation.

5.2   Identification

Because the controller design needs the transfer function of the system, it was identified in RobotStudio by step-changing plate angles and measuring position of the ball. The system has relatively linear behavior for small angles of the plate ($<10°$), so it was identified in this region. It was approximated by continuous-time transfer function (Eq. 2) and discretized afterwards.

$$G(s) = \frac{C}{s^2\,(Ts+1)} = \frac{C}{Ts^3 + s^2} = \frac{-0.1213}{0.05139s^3 + s^2} \tag{2}$$

where $G(s)$ is continuous transfer function with complex variable s, C is velocity gain and T is the time constant of the system.

5.3   Controller Design

The purpose of this paper is not to design the best controller in terms of quality, but just to evaluate abilities of ABB robotic system for given criteria. Thus the LQ 2DoF polynomial controller design from the previous work of authors [4] is chosen as a reference, mainly because of previous experiences and positive results in terms of implementation and robustness. The output of the controller for both coordinates is thus obtained using Eq. 3.

$$u_k = (1 - p_1)u_{k-1} + (p_1 - p_2)u_{k-2} + p_2 u_{k-3}$$
$$+ \ r_0 w_k - q_0 y_k - q_1 y_{k-1} - q_2 y_{k-2} - q_3 y_{k-3} \tag{3}$$

where u_{k-i} is output of the controller, y_{k-i} is the position of the ball in one coordinate, w_k is desired value and p_i, q_i and r_0 are individual calculated parameters shown in Eq. 4.

$$\begin{bmatrix} r_0 \\ p1 \\ p2 \end{bmatrix} = \begin{bmatrix} -0.0352 \\ -1.4072 \\ 0.4820 \end{bmatrix}, \quad \begin{bmatrix} q0 \\ q1 \\ q2 \\ q3 \end{bmatrix} = \begin{bmatrix} -150.2577 \\ 416.1380 \\ -383.1942 \\ 117.2787 \end{bmatrix} \tag{4}$$

5.4 Results

Results of ball stabilization with controller designed in previous chapter are shown in Figs. 4 and 5. They show stabilization of the ball in the center of the plate. After the ball is stabilized, it is randomly moved to another position which is repeated several times (even before stabilization in the last jump). Figure 6 shows the setup in RobotStudio virtual environment.

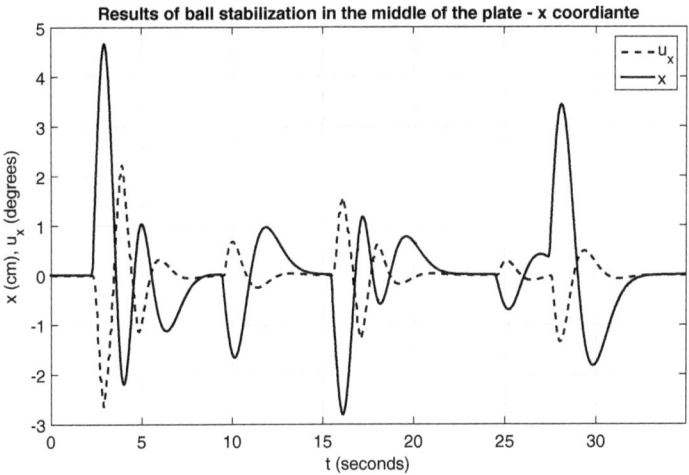

Fig. 4. Results for x coordinate

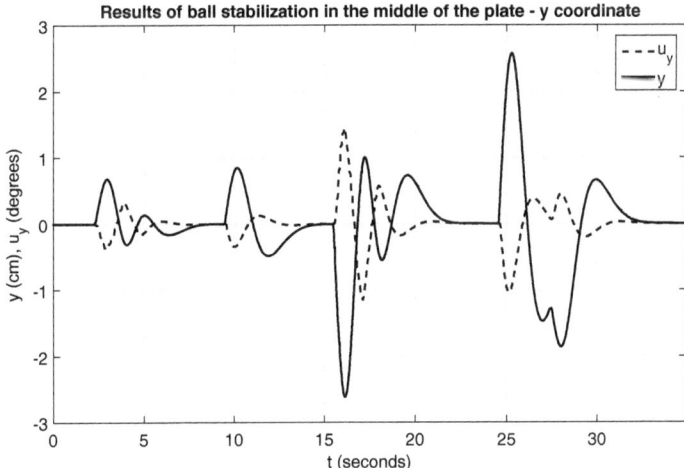

Fig. 5. Results for y coordinate

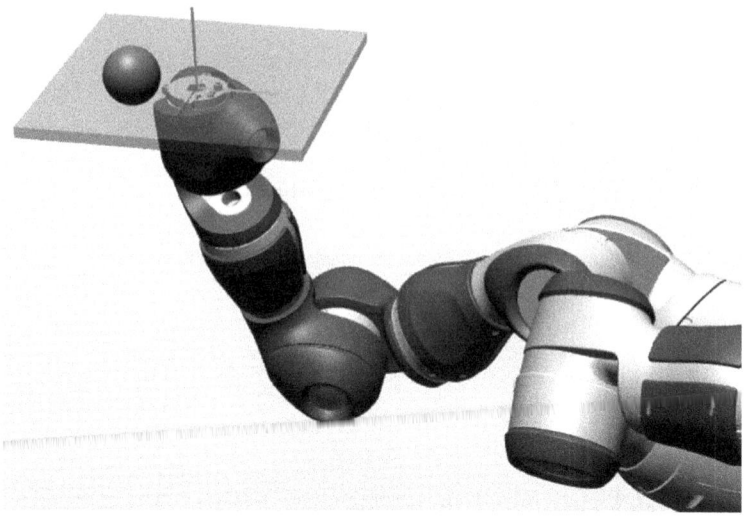

Fig. 6. Ball & Plate setup in RobotStudio environment

6 Conclusion

The overview of the project introduced main components of the real system and provided a brief feasibility study of the system with results proving that the project has a solid background in software and also hardware part. The optional goal is to use industrial-type hardware to comply with the usage of the industrial robotic manipulator. The ball stabilized in the frame of 5–10 s, which is not the best result regarding the quality of control, but it can be seen that the robot is fully capable of moving the plate in order to stabilize the ball. The only problem with the real setup is whether servomotors of the real robot will be able to sustain fast changes of angles and if not how to avoid this. Nonetheless, results presented in this paper are satisfying enough to continue with the project, design a better control strategy and begin with the implementation of the real system.

Acknowledgment. This article was created with support of the Ministry of Education of the Czech Republic under grant IGA reg. n. IGA/CebiaTech/2018/002.

References

1. Bohigas, O., Manubens, M., Ros, L.: A linear relaxation method for computing workspace slices of the stewart platform. J. Mech. Robot. **5**, 9 (2013). https://doi.org/10.1115/1.4007706. ISSN 1942-4302
2. Bobál, V., Böhm, J., Fessl, J., Macháček, J.: Digital Self-tuning Controllers. Springer-Verlag, London (2005)
3. Matušů, R., Prokop, P.: Algebraic design of controllers for two-degree-of-freedom control structure. Int. J. Math. Models Meth. Appl. Sci. **7**, 630–637 (2013)

4. Spaček, L., Bobál, V., Vojtěšek, J.: LQ digital control of Ball & Plate system. In: 31st European Conference on Modelling and Simulation ECMS 2017, pp. 427–432. European Council for Modelling and Simulation (2017)
5. Fryman, J., et al.: Safety of industrial robots: from conventional to collaborative applications. In: 7th German Conference on Robotics, ROBOTIK 2012, pp. 1-5, May 2012
6. ABB Robotics - IRB 14000 YuMi. http://new.abb.com/products/robotics/industrial-robots/yumi
7. Nokhbeh, M., Khashabi, D., Talebi, H.A.: Modelling and control of Ball-Plate system. Amirkabir University of Technology (2011)
8. ABB RobotWare Overview. http://new.abb.com/products/robotics
9. PROSYSTEMY Ltd. http://www.prosystemy.sk/

Simulink Model of a Coupled Drives Apparatus

Petr Chalupa$^{(\boxtimes)}$ ⓘ, František Gazdoš ⓘ, Michal Jarmar,
and Jakub Novák ⓘ

Tomas Bata University in Zlin,
nam. T. G. Masaryka 5555, 760 01 Zlín, Czech Republic
chalupa@fai.utb.cz

Abstract. The paper presents process of modelling of the CE108 Coupled Drives Apparatus developed by TecQuipment Ltd. This laboratory plant serves as a laboratory equipment representing a problem of modelling and control of producing continuous lengths materials, which is common in many industrial applications. A mathematical model based on the first principle modelling, which was created by designers of the plant, is used as a starting point. The characteristics of this initial model is compared with real-time measurements and the model is improved. Simulink representation of the model is designed and verified. The Simulink model is suitable for designing and verifying various control strategies not only for the CE108 Coupled Drives Apparatus but in general for similar industrial processes.

Keywords: Coupled Drives · First principle modelling · Simulink
Identification

1 Introduction

Almost all control design techniques are based on some kind of a model of a controlled plant [1]. The model provides necessary information regarding the plant and it can be used to investigate properties and behavior of the modelled plant without a risk of violating technological constraints of the real plant. There are two basic approaches of obtaining plant model: the black box approach and the first principles modelling (mathematical-physical analysis of the plant).

The black box modelling [2, 3] uses just input and output signals of the plant and do not take in account internal structure of the modelled plant and corresponding physical laws. It is possible to use the same identification algorithm for wide set of various controlled plants [4, 5] but resulting model is valid only for input signals it was derived from. On the other hand, the first principle modelling provides general models valid for whole range of plant inputs and states. Modelling is based on analyzing the modelled plant and combining physical laws [6]. Unfortunately, there are usually many unknown constants and relations when performing analysis of the plant. Therefore, some simplifications must be used to obtain reasonable results in more complicated cases. These simplifications must relate with the purpose of the model. The first principle modelling is also referred to as white box modelling.

© Springer International Publishing AG, part of Springer Nature 2019
J. Machado et al. (Eds.): HELIX 2018, LNEE 505, pp. 96–102, 2019.
https://doi.org/10.1007/978-3-319-91334-6_14

The CE108 Coupled Drives Apparatus, which was developed by TecQuipment Ltd., can serve as a laboratory model for many industrial processes involving tension and speed of interacting drive systems (e.g. production of paper, wire, textiles, metal sheet or plastic films).

The goal of this work is to obtain a Simulink model of the apparatus, which can be used for design of a control strategy. A previously published work [7] presents the first principle model in detail and provides basic characteristics of the model. Current paper is focused on designing an enhanced Simulink model, which is based on the first principle model improved by results of real-time measurements.

The paper is organized as follows. Section 2 presents the modelled system – CE108 Coupled Drives Apparatus. Section 3 is focused on creating the Simulink model of the whole system and Section 4 compares the resulting model with the real-time plant.

2   Coupled Drives Apparatus

A photo of the CE108 Coupled Drives Apparatus is presented in Fig. 1.

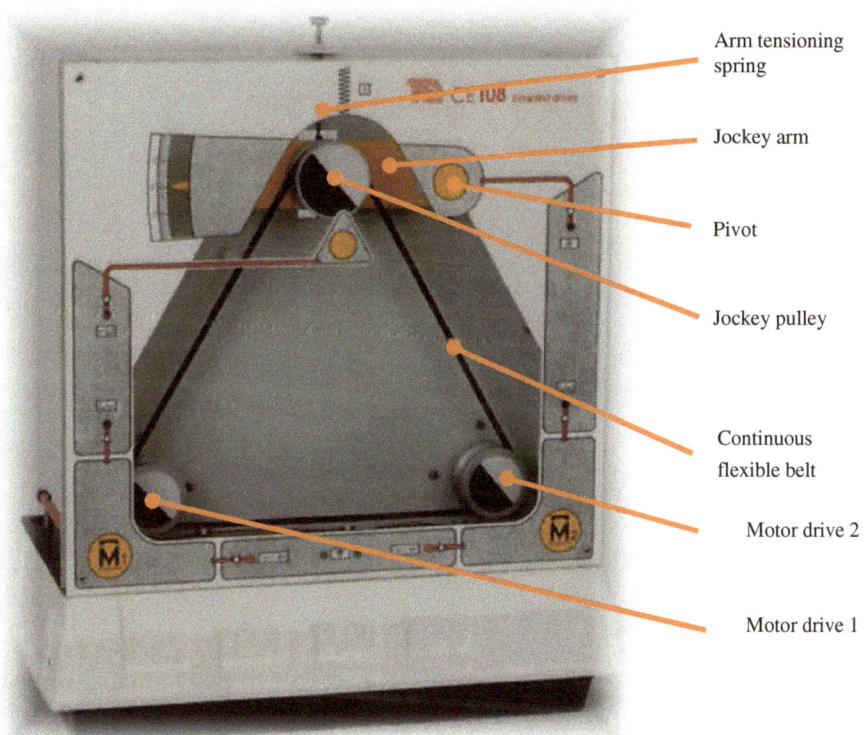

Fig. 1. CE108 Coupled Drives Apparatus

The apparatus consists of motors M1 and M2, which are used to control the speed and tension of the continuous flexible belt, two pulleys attached on the drive motor shafts and a jockey pulley. The jockey pulley is mounted on a swinging arm that is supported by a spring. The position of the arm corresponds to the tension in the drive belt [8, 9].

The CE108 Coupled Drives Apparatus is driven by two input signals – control voltages of motors M1 (u_1) and M2 (u_2). The apparatus provides four output signals in the form of voltage (−10 V to +10 V): speed of the belt, i.e. rotations of the jockey pulley; tension of the belt measured as an angle of the arm; rotations of the motor M1; rotations of the motor M2. The rotations of the motors can also be considered states of the apparatus.

In the most often used setup, the apparatus is considered a Two Inputs Two Outputs (TITO) system with control voltages as the inputs and belt speed and tension as the outputs

Many static and dynamic properties of the system were measured and results were presented in [7]. A typical static characteristic of belt speed is copied to this paper to demonstrate system behaviour (see Fig. 2).

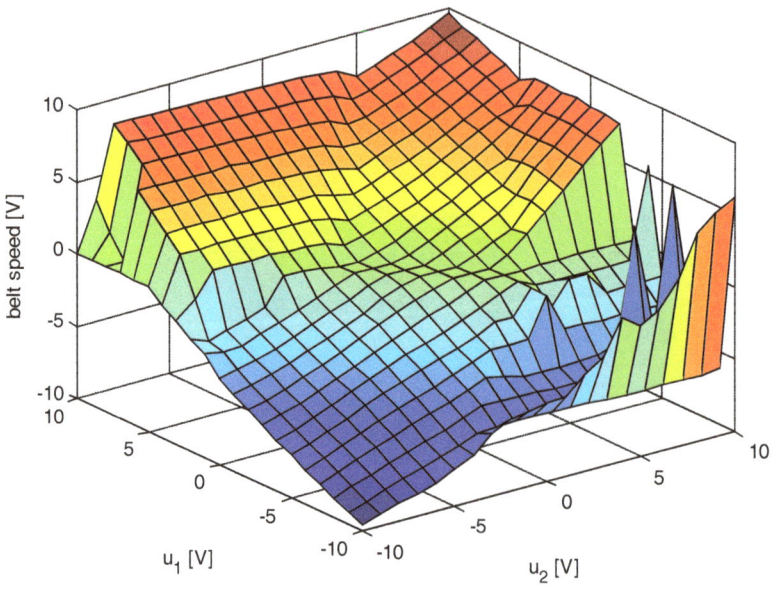

Fig. 2. Static characteristics of belt speed

As stated in [7], there are several principal differences between the linear first principle model and real-time apparatus:

- the CE108 apparatus is not linear in the whole range of its inputs;
- the steady outputs of the CE108 apparatus do not depend only on the steady inputs but also on the previous course of the inputs;

- the behaviour of the CE108 apparatus in the area where inputs have opposite sign is very far from deterministic;
- the steady outputs are not reached for the steady inputs in some cases but the system remains in undamped oscillations.

3   Creating Simulink Model

The process of creating a Simulink model of the whole CE108 plant is described in the following paragraphs. As stated in the previous chapter and as can be seen from Fig. 2, the system cannot be modelled by a single linear system. The input space was divided into four sections and a model was created for each section. The division into sections is presented in Fig. 3.

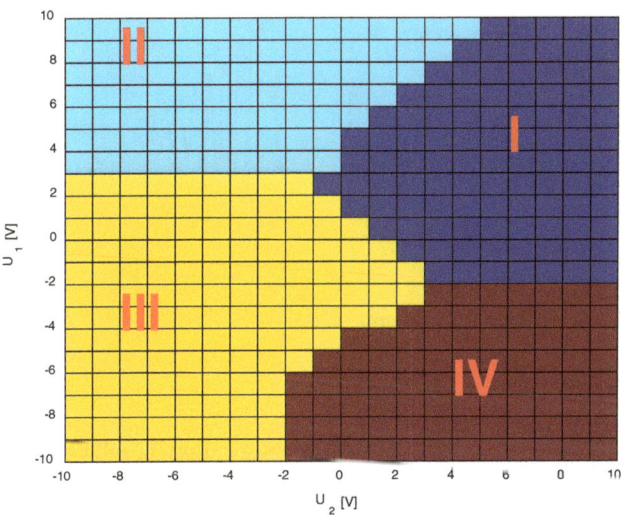

Fig. 3. Division of input space into sections

The division is valid for both outputs of the CE108 apparatus – belt speed and belt tension. The sub-models corresponding to each section are derived as follows:

- Section I – a sub-model based on first principle modelling with refined parameters,
- Section II – a sub-model obtained by experimental identification,
- Section III – a sub-model on first principle modelling with refined parameters ensuring better correspondence of the model and the real-time plant
- Section IV – a sub-model obtained by experimental identification.

The first principle modelling used for Section I and III is presented in [7] in detail. Off-line ARX identification was used to create models for Section II and IV.

3.1 Experimental Identification

The sub-models corresponding to Section II and IV were obtained by experimental identification. The basic idea of creating these sub-models consists in assumption that the output of the system depends on only one input parameter u_1. It should be noted that this is a simplification of the real behavior of the plant while both belt tension and belt speed have highly stochastic behavior in these sections.

The belt speed was modelled by a second order system in both Section II and IV but the parameters were identified separately. The belt tension was modelled by third order oscillatory system. Examples of comparison of step responses of the apparatus and the model are presented in Fig. 4. The courses correspond to step changes of u_1 and u_2 from zero (i.e. $u_1(0) = u_2(0) = 0$ V) to $u_1 = 7$ V and $u_2 = -3$ V. The change was applied just after the start of the measurements. The final state corresponds to Section II (see Fig. 3).

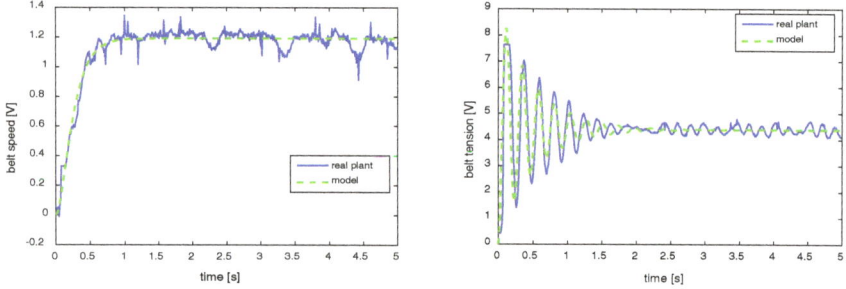

Fig. 4. Experimental identification of belt speed and belt tension

3.2 Simulink Model

A scheme of the Simulink model of the whole system is presented in Fig. 5. The scheme contains four sub-models corresponding to Section I–IV of the input space as well as a switching logic block that is responsible for passing token among the sub-models.

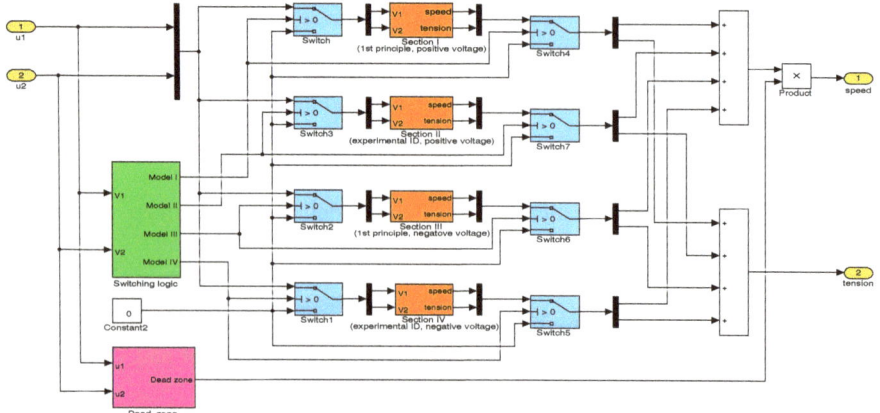

Fig. 5. Simulink model of the CE108 Coupled Drives Apparatus

Moreover, the Simulink model contains a Dead_zone block, which represent a dead zone observable in Fig. 2 between Section I and Section III, i.e. approximately around the line $u_1 = -u_2$.

The internal structures of the sub-models for Section III and Section IV are presented in Fig. 6. Details concerning original differential equations can be found in [7].

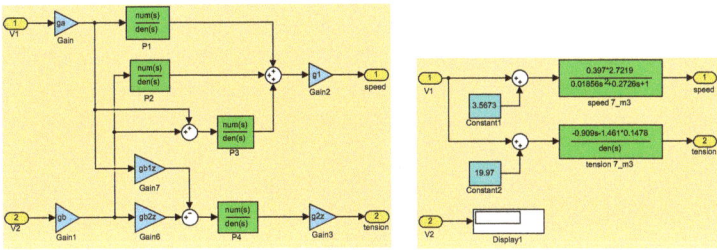

Fig. 6. Internal structure of sub-models for Section III (left) and Section IV (right)

4   Comparison of Simulink Model and Real-Time Plant

The Simulink model of the whole system presented in the previous chapter was compared with the real-time CE108 Coupled Drives Apparatus from several points of view. Comparison of static characteristics of belt speed is presented in Fig. 7 and comparison of static characteristics of belt tension is presented in Fig. 8.

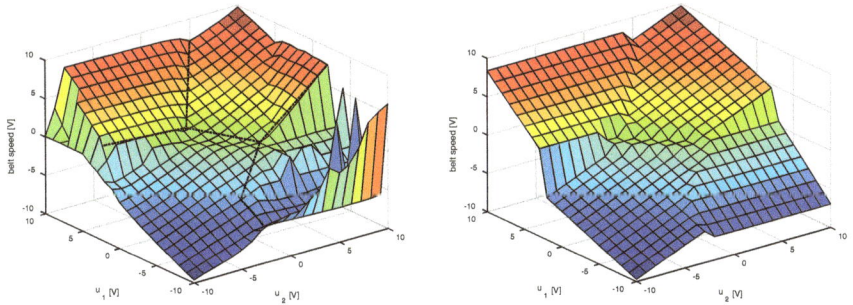

Fig. 7. Static characteristics of belt speed: real-time apparatus (left) and Simulink model (right)

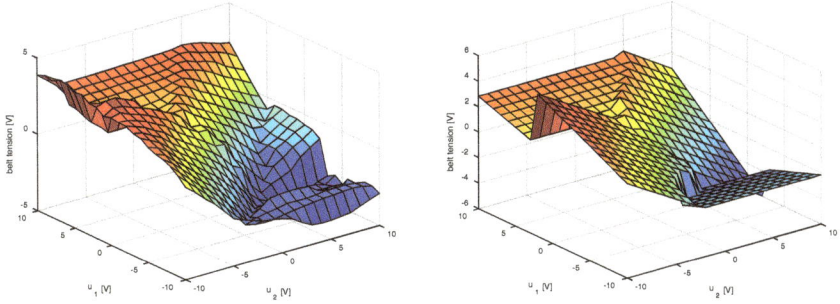

Fig. 8. Static characteristics of tension: real-time apparatus (left) and Simulink model (right)

5   Conclusion

The first principle model of the CE108 Coupled drives Apparatus was derived in previous works [7–9]. Significant differences between the first principle model and the real-time apparatus were observed especially in areas where input signals have opposite signs. Simulink model of the plant was created by combining four sub-models based on the input space sections. Resulting model represents the plant well including also dead zone of the belt speed. Both static and dynamic characteristics of the model and real-time plant were compared and a good accuracy was observed (see Figs. 4, 7 and 8). However, the model does not cover stochastic behavior of the real plant, which was observed – especially in sections with opposite signs of inputs. Fortunately, similar industrial systems do not operate in these sections and therefore the resulting Simulink model is a good starting point for the subsequent control system design.

Acknowledgements. This work was supported by the Ministry of Education, Youth and Sports of the Czech Republic within the National Sustainability Programme project No. LO1303 (MSMT-7778/2014).

References

1. Bobál, V., Böhm, J., Fessl, J., Macháček, J.: Digital Self-tuning Controllers: Algorithms Implementation and Applications. Springer, London (2005)
2. Liu, G.P.: Nonlinear Identification and Control – A Neural Network Approach. Springer, London (2001)
3. Ljung, L.: System Identification: Theory for the User, 2nd edn. Prentice Hall, Upper Saddle River (1999)
4. Mikleš, J., Fikar, M.: Process Modelling, Identification, and Control. Springer, Heidelberg (2007)
5. Codrons, B.: Process Modelling for Control: A Unified Framework Using Standard Black-Box Techniques. Springer, London (2005)
6. Himmelblau, D.M., Riggs, J.B.: Basic Principles and Calculations in Chemical Engineering. Prentice Hall, Upper Saddle River (2004)
7. Chalupa, P., Novák, J., Jarmar, M.: Model of coupled drives apparatus – static and dynamic characteristics. In: MATEC Web Conference, vol. 76, Article Number 02011. EDP Sciences (2016)
8. Hagadoorn, H., Readman, M.: Coupled Drives 1: Basics. http://www.control-systems-principles.co.uk/whitepapers/coupled-drives1.pdf. Accessed 21 Jan 2018
9. Hagadoorn, H., Readman, M.: Coupled Drives 2: Control and analysis. http://www.control-systems-principles.co.uk/whitepapers/coupled-drives2.pdf. Accessed 21 Jan 2018

Adaptive Control of Temperature Inside Plug-Flow Chemical Reactor Using 2DOF Controller

Jiri Vojtesek$^{(\boxtimes)}$ and Lubos Spacek

Faculty of Applied Informatics, Tomas Bata University in Zlin,
Nam. T.G. Masaryka 5555, 760 01 Zlin, Czech Republic
vojtesek@utb.cz
http://www.utb.cz/fai

Abstract. The tubular chemical reactor is a industrial equipment widely used in the chemical or biochemical industry for production of various kinds of products. The mathematical model of such system is described by partial differential equations that are solved numerically. This article presents simulation results of the mean reactant's temperature control inside the plug-flow tubular chemical reactor. The adaptive approach here is based on the recursive identification of the external linear model as a simplified mathematical representation of the originally nonlinear system. The control synthesis is based on the polynomial theory with the Pole-placement method and the spectral factorization. These methods are easily programmable and they also offers tuning of the controller. Used two degrees-of-freedom (2DOF) control structure divides the controller into two parts – the first in the feedback part and the second one in the feedforward part of the control loop.

Keywords: Adaptive control · 2DOF · Tubular chemical reactor
Recursive identification · Pole-placement method

1 Introduction

It is known that most of the processes in the industry has a nonlinear behaviour [1] which causes difficulties not only in the modelling and simulation but mainly in the control [2]. Tubular chemical reactors are from the parameter's point of view typical members of the system with distributed parameters [3] because besides the time variable, also the space variable must be taken into the consideration. The mathematical model of such system is then described by one or the set of partial differential equations (PDE).

A mathematical solution of the PDE is not trivial but we can simplified the computation for example with the use of the Finite difference method [4]. The set of PDE is then transformed to the set of ordinary differential equations (ODE) that are much better solvable for example with the Euler method, Runge-Kutta's methods or its modifications etc.

© Springer International Publishing AG, part of Springer Nature 2019
J. Machado et al. (Eds.): HELIX 2018, LNEE 505, pp. 103–109, 2019.
https://doi.org/10.1007/978-3-319-91334-6_15

It is common, that chemical reactors, not only tubular ones, must be cooled because of the exothermic character of the reaction inside. Tubular chemical reactors offers two types of the cooling varying in the direction of cooling – (I) the co-current or the (II) counter-current cooling. It was proofed for example in [5] that counter-current cooling has better cooling efficiency than co-current.

Control configuration in this work can be described as a system with two degrees-of-freedom (2DOF) [6]. This structure has one part in the feedback and the second one in the feedforward part of the control loop. The reason for using this configuration could improve the course of the output variable and reduce overshoots of the output variable.

Adaptive control [7] takes its philosophy from the nature where plants, animals or even human beings "adopts" their behaviour to the actual conditions or the environment. From the control point of view, this adaptation could be done for example with the change of parameters, structure etc. according to the actual state and conditions of the controlled system [8]. In our case, the adaptation is satisfied by the recursive identification of the External Linear Model (ELM) as a linear representation of the originally nonlinear system. Parameters are then recomputed according to the identified ELM's parameters because they are connected via the polynomial synthesis and The Pole-placement method.

All results in this work comes from simulations made by the mathematical package Matlab. The application to the real system is the part of the future work.

2   Hybrid Adaptive Control

It was already mentioned, that the adaptation is satisfied by the recursive identification of the ELM. As parameters of the controller are connected with parameters of ELM via computational relations, the controller can react and change its parameters according results of the identification.

The ELM could be for example in the form of the continuous-time transfer function (TF)

$$G(s) = \frac{Y(s)}{U(s)} = \frac{b(s)}{a(s)} \tag{1}$$

where $Y(s)$ and $U(s)$ are Laplace transforms of the input and the output variables. Parameters of unknown polynomials $a(s)$ and $b(s)$ are identified recursively during the control for example with the use of the Recursive Least-Squares method with some forgetting factor [12].

The on-line identification of the continuous-time (CT) ELM (1) could cause problems. Much better from the computational view is the recursive identification of the discrete-time (DT) ELM. Disadvantage of the DT ELM is the sensitivity of this model to the sampling period that could cause less accurate results.

The compromise between these two types can be found on so called "delta" models that are special types of the DT models in which both input and output variables are related to the sampling period. It was proofed for example in [9], that parameters of the delta-model approaches to the CT model for sufficiently small sampling period. This approach combines benefits of both models - accuracy of the CT model and better applicability of the DT model. That is why we called this approach "hybrid", because the ELM is considered as a CT one but the recursive identification is computed in discrete time intervals via delta models.

The ARX identification model for identification is then

$$y_\delta(k) = \boldsymbol{\theta}_\delta^T(k) \cdot \boldsymbol{\phi}_\delta(k-1) + e(k) \tag{2}$$

where $\boldsymbol{\theta}_\delta$ represents unknown vector of parameters, $\boldsymbol{\phi}_\delta$ is a data vector whose parts are known from the measurements and $e(k)$ is a general random immeasurable component.

The control scheme is based on the two degrees-of-freedom (2DOF) control configuration [6] schematic representation of which is displayed in Fig. 1.

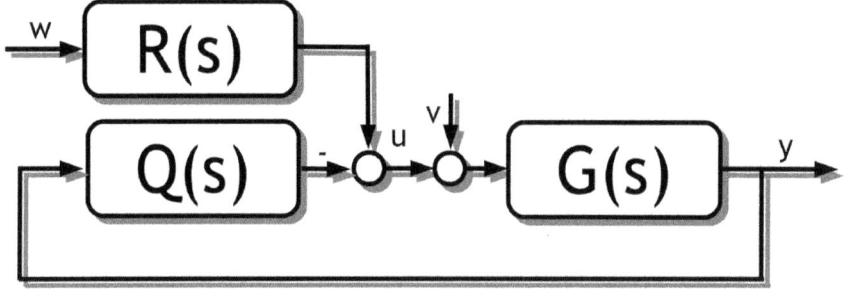

Fig. 1. Two Degrees-of-freedom (2DOF) control configuration

Block $G(s)$ represents ELM of the controlled system (1), $Q(s), R(s)$ are TF of the feedback and feedforward part of the controller, w is a reference signal (wanted value), v denotes an error, u is the input to the system (and also the output from the controller) and y is the output from the system. Transfer functions of the controller has then general form

$$Q(s) = \frac{q(s)}{p(s)}; \quad R(s) = \frac{r(s)}{p(s)} \tag{3}$$

for polynomials $p(s), q(s), r(s)$ satisfying condition $\deg p(s) \geq \deg q(s)$ and $\deg p(s) \geq \deg r(s)$ and $p(s) = s \cdot \tilde{p}(s)$.

Parameters of unknown controller's polynomials $\tilde{p}(s), q(s)$ and $r(s)$ are computed from the set of Diophantine equations [11]

$$a(s) \cdot s \cdot \tilde{p}(s) + b(s) \cdot q(s) = d(s), \quad t(s) \cdot s + b(s) \cdot r(s) = d(s) \tag{4}$$

The polynomial $t(s)$ in the second Diophantine equation is additional polynomial that is not used in transfer functions and the polynomial $d(s)$ on the right

side of Eq. (4) is optional stable polynomial that could be chosen for example
with the use of Pole-placement method:

$$d(s) = n(s) \cdot (s + \alpha)^{\deg d - \deg n} \tag{5}$$

where α is optional tuning parameter and polynomial $n(s)$ comes from the spectral factorization $n^*(s) \cdot n(s) = a^*(s) \cdot a(s)$. Advantage of this method can be
found in the interconnection of the polynomial $d(s)$ with the controlled system
via polynomial $a(s)$ in the denominator of the system's TF $G(s)$ (1).

Degrees of these polynomials are $\deg \tilde{p}(s) = \deg a(s) - 1$, $\deg q(s) = \deg a(s)$,
$\deg r(s) = 0$ and $\deg d(s) = 2 \cdot \deg a(s)$.

3   Plug-Flow Chemical Reactor

The system under the consideration is the tubular chemical reactor with plug-
flow inside. The construction is simple - it is, in fact, one main pipe with number
of individual pipes inside [3]. The reactant comes through the reactor inside
smaller 1 200 pipes and the reaction is cooled by the counter-current flow of the
cooling that surrounds individual pipes.

3.1   Mathematical Model

The mathematical model of this tubular chemical reactor with counter-current
cooling could be described by the set of PDE:

$$
\begin{aligned}
\frac{\partial c_A}{\partial t} + v_r \cdot \frac{\partial c_A}{\partial z} &= -k_1 \cdot c_A \\
\frac{\partial c_B}{\partial t} + v_r \cdot \frac{\partial c_B}{\partial z} &= k_1 \cdot c_A - k_2 \cdot c_B \\
\frac{\partial T_r}{\partial t} + v_r \cdot \frac{\partial T_r}{\partial z} &= \frac{h_r}{\rho_r \cdot c_{pr}} - \frac{4 \cdot U_1}{d_1 \cdot \rho_r \cdot c_{pr}} \cdot (T_r - T_w) \\
\frac{\partial T_w}{\partial t} &= \frac{4}{(d_2^2 - d_1^2) \cdot \rho_w \cdot c_{pw}} \cdot [d_1 \cdot U_1 \cdot (T_r - T_w) + d_2 \cdot U_2 \cdot (T_c - T_w)] \\
\frac{\partial T_c}{\partial t} - v_c \cdot \frac{\partial T_c}{\partial z} &= \frac{4 \cdot n_1 \cdot d_2 \cdot U_2}{(d_3^2 - n_1 \cdot d_2^2) \rho_c \cdot c_{pc}} (T_w - T_c)
\end{aligned}
\tag{6}
$$

The variable T in (6) is the temperature, d_1, d_2, d_3 represents diameters, ρ are
densities, c_p means specific heat capacities, U_1, U_2 stands for the heat transfer
coefficients, n_1 is a number of tubes and L represents the length of the reactor.
Index $(\cdot)_r$ is related to the reaction compound, $(\cdot)_w$ for the metal wall of the
pipes and $(\cdot)_c$ is used for the cooling liquid.

Variables v_r and v_c represents fluid velocities that are computed from

$$v_r = \frac{q_r}{f_r}; \quad v_c = \frac{q_c}{f_c} \quad \text{where} \quad f_r = n_1 \frac{\pi \cdot d_1^2}{4}; \quad f_c = \frac{\pi}{4} \left(d_3^2 - n_1 \cdot d_2^2 \right) \tag{7}$$

The nonlinearity of this system is mainly because of the reaction heat, h_r, and reaction velocities, k_i, that are computed via Arrhenius law:

$$h_r = h_1 \cdot k_1 \cdot c_A + h_2 \cdot k_2 \cdot c_B$$
$$k_i = k_{0i} \cdot \exp\left(\frac{-E_i}{R \cdot T_r}\right) \quad \text{for } i = 1, 2 \tag{8}$$

with k_{0i} as a pre-exponential factors, E_j are activation energies, R represents universal gas constant and h_i denotes activation energies. Fixed parameters of the reactor are not mentioned here due to the length of the contribution but they could be found in [5].

This system has five state variables - concentrations of compounds A and B $c_A(z, t)$, $c_B(z, t)$, the temperature of the reactant $T_r(z, t)$, the temperature of the metal wall $T_w(z, t)$ and the temperature of the coolant $T_c(z, t)$. There is also wide range of input variables, the one used here is the change of the flow rate of the coolant q_c mainly from the practical point of view. This flow rate can be found in the computation of the fluid velocity v_c as it is shown in (7). On the other hand, the mean reactant temperature inside the reactor was chosen as a controlled output. Both, control ($u(t)$) and controlled ($y(t)$) variable are then

$$u(t) = \frac{q_c(t) - q_c^s}{q_c^s} \cdot 100 \, [\%] \quad y(t) = \frac{\sum\limits_{z=1}^{N} T_r(z, t)}{N} \, [\text{K}] \tag{9}$$

where q_c^s denotes initial (steady-state) volumetric flow rate of the coolant, z is space variable that corresponds to the discretization mentioned above where the length of the reactor L is divided into N equivalent parts.

Detailed results of the steady-state and the dynamic analyses are again discussed in [5], but we can say that the observed dynamic system produces expected nonlinear behaviour and the working point could be chosen by the combination of the flow rates $q_r^s = 0.15 \, \text{m}^3 \cdot \text{s}^{-1}$ and $q_c^s = 0.275 \, \text{m}^3 \cdot \text{s}^{-1}$. The steady-state mean temperature of the coolant for this working point is then $T_{r,mean} = 333.17 \, \text{K}$. This value is also initial condition for dynamic analysis where step responses of controlled output $y(t)$ according to obtained results of the dynamic analysis shown in [5] could be described by the second order TF

$$G(s) = \frac{b_1 s + b_0}{s^2 + a_1 s + a_0} \tag{10}$$

This continuous-time TF is used as an ELM in the adaptive control.

3.2 Simulation of Adaptive Control

Once we have ELM of the controlled system in the form of TF (10), we can move on to the hybrid adaptive control described in Sect. 2. Degrees of the polynomials $\tilde{p}(s), q(s), r(s)$ and $d(s)$ are for the second order TF (10) $\deg \tilde{p}(s) = 1$, $\deg q(s) = 2$, $\deg r(s) = 0$ and $\deg d(s) = 4$ which means that TF (3) could be rewritten to

$$Q(s) = \frac{q_2 \cdot s^2 + q_1 s + q_0}{s \cdot (p_1 s + p_0)}; \quad R(s) = \frac{r_0}{s \cdot (p_1 s + p_0)} \tag{11}$$

and polynomial $d(s)$ on the right side of Diophantine equations (4) has form
$d(s) = (s + \alpha)^2 \cdot n(s)$.

Similarly, the data vector ϕ_δ and the estimated vector of parameters $\boldsymbol{\theta}_\delta$ in (2) are

$$\phi_\delta = [-y_\delta(k-1), -y_\delta(k-2); u_\delta(k-1), u_\delta(k-2)]^T; \boldsymbol{\theta}_\delta = [a'_1, a'_0, b'_1, b'_0]^T \quad (12)$$

where

$$y_\delta(k) = \frac{y(k) - 2y(k-1) + y(k-2)}{T_v^2}$$

$$y_\delta(k-1) = \frac{y(k-1) - y(k-2)}{T_v} \quad y_\delta(k-1) = y(k-2) \quad (13)$$

$$u_\delta(k-1) = \frac{u(k-1) - u(k-2)}{T_v} \quad u_\delta(k-1) = u(k-2)$$

The control with the adaptive 2DOF controller was tested on different changes of the reference signal $w(t)$ and various values of the tuning parameter $\alpha = 0.002$, 0.004 and 0.01. The sampling period was chosen as $T_v = 1\,\text{s}$ and the starting vector of parameters was $\boldsymbol{\theta}_\delta(0) = [0.1, 0.1, 0.1, 0.1]^T$. Results are presented in Fig. 2.

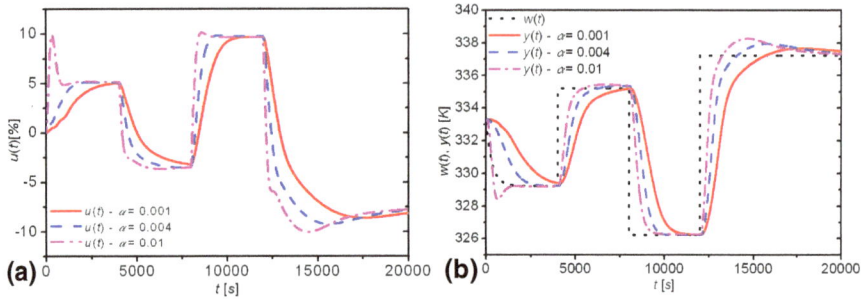

Fig. 2. Results of the adaptive control - (a) the course of the input variable $u(t)$ and (b) courses of the reference signal $w(t)$ and the output variable $y(t)$

Results of simulations displayed in Fig. 2 have shown that the proposed approach have very good control results and we can say that this type of 2DOF adaptive controller could be used for control of this type of systems or similar. We can also read from the graphs, that increasing value of the tuning parameter α results in the quicker output response but also in bigger overshoots of the output variable.

4   Conclusion

Proposed hybrid adaptive control in this paper is based on the recursive identification of the ELM that was designed as a second order transfer function with the relative order one with the knowledge from the dynamic analysis. The control system has two degrees-of-freedom (2DOF) configuration which was designed with the use of the polynomial synthesis and the Pole-placement method with Spectral factorization. These methods produce not only structure, but also relations for computation of the controller's parameters that depends on the ELM and as they are estimated during the control, parameters of the controller are recomputed too. The controller also provides selection of the course of the output variable via the choice of the parameter α that represents the position of the root in the Pole-placement method. It was shown that the increasing value of this parameter produces quicker output response but also overshoots of the output variable.

Acknowledgment. This article was created with support of the Ministry of Education of the Czech Republic under grant IGA reg. n. IGA/CebiaTech/2018/002.

References

1. Russell, T., Denn, M.M.: Introduction to Chemical Engineering Analysis. vol. xviii, 502 p. Wiley, New York (1972) ISBN 04-717-4545-6
2. Ingham, J., Dunn, I.J., Heinzle, E., Penosil, J.E.: Chemical Engineering Dynamics. An Introduction to Modelling and Computer Simulation, 2nd edn. VCH Verlagsgesellshaft, Weinheim (2000)
3. Dostal, P., Prokop, R., Prokopova, Z., Fikar, M.: Control design analysis of tubular chemical reactors. Chem. Papers **50**, 195–198 (1996)
4. Evans, L.C.: Partial Differential Equations. Graduate Studies in Mathematics, vol. 19, 2nd edn. American Mathematical Society, Providence (2010). ISBN 978-0821849743
5. Vojtesek, J., Dostal, P.: Adaptive control of the tubular reactor with co-and counter-current cooling in the jacket. In: 23rd European Conference on Modelling and Simulation, Madrid, pp. 544-550 (2009). ISBN 978-0-9553018-8-9
6. Grimble, M.J.: Robust Industrial Control Systems: Optimal Design Approach for Polynomial Systems. Wiley, Hoboken (2006). ISBN 978-0-470-02073-9
7. Astrom, K.J., Wittenmark, B.: Adaptive Control, 2nd edn. Prentice Hall, Reading (1994). ISBN 978-0201558661
8. Bobal, V., Bhm, J., Fessl, J., Machacek, J.: Digital Self-tuning Controllers: Algorithms. Implementation and Applications. Advanced Textbooks in Control and Signal Processing. Springer, London (2005). ISBN 1-85233-980-2
9. Stericker, D.L., Sinha, N.K.: Identifcation of continuous-time systems from samples of input-output data using the δ-operator. Control Theor. Adv. Technol. **9**, 113–125 (1993)
10. Mukhopadhyay, S., Patra, A.G., Rao, G.P.: New class of discrete-time models for continuos-time systeme. Int. J. Control **55**, 1161–1187 (1992)
11. Kucera, V.: Diophantine equations in control - a survey. Automatica **29**, 1361–1375 (1993)
12. Mikles, J., Fikar, M.: Process Modelling, Identification, and Control. Springer, Berlin (2007). ISBN 978-3-540-71970-0

CHP Production Optimization Based on Receding Horizon Strategy

Jakub Novak[⊠], Petr Chalupa, Viliam Dolinay, and Lubomir Vasek

Faculty of Applied Informatics, Tomas Bata University in Zlin, Nam TGM 5555,
Zlin, Czech Republic
jnovak@fai.utb.cz

Abstract. In the paper model of cogeneration plant with heat storage was developed for investigating the potential of implementing the heat storage. The model is used within the operation optimization to maximize the revenue from power sales. Receding horizon of 7 days is used for operational planning resulting in linear optimization problem. The method is used to evaluate the benefit of different sized storages using the real-life heat demand data for Malenovice district in Czech Republic and day-ahead electricity prices for a one-year horizon. The results show that the revenue from power sales can be significantly improved by shifting operation to the most commercially feasible points in time.

Keywords: Cogeneration · Heat storage · Linear programming

1 Introduction

A combined heat and power (CHP) plant simultaneously produces thermal and electrical energy, primarily to meet industry and household demand. The energy efficiency of the CHP plants is higher than the energy efficiency of the conventional thermal power plants, leading to decreased environmental impact and significant economic profitability. Due to the coupling of its heat and power output, CHP plant has limited flexibility. Installation of heat storage tank (HST) allows for heat accumulation and increase of the flexibility of the heat and electricity production. It enables to shift production of electricity and heat to hours where electricity prices are high especially during the days with low heat demand. Heat storage tanks also allows operation of the production units at more efficient loads. There are many studies considering CHP plant and HST optimization. Sliding window method was developed in [1] for optimization of the power production. The approach uses weather forecast for heat demand prediction and power prices for maximization of the revenue from power sales. Christidis et al. [2] optimized the design of heat storage devices together with the operation of a power plant supplying a large district heating network by formulating a mixed integer linear programming (MILP) problem that was solved using CPLEX software. MILP technique is also used in [3] for calculation of the investment potential of a heat storage. The effects of power prices uncertainties for different capacity of thermal storage are evaluated in simulation study [4]. The short-term operation planning of the CHP with heat storage is closely linked to the unit commitment problem.

© Springer International Publishing AG, part of Springer Nature 2019
J. Machado et al. (Eds.): HELIX 2018, LNEE 505, pp. 110–115, 2019.
https://doi.org/10.1007/978-3-319-91334-6_16

In this paper, we propose dynamic model with a 1-h sampling period which is used in the optimization scheme that maximizes the revenue from the power sales. The amount of consumed fuel must remain the same for the given prediction horizon as given by historical values and only the production of heat and electricity can be "shifted" in time and during the time of decreased production the demanded heat is obtained from the heat storage. The heat demand from the district heating system and the prices for the electrical power system are assumed to be known (historical data from 2016). An optimization model based on dynamic programming is used to calculate the optimal operational strategy for the operation of the turbine and the thermal storage. This is done for 1-week prediction horizon. The results are compared with a reference case with no heat storage, thus yielding the investment potential for the heat storage.

2   Mathematical Formulation of the Problem

Simple model of the turbine is required for the optimization in the short-term operational planning. Several assumptions have been made for the development of the CHP model. The dynamics of the turbine is not considered and it is possible to change the output of the turbine in the whole operating range (2–25 MW). Thermal losses from the heat storage are neglected. Historical data of heat demand, electrical power from the turbine and power prices for the day ahead electricity market are used in the simulations. The performance of the CHP with the heat storage tank is compared with the performance of CHP plant without the storage by using the historical data from the plant.

In order to compare these two variants, the same amount of electrical energy must be produced and thus also the same amount the fuel must be consumed during the optimization period of 1 year. The electrical energy production can only be shifted to the most economically favourable time but during the period of decreased power production the heat demand must be covered by the heat from the heat storage.

The optimized variable is the difference between the electrical power of the plant with heat storage and the electrical power given by the historical values ΔP_{MWe} [MW]. For given prediction horizon Np the amount of consumed fuel and produced power energy must be the same:

$$\sum_{i=1}^{N_p} \Delta P_{MWe}(i) = 0 \tag{1}$$

where i is the index for time step (hour). Constant power to heat ratio (ratio between the electrical power output and heat contained in the steam extracted from the turbine for district heating) is considered.

$$\Delta P_t(i) = k\Delta P_{MWe}(i) \tag{2}$$

where ΔP_t [MW] is the difference between the actual heat power and the historical heat power which corresponds to the difference of electrical power output. The value of the ratio k is constant for the whole operating region of the turbine. When the value of ΔP_t

is positive, the extra heat ΔP_t is used to charge the heat storage. In case of negative ΔP_t, the heat storage is discharged and the heat from it is used for district heating.

$$\Delta P_t(i) = \Delta P_{AKU}(i) \tag{3}$$

The symbol ΔP_{AKU} [MW] represents the change of energy in heat storage. The heat demand must be satisfied at any step so the heat must be either extracted from the turbine or discharged from the heat exchanger. The amount of energy in the heat storage $P_{aku}(i)$ [MWh] is constrained:

$$0 \le P_{aku}(i) \le Q_{max} \tag{4}$$

where the Q_{max} [MWh] is the capacity of the heat storage. The actual amount energy in the heat storage is given by the initial state of the heat storage and by integration of values of $\Delta P_{AKU}(l)$ upto the time sample l.

$$P_{aku}(i) = P_{aku}^{init} + \sum_{j=1}^{i} \Delta P_{aku}(i) \tag{5}$$

The electrical power output of the turbine is also constrained:

$$P_{MWe}^{min} \le P_{MWe}(i) + \Delta P_{MWe}(i) \le P_{MWe}^{max} \tag{6}$$

where P_{MWe}^{min} [MW] and P_{MWe}^{max} [MW] are the minimal and maximal values of the electrical power output of the turbine.

The charge/discharge rate is also constrained because of the design parameters of the heat station. The heat station is used to convert the heat from the steam extracted from the turbine to hot water for the district heating network or for the heat storage.

$$\Delta P_{AKU}(i) + P_{DH}(i) \le P_{HES}^{max} \tag{7}$$

where P_{HES}^{max} [MW] is the maximal heat power output of the heat station and $P_{DH}(i)$ is the heat load given by the historical values.

3   Linear Programming Model

The considered CHP plant Otrokovice that provides the heat for Malenovice district is equipped with a 25 MW backpressure turbine and a 25 MW condensations turbine. The 25 MW condensation turbine is used only for provision of ancillary services and is not included in the optimization. The hourly model of the CHP for maximization of the revenue from power sales computed at the day D is as follows:

$$max \sum_{i=1}^{N_p} \Delta P_{MWe}(i) * T_s * c_{MWh}(i) \tag{8}$$

Subjected to: $\sum_{i=1}^{N_p} \Delta P_{MWe}(i) = 0$

$$\Delta P_t(i) = k\Delta P_{MWe}(i)$$

$$\Delta P_t(i) = \Delta P_{AKU}(i)$$

$$0 \le P_{aku}(i) \le Q_{max}$$

$$P_{MWe}^{min} \le P_{MWe}(i) + \Delta P_{MWe}(i) \le P_{MWe}^{max}$$

$$\Delta P_{AKU}(i) + P_{DH}(i) < P_{HES}^{max}$$

where $c_{MWh}(i)$ [EUR/MWh] is the price of electricity for the day-ahead market at time step i. The electricity prices on the deregulated day-ahead market vary from hour to hour and electricity production in the CHP plant is more profitable during certain hours. T_s is the esampling period of 1 h.

The planning horizon, for the operational strategy of the plants with storage used in this paper, is the coming 7-day period. Using the receding horizon strategy the optimization is performed on the prediction horizon of 7 days (D + 1, D + 2,..., D + 7) but the results are used only for the following day. The time window is then shifted by 1 day and the optimization is repeated. The power prices for days D + 2 to D + 7 are assumed to be the same as in day D + 1 because they are not known at the time of optimization D. The planning strategy is based on the assumption that the future heat demand is known therefore the results represent the maximum revenue potential.

4 Computational Results

The considered CHP plant is equipped with a 25 MW backpressure turbine and a 25 MW condensations turbine. The 25 MW condensation turbine is used only for provision of ancillary services and is not included in the optimization. For district heating the heat station with the maximal power output of 50 MW is engaged. Total heat consumption of Malenovice district in year 2016 was 113,6 GWh. The parameters of the CHP plant together with historical values of D + 1 electricity prices c_{MWh}, heat loads P_{DH}, power outputs P_{MWe} define the constraints of the linear programming problem. Table 1 summarizes the values of the parameters for the CHP model used in this study. In the study the plant operates using a fixed power-to-heat ratio even though the actual CHP plant is more complex and flexible in its operation.

Table 1. Model parameters

Parameter	Value
Maximal electrical power output P_{MWe}^{max}	25 MW
Minimal electrical power output P_{MWe}^{max}	2 MW
Power to heat ratio k	3
Heat station maximal power P_{HES}^{max}	50 MW
Capacity of the heat storage Q_{max}	10, 25, 50, 75, 100, 125, 150, 175 MWh

The model is solved with different amounts of storage capacity, running from 10 to 150 MWh. The optimization results for a single winter and summer day and a 100 MWh thermal storage are presented in Figs. 1 and 2.

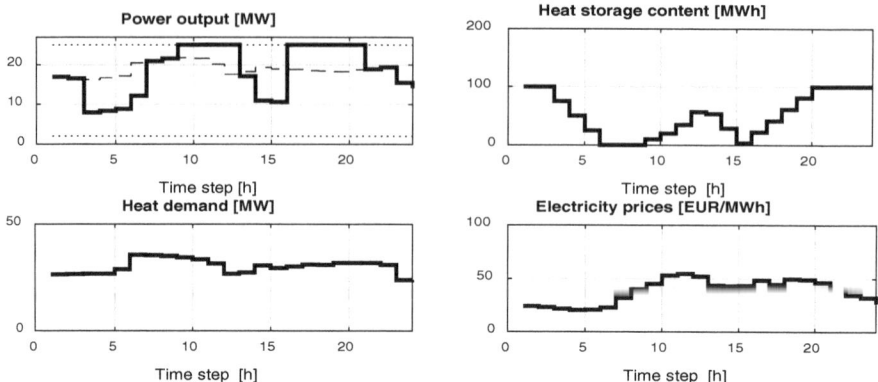

Fig. 1. Optimization results – January 5th 2016 (bold line – CHP plant with heat storage, dashed line – historical data)

It is evident that during winter when the heat demand is high the flexibility of control is increased and CHP plant can benefit from moving the production to times with high electricity prices and store the extra heat in the heat storage. During the summer the flexibility is decreased and thermal storage has marginal effect on the value of revenue from power sales. Thus operating the heat storage during summer is questionable because of the operating costs. The results for different capacities of the thermal storage are presented in Table 2.

The results show the increased revenue from power sales for increasing capacity of the thermal storage. Without the storage the revenues from power sales would be 2.47 million Euro using the power production values and day-ahead prices if no self-consumption is considered. The increase of profit is 3–10% according to the size of heat storage. The limiting factor is the low heat demand of the Malenovice district and the thermal power output of the heat station located in the CHP plant Otrokovice. Revenue from the heat storage investment depends on the capital costs of the heat storage technology.

5 Conclusion

In general, the feasibility of using additional thermal storage depends on the individual conditions, such as electricity and fuel prices, their variation over time, investments costs, national energy policy, taxes and the local energy demand. The increased flexibility of the plant with heat storage is used to benefit from significant variations in prices between peak and off-peak hours on the day ahead market for electricity. In order to evaluate the effects of adding the heat storage to a CHP plant a method based

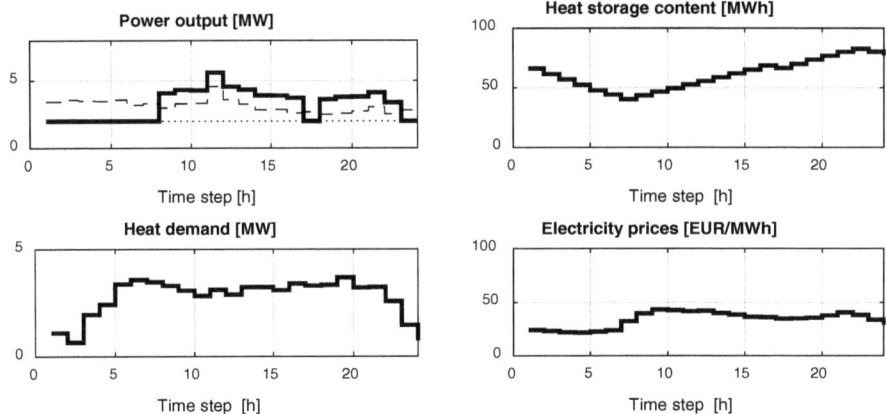

Fig. 2. Optimization results – May 25th 2016 - CHP plant with heat storage, dashed line – historical data)

Table 2. Revenue from power sales for different capacities of the thermal storage

Capacity [MWh]	10	25	50	75	100	125	150	175
Revenue [kEUR]	39,2	85,4	136,1	170,9	197,1	216,6	230,4	239,8

on receding horizon was developed. Historical data from Malenovice district and CHP plant Otrokovice were used to demonstrate the method and the feasibility study enables to evaluate the effects of adding the heat storage to a plant.

Acknowledgment. This work was supported by the Ministry of Education, Youth and Sports of the Czech Republic within the National Sustainability Programme project No. LO1303 (MSMT-7778/2014) and by Technology Agency of Czech Republic within the project TH02020979.

References

1. Fang, T., Lahdelma, R.: Optimization of combined heat and power production with heat storage based on sliding time window method. Appl. Energy **162**, 723–732 (2016)
2. Christidis, A., Koch, C., Pottel, L., Tsatsaronis, G.: The contribution of heat storage to the revenueable operation of combined heat and power plants in liberalized electricity markets. Energy **41**(1), 75–82 (2012)
3. Rolfsman, B.: Combined heat-and-power plants and district heating in a deregulated electricity market. Appl. Energy **78**, 37–52 (2004)
4. Schacht, M., Werners, B.: Impact of heat storage capacity on CHP unit commitment under power price uncertainties. In: Operational Research Proceedings, pp. 487–492. Aachen University, Germany (2014)

Decentralized Adaptive Suboptimal LQ Control in Microsoft Excel VBA

Karel Perutka[✉]

Tomas Bata University in Zlin,
Nám. T.G. Masaryka 5555, 76001 Zlin, Czech Republic, EU
kperutka@fai.utb.cz

Abstract. Suboptimal linear quadratic (LQ) control has been popular very much because of its very good results in several tasks of control. It allows us to implement it on many systems even on unstable systems which it stabilizes. Adaptation enlarges the area of the usage especially in the case when standard controllers with fixed parameters gives unsatisfactory results. In this paper, the main attention is dedicated to the usage of the modified instrumental variable technique as the identification part of the self-tuning controllers, and to the implementation in Microsoft Excel. So this approach was verified by simulation in Microsoft Excel Visual Basic for Applications (VBA) on two input two output (TITO) systems.

Keywords: Adaptive control · Instrumental variable · Linear quadratic control
Microsoft Excel VBA · Self-tuning control · Suboptimal control

1 Introduction

Suboptimal linear quadratic control, which was introduced in [1], has been popular very much because of its very good results in several tasks of control. The systems can be non-smooth, and for instance they can be time delayed, too. It is also possible to choose the value of the overshooting of the output signal and the speed of reaching the set point without overshooting or automatically set the fastest achievement of the set point without overshooting [2]. By the way whereof it implies the approach to control of single input single output (SISO) systems sketched in short above has to be and is very successful.

Adaptation enlarges the area of the usage of this approach [3, 4], for example the adaptive-tuning of extended Kalman filter used for small scale wind generator control [5]. Nonlinear systems might be the examples, for example the self-tuned local feedback gain based decentralized fault tolerant control for a class of large-scale nonlinear systems [6]. The fundamental principle of adaptation is based on the fact derived from natural systems. The behavior of natural systems and these systems themselves are changing in accordance with the change of the living environment. Before the natural system changes itself the changes in the living environment have to be recognized, identified. Therefore, the appropriate identification algorithm is crucial for the behavior of the controlled system. The usage of self-tuning controller is common in practice, for example for temperature control at polymerization reactor [7], speed control of

© Springer International Publishing AG, part of Springer Nature 2019
J. Machado et al. (Eds.): HELIX 2018, LNEE 505, pp. 116–123, 2019.
https://doi.org/10.1007/978-3-319-91334-6_17

electrical motor drives [8], PID control in pursuit of plug and play capacity [9], design of an implicit self-tuning PID controller based on the generalized output [10], or at velocity feedback for vibration control on the flexible structure [11]. In this paper, the main attention is dedicated to high order SISO systems. Nevertheless, this approach can be successfully extended on other systems in various types of control such as autonomous decentralized control of multi input multi output (MIMO) systems [12]. Self-tuning MIMO disturbance feedforward control for active hard-mounted vibration isolators is nice example MIMO systems control [13]. Modified instrumental variable method based on the linear transformation of the action and output signal using reinitialized partial moments is used as the identification algorithm, for more details see [14]. The principle of the method of the reinitialized partial moments (RPM) reposes on the calculation of the integrals of the input-output signals stabilized using the function which is given in advance. It has one big advantage – the terms of the initial conditions do not appear in the phase of the estimation of the model parameters. In our case, continuous-time suboptimal linear quadratic tracking controller controls the continuous-time system. The usage of Microsoft Excel in the area of control became popular in the last years, for example for data acquisition and real-time control [15]. Simulation results, as can be seen thereinafter, show us very effective control of high order system using the suboptimal LQ tracking method ant it is one of the ways how high order SISO system can be excellently controlled. The implementation of the selected method of control in MS Excel VBA was the main goal of the paper.

2 Theoretical Background

2.1 Description of TITO System

Let us describe this approach on two input two output (TITO) system because it is simpler to understand and there are no problems to extend it up to higher I/O systems. The internal structure of the system is shown in Fig. 1.

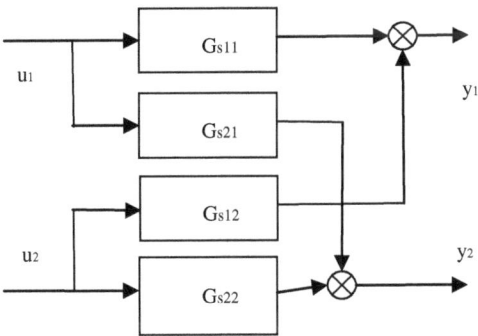

Fig. 1. Two input – two output system – the "P" structure

The transfer matrix of the system is

$$G = \begin{bmatrix} G_{s11} & G_{s12} \\ G_{s21} & G_{s22} \end{bmatrix} \tag{1}$$

It is possible to assume that the system is described by the matrix fraction

$$\mathbf{G} = \mathbf{A}^{-1}\mathbf{B} = \mathbf{B}_1\mathbf{A}_1^{-1} \tag{2}$$

See [12] for more details.

2.2 Decentralized Control

The system has the "P" structure and it is decomposed de facto to 2 single input single output systems (SISO) G_{s11} and G_{s22} controlled by 2 SISO controllers. The corresponding transfer functions of these subsystems can be written as

$$G_{s11}(s) = \frac{G_{s21}(s)G_{s12}(s)G_{r22}(s)}{1 + G_{s22}(s)G_{r22}(s)} \tag{3}$$

$$G_{s22}(s) = \frac{G_{s21}(s)G_{s12}(s)G_{r11}(s)}{1 + G_{s11}(s)G_{r11}(s)} \tag{4}$$

See [12] for more details.

2.3 Suboptimal Linear Quadratic Tracking

Suboptimal linear quadratic tracking was introduced in [1]. Let us consider Fig. 2.

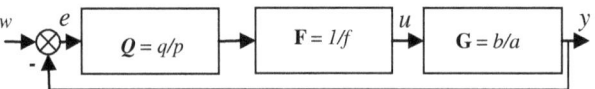

Fig. 2. System with feedback controller

Let us minimize the quadratic functional

$$J = \int_0^\infty \left\{ \mu e^2(t) + \varphi u^2(t) \right\} dt \tag{5}$$

where $\mu \geq 0, \varphi > 0$ are the penalty constants. The controller synthesis method giving the suboptimal tracking of the reference signal for a group of reference signals is needed. The spectral factorization of the equations

$$(as)^*\varphi as + b^*\mu b = g^*g, n^*n = a^*a \tag{6}$$

gives the coefficients of g, n which are used to solve the Diophantine equation

$$asp + bq = gn \tag{7}$$

from which the parameters of the controller are obtained. There was introduced the regression method for auto-tuning the penalization constants of the quadratic functional [2]. Let us suppose the system transfer function in the form

$$G_S(s) = \frac{b_0}{s^4 + a_3 s^3 + a_2 s^2 + a_1 s + a_0} \tag{8}$$

Then the spectral factorization gives the coefficients that have to be evaluated numerically. In accordance with the system transfer function hereinbefore mentioned the controller has the form

$$G_R(s) = \frac{q_4 s^4 + q_3 s^3 + q_2 s^2 + q_1 s + q_0}{s(p_4 s^4 + p_3 s^3 + p_2 s^2 + p_1 s + p_0)} \tag{9}$$

See [1] for more details.

2.4   Identification

The instrumental variable method is well-known and therefore it will not be described, for more details see [14]. However the used method of linearization is not so known and will be briefly described, see [14] for more details. Let us consider the system represented by the linear model of the nth order. The input-output data are transformed according to the following: the signals are changed using the function $p_i(\tau)$ and then integrated in the interval $[0, \hat{t}]$ where t is the parameter of the synthesis of the method yielding the following algebraic equation

$$y(t) - \alpha_n^y(t) - \sum_{i=0}^{n-1} a_{n-i} \alpha_i^y(t) = \sum_{i=0}^{m} b_{m-i} \alpha_i^u(t) + \varepsilon_0(t) \tag{10}$$

where

$$\alpha_i^u(t) = \int_0^{\hat{t}} p_i(\tau) u(t - \hat{t} + \tau) d\tau \tag{11}$$

$$\varepsilon_0(t) = \int_0^{\hat{t}} p_0(\tau) v(t - \hat{t} + \tau) d\tau \tag{12}$$

for which the function $p_i(\tau)$ is

$$p_0(\tau) = \frac{\tau^n (\hat{t} - \tau)^{n-1}}{(n-1)\hat{t}^n} \tag{13}$$

$$p_1(\tau) = -\frac{dp_0(\tau)}{d\tau}; \ldots; p_i(\tau) = -\frac{dp_{i-1}(\tau)}{d\tau} = (-1)i\frac{d^i p_0(\tau)}{d\tau^i}; i = 1\ldots n \qquad (14)$$

The integration is realized in the interval of the lengths $\hat{\imath}$ in the all observation horizon $[0, T]$. Because N samples of input-output data signal are available and when we noted $\hat{k} = \hat{\imath}/T_e$ we have $N - \hat{k}$ measurements for the estimation. It is now possible to obtain the following matrix relation (t varies from $\hat{\imath}$ to T), k varies from $\hat{k}$ to $N - 1$)

$$Y = \Gamma\Theta + \Lambda + V_0 \qquad (15)$$

where

$$Y = \left[y(\hat{k}) \cdots y(N-1)\right]^T \qquad (16)$$

$$\Gamma = \begin{bmatrix} \alpha_0^y(\hat{k}) & \cdots & \alpha_{n-1}^y(\hat{k}) & \alpha_0^{yu}(\hat{k}) & \cdots & \alpha_m^u(\hat{k}) \\ \vdots & \ddots & \vdots & \vdots & \ddots & \vdots \\ \alpha_0^y(N-1) & \cdots & \alpha_{n-1}^y(N-1) & \alpha_0^u(N-1) & \cdots & \alpha_m^u(N-1) \end{bmatrix} \qquad (17)$$

$$\Lambda = \left[\alpha_n^y(\hat{k}) \cdots \alpha_n^y(N-1)\right]^T \qquad (18)$$

$$\Theta = [a_1 \cdots a_n \ b_0 \cdots b_m]^T \qquad (19)$$

$$V_0 = \left[\varepsilon(\hat{k}) \quad \cdots \quad \varepsilon(N-1)\right]^T \qquad (20)$$

3 Simulation Example

Let us have the system described by

$$G_{sij} = \frac{K_{ij}}{(0.4s + 1)(0.8s + 1)}; \mathbf{K} = \begin{pmatrix} 2.17 & 0.34 \\ 0.11 & 1.79 \end{pmatrix} \qquad (21)$$

Whole program defining the data, computing the solution and drawing the results, was written in Microsoft Excel Visual Basic for Applications (VBA) editor. The reason was simple, to work with the software which is available in almost every computer. For example, the part of the source code, the selected part of the main computational function of the program created in MS Excel VBA, is shown below

```
Function mainComputationCycleLQ(N As Integer, t As Dou-
ble, mi_lq1 As Double,fi_lq1 As Double,b0_1 As Dou-
ble,a3_1 As Double,a2_1 As Double,a1_1 As Double,a0_1 As
Double,mi_lq2 As Double,fi_lq2 As Double,b0_2 As Dou-
ble,a3_2 As Double,a2_2 As Double,a1_2 As Double,a0_2 As
Double, zR1 As Double, outR1 As Double)
Dim k, M As Integer
Dim A1(0 to 9) As Double
Dim A2(0 to 9) As Double
Dim A3(0 to 1) As Double
Dim A4(0 to 1) As Double
M=N-1;
For k=0 To M Step 1
'controllers parameters computation:
A1=suboptLQ(mi_lq1,fi_lq1,b0_1,a3_1,a2_1,a1_1,a0_1);
p4_1=A1(0); p3_1=A1(1); p2_1=A1(2); p1_1=A1(3);
p0_1=A1(4); q4_1=A1(5); q3_1=A1(6); q2_1=A1(7);
q1_1=A1(8); q0_1=A1(9);
A2=suboptLQ(mi_lq2,fi_lq2,b0_2,a3_2,a2_2,a1_2,a0_2);
p4_2=A2(0); p3_2=A2(1); p2_2=A2(2); p1_2=A2(3);
p0_2=A2(4); q4_2=A2(5); q3_2=A2(6); q2_2=A2(7);
q1_2=A2(8); q0_2=A2(9);
'action signals computation:
A3=rkf4r_c2j3contr(t(k),zR1(k),dt,p4_1,p3_1,p2_1,p1_1,p0_
1,q0_1,q1_1,q2_1,q3_1,q4_1,e1);
...
Next
End Function
```

The results of simulation are shown below. The control action was not limited considering the given time constants of the system and long settling time. The relatively long settling time was caused by two weight constants of the controller. The usage of Excel for adaptive control in conjunction with possible hardware input/output to real systems is the main advantage of this approach (Figs. 3 and 4).

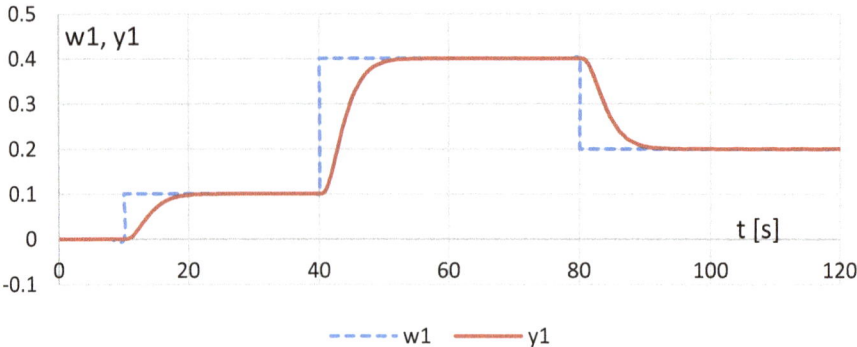

Fig. 3. History of control for 1st controller

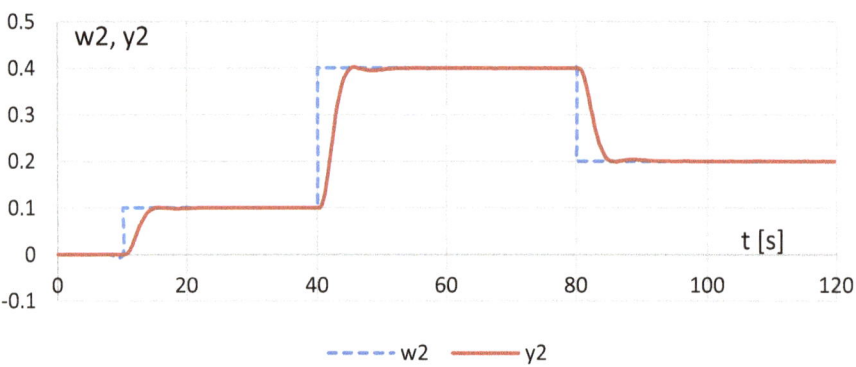

Fig. 4. History of control for 2nd controller

4   Conclusions

In this paper, the approach of control of TITO system was presented. The results show that this algorithm joining the instrumental variable with reinitialized partial moments and linear suboptimal quadratic tracking together in Microsoft Excel Visual Basic for Applications is suitable for control of multivariable systems. The main goals proposed for this paper were achieved.

References

1. Dostál, P., Bobál, V.: The suboptimal tracking problem in linear systems. In: Proceedings of the 7th Conference on Control and Automation, pp. 667–673. Haifa, Israel (1999)
2. Perutka, K.: Method of setting penalization constants in suboptimal linear quadratic tracking method. In: Proceedings of the 26th ASR 2001 Seminar on Instruments and Control. Ostrava, Czech Republic (2001)
3. Åström, K.J., Wittenmark, B.: Adaptive Control. Addison-Wesley, MA (1989)

4. Bobál, V., Böhm, J., Prokop, R., Fessl, J.: Practical Aspects of Self-Tuning Controllers: Algorithms and Implementation. VUTIUM Press, Brno University of Technology, Brno (1999)
5. Al-Ghossini, H., Locment, F., Sechilariu, M., Gagneur, L., Forgez, C.: Adaptive-tuning of extended Kalman filter used for small scale wind generator control. Renew. Energy **85**, 1237–1245 (2016)
6. Zhaoa, B., Lib, Y., Liuc, D.: Self-tuned local feedback gain based decentralized fault tolerant control for a class of large-scale nonlinear systems. Neurocomputing **235**, 147–156 (2017)
7. Vasanthi, D., Pranavamoorthy, B., Pappa, N.: Design of a self-tuning regulator for temperature control of a polymerization reactor. ISA Trans. **51**, 22–29 (2012)
8. Zaky, M.S.: A self-tuning PI controller for the speed control of electrical motor drives. Electr. Power Syst. Res. **119**, 293–303 (2015)
9. Mendes, J., Osório, L., Araújo, R.: Self-tuning PID controllers in pursuit of plug and play capacity. Control Eng. Pract. **69**, 73–84 (2017)
10. Ashida, Y., Wakitani, S., Yamamoto, T.: Design of an implicit self-tuning PID controller based on the generalized output. IFAC PapersOnLine (50-1), pp. 13946–13951 (2018)
11. Zillettia, M., Elliott, S.J., Gardonio, P.: Self-tuning control systems of decentralised velocity feedback. J. Sound Vib. **329**, 2738–2750 (2010)
12. Bobál, V., Perutka, K., Dostál, P.: Decentralized adaptive continuous-time control of coupled drives apparatus. In: Proceedings of the IFAC Workshop on Adaptation and Learning in Control and Signal Processing, pp. 57–62. Cernobbio-Como, Italy (2001)
13. Beijen, M.A., Heertjes, M.F., Van Dijk, J., Hakvoort, W.B.J.: Self-tuning MIMO disturbance feedforward control for active hard-mounted vibration isolators. Control Eng. Pract. **72**, 90–103 (2018)
14. Mensler, M.: Analyse et étude comparative de méthodes d´identification des systèmes à representation continue: Développmnet d´une boîte à outils logicielle. CRAN, Nancy, France (1999)
15. Nourdine, A.: Data acquisition and real-time control using spreadsheets: Interfacing Excel with external hardware. ISA Trans. **49**, 264–269 (2010)

Formal Verification: Focused on the Verification Using a Plant Model

Joel Galvão$^{(\boxtimes)}$, Cedrico Oliveira, Helena Lopes, and Laura Tiainen

Mechanical Engineering Department, University of Minho, Guimarães, Portugal
jmrgalvao@hotmail.com

Abstract. The main goal of this paper is present a review and discussion about the option of using plant models in formal verification techniques. Relevant works in the field considering different approaches are reviewed and the importance of choosing the level of detail correctly is discussed. Although exists few works about this topic, the studies revealed the necessity and importance to consider the plant model in formal verification.

Keywords: Formal verification · Real time modelling · Model checking
Plant model

1 Introduction

In the 60's the industrial control systems [1] were made using electromechanical devices such as relays or drum switches. The need to design and develop new and more flexible systems arises from industry. This specification combined with the today's microprocessors make it possible to control large and complex organizations and enable making changes through alterations in the code.

To analyse a system, simulation can be used to test the program by sub-system part testing [2–6]. Today's challenge is that industries need a more powerful tool to ensure that all control states are tested.

There is a growing interest in using formal verification methods [7–9] is within software applications for industrial controllers. Simply because it is possible to test if the system reach deadlock (on state without any transposable transition), testing all the possible system behaviour evolution scenarios and ensure that the project requirements are fulfilled without having it done physically. This is very useful because doing real time verification [3] can be dangerous to the operator, it can lead to damaging of the machines and wasting materials. However, sometimes it takes a large amount of time to get these results and need specialized personal to create the models. Also a correct understanding of the logics used to express the properties of a system's behaviour is needed in order to guarantee that properties correctly encode the intent of the verification process.

To do formal verification we can use model checking [10] techniques (see Fig. 1), that are basically a technique where same mathematic language models are applied to the controller and process. The system can then be simulated and properties examined with the help of algorithms created to the effect.

© Springer International Publishing AG, part of Springer Nature 2019
J. Machado et al. (Eds.): HELIX 2018, LNEE 505, pp. 124–131, 2019.
https://doi.org/10.1007/978-3-319-91334-6_18

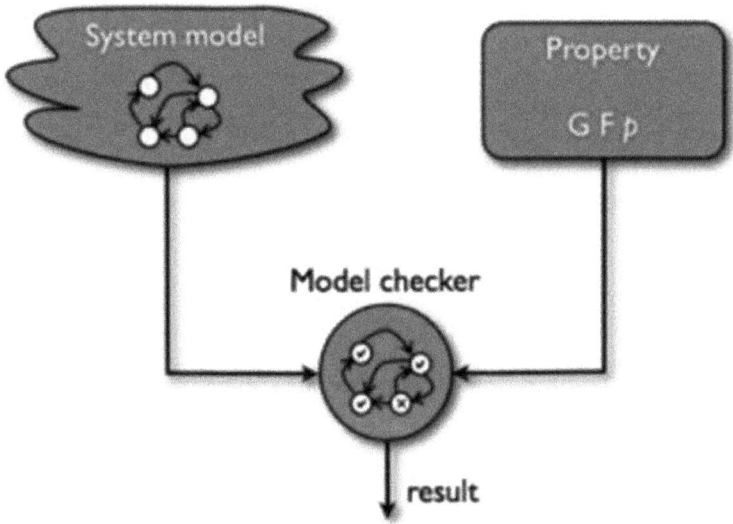

Fig. 1. The overview of the model checking method [8]

It is also possible to use theorem-proving [11] methods which are restricted by a given world model. They are used for giving answers related to which operations are applicable and if the goal set was satisfied or not. The overall performance of the system can be improved by employing separate strategies for two activities: the theorem proving what occurs within a world model, and the search through the space of models.

The structure of the programs can follow two different standards: the IEC 61131 [12] or the IEC 61499 [13]. The third part of IEC 61131 [14] establishes definitions about the programming aspects of industrial controllers like logical programming blocks and programming languages in a single unit. The IEC 61499 [13] standard system, in which the intelligence is embedded into a number of software components, can be distributed across networked devices.

Some important work is done by Frey and Litz, the model of the control design process in PLC-programming [15]. Their work focused on formal methods for validation and verification PLC programs. Other important works were proposed by Vyatkin and Hanisch [16]. They developed a tool that follows IEC 61499 standard and provides a formal and model-based verification of the execution control of function blocks. Also, Roussel and Denis [17] presented a theorem-proving as a method to express and handle Ladder Diagram programs with specific algebra to verify the safety properties of these programs. The theorem-proving method uses a formal framework to represent Boolean variables states, events and physical delays. In 2006, Machado [18, 19] showed the impact of the use of a plant model on the formal verification. Through a study-case, it was showed that model and non-model approaches can complement each other. It considers an approach that works with the standardized source code of controllers and can be applied to distributed systems taking into consideration other issues besides the pure logic control.

Same authors like Kunz and his team [20] propose a technique of simulation by model-in-the-loop [21] and formal verification performed in the same environment. They show that the two techniques can complement each other. In the first phase the system is simulated and basic mistakes are found and corrected, in a second phase the system is test against the model checker tool to test if the system already simulated corresponds to what is needed in all controller states. If not, a situation path as result is given to the user showing where is something wrong.

This paper is composed of a critical review of articles that are considered relevant for the thematic of formal verification considering plant model. Also, the methods used for each of present methodology are discussed. The second section discusses relevant features of formal verification using the plant model to ensure dependability and security of program in mechatronic systems. In the third section is discussed the importance of correctly expressing the properties, the impact of the abstraction level and granularity in the formal verification technique.

2   Literature Review

Carpanzano and his working team [19], applied a method for specifying the behaviour of diverse parts of robotic cell. Basically, for each part of the robotic cell, its behaviour is modulated and verified the formal language TRIO [20] and the Zot [21] model checker. Each element is modulated as a unit with its behaviour using a state flow technique. The program structure follows the IEC 61499. There was no really distinction between the program part and the plant model part.

On the other hand, Enoiu et al. [5] present a method of converting programs written in FBD (function blocks diagram) to timed automata [22]. They considered models for the base elements of a program written in FBD, adding the synchronic of the internal PLC cycle (scan cycle) and the environment around it: the plants. It follows the interaction between the different models and finally the models of the base elements of a code in FBD, namely the functions and functional blocks described in Fig. 2. The properties are written in Timed Computation Tree Logic [22] and the tool used in UPPAAL [23]. This research focuses mainly on the program structure and does not show how the plant is modelled in this approach.

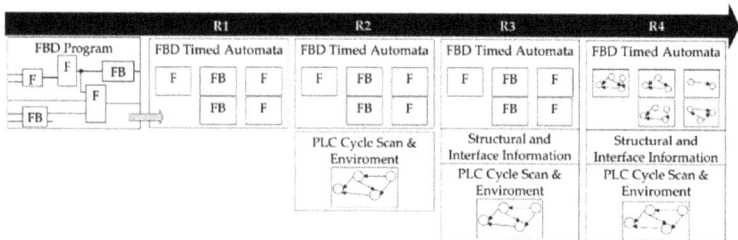

Fig. 2. Function Block Diagram to Timed Automata Transformation Process [5].

Other work is present by Pacheco and his team [23] where the development of the plant models for formal verification is discussed. This plant model approach using a modular approach was first present by Machado et al. [19]. The work of Pacheco attempts to authenticate the results using a simulation bench and proposes a model for a door of an elevator (Fig. 3), according to the plant model proposed by [24]. The model is posed of four states. The first state represents a closed door (P1), the second an opening door (P2), the third (P3) an open door and the fourth state (P4) is a closing door.

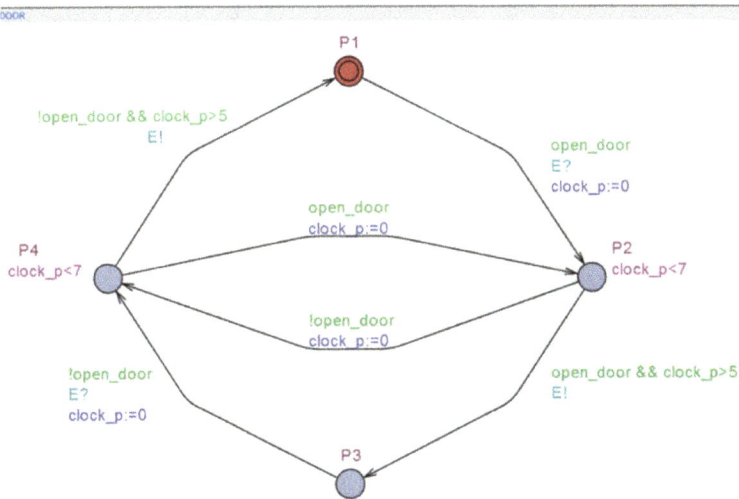

Fig. 3. Timed automata model of an elevator door [23]

The same modular technique is used to modulate a parking lot by Machado and Galvão [25] and also by Canadas and his working team [3] to do simulation.

Other work is proposed by Buzhinsky and Vyatkin [26] presents a model for pneumatic cylinder according to Fig. 4.

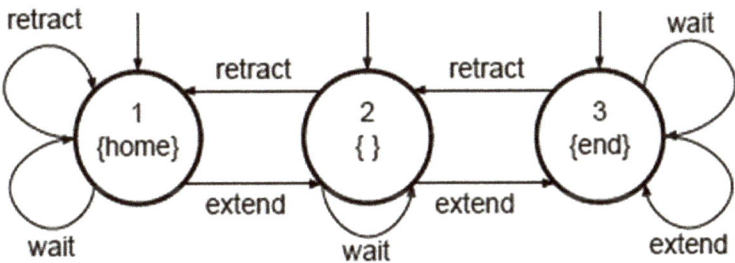

Fig. 4. Plant model representation of [26] in NCES

This paper shows that there is still a gap for formal verification considering the plant model and they try to solve it. They use net condition/event systems (NCESs) [27] and evaluated the model using checker tools such as the visual verifier ViVe/SESA [28] formerly implemented for this study. Other approach using the same tools was developed by Sorouri, Patil, Salcic and Vyatkin [29] to test the decentralized controller Function Blocks discrete-event dynamics of the plant software composition method for automated machines that exploit their mechatronic modularity. They follow IEC 61499 Function Blocks' (FBs) architecture. As case study, the proposed approach is used in pick-and-place manipulator with decentralized control synthesized (Fig. 5).

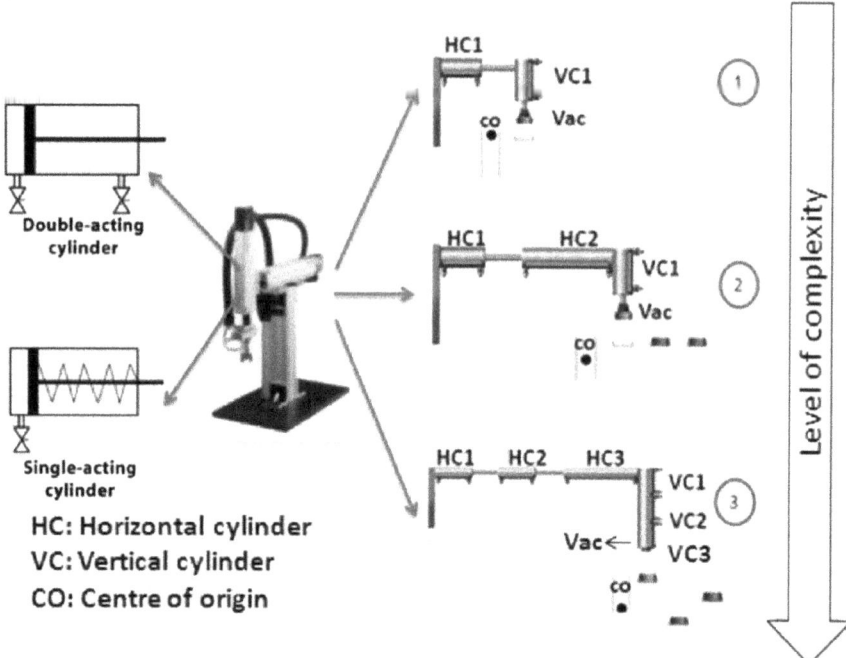

Fig. 5. Technique proposed by [29].

Basically, for each desired behaviour of a certain group of machines can be composed of actions of its mechatronic mechanisms, as well as completely dispersed scheduling and process control.

3   Properties, Abstraction Level and Granularity

Writing appropriate properties, in a logic suitable for verification, is a skillful process. Errors in this step of the process can create serious problems since a false sense of safety is gained from the analysis. [8] There are some works proposed by reference [5] that are focus mainly in the program itself and do not consider the physical part in detail.

In other cases like Carpanzano and his working team [30] the behaviour of each machine is modulated separately and his interaction tested, there was no real separation between the physical and control part.

In 2006, Machado et al. studied the impact of detail modelling of the plant using a modular approach. In fact, Machado and Galvão [25] proposed a modular approach that made possible to do formal verification, and show that is very useful because is possible to reutilize the models like was discuss also in [29]. But modulating the interaction between different models is hard to obtain in a realistic way as also discussed by Machado and Galvão [25]. In fact, there still is a gap in formal verification concerning the use of the plant models as discussed [26], but it is important to consider the model construction in several different ways taking into account the level of abduction and the granularity chosen. The choice of the degree of abstraction and the granularity [18] of the models has a direct effect on the obtained results in the use of the analysis techniques. Some aspects, which may be referred to, are:

- Computer resources and time available - Reduced granularity implies more robust computing resources and longer time periods needed to obtain results;
- Properties to be tested - If the user wants to prove simple properties of the models (for example, proving whether a sensor is "switched on" or "off" when certain events occur), models with fewer states should be used. However, when it is desired to prove more complex properties, such as the advancement or retreat of a pneumatic cylinder located in its intermediate position, for the controller to give a certain order, more complex models with more states should be used.

Is was showed that there are few works concerning the thematic of formal verification considering the plant model. But with the growing interest of the "fusion" of the physical and cyber world caused by the industry 4.0 [31], the need for reliability is extremely important [32] because the impact of error could be catastrophic and consider the plant model is unavoidable as discussed in [33].

4   Conclusions

Relevant literature has been analysed, regarding the important achievements in the last years, concerning use of formal verification. In fact, with the exponential increase of industry 4.0 leading to an increasing merger between the physical environment and the virtual environment the need for techniques such as formal verification are imperative as a way of increasing the dependability of the applied control systems.

To effectively apply this technique as demonstrated above, it is necessary to develop models with the correct degree of granularity for the properties that you want to test and considering the degree of abstraction in which you are working.

Application of a modular mechatronic approach is considered as one of the hypotheses to solve problem since it allows to reuse models, however it is complex to ensure that the different sub models interact with each other in a realistic way.

References

1. Considine, D.M., Considine, G.D.: Standard Handbook of Industrial Automation. Chapman & Hall, New York (1986)
2. Baresi, L., Carmeli, S., Monti, A., Pezzè, M.: PLC programming languages: a formal approach. In: Proceedings of Automation 1998 (1998)
3. Canadas, N., Machado, J., Soares, F.: Simulation of cyber physical systems behaviour using timed plant models. Mechatronics, **0–1** (2017). https://doi.org/10.1016/j.mechatronics.2017.10.009
4. Chioran, D., Machado, J.M.: Design of a mechatronic system for application of hardware-in-the-loop simulation technique (2011)
5. Enoiu, E.P., Sundmark, D., Pettersson, P.: Model-based test suite generation for function block diagrams using the uppaal model checker. In: 2013 IEEE Sixth International Conference on Software Testing, Verification and Validation Workshops (ICSTW), pp 158–167 (2013)
6. Barth, M., Fay, A.: Automated generation of simulation models for control code tests. Control Eng. Pract. **21**, 218–230 (2013)
7. Zhang, Y., Dong, Y., Hong, H., Zhang, F.: Code formal verification of operation system. Int. J. Comput. Netw. Inf. Secur. **2**, 10–18 (2010)
8. Campos, J., Machado, J.: A specification patterns system for discrete event systems analysis. Int. J. Adv. Robot. Syst. **10**, 315 (2013). https://doi.org/10.5772/56412
9. Meenakshi, B.: Formal verification. Resonance **10**, 26–38 (2005)
10. Alur, R., Courcoubetis, C., Dill, D.: Model-checking in dense real-time. Inf. Comput. **104**, 2–34 (1993)
11. Fikes, R.E., Nilsson, N.J.: Strips: A new approach to the application of theorem proving to problem solving. Artif. Intell. **2**, 189–208 (1971). https://doi.org/10.1016/0004-3702(71)90010-5
12. Ćengić, G., Åkesson, K.: On formal analysis of IEC 61499 applications, Part A: Modeling. IEEE Trans. Industr. Inf. **6**, 136–144 (2010)
13. Vyatkin, V.: IEC 61499 Function Blocks for Embedded and Distributed Control Systems Design. ISA-Instrumentation, Systems, and Automation Society, Oneida (2007)
14. John, K.-H., Tiegelkamp, M.: IEC 61131-3: Programming Industrial Automation Systems: Concepts and Programming Languages, Requirements for Programming Systems, Decision-Making Aids. Springer Science & Business Media, Heidelberg (2010)
15. Frey, G., Litz, L.: Formal methods in PLC programming. In: 2000 IEEE International Conference on Systems, Man, and Cybernetics, Nashville, USA, pp. 2431–2436 (2000)
16. Vyatkin, V., Hanisch, H.-M.: Verification of distributed control systems in intelligent manufacturing. J. Intell. Manuf. **14**, 123–136 (2003)
17. Roussel, J.-M., Denis, B.: Safety properties verification of ladder diagram programs. J. Eur. des Systèmes Autom. **36**, 905–917 (2002)
18. Machado, J.M.: Influence de la prise en compte d'un modèle de processus en vérification formelle des Systèmes à Evénements Discrets. Universidade do Minho (2006)
19. Machado, J.J.B., Denis, B., Lesage, J.-J., et al.: Logic controllers dependability verification using a plant model. In: Proceedings of the 3rd IFAC Workshop on Discrete-Event System Design, DESDes 2006, Rydzyna, Poland, 26–28 September 2006
20. Kunz, G., Machado, J., Perondi, E.: Using timed automata for modeling, simulating and verifying networked systems controller's specifications. Neural Comput. Appl. **28**, 1031–1041 (2017)

21. Plummer, A.R.: Model-in-the-loop testing. Proc. Inst. Mech. Eng. Part I J. Syst. Control Eng. **220**, 183–199 (2006). https://doi.org/10.1243/09596518JSCE207
22. Alur, R., Dill, D.: Automata for modeling real-time systems. In: Proceedings of Seventeenth International Colloquium on Automata, Languages, and Programming, pp. 322–335 (1990)
23. Pacheco, R., Gonzalez, L., Intriago, M.: Issues to be considered on obtaining plant models for formal verification purposes. In: IOP Conference Series: Materials Science and Engineering, vol. 147 (2016). https://doi.org/10.1088/1757-899X/147/1/012050
24. Machado, J., Denis, B.: A generic approach to build plant models for DES verification purposes, pp. 407–412 (2006)
25. Machado, J., Galvão, J., Fernandes, A.: Formal verification considering a systematic modeling approach for function blocks. J. Braz. Soc. Mech. Sci. Eng. **39**, 4107–4113 (2017). https://doi.org/10.1007/s40430-017-0893-7
26. Buzhinsky, I., Vyatkin, V.: Plant model inference for closed-loop verification of control systems : initial explorations. In: 2016 IEEE 14th International Conference on Industrial Informatics (INDIN), pp. 736–739 (2015)
27. Vyatkin, V.V.: Net condition/event systems (NCES) (2005). http://homepages.engineering. auckland.ac.nz/ ~ vyatkin/nces/net_condition_event_systems.htm. Accessed 18 Jun 2015
28. Vyatkin, V., Starke, P., Hanisch, H.-M.: ViVe and SESA model checkers (2007). http:// homepages.engineering.auckland.ac.nz/ ~ vyatkin/tools/modelchekers.html. Accessed 19 Mar 2015
29. Sorouri, M., Patil, S., Salcic, Z., Vyatkin, V.: Software composition and distributed operation scheduling in modular automated machines. IEEE Trans. Industr. Inf. **11**, 865–878 (2015). https://doi.org/10.1109/TII.2015.2430836
30. Carpanzano, E., Ferrucci, L., Mandrioli, D., et al.: Automated formal verification for flexible manufacturing systems. J. Intell. Manuf. **25**, 1181–1195 (2014)
31. Lasi, H., Fettke, P., Kemper, H.G., et al.: Industry 4.0. Bus. Inf. Syst. Eng. **6**, 239–242 (2014). https://doi.org/10.1007/s12599-014-0334-4
32. Kunz, G., Machado, J., Perondi, E., Vyatkin, V.: A formal methodology for accomplishing IEC 61850 real-time communication requirements. IEEE Trans. Industr. Electron. **64**, 6582–6590 (2017). https://doi.org/10.1109/TIE.2017.2682042
33. Khaitan, S.K., McCalley, J.D.: Design techniques and applications of cyber physical systems: a survey. IEEE Syst. J. **9**, 350–365 (2015). https://doi.org/10.1109/JSYST.2014. 2322503

Mechatronics Design, Medical Devices and Wellbeing

Mechatronic Design and Control of a Robotic System for Inspection Tasks

Pierluigi Rea$^{(\boxtimes)}$ and Erika Ottaviano

DICeM: Department of Civil and Mechanical Engineering,
University of Cassino and Southern Lazio,
via G. Di Biasio 43, 03043 Cassino, FR, Italy
{rea, ottaviano}@unicas.it

Abstract. Mobile robots are being widely used in various security and inspection applications. This contribution describes a mechatronic design of a mobile robot with the function of inspection system in indoor environment. This robot is tele operated and can be remotely controlled by mobile phone or tablet. It is equipped by the so-called internal and external sensors, which are efficiently managed by the designed mechatronic control scheme. The development of a paradigmatic robotic system for inspection and security tasks is the main objective of this paper to demonstrate the feasibility of mechatronic solutions for inspection of indoor sites. The development of such systems will be exploited in the form of a tool-kit to be flexible and installed on a mobile system, in order to be used for inspection and monitoring, introducing high efficiency, quality and repetitiveness in the addressed sector. The interoperability of sensors with wireless communication constitutes a smart sensors tool-kit and a smart sensor network with powerful functions to be used efficiently for inspection purposes.

Keywords: Hybrid leg-wheel locomotion · Inspection · Interoperability
Low-cost monitoring · Experimental tests

1 Introduction

Inspection and home security robotics are becoming a reality in recent years. Moreover, the advent of robots is an intrinsic part of everyday life and has gained immense popularity in recent years [1–3]. Soft computing and artificial intelligence are successfully applied to machinery control, robot manipulation and engineering applications. With the integration of nonlinear system and communication technology, smart and secure industry and home is no longer an idea of science fiction. Therefore, a robot may play an important role in such environments. Indoor inspection, surveillance and home safety are becoming critical issues at this time to organizing a smart and secure place to live and/or work [4, 5].

Advances in Closed Circuit TV (CCTV) technology is turning video surveillance equipment into the most valuable loss prevention tool. Such technology is a safety and security tool, which has been made available for either industrial, commercial or residential applications. The use of surveillance systems can alert users for threatening situations worsen, as well as provide them with an important record of events, inspect

© Springer International Publishing AG, part of Springer Nature 2019
J. Machado et al. (Eds.): HELIX 2018, LNEE 505, pp. 135–142, 2019.
https://doi.org/10.1007/978-3-319-91334-6_19

part of a production plant or inside an apartment for verification of structural and/or electrical components. However, in the conventional surveillance systems, only cameras and infrared cameras are used and their positions are fixed. As a result, a user may have problems viewing all video streams of cameras at the same time for tracking a moving object through a mobile device due to the object dynamics and limited network bandwidth. To solve these problems, devices have been developed for automatic inspection; these systems are equipped with sensors allowing the exploration of a building, as reported in [6]. Inspection and monitoring systems are only apparently tools easy to use and manage, in fact, they hide drawbacks such as high purchase and maintenance costs as well as significant financial commitment related to data management and processing. Those factors may greatly influence the wide spreading of those systems, in order to enhance the use of these technologies, new solutions have been explored dealing with the concept of robotic and automatic survey using low cost technology [7] More specifically, the use of a robotic platform may drastically reduce the time and cost needed for a relief, if compared to a classical approach. Moreover, the use of a low-cost technology both for the mechanical design of the mobile robot and the onboard sensors, allows the wide spreading of the robotic system and substitution in the case of damages, or if the robot is lost.

2   Requirements and Solutions for the Mechatronic Design of Surveillance and Inspection Robotic Systems

The tasks of surveillance and automatic inspection need careful consideration to identify basic requirements and related solutions, as it is schematically represented in Fig. 1. First, basic requirements related to mobility issues deals with the site to inspect, i.e. buildings, industrial plants or any other indoor environment. They have in common to require a small sized mobile robot that can travel across a large variety of scenarios.

Mobile robots are usually classified according to the type of locomotion. Walking machines are well suited for unstructured environment because they can insure their stability in a wide range of situations, but they are mechanically complex and require high power and control efforts. On flat surfaces, they demonstrate their main limitation being low speed motion and high power consumption if compared with the other solutions [8]. Wheeled robots are the optimal solution for well-structured environment mainly flat and regular terrain. In off-road, their mobility is often very limited and highly depends on the type of environment and the dimension of the obstacles. Hybrid mobile robots combine advantages of both locomotion types, therefore in the recent past they are preferred for a large variety of scenarios and applications. According to the task and overall costs, three navigation modes are possible, pure tele-operation, safeguarded teleoperation, and autonomous navigation. The choice depends by the application and the environment. The sensorization is strictly related to the navigation type and the level of sophistication of the inspection. It is possible to classify internal and external sensors. The internal ones give the robot mobility control and navigation capabilities. They are proximity sensors, encoders, and GPS, accelerometers, gyroscopes, magnetic compasses, tilt and shock sensors. External sensors are related to the specific task, for the inspection and surveillance sensors such as cameras, thermal

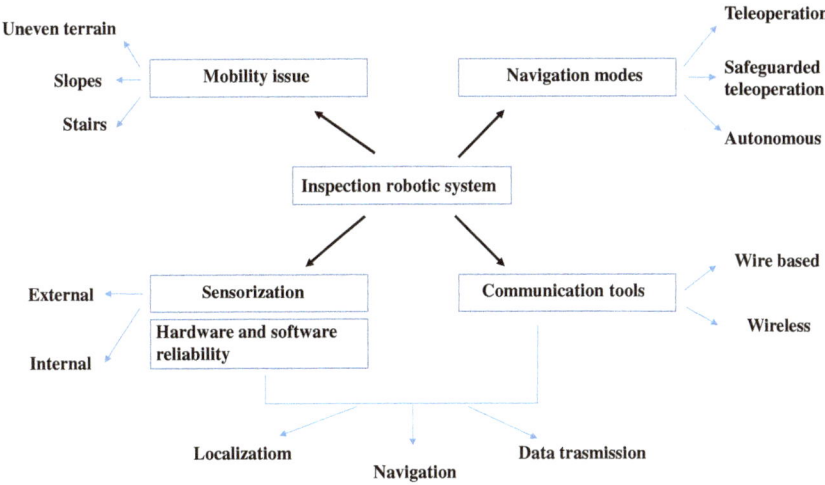

Fig. 1. Flow-chart of the main requirements and solutions for a mechatronic design of inspection robotic systems

cameras, laser, light, temperature, gas, smoke, oxygen, humidity, listening and ultra-sound may be considered. Hardware and software reliability deals with the end-user/application of the robot. In fact, this issue has to be clarified at very early stage of the design since it is related to operation, maintenance, failure prevention. Last but not the least; communication tools are essential either for localization, navigation and data transmission. In addition to wired and wireless communication, the new trend is the interoperability with any automatic and robotic system with other devices, such as home automation and security system in industrial plants according to industry 4.0.

3 The Mechatronic Design of Inspection Robotic System for Indoor Applications

The mechatronic design of the inspection robotic system can be schematically made by two main parts; one is devote to the robot mobility and operation modes, and the other one is responsible to manage the external sensor suite, as shown in Fig. 2. According to the scheme reported in Fig. 2, (1) is the tablet or mobile phone used for the robot motion control and navigation by means of the two sliders represented in the zoomed view in (7), and the program is written with the DENSION WIRC software. The USB WiFi router type TP-LINK Model TL-WN821N is (2) and allows the tablet to access and connect to the WIRC hardware. The webcam CAM (3) is the Logitech U0024 type. The DENSION WIRC hardware (4) with four digital Inputs, four Digital Outputs and eight channels, is used to control the Servo Motors of the robot. The hardware in (4) is connected to and command the Arduino board (5), which drives via relay (6) the robot's actuators. The target (8) is displayed on the tablet or the mobile phone. The overall mechatronic system architecture for the robot navigation is represented in (9).

Therefore, when the robot in Fig. 3 moves in tele operated mode, the internal sensor suite is used to help the user to understand the environment and guide the robot to the path or close a place to inspect. In particular, (2) is an electronic board equipped with accelerometer, gravity and gyroscope sensors, GPS sensor, magnetic field and acceleration sensors. The front camera view for navigation is displayed as (3). External sensors are used for inspection and monitoring tasks. More specifically, in Fig. 3, (1) is the FLIR ThermaCAM S40 is the thermal infrared camera for videos, The 3D scan is the Xbox Kinect shown in (4), provided with an infrared sensor, and two additional micro cameras. An overview of possible industrial and non-industrial applications is reported in [9, 10]. The overall layout for the system is shown in Fig. 4, in which the main components may be recognized, mainly the robot in (2) with external and internal sensors, operating (3) and monitoring (5) (6) systems and power supply (6); (1) is the target. Figure 5 shows the control architecture to operate the robot and a representation of the interoperation with the equipment on-board.

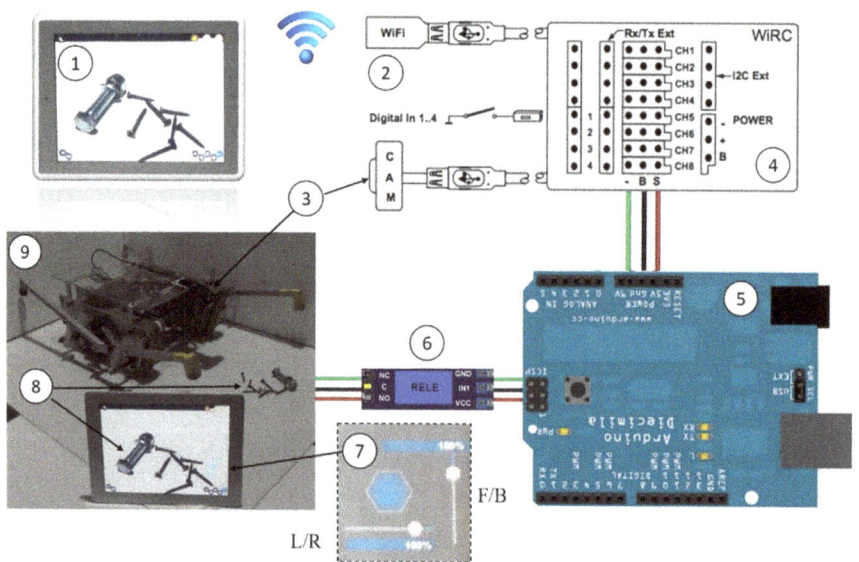

Fig. 2. Mechatronic architecture of the control for the THROO robot.

Table 1 summarizes the main features of the proposed system. The integration of sensors and its management has been a subject of research activity in different specific domains [11–13]. An industrial network laboratory prototype has been proposed in [14] in which several kits have been implemented.

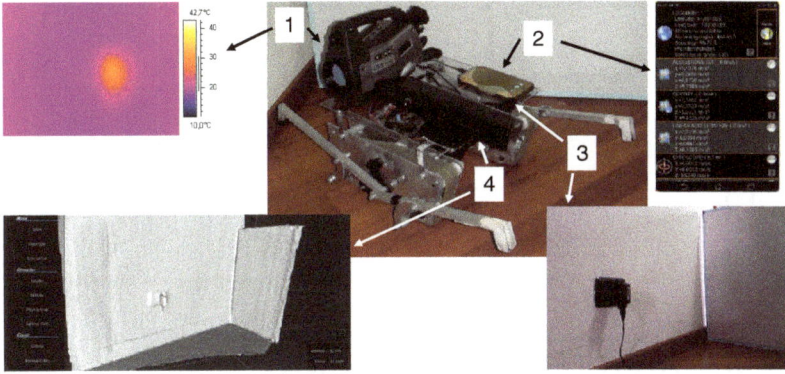

Fig. 3. Sensor installation on the inspection robot.

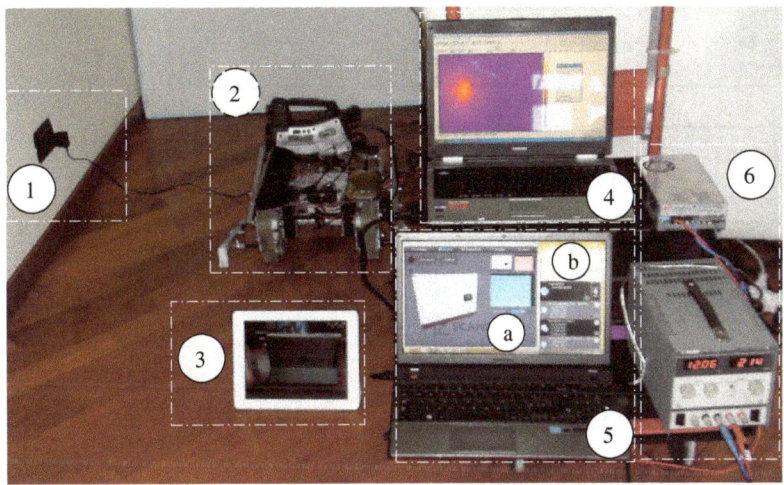

Fig. 4. Overview of the proposed mechatronic/robotic system.

Figure 6 shows the robot operation during an indoor inspection. In particular, thermal detection of an electrical component is carried out. The interoperability of the sensors on board with navigation sensorization is managed by the control board and the WIRC Controller.

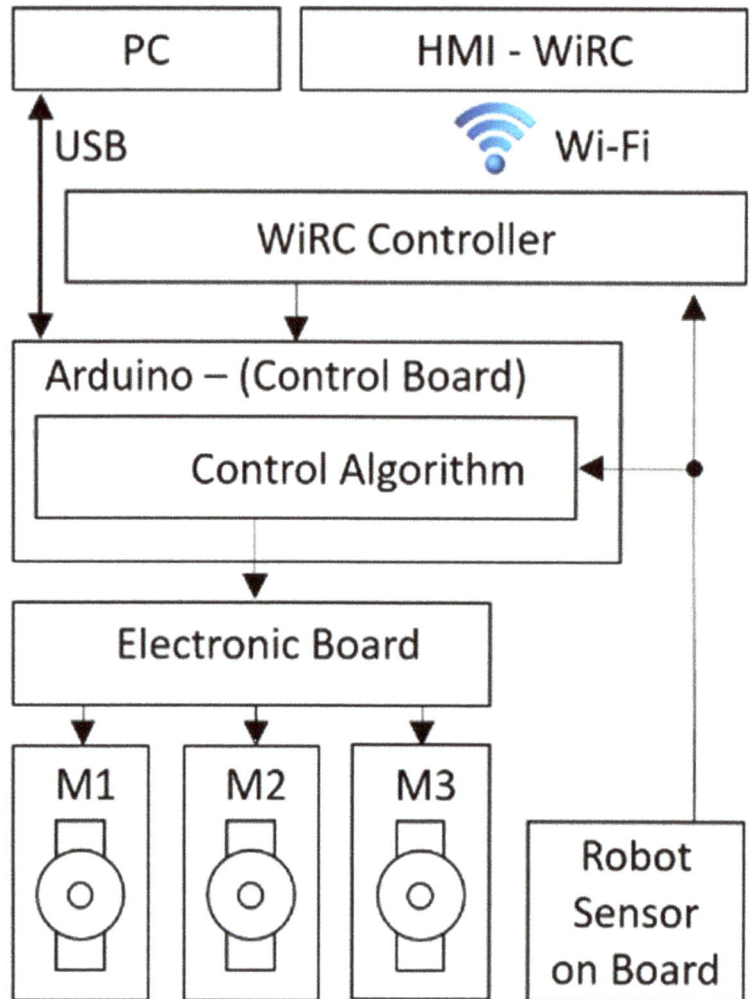

Fig. 5. A scheme for the control architecture.

Table 1. System specifications.

Parameter description	Specification		
Robot system Hybrid mobile robot THROO	Size (LxHxW)	300(550)x140x400mm	
	Mass	4.5 kg no batteries	
	Max speed	Up to 0.5 m/s	
	Actuation	DC 24V 12 Nm 24 W	
	DOFs	2 (track) 1 (legs)	
	Max step size	100 mm	
Internal sensors	Range	Resolution	power
PSH Accelerometer (Intel Inc.)	0...39.227	0.01 (0.024%)	0.006mA
PSH Gyroscope sensor (Intel Inc.)	0...34.907	0.002 (0.005%)	6.1mA
PSH Gravity sensor (Intel Inc.)	0...19.613	0.005 (0.024%)	0.006mA
PSH Magn. field sensor (Intel Inc.)	0...800	0.5 (0.062%)	0.1mA
PSH Lin. Accel. sensor (Intel Inc.)	0...19.613	0.005 (0.024%)	0.006mA
External sensors	Model		
Thermal camera (FLIR)	ThermaCAM S40		
Front camera (Logitech)	U0024 type		
3D scan (Xbox)	Kinect		
Communication USB WiFi router	TP-LINK Model TL-WN821N		
Control station No. monitors	2		
No. computers/CPU	2		

Fig. 6. Photo sequence of an experimental test of data acquisition.

4   Conclusion

The mechatronic system and control proposed in this paper constitutes a solution for a broad range of scenarios spacing from home-security, inspection of industrial sites, brownfields, historical sites or sites dangerous or difficult to access by operators. First experimental tests are reported to show the engineering feasibility of the system and interoperability of the mobile hybrid robot equipped with sensors that allow real-time multiple acquisition and storage. The robot is equipped with external and internal sensors e.g., gyroscope, accelerometer, inclinometer, thermal camera, 3D motion capture system. First experimental results are reported in this paper.

Acknowledgement. The research leading to these results has received funding PRASG: University Research Projects "Starting Grant" from the University of Cassino and Southern Lazio. The first author would like to thank the Italian Ministry of Education, Universities and Research (MIUR) through the funded program PRASG to obtain results of this research activity.

References

1. Balaguer, C., Montero, R., Victores, J.G., Martínez, S., Jardón, A.: Towards fully automated tunnel inspection: a survey and future trends. In: The 31st International Symposium on Automation and Robotics in Construction and Mining ISARC, Sydney (2014)
2. Castelli, G., Ottaviano, E., Rea, P.. A Cartesian cable-suspended robot for improving end-users' mobility in an urban environment. Rob. Comput. Integr. Manuf. **30**(3), 335–343 (2014)
3. Rea, P., Ottaviano, E., Conte, M., D'Aguanno, A., De Carolis, D.: The design of a novel tilt seat for inversion therapy. Int. J. Imaging Rob. **11**(3), 1–10 (2013)
4. Tseng, C.-C., Lin, C.-L., Shih, B.-Y., Chen, C.-Y.: SIP-enabled surveillance patrol robot. Rob. Comput. Integr. Manuf. **29**, 394–399 (2013)
5. Borja, R., de la Pinta, J.R., Álvarez, A., Maestre, J.M.: Integration of service robots in the smart home by means of UPnP: a surveillance robot case study. Rob. Auton. Syst. **61**, 153–160 (2013)
6. Rea, P., Ottaviano, E.: Design and development of an inspection robotic system for indoor applications. Rob. Comput. Integr. Manuf. **49**, 143–151 (2018)
7. Rea, P., Pelliccio, A., Ottaviano, E., Saccucci, M.: The heritage management and preservation using the mechatronic survey. Int. J. Arch. Heritage **11**(8), 1121–1132 (2017)
8. Siegwart, R., Nourbakhsh, I.R.: Introduction to Autonomous Mobile Robots. MIT Press, Cambridge (2004)
9. Figliolini, G., Rea, P.: Overall design of Ca.U.M.Ha. robotic hand for harvesting horticulture products. Robotica **24**(3), 329–331 (2006)
10. Ottaviano, E., Ceccarelli, M., Castelli, G.: Experimental results of a 3-DOF parallel manipulator as an earthquake motion simulator. In: Proceedings of the ASME Design Engineering Technical Conference, Salt Lake City, 2 A, 215-222, Code 64323 (2004)
11. Figliolini, G., Rea, P.: Ca.U.M.Ha. robotic hand (Cassino-Underactuated-Multifinger-Hand). In: IEEE/ASME International Conference on Advanced Intelligent Mechatronics, Zurich (2007). Article no. 4412562
12. Thomas, F., Ottaviano, E., Ros, L., Ceccarelli, M.: Performance analysis of a 3-2-1 pose estimation device. IEEE Trans. Rob. **21**(3), 288–297 (2005)
13. Sorli, M., Figliolini, G., Pastorelli, S., Rea, P.: Experimental identification and validation of a pneumatic positioning servo-system. In: Power Transmission and Motion Control, Bath, pp. 365–378 (2005)
14. Leão, C.P., Soares, F.O., Machado, J.M., Seabra, E., Rodrigues, H.: Design and development of an industrial network laboratory. Int. J. Emerg. Technol. Learn. **6**(2), 21–26 (2011)

Influence of the Indentation Speed on Viscoelastic Behavior of the Human Finger

Vlad Cârlescu[1(✉)], Dumitru N. Olaru[1(✉)], Gheorghe Prisăcaru[1],
Cezar Oprişan[1], Radu Ştefan Ştirbu[1], and José Machado[2]

[1] Department of Mechanical Engineering, Mechatronics and Robotics, The
"Gheorghe Asachi" Technical University of Iasi, Iasi, Romania
carlescu.vlad@yahoo.com, dolaru@mail.tuiasi.ro
[2] Department of Mechanical Engineering, MEtRICs Research Center, University
of Minho, Guimarães, Portugal
jmachado@dem.uminho.pt

Abstract. In this paper the authors presented a methodology to study the viscoelastic behavior of the human finger tissue by using the indentation process with a steel ball and indentation speed between 0.02 mm/s to 4 mm/s. Considering that at very low speeds the viscoelastic effects can be neglected, an effective Young's modulus of the human finger as function of the indentation force was determinate based on the Hertzian equation and validated with experimental data. The influence of the indentation speed on both, the indentation force and depth, was studied and a mathematical model was proposed. The authors observed that the normal force resulting from the indentation process increase with speed due to an important damping effect in the finger tissue. Also, based on the Hertzian contact model the contact pressure during the indentation process has been calculated.

Keywords: Human finger · Young's modulus · Indentation speed
Viscoelastic behavior

1 Introduction

Human skin acts as an essential physical and chemical barrier against external environment. In daily life, human skin often comes into contact with many different materials and rubs against external surfaces that involve gripping, feeling and manipulating objects which in some cases can cause irritation and friction injuries. Therefore, human skin tribology is a very diverse and complicated area of study [1, 2].

One of the most important mechanical parameter of the human skin is the Young's elastic modulus and examination of the literature gives a wide spread of data. This is because the human skin is a complex material including three layers with different mechanical properties: epidermis, dermis and subcutaneous fat. Techniques such as tensile testing [3, 4], indentation [5–8] and suction methods [9] are used. Data varies across the different measurement techniques for the same skin sample. The effective Young's modulus in skin being loaded by a probe in a tribological test will also vary with applied load and also with probe diameter [10]. Derler and Gerhardt [11] show

© Springer International Publishing AG, part of Springer Nature 2019
J. Machado et al. (Eds.): HELIX 2018, LNEE 505, pp. 143–150, 2019.
https://doi.org/10.1007/978-3-319-91334-6_20

that elastic moduli of human skin in vivo varying over 4–5 orders of magnitude from 0.004 MPa to 57 MPa, depending on the measurement technique, anatomical site, skin hydration level, age, individual person. Zahouani et al. [12] proposed a Kelvin–Voigt viscoelastic model, and solved the differential equation for small vibration of the human skin induced by a variable displacement applied to the skin surface with a spherical indenter. The authors obtained both the stiffness and the damping properties of the skin. Notably, while a great deal of data exists for forearm skin and ex vivo skin and different skin layers, little data exists for skin substitutes, especially engineered skin [2, 13]. Thus, mechanical and tribological properties and tactile sensing are prerequisites for implementation in robots that deals with dexterous tasks [14–17].

In this paper, the authors used a methodology previously presented [7, 8] to determine an effective elastic modulus of the finger tissue by indentation with a steel ball. The response of the finger tissue by indentation with various speeds (from 0.02 mm/s to 1 mm/s) has been evidenced experimentally and a general equation of the normal force as function of indentation speed and indentation depth has been obtained for the contact between a steel ball and the human finger.

2   Experimental Procedure and Equipments

The experiments were realized using the Tribometer CETR UMT-2 (Fig. 1) equipped with a force sensor (20 N) having the possibility to determine the normal force F_z applied on the middle finger from the left hand of a 34 year old man subject by a steel ball. The experiments were realized in a displacement-controlled technique imposing to the carriage a maximum vertical displacement of 4 mm. The steel ball with a diameter of 7.983 mm was placed in contact with finger and indentation measurements were performed at different indentation speed between 0.02 mm/s and 4 mm/s. The experimental parameters such as displacement, time and speed were previously imposed serving as input data and the force resulting from the indentation process was automatically recorded.

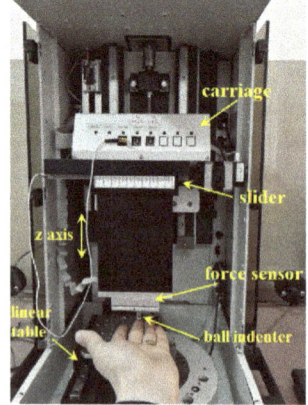 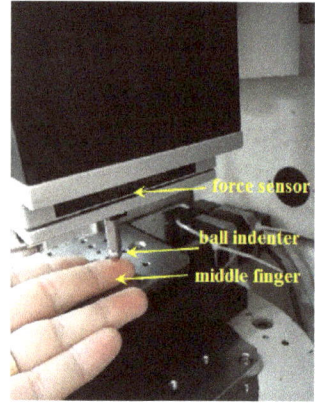

Fig. 1.   The Tribometer CETR UMT-2 used for indentation tests

3   Experimental Results

The experimental force-depth curves recorded at each indentation speed are presented in Fig. 2. The negative values of the applied normal force F_z corresponds to the coordinate system of the Tribometer. The experimental results obtained in the ball-finger indentation process are similar to those previously reported by the authors [8] when a steel cylinder having 7 mm diameter was used. From the Fig. 2 it can be observed that the force resulting from the indentation process increase with indentation speed as a result of the increasing in the rigidity of the finger tissue. Also, a viscoelastic behavior of the finger tissue is evidenced and the authors intend to obtain a general equation of the normal force F_z as function of indentation depth and indentation speed.

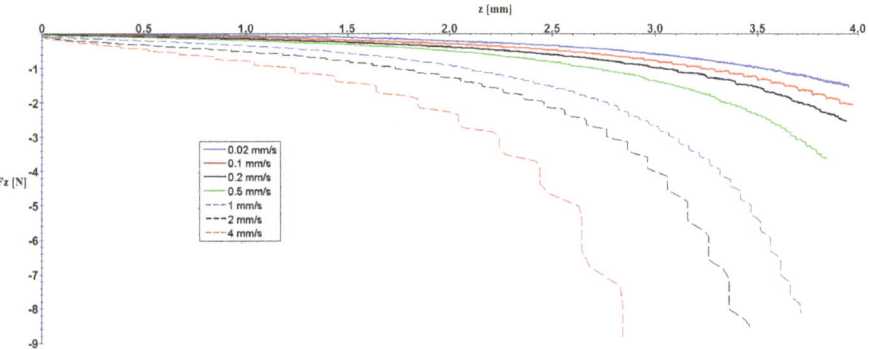

Fig. 2. Variation of the normal force (F_z) versus the indentation depth (z) at different speeds

3.1   Determination of the Effective Elastic Young's Modulus E_f

The effective Young's modulus of the finger's tissue was determine according to the methodology previously reported by the authors [7, 8]. The methodology is based on the Hertzian equation of the deformation δ realized by a ball indenter on the finger's surface:

$$\delta = \delta^* \left\{ \frac{3 \cdot F_z}{2 \sum \rho} \cdot \left[\frac{1 - v_b^2}{E_b} + \frac{1 - v_f^2}{E_f} \right] \right\}^{2/3} \cdot \frac{\sum \rho}{2} \tag{1}$$

where F_z is the normal force, E_b, E_f, v_b and v_f are the Young's modulus and Poisson's ratio of the steel ball and the finger, respectively. Curvature (sum $\Sigma \rho$) and geometrical dimensionless parameter δ^* have been determined according to the geometry of the finger (R_x and R_y) and the diameter of the ball [8]. We have adopted the following values: transversal radius $R_x = 8$ mm, longitudinal radius $R_y = 30$ mm, ball diameter $d = 7.983$ mm and $E_b = 2.1 \cdot 10^5$ MPa. Considering that the Young's modulus of the human finger is less than 1 MPa and imposing $v_f = 0.5$, Eq. 1 becomes:

$$\delta = 0.357 \cdot (F_z/E_f)^{2/3} \qquad (2)$$

where F_z is given in N, E_f in MPa and δ in mm. According to the Tribometer's parameters, the indentation depth z corresponds with the elastic deformation δ. The effective Young's modulus was determinate by curve fitting of the experimental results with Eq. 2 in the vicinity of the various normal forces F_z at the indentation speed of 0.02 mm/s. We have assumed that at very low indentation speeds the viscous effects can be neglected and elastic effect is dominant. The variation of the effective Young's modulus of the finger E_f as function of the normal force F_z is illustrated in Fig. 3 and a linear equation has been obtained:

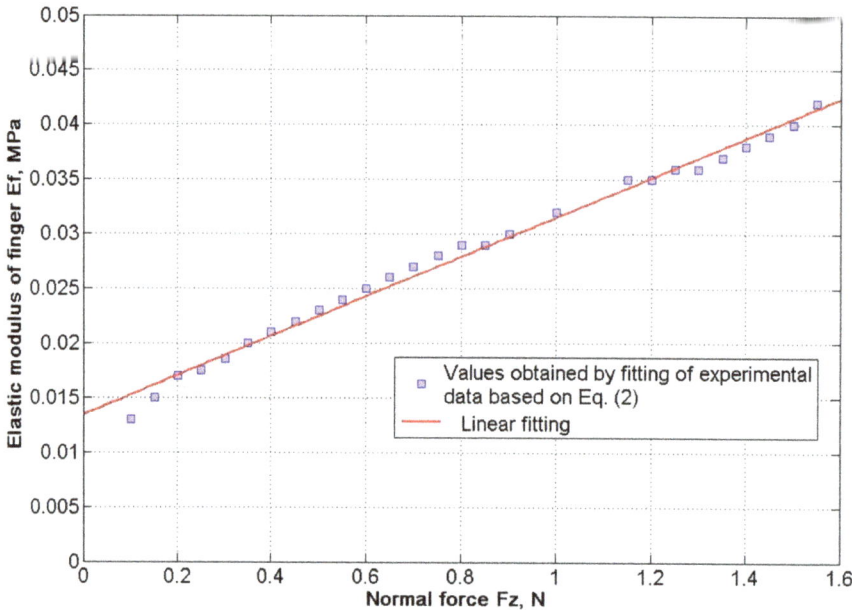

Fig. 3. Young's modulus E_f versus normal force F_z at indentation speed of 0.02 mm/s

$$E_f(F_z) = 0.0181 \cdot F_z + 0.0134 \qquad (3)$$

A similar linear dependence between E_f and F_z was previously reported in [8]. From Fig. 3 it can be observed that the effective Young's modulus of the finger varied between 0.013 MPa to 0.045 MPa similar to those reported in literature [11].

Figure 4 illustrate the dependence of the indentation depth δ as function of the normal force F_z obtained by indentation experiments and calculated according to Eqs. 2 and 3. A very good correlation has been obtained.

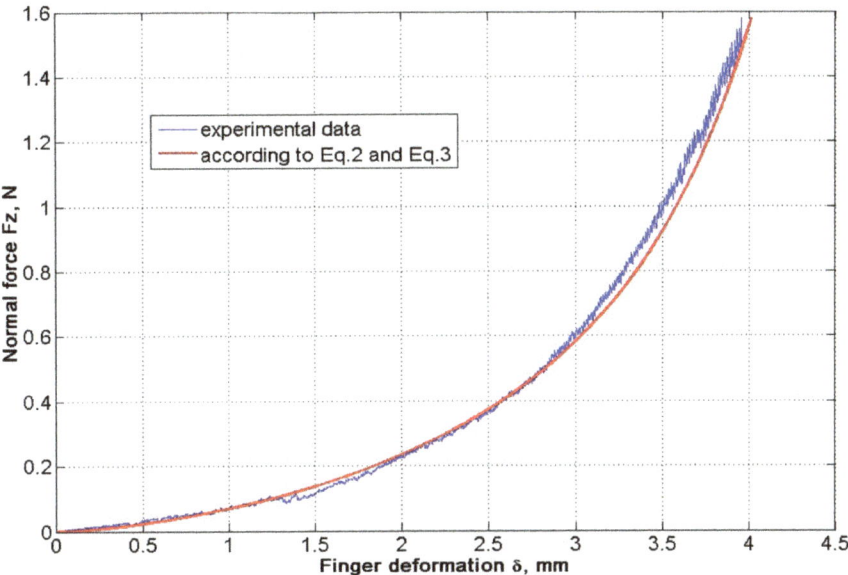

Fig. 4. Experimental and theoretical dependence of finger's deformation with normal force F_z

3.2 Influence of the Indentation Speed on the Deformation and Normal Force

The influence of the indentation speed on the finger deformation and normal force is illustrated in Fig. 5. It can be observed that the normal force resulting from the indentation process of the finger increase with speed from 1.5 to 9 N while the deformation of finger decrease. Based on the same mathematical model developed in [8] for the cylinder-finger indentation, the authors propose that normal force F_z can be expressed as a function of the indentation speed and indentation depth δ, by following general equation:

$$F_z(v, \delta) = F_z(v) \cdot e^{\delta/\delta(v)} \tag{4}$$

where $F_z(v)$ and $\delta(v)$ are the force function and indentation depth function depending of the indentation speed.

The force and indentation depth functions, $F_z(v)$ and $\delta(v)$, respectively, were obtained by curve-fitting the values from Fig. 5 and the following equations are expressed:

$$F_z(v) = 0.0981 \cdot v^{0.459} \tag{5}$$

$$\delta(v) = -0.05 \cdot v + 0.958 \tag{6}$$

where indentation speed v is given in mm/s.

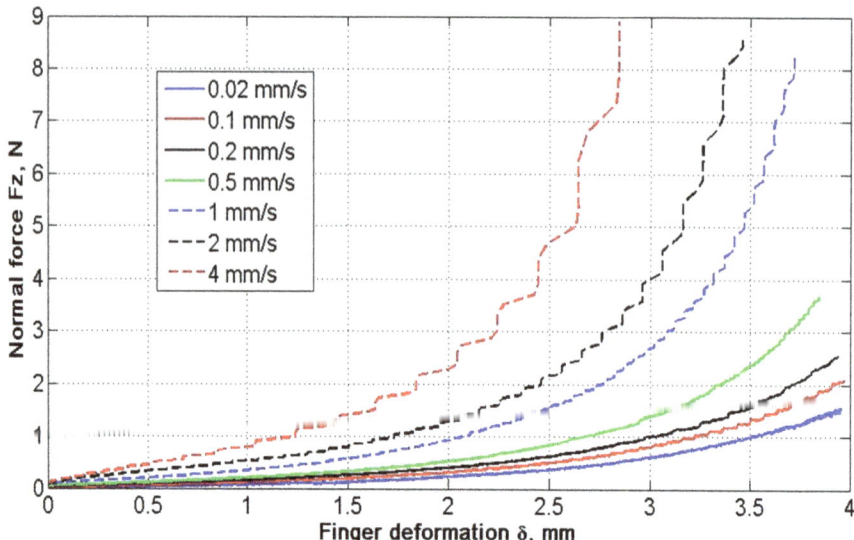

Fig. 5. Variation of the normal force F_z as function of indentation depth and speed

3.3 Contact Pressure in a Ball-Finger Indentation Process

The contact pressure during the indentation process has been considered as a Hertz pressure distribution on contact ellipse [18]. The maximum contact pressure p_{max} has been obtained by Hertzian equation:

$$p_{max}(v, \delta) = 3 \cdot F_z(v, \delta)/2 \cdot \pi \cdot a(\delta) \cdot b(\delta) \qquad (7)$$

where $a(\delta)$ and $b(\delta)$ are the semi-major and semi-minor axis of the contact ellipses depending only on the indentation depth δ. To determine the contact surfaces between the ball and the finger a two bodies method and their intersections from CATIA V5 was used. The two semi axis of the projected contact ellipses were calculated for the new obtained body. The maximum contact pressure calculated for the tested conditions varied between 0.045 MPa and 0.35 MPa with increasing values by increasing the indentation speed.

4   Conclusions

In this paper the authors determined the effective Young's modulus for a middle human finger indented with a steel ball by using the Hertz contact theory. Supposing that viscous effects can be neglected at low indentation speeds, values for effective Young's modulus between 0.013 to 0.045 MPa have been obtained. Viscoelastic behavior of the human finger tissue was studied in terms of increasing the indentation speed from 0.02 to 4 mm/s. A deformation of 4 mm was imposed to the finger and an important increase of the normal forces at different indentation speeds has been observed. Thus, by increasing of the indentation speed an important damping effect in the finger tissue appears. Based on the experimental results the authors found a force function and indentation depth function depending of the indentation speed. Also, the variation of the ball-finger maximum contact pressure as function of indentation speed was determinate. The obtained equations characteristic to the viscoelastic behavior of the human finger tissue can be considered when studying both the prehension mechanism and tactile perception of human hand.

References

1. Thieulin, C., Pailler-Mattéi, C., Djaghloul, M., Abdouni, A., Vargiolu, R., Zahouani, H.: Wear of the *stratum corneum* resulting from repeated friction with tissues. Wear **376–377**, 259–265 (2017)
2. Bostan, L.E., Taylor, Z.A., Carré, M.J., MacNeil, S., Franklin, S.E., Lewis, R.: A comparison of friction behavior for ex vivo human, tissue engineered and synthetic skin. Tribol. Int. **103**, 487–495 (2016)
3. Park, A.C., Baddiel, C.B.: Rheology of stratum corneum I: a molecular interpretation of the stress-strain curve. J. Soc. Cosmet. Chem. **23**, 3–12 (1972)
4. Park, A.C., Baddiel, C.B.: The effect of saturated salt solutions on the elastic properties of stratum corneum. J. Soc. Cosmet. Chem. **3**, 471–479 (1972)
5. Geerligs, M., van Breemen, L., Peters, G., Ackermans, P., Baaijens, F., Oomens, C.W.J.: In vitro indentation to determine the mechanical properties of epidermis. J. Biomech. **44**, 1176–1181 (2011)
6. Pailler-Mattei, C., Bec, S., Zahouani, H.: In vivo measurements of the elastic mechanical properties of human skin by indentation tests. Med. Eng. Phys. **30**, 599–606 (2008)
7. Oprişan, C., Cârlescu, V., Barnea, A., Prisacaru, G., Laru, D,N., Plesu, G.: Experimental determination of the Young's modulus for the fingers with application in prehension systems for small cylindrical objects. In: IOP Conference Series: Materials Science and Engineering 147, p. 012058 (2016)
8. Cârlescu, Vlad, Olaru, Dumitru N., Prisacaru, Gheorghe, Oprisan, Cezar, Machado, José: Sensors and actuators on determining parameters for being considered in selection of elastomers for biomimetic hands. Sensors **17**, 1190 (2017)
9. Hendriks, F.M., Brokken, D., van Eemeren, J., Oomens, C.W.J., Baaijens, F.P.T., Horsten, J.B.A.M.: A numerical-experimental method to characterize the non-linear mechanical behavior of human skin. Skin Res. Technol. **9**, 274–283 (2003)
10. Kwiatkowska, M., Franklin, S.E., Hendriks, C.P., Kwiatkowski, K.: Friction and deformation behaviour of human skin. Wear **267**, 1264–1273 (2009)
11. Derler, S., Gerhardt, L.-C.: Tribology of skin: review and analysis of experimental results for the friction coefficient of human skin. Tribol. Lett. **45**, 1–27 (2012)
12. Zahouani, H., Boyer, G., Pailler-Mattei, C., Ben Tkaya, M., Vargiolu, R.: Effect of human ageing on skin rheology and tribology. Wear **271**, 2364–2369 (2011)
13. Morales, H.M.: Mimicking the Tribo-Mechanical Performance of Human Skin: A Scale – Dependent Approach Based on Poly(Vinyl Alcohol) Hydrogel. Ph.D. thesis, University of Twente (2016)
14. Liu, X.: Understanding the effect of skin mechanical properties on the friction of human finger-pads. Ph.D. thesis, Department of Mechanical Engineering, University of Sheffield (2013)

15. Jockusch, J., Walter, J., Ritter, H.: A tactile sensor system for a three-fingered robot manipulator. In: Proceedings of the International Conference on Robotics and Automation (ICRA). IEEE vol. 4, pp. 3080–3086 (1997)
16. Wettles, N., Santos, V.J., Johansson, R.S., Loeb, G.E.: Biomimetic tactile sensor array. Adv. Rob. **22**, 829–849 (2008)
17. Figliolini, G., Rea, P.: Overall design of Ca.U.M.Ha. robotic hand for harvesting horticulture products. Robotica **24**(3), 329–331 (2006)
18. Harris, T.A., Kotzalas, M.N.: Rolling Bearing Analysis. Essential Concepts of Bearing Technology. CRC Press (2007)

Mechatronic Design of a Low-Cost Control System for Assisting Devices

Erika Ottaviano, Pierluigi Rea$^{(\boxtimes)}$, and Alessandra Grandinetti

DICeM: Department of Civil and Mechanical Engineering,
University of Cassino and Southern Lazio,
via G. Di Biasio 43, 03043 Cassino, FR, Italy
{ottaviano,rea}@unicas.it

Abstract. In this paper, we focus our interest in the development of low-cost control systems of mechatronic devices for the homecare of the elderly and people with reduced mobility with the goal of assisting people in daily-life activities. The development of such systems will be exploited in the form of a tool kit to be flexible and applied either to assistive systems, in order to aid movements with controlled position, and in future force/torque and acceleration, introducing reliability, repetitiveness at a relatively low-cost in the related sector. In particular, the development of a low-cost control system design allows the applicability and interconnection of devices and solutions for the home-care. Our aim is to test low-cost technologies for the homecare of elderly and people with motor impairments trying to reduce significantly the overall costs to facilitate wider use of assisting and rehabilitation systems for the home-care. In particular, a Sit-To-Stand (STS) assisting device is considered as a paradigmatic example. Numerical simulations and results of an experimental test-bed are discussed in this paper.

Keywords: Mechatronics · Control system · Assisting device
Simulation

1 Introduction

Recent studies report that the number of elderly and people with motion disabilities that cannot perform Activities of Daily Life (ADL) without the help of other people is increasing. In order to face this problem, nurses, caregivers or relatives are involved for continuous aiding at home, alternatively, a number of commercial solutions have been developed to help the execution of some tasks. Among several ADL, such as walking, feeding, bathing, toileting, we are interested in the Sit-To-Stand (STS), which is one of the most frequent and, despite its apparent simplicity, it is a mechanically demanding functional task undertaken daily, requiring a strong coordination between posture and movement. Some commercial solutions were developed [1]. Most of these devices act like end-effector robot, i.e. they provide assistive forces only at one body segment, while the trunk is either rigidly supported by a back support of a seat or through use of the arms. All these approaches may or surely result in unnatural kinematics.

© Springer International Publishing AG, part of Springer Nature 2019
J. Machado et al. (Eds.): HELIX 2018, LNEE 505, pp. 151–157, 2019.
https://doi.org/10.1007/978-3-319-91334-6_21

The purpose of this work is to propose a mechatronic design for the control of the actuation of a novel assisting device designed for the STS, as evolution of the ones developed by the authors [2, 3] that can be either used as rehabilitation device, as the one proposed in [4]. The mechatronic design is particularly important when the motion or force have to be controlled [5, 6]. Simulation results are reported to show the feasibility of the solution together with an experimental test bed of the mechatronic design.

2 The Design of a 2-DOF Mechanism for the Sit-to-Stand Application

The so-called "natural" STS movement is composed by a translation in two directions with a rotation of the trunk, as described in [7]. It can be defined as a balanced movement of the body's center of mass from a seated position to the standing one [8] being a transition between two stable postures [9]. Several factors can influence the STS, e.g. the anthropometric data, age, environment, strategies (velocity, upper limb configuration) and objective (assistance, rehabilitation). Therefore, the design of a support mechanism, which is devoted to generate the requested motion and support the body of the individual during the STS is a crucial element in the design of the device [10]. Among several solutions, we have chosen to consider a 2-DOF mechanism shown in Fig. 1 in order to reproduce any trajectory in the sagittal plane having a fixed orientation of the trunk. The combination of the motion laws for the two actuators will produce then any trajectory in a region of the sagittal plane. Experiments and numerical simulations were conducted and aimed at clarifying the effects of the motion law of the actuation and speed for the execution of the STS movement supported by the assisting device. CAD based modelling allows the realization of realistic simulations that are

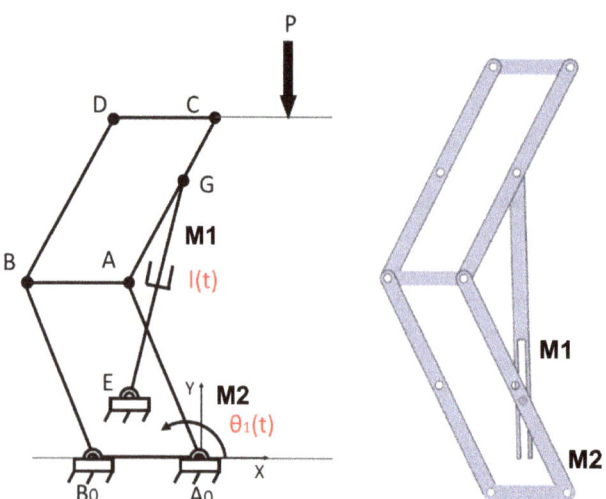

Fig. 1. Kinematic scheme and planar sketch of the 2-DOF STS mechanism.

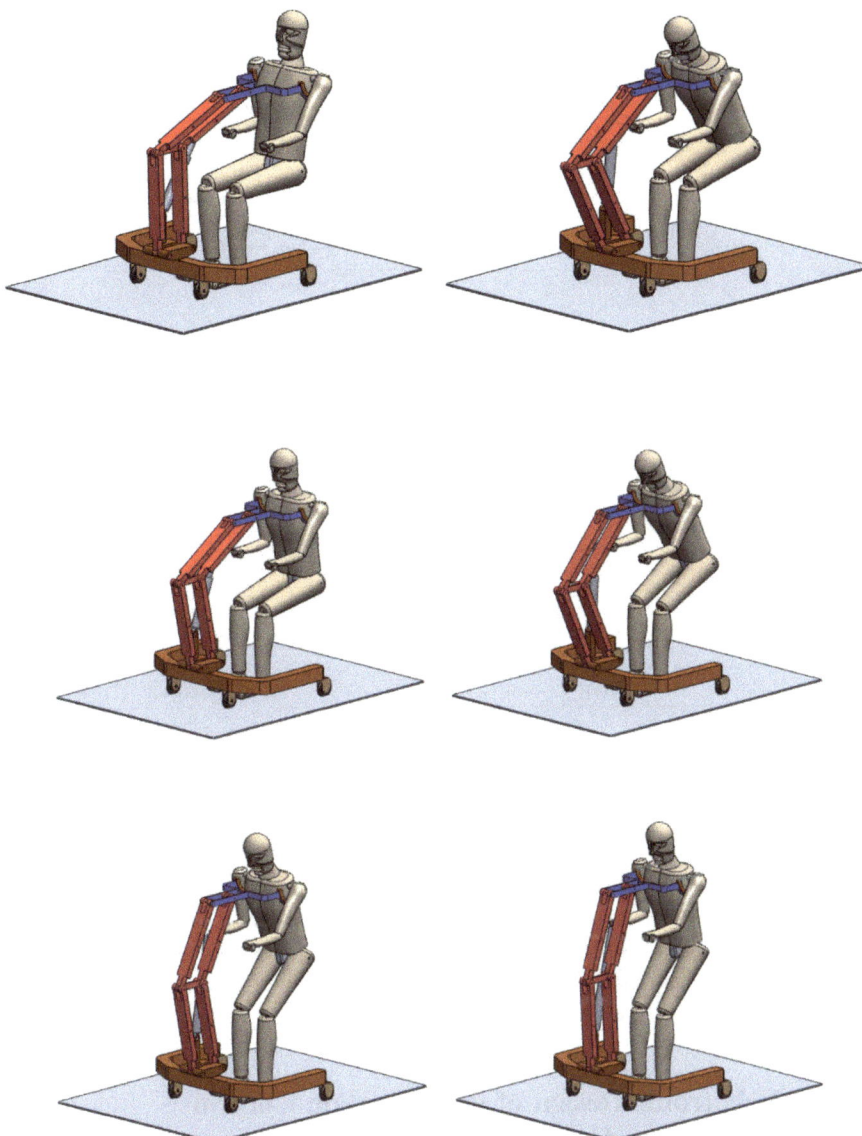

Fig. 2. Motion sequence of the STS with the proposed 2-DOF mechanism.

effectively used at the design stage [11]. Figure 2 shows the motion sequence of the device during the STS motion interacting with a realistic model of the human body, [12]. Figure 3 reports the motion laws of the two actuators M1 and M2 and their velocity profiles.

During the simulation, the velocity and acceleration components in the sagittal plane have been also displayed to monitor the effect of the STS device on a reference

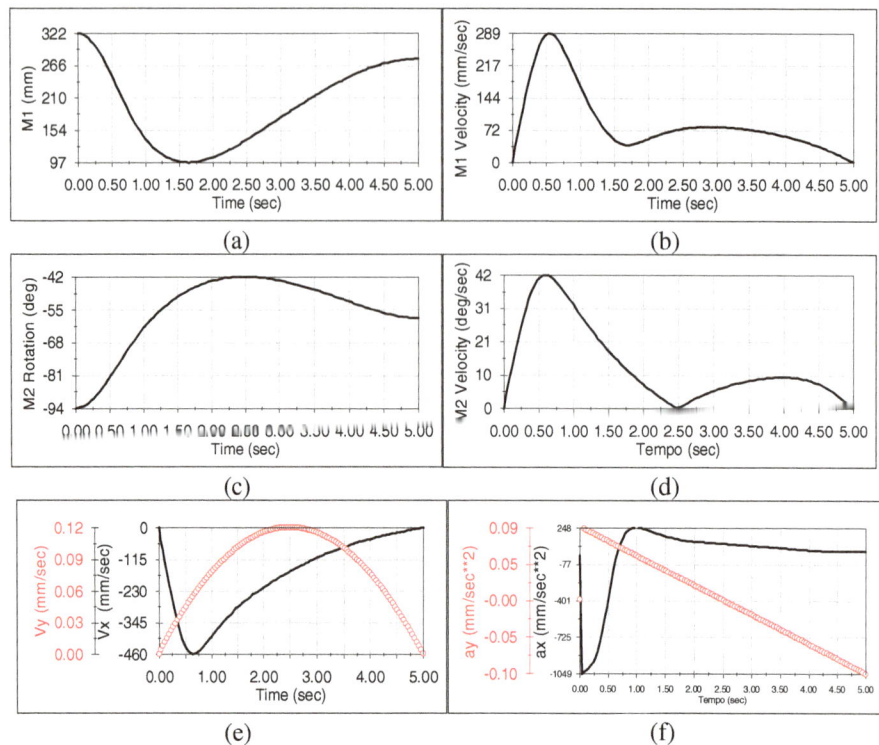

Fig. 3. Results for the simulation in Fig. 2: (a) displacement the linear actuator M1; (b) velocity profile of M1; (c) angular displacement of the rotational actuator M2; (d) velocity profile of the M2; (e) point S velocity; (f) point S acceleration.

point on the human body model. In particular, a reference point S has been chosen on the center of the interface of the mechanism that supports the armpit.

This point has been chosen to monitor the effects of the movement on the body, kinematics can be used to drive the motors M1 and M2.

3   The Mechatronic Design of the Actuation Control for the Sit-to-Stand Application

Figure 4 shows a general scheme for the architecture of the control system of the assisting device. It is composed by three parts. The first one deals with the acquisition/monitoring task, in which the motion is captured and recorded by a MoCap system. Alternatively, it can be simulated by CAD based design with a model interacting with a human body model, [12]. The second part of the control deals with the kinetostatic/dynamic model of the system. Data obtained by the acquisition are used into the math model to get the motion law of the actuation system, which is sent to Arduino to drive the actuation on a scaled prototype, as reported in the illustrative

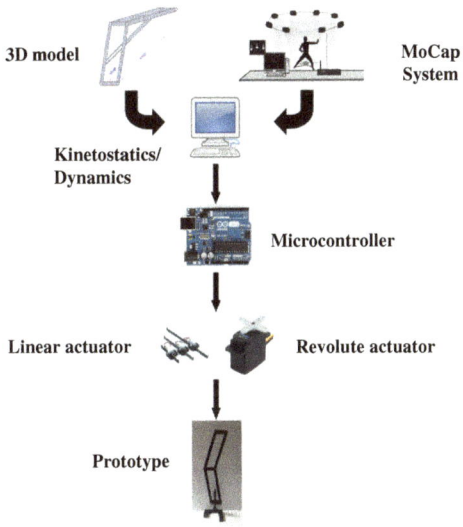

Fig. 4. A scheme for the mechatronic architecture for the control of the system.

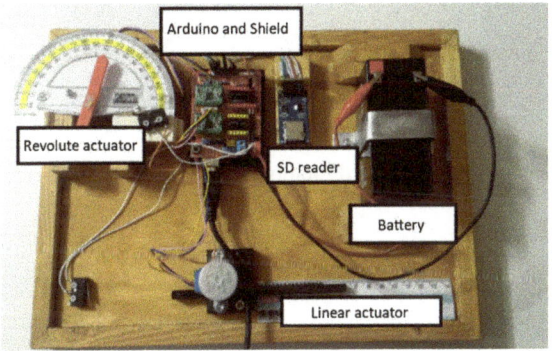

Fig. 5. Experimental test-bed for the actuation control.

example here or alternatively is sent to a real prototype. Figure 5 shows the experimental test-bed that is realized as illustrative example to demonstrate the effectiveness of the proposed procedure. Arduino reads the motion laws for the actuation from a memory card, and then it commands the two stepper motors, one revolute one linear to move the prototype. Figure 6 reports a flowchart for the programming that is used for Arduino. It is worth noting that the same procedure and results can be extended to a real scale prototype. Figure 7 shows a 1:10 prototype obtained by 3D printing technique.

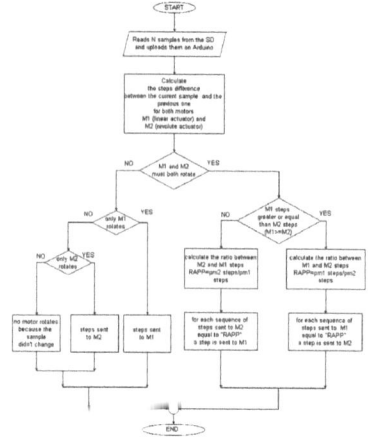

Fig. 6. Flow-chart for the programming.

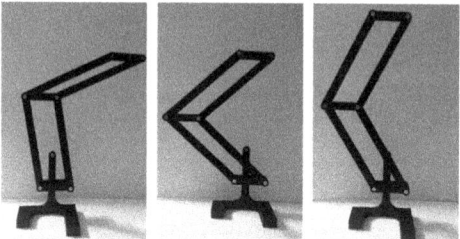

Fig. 7. Motion sequence of first scaled prototype obtained with a 3D printing technique.

4   Conclusion

In this paper, we presented a mechatronic design for the actuation control of STS devices with the aim to provide solutions for the homecare of elderly and people with reduced mobility. The mechatronic solution has the aim to test low-cost technologies trying to reduce significantly the overall costs to facilitate interconnection and wider use of assisting systems. Further development is design of a brake system or an irreversible mechanical transmission for security issues.

Acknowledgement. The research leading to these results has received funding PRASG: University Research Projects "Starting Grant" from the University of Cassino and Southern Lazio. The second author would like to thank the Italian Ministry of Education, Universities and Research (MIUR) through the funded program PRASG.

References

1. Krishnan, R.H., Pugazhenthi, S.: Mobility assistive devices and self-transfer robotic systems for elderly, a review. Intel. Serv. Robot. **7**, 37–49 (2014)
2. Rea, P., Ottaviano, E., Castelli, G.: Procedure for the design of novel assisting devices for the sit-to-stand. J. Bionic Eng. **10**(4), 488–496 (2013). ISSN 1672-6529
3. Rea, P., Ottaviano, E.: Functional design for customizing sit-to-stand assisting devices. J. Bionic Eng. **15**(1), 83–93 (2018). https://doi.org/10.1007/s42235-017-0006-4
4. Rea, P., Ottaviano, E., Conte, M., D'Aguanno, A., De Carolis, D.: The design of a novel tilt seat for inversion therapy. Int. J. Imaging Robot. **11**(3), 1–10 (2013)
5. Figliolini, G., Rea, P.: Overall design of Ca.UMHa. robotic hand for harvesting horticulture products. Robotica **24**(3), 329–331 (2006)
6. Ottaviano, E., Ceccarelli, M., Paone, A., Carbone, G.: A low-cost easy operation 4-cable driven parallel manipulator, Barcelona, pp. 4008–4013 (2005). Article no. 1570734.
7. Marlene, J.A., Cooper, J.: Biomechanics of Human Movement. Brown & Benchmark (1995)
8. Roebroeck, M.E., Doorenbosch, C.A.M., Harlaar, J., Jacobs, R., Lankhorst, G.J.: Biomechanics and muscular activity during sit-to-stand transfer. Clin. Biomech. **9**(4), 235–244 (1994)
9. Tully, E.A., Fotoohabadi, M.R., Galea, M.P.: Sagittal spine and lower limb movement during sit-to-stand in healthy young subjects. Gait Posture **22**(4), 338–345 (2005)
10. Ceccarelli, M., Carbone, G., Ottaviano, E.: Multi criteria optimum design of manipulators. Bull. Pol. Acad. Sci. Tech. Sci. **53**(1), 9–18 (2005)
11. Castelli, G., Ottaviano, E., Rea, P.: A Cartesian cable-suspended robot for improving end-users' mobility in an urban environment. Rob. Comput.-Integr. Manuf. **30**(3), 335–343 (2014)
12. Castelli, G., Ottaviano, E.: Modelling, simulation and testing of a reconfigurable cable-based parallel manipulator as motion aiding system. Appl. Bionics Biomech. **7**(4), 253–268 (2010). https://doi.org/10.1080/11762322.2010.512733

Boccia Court Analysis for Promoting Elderly Physical Activity

Alexandre Calado, Pedro Leite$^{(\boxtimes)}$, Filomena Soares, Paulo Novais,
and Pedro Arezes

Algoritmi Centre, University of Minho, Guimarães, Portugal
alexandreluiscalado@gmail.com,
a66161@alunos.uminho.pt, fsoares@dei.uminho.pt,
pjon@di.uminho.pt, parezes@dps.uminho.pt

Abstract. Physical inactivity is one of the leading risk factors for global mortality. Older adults, in particular, are more probable to suffer the consequences of physical inactivity, since it is one of the most sedentary age groups. On the other hand, engaging physical activity can have various benefits for the prevention of several diseases and functional loss prevention, therefore, it is critical to encourage its regular practice amongst the elderly. Boccia is a simple precision ball sport that is easily adaptable for individuals with physical limitations, which makes it a perfectly good game for this circumstance. The present paper proposes a ball-detection based system for monitoring the Boccia court, compute the current game score and display it on a user interface. The future goal of such system will be to motivate the elders to participate more frequently in the Boccia game and make the overall game experience more enjoyable. The proposed system was tested with twenty video recordings of different simulated game situations. Overall, the obtained results were encouraging, having only one incorrect game score being computed by the developed algorithm.

Keywords: Boccia · Object detection · Sports analysis · Physical inactivity

1 Introduction

According to the World Health Organization [1], physical inactivity can be defined as an absence of bodily movement produced by skeletal muscles that require energy expenditure. It has been identified as the fourth leading risk factor for global mortality, which represents 6% of deaths globally and it is also associated with increased risk of type 2 diabetes mellitus, cardiovascular diseases, obesity, breast and colon cancers and other chronic diseases [1, 2].

Matthews et al. [3] identified older adults (aged $\geq$ 60 years) to be the most sedentary group in the United States. It has been shown in previous studies that this age group can have substantial benefits from the practice of physical activity such as preventing functional loss [4], reducing the risk of falling [5], controlling blood pressure [6], improving bones and joint health [2] and even maintaining mental health [7]. However, 50% of sedentary adults have no intention of starting an exercise plan and from those who are engaged in physical activity, only 30% of older man and 15%

© Springer International Publishing AG, part of Springer Nature 2019
J. Machado et al. (Eds.): HELIX 2018, LNEE 505, pp. 158–164, 2019.
https://doi.org/10.1007/978-3-319-91334-6_22

of older woman perform it regularly [8]. Facing these statistics, along with the current increase in older population, it is paramount to find innovative solutions for encouraging the elderly to engage in physical activity on a regular basis.

Boccia is a simple precision ball sport that resembles *pétanque*. For the context of this paper, it has the main advantage of being easily adaptable according to the age and limitations of the players. Besides, Boccia can be played as a team sport which promotes social interaction and encourages the individual to participate more often.

This paper proposes a system that monitors the Boccia game scenario through the use of a camera and, by computing an algorithm based on object detection, it returns the game score, in real time, to a User Interface (UI) that can be consulted by all of the players during the game.

The proposed system will be used during a Boccia game in a nursing home environment with the objective of making the game experience more enjoyable for the player and promoting the Boccia practice amongst the elderly.

In a later stage of this project, this system will be used synchronously with the recording of acceleration data of the player's arm, along with the Kinect to determine angles and analyse movements during the game [9, 10]. Afterwards, all of the extracted data will be automatically processed to help enhance the player's performance by suggesting improvements in the movement of the ball. This data will also be available to the caregiver to detect physical or cognitive declines in the individual.

The paper is structured as follows: in Sect. 2 a state-of-art about object detection and tracking applied to sports is presented, along with a brief description of Boccia and the respective scoring rules. Section 3 presents the proposed system's architecture. Section 4 described the used methodology to test the proposed system and respective results. Finally, in Sect. 5, conclusions and future work are addressed.

2 Background

To the best of the authors' knowledge, there are no studies referencing Boccia as a context for object detection or tracking, nonetheless, this type of techniques has been applied with success to several other sports.

For instance, [11] used six monochrome cameras to track a tennis ball in 3D, based on its motion, intensity and shape. The system was successfully used during the international television broadcasts of tennis matches and provided the enabling of virtual replays, game statistics and other interesting features. Basketball has also been a target of various studies associated with ball tracking. Wu et al. [12] successfully tested a ball tracking algorithm on videos from a basketball tournament. This algorithm consisted in detecting the ball based on its colour and shape and, if the ball was indeed detected, the actual tracking would start in the following frame. Chen et al. [13] proposed a more complex approach by applying a physics-based algorithm for predicting the trajectory of a basketball in 3D. The algorithm was similar to the one developed by [12], however, it exploited the 2D shooting trajectory, along with the detection of the court lines, to reconstruct the 3D ball trajectory and infer the shooting location, which can be very useful in the context of game analysis. An analogous approach was developed for volleyball [14].

Curling, similarly to Boccia, is a strategy-based sport that depends of the curling stone's position, therefore, it is relevant to automatically annotate it by using object tracking algorithms, as observed in [15]. In this work, mean-shifting tracking was applied by using the detection of colour and edges of the curling stone. In certain situations of the game, the stone may become occluded by one of the players, which can lead to tracking interruption. To overcome this problem, when the curling stone becomes occluded, the algorithm maintains the tracking by using Kalman prediction [16].

Overall, the use of object detection and tracking applied to sports can provide the spectator with innovative ways of experiencing sport, such as highlighting important events, along with the gathering of richer data for statistical analysis. It can also provide valuable assistance for training, through tactics analysis, and relevant information for referee decision [17].

2.1 The Boccia Game

Boccia is a ball precision game that became a Paralympic sport in 1984. It can be played individually or by teams, which will be the focus of this paper. The game is played with one white ball, which is called the *jack*, along with six blue and six red balls. Each of the six different coloured balls is given to a team composed by three players, the red or blue team, respectively. Since the main feature of the proposed algorithm is returning the game score in real time, this section will focus on the Boccia rules regarding scoring.

The game is divided into six segments, which are called "ends". Throughout each end, each team throws their respective six balls as close as possible to the *jack*. As each end finishes, the score is annotated as it follows:

- The team that placed the ball closer to the *jack* will earn one point for each ball placed at a shorter distance from the *jack* than the closest opponent's ball to the *jack*.
- If two or more balls of different colour are equidistant from the *jack*, then each team earns a point per ball.

After all the six ends have been played, the points from each end are summed and the team that finished with a higher number of points is declared the winner. In case of a tie, extra ends will be played and the first team to win one will be considered the winner. The interested reader should refer to [18] for further knowledge about the current international Boccia rules.

For promoting physical activity, Boccia features the main advantage of being easily adaptable according to limitations and age of the player. The rules and size of the field can also be readapted for a nursing home, which makes Boccia a very accessible sport for older adults [19].

3 Proposed System

3.1 Architecture

The proposed system (Fig. 1) relies on three main components: object detection device, processing unit and interface.

Fig. 1. System architecture.

The selected device was a Microsoft LifeCam VX-1000 computer webcam, which provides a video stream, in real-time, to be processed by the algorithm in the computer, which was developed using Python programming language. In each of the video stream frames, the algorithm automatically detects the balls included in the camera's field-of-view (FOV) according to their colour: white, red or blue. After the white ball is detected, the distances between any blue or red ball covered by the FOV and the *jack* are calculated in pixels by using the centroid coordinates of each of the contours segmenting each ball, which are also calculated in pixels. Based on these values, the algorithm computes the score of the current game scenario according to the Boccia rules.

Finally, the connection between the algorithm and the UI is enabled by using TCP/IP communication. Each second, the current game score is sent by the algorithm to the UI which is displayed on a screen.

4 Preliminary Results

The proposed system was tested in a nursing home in S. Torcato, Guimarães, in Portugal, in a wide room that was used by the elders to play Boccia with adapted rules. As referenced by the caretaker, every Boccia game would start with the *jack* placed at an arbitrary distance from the players, depending on the individual's limitations. Furthermore, no court lines are considered during gameplay, and players throw the Boccia balls while sitting on a chair.

Since the system is based on object detection, it was preponderant to position the camera facing the court in a way that the limited 2D perspective would not interfere with the distance calculation between each Boccia ball and the *jack*.

Another important factor to have into account while positioning the camera is the luminosity. The level of luminosity and light angle have a direct effect on how the algorithm detects colours, which can lead to false positives or even the detection of incorrect colours. Considering the player's perspective when facing the court, there was a window with a curtain on the right side and lights on the ceiling, directly above the court. The authors considered these to be the elements that could be adjusted for optimizing the luminosity level. After testing the system under different light conditions, it was decided that the testing should be performed with the ceiling lights on and the window curtain completely closed. Again, considering the player's perspective facing the court, the system was tested using two camera locations:

- **Camera 1:** placed on left side of the court, facing the window, at a height of approximately 2, 15 m (using a table and a tripod);
- **Camera 2:** placed on the right side of the court, facing the left side wall, at a height of approximately 2,12 m (using a table and a tripod);

For each of the camera locations, ten game situations depicting different game scores were simulated using the *jack* and just three blue balls and three red balls. Each of these game situations was recorded for 20 s and all the Boccia balls were positioned at an arbitrary distance from the *jack* and included in the camera's FOV before each recording. After all the recordings were performed, the algorithm parameters were adjusted to optimize colour detection and eliminate false positives. Finally, the algorithm was tested in each of the recordings and the computed game score was compared to the real game score for each situation.

Overall, the obtained results were encouraging. Considering position 1, the algorithm computed an incorrect game score for only one of the ten recordings. On the other hand, for position 2, all of the ten computed game scores were correct. Figure 2 depicts one of the game situations tested by the algorithm plus the corresponding game score and the current end being played, both displayed in the UI. There is a table on each side of the UI that keeps track of the points scored in the end of each of the six ends by the blue and red team, respectively. The total score of the respective team is shown at the bottom of each of these tables. Finally, in the bottom center of the UI, there are two buttons. The left button should be clicked on when each end finishes to store the score in the table. Moreover, the right button should be clicked on when the players desire to restart the game, which will reset all the points stored in the score tables to zero.

5   Final Remarks

The present paper proposed a system based on a ball detection algorithm to monitor the court of a Boccia game and automatically compute the game score to be displayed in a UI. The objective of such system is to motivate the elderly to engage more frequently in the game of Boccia and therefore promoting physical activity.

The system was tested by using the developed algorithm on recordings of Boccia game situations simulated in a space that belonged to a nursing home, where the elders usually played the game with adapted rules. Due to luminosity conditions, videos were recorded by using two distinct locations of the camera.

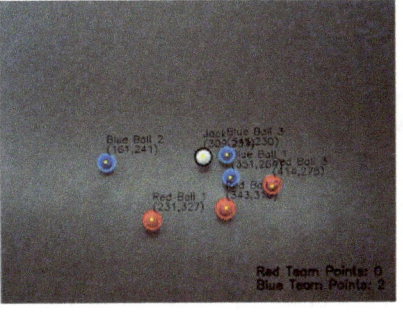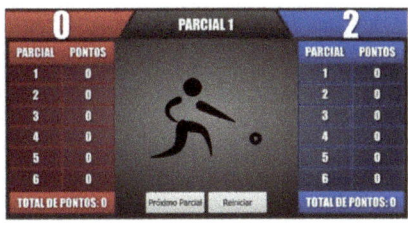

Fig. 2. At the left: Example of one of the recorded game situations, using camera position 2. At the right: The UI, in Portuguese, showing the corresponding game score of the current end ("*parcial*") being played.

Considering the total of twenty recorded videos, the algorithm returned an incorrect score for only one of the game situations recorded with camera in position 1. This error might have been due to the non-optimal camera location, which could have lead the limited 2D perspective to interfere with the computation of the distance between the Boccia balls and the *jack*. Positioning the camera in the ceiling, directly above the court could be one possible solution for this issue. However, from a general point of view, the results from the performed system test were encouraging.

For future reference, the used camera should be replaced by a different model, with a broader FOV (>55°) in order to cover all of the court's area. Furthermore, it is inconvenient to manually adjust the algorithm parameters, which depend of the surrounding environment, to acquire better results. Future work should also focus on developing an automatic calibration method for these parameters in order to make the system more easily adaptable to scenarios with different lighting conditions. Regarding future work, it is planned to test the proposed system, in real-time, during a Boccia game played by the elders living in the nursing home. It is also important to evaluate the usability of the system, which includes not only its effectiveness and efficiency, but also the resulting users' satisfaction towards the use of the system to understand if it can be a factor that influences participation in the game.

Acknowledgements. This article is a result of the project Deus ex Machina: NORTE-01-0145-FEDER-000026, supported by Norte Portugal Regional Operational Programme (NORTE 2020), under the PORTUGAL 2020 Partnership Agreement, through the European Regional Development Fund (ERDF).

References

1. W.H.O.: Global recommendations on physical activity for health, World Health Organization, Geneva (2010)
2. Lee, I.-M., Shiroma, E.J., Lobelo, F., Puska, P., Blair, S.N., Katzmarzyk, P.T.: Impact of physical inactivity on the world's major non-communicable diseases. Lancet **380**, 219–229 (2012)

3. Matthews, C.E., Chen, K.Y., Freedson, P.S., Buchowski, M.S., Beech, B.M., Pate, R.R., et al.: Amount of time spent in sedentary behaviors in the United States, 2003–2004. Am. J. Epidemiol. **167**, 875–881 (2008)
4. Stessman, J., Hammerman-Rozenberg, R., Cohen, A., Ein-Mor, E., Jacobs, J.M.: Physical activity, function, and longevity among the very old. Arch. Intern. Med. **169**, 1476–1483 (2009)
5. Gillespie, L.D., Robertson, M.C., Gillespie, W.J., Sherrington, C., Gates, S., Clemson, L.M., et al.: Interventions for preventing falls in older people living in the community. Cochrane Database of Syst. Rev. (2012)
6. Westhoff, T.H., Franke, N., Schmidt, S., Vallbracht-Israng, K., Meissner, R., Yildirim, H., et al.: Too old to benefit from sports? The cardiovascular effects of exercise training in elderly subjects treated for isolated systolic hypertension. Kidney Blood Pressure Res. **30**, 240–247 (2007)
7. Salguero, A., Martínez-García, R., Molinero, O., Márquez, S.: Physical activity, quality of life and symptoms of depression in community-dwelling and institutionalized older adults. Arch. Gerontol. Geriatr. **53**, 152–157 (2011)
8. Schutzer, K.A., Graves, B.S.: Barriers and motivations to exercise in older adults. Prev. Med. **39**, 1056–1061 (2004)
9. Silva, V., Ramos, J., Soares, F., Novais, P., Arezes, P., Figueira, C., et al.: A wearable and non-wearable approach for gesture recognition – Initial results. In: The 9th International Congress on Ultra Modern Telecommunications and Control Systems, pp. 185–90 (2017)
10. Figueira, C., Silva, J., Santos, A., Sousa, F., Silva, V., Ramos, J., et al.: iBoccia: monitoring elderly while playing Boccia gameplay. ICINCO **1**, 670–675 (2017)
11. Pingali, G., Opalach, A., Jean, Y.: Ball tracking and virtual replays for innovative tennis broadcasts. In: Proceedings-International Conference on Pattern Recognition, vol. 15, pp. 152–156 (2000)
12. Wu, L., Meng, X., Liu, X., Chen, S.: A new method of object segmentation in the basketball videos. In: Proceedings - International Conference on Pattern Recognition, vol. 1, pp. 319–22 (2006)
13. Chen, H.T., Tien, M.C., Chen, Y.W., Tsai, W.J., Lee, S.Y.: Physics-based ball tracking and 3D trajectory reconstruction with applications to shooting location estimation in basketball video. J. Vis. Commun. Image Represent. **20**, 204–216 (2009)
14. Chen, H.T., Tsai, W.J., Lee, S.Y., Yu, J.Y.: Ball tracking and 3D trajectory approximation with applications to tactics analysis from single-camera volleyball sequences. Multimedia Tools Appl. **60**, 1–27 (2011)
15. Kim, J.: Curling stone tracking by an algorithm using appearance and colour features. In: Proceedings of the World Congress on Electrical Engineering and Computer Systems and Science (EECSS), pp. 1–6 (2015)
16. Kalman, R.E.: A new approach to linear filtering and prediction problems 1. J. Fluids Eng. **82**, 35–45 (1960)
17. Wang, J.R., Parameswaran, N.: Survey of sports video analysis: research issues and applications. In: School of Compute Science and Engineering the University of New South Wales, vol. 113, pp. 115–118 (2006)
18. BISFed: BISFed International Boccia Rules (v.2) (2017)
19. Silva, V., Ramos, J., Soares, F., Novais, P., Arezes, P., Sousa, F., Silva, J.O., Santos, A.: iBoccia: a framework to monitor the Boccia gameplay in elderly. In: Lecture Notes in Computational Vision and Biomechanics, vol. 27 (2018)

Development and Optimization of a New Suspension System for Lower Limb Prosthesis

Andreia S. Silveira⑩, Patrícia A. Senra⑩, Eurico Seabra$^{(\boxtimes)}$⑩,
and Luís F. Silva⑩

Department of Mechanical Engineering, School of Engineering,
University of Minho, 4800-058 Guimarães, Portugal
eseabra@dem.uminho.pt

Abstract. The increasing rate of lower limb amputations reinforces the need to develop a new suspension system that provides a better quality of life for the lower limb amputees. This study aimed to present a novel suspension system that improves amputee's satisfaction in terms of donning and doffing process of the prosthetic lower limb. The design of the proposed suspension system was developed following the design methodology, to establish the amputee's needs, objectives, functions, requirements and specifications in order to optimize the final solution. The final solution is a combination of a guiding and fixation mechanisms that improve the donning and doffing process by driving the serrated pin to the fixation system. The proposed suspension system is a good alternative to improve the quality of life of amputees with lower activity level on the daily basis.

Keywords: Lower limb prosthesis · Suspension system
Amputee's satisfaction · Mechanical design
Guiding system and fixation system

1 Introduction

Lower limb amputees perform daily the donning and doffing process to apply the prosthetic limb. The suspension system of prosthesis allows the firm attachment between the residual limb and the prosthetic limb and prevents excessive translation, rotation, and vertical movements between the residual limb and the socket [1, 2].

Suspension systems have a fundamental role in the adaptation of the amputee to the lower limb prosthesis in order to replace the lost limb functions. Several prosthetic suspension systems are commercial available, including pin/lock systems, lanyard system, straps and hinges, suction systems and magnetic system [3, 4]. Depending on the suspension system used, a different donning and doffing method is required, some of them are more time-consuming and require more hand strength than others. Each suspension system has disadvantages and advantages depending on the type of user [2, 5, 6].

The selection between the commercial solutions requires a careful evaluation in order to choose the suspension system that best fits the amputee's needs, since a poor suspension can cause skin problems, pain, gait instability, shear stress and volume loss of the residual limb [2, 7].

© Springer International Publishing AG, part of Springer Nature 2019
J. Machado et al. (Eds.): HELIX 2018, LNEE 505, pp. 165–171, 2019.
https://doi.org/10.1007/978-3-319-91334-6_23

To select the most suitable suspension system, it is important to understand the overall satisfaction of amputees using the selected suspension system [3, 8]. Several studies evaluated the satisfaction and functionality of the suspension systems available in the current market using a Prosthetics Evaluation Questionnaire (PEQ) [5, 7, 9]. PEQ rates the participant's feedback about the satisfaction in different domains (fitting, walking on diverse surfaces, appearance, donning and doffing and sitting) and perceives problems such as pistoning, sweating, skin irritations, residual limb pain, swelling, smell and sounds [1, 10].

Some studies pointed out the ease of donning and doffing as an important factor to amputee's satisfaction since the donning and doffing technique differs according to the suspension used. Also, an impaired hand function increases the risk of skin problems and malfunction suspension [8, 11, 12]. Some suspension systems like suction systems requires a proper hand function during the donning and doffing process for a safety and correct suspension [3, 0].

Some studies also highlighted the preference of amputees for the pin/lock system due to the easy donning and doffing process using a serrated pin attached to the distal end of the liner to suspend the prosthesis [13–15]. On the other hand, the study by Eshraghi et al. (2012) suggested some difficulties of donning and doffing with pin/lock system because some patients experienced a bit of struggle when aligning the pin with the locking system during donning process [8].

The authors claim that the suspension systems available in the current market, do not provide the quality of life that the lower limb amputees deserve, since they do not ensure the total safety and enough satisfaction to use the prosthesis in long term. These limitations are associated with the difficulties during the donning and doffing process, especially for amputees with lower activity levels. Therefore, this paper proposes a novel approach to solve some of the current limitations of the pin/lock systems in order to improve the quality of life of lower limb amputees and, at the same time, to increase amputee's satisfaction with the prosthesis. It is presented a new suspension system that ensures an effective serrated pin insertion, while providing comfort and an easy process of donning and doffing the prosthesis.

2   Suspension System Design

The proposed suspension system was developed following the design methodology, to understand the amputees' needs, define the statement-problem, create several alternative solutions and prototyping the selected solution.

2.1   Design Concept

The main user was defined as an individual with lower level of activity and manual dexterity, that also presents difficulties in terms of adaptation and learning, since the literature reports that lower limb amputations are increasing due to the incidence of vascular diseases, such as diabetes, which especially affects the older population [16].

It was essential to develop a simple system that allows the reduction of the existing barriers during the adaptation to the prosthesis and improves amputee's satisfaction in terms of donning and doffing process. Therefore, the mechanical design was driven by the need to develop a simple and functional mechanism that ensures that the serrated pin is inserted and secured correctly into the housing without colliding at its ends. It was also intended to be a universal mechanism that could be easily adapted to the different types of pin/lock systems available in the current market.

The other parameters that were also taken into account during the conceptual design phase were safety, comfort, ergonomics, easy to assembly and maintenance, aesthetic appearance and costs.

2.2   Conceptual Design Solution

The proposed design, illustrated in Fig. 1, presents a guiding and fixation systems inside the mounting case, a serrated pin and a housing. It has two mainly functions: guide the serrated pin into the housing and establish a firm attachment between the residual limb and the prosthesis by fixating the serrated pin.

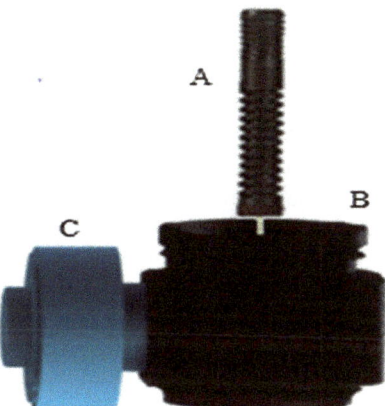

Fig. 1. The proposed suspension system: (A) serrated pin, (B) housing and (C) mounting case with the guiding and fixation systems inside.

This solution proposes to guide the serrated pin into the housing with a retractable mechanism that winds and unwinds a wire connected to the serrated pin, with the energy release from a power spring. To achieve the fixation of the serrated pin with the pinion gear, it is proposed a ratchet mechanism, as the fixation system, to prevent the rotation in counterclockwise direction of the pinion gear.

Figure 2 shows an exploded view of all the components of the proposed suspension system. The shaft, in which the ratchet gear is mounted, is fastened to the end of the pinion gear. The power spring is assembled inside the reel, that it is mounted on a shaft. The wire is wound around the spool, that it is mounted on the shaft portion of pinion gear. Each pawl has two openings, the inferior opening receives the pin of the button and the superior opening engages with the pin of the reel. The pawls also have a middle opening to receive one end of the torsion spring. Both torsion springs are connected to the respective pawls at one end and has its opposite end fitted into the inner wall of the mounting case. The button includes two pins and a central opening to support the end of the shaft. The reel has two pins that hold the pawls and act as the center of the rotation of the pawls. All the components are assembled inside the mounting case that it is threaded to the receiving compartment of the housing with the pinion gear head positioned at the axial hole of the housing.

Fig. 2. Exploded view of the proposed model with the components: (A) button, (B) pawls, (C) ratchet gear, (D) torsion springs, (E) reel, (F) shaft, (G) mounting case, (H) spool with the wire, (I) pinion gear and (J) power spring.

To connect the residual limb to the prosthesis, the patient must first rotate the button to unlock the rotation of the ratchet gear backwards in order to pull out the wire. The pulling of the wire causes the rotation of the pinion gear in counterclockwise direction and the power spring is fully wound around the shaft. The wire connects with the serrated pin, which is attached to the distal part of the residual limb, through a nut. Then, the patient rotates again the button to the first position, where the pawls engage with the teeth of ratchet gear. Subsequently, the patient lightly moves the residual limb toward the socket direction. In turn, the power spring goes back to its initial position and its energy, stored during the pulling, is enough to wind the wire around the spool by the rotation of the shaft in the clockwise direction. This succession of events allows the serrated pin to be correctly guided into the housing. At this point, the serrated pin engages with the teeth of the pinion gear and the pinion gear rotates in the clockwise direction until the serrated pin is totally inserted inside the housing. Yet, if in some instance the serrated pin attempts to move in the release direction, both pawls lock up the rotation of the ratchet gear. In this way, the proposed solution guarantees that the serrated pin is fixed with the teeth of pinion gear and cannot be released until the patient desires it.

To remove the residual limb from the prosthetic limb, the patient just needs to rotate again the button to release the serrated pin. The pawls are disconnected from the ratchet gear with the rotation of the button to unlock the pinion gear rotation in the counterclockwise direction with the release of the serrated pin.

3   Prototyping and Validation

The final solution was prototyped to properly evaluate the viability and the possible improvements of the conceptual idea. A low-cost prototype model was created. The components were produced using 3D printing and they were assembled to create the prototype shown in Fig. 3.

Fig. 3. Prototype model of the new proposed suspension system.

A detail analysis was carried out to verify if the proposed solution follows all the established requirements and specifications. Safety and functional tests were conducted to evaluate the viability of the solution, and, if necessary, to identify possible improvements.

The safety analysis consisted on detecting possible failures and evaluating its effects for the system and patient in order to identify required corrective actions to prevent failures and to assure the reliability of the suspension system. Table 1 specifies possible failures, as well as the recommended actions to prevent these failures. They were classified as critical, major and minor failures. The critical failure is addressed when the user safety is at risk, and it is necessary to undertake improvement actions on the designed solution. The major failure affects the performance of the product but does not affect the user safety and again improvement actions must be undertaken. Minor failures do not affect product performance or put the user at risk and there is no need for improvement actions on the designed solution.

From the evaluation conducted on the prototype it was observed that it was possible to adapt the proposed solution to the different types of pin/lock systems, and, in turn, it can be applied in patients that already have the prosthesis with the pin/lock system.

It was also verified that there are still some improvements needed to optimize the proposed system. Besides the suggestions in Table 1, the ratchet system still needs to be improved, so that the serrated pin can easily enter the housing without any great effort to be developed by the user.

Due to the limitations imposed by the prototype of Fig. 3, it is still necessary further research to full evaluate the performance of the proposed system. Subsequently, in the future, it is intended to produce a new prototype to be tested on patients. The new system will be compared with prosthetic suspension systems available in the current

Table 1. Potential failures for the proposed suspension system.

System	Potential failure	Type of failure	Cause	Effect	Recommended action
Guiding	Spring failure	Major failure	Over extension of the spring coils	Serrated pin is not guided	Careful spring design
	Wire break	Major failure	High pulling force	Serrated pin is not guided	Select a wire with higher tensile strength
	Nut disconnection	Major failure	Corrosion and debris	Serrated pin is not guided	Coating the serrated pin and ensure it is daily cleaned
	Button blocking	Major failure	Debris	The user cannot pull out the wire	Place a spongy layer on the entrance of the housing
Fixation	Ratchet gear rotates backwards	Critical failure	Pawls are not correctly aligned with the ratchet gear	Release of the serrated pin	Design a ratchet gear with bigger number of teeth
	Button blocking	Major failure	Debris	The user cannot remove the serrated pin	Place a spongy layer on the entrance of the housing
	Serrated pin disengagement	Critical failure	Corrosion and debris	Poor suspension	Coating the serrated pin and ensures it is cleaned daily

market, in terms of patient's satisfaction to determine the real benefits and advantages of the proposed system. Besides that, further tests will also be carried out in terms of pistoning and gait analysis.

4   Conclusions

This study proposed a new suspension system that revealed to be a good alternative for amputees with lower activity level and bad hand function, since it facilitates the donning and doffing processes, and, at the same time, ensures a firm attachment between the residual limb and the prosthesis. The suspension system was developed to improve the quality of life of amputees and increase amputee's satisfaction. Not only it provides a better suspension but also promotes a frequent use of the prosthesis. However, further tests are still needed to determine the real benefits and advantages of the novel suspension system herein proposed.

References

1. Andrysek, J.: Lower-limb prosthetic technologies in the developing world: a review of literature from 1994–2010. Prosthet. Orthot. Int. **34**(4), 378–398 (2010)
2. Eshraghi, A., Azuan, N., Osman, A., Karimi, M., Ali, S.: Pistoning assessment in lower limb prosthetic sockets. Prosthet. Orthot. Int. **36**(1), 15–24 (2012)
3. Gholizadeh, H., Abu, N.A.A., Eshraghi, A., Ali, S., Yahyavi, E.S.: Satisfaction and problems experienced with transfemoral suspension systems: a comparison between common suction socket and seal-in liner. Arch. Phys. Med. Rehabil. **94**(8), 1584–1589 (2013)
4. Gholizadeh, H., Osman, N.A.A., Eshraghi, A., Ali, S.: Transfemoral prosthesis suspension systems a systematic review of the literature transfemoral prosthesis suspension systems. Am. J. Phys. Med. Rehabil. **93**(9), 809–823 (2014)
5. Baars, E.C.T., Geertzen, J.H.B.: Literature review of the possible advantages of silicon liner socket use in trans-tibial prostheses. Prosthet. Orthot. Int. **29**(1), 27–37 (2005)
6. Gholizadeh, H., Azuan, N., Osman, A., Eshraghi, A., Sævarsson, S.K., Abu, W., Wan, B., Pirouzi, G.H.: Transtibial prosthetic suspension: less pistoning versus easy donning and doffing. J. Rehabil. Res. Dev. **49**(9), 1321–1330 (2012)
7. Safari, M.R., Meier, M.R.: Systematic review of effects of current transtibial prosthetic socket designs—Part 1: Qualitative outcomes. J. Rehabil. Res. Dev. **52**(5), 491–508 (2015)
8. Eshraghi, A., Osman, N.A.A., Karimi, M.T., Gholizadeh, H., Ali, S., Wan Abas, W.A.B.: Quantitative and qualitative comparison of a new prosthetic suspension system with two existing suspension systems for lower limb amputees. Am. J. Phys. Med. Rehabil. **91**(12), 1028–1038 (2012)
9. Board, W.J., Street, G.M., Caspers, C.: A comparison of trans-tibial amputee suction and vaccum socket conditions. Prosthet. Orthot. Int. **25**, 202–209 (2001)
10. van de Weg, F.B., van der Windt, D.A.W.M.: A questionnaire survey of the effect of different interface types on patient satisfaction and perceived problems among trans-tibial amputees. Prosthet. Orthot. Int. **29**(3), 231–239 (2005)
11. Ali, S., Osman, N.A.A., Naqshbandi, M.M., Eshraghi, A., Kamyab, M., Gholizadeh, H.: Qualitative study of prosthetic suspension systems on transtibial amputees' satisfaction and perceived problems. Arch. Phys. Med. Rehabil. **93**(11), 1919–1923 (2012)
12. Gholizadeh, H., Azuan, N., Osman, A., Eshraghi, A., Ali, S., Arifin, N.: Evaluation of new suspension system for limb prosthetics. Biomed. Eng. Online, pp. 1–13, 2014
13. Ali, S., Osman, N.A.N., Mortaza, N., Eshraghi, A., Gholizadeh, H., Abas, W.A.B.B.W.: Clinical investigation of the interface pressure in the trans-tibial socket with Dermo and Seal-In X5 liner during walking and their effect on patient satisfaction. Clin. Biomech. **27**(9), 943–948 (2012)
14. Klute, G.K., Berge, J.S., Biggs, W., Pongnumkul, S.: Vacuum-assisted socket suspension compared with pin suspension for lower extremity amputees: effect on fit, activity, and limb volume. Arch. Phys. Med. Rehabil. **92**(10), 1570–1575 (2011)
15. Coleman, K.L., Boone, D.A., Laing, L.S., David, E., Smith, D.G.: Quantification of prosthetic outcomes: Elastomeric gel liner with locking pin suspension versus polyethylene foam liner with neoprene sleeve suspension. J. Rehabil. Res. Dev. **41**(4), 591–602 (2004)
16. Ziegler-graham, K., Mackenzie, E.J., Ephraim, P.L., Travison, T.G., Brookmeyer, R.: Estimating the prevalence of limb loss in the United States: 2005 to 2050. Arch. Phys. Med. Rehabil. **89**(3), 422–429 (2008)

Impact of UTAUT Predictors on the Intention and Usage of Electronic Health Records and Telemedicine from the Perspective of Clinical Staffs

P. Venugopal[1], S. Aswini Priya[1], V. K. Manupati[2],
M. L. R. Varela[3(✉)], J. Machado[4], and G. D. Putnik[3]

[1] Department of Technology Management, School of Mechanical Engineering,
VIT, Vellore, India
[2] Department of Manufacturing, School of Mechanical Engineering,
VIT, Vellore, India
manupativijay@gmail.com
[3] Department of Production and Systems, School of Engineering,
University of Minho, Guimarães, Portugal
{leonilde,putnikgd}@dps.uminho.pt
[4] Department of Mechanical Engineering, School of Engineering,
University of Minho, Guimarães, Portugal
jmachado@dem.uminho.pt

Abstract. Technology adoption play significant role for good healthcare services. It is very important to understand and identify the requirements and perceptions of hospital employees for the implementation of technology in their work place. This study determines the impact of performance expectancy, effort expectancy and social influence on behavioural intention as well as the impact of facilitating conditions on technical/clinical staff's perspective. The structured questionnaire is administered to 770 clinical staffs on the usage of telemedicine and electronic health records in hospitals. A valid sample of 568 was returned back for further analysis. Regression analysis using AMOS 20 is performed to examine the effect of the constructs. Findings revealed that performance expectancy, effort expectancy, and social influence have a significant impact on behavioural intention and facilitating condition also significantly impacts behavioural intention which in turn impacts usage behaviour of electronic health records and telemedicine. The limitations and future research are suggested and delineated.

Keywords: Electronic health records · Telemedicine · Performance expectancy
Effort expectancy · Social influence · Facilitating conditions
Behavioural intention

1 Introduction

Medicine among Primitive Peoples was evident of surgery in skulls from the Stone Age. Primitive people used different experiments for their injuries, for example Australian Aborigines covered broken arms in clay, which hardened in the hot sun.

© Springer International Publishing AG, part of Springer Nature 2019
J. Machado et al. (Eds.): HELIX 2018, LNEE 505, pp. 172–177, 2019.
https://doi.org/10.1007/978-3-319-91334-6_24

Cuts were covered with fat or clay and bound up with animal skins or bark. Whereas Ancient Egyptian Medicine was about 3000 BC the curtain rises on Egyptian civilization. Egyptian doctors used a huge range of drugs obtained from herbs and minerals. They were drunk with wine or beer or sometimes mixed with dough to form a 'pill'. Secondly, Ancient Greek Medicine had its roots in modern medicine are in ancient Greece. People who were ill-used to offer sacrifices to the god and sleep overnight in the temple. Greek doctors developed a rational theory of disease and sought cures. However one did not replace the other. Finally, the Roman Medicine conquered Greece and afterward, doctors in the Roman Empire were often Greeks.

In India after agriculture healthcare is the largest sector offering more employment and generating more revenue. The Indian healthcare is divided into two categories one is public/government and second one is private. Government sector healthcare Institutions are very limited in number in few cities, primary healthcare centres in rural areas whereas most of the private sector healthcare institutions are in urban areas. There are well trained medical/healthcare professions and the cost wise it is very low when compared to US and Western countries. The present Indian healthcare market size is US$100 billion and anticipated to grow up to US$ 280 billion by 2020 with 22.9 compound annual growth rate. According to department of industrial policy and promotion, Indian hospitals and diagnostic centres are drawing the Foreign Direct Investment too. To promote health industry in India, Social Endeavour for Health and Telemedicine was introduced to rural population through information technology accessibility, improving the skills and by introducing the technologies for the success of 'Digital India'. Another initiative by Government of India is the agreement between Sweden and India for improving the quality of health through the technology innovation and research and development. Prime Minister of India, initiated E-health services with an objective to offer effective and economical healthcare services to all people through electronic health records, for instance to get online appointments.

2 Problem Contextualization

As on March 2011, Indian population was Rs. 121 crore, out of which Rs. 83.3 crore (68.84%) live in rural areas while 37.7 crore (31.6%) live in urban areas, as per census of India. The access of healthcare services is not yet achieved fully in many developing countries, even though it is guaranteed for all people throughout the world. Particularly in rural area of developing countries are unable to access health care services. As per World Health Organization, a pregnant women should have at least four times Anatanal care (ANC) by a trained health provider. According to the Deloitte report of 2015, the estimation of expenditures allotted to this sector is five percent of Gross Domestic Product (GDP) in 2013 and it is anticipated to remain stable till 2016. In terms of infrastructure, India has only one bed for every 1050 patients and yet it accounts for 100000 beds at present decade with an investment of about $50 billion. And it also lacks in qualified medical professionals to efficiently diagnose the diseases through the proper delivery of services. The report states that doctor-patient ratio is considerably lower in India than World Health Organization statistical report the ratio of India is about 0.7 doctors and 1.5 nurses per 1000 people which is comparatively lower than

2.5 doctors and nurses per 1000 people. As there is a scarcity of doctors and nurses in India, opting for traditional method of services is seem to be not effective and difficult for patients to access to it. The problem also includes traditional consultation process which is very lengthy and time consuming in urgency cases, long-distance travel times, improper and inefficient delivery of services and non-technology up gradation in the past. All these factors contributed to the significant adoption of telemedicine in India. Kai Zheng et al. (2005) studied about the behaviour of receipt and espousal of the clinical system from the perspective of clinicians for continual sickness and precautionary care management. The technology tested is useful for taking the appointment of the doctor; information of the patient can be browsed, reminding the doctors' directions, for recording the significant symptoms. It was found that there was a larger confrontation. Zaidi (2008) measured the clinicians' perceptions and usefulness of a web-based antibiotic approval system and the future need to be carried on the adoption of other technologies.

3   Methodology

The hospital employees' especially technical/clinical staff is the sampling units for the study. In particular technical/clinical staff from Vellore District in Tamilnadu has been selected as the sampling unit for the study. Healthcare sector has been chosen for the study purpose of the research. Convenience sampling was adopted to select the clinical staffs from the identified hospitals of Vellore. Convenience sampling would guarantee the availability of the respondents could be included in the sample. A sample of 770 clinical staff was contacted for elucidating their responses, out of which 568 clinical staff willingly filled up the questionnaire after filling the missing values with mean series all the respondents have been chosen for the final study. In the present research, compiles of both primary as well as secondary data sources. The secondary data sources include a database of referred journals from Emerald publishers, EBSCO, IEEE, MIS quarterly etc. other secondary data sources like government reports, conference proceedings and other sources.

The primary data associated to the demographic and the perceptions of respondents on performance expectancy, effort expectancy, social influence, facilitating conditions, behavioural intention and usage behaviour of electronic health records and telemedicine were collected by administering through structured questionnaire with technical/clinical staff of Vellore District – Vellore, Ambur, Ranipet, Walaja etc. Once the data collection is completed, the data was coded, cleaned, labelled and verified with regard to the missing values. The corrected data was taken for analysis. In this research, SPSS was utilized to test the reliability and validity of the instrument and to examine the perception of hospital employees whereas to examine the effects of performance expectancy, effort expectancy, and social influence constructs influence on behavioural intention measured is using Regression in AMOS 20. The usage behaviour determinants of behavioural intention and facilitating condition effect also measured using Regression in AMOS. The weights of items of the behavioural intention determinants and usage behaviour determinants were taken and regression formula was applied to constructs.

4 Results and Analysis

Table 1 shows the demographic profile of Clinical Staffs in the healthcare services. Out of five hundred and sixty-eight appropriate responses, 29.9% (170) were female clinical staffs and 70.1% are male clinical staffs (398), belongs to the age group of between 20–25 years (100%). The experience/service in the medical field reported is 97.9% (556) with 0–5 years, 2.1% (12) with 11–15 years respectively.

Table 1. Socio-demographic profile of the participant.

Demographic details of clinical staff's		Number of participants (568)	
		Frequency	Percent
Gender	Male	398	70.1
	Female	170	29.9
Age	20–25	568	100
	26–30	–	–
	31–35	–	–
	36–40	–	–
	41–45	–	–
	46–50	–	–
	Above 50	–	–
Experience	0–5	556	97.9
	6–10	–	–
	11–15	12	2.1
	16–20	–	–
	21–25	–	–
	26–30	–	–
	Above 31		–

Figure 1 shows the regression and correlation analysis for the clinical staff which has been used for further analysis. The above formula shows that one unit change in Performance Expectancy will result in 0.65 unit change in Behavioural Intention, one unit change in Effort Expectancy will result in 0.21 unit change in Behavioural Intention and one unit change in Social Influence will result in 0.08 unit change in Behavioural Intention. It was found that all the three factors have a positive effect on Behavioural Intention to accept and use electronic health records and telemedicine. The R-Square value for Behavioural Intention is 0.48; it shows that 48% of variance or performance in Behavioural Intention is together explained by these three Factors. It clearly shows that by bringing positive perception on the three factors can alone cause 48% Behavioural Intention to use electronic health records and telemedicine and among the three factors Performance Expectancy seems to be more important.

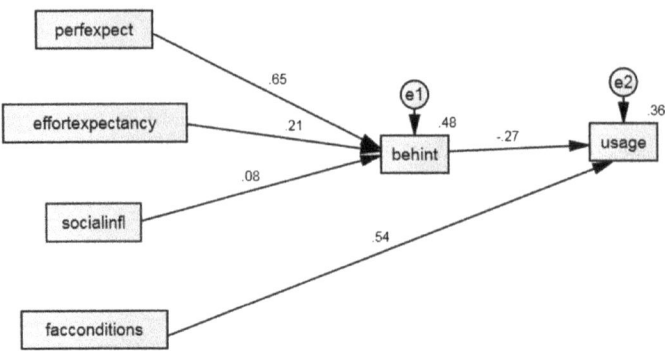

Fig. 1. Testing model. Source: Primary Data - Testing Model to determine the direct effect of clinical staff's perceptions on performance expectancy, effort expectancy and social influence upon their behavioural intention as well as the effect of behavioural intention and facilitating conditions on usage behaviour. [Behavioural Intention = (0.65 * Performance Expectancy) + (0.21 * Effort Expectancy) + (0.08 * Social Influence)]

$$[\text{Usage Behaviour} = (0.54 * \text{Facilitating Conditions}) + (-0.27 * \text{Behavioural Intention})]$$

The above formula shows that one unit change in Facilitating Conditions will result in 0.54 unit change in Usage Behaviour and one unit change in Behavioural Intention will result in −0.27 unit change in Usage Behaviour. It was found facilitating conditions have a positive effect and behavioural intention have a negative effect on causing Usage Behaviour or likely to use behaviour. The R-Square value for Usage Behaviour is 0.36; it shows that 36% of variance or performance in Usage Behaviour is together explained by these three Factors. It clearly shows that by bringing positive perception on these two factors facilitating conditions can alone cause 54% of Usage Behaviour on electronic health records and telemedicine and among the two factors facilitating conditions seems to be more important. Hence the hypothesis of H02 is accepted. All the variables are statistically significant at 5% level ($p < 0.05$).

5   Conclusion

The principal purpose of the study is to validate the dimensions that portray the behavioural intention and use behaviour or likely to use in the context of electronic health records and telemedicine technologies. Technology adoption can significantly put in hospitals to improve good healthcare service. It is very important to understand and identify the requirements and perceptions of hospital employees for the implementation of technology. The patients need not wait more for a lean period through technology adoption, electronic health records will help the doctors by providing the entire health information of the patient and telemedicine will helps the patients in avoiding number of visits and offering healthcare service in remote places through telemonitoring. The technology adoption and technology acceptance and use from the perspective of information technology measured different dimensions both qualitatively and quantitatively. Further the authors have also suggested for further measurement of technology acceptance and use from other contexts.

Acknowledgements. This work is supported by Fundação para a Ciência e Tecnologia (FCT) with ref. PEst2015-2020.

References

1. AlDossary, S., Martin-Khan, M.G., Bradford, N.K., Smith, A.C.: A systematic review of the methodologies used to evaluate telemedicine service initiatives in hospital facilities. Int. J. Med. Inf. **97**, 171–194 (2017)
2. Bickmore, T., Vardoulakis, L., Jack, B., Paasche-Orlow, M.: Automated promotion of technology acceptance by clinicians using relational agents. In: Intelligent Virtual Agents, pp. 68–78. Springer, Heidelberg, January 2013
3. Chiang, M.F., Boland, M.V., Margolis, J.W., Lum, F., Abramoff, M.D., Hildebrand, P.L.: Adoption and perceptions of electronic health record systems by ophthalmologists: an American Academy of Ophthalmology survey. Ophthalmology **115**(9), 1591–1597 (2008). American Academy of Ophthalmology Medical Information Technology Committee
4. Gheorghe, M., Petre, R.: Integrating data mining techniques into telemedicine systems. Informatica Economica **18**(1), 120–130 (2014)
5. Jeyakodi, T., Herath, D.: Acceptance and use of electronic medical records in Sri Lanka. Sci. Res. J. (SCIRJ) **4** (2016)
6. Katzenstein, J., Yrle, A.C., Chrispin, B., Hartman, S., Lundberg, O.: Telemedicine: an innovative technique in healthcare delivery. Acad. Health Care Manag. J. **8**(1/2), 49 (2012)
7. Narattharaksa, K., Speece, M., Newton, C., Bulyalert, D.: Key success factors behind electronic medical record adoption in Thailand. J. Health Organ. Manag. **30**(6), 985–1008 (2016)
8. Tavares, J., Oliveira, T.: Electronic health record portal adoption: a cross country analysis. BMC Med. Inform. Decis. Mak. **17**(1), 97 (2017)
9. Ward, R., Stevens, C., Brentnall, P., Briddon, J.: The attitudes of health care staff to information technology: a comprehensive review of the research literature. Health Inf. Libr. J. **25**(2), 81–97 (2008)
10. Zaidi, S.T.R., Marriott, J.L., Nation, R.L.: The role of perceptions of clinicians in their adoption of a web-based antibiotic approval system: do perceptions translate into actions? Int. J. Med. Inf. **77**(1), 33–40 (2008)
11. Zurovac, J., Dale, S., Kovac, M.: Perceptions of electronic health records and their effect on the quality of care: results from a survey of patients in four states. Mathematica Policy Research (2012)

Smartphone-Based Solution for Pedestrian Detection and Communication with a Driver Assistance System

Pedro Sousa, André Correia, and Sara Paiva$^{(\boxtimes)}$

ARC4DigiT, Applied Research Center for Digital Transformation,
Instituto Politécnico de Viana do Castelo, Viana do Castelo, Portugal
`sara.paiva@estg.ipvc.pt`

Abstract. This paper presents an on-going project that consists of a smartphone-based solution for detecting pedestrians in crosswalks and communicating that information for a driver assistance system. The solution assumes a pedestrian using a smartphone, beacons in traffic lights and a driver with an android app, both apps communicating with a central database. The pedestrian app warns the user of crosswalks nearby and notify drivers if crosswalks are approaching and also if there are people in it. We also present the on-going tests that are being carried out to conclude the usefulness of the system in a real.

Keywords: Pedestrian detection · Notification of drivers
Driver assistance system · Crosswalks · Beacons

1 Introduction

Detection of pedestrians in crosswalks is an area of continuous research as it falls in the important scope of citizen urban security where plenty of actions and measures are carried out annually by local councils. The need for solutions in this field is supported by the big number of deaths by trampling and solutions to minimize this number are certainly embraced by everybody. Some causes of these accidents include distraction of pedestrians, improper use of smartphone when entering crosswalks, drivers' distraction or fatigue, among others. In this paper, we present the design of a solution for warning drivers of the presence of crosswalks and pedestrians in it, through a mobile android application. This requires an algorithm for detecting the presence of pedestrians, which is achieved by a mobile app being used by the pedestrian, Bluetooth low energy technology in traffic lights and signal processing. The rest of this paper is organized as follows: in the next section we will present some related work. Next, we explain the architecture and the overview of the designed solution. As this is an on-going work, we will present the scenarios in which the solution is currently being tested and what type of conclusions we will be able to have in the end of this study.

© Springer International Publishing AG, part of Springer Nature 2019
J. Machado et al. (Eds.): HELIX 2018, LNEE 505, pp. 178–184, 2019.
https://doi.org/10.1007/978-3-319-91334-6_25

2   Related Work

The current literature presents us with many studies and results of solutions for crosswalk and pedestrian detection and driver assistance systems. One related work is presented in [1] where authors present a work in progress that tackles the problem of crosswalk detection and self-localization, enabling blind and visually impaired users to acquire 360° image panoramas while turning in place on a sidewalk. Another work refers to the importance of pedestrian detection as part of a nurban pedestrian safety system [2]. Another approach that uses laser feature extraction is presented by Hernandéz et al. [3]. Sukuzi et al. base their solution [4] on a sensor fusion of a monocular camera and a millimeter wave radar. Other proposed works use traffic light detection [5], 3D density maps [6], bionic eyeglass [6], maximally stable extremal regions (MSER) and extended random sample consensus (ERANSAC) [7], RGBD cameras [8]. Regarding the integration of detection system with driver assistance systems, a work is presented by Anselm Haselhoff and Anton Kummert [9]. Another work is presented in [10] where authors use the advances in camera-based driver assistance for the assistance of blind and visually impaired people by developing a concept for the transfer of object detection algorithms in the traffic domain.

3   Implementation

3.1   Overview

The solution we designed has in mind minimizing the number of accidents in cross-walks. Two different perspectives exist: the pedestrian and the driver. Regarding pedestrians, we currently assist plenty of them entering crosswalks distracted by using their smartphones so one of the functionalities we implemented is a notification for the user, whenever he is using the phone and a crosswalk is near (not necessarily in it). This is achieved by a background service that detects proximity of beacons placed in traffic lights near crosswalks. Regarding drivers, we inform them of nearby crosswalks approaching using the GPS coordinates of the moving car. On another hand, we also inform if there are pedestrians in it. We obtain this information with an algorithm that detects if a pedestrian is inside a crosswalk that mainly uses beacons signal intersection.

In the next sections, we will explain in detail each of the developed components of the system.

3.2   Architecture and Components

The architecture of the system is composed of several components, as illustrated in Fig. 1. As part of the architecture, we have a pedestrian that carries an android smartphone with an app that mainly consists of a background service that starts with the reboot of the system. The background process looks for beacons and takes three actions: if the smartphone is being used and the user is near a crosswalk, the user gets notified to be careful or to turn off the phone; if the user is inside the crosswalk, notify a central database (so later the driver can receive this information); if the user is outside the crosswalk, also notify the central database. The driver has another app and one of the functionalities for the driver is to get notified of a nearby crosswalk approaching, using his current GPS coordinates obtain by the satellite. Whenever a crosswalk is detected to be approaching, the app will check the central database to see if any pedestrians were notified to be in it, in order to inform the driver.

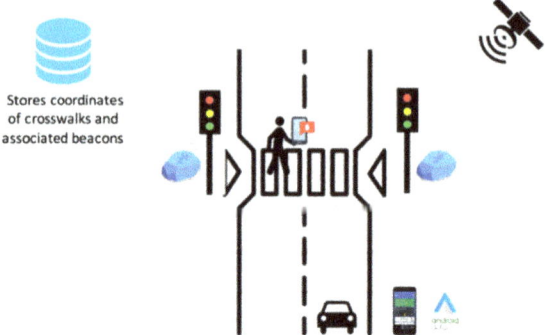

Fig. 1. System overview and its components

3.3 Implementation

Data Model. The supporting data model for this project has information about crosswalks and their location. One of the features of the system will be to locate nearby crosswalks given a car location. Considering a city has thousands of crosswalks, it would not be efficient to compare the current car position with all crosswalks. So, we

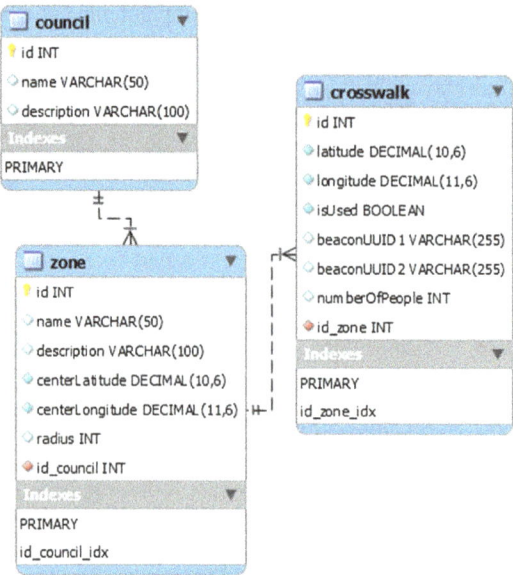

Fig. 2. Supporting data model

created a notion of zone, with a center point and a radius, that is part of a council of a given city. Crosswalks are then created as being part of a zone. The final data model that supports the system is shown in Fig. 2.

Driver App

Detecting the Zone

The first thing the driver app needs to do is, based on the car current position, calculate the zone he is in. The zones (central point and radius) are obtained once when the app starts and stored locally to avoid too much network connection. When a new car location is obtained we calculate the zone the car is in which is given by the zone that has a distance inferior to 500 m to the car location. We defined that all zones have a 500-m radius and there are no intersections between zones.

Retrieving Crosswalks

After the zone is calculated, we obtain the crosswalks in it and the driver is presented with a map that shows him the nearby crosswalks (Fig. 3). A sound notification is also produced so the user avoids looking frequently to the app.

Fig. 3. UX prototype to present crosswalk information to drivers

Fig. 4. UX prototype to notify of the presence of pedestrian in crosswalks

Retrieving People

When a crosswalk is approaching, we obtain from the central database if there are pedestrians in it and also inform the user graphically (Fig. 4) and also with sound.

Pedestrian App

Assumptions

For the development of the pedestrian app features, we assume a beacon in each traffic light of a crosswalk. The beacon has information about its type (in this case crosswalk) and the crosswalk identification, as it is stored in the database. In a IoT real scenario, the smartphone would certainly receive information of other beacons, such as in cars, bus stops, etc., and our system is only interested in processing beacons from crosswalks.

Estimote SDK

For capturing beacon signal we used the Estimote SDK, namely the class *BeaconManager* which is a gateway to almost all interactions with Estimote Beacons. Next, we created the necessary *BeaconRegions*, for all the beacons we wish to monitor. In our case, we configured two for each crosswalk we need to monitor.

Determine if Crosswalk Is Nearby

As soon as the regions are created, the app starts to monitor them in background. The method *onEnteredRegion* is triggered when the app enters in the range of a given region. At this point, we verify it is a crosswalk type beacon and inform the user, via a notification to the screen and with voice that a crosswalk is approaching, and he should turn off the phone to enter it. Of course, this notification is only sent is the user is using the phone. We verify this using the class *KeyguardManager* in concrete with the method *inKeyguardRestrictedInputMode* that returns a boolean value depending if keyguard screen is showing or in restricted key input mode.

Determine if Someone Is in the Crosswalk

To conclude if someone is inside the crosswalk we must detect, at the same time, signals from the two beacons of the traffic light crosswalk, which corresponds to the user being in the intersection of the two signals. When the *onEnteredRegion* method is triggered we keep track of the identification of the beacons and, as soon as we have two signals detected, we consider the user is inside the crosswalk and notify a central database of the existence of one more person inside the crosswalk. This information will then be used by the driver app.

Determine if Someone Left the Crosswalk

This is the final functionality regarding the pedestrian app and it has the goal of decreasing the number of people in a specified crosswalk. The SDK Estimote has a limitation of approximately 20 s to notify the application that a given beacon is no longer on the range. This time is bigger than the desired for our case, so we implemented an alternative way. The pedestrian app monitors the signals coming from the two beacons of the crosswalk and, according to the signal strength received, a distance to the traffic light is calculated and conclusion about the user being out of the crosswalk can be made. In this case, the central database is informed.

4 Tests

In this section, we describe in what the several tests consist of and which conclusions we aim to gather. The test will be performed in the School of Technology and Management of the Viana do Castelo Polytechnic Institute.

4.1   Pedestrian Tests

Test #1: At a normal walking velocity from the three different directions in Fig. 5, is the user notified at time of C1 crosswalk approaching?

Goal: are the notifications relevant and can they achieve the mission of avoiding distractions of pedestrians when entering crosswalks?

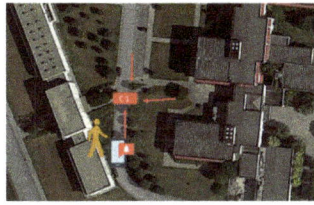
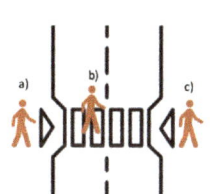

Fig. 5. Test scenario for testing if notifications of crosswalks approaching are sent in time.

Fig. 6. Test scenario to conclude if technology is adequate to differentiate if a pedestrian is inside or outside a crosswalk

Fig. 7. Test scenario to conclude if drivers are notified in advance of pedestrians and crosswalks

Test #2: using beacon signal intersection, is it possible to accurate state if the user is inside or outside a crosswalk? What is the result achieved in positions (a), (b) and (c) of Fig. 6?

Goal: are beacons a correct and appropriate technology for identifying pedestrians inside a crosswalk?

Test #3: what is the behavior of the system with 6 to 10 people entering and leaving the crosswalk?

Goal: are beacons a correct and appropriate technology for identifying pedestrians inside a crosswalk, considering several people in it?

4.2   Driver Tests

Test #4: At a velocity of 20/40/60 km/h, what is the average distance x to the crosswalk that the driver is notified of its existence (as shown in Fig. 7)?

Goal: does the system notify drivers of the existence of crosswalks with the necessary advance?

Test #5: At a velocity of 20/40/60 km/h, what is the average distance x to the crosswalk that the driver is notified of the existence of pedestrians (as shown in Fig. 7)?

Goal: does the system notify drivers of the existence of pedestrians in crosswalks with the necessary advance?

5 Conclusions

In this paper, we presented a solution to notify drivers of the presence of crosswalks as well as pedestrians in it. This is a field where contributions are welcome as there are still a big number of deaths caused by trampling. The design of the solution included two android applications - for the pedestrian and for the driver - and beacons in traffic light of crosswalks. As an additional functionality, and because there are many pedestrians that use their phone when entering crosswalks, we make that verification and advise the user to turn off the phone if he is near a crosswalk and the phone is in use. We present with detail the implementation and algorithm developed to achieve this behavior and be able to deliver notification to drivers. As this is an-going project, we present the test cases that are currently being carried out and what type of conclusion we wish to achieve.

References

1. Priebe, S.J., Keenan, J.M., Miller, A.C.: NIH Public Access. In: IEEE International Conference on Multimedia and Expo Workshops, pp. 581–586 (2013)
2. Sichelschmidt, S., Haselhoff, A., Kummert, A., Roehder, M., Elias, B., Karsten, B.: Pedestrian crossing detecting as a part of an urban pedestrian safety system. In: IEEE Intelligent Vehicles Symposium, pp. 840–844 (2010)
3. Hernández, D.C., Filonenko, A., Seo, D., Jo, K.-H.: Crosswalk detection based on laser scanning from moving vehicle. In: IEEE International Conference on Industrial Informatics, pp. 1515–1519 (2015)
4. Suzuki, S., Raksincharoensak, P., Shimizu, I., Nagai, M., Adomat, R.: Sensor fusion-based pedestrian collision warning system with crosswalk detection. In: IEEE Intelligent Vehicles Symposium, 2010, no. Itarda 2007, pp. 355–360 (2010)
5. Choi, J., Ahn, B.T., Kweon, I.S.: Crosswalk and traffic light detection via integral framework. In: 9th Korea-Japan Joint Workshop on Frontiers of Computer Vision, pp. 309–312 (2013)
6. Llorca, D.F., Parra, I., Quintero, R., Fernández, C., Izquierdo, R., Sotelo, M.A.: Stereo-based pedestrian detection in crosswalks for pedestrian behavioural modelling assessment. In: 11th International Conference on Informatics in Control, Automation and Robotics, pp. 102–109 (2014)
7. Zhai, Y., Cui, G., Gu, Q., Kong, L.: Crosswalk detection based on MSER and ERANSAC. In: IEEE Conference on Intelligent Transportation Systems, pp. 2770–2775 (2015)
8. Wang, S., Tian, Y.: Detecting stairs and pedestrian crosswalks for the blind by RGBD camera. In: IEEE International Conference on Bioinformatics and Biomedicine Workshops, pp. 732–739 (2012)
9. Haselhoff, A., Kummert, A.: On visual crosswalk detection for driver assistance systems. In: IEEE Intelligent Vehicles Symposium, pp. 883–888 (2010)
10. Jakob, J., Tick, J.: Concept for transfer of driver assistant algorithms for blind and visually impaired. In: IEEE 15th International Symposium on Applied Machine Intelligence and Informatics, pp. 241–246 (2017)

Rehabilitation and Product Design: Towards the Inclusion of People with Disabilities Through Interdisciplinary Collaboration

Fausto O. Medola[1]($\boxtimes$), Frode E. Sandnes[2],
Ana Claudia T. Rodrigues[3], Luis C. Paschoarelli[1],
and Luciana M. Silva[3]

[1] Sao Paulo State University (UNESP), 17033-360 Bauru, SP, Brazil
fausto.medola@faac.unesp.br
[2] OsloMet – Oslo Metropolitan University, Oslo, Norway
[3] SORRI Bauru Rehabilitation Center, 17033-130 Bauru, SP, Brazil

Abstract. In low-GDP countries, such as Brazil, many older people and people with disabilities rely on the public health system for access to assistive devices, which are important to support them to live healthy and dignified lives. However, many problems limit their access to specialized prescriptions and appropriate devices that satisfy their needs and expectations. This paper explores the challenges, contributions and results from an interdisciplinary collaboration between two different yet related areas of knowledge – Product Design and Rehabilitation – in education, research and innovation in assistive technologies. The combination of product design education and rehabilitation services has mutual benefits from the perspective of research and innovation, thus contributing for the community. Here, we report on the process and benefits resulting from four years of interdisciplinary collaboration on assistive products for the aging population and people with disabilities.

Keywords: Assistive technologies · Product design · Rehabilitation

1 Introduction

The World Health Organization (WHO) estimates that one billion people in the world experience disability [1]. While there have been important achievements towards the promotion of access for rehabilitation services and technologies, still there are several problems that, ultimately, limit people with disabilities of having equal opportunities and living dignified lives.

Recent WHO guidelines on promoting inclusion of people with disabilities emphasize improving the variety and quality of products supporting people's independence and quality of life. A list of the 50 most prioritized assistive devices for promoting inclusion and allowing a healthy, productive and dignified life has been proposed [2]. The list highlights the importance of providing the users access to more diverse, better quality and customable assistive devices, as well as to specialized prescription, maintenance and follow-up.

© Springer International Publishing AG, part of Springer Nature 2019
J. Machado et al. (Eds.): HELIX 2018, LNEE 505, pp. 185–191, 2019.
https://doi.org/10.1007/978-3-319-91334-6_26

Most of the Brazilian population rely on the public health system for rehabilitation and assistive technology services. Data from the SORRI BAURU Rehabilitation Center reveals that most of the assistive devices are provided via the public health system [3]. It has been pointed out that the AT product list provided by the Brazilian government is restricted in terms of the variety of the products and the costs supported for each device. As a result, those who depend on the public system to acquire an assistive device will have limited options and, most of the times, will not receive the most appropriate device for their specific needs.

The design of products for people with disabilities have unique characteristics, requiring skills and a new understanding of what it means to be human [4]. The interaction between the user and the assistive device transcends the context of only practical issues, as the significance of such products have also been related to acceptance and stigma issues [5].

Disability and personal identity are closely linked. Thus, a disability is a human condition that cannot be addressed by technology alone; form and function are equally important. An assistive device should meet the users' needs while supporting personal expression. Assistive technology that works satisfactorily but leaves its user feeling diminished is a failure. Taking a holistic approach when addressing assistive technologies is therefore important [4].

In low-GDP countries, demographic data reveals the tendency of an ageing society, which is already established in high-GDP nations. The ageing process affect people's physical, sensorial and cognitive abilities, leading to functional difficulties in many activities of daily life. Therefore, living longer brings together the challenge of maintaining independence, social participation and quality of life. An integrated and interdisciplinary-based approach on health, rehabilitation and technology is perhaps the strategy most likely to have success on promoting independence and quality of life for the wide range of people who will experience disability.

From a pedagogical perspective, developing students' empathy for the needs of people with disabilities, as well as sensitivity for the need of a holistic approach to assistive technology design has been a challenging but necessary target in Design education. This paper presents the framework of an interdisciplinary approach in the areas of Product Design and Rehabilitation towards the improvement of Assistive Technology research and development.

2   Interdisciplinarity in Assistive Technologies

When addressing complex problems from the perspective of the specificities and individualities of each discipline, it might be difficult to figure out how to connect into practical approaches and sciences from different areas of knowledge. A full view of the situation may reveal the potential of such areas in complementing each other, fulfilling gaps and empowering education, research and innovation.

Product Design education addresses a variety of issues that are organized in thematic axes mostly related to technology, humanity, science and expression. While exploring human characteristics and needs is well addressed in disciplines such as Ergonomics, Human-Computer Interaction, among others, Product Design students often lack knowledge about health sciences related to people with disabilities. For example: to design a wheelchair for individuals with paraplegia, knowing that there is an impairment in the lower limbs movement is not enough to have a full picture of all the related issues, characteristics and needs of the users. Therefore, health sciences represent important knowledge about how to properly design assistive technologies and products for independent living.

On the other hand, rehabilitation professionals are very effective in – among many other skills - detecting what the patient really needs and, many times, they recognize the features that a product must have to best meet the patients' needs. However, health and rehabilitation professionals are not trained and familiar with how to design and make a prototype, especially in industrial levels, even though physical and occupational therapists, among other rehabilitation professionals, have the skills to make customized devices, tools and adaptations to benefit the users' independence in daily activities.

An important contribution for the comprehension of the complexity and multi-factorial characteristic of the challenges for the inclusion of the people with disabilities was provided by the International Classification of Functioning, Disability and Health (ICF), established in 2001 as the WHO framework for describing and measuring health and disability [6]. This holistic approach substitutes the negative focus of the disability by a positive perspective focused on the users' potentials. According to the ICF model, people's functionality and limitations are determined by the environmental contexts [7], and the assistive devices work as facilitators - improving independence and autonomy – for the activities of daily living (ADL). This holistic view provides a common platform for the interdisciplinary collaboration between Design, Rehabilitation and related areas, emphasizing the need of considering not only the disability itself, but also people's characteristics, needs and potentials, integrated with environmental context (Fig. 1).

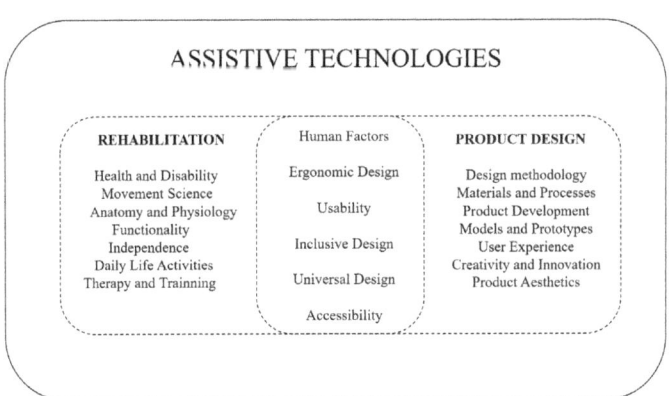

Fig. 1. Specific and shared knowledge: Design and Rehabilitation based collaboration on Assistive Technologies.

3   Product Design and Rehabilitation: Education, Research and Innovation

The collaboration between Product Design (PD) at UNESP and SORRI BAURU Rehabilitation Center was built up based on the mutual understanding that only an interdisciplinary-based approach would meet the needs and fill gaps from each specific area: design and rehabilitation. From the academic side of PD, there was a consensus that theoretical classes did not provide a complete understanding of the users' characteristics, needs and expectations, and that this knowledge is essential for the design of products for people with disabilities. From the perspective of SORRI rehabilitation professionals, the challenge was to apply practical observations and the experience with the user into a reproducible and replicable design process. Combining rapid prototyping technologies with customized orthotic/prosthetic process was also a recognized challenge for the rehabilitation approach.

The interdisciplinary collaboration between SORRI and Product Design School at UNESP was based on an integrated structure of three main axes: Education, Research and Innovation (Fig. 2).

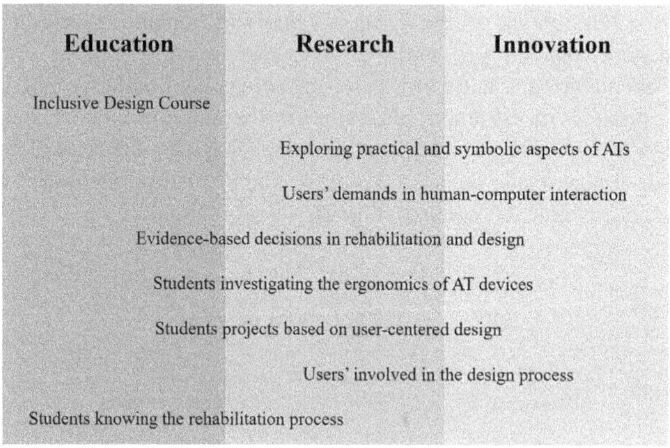

Education	Research	Innovation

Inclusive Design Course

Exploring practical and symbolic aspects of ATs

Users' demands in human-computer interaction

Evidence-based decisions in rehabilitation and design

Students investigating the ergonomics of AT devices

Students projects based on user-centered design

Users' involved in the design process

Students knowing the rehabilitation process

Fig. 2. UNESP Product Design and SORRI Rehabilitation: initiatives on interdisciplinary collaboration.

Assistive Technology Research and Development initiatives have focused on improving users' independence, satisfaction and quality of life. In the context of wheelchair mobility, the benefits of practicing adapted sports [8], as well the implementation of power assistance in manual wheelchairs [9, 10], handrim design [11] and the influence of axle position and the use of accessories on manual propulsion biomechanics [12, 13] have been reported. Additionally, computer usage by people with tetraplegia [14] and text entry optimization [15, 16] have also been addressed. A research and innovation-based study described the involvement of the users' in the process of designing a new walker as assistive mobility device for children with cerebral palsy [17]. The challenges and opportunities of the mobility aids for the elderly have been discussed in the context of the Brazilian market [18].

Parallel to the research studies, initiatives on education in the scope of Rehabilitation and Product Design were taken. At UNESP, a new course was proposed for the Bachelor Program of Design, named "Inclusive Design", aiming to provide the students a view on the many aspects related to disability, social inclusion, accessibility and assistive technologies, and how this knowledge can be implemented in the design process from the perspective of inclusivity. In this course, the students are encouraged to center the design process in the user, exploring their characteristics, needs, expectations, to design proposals with the aim of enhancing function, independence, social participation and quality of life. In this course, the students' work in group in the development of solutions based on the demands from SORRI, which has strengthened the students' engagement on the project. Students' visit to SORRI Center, lectures gave by SORRI staff and the contact with the SORRI rehabilitation team, patients, families and caregivers were experiences provided to support the students during the Inclusive Design course. Additionally, innovation has been an important target of the UNESP-SORRI collaboration. From the research-education platform, innovation-based projects have been carried out, resulting in Product Design bachelor's final projects. Working together with SORRI staff and other health and rehabilitation institutions from Bauru city, UNESP students have developed projects based on patients' needs and

Fig. 3. Product Design students' projects based on exploring, designing and testing with subjects. Reproduced with kind permission by Aline Darc P. dos Santos, Beatriz Martino Matos and Ana Elisa Franchini.

expectations, such as: customized upper limb prosthesis made prototyped with 3D printing, jewelry-look (ring) orthosis for fingers deformity due to rheumatoid arthritis), multisensorial experience toy, mobile sink for institutionalized elderly and tableware kit for elderly people with motor difficulties on the hands (Fig. 3).

4   Conclusion

The contributions of a fruitful collaboration between Product Design and Rehabilitation are explored herein. The integrated platform based on education-research-innovation inspired the academics from Computer Science and Product Design at Oslo Metropolitan University (OsloMet, Norway) and Product Design at UNESP in proposing an international collaboration on Assistive Technologies and Produoto for Independent Living. In January 2017, the 4-years project "Collaborative Design and Rapid Prototyping of Assistive Technologies and Products for Independent Living" (SIU – Norwegian Centre for International Cooperation in Education, Project Number UTF-2016-long-ter/10053) started, which has SORRI BAURU as a partner. The aims, scope and strategies of the project have been previously described [19]. Educational, research and innovation initiatives from this project are being currently developed and will be reported in future studies.

References

1. World Health Organization. Health statistics and information systems: Global Health Estimates (GHE) (2011). http://www.who.int/mediacentre/news/releases/2011/disabilities_20110609/en/
2. World Health Organization. Priority Assistive Products List: Improving access to assistive technology for everyone, everywhere (2016). http://www.who.int/phi/implementation/assistive_technology/EMP_PHI_2016.01/en/
3. Sorri Bauru. Relatorio Anual – 2016 (2016). http://sorribauru.com.br/site/conteudo/205627-relatorios-anuais.html?menu_id=27
4. Reiser, S., Bruce, R., Martin, J., Skidmore, B.: Making together: an interdisciplinary inter-institutional assistive-technology project. IEEE Comput. Graph. Appl. **37**(5), 9–14 (2017)
5. Lanutti, J.N.L., Medola, F.O., Gonçalves, D.D., Silva, L.M., Nicholl, A.R.J., Paschoarelli, L. C.: The significance of manual wheelchairs: a comparative study on male and female users. Procedia Manuf. **3**, 6079–6085 (2015)
6. World Health Organization. How to use the ICF: A practical manual for using the International Classification of Functioning, Disability and Health (ICF). Exposure draft for comment. WHO, Geneva (2013)
7. Buchalla, C.M.: A Classificação Internacional de Funcionalidade, Incapacidade e Saúde. Acta Fisiátrica **10**, 29–31 (2003)
8. Medola, F.O., Busto, R.M., Marçal, A.F., Achour Junior, A., Dourado, A.C.: Sports on quality of life of individuals with spinal cord injury: a case series. Revista Brasileira de Medicina do Esporte **17**, 254–256 (2011)

9. Medola, F.O., Purquerio, B.M., Elui, V.M.C., Fortulan, C.A.: Conceptual project of a servo-controlled power-assisted wheelchair. In: 5th IEEE RAS/EMBS International Conference on Biomedical Robotics and Biomechatronics, pp. 450–454 (2014)
10. Lahr, G.J.G., Medola, F.O., Sandnes, F.E., Elui, V.M.C., Fortulan, C.A.: Servomotor assistance in the improvement of manual wheelchair mobility. Stud. Health Technol. Inform. **242**, 786–792 (2017)
11. Medola, F.O., Paschoarelli, L.C., Silva, D.C., Elui, V.M.C., Fortulan, C.A: Pressure on hands during manual wheelchair propulsion: a comparative study with two types of handrim. In: European Seating Symposium, Dublin, pp. 63–65 (2011)
12. Bertolaccini, G.S., Carvalho Filho, I.F.P., Christofoletti, G., Paschoarelli, L.C., Medola, F. O.: The influence of axle position and the use of accessories on the activity of upper limb muscles during manual wheelchair propulsion. Int. J. Occup. Safety Ergon. 1–5 (2017). Epub ahead of print
13. Bertolaccini, G.S., Sandnes, F.E., Paschoarelli, L.C., Medola, F.O.: A descriptive study on the influence of wheelchair design and movement trajectory on the upper limbs' joint angles. In: International Conference on Applied Human Factors and Ergonomics, pp. 645–651. Springer (2017)
14. Medola F.O., Lanutti J., Bentim C.G., Sardella A., Franchinni A.E., Paschoarelli L.C.: Experiences, problems and solutions in computer usage by subjects with tetraplegia. In: Marcus, A. (eds.) Design, User Experience, and Usability: Users and Interactions. DUXU 2015. LNCS, vol 9187. Springer (2015)
15. Sandnes, F.E., Medola, F.O.: Effects of optimizing the scan-path on scanning keyboards with QWERTY-layout for english text. Stud. Health Technol. Inform. **242**, 930–938 (2017)
16. Sandnes, F.E., Medola, F.O.: Exploring russian tap-code text entry adaptions for users with reduced target hitting accuracy. In: DSAI 2016, Proceedings of the 7th International Conference on Software Development and Technologies for Enhancing Accessibility and Fighting Info-exclusion, pp. 33–38. ACM (2016)
17. Nicholl, A.R., Busnardo, R.G., da Silva, L.M., Rodrigues, A.C., Luz, F.R., Bentim, C.C., Medola, F.O., Paschoarelli, L.C.: Development of the SORRI-BAURU posterior walker. Stud. Health Technol. Inform. **217**, 1003–1008 (2015)
18. Medola, F.O., Bertolaccini, G.S., Boiani, J.A.M., Silva, S.R.M.: Mobility aids for the elderly: challenges and opportunities for the Brazilian market. Gerontechnology **15**, 65–97 (2016)
19. Sandnes, F.E., Medola, F.O., Berg, A., Rodrigues, O.V., Mirtaheri, P, Gjøvaag, T.: Solving the grand challenges together: a Brazil-Norway approach to teaching collaborative design and prototyping of assistive technologies and products for independent living. In: Proceedings of the Engineering & Product Design Education – EPDE2017. The Design Society (2017)

3D Printing as a Design Tool for Wearables: Case Study of a Printed Glove

Luisa M. Arruda$^{(\boxtimes)}$ ⓘ and Helder Carvalho ⓘ

Centro de Ciência e Tecnologia Têxtil, University of Minho, Campus Azurém,
Guimarães, Portugal
luisamendesarruda@gmail.com, helder@det.uminho.pt

Abstract. In the research work herein described, the body is analyzed from the perspective of the integration of mechanical and electronic resources, both seen as communicative systems. In this sense, the body becomes a project design for both fashion design and engineering, and therefore requires technical specificities of these wearable devices attached to them. Specifically, this paper described the development of the prototype of a glove, produced with 3D printing technology (FDM). The glove is the first step in the development of prostheses that are integrated into garments. In this work, the 3D printing method, its limits and capabilities are evaluated, a draw printing materials are studied concerning print quality and user comfort. As a conclusion, we pointed out the need for a constant search for flexible filaments more appropriate for garments, especially with regard to movements of opening and closing of the hands. We also confirmed the effectiveness of the 3D printing technique as a tool for Designers to quickly and inexpensively visualize the initial shape of their products and thus be able to make changes in a more appropriate way.

Keywords: 3D printing · Design · Wearables

1 Introduction

In an analysis of the garments developed by the engineer and artist Flavio de Carvalho, we identify the evolutionist thought of Charles Darwin that clarifies the associations of biology to reflections on the evolution of the garments [1]. As well as being based on the theory of the utopian body of Michel Foucault, Carvalho thought of the garments as a blossoming of utopias, ghosts or body desires, projecting them to another space, garment as prosthesis [2]. This discussion makes it possible to think about the technological body, in which its traditional limits are questioned. It is not just a garment, but also an extension, prostheses that penetrate and line the human body through the biotechnological sciences.

Coexisting with the concept described, advances in miniaturization of electronic devices have generated a growing interest in Wearable Technologies. It is a fact that wearable systems are non-invasive devices that allow clinicians to monitor individuals for long periods of time [3]. This happens because the contemporary body admits the projection of these tools both internally and close to the skin. This is a path that theorists call confirmation of self-regulatory systems [4].

© Springer International Publishing AG, part of Springer Nature 2019
J. Machado et al. (Eds.): HELIX 2018, LNEE 505, pp. 192–198, 2019.
https://doi.org/10.1007/978-3-319-91334-6_27

These considerations, added to the advances of the Additive Manufacturing (AM), provide a conceptual extension of 3D printing, resulting in processes with potential to offer new, advanced products related to wearables. However, addressing the issue of these new products to be worn by humans, requires parameterizing the body as support in many dimensions: shapes, sizes, thermophysiological comfort, psychological comfort, etc. Whether it is following its forms, or building new volumes, clothing necessarily presents itself aggregated to body culture [4].

Therefore, based on the assumption that garments are prosthetics of the body, and motivated by the demand of wearable devices with monitoring functionalities, we have chosen for the execution of the 3D printing tests, the modeling of a medical glove. In addition, it has been chosen because it is a garment that integrates – coats -the part of the body referring to the upper limb, the forearm and the hand, and therefore, its specific movements require great flexibility of the material. In recent studies, we have identified the patent of two gloves: a glove with compass and thermometer (CN201088158 Y) [5], and in 2017, the medical glove to prevent infection, with bacteria thermometer (CN106983498 A) [6].

2   Materials and Methods

2.1   3D Printing

3D printing is a form of AM technology, in which products or prototypes are constructed by depositing materials, layer by layer, through a series of transverse slices.

Today, more than 100 different types of 3D printers are available on the market, many of which are desktop printers, featuring relatively low and affordable prices. There are also industrial-scale 3D printers with higher prices, however, capable of processing a wide range of materials on a larger scale. In the same way, several techniques are used by these printers. We opted for the Fused Deposition Modeling technique (FDM), because this is one of the most used AM techniques in garments [8]. It consists in the use of a thermoplastic filament, which is introduced into an extrusion head, and is heated to a semi-liquid state, before being extruded and deposited in thin layers by the nozzle [9]. Since we assume the choice of flexible thermoplastic printing material, it is, therefore, a suggested technique for its use.

In order to be able to choose the most suitable filament for the glove, the two most important parameters are high resistance to bacteria and high flexibility. The glove should make possible the following movements: flexion and extension; adduction and abduction; internal and external rotation; and circumference. In our search, we found a flexible thermoplastic elastomeric filament (TPE), based on polyurethane and other additives: *Filaflex Original 82A,* from the Spanish company *Recreus*. Table 1 presents the physical properties of the filament as well as the printing properties provided by the manufacturer.

The printer used was the UP Plus 2, 3DP-14-4A, from UP! For initial printing tests, we searched for models that were similar to a glove. On the *Open Bionics* website, we found a one-hand file that was suggested for printing on flexible materials (*"New flexy hand for filaflex"*). From this file only the thumb was printed, with three different

Table 1. Printing and properties of Filaflex Original 82A. Source: *Recreus*.

Physical properties	Typical value	Test method
Density	1,14 g/cm^3	ISO 1183-1-A
Hardness shore A	82	ISO 7619-1
Elongation at break	665%	ISO 37
Tensile strength	42 Mpa	ISO 37
Abrasion loss	23 mm^3	ISO 4649-A
Flammability rating	HB	UL 94
Printing properties	Value	
Printing temperatures	225–235 °C	
Printing speed	20–40 mm/s	
Hot-Bed temperature	0 °C	
Optimal layer height	0,2 mm	
Minimal Nozzle diameter	0,3 MM (0,4 mm or higher recommended)	
Retraction parameters	3,5–6,5 mm (speed 20–120 mm/s)	

filaments: acrylonitrile butadiene styrene (ABS), polylactic acid (PLA), and TPE. These experiments were designed to explore the skills of the printer and the resulting properties of the models. For this purpose, we used the following recommended temperatures: For ABS, the extruder temperature was 240 °C and the bed temperature 110 °C; for PLA, the extruder temperature was 195 °C, and the bed temperature 60 °C. To print TPE, the extruder temperature was set to 230 °C, the bed was not heated.

The results printing with the flexible filament on our desktop printer were not satisfactory, as will be explained later. We verified that the adversity resulted from the specificity of the filament. Using this filament required setting more printing parameters than the UP software allows. Thus, it was necessary to use another printer. The company *Xpim* in the city of Braga/Portugal had experience with flexible filaments, and they were able to print in larger sizes (it was possible to print the whole glove). A POM printer was used for the final prototype. Table 2 details its technical specifications. In this course, it was also necessary to change the type of the flexible filament. Although we did the initial tests with Filaflex, the prototype was printed with Ninjaflex Cheetah, as this is the flexible filament normally used by *Xpim*. This material has a shore hardness scale of 85, suitable for flexible materials.

The software used for slicing was Cura, as it allows a more detailed configuration of the printing parameters. The bed was prepared with Kapton Polyimide tape and water-soluble PVA support material was used.

2.2   Modeling

The proposed three-dimensional model for the respective study is a glove. Measurements of a male arm were made using a measuring tape. The resulting values are: forearm width (35,5 cm); wrist width (21 cm); length between wrist and end of the middle finger (19 cm); length of the forearm to end of the middle finger (35 cm). Based

Table 2. Specifications used in the final prototype and the print parameters.

POM printer technical specifications		Printing parameters	
Dimensions	95 × 95 × 170 cm	Diameter of the filament	2,95 mm
Print volume	70 × 60 × 60 cm	Print runtime	10 h
Resolutions	100–900 μ	Print temperature	230 °C
Extrusion °C	Up to 350 °C	Bed temperature	45 °C
Bed temperature	Up to 120 °C	Print speed	50 mm/s
Energy consumption	240 V, 9 A, 50–60 Hz	Layer height	0,25 mm
		Percentage of compactness	100%

on these measures, we started modeling using the *SolidWorks* 3D Computer Aided Design (CAD) software. Due to the characteristics of the three-dimensional model, we observed that it would be more appropriate to change the modeling process for software that has a cognitive feature similar to the *moulage* process. In Fashion Design, *moulage* is a three-dimensional modeling technique of clothing in which the fabric is molded, pinched, scratched and cut straight onto a mannequin [7]. In this sense, we found the sculpt feature in *Fusion 360* software, and we opted for this simple and intuitive functionality.

In the sequence, with the same method of the *moulage* - in which modeling is done on a mannequin of the human body - we searched on the website cgtrader.com, for the base of a male arm with extension *Object File Wavefront 3D* (OBJ). From then on, we molded the glove with the pre-established measures. Subsequently, we defined the thickness of the model (2 mm). Figure 1 presents the rendered 3D model of the glove.

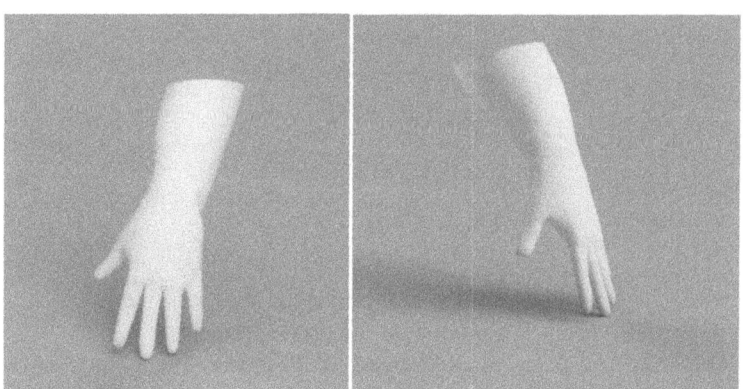

Fig. 1. Rendered glove.

3 Results

As far as the initial tests with the three filaments (ABS, PLA and Filaflex) are concerned, satisfactory prints were achieved with the first two. However, being rigid materials, they did not meet the comfort requirements for a wearable part.

In the first printing test with the Filaflex filament, we followed the parameters provided by the supplier. However, the model showed some discontinuities in parts of the print. In addition, the matte transparent filament turned glossy when printed, which is indicative of problems with temperature.

In the second printing attempt with Filaflex we changed to the black filament. Extrusion temperature was to 265 °C, because we noticed that the printer always printed below the temperature set. We observed that, regarding color, the same issue occurred: the black matte filaments became glossy as they went through the extrusion process. Still, print quality was better, but we noticed that the filaments were not bonding to each other. They were forming a kind of a lace (Fig. 2). New attempts provided the same unsatisfactory results.

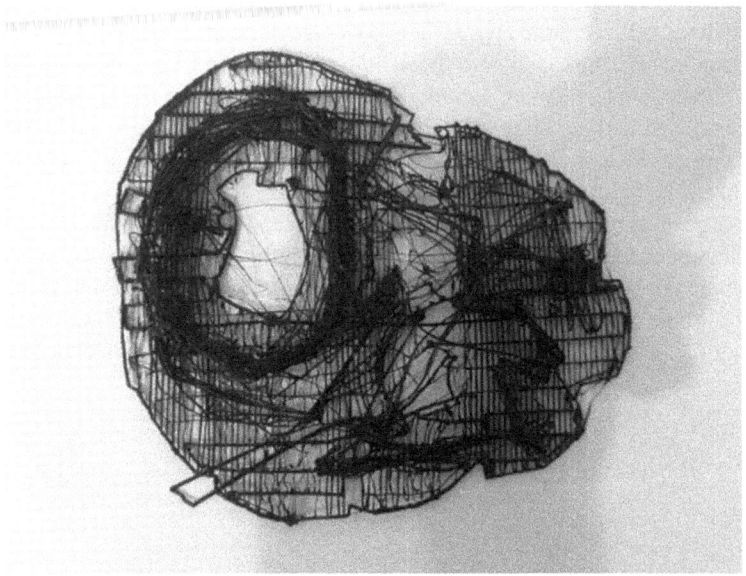

Fig. 2. Glove print test

The next step was searching for the causes of these problems, namely identifying if it was the 3D model, the slicing process or the print parameters.

In this sense, the first change was the choice of the printer. We found at *Xpim* a way to test the print of the prototype again. *Xpim* printed the modelled glove with NinjaFlex flexible filament twice.

In the first attempt, the result was not satisfactory at the fingers, because their interior was filled by strings of filament. This happened because, when travelling from one side of the perimeter to the other, the nozzle was not able to stop the filament flow, leaving remains of fused filament behind. To solve this problem the initial speed of 50 mm/s was reduced to 60% at the fingers.

In the next attempt, the interior of the fingers was not filled anymore, but small voids appeared at the perimeter. This meant that the nozzle was not able to produce filament quickly enough when it reached the perimeter of the finger. Reducing the "retraction" parameter allowed mitigating this problem, and a satisfactory result was obtained (Fig. 3).

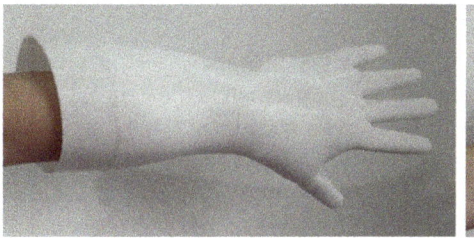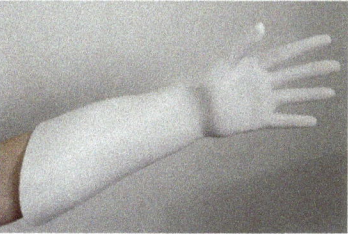

Fig. 3. Result of the glove, dressed on one arm.

4 Discussion

The prototype presented, a glove, acts as a garment, so it is a direct extension of the surface of the body. In this perspective, with regard to the adopted modeling methodology, in which there is a correspondence of the concepts of *moulage* in the 3D modeling of the piece, we observe that the anatomy of the glove fulfills the dressing requirement. In this sense, the choice of Fusion 360 software was successful.

Regarding the filaments, it was possible to use the ABS and PLA filaments following the specifications of the suppliers regarding temperature and printing speed. They also were easier to handle, adjusting very few parameters, and in fact have good print quality even using a desktop printer.

However, Filaflex flexible filament requires a greater specificity of parameters, and a slower printing speed (the slower, in this case, the better the quality). Printing results with this filament have confirmed that we still need to experience 3D printing speeds and feeds according to the specificity of each material and the printer model used. Although there are indications from manufacturers, these standards are not adequately addressed, which proves to be an area of constant research [10].

With Ninjaflex, the best result within the given time frame could be achieved. However, its flexibility does not meet the comfort requirement that a garment demands, nor does it possess the necessary breathability. The rotational movement of the pulses is maintained, however, the flexibility of the resulting surface is not sufficient to perform the movement of opening and closing the hand.

5 Conclusions

The observations made in this work were only possible due to the prototyping of the piece by the 3D technique, which proved feasible as to the exploratory character that the design creation process demands. To this extent, this study also intends to provide insights into the fabrication of garment pieces as prostheses using AM.

Regarding the design of the glove, the analysis of the model used shows that further study and modeling of the arm shape is necessary. For future work, in order to integrate the garment into the prosthesis, it is necessary to improve design and aesthetics of the proposed object to make it functional and based on ergonomic concepts. Is also important to carry on searching for and studying flexible materials that allow comfort and the necessary movements of the users.

As an initial study, we perceive as positive the findings made in relation to choosing of the *Fusion 360* modeling software, as a correlation to the garment *moulage* technique. This also narrows the relationship between fashion and AM, and confirms 3D printing as a design tool.

Acknowledgements. This work is financed by FEDER funds through the Competitivity Factors Operational Programme - COMPETE and by national funds through FCT – Foundation for Science and Technology within the scope of the project POCI-01-0145-FEDER-007136.

References

1. Carvalho, F.: A moda e o novo homem: dialética da moda. Azougue Editorial, Rio de Janeiro (2010)
2. Foucault, M.: O corpo utópico, as heterotopias. 1nd edn. N-1 edições, São Paulo (2013)
3. Bonato, P.: Wearable Sensors/Systems and their impact on Biomedical engineering: an overview from the Guest Editor. IEEE Eng. Med. Biol. Mag. (2003). https://doi.org/10.1109/MEMB.2003.1213622
4. Avelar, S.: Moda, globalização e novas tecnologias. Estação das Letras e Cores, São Paulo (2011)
5. 航张: Glove with compass and thermometer. CN 201088158Y, 23 July 2008
6. 董攀: Glove type thermometer for preventing doctor from infecting bacteria. CN 106983498 A, 28 July 2017
7. Grave, M.F.: A modelagem sob a ótica da ergonomia. Zennex, São Paulo (2004)
8. Yap, Y.L., Yeong, W.Y.: Additive manufacture of fashion and jewellery products: a mini Review. Virtual Phys. Prototyp. **9**(3), 195–201 (2014). https://doi.org/10.1080/17452759.2014.938993
9. Silva, D.N., Broega, A.C., Menezes, M.D.S.: Uma abordagem ao conforto nos produtos vestíveis impressos em 3D, 13º Colóquio de Moda, Bauru, São Paulo, 11–15 de October 2017
10. Petrick, I.J., Simpson, T.W.: 3D printing disrupts manufacturing: how economies of one create new rules of competition. Res. Technol. Manag. **56**(6), 12–16 (2013). https://doi.org/10.5437/08956308x5606193

Design, Development and Construction of a New Medical Wrist Rehabilitation Device: A Project Review

Eurico Seabra$^{(\boxtimes)}$ ⓘ, Luís F. Silva ⓘ, Ricardo Ferreira ⓘ, and Valdemar Leiras ⓘ

Department of Mechanical Engineering, School of Engineering, University of Minho, 4800-058 Guimarães, Portugal
eseabra@dem.uminho.pt

Abstract. This paper reviews in detail the design and development procedures for the construction of medical device for the rehabilitation of the wrist. The construction of the actual prototype was carried out with off-the-shelf components. The objective was to develop a functional and low cost device. The system is divided in two main components, one for the rehabilitation of the wrist and the other one to perform proprioception exercises, providing relaxation of the patient's pain due to injuries on the wrist. All the hardware is assembled in one portable support, capable to make vertical and horizontal adjustments. Both systems are fully controlled through a software specially developed to be used by physiotherapists and patients, allowing also the registration of the rehabilitation evolution.

Keywords: Biomechanics · Wrist · Rehabilitation · Proprioception
Control system

1 Introduction

Fracture of the wrist is one of the most complicated pathologies, because it is associated with a wide range of movements; in a healthy pulse three main types of movements can be considered: flexion and extension, adduction and abduction, pronation and supination. This type of injury has a large incidence in adulthood, and occurs mostly in women, since osteoporosis increases the brittleness of bones, and in case of impact there is a great susceptibility to bone breakage. In relation to younger individuals, this type of fracture is mainly due to sports injuries [1, 2].

Therefore this paper will briefly review the project undertaken by the authors to study, design, develop and construct a new medical device capable to help and to carry out the rehabilitation procedure of the wrist. Over the next topics, some of its development phases and obtained results will be presented.

© Springer International Publishing AG, part of Springer Nature 2019
J. Machado et al. (Eds.): HELIX 2018, LNEE 505, pp. 199–205, 2019.
https://doi.org/10.1007/978-3-319-91334-6_28

2 The Wrist Rehabilitation Device

This project began with the adaptation of a device already available in the market called Powerball®, which was designed to perform fitness exercises on the wrist, forearm and shoulder. Through the modification of this device, the Bioball was created (see Fig. 1). This new device uses an A28L brushless outrunner motor, powered by a 12 V – 2 A power source through a MAG8 electronic speed controller, with an eccentric mass coupled to its shaft for creating vibration, which is useful for proprioception and relaxation purposes. Removing the ball from its fixed mechanism, it is possible to engage a handle which allows massages on the wrist with controlled frequencies.

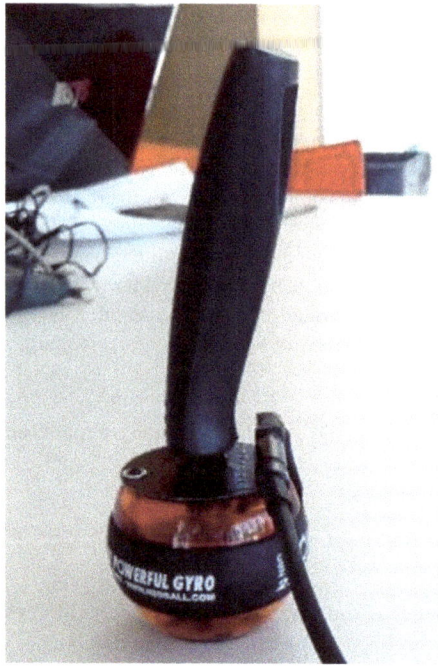

Fig. 1. The Bioball with its handle for relaxation and proprioception purposes.

It also uses a Superior Electric M062-LE04 stepper motor, NEMA 23 [3], powered by the same power source and controlled through a TB6600 driver; this motor is coupled to an eccentric bar having at its end the Bioball. Combining the position adjustment of the support, this device enables the reproduction of the movements associated with the wrist. The control of both motors and respective drivers is carried out by an Arduino controller, through a special designed software created in LabVIEW (by National Instruments™) [4], or directly through the hardware present on the support of the device, with different push buttons and an emergency button.

3 The Wrist Rehabilitation Device

This multifunctional device is divided in two main subsystems: one for relaxation sessions and proprioception tests, and another one for rehabilitation purposes through the reproduction of the movements of the wrist.

3.1 Relaxation and Proprioception

Through vibrations between 30 Hz and 50 Hz it is possible to promote the relaxation of the affected regions, ensuring the well-being of the patient [5–7]. As shown in Fig. 1, it is possible to attach a handle to the Bioball; the therapist can now start the relaxation sessions without being directly affected by the induced vibration, allowing the use of the device for long periods of time.

Depending on the type of injury, usually a loss of sensitivity is experienced on the wrist, which also might affect the ability and control of the respective movements. This means that the patient has its own proprioception capabilities reduced at the rehabilitation stage. With the vibration induced by the Bioball (which is controlled by a specially designed software) it is possible to perform proprioception exercises and to control the patient's improvement with this component of the rehabilitation.

3.2 Rehabilitation Through the Reproduction of the Movements of the Wrist

This device enables two types of rehabilitation: active and passive. For active rehabilitation the patient creates resistance to the movement generated by the device, exercising therefore the muscles on the wrist, forearm and hand. In passive rehabilitation the device helps the patient to perform each exercise. This procedure helps to increase the amplitude of each movement of the wrist; the data corresponding to each one of the exercises can be computed by the developed software.

As mentioned, the movements of the wrist are divided in three main groups: flexion and extension, adduction and abduction, pronation and supination. Each one of these movements is associated to a limit angle when it comes to a healthy wrist [8]; these values are presented in Table 1:

Table 1. Angles associated to each movement of a healthy wrist.

Movement type	Angle
Flexion	85°
Extension	85°
Adduction	45°
Abduction	15°
Pronation	75°
Supination	85°

A maximum amplitude for each movement must be respected to avoid any kind of injuries to the patient. For this purpose the device allows full control over the amplitude movements that will be reproduced. For ensuring the patient security during the exercises, low rotation speeds of about 1.88 rpm are also used. For the sake of completeness, the device only enables the adjustment of the rotation speed to a maximum of 10 rpm.

4   Operation Modes

Due to the different anatomy of the patients, the support of the device can be adjusted vertically and horizontally to provide a more comfortable use (see Fig. 2). It is possible to place the device on a table, writing desk or any other daily use object capable to provide a proper fastening of the device.

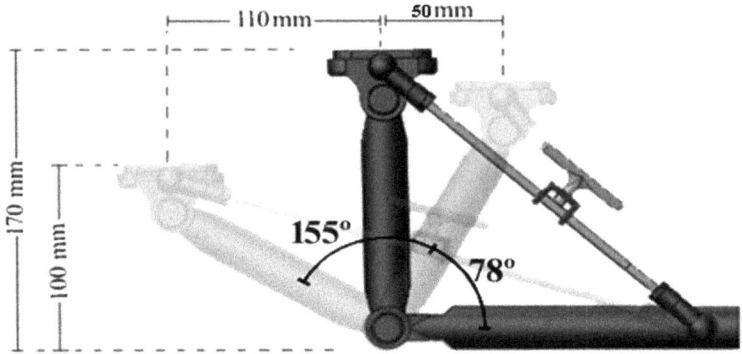

Fig. 2. Limit adjustment positions of the support to provide a comfortable use.

The support of the device enables different grabbing positions of the Bioball to move the wrist around the motor axis, simulating all the natural movements of the wrist. As shown in Fig. 3, for each movement the patient must grab the Bioball in a different way, and the position of the support should also be adjusted to carry out the desired exercises.

In Fig. 3 it is possible to distinguish three configurations of the support and grabbing positions (1, 2 and 3) for the reproduction of the movements of the wrist. The operation modes, related with each support configuration (1, 2 and 3), represent, respectively, the adduction and abduction, flexion and extension, supination and pronation movements. In the case of pronation and supination, due to the nature of these movements, it is possible to couple the Bioball directly to the stepper motor shaft; in this situation, the eccentric bar will not be used. The adjustment of the support is needed in this rehabilitation exercise, so the Bioball can be placed horizontally.

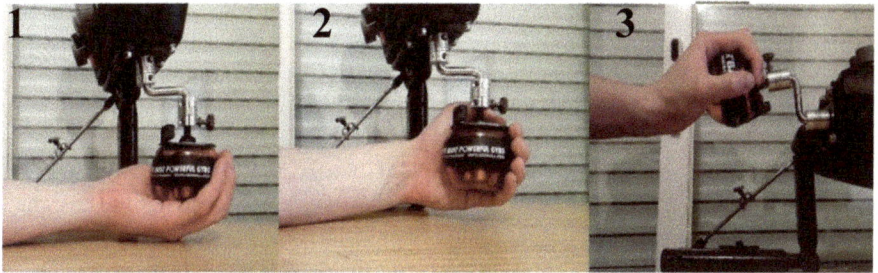

Fig. 3. Different arrangements of the support and grabbing position for each rehabilitation movement.

5   Results

The validation of this device was accomplished by physiotherapists at the Hospital of Braga (in Portugal), based on their own experience in the rehabilitation of patients with wrist pathologies.

Force analysis and power consumption tests with speed variation were performed with the developed device, evaluating the torque/speed/power ratio and determining the

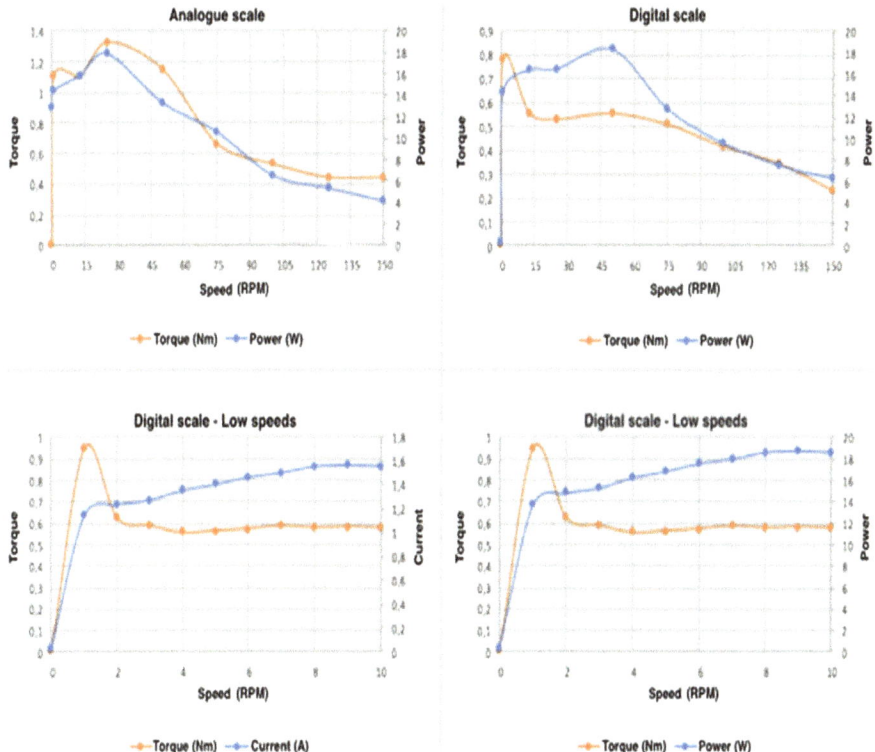

Fig. 4. Stepper motor torque variations with speed.

optimum working ranges for rehabilitation purposes. Some of the obtained results are shown in Fig. 4. For active rehabilitation, the stepper motor used to reproduce the movements of the wrist is capable to produce a maximum torque of 1.32 Nm [9].

As shown in Fig. 4, the highest values for torque are achieved around 1 rpm, decreasing for higher values of speed.

The developed software provides a friendly interface to the user, which is an important factor for unspecialized users. Figure 5 shows the front panel of the designed software. The user has always the possibility to save the data regarding the performed exercises. If the data is saved, a file is created with the date, hour and characteristics of the exercises, allowing, if necessary, a future evaluation of the rehabilitation progress of the patient (see Fig. 6); this figure shows, as an example, the gain of amplitude obtained for each group of movements during the testing phase of the device [10].

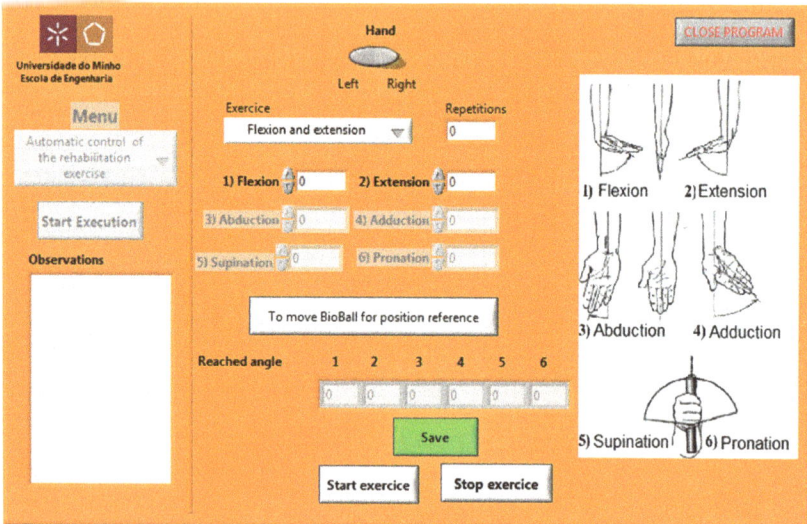

Fig. 5. The front panel of the designed software: user interface.

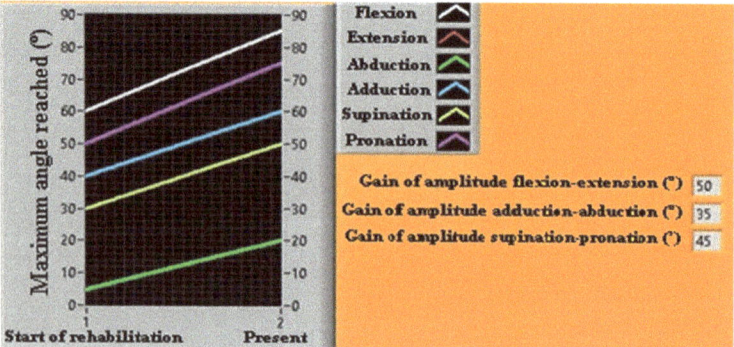

Fig. 6. Output results showing the evolution of the angle achieved for each rehabilitation movement.

6 Conclusions

Existing controlled devices are large in size, do not allow any type of portability and are oriented to address only one function. Due to the degree of complexity of these equipments a specialized operator is usually required.

The proposed device presents several solutions for the different steps of recovery/rehabilitation of an injured wrist. It is possible to obtain the gain of amplitude for each group of movements, as well as the increase of amplitude that the patient is capable to perform for each singular movement.

According to the first preliminary results obtained, the proposed prototype has proven to be useful for rehabilitation purposes. Nevertheless further tests should be performed in the near future to determine its real benefits on patients with wrist injuries.

Acknowledgments. The authors gratefully acknowledge the support given by the staff of the Department of Occupational Therapy of the Hospital of Braga (Portugal) during the development of the prototype.

References

1. Lana, L.: Avaliação clínica da propriocepção. Rio de Janeiro (2013)
2. Angermann, P., Lohmann, M.: Injuries to the hand and wrist - a study of 50,272 injuries. J. Hand Surg. Br. Eur. **18**(5), 642–644 (1993)
3. Oriental motor. www.orientalmotor.com. Accessed 20 Nov 2017
4. Elliott, C., Vijayakumar, V., Zink, W., Hansen, R.: National instruments LabVIEW: a programming environment for laboratory automation and measurement. SAGE J. **12**(1), 17–24 (2007)
5. Brewin, M.P., Bexon, C.J., Tucker, S.C.: A case report on the use of vibration to improve soft tissue extensibility after major trauma. J. Hand Ther. **30**, 367–371 (2017)
6. Camerota, F., Celletti, C., Di Sipio, E., De Fino, C., Simbolotti, C., Germanotta, M., Mirabella, M., Padua, L., Nociti, V.: Focal muscle vibration, an effective rehabilitative approach in severe gait impairment due to multiple sclerosis. J. Neurol. Sci. **372**, 33–39 (2017)
7. Christiansen, B.A., Silva, M.J.: The effect of varying magnitudes of whole-body vibration on several skeletal sites in mice. Ann. Biomed. Eng. **34**(7), 1149–1156 (2006)
8. Lacerda, J., José, P., Costa, L., Dias, N., Marques, T.: Projeto de Dispositivos Médicos e Reabilitação. Universidade do Minho, Braga (2010)
9. Leiras, V.: Design of a control system for a medical wrist rehabilitation device. Master thesis in Mechatronic Eng, School of Engineering, University of Minho, Guimarães (2017)
10. Ferreira, R.: Development and construction of a medical device for the rehabilitation of the wrist. Master thesis in Mechanical Eng., School of Engineering, University of Minho, Guimarães (2017)

Comparing Childhood Hypertension Prevalence in Several Regions in Portugal

M. Filomena Teodoro[1,2(✉)], Carla Simão[3,4], Margarida Abranches[5],
Sofia Deuchande[6], and Ana Teixeira[7]

[1] CINAV, Portuguese Naval Academy, Portuguese Navy, Base Naval de Lisboa,
Alfeite, 2810-001 Almada, Portugal
maria.alves.teodoro@marinha.pt
[2] CEMAT - Center for Computational and Stochastic Mathematics,
Instituto Superior Técnico, Lisbon University, Avenida Rovisco Pais, n. 1,
1048-001 Lisboa, Portugal
[3] Medicine Faculty, Lisbon University, Av. Professor Egas Moniz,
1600-190 Lisboa, Portugal
[4] Pediatric Department, Santa Maria's Hospital, Centro Hospitalar Lisboa Norte,
Av. Professor Egas Moniz, 1600-190 Lisboa, Portugal
[5] Pediatrics Department, D. Estefânia's Hospital, R. Jacinta Marto,
1169-045 Lisboa, Portugal
[6] Pediatrics Department, Cascais's Hospital,
Av. Brigadeiro Victor Novais Gonçalves, 2755-009, Alcabideche Cascais, Portugal
[7] Pediatrics Department, São João's Hospital, Alameda Prof. Hernani Monteiro,
4200-319 Porto, Portugal

Abstract. The pediatric high blood pressure has severe the risk factors and it's prevention is mandatory. To evaluate the pediatric arterial hypertension caregivers Knowledge, in [1,2] was done a preliminary study of an experimental and simple questionnaire with 5 questions previously introduced in [3]. The analysis of an improved questionnaire applied to children caregivers and filled online was completed in [4,5]. In the present work, we obtain estimates about the childhood hypertension prevalence in several regions of Portugal. As preliminary approach, we perform an analysis of variance. The results evidences significant differences of high blood pressure prevalence between girls and boys; also the children's age is a significant issue to take into consideration.

Keywords: Pediatric hypertension · Caregiver · Questionnaire
Statistical approach · Analysis of variance

1 Introduction

The high arterial blood pressure (HABP) is a condition which, although traditionally considered a disease of adults, may increase during the pediatric age and in most cases, silently. The diagnostic criteria for Pediatric Hypertension (PH) have as their main reference the normal distribution of blood pressure (BP) in

© Springer International Publishing AG, part of Springer Nature 2019
J. Machado et al. (Eds.): HELIX 2018, LNEE 505, pp. 206–213, 2019.
https://doi.org/10.1007/978-3-319-91334-6_29

healthy children [6] and based on the concept that the pediatric BP increases with age and with body mass [7]. Thus, taking into account that there is a strong correlation between BMI and BP levels [7], the PH has become highly prevalent among children and teens [8] due to a parallel growth with the epidemic of childhood obesity [9], the which suggests that obesity is a major risk factor in the development of pediatric hypertension [8,10].

Under the aim of a preliminary study about PH caregivers (parents or their legal representatives) acquaintance, a simpler questionnaire composed by five questions was introduced in [3]. To complete such work, different statistical techniques were performed in [1,2,11], namely logit modeling, exploratory factorial analysis and MANOVA, following e.g. [12,13]. In [14], a redesign of such questionnaire was built (including fifteen questions) and was applied to a different target population. This work was continued in [4,5] using statistical multivariate techniques. In addition, in both questionnaires was included a question to check if the regular measurement of the BP, according to current European recommendations, was usually fulfilled.

A study on prevalence of pediatric hypertension at national level promoted by the Portuguese Society of Pediatrics (SPP) and Pediatric Hypertension Group (PHG) is ongoing. The medical individual characteristics of children and adolescents are observed by a medical team and a questionnaire was designed to be answered by caregivers of children and Portuguese teenagers where sociodemographic and familiar health history details are inquired. The data collection is still ongoing over several regions of Portugal (main land and islands).

In this article the objectives are to characterize the BP profile of a pediatric population of school age and to evaluate the prevalence of PH and normal-high BP and analyze the relation between PH / normal-high BP and age, sex, race and geographical origin. It was performed a preliminary approach, which will be used as reference when the designed sample at national level is complete.

This article is comprised of an introduction and results and final remarks Sects., Sect. 2 containing the description of data and methodology. The empirical application can be found in Sect. 3.

2   Data and Methodology

We have performed and observational, prospective, transverse pilot study (Tracking) in May 2017, in public and private schools from Northern to South of Portugal (Aveiro, Braga, Lisboa).

A sample, almost 5 hundred of observations, was collected by the PHG during the Hypertension Day activities, during an action of disclosure and tracking of PH. The target population was the set of students from first year until 12nd year of schooling in three different groups of schools in Lisbon, Aveiro and Braga. These groups of schools have a population of students, aged between 5 of and 18 years old. The data collection occurred in May 2017, 25th, after adequate approval from Ministry of Education, General Direction of Health, Data Protection Committee and school Executive Committee. It was measured the blood

Table 1. Classification of pediatric hypertension (adapted from [9]).

Class	SBP and/or DBP percentile
Normal	<90
Normal-high	≥90 until <95 $mmHg$
	Teenagers: AP > 120/80 $mmHg$, including percentile <90
HBP- class 1	95 ≤ Class 1 ≤ 99 + 5 $mmHg$
HBP - class 2	>99 + 5 $mmHg$

pressure by oscillometric method (up to 3 measurements) and annotated some medical and socio-demographic characteristics for each student. The age was classified in three intervals: I (5 10 years); II (11–15 years); III (16–18 years). To characterize of the tensional profile was used the criteria defined SEH 2016. The classification of pediatric hypertension is presented in Table 1 (adapted from [9]).

The blood pressure measurements and remaining information are analyzed statistically following [13]. This sample is not representative of the Portuguese population, but can be considered to get some preliminary results that can be used as an alert to the PBP problem and also as an incitement to complete the sample collection stage of ongoing PBP study.

3   Empirical Application

The first step organizes the data and get some simple measures by descriptive statistic techniques. In Fig. 1 we can find the characterization of students per age, gender, origin and race. The non-Caucasian students were few when compared with the number of remaining students, so the race was not considered as explanatory variable. In a preliminary data analysis and taking into account the non-quantitative nature of some involved variables, were calculated some measures of association, nonparametric Spearmann correlation coefficient, and performed several tests, namely nonparametric test of Friedman, median test, rank test, signal test.

There are 222 girls and 249 boys. The difference of between girl and boys proportion is not significant ($p - value = 0.4834$).

In Fig. 2, we can observe that, in sample, 85% of child have normal level of BP; 7% of students have HBP and 8% of students have pre-HBP. The blood pressure measurements and remaining information are analyzed statistically. This sample is not representative of the Portuguese population, but it can be considered to get some results to be used as an alert and valuable contribution of ongoing study. The most reasonable is to calculate the % of hypertensives by gender and to test whether the percentage of hypertension occurrence in boys is higher than in girls (7.22% versus 5.24%). The $p - value$ of the test is 0.4767, this difference is not statistically significant.

For normal-high BP, the percentage of hypertension occurrence by gender was calculated and a test was done to evaluate whether this percentage in boys

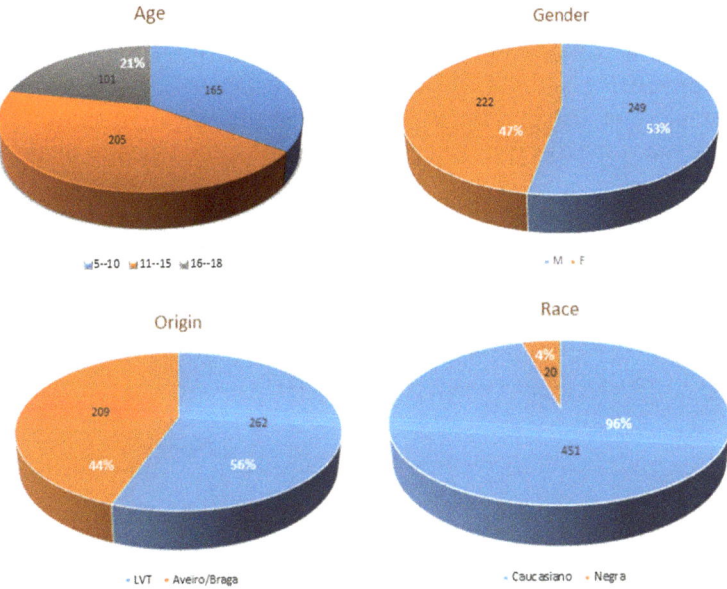

Fig. 1. Sample characterization (age, gender, origin, race).

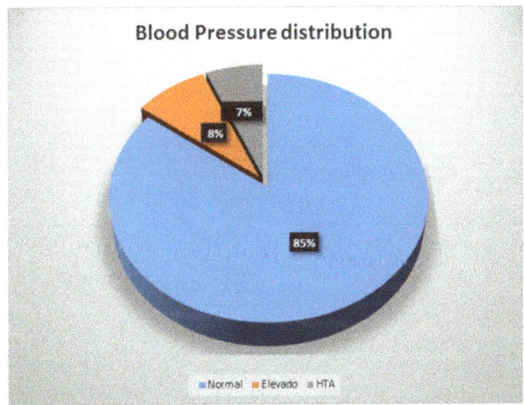

Fig. 2. HBP distribution in sample (471 children).

is higher than in girls (10.04% versus 6.30%). This difference, while awakening some attention to us, is not statistically significant for a significant level $\alpha = 5\%$, for $\alpha = 10\%$ the test is significant ($p - value = 0.0637$). This means that one should take some attention with the difference of HBP values obtained between boys and girls.

Considering the geographical region origin as an explanatory variable, the BP values did not conduced to statistically different results per region. Detailing, when it is performed the difference of HBP and pre-HBP tests considering North versus South, there were not statistically significant differences. For HBP, the difference between North and Lisboa conduced to a test with a $p - value = 0.4527$, the pre-HBP conduced to a $p - value = 0.2859$.

The box-plots of BP per age and per age group are represented in Fig. 3 (top and bottom respectively). It is visible that BP value tends to increase with the age.

Taking into consideration the age group defined previously, the differences between the 1st and 2nd groups, the 1st and 3rd groups and between the 2nd and 3rd groups gave different results. The tests were done separately in each situation. The values of the $p - values$ are as follows: testing the null hypothesis

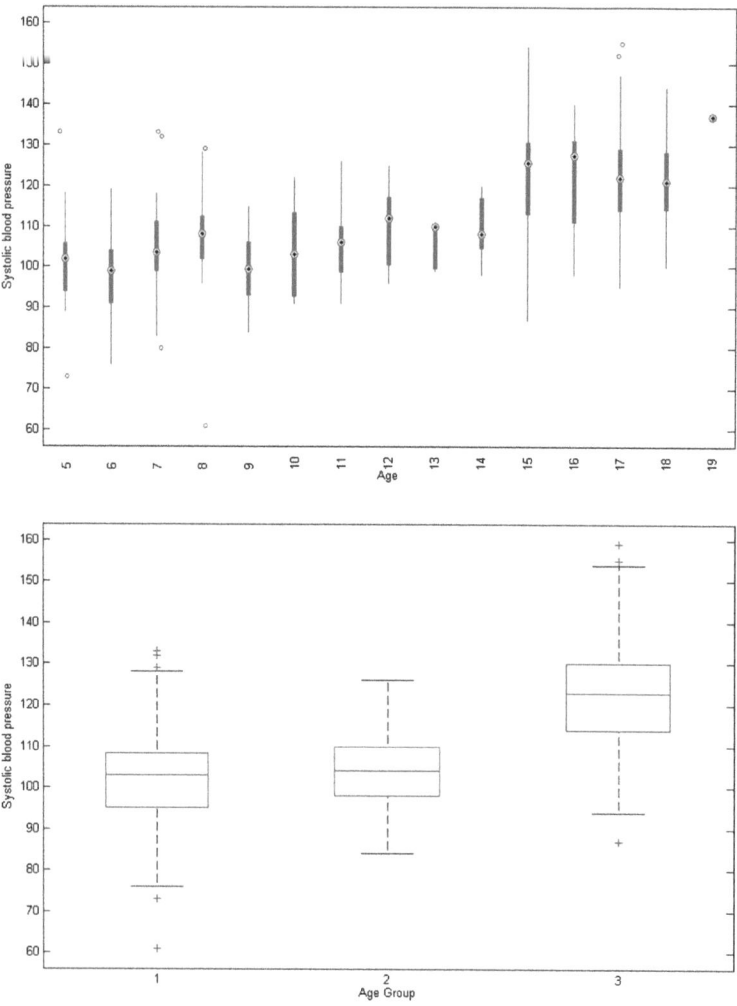

Fig. 3. Distribution of BP per age (top) and per age group (bottom).

that the older students have a higher HBP against the students of group 1 or group 2, the $p-values$ obtained are, respectively, 0.0284 and 0.0302. It means that in terms of the groups 1 and 2 can be agglutinated, there is no significant difference. The group 3 is the one that presents higher and more significant values. We can identify these facts in Fig. 3, where we can observe the BP distribution per age and per age group.

Table 2. ANOVA results: Univariate F Tests *p-value*. Design: Intercept + Gender + Ed.Level+ Gender * Ed.Level.

	p-value
Age Group	0.0000
Gender	0.0208
Age Group*Gender	0.0038

We have performed an analysis of variance (ANOVA) two factors so we could analyze gender and age variables influence on BP. The residual analysis was complete. Initially, the homogeneity variance Levene's Test does reject the null hypothesis (for usual significance levels) ($p-value = 0.0213$) so the data was transformed so we can go on with the analysis. The univariate F test to evaluate the effect of gender and age relatively to BP, can be found in Table 2. Notice that gender and age groups have significant effects to BP ($p-value = 0.0208$ and $p-value = 0.000$); also gender and gender*age level have significant effects ($p-value = 0.0037$). The pairwise comparison using the estimated marginal means, concludes that, for a boy, BP has a mean greater than for a girl

Table 3. ANOVA design: Intercept + Gender + Ed.Level+ Gender * Ed.Level. Label for each pair of treatments (on left). BP multiple comparison (Scheffé). Yellow means significant differences considering a 5% significance level (on right).

Label	Age Group	Gender
1	1	Male
2	2	Male
3	3	Male
4	1	Female
5	2	Female
6	3	Female

Groups label		Lower bound	Center	Upper bound
1	2	-9.505	-0.603	8.300
1	3	-32.467	-24.488	-16.510
1	4	-6.611	1.853	10.317
1	5	-11.905	-2.269	7.368
1	6	-22.544	-13.932	-5.320
2	3	-32.374	-23.886	-15.398
2	4	-6.491	2.455	11.401
2	5	-11.728	-1.666	8.396
2	6	-22.415	-13.329	-4.243
3	4	18.314	26.341	34.368
3	5	12.965	22.220	31.475
3	6	2.373	10.557	18.739
4	5	-13.796	-4.121	5.555
4	6	-24.441	-15.785	-7.128
5	6	-21.469	-11.663	-1.857

$(p - value = 0.033)$. When we consider the model with interaction, the same is verified, for a boy the BP mean is greater than for a girl, when both have similar age. We can conclude the same for the age factor, older students have BP mean greater than younger students, for the same gender.

We also performed multiple comparisons (Sheffé) accordingly with the labels in Table 3 (on left). The simultaneous confidence intervals (Sheffé) for each treatment difference considering 95% significance level, can be found in Table 3 (on right). The significant differences are point out in yellow. The results are in accordance with the remaining analysis.

4   Results and Final Remarks

While the data collection at national level is not complete, a smaller sample, with 471 observations, collected by the Pediatric Hypertension Group during the Hypertension Day activities, is considered to be analyzed statistically. This sample is not representative of the Portuguese population, but can be considered to get some indicative results as an alert to the need and value of ongoing national study. It is performed a descriptive analysis and applied an analysis of variance. Relationships between socio-demographic variables and blood pressure are evaluated. The statistical approach, estimating a model with relevant information using more elaborate techniques such as generalized linear models is still going on. Some important issues about pediatric hypertension were obtained, e.g. gender and age are related with pediatric high blood pressure, reinforcing the importance and need of a caregiver's knowledge improvement and prevention measures implementation about pediatric high blood pressure.

Acknowledgements. This work was supported by Portuguese funds through the FCT, *Center for Computational and Stochastic Mathematics* (CEMAT), University of Lisbon, Portugal, project UID/Multi/04621/2013, and *Center of Naval Research* (CINAV), Naval Academy, Portuguese Navy, Portugal.

References

1. Teodoro, M.F., Simão, C.: Perception about pediatric hypertension. J. Comput. Appl. Math. **312**, 209–215 (2017)
2. Teodoro, M.F., Simão, C.: Completing the analysis of a questionnaire about pediatric blood pressure. Trans. Biol. Biomed. World Sci. Eng. Acad. Soc. **14**, 56–64 (2017)
3. Costa, J.R.B.: Hipertensão arterial em idade pediátrica: que conhecimento têm os prestadores de cuidados sobre esta patologia?. Master Thesis, Medical Faculty, Lisbon University (2015)
4. Teodoro, M.F., Romana, A., Simão, C.: An issue of literacy on pediatric hypertension. In: Simos, T., et al. (eds.) Computational Methods in Science and Engineering 1906, 110006. AIP, Melville (2017)
5. Teodoro, M.F., Simão, C., Romana, A.: Questioning caregivers about pediatric high blood pressure. In: Gervasi, O. et al., (eds.): Computational Science and Its Applications – ICCSA 2017. LNCS 10408(V), 1–10. Springer, Heidelberg (2017)

6. Andrade, H., Antonio, N., Rodrigues, D.: Hipertensão arterial sistémica em idade pediátrica. Revista Portuguesa de Cardiologia **29**(3), 413–432 (2010)
7. Lurbe, E., Cifkovac, R.F.: Management of high blood pressure in children and adolescents: recommendations of the European society of hypertension. J. Hypertens. **27**, 1719–1742 (2009)
8. Muntner, P., He, J.: Trends in blood pressure among children and adolescents. J. Am. Med. Assoc. **291**, 2107–2113 (2004)
9. National high blood pressure education program working group on high blood pressure in children and adolescents. The fourth report of the diagnosis, evaluation and treatment of high blood pressure in children and adolescents. Pediatrics 114, 555-576 (2004)
10. Stabouli, S., Kotsis, V.: Adolescent obesity is associated with high ambulatory blood pressure and increased carotid intimal-medial thickness. J. Pediatr. **147**, 651–656 (2005)
11. Teodoro, M.F., Simão, C.: Notes about pediatrics hypertension literacy. Trans. Biol. Biomed. World Sci. Eng. Acad. Soc. **14**, 89–97 (2017)
12. Anderson, T.W.: An Introduction to Multivariate Analysis. Wiley, New York (2003)
13. Tamhane, A.C., Dunlop, D.D.: Statistics and Data Analysis: from Elementary to Intermediate. Prentice Hall, New Jersey (2000)
14. Romana, A.: Hipertensão Arterial em Pediatria. Um estudo observacional sobre a literacia dos cuidadores, Master Thesis, Medical Faculty, Lisbon University (2017)

LIGHTness: Interactive Luminous Ballet Outfit

Virginia Cardoso, Rachel Boldt, Hélder Carvalho$^{(\boxtimes)}$ (iD),
and Fernando Ferreira (iD)

Centre for Textile Science and Technology,
University of Minho, Guimarães, Portugal
virginiavianadesigner@gmail.com,
rachelsagerb@gmail.com, {helder,fnunes}@det.uminho.pt

Abstract. The priorities of dressing have broadened and become more complex over the years, creating challenges in developing new products. The techno-logical evolution allied with its democratization has provided new possibilities for experimentation in the area of fashion and clothing in order to meet the current complex expectations. The article presents the experimental develop-ment of a transdisciplinary fashion product, which proposes the expansion of the communicability of dance through technological stage design. According to the metalinguistic concept allied to the computational logic pre-established in a project plan, the proposed product has, attached to a ballet costume, sensors to capture movements and transform them into light and colors with different intensities, through direct interaction with the movement direction and intensity. As a result, the LIGHTness project explores the forms, interactivity and light effects of ballet, emphasizing the dichotomy between strength and lightness in this type of dance.

Keywords: Ballet · Technological stage design · Fashion product
Interactive textile · Smart wearable

1 Introduction

In the field of fashion design, several professionals use clothing and its ability to express and communicate as an object of study. From gala dresses to jeans, all clothes have the potential to communicate an idea and assist in projecting the image of the wearer.

In arts, as well as in fashion, the communication of subjective aspects is, in many cases, non-verbal. For example, the prevalence of cold colours in horror movies, the accelerated rhythm of cheerful music, the fluidity and slowed velocity of the ballerina in a romanticized spectacle, are part of the characteristics of this interlocution. How-ever, these elements do not constitute a dialogue in an autonomous way, but rather they constitute a set of elements that, when combined favourably, allow creating a desired climate.

This article presents the experimental development of a transdisciplinary fashion product that proposes to unite art and technology. The project results from a challenge

© Springer International Publishing AG, part of Springer Nature 2019
J. Machado et al. (Eds.): HELIX 2018, LNEE 505, pp. 214–220, 2019.
https://doi.org/10.1007/978-3-319-91334-6_30

posed to students of the MSc course in Design and Marketing at University of Minho, Portugal. The task sought to approach design students to the possibilities of technology by designing an interactive textile product. Specifically, the assignment was to create an interactive textile product using sensors and light. The authors focused on producing an interactive ballet dress.

Costumes and clothes are considered one of the most important elements of success in a spectacle [1]. Besides the skills of the dance professionals, the costumes and the scenarios build the atmosphere of the show, helping to construct and enhance the emotion and beauty presented by the dance.

In the universe of dance, the dancers' movements and their dramatization are the key factors to construct the information that will be communicated within the plot. The viewer should perceive a natural ease of the dancer in performing the moves, which is achieved by expressing softness and lightness in executing the movements in a choreography. However, it is known that muscular strength is fundamental for the dancer to develop a good movement execution [2]. Citing [3]: "The performance of the movements supposes technical perfection and, despite the harsh characteristics, the ballet tries to exalt lightness, delicacy and beauty, which are attributes associated with the feminine gender."

Focusing on exploring the dichotomy between the light and fluid beauty presented by the ballerinas and their strength and commitment used during the execution of ballet steps, the concept was developed as "lightness and strength", later renamed as "LIGHTness, wearing dance". This concept, applied to clothes, has the objective of materializing through lighting the moments in which strength and intensity are present in the performance.

2 Background and State-of-the Art

2.1 Smart Wearables

Wearable technologies, e-textiles, electronic textiles and smart fabrics are terms that indicate the application of electronic and other technologies, aiming to offer additional functions to worn objects, e.g. clothing and accessories [4]. The advance of researches in this area is growing and has close relationship with the wide development of computing, its miniaturization, democratization and multiple applications at individual and customized levels.

Among the research topics in this area it is possible to notice the wide range of applications, from medical, rehabilitation, enhancement of military articles, traffic safety, pure technology experiments, such as experimental laboratories [5] and projects of an artistic-conceptual objective in the area of fashion. Previous projects resulting from work in the MSc in Design and Marketing at University of Minho approach the interactivity between wearables and environment in daily life and in arts [6, 7].

2.2 Interactive Art Projects in Ballet and Dance

Several commercial products exist in luminous dresses for dance and ballet. The company Etereshop proposes "interactive tutus" and "wing dresses" [8]. Whilst some of these products display a static LED light pattern, others can be pre-programmed or controlled remotely. A direct interaction between the dance movements and the dress is not present. Still, beautiful spectacles can be observed [9].

Another example of a ballet show using a pre-programmed LED ballet tutu was choreographed by Nicolai Kabaniaev. In this case, a synchrony between dance, music and light effects exists, achieved through careful preparation and rehearsal [10].

The project "E-Traces" by Lesia Trubat uses ballet pointe shoes with sensorized tips that allow dancers to translate their movements into digital images in a mobile application [11]. In this case, true interaction between dance and effect exists.

In the project herein described, the intention was to transform the dance movements into light effects that emphasize the dancer's strength and elegance in real-time, in a truly interactive effect that adds beauty and complexity to the spectacle.

3 Objectives

As pointed out before, the authors intended to explore the dichotomy between lightness and fluidity presented during ballet performance and the strength, the effort required to produce this perception.

To implement this concept it was proposed that the movements should be materialized through light emitting devices embedded discretely in the ballet dress or accessories.

Fundamental requirements for the final product are a non-intrusive integration of the electronic components and light emitting devices in the ballet dress, sufficient autonomy for the performances and an attractive and innovative aesthetical design.

4 Methodology

The work was divided into three main stages. The first one was exploratory, researching academic, commercial and other sources for previous projects that could be inspiring and guiding to the current implementation, or other that could be translated to the ballet universe. In this phase, also an extensive search for materials, technologies and products enabling the development of the desired product was carried out.

In the second phase, the work focused on the thematic selection for the product to be addressed. Starting from the potential of the textile base material of the project, the possible application to the scenic dance spectacles was considered. A data search and analysis in the contexts of show business, productive partnerships and ballet professionals was carried out.

Finally, a design methodology was applied for product development. At this point, the concept and morphological orientations of the product were elaborated, as well as the generation of alternatives, analysis of the technical feasibility, selection and development of the project. The result was a construction of the concept project and finally the implementation and testing of the prototype.

5   Results and Discussion

The final concept of the project was to test the aesthetic and communicative possibilities of apparel by interpreting and displaying effort of the dancer during the performance. For this purpose, the interactive product should represent the dancer's movement direction and intensity using light movement, intensity and colours. A second feature defined for the product was to detect force applied on the ballet shoes, when the dancer is in the difficult "en pointe" position. This effort should be emphasized and rewarded by additional light effects.

The developed product is divided into three fundamental parts: the **costume**, the central **electronic platform** and the **sensing pointe shoes**.

5.1   The Costume

At a conceptual-aesthetic level, the ballet costume construction was based on structures that offer a sense of strength, such as combat armour, and were built with fluid materials that show lightness, such as tulle fabric and a light plain weave, in white colour. Another element used were fabrics capable of assigning a diffusion effect to the light from the LEDs. Figure 1 shows the schematic design of the developed concept, its hierarchy and the material selection.

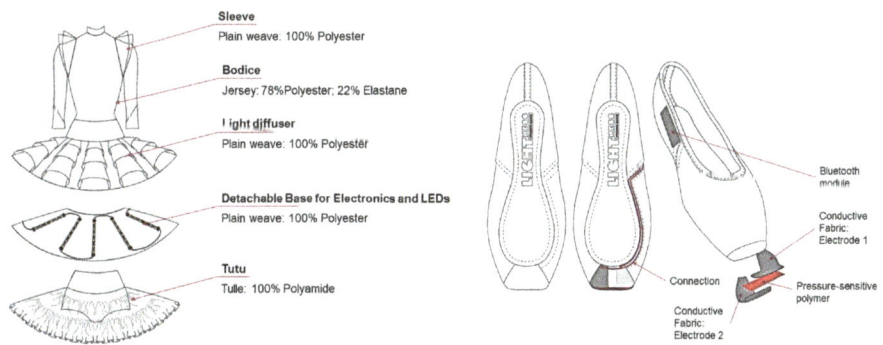

Fig. 1. Ballet costume and shoes, schematic drawings and details.

6   Electronic Platform

To implement the electronic platform, independent devices were developed to gather most of the electronic elements. This optimises fundamental factors such as portability; stability; safety and easy removal were observed to assist in the outfit's maintenance and storage.

The platform consists of a central controller module, two peripheral modules at the pointe shoes and LED strips. All of the modules are based on the Arduino Nano prototyping platform. The central controller includes a three-axis MMA7361L analogue accelerometer module and uses the nRF24L01 + communication module to receive force information from the shoe modules. WS2813 programmable LED strips are distributed evenly over its textile support. Segments of 6 LEDs are placed vertically on the support and linked by specific connectors. The circuit is fed by a 3.7 V LiPo battery combined with a DC-DC converter to boost voltage to 5 V.

7 Pointe Shoes

The classic ballet pointe shoes are fundamental to execute a good performance, since they add beauty to the dance and their physical construction provides support for the execution of the techniques, especially for the "en pointe" position.

The pressure sensors used are piezoresistive sensors based on Velostat. Velostat is a flexible piezoresistive polymer film. To incorporate the sensors in the shoes, it was necessary to make small modifications to the traditional ballet shoes. This was made at the external part of the shoe, in order not to interfere at the functional level and comfort, Fig. 1 shows the schematic representation of the shoe. The sensor was added to the pointe and was then covered with fabric similar to the one of the shoe by gluing. Two wires connect the sensor to a small box containing an Arduino Nano with the nRF24L01 + communication module, 3.7 V LiPo battery and DC-DC converter.

8 Interactivity

In order to materialize the objectives, a chromatic variation linked to luminous effects representing the physical effort and force was proposed. The chromatic variation ranging from cold colours, at moments where less movement is present, to warmer colours at moments of greater effort was implemented using the signals of the accelerometer. Moreover, the spatial information provided by the combination of the signals in the three axes of the accelerometer was used to light the LEDs in a way that can represent a sense of directionality of the movement (jumps, left and right rotations and combinations). Finally, the "en pointe" position is detected by the pressure sensor in the shoes and produces an intense glitter effect. Although it was intended to quantify the pressure and make the glitter's intensity proportional to it, in a first version of the dress the same glitter effect is produced whenever the dancer is in the position.

9 Presentation

The proposed costume has an experimental, objective, poetic and metalinguistic character. In addition, it is intended to express the dichotomy between the force employed by the ballerina and the visual result of the ballet steps during the act of dancing (Fig. 2).

Fig. 2. Visual effects

As an experimental project, the product was developed using electronic components available on the market. Therefore, the result is a functional prototype that can be significantly enhanced regarding miniaturization and integration of the material.

In a first test by a professional ballerina, the glitter effect associated to the "en pointe" position resulted as a pleasant surprise in what concerns an actual reward and enhancement of this noble and difficult technique. The display of movement was poor due to a lack of sensitivity to the movement. This was not due to the accelerometer, but to the thresholds and levels defined in the controller's firmware.

In a second moment, after adjustment of these values, a true sense of directionality and intensity could be observed in the light effects produced. A problem was observed with the connectors used to link the individual LED strip segments, at which bad contacts generated by movement would sporadically generate pixel display errors. This was later solved by applying tin solder to the contacts at the connectors.

10   Conclusion

In the art scene, the aesthetic increase as a conceptual reinforcement is constantly used. The development of the costume presented, even with an experimental character, highlights the importance of the spectacle, emphasizing the search for new platforms that could communicate and connect with the public by showing, for instance, the dancers' feelings. With this proposal, we believe that the body language and power of dance communication are increased, because it surpasses the communication of a traditional costume, presenting a luminous interactive aesthetics resulting from choreography. Beyond the sheer representation of force during the dance, the luminous answers may allow people to implement different concepts, being able to collaborate aesthetically and conceptually in different spectacles.

In the universe of smart wearables, the project shows itself to be innovative and with potential for commercial relevance. The devices that capture the interaction, especially the pressure sensors, are perfectly integrated into the costume without interfering with the primary performance.

In addition to the construction of an interactive ballet outfit, the project presents new possibilities for research and development of interactive electronic products in the ballet area. For example, pointe shoes with a pedagogical character that help trainees by indicating the correct position of the feet using their ability to perceive shoe position and pressure distribution of the tip in contact with the floor. Besides incremental improvements to the current prototype (especially regarding miniaturization and weight reduction), the use of the shoe sensor for assessment of correct positions in dance training is a subject with potential for future work.

Acknowledgements. Gratitude to Marlene Cardoso for her contribution in the initial phase of the project, to the Bailado Academy for the tests, to Marlon Carvalho and Patrícia Azambuja for photography and to the Center for Textile Science and Technology, University of Minho.

This project is financed by FEDER funds through the Programa Operacional Factores de Competitividade - COMPETE and by National Funds through FCT Foundation for Science and Technology under the project POCI-01-0145-FEDER-007136.

References

1. Puccini, C.C.: A Importância do Figurino na Construção dos Protagonistas de Vem Dançar Comigo 1 The Importance of the Costume on the Main Character´s Construction of Stricly Ballroom. 4o Colóquio de Moda (2008). http://www.coloquiomoda.com.br/anais_ant/anais/4-Coloquio-de-Moda_2008/35755.pdf. Accessed 13 Mar 2018
2. Candiotto, V.M.: Flexibilidade e força, componentes importantes no treinamento de bailarinas. Universidade do Extremo Sul Catarinense – UNESC (2007). http://docplayer.com.br/69009501-Flexibilidade-e-forca-componentes-importantes-no-treinamento-de-bailarinas.html. Accessed 14 Mar 2018
3. Gonçalves, M.C., Turelli, F.C., Vaz, A.F.: Corpos, dores, subjetivações: Notas de pesquisa no esporte, na luta, no balé. Movimento **18**(3), 141–158 (2012). http://seer.ufrgs.br/Movimento/article/viewFile/27166/21143. Accessed 14 Mar 2018, apud zur LIPPE (1988)
4. WEAR Sustain Glossary, Wear Sustain Project. https://wearsustain.eu/wp-content/uploads/2017/03/WEAR-Glossary-Call-2.pdf. Accessed 14 Mar 2018
5. Karpati, J.E., Chromosonic (2014). http://chromosonic.tumblr.com/post/85225941672/prostheticknowledge-chromosonic-experimental. Accessed 15 Mar 2018
6. Paiva, A., Catarino, A.P., Cabral, I., Carvalho, H.: Noise: sound reactive fashion. In: Proceedings of the 15th AUTEX World Textile Conference 2015, Bucharest, Romania, 10–12 June (2015)
7. Santos, M., Catarino, A., Cabral, I., Carvalho, H.: Wearable platform: the visual artistic manifest in interface with design, fashion and technology. In: Proceedings of the Third International Congress on Fashion and Design CIMODE 2016, Buenos Aires, Argentina, pp. 2282–2295 (2016)
8. Etereshop, Smart LED Ballet Tutu -164 LEDs. http://www.etereshop.com/product/smart-led-ballet-tutu-164-leds/. Accessed 07 Feb 2018
9. Contraband Events International, LED Ballerinas - Modern Ballet Show. https://www.youtube.com/watch?v=PGyPAXxa3U0. Accessed 07 Feb 2018
10. Polakoff, M., Kabaniaev, N.: Interactive LED ballet tutu. https://www.youtube.com/watch?v=i_G5EJ83eM8&t=83s. Accessed 07 Feb 2018
11. Trubat, L.: E-Traces. http://www.lesiatrubat.com/, Accessed 07 Feb 2018

New Vibratory Device for Wrist Rehabilitation

Ana Oliveira⬤, João Freitas⬤, Eurico Seabra⬤, Luís F. Silva$^{(\boxtimes)}$⬤,
and Hélder Puga⬤

Department of Mechanical Engineering, School of Engineering,
University of Minho, 4800-058 Guimarães, Portugal
lffsilva@dem.uminho.pt

Abstract. Wrist injuries are very common in most of the population, specially bone fractures, but also other pathologies such as tendinitis and neurological diseases. When the wrist is injured, their flexion-extension and radial-ulnar deviation and pronation-supination movements of the forearm are compromised and interdict most daily activities. Therefore, it is important to recover the normal functionality of the wrist, namely, the range of motion. In this paper, a new vibratory device for wrist rehabilitation is proposed. This device aims to overcome the main problems of the existing technologies, such as their price and complexity. For this purpose, two models were developed. In this work numerical simulations were performed, as well as experimental tests on patients and on the device itself. The devices promoted an increase of range of motion of the wrist and forearm at appropriate frequencies. Thus, they constitute an important complement to the rehabilitation process of patients with wrist injuries.

Keywords: Wrist · Medical device · Rehabilitation · Vibration

1 Introduction

The wrist is an important region of the human body that connects the forearm to the hand, being essential for the accomplishment of numerous basic daily activities. The wrist joint is responsible for flexion-extension and radial-ulnar deviation. The forearm is responsible for pronation-supination movements [1]. There are numerous pathologies that can affect the normal performance of the wrist, such as tendinitis, fractures and neurological diseases. Thus, when the wrist is injured, in addition to the inherent pain, its movements can also be compromised. Existing rehabilitation devices do not allow all movements of the wrist and forearm, and their replacement is necessary when the movement is partially recovered. In addition, they are not mainly aimed at wrist movements, being more focused on the grip movement. The latest developments are based in vibratory therapies for the aid of wrist rehabilitation. The application of vibratory therapies helps in the recovery of traumatic injuries to the wrist and forearm, with improved soft tissue extensibility, active wrist extension, and the composite extension of all fingers [2]. In addition, vibratory therapies can function as a form of biophysical stimulation and aid individuals with osteoporosis [3–5]. This type of stimulation has also been shown to accelerate the healing process of fractures [6].

© Springer International Publishing AG, part of Springer Nature 2019
J. Machado et al. (Eds.): HELIX 2018, LNEE 505, pp. 221–228, 2019.
https://doi.org/10.1007/978-3-319-91334-6_31

In this paper, a new active vibratory device is proposed, which allows complete rehabilitation of the forearm and the wrist through pain relief, amplitude gain and muscle strengthening. This device is portable and capable of assisting all of the wrist and forearm movements. A simulation study was also carried out to predict the vibration modes produced by the geometries proposed.

2 Methodology

The device must have an ergonomic cylindrical handle and an attractive shape. In addition, it should be designed according to human anthropometric measurements [7].

2.1 Geometric Models and Devices

For this study, two geometric models have been proposed. The first model, Fig. 1(a), consists of a tube with two motors and offset masses, one at each end and joined by the respective motor shaft. In this model, the motors are aligned by the axis in the cylindrical body (diameter = 26 mm; length = 100 mm), with offset masses with 10 mm of diameter and 5 mm of length. The second model, Fig. 1(b), presents a dumbbell shape and the motors are no longer aligned by the same axis as the body of the device, but in a perpendicular direction. For this model, the main body's cylinder has a diameter of 28 mm and a length of 100 mm, in each of its ends there is also a cylinder with 30 mm of diameter and 55 mm of length, were the motors and masses are stored. In this case, offset masses with 12 mm in diameter and 8 mm in length were used, likewise, two physical prototypes were constructed based on the geometrical models created.

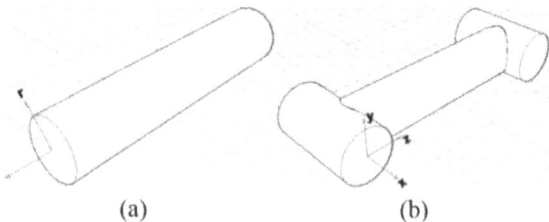

(a)   (b)

Fig. 1. Geometric models: (a) first prototype and (b) second prototype, with respective axes.

2.2 Numerical Simulation

In order to optimize the devices, a static simulation was performed in COMSOL Multiphysics® 5.1 software (COMSOL inc., 2015), which sought to find the ideal geometry for devices, taking into account the modes of vibration and the natural frequencies of the different geometries. For the numerical simulations simplified models of the first and second geometries were considered, the material chosen for these simulations was aluminum 6063 (density = 2700 kg/m^3; Young's modulus = 69 GPa;

Poisson's ratio = 0,33). In these models, the offset masses were not considered for the numerical simulation, due to their small dimensions relative to the whole body. Alternatively, only two larger masses were considered at the ends of the models, which would include the motors, shafts and respective offset masses. For the simulations, the following conditions were considered: for its boundary conditions the body was free in space, with no external fixation point; the connections between the body and the motors are rigid and its initial conditions were considered as static (acceleration = 0 m/s^2; velocity = 0 m/s; displacement = 0 m). The nodes generated by the mesh were tetrahedral: the first prototype mesh had 40490 nodes, and the second prototype mesh had 18864 nodes.

2.3 Experimental Tests on Patients

Both prototypes were tested in nine patients with different clinical conditions (see Table 1). Each patient underwent three sessions on consecutive days of fifteen minutes of therapy with each one of the devices. Thus, five patients were first submitted to the first, while the other four were initially submitted to the second prototype. After three days of therapy with one of the prototypes, patients rested for two days and were subsequently submitted to three more therapy sessions on consecutive days with the other prototype.

Table 1. Medical conditions of the patients.[a]

	Gender	Age (years)	Pathologies	Side	Surgery	Physiotherapy sessions	Pain	Background
P1	F	72	CT	L	Yes	23	Yes	–
P2	F	31	DQ	L	No	1	Yes	TT
P3	F	64	DQ	R	No	72	Yes	CTS
P4	M	55	TI	L	No	28	Yes	–
P5	M	25	CT	L	Yes	25	Yes	–
P6	F	83	T	R	Yes	15	Yes	RA, A, CF
P7	M	34	DQ	R	No	9	Yes	–
P8	F	46	SFS	L	No	15	Yes	–
P9	F	59	CF	L	No	25	Yes	–

[a]A – Arthrosis; CF – Colles fracture; CT – Cut in tendons of the wrist; CTS – Carpal tunnel syndrome; DQ – De Quervain's Tenosynovitis; F – Female; L – Left; M – Male; R – Right; RA – Rheumatoid arthritis; T – Tendinitis of the wrist; TI – Traumatic impact; TT – Tendinitis of the thumb.

2.4 Measurements of Accelerations

Measurements of their accelerations were made in each one of the devices. To achieve this, a variable power supply was required to power the motors in the device. Each device had an accelerometer that was connected to the signal conditioner, which in turn was connected to a computer through a data acquisition board.

In the first model, the measurements were performed on only one axis, and in the second model, the measurements were performed in the three axes (see Fig. 1).

3   Results and Discussion

This section presents the results obtained in the numerical simulation, the joint gain obtained in patients and the results of the measurements of the accelerations.

3.1   Numerical Simulation

The models proposed were simulated for their eigenfrequencies iterating over several lengths of the geometry of the bodies between 100 to 160 mm. The first modes of vibration can be found on Fig. 2(a) for the first model, and on Fig. 2(b) for the second model, On Table 2 can be found the lengths of the bodies for each model as well as each respective eigenfrequency; all of these frequencies correspond to the mode of vibration shown in Fig. 2. These results suggest a decrease of the eigenfrequencies as the length of the bodies increases. Also, the second model reveals smaller eigenfrequencies for every iteration of the simulation. The first mode of vibration for this device is desirable, because it produces a flexion motion that can be aligned with the radial/ulnar deviation and the pronation/supination movements of the wrist.

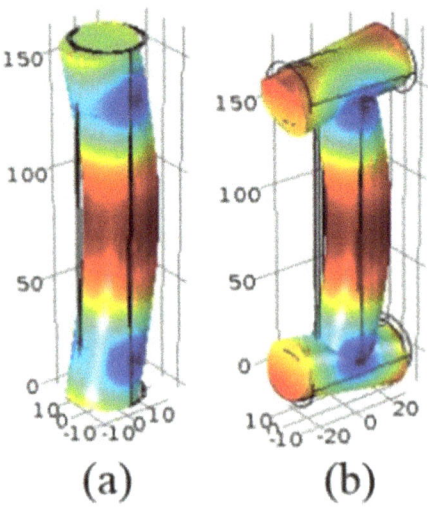

Fig. 2. First mode of vibration: (a) for the first model and (b) for the second model.

3.2   Study in Patients

Table 3 shows the joint gain obtained at the end of the six sessions of therapy with both prototypes and the joint gain resulting from a single session of vibratory therapy. It also shows that at the end of the therapy with both prototypes a significant joint gain was

Table 2. Results of the simulations.

First model – First mode of vibration				
Length (mm)	100	120	140	160
Eigenfrequency (Hz)	8091,7	6032,7	4631,0	3646,5
Second model – First mode of vibration				
Length (mm)	100	120	140	160
Eigenfrequency (Hz)	4168,9	3483,6	2930,2	2481,7

obtained in all movements of the wrist and forearm for both genders. In female subjects, greater joint gains were observed than in male subjects. In pronation and supination movements in male subjects, the joint gain obtained was null since all patients already had this movement fully recovered. The results obtained by the vibration therapy of this study were much higher than the results obtained in other works for both the female and the male genders [8]. Other authors who also used a vibratory therapy achieved a greater joint gain for the extension of the wrist [2]. However, these authors used a therapy with longer exposure time and with resonant frequencies coincident with the resonance frequencies of the hand and arm, which should be avoided. However, it must be taken into account that the gains of amplitudes obtained are also dependent of the physiotherapy to which the patients were submitted. In this way, the joint gain obtained after a single session of vibratory therapy was measured, without the interference of other variables. In these conditions, there is a very significant joint gain resulting from a single session of vibration therapy. A great dispersion of results can also be observed for flexion, extension, pronation and supination (see Table 3).

The joint gain obtained with each one of the prototypes was also studied. Table 4 shows these results. The results obtained for both prototypes are similar; however, it suggests a little tendency for greater joint gains with the second prototype. In this case, there is also a great dispersion of results for flexion, extension, pronation and supination (see Table 4).

3.3 Accelerations

The motors of the devices were supplied with a voltage of 4 V as the accelerations were measured. In this way, Table 5 presents the peak frequencies, found for this voltage for each one of the prototypes, according to the measured axis and the gender of the individual.

For proper wrist rehabilitation, the hand and arm resonance frequencies between 20 Hz and 70 Hz should be avoided [9, 10]. In this study, the frequencies did not match with this resonance frequencies range, which is a positive aspect of the work. However, the obtained frequencies are inferior to those predicted by the simulation. In addition, there are no significant differences between the natural frequencies of both prototypes, contrary to that predicted by the numerical simulation. These differences may result from the material used and the simplifications adopted in the simulations.

Table 3. Total joint gain and joint gain after a single session, in parenthesis are the values for the patients that did not have a fully recovered movement.

	Movement	Female		Male	
		Mean	Standard deviation	Mean	Standard deviation
Total joint gain	Flexion	17,3°	8,6°	14,7°	11,9°
	Extension	16,5°	10,2°	12,3°	9,3°
	Radial deviation	4,3° (5,2°)	3,1° (2,5°)	5,0°	2,0°
	Ulnar deviation	8,5°	2,8°	10,0°	3,0°
	Pronation	9,2° (27,5°)	20,1° (31,8°)	0,0°	0,0°
	Supination	11,2° (13,4°)	13,6° (13,9°)	0,0°	0,0°
Joint gain after a single session	Flexion	4,8°	4,7°	4,3°	0,6°
	Extension	5,3°	5,8°	2,3°	0,6°
	Radial deviation	2,0° (2,4°)	1,8° (1,7°)	2,0°	0,0°
	Ulnar deviation	3,3°	1,2°	2,3°	1,5°
	Pronation	2,3° (7,0°)	4,8° (7,1°)	0,0°	0,0°
	Supination	3,2° (3,8°)	4,0° (4,1°)	0,0°	0,0°

Table 4. Joint gains for each prototype, in parenthesis are the values for the patients that did not have a fully recovered movement.

	Movement	Joint gain			
		Female		Male	
		Mean	Standard deviation	Mean	Standard deviation
Prototype 1	Flexion	9,3°	5,0°	7,0°	2,6°
	Extension	11,5°	6,6°	4,3°	0,6°
	Radial deviation	3,2° (3,8°)	2,7° (2,5°)	3,3°	0,6°
	Ulnar deviation	4,0°	1,7°	5,3°	2,5°
	Pronation	3,8° (11,5°)	8,9° (14,8°)	0,0°	0,0°
	Supination	5,2° (6,2°)	6,4° (6,6°)	0,0°	0,0°
Prototype 2	Flexion	10,3°	8,2°	11,3°	8,1°
	Extension	8,0°	5,9°	9,7°	8,1°
	Radial deviation	2,8° (3,4°)	2,0° (1,7°)	3,3°	0,6°
	Ulnar deviation	6,3°	2,3°	7,3°	3,5°
	Pronation	5,3° (16,0°)	11,2° (17,0°)	0,0°	0,0°
	Supination	7,0° (8,4°)	7,7° (7,7°)	0,0°	0,0°

Table 5. Peak accelerations found.[a]

	Prototype 1		Prototype 2					
	F	M	X axis		Y axis		Z axis	
			F	M	F	M	F	M
Frequency (Hz)	84,0	82,2	83,2	79,6	78,4	80,2	84,0	82,2

[a]F – Female; M – Male.

4   Conclusions and Future Work

In this study it was possible to verify that the vibratory therapy promotes the joint gain of all the wrist movements, being the results slightly higher for the second prototype. There were differences between what the simulations predicted and what was obtained in practice in terms of the eigenfrequency of the devices. In addition, it was verified that the natural frequencies obtained are low and do not match with the resonance frequencies range of the hand and the arm, which is an essential aspect for the patients' health. Future work should contemplate the construction of a new prototype, similar to the second prototype, in materials suitable for construction. Besides that, new simulations of a dynamic nature and in more detail should also be carried out. In addition, new tests should also be performed on a larger number of patients, in order to offer greater support to the proposed solution.

References

1. Garratt, B.J.: The development of a wrist rehabilitation device for movement therapy. Bachelor of Engineering (2009)
2. Brewin, M.P., Bexon, C.J., Tucker, S.C.: A case report on the use of vibration to improve soft tissue extensibility after major trauma. J. Hand Ther. **30**(3), 367–371 (2017). https://doi.org/10.1016/j.jht.2016.11.012
3. Gilsanz, V., Wren, T.A., Sanchez, M., Dorey, F., Judex, S., Rubin, C.: Low-level, high-frequency mechanical signals enhance musculoskeletal development of young women with low BMD. J. Bone Miner. Res. **21**(9), 1464–1474 (2006). https://doi.org/10.1359/jbmr.060612
4. Rubin, C., Judex, S., Qin, Y.X.: Low-level mechanical signals and their potential as a non-pharmacological intervention for osteoporosis. Age Ageing **35**(suppl_2), ii32–ii36 (2006). https://doi.org/10.1093/ageing/afl082
5. Rubin, C., Recker, R., Cullen, D., Ryaby, J., McCabe, J., McLeod, K.: Prevention of postmenopausal bone loss by a low-magnitude, high-frequency mechanical stimuli: a clinical trial assessing compliance, efficacy, and safety. J. Bone Miner. Res. **19**(3), 343–351 (2004). https://doi.org/10.1359/JBMR.0301251
6. Leung, K.S., Shi, H.F., Cheung, W.H., Qin, L., Ng, W.K., Tam, K.F., Tang, N.: Low-magnitude high-frequency vibration accelerates callus formation, mineralization, and fracture healing in rats. J. Orthop. Res. **27**(4), 458–465 (2009). https://doi.org/10.1002/jor.20753

7. Tilley, A.R.: The Measure of Man and Woman: Human Factors in Design. The Whitney Library of Design, New York (1993)
8. Hsieh, W.M., Hwang, Y.S., Chen, S.C., Tan, S.Y., Chen, C.C., Chen, Y.L.: Application of the Blobo bluetooth ball in wrist rehabilitation training. J. Phys. Ther. Sci. **28**(1), 27–32 (2016). https://doi.org/10.1589/jpts.28.27
9. Duarte, M.L.M., Misael, M.R., Freitas Filho, L.E.: Experimental evaluation of vibration comfort for a residential environment. In: IMAC-XX: Conference and Exposition on Structural Dynamics, pp. 1376–1377. Society of Photo-Optical Instrumentation Engineers (2002)
10. Mester, J., Kleinöder, H., Yue, Z.: Vibration training: benefits and risks. J. Biomech. **39**(6), 1056–1065 (2006). https://doi.org/10.1016/j.jbiomech.2005.02.015

Design of a Smart Garment for Cycling

André Paiva[2], Daniel Vieira[1], Joana Cunha[2],
Hélder Carvalho[2(✉)], and Bernardo Providência[1]

[1] Landscapes, Heritage and Territory Laboratory,
University of Minho, Guimarães, Portugal
mail@danielvieira.pt,
providencia@arquitectura.uminho.pt
[2] Centre for Textile Science and Technology,
University of Minho, Guimarães, Portugal
dmpaiva.s@gmail.com, {jcunha,hcarvalho}@det.uminho.pt

Abstract. Given the premise that cycling encourages a healthy lifestyle and promotes wellbeing, this paper describes the design process for the development of a cycling garment, with embedded electrodes for heart rate measurement to further widen the possibilities of health and performance monitoring, in a practical and unobtrusive way. Electrodes were produced and tested, using different textile materials and techniques. Signal measurement is based on the BBB Y8YBH20 heart rate monitor and performance tests were done using the Polar Beat mobile app. Finally, a prototype of the garment is presented.

Keywords: E-textile · Wearables · Smart textiles · Cycling wear
Design

1 Introduction

Information technologies (ICT), portable technologies and smart textiles are changing the way we think and design new concepts and paradigms for the area of sports and well-being. With the advent of new types of textiles, such as e-textiles (i.e.: for heart or muscle activity monitoring), the textile becomes a platform to integrate new technologies and increment functionalities. By adding computational technologies, the textile becomes "smart", providing the user information in a ubiquitous way, promoting a new form of interaction with the textile.

This paper describes the design process of a skinsuit for road cycling, triathlon, track cycling, mountain cycling and cyclocross, where e-textile technology for heart rate analysis was explored. For this, we had to understand: (i) the purpose of the concept; (ii) the used technology; (iii) the context in which the product is used.

© Springer International Publishing AG, part of Springer Nature 2019
J. Machado et al. (Eds.): HELIX 2018, LNEE 505, pp. 229–235, 2019.
https://doi.org/10.1007/978-3-319-91334-6_32

2   State of the Art

2.1   Exploring the Technology: Textile Electrodes and Sensors

The study of textile sensors for vital signs and biometrics has been driven by the idea of building remote monitoring systems that can be continuously connected to the body without compromising comfort, just like clothing, adding new functionalities.

In electrocardiography (ECG), surface electrodes are used, the same type as used in surface EMG and EEG. Several textile materials and techniques have been used in order to produce textile electrodes. Some examples include knitting [1, 2], weaving [3, 4], and embroidery [5], using conductive yarns; printing with conductive ink [6], coating with conductive rubber [7].

Yao [8] have developed an X-band with three conventional silicon electrodes for continuous monitoring of cardiac activity. Zięba et al [9] created a sleeveless seamless shirt that includes a knitted stretch sensor to measure breathing rate. Manero et al. [10] have presented a prototype of running legging with three pairs of embroidered electrodes (and circuits) for continuous monitoring of muscular activity in the quadriceps. Ma et al. [11] proposed a shirt with textile electrodes for ECG and EMG and a textile breathing sensor, as well as conventional accelerometers and temperature sensor. Frydrysiak and Tesiorowski [12] proposed a remote monitoring system, which includes textiles sensors attached to a shirt for measurement of heart rate, breathing rate and temperature.

A few commercial brands have been investing in smart clothing, mainly for the sports market. Hexoskin [13] has introduced in the market a shirt that combines textile and conventional sensors to provide data about heart rate and variability, breathing rate, oximetry and activity parameters. AiQ Smart Clothing developed *Bioman*, a shirt that has embedded sensors to monitor heart rate, breathing rate and skin temperature [14]. Athos® gave form to a compression shirt and shorts for sports that makes use of printed sensors with conductive rubber to provide information about heart rate, respiration and muscle activity [15]. Myontec produced cycling shorts with textile electrodes sewn to the shorts to detect electrical signals from quadriceps and hamstrings muscles [16].

So far, a truly invisible integration of technology does not yet seem to be possible, or at least, none is known to be in the market. Nonetheless, all visible technology is designed in order to look pleasantly (and even unnoticed), as if it is part of the garment and not just some strange object.

3   Materials and Methods

The project follows a design process suggested by the authors that requires the understanding: (i) the purpose of the concept; (ii) the used technology; (iii) the context in which the product is used. Various materials have been explored to make textile electrodes, in order to find a suitable material and method to integrate ECG electrodes in the garment. Performance tests were done using the Polar Beat mobile app. To make a prototype of the suit (size S), patterns were designed and cut, first for a partial prototype and later for a full prototype of the garment. After each fitting, a new prototype was done to correct the size measurements.

3.1   Design Process

A product was designed that responds to contemporary standards for cycling apparel and technology, combined in a fashionable way. The designed product is a skinsuit (Fig. 1), that integrates ECG electrodes and a heart monitoring device to continuously monitor heart rate, integrated in a way that it is almost invisible, therefore without compromising aesthetics. A polyamide/elastane jersey fabric was chosen for the cycling garment, which is smooth and provides some compression, keeping the electrodes in contact with the body.

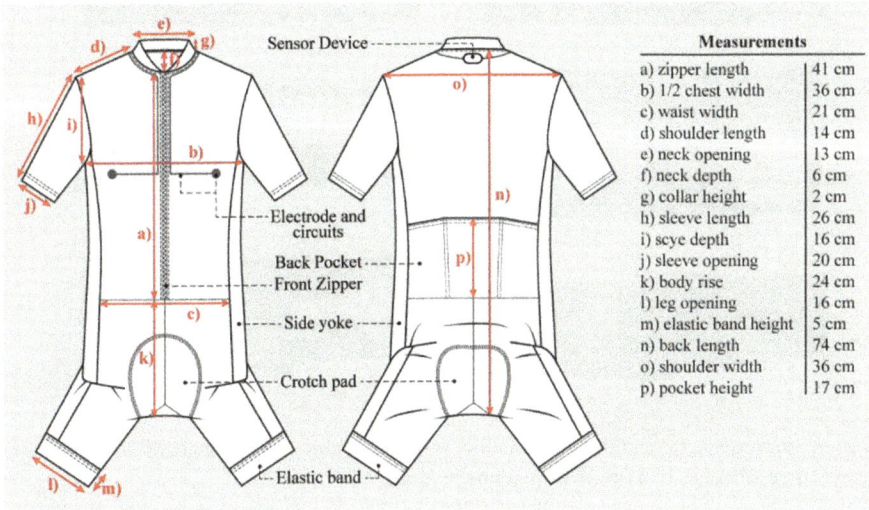

Measurements	
a) zipper length	41 cm
b) 1/2 chest width	36 cm
c) waist width	21 cm
d) shoulder length	14 cm
e) neck opening	13 cm
f) neck depth	6 cm
g) collar height	2 cm
h) sleeve length	26 cm
i) scye depth	16 cm
j) sleeve opening	20 cm
k) body rise	24 cm
l) leg opening	16 cm
m) elastic band height	5 cm
n) back length	74 cm
o) shoulder width	36 cm
p) pocket height	17 cm

Fig. 1. Detailed drawing of the skinsuit, with specifications for size S

The garment was designed with raglan sleeves, but the position of the ECG device and connections' shape made it impossible, because of the scye seams. It was then decided to use a conventional sleeve, so the shoulder seam could be removed. The electrodes would be put near the frontal center, but given the changes in body shape, the electrodes would not have a good electrical contact. Therefore, it was settled that they would be moved closer to the side, below the chest, where the body shape is more uniform. The heart rate monitoring device (BBB Bluepulse Heart Rate Sensor) was put in the back, near the collar, so it would not interfere with movements.

3.2   Choosing the E-textile

Before building a prototype, several conductive materials – woven fabric (Statex Bremen), knitted fabric (Statex Silverell 1), silver ink (Dupont PE828) and conductive silicone (ELASTOSIL® LR 3162 A/B) – were studied, to check their performance as ECG electrodes. For that, elastic bands were produced in the fabric selected for the cycling skinsuit and electrodes were embedded using the appropriate technique for each material (Fig. 2). Thermoplastic polyurethane (TPU) was used on the connections and on the border of the electrodes to isolate the connections from the body and to attach the electrodes to the fabric, maintaining the electrodes visible. Snaps were attached to the end of the connection to plug the monitoring device. The snaps were covered by a piece of fabric, so they do not contact the skin.

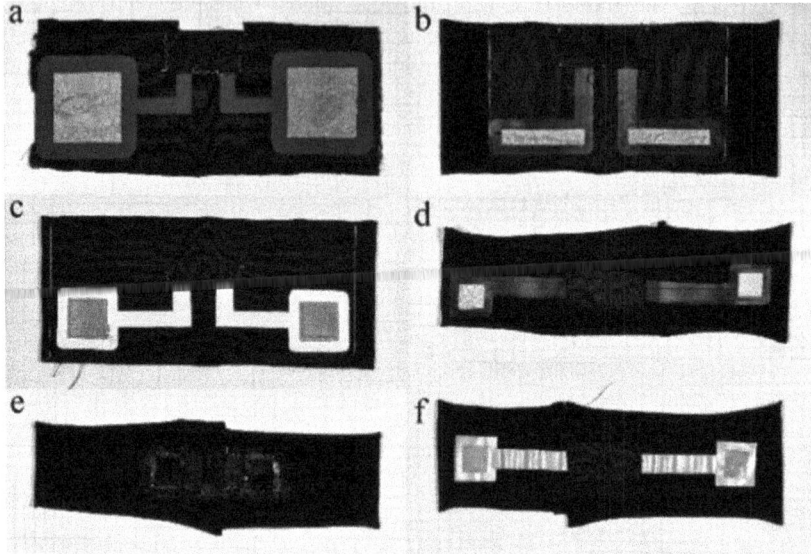

Fig. 2. Stretch bands: (a) and (b) woven fabric; (c) woven fabric with sponge; (d) knitted fabric; (e) conductive silicone; (f) printed with conductive ink

In the conductive fabric version (Fig. 2a, b and c) the fabrics are sewn to the elastic band and in the knitted version (Fig. 2d), the conductive fabric is attached with adhesive. In the electrodes of Fig. 2c, a 3 mm sponge was put inside the electrode to create some volume. The conductive silicone comprises two components, mixed together in a ratio of 1:1. The mixture was coated onto the knitted fabric (Fig. 2e) and cured at 160 °C, for 10 min. For the printed version (Fig. 2f), the ink was applied using a brush and it was then cured at 80 °C for 20 min. The connections were isolated with TPU film, except in the case of the silicone, where this was not needed, since the fabric used to cover the snap also covered the short silicone circuit.

For heart rate measurement, the Polar Beat app was used, which provides real-time information about heart rate and parameters related to cycling (speed, distance...).

3.3   Prototyping

According to the design specification, a prototype was prepared for size S (small). At first, a prototype of the upper garment was built to check for measurements of the torso, which would then define the measurements of the rest of the body. This is an easy way to check if measurements are correct without building the whole garment. The patterns for the bottom were designed and a new prototype was made, comprising the upper and bottom parts, adding the zipper in the front, but keeping aside other details (pockets...).

Finally, after measures were checked, a prototype of the whole garment, considering all details (except the e-textile) was assembled, in order to understand its construction, but also to check the fitting (Fig. 3).

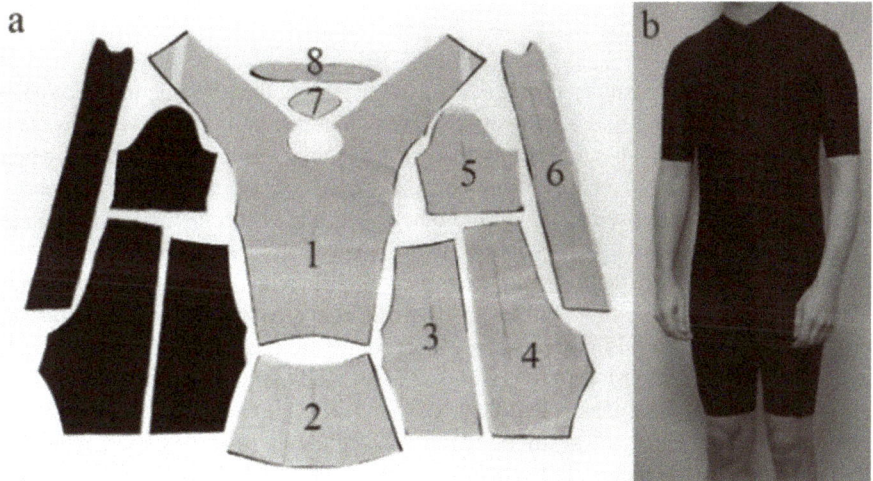

Fig. 3. Garment prototype: (a) Patterns cut (1) torso; (2) back pocket; (3) front leg; (4) back leg; (5) sleeve; (6) side yoke; (7) crotch yoke; (8) collar; (b) fitting

4   Results

With all the electrodes, except the conductive silicone based electrodes, the mobile app showed a heart rate, when wetted. With dry electrodes, it was very hard to get a reading. It was only possible to get a reading when applying high pressure on the electrodes. In dry state, the electrodes with the sponge (elastic band *c*) gave better readings, because the sponge promotes better contact with the skin, but even in this version the readings were very unstable. It is easier to get a response with wet electrodes, as can be seen from Fig. 4, which shows an example of the heart rate shown in the app, using elastic band *c*, where four stages of exercise can be view: (a) static position, (b) walking, (c) running and (d) relaxing.

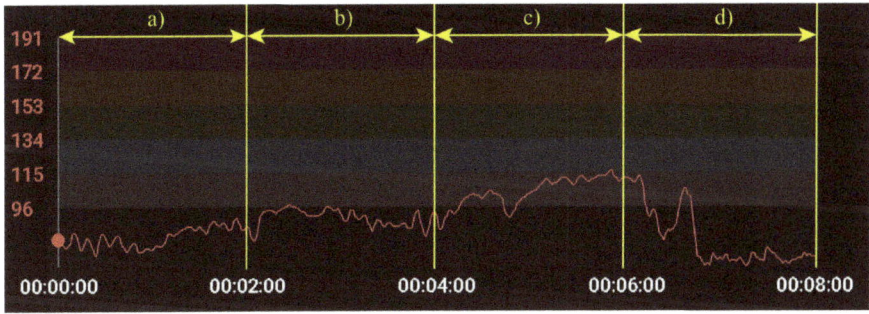

Fig. 4. Heart rate graphic acquired from the app, using the elastic band *c*

The electrodes made with conductive fabric (Bremen) do not have elasticity, unlike its counterparts, the knitted, printed or coated electrodes, which make these last ones more suitable for a garment fitted to the body and, therefore, needing to be stretched.

5 Discussion

If a garment with good elasticity is required, and considering that the silicone electrodes don't provide any reading, the best electrodes are the knitted and printed versions. The printed method resulted in connections that increase their electrical resistance when deformed and do not recover to the initial value. This may occur because of spaces created in the print after stretching. On the other hand, the electrodes with the sponge provided the best response. The volume of this electrode assembly compresses the electrodes against the skin, therefore, improving the contact with the skin and the reading.

There is a relevant design difference between using conductive ink and fabric. When printing on a thin fabric, the ink is visible on the other side of the fabric, which raises aesthetic concerns. Moreover, given this issue, the connections need to be isolated from both sides. This means that they will be visible and the designer will have to design a pattern that meets the aesthetic requirements. On the other hand, attaching the knitted fabric with a dielectric TPU adhesive will not only hide the circuits inside the garment, but will also isolate it from the outside, making it invisible.

The electrodes produced were tested in stretch bands, which comprise short connections along with the electrode. They may not have the same performance in the suit, given that the connections follow a longer track. A prototype of the suit with embedded electrodes for heart rate monitoring is being prepared for testing, following the above results.

6 Conclusion

Developing appealing and functional new products with e-textiles asks for a multi-disciplinary approach encompassing design and engineering expertise that changes the way a product is thought and designed.

In this paper, the design process of a cycling smart garment was presented. Several approaches to using e-textiles as electrodes to measure heart rate were reported and a prototype of a cycling suit was made. This kind of work involves expertise in fashion and clothing design, as well as textile and electronics engineering, requiring close cooperation in a team.

The next step will be to produce a prototype of the same suit, but this time with the electrodes embedded. Future work may include exploration of other ways of integrating technology into clothing.

Acknowledgements. This work is financed by FEDER funds through the Competitivity Factors Operational Programme – COMPETE, by national funds through FCT – Foundation for Science and Technology within the scope of the project POCI-01-0145-FEDER-007136 and by TSSiPRO-NORTE-01-0145-FEDER-000015 funded by the regional operational program NORTE 2020, under the PORTUGAL 2020 Partnership.

References

1. Carvalho, H., Catarino, A., Rocha, A., Postolache, O.: Health monitoring using textile sensors and electrodes: an overview and integration of technologies. In: IEEE International Symposium on Medical Measurements and Applications (MeMeA), Lisbon (2014)
2. Paiva, A., Catarino, A., Carvalho, H., Postolache, O., Postolache, G.: Development of dry textile electrodes for electromyography. In: 9th International Conference on Sensing Technology (ICST), Auckland, pp. 447–451 (2015)
3. Pylatiuk, C., Müller-Riederer, M., Kargov, A., Schulz, S., Schill, O., Reischl, M., Bretthauer, G.: Comparison of surface EMG monitoring electrodes for long-term use in rehabilitation device control. In: IEEE International Conference on Rehab. Robotics, Kyoto, pp. 300–304 (2009)
4. Löfhede, J., Seoane, F., Thordstein, M.: Textile electrodes for EEG recording - a pilot study. Sensors **12**, 16907–16919 (2012)
5. Puurtinen, M., Komulainen, S., Kauppinen, P., Malmivuo, J., Hyttinen, J.: Measurement of noise and impedance of dry and wet textile electrodes, and textile electrodes with hydrogel. In: Proceedings of the 28th IEEE Engineering of Medicine and Biology Society Annual International Conference, pp. 6012–6015 (2006)
6. Tao, D., Zhang, H., Wu, Z., Li, G.: Real-time performance of textile electrodes in electromyogram pattern-recognition based prosthesis control. In: Proceedings of the IEEE-EMBS International Conference on Biomedical and Health Informatics, pp. 487–490 (2012)
7. De Rossi, D., Rocha, A.M., Abreu, M.J., Fardin, D., Da Silva, J., Ferreira, J., Tavares, V., Correia, M., Dias, R.: E-legging for monitoring the human locomotion patterns. J. Text. Eng. **59**(6), 153–158 (2013)
8. Yao, M.: Real-time monitoring of heart rate using wellness belt equipped with electrocardiogram sensors. In: Proceedings of the 35th Chinese Control Conference, Chengdu (2016)
9. Zięba, J., Frydrysiak, M., Blaszczyk, J.: Textronic clothing with resistance textile sensor to monitoring frequency of human breathing. In: IEEE International Symposium on Medical Measurements and Applications Proceedings (MeMeA), Budapest (2012)
10. Manero, R.B.R., Shafti, A., Michael, B., Grewal, J., Fernandez, J.L.R, Althoefer, K.: Wearable embroidered muscle activity sensing device for the human upper leg. In: 2016 IEEE 38th Annual International Conference on the Engineering in Medicine and Biology Society (EMBC), Orlando, pp. 6062–6065 (2016)
11. Ma, Y.C., Chao, Y.P., Tsai, T.Y.: Smart clothes - prototyping of a health monitoring platform. In: IEEE 3rd International Conference on Consumer Electronics, Berlin (2013)
12. Frydrysiak, M., Tesiorwski, L.: Wearable care system for Elderly people. Int. J. Pharma Med. Biol. Sci. **5**(3), 171–177 (2016)
13. Carre Technologies Inc.: Getting Started Guide for iPhone and OS X Users. Carre Technologies Inc. (2013)
14. Krishnamurthy, G.: AiQ's BioMan Biomonitoring Shirt and Other Smart Clothing Technology (2013). https://www.medgadget.com. Accessed 6 Jan 2018
15. Berger, J.: The Most Interesting Wearable in Fitness: With Athos, Has the Personal Trainer Met its Match? (2015). http://www.gearpatrol.com. Accessed 6 Jan 2018
16. Pesola, A.: Reduced muscle Activity, sedentary time and cardio-metabolic benefits: effectiveness of a one-year family-based cluster randomized controlled trial. Ph.D. thesis, Faculty of Sport and Health Science, University of Jyväskylä, Jyväskylä (2016)

Sleepless Night

A Lighting Panel for Kids

Catarina Fernandes, Gizelle Carvalho, Lettícia Souza,
and Helder Carvalho(✉) 📍

University of Minho, Guimarães, Portugal
catasf@hotmail.com, letticia.pink@hotmail.com,
gizellecarvalho16@yahoo.com.br, helder@det.uminho.pt

Abstract. Creating this product started with the challenge of developing a textile based product that incorporated design, human interaction and light. Very early on during the design process, a gap in the kid's segment was identified leading to the investigation into children's illumination products and the idea to develop a new form of night-lights. Looking at existing technology and products, the idea became clear: to develop an interactive wall panel that reacted according to specific stimuli. By carrying out questionnaires and interacting with children in person the specific stimuli were identified; the panel would react to sound and movement. To build the panel electronic equipment such as LEDs, a microprocessor, sound sensors and movement sensors were used, along with MDF boards for the structure and specific fabrics for appearance. Due to a thorough investigation, specific colours and shapes were identified. Consequently, the outcome was a product we believe provides interaction, well-being and safety to the young target users, along with providing the parents with a calm and peaceful environment for their loved ones to sleep.

Keywords: Kids · Sleep · Light · Textile · Design · Interactive

1 Introduction and Objectives

The initial idea for this project arose after understanding that children's sleep cycles are very different from adults and by identifying their fear of darkness and loneliness when having to sleep in their own beds. Thus, the need was identified for an interactive object that would ease the transition from the parents' bedroom to their own bedroom, as well as keep the child calm if they were to wake up earlier than expected. A focal point was to understand the children's routine, their sleeping habits and their likes and dislikes as well as understanding what are the parents' views and difficulties during the transition, along with identifying what factors could contribute to a better nighttime experience for the family.

Based on Oliveira, a baby's or child's sleep cycle is very different from that of adults. Babies have two states of sleep: the active (EM) and the quite state. During the first nine months of life, each cycle lasts about 50 to 60 min. Both states of sleep are very light when compared to an adult, which is the reason why babies wake up so

© Springer International Publishing AG, part of Springer Nature 2019
J. Machado et al. (Eds.): HELIX 2018, LNEE 505, pp. 236–243, 2019.
https://doi.org/10.1007/978-3-319-91334-6_33

easily. As the child grows the sleep cycles lengthen, nevertheless they do not last longer than 100 min until schooling age is reached [1].

The objectives derived from the research became clearer as the initial studies progressed. The main objective was to develop a product that would propose a balance between creating a calm environment, entertainment and a fun and safe human interaction. It was also crucial to focus on the aesthetics of the product, as it had to be functional but also assume a more decorative function during the day. It was important that the product would be able to transition from a simple nightlight to a more complex piece of decoration allowing the panel to be used even after the child has grown. According to Kazazian everyday objects must change radically to become more efficient, satsifying the need for products with longer life cycles and that are increasingly more sustainable [2].

Consequently, functionality was incorporated with design to develop an interactive artefact that combined electronics with visually attractive textiles, colours and shapes where sustainability was always considered. Product use was also taken into consideration, with the product being activated only when the child is awake, by inclusion of sound and movement sensors used to detect activity.

2 Colors, Shapes and Safety

Colour can be considered a form of non-verbal communication, suggesting a spectrum of emotions varying from culture to culture, for example white. In the western cultures, white is associated with peace, fertility or religion while for other cultures i.e. Hindu's, associate white with death.

França, and Spinillo explain that colour is one of the principal acts towards communication with children and it has a pivotal role in a purchasing environment. Generalizing, children like colours such as reds, yellows and greens, in this order of preference. A study showed that children have a preference towards warm colours; these portray strength, vigour, energy, joy, emotion, sentiment, aggressiveness and even nervousness [3].

Form, just like colour, has visual impact and influence. Again, França, and Spinillo, explain that shape can also be associated with affectionate values and these can derive from three basic shapes: the square, the circle and the triangle. Rounded shapes are associated with positive sensations like protection, peace and cold colours like blue. Similarly, the square transmits sensations of tranquillity and stability [3].

Concerning safety, it was identified that the interactive panel could not have small visible pieces that could easily be broken off and swallowed [4]. It was also indisputable that the existence of visible electrical cables and strings had to be avoided. APSEI adverts that all electrical installations must be, at all times, in good condition and that no electrical cables can be stuck under household appliances or furniture, nor should these have knots along them [5].

As this product is electricity dependent and generates heat it was crucial that the materials used were flame retardant. The textiles used should also have hypoallergenic properties.

3 Development

3.1 Research Methods

The first step was to adapt the product to an infant market segment. To improve our knowledge on infants and their needs, it was necessary to interact with them in first hand. For this purpose, a visit to Fraterna was organized. Fraterna is a community centre dedicated to solidarity and social integration in Guimarães, Portugal, where children between the ages of 1 and 5 attend the day centre. At the institution, we held an informal discussion with both the students and teachers to better our understanding of children's sleep routine and preferences regarding colours and shapes.

It was crucial to identify colours that the children liked but also ones that transmitted a sense of tranquillity and calmness to facilitate naptime.

A questionnaire was also developed and distributed to parents or guardians of children between the ages of 1 and 5. The results aided in the comprehension of the sleeping habits, likes and dislikes of children in general and how much parents would be willing to spend on this kind of product.

3.2 Prototype

The prototype consisted of three parts; the basic structure, the electrical circuits and the textile-based cover (see Fig. 1).

The basic structure, first layer, was made in MDF, Medium Density Fibreboard, which served as the support system for the electrical circuits. It was imperative that the circuits were fixed in a permanent position to prevent damaging or loose bits that could easily be swallowed. The second layer is a small frame like structure used to create distance between the LED's and the fabric layer, which is the third layer.

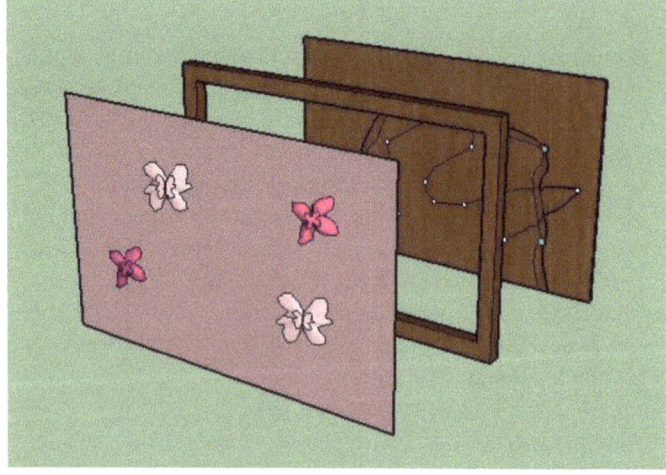

Fig. 1. Prototype mock-up

The electronics used an Arduino Uno microcontroller, WS2813 individually addressable LEDs, and a sound sensor board (see Fig. 2). For movement detection, a Velostat-based force sensor was used. Velostat is a piezoresistive polymer sheet with which it is straightforward to construct simple force sensors. For motion detection, the Arduino firmware monitors the force signal to detect variations.

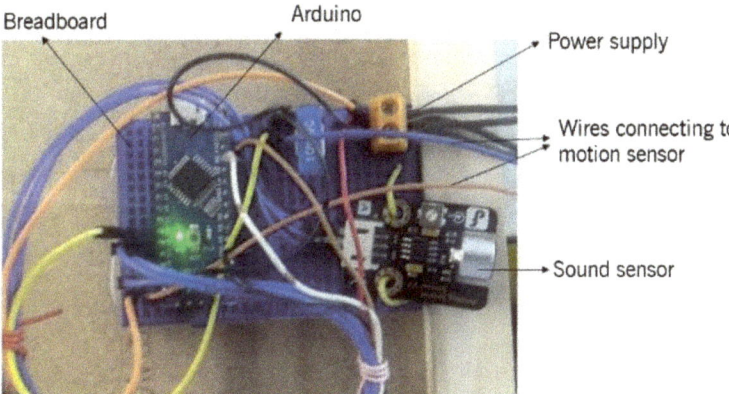

Fig. 2. Electrical components

Besides the Velostat layer, two conductive fabric layers are necessary to serve as electrodes, one at each side of the Velostat. Connections were hand sewn with conductive thread in such a way that the electrodes would not touch directly (see Fig. 3a).

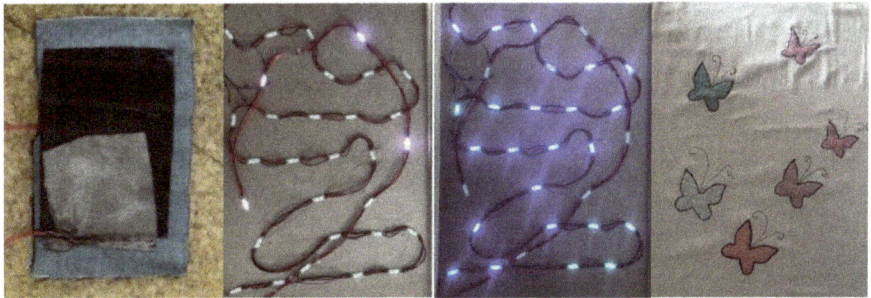

Fig. 3. From left to right (a) Force sensor, (b) Illuminated circuit activated by the sound sensor, (c) Illuminated circuit activated by the force sensor, (d) Front view of the panel

The assembled Velostat piece was covered with a polyester fabric and placed in the child's mattress to detect movement. The sound sensor was placed in a fixed position inside the panel in order to detect the child crying.

The LEDs are attached to the MDF panel and arranged according to the butterfly and star designs. 35 LEDs were used in total, where 6 of them represent the butterflies, activated by sound (see Fig. 3b) and the remaining 29 represent the starry sky, activated movement (see Fig. 3c). The LEDs are connected to the Arduino microprocessor that is attached to a breadboard located in the bottom right corner of the panel.

The textile layer is composed of a raw cotton base where butterfly designs were embroidered. The butterflies are composed of polyester wadding and Trevira CS, 100% polyester (see Fig. 3d). This fabric was developed by Philips® for lighting panels [6].

The fabric possesses anti-flame finishing and light diffusing characteristics essential in meeting the objectives and specifications of this project.

The wadding used is also 100% polyester and serves as a filling, aiding in the diffusion of light throughout the butterfly design (see Fig. 4).

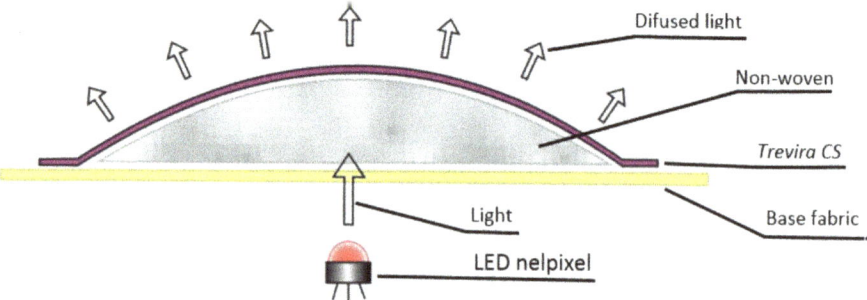

Fig. 4. Cross-sectional view of the butterfly design

The butterflies are illuminated assuming the colour of the fabric behind the white light emitted by the LEDs. Two butterflies light up at a time, from the bottom of the panel to the top, to create the impression of movement and the butterflies flying away. The stars light up in a random way, considering both the LEDs that are lit, as well as the colour in which they are lit, emitting a slow light that fades in and out to create the impression of a starry sky.

The outer case, where the butterfly designs are sewn on, is fully composed of textile and thread and held in place in the panel by elastic ribbon fixed around it. This allows for an easy removal, easing cleaning. The case covers and protects the electric circuits and allows easy access to the components in case any maintenance is necessary.

The panel is to be firmly wall mounted out of the child's reach, as the interaction is solely visual.

4   Results and Discussion

By detecting the sound of a child's cry and instantly lighting up the butterflies as a form of entertainment, as well as reacting to the baby's movement in bed, the initial intent of interaction of this project was achieved (see Fig. 5).

Fig. 5. Lighting effects created by the working prototype.

Even though the prototype uses wires to connect to the force sensor, which was identified during the research as something to avoid, this should not happen in the final product. The prototype was intended to showcase the idea through a working model.

For the actual product, wireless technology would be used, such as Bluetooth or Wi-Fi, which would relive the panel of the exposed wiring, enhancing safety. Furthermore, the panel was subjected to tests to analyse its functions and how both children and their parents reacted. At first, the lighting panel was tested by a five-year-old girl to test the sound system (Fig. 5).

The panel was the exposed during Gnoc Noc exhibition in Guimarães (2017), in a small room with dim lighting. For the exhibition, the Velostat sensor was placed in a pillow that the children could press and instead of crying we asked the audience to clap (Fig. 6).

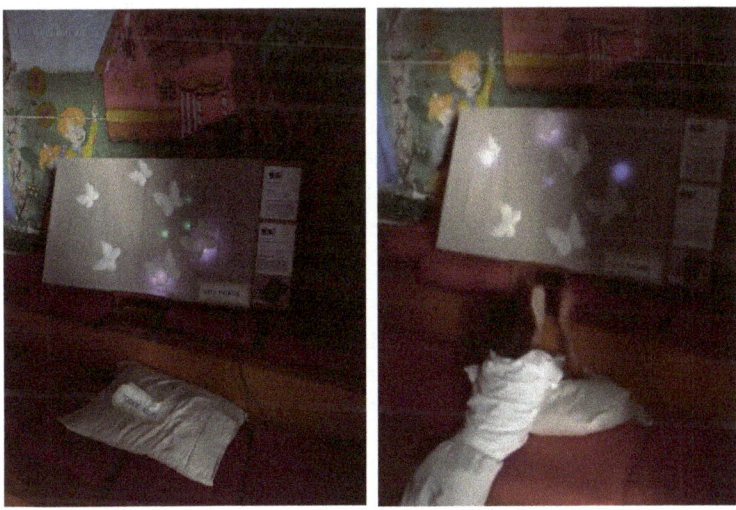

Fig. 6. Noc Noc exhibition in Guimarães.

During Gnoc Noc the product was tested by children between the ages of 4–10. The event lasted two days and on average 25 children per day tested the panel for 8–12 min each.

The results achieved were satisfactory as both children and parents demonstrated great enthusiasm when interacting with the panel.

5   Conclusions

In this project, using methodologies from both design and technology areas, it was possible to develop a product with practical and commercial interest. Using knowledge from multiple areas of study, creativity and innovation, a product that can enhance the home environment and everyday family life was developed.

We believe this product to be more than a simple decoration artefact due to its functionality. It became more than a simple piece of decoration, as it is capable of identifying sound and movement, and translating them into aesthetically pleasing luminous effects, presenting itself differently each time, through either a shift in hue or the lighting sequence.

Once complete, the prototype was subjected to testing which leads us to conclusions regarding how the lighting panel could be improved. The most obvious improvement would be the inclusion of wireless communication. This would not provide a safer product, considering the link between force sensor and controller. Moreover, a mobile phone app could allow the panel to be remotely controlled, or even to emit alerts when the child wakes up or cries.

There is also the need to identify or develop a different fabric to use other than the Trevira CS, due to its cost. The idea would be to reduce costs of production without compromising the function, aesthetics or safety of the product.

Nevertheless, the results were very satisfactory. Further development involving teamwork between designers and technologists can improve and enhance this product and certainly lead to ideas for new products in which the synergies between these two areas can be fully explored.

Acknowledgements. This work is financed by FEDER funds through the Competitivity Factors Operational Programme - COMPETE and by national funds through FCT – Foundation for Science and Technology within the scope of the project POCI-01-0145-FEDER-007136.

References

1. Oliveira, K.: Entenda como funciona o sono do bebê (2015). https://pediatriadescomplicada.com/2015/03/16/entenda-como-funciona-o-sono-do-bebe. Accessed 16 Nov 2016
2. Kazazian, T.: Haverá a idade das coisas leves – Design e Desenvolvimento Sustentável, p. 10. Senac, São Paulo (2005)
3. França, M.S., Spinillo, C.G.: Cores e formas na estamparia infantil: a criança como usuário participativo. In: 7 Congresso Brasileiro de Pesquisa e Desenvolvimento em Design, Curitiba (2016)

4. Silva, F.P., Nunes, V.A.V.: A questão da segurança no vestuário infantil. In: VII Colóquio de Moda, Maringá (2011)
5. APSEI: Segurança em casa. https://www.apsei.org.pt/areas-deatuacao/cidadao/seguranca-em-casa. Accessed 21 Mar 2017
6. Philips - Luminous textile with Kvadrat Soft Cells. http://www.light-ing.philips.com/main/products/luminous-textile. Accessed 03 Jan 2017

Design of a Long Sleeve T-Shirt with ECG and EMG for Athletes and Rehabilitation Patients

André Paiva[1] , André Catarino[1] , Hélder Carvalho[1($\boxtimes$)] ,
Octavian Postolache[2] , Gabriela Postolache[3] ,
and Fernando Ferreira[1]

[1] Centre for Textile Science and Technology, University of Minho, Guimarães,
Portugal
dmpaiva.s@gmail.com,
{whiteman,helder,fnunes}@det.uminho.pt
[2] Institute of Telecommunications, Instituto Superior Técnico, Lisbon, Portugal
octavian.postolache@gmail.com
[3] Universidade Atlântica, Oeiras, Portugal
gabrielap@uatlantica.pt

Abstract. Considering the importance of strength training in sports injury prevention and rehabilitation, an e-textile system that assists the athlete's training and rehabilitation was designed. This paper reports the construction of a prototype of a T-shirt with embedded textile electrodes to monitor cardiac and muscle activity. Signal tests were conducted and are presented. It is the authors' intention to study and integrate other textile sensors that may provide useful information to the user.

Keywords: e-textile · Smart clothing · Wearables · Strength training
Rehabilitation

1 Introduction

Physical activity has proven benefits for the prevention and treatment of health diseases [1], and contributes to prevent sports injuries [2]. Strength training exercises are often prescribed for athletes training, prevention and rehabilitation [3]. Studies have shown positive effects of physical activities, for instance, on people with osteoarthritis of the knee [4], hamstring injuries [5] and low back pain [6], and it has been proved that strength training can reduce the number of sports injuries to one-third [2].

One of the issues related with athletes' injuries is early detection. Athletes may get injured and keep it secret, fearing discrimination, which means that some injuries may seem to have been developed in the weight room, but it most likely occurred in the field and only worsened in the weight room [6].

A system that provides EMG data in real time may help detecting the existing injury and the athlete may have an adequate intervention in due time. In the present study, a system able to collect biometric data in real-time, for sports injury prevention

© Springer International Publishing AG, part of Springer Nature 2019
J. Machado et al. (Eds.): HELIX 2018, LNEE 505, pp. 244–250, 2019.
https://doi.org/10.1007/978-3-319-91334-6_34

and rehabilitation, is proposed. In this paper, the design and development of a proto-type of a T-shirt with knitted electrodes for electromyography (EMG) and electro-cardiography (ECG), and respective preliminary tests are reported.

2   Related Work

ECG is used to monitor cardiac activity, while EMG relates to muscle activity, both important parameters to be measured during rehabilitation [7]. In both techniques, surface electrodes can be used. Textile electrodes have been widely reported in liter-ature. Weaving [8], knitting [9] and embroidering [10] with conductive yarns represent few examples of attempts to replace conventional electrodes with the textile counter-part. In a previous study, it has been suggested that tuck loop-based flat knitted elec-trodes provide better response than the float loop-based counterpart [11]. Capacitive textile electrodes that do not need to be in contact with the skin for ECG [12] and EMG [13] have also been reported.

In the design and commercialization of clothes and garment accessories for humans, biomedical system has nowadays a great importance [14], and several examples proved this trend. Szczęsna et al. [15] developed a prototype of a waistband with textile electrodes for monitoring of cardiac activity. The T-shirt presented by Lage et al. [16] includes seamless knitted electrodes for ECG. Manero et al. [17] used embroidery to embed EMG electrodes in running leggings, to monitor quadriceps muscle activity.

3   Materials and Methods

A long sleeve knitted T-shirt with ECG and EMG knitted electrodes was designed and produced. It comprises four ECG electrodes – LA and RA under the armpit (Lead I), in the chest horizontal line, and the RL and LL in the waistline (Lead II), between the front center and the lateral – and EMG electrodes for *biceps brachii*, *extensor carpi radiali longus*, and *flexor carpi ulnaris*.

The knitted fabric was produced in a seamless knitting machine, which produced knitted fabrics in the form of a tube that can have the size of the torso, for instance, eliminating the need of side seams. For the sleeves, a separated tube was produced and the sleeves were then cut and sewed. The T-shirt was produced in two layers of fabric, the inner layer containing the electrodes and connections and the outer layer being totally black, giving the looks of a conventional garment and protecting the conductive layer. Polyamide/elastane (PA/EA) composes the nonconductive part and a silver coated PA (ELITEX 110/f34) builds the conductive areas.

Garment patterns were cut according to the desired shape and taking into consid-eration the position of the electrodes, and the pieces were then put together to give shape to the T-shirt. A *Shimmer3 ExG* device was connected to the T-shirt, which is an electronic device that allows the measurement of physiological signals, namely, 4-lead ECG (LA-RA, LL-RA, LL-LA and Vx-WCT) and 2-channel EMG. Metallic snaps, compatible with *Shimmer*'s channels, were inserted in the extremities of the connections. A zipper was placed in the outer layer, to give access to the connection points, therefore, being able to connect the electrodes to the *Shimmer* electronic device. To isolate the circuits from the skin, knitted fabric was cut in shapes appropriate to cover the conductive areas (except the electrodes, which need to be in contact with the skin). This was done to produce a prototype, for a final version it is intended to use a thermoplastic polyurethane film to cover the circuits.

Preliminary tests were carried out with dry and wet electrodes, as well as with and without pressure on the electrodes, thus comprehending four different conditions. An elastic band was used to apply pressure on the electrodes. No skin preparation was carried out. The ECG tests were made with a subject seated in a static position, and simulating some strength training exercises (without the weights) that would involve the movement of the arms and shoulder, in order to understand the effect that moving the arms has on the ECG signal. The selected exercises were (i) *high-pulley curls*, and (ii) *seated front presses*, performed according to the book of Delavier [18]. For sEMG tests, only the biceps electrodes were tested, which were carried out with the subject performing curls with a 5 kg dumbbell. The *Shimmer* device was set for frequency = 512 Hz (ECG) and 1024 Hz (sEMG), gain = 4 (ECG) and 12 (sEMG), and resolution = 24 bits (both).

4   Results

4.1   The Prototype

The prototype's dimensions and patterns were specified and the garment was produced using seamless knitting technologies by a knitwear manufacturer. It was soon obvious that a great difficulty exists in this type of material (knitted, elastic) to meet design specifications. The produced sleeve didn't fit the required measurements and it was not possible to have the flexor and extensor electrodes in place. The size of the prototyped sleeve is also larger than the one initially designed, which makes a loose fit, instead of the intended skinny fit that keeps the electrodes in contact with the body. The sleeve was then cut according to the desired measures, only with the biceps electrodes in the correct position.

Figure 1 shows pictures of the prototyped T-shirt. It can be seen from Fig. 1c that the electrodes and connection tracks (covered by jersey fabric) are in the inside of the T-shirt, being invisible at the outer side (Fig. 1b and d). After dressing the garment (Fig. 1d), it was noticed that the sleeve is too tight over the *brachyorradialis* muscle and the collar is too high, creating discomfort.

4.2   Performance Tests

ECG Tests. Figure 2 shows the ECG response on the LA-RA lead with the subject seated and still, for dry and wet electrodes. As can be observed, dry electrodes (Fig. 2a and b) produce a lot of noise and when applying pressure only a small signal is observed (Fig. 2c), with significant low-frequency fluctuation. When the electrodes are wetted with water (Fig. 2c and d) the signal improves notably, even without pressure on the electrode (Fig. 2c).

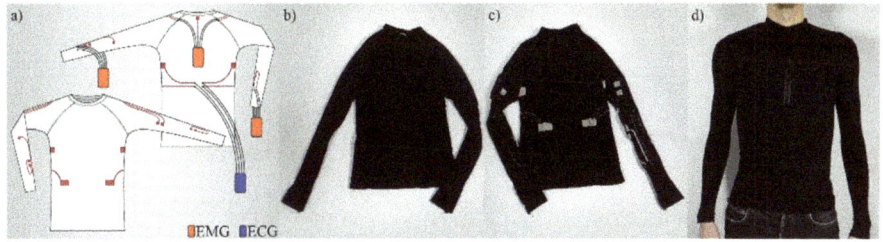

Fig. 1. Prototype of the T-shirt

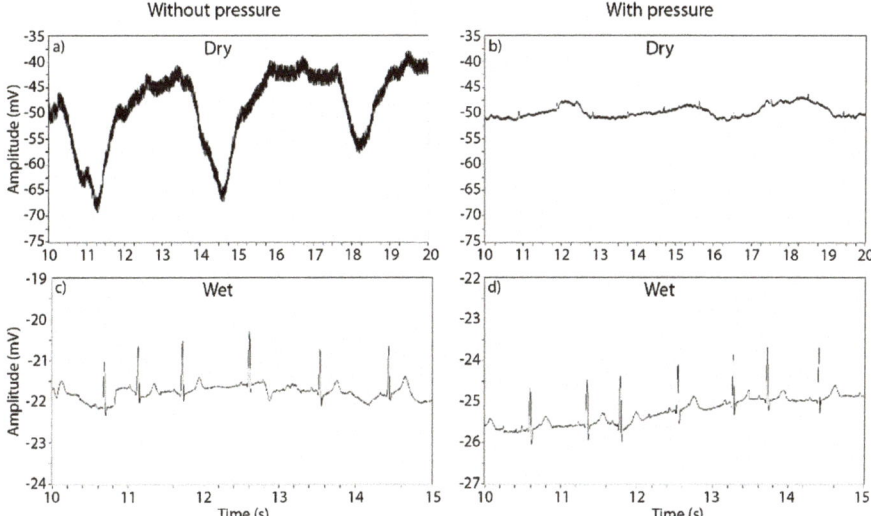

Fig. 2. ECG signals with the person seated, derivation LL-RA

The following tests were conducted with wet electrodes. Figure 3 shows the ECG signal for exercise (i). The signal is smaller without pressure applied, near the noise level, while when applying pressure, cardiac signals are better distinguished from noise.

Figure 4 shows the signal when the subject is performing exercise (ii), with and without pressure (Fig. 4b and a, respectively). It can be noticed that, with pressure, the signal is more stable than its counterpart. Differences in the position of the arms can also be noticed. In Fig. 4a, the signal can be divided into three exercise moments: (1) arms down, (2) arms moving up, (3) arms up and (4) arms moving down. Moment (3) shows more noise than moment (1). In Fig. 4b, these moments are barely distinguished – (1) arms down, and (2) arms up – but the movement of the arm doesn't produced the peak seen in Fig. 4a.

In the first tests, using this specific T-shirt design, sEMG signals were very weak.

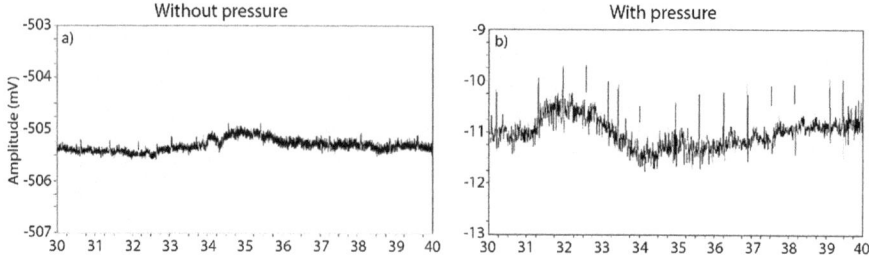

Fig. 3. ECG signal for exercise (i), LA-RA derivation of ECG with more and less contact of electrodes with skin

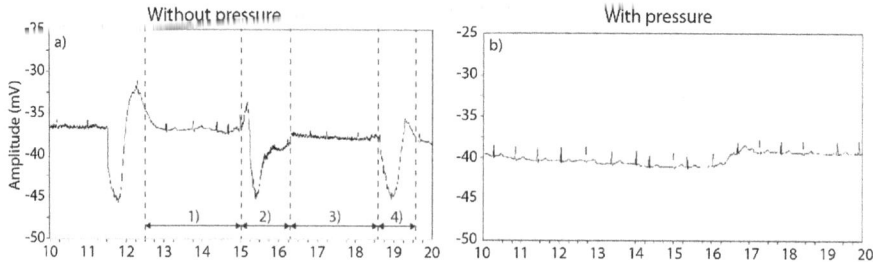

Fig. 4. ECG signal for exercise (ii), derivation LA-RA

5   Discussion

Wet electrodes showed better results than dry electrodes, as expected, since skin-electrode impedance is much lower, just as the with the conductive gels used in conventional exams. This is not an issue if designing sportswear, given that sweat will wet the electrodes. Before starting exercising, the athlete may pulverize the electrodes with water, but this may not be the case in an everyday scenario. Although we may produce some sweat during the day, the skin is usually dry. For a scenario like this (i.e.: monitoring of home patients), the system must operate correctly with dry electrodes.

As stated before, the electrode pattern was not produced with the required precision, but it is possible to produce the electrode pattern with precision. Seamless knitting machines use specific programs to draw knitting pattern, in which measures aren't defined using metric measures (such as cm or inch), but loops. Outside these programs, designers, pattern cutter and tailors work with metric measures, such as cm or inch. Which means that, before drawing in the specified program, measures in loops need to be taken. The question is: how does one do it? A possible proposal is the following: (1) design and produce the electrode patterns and clothing with the desired measures; (2) produce the knitted fabric with the desired specifications (i.e.: elasticity); (3) take the contexture; (4) convert metrics into loops; (5) use the loops to draw the desired patterns in the seamless program.

The noise present in the ECG signal when exercising results from the shirt being stretched off and losing contact with the body under the armpit, when the arms are lifted. Since the LA and RA electrodes are too close to the armpit, they lose contact with the skin during movements, which means that more favorable positions for the electrodes will need to be found.

Several effects can justify the weak sEMG signal, such as low contact between electrodes and the skin (either the muscle electrodes or the reference electrode), the knitted structure or ambient noise. At least two factors can be excluded: (1) the incorrect distance between the electrodes, since they have the required distance; (2) the conductivity of the tracks, since the resistance between electrode and connection point is less than 10 ohms. Considering that previous studies had resulted in functional EMG sensing [11], further experiments will be carried out to improve this feature.

6 Conclusion

A way of integrating electronic systems into clothing, hidden inside a long sleeve knitted shirt, therefore maintaining the aesthetics of a conventional garment, has been demonstrated. Smart clothing, which mixes apparel and technology, needs to be designed considering not only functional requirements and operability, but also the way the individual expresses himself in society.

Improvements need to be done in order to obtain a good sEMG response, as well as to improve the contact of the ECG electrodes, making sure that the movement artefacts do not affect the signal. This should be achieved, ideally, with the electrodes in dry state. The solution may include other textile structures, materials and ways to create pressure.

Other textile sensors can also be of interest in a garment that is designed to support physical training and rehabilitation, such as temperature, breathing or bending sensors. This shall also be studied and embedded in a future prototype.

Acknowledgements. This work was supported by Project "Deus ex Machina", NORTE-01-0145-FEDER-000026, funded by CCDRN, through Sistema de Apoio à Investigação Científica e Tecnológica (Projetos Estruturados I&D&I) of Programa Operacional Regional do Norte, Portugal 2020 and by project TailorPhy, PTDC/DTP-DES/6776/2014.

References

1. O'Shea, S.D., Taylor, N.F., Paratz, J.: Peripheral muscle strength training in COPD. Chest **126**(3), 903–914 (2004)
2. Lanersen, J.B., Bertelson, D.M., Anderson, L.B.: The effectiveness of exercise interventions to prevent sports injuries: a systematic review and meta-analysis of randomized controlled trials. J. Sports Med. **48**, 871–877 (2014)
3. Carpinalli, R.N., Otto, R.M.: Strength training: single versus multiple sets. Sports Med. **2**, 73–84 (1998)

4. Lange, A.K., Vanwanseele, B., Singh, M.A.F.: Strength training for treatment of osteoarthritis of the knee: a systematic review. Arthritis Rheum. (Arthritis Care & Research) **59**(10), 1488–1494 (2008)
5. Erickson, L.N., Sherry, M.A.: Rehabilitation and return to sport after hamstring strain injury. J. Sport Health Sci. **6**, 262–270 (2017)
6. Kraemer, W.J., Dziados, J.: Medical aspects and administrative concerns in strength training. In: Kraemer, W.J., Keijo, H. (eds.) Handbook of Sports Medicine and Science: Strength Training for Sport, pp. 163–175. Blackwell Science Ltd., Oxford (2002)
7. Patel, S., Park, H., Bonato, P., Chan, L., Rodgers, M.: A review of wearable sensors and systems with application in rehabilitation. J. NeuroEng. Rehab. **9**(21), 1–17 (2012)
8. Pylatiuk, C., Müller-Riederer, M., Kargov, A., Schulz, S., Schill, O., Reischl, M., Bretthauer, G.: Comparison of surface EMG monitoring electrodes for long-term use in rehabilitation device control. In: IEEE International Conference on Rehabilitation Robotics, Kyoto, pp. 300–304 (2009)
9. Carvalho, H., Catarino, A., Rocha, A., Postolache, O.. Health monitoring using textile sensors and electrodes: an overview and integration of technologies. In: IEEE International Symposium on Medical Measurements and Applications (MeMeA), Lisbon (2014)
10. Puurtinen, M., Komulainen, S., Kauppinen, P., Malmivuo, J., Hyttinen, J.: Measurement of noise and impedance of dry and wet textile electrodes, and textile electrodes with hydrogel. In: Proceedings of the Annual International Conference of 28th IEEE Engineering of Medicine and Biology Society, pp. 6012–6015 (2006)
11. Paiva, A., Catarino, A., Carvalho, H., Postolache, O., Postolache, G.: Development of dry textile electrodes for electromyography. In: 9th International Conference on Sensing Technology (ICST), Auckland, pp. 447–451 (2015)
12. Boehm, J., Yu, X., Neu, W., Leonhardt, S., Teichmand, D.: A novel 12-lead ECG T-shirt with active electrodes. Electronics **5**(75), 1–15 (2014)
13. Linz, T., Gourmelon, L., Langereis, G.: Contactless EMG sensors embroidered onto textile. In: Linz, T., Gourmelon, L., Langereis, G. (eds.) 4th International Workshop on Wearable and Implantable Body Sensor Network (BSN 2007). IFMBE Proceedings, vol. 13. pp. 29–34. Springer, Berlin (2007)
14. Toda, M., Junichi, A., Sakurawa, S., Yanagihara, K., Kunita, M., Iwata, K.: Wearable biomedical monitoring system using textile. In: 10th IEEE International Symposium on Wearable Computers, Montreaux (2006)
15. Szczęsna, A., Nowak, A., Grabier, P., Rozentryt, P., Wojciechowska, M.: Wearable sensor vest design study for vital biomedical engineering. In: Grik, M., Tkacz, E., Paszenda, Z., Piętka, E. (eds.) Innovation in Biomedical Eng, vol. 526, pp. 330–337. Springer, Cham (2017)
16. Lage, J., Catarino, A.P., Carvalho, H., Rocha, A.M.: Smart shirt with embedded vital sign moisture sensing. In: SPWID 2015: The 1st International Conference on Smart Portable, Wearable, Implantable and Disability-Oriented Devices and System, Lisbon, pp. 25–30 (2015)
17. Manero, R.B.R., Shafti, A., Michael, B., Grewal, J., Fernandez, J.L.R, Althoefer, K.: Wearable embroidered muscle activity sensing device for the human upper leg. In: 2016 IEEE 38th Annual International Conference on the Engineering in Medicine and Biology Society (EMBC), Orlando, pp. 6062–6065 (2016)
18. Delavier, F.: Strength training anatomy. In: Human Kinetics, Montreal (2006)

Real-Time Data Movements Acquisition of Taekwondo Athletes: First Insights

Pedro Cunha[1(✉)], Vítor Carvalho[2,3(✉)], and Filomena Soares[1(✉)]

[1] Department of Industrial Electronics, Algoritmi Research Centre, Minho University, Braga, Portugal
id5514@alunos.uminho.pt, fsoares@dei.uminho.pt
[2] School of Technology, IPCA, Barcelos, Portugal
vcarvalho@ipca.pt
[3] Algoritmi Research Centre, University of Minho, Braga, Portugal

Abstract. In the last two decades, Taekwondo is a combat sport who has gained popularity, being inducted as an official Olympic Sport since 2000. Lately, more than 4500 federated athletes existing in Portugal have achieved excellent results in national and international competitions recognized by the International Olympic Committee, Portugal Olympic Committee, European Taekwondo Union and World Taekwondo Federation. In this paper, it was performed a study to establish the reliability and accuracy of the Microsoft Kinect 2 in real-time image data collection of the Taekwondo athletes movements. This work will reinforce the research on motion-analysis, targeting the improvement of the athlete's performance and the establishment of a new level of training and competition, contributing to sport modernization and development.

Keywords: Kinect · Martial arts · Motion analysis · Real-time Taekwondo

1 Introduction

Ubiquitous computing has been suffering a fast-growing with applications in many areas. One of the fields where this growth is most visible is in Sports, where it is used from the training practices to automatic score and refereeing [1]. This reality makes possible to monitor athletes in various aspects, for example, by measuring accelerations and forces, recognizing and tracking movements, quantifying the effort applied and many other functionalities. Grand Master David Chung Sun Yong introduced Taekwondo in Portugal in 1974 [2]. Since then, it has gained popularity over the last two decades being introduced as an Olympic Sport in the year 2000. The number of federated athletes has been increasing, reaching currently more than 4500 athletes in Portugal [3]. Besides the popularity of the sport, training methods have not accompanied the technological development. The current methods used by coaches for evaluating the athlete's performance are mostly made manually. This process becomes time consuming and it is not efficient, since the coach must analyze each team trainee

© Springer International Publishing AG, part of Springer Nature 2019
J. Machado et al. (Eds.): HELIX 2018, LNEE 505, pp. 251–258, 2019.
https://doi.org/10.1007/978-3-319-91334-6_35

video at a time to obtain relevant feedback that could be used to help athletes to improve their performance.

With the evolution of technology in the field of the image acquisition and in data processing power, hardware solutions that fit both features have come up. One of these examples was the Microsoft Kinect 1 and later the updated Microsoft Kinect 2 [4]. Standing out for being a portable, inexpensive and markers free when to determine anatomical landmarks.

Considering that, in this paper the main objective is to understand if it is possible to establish the Microsoft Kinect 2 as a reliable and efficient tool to acquire valid real-time data images from the athletes' movements during the practice of Taekwondo martial art in the training environment

This paper is organized in five sections. In Sect. 2, it is presented the state of the art; in Sect. 3 is described the planned methodology; in Sect. 4 data analysis is presented; in Sect. 5 the final remarks are enunciated.

2   State of the Art

Evaluating the performance of the athletes is a constant challenge for coaches at any level of sport. In case of Taekwondo martial art there is a higher difficulty level when it is necessary to evaluate the evolution of the athlete's performance. For this, the coach uses the visual verification of the movement's execution and the corresponding speed. Motion analysis is a key issue for trainers as they can improve technical skills by correcting the trainee's motion of the body when performing the movement.

The need for motion analysis extends to several areas of application, such as health, work and industry, among others. Regarding health, Mobini et al. [5] presented a study of a hand acceleration measurement for rehabilitation applications, using Microsoft Kinect. Extracting velocity and acceleration from Kinect position data, evaluating the accuracy of the results. Applying for different methods, experimentally comparing the performance results with clinical optical system, Vicon, and inertial measurement unit, Xsens, as references. Regarding sports, Hachaj et al. [6] evaluated the effectiveness of Kinect and Kinect 2 for specialized actions karate techniques, namely Oyama. The study indicates Microsoft Kinect 2 as a more accurate calculation of legs joints positions when compared with its predecessor.

The accuracy of the data collection has a major importance in the validation of results. Taking that into account, in [7–9] the researchers studied the Kinect 3D map accuracy and its capacity in estimating the position of joints when using its software tool [10–13]. Still in the study carried out in [13], the Kinect accuracy in estimating static joint positions is about 4–7 cm.

Considering the real-time approach Pinto et al. [14] presented a system that uses Microsoft Kinect and image processing to recognize and record in real-time the number of occurrences of a movement in Taekwondo training environment. The recognition of the movements was made through the calculation of the angles between human body joints and comparing them with the correct values of each movement previously saved in a database.

3 Methodology

The goal of the national Portuguese Taekwondo team coach is to bring the team to the upper positions. This effort is perceptible by the outstanding results obtained in European and World competitions, also as in the Olympic Games [15, 16].

The main goal of the presented study was to compare and verify the accuracy and reliability of the Microsoft Kinect 2 compared to a well-established system. To carry out this study we opted to perform a transversal action research with a qualitative method approach.

The use of valid data is important for the accuracy of the developed system. Considering that, as initial stage it was decided to collect data of the movements of the Taekwondo athletes using a tested commercial hardware and software that served as reference. According to the requirements of the data to be collected and the normal conditions of its acquisition, the XSens MVN Link was chosen. Following this idea, a data collection session of Taekwondo athlete movements using Xsens MVN Link, MVN Analyze [17] and Microsoft Kinect 2 was performed at the Digital Games Lab in Polytechnic Institute of Cávado and Ave (IPCA), Barcelos, Portugal. The data collection session had the participation of a Braga Taekwondo athlete who made several movements used in the practice of their sport. Five repetitions of each of twenty leg and arm movements were performed.

The MVN Link hardware consist in a lycra suit with 17 wireless motion trackers attached in predefined positions that allow full-body motion measurements. The data from the motion trackers is sent to a computer where the MVN Analyze software provides an interface for recording and playback the data using a scalable biomechanical model. It displays an avatar in real-time of the MVN Link performed movements.

The MVN Analyze software divides the human body into twenty-three segments (Fig. 1) and the analysis of movement is performed by reading and calculating the position and orientation of each one of the segments [18], providing 120 values per second for every segment.

The other option for data acquisition was the Microsoft Kinect 2 sensor along with SDK 2.0. The Microsoft Kinect V2 hardware is composed by a depth sensor, an infrared sensor, four microphones and a 1080 p camera. It allows to collect body and hands position and orientation data, in real-time, in a three-dimensional environment. [19–21].

The data provided are relative to the points relating to the twenty five joints of the human body recognized by the Kinect SDK (Fig. 2). This version of the Kinect provides 30 values per second for any one of the joints, the same value of the integrated camera frame rate.

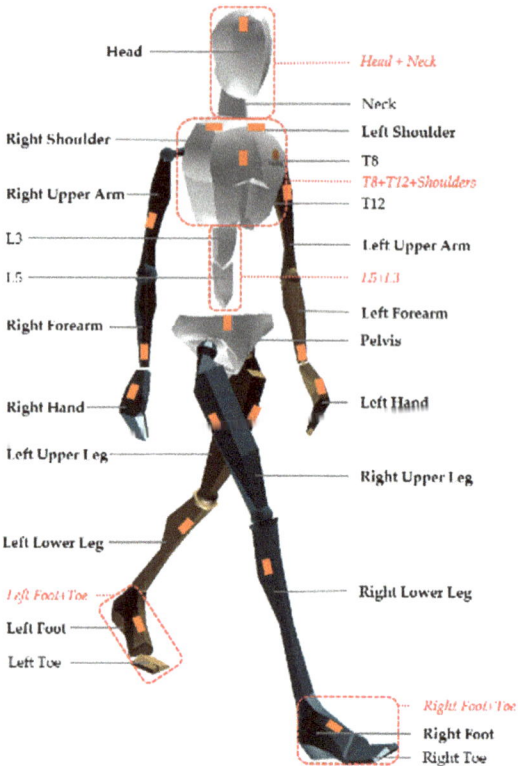

Fig. 1. MVN Analyze human body segments division [18]

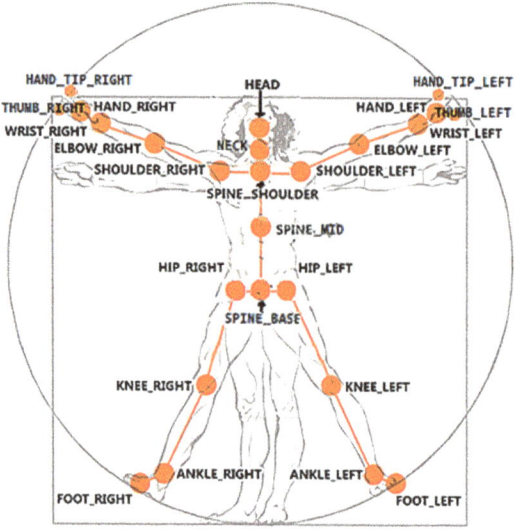

Fig. 2. Microsoft Kinect 2 skeleton positions relative to the human body

4   Data Analysis

In the evaluation of the Taekwondo athletes' performance, both technique and velocity are needed to determine if there is an evolution and improvement comparing with previous evaluations. In this paper, the study carried out intend to define the method and the hardware able to carry out this task effectively and at low cost.

The data structure of both devices, although similar in number of elements, is different because MVN Analyze reads the different segments of the human body and Microsoft Kinect 2 reads the different joints of the human body. Considering this, to obtain similar readings the hand was chosen because of its position values proximity between both equipment.

Due to logistic reasons (space and electric power sources localization in the set-up room), the computers to run the MVN Analyze Software and to acquire the data from Microsoft Kinect 2 were positioned in different places. This fact along with the different data reading of the position coordinates in a tridimensional environment lead to different position reading values.

The data collected with MVN Analyze and Microsoft Kinect 2 were analyzed using MATLAB software. The movement chosen to perform the analysis of the data, in order to compare both equipment reading values was the Chi Jireugi (Lift Punch), an arm movement.

The individual values of the three axis (Fig. 3) as well as the three-dimensional movement path of the right hand (Fig. 4) were obtained. After analyzing the data and by interpreting the plots obtained, either for XSens MVN Analyze data or for data from Microsoft Kinect 2 was possible to realize that both have similar curves.

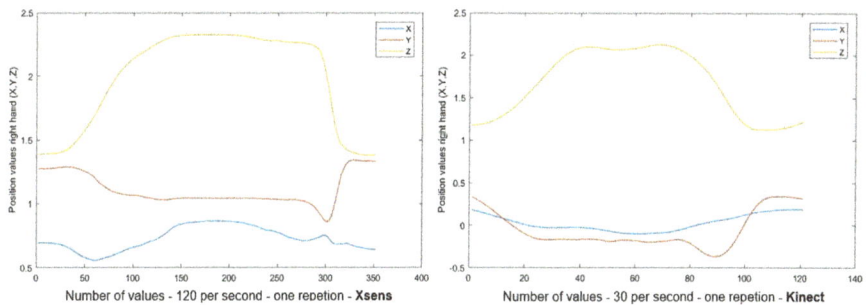

Fig. 3. XSens MVN Analyze and Kinect 2 three axes position values of right hand in one Chi Jireugi (Lift Punch) movement plot

When comparing the plots of three axis of both equipment (Fig. 3) the curves of each axes present similar reading values throughout the execution of the movement. This similarity is more perceptible in the Z and Y axes, where exist more noticeable displacement values of the right hand during the movement execution. Therefore, these will be the most relevant values to perform an analysis of the movements performed,

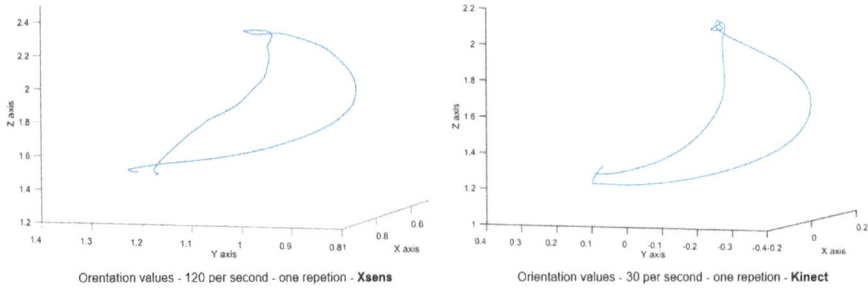

Fig. 4. XSens MVN Analyze and Kinect 2 three-dimensional position values of right hand in one Chi Jireugi (Lift Punch) movement plot

interpreting and identifying which are these movements, according to reference values collected previously.

In the three-dimensional plots comparison (Fig. 4), is presented the right hand trajectory during the of the Chi Jireugi movement performance, according to the values obtained by both equipment. In the plots, it is possible to realize that the data collected on both equipment have similar profiles resulting in similar curves. Since the beginning and along of the right hand movement, in both plots we can see the proximity of the trajectory of the movement.

5 Final Remarks

The accuracy of the data collected is of great importance for the correct performance of its interpretation. Just as the determination of an economically more affordable solution in the collection of these data and less invasive for the athlete is important for the development of a technology that can be used in the practice of Taekwondo.

This study aimed to verify the reliability and accuracy of Microsoft Kinect 2 in the collection of image data of the movements performed by Taekwondo athletes during training. The study consisted of a comparison between the Xsens MVN Analyze, a state-of-the-art system built and marketed for the acquisition of biomechanical data from the human body and the Microsoft Kinect 2, a motion sensor with similar features but at a more affordable cost.

There was collected data values from several movements used in the practice in the Taekwondo, being chose for analysis the Chi Jireugi movement. After the analysis of the data it was possible to verify that both equipment presented similar data collection values, resulting in similar interpretations of the data values for the same movement execution and trajectory.

Thus, it allows to conclude that Microsoft Kinect 2 fulfills the necessary requirements for the acquisition of data in real-time of the movements performed by the athletes of Taekwondo in the training environment.

Acknowledgments. The authors would like also to express their acknowledgments to COM-PETE: POCI-01-0145-FEDER-007043 and FCT – Fundação para a Ciência e Tecnologia within the Project Scope: UID/CEC/00319/2013. Pedro Cunha thanks FCT for PhD scholarship SFRH/BD/121994/2016.

References

1. Baca, A., Dabnichki, P., Heller, M., Kornfeind, P.: Ubiquitous computing in sports: a review and analysis. J. Sports Sci. **27**(12), 1335–1346 (2009)
2. E. H. A. Costa: História do Taekwondo. http://www.dojangabiliocosta.com/home. Accessed Feb 2018
3. Instituto Português do Desporto e Juventude, I.P. http://www.idesporto.pt/conteudo.aspx?id=103. Accessed Feb 2018
4. Microsoft | Developer Network: Kinect para Windows. https://developer.microsoft.com/pt-pt/windows/kinect. Accessed Feb 2018
5. Mobini, A., Behzadipour, S., Foumani, M.S.: Hand acceleration measurement by Kinect for rehabilitation applications. Sci. Iran. **24**(1), 191–201 (2017)
6. Hachaj, T., Ogiela, M.R., Koptyra, K.: Effectiveness comparison of Kinect and Kinect 2 for recognition of Oyama karate techniques. In: Proceedings of 2015 18th International Conference on Network-Based Information Systems, NBiS 2015, pp. 332–337 (2015)
7. Khoshelham, K.: Accuracy analysis of kinect depth data. In: ISPRS Workshop Laser Scanning (2011)
8. Dutta, T.: Evaluation of the KinectTM sensor for 3-D kinematic measurement in the workplace. Appl. Ergon. **43**(4), 645–649 (2012)
9. Smisek, J., Jancosek, M., Pajdla, T.: 3D with Kinect. In: Consumer Depth Cameras for Computer Vision, pp. 3–25. Springer (2013)
10. Pfister, A., West, A.M., Bronner, S., Noah, J.A.: Comparative abilities of Microsoft Kinect and Vicon 3D motion capture for gait analysis. J. Med. Eng. Technol. **38**(5), 274–280 (2014)
11. Livingston, M.A., Sebastian, J., Ai, Z., Decker, J.W.: Performance measurements for the Microsoft Kinect skeleton. In: 2012 IEEE Virtual Reality Workshops (VR). IEEE (2012)
12. Obdrzalek, S., Kurillo, G., Ofli, F, et al.,: Accuracy and robustness of Kinect pose estimation in the context of coaching of elderly population. In: 2012 Annual International Conference of the IEEE on Engineering in Medicine and Biology Society (EMBC). IEEE (2012)
13. Mobini, A., Behzadipour, S., Saadat Foumani, M.: Accuracy of Kinect's skeleton tracking for upper body rehabilitation applications. Disabil. Rehabil. Assistive Technol. **9**(4), 344–352 (2014)
14. Pinto, T., Faria, E., Cunha, P., Soares, F., Carvalho, V., Carvalho, H.: Recording of occurrences through image processing in Taekwondo Training: first insights. In: Tavares, J., Natal Jorge, R. (eds.) VipIMAGE 2017, ECCOMAS 2017. Lecture Notes in Computational Vision and Biomechanics, vol. 27. Springer (2017)
15. Federação Portuguesa de Taekwondo: Resultados de excelência desportiva. http://fptkd.com/index.php/component/k2/item/308-resultados-de-excelencia-desportiva. Accessed Feb 2018
16. Comité Olímpico Portugal: Rui Bragança triunfa no Taekwondo. http://comiteolimpicoportugal.pt/rui-braganca-triunfa-no-taekwondo/. Accessed Feb 2018
17. XSENS: Xsens MVN. https://www.xsens.com/products/xsens-mvn-analyze/. Accessed Feb 2018
18. XSENS: Xsens MVN. https://www.xsens.com/software/mvn-analyze/. Accessed Feb 2018

19. Freedman, B., Shpunt, A., Machline, M., Arieli, Y.: Depth mapping using projected patterns, WO Patent 2,008,120,217 (2008)
20. Shpunt, A.: Depth mapping using multi-beam illumination, WO Patent 2,008,087,652 (2008)
21. Spektor, E., Mor, Z., Rais, D.: Integrated processor for 3D mapping, WO Patent 2,010,004,542 (2010)
22. Microsoft | Developer Network: JointType Enumeration. https://msdn.microsoft.com/en-us/library/microsoft.kinect.jointtype.aspx. Accessed Feb 2018

"Barty" - A Serious Game to Fight Childhood Obesity: First Insights

Fátima Gonçalves[1(✉)], Vítor Carvalho[1,3(✉)],
and Demétrio Matos[2(✉)]

[1] 2Ai-School of Technology,
Polytechnic Institute of Cávado and Ave, Barcelos, Portugal
al1481@alunos.ipca.pt, vcarvalho@ipca.pt
[2] School of Design: ID+,
Polytechnic Institute of Cávado and Ave, Barcelos, Portugal
dmatos@ipca.pt
[3] Algoritmi Research Centre, University of Minho, Guimarães, Portugal

Abstract. According to World Health Organization, childhood obesity, classified as the epidemic of the 21st century, is reaching alarming values all over the world. Mobile technologies, especially smartphones and tablets, are an important part of a children's life. Many studies believe that these technologies can be used in a healthy way, helping children to grow, so caregivers allow its use, on the hope that can help at their kid's development. With educational games kids learn better and faster, because they feel motivated. To contribute to the fight against this epidemic, a mobile Android application has been developed to educate and motivate children to achieve healthy lifestyles, as well as allowing caregivers to follow and be part of the process, giving them tools to contribute on the improvement on the life quality of the children.

Keywords: Obesity · Childhood · Serious game · Mobile · Feed day

1 Introduction

Family context is the biggest influencer in children's physical and psychosocial development, mainly because it's their first learning environment.

Children's food habits are modeled by parents, family members, friends and by external factors such as various food advertisements. Increase of excessive weight and obesity at childhood, led World Health Organization – WHO [1] to declare it as the epidemic of the century.

Studies indicate that Portugal is one of the countries affected by this problem and one of the ones that will progress negatively in the future. Children's bad food habits and lack of physical activities will influence their grown up life, creating a bigger problem for the generations to come [2].

Games can be used in a casual or serious way. Playful activities are one of the most effective methodological resources to commit children in learning. This is only possible because playing is inherent to them, constituting their main way of learning [3]. What distinguishes a serious game from the others is their educational background, carefully

© Springer International Publishing AG, part of Springer Nature 2019
J. Machado et al. (Eds.): HELIX 2018, LNEE 505, pp. 259–266, 2019.
https://doi.org/10.1007/978-3-319-91334-6_36

studied and structured so that fun is not its top priority. However, this doesn't mean that they are not capable of creating entertainment and get the attention of their target audience [4]. That said, a serious game is nothing less than a learning process delivered through entertainment.

Playing is the children's job. This will allow them social interaction, as well as development of their cognitive and motor skills. Parents are children's first teachers, guiding and stimulating them since young age [5]. Playing with them while smiling, expecting children to smile back, it's just one of the simplest examples that demonstrate small challenges kids face. This will enable them the development of human interaction which will be very useful during their growth.

This project intends to create a mobile app for Android devices, leading parents and their children to play together and make their relation stronger, evaluating the kid's food day through a moment of teaching and reflection from the game.

After installing the app user must record their age and gender, among other relevant data. With this, it will be possible to change the game's behavior, to achieve the best result for the children. One example of this is the estimated calories.

The game is composed by a series of daily challenges and has a mascot to help on the interaction with the children. This way, they will keep motivated to play every day, allowing them to consolidate their knowledge through progressive challenges [6]. With these daily tasks, it's possible to collect their food day, and through it, change and adapt the mascot behaviors and health status. By having better food habits, they will improve the mascot health and status. This analogy between the children's food day and the mascot status leads children to be more careful and aware of the short and long-term consequences of their food habits.

This paper is organized in 4 sections. Section 1, presents the introduction. Section 2, presents the state of the art. Section 3, presents the game. Section 4, presents final remarks.

2   State of the Art

To fight the epidemic of the century new approaches are needed. The video game industry, that generates more money than Hollywood productions, are much appreciated by children. Therefore, it is considered that games can be a good tool in the prevention or treatment of childhood obesity. Games for Health - G4H is a new area of research and development that uses game entertainment to achieve health related goals [7].

In 2013 the number of smartphones already represented more than half of the global mobile phone market [8]. In this way, the development of new applications for these devices offers an excellent opportunity to expand the access to information regarding the health of its user, promoting the implementation of health programs, especially among children and young people, who accept easily the new technologies. With a large market, is possible to use new approaches that can have a real-world impact. Combining the tendency of young people for new technologies and the need to encourage them to change their behavior this can be possible [9].

A serious game consists of a game that aims to educate through various ways, promoting learning as well as behavioural changes. They are a new and innovative way of educating, which is applied in many areas, such as health, helping children and adults to treat chronic diseases such as diabetes, asthma or cancer, or other psychotherapeutic treatments. The serious game on the web platform *DigesTower*, developed by Dias et al. [10], intends to study and evaluate the use of such a game.

Kids in school age are *DigesTower's* target audience. The game uses the human digestive system in the background, and is classified as the tower defence game. When the main character, Elise, is hungry, she heads to the freezer to choose anything to eat, and the game starts. The food chosen by the user enters the digestive tract at regular intervals, and the enzymes digest each food into the correct organ. In each step, there is a brief explanation of the specific digestion functions performed by each organ. At the end of the level the user receives the feedback of the health status and is informed of his score in the game.

As conclusions of the tests made on *DigesTower*, it was possible to verify that is a good strategy to change the children mindset, to have a healthier diet and lifestyle, with the goal of not being overweight or even obese.

To help children and young people, that are a vulnerable segment of the population, increasing the consumption of fruits, juices and vegetables - FJV, at home and outside and believing that their consumption reduces the risk of obesity and certain types of cancer, the web game *Squire's Quest I & II* were created: Implementation Intentions and Children's Fruit, Juice and Vegetable (FJV) Consumption in Baylor College of Medicine by Thompson et al. [11], in the state of Texas. Two studies were performed regarding the game, and had the following conditions: the participants must be attendants of the elementary school, need to speak and write in English, must have access to a computer and an associated tutor who also participates and supports the children in the studies. The game consisted of choosing and eating food from a pre-set menu, asking the tutors to buy it if it was not available in their homes. When a food is not to your liking, becoming an obstacle in your progression in the game, a resolution plan is defined to help the participant to eat better and rise to the next level.

The first study made to *Squire's Quest I* began in 2003, and aimed to register and evaluate two groups of children. in the first group the FJV consumption was analysed over 8 days; in the second group the consumption was analysed for 4 days. After the evaluations, they were asked to play the game, and to record the food consumption on the following 4 days. In the end, it was possible to verify that the children of the second group increased their consumption of FJV compared to those of the first group, because the children of the first group were object of a more intensive testing and got bored, leading to a general disinterest on the game.

The second study, carried out in version II, was performed between 2009 and 2011, and its main purpose was to test the effects of creating FJV consumption goals attributed to participants while they were playing *Squire's Quest*. In this study, participating children were to play all 10 levels, while tutors were given a newsletter for each of those levels, as well as web access where they could collect information on healthy foods and participate in data collection activities. A small subset was randomly selected to participate in an interview about the study and its effect on the home food environment. As conclusion of this study, there was an increase in FJV consumption in each participant.

Nutri Ventures is an educational project with the support of several nutritionist associations that guarantee and validate the developed content. Their main goal is to change the eating habits of children worldwide, seeking through entertainment, tools that make children's nutrition healthier and fun. For this, they have games in which food is a central part of the story, leading children to create good relationships with all foods. To spread the message, the company created a child television show with musical contents, as well as a digital platform where children can play adventures of nutrition [12].

On the child growth phase, it's extremely important that she has an healthy eating and therefore is essential to choose safe foods regarding quality and hygiene and they need to meet all the necessary and essential nutrients.

The energy requirements in this group age are entirely dependent on the needs required for the organism normal needs, for the growth and by the energy expenditure that is inherent for the physical activity. On average, the needs range from 1.300 calories for children with three years old and 1.700 for children with the age of six. It is also worth to note that, in average, female children need less 150 calories when compared to male children, because they have lower energy needs [13].

The distribution of the total daily caloric value should focus on the following percentages of macronutrients, 60% carbohydrates, 27% fat and 13% protein, divided as follows during the day: breakfast 15%, morning snack 5%, lunch 35%, afternoon snack 15% and dinner 30% [13].

From the research carried out, it was not possible to find a mobile application that covered the gaming side as well as the analysis of the evolution of the user that the caregivers could follow, so that they could avoid or combat the child's health problems. The mobile application was created on Android, named *Barty*, and will help children and caregivers to educate and fight childhood overweight and obesity.

3   The Game

Symbols were created and transformed along several generations to be part of their religion or visual art. They become a tool for understanding the world. Erich Fromm classifies the symbols into two categories: conventional or accidental and universal. The conventional ones are used for informational purposes, while the universal ones have a wide and deeper meaning. Before the discovery of the main function of the heart, it was seen as the center of life, courage and reason, being the most universal of the existing symbols [14].

The figure of the heart has evolved and the rounded representation in the shape of a peach appears for the first time, with a groove in the upper part, in an anatomy work of Vigevano in 1347. At present days, the symbol of the heart isolated or crossed by an arrow of Cupid is the universal symbol of love [14].

With the believing that a symbol that represents love and life, would be the best way for the interaction with parents and children, *Barty* was created, Fig. 1.

This central character of the game takes the form of a heart and interacts with parents and children so that they live life with love and a healthy body. The name *Barty* has originated in the union of the words Body and Hearty.

Fig. 1. Mascot - *Barty*

The representation of this symbol was not intended to have a high level of abstraction so that the meaning was not explicit, but also not to include a great level of detail that could lead to an unpleasant symbol and could also lose the meaning of love. In this context, we followed a compromise between abstraction and reality, following the ideals of illustrator and graphic designer *Christoph Niemann* who developed *The Abstract-o-Meter* [15] represented in Fig. 2.

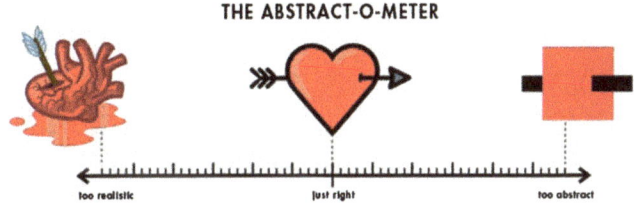

Fig. 2. The abstract-o-meter by Christoph Niemann [15]

The application consists of five main sections, Fig. 3:

The user information and caregiver responsible for the users of each application is stored in the android internal database through SQL Lite, following the structure defined in Fig. 4.

For each user who wants to register, they are asked to enter their name and photo so that it is easily identified in the users list. The date of birth and gender are required so that the calorie values on the child's feed day can be adapted to their age group and gender. Weight and height registration are important for evaluating user evolution.

The caregiver association in each installation of the application is important because only with this it will be possible to draw conclusions about the evolution of all responsible users and caregivers. The password allows only the caregiver to have access to sensitive (personal) information about all users, and the ability to edit or remove users. In the information only accessible by the caregiver it's still possible to make an analysis of the child's evolution in a given game, for example, how many

1. List of configured users;

2. Control of information accessible by caregivers;

3. Presentation of the character varying according to the level and state defined in the user;

4. Challenge(s) presented to users;

5. Addition of a new user.

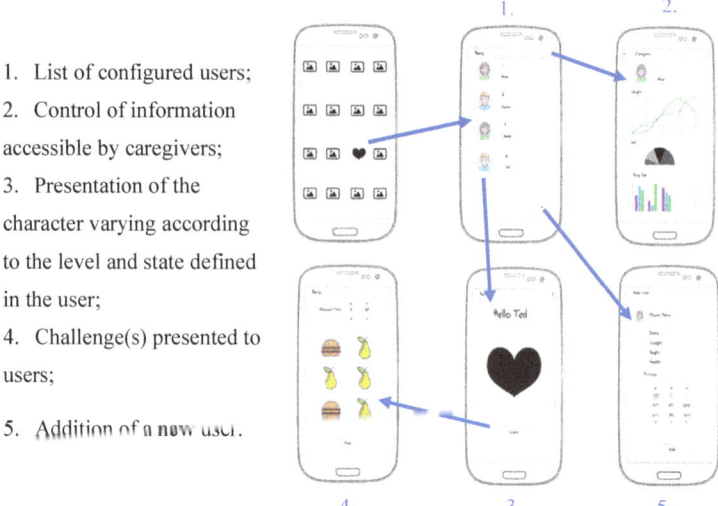

Fig. 3. Game mockup

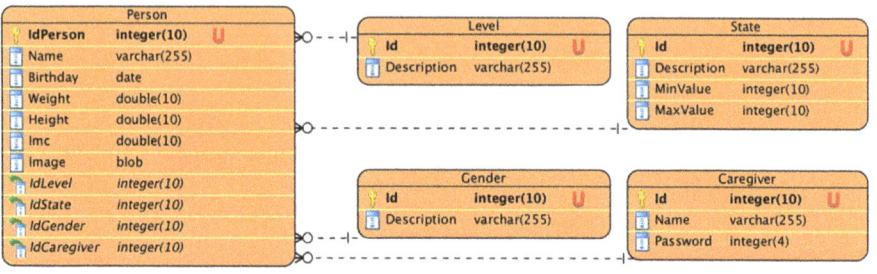

Fig. 4. Database structure for person and caregiver

movements he needs to overcome a given challenge or even the time needed to finish it, being able to evaluate his performance over time.

The registration of new users is available without any type of control, because it is not intended to limit access to the application.

The levels will have the following definition: beginner, intermediate and advanced, which will influence the way *Barty* is dressed.

On the other hand, each level is associated to four states, every quarter of 100%, which influence the physical and colour aspect of *Barty*:

0–25% - green colour, fat and sad	25–50% - yellow colour, slimmer and listless
50–75% - orange colour, athletic shape and joyful	75–100% - red colour, athletic shape and very cheerful

The main page of the game *Barty* appears with the corresponding aspect with the level and state of the user selected. In this screen, it's possible that the user initiates new challenges that, when finalized, can affect the level and state of the mascot.

To motivated the child to use the application, the educational side cannot be very explicit, the games should be catchy and *Barty* must interact with the user, stimulating him to play new challenges with increasing complexity. Taking care about the age range for which the game is intended, audio should be used to voice the mascot, as well as for successful or failed game actions, or other touching options.

The database structure that will store the information regarding the proposed challenges has not yet been defined, but should contain the designation of the layout of each game available in the application, the level of difficulty and other information that can be considered relevant during implementation.

4 Final Remarks

Due to the relevance of preventing childhood obesity, a serious game is being developed with the purpose to give the child, a better understanding on food habits, by guiding them on good food habits. This game will be tested in children between 3 and 6 years old.

As future work it is necessary to define the best challenges for the children, implement information management accessible by caregivers and include audio to help younger children, as well as performing real world tests to validate the game objective.

Acknowledgments. The authors would like to express their acknowledgements to the nutritionists Ana Miranda and Mariana Afonso for the follow-up on the development stage of this application, which led to the use of the best practices of nutrition, and to COMPETE: POCI-01-0145-FEDER-007043 and FCT – Fundação para a Ciência e Tecnologia within the Project Scope: UID/CEC/00319/2013.

References

1. World Health Organization: Report of the commission on ending childhood obesity. World Health Organization, Geneva (2016)
2. do Vale Leiras, E.M.: Comportamento alimentar da criança: a influência materna (2015)
3. Ramos, F.P., da Silva Santos, L.A., Reis, A.B.C.: Educação alimentar e nutricional em escolares: uma revisão de literatura. Cadernos de Saúde Pública **29**(11), 2147–2161 (2013)
4. Djaouti, D., Alvarez, J., Jessel, J.-P., Rampnoux, O.: Origins of serious games (2011)
5. Anderson-McNAmee, J.K., Bailey, S.J.: The importance of play in early childhood development. Mont. State Univ. Ext. MontGuide (2010)
6. Wang, L.-C., Chen, M.-P.: Innovations in education and teaching international (2017)

7. Baranowski, T.: Might video games help remedy childhood obesity? Child. Obes. **11**(4), 331–334 (2015)
8. Pereira, J.P.: Smartphones em Portugal ultrapassaram telemóveis tradicionais – PÚBLICO (2014). https://www.publico.pt/2014/02/20/tecnologia/noticia/smartphones-em-portugal-ultrapassaram-telemoveis-tradicionais-1624400. Accessed 30 Dec 2016
9. Fitton, D., Bell, B.: Working with teenagers within HCI research: understanding teen-computer interaction, pp. 201–206 (2014)
10. Dias, J.D., et al.: Serious game development as a strategy for health promotion and tackling childhood obesity. Rev. Lat. Am. Enfermagem **24**, e2759 (2016)
11. Thompson, D., et al.: Creating action plans in a serious video game increases and maintains child fruit-vegetable intake: a randomized controlled trial. Int. J. Behav. Nutr. Phys. Act. **12**, 39 (2015)
12. Nutriventures. http://www.nutri-ventures.com/pais/pt. Accessed 26 Aug 2017
13. Nunes, E., Breda, J.: Manual para uma alimentação saudável em jardim de infância. Lisboa
14. Prates, P.R.: The heart as symbol. História, Ciências, Saúde – Manguinhos, vol. 12, pp. 1023–1031 (2005)
15. Abstract-O-Meter|Christopher Simmons. http://minesf.com/resources/cca/2010/01/25/abstract-o-meter/. Accessed 30 Sept 2017

Mechatronic System for the Promotion of Physical Activity in People with Motor Limitations: First Insights

Leandro Pereira[1], José Machado[1,2], Vítor Carvalho[3,4(✉)],
Filomena Soares[4], and Demétrio Matos[5]

[1] Mechanical Engineering Department, University of Minho,
Campus de Azurem, Guimarães, Portugal
a75901@alunos.uminho.pt, jmachado@dem.uminho.pt
[2] MEtRICs Research Centre, University of Minho, Campus de Azurem,
Guimarães, Portugal
[3] 2Ai Lab – IPCA-EST, Campus do IPCA, Barcelos, Portugal
vcarvalho@ipca.pt
[4] Algoritmi Research Centre, University of Minho, Campus de Azurem,
Guimarães, Portugal
fsoares@dei.uminho.pt
[5] ID+ – IPCA-ESD, Campus do IPCA, Barcelos, Portugal
dmatos@ipca.pt

Abstract. The increase in the elderly population takes along concerns to health professionals. A way of helping health professionals and the elderly may be through the development of systems that allow people to maintain their mobility regardless of age and health problems that may arise, such as: obesity, Parkinson's, hypertension, arthritis, diabetes and even the passive symptoms of stroke. This article has two main objectives, the first is the study and analysis of the available devices that help people with mobility problems and the second is the proposal of a system that permits the accomplishment of physical activity and rehabilitation in a pleasant way.

Keywords: Elderly · Fitness games · Serious games · Rehabilitation
Exergames

1 Introduction

Demographic trends point to a sharp increase in the number of elderly people worldwide and in Portugal. According to the Portuguese census of 2011, the percentage of young people decreased from 16% to 15% between 2001 and 2011, while the elderly population increased from 16% to 19%. Consequently, the Portuguese aging index increased from 102 to 128 between 2001 and 2011 (INE, 2012) [1]. This increase of the elderly and in particular of those with mobility problems is a situation that brings concerns to society and, in particular, to health professionals. One way of helping the health professionals and people with motor limitations can be through the use of games oriented to the physical activity and rehabilitation [2].

© Springer International Publishing AG, part of Springer Nature 2019
J. Machado et al. (Eds.): HELIX 2018, LNEE 505, pp. 267–274, 2019.
https://doi.org/10.1007/978-3-319-91334-6_37

One of the most common reasons for hospitalization of the elderly are falls; they occur during daily activities, and frequently they are due to the degradation of the mechanisms necessary for maintaining balance. These falls can cause injuries and consequently loss of autonomy. The primary basis for the prevention of falls in the elderly is physical activity [3]. In this way, physical activity is defined as a factor related to the increase of quality of life and a healthy and effective behavior for prevention of diseases. Physical activity can also have effects in reducing anxiety, stress and depression, helping to maintain mental health and ensure physical vitality [4]. Another advantage of exercise is the improvement of cardiovascular function and the reduction of risks associated with chronic diseases [5].

Not only aging, but also chronic diseases such as obesity, Parkinson's, hypertension, arthritis, diabetes and even symptoms resulting from strokes can force people to gradually abandon their independence, making it more difficult to performing daily tasks such as: bathing, dressing and walking. As many of the above diseases require rehabilitation, some researchers believe that active games (games involving physical activity) can increase adherence to treatment [6].

In order to achieve the main goals of this paper, the respective organization is as follows: a brief historical review of related systems is presented in Sect. 2; Sect. 3 presents some interactive devices being used nowadays for those purposes; in Sect. 4 a conceptual solution is illustrated, solving some gaps that have been found, and, finally, Sect. 5 presents some final remarks.

2   Historical Review of the Exergames

The first active game that appeared was Foot Craz in 1987. This game consisted of a small cushion with five buttons that responded to the touch; a year later, Nintendo Power Pad was released, which was similar to the previous game but larger. Dance Dance Revolution came out in 1997 and consisted of a dance floor with four arrows: up, down, left and right; these arrows were pressed with the feet [7]. In the decade of 2000 appeared the fashion devices as camera EyeToy (a camera that allows an interaction with the game according to the movements of the body) for PlayStation 2 and Nintendo Wii Fit.

Besides these, there were still other important devices launched. Around 1986 appeared the Computrainer that consisted of a static bike designed as a motivational training tool. The Computrainer allowed users to navigate through a virtual landscape, while monitoring data such as power and speed [11]. In the 1990s, several devices of the same type emerged, where the most sophisticated was Tectrix VR Bike originally developed by CyberGear Inc. This system allowed users to pedal in various virtual environments. Currently, the VR Bike software is distributed alongside Trixter Bikes [12].

Other important developments in active game arena came with the release of the well-known Wii console. The Wii console was officially announced at the annual E3 gaming fair 2005 and launched in late 2006. The console had great success because it used a wireless controller equipped with an accelerometer capable of detecting movements in 3D [13].

3   Interactive Devices in the Promotion of Physical Activity

3.1   Physioland

Physioland presents a serious game developed with the objective of enabling the accomplishment of physiotherapy thus helping health professionals. This game presents a medieval concept in which the player will have to perform tasks with therapeutic goals. The overall architecture of the game is based on five main components: the game, the hardware, the peripherals, the remote database and the player.

This serious game also has an auxiliary structure that serves as a support to the patient, when he/she is performing the physiotherapeutic exercises [8].

3.2   ReaKing

Another virtual reality game is ReaKinG; the goal of this game is to get people to perform physical rehabilitation exercises, using a Kinect camera. One of the goals of this game is to make aging more fun and at the same time promote the level of independence. The ReaKinG consists of two types of games (aerobic games and strength games) that can be selected according to the user's preference.

In aerobic games, the player will find objects on the path, and the goal is to pick up the coins that appear on the way. Each coin is placed in different strategic positions so that the user can perform different movements (abduction, adduction, rotation, among others). In strength games, the patient should practice some exercises, as if he/she was in the gym [9].

3.3   Smartfloor

The Smartfloor is a platform game that is designed to combat childhood obesity. It was developed in a modular way, with 36 steel panels (50 cm^2) and load cells to measure the deformation located in each of its 4 corners. The panel pairs are connected to a circuit board that amplifies, filters and scans the 8 analog signals. The system also contains a microprocessor that calculates the total force and position of the pressure center for each panel. The device is like the Nintendo Wii Fit concept but allows the user to measure motion in a larger area [10].

3.4   PhysioSensing

The PhysioSensing is a pressure platform similar to the previous one. This platform was developed for physiotherapy and rehabilitation activities, allowing to evaluate balance, stability and posture. In summary, it can be said that this equipment has several functionalities, among which: balance/stability and load transfer. Physiosensing also allows the mapping of plantar pressure through its 1,600 sensors, and the execution of physiotherapeutic games in which the patient is challenged to be a character who manages an obstacle course, where each obstacle is programmed to represent a therapeutic situation [14].

3.5   Twall

Twall is an interactive sport that consists in a wall that uses light pulses to generate motion sequences in a specific way. In this game, the movements are stimulated using illuminated buttons that are turned off by touch. It can be used for resistance training, flexibility and response time. In addiction it is possible to adjust Twall to the training area, the radius of action and the area of visual perception. Another advantage offered by this product is the possibility of including cognitive tasks and being fully operational without using a computer [15].

3.6   GymTop USB

The GymTop USB is a board with visual biofeedback technology, which can be used for balance training, diagnosis of anatomical problems, resistance training and body-building of the legs and torso. The software consists of 12 training programs very intuitive and exciting. This system also presents the ability to generate a clinical report for each exercise performed [16].

3.7   Trixter VR

Trixter VR was developed to introduce an entirely new dimension to indoor cycling. This stationary bike offers a large quantity of virtual realities, where it is possible to carry out local and even international competitions. This bike offers additional features and innovative data loading technology to keep users interested.

   This bike includes a classic mode (series of predefined profiles) and real video tours (where the player can pedal through the big cities and touring different countries.). Cyclists can pedal alone or compete with other local or worldwide users. For extra motivation, there is a point system displayed on the screen, allowing cyclists to unlock the new skill levels as they travel [12].

3.8   Comparison of the Equipment for Physical Activity Promotion

By analyzing the products described previously, it is possible to verify the existence of a great variety of products in the market that allow the performance of physical exercise in an iterative way. Each of these products have different functionalities. The advantages and disadvantages of these products are analyzed below.

   Physioland presents a virtual reality game, its advantages are: the presence of an auxiliary structure (that serves as a support to the patient) and its ability to adjust to any person through the calibration (necessary procedures for applying the anthropometric data of the player to the game) of the game; on the other hand, it can be said that this product has the disadvantage of requiring some space for its installation.

   ReaKing presents another virtual reality game that, comparing with Physioland, it has the disadvantage of not including the auxiliary structure.

   SmartFloor presents a platform game very similar to the Wii console. This system has the following advantages: compact, portable and it can be used by anyone without adjustment or calibration; on the other hand, this system presents the disadvantage of not being very safe or comfortable to be used by people with motor limitations.

Physiosensing, compared to Smartfloor, is more oriented towards the rehabilitation and physiotherapy, offering more functions and applications oriented to this domain.

Twall is more used for practicing sport, as it is relatively compact and portable and it is capable of being used without a computer.

GymTop USB is more focused on rehabilitation, allowing to generate clinical reports for each activity performed.

Trixter VR presents a virtual reality bike with great adjustability and versatility, but it does not present adequate ergonomics.

In summary it can be verified that despite the variety of interactive equipment that motivates the people for the physical exercise there are few that also promote a level of comfort and safety for people with motor limitations.

4 Proposed Solution

The proposed mechatronic system involves the creation of a product that allows practicing physical exercise, especially those with mobility and locomotion problems. This model is also capable of adjusting ergonomically to any person and change required load for the upper and lower limbs. In Fig. 1 it is presented the proposed 3D CAD conceptual model of the intended system.

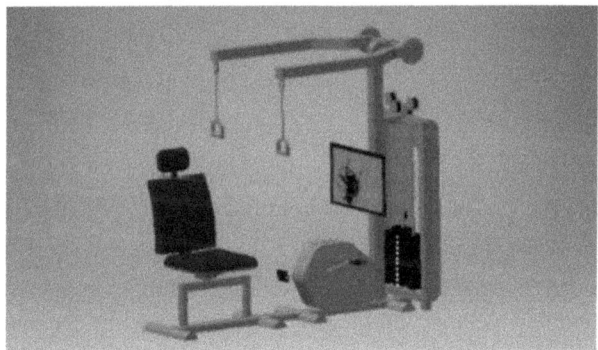

Fig. 1. Proposed 3D CAD conceptual model

This model consists of a stationary bicycle, very similar to the bicycles in the gyms, but with a didactic component. For the creation of the didactic part, it was chosen the creation of a virtual game that will allow the patient's adhesion by stimulating him/her to exercise in a healthy way. The game to be developed consists on a serious game. In this case, as the patient pedals and moves the arms, he/she is performing the game and the tasks related to physical exercise, rehabilitation and wellbeing.

The game chosen to integrate the system is a game that simulates the use of a paramotor. A paramotor, also known as parafly is considered an adaptation of the paraglider. Flying paragliders require a minimum wind speed, while in the paramotor the wind speed is generated thanks to the speed provided by the motor to its blades.

In the case of the game, the speed that is provided by the paramotor will be simulated by the pedaling speed of the patient. This causes the user to go up and down in the game scenario according to the speed. On the other hand, direction (left or right) will depend on the position of the patient's arms.

To create the game, it is necessary the acquisition of the following inputs: the detection of the pedal speed, the detection of the position of the arms of the structure and the detection of the load applied to the pedal and the arms. To obtain this data it will be necessary to use encoders and sensors to measure position and speed. Another relevant aspect of the game will be the placement of objects along the virtual path that should be intercepted by the user (for example, coins). One of the objectives of this model is that it is as compact as possible. In this way it was chosen to divide the system in two modules (chair and main structure). This division in modules, besides allowing the disassembly and division of the equipment will allow the adaptation to any chair, enabling the use of wheelchairs, as shown in Fig. 2.

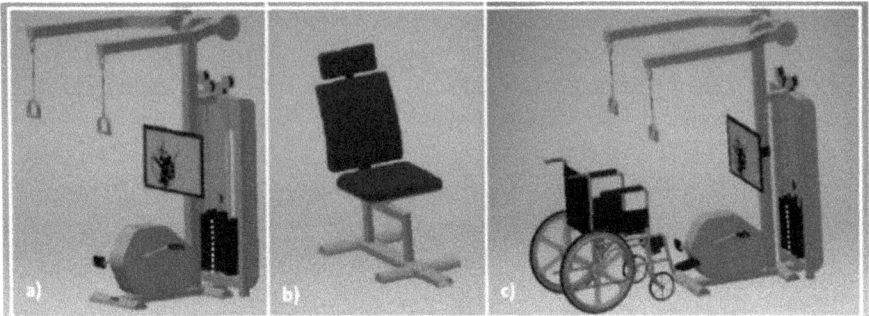

Fig. 2. Division of the system in modules and possibility of using a wheelchair (a: main structure module; b: chair module; c: use of a wheelchair) [17]

It is also intended that the chair module will bring comfort and safety to the patient, making it an important complement to the main structure of the model. Another advantage of using the chair module is its ability to adjust the system to different people, in this case the chair will have the ability to rotate, adjust the height, adjust the distance to the pedal and adjust a slope of the backrest.

5   Final Remarks

Currently there are several equipment that intend to promote physical activity in an interesting way. Some of this equipment have the objective of improving the physical condition of elderly and people with motor limitations through physiotherapy and rehabilitation exercises. However, according to the analysis performed it can be verified that despite the existence of the diversity of equipment, few present the level of comfort and safety necessary to be used by people with motor difficulties. In this way, the proposed solution presents an added value, since it will provide the achievement of exercise in an iterative, safe and comfortable way.

In the future it is intended to optimize the illustrated model, design its control system and verify its usability, safety and comfort, by building a concept proof's physical prototype.

Acknowledgements. The authors would like to express their acknowledgments to COMPETE: POCI-01-0145-FEDER-007043 and FCT – Fundação para a Ciência e Tecnologia within the Project Scope: UID/CEC/00319/2013.

References

1. Abreu, M., Caldevilla, N.: Attitudes toward aging in Portuguese nursing students. Procedia Soc. Behav. Sci. **171**, 961–967 (2015). https://doi.org/10.1016/j.sbspro.2015.01.215
2. Martins, T., Carvalho, V., Soares, F.: Web platform for serious games' management. Procedia Procedia Comput. Sci. **64**, 1115–1123 (2015). https://doi.org/10.1016/j.procs.2015.08.571
3. Szerdiová, L., Šimšík, D., Galajdová, A., Onofrejová, D.: Evaluation of seniors gait training with mechatronic device. Procedia Eng. **96**, 444–453 (2014). https://doi.org/10.1016/j.proeng.2014.12.114
4. Lok, N., Lok, S., Canbaz, M.: The effect of physical activity on depressive symptoms and quality of life among elderly nursing home residents: randomized controlled trial. Arch. Gerontol. Geriatr. **70**, 92–98 (2017). https://doi.org/10.1016/j.archger.2017.01.008
5. May, R.: Exercise in the frail elderly. Saint Louis University (2002)
6. Zeng, N., Pope, Z., Eun, J., Gao, Z.: A systematic review of active video games on rehabilitative outcomes among older patients. J. Sport Health Sci. **6**(1), 33–43 (2017). https://doi.org/10.1016/j.jshs.2016.12.002
7. Bogost, I., St, C.: The rhetoric of exergaming. The Georgia Institute of Technology, 10 (2005). https://doi.org/10.1093/llc/fqn029
8. Martins, T.: Desenvolvimento de um jogo sério para fisioterapia, monitorização e motivação de pacientes com doenças neurológicas. Universidade do minho (2017)
9. Pedraza-hueso, M., Martín-calzón, S., Díaz-pernas, F.J., Martínez-zarzuela, M.: Rehabilitation using kinect-based games and virtual reality. Procedia Procedia Comput. Sci. **75** (Vare), 161–168 (2015). https://doi.org/10.1016/j.procs.2015.12.233
10. Heller, B., Senior, T., Wheat, J.: The Smartfloor: a large area force-measuring floor for investigating dynamic balance and motivating exercise. Procedia Eng. **72**(2011), 226–231 (2014). https://doi.org/10.1016/j.proeng.2014.06.040
11. Dinh, T.: RacerMate CompuTrainer Pro Review (2013). http://www.roadbikereview.com/reviews/racermate-computrainer-pro-review. Accessed 27 Jan 2018
12. Pulsefitness: New ground breaking features on Trixter VR (n.d.). http://www.trixter.net/. Accessed 28 Jan 2018
13. Iwata, S.: Media briefing speech at E3 (2006). https://web.archive.org/web/20060604061842/http://www.nintendo.co.jp/n10/e3_2006/speech/english.html. Accessed 27 Jan 2018
14. Sensisng Future: Physiosensing balance and pressure plate (n.d.). https://www.physiosensing.net/. Accessed 27 Jan 2018

15. Motion Fitness: TWALL targeted training with a fun factor (n.d.). https://www.motionfitness.com/tWALL-s/328.htm. Accessed 27 Jan 2018
16. Sensing Future: GymTop USB (n.d.). https://www.sensingfuture-store.com/product-page/gymtop. Accessed 27 Jan 2018
17. Tanay: Wheelchair (catia model) (2016). https://grabcad.com/library/wheelchair-23. Accessed 5 Feb 2018

Cyber-Physical Systems, IoT and Industry 4.0

Conceiving the Cyber-Physical Systems with Object Enhanced Time Petri Nets

Tiberiu S. Letia$^{(\boxtimes)}$ and Dahlia Al-Janabi

Technical University of Cluj Napoca, 400114 Cluj Napoca, Romania
Tiberiu.Letia@aut.utcluj.ro, dahliajanabi@gmail.com

Abstract. The conceiving of the Cyber-Physical Systems (CPSs) requires models that can sustain the specification, the synthesis, the design, the verification, the implementation and the testing. The newly introduced models consisting of Object Enhanced Time Petri Nets (OETPNs) can be successfully used to meet this requirement. Changing the unique kind of tokens in the classical Petri Nets (PNs) with the software passive or active objects opens the possibility to describe the object dynamic instantiation and distributed concurrent tasks that are moved through a computer net with a dynamical structure. The moving tasks can be loaded with agents missions obtaining thus a so called dynamic multi-agent system. The need to move Java compiled codes to another computer, in the absence of the source classes at the destination, is fulfilled by changing the problem into one with tasks instantiated with different parameters.

Keywords: Cyber Physical System · Control synthesis
Object Petri Nets · Software engineering

1 Introduction

The current research aims to diminish the obstacle mentioned in [1] to efficiently achieve CPSs based on rigorous modeling of the interaction among different components and between the physical and the cyber sides. The software is required for the monitoring and the control of processes involving feedback loops where physical parts affect the computation and vice versa.

The modeling of the linking of physical devices (i.e. things) endowed with continuous time behaviors, and the software objects behaving in a discrete-time manner (with possible asynchronous reactions) can be performed with low accuracy. Both physical things and software objects will further be named as objects and their different particular features are considered as a result of their class (i.e. types) instantiation. These objects behave concurrently and interact with each other through communication channels having drivers that implement protocols to make compatible the different types of information they manage.

Distributed CPSs can include dynamic instantiations or destructions of software passive or active objects, and dynamic entrances and exits of the physical

© Springer International Publishing AG, part of Springer Nature 2019
J. Machado et al. (Eds.): HELIX 2018, LNEE 505, pp. 277–283, 2019.
https://doi.org/10.1007/978-3-319-91334-6_38

passive and active objects. Both kinds of objects can move through a given net consisting of physical (including hardware) and software parts.

Besides the multi-tasking feature, many of the CPSs have to meet real-time constraints that increase their model synthesis and verification. This leads to an NP-complete problem that is very difficult to solve and often evolutionary methods are involved [2]. The current research leads to the conclusion of the use of an interactive approach where the heuristic methods are integrated with the human expert activities. Due to the system complexity, the verification should combine formal verifications with computer simulations using benchmarks.

Some devices connected in CPSs are very complex and include software programs concurrently executing different tasks that have often to be remotely updated. This activity is usually performed by stopping the device, the sending of the new executable codes and its restarting. All the device tasks are stopped even if only one of them is updated. The current proposed approach allows the task changing or migration without the need to stop all the activities of a running device.

Multi-agent systems are often included in the CPSs for resilient and safety purposes [3]. Due to the fact that CPSs become more and more complex, the agents have to perform more complex tasks besides the regular monitoring and control reactions, involving activities like planning, negotiating etc. [4]. The agents are included in moving vehicles or other different kinds of devices, but also software agents move through dynamic networks fulfilling real-time requirements. The moving agents have to adapt to their new environments continuing to perform their assigned missions.

2   Related Works

For a very long time Petri nets (PNs) have been known for their capabilities to describe concurrent dynamic systems and to verify their properties. A good review of PN history can be found in [5]. There are many points of view for using the PNs in software engineering, control engineering, etc. A general conclusion is that the PNs can be successfully applied for approaching almost any kind of dynamical systems where the notion of state is compulsory and the description of their evolution is required.

Some developments of PNs concern the unification of the PN features leading to the so called Unified Enhanced Time Petri Nets where the tokens representing real numbers are processed by arithmetical or logical operations (including the fuzzy logic) [6]. These kinds of models were further developed by changing the previous tokens with software object types [7].

The ordinary PNs are appropriate for modeling and verification of simple flow of executions. The Object Petri Nets (OPNs) were conceived to cover the gap between flow of executions and Object Oriented Programing (OOP) that concerns the interactions between objects [12]. Some of the proposals extend the Colored Petri Nets to get the OPNs as [8]. One trend to develop OPNs focuses on the mathematical analysis and tries to keep the OPNs as simple as possible to facilitate the properties demonstrations [14,15]. The other trend concerns the enhancing the OPNs as much as possible with object-oriented notions and properties, but this increases the difficulties to prove different properties, meanwhile increasing the model's description power [9–11,13].

A PN based object-oriented modeling language that can be seen as an object-oriented extension for high-level PNs or as a method providing formal concurrent semantics for object-oriented models was conceived [16]. This was used for the analysis of a protocol performance.

3   Object Enhanced Time Petri Nets (OETPNs)

The OETPN models are defined In reference [7]. The OETPN *marking* is a vector $\mathbf{M}$ with the elements as references to objects if the corresponding places have tokens or ϕ (i.e. null) if they do not. A transition t is *enabled* or *admissible* from the current marking $\mathbf{M}$ if there is a fulfilled guard in λ_t. It is assumed that only one condition can be *true* from any given marking. If more conditions can be simultaneously fulfilled for a transition, the first found is taken into account. Similarly, it is supposed that there are no conflict transitions for any marking. If there are conflict transitions the transition with the lowest index wins the election.

The *execution* of an enabled transition has two phases. In the first one, the tokens from the transition input place set are moved to a temporary marking vector and a clock (i.e. counter) is set with the transition delay. The counter is decreased at each clock *tic* and when it reaches zero, the transition execution enters the second phase. The tokens from the temporary marking are used by the mapping to determine the tokens that will be set in the transition output places. The new tokens are set in the corresponding places if they do not contain active objects. If active objects exist there, they have to be stopped before their changing occurs. If a new token has the attribute *active* true, the new task is started. Details of OETPN models and the simulation algorithm are given in reference [7].

4   Example of Application

OETPNs are capable to model object-oriented distributed applications containing multiple heterogeneous agents that move through a heterogeneous computer network. Figure 1 presents a component diagram that links two OETPNs denoted by $OETPN_A$ and $OETPN_B$ that are implemented on a platform and communicate as peer to peer through the TCP/IP protocol.

$OETPN_A$ works as supervisor for $OETPN_B$ that controls a power station with two generators represented schematically in Fig. 2. $OETPN_B$ implements a level controller and two generator controllers that control the frequency by control signal u_{k1} (i = 1, 2) and the voltage by u_{i2}. The notation h denotes the lake level with Z (zero), L (low), M (medium) and H (high) as marks.

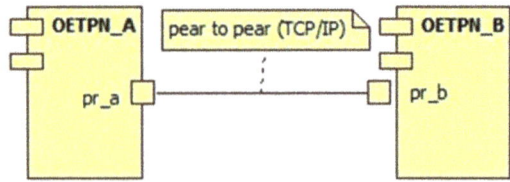

Fig. 1. Example of two interconnected OETPN models.

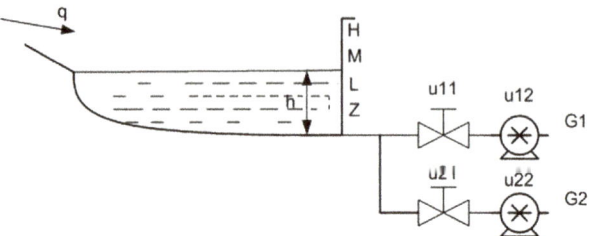

Fig. 2. Hydro-power station scheme.

The level controller has the specification:

- $IF(h_c > h_a)\&(h_c > L)\&(G_1 stopped) THEN start G_1$
- $IF(h_c > h_a)\&(h_c > M)\&(G_2 stopped) THEN start G_2$
- $IF(h_c < h_a)\&(h_c < M)\&(G_2 active) THEN stop G_2$
- $IF(h_c < h_a)\&(h_c > L)\&(G_1 active) THEN stop G_1$

where h_c represents the current level and h_a the anterior level.

Figure 3 shows the parent $OETPN_B$ that is composed of two branches. The upper branch reads the level h_c through the place p_3 linked to an input channel, compares it with the anterior value h_a stored in p_5, reads the generator states thorough p_7 and p_8 and starts the corresponding generator through the output channels linked to p_{10} and p_{11}. The lower branch has the role to set or reset a new controller for the generators sent by $OETPN_A$. It reads the controller child $OETPN_C$ and the destination by p_{13} that is linked to the communication channel.

A child OETPN model corresponding to the generator controller is presented in Fig. 4. It can be started or stopped using the places p_2 and p_{10} respectively. It reads the generator output values (y_1, y_2), the reference values (r_1, r_2) and applies the control signals (c_1, c_2). The update of the child $OETPN_C$ with another one represented in Fig. 5 is required with the aim to improve the control performances. The new child takes into account besides the current error e_c, the anterior error e_a. Before replacing the old controller with a new one, the previous one has to be stopped if it is working (i.e. running).

A reasonable requirement of PNs is to avoid the conflicts consisting in the simultaneous admissibility of transitions under the same input places. In the case of OETPN models the conflicts that exist at the PN level can be removed by adequately conceiving of the guard conditions.

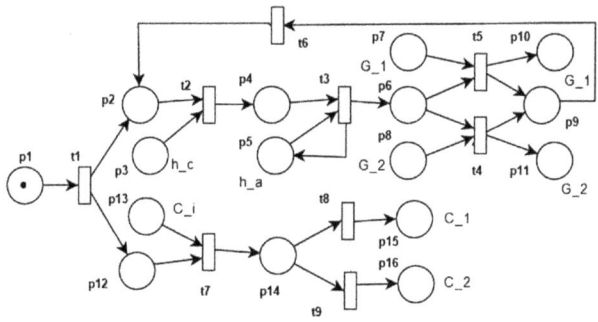

Fig. 3. $OETPN_B$ parent.

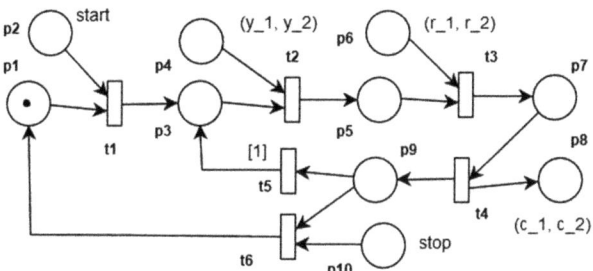

Fig. 4. Controller child $OETPN_C$ model.

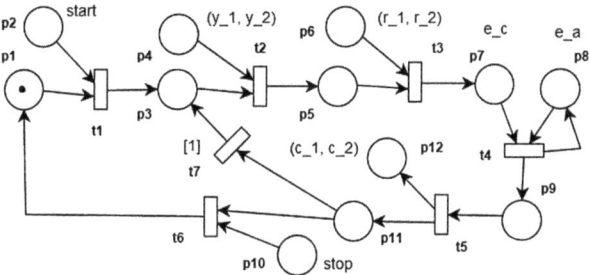

Fig. 5. Second controller child $OETPN_C$ model.

5   Conclusions

The newly introduced OETPN models allow joining the control engineer approach with the software engineer approach. They facilitate the description of the physical devices' behaviors and their connection to controllers. Any physical device can be described by an OETPN model due to the fact that there is no restriction to the kind of information or behavior. For complex applications where a large number of elements are included can be used components that include OETPN models. More than this, the OETPN models can include other OETPN models or any number of passive objects. The hierarchical construction is sustained by OETPN as has been shown above.

The interfaces between OETPN models are clearly defined and these can be used for specification, design, implementation, and verification or testing.

OETPN models include the information types requested for software implementation modeling and even the instantiation process is described. This leads to the use of more complex structured information that increases the control engineer thinking power. On the other side, having clearly defined the types, the software engineer is helped to design the applications and their implementation avoiding any confusion between control and software engineer.

The same kind of models allow the description of the concurrency including the parallel execution, the synchronization, the asynchronous event signaling and the input of information from the outside or sending the information through output channels. The control engineer has no concern for the possibility to implement his specification. All the rules involved by control specifications are clearly added to transition, so the verification that the software implements the control requirements is easily performed.

The introduced ϕ symbol allows the description and the implementation of rules for the case of the lack of the expected information. The exceptions involved by OOP that disturb the regular executions and that are not easily introduced in control specification or in software implementation are strongly solved here allowing the definition of rules for the case when some devices or operations do not behave properly. This also allows and sustains the analysis of state consistency of the physical devices and software components.

As a final conclusion, the proposed approach allows the contraption of a framework for the development of CPSs that is appropriate for a large variety of distributed applications.

References

1. Nuzzo, P., Sangiovanny-Vincentelly, A., Bresolin, D., Geretti, L., Villa, T.: Methodology with contracts and related tools for the design of Cyber-physical systems. Proc. IEEE, **103**(11) (2015). https://doi.org/10.1109/JPROC.2015.2453253
2. Sapiecha, K., Ciopinski, L., Deniziak, S.: Synthesis of self-adaptive supervisors of multi-task real-time object-oriented systems using developmental genetic programming. In: Recent Advances in Computational Optimization. Studies in Computational Intelligence, vol. 610. Springer (2016). https://doi.org/10.1007/978-3-319-21133-6-4

3. Zhu, Q., Bushanell, L., Basar, T.: Resilient distributed control of multi-agent systems. Lecture Notes in Control and Information Sciences, vol. 449. Springer (2013). https://doi.org/10.1007/978-3-319-01159-2-16

4. Calvaresi, D., Appoggetti, K., Lustrissimini, L., Marinoni, M., Sernani, P., Dragoni, A.F., Schumacher, M.: Multi-agent systems negotiation protocols for cyber-physical systems: results from a systematic literature review. In: International Conference on Agents and Artificial Intelligence, Madeira (2018)

5. Silva, J.R., Miralles, J.A.S.P., Salmon, A.O., Gonzalez del Foyo, P.M.: Introducing object-oriented in unified Petri net approach. In: ABCM Symposium Series in Mechatronics, vol. 4, pp. 451–459 (2010)

6. Letia, T.S., Kilyen, A.O.: Unified enhanced time Petri net models for development of the reactive applications. In: Proceedings of EBCCSP: Madeira, IEEE Xplore (2017). https://doi.org/10.1109/EBCCSP.2017.8022831

7. Letia, T.S., Al-Janabi, D.: Object enhanced time Petri net models. In: Proceedings of IEEE Conference on AQTR 2018, Cluj-Napoca, May 2018. (to be appeared)

8. Lakos, C.: From coloured Petri nets to object Petri nets. LNCS, vol. 935, pp. 278–297 (1995)

9. Buchs, D., Guelfi, N.: A formal specification framework for object-oriented distributed systems. IEEE Trans. Soft. Eng. **26**(7), 635–652 (2000)

10. Kohler-Bussmeier, M.: A survey of decidability results for elementary object systems. In: Farwer, B. (ed.) LAM 2010, vol. 1(1), pp. 18-35 (2014)

11. Valk, R.: Object Petri nets using the nets-within-nets paradigm. In: ACPN 2003. LNCS, vol. 3098, pp. 819–848. Springer (2004)

12. Miyamota, T., Kumagai, S.: A survey of objects-oriented Petri nets and analysis methods. IEEE Trans. Fund. Electron. Comput. Sci. (2015). https://doi.org/10.1093/ietfec/e88-a.11.2964

13. Koci, R., Janousek, V.: Object oriented Petri nets in software and deployment. IJSSST **10**(3), 31–43 (2013)

14. Lamazova, I.: Nested Petri nets-a formalism for specification and verification of multi-agent distributed systems (2000)

15. Kohler, M., Rolke, H.: Properties of object Petri nets. In: Cortadella, J., Reisig, W., (eds.) International Conference on Application and Theory of Petri Nets 2004. LNCS, vol. 3099, pp. 278–297. Springer (2004)

16. Vale de Azevedo Guerra, F., Abrantes de Figueiredo, J.C., Guerrero, D.D.S.: Protocol performance analysis using a timed extension for an object oriented Petri net language. Electronic Notes in Theoretical Computer Science, vol. 130, pp. 187–209. Elsevier (2005)

Mechatronic System Using a Straight-Line Motion Mechanism for AGV Application

João Sousa[1(⊠)], Joel Galvão[1], José Machado[1], João Mendonça[1],
Toni Machado[2], and Pedro Vaz Silva[2]

[1] MEtRICs Research Centre, University of Minho, Guimarães, Portugal
jsousa@dem.uminho.pt
[2] Bosch Car Multimedia Portugal S.A., Braga, Portugal

Abstract. This paper describes the design of an exact straight line motion mechanism, commonly known as a Scott-Russell mechanism (SRM), for industrial application. It is applied in an internal materials movement system using autonomous guided vehicles (AGVs) for supporting logistics processes along the supply chain, aiming to replace the current manual operations. The proposed mechanism, analyzed regarding its kinetic and kinematic aspects using SolidWorks Motion, allows the system to discard electric current during motion and to be actuated when docked in the reception station using a linear solenoid. Additionally, the proposed mechanism can be attached below each trolley level's base, occupying minimal space and avoiding interference with larger Returnable Packages handling motions. Furthermore, it can be used in similar applications where a small-occupied volume mechanism is necessary and primarily, where a straight-line motion is required using a passive actuator to perform one of the motions. The mechanical and electrical application requirements are identified along with the modelling phases of the mechanism. The electrical diagram is defined and a compression spring and a linear solenoid were selected using kinematic and kinetic simulation with CAD/CAE software Solidworks.

Keywords: Mechanism · Kinematics · Mechatronics · AGV · Trolley

1 Introduction

The use of autonomous guided vehicles (AGVs) in factories shop floors are becoming more common and Bosch BrgP does not want to fall behind. Bosch's internal logistics is founded on a Just-In-Time (JIT) system to supply raw materials on time whenever required by the production plan. The concept of milk-run (a cyclic method to supply materials within the production area) logistics system comprises a transportation system in which all input and output material requirements of several assembly stations are covered by vehicles that cycles by all stations and moves accordingly to a pre-defined timetable.

The produced car multimedia items in Bosch BrgP require different type of raw materials provided to the assembly lines with internal milk-runs. Vehicles used to perform milk-runs are illustrated in Fig. 1. When loaded, those vehicles perform a predefined route, unloading raw materials and loading empty returnable packages

© Springer International Publishing AG, part of Springer Nature 2019
J. Machado et al. (Eds.): HELIX 2018, LNEE 505, pp. 284–291, 2019.
https://doi.org/10.1007/978-3-319-91334-6_39

Fig. 1. Vehicle used in the Bosch BrgP for supporting the raw materials flow with attached trolleys carrying Returnable Packages (RPs) next to a dynamic rack

Fig. 2. Trolley transferring all trolley's RPs to the reception station

(RP) using the towed trolleys, in a predefined set of locations (buffers next to the production lines) named, in this study, as supermarkets.

A supermarket comprises a set of dynamic racks to temporarily store non-bulky material and a set of static trolleys to store bulky material. Each channel of a dynamic rack has a label that identifies the type of packages/material to store and help the milk-run operator to identify delivery points.

The currently being developed handling system comprises the following sub-systems: supermarket (comprising a delivery station, a reception station and the dynamic rack) and the vehicle with towed trolleys as overviewed by Fig. 2. The objective of the handling system is to autonomously transfer RPs from the trolleys to the dynamic racks of the reception station and to receive RPs from the delivery station.

The main goal is to develop an internal materials handling system using AGVs for supporting logistics processes along the internal supply chain, aiming to replace current manual processes and to improve overall system performance. The autonomous vehicles implementation will improve and optimize current processes allowing Bosch BrgP more capable of implementing industry 4.0's key processes.

2   Application Overview

The trolley structure has a structural profile frame. Each level was conceived with an array of passive anti-static polymeric rollers, enabling a smooth transfer of the RPs, both in human operated manual mode and automatic mode from the supermarket side. The structure of a trolley can be overviewed by Fig. 3. In order to completely deliver and receive RPs from the reception station and the delivery station, three systems are required: the lateral stopping mechanism, the returnable packages translation mechanism and the electrical control circuit.

The RPs lateral stopping mechanism is only used on one side of the trolley being a static lateral stopping mechanism present on the other side. It has function of inhibiting the RPs from falling from each trolley's level when vehicle with coupled trolleys is in motion and assist the reception and delivery of RP's to the reception station and the delivery station.

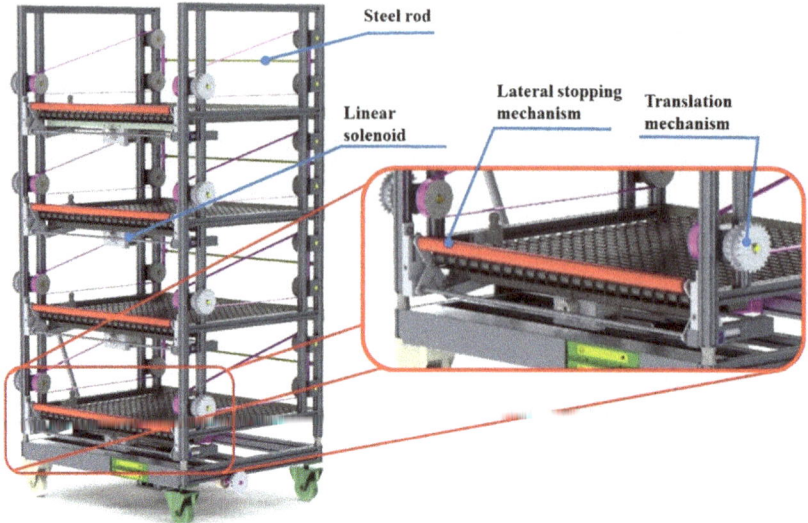

Fig. 3. Perspective view of the trolley highlighting the lateral stopping mechanism

The returnable packages translation mechanism comprises an electric motor with gearbox and gear transmission (fixed under the lower level) that transmits rotational motion to the lateral sprockets in each different level using chain rollers (not represented in Fig. 3). The rotating sprockets are connected to the pulleys by a common shaft leading to the rotation of the timing belts. A steel rod is then connected to each pair of timing belts (a pair for each level) by an insert connected to the timing belt by heat treatment.

The rotation of each pair of timing belts will then lead to the translation of the returnable packages by pushing them with the steel rod.

The RPs translation mechanism (Fig. 4) has the functions to remove all RPs from each level and to provide constant linear velocity to the RPs.

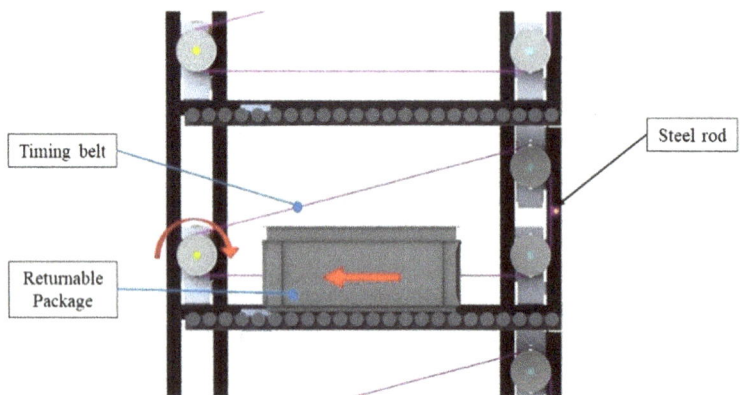

Fig. 4. Lateral view of the returnable packages translation mechanism

The handling system design requires that the power source for actuating devices should be electric, although no control board should be applied. With no control board, there is no ability to control a linear actuator to act in both directions of motion. Nevertheless, an electric control circuit, responsible for controlling all electromechanical components in the trolley, will later be described. While moving, the trolley does not have access to electric power and only when docked in the supermarket, all trolley's electromechanical components have access to electric power.

Additionally, for safety and security reasons, and to avoid problems when interacting with the supermarket's moving devices, all trolley's moving parts should not cross the structural frame's limits.

From the requirement analysis, a straight-line motion mechanism is required. Additionally, the mechanism should be attached to a passive roller that will ease the RPs removal from the trolley when vehicle is at the supermarket.

The returnable packages lateral stopping mechanism comprises an articulated mechanism actuated by a linear solenoid. Each level has its own stopping mechanism so in case of malfunction of a particular level; the electric motor will not be actuated, and RPs will not be delivered. The articulated mechanism has two possible positions: one being fully actuated and other when not actuated.

When fully actuated, by means of a linear solenoid, the mechanism lowers the passive roller to the same height of the level rollers, aiding the translation of the RPs from the trolleys to the dynamics racks because the stopping mechanism's roller assumes the same function of the other passive rollers. When not actuated, the mechanism will then return to the original position. The return movement is enabled by compression springs. The original position will inhibit the RPs from falling from the trolley when in motion. On the opposite side, a static stopping mechanism is used.

Additionally, a moving steel rod will be used for the removal of RPs from the trolley to the reception station. The moving steel's default stopping position will be guaranteed by a limit switch (one for each level).

3 Mechanism Design

The lateral stopping mechanism was modelled using Solidworks CAD/CAE software as Fig. 5 overviews. Designed mechanism uses an exact straight line motion mechanisms [1, 2], particularly a Scott-Russel type mechanism where joints X and Y are designed to be in the same part, with slider crank linkages. An additional bar was designed, linked to joint Y in order to extend the mechanism so a linear actuator can be applied. The required displacement of variable "a" is 30 mm and the input displacement is 30 mm (Fig. 6).

Nevertheless, this kind of mechanism has the ability of amplifying the output displacement [3–6]. So, a smaller input displacement could be achieved with a variant of proposed mechanism.

Where:

- F_R: Force of passive roller;
- F_{Sp}: Force of spring;
- F_{Sl}: Force of linear solenoid.

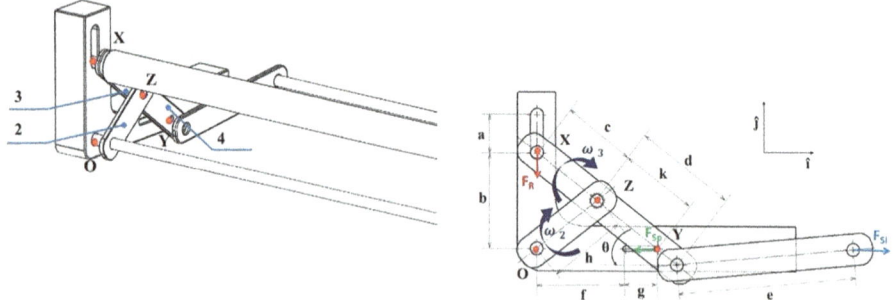

Fig. 5. Solidworks model of the SRM.   **Fig. 6.** Schematic diagram of the mechanism

The location of points Y and X with respect to the origin O are given by:

$$Y = (h+k)\cos(\theta)\hat{\imath} + (h-k)\operatorname{sen}(\theta)\hat{\jmath} \tag{1}$$

$$X = (h-c)\cos(\theta)\hat{\imath} + (h+c)\operatorname{sen}(\theta)\hat{\jmath} \tag{2}$$

and the motion of the same points is given by:

$$\nabla Y = -(h+k)\operatorname{sen}(\theta)\hat{\imath} + (h-k)\cos(\theta)\hat{\jmath} \tag{3}$$

$$\nabla X = -(h-c)\operatorname{sen}(\theta)\hat{\imath} + (h+c)\cos(\theta)\hat{\jmath} \tag{4}$$

Since the linear solenoid can only trigger in one direction, other means should be applied in order to guarantee that the lateral stopping mechanism could have two discrete positions as previously mentioned.

With the trolleys not being energized during motion, the linear solenoid can only actuate when the trolleys are docked. In this context, a compression spring should be applied to guarantee that the passive roller is in its upward position during trolley's motion.

Proposed mechanism comprises a linear solenoid with 50 N delivery force capacity, with an "almost" constant force during the 30 mm required displacement.

3.1   Spring Selection

Using Solidworks Motion's simulation, the required force for the compression spring is calculated. In Fig. 7, the required force, using a total estimated mass of 2.5 kg for the passive roller is plotted. As seen in Fig. 7, roughly 32 N are required in order to elevate the passive roller. Figure 8 shows the position of the compression spring (one on each side of the mechanism).

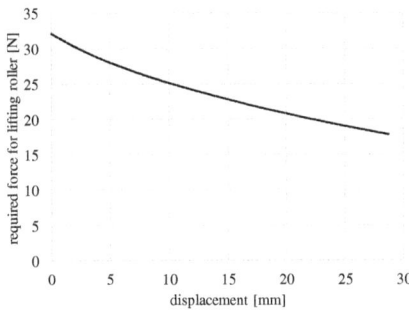

Fig. 7. Required force for lifting roller.

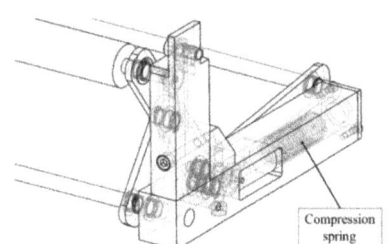

Fig. 8. Detail of the lateral stopping mechanism

To balance the motion of the mechanism, a single solenoid is considered, but two compression springs are used, splitting the required force.

The compression spring is designed with calculations based on data from standards EN 13906-1 and DIN EN 15800 for helical compression cylindrical springs [7] and has the following specifications:

- Maximum working loading: 20 N;
- Minimum working loading: 15 N;
- Spring constant: 0.17 N/mm.

4   Electric Requirements

As mentioned, no control board should be applied, and only when docked in the supermarket trolley's electromechanical components have access to electric power.

Electric power will be delivered via charging contacts, mounted in the reception station and in each trolley.

For controlling the motor actuation of both the lateral stopping mechanism and the translation mechanism, a series of sensors are applied, namely, limit switches and relays. Figure 9 shows the electrical diagram required for electromechanical devices in each trolley.

For the returnable packages translation mechanism, each level has a limit switches. The purpose is to stop the motor from actuating when the steel rod passes its highest point. The reason is that, when the bar is at its highest point, there is enough room for introducing RPs in each trolley's level. For the lateral stopping mechanism, again, a limit switch is used for each level. The aim is to detect when every solenoid is actuated, corresponding to the moment when the passive roller, in the lateral stopping mechanism, is at its lowest position, allowing the RP's from being extracted or introduced in the trolley. Only when every solenoid is actuated, the motor from the translation mechanism can be actuated for extracting the RPs, and the actuators in the delivery station can be actuated for delivering RPs to the trolleys. Additionally, a light emitter will be applied in the trolley in order to inform the reception station that the lateral

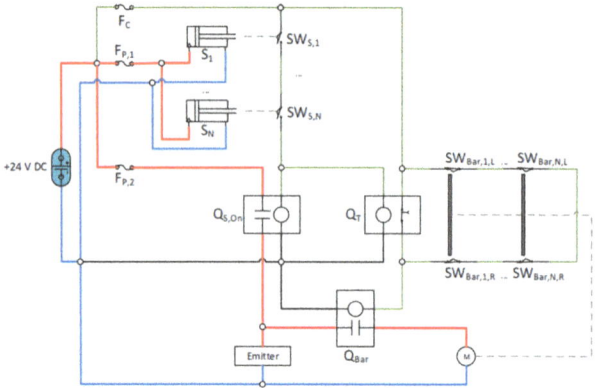

Fig. 9. Electrical diagram for each trolley

stopping mechanism is actuated and the translation mechanism is also actuated (which confirms that RPs are being delivered).

5 Conclusions

Present paper describes the mechatronic design of a straight linear motion planar mechanism to be applied in a novel materials handling system using AGVs along with its electric requirements to control a lateral stopping mechanism and a translation mechanism. Solidworks Motion is used to analyze the mechanism concerning its kinetics and kinematics characteristics, in order to identify key displacements and forces involved in the motion. A linear solenoid and compression springs where designed using simulation data in order to enable the functional requirements. The application of a SRM respected proposed requirements and enabled the use of a linear actuator to support the lowering of the passive roller. This system is now in the production step and will soon be implemented in Bosch BrgP shop floor.

Acknowledgment. This work is supported by: European Structural and Investment Funds in the FEDER component, through the Operational Competitiveness and Internationalization Programme (COMPETE 2020) [Project n° 002814; Funding Reference: POCI-01-0247-FEDER-002814].

References

1. Rao, J.S.: Kinematics of Machinery Through HyperWorks. Springer, Dordrecht (2011)
2. Sclater, N.: Mechanisms and Mechanical Devices Sourcebook, 5th edn. McGraw-Hill, New York (2011)
3. Wu, Z., Li, Y.: Design of control strategy for a novel compliant flexure-based microgripper with two jaws. In: Volume 5A: 39th Mechanisms and Robotics Conference. ASME, p. V05AT08A016 (2015)

4. Ai, W., Xu, Q.: New structural design of a compliant gripper based on the Scott-Russell mechanism. Int. J. Adv. Robot. Syst. **11**, 192 (2014). https://doi.org/10.5772/59655
5. Pinskier, J., Shirinzadeh, B., Clark, L.: Design & optimization of a compact, large amplification XY flexure-mechanism. In: International Conference on Manipulation, Automation and Robotics at Small Scales, MARSS 2017 – Proceedings (2017)
6. Chen, C.M., Hsu, Y.C., Fung, R.F.: System identification of a Scott-Russell amplifying mechanism with offset driven by a piezoelectric actuator. Appl. Math. Model. **36**, 2788–2802 (2012). https://doi.org/10.1016/j.apm.2011.09.064
7. Budynas, N.: Shigley's Mechanical Engineering Design, 8th edn. Analysis 1059 (2006)

Industrial IoT Devices and Cyber-Physical Production Systems: Review and Use Case

Eva Masero Rúbio[1], Rogério Pais Dionísio[2],
and Pedro Miguel Baptista Torres[2(✉)]

[1] Allbesmart, Lda, Avenida do Empresário,
nº 1, 6000-767 Castelo Branco, Portugal
[2] Escola Superior de Tecnologia, Instituto Politécnico de Castelo Branco,
6000-767 Castelo Branco, Portugal
pedrotorres@ipcb.pt

Abstract. The present paper describes the state of the art related to IIoT Devices and Cyber-Physical systems and presents a use case related to predictive maintenance. Industry 4.0 is the boost for smart manufacturing and demands flexibility and adaptability of all devices/machines in the shop floor. The machines must become smart and interact with other machines inside and outside the industries/factories. The predictive maintenance is a key topic in this industrial revolution. The reason is based on the idea that smart machines must be capable to automatically identify and predict possible faults and actuate before they occur. Vibrations can be problematic in electrical motors. For this reason, we address an experimental study associated with an automatic classification procedure, that runs in the smart devices to detect anomalies. The results corroborate the applicability and usefulness of this machine learning algorithm to predict vibration faults.

Keywords: Cyber-physical systems · Industrial IoT · Smart Factories
Machine learning · Predictive maintenance

1 Introduction

The Internet of Things (IoT) is a novel paradigm that is rapidly gaining ground in the scenario of modern wireless telecommunications. The basic idea of this concept is the pervasive presence around us of a variety of things or objects – such as Radio-Frequency IDentification (RFID) tags, sensors, actuators, mobile phones, etc. – which, through unique addressing schemes, can interact with each other and cooperate with their neighbours to reach common goals [1, 2].

Information and communication systems are invisibly embedded in the environment around us. In this generation, there are enormous amounts of data which must be stored, processed and presented in a seamless, efficient and easily interpretable form. Cloud computing can provide the virtual infrastructure for such utility computing which integrates monitoring devices, storage devices, analytics tools, visualization platforms and client delivery [3]. The cost-based model that Cloud computing offers

© Springer International Publishing AG, part of Springer Nature 2019
J. Machado et al. (Eds.): HELIX 2018, LNEE 505, pp. 292–298, 2019.
https://doi.org/10.1007/978-3-319-91334-6_40

will enable end-to-end service provisioning for businesses and users to access applications on demand from anywhere [4].

While the IoT affects among personal electronic devices, healthcare, or smart homes, the Industrial Internet of Things (IIoT) refers to industrial environments.

The IoT is expected to offer promising solutions to transform the operation and role of many existing industrial systems such as transportation systems and manufacturing systems. E.g., when IoT is used for creating intelligent transportation systems, the transportation authority will be able to track each vehicle's existing location, monitor its movement, and predict its future location and possible road traffic [5].

This document presents the current trends in IIoT research propelled by applications and the need for convergence in several interdisciplinary technologies. Specifically, it presents:

- Overall IIoT vision, cyber-physic systems and some possible platforms to implement it (Sect. 2)
- Communication Protocols (Sect. 3)
- OPC Servers (Sect. 4)
- Use Case using NI LabVIEW software for developing systems and data Acquisition with NI DAQ 6008 (Sect. 5).

2 IIoT and Cyber-Physical Systems

Nowadays, embedded, mobile, and cyber-physical systems (CPS) are ubiquitous and used in many applications, from industrial control systems, modern vehicles, to critical infrastructure. Current trends and initiatives, such as Industry 4.0 and Internet of Things promise innovative business models and novel user experiences through strong connectivity and effective use of next generation of embedded devices [6].

Industrial and IP-enabled low-power wireless networking technologies are converging, resulting in the IIoT. On the one hand, low-power wireless solutions are available today that answer the strict reliability and power consumption requirements of industrial applications. On the other hand, a range of standards have been published to allow low-power wireless devices to communicate using the Internet Protocol (IP) [7].

Cyber-physical systems are the indispensable technological link for the merging of real and virtual worlds. Cyber-physical systems consist of objects featuring embedded software and electronics that are connected with each other or via the Internet to form a single networked system. They include sensors and components for moving or controlling a mechanism or system, so called actuators linking the CPS to the outside world. The sensors enable the systems to acquire and process data. The data is subsequently made available to network-based services that use actuators to directly impact measures taken in the real world (Fig. 1). This leads to the merging of the physical world and cyberspace into the Internet of Things [8].

The integration of shop floor devices into IIoT platforms, with the concept of Smart-Object (SO), which will enable processing of data to aid decision-making, not only with the information of one device, but also with several devices, forming a cyber-physical system.

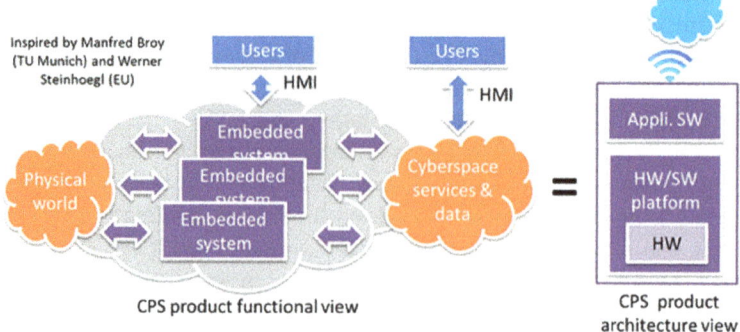

Fig. 1. Architecture of a CPS [9].

CPSs are used in industrial production to build up Internet-based architectures that facilitate remote control of stand-alone production systems. This system of sub-systems is composed of application software running on a hardware/software platform, including specific hardware components (Fig. 2).

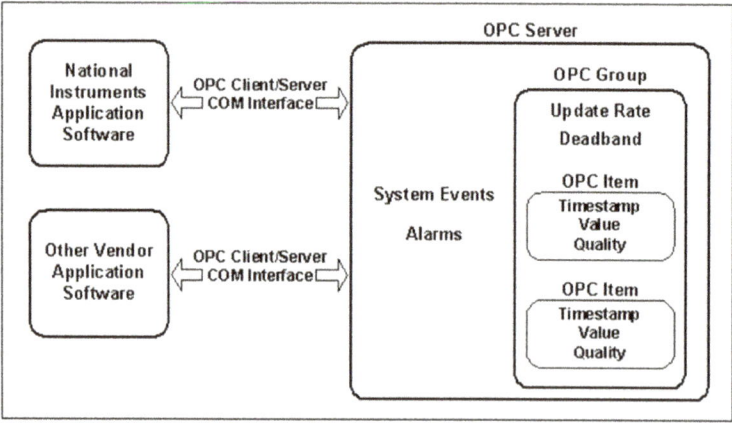

Fig. 2. An example implementation of the OPC specification [11].

3   OPC Servers

OPC is a standard interface between numerous data sources, – such as programmable logic controllers (PLCs), remote terminal units (RTUs), and sensors on a factory floor – to HMI/SCADA applications, application tools, and databases. With OPC, device-side server and application software can communicate without duplicating device driver development and providing support for hardware feature changes [10].

OPC servers provide a method for many different software packages to access data from a process control device. We have an OPC Server and one or more OPC Clients that communicate with the server to write or read data. In this document, we will use MatrikonOPC Server.

3.1 Matrikon OPC Server

MatrikonOPC Simulation Server is a free utility used to help test and troubleshoot OPC applications (clients) and connections. Testing applications on "live" OPC servers may result in loss of actual production data. The MatrikonOPC Simulation Server creates a simulated environment so that in the event of a problem, no real process data is lost.

4 Industrial Communications

Industrial communications are at the heart of automated systems. Numerous defined standards exist (RS-485, AS-I, PROFIBUS, PROFINET), and these standards continue to grow as new industrial Ethernet protocols emerge [12].

The vertical integration of management execution systems with factory floor equipment has led to the convergence of the Ethernet TCP/IP protocol with industrial field busses. Several industrial Ethernet protocols have emerged, including Profinet, Ethernet/IP, ModbusTCP/IP, EtherCAT®, and Ethernet Powerlink. Each protocol addresses very specific needs, so requirements on the hardware platform vary accordingly.

4.1 Rs-485

RS-485 is a standard defining the electrical characteristics of drivers and receivers for use in serial communications systems. RS-485 allows multiple devices (up to 32) to communicate at half-duplex on a single pair of wires, plus a ground wire (more on that later), at distances up to 1200 m (Fig. 3). Data is transmitted differentially on two wires twisted together, referred to as a "twisted pair." The use of a balanced line means RS485 has excellent noise rejection and is ideal for industrial and commercial applications [13].

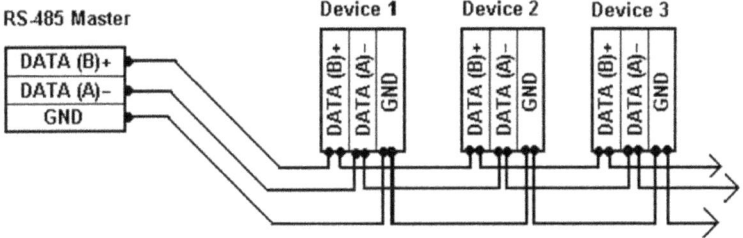

Fig. 3. Wire RS-485 connections [13].

4.2   EtherCAT®

EtherCAT® (Ethernet for Control Automation Technology) is an open industrial real-time and deterministic Ethernet Fieldbus standard that is widely regarded as the best industrial network to date [14]. EtherCAT® is based on the principle of 'Ethernet on the fly' processing, a driving factor for it's incredibly fast cycle times for industrial automation applications. It supports a wide variety of standard application layer implementations, including: CANopen over EtherCAT (CoE), Ethernet over EtherCAT (EoE), File-access over EtherCAT (FoE), Vendor-specific over EtherCAT (VoE) and Servo Profile over EtherCAT (SoE) mailbox protocols. These previous communication profiles allow multiple vendors to make fully compliant devices with the same application implementations [15].

- CANopen over EtherCAT (CoE): This protocol is used to configure slaves and exchange data objects on application level.
- Ethernet over EtherCAT (EoE): The EtherCAT master implements the Ethernet over EtherCAT mailbox protocol to enable the tunnelling of Ethernet frames to special slaves, that can either have physical Ethernet ports to forward the frames to, or have an own IP stack to receive the frames.

The process management level has special communication requirements that differ slightly from the requirements addressed by the EtherCAT Device Protocol, described previously. Machines or sections of a machine often need to exchange status information and information about the following manufacturing steps with each other. Additionally, there is usually a central controller that monitors the entire manufacturing process, which provides the user with status information on productivity, and assigns orders to the various machine stations. The EtherCAT Automation Protocol (EAP) defines interfaces and services for:

- Exchanging data between EtherCAT master devices (master-master communication)
- Communication to Human Machine Interfaces (HMI)
- A supervising controller to access devices belonging to underlying EtherCAT segments (Routing)
- Integration of tools for the machine or plant configuration, as well as for device configuration.

Specifically, in this chapter, we focus on EtherCAT® industrial protocol with Arduino [15]. The union of the world Arduino and the world EtherCAT® will allow you to create innumerable devices for the automation in a simple, rapid and economic way. Some of the possible applications are:

- I/O devices both analogical and digital with possibility of pre-elaboration and elaboration of the signals as averages, linearization, filters, among other functions.
- Generators of PWM signals to pilot motor and other.
- Serial gateways RS-232, RS-485, CAN, Modbus TCP/IP, from and toward EtherCAT®.

5 Use Case

One of the important concepts related to smart devices is the ability to implement machine learning techniques, automatically identify machine anomalies or predict future events. In this work, the classification of vibration gravity was implemented to predict anomalies in electric motors. This event identification and classification automatically assists maintenance technicians and engineers to predict future problems and act in advance.

Machine learning is a kind of artificial intelligence that provides systems with the ability to learn without being explicitly programmed. Machine learning focuses on developing computer programs that can be changed, when exposed to new data. The algorithms use data to detect patterns and adjust program actions accordingly. Typically, these algorithms can be categorized as being supervised, unsupervised and reinforcement learning. Supervised algorithms can apply what has been learnt in the past to new data. Unsupervised algorithms can draw inferences from datasets. In reinforcement learning, the algorithm learns a policy of how to act given an observation of the world. Every action has some impact in the environment, and the environment provides feedback that guides the learning algorithm [16].

Supervised learning includes two categories of algorithms, classification for categorical response values, where the data can be separated into specific "classes" and regression for continuous-response values. The vibration's motor severity was classified through the k-nearest neighbor's algorithm (k-NN). The standard ISO 2372 it was used in the k-NN supervised learning procedures. This standard indicates a suggested classification, according to vibration's severity range limits (velocity) and machines classes, in four groups: A = Good, B = Satisfactory, C = Unsatisfactory and D = Unacceptable.

The classification model is trained considering 4000 samples cases, with evident occurrences of the four classes. Before classification, we need to identify the events. For the example of vertical vibration, the system detects an upper peak of vibration velocity, depicted in Fig. 4. Based on the magnitude of the peaks, the system can automatically classify the vibrations. In this way, it is possible to classify vibration's severity in one of these four groups. As the value of the peak is 3,9 mm/s and the class

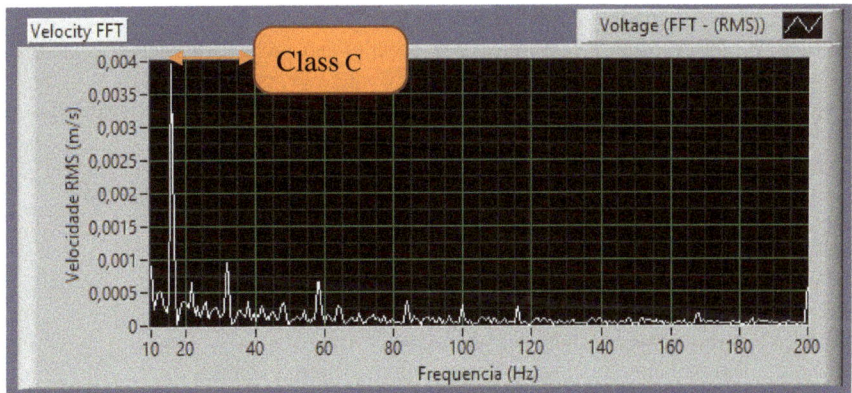

Fig. 4. User interface: velocity's FFT, identification and classification result.

of motor is Class I (<20 HP), the vibration's severity is the Class C, that is, the vibrations are 'Unsatisfactory' – according to ISO 2372 such as it was indicated in the previous data analysis.

6 Conclusion

This paper gives an overview of the technologies currently available to implement a CPS, describing the architecture, communication protocols and information servers. In the study case, the CPS analyse and record the vibration data of an electrical motor to identify the vibration's severity and to implement a predictive maintenance using decision tree classifiers according to machine learning fundamentals. The proposed implementation successfully detected the vibration severity fault and allowed to demonstrate the concept of Cyber-Physical systems.

Further developments of this work will include the deployment and testing of CPSs using the proposed techniques in real industrial environments, with harsh environmental conditions and a diversity of production machines in the shop floor.

Acknowledgments. This   research   was   supported   by   project   PRODUTECH-SIF (COMPETE2020).

References

1. Liu, Y., Peng, Y., Wang, B., Yao, S., Liu, Z.: Review on cyber-physical systems. IEEE/CAA J. Autom. Sinica **4**(1), 27–40 (2017)
2. Atzori, L., Lera, A., Morabito, G.: The Internet of Things: a survey. Comput. Netw. 1–19 (2010)
3. Silva, R., Reis, J., Neto, L., Gonçalves, G.: Universal parser for wireless sensor networks in industrial cyber physical production systems. In: 2017 IEEE 15th International Conference on Industrial Informatics (INDIN), pp. 633–638, July 2017
4. Gubbi, J., Buyya, R., Marusic, S., Palaniswami, M.: Internet of Things (IoT): a vision, architectural elements, and future directions. 1–19 (2013)
5. Da Xu, L., He, W., Li, S.: Internet of things in industries: a survey. IEEE Trans. Ind. Inform. **10**(4), 2233–2243 (2014)
6. Sadeghi, A., Wachsmann, C., Waidner, M.: Security and privacy challenges in industrial internet of things. In: IEEE Design Automation Conference, pp. 1–6 (2015)
7. Dujovne, D., Watteyne, T., Vilajosa, X., Thubert, P.: Deterministic IP-enabled industrial internet (of things). IEEE Commun. Mag. (2015)
8. Cyber physical system. https://www.hbm.com. Accessed 21 Jan 2018
9. Architecture of a Cyber physical system. www.eurocps.org. Accessed 21 Jan 2018
10. OPC servers. http://home.hit.no. Accessed 21 Jan 2018
11. OPC specification. http://ni.com. Accessed 21 Jan 2018
12. Industrial communication. http://www.atmel.com. Accessed 21 Jan 2018
13. RS-485 protocol. www.bb-elec.com. Accessed 21 Jan 2018
14. EtherCAT. https://www.beckhoff.com. Accessed 21 Jan 2018
15. EtherCAT and Arduino. http://www.bausano.net. Accessed 2 Oct 2017
16. Safavian, S.R., Landgrebe, D.: A survey of decision tree classifier methodology. IEEE Trans. Syst. Man Cybern. **21**(3), 660–674 (1991)

Scientific and Methodological Approach for the Identification of Mathematical Models of Mechanical Systems by Using Artificial Neural Networks

Ivan Pavlenko[1], Justyna Trojanowska[2(✉)], Vitalii Ivanov[1],
and Oleksandr Liaposhchenko[1]

[1] Sumy State University, 2 Rymskogo-Korsakova St., Sumy 40007, Ukraine
[2] Poznan University of Technology, Chair of Management
and Production Engineering, 3 Piotrowo Street, 60-665 Poznan, Poland
justyna.trojanowska@put.poznan.pl

Abstract. The article is aimed at developing the scientific and methodological approach of using artificial neural networks (ANN) for solving applied problems in the field of mechanical engineering. This approach is based on the comprehensive implementation of ANN with the modern methods of numerical analysis (e.g., the finite element method) and analytical methods of the research with the use of mathematical modeling of the dynamic state for mechanical systems. Conceptual schemes for the implementation of the abovementioned approach are proposed for solving a number of interdisciplinary problems, such as investigation of the dynamics for rotary machines and hydroaeroelastic interaction of gas-liquid mixtures with deformable structural elements, as well as the dynamic analysis of fixtures. The main advantages of the proposed approach in comparison with the traditional regression analysis are the ability to learn and improve the ANN architecture, and to solve nonlinear problems of the parameters' identification for mathematical models by using data of the results of physical experiments and numerical simulations. This approach allows refining parameters of the linear and nonlinear mathematical models describing the complicated mechanical and hydro-mechanical interactions under the impossibility of determination of an absolutely precise solution of the equations describing the process, as well as the incompleteness of the initial data.

Keywords: ANN architecture · Estimation of parameters
Finite element method · Nonlinear characteristics · Numerical simulation
Regression analysis

1 Introduction

Artificial neural networks (ANN) were originated as an attempt for modeling the information processing capabilities of nervous systems. And now they can be considered as the approach for solving applied computational problems, such as using e-sourcing platform for search of potential suppliers [1], in an application in the control system of an automated people mover [2] or in production process optimization [3–5].

© Springer International Publishing AG, part of Springer Nature 2019
J. Machado et al. (Eds.): HELIX 2018, LNEE 505, pp. 299–306, 2019.
https://doi.org/10.1007/978-3-319-91334-6_41

A comprehensive description of ANN models and approaches for their creation are presented in the works [6–8]. This point of view allows concluding, that ANN can be explained as an extremely simplified model of the brain. It widely used for solving the problems of the classification (pattern recognition, feature extraction, image matching), noise reduction, and prediction (or extrapolation) due to the following advantages over the traditional regression procedure can be stated:

- ANN figures out how to perform its function;
- determination of the function based only upon the sample inputs;
- ability to generalize, i.e. to produce reasonable outputs for the corresponding inputs.

However, ANN remains an almost unexplored for solving applied problems in the field of engineering mechanics due to the absence of the unified scientific and methodological approach for its implementation. Moreover, there is no single ANN notation that is universally accepted. Research works in many diverse fields (e.g. engineering, physics, mathematics) tend to use vocabulary peculiar to their specialty. As a result, many authors made to seem ANN more complicated than they actually are [9].

Due to the abovementioned, this paper is devoted to generalize the procedure of solving a wide range of applied research problems in the field of mechanical engineering and related fields of science on the following examples:

- estimation of eigenvalues for dynamic systems;
- solving non-linear problem of the identification of parameters of mathematical models for dynamic systems;
- analysis of the dynamic state of mechanical systems;
- estimation of parameters for mechanical and hydro-mechanical systems.

It should be noticed, that all the research problems can be conditionally divided into two categories:

- conducting the directional research by using ANN as the generalized regression procedure;
- solving the inverse problem for the identification of unknown parameters of reliable mathematical models.

These problems can be solved by assuming that ANN is generally able to train itself by initial data (empirical or analytical) [10]. There is a widened set of publications available about the application of different kind of approaches for solving collaborative projects in the area of production engineering, for instance [11–13].

2 Materials and Methods

ANN can be useful for solving directional research due to its ability to approximate obtained experimental results. The cooperation of many "neurons" of ANN performs the evaluated parameters. First significant attempts to implement ANN in the field of mechanical engineering were made in the work [14] for the case of continuous mechanical systems. Another example of the application of ANN in civil engineering is demonstrated within the research paper [15] for solving complex engineering problems for the construction of design processes. ANN can also be successfully used for prognostic modelling in different engineering fields.

The design schemes of the application of ANN for the directional research and for solving the inverse problem are presented on Fig. 1.

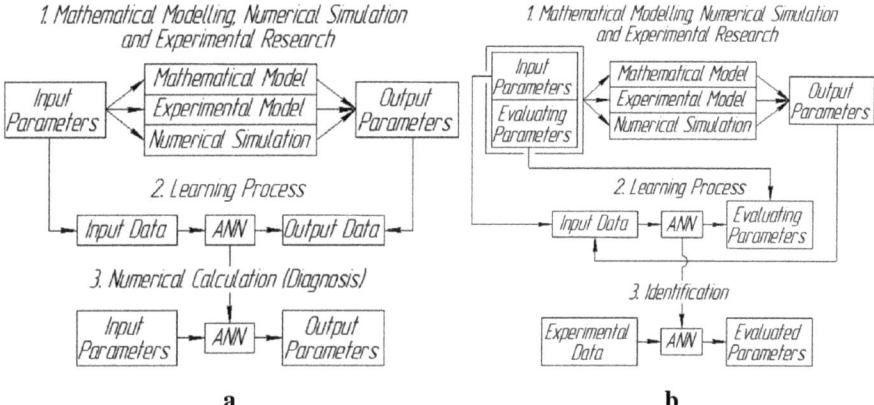

Fig. 1. Design scheme of the application of ANN for the directional research (a) and for solving the inverse problem (b)

The directional research (Fig. 1a) is used for numerical simulation of mechanical systems. However, the unknown parameters for the comprehensive interdisciplinary problem can be evaluated only by using the inverse research (Fig. 1b).

3   Results and Discussion

3.1   Estimation of Critical Frequencies of the Rotor

The mathematical model of the rotor dynamics for centrifugal machines is proposed on the work [16]. This model can be realized by using ANN for determining critical frequencies in the case of non-linear mathematical model. The design schemes of the application of ANN for the determination of critical frequencies of the rotor, as well as for the evaluation of bearing stiffness coefficients are presented on Fig. 2.

This methodology has been realized for the multistage centrifugal compressor 295GC2-190/44-100 M by using ANN for input data (coefficients of the nonlinear stiffness), three hidden layers (the total number of neurons – 31), output data (critical frequencies) for the following settings: learning rate – 0.001; transfer function – hyperbolic tangent; total number of training cycles – $1 \cdot 10^{6}$; target error – $1 \cdot 10^{-5}$; initialization method of threshold – random; initialization of weight factor – random; update interval – 500 cycles. Maximum relative error of the evaluated results is 0.9%.

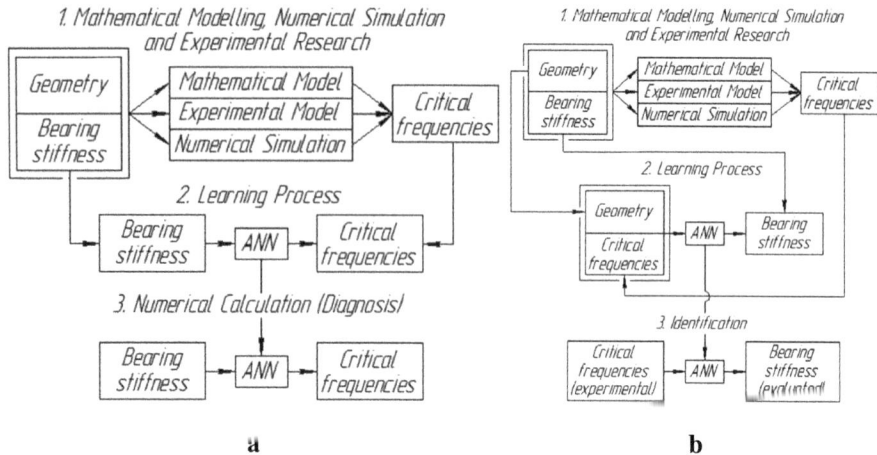

Fig. 2. Design scheme of the application of ANN for the determination critical frequencies of the rotor (a) and evaluation of bearing stiffness coefficients (b)

3.2 Dynamic Balancing of the Rotor

The mathematical model of the rotor balancing procedure for rotors supported on ball bearings of centrifugal machines is proposed on the work [17] with the use of the combined application of finite element analysis. This model can be realized by using ANN for determining the system of residual imbalances by the experimental data of measured displacements of the rotor.

The design schemes of the application of ANN for the dynamic analysis of the rotor under the system of imbalances and for rotor balancing are presented on Fig. 3.

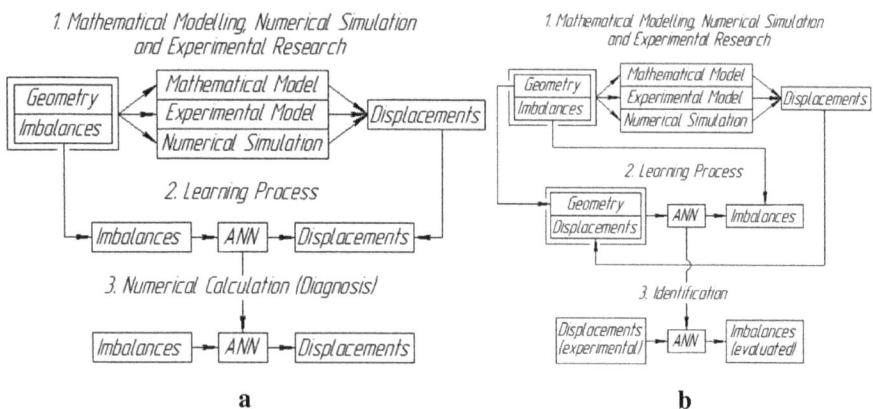

Fig. 3. Design scheme of the application of ANN for the dynamic analysis of the rotor (a) and rotor balancing (b)

3.3   Dynamic Analysis of the Mechanical System "Fixture – Workpiece"

General approach for the numerical simulation of the system "fixture–workpiece" is proposed in the work [18] on the example of lever machining. For this field of study, ANN can be useful for determining unknown system of cutting and clamping forces.

The design schemes of the application of ANN for the dynamic analysis of the mechanical system "fixture – workpiece" under the clamping and cutting forces, as well as for the evaluation of these forces are presented on Fig. 4.

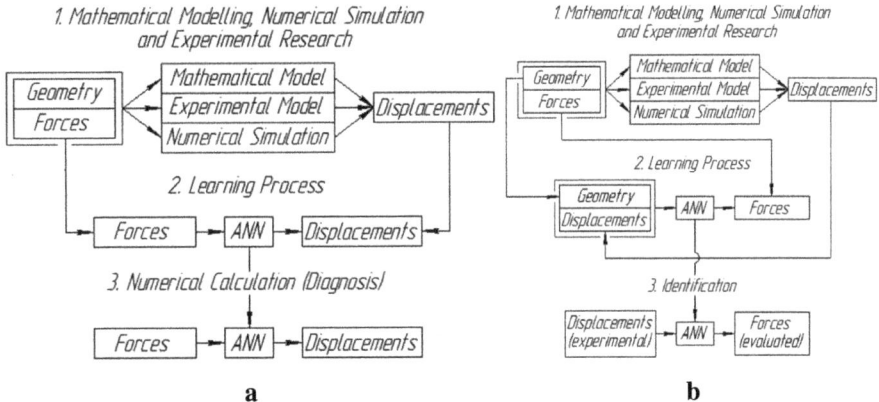

Fig. 4. Design scheme of the application of ANN for the dynamic analysis of the mechanical system "fixture – workpiece"

3.4   Estimation of the Parameters of the Inertial Dynamic Separation Device

The mathematical model of crossed movement and gas-liquid flow interaction with captured liquid film in the inertial-filtering separation channels is proposed in the paper [19], as well as the appliance of the inertial gas-dynamic separation of gas-dispersion flows in the curvilinear convergent-divergent channels for the improvement of the compressor equipment is stated in the paper [20]. ANN can be implemented to these models for solving the problem of the identification of parameters of the mathematical model of the dynamic state for abovementioned hydro-mechanical systems.

The design schemes of the application of ANN for the dynamic analysis of the hydro-mechanical system "multiphase flow – flexible element" for the inertial dynamic separation device, as well as for the evaluation of the added mass for the mathematical model of the interaction between the multiphase flow and flexible dynamic elements are presented on Fig. 5.

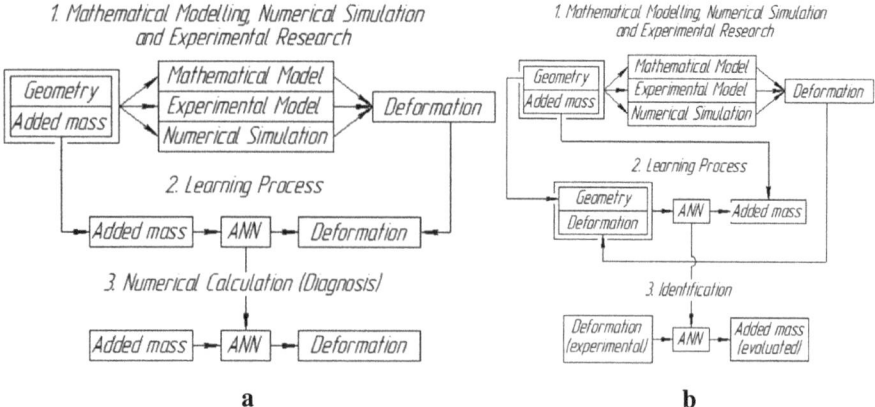

Fig. 5. Design scheme of the application of ANN for the dynamic analysis of the hydro-mechanical system "multiphase flow – flexible element" (a) and for the evaluation of the added mass (b)

4   Conclusions

The scientific and methodological approach of using ANN for solving applied problems in the field of mechanical engineering is proposed. The corresponding design schemes are presented for the identification of mathematical models for the following mechanical and hydro-mechanical systems: "rotor – bearing supports", "fixture – workpiece", "multiphase flow – flexible elements". This approach allows solving the direct and inverse interdisciplinary problems on the basis of the comprehensive implementation of ANN, numerical simulation and analytical methods. The main advantages of the proposed approach in comparison with the linear and nonlinear regression analysis are stated. Additionally, the proposed methodology can be used for solving both for the comprehensive mechanical engineering and interdisciplinary problems.

All the above mentioned design schemes should be implemented within the further research for solving complicated problems in the field of mechanical engineering.

Acknowledgements. All the achieved results were obtained within the following research projects commissioned by the Ministry of Education and Science of Ukraine: "Investigation of rotor dynamics for turbopump units and reciprocating compressor plants" (No. 0117U004922); "Development and implementation of energy efficient modular separation devices for oil and gas purification equipment" (No. 0117U003931, No. 15.01.06–01.17/20.ZP).

References

1. Fuks, K., Kawa, A., Wieczerzycki, W.: Improved e-sourcing strategy with multi-agent swarms. In: Computational Intelligence for Modelling Control & Automation, pp. 488–493 (2008)
2. Kunz, G., Machado, J., Perondi, E.: Using timed automata for modeling, simulating and verifying networked systems controller's specifications". Neural Comput. Appl. **28**(5), 1031–1041 (2017)
3. Santos, A.S., Varela, M.L.R., Putnik, G.D., Madureira, A.M.: Alternative approaches analysis for scheduling in an extended manufacturing environment. In: Proceedings of the Nature and Biologically Inspired Computing, pp. 97–102 (2014)
4. Sika, R., Rogalewicz, M.: Methodologies of knowledge discovery from data and Data Mining methods in mechanical engineering. Manag. Prod. Eng. Rev. **7**(4/2016), 97–108 (2016)
5. Dostatni, E., Diakun, J., Grajewski, D., Wichniarek, R., Karwasz, A.: Multi-agent system to support decision-making process in ecodesign. In: Herrero, Á., Sedano, J., Baruque, B., Quintián, H., Corchado, E. (eds): Advances in Intelligent Systems and Computing, vol. 368, 463–474, Springer, Cham (2015)
6. Rojas, R.: Neural Networks. Springer, Heidelberg (1996)
7. Kriezel, D.: A Brief Introduction to Neural Networks (2008). http://www.dkriesel.com/science/neural_networks. Accessed 2017/12/10
8. Ratnayake, R.M.C., Antosz, K.: Development of a risk matrix and extending the risk-based maintenance analysis with fuzzy logic. Procedia Eng. **182**, 602–610 (2017)
9. Hagan, M.T., Demuth, H.B., et al.: Neural network design. 2nd edn. Pws publishers, Boston. http://hagan.okstate.edu/nnd.html. Accessed 10 Dec 2017
10. Zhao, L., Li, W., et al.: Artificial neural networks based on fractal growth. Adv. Autom. Robot. **2**, 323–330 (2011)
11. Ferraz, A., Brito, J., Carvalho, V., Machado, J.: Blood type classification using computer vision and machine learning. Neural Comput. Appl. **28**, 2029–2040 (2017)
12. Putnik, G.D., Ferreira, L., Shah, V., Putnik, Z., Castro, H., Cruz-Cunha, M.M., Varela, L.: Effective service dynamic packages for ubiquitous manufacturing system. Virtual and Networked Organizations, Emergent Technologies and Tools, pp. 207–219. Springer, Heidelberg (2011)
13. Varela, M.L.R., Ribeiro, R.A.: Distributed manufacturing scheduling based on a dynamic multi-criteria decision model. In: Zadeh, L., Abbasov, A., Yager, R., Shahbazova, S., Reformat, M. (eds): Recent Developments and New Directions in Soft Computing. Studies in Fuzziness and Soft Computing, vol. 317, pp. 81–93. Springer, Cham (2014)
14. Kapania, R.K., Liu, Y.: Applications of artificial neural networks in structural engineering with emphasis on continuum models. Virginia Polytechnic Institute and State University, Blacksburg (1998)
15. Lazarevska, M., Knezevic, M., et al.: Application of artificial neural networks in civil engineering. Tehnicki vjesnik **21**(6), 1353–1359 (2014)
16. Pavlenko, I.: Static and dynamic analysis of the closing rotor balancing device of the multistage centrifugal pump. Applied Mechanics and Materials, vol. 630, pp. 248–254. Trans Tech Publications, Zurich (2014). DOI: https://doi.org/10.4028/www.scientific.net/AMM.630.248
17. Pavlenko, I.V., Simonovskiy, V.I., Demianenko, M.M.: Dynamic analysis of centrifugal machines rotors supported on ball bearings by combined application of 3D and beam finite element models. IOP Conference Series: Materials Science and Engineering, vol. 233, pp. 1–8, 012053. https://doi.org/10.1088/1757-899x/233/1/012053

18. Ivanov, V.O., Mital, D., et al.: Numerical simulation of the system "fixture–workpiece" for lever machining. Int. J. Adv. Manuf. Technol. **91**(1–4), 79–90 (2017). https://doi.org/10.1007/s00170-016-9701-2
19. Liaposhchenko, O., Pavlenko, I., Nastenko, O.: The model of crossed movement and gas-liquid flow interaction with captured liquid film in the inertial-filtering separation channels. Sep. Purif. Technol. **173**, 240–243 (2017)
20. Liaposhchenko, O.O., Sklabinskyi, V.I., et al.: Appliance of inertial gas-dynamic separation of gas-dispersion flows in the curvilinear convergent-divergent channels for compressor equipment reliability improvement. IOP Conference Series: Materials Science and Engineering, vol. 233, pp. 1–8, 012025 (2017)

A Formal Approach for Railroad Traffic Modelling Using Timed Automata

Camelia Avram[1](✉), Karolina Bezerra[2], Dan Radu[1], Jose Machado[2],
and Adina Astilean[1]

[1] Technical University of Cluj Napoca, Cluj Napoca, Romania
camelia.avram@aut.utcluj.ro
[2] Department of Mechanical Engineering, University of Minho, Braga, Portugal

Abstract. The present paper proposes a new modelling framework well-suited to describe in detail multiple modelling features and interactions among different components of railway transport systems. Timed automata and Uppaal software tool were chosen because allow not only the introduction of additional features but also the construction and the verification of timed models. Another important advantage of the presented formalism is given by the possibility to operate modifications on the simulated maps only by changing the initialization variables and without changing the UPPAAL model. Different templates were defined to represents the complexity of railroad train station, two types of structures being taken into consideration: infrastructure – *road, switches, traffic lights, platforms*, and moving vehicles – *engines, cars*. To model and analyze the two types of structures, a two layered model based on Timed Automata was considered; one represents the components of the infrastructure and the relations between them and the second one represents the mobile components of the railroad traffic. The interactions between layers were defined using a set of rules implemented in "C" programming language. A case study was considered for simulation purposes, some possible scenarios being presented. A list of queries was generated in order to verify various properties. The proposed framework proved to be capable to model railway complex structures and to verify their behavior.

Keywords: Railroad traffic control · Timed automata model
Computer simulation · Formal verification

1 Introduction

The constant increasing of freight transport volume and number of passengers using rail transport lead to new, various problems which must be solved in this field. The structural characteristics, organizational mechanisms and the wide range of requirements of railway traffic are the main reasons for which many different approaches are needed to solve the specific problems of the domain. The modernization and innovations efforts imposed both by the exigencies of customers and the increased complexity of the system is supported by actual research orientations grouped around the main problems of different sectors.

© Springer International Publishing AG, part of Springer Nature 2019
J. Machado et al. (Eds.): HELIX 2018, LNEE 505, pp. 307–314, 2019.
https://doi.org/10.1007/978-3-319-91334-6_42

A special attention was given to the optimization of train movement many studies trying to find new possibilities to improve the efficiency of usage of railway infrastructure. Many research papers focuses on a rescheduling of trains timetables, based on new strategies and using different methods and algorithms [1] Analysis methods were developed to highlight the interactions among a multitude of components and the specific factors influencing the systems performances [2, 3]. The determination of optimal train stopping patterns [4], the analysis of delay effects on the passengers and freight traffic and the elaboration of new models for simulation purposes [5] are only some of other related research themes corresponding to actual orientations. Factors which influence delay propagation were identified and analyzed and some of their interdependencies were highlighted [6].

Aspects related to operational activities carried out in railway stations were also approached in actual research studies, playing an important role in processes aiming to improve service quality and to increase considered performances indicators. The management of railway stations is a distinct issue because it implies various specific aspects and requirements. Some of the most important refer to passengers and train movements, platforms usage, traffic routing and priorities, rolling stocks scheduling, security and safety assurance.

In the conditions in which the allocation of resources in railway stations is characterized by a pronounced dynamicity, standard specific rules, both for passengers and freight transport, have to be respected. Many proposed passengers transfer solutions are based on studies analyzing passengers' behaviors [7] and estimations of travelers' flows [8].

A special place is occupied by the scheduling of activities in marshalling yards [9] such as arrivals, verifying sorting, grouping of rail cars and the connections with the outbound trains. The performances of these operations influence in great measure the global time efficiency of the corresponding stations.

Though many modelling and simulation techniques were proposed to offer support for operational problems and the scheduling of activities in railway networks [10–14], still there are a lot of aspects which must be integrated and analyzed to improve the efficiency of different approaches.

Taking into account the above enumerated aspects, the present paper proposes a new modelling framework well-suited to describe in detail multiple modelling features and interactions among different components of railway transport systems. Timed automata and Uppaal software tool were chosen not only because allow the introduction of additional features but also because allow the construction and the verification of timed models.

2 Railroad Train Station Modelling

2.1 Modelling Aspects of Railroad Traffic

In order to accurately model the railroad traffic, some aspects should be taken into consideration as follows:

– Modelling different types of vehicles and their behavior;

- Modelling the infrastructure (railroad with different sizes);
- Modelling switches and traffic lights;
- Modelling speed and acceleration of vehicles;
- Modelling entry and exit points for traffic (possible intersections with road traffic);
- Modelling sensors for exact location on each traffic cell, balks (located in front of the engine), switches and crossings (to control the road traffic around the entering and exiting points of the train stations);
- Modelling platforms and railroad traffic line cells;
- Modelling pedestrians (only to specify the location for each car according to traveling tickets).

To accomplish the above mentioned objectives, a modular approach using timed automata was considered to develop a new methodology for modelling and verification purposes. It is appropriated to define modular models for each vehicle, for each railroad for each traffic line, for each switch, for each incoming and outgoing traffic and for each physical component.

The railroad traffic line cell was defined as the space occupied by a vehicle and the safety distance around (front and back) according to average speed (the safety distance can vary in direct correlation with the travelling speed).

In order to deal with this complexity, it was decided to create a model based on a matrix that will describe in an accurate way, the behavior of the considered railroad station and traffic participants. Matrices and vectors were used to model all the parts of this complex systems.

This methodology features the following main characteristics:

- The physical environment is a one dimensional grid of rectangular cells;
- It is a single cell model because each elements will occupy only one railroad's cell for each time iteration;
- The time is a stochastic feature and its choice is completely non-deterministic;
- This model can be easily extended;
- Modelling the railroad station.

Different templates were defined to represent the complexity of railroad train station (two types of structures are taken into consideration: infrastructure – *road, switches, traffic lights, platforms*, and moving vehicles – *engines, cars*).

The formalism adopted to model the system was Timed Automata (TA). UPPAAL environment was used for performing the simulation and as model checker.

To model and analyze the above mentioned types of structures, a two layered model based on Timed Automata was considered; one layer represents the components of the infrastructure and the relations between them (Fig. 1) and the second one represents the mobile components of the railroad traffic (Fig. 2). The interactions between layers are defined by a set of rules which are implemented in "C" programming language.

The circulation of the individual vehicles in a traffic flow is described by a set of rules that reflects the movements of trains and the switching lane behavior, evolving in time and space.

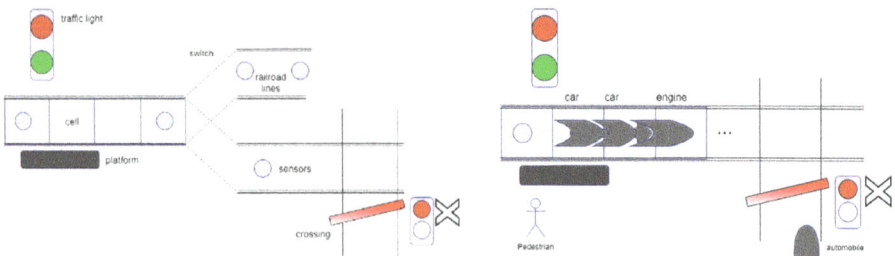

Fig. 1. The first layer of the structure   **Fig. 2.** The second layer of the structure

The characteristics of the defined structures include:

- *Railroad lanes* – code, section, length, number of cells;
  Switch – code of the connected railroads lanes, code of the switch;
- *Traffic signs* – semaphore for railroad traffic, semaphore for road traffic;
- *Platform* – are located alongside of the railroad lanes;
- *Pedestrian* – located on the platform;
- *Automobile* – just as on/off traffic;
- *Sensors* – railroad line, balks detection.

The use of this formalism to represent the railroad traffic allows us to have a general representation and in the same time to model complex structures. The modification of the simulated map is possible only by changing the initialization variables and without changing the UPPAAL model.

The intersection between modelling layers:

- *First layer* – define components characteristics, variables, constants, functions to express the evolution of matrices and lists;
- *Second Layer* – define components characteristics, variables, constants, functions defined in C language;

The two layers interaction is realized by specific defined functions and queries. The interactions between the two layers and the behavior of the two layers are validated, verified and several charts and analysis are generated.

3   Application of the Methodology – Case Study

3.1   Presentation of Different Situations

The case study considered for simulation purposes involved a small group of railroads, switches, traffic lights and sensors; It contains the maximum possible of traffic-elements, and is presented in Fig. 3.

This section details the physical environment of the simulation, describing the railroad (number, direction and length of the railroads segments), the location of the pedestrian crossings, platforms, goods train segments, sensors, freight tram stations,

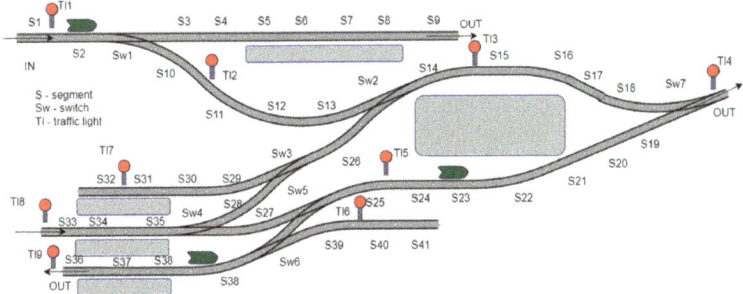

Fig. 3. Representation of the Entire railroad network considered in this simulation

tram railways, the intersections between roads, and the way traffic lights for road, rail, and pedestrian traffic operate.

For each element, various pieces of information were collected and structured, such as:

Length; Link between segments (previous and following segments); Semaphore; Sensors (two – located at the beginning and at the end of the segment); Pedestrian platform location (GPS coordinates); Type of railroad vehicle allowed.

The UPPAAL model is based on templates for: segments, traffic lights, and railroad vehicles. In the global declaration first is declared the number of vehicles, segments, traffic lights and switches using the function "typedef".

This variable defines the maximum length of the considered railroad (maximum number of cells). With this new variable and with the variable number of segments ("noS") was created the matrix "idexSC". The matrix is a map of the railroad cell coordinates. Each line corresponds to a segment ID of which first line ID is 0 and of which last line ID is the last segment's ID. The column number 1 corresponds to the number of cells of each segment and the following columns are filled with −1 which means that the segment's cell is empty. Therefore it is easy to extend the map and to implement other features in subsequent models.

3.2   Simulation and Verification of Results

To implement the considered map, 41 railroad segments, 9 traffic lights, 6 switches and 20 railroad vehicles (consisting of one engine and at least 5 cars) were declared. Afterwards, the traffic elements and the pre-defined routes for some of the trains were implemented.

For the formal verification of the model it is enough to know that, in UPPAAL version, the logic A is the universal quantifier on paths (for any path...), E is the specific quantifier on paths (there is a path...), [] is the universal quantifier over states in a path (for any state...) and <> is the specific quantifier over states in a path (there is a state...).

To test the correct behavior of this model, in the option Verifier of UPPAAL were implemented several queries. The initial values and variables taken into consideration for the formal verification of the map are presented in Fig. 3.

A list of queries was generated in order to verify various properties. Some of the queries were built in order to double check some important railroad traffic behaviors. The response of the query could either be "yes" or "no". Based on these responses, a report of a traffic simulation scenario can be generated.

The following properties can verifies and validates:

- Properties that need to be verified: greater than a minimum value
  - 1–2. E[]/E<> forall (i : idV) vehicle(i).time_per_road(idS) > 0.5 * indexSC[p][q]
- Properties that need to be verified: smaller than a maximum value
  - 1–2. E[]/E<> forall (i : idV) vehicle(i).time_per_road(idS) < 27 * indexSC[p][q]
- Properties that need to be verified: greater than a minimum value and smaller than a maximum value
  - 1–2. E[]/<> (forall (i : idV) vehicle(i).time_per_road(idS) > 0.5 * indexSC[p][q]) && (forall (i : idV) vehicle(i). time_per_road (idS) < 27 * indexSC[p][q])
- Properties that need to be verified: smaller than a maximum value, greater than a minimum value, greater than a minimum value and smaller than a maximum value (vehicle)
  - 1–2. E[]/<> vehicle(idV).time_per_road(idS) > 0.5 * indexSC[p][q]
  - 3–4. E[]/<> vehicle(idV).time_per_road(idS) < 27 * indexSC[p][q]
  - 5–6. E[]/<> (vehicle(idV).time_per_road(idS) > 0.5 * indexSC[p][q]) && (vehicle(i).time_per_road(idS) < 27 * indexSC[p][q])
- Properties that need to be verified: if the vehicles leave the road
  - 1. E forall (i : idV) vehicle(i).response = −1

In these queries inserted in UPPAAL, the maximum values for validation variables taken into consideration are the values presented at the beginning of the simulation scenario. The minimum time is the result of the multiplication of 0.5 time units by the number of cells (from each segment), contained in the first column of the process definition matrix. The maximum time is the multiplication of 27 time units by the number of cells, which is contained in the first column in the process definition matrix.

The cellular automata allow the observation of different phenomena, managing to break down their components in individual variables. They allow understanding how local changes affect the whole grid of cells. Due to their elementary structure are suitable for a modular approach and the resolution (level of detail) and the obtained system size (the network size that needs to be covered) are adequate for the proposed model.

In the context of the formal verification, this model presents some possible scenarios than can occur, and the obtained results allow for the maximum level of permissiveness. It was demonstrated that UPPAAL software is capable to deal with low/medium complexity models, but for the implementation in high complexity models it is largely limited. In the presented approach its limits were taken into consideration and its application proved to be a success in the specified context.

4   Conclusions

The main goal of this work was the creation of a new systematic approach for complex railroad traffic sceneries. The implemented structure can be further used for modeling, analysis and verification purposes of various railroad traffic situations. A series of aspects of this work can be developed and explored deeper, providing room for new perspectives in this domain.

An important part of the future work will be focused on the scheduling of activities in marshaling yards, a special attention being given to the real-time characteristics of the considered applications.

References

1. Caimi, G., Kroon, L., Liebchen, C.: Models for railway timetable optimization: applicability and applications in practice. J. Rail Trans. Plan. Manag. 6(4), 285–312 (2017)
2. Schittenhelm, B., Landex, A.: Danish key performance indicators for railway timetables. In: Proceedings from the Annual Transport Conference at Aalborg University, pp. 1–28 (2016)
3. Goverde, R.M.P., Odijk, M.A.: Performance evaluation of network timetables using PETER. In: Allan, J., Andersson, E., Brebbia, C.A., Hill, R.J., Sciutto, G., Sone, S. (eds.) Computers in Railways VIII. WIT Press, Southampton (2002)
4. Lin, D.-Y., Ku, Y.-H.: Using genetic algorithms to optimize stopping patterns for passenger rail transportation (2014)
5. Ricci, S., Tieri, A.: A Petri nets based decision support tool for railway traffic conflicts forecasting and resolution. Department of Hydraulics, Transport and Roads, University Rome, Italy (2017)
6. WIT Transactions on State of the Art in Science and Engineering, vol. 40. WIT Pres (2010)
7. Nagy, E., Csiszár, C.: Analysis of delay causes in railway passenger transportation. Periodica Polytech. Transp. Eng. 43(2), 73–80 (2015)
8. Virgona, A., Kirchner, N., Alempijevic, A.: Sensing and Perception Technology to Enable Real Time Monitoring of Passenger Movement Behaviours Through Congested Rail Stations. Australasian Transport Research Forum, Sydney (2015)
9. Ahna, Y., Kowadab, T., Tsukaguchia, H., Vandebona, U.: Estimation of passenger flow for planning and management of railway stations. Transp. Res. Procedia 25, 315–330 (2017). World Conference on Transport Research
10. Zhou, W., Yang, X., Qin, J., Deng, L.: Optimizing the long-term operating plan of railway marshalling station for capacity utilization analysis. Sci. World J. (2014)
11. Caprara, A., Fischetti, M., Toth, P.: Modeling and solving the train timetabling problem. Oper. Res. 50(5), 851–861 (2002)
12. Dorfman, M.J., Medanic, J.: Scheduling trains on a railway network using a discrete event model of railway traffic. Transp. Res. Part B: Methodol. 38(1), 81–98 (2004)
13. Schobel, A., Scholl, S.: Line planning with minimal traveling time. In: Kroon, L.G., Mohring, R.H. (eds.) 5th Workshop on Algorithmic Methods and Models for Optimization of Railways, Dagstuhl, Germany (2006)

14. Li, W., Zhu, W.: A dynamic simulation model of passenger flow distribution on schedule-based rail transit networks with train delays. J. Traffic Transp. **3**(4), 364–373 (2017)
15. Jiang, Z., Xie, C., Ji, T., Zou, X.: Dwell time modelling and optimized simulations for crowded rail transit lines based on train capacity. Traffic Transp. **27**(2), 125–135 (2015)
16. UPPAAL. http://www.uppaal.org/. Accessed 20 Mar 2018

An Analysis of Simulation Models in a Discrete Manufacturing System Using Artificial Neural Network

Sławomir Kłos[(⊠)] and Justyna Patalas-Maliszewska

Faculty of Mechanical Engineering, University of Zielona Góra,
Licealna 9, 65-417 Zielona Góra, Poland
s.klos@iiz.uz.zgora.pl, j.patalas@iizp.uz.zgora.pl

Abstract. Computer simulation is a very important method for studying the behaviour of discrete manufacturing systems. This paper presents the results of simulation research and how buffer capacity, allocated in a manufacturing system, influence the throughput and work-in-progress. The simulation model of the manufacturing system is prepared using Tecnomatix Plant Simulation Software. The impact of individual buffers on the effectiveness of the system is analysed using the artificial neural networks module included in the software package. Simulation experiments were prepared for different capacities of intermediate buffers, located between manufacturing resources as input parameters with the throughput *per* hour and the average life span of products as the output parameter. A methodology for improving of the effectiveness of the system is proposed.

Keywords: Manufacturing systems · Computer simulation
Artificial neural networks · Buffer allocation · Throughput · Work-in-progress

1 Introduction

Computer simulation is a powerful method used for the design and study of manufacturing processes in industry. The construction of simulation models of manufacturing systems facilitates evaluation of the effectiveness, the work-in-progress, the demand for resources, the buffer allocation capacity and the time reduction of a cycle [1]. Advanced computer simulation tools enable analysis of the general performance of manufacturing systems, scheduling methods, support for facility layouts, automated material handling and the design of automated guided vehicle systems and so forth.

By doing this, we can reduce manufacturing costs and improve a system's throughput. One very important advantage of computer simulation is the possibility of creating a model of a manufacturing system whose behavior could be said to be very close to a real system. Modelling manufacturing system processes and production processes is time-consuming and needs expert knowledge which increases the costs of the simulation method. In order to reduce modelling time, some simplifications to the manufacturing system are often introduced which do not significantly affect the behavior of the system. In the paper, an analysis of a simulation model of a discrete, parallel serial manufacturing system is presented, using artificial neural networks. The simulation

© Springer International Publishing AG, part of Springer Nature 2019
J. Machado et al. (Eds.): HELIX 2018, LNEE 505, pp. 315–322, 2019.
https://doi.org/10.1007/978-3-319-91334-6_43

model is prepared using Tecnomatix Plan Simulation v.12, where the module of artificial neural network is implemented. The simulation experiments are prepared for different allocation of the buffer capacities within the system. For analysis of the impact of the buffer capacities on throughput and the work-in-progress - that is, the average lifespan of products and the time that the product resides in the system-artificial, neural networks are used.

1.1 Literature Overview

Simulation has been successfully utilised in a great deal of research related to the design and operation of manufacturing systems. Computer simulation is a highly effective tool for visualising, understanding and analysing the dynamics of manufacturing systems. Due to its complexity and importance, the buffer allocation problem has been studied widely and numerous publications are available in the literature [2]. Demir et al. [3] have presented a comprehensive survey on the buffer allocation problem in production systems. Shi and Gershwin [4] have presented an effective algorithm for maximising profits through optimisation of the buffer size in production lines. They considered both the cost of buffer space as well as the average inventory cost with distinct cost coefficients for different buffers. To solve the problem, a corresponding, yet unconstrained problem was introduced and a nonlinear programming approach was adopted. An analysis of mathematical models, describing the buffer allocation problem (BAP) was made by Papadopoulos et al. [5]. They presented the results of experiments using a search algorithm in association with an evaluative algorithm in order to obtain the mathematical optimum of the specified objective function. Tsadiras et al. [6] used an artificial neural network in order to create a decision support system for finding a sufficiently acceptable solution to the problem of the allocation of the production line buffer. Qudeiri et al. [7] used a genetic algorithm for studying the design of the serial-parallel production line. They tried to find the nearest optimal design for a serial-parallel production line which would maximise production efficiency by optimising the buffer size between each pair of work stations, the machine numbers in each of the work stations and between machine types. Fernandes and Carmo-Silva [8] presented a simulation study of the rôle of sequence-dependent set-up times in decision making, at the order release level of a workload-controlled, make-to-order, flow-shop. They indicated that the local strategy, which had traditionally been adopted in practice - and in most of the studies dealing with sequence-dependent set-up times - had not always given the best results. Matta [9] has presented mathematical programming representations for the simulation-based optimisation of buffer allocation in flow lines. Can and Heavey [10] provided a comparative analysis of genetic programming and artificial neuron network for the meta-modelling of discrete-event simulation models in manufacturing systems. Krenczyk et al. have presented a semi-automatic simulation model generation of virtual dynamic networks for production flow planning [11]. Kujawińska et al. [12] proposed a decision support tool, based on soft modelling in order to determine the allowances for the variability of consecutive technological processes caused by the material, the machine, the operator, the measurement system and other random factors. Another example of a decision support system for modeling a technological process using neural networks has been presented in [13, 14].

1.2   Problem Specification

In this paper, a simulation model of a discrete manufacturing system is prepared using Tecnomatix Plant Simulation Software. The system includes manufacturing resources and buffers. In the system, different products can be manufactured and the production programme is divided into batch sizes. The processing and set-up times are defined using lognormal distribution. For the system, 100 simulation experiments are randomly generated as input values, for the different allocations of buffer capacity, with the average throughput and lifespan of products being the output values. Based on the results of these simulation experiments, a set of learning data for artificial neural network is prepared. In order to analyse the behaviour of the simulation model, a two-layer, artificial neural network is prepared, using the standard, Tecnomatix Plant Simulation function. The main problem considered in the paper can be formulated as follows: "Given a simulation model of a discrete manufacturing system, how may the impact of the buffer capacity allocation on the average throughput and lifespan of products, using artificial neural networks, be analysed? In the next chapter, the simulation model of the discrete manufacturing system is described and the structure of the artificial neural network is presented. The third chapter contains the results of the simulation research and an analysis of the behaviour of the manufacturing system, using artificial neural networks. In the final chapter, conclusions and directions for further study are presented.

2   Simulation Model of the Manufacturing System

The model of the discrete manufacturing system was prepared on the basis of a real example of a manufacturing system for small, mass-produced spare-parts. The model and simulation experiments are implemented using Tecnomatix PLM Simulation Software. The simulation model is presented in Fig. 1. The manufacturing resources arc denoted as M_{11}, M_{21}, M_{12}, M_{21}, M_{32}, M_{13}, M_{23} and the buffers as B_1, B_2, B_3, B_4, B_5, B_6. It was assumed that the efficiency of manufacturing resources is approximately 95%. The processing and set-up times are based on a lognormal distribution. A lognormal distribution is a continuous distribution in which a random number has a natural logarithm corresponding to a normal distribution. The realisations are non-negative, real numbers. Operation times are presented in Table 1. In the system, four products (A, B, C, D) are manufactured, based on the following sequence of the size of the production batches: 100 A, 300 B, 80 C, 120 D, 60 A, 200 B, 150 C, 80 D. Four different products are manufactured and any change of batch requires a considerable increase in set-up time. For the processing values and set-up times, a set of 100, random buffer capacity experiments is generated. The density of the lognormal distribution Lognor (σ, μ) is calculated as follows [15]:

$$f(x) = \frac{1}{\sigma_0 x \sqrt{2\pi}} \cdot \exp\left[\frac{-\ln(x - \mu_0)^2}{2\sigma_0^2}\right] \tag{1}$$

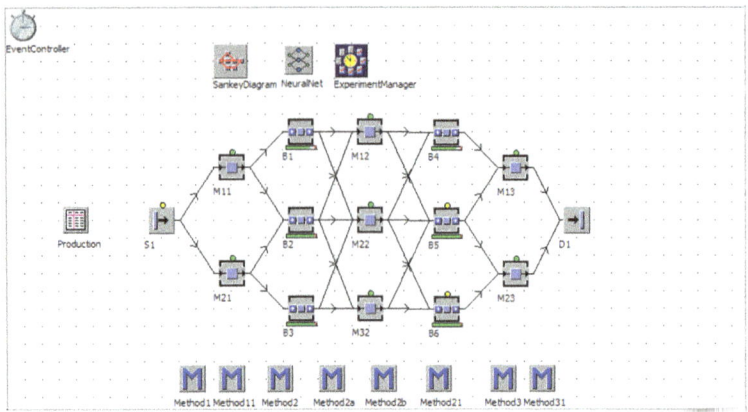

Fig. 1. Simulation model of the manufacturing system

where σ and μ are respectively mean and standard deviations are defined as follows:

$$\mu = \exp\left[\mu_0 + \frac{\sigma_0}{2}\right] \tag{2}$$

$$\sigma^2 = \exp(2\mu_0 + \sigma_0^2) \cdot (\exp(\sigma_0^2) - 1) \tag{3}$$

The maximum of the density function is defined as:

$$\exp(\mu_0 - \sigma_0^2) \tag{4}$$

The input value for the experiments were the buffer capacity values, ranging from 1 to 20, while as an output value, the average lifespan of the products, put through the system, was analysed. The results of the experiments are presented in Fig. 2. The greatest throughput value for the experiments is 13,23 and the smallest lifespan value is 2:09:58.

Table 1. The matrix of operation and set-up times

Manufacturing resources	Operation times		Set-up times	
	σ^2	μ	σ^2	μ
M_{11}	480	20	840	200
M_{21}	480	20	840	200
M_{12}	500	30	840	400
M_{21}	300	30	840	400
M_{32}	500	30	840	400
M_{13}	480	20	840	100
M_{23}	480	20	840	100

The results of simulation experiments are used as learning data for a two-layer, artificial neural network. The first layer includes 9 neurons while the second layer has 3 neurons. The activation function is a hyperbolic tangent function, used with an activation magnitude of 0,9; the portion of training data is 90%. Training is conducted with 300 training steps. The best mean error for the structure of artificial neural network is equal to 2,4%. In the next chapter, an analysis of the behaviour of the model of a discrete manufacturing system, using artificial neural networks, is presented.

3 Analysis of the System Using Artificial Neural Networks

The impact of each individual buffer, within a six buffer system (B_1, B_2, …, B_6) on throughput and lifespan may be analysed using artificial neural networks. In Fig. 3, examples of the relation between buffer capacity B_5 and the value of the system throughput is presented. The chart show that the best results are obtained for the capacity from 9–11. Based on an analysis of the training results for the different buffer capacities, a set of buffer capacities is prepared, in order to guarantee a high throughput along with a relatively low, product lifespan.

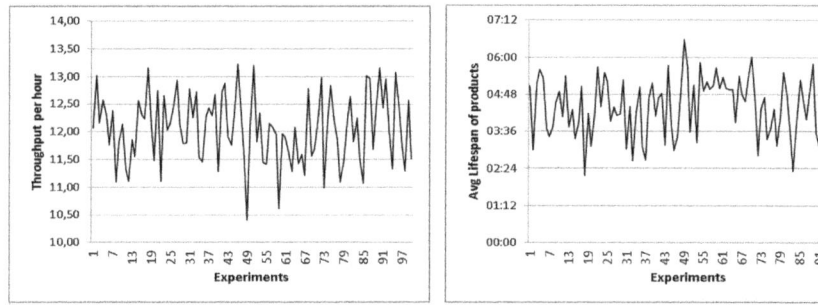

Fig. 2. The results of simulation experiments: throughput per hour and the lifespan of products

Table 2 shows the throughput and lifespan results of the simulation experiments; these are, respectively, T_S and L_S while the same values, calculated by artificial neural networks, are denoted as T_N and L_N; errors ΔT and ΔL are presented in the last two columns.

The error ΔT is about 7% but the ΔL error is from 3%–115%. The greatest throughput value is obtained for a buffer capacity equal to 3, 4 and 5 (TS = 13,78). The smallest lifespan value is obtained for the single capacity of all buffers (LS = 0:51:17).

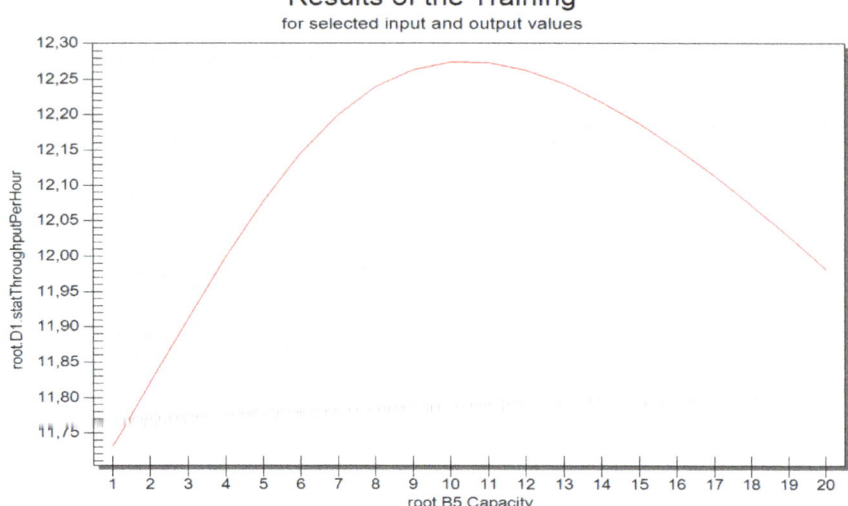

Fig. 3. The values of flow indices for the different variants of operation times

Table 2. Comparative analysis of simulation experiments and artificial neural network results

	B_1	B_2	B_3	B_4	B_5	T_S	L_S	T_N	L_N	ΔT	ΔL
Exp 01	1	1	1	1	1	13,62	**00:51:17**	12,73	01:50:06,8	0,07	1,15
Exp 02	2	2	2	2	2	13,67	01:05:27	12,75	01:57:51,1	0,07	0,80
Exp 03	3	3	3	3	3	**13,78**	01:12:49	12,77	02:07:11,3	0,07	0,75
Exp 04	4	4	4	4	4	**13,78**	01:16:57	12,79	02:18:23,3	0,07	0,80
Exp 05	5	5	5	5	5	**13,78**	01:23:09	12,80	02:31:42,8	0,07	0,82
Exp 06	10	10	10	10	10	13,58	02:48:53	12,77	04:11:35,6	0,06	0,49
Exp 07	15	15	15	15	15	13,34	04:45:16	12,57	05:46:22,0	0,06	0,21
Exp 08	20	20	20	20	20	13,15	06:08:29	12,42	06:19:43,9	0,06	0,03

4   Conclusions

In the paper, an analysis of the simulation model of a discrete manufacturing system, using artificial neural networks, is presented. The simulation model is created using Tecnomatix Plant Simulation Software, while in order to build the artificial neural networks model, the standard module of the software is used. The learning set is prepared based on the results of 100 simulation experiments. The following general conclusion can be formulated:

- it is possible to analyse the impact of individual buffers on the behaviour of the system, using artificial neural networks,
- estimating throughput, using artificial neural networks, gives relatively good results (about 7%),

- estimating the average lifespan of products, using artificial neural networks is prone to serious error, maybe as much as 115%,
- using artificial neural networks, allocating buffer capacity *guaranteeing* an acceptable level of system throughput and product lifespan, is possible.

Further research will address more complex manufacturing system structures as well as the impact of the effectiveness of manufacturing resources and maintenance on the throughput and work-in-progress of a manufacturing system.

References

1. Gangala, C., Modi, M., Manupati, V.K., Varela, M.L.R., Machado, J., Trojanowska, J.: Cycle time reduction in deck roller assembly production unit with value stream mapping analysis. In: Recent Advances in Information Systems and Technologies, WorldCIST 2017. Advances in Intelligent Systems and Computing, vol. 571, pp. 509–518. Springer, Heidelberg (2017)
2. Seleima, A., Azaba, A., Geddawy, T.: Simulation methods for changeable manufacturing. In: 45th CIRP Conference on Manufacturing Systems 2012, Procedia CIRP 3, pp. 179–184 (2012)
3. Demir, L., Tunali, S., Eliiyi, D.T.: The state of the art on buffer allocation problem: a comprehensive survey. J. Intell. Manuf. **25**, 371–392 (2014)
4. Shi, C., Gershwin, S.B.: An efficient buffer design algorithm for production line profit maximization. Int. J. Prod. Econ. **122**, 725–740 (2009)
5. Papadopoulos, C.T., O'Kelly, M.E.J., Tsadiras, A.K.: A DSS for the buffer allocation of production lines based on a comparative evaluation of a set of search algorithms. Int. J. Prod. Res. **51**, 4175–4199 (2013)
6. Tsadiras, A.K., Papadopoulos, C.T., O'Kelly, M.E.J.: An artificial neural network based decision support system for solving the buffer allocation problem in reliable production lines. Comput. Ind. Eng. **66**, 1150–1162 (2013)
7. Qudeiri, J.A., Yamamoto, H., Ramli, R., Jamalim, A.: Genetic algorithm for buffer size and work station capacity in serial–parallel production lines. Artif. Life Robot. **12**, 102–106 (2008)
8. Fernandes, N.O., Carmo-Silva, S.: Order release in a workload controlled flow-shop with sequence-dependent set-up times. Int. J. Prod. Res. **49**(8), 2443–2454 (2011)
9. Matta, A.: Simulation optimization with mathematical programming representation of discrete event systems. In: Proceedings of the 2008 Winter Simulation Conference, pp. 1393–400 (2008)
10. Can, B., Heavey, C.: A comparison of genetic programming and artificial neural networks in metamodeling of discrete-event simulation models. Comput. Oper. Res. **39**(2), 424–436 (2012)
11. Krenczyk, D., Skolud, B., Olender, M.: Semi-automatic simulation model generation of virtual dynamic networks for production flow planning. IOP Conf. Ser.: Mater. Sci. Eng. **145**, 042021 (2016)
12. Kujawińska, A., Diering, M., Żywicki, K., Rogalewicz, M., Hamrol, A., Hoffmann, P., Konstańczak, M.: Methodology supporting the planning of machining allowances in the wood industry. In: SOCO Leon, pp 338–347. Springer, Heidelberg (2017)
13. Rojek, I.: Technological process planning by the use of neural networks. Artif. Intell. Eng. Des. Anal. Manuf. **31**(1), 1–15 (2017)

14. Ivanov, V., Mital, D., Karpus, V., Dehtiarov, I., Zajac, J., Pavlenko, I., Hatala, M.: Numerical simulation of the system "fixture – workpiece" for lever machining. Int. J. Adv. Manuf. Technol. **91**(1), 79–90 (2017)
15. Tecnomatix Plant Simulation version 12.2.0 on-line documentation

Urban Transport System Enhanced Through IoT Infrastructure and Mobile Services Provided to Citizens

Ricardo Araújo and Sara Paiva[✉]

ARC4DigiT, Applied Research Center for Digital Transformation,
Instituto Politécnico de Viana do Castelo, Viana do Castelo, Portugal
sara.paiva@estg.ipvc.pt

Abstract. In this paper, we present an on-going project that aims to promote transport sustainability, urban mobility and citizens accessibility in Viana do Castelo. We contextualize and describe the current transport system and identify the improvements that could be made with an IoT infrastructure on it, namely the services that could be provided to citizens and their benefits. We also present the designed solution and implementation details as well as the tests that are being carried out with the developed prototype.

Keywords: Urban mobility · Transport sustainability · IoT · Accessibility
Citizens

1 Introduction

Smart cities are commonly referred to as cities that use Information and Communication Technologies at their service and to improve the life of their citizens. The application of new technologies to cities represents, in its majority, what is currently called of Internet of Things (IoT), which is an emergent reality, gaining more and more momentum. Cities, to be smart, will depend on a wide variety of sensors in almost all types of equipment, places and objects across cities and the gathered data will allow to improve monitoring and tracking, create alert systems, take preventive actions, among other functionalities. In a not very distant future, a smart city will be able to anticipate citizens' problems and provide a better service to them. Urban mobility represents the conditions that are created to citizens, so they can easily move inside the city and between its major places, using cars and public transportations. Urban mobility plans are being developed by municipalities to assure that these conditions are provided. Several causes exist so these plans assume a great deal of importance: the growth of the number of people within cities; the increasing number of roads being built; the different and growing number of transports within the city; the need people have of moving within the city several times during the day because of their jobs and personal activities. One of the aspect urban mobility plans refer to is the promotion of an efficient public transport system that promote accessibility of citizens. And this leads us to the concept of transport sustainability which refers to the application of new technologies to minimize the loss of time and improve citizens satisfaction [1]. Next generations

© Springer International Publishing AG, part of Springer Nature 2019
J. Machado et al. (Eds.): HELIX 2018, LNEE 505, pp. 323–329, 2019.
https://doi.org/10.1007/978-3-319-91334-6_44

quality of life, namely on levels of pollution, should not be compromised by what we are currently doing towards achieving transports sustainability. Some of the actions being promoted by cities around the world include the creation of bicycle paths or fees to enter the center of the city with their own cars which must be accompanied by a good, working and efficient transport system. These thematic has also been important, over the years, for the Europen Comission (EC) who adopted, in 2009, the Action Plan on Urban Mobility [2] to encourage authorities to define measures to achieve their goals on this field. The same commission published, in 2011, the white paper "Roadmap to a Single European Transport Area - Towards a competitive and resource efficient transport system" [3] that advises cities to develop plans to promote sustainable urban mobility plans. Another initiative dates from December 2013 when the EC adopted the Urban Mobility Package "Together towards competitive and resource-efficient urban mobility" [4].

In this paper, we present one of the applications IoT can play in smart cities, in its urban mobility and in transport sustainability. The presented prototype was developed in Viana do Castelo, a city on the North of Portugal, and consists of a mobile solution that takes advantage of an IoT infrastructure and geo-referentiation to provide useful services to citizens and contribute to evolve to a sustainable city and transport system.

This paper is structured as follows: in the next section, we present some related work on urban mobility and transport sustainability. After that, we present the concrete problem we are addressing in the city of Viana do Castelo and state the functionalities and benefits this city would have in adopting an IoT infrastructure. We explain the implementation stage and developed prototype; and finally, present conclusions of this work.

2   Related Work

The current literature presents us with may studies regarding the use of IoT to contribute to the creation of smart cities. The domains that can benefit from technology applied to cities are plenty and include smart transport, smart tourism and recreation, smart health, ambient-assisted living, crime prevention and community safety, governance, monitoring and infrastructure, disaster management, environment management, refuse collection and sewer management, smart homes and smart energy [5]. Picking up on the transport category, Dong et al. present [6] a framework of future Innovative Urban Transport (IUT), to support next generation urban transport. Another example is presented in [7] which refers to the design of an intelligent urban transportation system in Jamaica based on the Internet of Things. One other project focuses on solving the problems conductors face in collecting the fare from the passengers [8]. Handte et al. present an interesting study regarding the application of IoT to urban contexts and they present the Urban Bus Navigator [9], an IoT enabled navigation system for urban bus riders. GHOST is a location-based service and its goal is to implement an Intelligent Transport System for the public transport which can exploit the geo-referenced information of urban elements along the bus lines, monitoring them in a smart, continuous and autonomous way [10]. Finally, another example that can be provided highlights the opportunity to take advantage of emerging technologies, like an open source data platform and its application to the transport domain [1].

3 Problem Statement

The main motivation for developing this project was to improve the services to citizens in a city that is actively working on transport sustainability.

This project works with two types of transports: electrical bus (we will call it from now on E-BUS) and the urban bus (from now on U-BUS). The E-BUS type of transportation consists of two buses that circulate via the main arteries of the city from 9 a.m. to 6 h 30 p.m. The bus, that has the colors of the city – yellow and black - is shown in Fig. 1.

This bus goes by the main places of the historical center such as the city hall, finances, health center, shopping, interface, funicular, etc. The entire route is shown in Fig. 2. One important aspect of this bus is that it has no pre-determined stops. As long as a citizen rings the bell or signals from the road, the driver stops. The other type of transport, the U-BUS, intends to drive people from the center of the city to the surrounding areas such as Meadela, Carreço, Areosa, Perre, etc. These buses have pre-determined stops in one of the plenty bus stops around the city and surroundings. Considering this behavior of the two types of transport, we detect some enhancements in the transport system services that could be made available for citizens. In the case of the E-BUS, citizens never know where the bus is and how long it will take to get to where the person is at the moment. This is the main service to provide to citizens regarding E-BUS. The U-BUS is a little different in its behavior because of the stops but currently we have no information being provided to citizens. So, one of the things we detect as a problem is citizens not knowing where the buses are nor how long they will take to get to a certain stop. Another important feature for citizens would be to know the bus stop they need to go to get to a given destination, get directions to that bus stop and know how long the bus will take to get there.

Fig. 1. Electrical bus that continuously circulates in the historical center of Viana do Castelo

Fig. 2. Circular route made by E-BUS going through the main arteries of the city

4 Proposed Solution

Based on the problem statement priory mentioned, we designed a solution that we will present over the next sections.

4.1 Overview

The solution to enhance the transport system in Viana do Castelo and promote urban mobility is based on sensors, bus stops geo referencing and mobile apps. The main functionalities that will be available to citizens via a mobile application are:

- In a specific point of the city, know how long the E-BUS will take to get to the current point
- Anytime, anywhere check the real time moving position of E-BUS in a map
- In a specific point of the city, know the stop to go to in order to get to a given destination (ex.: I want to go to Meadela. To what bus stop should I go to?)
- Get directions from any point of the city to a given bus stop
- Know how long a U-BUS will take to get to a given bus stop
- Anytime, anywhere check the real time moving position of U-BUS in a map
- Get location-aware information in bus stops

These functionalities will bring citizens the following benefits:

- Better schedule of their time
- Awareness regarding bus delay time
- Reduction of time spent in bus stops
- Possibility of spending more time finishing a given task (work or personal), by knowing approximate hours bus arrives
- Less levels of stress (with the awareness of real time bus schedules)

4.2 Architecture and Components

The architecture that we designed to deliver the functionalities we intend is represented in Fig. 3. It mainly relies on beacons in bus stops and on the buses ability to send their GPS position to a central database. With this infrastructure, we can deliver high quality end-services to citizens to promote urban accessibility and enhance the transport sustainability.

Fig. 3. Architecture designed and proposed for the solution of enhancing transport sustainability in Viana do Castelo

4.3   Supporting Data Module

The remote database is responsible for storing information to be managed by the back-office and that are fundamental for the apps to work. The structure is shown in Fig. 4 and it mainly consists of buses, bus stops and routes. A bus has a type (electrical or urban). The route of the electrical bus is composed of a set of ordered points and the route of the urban bus is composed of a set of bus stops.

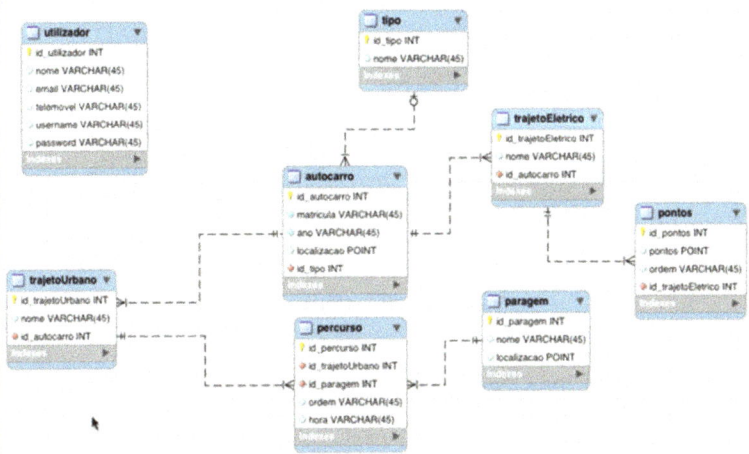

Fig. 4. Supporting data model of the system

4.4   Back-Office

The developed back-office has the purpose of being used by a legal authority, such as the City Hall, in order to enter information to be kept in the data model shown in the previous section. The functionalities are mainly about managing bus stops, routes, schedules and associated buses.

4.5   Mobile Application

In order to deliver the functionalities mentioned in the overview section of this chapter, we developed an android application that provides the services to citizens. The main screen (Fig. 5(a)) shows a map with the real-time position of electrical and urban buses and also the user´s position. The user can click on one bus and he is presented with information such as (Fig. 5(b)): how long does the bus will take to get to the bus stop closer to the user´s position and the bus route (next stops). If the user enters a desired destination in the search box on the top, he is presented with the bus stop that he should go to, how long he takes to get to that bus stop according to its current location, directions to get there, nearest bus that goes to that location and how long it takes to get to the bus stop (Fig. 5(c)). If the user clicks a bus stop in the map (Fig. 5(d)), he will be able to see the buses that will stop in this bus stop, at what time and its destination. This screen is automatically shown if the user is using the app and he is near a bus stop, which is detected by capturing beacon signal that is part of the infrastructure we assumed for the design of this solution.

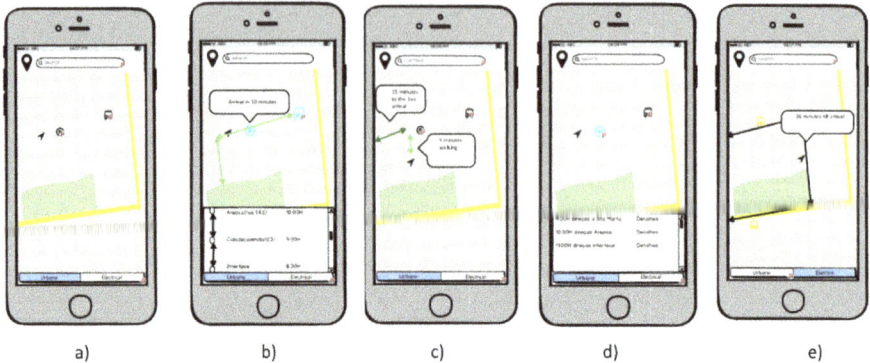

a)   b)   c)   d)   e)

Fig. 5. Mobile app prototype UX

Regarding electrical buses, Fig. 5(e) shows a map with the route these buses make, the buses currently on the road (typically there are two), the real-time position of each one and the estimate time for their arrival to the user's current position (these buses stop everywhere so there is no bus stop to consider).

5   Discussion

The solution and technologies adopted to solve the lack of information currently provided to citizens of the city of Viana do Castelo revealed to be adequate. Beacons are assumed to be in bus stops and buses sending their GPS position to a central database that is queried by the mobile app to deliver end services to citizens. The city lacks an IoT infrastructure that allows the gathering and processing of data and its presentation to citizens in an easy and useful way. Tests made in the field showed that the calculation of expected time of arrival and distance of the bus to the citizen current position is correctly made, with and without traffic. We take into consideration the hour and GPS positions to calculate velocity of the bus and also the distance to the user's position. The proposed solution, therefore, allows to solve the concrete problems identified in the urban mobility of the city of Viana do Castelo also promoting transport sustainability.

6 Conclusions

In this paper, we presented a proposal for an IoT infrastructure for the public transportation reality of the city of Viana do Castelo, in the north of Portugal, with the main purpose of addressing urban mobility and transport sustainability. The city has two types of transports: electrical buses make a circular route around the historical center all day long and have no predetermined stops; urban buses go to the periphery of the city and have pre-determined stops. We presented the architecture of the proposed solution, UX designs and the main benefits the developed mobile application can have to citizens and to enhance the city urban mobility.

References

1. Soriano, F., Samper, J., Martinez, J., Cirilo, R., Carrilo, E.: Smart cities technologies applied to sustainable transport. Open data management. In: 8th Euro American Conference on Telematics and Information Systems (EATIS (2016)
2. European Commission: Action Plan on Urban Mobility COM (2009). http://ec.europa.eu/transport/themes/urban/urban_mobility/action_plan_en.htm
3. European Comission: White Paper Roadmap to a Single European Transport Area – Towards a competitive and resource efficient transport system (2011). https://ec.europa.eu/transport/themes/strategies/2011_white_paper_en. Accessed 27 Nov 2017
4. European Comission: Urban Mobility Package: Together towards competitive and resource-efficient urban mobility. (2013). https://ec.europa.eu/transport/themes/urban/urban_mobility/ump_en. Accessed 27 Nov 2017
5. Dlodlo, N., Gcaba, O., Smith, A.: Internet of things technologies in smart cities. In: IST-Africa Conference (2016)
6. Dong, X., Zhou, J., Hu, B., Riekki, J., Xiong, G., Wang, F., Zhu, F.: Framework of future innovative urban transport. In: Conference on Intelligent Transportation Systems, pp. 19–23 (2016)
7. Deans, C.: The design of an intelligent urban transportation system in Jamaica based on the internet of things. In: IEEE SoutheastCon, pp. 1074–1077 (2015)
8. Mrityunjaya, D.H., Kumar, N., Ali, S., Kelagadi, H.M.: Smart transportation. In: International Conference on IoT in Social, Mobile, Analytics and Cloud, pp. 1–20 (2010)
9. Handte, M., Foell, S., Wagner, S., Kortuem, G., Marron, P.J.: An internet-of-things enabled connected navigation system for urban bus riders. IEEE Internet Things J. 3(5), 735–744 (2016)
10. Tadic, S., Kavadias, C., Favenza, A., Tsagaris, V.: GHOST: a novel approach to smart city infrastructures monitoring through GNSS precise positioning. In: IEEE 2nd International Smart Cities Conference: Improving the Citizens Quality of Life, ISC2 2016, no. 641495 (2016)

Organizational Strategies Induced by the Fourth Industrial Revolution: Workforce Awareness and Realignment

Marcelo Gaspar[1(✉)], Jorge Julião[2], and Mariana Cruz[2]

[1] Escola Superior de Tecnologia, Instituto Politécnico de Castelo Branco,
Castelo Branco, Portugal
calvete@ipcb.pt
[2] Católica Porto Business School,
Universidade Católica Portuguesa, Porto, Portugal

Abstract. On the midst of this new Industrial Revolution, a disruptive change of technological, economic and social systems is expected to take place in a near future. In this context, significant changes are forecast to impact the employment landscape over the coming years. New challenges and new opportunities will arise in this digital landscape, demanding from future workforce the ability to adapt and grow in such ground-breaking environments. The objective of this research is to understand the main people-focused factors related to the shaping of the fourth industrial revolution, as well as to assess Portuguese local workforce awareness and realignment towards the Industry 4.0 paradigm.

Keywords: Industry 4.0 · Organizational strategies · Workforce skillset

1 Introduction

The *Fourth Industrial Revolution*, also known as Industry 4.0, is building on the third, as it is characterized by the use of cyber-physical systems, which result from the fusion of technologies that are blurring the lines between the physical, the digital, and the biological spheres [1]. Based on what occurred for previous industrial revolutions, new organizational strategies will also be induced by the fourth, as work context and social interactions are foreseen to be soon drastically transformed. Thus, current debate focuses on the organizational strategies induced by this fourth industrial revolution, namely on the workforce awareness and realignment regarding their integration and cooperation with such ground-breaking digital technologies and the related significances [2].

1.1 Motivation

The digital age, resulting from the fourth industrial revolution, will need new work organisation and new work design to effectively prepare and adapt to the impact that such challenges will comprise. Thus, in a (near) future characterised by increasing automation and real-time oriented control systems, it is key to ensure that people's jobs are good, safe and fair [3].

Hence, the importance of identifying and promoting solutions regarding the shaping of the new workers' skillset, as well as to perceive the way companies are preparing and adapting to these (near) future challenges. To this end, current investigation aims at contributing to the understanding of how the workforce awareness and realignment is effectively taking place, regarding their integration and cooperation with the new work paradigm.

1.2   Research Question

Following two previously pilot experiences that were carried-out focusing on the assessment of a Portuguese Region local workers' perceptions regarding their intercultural communication skills needed to adapt to the Industry 4.0 model [4, 5], it was realized that further research was needed to contribute to understanding the main people-focused factors related to the shaping of the Portuguese fourth industrial revolution workforce.

In this third pilot experience, focus will be put in the understanding of the same Portuguese region recently unemployed skilled workers' perceptions towards the challenges encompassing the new industrial revolution. Namely, the aim of current research is to contribute to the understanding of two combined research questions:

- Are (Portuguese) recently unemployed skilled workers aware of the Industry 4.0 paradigm and its implications?
- Are (Portuguese) recently unemployed skilled workers driven to realign their skillset towards the Industry 4.0 paradigm?

2   Methodology

Using both quantitative and qualitative approaches, this study combines different types of records and methods of analysis in order to provide an exploratory account of the envisaged findings, collecting and assessing the interviewees' feedback whilst paying attention to their' views and practices.

2.1   Focus Group

This study focused on a group of recently unemployed skilled workers applying for a job at a Portuguese industrial company. The company is located at an inland Portuguese region, currently classified as low density and peripheral. This company employs over 200 workers and exports over 70% of their annual production to the global market. The company also integrates a national industrial group that contributes to the Portuguese world leadership in their niche market of products.

This exploratory research was based on focus group discussions, individual interviews and dedicated questionnaires and took place during a two-month period (end of

2017 and beginning of 2018), in which the participants were undergoing a preparatory period of training to apply to the referred industrial company.

Cumulatively to the individual and group interviews, a self-developed survey questionnaire was used to identify the perception of these recently unemployed skilled workers and their points of view on the Industry 4.0 paradigm. The information for the development of the survey questionnaire was based on structured and semi-structured interview guidelines. These guidelines were supported on previous studies focusing on needs, analysis, feedback from informal in-depth interviews and discussions with local workforce, as well as on the experiences of the researchers.

3 Findings

A recent global Industry 4.0 survey [6] conducted amongst over 2 000 respondents of more than 26 countries (including Portugal) emphasises that, on what regards to the new industrial revolution paradigms, the focus has to be put into people and culture to drive transformation. This study concludes that while investing in the right technologies is important, ultimately success or failure will depend mainly on a broad range of people-focused factors.

Thus, in order to contribute to the further understanding of these human-related factors, current investigation focuses on a group of Portuguese recently unemployed skilled workers and their underlying perceptions and attitudes towards the recent work paradigms related to the new industrial revolution.

3.1 Participants Characterization

To encompass the respondents' perceptions and points of view in this Industry 4.0 investigation, a preliminary characterization of the group is needed to support, put into perspective and contextualise their underlying.

Thus, a brief outline of their personal attributes, *e.g.*, their age and gender, as well as their personal qualifications may help situating the group towards the maturity level and professional expectations. Additionally, as the group has a common attribute of being recently unemployed, their previous work experiences, as their personal work-related abilities may also allow inferring about the participants' attitudes and perspectives towards new and future work opportunities.

Age and Gender
Even though all participants were categorised as recently unemployed, their age range and their previous work experience was very varied. Hence, the age range of the participants varied from the mid-twenties to the early fifties. On what refers to the gender of the participants, the group presented a 41% male and 59% female distribution.

Academic Qualifications
All participants referred having higher education qualifications, namely bachelor, licentiate or master degrees. The minority of the individuals (6%) referred having a 3-year bachelor degree. On what refers to those holding a licentiate (65%), no

distinction was made towards identifying those having a 3-year degree and the ones having a 5-year degree qualification. The individuals with the 5-year degree had their graduation previous to the Higher Education Bologna reorganization process, which took place in the 2006/2007 academic year.

Finally, the participants referred having a master degree (29%) have attended a 2nd cycle of studies additionally to their previous degrees. All participants mentioned that the specialization was important to their performance in the future work environment they were applying to the above mentioned industrial company.

Previous Work Experience

On what refers to the participants' previous work experiences, the respondents mentioned that all had previous work experience with varied time-periods. 58% of the enquired participants claimed having performed mainly high-level technical tasks in their previous jobs, whilst the remaining 42% referred doing mainly regular, non-skilled tasks, during their professional background.

Personal Work-Related Abilities

When asked about «what characteristics you consider to be essential in defining your profile to a potential employer?», the respondents mentioned mainly work related characteristics, such as being a «hard worker», always «arriving on time», being «eager to learn», amongst similar others.

3.2   ICT Tools and Skills

As above referred, the participants underwent a preparatory period of training on the process of applying to the local industrial company. The preparation comprised regular face-to-face training sessions and blended learning classes using virtual learning environments. Online tasks were custom-designed in order to improve the trainees' engagement with these different forms of participation and to enhance their learning outcomes. This latter aspect is key; as current industrial companies need digitally skilled workers to make sure that they contribute to drive the change that is needed regarding the new Industry 4.0 paradigm.

Personal Computers *vs.* Computers at Work

Nowadays, industrial companies need to develop a robust digital culture that challenges their co-worker to continuously improve upon their online communicational and functional skills.

When talking about their previous work experiences, the participants were asked if they had a computer allocated to them. Thus, almost all respondents (95,7%) confirmed having had a computer at their workplace. However, when asked about such computers having online communication accessories, more than half (63,6%) referred having loudspeakers and less than half (45,5%) stated they also had a microphone and a camera. This allows concluding that for the majority of the respondents, online communications were not a demanded task at their previous work experiences.

Bring Your Own Device

During the training period, the candidates were asked to bring a mobile communication device to participate on the dedicated online learning environment tasks. To that end, all participants brought their own mobile devices to the training classes: 74% brought laptops, 17% used their smartphones and 9% brought a tablet. All of the job candidates participated with success on the online tasks. Nonetheless, even though all of the trainees had brought their own mobile communication device during the company training period, when questioned about «are you available to bring your own computer to the workplace?», not all of the participants agreed to do so. Globally, it was observed that almost all of the female participants (84,6%) showed to be available to bring their own computer to the workplace. whilst only half of the male answered positively to that question. On the opposite, less than 8% of women stated not being available to bring their own computer to work, whilst that percentage increased up to 30% for men.

In a Nutshell
Current industrial companies need to attract, retain, and train digital natives and other employees who are comfortable working in a dynamic ecosystem environment [6]. Thus on what concerns to the participants' digital tools and skills, it is possible to identify that, even though almost all of the participants owned a personal computer with online Web 2.0 communication tools and skills, the same availability to use (and bring) such skills and tools to the work place decreased significantly.

3.3 Industry 4.0 Related Skills

According to the World Economic Forum report [1], the future of jobs, skills and workforce strategy will soon be drastically changed. The ability to correctly interpret and perceive these changes will allow current and future workforce gain a higher level of awareness to prepare and adapt and allow the organizational alignment with such work paradigm change.

Thus, amongst the set of the future skills mentioned on that report, the relational, intercultural & communicative, collaborative and ICT skills can be highlighted as core work-related skills to invest-on, and to prepare accordingly to the envisaged work demands. To encompass the participants' perceptions and points of view towards such set of skills, a brief survey was proposed to the group, highlighting the main skills identified in the above mentioned report.

Digital Skills
Even though the WEF [1] report refers to the ICT skills as mere basic skills, to perceive the awareness of current study participants' towards such skills, a quick questionnaire was designed using a dedicated Likert scale [7].

The observation of the results allowed concluding that three-fourths of the participants showed to be aware of their digital skills, with approximately 29% referring being fully aware of those skills. However, a group of one-fourth of the participants showed that they are still unaware of the importance of that skills have to their future jobs, particularly when considering that they were undergoing a job application process at the moment of the survey.

Collaborative Work Skills
The WEF [1] report includes the social skills of coordinating with others in the set of cross-functional skills needed to encompass to the Industry 4.0 work-related paradigm.

Most of the participants seemed to be aware (over 80%) of the importance of having collaborative work skills, with a group of approximately 2 out of 10 not being aware of their own collaborative and cooperative skills. Again, to what was mentioned on the above comment on the digital skills, this is concerning considering that they were undergoing a job application process when they participated in this quick survey.

Intercultural and Communicative Skills
When considering the design principles for Industry 4.0 scenarios, intercultural and communicative skills are highlighted as one of the main principles to guarantee the interoperability of such work paradigm [8]. Thus, the ability to communicate amongst workers using digital resources to promote effective and efficient cooperation networks is a key issue on which to plan in-service linguistic programs that should also take into consideration the basic skills workers already possess, their perceptions on how they learned foreign languages and suggestions on Web 2.0 and Web 3.0 technologies to enhance workers' performance. Thus, a quick survey was proposed to the participants regarding their awareness related to their intercultural and communicative skills in a working scenario.

The results showed that, as what occurred for previous categories, most of the participants seemed to be aware (over 80%) of the importance of the collaborative work skills, with a group of approximately 2 out of 10 not being aware of their own intercultural and communicative skills. This further contributes to the fact that a significant number of participants seem unaware of the role of such skills to their professional profile and the effect on their future job.

Relational Skills
The social relational skills, such as emotional intelligence or the negotiation abilities were highlighted on the Future of Jobs report [1] as cross-functional skills in an Industry 4.0 scenario. On what concerns the social skills, here classified as relational skills, the participants were asked about their personal perception towards such abilities. Thus, when compared to the previous categories' surveys, the results varied significantly. Even though the majority of the group still refers being aware about their relational skills, the group of participants unaware of such set of skills increased to almost the double of previous categories (from 18% to 36%). Again, as what was referred for previous categories, a significant group of participants seemed to be unaware of the significance of the relational skills related to their personal performance in a future work scenario, with more than 3 out of ten respondents to assume that lack of skills.

4   Summary and Conclusions

The purpose of this study was to further contribute to the understanding of the main people-focused factors related to the shaping of the fourth industrial revolution in modern business environments that affect the new skillset alignment needed to fully implement and adapt to the envisaged Industry 4.0 paradigm. Considering this set of skills, the aim of the research was mainly to assess local Portuguese workforce awareness and realignment, namely assessing the viewpoints and perceptions of recently unemployed skilled workers. This study allowed concluding that the majority of the group interviewees' referred being aware of their main Industry 4.0 related skills. Nonetheless, a significant number on the group of participants seemed to be unaware of the significance of such skills when related to their personal performance in a work scenario

As future developments of this study, further research is needed to complement the findings of this exploratory experiment. Additionally, the viewpoints of the companies related to the same issue may also contribute to the understanding of local Portuguese workforce awareness and realignment towards the Industry 4.0 paradigm.

References

1. World Economic Forum. The Future of Jobs Employment, Skills and Workforce Strategy for the Fourth Industrial Revolution (2016)
2. Dombrowski, U., Wagner, T.: Mental strain as field of action in the 4th industrial revolution. Proc. CIRP **17**, 100–105 (2014)
3. Henning, K., Wolfgang, W., Johannes, H.: Recommendations for implementing the strategic initiative INDUSTRIE 4.0. Final report of the Industrie 4.0 WG, p. 82, April 2013
4. Régio, M.M.A., Gaspar, M.R.C., Farinha, L.M.C., Morgado, M.M.A.P.: Forecasting the disruptive skillset alignment induced by the forthcoming industrial revolution. Rom. Rev. Precis. Mech. Opt. Mechatron. **2016**(49), 24–29 (2016)
5. Régio, M., Gaspar, M., Farinha, L., Morgado, M.: Industry 4.0 and telecollaboration to promote cooperation networks: a pilot survey in the Portuguese region of Castelo Branco. Int. J. Mechatron. Appl. Mech. **2017**(1), 243–248 (2017)
6. Reinhard, G., Jesper, V., Stefan, S.: Industry 4.0: building the digital enterprise. In: 2016 Global Industry 4.0 Survey, pp. 1–39 (2016)
7. Wadgave, U., Khairnar, M.R.: Parametric tests for Likert scale: for and against. Asian J. Psychiatr. **24**, 67–68 (2016)
8. Hermann, M., Pentek, T., Otto, B.: Design principles for industrie 4.0 scenarios. In: Proceedings of Annual Hawaii International Conference on System Sciences, pp. 3928–3937, March 2016

Innovations in Industrial Context and Advanced Manufacturing

A Dynamic Selection of Dispatching Rules Based on the Kano Model Satisfaction Scheduling Tool

L. Ferreirinha[1], S. Baptista[1], A. Pereira[1], A. S. Santos[1,2], J. Bastos[1],
A. M. Madureira[1], and M. L. R. Varela[2(✉)]

[1] School of Engineering, Polytechnic Institute of Porto (ISEP/IPP),
Porto, Portugal
`l.ferreirinha96@gmail.com`, `sara.raquel.mb@gmail.com`,
`gimirra@hotmail.com`, {`abg,jab,amd`}`@isep.ipp.pt`
[2] Department of Production and Systems,
University of Minho (UM), Braga, Portugal
`leonilde@dps.uminho.pt`

Abstract. Production scheduling is a function that can contribute strongly to the competitive capacity of companies producing goods and services. Failure to stagger tasks properly causes enormous waste of time and resources, with a clear decrease in productivity and high monetary losses. The efficient use of internal resources in organizations becomes a competitive advantage and can thus dictate their survival and sustainability. In that sense, it becomes crucial to analyze and develop production scheduling models, which can be simplified as the function of affecting tasks to means of production over time. This report is part of a project to develop a dynamic scheduling tool for decision support in a single machine environment. The system created has the ability, after a first solution has been generated, to trigger a new solution as some tasks leave the system and new ones arrive, allowing the user, at each instant of time, to determine new scheduling solutions, in order to minimize a certain measure of performance. The proposed tool was validated in an in-depth computational study with dynamic task releases and stochastic execution time. The results demonstrate the effectiveness of the model.

Keywords: Dynamic production scheduling · Single machines
Decision support tool

1 Introduction

With the increase in complexity and the need for flexibilization of productive systems, new heuristics have been developed that have the capability of producing better results for scheduling problems [1, 2]. The scheduling affects the resources of the activities and determines in which sequence the jobs must be executed, in order to optimize a given measure of performance [3, 4].

© Springer International Publishing AG, part of Springer Nature 2019
J. Machado et al. (Eds.): HELIX 2018, LNEE 505, pp. 339–346, 2019.
https://doi.org/10.1007/978-3-319-91334-6_46

In view of the above, it is necessary to analyze more agile and flexible methods to solve these problems, integrating an analysis that does not only contemplate the satisfaction of the interests of the client, but also the interests of the company.

This paper presents a decision-support tool for scheduling on single machines, not only in a static environment but also in a dynamic one. In a static environment, the tool allows an analysis of several measures of performance simultaneously, which leads to a greater balance of the interests of the stakeholders [5, 6]. Regarding dynamic scheduling, the tool was developed with the purpose of not requiring any interaction with the user, the software itself alternates between two priority rules according to the objectives in question. This article will only discuss the dynamic environment. To validate the tool, a vast set of stochastic jobs from normal distributions were generated.

The remaining sections of this paper are organized as follows: in Sect. 2 the problems of production scheduling are approached. The tool developed is presented in Sect. 3. In Sect. 4 is where the paper finally presents some conclusions and provides some ideas for future work.

2 Scheduling Problems

The production scheduling is preponderant for the survival of a company and can be defined as a decision function that distributes the resources available through the operations over a period of time [1, 7].

In scheduling problems, two types of feasibility constraints are usually found: the capacity limits of the machine and the technological constraints that condition the production order. The scheduling problems aim to answer two questions [8]: What resources to use to perform a given job and when to perform each job. Regarding the first question it allows the allocation of jobs while the second question defines the sequence of jobs [9]. When there are alternative processors that can perform a job, we talk about affectation. As the processors finish executing jobs, the dispatching function chooses and forwards, iteratively, the jobs for execution. The sequencing is a job ordering function for each station. Sometimes sequencing is approached with the problem of affectation [3].

2.1 Scheduling on Single Machines

Production scheduling on single machines is a one-operation environment consisting of a single processor available to perform each job. This type of scheduling is fundamentally a sequencing problem, having as the only objective the ordering of jobs to be produced in a single processor (Fig. 1).

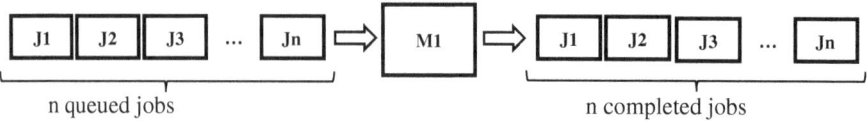

Fig. 1. Single machine environment.

Despite its simple approach, it is very common to find a single machine feature inserted in a more complex production environment, such as the "shop" environment [10, 11]. Often, the breakdown of more complex environments leads to single machine environments, allowing them to be solved in a simpler and faster way [12, 13].

2.2 Production Scheduling Methods

In a given problem the evaluation of each solution is only possible when the solution space is relatively small. As the size of the problem increases the resolution becomes intractable. The methods for solving this type of problem can be divided into [7]: Enumerative methods, where the optimal solution is found through implicit enumeration and comparison of all possible solutions. Such methods refer to dynamic programming, constraint programming, and Branch and X optimization methods; Heuristic methods which allow us to find satisfactory solutions efficiently. It is possible to identify local search heuristics, meta-heuristics, and constructive heuristics. The latter refers to optimization techniques that start from an empty solution and sequentially construct solutions without considering the impact of decisions in later stages; And finally, the priority rules, which sorts the jobs in a sequence and determine in what order they should be executed [8, 13–15].

2.3 Dynamic Scheduling

In industrial environments is rare to schedule in a static environment, typically new tasks are launched during the implementation of a scheduling solution, making it immediately obsolete. In such scenario, it is necessary to adapt the scheduling solutions in order to incorporate the jobs that are arriving at the shop floor. In this type of scheduling, it is assumed that all tasks are not known at the beginning of the problem, that is, new tasks can be released during the implementation of a given scheduling solution.

A dynamic scheduling problem is handled through purely reactive models that respond to the launch of new tasks. Generally, dynamic scheduling reorders a queue of jobs, which have not yet been executed, whenever a new task is released [7, 9, 16, 17].

Hyper-heuristics have become increasingly popular, due to their ability to solve real-world optimization problems, such as scheduling problems in dynamic environments. A hyper-heuristic is a heuristic that seeks to automate the processes of selection, combination, generation or adaptation of several simpler heuristics. The most important characteristic of a hyper-heuristic is that they search a space of heuristics instead of directly searching the space of solutions [18, 19].

Many papers have addressed the application of different heuristics, including hyper-heuristics in either dynamic environments and in more complex scheduling problems, such as the hybrid flow shop (HFS) and job shop problems [21–24]. For example, in [21] it is proposed a new hybrid Dispatching Rule Based Genetic Algorithms (DRGA) which searches for the best sequence of dispatching rules and the number of operations to be handled by each dispatching rule simultaneously. It was used to solve different variants of the multi-objective job shop problem. In [19] the authors designed a framework that uses the genetic programming hyper-heuristic techniques to combine Palmer's and Gupta's algorithms, in order to obtain new and better heuristics to solve a flow shop scheduling problem. In [20] the authors propose a Genetic Algorithm (GA) for a Hybrid Flow Shop problem, which combines a meta-heuristic with a hyper-heuristic. What differentiates the present article from those presented above is that the developed tool decides between two priority rules based on a real-time monitoring of two performance criteria, while the majority do research based on a metaheuristic (GA).

3   Dynamic Selection of Dispatching Rules Based on the Kano Model Satisfaction Scheduling Tool (DSDR-KMS-ST)

The developed tool allows the analysis of scheduling problems in single machines. This application, apart from allowing quick access to the insertion, removal, edition, and visualization of jobs, also gives the user the possibility to choose their environment, static or dynamic.

3.1   Dynamic Environment

The dynamic environment was designed in such a way that the user has no interaction with the tool except for the definition of the objectives, where the goals are classified through the degree of satisfaction of the Kano's Model. The DSDR-KMS-ST was designed to schedule the work in progress whenever new tasks enter the system. As soon as the objectives are defined, the software itself schedules the jobs autonomously by alternating between two priority rules. It should be noted that the DSDR-KMS-ST presented here intends to portray a test of the Kano's Model concept with two performance measures. In this initial phase, the user has the possibility to set the maximum tardiness value as well as minimize the average flow time. What underlies here is that the user does not want their maximum tardiness to exceed the set value and, at the same time, wants the average flow time to be as small as possible. The maximum tardiness represents an obligatory attribute in the Kano's model, so if it is not fulfilled it results in an extreme dissatisfaction of the "client". As for the average flow time, it represents a proportional attribute, since it corresponds to a degree of satisfaction proportional to the degree of performance of the attribute, i.e., the smaller the average flow time, the higher the satisfaction of the "client" [21, 22]. The DSDR-KMS-ST also shows several useful performance indicators, at each instant, informing the user and giving him a chance to evaluate the performance of the system (Fig. 2).

3.2   Computational Study

To validate the operation of the DSDR-KMS-ST, normal distributions were used to generate 100 jobs with stochastic attributes (Table 1).

It is known that the minimization of the maximum tardiness in single machines is achieved through the EDD rule, and the minimization of the average flow times is achieved through the SPT rule. Thus, an algorithm was developed to alternate between these two priority rules as time passes and new tasks are released. Therefore, whenever

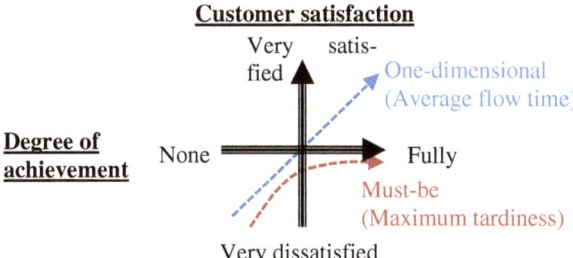

Fig. 2. Kano's model [22].

Table 1. Distributions used.

	[1–20]	[21–40]	[41–60]	[61–80]	[81–100]
rj	N(75,35)	N(175,45)	N(275,45)	N(375,45)	N(475,45)
dj [N(μ,σ) + rj + pj]	N(150,2)	N(215,2)	N(300,2)	N(400,2)	N(525,2)
pj	N(10,2)				
wj	N(10,3)				

the maximum tardiness does not exceed the defined value, the SPT rule is used to minimize the average flow time. If the algorithm detects that, through the SPT, the maximum tardiness is higher than the stipulated value, it uses the EDD to minimize the maximum tardiness.

3.3   Results

After the generation of the 100 jobs it was establish that the maximum value of the requested tardiness would be zero. The obtained results are shown in Figs. 3 and 4. It should be noted that in the example used, a margin value was defined for the maximum tardiness value. That is, instead of the tool only turning to the EDD when the maximum tardiness is higher than the set value, it uses this rule when the margin value is achieved or overpassed. Such a margin value can have a significant impact on the final solution because if it is too high, minimizing the average flow time is compromised since EDD is used very early. If it is too small, the system may not be able to correct in time the maximum tardiness hitherto generated. For this example, it was defined that the margin value would be equal to twice the average of the processing time.

Figure 3 compares the average flow time and the work in progress between the model and the system based only in EDD. As we can see, the average flow time in the system based on EDD tends to get higher faster than the DSDR-KMS-ST. As for the work in progress, the same thing happens, the system based only in EDD tends to create a bigger WIP than the DSDR-KMS-ST. Figure 4 shows the effectiveness of the tool, where the maximum tardiness was always less than zero and the average flow time was minimized whenever possible, reaching 718.45 in the final phase (in the system based only in EDD reached 747,65). The SPT rule was used 56 times, gray on the performance chart, and the EDD rule was used 33 times, represented in red on the same chart. There are 33 tasks in progress and 67 have been processed.

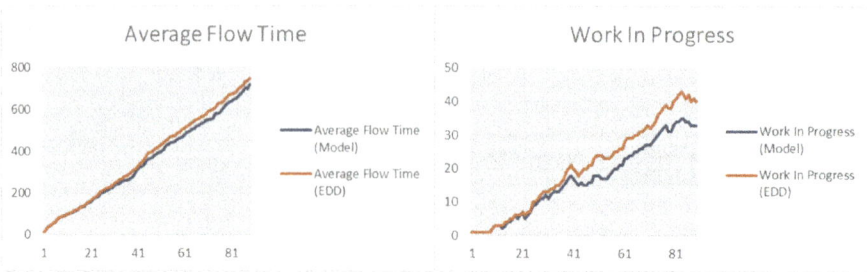

Fig. 3. Obtained results.

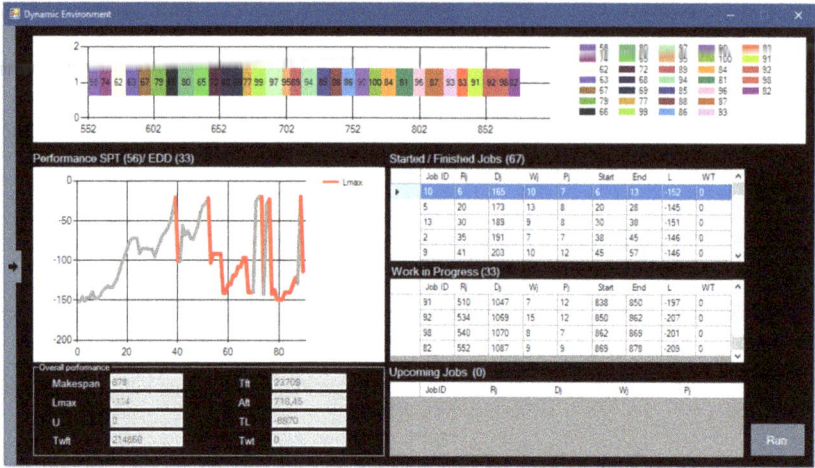

Fig. 4. DSDR-KMS-ST.

4 Conclusions

The DSDR-KMS-ST intends to demonstrate an approach to the dynamic scheduling production without user intervention. The software itself, according to the defined objectives, and as new tasks enter the system, alternates between two rules of priority in order to comply with the objectives. This tool also has the autonomy to demonstrate, at each instant, several performance indicators to keep the organization informed. Regarding future work, it is intended that it can classify more performance criteria by its degree of satisfaction and that the tool dynamically switches between dispatching rules to optimize the customer satisfaction function. It is also intended that, in the future, the tool should be link to the MRP system, which allows it to know when new tasks will enter the system. It could also be useful to analyze the impact that the margin may have on the performance of the tool.

References

1. Artiba, A., Elmaghraby, S.E.: The Planning and Scheduling of Production Systems: Methodologies and Applications, 1st edn. Springer, Heidelberg (1996)
2. Kalinowski, K., Zemczak, M.: Preparatory stages of the production scheduling of complex and multivariant products structures. Adv. Intell. Syst. Comput. **368**, 475–483 (2015)
3. Varela, M.L.R.: Uma contribuição para o escalonamento da produção baseado em métodos globalmente distribuídos. Doctoral dissertation, Universidade do Minho (2007)
4. Krenczyk, D., Skolud, B.: Computer aided production planning - SWZ system of order verification. In: IOP Conference Series: Materials Science and Engineering, vol. 95, p. 012135 (2015)
5. Shabtay, D., Steiner, G.: The single-machine earliness-tardiness scheduling problem with due date assignment and resource-dependent processing times. Ann. Oper. Res. **159**(1), 25–40 (2008)
6. Trojanowska, J., Varela, M.L.R., Machado J.: The tool supporting decision making process in area of job-shop scheduling. In: Rocha, Á., Correia, A., Adeli, H., Reis, L., Costanzo, S. (eds.) Recent Advances in Information Systems and Technologies. WorldCIST 2017. Advances in Intelligent Systems and Computing, vol. 571, pp. 490–498. Springer, Heidelberg (2017)
7. Serra e Santos, A.: Análise do Desempenho de Técnicas de Otimização no Problema de Escalonamento. Tese de Mestrado, Instituto Superior de Engenharia do Porto (2015)
8. Baker, K.R., Trietsch, D.: Principles of Sequencing and Scheduling, 1st edn. Wiley, Hoboken (2009)
9. Pinedo, M.L.: Scheduling Theory, Algorithms, and Systems, 4th edn. Springer, Heidelberg (2010)
10. Madureira, A., Pereira, I., Pereira, P., Abraham, A.: Negotiation mechanism for self-organized scheduling system with collective intelligence. Neurocomputing **132**, 97–110 (2014)
11. Varela, L.R., Pinto, T.: Comparing extended neighborhood search techniques applied to production scheduling. Methods **6**(10), 11–12 (2010)
12. Madureira, A., Pereira, I.: Intelligent bio-inspired system for manufacturing scheduling under uncertainties. In: Hybrid Intelligent Systems (HIS), pp. 109–112 (2010)
13. Blazewicz, J., Ecker, K.H., Pesch, E., Schmidt, G., Weglarz, J.: Handbook on Scheduling: From Theory to Applications. Springer, Heidelberg (2007)
14. Talbi, E.G.: Metaheuristics: From Design to Implementation, vol. 74. Wiley, Hoboken (2009)
15. Pfund, M., Fowler, J.W., Gupta, J.N.: A survey of algorithms for single and multi-objective unrelated parallel-machine deterministic scheduling problems. J. Chin. Inst. Ind. Eng. **21**(3), 230–241 (2004)
16. Terekhov, D., Down, D.G., Beck, J.C.: Queueing-theoretic approaches for dynamic scheduling: a survey. Oper. Res. Manag. Sci. **19**(2), 105–129 (2014)
17. Varela, M.L.R., Ribeiro, R.A.: Distributed manufacturing scheduling based on a dynamic multi-criteria decision model. In: Recent Developments and New Directions in Soft Computing, pp. 81–93 (2014)
18. Vázquez-Rodríguez, J.A., Petrovic, S.: A new dispatching rule based genetic algorithm for the multi-objective job shop problem. J. Heuristics **16**(6), 771–793 (2010)
19. Nugraheni, C.E., Abednego, L.: On the development of hyper heuristics based framework for scheduling problems in textile industry. Int. J. Model. Optim. **6**(5), 272 (2016)

20. Rodríguez, J.A.V., Salhi, A.: A robust meta-hyper-heuristic approach to hybrid flow-shop scheduling. In: Evolutionary Scheduling, pp. 125–142 (2007)
21. Högström, C., Rosner, M., Gustafsson, A.: How to create attractive and unique customer experiences: an application of Kano's theory of attractive quality to recreational tourism. Mark. Intell. Plann. **28**(4), 385–402 (2010)
22. Löfgren, M., Witell, L., Gustafsson, A.: Theory of attractive quality and life cycles of quality attributes. TQM J. **23**(2), 235–246 (2011)
23. Reddy, M.S., Ratnam, C., Agrawal, R., Varela, M.L.R., Sharma, I., Manupati, V.K.: Investigation of reconfiguration effect on makespan with social network method for flexible job shop scheduling problem. Comput. Ind. Eng. **110**, 231–241 (2017)
24. Varela, M.L.R., Trojanowska, J., Carmo-Silva, S., Costa, N.M.L., Machado, J.: Comparative simulation study of production scheduling in the hybrid and the parallel flow. Manag. Prod. Eng. Rev. **8**(2), 69–80 (2017). https://doi.org/10.1515/mper-2017-0019

Inventive Methods in Designing an Environmentally Friendly Household Appliance

Ewa Dostatni[1], Izabela Rojek[2(✉)], Paulina Szczap[1],
and Marta Tomczuk[1]

[1] Department of Management and Production Engineering, Poznan University of
Technology, Poznan, Poland
ewa.dostatni@put.poznan.pl, {paulina.szczap,
marta.tomczuk}@student.put.poznan.pl
[2] Institute of Mechanics and Applied Computer Science, Kazimierz Wielki
University, Bydgoszcz, Poland
izarojek@ukw.edu.pl

Abstract. The paper looks at a concept of an eco-friendly household appliance. Owing to an innovative approach to the design process, the functionality of an existing product has been altered. Inventive creation methods have been applied to involve respondents in the search for innovative solutions. A new approach to designing products, eco-design is based on identifying the environmental impact of a product and incorporating it into early stages of the design process. Eco-innovation consists in creating innovative goods, services, products, processes, systems or procedures at competitive prices, aimed at meeting human needs and ensuring better quality of life with minimum possible consumption of natural resources (materials, energy, space) per manufacturing unit and minimum possible release of toxic substances. Selected inventive creation methods used in the study, such as brainstorming, stimulation, crumbling, play on words, and superposition, are described. The results of the study have been used to design the concept of an eco-innovative product – a small refrigerator for storing medicines and cosmetics. An ecological analysis of the product has been performed and its price estimated and compared to the market price of a similar product.

Keywords: Eco-design · Eco-innovation · Inventive creation method
Household appliance

1 Introduction

Eco-design combines technical feasibility and environmental requirements [1–5]. A new approach to designing products, eco-design consists in identifying the environmental footprint of a product and incorporating it into the design process as early as the initial stages of product development [6–11]. Eco-design can also be part of eco-innovation. To be able to refer to eco-innovation, we need to define innovation in the first place. Innovation consists in manufacturing new products or improved versions

© Springer International Publishing AG, part of Springer Nature 2019
J. Machado et al. (Eds.): HELIX 2018, LNEE 505, pp. 347–353, 2019.
https://doi.org/10.1007/978-3-319-91334-6_47

of existing products. Eco-innovation, on the other hand, involves designing innovative goods, services, products, processes, systems or procedures at competitive process, aimed to meet human needs and boost the quality of life for consumers while ensuring minimum possible consumption of natural resources (materials, energy, space) per manufacturing unit and minimum release of toxic substances. Having a special role in creative problem solving, inventive creation methods are applied in seeking to develop new solutions, ideas and products [12].

The paper presents application of eco-design methodology in the development of a household appliance using inventive creation methods. The authors took an innovative approach to the design process with the aim to define functionality of the product, and used inventive methods to involve respondents in the search for innovative solutions. The concepts of eco-design, eco-innovation and science of inventive creation are explained. The product under analysis was designed with the use of selected inventive creation methods. Research into functionality and environmental friendliness of the designed product was conducted with participation of respondents. The concept and design of an "eco-product" were developed, the ecological analysis of the product was conducted, and the costs of manufacturing of the new product was estimated.

2   Literature Review

There are many works that have been carried out about frameworks and platforms for supporting decision making regarding product development, along with underlying supply chain and even though virtual organizations specification or creation for this purpose, for instance [13–17]. Regulations implemented in the European Union require that designers and engineers take appropriate actions related to ecodesign [18–21]. However, insufficient knowledge of ecodesign often limits the creative approach to such engineering solutions [22]. The tools supporting ecodesign include also solutions dedicated e.g. specifically to product designers, based on the automation of the design process taking into account environmental concerns. There are also a number of papers in the literature on the design of technological processes using more or less sophisticated IT tools.

3   Research Methodology

3.1   Inventive Creation Methods in Eco-Design

The object of the study discussed in this paper is a household appliance – a refrigerator. It was assumed that the refrigerator would be compact in size, made of eco-friendly materials which are completely or partly recyclable, and that it would be used for storing medicines and cosmetics. The selection of research methods necessary to design an innovative household appliance started with an analysis of all the existing inventive creation methods. Inventive creation methods assist in creative problem solving, hence it was important to split the entire cycle into the following phases [23]: understanding and defining the problem, collecting and analysing information, generating ideas, analysing and evaluating ideas, finding and defining alternative solutions, and acting on the selected solution. At the design stage of the study, focus was put mainly on three phases: collecting and analysing data, generating ideas, and analysing and evaluating ideas. Due to the defined purpose of the research, its planning of research focused mainly on the phases of: gathering and analysing information, searching for ideas as well as analysing and evaluating ideas and creating solution designs. The following inventive methods were selected for the research: brainstorming, stimulation and crushing method, game method with words and superposition.

3.2   Inventive Study

Based on selected inventive methods, appropriate surveys have been prepared. The research was carried out in two groups. The first group consists of 60 students from the field of Management and Production Engineering, studying at the last semester of the master's studies. The second group of 50 people were students of the last semesters of engineering studies in the field of Mechanics and Machine Design. The research was carried out in January 2017. The study, designed as a questionnaire, was divided into two stages, which were conducted by means of different methods. The entire study was based on one key method, which was brainstorming. It was assumed that the study group would be split into several smaller groups of three to four individuals. The part of the results of questionnaires are shown in Table 1.

Table 1. The part of the results of questionnaires

Equipment/appearance	Materials	Ecological solutions
Different temperature in selected zones	Ultrasonic aggregate instead of gas aggregate	Solar energy as a source of energy
Recoiling- wheels	PVC composite with wood	Elimination of freons

Stage 1. At this stage, the methods of play on words (associations) and superposition were applied. It was assumed that the study group would be familiar with the subject matter of the study, i.e. the design of a household appliance – a refrigerator, but unfamiliar with the initial concept of the product (compact in size, eco-friendly).
Stage 2. At this stage, the methods of stimulation and crumbling were used. The study group was familiarised with the initial concept of the refrigerator as a compact size household appliance featuring eco-friendly solutions. Stage 2 consisted of a set of questions and was aimed at obtaining specific suggested solutions which could be drawn on during the development of the product concept.

The play on words method applied at Stage 1 utilized the phrase "association – a refrigerator in 20 years' time", and a set of prompt questions. One study group focused more on accessories and physical appearance of the refrigerator, while the other group paid more attention to the technical issues, such as materials and eco-friendly solutions. The criteria taken into consideration by the respondents included technical feasibility of the solution and its eco-friendliness. The feedback provided by the respondents was consistent with the initial assumptions made for the product concept.

3.3   Development of the Product Concept on the Basis of the Study Results

The refrigerator has been designed in the shape of a 60 × 40 × 35 cm cuboid with a door made of smart glass [24]. The insulation is made of polyurethane foam – its excellent thermal insulating properties make it possible to fit the refrigerator with a small-size refrigerating unit. To ensure that the refrigerator casing be eco-friendly and biodegradable, it is made of chipboard – a wood derivative. On the outside, the chipboard is covered with veneer made of natural wood, in the colour selected by the customer. The refrigerator features LED lighting to minimise energy consumption. Edges of the casing are slightly rounded to increase user safety. The interior is made from a blend of polystyrene with silver. Isobuthane (R600a), recommended for application in household refrigerators, is used as a coolant [25]. To expand its functionality, the refrigerator features adjustable temperature zones within a range of 1 °C to 7 °C. Two refrigerating units are fitted to enable simultaneous storage of products at different recommended storage temperatures. The modular design of the interior comprises a set of containers of various sizes, made of transparent polystyrene. The underlying concept of the design was to ensure that the maximum number of components would be recyclable and easily separable from others. Once the assumptions were defined, the structure of the refrigerator was developed (see Fig. 1).

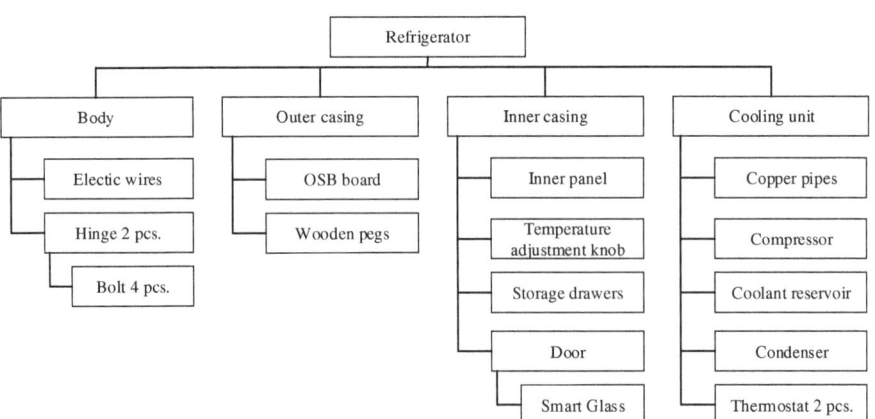

Fig. 1. The structure of the refrigerator.

4   Results of Study

4.1   Ecological Analysis of the Refrigerator

The process of eco-design of a product life cycle is based on the following principles [1]:

- seeking environmental efficiency,
- thrifty use of resources and use of available renewable resources,
- improving the product durability,
- designing for future reuse of the product,
- designing for recyclability of materials,
- designing for disassemblability and minimization of toxicity,
- environmentally friendly manufacturing,
- reducing the environmental impact of the product during its use,
- disposal of non-recyclable materials.

Ecological criteria of a product are defined through an analysis of the principles listed above, and specified in more detail in the design process. The criteria are assigned levels at which they are met, and a question checklist is drawn up. The checklist is then used to determine whether the ecological criteria have been met, and if yes – at what level. A section of a qualitative analysis of meeting the ecological criteria is presented in Table 2. Levels A, B, and C identify weaknesses of the analysed solution. Level A is assigned when all the criteria are met. Level B means that the criteria are met in an acceptable degree, while level C indicates an urgent need for alteration to introduce eco-friendly alternatives [1].

Table 2. A selected section of ecological analysis of a product on the basis of a question checklist (developed on the basis of [1]).

The principle of thrifty use of resources is implemented on the basis of the following criteria:	
Use of recycled materials:	**A – high content of recycled material (70-100%)**
	B – medium content of recycled material (up to 30%)
	C – low content of recycled material (less than 30%)

Table 2 is a fragment of the checklist (ready), which was not developed by us only by [1], he proposed such divisions. Our designed product has been subjected to a recycling assessment using a system to perform the analysis and we have received that we can recover 75% of materials for reuse, so it falls under criterion A.

4.2   Estimation of the Cost of the Designed Product

The cost of the designed refrigerator was estimated on the basis of market prices of materials used in the manufacturing. The cost of materials was increased by the labour costs, costs of operation of the machines used, and other costs incurred in the manufacturing process. The price of a similar product available from a household appliance retailer is ca. PLN 480 [26], whereas the cost of the refrigerator presented in this paper has been estimated at ca. PLN 790. The difference is generated by the eco-friendly materials, which are more expensive than the materials being in common use, as well as by the innovative solutions applied, such as smart glass or wooden casing. However, the underlying reason for the difference is that the cost has been estimated for a prototype. In series production, the cost of the refrigerator would be much closer to the price of a comparable product available on the market.

5   Conclusion

Creating an innovative household appliance is a challenging task, even more so considering high competitiveness of the market. Manufacturers are in a constant race for more modern and more eco-friendly products, leaving the competition behind.

Eco-design offers a new approach to product design. It draws on identifying the room for improvement of environmental friendliness of the product and implementing the findings at early stages of the design process. The design of the refrigerator discussed in this paper is based on the principles of eco-design, with emphasis on eco-friendly, biodegradable and recyclable materials.

The surge in interest in more eco-friendly and innovative products in the modern society opens the door to designing new products, such as the refrigerator presented above. Its price being higher than that of a comparable, yet less eco-friendly, traditional product available on the market, the refrigerator is more environmentally friendly and less expensive in operation owing to lower energy consumption.

Acknowledgments. This research was supported in part by: 02/23/DSPB/7716 financed from a research grant awarded by the Ministry of Science and Higher Education.

References

1. Adamczyk, W.: Ecological Problems of Product Quality [in Polish]. Scientific Publisher PTTŻ, Cracow (2002)
2. AENOR: Environmental Management of Design and Development Process. Design for Environment, UNE 150301 Standards (2003)
3. Gehin, A., Zwolinski, P., Brissaud, D.: Integrated design of product lifecycles-the fridge case study. CIRP J. Manuf. Sci. Technol. **1**(4), 214–220 (2009)
4. Low, J.S.C., Lu, W.F., Song, B.: Methodology for an integrated life cycle approach to design for environment. In: Su, D., Zhu, S. (eds.) Key Engineering Materials, vol. 572, pp. 20–23 (2014)
5. Karwasz, A., Trojanowska, J.: Using CAD 3D system in ecodesign—case study. In: Golinska-Dawson, P., Kolinski, A. (eds.) Efficiency in Sustainable Supply Chain, Part II, pp. 137–160. Springer, Heidelberg (2017)
6. Dostatni, E., Diakun, J., Grajewski, D., Wichniarek, R., Karwasz, A.: Functionality assessment of ecodesign support system. Manag. Prod. Eng. Rev. **6**(1), 10–15 (2015)

7. Grajewski, D., Diakun, J., Wichniarek, R., Dostatni, E., Buń, P., Górski, F., Karwasz, A.: Improving the Skills and knowledge of future designers in the field of ecodesign using virtual reality technologies. In: International Conference Virtual and Augmented Reality in Education Proceedings, Monterrey, Mexico, pp. 348–358 (2015)

8. Dostatni, E., Diakun, J., Grajewski, D., Wichniarek, R., Karwasz, A.: Multi-agent system to support decision-making process in ecodesign. In: Herrero, Á., Sedano, J., Baruque, B., Quintián, H., Corchado, E. (eds.) Advances in Intelligent Systems and Computing, vol. 368, pp. 463–474. Springer, Cham (2015)

9. Karwasz, A., Dostatni, E., Diakun, J., Grajewski, D., Wichniarek, R., Stachura, M.: Estimating the cost of product recycling with the use of ecodesign support system. Manag. Prod. Eng. Rev. 7(1), 33–39 (2016)

10. Dostatni, E., Rojek, I., Hamrol, A.: The use of machine learning method in concurrent ecodesign of products and technological processes. In: Hamrol, A., Ciszak, O., Legutko, S., Jurczyk, M. (eds.) Advances in Manufacturing, pp. 321–330. Springer, Heidelberg (2018)

11. Czapczuk, A., Dawidowicz, J., Piekarski, J.: Artificial intelligence methods in the design and operation of water supply systems. ROCZNIK OCHRONA SRODOWISKA 17, 1527–1544 (2015)

12. Martyniak, Z.: Introduction to the Invectiveness [in Polish]. University of Economics, Cracow (1997)

13. Arrais-Castro, A., Varela, M.L.R., Putnik, G.D., Ribeiro, R.A., Dargam, F.: Collaborative negotiation platform for networked organizations using dynamic multi-criteria decision model. Int. J. Decis. Support Syst. Technol. 7(1), 1–14 (2015)

14. Arrais-Castro, A., Varela, M.L.R., Putnik, G., Machado, J.: Collaborative platform for virtual organization synthesis using a multi-criteria decision model. Int. J. Comput. Integr. Manuf. 30(4–5), 483–500 (2018)

15. Castro, A.A., Varela, M.L.R., Carmo-Silva, S.: An architecture for a web service based product configuration information system. Commun. Comput. Inf. Sci. 110, 20–31 (2010)

16. Bezerra, K., Machado, J., Carvalho, V., Castro, M., Costa, P., Matos, D., Soares, F.: Bath-ambience - a mechatronic system for assisting the caregivers of bedridden people. Sensors 17(5), 1156 (2017)

17. Florian, H., Mocanu, A., Vlasin, C., Machado, J., Carvalho, V., Soares, F., Astilean, A., Avram, C.: Deaf people feeling music rhythm by using a sensing and actuating device. Sens. Actuators A. Phys. 267, 431 442 (2017)

18. Azevedo, S.G., Carvalho, H., Machado, V.C.: The influence of Green practices on supply chain performance: a case study approach. Transp. Res. Part E 47(6), 850–871 (2011)

19. Buyukozkan, G., Cifci, G.: A novel hybrid MCDM approach based on fuzzy DEMATEL, fuzzy ANP and fuzzy TOPSIS to evaluate green suppliers. Expert Syst. Appl. 39, 3000–3010 (2012)

20. Lin, R.J.: Using fuzzy DEMATEL to evaluate the green supply chain management practices. J. Clean. Prod. 40, 32–39 (2013)

21. Govindan, K., Khodaverdi, R., Vafadarnikjoo, A.: Intuitionistic fuzzy based DEMATEL method for developing green practices and performances in a green supply chain. Expert Syst. Appl. 42(20), 7207–7220 (2015)

22. Yang, C.J., Chen, J.L.: Forecasting the design of eco-products by integrating TRIZ evolution patterns with CBR and simple LCA methods. Expert Syst. Appl. 39(3), 2884–2892 (2012)

23. Antoszkiewicz, J.: Heuristic methods. Creative problem solving [in Polish] Państwowe Wydawnictwo Ekonomiczne, Warsaw (1990)

24. Profiglass Homepage. http://www.profiglass.com.pl/. Accessed 27 Dec 2016

25. Linde-gaz Homepage. http://www.linde-gaz.pl/. Accessed 27 Dec 2016

26. Mediaexpert Homepage. http://www.mediaexpert.pl/. Accessed 10 Jan 2017

Process Mapping in a Prototype Development Case

José Dinis-Carvalho[⊠], Diana Santos, Mariana Menezes, Melissa Sá,
and Joana Almeida

Production and Systems Department, University of Minho, Braga, Portugal
dinis@dps.uminho.pt,
diana_ribeiro_santos@hotmail.com,
mariana_meneses_21@hotmail.com,
melissa.c.sa.94@gmail.com, joanambpa@gmail.com

Abstract. Process mapping in indirect areas is a powerful tool to increase company awareness of the complete process and it allows to identify problems and improve opportunities leading to the reduction of lead time and waste as well as improving productivity and competitiveness. In this article a team of students carried out a project in a company aiming to reduce the lead time in the prototype development using a process mapping tool called VSDia (Value Stream Design in Indirect Areas). This tool allowed the identification of several improvement opportunities leading to a reduction of the lead time from 27 to 17 days as well as gains in productivity. This activity also created in the personnel a new awareness of the whole process, the particularities of the process steps and the impact of each process step in other steps of the process.

Keywords: Process mapping in indirect areas · Lean thinking

1 Introduction

The identification of the value and non-value adding steps that are currently required to produce a product or a family of products in a production unit is one of the key principles of the lean thinking [1]. These principles were identified based on the production approach developed by Toyota under the name of Toyota Production System [1, 2]. A very commonly used tool to describe the value and non-value adding process steps (process mapping) in manufacturing environments where lean principles are applied is the Value Stream Mapping (VSM) methodology [3, 4]. In fact the VSM is very often referred as the key methodology in the process of implementing lean and continuous improvement. As the lean principles and concepts started to be implemented in non-manufacturing environments such as offices and services, the VSM was also a tool adopted. The problem is that the characteristics of non-manufacturing environments (ex: indirect areas) are not effectively represented using VSM since the processing steps are more focused in processing information with very complex routes. Other processing mapping tools were adapted by lean practitioners in different companies when the representation of information processing was necessary. Some used simple tools such as large boards where post-it were placed representing process steps and lines to represent information exchange between process steps.

© Springer International Publishing AG, part of Springer Nature 2019
J. Machado et al. (Eds.): HELIX 2018, LNEE 505, pp. 354–360, 2019.
https://doi.org/10.1007/978-3-319-91334-6_48

This article reports a project that was carried out by an Industrial Engineering team of students in a company dedicated to the production of wiring systems for heavy machinery. The company identified as a problem the long lead time required in the wiring system prototype process because it can be a competitive factor in the market. The objective of the project was to identify opportunities to reducing that lead time.

2 Methodology

Value Stream Design in Indirect Areas (VSDia) [5, 6] is a simple and visual tool to represent broadly a process from the beginning to the end. A process is an organized group of related activities that work together to create a result of value [7]. In fact, the process mapping is meant to examine the current state of a process and, through its graphical representation becomes apparent which activities add value or not to the product. Thus, it is possible to identify opportunities for improvement and designing the future state of the process [8]. The methodology applied in this work follows four steps that must be performed sequentially [6]:

i. **Preparation** - In this step the essential conditions are prepared, in particular a room and the material used to map the current state of the process. In addition, the relevant stakeholders are invited to the meeting at which the expert method clarifies the objectives thereof, sensitizing them to be honest and transparent in the description of tasks.

ii. **Value flow analysis** - The method applied in this step uses several elements, as can be seen in Fig. 1 and a brief description of each element is presented below.

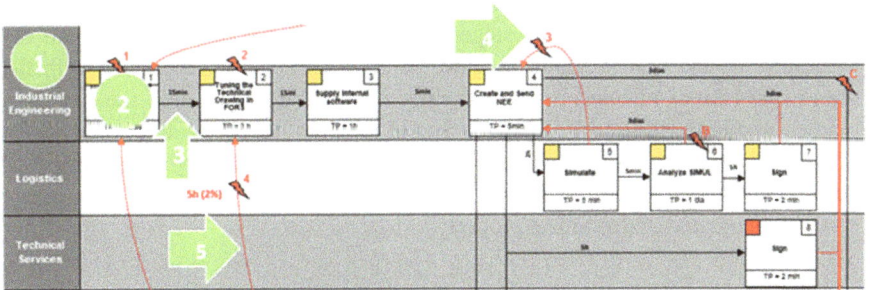

Fig. 1. Mapping Processes

The element 1 refers to *tracks* or lanes - each track represents an entity or department intervening in the process.

The element 2 refers to *process boxes* – each process box contains information about a task carried out by a particular entity and consisting of the following fields: Sequential numbering of the tasks; task description; task processing time; type of value for the customer. The value for the customer may take one of three colours depending on the value the task is perceived by the customer as well as by the team in charge of the process mapping. Green colour will be assigned to the value-added tasks, yellow to the tasks that add no value but that are necessary and red to those which do not add any value.

The element 3 refers to *connectors* – they establish a connection between processes. It may include information regarding the transition time.

The element 4 refers to *flashes* - signs that represent problems identified throughout the process. These are numbered and, at the same time, a list with detailed description of the problems and possible improvement solutions is created.

The element 5 refers to *queries* – they arise whenever it is necessary to go back in the process to complete information that is missing or even correct errors, avoiding the continuation of the process. The queries are represented by dashed red arrows, being accompanied by a flash. In addition, they should also contain information on the frequency that they occur.

iii **Value stream design**   Refers to the mapping of the future improved process, aiming the following lean vision: 100% of value-added activities, continuous flow of information, parallel processing, levelling cadence and workload, customer orientation, ideal capacity, minimum number of connections between entities, elimination of rework (Perfect Quality), customer shaped information and elimination of waiting. Also according to [9], mapping the future state with the suggested changes should take into account three aspects:

(a) Customer Requirements: Customer needs should be well understood;

(b) Simplicity and transparency: the process should be as simple as possible for employees in order to provide the desired output to the customer (internal/external);

(c) Elimination of all mental barriers: all suggestions should be respected even less hypothetical, since they can be beneficial for combination with others.

iv. **Implementation** – In this last stage standards are developed to implement the measures of the new working method. In order to ensure compliance with the rules, the entities affected by the changes will be trained. Finally, the improvements resulting from this method should be monitored through a checklist designed to determine if the desired performance is achieved. This project is complete when the performance measures and suggestions for improvement are in fact implemented and running properly.

3 Application

In a first phase, the student team gathered the necessary material for the construction of the map, namely the post-it notes for the tasks, the flashes to identify the problems and improvement opportunities, the paper with large area where the map will be drawn, and pens. Thereafter the paper was placed on a room wall so that the participants would be able to easily visualize the process. To finalize the preparation of the session, a panel was also used to enumerate the flashes and queries describing all the problems encountered in the course of the mapping process. The existence of this panel with the problems described, enhances the visual concept of the tool used, giving opportunity for participants to suggest immediate improvement proposals.

The process mapping event was then scheduled and at the planned time all the following representatives of all entities involved in this process met in the room: Industrial Engineering, Logistics, Technical Services, Quality, Segment 1, Segment 5, Production Engineering, FST (Planning Department), and Packaging and Shipping. This session was led by three students, one responsible for the registration of the problems encountered during the process on the panel and two responsible for leading the mapping process.

Completed the process mapping session, it was possible to understand in detail all the stages of production, giving opportunity to identifying problems and solve them in order to improve the efficiency of the processes. The end result of the session can be viewed in Fig. 2 showing the final map. This current state map also shows the part related to office operations on the left side of the image and then on the right side the operations that are mainly physical operations. All the problems and improvement opportunities identified were marked on the map with a red flash symbol as shown Fig. 2.

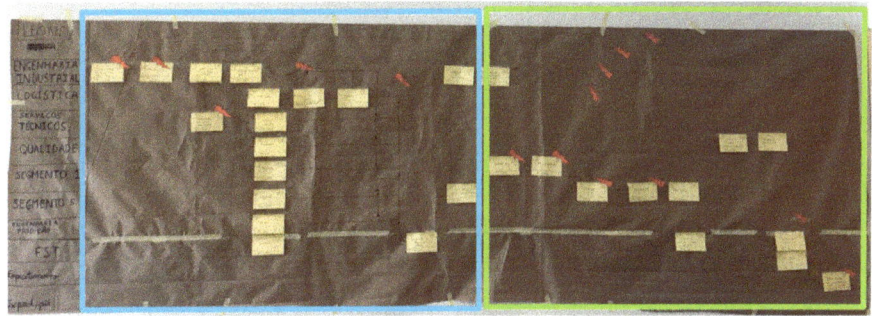

Office operations   **Production operations**

Fig. 2. Process map of the current state.

4   Major Improvements

After the analysis of the entire map, the discussing was centred in the decision on what problems should be solved first. All problems had to be ranked based on its effect in performance improvement and on the complexity of possible solutions.

4.1   Opportunity 1 – Reduction of the Number Signatures

One of the problems was related to the fact that one document (prototyping order) was sent to be signed in several departments, but in reality, this document just needed the signatures of two departments (Industrial Engineering and Logistics), the remaining departments (Technical Services, Quality, Segment 1, Segment 5 and Production Engineering) only needed to be informed about the existence of that document.

4.2   Opportunity 2 – Reduction of the Lead Time in Cable Cutting Process

Another interesting opportunity for improvement was that the lead time for cable cutting orders was of one week. For optimization purposes all orders wait until Fridays to be sent to the cable cutting area. The developed proposal is to send wire cutting orders every day as needed, and making this requires a daily meeting in the cable cutting area in order to optimize the cutting sequences to reduce setup times.

4.3   Opportunity 3 – Reduction of the Lead Time in Manual Test

After the assembly of the prototype all wires (several hundreds of wires) must go through a manual test. The worker must verify if the components of each one of the wires in the prototype follow the customer specifications, The operator identifies the reference code of each wire, then search that code in a paper database and compares each wire specification with each wire in the prototype. This operation can take up to 16 h of work for a single prototype. In order to reduce the operation time, the proposal includes the introduction of a code scanner as well as a new system that displays in a monitor the respective component characteristics to simplify the task and reduce the operation time.

4.4   Opportunity 4 and 5 – Code Generation for New Components

In the first stages of the process, the industrial engineering department must, very often, generate reference codes for new components. Since this operation takes considerable amount of work and time because it requires communication with the customer to clarify details, the process is simplified by creating temporary codes so the process proceeds to next steps. The problem is that only at later stages when tests quality checks are required, the definitive reference codes are needed, the industrial engineering personnel are asked to provide the definite codes. The new procedure is that the industrial engineering department although letting the process continuing to next stages must start immediately the new reference coding process before the quality check requiring them.

4.5   Other Improvements

Apart from the problems and opportunities identified in the process mapping activities the team also identified other improvement opportunities by observing the production area. These opportunities did not have impact in the prototyping lead time but have impact in the productivity of this area. One of the problems was related to the lack of organization in the picking area, which led the worker to search for references in a large disorganized shelf. To solve this problem the boxes with wires are arranged in columns properly identified taking into account the sequence of operations to be performed in the pre-confection area. To make this effective organization, the software used by the company should be able to directly locate the box, by adding a parameter "column".

Another problem was identified in the pre-confection area (responsible to assemble components in wires), FIFO was not followed, due to the fact that the boxes were placed randomly on the workstation and boxes with small batches tended to be left to the next shift since the setups are the less wanted operations by workers. In order to eliminate this problem, the logistics should sequence the boxes according to the daily plan provided by this planning system. The sequence takes into account the demand of wires in the assembly line as well as minimizing the number of setups when possible. This plan also takes into consideration a lead time restriction of two days in this area.

5 Results

The process mapping activities resulted to be a powerful tool in creating awareness among different department's personnel regarding the whole process of prototyping developments. The process mapping session helped to understand others problems that were not understood before. People become aware that some problems can be easily solved by simple understand them. This process mapping session resulted in the identification of several improvement opportunities that led to an immediate lead time reduction from 27 to 21 days in the prototyping development. This reduction derives from the elimination of some signatures that were required for prototyping development order (Opportunity 1) and by decreasing the cutting planning time from 5 to 3 days (Opportunity 2).

The other improvements do not have results yet because the implementations are not finished but the estimation is that after fully implementation of Opportunity 3 the lead time can be reduced in one day while the productivity can be doubled. With the new procedure (Opportunity 4 and 5) where the industrial engineering personnel start the generation of new codes before they are asked can lead to an average reduction of 2 days in the lead time.

Finally, the improvements in the picking areas as well as implementation of FIFO lanes in the pre-confection will contribute to a lead time reduction of at least one day in average. The final expected average lead time value in the prototyping process is 17 days, representing 37% reduction. This reduction represents an impressive improvement in the company's global performance since the company developed 651 different prototypes in 2017.

6 Conclusion

The objective of this work was to carry out a process mapping activity in the prototype development process in a wire systems company. The activity forced to put all people responsible for each step of the process in the same room building the map. The clarification of the different process steps, its problems and its implication in the whole process, represented a major gain in the organization. This global awareness help in the identification of improvement opportunities as well as the identification of solutions. The solutions already implemented as well as the solution being implemented will lead to a 37% reduction of the lead time which is of major importance in the market. From the testimonies collected among different players in the company it became clear that no one had a clear picture of the quantity of steps involved in the prototyping generation neither the complexity of some steps or the effect of each step in the whole chain. Some players reported that although feeling that the lead time was large, they did not think it was as large as 27 days. Some people reported that they not know all the stages of the process, or even know of its existence in the company. This activity helped in this process of awareness, and the improvements proposed were well accepted and understood by all, due to the fact that now the whole process is known as well as its problems.

Acknowledgements. This work has been partially supported by projects COMPETE-POCI-01-0145-FEDER-007043 and FCT-UID-CEC-00319-2013, from Portugal.

References

1. Ohno, T.: Toyota Production System (1988)
2. Monden, Y.: Total framework of the toyota production system. In: Toyota Production System: An Integrated Approach to Just-In-Time (1998)
3. Rother, M., Shook, J.: Learning to see: value stream mapping to add value and eliminate muda (1999)
4. Jones, D., Womack, J.: Seeing the Whole. Lean Enterprises Institute Inc, Cambridge (2002)
5. Braga, I.: Aplicação do VSDiA para melhoria do processo de cotação na indústria eletrónica. University of Minho (2012)
6. Silva, A.: Mapeamento do processo de pedido de alterações e implementação de scrum. University of Minho (2016)
7. Browning, T., Fricke, E., Negele, H.: Key concepts in modeling product development processes. Syst. Eng. **9**, 104–128 (2006)
8. Nyemba, W.R., Mbohwa, C.: Process mapping and optimization of the process flows of a furniture manufacturing company in Zimbabwe using machine distance matrices. Proc. Manuf. **8**, 447–454 (2017)
9. Tapping, D., Shuker, T.: Value stream management for the lean office: eight steps to planning, mapping, and sustaining lean improvements in administrative areas (2003)

Minimization of Fuel Consumption in a Logistic Company: Implementation of Six Sigma and Drivers' Skills Management with the Use of Fuzzy Logic

Dorota Stadnicka$^{(\boxtimes)}$ and Antosz Katarzyna

Faculty of Mechanical Engineering and Aeronautics,
Rzeszow University of Technology, Rzeszów, Poland
{dorota.stadnicka,katarzyna.antosz}@prz.edu.pl

Abstract. Considering logistics management plays a very important role in costs generating. There are different kinds of costs. One significant kind of costs is connected to fuel consumption. Logistic companies try to implement different solutions in order to decrease this cost. This work presents the implementation of a six sigma project to identify important factors which influence fuel consumption. Based on the performed analysis, we know that one of the most important factors is a human factor. Therefore, in the second part of the paper the authors present an analysis which explains how, with the use of fuzzy logic, a company can identify drivers who should be trained to improve their skills. As a result it allows to minimize fuel consumption.

Keywords: Logistic costs · Fuzzy logic · Skills management · Six sigma

1 Introduction

There are many factors which influence the costs of logistic processes. One of the main factors indicated in the literature is the cost of fuel consumption [1]. Fuel consumption depends e.g. on the type of vehicles, delivery routes, tonnage and the skills of the drivers [2–4]. An important element of drivers' competences is the way of their training. Some publications show how the drivers training influences the decrease of fuel consumption [5]. Many publications highlight the advantages of the eco drivers' training. For example, in the paper [6], the influence of eco driving training on the reduction of city and highway fuel consumption is presented. The paper [7] analyses the bus operations real data for the potential eco-driving benefits and proposes an eco-driving algorithm for transit buses. Companies are looking for methods which can help to increase their effectiveness. The methods connected with lean manufacturing (LM) or Six Sigma (SS) are known very well in manufacturing companies, but still they are not so popular in other kind of companies e.g. logistic companies. In the literature we can find some examples of the implementation of these methods in logistics. For example, SS and LM were implemented to solve problems with transportation and health industries in French [8] or in Chinese [9] companies. In the work [10], the author presents the possibilities of SS and LM methods to rationalize and

© Springer International Publishing AG, part of Springer Nature 2019
J. Machado et al. (Eds.): HELIX 2018, LNEE 505, pp. 361–368, 2019.
https://doi.org/10.1007/978-3-319-91334-6_49

automate simultaneously the production and logistic processes in order to improve the quality, to reduce waste and increase agility. In the paper [11], the DMAIC (Define, Measure, Analyse, Improve, Control) methodology was applied in a logistic service.

A literature analysis shows that, in the area of logistics, many modern management methods are used in the transport area. However, it is not so easy to decide on what exactly should do a logistic company which wants to decrease fuel consumption. The following questions are posed: What are the factors which influence fuel consumption in a certain company? How to identify them? What kind of actions should be taken to obtain improvement? The companies need to use a system to support a decision-making process [12]. In the works [13, 14] the examples of such systems used in the maintenance processes are presented. In the work [15] the author emphasizes the importance of a decision-making process in the logistic and supply chain management.

In this work, the authors propose to implement SS methodology which can help a company to identify important factors influencing fuel consumption on the basis of the collected data concerning the logistic process. Then, the authors propose a method which can support a decision-making process. In Sect. 2 the methodology and goal of the work is presented. Section 3 describes the problem which was defined in a case study company and presents the use of SS methodology. Section 4 presents the development of a method of supporting a decision-making process which uses fuzzy logic. The last section of this paper concludes the work and presents its limitations as well as the future research.

2 The Research Goal and Plan

The main goal of the research was to build a tool, which can support a decision-making process concerning drivers' training in a logistic company. One of the main problems in logistic companies is fuel consumption. The conducted research aimed to identify the factors which influence fuel consumption in order to allow companies to take decisions which help to minimize fuel consumption [16]. In the research, a case study was analysed. The data from a logistic company in which a high fuel consumption problem exists were analysed. The work consists of two stages. The goal of the first stage of the research was to identify important factors which influence fuel consumption. In this stage the problem was described together with the goal of the SS project and the scope of the project. Then, the transport process was analysed in order to identify potential factors which can influence fuel consumption. The data concerning these factors were collected in the period of four months. The data were used in a statistical analysis to identify important factors. The second stage was to propose a tool for supporting a decision-making process. In this stage the selected factors were analysed. The model presenting how the factors influence fuel consumption was developed. Then, on the basis of the important factors, a tool to support a decision-making process was proposed and implemented in a case study company.

3 Six Sigma Project

3.1 Problem Definition, Goal and Scope of the Project

The problem of high fuel consumption, which causes high costs of the logistic processes in a company, was a subject of the SS project. Till now, the case study company have tried to minimize the costs concerning fuel, for example by purchasing fuel in a carefully chosen net of petrol stations. However, the market still forces the company to minimize the transport costs. Therefore. the company decided to start the SS project. The main goal of the SS project was to discover the factors which influence fuel consumption and to undertake improvement actions in order to minimize fuel consumption. The company has a wide transport fleet in the form of over 100 sets of tractor + semi-trailer and realizes different kinds of transport processes inside the country as well as to the selected European countries. However, for this SS project purpose the transports realized on the same route and only on international routes were chosen. The trucks use the same routes and roads and refuel at the same petrol stations.

3.2 SIPOC Development, Data Collection and Analysis

A SIPOC diagram was developed to present the transport process. Petrol stations are suppliers (S) which deliver fuel of a certain quality (I – Inputs – x1), what can influence fuel consumption (O – Outputs) in a transport process (P) realized for different clients (C). Other factors which can have influence on the process are: truck brand (x2), driver (x3), route (x4), emission standard (x5), tonnage (x6) and distance (x7). The data were collected from 1 FEB, 2017 to 31 MAY, 2017 for 100% of the realized transports. In this period of time 122 transports were realized. The data concerning the presented factors (x2–x7) and the output (y) were collected. Only the fuel quality wasn't monitored as the drivers refuel in the same net of petrol stations. The Anderson-Darling normality test for the average fuel consumption (AFC) (y) was used. The gathered data has normal distribution with P-value = 0.141 > 0.05. The collected data were the object of a further analysis. In order to identify important factors which influence fuel consumption, the following statistical tests were performed. One-way ANOVA statistical test for x2, x3, x4 and x5 and regression analysis for x6 and x7 were used to calculate P-values. On the basis of the statistical test results, it can be said that the factors which have influence on fuel consumption are: a driver (x3; P-value = 0.001), emission standards of the trucks (x5; P-value = 0) and tonnage (x6; P-value = 0.007), what is in accordance with previous research [3]. Equation (1), being the result of a regression analysis, presents how tonnage (T) influences the AFC. The higher tonnage fuel consumption increases. However, it's still more justified to accept higher AFC because of the higher tonnage, than sending another truck. AFC depends on emission standards in the trucks. For Euro 3 and Euro 4 AFC is much higher than for Euro 5 and Euro 6. It means that newer trucks consume less fuel. Additionally, drivers themselves have influence on fuel consumption.

$$AFC = 28.1 + 0.144T \tag{1}$$

They may possess different skills what influences their style of driving. The works [6, 7] show that training is an important element in developing driving skills. Therefore, in order to minimize the costs concerning fuel consumption, drivers should be trained to apply eco-driving into practice. However, drivers are already trained and they have obtained special permissions to drive trucks. Thus, any additional training of all drivers would also create costs, particularly when a company often employs new drivers. Therefore, the question is how a company should manage the drivers' skills to decrease fuel consumption and not to increase the trainings costs much. Thus, a company will have to decide which drivers should be trained. To answer this question the authors propose a method which will support a decision-making process. The authors of this publication have not reached any works which would suggest any intelligent system that could support the process of the drivers' skills management in enterprises. Therefore, they propose to use fuzzy logic in a decision-making process and which can support the drivers' skills management.

4   Development of a Method to Support a Decision-Making

The drivers may be of different ages (DA – age of a driver), they can have a driving licence and other specific permissions (Cat. C, Cat. CE) for a longer or shorter time (YDL – number of years of possessing a driving license), they can have different numbers of permissions (NPE), they can work longer or shorter in the case study logistic company. These data were collected for each of 56 drivers. From all the mentioned factors, the authors wanted to choose these factors which have influence on AFC. Therefore, the regression analyses were made, P-values were calculated and an regression Eq. (2) were presented. On the basis of the regression analysis results, it can be said that only DA (P-value = 0.007), YDL (P-value = 0.026) and NPE (P-value = 0.018) have influence on AFC there before these factors were taken into consideration.

$$AFC = 26.1 + 0.194\,DA - 0.17YDL + 0.0078YCC - 0.0105YCE + 0.1NPE \tag{2}$$

For the purpose of the development of a tool supporting a decision-making process, the continuous data concerning factors having influence on AFC were transformed into attribute data (Low, Medium, High). Furthermore, the levels of AFC were established. It is presented in Table 1. The values in the Table 1 were developed based on the experience of the authors and experts from the company. Then, from the factors mentioned in Table 1 two (YDL and NPE) were chosen to develop a matrix of fuel consumption (Table 2) which will be used in the proposed method. The Table 2 was developed on the base of Table 1 and analysis of collected data from the case study company. In order to explain the obtained results, it can be said that drivers with more permissions were trained more times. Thus, there is a higher probability that eco-driving rules were included in the previous trainings, therefore AFC decreases. High AFC was received for the drivers with a low NPE and a low YDL. It can be said that they are not experienced yet. Nonetheless, we can also see that high AFC was received for the drivers with a low NPE and a high YDL. This can be explained by the fact that they have bad habits and they haven't been probably trained on eco-driving rules.

Table 1. Data concerning drivers transformed into attribute data and levels of AFC [l/100 km]

Level	Low	Medium	High
DA	DA < 40	$40 \leq DA \leq 50$	>50
YDL	$YDL \leq 15$	15 < YDL < 26	$YDL \geq 26$
NPE	$NPE \leq 4$	4 < NPE < 9	$NPE \geq 9$
Levels of AFC	AFC < 30.54	$30.54 \leq AFC \leq 31.47$	AFC > 31.47

Therefore, at first, these two groups of drivers should be trained on eco-driving rules. Additionally it is worth to emphasize that, the drivers were trained in different courses before therefore we can't be sure that all of them were not trained on eco-driving rules well. They have also different kinds of permissions, what is also connected with the course program. Therefore, in order to choose the drivers who should be trained, a fuzzy logic analysis was implemented. In a Mamdami fuzzy interference system (FIS), which was used in the presented case study two inputs were used: YDL and NPE. The output is AFC. The fuzzy inference process consists of three stages: Fuzzification, Rules evaluations and Defuzzification. The Matlab R2012a fuzzy logic designer was used. Because both, the inputs as well as the output were transformed into attribute data on the three levels (Table 1) and, in order to minimize the difference concerning the mathematical modelling and the practical application, the triangular membership functions (TriMF) were used. Based on the AFC matrix (Table 2), the fuzzification rules were established and used in the further analysis.

5 Fuzzy Logic Implementation in Decision-Making Process Concerning Drivers' Skills Improvement

In order to demonstrate how the developed Mamdami FIS works, the following example was analysed with the use of Matlab Fuzzy Toolbar (R2012a) simulator. When a company signs an agreement with a new driver, it checks a YDL and NPE. These data can be used to decide whether this driver should be trained on eco-driving rules. To present how the described logic works, a driver having 8 permissions (NPE) and 20 years of possessing a driving licence (YDL) was taken into consideration. With the use of these data, the potential AFC could be calculated, and it is 28.2 [L/100 km] (Fig. 1). Based on Fig. 1, we can see that mainly rule 5 works. The value obtained for AFC equals 28.2 what means that the AFC for this driver should be low, and it is not necessary to provide him with an additional training. The data were also analysed to see how the age of drivers influences AFC. The authors discovered that in 70% of cases, when the lowest AFC was obtained, the driver was younger than 50

Table 2. Matrix of AFC depending on YDL and NPE; L – low, M – medium, H – high.

NPE	YDL		
	YDL ≤ 15 (L)	15< YDL< 26 (M)	YDL ≥ 26 (H)
NPE ≤ 4 (L)	AFC>31.47	30.54≤AFC≤31.47	AFC>31.47
	31.61	31.09	31.68
4 < NPE< 9 (M)	AFC<30.54	AFC< 30.54	30.54≤AFC≤31.47
	29.79	30.18	31.12
NPE ≥ 9 (H)	AFC<30.54	30.54≤AFC≤31.47	30.54≤AFC≤31.47
	30.23	30.63	30.56

years old. Therefore, the authors propose to provide additional training for the drivers older than 50 years old, in cases presented in Table 3. The values in the Table 3 were developed based on the experience of the authors and experts from the company as well as on the base of data analysis. The drivers employed in the case study companies, who were additionally trained on eco-driving rules, were able to obtain lower fuel consumption.

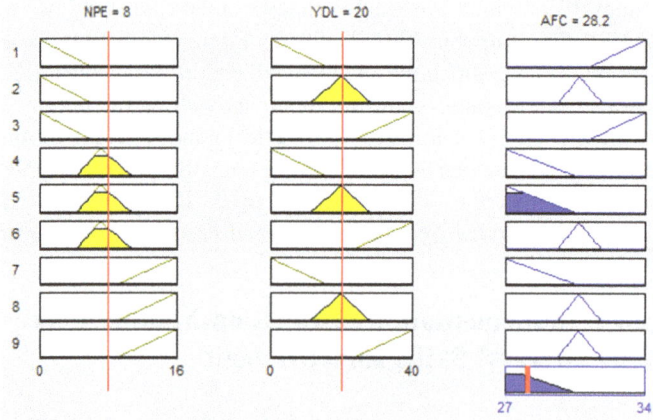

Fig. 1. Calculations of AFC

6 Conclusions

The matrix was created on the basis of the real data obtained from the logistic company, which took into account the data concerning 56 drivers and all the routes done by them in the selected period of time. The higher number of NPE, the more time a driver spent on training. As a result he/she could remember eco-driving rules better what allows for saving fuel. The higher number of YDL means that a driver was trained much time ago when eco-driving rules might not be included in the driver's training.

Table 3. Decision matrix for drivers' training

AFC	DA < 50		DA ≥ 50
Low	not necessary	≥	not necessary
Medium	not necessary		recommended
High	recommended		recommended

Additionally, a driver can also forget the training. During a long time of driving, they may have created their own rules and own habits in driving. However, while they take a next training to obtain a new permission, the eco-driving rules are reminded again and again then. This knowledge put into practice causes the decrease in fuel consumption.

The paper suggests the implementation of fuzzy logic as a tool supporting the management of the drivers' skills. On the basis of the analysis, the a manager takes a decision concerning additional trainings for the drivers in order to improve their eco-driving skills. The proposed method has limitations which are connected to the fact that the method was developed on the basis of the data from one logistic company, taking into account a certain group of drivers. Additionally, only international transports realized with use of the same routes were taken into consideration. Therefore, in the future work, additional data from the case study company as well as the data from other companies, in order to ensure that the developed method can be applied in any logistic company or to propose a way to adopt the presented method to different conditions can be analysed.

References

1. Li, L., Wang, X., Song, S.: Fuel consumption optimization for smart hybrid electric vehicle during a car-following process. Mech. Syst. Sig. Process. **87**, 17–29 (2017)
2. Makarova, I., Shubenkova, K., Pashkevich, A.: Logistical costs minimization for delivery of shot lots by using logistical information systems. Proc. Eng. **178**, 330–339 (2016)
3. Dindarlo, S.R., Siami-Irdemoosab, E.: Determinants of fuel consumption in mining trucks. Energy **112**, 232–240 (2016)
4. Toledo, G., Shiftan, Y.: Can feedback from in vehicle data recorders improve driver behavior and reduce fuel consumption? Transp. Res. Part A: Policy Pract. **94**, 194–204 (2016)
5. Díaz-Ramirez, J., Giraldo-Peralta, N., Flórez-Ceron, D., Rangelb, V., Mejía-Argueta, C., Huertas, I., Bernal, M.: Eco-driving key factors that influence fuel consumption in heavy-truck fleets: A Colombian case. Transp. Res. Part D: Transp. Environ. **56**, 258–270 (2017)
6. Barl, P., Gilbert-Gonthier, M., Lopez, M.A., Castroa, L., Moreno, M.: Eco-driving training and fuel consumption: Impact, heterogeneity and sustainability. Energy Econ. **62**, 187–194 (2017)
7. Xu, Y., Li, H., Liu, H., Rodgers, M.O., Guensler, R.L.: Eco-driving for transit: an effective strategy to conserve fuel and emissions. Appl. Energy **194**, 784–797 (2017)

8. Alhuraish, I., Robledo, C., Kobi, A.: Assessment of lean manufacturing and six sigma operation with decision making based on the analytic hierarchy process. IFAC-PapersOnLine **49**(12), 59–64 (2016)
9. Yuan, Z., Yan, X., Xuan, Z.: Research on the significance of six-sigma's implementation in logistics corporation. In: Proceedings of IEEE International Conference on Automation and Logistics, vol. 1(6), pp. 2089–2098. IEEE explorer (2008)
10. Nicoletti, B.: Lean and automate manufacturing and logistics. In: Rannenberg, K. (ed.) Advanced in Production Manufacturing Systems APMS 2013, PT II, Book Series. IFIP Advances in Information and Communications Technology, vol. 415, pp. 278–285, Springer, Cham (2013)
11. Gutierrez-Gutierrez, L., Leeuw, S., Dubbers, R.: Logistics services and lean six sigma implementation: a case study. Int. J. Lean Six Sigma **7**(3), 324–342 (2016)
12. Ranisavljević, P., Spasić, T., Mladenović-Ranisavljević, I.: Management information system and decision making process in enterprise. Econ. Manag. Inf. Technol. **1**(3), 184–188 (2012)
13. Antosz, K., Stadnicka D.: An intelligent system supporting a forklifts maintenance process. In: Burduk, A., Mazurkiewicz, D. (eds.) Intelligent Systems in Production Engineering and Maintenance – ISPEM 2017. ISPEM 2017. Advances in Intelligent Systems and Computing, vol. 637, pp. 13–22. Springer, Cham (2018)
14. Ratnayake, R.M.C., Antosz, K.: Risk-based maintenance assessment in the manufacturing industry: minimisation of suboptimal prioritisation. MPER – Manag. Prod. Eng. Rev. **8**(1), 38–45 (2017)
15. Yazdani, M., Zarate, P., Coulibaly, A., Zavadskas, E.K.: A group decision making support system in logistics and supply chain management. Expert Syst. Appl. **88**(1), 376–392 (2017)
16. Kurzeja, T.: Analysis of factors influencing on fuel consumption in a logistic company. Unpublished work under supervising of Dorota Stadnicka, Rzeszow (2017). (in Polish)

Optimization of the Medium-Term Production Planning in the Company-Case Study

Perłowski Ryszard, Antosz Katarzyna[(✉)], and Zielecki Władysław

Faculty of Mechanical Engineering and Aeronautics,
Rzeszow University of Technology, Rzeszow, Poland
{rpztmiop, katarzyna.antosz, wzktmiop}@prz.edu.pl

Abstract. Contemporary enterprises of sustainable development, try to optimize the use of resources. Medium-term planning is one of the main elements of the MRP II production management systems, in which the key result of planning is the development of a production plan. That plan enables to meet the forecast demand for products and services on the basis of the resources engaged for this purpose in the accepted term. The aim of this article is to indicate the possibility of increasing the accuracy of informal methods in production planning, in particular graphic-table techniques with the use of an IT tool. The work discusses decision-making options and strategies for balancing projected demand with available production capacities.

Keywords: Medium-term production plan · Optimization
Aggregate planning

1 Introduction

The most important goal that every company sets for itself is to maximize profits, while minimizing production costs. Therefore, the main goal of a production flow control is to manufacture products in quantities and on dates corresponding to the adopted sales plans [1, 2]. It is also associated with shorter product life cycles, while producing more product variants as a result of more variable customer requirements [3]. Such market conditions require, above all, an efficient planning process that affects the timeliness of deliveries to the customer [4]. Actions aimed at this can be achieved by controlling the main parameters of the production flow, such as quantity and time. Each individual operation should be planned and performed in accordance with the dates allowing for the execution of the sales plan. All these activities, at the same time, should take into account the trade-off between the overloading of employees, the operation of production equipment, the use of space and minimal involvement of financial resources. Achieving such a compromise results in minimal costs borne by the company, which certainly is for the benefit of the company. Optimal production planning and scheduling is a big challenge for manufacturing companies. In practice, the planning problem is usually more complex due to many limitations of tasks and machines, more complex structures of production systems and various target criteria [5]. In recent years, a lot of new research has been carried out on planning, new concepts and techniques of repetitive production operations. As a result of innovations, such as computer-integrated production/management (CIM) and just-in-time (JIT), new processes are

© Springer International Publishing AG, part of Springer Nature 2019
J. Machado et al. (Eds.): HELIX 2018, LNEE 505, pp. 369–376, 2019.
https://doi.org/10.1007/978-3-319-91334-6_50

created in today's companies that capture the benefits of repeatable production and continuous flow generation [6]. There are also two-level methods of production planning and task planning [7], the optimal planning strategy [8] is determined, the methods of Aggregate Production Planning (APP) [9] are successfully used. For APP method many planers use linear programming and simulations methods. In the work [10] stochastic multi-objective optimisation techniques is used for APP. The authors in the work [11] used stochastic nonlinear programming models for APP. Leung in the work [12] to handle APP with stochastic demand and cost parameters proposed using stochastic linear programming method. Moreover there are many publications such work [13] in which different kinds of robust optimisation techniques to study APP is used. Additionally in last time the in publications i.e. [14] using the genetic algorithm in APP is observed. Unfortunately, despite the many tools available, the planning process is still poorly implemented. The main reasons are poor planning, poorly selected tools and planning techniques or ignorance of the employees implementing the planning process [15]. Additionally these methods are very complicated and not so often useful in the industry.

The aim of this article is to present a developed simple and useful solution that supports the process of medium-term planning (MTP) in a production company. In the second part, the role, objectives and methods of MTP were briefly characterized. Section 3 presents the methodology and purpose of the work. Section 4 presents the problem of optimizing planning using an IT tool. The last part of this article concludes the work and presents limitations as well as future research.

2   Medium-Term Planning (MTP)

Medium – term planning is strategic planning. A typical planning horizon is 1–2 years. It is the link binding and coordinating the strategic business objectives with the operational objectives of production (Fig. 1).

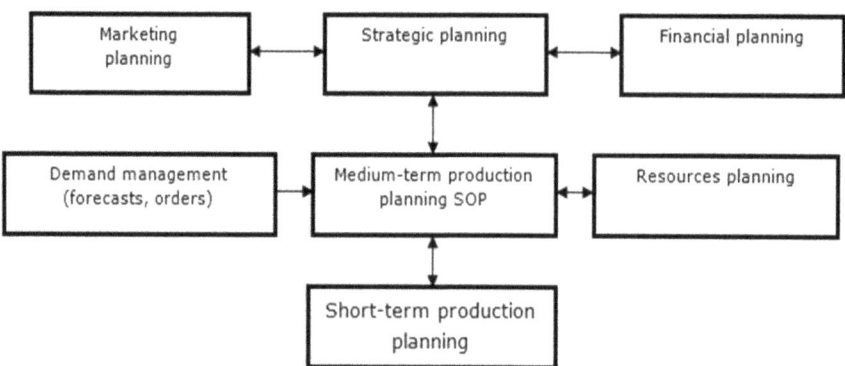

Fig. 1. Links within medium-term production planning

The key result of planning is the development of a production plan responding to the forecast demand for products and services on the basis of the resources engaged for this purpose in the accepted time. While preparing a medium-term plan, the following are used: a strategic plan, forecast, available production size, adopted decision options and the costs of their implementation and an adopted planning strategy [9]. By limiting the planning units by grouping them, the S&OP plan is defined as an aggregated production plan. Such a plan concerns mostly the product family distinguished e.g. on the basis of the construction and technology similarities. Aggregated planning is related to the forecast demand which is often variable in the form of upward or downward trends as well as in seasonal fluctuations. Taking account of the size of the forecast demand in relation to the available production size, this problem may be considered in terms of static or dynamic (Fig. 2). For APP we can use three basic strategies: even production - (Level Strategy - LS), production for demand - Chase Strategy (CHS), variable production – Hybrid/Mixed Strategy (MS) [9].

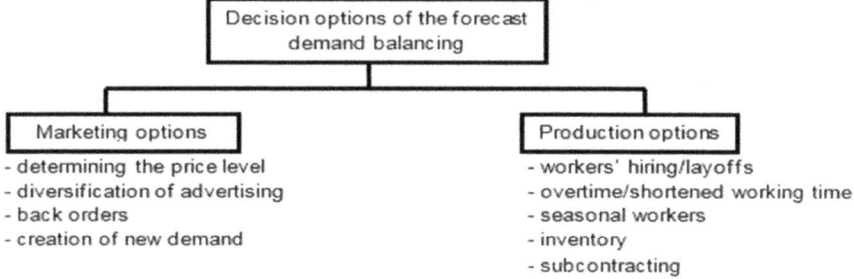

Fig. 2. Basic decision options of the forecast demand balancing

MTP may be realized with a number of methods. Generally, they can be divided into informal (manual) and formal (mathematical) methods which are often IT supported. In practice, the informal approach is most commonly used. It involves the approximation method and the trial-and-error approach. This way, further plan alternatives are created and, next, the plan which guarantees the acceptable costs is chosen. The most common technique, based on the informal approach, is a graphic-tabular technique which uses graphs and design tables [16]. Second method of enabling the MTP involves mathematical methods [17]. The growing complexity of decision processes in a company as well as of relations between them make the informal methods inaccurate and, in consequence, increase the negative results of the mistakes made.

3   The Goal and Methodology

The goal of the research was to build a tool supporting the mid-term planning process in a production company. One of the main problems of the company is the problem of production size optimization. The problem is important because in the case of a poorly selected production size, the company can generate back-orders or create unnecessary inventories. Moreover, both: the back – orders and unnecessary inventories generate unnecessary costs for the company. The proposed solution uses a method combining elements of graphical and tabular techniques with the elements of mathematical methods in the form of a spreadsheet, which can easily and quickly support the planner in MTP. Using the developed spreadsheet and the Solver add-on makes it possible to solve optimization problems, especially linear ones, with a small number of decision variables (up to 200). The Solver module is a standard addition to a spreadsheet that can be used to solve optimized linear and non-linear tasks, as well as the whole-object tasks, including the tasks with binary variables. Planning simulations were carried out for the selected company using the APP. Costs were determined for each strategy and production optimization measures were carried out.

4 Optimization Modelling – Case Study

4.1 Definition of Inputs

The analysed company respects the principles of sustainable development, optimizing the use of available resources, planning to use resources in accordance with the forecasted demand. The presented example concerns the development of the S&OP plan for a production company. The company manufactures rubber products for the automotive industry. Using the Glenday Sieve (GS) method, a group of products was selected for which it was appropriate to apply a level strategy, including production costs, inventories and unrealized orders. Monthly forecasts for the selected product were used to conduct the simulation. The company's production size allows for the production of 1300 items per month. At the same time, the company declares the possibility of increasing it to 1800 pcs/month. The optimization activities for the selected company were carried out in two stages: selecting the volume of production manually in the prepared form and Conducting optimization activities using the Solver add-on.

The above activities were carried out for three strategies: variant I- fixed volume of production (Level Strategy), variant II - strategy for chasing demand (Chase Strategy), variant III - mixed production (Hybrid/Mixed Strategy). Before starting the planning, the person servicing the sheet (planner) introduces the forecast demand in the adopted plan horizon (column - demand forecast), determines the assumed production size (column - production) and assumes appropriate unit costs for individual strategies (Fig. 3).

4.2 Analysis of Variants

In Level Strategy we assume a constant production size of 1300 units. The increased demand is balanced by inventories created in periods when the demand is lower than the assumed production size (periods 1–2, Fig. 3).

Appropriate sheet formulas automatically balance the demand with the size of the inventory. However, the planned variant of the plan will show significant shortages in the form of back-orders (periods 3–12). Using the Solver it is possible to determine the

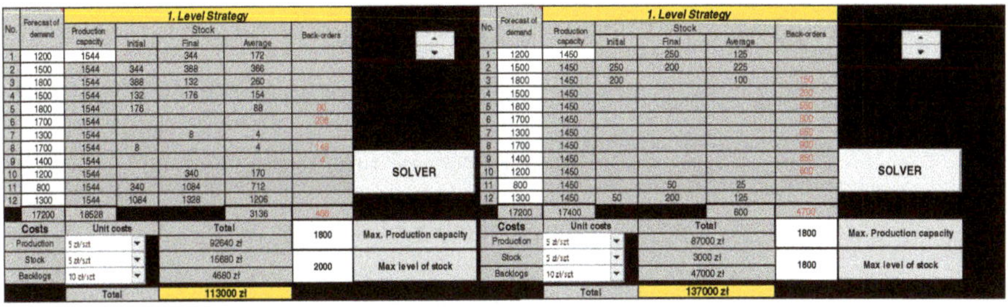

Fig. 3. Version of the plan according to the LS (a) without limitations of the stock level, (b) with the limit of the stock level up to 250 items.

optimal production size, balancing analysed costs (objective function - the minimum total costs). The batch size to the maximum is declared by the company (1800 pieces/month). In this case, the minimum total cost is reached for the level of 1544 items/month. The company, however, has a limited space for storing inventories - a maximum of 250 items/month. By adding further imposed constraints, Solver will set an optimal monthly production batch size to 1450 pcs. The proposed sheet has also the option of cost balancing. The CHS adjusts the level of production size according to the size of the forecast demand. If the size of demand increases, the production size also increases and vice versa. In this case, the total cost will be much higher compared to the MTP. It is particularly time-consuming to develop a plan according to Hybrid/Mixed Strategy, where the number of possible plan variants is many times higher than in previous strategies. With a constant level of production capacity, it is possible to shorten working hours or extend them (overtime), and with an insufficient number of hours, the use of subcontracts to balance changes in demand. There is also a possibility of reducing the production level in the periods of lower demand. The plan may be developed in two ways. The first method consists in "manual" modification of decision options (nominal output volume, overtime hours and subcontracts). Each change is immediately verified by the calculated total costs of this strategy. For an experienced planner, it is possible to get a satisfactory result after few dozen attempts. The other method uses the Solver add-in. In order to find a plan that assures minimal costs of the strategy, the parameters of the Solver module should be appropriately defined. An objective function (objective cell), which is mathematically described as a correlation (1) – in this case the total costs of the strategy – should reach the minimum value. The decision variables $x1 \div x48$ (cells changed) include accordingly: the size of the nominal production, overtime, subcontracting and inventory level. In this case, the production size in the nominal time equals to 1450 items per month (2), and the overtime production as well as the number of subcontracts shouldn't go above the assumed 75 items per month and 150 items per month accordingly (3, 4). In order to achieve a reasonable plan, the condition of the total values of decision variables (5) and the conditions describing a limited production size must be entered. The function of the objective:

$$f(x) = a(x_1 + L + x_{12}) + b(x_{13} + L + x_{24}) + c(x_{25} + L + x_{36}) + d(x_{37} + L + x_{48}) \rightarrow \min \quad (1)$$

$$x_1 \div x_{12} \leq 1450 \text{ – limited production size} \quad (2)$$

$$x_{13} \div x_{24} \leq 75 \text{ – limited overtime production} \quad (3)$$

$$x_{25} \div x_{36} \leq 150 \text{ – limited number of subcontracts} \quad (4)$$

$$x_1 \div x_{36} \text{ – whole numbers} \quad (5)$$

where $x_{i,j}$ is decision variables: x_1 ... x_{12} – production size in nominal time, x_{12} ... x_{24} – production size in overtime, x_{25} ... x_{36} - production size in subcontracts, x_{37} ... x_{48} - average inventory level, a – unit cost of nominal production, b – unit cost of production in overtime, c – unit cost of production in subcontracts, d – unit cost of holding inventory. In addition, the accepted condition is a limitation in relation to the level of stocks, securing both against lack in the stock and overstock (min > = 0, max <= 250).

After entering all the parameters, the Solver module will generate an optimal aggregate plan of the production rate changes which guarantees the lowest costs. At the same time, these costs are the lowest from among the analysed production strategies (Fig. 4).

No.	Forecast of demand	1. Level Strategy					2. Chase Strategy			3. Hybrid Strategy					
		Capacity	Stock			Back-orders	Capacity	Inc. Cap. Zdoln	Decrea. Cap. Zdoln	Capacity			Inventories		
			Initial	Final	Average					Nominal.	Overtime	Subcon.	Initial	Final	Average
1	1200	1544		344	172		1200		344	1450				250	125,0
2	1500	1544	344	388	366		1500	300		1450	50		250	250	250,0
3	1800	1544	388	132	260		1800	300		1450	75	25	250		125,0
4	1500	1544	132	176	154		1500		300	1450	75	125		150	75,0
5	1800	1544	176		88	80	1800	300		1450	75	150	150	25	87,5
6	1700	1544				236	1700		100	1450	75	150	25		12,5
7	1300	1544		8	4		1300		400	1450	25			175	87,5
8	1700	1544	8		4	148	1700	400		1450	75		175		87,5
9	1400	1544				4	1400		300	1400					
10	1200	1544		340	170		1200		200	1200					
11	800	1544	340	1084	712		800		400	800					
12	1300	1544	1084	1328	1206		1300	500		1300					
	17200	18528			3136	468		1800	2044	16300	450	450			850,0

Costs	Unit costs	Total	Costs	Unit costs	Total	Total	Unit costs		Total
Production	5 zł/szt	92640 zł	Production	5 zł/szt	86000 zł	Production	5 zł/szt	Production	81500 zł
Inventory	5 zł/szt	15680 zł	Increasing Cap.	30 zł/szt	54000 zł	Ovetrtime Prod.	10 zł/szt	Ovetrtime Prod.	4500 zł
Back - orders	10 zł/szt	4680 zł	Decrasing. Cap.	30 zł/szt	61320 zł	Subcontracts Prod.	20 zł/szt	Subcontracts Prod.	9000 zł
						Inventories	5 zł/szt	Inventories	4250 zł
	Total	113000 zł		Total	201320 zł			Total	99250 zł

Fig. 4. View of an aggregate planning spreadsheet.

5 Conclusions

A production planning process is an important element of the enterprise's operation. The main purpose of this process is to manufacture products in a number that meets the customer's requirements in accordance with the agreed delivery date. The implementation of the customer's orders involves the use of the resources of the company. Then, it is important for the company which properly plans its activities not to be exposed to generating unnecessary costs. Rational production planning is possible with the help of a variety of organizational and analytical techniques and methods that allow you to reduce or eliminate the activities that do not add value to your products or bring profits. The presented technique is an effective tool supporting planning in an enterprise. The company's management can quickly check which scheme of resource allocation (production capacity) will allow them to meet the demand and minimize costs. The adoption of the final variant of the plan must be taken in conjunction with other information, e.g. Does the company have warehouse space to store a certain amount of stock? Is it possible to change production capacities quickly? Are savings big enough to overcome these difficulties? The presented technique indicates a way to quickly and efficiently determine a plan in order to minimize costs. It is also the starting point for making a final decision about the shape of the plan that belongs to man.

The presented work has limitations related to the fact that the method was developed on the basis of the data from one company. Therefore, in the future work, the authors would like to collect additional data from the case study company as well as the data from other companies, in order to ensure that the developed method can be used in any company or to propose a method to adopt the presented method for different conditions.

References

1. Tuomikangas, N., Kaipia, R.: A coordination framework for sales and operations planning (S&OP): synthesis from the literature. Int. J. Prod. Res. **154**, 243–262 (2014)
2. Krolczyk, J., Krolczyk, G., Legutko, S., Napiorkowski, J., Hloch, S., Foltys, J., Tama, E.: Material flow optimization – a case study in automotive industry. Tehnicki Vjesn. Tech. Gaz. **22**(6), 1447–1456 (2015)
3. Koren, Y.: The Global Manufacturing Revolution: Product–Process–Business Integration and Reconfigurable Systems. Wiley, Hoboken (2010)
4. Koren, Y., Shpitalni, M.: Design of reconfigurable manufacturing systems. J. Manuf. Syst. **29**(4), 130–141 (2010)
5. Trojanowska, J., Varela, M.L.R., Machado, J.: The tool supporting decision making process in area of job-shop scheduling. In: Rocha, Á., Correia, A., Adeli, H., Reis, L., Costanzo, S. (eds.) Recent Advances in Information Systems and Technologies, WorldCIST 2017. Advances in Intelligent Systems and Computing, vol. 571, pp. 490–498. Springer, Cham (2017)
6. Kreipl, S., Michael, P.: Planning and scheduling in supply chains: an overview of issues in practice. Prod. Oper. Manag. **13**(1), 77–92 (2004)
7. Cho, H.-M., Jeong, I.-J.: A two-level method of production planning and scheduling for bi-objective re-entrant hybrid flow shops. Comput. Ind. Eng. **106**, 174–181 (2017)
8. Chiang, C.H., Min-Hsiu, F., Yeu-Shiang, H.: Production planning of new and remanufacturing products in hybrid production systems. Comput. Ind. Eng. **108**, 88–99 (2017)
9. Abassa, S., Gomma, M.A., Elsharawy, G.A., Elsaid, M.S.: Generalized production planning problem under interval uncertainty. Egypt. Inf. J. **11**(11), 27–31 (2010)
10. Rakes, T., Franz, L., Wynne, A.: Aggregate production planning using chance-constrained goal programming. Int. J. Prod. Res. **22**(4), 673–684 (1984)
11. Vörös, J.: On the risk-based aggregate planning for seasonal products. J. Prod. Econ. **59**(1–3), 195–201 (1999)
12. Leung, S.C.H., Wu, Y., Lai, K.: A stochastic programming approach for multisite aggregate production planning. J. Oper. Res. Soc. **57**(2), 123–132 (2013)
13. Niknamfar, A.H., Akhavan Niaki, S.T., Pasandideh, S.H.R.: Robust optimization approach for an aggregate production–distribution planning in a three-level supply chain. Int. J. Adv. Manuf. Technol. **76**(1), 623–634 (2015)

14. Savsani, P., Banthia, G., Gupta, J., Ronak, V.: Optimal aggregate production planning by using genetic algorithm. In: Proceedings of the International Conference on Industrial Engineering and Operations Management, IEOM 2016, pp. 863–874. IOEM Society Bangladesh (2016)
15. Gansterer, M.: Aggregate planning and forecasting in make-to-order production systems. J. Prod. Econ. **170**, 521–528 (2015)
16. Bozarth, C., Handfield, R.B.: Introduction to Operations and Supply Chain Management. HELION, Gliwice (2007). (in polish)
17. Sysło, M., Narsingh, D., Kowalik, J.S.: Discrete Optimization Algorithms. PWN, Warszawa (1995). (in polish)

Methodology of Estimating Manufacturing Task Completion Time for Make-to-Order Production

Krzysztof Żywicki, Filip Osiński[(⊠)], and Radosław Wichniarek

Chair of Management and Production Engineering,
Poznan University of Technology, Poznan, Poland
{krzysztof.zywicki,filip.osinski,
radoslaw.wichniarek}@put.poznan.pl

Abstract. Timely delivery of products designed to individual requirements of customers is increasingly difficult for manufacturers. It results from difficulties in effective development of production schedules and material flow control. The reason is, among other things, the lack of real data on standard times required to complete individual manufacturing processes. This article presents the assumptions on the method of estimating standard times for highly individualized products. The proposed method has been incorporated into computer software that was verified at an actual production company that manufactures electric heating elements.

Keywords: Standardizing time · Preparing production
Production management

1 Introduction

Individual requirements of customers, particularly in the scope of construction or configuration of finished products are currently one of the main factors to influence functioning of the manufacturing systems. This affects a lot of decision-making processes concerning, among other, the technology of manufacturing or production planning and management. Therefore, effective design of products to fulfil individual requirements of customers (for example the implementation of mass customization strategy) and effective production management are necessary components of a flexible manufacturing system consistent with the concept of smart factory [1–6].

Structural and technological data have direct influence upon manufacturing task completion time and use of manufacturing resources. This results from standard times of completion of individual manufacturing stages enabling to develop a production schedule [7–13].

Methods of standardizing work time can be divided into two basic categories connected with each method's intended purpose. Summary methods apply for discrete manufacturing, variant manufacturing or small-scale manufacturing, whereas in the case of mass production or serial production, analytical methods are used.

© Springer International Publishing AG, part of Springer Nature 2019
J. Machado et al. (Eds.): HELIX 2018, LNEE 505, pp. 377–383, 2019.
https://doi.org/10.1007/978-3-319-91334-6_51

Summary methods of standardizing work time are based on determining work time for the entire task, without analysing its individual components or specific operations. This group of methods includes:

- estimation method – determining work time by an employee who, on the basis of his/her own experience, defines the amount of time required to perform a given manufacturing task
- statistical method – using historical data on time required to manufacture products or analogous operations
- comparative method – using historical data for similar manufacturing operations, taking into consideration modification of given operation's time by specific factors, such as, for instance, dimensions, area, and weight of the product or workpiece under processing.

Manufacturing of customized products is a great challenge for production scheduling and management. They have direct impact on the efficiency of resources and achievement of goals connected with manufacturing costs. That is why systemic solutions as regards production scheduling, monitoring material flow, analysis and decision-making are required. Since there are a lot of factors to be taken into account when preparing a production schedule, it is necessary to implement methods that allow for introducing dynamic adjustments to it [14, 15].

Bearing in mind the diversity of customers' requirements, the adopted scheduling methods must take into consideration the possibility of manufacturing goods with application of different manufacturing resources. This enables fast response to new orders or if new factors appear that makes it impossible to use the resources. That is the reason for growing importance of close integration between scheduling and production preparation. Additionally, the scheduling process should be performed on the basis of feasible material flow variants. This allows for determining the most efficient material flow with reference to the assumed criteria, e.g. using production resources, the shortest delivery time or minimizing stock [16–19].

In this context, calculating standard times for performance of manufacturing processes is of increasing importance. The degree of accuracy in determining standard times translates directly into accuracy of production scheduling and management. On the other hand, however, the amount of labour required to define the standard times is equally important, particularly taking into account the degree of diversity and low repeatability of manufactured products.

The article provides assumptions of the method of estimating manufacturing completion times for production of individual structural product designs.

2 Methodology of Standardizing Time of Completion of Manufacturing Orders

The suggested methodology of standardizing time is based on application of the following combined methods of calculation: a summary method (statistical and comparative) and an analytical-measuring method. Those two methods combined allow for reducing errors resulting from both numerous variants of products and differences in production batch size.

The application of the methodology is based on the following input data:

- group of products – the criterion of structural and technological similarity + structural variants,
- technological processes – specific technological operations,
- critical structural features for product groups.

The assumption of the calculation methodology is estimating standard times based on the knowledge about already performed manufacturing tasks that includes the actual times of completion of individual technological operations.

The following actions should be performed for each new manufacturing task subject to time standardizing:

- assign a product group,
- determine a structural variant,
- define values of critical features.

Defining the above data, allows to calculate standard times of completion of consecutive technological operations. Assigned data are available to customer service from the very beginning of the quoting process. It allows for very quick determination of labor consumption and completion time for the offered products.

Time required to complete a technological operation is determined for one piece of product and for the number of pieces included in the manufacturing order. Time required to complete an operation for one piece of the product is calculated analytically on the basis of data from already completed operations under other manufacturing orders. The basis for this is the correlation between the values of critical features and the actual time required to complete the operation. The use of historical data in connection with their constant update with the latest execution times, also allows to update the obtained results, due to the increasing efficiency of processes and thus shortening production times.

3   Computer Software for Estimating Time Required to Complete a Manufacturing Orders

The elaborated method of calculation has been implemented in computer software supporting estimation of times required to complete manufacturing tasks. The software contains a technological knowledge database, a module for defining manufacturing orders and algorithms used for estimating times required to complete given operations. The technological knowledge database includes:

- structure of technological processes – a list of operation for individual product families,
- specific factors affecting standard time required to complete a given technological operation,
- times required to complete manufacturing orders.

Defining new manufacturing orders subject to time estimation consists in selecting a product family which is a basis for generating consecutive technological operations. Then, factors and their values that affect the time required for completion are determined (Fig. 1).

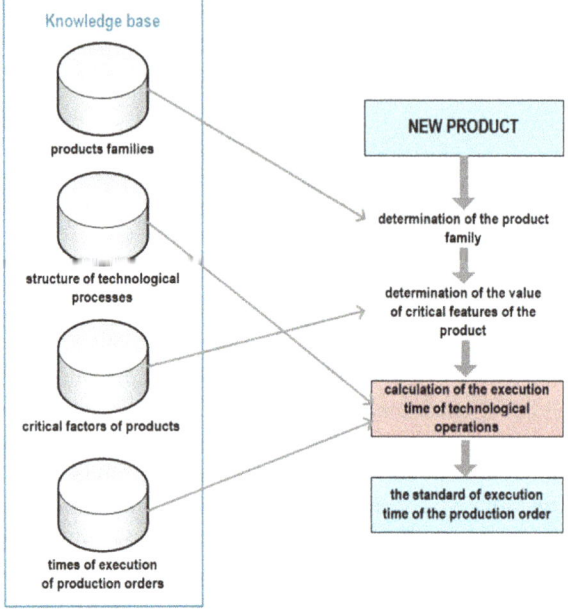

Fig. 1. Program structure for calculating the execution of production orders

Next, the algorithm selects the most similar manufacturing process and calculates estimated time required to complete a given manufacturing order. As a result, time required to complete consecutive technological operations which are used as input data in production scheduling is determined.

4 Examples of Application

The method of estimating standard time implemented in the software was validated at a company that produces electrical heating elements. The plant manufactures heating elements for all types of devices that require increasing temperature and maintaining it at a defined level. Under the elaborated methodology of standardizing manufacturing time the manufactured products were divided into 7 main groups in which individual structural variants were distinguished. For each group of products critical features that directly affect the time required to complete technological operations were determined. These included: dimensions, shape, power connection type, presence of temperature sensors, winding type, heating area, heating power or presence of radiator. Such a high number of variants of finished products make it necessary for each order to be treated individually. A simple estimation of time required to manufacture a given product becomes impossible. Furthermore, the manufacturing is in most part based on manual work in workshop conditions.

For such prepared input data the planned and actual times were verified. Planned time required to perform the operations were evaluated both in relation to the total time of manufacturing and in relation to time required to perform individual operations for one workpiece (unit time). Total manufacturing time forecasted by the software pursuant to the suggested method was on average ca. 5% longer than the actual manufacturing time. Much greater differences occurred for individual technological operations. The greatest differences were found in core preparation and insulation processes (the estimated time was 38% longer for core preparation and 20% shorter for insulation, than the actual time of completion). The reasons for such differences are the specificity of the operations. The first of them – core preparation – is based on manual assembly of core elements, and substantially it depends on the availability of finished components or the necessity of performing additional processing on individual components. The second operation – insulation – requires extremely precise manual operations (Fig. 2).

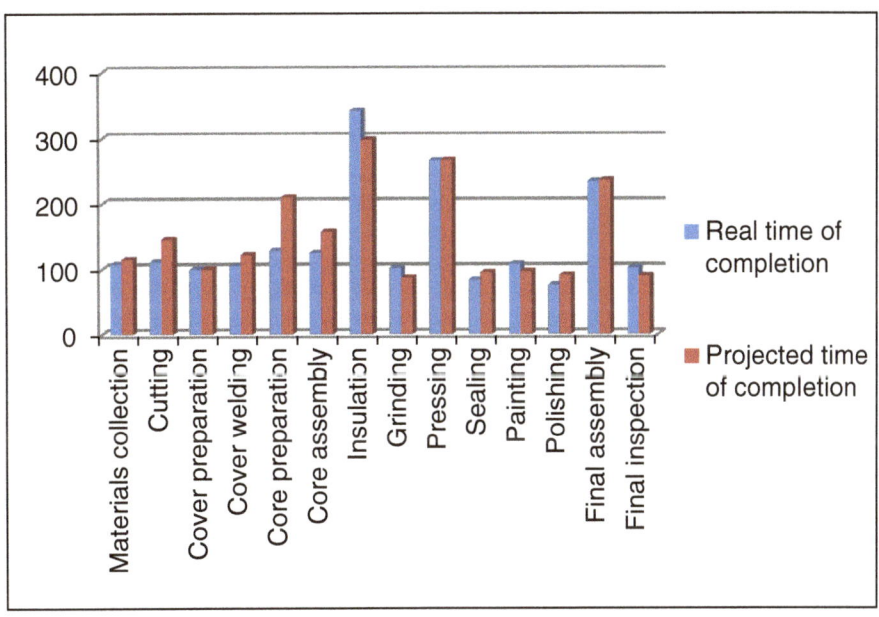

Fig. 2. The results of the methodology validation

5   Conclusion

The main challenges for manufacturers nowadays are clients' requirements as regards diversity of product structures. The situation calls for an ongoing adjustment of production, mainly by flexible planning and management. Data on times required to complete individual manufacturing tasks are the basic requirement of effective production management. However, structural diversity of products makes defining real standard times more and more complicated and laborious. Thus, application of calculation or measuring methods does not guarantee that the same will be determined correctly. The article proposes a methodology for estimating times required to perform manufacturing tasks based on a method of analysis and measurement. It estimates times required to complete manufacturing tasks based on the products' critical features. The methodology was verified against manufacturing of electrical heaters and showed similarities between standard times and actual times that were required to complete manufacturing orders.

Acknowledgments. The presented results of the research, carried out under the theme No. 02/23/DSPB/7716, were funded with a grant to science granted by the Ministry of Science and Higher Education.

References

1. Ivanov, D., Dolgiu, A., Sokolov, B., Werner, F., Ivanova, M.: A dynamic model and an algorithm for short-term supply chain scheduling in the smart factory industry 4.0. Int. J. Prod. Res. **54**(2), 386–402 (2016)
2. Górski, F., Zawadzki, P., Hamrol, A.: Knowledge based engineering as a condition of effective mass production of configurable products by design automation. J. Mach. Eng. **16**, 5–30 (2016)
3. Kai-Frederic Seitza, K.F., Nyhuisa, P.: Cyber-physical production systems combined with logistic models – a learning factory concept for an improved production planning and control. Proc. CIRP **32**, 92–97 (2015). The 5th Conference on Learning Factories 2015
4. Zhang, Y., Xie, F., Dong, Y., Yang, G., Zhou, X.: High fidelity virtualization of cyber-physical systems. Int. J. Model. Simul. Sci. Comput. **4**(02), 1340005 (2013)
5. Shrouf, F., Ordieres, J., Miragliotta, G.: Smart factories in industry 4.0: a review of the concept and of energy management approached in production based on the Internet of Things paradigm. In: 2014 IEEE International Conference on Industrial Engineering and Engineering Management (IEEM). IEEE (2014)
6. Żywicki, K., Zawadzki, P., Hamrol, A.: Preparation and production control in smart factory model. In: Rocha, Á., Correia, A., Adeli, H., Reis, L., Costanzo, S. (eds.) Recent Advances in Information Systems and Technologies, WorldCIST 2017. Advances in Intelligent Systems and Computing, vol. 571. Springer, Cham (2017
7. Żywicki, K., Zawadzki, P.: Fulfilling individual requirements of customers in smart factory model. In: Hamrol, A., Ciszak, O., Legutko, S., Jurczyk, M. (eds.) Advances in Manufacturing. Lecture Notes in Mechanical Engineering. Springer, Cham (2018)
8. Verhagen, W.J.C., Bermell-Garcia, P., Van Dijk, R.E.C., Curran, R.: A critical review of Knowledge-Based Engineering: an identification of research challenges. Adv. Eng. Inf. **26** (1), 5–15 (2012)
9. Choi, J.W., Kelly, D., Raju, J., Reidsema, C.: Knowledge-based engineering system to estimate manufacturing cost for composite structures. J. Aircr. **42**(6), 1396–1402 (2005)
10. Dostatni, E., Diakun, J., Grajewski, D., Wichniarek, R., Karwasz, A.: Functionality assessment of ecodesign support system. Manag. Prod. Eng. Rev. **6**(1), 10–15 (2015)

11. Grajewski, D., Diakun, J., Wichniarek, R., Dostatni, E., Buń, P., Górski, F., Karwasz, A.: Improving the skills and knowledge of future designers in the field of ecodesign using virtual reality technologies. In: Proceedings of International Conference Virtual and Augmented Reality in Education, Monterrey, Mexico, pp. 348–358 (2015)
12. Kalinowski, K., Zemczak, M.: Preparatory stages of the production scheduling of complex and multivariant products structures. Adv. Intell. Syst. Comput. **368**, 475–483 (2015)
13. Krenczyk, D., Skolud, B., Olender, M.: Semi-automatic simulation model generation of virtual dynamic networks for production flow planning. IOP Conf. Ser. Mater. Sci. Eng. **145**, 042021 (2016)
14. Varela, M., Trojanowska, J., Carmo-Silva, S., et al.: Comparative simulation study of production scheduling in the hybrid and the parallel flow. Manag. Prod. Eng. Rev. **8**(2), 69–80 (2017). Accessed 6 Jan 2018
15. Trojanowska, J., Varela, M.L.R., Machado, J.: The tool supporting decision making process in area of job-shop scheduling. In: Rocha, Á., Correia, A., Adeli, H., Reis, L., Costanzo, S. (eds.) Recent Advances in Information Systems and Technologies, WorldCIST 2017. Advances in Intelligent Systems and Computing, vol. 571. Springer, Cham (2017)
16. Ivanov, D., Dolgui, A., Sokolov, B.: A Dynamic approach to multi-stage job shop scheduling in an industry 4.0-based flexible assembly system. In: Lödding, H., Riedel, R., Thoben, K.D., von Cieminski, G., Kiritsis, D. (eds.) Advances in Production Management Systems. The Path to Intelligent, Collaborative and Sustainable Manufacturing, APMS 2017. IFIP Advances in Information and Communication Technology, vol. 513. Springer, Cham (2017)
17. Kaihara, T., Kokuryo, D., Fujii, N., Hirai, K.: A proposal of production scheduling method considering users' demand for mass customized production. In: Lödding, H., Riedel, R., Thoben, K.D., von Cieminski, G., Kiritsis, D. (eds.) Advances in Production Management Systems. The Path to Intelligent, Collaborative and Sustainable Manufacturing, APMS 2017. IFIP Advances in Information and Communication Technology, vol. 513. Springer, Cham (2017)
18. Mehrsai, A., Figueira, G., Santos, N., Amorim, P., Almada-Lobo, B.: Decentralized vs. centralized sequencing in a complex job-shop scheduling. In: Lödding, H., Riedel, R., Thoben, K.D., von Cieminski, G., Kiritsis, D. (eds.) Advances in Production Management Systems. The Path to Intelligent, Collaborative and Sustainable Manufacturing, APMS 2017. IFIP Advances in Information and Communication Technology, vol 513. Springer, Cham (2017)
19. Klos, S., Skrzypek, K., Dąbrowski, K.: ERP-based innovation management system for engineering-to-order production. In: Innovation Management, Development Sustainability, and Competitive Economic Growth - Vision 2020. International Business Information Management Association (IBIMA), pp. 3007–3016 (2016)

Quality Management in Training Companies

Ana Fernandes[1] ⓘ, Henrique Vicente[1,2] ⓘ,
Margarida Figueiredo[1,3] ⓘ, Jorge Ribeiro[4] ⓘ, and José Neves[2(✉)] ⓘ

[1] Departamento de Química, Escola de Ciências E Tecnologia,
Universidade de Évora, Évora, Portugal
anavilafernandes@gmail.com, {hvicente,mtf}@uevora.pt
[2] Centro Algoritmi, Universidade do Minho, Braga, Portugal
jneves@di.uminho.pt
[3] Centro de Investigação em Educação e Psicologia,
Universidade de Évora, Évora, Portugal
[4] Escola Superior de Tecnologia e Gestão, ARC4DigiT – Applied Research
Center for Digital Transformation, Instituto Politécnico de Viana do Castelo,
Viana do Castelo, Portugal
jribeiro@estg.ipvc.pt

Abstract. This study was carried out in training companies and aims to evaluate customer satisfaction. It focusses at the *Organizations' Quality-of-Management* (*QoM*) that is in itself a major competitive advantage to differentiate them. Indeed, the universe of discourse is set in order to consider not only the complex relationships among the entities that populate it, but also to take into account its inner structure, where incomplete, unknown or even self-contradictory information or knowledge are present. One's goal is at the development of a comprehensive and integrated computational model to ensure the *Organizations' Performance* and its *QoM* in order to fulfill customer's requirements. It is based on a *Logic Programming* approach to *Knowledge Representation* and *Reasoning* and grounded on an *Artificial Neural Networks* approach to computing.

Keywords: Artificial Intelligence · International Standard ISO 9001
Quality-of-Management · Logic Programming
Knowledge Representation and Reasoning · Artificial Neural Networks

1 Introduction

Despite conflicting views and relative uncertainty about where *Artificial Intelligence* (*AI*) will go in the future, the technology is already around us. More and more often in business operations and even in everyday life it does not seem to slow down a bit. Among them, *Quality-of-Management* (*QoM*) faces the biggest change as intelligent systems make their way in. Indeed, the implementation at the organization level of a *QoM* structure according to some standards, namely *ISO 9001*, may ensure its performance and the quality of its services/products [1]. In fact, the use of *ISO 9001* in organizations may enable continuous improvement of *QoM*, leading to increased customer satisfaction and the opportunity to enter new markets [2]. Two management indicators (i.e., complaints and customer satisfaction) were used that allow us to

© Springer International Publishing AG, part of Springer Nature 2019
J. Machado et al. (Eds.): HELIX 2018, LNEE 505, pp. 384–390, 2019.
https://doi.org/10.1007/978-3-319-91334-6_52

address issues such as complaints from multiple trainees, quality support materials and organization's training, as set in Sect. 3. The paper therefore presents a formal method to assure *QoM*, grounded on a *Logic Programming (LP)* approach to *Knowledge Representation* and *Reasoning (KRR)* [3, 4], and complemented by a computer framework centered on *Artificial Neural Networks (ANNs)* [5]. *ANNs* can be described as a connected structure of basic computing units (artificial neurons or nodes) with learning capabilities. The neural multi-layer feed-forward network architecture is one of the most popular *ANN* structures. It consists of two or more layers of artificial neurons, including an input layer, an output layer and a number of hidden layers with a certain number of active neurons connected by modifiable weights. The number of nodes in the input level determines the number of independent variables, and the number of nodes in the output level indicates the number of dependent variables [5]. Several studies have shown how *ANNs* can successfully be used to model data and capture complex relationships between inputs and outputs [5, 6].

This paper is divided into five sections, where the former one stands for an opening to publicize the problem to be addressed, followed by a section where one's approach to *KRR* is the object of attention. A methodology for problem solving is presented in the third section while the fourth one introduces the case study and a possible solution to the problem using *ANNs*. Finally, a conclusion is set and guidelines for future work are outlined.

2 Knowledge Representation and Reasoning

Knowledge Representation and *Reasoning (KRR)* aims at the understanding of the information's complexity and the associated inference mechanisms. Automated reasoning capabilities enables a system to fill in the blanks when one is dealing with incomplete information, where data gaps are common, i.e., the fundamentals and the attributes of the logical functions go from discrete to continuous, allowing for the representation or handling of unknown, vague, or even self-contradictory information/knowledge. In this work a data item is to be understood as find something smaller inside when taking anything apart, i.e., it is mostly formed from different elements, namely the *Interval Ends* where their values may be situated, the *Quality-of-Information (QoI)* they carry, and the *Degree-of-Confidence (DoC)* put on the fact that their values are inside the intervals just referred to above [7]. These are just three of over an endless element's number. Undeniably, one can make virtually anything one may think of by joining different elements together or, in other words, viz.

- What happens when one splits a data item? The broken pieces become data item for another element, a process that may be endless; and
- Can a data item be broken down? Basically, it is the smallest possible part of an element that still remains the element.

Therefore, the proposed approach to this issue, put in terms of the logical programs that elicit the universe of discourse, will be set as productions of the type, viz.

$$
predicate_{1 \leq i \leq n} - \bigcap_{1 \leq j \leq m} clause_j(((\left[A_{x_1}, B_{x_1}\right](QoI_{x_1}, DoC_{x_1})), \cdots \\
\cdots, (\left[A_{x_m}, B_{x_m}\right](QoI_{x_m}, DoC_{x_m}))) :: QoI_j :: DoC_j
$$

(1)

n, $\cap$, m and A_{x_m}, B_{x_m} stand for the cardinality of the predicates' set, conjunction, predicate's extension, and the interval ends where the predicates attributes values may be situated, respectively. The metrics $\left[A_{x_m}, B_{x_m}\right]$ QoI and DoC show the way to data item dissection [7, 8], i.e., a data item is to be understood as the data's atomic structure. It consists of identifying not only all the sub items that are thought to make up an data item, but also to investigate the rules that oversee them, i.e., how $\left[A_{x_m}, B_{x_m}\right]$, QoI_{x_m}, and DoC_{x_m} are kept together and how much added value is created.

3   Methods

3.1   Data Collection

In order to collect data a questionnaire was designed specifically for this study. The questions in the questionnaire were divided into three sections, where the former one sets the questions about the *Trainees' Complaints* (Table *Trainees' Complaints*, Fig. 1). Following are the questions on the opinion of the trainees on the *Quality of the Support Materials* (Table *Quality of Support Materials*, Fig. 1). The last one contains questions related to the apprentices' opinion on the *Training Organization* (Table *Training Organization*, Fig. 1). Qualitative values were also used to assess the various problems. For example, in the previous section of the questionnaire, a scale was specified in terms of *none*, *very few*, *few* and *often*. In the remaining sections the scale used was given in terms of *none*, *very low*, *low*, *medium*, *high* and *very high*.

3.2   Feature Extraction

The feature extraction's process focused on the more relevant issues involved in each topic affecting the trainees' satisfaction. The answers obtained in the above-mentioned questionnaire were stored in the respective tables as shown in Fig. 1. In order to quantify the qualitative information, the method described in [9] was followed.

4   Case Study

4.1   Sample Characterization

The questionnaire was applied to a cohort of 227 apprentices with a mean age of 24.8 years at the age of 19 to 38 years. The gender distribution was 44.9% and 55.1% for men and women, respectively. In terms of trainee status, 32.6% of respondent had already the training, 17.2% have left, and 50.2% attended the training.

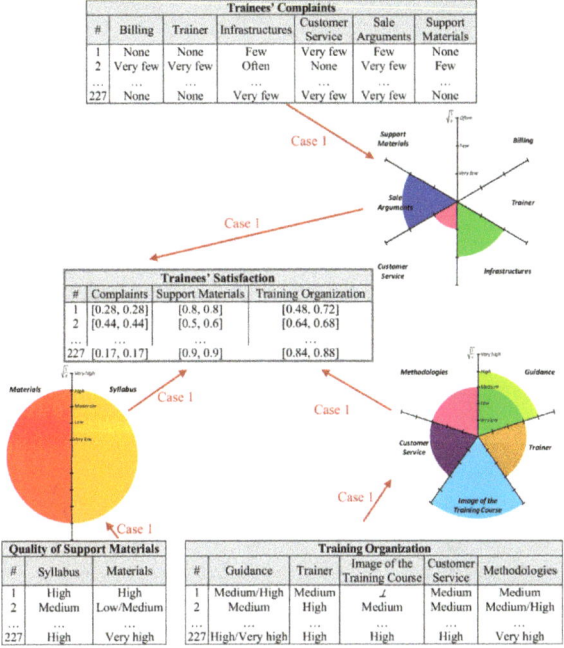

Fig. 1. A fragment of the *trainees' satisfaction* knowledge base.

4.2 Computational Model

In order to develop an intelligent system for trainees' satisfaction screening a knowledge database was build up. The mentioned database is provided in the form of extensions of the relations (or tables) illustrated in Fig. 1. Although the bulk of the data is known, there are some items that are unknown, be in a set or belong to an interval (e.g., in case 1 the *Image of the Training Course* is unknown, represented by the symbol ⊥, while the *Guidance* varies between *Medium* and *High*). Thus, the objective function that maps system behavior may now be set in terms of the extension of the predicate *trainees' satisfaction (ts)*, viz.

$$ts : Com_{plaints}, Support M_{aterials}, Training O_{rganization} \rightarrow \{0, 1\} \qquad (2)$$

where 0 and 1 denote *true* and *false*, respectively. The extension of the predicate *ts*, regarding to a term (trainee) characterized by feature vector $Com_{plaints} = [0.17, 0.17]$, $Support M_{aterials} = \perp$, $Training O_{rganization} = [0.72, 0.84]$, is now given in the form [7], viz.

$\{$

$$\neg\, ts\left(\left([A_{Com}, B_{Com}](QoI_{Com}, DoC_{Com})\right),\ \left([A_{SM}, B_{SM}](QoI_{SM}, DoC_{SM})\right),\right.$$

$$\left.\left([A_{TO}, B_{TO}](QoI_{TO}, DoC_{TO})\right)\right)$$

$$\leftarrow not\ ts\left(\left([A_{Com}, B_{Com}](QoI_{Com}, DoC_{Com})\right),\ \left([A_{SM}, B_{SM}](QoI_{SM}, DoC_{SM})\right),\right.$$

$$\left.\left([A_{TO}, B_{TO}](QoI_{TO}, DoC_{TO})\right)\right)$$

$$ts\left(\left([0.17, 0.17](1_{[0.17,0.17]}, 1_{[0.17,0.17]})\right), \left([0, 1](1_{[0,1]}, 0_{[0,1]})\right),\right.$$

$$\underbrace{\left.\left([0.72, 0.84](QoI_{[0.72,0.84]}, 0.99_{[0.72,0.84]})\right)\right)}_{attribute's\ values\ ranges\ once\ normalized\ and\ respective\ QoI\ and\ DoC\ values} :: 1 :: 0{,}67$$

$$\underbrace{[0, 1] \qquad\qquad\qquad [0, 1] \qquad [0, 1]}_{attribute's\ domains\ once\ normalized}$$

$\} :: 1$

Program 1. The feature vector's logical form.

The computational model is now set in terms of an *ANNs* approach to computing as it was referred to above. *ANNs* were selected due to their dynamic properties such as adaptability, robustness and flexibility. Figure 2 shows how the normalized values of the *Interval Ends* and the respective *DoC* and *QoI* values operate as inputs to the *ANN*. The result is the evaluation of the trainees' satisfaction as well as a measure of the trust that can be expected from such a prediction. The *ANN* topology consists of an input layer with 3 nodes (corresponding to 12 nodes in the preprocessing one), a hidden layer with 4 nodes and a 2-node output layer.

A set with 227 records were used to test the model. The dataset was divided in exclusive subsets through the ten-folds cross validation [3]. In order to guarantee the statistical significance of the attained results 30 (thirty) experiments were applied in all tests. The back-propagation algorithm was used in the learning process of the *ANN*. As the activation function in the pre-processing layer it was used the identity one, while in the other layers was passed down the sigmoid. Table 1 present the confusion matrix (the values denote the average of the 30 experiments) for proposed model. A perusal of

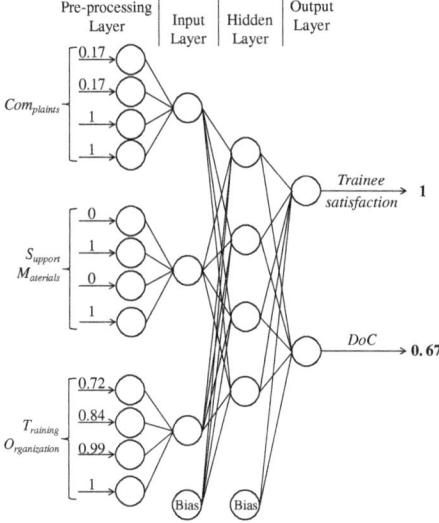

Fig. 2. *ANN's* topology.

Table 1 shows that the model accuracy was 96.9% (220 instances correctly classified in 227). Such results seem to suggest that the *ANN* model exhibits a good performance in the screening of trainees' satisfaction.

Table 1. The confusion matrix regarding proposed model.

Output	Model output	
	True (1)	False (0)
True (1)	TP = 156	FN = 2
False (0)	FP = 5	TN = 64

5   Conclusions

This work presents an *Artificial Intelligence* (*AI*) based system for assessing trainees' satisfaction, being the information acquired according the requirements of *International Standard ISO 9001*. Indeed, quality professionals need to expand their skill set to include more technical knowledge in order to take full advantage of these new technologies and problem solving processes. It is grounded on a formal framework based on *Logic Programming* for *Knowledge Representation and Reasoning*, complemented by an *Artificial Neural Network* stance to computing. In fact, this is one of the added values resulting from the complementary between *Logic Programming* and *ANNs* remarkable information processing capabilities, such as nonlinearity, high parallelism, robustness, error and fault tolerance, learning ability, generalization, and the possibility of handling incomplete information or knowledge.

Acknowledgments. This work has been supported by COMPETE: POCI-01-0145-FEDER-007043 and FCT – Fundação para a Ciência e Tecnologia within the Project Scope: UID/CEC/00319/2013.

References

1. Portuguese Institute of Quality: NP EN ISO 9001:2015 – Quality management systems – requirements. Portuguese Institute of Quality Edition, Caparica (2015) (In Portuguese)
2. Christian, D., Drilling, S.: Implementing Quality in Laboratory Policies and Processes – Using Templates, Project Management and Six Sigma. CRC Press, Boca Raton (2010)
3. Neves, J.: A logic interpreter to handle time and negation in logic databases. In: Muller, R., Pottmyer, J. (eds.) Proceedings of the 1984 annual conference of the ACM on the 5th Generation Challenge, pp. 50–54. Association for Computing Machinery, New York (1984)
4. Neves, J., Machado, J. Analide, C., Abelha, A., Brito, L.: The halt condition in genetic programming. In: Neves, J., Santos, M.F., Machado, J. (eds.) Progress in Artificial Intelligence. LNAI, vol. 4874, pp. 160–169. Springer, Berlin (2007)
5. Haykin, S.: Neural Networks and Learning Machines. Prentice Hall, New York (2009)
6. Vicente, H., Dias, S., Fernandes, A., Abelha, A., Machado, J., Neves, J.: Prediction of the quality of public water supply using artificial neural networks. J. Water Supply Res. Technol. – AQUA **61**, 446–459 (2012)
7. Fernandes, F., Vicente, H., Abelha, A., Machado, J., Novais, P., Neves, J.: Artificial neural networks in diabetes control. In: Proceedings of the 2015 Science and Information Conference (SAI 2015), pp. 362–370. IEEE Edition (2015)
8. Silva, A., Vicente, H., Abelha, A., Santos, M.F., Machado, J., Neves, J., Neves, J.: Length of stay in intensive care units – a case base evaluation. In: Fujita, H., Papadopoulos, G.A. (eds.) New Trends in Software Methodologies, Tools and Techniques, Frontiers in Artificial Intelligence and Applications, vol. 286, pp. 191–202. IOS Press, Amsterdam (2016)
9. Fernandes, A., Vicente, H., Figueiredo, M., Maia, N., Marreiros, G., Neves, M., Neves, J.: A case-base approach to workforces' satisfaction assessment. In: Research and Practical Issues of Enterprise Information Systems, Lecture Notes in Business Information Processing, vol. 268, pp. 191–206, Springer, Cham (2016)

Improving Internal Transport

Anna Karwasz[✉] and Daria Skuza

Faculty of Mechanical Engineering and Management,
Poznan University of Technology, Piotrowo 3, 60-965 Poznan, Poland
anna.karwasz@put.poznan.pl,
daria.skuza@student.put.poznan.pl

Abstract. The paper presents possibilities for improving internal transport in a manufacturing company. The authors provide a general analysis of logistic processes as well as questions relating to production logistics, using Lean Management tools like supermarket or electric logistic train. They analyse the current state of internal transport and present changes and possible improvement concepts. Internal transport, which is responsible for the flow of materials, movement of raw materials, semi-finished products and finished products within a company, has a significant impact on production processes. It can be improved thanks to the application of simple Lean Management tools, which will make it possible to eliminate all kinds of waste. Using the solutions presented in the paper, the analysed company may expect to continue to eliminate waste and strengthen its position among the producers of wire bunches and system panels used in household appliances.

Keywords: Internal transport · Production logistics · Lean Management
Kanban · Supermarket · Electric logistic train

1 Introduction

A changing environment, customers' requirements and increasing competition force companies to introduce corrections into many areas of their operation, corrections improving the services provided [1]. Various solutions are sought to manufacture products more quickly and more cheaply while maintaining the right quality [2]. The production system of any company cannot function well without well-organised logistics the main objective of which is effective and efficient materials management. This makes it possible to maintain the continuity of production processes and fulfil customers' orders on time. This is an area that benefits from reorganisations aimed at improving the flow of raw materials, components and finished products, as it has a direct impact on the maximising of the income from all of the company's operations.

In order to achieve visible benefits in internal transport, it is not necessary to invest substantial amounts in technological innovations. An effective and, at the same time, simple way to improve the current state is to apply Lean Management solutions. Their implementation, not only in this area, makes it possible to systematically eliminate waste, such us excessive inventory, ineffective use of production resources, means of transport and personnel, or unplanned down times generating costs that add no value for the customer.

© Springer International Publishing AG, part of Springer Nature 2019
J. Machado et al. (Eds.): HELIX 2018, LNEE 505, pp. 391–397, 2019.
https://doi.org/10.1007/978-3-319-91334-6_53

2   The Logistics of Internal Transport

Production logistics is a subsystem of logistics linking supply and distribution, a subsystem associated with the organisation of the production system with its nearest warehouse and transport environment [3, 4]. Production logistics encompasses physical flow and storage of materials as well as information and decision stream, including organising, planning and decision making as well as control of flow intensity. Its main tasks include [5]:

- Internal transport of materials, semi-finished products, raw materials, waste and finished products.
- Managing work-in-progress inventory as well as process inventory.
- Operations within technological production processes.

An area directly linked to production logistics is internal transport [6]. This transport services material flow within the company, from the entry of raw materials and intermediate products to the supply warehouse to the transfer of finished products to the finished product warehouse [5]. From an economical point of view, this type of transport should be characterised by the shortest route to destination, which is associated with the shortest time needed to cover it, maximum use of the available means of transport and, at the same time, lowest possible degree of their wear and tear.

Inventory maintenance, product circulation and protection require the application of various technical means that make up a logistic infrastructure. Its choice has a major impact on the speed of flow, maintenance of appropriate use value of products, efficiency of logistics processes as well as on production processes themselves [7, 8]. The internal transport infrastructure comprises [3]:

- Means of transport – trucks, cranes, conveyors, loaders, manipulators.
- Means of storage – joists, racks, trestles, stands, clamps.
- Auxiliary means – pallets, pallet superstructures, containers, tanks, lifting slings.

2.1   The Idea of Internal Logistics in the Lean Management Concept

Lean Management provides for a reduction in the costs and increase in the productivity of resources by eliminating non-added value actions [9]. Initially, Lean Management was only about production processes; today its application has been extended to include all spheres of companies' operations, including logistics and processes associated with it.

The concept of Lean is based primarily on eliminating waste (*muda* in Japanese) from a company [10]. Waste is any activity requiring resources and work but not creating any added value for the customer. A desired state in processes is a point in which the customer ordering a product will pay only for its manufacture and not for the functioning of the enterprise [11]. The precursor of the Lean method, Taiichi Ohno, identified seven types of waste, which acquire a different dimension in the supply chain. They are: overproduction, waiting, transport, unnecessary motion, excessive inventory, space, defects and their repair [12]. The implementation of the Lean concept requires specific tools, often defined as the Lean Toolbox, which make it possible to eliminate various types of waste [11]. Among all the available tools, many are directly or indirectly associated with logistics processes in an enterprise:

- 5S method – an organised model of action the objective of which is to shape the work environment in an appropriate manner and set its standards. It provides for the creation and improvement of efficient and clean work places [13, 15].
- Kanban – a tool for efficiently managing material and information flow in a process. Controlling the inventory by coordinating the transfer of components between all production processes, as a result of which material excess and shortages are avoided [16–18].
- Supermarket – is used to store a specific amount of materials, finished products and work in progress in order to ensure production continuity and delivery to the customer on time. It is usually located near the production process [13, 19, 20].
- Logistic train – a pull truck with carts. A characteristic feature of this solutions is increased frequency of deliveries, transport synchronisation as well as delivery and pick-up of components with a minimal amount of transport means (limiting the use of forklift trucks and pallet trucks on the shop floor) thanks to the possibility of manipulating the number of trailers as required [21, 22].
- Spaghetti chart – a visualisation tool for defining the physical movement of materials or people. It makes it possible to analyse the actual routes covered by plotting, usually manually, the movement of a given element on a plan of the area in question [23].

3   Characteristics of the Research Object

The analysed company manufactures wire bunches and conductors, using IDC techniques (connecting blocks to insulated wires without removing the insulation before the connection) and CRIMP (creation a connection by inserting a terminal with its insulation removed into the connector and then mechanically squeezing it around the wire) as well as assembly of system panels for many household appliances customers. The company has four main production departments:

- K+V (IDC+CRIMP) – the main department – provides the company with components to produce wire bunches. IDC – wire cutting and insertion of connectors. CRIMP – manufacturing crimp connections.
- TAMPODRUK (tampography) – preparing clichés for overprinting on strips, drawers and keys. Preparing the inks used in tampography. Creating overprints and welding panel displays.
- SYSTEMY (systems) – manufacturing subassemblies and bunches for system panels. Conductors produced in the K+V department as well as overprinted panels with drawers are delivered to the process line. In addition, there are individual work places where bunches, displays and knobs are manufactured without connecting assembly.
- WIĄZKI (bunches) – manufacturing various kinds of wire bunches on assembly benches and/or turntables for other types of the customer's household appliances.

4   Current State Analysis

Internal transport is responsible for supplying materials, semi-finished products and packaging to work stations and transferring the finished products to the output warehouse in order to ensure production continuity and on-time delivery.

The site comprises two production shop floors connected by corridors. There are transport routes indicated on the shop floors alongside which employees responsible for the logistics move using intermittent transport means, as well as tracks for pedestrians. On a considerable part of the shop floor the two paths run alongside each other, which is why in particularly dangerous areas they are separated by special barriers; zebra crossings are placed where the two paths cross. The width of the transport routes makes it possible to move goods both by means of pallet trucks and fully loaded electric logistic train. The logistic train has its parking place in one of the corridors connecting the two shop floors.

Goods are moved with the help of various means of transport like forklift trucks, hand lift trucks, electric lift trucks.

4.1   Material Flow

The flow of products within the company starts with the arrival of materials and their reception in the warehouse. This takes place in the input section of the warehouse. The goods are verified, entered into books and placed on appropriate racks in line with the warehousing system in place. The company receives components like connection blocks, wires, contacts, electronic parts, injection moulded plastic parts, packaging.

The various parts needed in production are ordered by shop floor employees responsible for specific departments by means of code scanners. After receiving an

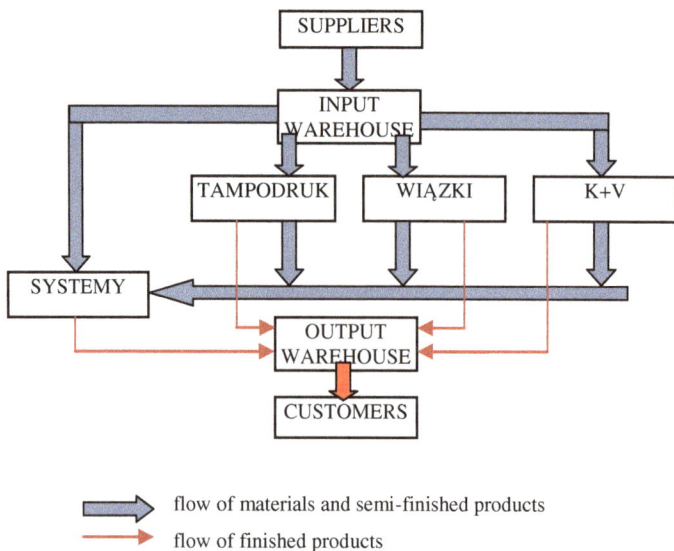

Fig. 1. Flow of all materials within the company [14].

order the warehouseman prepares the materials needed, transfers them from the input warehouse to the right production warehouses and releases them outside the warehouse door from the side of the shop floor. In the case of an urgent order, all these tasks are to be completed within half an hour; in all other cases the delivery time is one hour. From there the materials are picked up and then delivered to racks situated in the relevant departments or specific work places. Next, products are transferred to semi-finished product warehouses for the next production stage or to the finished products warehouse, where they are prepared to be shipped to the customer. A general flow of materials within the company is presented in Fig. 1.

5   Identification of Waste

The following waste was identified on the basis of observations:

- Too many parts delivered to the TAMPODRUK department.
- Electric logistic train used improperly owing to a lack of standardised work for the device and too long loading and unloading times.
- Electric logistic train equipment – poor trailer structure.
- Damage to the transported material.
- Too much time spent on ordering the materials, searching for the materials, their sorting, cleaning the work place.
- Incorrectly designated accumulation bins. The placement of some pieces of the shop floor equipment hinders the delivery of components to the work place and pick-up of finished products.

6   Improvement Processes

The following improvement processes were implemented on the basis of the analysis conducted in the company:

- A different kind of transport carts was introduced to eliminate multiple lifting and lowering of the hand lift truck. Unused platform trucks were used to verify the correctness and effectiveness of the new concept.
- Loading and unloading are to take place already in the warehouse areas, which will reduce the risk of material damage.
- Temporary stops for the electric logistic train were designated together with a loading zone and a rule was introduced whereby after filling the packaging unit, the packaging employee pushes the cart with the finished product to this zone.
- New equipment for the logistic train made it possible to reduce the loading time and transport time of the finished product from the shop floor to the output warehouse, and the time for preparing empty packaging for transport from 145.70 s to 87.52 s, i.e. by nearly 40%.

- The number of tasks performed by the driver was reduced from 26 to 14, which improved work ergonomics by eliminating processes involved in the handling of the hand lift truck.
- Creation of a supermarket in TAMPODRUK for injection moulded panels used to create the manufacturer's overprints, and further manual assembly.
- Reduction of the raw materials inventory to a level that is safe for the production processes and does not cause problems with inventory backlog. There was an improvement in the material pick-up system – the first material to be picked up is the one delivered first.
- Introduction of 5S measures.

7 Conclusions

Internal transport, which is responsible for the flow of materials, movement of raw materials, semi-finished products and finished products within a company, has a significant impact on production processes. It can be improved thanks to the application of simple Lean Management tools, which will make it possible to eliminate all kinds of waste.

Using the solutions presented in the paper, the analysed company may expect to continue to eliminate waste and strengthen its position among the producers of wire bunches and system panels used in household appliances.

Acknowledgments. The presented results of the research, carried out under the theme No. 02/23/DSPB/7716, was funded with a grant to science granted by the Ministry of Science and Higher Education.

References

1. Kujawińska, A., Vogt, K., Hamrol, A.: The role of human motivation in quality inspection of production processes. In: Advances in Intelligent Systems and Computing, vol. 490 (2016)
2. Kujawińska, A., Vogt, K.: Human factors in visual quality control. Manag. Prod. Eng. Rev. **6**(2), 25–31 (2015)
3. Szymonik, A.: Production Logistics. Processes, Systems, Organisation. Difin, Warsaw (2012)
4. Kawa, A.: Supply chains of cross-border e-commerce. In: Król, D., Nguyen, N., Shirai, K. (eds.) Advanced Topics in Intelligent Information and Database Systems, ACIIDS 2017. Studies in Computational Intelligence, vol 710, pp. 173-183. Springer (2017)
5. Niziński, S., Żurek, J.: General Logistics. Wydawnictwa Komunikacji i Łączności WKŁ, Warsaw (2011)
6. Dohn, K.: Organisation of internal transport processes – case studies. Warsaw University of Technology Research Papers. Transport, no. 70, Oficyna Wydawnicza Politechniki Warszawskiej, Warsaw (2009)
7. Skowronek, C., Sarjusz-Wolski, Z.: Logistics in an Enterprise. Polskie Wydawnictwo Ekonomiczne, Warsaw (2012)

8. Ivanov, V., Mital, D., Karpus, V., Dehtiarov, I., Zajac, J., Pavlenko, I., Hatala, M.: Numerical simulation of the system "fixture – workpiece" for lever machining. Int. J. Adv. Manuf. Technol. **91**(1), 79–90 (2017)
9. Kasperek, M.: Lean logistics – an as-is analysis. Mater. Manag. Logist. (5), 2–10 (2013)
10. Mazur M., Ulewicz R.: Improving internal transport using the lean concept – a case study. Organ. Overv. (7), 6–13 (2015)
11. Szatkowski, K.: Modern Production Management – Process Approach. PWN, Warsaw (2014)
12. Lopes, J.J., Varela, M.L.R., Trojanowska, J., Machado, J.: Production flow improvement in a textile industry. In: Rocha, Á., Guarda, T. (eds.), Advances in Intelligent Systems and Computing, vol. 721, pp. 224–233. Springer (2018)
13. Tarała, M. (ed.): Lean Manufacturing. Improving Production. Oficyna Wydawnicza Politechniki Rzeszowskiej, Rzeszów (2015)
14. Skuza, D.: Improving internal transport in a selected production facility, thesis, Faculty of Mechanical Engineering and Management, Poznań University of Technology, Poznań (2017)
15. Kornicki, L., Kubik, S.: 5S for Operators, 5 Pillars of Work Place Visualisation. ProdPublishing, Wrocław (2008)
16. Kornicki, L., Kubik, S.: Kanban on the Shop Floor. ProdPublishing, Wrocław (2009)
17. Kubik, S.: Just-in-Time for Operators. ProdPublishing, Wrocław (2010)
18. Bendkowski, J., Matusek, M.: Production logistics. In: Practical Aspects. Part II. Tools, Methods, Systems. Wydawnictwo Politechniki Śląskiej, Gliwice (2013)
19. Marchwinski, C., Shook, J., Schroeder, A.: Lean Lexicon. Illustrated Dictionary of Lean Management Terminology. Wydawnictwo Lean Enterprise Institute Polska, Wrocław (2010)
20. Rewers, P., et al.: Production leveling as an effective method for production flow control – experience of polish enterprises. Proc. Eng. **182**, 619–626 (2017)
21. Piasecka-Głuszak, A.: Lean Management in the internal logistics of enterprises in Poland – survey results, Katowice University of Economics Research Papers, no. 249 (2015)
22. Trojanowska, J., Kolinski, A., Galusik, D., Varela, M.L.R., Machado, J.: A methodology of improvement of manufacturing productivity through increasing operational efficiency of the production process. In: Hamrol, A., Ciszak, O., Legutko, S., Jurczyk, M. (eds.) Advances in Manufacturing. Lecture Notes in Mechanical Engineering, pp. 23–32. Springer (2018)
23. Wilson, L.: How to Implement Lean Manufacturing. McGraw-Hill Professional, New York City (2009)

Comparison of Manual Assembly Training Possibilities in Various Virtual Reality Systems

Paweł Buń(✉), Filip Górski, and Nikola Lisek

Chair of Management and Production Engineering,
Poznań University of Technology, Piotrowo 3 STR, 60-965 Poznan, Poland
{pawel.bun, filip.gorski}@put.poznan.pl,
nikola.lisek@student.put.poznan.pl

Abstract. The paper presents results of preliminary studies on possibilities of using low – cost devices to train operators of production line on how to perform a simple assembly task in Virtual Reality. Two different approaches to manipulate virtual objects without haptic feedback were compared and tested. The overall time and successful rate was measured for all users in both systems. The results lead to conclusion that smartphone based training applications in VR have a long way to go, but with increasing computing power seems like a promising direction.

Keywords: Virtual reality · Interaction · Tracking system · Industrial training

1 Introduction

Both the hardware and software for Virtual Reality (VR) technology developed significantly in recent years. Rising interest in the VR branch from the companies related to digital entertainment, communication and visualization enabled development of low – cost interaction devices, allowing not only to obtain near-photorealism in visualization (Head Mounted Devices), but also allowing user directly interact with elements inside a visualization (systems for tracking and gesture recognition). Soon after that, software previously used to create games was adopted to enable development of VR applications that use these low – cost devices. The VR application should be understood as programmatically and logically enclosed unity, containing virtual three-dimensional models of objects, placed in an appropriate environment, with ensuring of interaction and immersion to the user [1]. Application that fully uses the potential of new interaction devices can make users feel physically located within the virtual world or feel a sense of presence. Producing a sense of presence sets VR apart and takes traditional computing interfaces to the next level [2].

Besides entertainment, VR can be used as a specialized engineering tool [3, 4] for medical education [5, 6], engineering education [7, 8] or advanced training and simulation systems [9].

Learning transfer is a key issue for virtual learning environments [10]. Main aim of the presented work is to create cost-effective training systems that ensure a level of immersion in VR sufficient to develop specific indispensable reflex actions in a controlled environment. To achieve that, the authors are looking for the most effective

© Springer International Publishing AG, part of Springer Nature 2019
J. Machado et al. (Eds.): HELIX 2018, LNEE 505, pp. 398–404, 2019.
https://doi.org/10.1007/978-3-319-91334-6_54

combination of hardware and software, that allows user to interact freely with Virtual Environment (VE). This paper presents preliminary studies on possibilities of using low – cost devices to train operators of production line on how to perform a simple assembly task. Two different systems were compared in the same simple assembly process application – a stationary (PC-based), cabled HMD solution with controllers and a mobile (Android phone based) HMD coupled with a large-area tracking system and a data glove.

2 Materials and Methods

2.1 Object Manipulation in VR

A defining feature of virtual reality (VR) is capability of manipulation of virtual objects interactively, rather than simply viewing a passive environment [11]. Manipulation in immersive virtual environments is difficult, partly because users must do it without the haptic contact with real objects. They rely on it in the real world to orient themselves and their manipulanda [12]. Currently, the most popular method of interacting with virtual environment is use of motion tracking systems [13] and gesture recognition using hardware such as data gloves or a Myo device, which can read muscle tone in user's arm by means of electromyography (EMG) [14].

2.2 Used Hardware and System Structure

In the course of the research, the authors wanted to compare two different approaches to manipulate virtual objects without haptic feedback. First of the two compared solutions is based on a consumer HMD solution - Oculus Rift CV1 (parameters in Table 1) with dedicated controllers (Oculus Touch), the second one is based on a smartphone with Android and 5DT Data Glove for manipulation, as well as Google Cardboard compatible HMD (Kruger & Matz Immerse, 120 FOV). For tracking of user's hand and head, PPT X large-area system was used, by WorldViz company.

Table 1. Parameters of Oculus Rift CV1

Parameter name	Value
Resolution	2160×1200
FOV (Field of View)	110
Communication	USB for tracking and control, display by HDMI
Interaction	Controller - Oculus Touch
Tracking	IR LED sensor – 3 DOF positional tracking, 1.5×1.5 m tracking space, 3 DOF rotational tracking

An Android smartphone Samsung Galaxy S7 (QHD screen resolution) was used as a display device in the second, low-cost solution, it is also used as a processing center and image source.

The big advantage of the 5DT DataGlove glove is its low weight and comfort of wearing - it is made of lycra and it adapts to any hand size. Finger positions (and, consequently, gestures) are recognized by means of optical fibers, placed on each finger. The system is capable of recognizing various gestures, such as fist, index finger point or others. Finger positions are detected on the basis of amount of light reaching a phototransistor placed at the end of each fiber (the glove has 14 in total). As the finger bends, the intensity of light reaching the end of the fiber decreases. Information is sent to a computer by a connector placed in the wrist part, where appropriate software recognizes user's gesture and adjusts visualization accordingly. The system recognizes at least 16 gestures, although they can be custom defined, as any state of fingers can be defined as a gesture.

The diagram of the proposed system is presented in the Fig. 1.

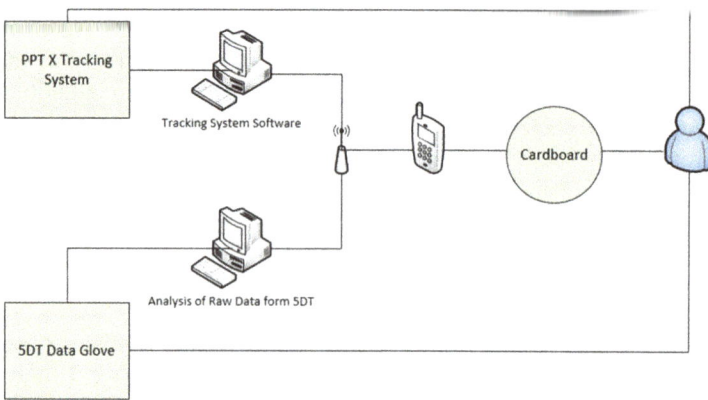

Fig. 1. Interaction system in the mobile VR system

2.3   Used Software

The application for both hardware systems was created using the Unity 3D software. The Unity is a long-known game engine, originally used mostly for building interactive 3D and 2D applications. Since it supports Oculus Rift, HTC Vive as well as Google Cardboard natively, it has become a very popular platform that allows building applications for most of the currently available commercial VR hardware, both PC-based and mobile.

For the purpose of the mobile (Android, Cardboard based) version, the virtual hand model with colliders on each finger was created. The virtual hand (VH) tracking is realized by a marker from the PPT X system, placed on user's hand. Each user's hand movement is reflected in the application as virtual hand movement, in real time. In the application, the option of recognizing user gestures is also necessary to grab and drop objects in virtual environment which is why the VH cooperates with 5DT DataGlove (Fig. 2). A number of custom scripts for data sending via UDP were prepared and implemented. For the Oculus version, standard assets from the Oculus Integration pack were used – it was definitely easier to prepare the application, as an integrated hardware system (Rift CV1 with controllers) was used.

Fig. 2. Model of the virtual hand with colliders on fingers and the control script node

Unity gives possibility of physics simulation. Most of the objects that are supposed to be affected by the physics engine should have a RigidBody component on them. It is possible attach one RigidBody object to another using a Fixed Joint component. Due to the fact that the virtual hand driven by a user was supposed to collide with its environment, it was decided that this kind of connection between hand and blocks will be the most suitable.

2.4   Virtual Environment

The main aim of the research was to check if proposed solutions can be used to train operators in Smart Factory Laboratory (an example of an elastic production line, realizing the Industry 4.0 concept [14]) and which one could be possibly more effective.

The production system, which is an equipment of a SmartFactory laboratory at Poznan University of Technology, consists of an automated production line. It consists of four transporting loops, by which there are production stands (six in total, as the line is elastic it could be less). Transport of products is realized by small pallets, that can be directed to any production stand. Their automated identification in the system is realized with use of the RFID technology. The whole production system is controlled by a customized computer program, named 4Factory [15]. System can be also used to simulate material flow [16].

The model prepared in 3D CAD software has been transferred to the 3ds Max environment, where a number of operations have been carried out to improve the quality of visualization. The laboratory model prepared in this way was imported to Unity 3D. To maintain high framerate in the mobile version, it was necessary to replace selected elements of the environment with geometric primitives and generally decrease quality of generated image.

The following interaction functions were programmed in the virtual model:

- virtual walk – realized by analogue stick in the Oculus version and on user's own feet (free walk) in the mobile version,
- object interaction – interactable objects can be grabbed by pushing the grasp button in the Oculus Touch controllers or by making the fist gesture in the mobile version,
- assembly – user can assemble a simple object out of available parts,
- animation – user can put assembled objects on a pallet and make it move.

3   Test Procedure and Results

The application was tested by Production Engineering students (10 persons). They compared possibilities of interaction with the virtual environment using the mobile version (large area tracking, Data Glove, Android device) and the PC version (Oculus CV1 with Oculus Touch).

The task was to assemble a simple product by dragging elements in to right spots, put it on a pallet (as seen in Fig. 3 - right) and press the right button in order to start animation of moving it to another stand. The user was presented in a pictorial form of how a complex product should look. At the moment of grabbing of any part, a shadow was visible where it should be placed, giving a hint. The overall time was measured for all users in both systems.

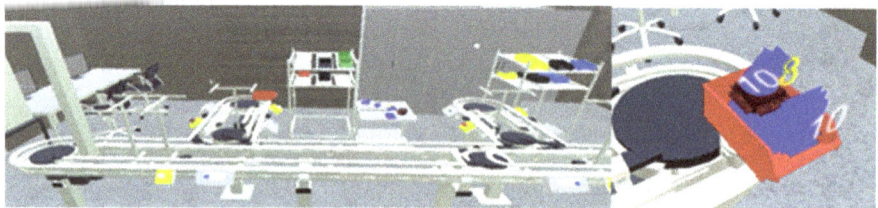

Fig. 3. Left - SmartFactory model after import to Unity 3D environment and geometry optimization, right - assembled product placed on a pallet. Quality settings – low.

The test results are presented in Table 2. Red cells represents failed attempts at assembly in reasonable time. In most of the cases, the main reason for that were glitches in the physics system, causing blocks to scatter. It is worth to mention that it happened only in the Android version, despite using exactly the same logic in both systems.

Table 2. Time needed for the assembly task in both versions

User		1	2	3	4	5	6	7	8	9	10	average	median
Oculus	Time [s]		136	73	70	23	54	57	68	92	53	69.55	68
Android	Time [s]	45	494					140	89	74	114	159.33	101.5

Main problem with the application in the mobile version were glitches in physics, that randomly occurred while trying to attach blocks to the assembly. As mentioned above, while the block was placed in right position the new FixedJoint was created between block and rest of the product, at the same time the FixedJoint between virtual hand and block should be removed. Apparently it did not happen. Two opposing forces applied in two points triggered a torque, after the disappearance of one of the forces (which should happened immediately after joining blocks), the remaining joints were torn apart and the components scattered all over the virtual environment. Perhaps the computing unit of the mobile device was not able to correctly interpret the command to immediately remove the joint between the virtual hand and the block, or it happened too late due to lower framerate. The long time of the assembly operation performed by the person no. 2 (in the Android version) resulted mainly from the persistence – despite continuous problems, the person wanted to finish the operation (unlike participants no. 3–6, who resigned at some moment). The person regularly and inadvertently let the blocks drop (as a result of the loosening of the hands) and fall from the table. Grabbing small elements from the ground level was problematic and took significant amount of time.

The times in the PC version were much shorter and there were almost no failed attempts. The users praised better graphics and fluency in the PC version, not minding the cables and physical controllers. The Android version was rated lower because of poor quality graphics, less framerate and non-repeatable gestures (it was difficult to perform the grabbing movement using the glove), despite advantages of being totally wireless and free of any extra physical controllers.

4   Conclusions

The VR solutions are a feasible option for the assembly process training. The users marked both presented versions as possible to implement as a real training solution before starting work on a production line. Still, it is noteworthy that features such as free movement over large space and lack of cables (the Android version) are much less important than overall fluency and high graphics quality when it comes to learning of manually performed operations.

The preliminary results lead to conclusion that smartphone based training applications in VR have a long way to go, due to the immersion effect broken by physics errors and poor quality. The PC based application with Oculus Rift CV1 seems much more promising direction at this point. In future, the authors will try to improve the hardware aspects in the mobile version, as well as try different tasks (such as vehicle operation, quality control and others). The next step to be taken is the validation of the solution - checking whether the person after virtual training using the proposed hardware solutions is able to correctly assemble the product or perform other tasks, e.g. perform activities related machine setup in a real stand.

Acknowledgements. The presented research results, were funded with grants for education allocated by the Ministry of Science and Higher Education in Poland **02/23/DSPB/7716**.

References

1. Górski, F., et al.: Effective design of educational virtual reality applications for medicine using knowledge-engineering techniques. Eurasia J. Math. Sci. Technol. Educ. **13**(2), 395–416 (2017)
2. Berg, L.P., Vance, J.M.: Industry use of virtual reality in product design and manufacturing: a survey. Virtual Real. **21**(1), 1–17 (2017)
3. Wu, Y., et al.: The virtual reality applied in construction machinery industry. In: Lecture Notes in Computer Science, vol. 8022, pp. 340–349 (2013)
4. Grajewski, D., et al.: Improving the skills and knowledge of future designers in the field of ecodesign using virtual reality technologies. Proc. Comput. Sci. **75**, 348–358 (2015)
5. Hamrol, A., et al.: Virtual 3D Atlas of a human body – development of an educational medical software application. Proc. Comput. Sci. **25**, 302–314 (2013)
6. Buń, P., et al.: Educational simulation of medical ultrasound examination. Proc. Comput. Sci. **75**, 186–194 (2015)
7. Martin-Gutierrez, J., et al.: Improving the teaching-learning process of graphic engineering students through strengthening of their spatial skills. Int. J. Eng. Educ. **31**(3), 814–828 (2015)
8. González, M.A., et al.: Virtual worlds. Opportunities and challenges in the 21st century. Proc. Comput. Sci. **25**, 330–337 (2013)
9. Falah, J., et al.: Development and evaluation of virtual reality medical training system for anatomy education. In: Studies in Computational Intelligence, vol. 591, pp. 369–383 (2014)
10. Ganier, F., et al.: Evaluation of procedural learning transfer from a virtual environment to a real situation: a case study on tank maintenance training. Ergonomics **57**, 828–843 (2016)
11. Bowman, D.A., Larry, F.H.: An evaluation of techniques for grabbing and manipulating remote objects in immersive virtual environments. In: Proceedings of the 1997 Symposium on Interactive 3D Graphics. ACM (1997)
12. Mine, M.R., et al: Moving objects in space: exploiting proprioception in virtual-environment interaction. In: Proceedings of the 24th Annual Conference on Computer Graphics and Interactive Techniques, pp. 19–26. ACM Press/Addison (1997)
13. Welch, G., Foxlin, E.: Motion tracking: no silver bullet, but a respectable arsenal. IEEE Comput. Graph. Appl. **22**, 24–38 (2002)
14. Buń, P., et al.: Possibilities and determinants of using low-cost devices in virtual education applications. Eurasia J. Math. Sci. Technol. Educ. **13**(2), 381–394 (2017)
15. Żywicki, K., et al.: Virtual reality production training system in the scope of intelligent factory. In: International Conference on Intelligent Systems in Production Engineering and Maintenance, pp. 450–458. Springer, Cham (2017)
16. Krolczyk, J., et al.: Material flow optimization – a case study in automotive industry. Teh. Vjesn. – Techn. Gaz. **22**(6), 1447–1456 (2015). https://doi.org/10.17559/TV-20141114195649

A System for Identifying Key Tacit Knowledge Resources Within Manufacturing Companies

Justyna Patalas-Maliszewska[(✉)] and Sławomir Kłos

Institute of Computer Science and Production Management,
University of Zielona Góra, Zielona Góra, Poland
{J.Patalas, S.Klos}@iizp.uz.zgora.pl

Abstract. Tacit knowledge is that special, expert knowledge which is always referred to within a company; it is also very difficult to identify it, much less describe it. In this paper, the TKnowS approach for the effective identification of Key Tacit Knowledge Resources, in manufacturing companies, is constructed and implemented in the form of a web-application, based on a case study. The TKnowS approach includes: (1) the definition of groups of workers according to processes which have been completed, using the Algorithm - k-means Clustering and Distance Method and the Euclidean Distances Method, (2) determining the competences of each worker, (3) determining the importance of each competence for each group of workers using the FAHP Method.

Keywords: Tacit knowledge · Identification of tacit resources
Euclidean Distances Method · FAHP Method · Manufacturing company

1 Introduction

Knowledge, by its very nature, may be divided into tacit knowledge and explicit knowledge. Knowledge, residing within employees, is referred to as tacit knowledge and is difficult to codify, identify or explain [1]. When defining tacit knowledge in a manufacturing enterprise, the experts, within that enterprise, should be recognized as such and relied on to perform that role. In a knowledge-based economy, an important role is played these experts who create new groups of employees, that is, knowledge workers; these employees then go on to co-create the value of the intellectual capital of the company. Manufacturing companies need to introduce models [2] for identifying knowledge resources arising from ongoing changes within a company and its environment. Knowledge sourcing refers to the activities of those workers who are engaged in tracking down the expertise, experience and opinions of others [3].

In this paper, special attention is paid to presenting our approach to the identification of internal knowledge resources, namely the identification of Key Tacit Knowledge Resources (TKnowS).

In the first stage of our TKnowS approach, the processes that are carried out by workers in a company are strictly described, using the Algorithm - k-means Clustering and Distance Method and the Euclidean Distances Method, in order to define the groups of workers according to completed processes.

© Springer International Publishing AG, part of Springer Nature 2019
J. Machado et al. (Eds.): HELIX 2018, LNEE 505, pp. 405–410, 2019.
https://doi.org/10.1007/978-3-319-91334-6_55

In the second stage the following groups of competences for workers, viz., Technical Competences (TC), Organisational Competences (OC), Behavioural Competences (BC), Social Competences (SC), and Personal Competences (PC) [4, 5] are distinguished. Then, by using the FAHP Method (the Fuzzy Analytic Hierarchy Process) it is possible to determine the relative predominance of a particular factor from the immeasurable elements for each group of workers according to completed processes.

2 An Approach to the Identification of Key Tacit Knowledge Resources in a Manufacturing Company

Knowledge sourcing often relies on IT systems and tools in today's organisations [6]. Due to the fact that knowledge sourcing is, in essence, learning from others in work settings, our problem in this paper is how to identify those experts named as Key Tacit Knowledge Resources in manufacturing companies and how to select those employees with the appropriate knowledge to undertake new projects in companies. The proposed approach, presenting processes completed by workers, provides an opportunity to evaluate expert knowledge within a company. Therefore the following stages are involved in our approach (see Fig. 1).

We proposed web-based questionnaires for each employee in the manufacturing companies that we surveyed, based on processes which they had completed in the usual manner. Each worker was required to select the processes, carried out by himself. Where processes were carried out more than 80% of the time, in a given month, by a particular worker, then that particular process was entrusted to that particular worker [7]. This worker was also required to actually select the tool used for completing the operation.

The Algorithm k-Means Clustering Method and the Euclidean Distances Method were then used to classify the workers according to the processes completed. The object of this clustering is the division of objects into groups called 'clusters' [8].

In Stage 2 of our approach, we proposed five, web-based questionnaires for each employee in the manufacturing companies surveyed, in order to obtain values for each of the defined competences, viz., Technical Competences (TC), Organisational Competences (OC), Behavioural Competences (BC), Social Competences (SC) and Personal Competences (PC). The values of each worker's competence, in a company, will be obtained through on-line interviews and/or tests. Using an algorithm for the answers to the questionnaires for each competence, it is possible to determine the values of the competences for each employee.

In Stage three, the FAHP Method (the Fuzzy Analytic Hierarchy Process) was implemented and used. By using FAHP, it is possible to determine the relative dominance of each competence for each group of workers, according to the manner in which jobs were completed (q.v. Stage 1) [9].

Key Tacit Knowledge Resources, according to completed processes in a manufacturing company, can now, finally, be defined. This approach, presented in the form an IT tool, was implemented and is presented in the next section.

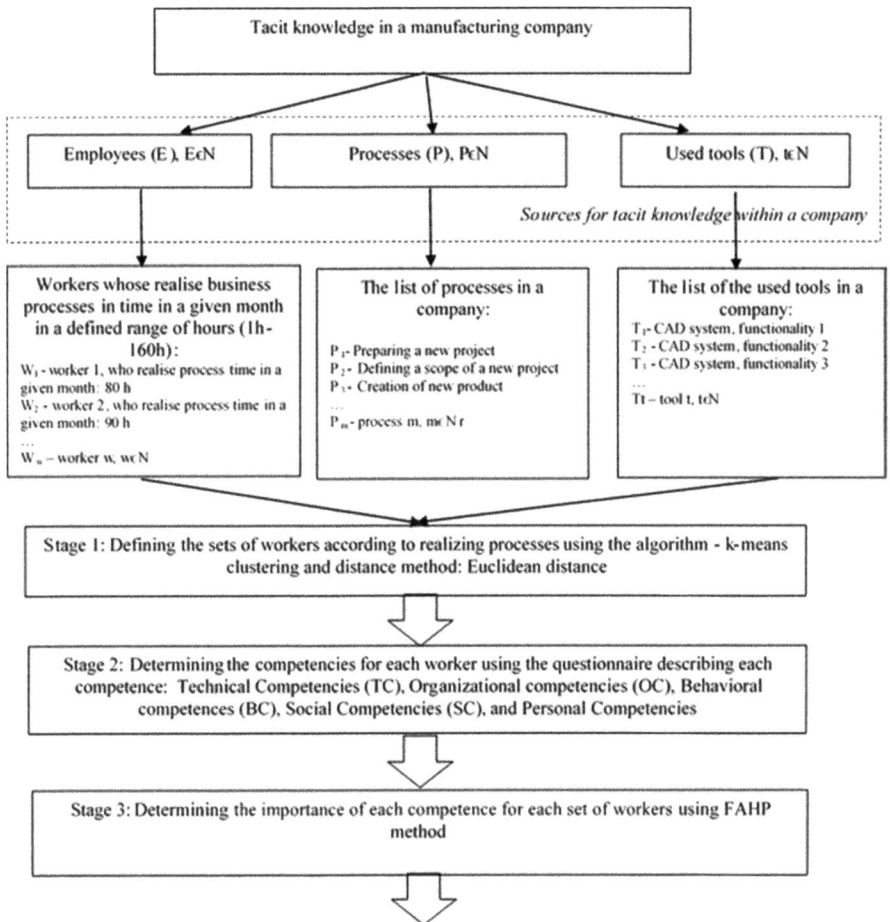

Fig. 1. An approach to the identification of Key Tacit Knowledge Resources in a manufacturing company

3 A System for the Identification of Key Tacit Knowledge Resources in a Manufacturing Company

To distinguish groups of workers according to processes completed within a company, a knowledge web-questionnaire is defined. Figure 2 presents an extract from the web-questionnaire wherein the time taken to complete processes in a given month is defined in hours, such as from 1 h to 160 h.

Currently, 69 different processes are defined in the Research and Development Dept. of a manufacturing company. With regard to defining processes, each worker is classified using the Algorithm - k-means Clustering and Distance Method and the Euclidean Distances Method with Statistica, ver. 13.3. In our example, we assembled groups of worker based on the results of 20 employees in the manufacturing company.

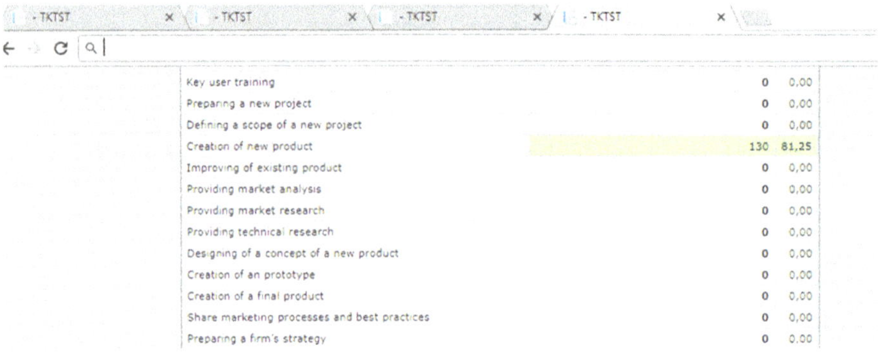

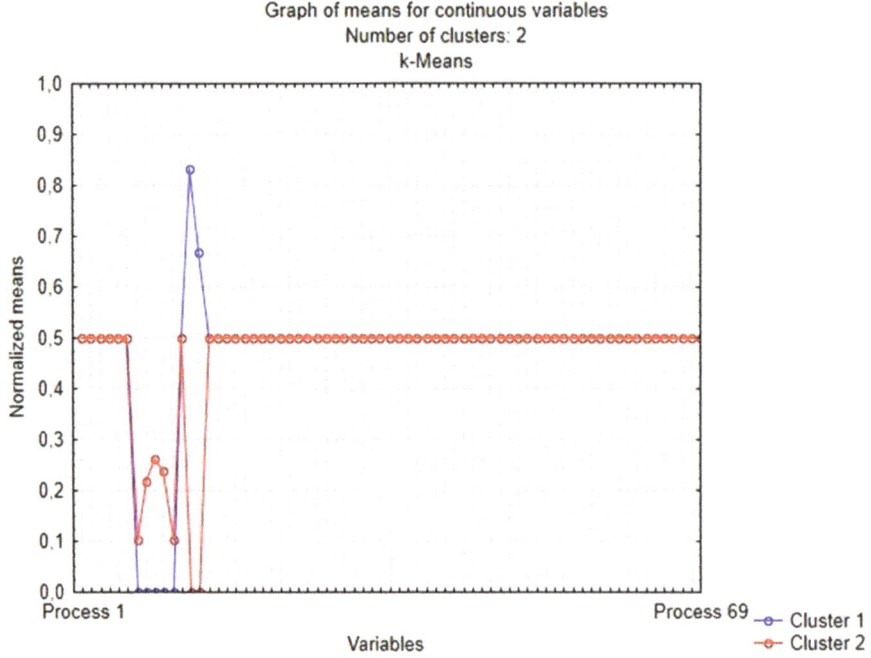

Fig. 2. An extract from the web-questionnaire wherein the time taken to complete processes in a given month is defined in hours, such as from 1 h to 160 h

Fig. 3. Groups of workers, according to completed processes, using the Algorithm - k-means Clustering and Distance Method and the Euclidean Distances Method

It was assumed that two groups of workers would be assembled. Figure 3 presents the results of the research.

For the first group of workers, the following processes were defined and carried out by a particular worker, viz., improvement to existing products and the provision of market analysis. For the second group of workers, this was the provision of both technical and market research, the conceptualization of new products, the creation of the prototype and the creation of the final product. So, only two workers are members of the first cluster (worker 16; worker 17; worker 18) the other employees are members of the second cluster.

According to Stage 2, in Fig. 1, the defined and implemented knowledge web-questionnaire enabled values for each competence to be obtained, according to selected groups of workers. In our study, we defined the tests for the second cluster workers. Each worker from this cluster completed a defined survey form for each competence, viz., for Technical Competences (TC), for Organisational Competences (OC), for Behavioural Competences (BC), for Social Competences (SC) and for Personal Competences (PC) (see Fig. 4 for an example describing TC).

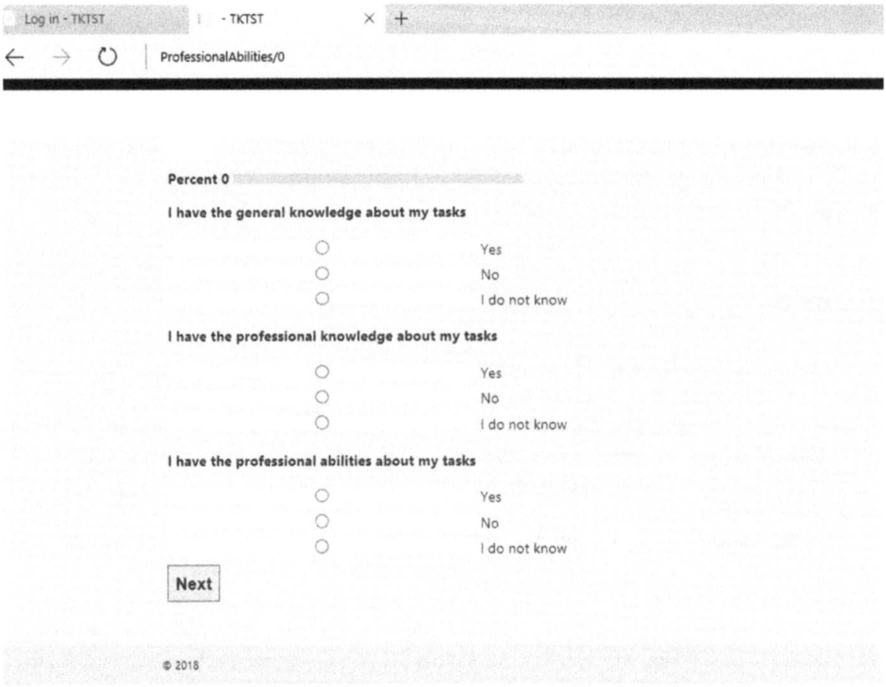

Fig. 4. An extract from the web questionnaire for employees, exemplifying Technical Competences

In the third stage of the proposed approach, managers determine the importance of each competence - TC, OC, BC, SC, PC for the second cluster of workers, according to the rules in the FAHP Method. Based on the fuzzy scale of preferences, the fuzzy weightings matrix of competences for the second cluster was defined.

In our case study, we can observe the most-important competence is social competence within the manufacturing company, for the following, processes, completed by the workers, the provision of market research, the provision of technical research, the conceptualisation of new products, the creation of the prototype and the creation of the final product.

Finally, the manager of a manufacturing company can determine the Key Tacit Knowledge Resources from the second cluster of workers, based on the values of the web-questionnaires for those employees, exemplifying and defining the importance of each competence, using the FAHP Method. Employees with the best results will be defined as TKnowS within a manufacturing company.

4   Conclusions

Knowledge in a manufacturing company is a resource, the appropriate development of which may determine the further development of that company. In this paper, we have argued that identification of Key Tacit Knowledge Resources would improve the effectiveness of a manufacturing company. In order to elaborate the application, an approach was built, which presents the principle of how a system works. The use of this application, as developed, may provide managers with the ability to determine critical tacit knowledge resources. Our case study presents a system for the identification of Key Tacit Knowledge Resources within a manufacturing company and will be developed further in our future work.

References

1. Lagerstrtöm, K., Andersson, M.: Creating and sharing knowledge within a transnational team – the development of a global business system. J. World Bus. **38**, 84–95 (2003)
2. Kujawińska, A., Diering, M., Rogalewicz, M., Żywicki, K., Hetman, Ł.: Soft modelling based methodology of raw material waste estimation. In: Burduk, A., Mazurkiewicz, D. (eds.) ISPEM 2017. Advances in Intelligent Systems and Computing, vol. 637, pp. 407–417. Springer, Heidelberg (2017)
3. Gray, P.H., Meister, D.B.: Knowledge sourcing effectiveness. Manag. Sci. **50**(6), 821–834 (2004)
4. Alam, M., Gale, A., Brown, M., Khan, A.I.: The importance of human skills in project management professional development. Int. J. Manag. Proj. Bus. **3**(3), 495–516 (2010)
5. Hecklaua, F., Galeitzkea, M., Flachsa, S., Kohlb, H: A holistic approach to human-resource management in Industry 4.0. In: Procedia CIRP, 6th CLF 6th CIRP Conference on Learning Factories, vol. 54 (2016)
6. Lin, C.Y., Kuo, T.H., Kuo, Y.K., Ho, L.A., Kuo, Y.L.: The KM chain - empirical study of the vital knowledge sourcing link. J. Comput. Inf. Syst. **48**(2), 91–99 (2007)
7. Patalas-Maliszewska, J., Krebs, I.: A model of the tacit knowledge transfer support tool: CKnow-board. In: Dregvaite, G., Damasevisius, R. (eds.) ICIST 2016. Communications in Computer and Information Science, vol. 639, pp. 30–41. Springer, Heidelberg (2016)
8. Cornuéjols, A., Wemmert, C., Gançarski, P., Bennani, Y.: Collaborative clustering: why, when, what and how. Inf. Fus. **39**, 81–95 (2018)
9. Nydick, R.L., Hill, R.P.: Using the analytic-hierarchy process to structure the supplier-selection procedure. Int. J. Purch. Mater. Manag. **28**(2), 31–36 (1992)

Classification of Products
in Production Levelling

Paulina Rewers[✉], Filip Osiński, and Krzysztof Żywicki

Faculty of Mechanical Engineering and Management,
Poznan University of Technology, Piotrowo 3, 60-965 Poznan, Poland
{paulina.rewers,filip.osinski,
krzysztof.zywicki}@put.poznan.pl

Abstract. The article focuses on the issue of production levelling (jap. Heijunka). The article contains an analysis of literature and presents the author's methodology of production levelling. The methodology consists of 5 steps. The first stage is the selection of products for which the production levelling will be carried out. The article presents the author's scheme for defining product groups, consisting of two main steps, from which 6 product groups are selected. Criteria that were taken into account in the creation of the scheme are primarily the rotation of products, sales of products (volume and value) and the minimum, economic production batch. The example of the use of product classification in an enterprise from the automotive industry is also presented. Separated product groups accounted for 28% of all manufactured products and approximately 88% of the company's total turnover.

Keywords: Products classification · Production levelling
Lean Manufacturing

1 Introduction

If production companies want to be competitive on the market, they have to adjust their production to growing customer demands. Nowadays clients want to buy products at lowest price possible, of the highest quality and in the shortest time possible [1, 2]. In order to meet their requirements, companies reach more frequently for Lean Manufacturing (LM) methods and tools, which not only eliminating waste during production, but also enhancing productivity, cutting down the costs and shortening production time [3]. The LM concept is based on Toyota Production System (TPS) [4, 5], which accounts for core production management methods modification in order to increase the actions' efficiency. A lot of models have been developed to explain the idea and to present the principles of achieving the assumed goals. One of the most important authors in this regard is Womack. Moreover, apart from grand rules of designing a value stream, he also formulated requirements that the processes should meet and actions presented. According to Womack and Jones each action should be [6]:

- valuable – it should create an added value for which the customer is willing to pay,
- capable – tasks should be performed always in the same manner, at the same time and with the same result,

© Springer International Publishing AG, part of Springer Nature 2019
J. Machado et al. (Eds.): HELIX 2018, LNEE 505, pp. 411–417, 2019.
https://doi.org/10.1007/978-3-319-91334-6_56

- available – it should be in compliance with the demand volume, enabling continuous course of process,
- flexible – it should permit a swift transition from production of one type of product to another.

The individual actions should be connected in such manner to ensure that the process is:

- flow – without any downtimes,
- pull – in compliance with current needs of direct consumer,
- level - by means of even distribution of production, in the shortest periods of time possible (days, shifts and even hours).

Individual assumptions are realized by application of proper Lean Manufacturing tools and methods, such as: 5S, SMED, Poka-Yoke, Just in Time, production levelling [7, 8].

The article presents an original methodology of production levelling. The emphasis was put on its first stage, which is product classification.

2 Production Levelling

The concept of production levelling (Japanese "Heijunka") was developed by Toyota in 1950s. According to Toyota's management concept, Heijunuka, combined with work standardization and the Kaizen approach are the foundation of Toyota Production System (TPS) [9, 10].

The authors of the article consider production levelling a method of planning oriented to even flow of goods from production. It consists of defining the sequence and the size of the flow of goods from the process in a way to ensure that the current demand was realized from stocks and does not cause any sudden changes in the production plan [11].

In literature there are a lot of interpretations of production levelling. The most significant include:

- a method of enhancing production capacity [12],
- a method of reducing stock levels [13],
- a method of increasing competitiveness [14],
- a method of eliminating discrepancy in production [15],
- a method of improving flexibility, reducing costs and improving the customer service quality [16–18],
- a method of make to stock (MTO) [19].

The article's authors emphasize that production levelling studies available in the literature are quite general, hence are very difficult to implement in real conditions. That is why there is a need to develop the subject and adjust the Heijunka assumptions to conditions prevailing in production companies.

3 Production Levelling Methodology

The authors of the article have been working diligently on the development of levelling production. The work resulted in original production levelling methodology described in papers [20–22]. The methodology is made up of five stages shown in Fig. 1.

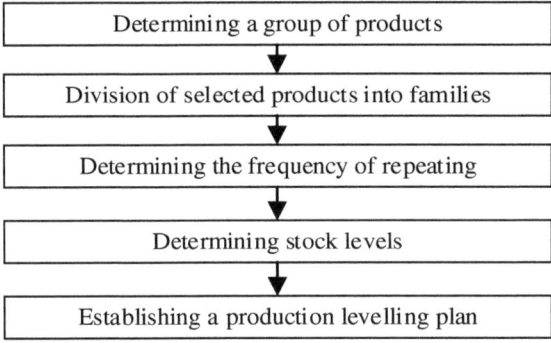

Fig. 1. Production levelling methodology, source: [22]

The first stage is selecting a product group for which production levelling is going to be performed. In production levelling the proper selection of products is crucial, as it determines further actions, and consequently the development of production plan. A production levelling plan is developed for key products manufactured by the company. A lot of criteria should be taken into account, for instance: the volume and frequency of orders, the value of orders, customer's demand, the time of product passage through the process, different variants of products, economic batch size, etc. It is very difficult to show the most important elements because as the company's specificity. Moreover, taking into account all of the criteria is connected with the necessity of applying advanced methods and techniques of product classification.

4 Classification of Products in Production Levelling

Production companies usually offer a wide range of products, therefore so many different variants it is very difficult to clearly indicate the products for which a production levelling plan will be developed. In many cases criteria such as sale volume, rate of product rotation, availability of raw materials or economic order quantity should be taken into account.

Considering the listed criteria, the authors suggest preforming two stages from which six product groups will be selected (Fig. 2). The first stage consists of selecting product groups which are not under regular production. They include:

- group 1 – products, which are being implemented or improved. After introducing them to regular sale, they should be re-classified to other groups.

- group 2 – products withdrawn from regular production. Production of these goods is ceased and the sale is continued while stock lasts.
- group 3 – special products, usually single. Tailor-made from customer's individual order.

The second stage – dividing the rest of products (which are not in groups 1, 2 and 3) as regards the rate of rotation. Three groups are distinguished here:

- group 4 – fast-moving products – manufactured and sold regularly. The products are classified is this group if their medium-term sale (e.g. per month) is greater than the minimum batch size.
- group 5 – medium-moving products – these are sold less frequently than fast-moving products but more frequently than slow-moving ones.
- group 6 – slow-moving products for which sale in a definite unit of time (e.g. per year) is lower than minimum batch size.

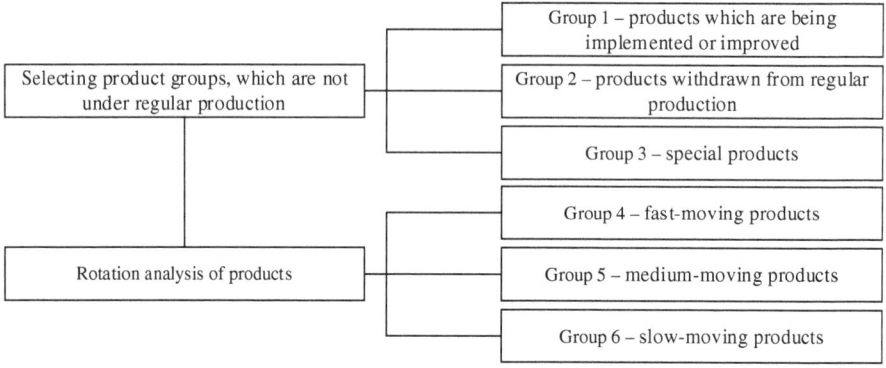

Fig. 2. Diagram of products classification

Classifying products to individual groups should be updated at least once per year. This period depends mostly on the seasonality of product sales. The presented products classification will work in enterprises producing mainly according to make to stock (MTO).

Firstly, group 4 products should be included in production levelling. Fast-moving products are the most desired ones. These are often key products for a company (bringing most profit from sales). Thanks to continuous sequence and number of manufacturing of those products, irrespective of fluctuations in demand, one can be certain that all customers will receive the ordered products on time and in quantity as required. Constant production of these products will also allow to reduce inventories and production lead time. Subsequently, group 5 and 6 can be also levelled.

5 Exemplary Application of Product Classification

The analysed company manufactures and distributes parts for automotive industry. Ca. 90% of production are spare parts. The rest of production is sold on the primary market (1%) and delivered exclusively to certain institutional customers (9%). Overall the 4334 different types of products are manufactured.

The company's basic aim is shipment of 90% of spare parts within 24 h from receiving customer order. The aim have to be fulfilled even though the entire production lead time is longer and lasts from 5 to 21 business days. The aim will be reached by the production levelling. The further part of the section describes products classification in the analysed company.

In the first step, products were assigned to three groups - 1, 2 and 3. The first group contains 104 products, the second – 242 products and the third – 2775 products. All these products represent 72% of the manufactured goods. Next, the fourth (fast-moving products), the fifth group (medium-moving products) and the sixth 6 (slow-moving products) were separated.

It was assumed that fast-moving products are the products which average monthly sales are greater than minimum economic batch quantity (40 pcs). The group includes 68 products. It was also assumed that the maximum stock of the fast-moving products is equal to average two-months' sales and the minimum stock is equal to 50% of the maximum one. Such a stock levels enable to ensure continuous products availability. Products of the fourth group are manufactured regularly, irrespective of the ordered quantities.

All products, which monthly sale is not classified to the fast-moving group are the medium-moving goods (the frequency of the product production is lower than one per month). However, their annual sales volume exceeds the economic order quantity. 211 products were classified into this group. The maximum stock level of the products was set at the level of four-months' sales and the minimum stock level was set at 25% of the maximum. When the minimum level is reached, demand of the products is requested. A planner places the products in the production plan and informs the sales department about the planned date of finishing the production.

The sixth group includes products which annual sales volume is lower than the minimum production batch (less than 40 pcs. per year). 934 products were classified into this group. The maximum stock level of the products group is equal to annual sales. On the other hand the minimum stock level is determined on the basis of market information and forecasts concerning further demand of the product.

Products of the groups 4, 5 and 6 are a total of 28% of all the manufactured products, which represents 88% of the company's total sales revenues. Due to products classification the production plan is stable and there is no delay in shipment of products to customers.

6 Conclusions

Production levelling is a goal pursued by a great many enterprises. Mostly it ensures the production balance and low stock levels, while at the same time ensuring timely deliveries to the customer. The production levelling methodology leads to development of production levelling plan. It is important stage is determining products for which a production levelling is going to be developed. Due to a great number of factors which have to be taken into account, including inter alia, the frequency and quantity of orders, the selection is not simple. Thanks to production levelling it is possible to reassure the production of goods which are most important for a company, which enables the company to focus on planning production of medium- and slow-moving goods.

The article presents an original scheme for classification of products. It is made up of the following two main steps:

- selecting groups of products which are not under regular production,
- dividing products into 6 groups by rate of rotation.

The criteria that were taken into account when developing the scheme include, inter alia: the products' rate of rotation, product sales and minimum production batch.

In the first place, group 4 is selected for production levelling, since the products included in that group are characterized by the highest rate of rotation and sales volume. Usually these are key products for a company – profit from sales of such products is the highest.

Exemplary application of the product classification is also shown. Groups 4, 5 and 6 represented 28% of all the manufactured goods, which corresponded to ca. 88% of the company's overall turnover. The presented classification is in its development phase. The authors predict its further development and implementation in other companies.

Acknowledgements. The presented results of the research, carried out under the theme No. 02/23/DSPB/7716, were funded with a grant to science granted by the Ministry of Science and Higher Education.

References

1. Hamrol, A.: Strategies and Practices of Efficient Operation. LEAN SIX SIGMA and other (in Polish). PWN, Warszawa (2015)
2. Zawadzki, P., Żywicki, K.: Smart product design and production control for effective mass customization in the Industry 4.0 concept. Manag. Prod. Eng. Rev. **7**(3), 105–112 (2016)
3. Trojanowska, J., Żywicki, K., Varela, M.L.R., Machado, J.: Improving production flexibility in an industrial company by shortening changeover time: a triple helix collaborative project. In: Peris-Ortiz, M., Ferreira, J., Farinha, L., Fernandes, N. (eds.) Multiple Helix Ecosystems for Sustainable Competitiveness. Innovation, Technology and Knowledge Management, pp. 133–146. Springer, Cham (2016)
4. Hopp, W.J., Spearman, M.L.: Factory Physics, 1st edn. Irwin, Chicago (1996)
5. Rinehart, J.: After lean production: evolving employment practices in the world auto industry. Am. J. Sociol. **104**(4), 1212–1214 (1997)

6. Womack, J.P., Jones, D.T.: Lean Thinking: Banish Waste and Create Wealth in Your Corporation. Simon & Schuster UK Ltd., London (2003)
7. Ohno, T.: Toyota Production System. Productivity Press, Portland (1988)
8. Rother, M., Shook, J.: Learning to See: Value Stream Mapping to Add Value and Eliminate Muda. The Lean Enterprise Institute, Cambridge (2003)
9. Monden, Y.: Toyota Production System. Industrial Engineering and Management Press, Institute of Industrial Engineers, Norcross (1983)
10. Van Staden, M., Du Plessis, W.: Being the lean dream. In: CIE42 Proceedings, 16–18 July 2012, Cape Town, South Africa (2012)
11. Rewers, P., Żywicki, K., Hamrol, A., Chabowski, P.: Methodology of conduct in providing creation of repeatable production plan (in Polish). Bus. Manage. **3**, 22–30 (2016)
12. Hüttmeir, A., Treville, S., Ackere, A., Monnier, L., Prenninger, J.: Trading off between Heijunka and just-in-sequence. Int. J. Prod. Econ. **118**, 501–507 (2009)
13. Coleman, J.B., Vaghefi, M.: Heijunka: a key to the Toyota production system. Prod. Inventory Manag. J. **34**(4), 31–35 (1994)
14. Teece, D.J., Pisano, G., Shuen, A.: Dynamic capabilities and strategic management. Strateg. Manag. J. **18**(7), 509–533 (1997)
15. Andel, T.: Accentuate Heijunka, eliminate junk, supply chain flow. Mater. Handling Eng. **54**(8), 77 (1999)
16. Korytkowski, P., Grimaud, F., Dolgiu, A.: Exponential smoothing for multi-product lot-sizing with Heijunka and varying demand. Manag. Prod. Eng. Rev. **5**(2), 20–26 (2014)
17. Ivanov, V., Vashchenko, S., Rong, Y.: Information support of the computer-aided fixture design system. In: Proceedings of 12th International Conference on ICTERI 2016, Kyiv, Ukraine, 21–24 June (2016)
18. Ivanov, V., Mital, D., Karpus, V., et al.: Numerical simulation of the system "fixture – workpiece" for levers machining. Int. J. Adv. Manuf. Technol. **91**(1), 79–90 (2017). https://doi.org/10.1007/s00170-016-9701-2
19. Araujo, L.F.D., Queiroz, A.A.D.: Production leveling (Heijunka) implementation in a batch production system: a case study. In: International Federation for Information Processing, pp. 105–112 (2010)
20. Żywicki, K., Rewers, P., Bożek, M.: Data analysis in production levelling methodology. In: Rocha, Á., Correia, A., Adeli, H., Reis, L., Costanzo, S. (eds.) Recent Advances in Information Systems and Technologies. WorldCIST 2017. Advances in Intelligent Systems and Computing, vol. 571, pp. 460–468. Springer (2017)
21. Rewers, P., Hamrol, A., Żywicki, K., Kulus, W., Bożek, M.: Production leveling as an effective method for production flow control—experience of polish enterprises. Procedia Eng. **182**, 619–626 (2017)
22. Rewers, P., Trojanowska, J., Diakun, J., Rocha A., Reis L.P.: A study of priority rules for a levelled production plan. In: Hamrol, A., Ciszak, O., Legutko, S., Jurczyk, M. (eds.) Advances in Manufacturing. Lecture Notes in Mechanical Engineering. Springer (2018)

Polish SME Energy Efficiency in the Years 2014–2016

Filip Osiński and Łukasz Grudzień[✉]

Poznan University of Technology,
pl. M. Skłodowskiej-Curie 5, 60-965 Poznań, Poland
{filip.osinski,lukasz.grudzien}@put.poznan.pl

Abstract. The use of energy and its various sources is becoming an increasingly important issue in the management of an industrial enterprise. The growing importance of this element is related to its significant share in the costs of running a business, a high impact on the natural environment and the increasing number of laws regulating this subject. Also International Standardization Organization noticed the problem of power consumption in companies and published of the ISO 50001 energy management standard. Presented research shows the state of energy management and awareness in SME enterprises in Poland in the years 2014–2016.

Keywords: Energy efficiency · Energy management system · Energy costs

1 Introduction

The issue of energy efficiency is widely described in the literature in relation to the production process [1–6]. Due to the increasing costs of all energy forms, energy efficiency has become one of the most important problems in enterprise management. Improving energy efficiency has a positive impact on the organization's productivity [7]. This is particularly evident in the case of industrial enterprises, where energy costs absorb 15–30% of all production costs [8]. The generation of energy is also associated with a huge threat to the environment. Due to the growing public awareness, entrepreneurs try to manage the energy consumption of their companies as part of improving the company's image. Energy efficiency is not only part of the general trend of caring for the environment, especially in the EU, but it is also an element of the wider issue of corporate social responsibility (CSR) [9]. Obtaining good energy efficiency is not only a manifestation of concern for reducing the company's own costs, but also shows that the interests of society are widely understood by companies. The reduction in energy intensity is directly related to the reduction in the consumption of raw materials and utilities in the processes carried out by enterprises as well as the reduction in the amount of greenhouse gas emissions [10]. This problem was recognized by the EU commission, which is trying to speed up the process by introducing certain legal regulations. An example of them can be the 2020 climate & energy package [11]. It assumes a reduction of greenhouse gas emissions by 20% with a simultaneous increase in energy efficiency by 20% and a share of renewable energy sources (RES) in the total energy production at the level of 20%. These are the EU's main objectives for

© Springer International Publishing AG, part of Springer Nature 2019
J. Machado et al. (Eds.): HELIX 2018, LNEE 505, pp. 418–424, 2019.
https://doi.org/10.1007/978-3-319-91334-6_57

achieving sustainable growth by 2020. The importance of the problem of energy efficiency was also noticed by the International Organization for Standardization, issuing in 2011 a standard containing requirements for the construction of the energy management system ISO 50001 [12].

2 Energy Efficiency in Enterprises

The energy consumed in enterprises should be understood as all media that are necessary for both production processes and other processes related to, for example, maintenance or broadly defined infrastructure, e.g. electricity, heat, fuels, cooling water, compressed air, etc. In addition, due to the source, two basic types of energy can be distinguished [13]. The first is primary energy, it means the energy used directly from energy resources (e.g. coal, oil, natural gas). The other type, secondary energy, is created from other energy sources, to enable its industry and residential use. An example of this is the electricity generated by the combustion of fuels in a power plant or as a result of the wind turbines work. Energy efficiency is, in turn, the ratio of the size of the applied effect, i.e. the effect obtained as a result of energy supply to a given facility, technical device or installation, in particular: mechanical work, thermal comfort, lighting for a given facility, technical device or installation, under typical conditions their use or operation, to the amount of energy consumed by this object, technical device or installation, or as a result of the service required to obtain this effect [14]. In practice, the greater the efficiency of a given device, installation, medium, the greater savings for the company and, consequently, greater opportunities to be competitive. According to the International Energy Agency, energy efficiency is a process of managing and limiting the growth of energy consumption [15]. Activities defined as energy-efficient consist in providing a higher level of services with the same amount of energy or the same level of services with less energy. To determine energy efficiency it is necessary to determine the level of input energy consumption and to determine the amount of energy to be saved.

Audit of energy efficiency is one of the most important elements of energy management. It is a study containing an analysis of energy consumption and defining the technical condition of a facility, devices and installations, containing a list of projects aimed at improving the energy efficiency of these facilities, equipment or installations, as well as assessing their economic viability and possible energy savings [16]. The audit is an element required both in the light of EU regulations and the ISO 50001 standard.

Based on the data collected during the audit, it is possible to determine the actual level of energy savings in a given company. Energy efficiency assumes how much energy is to be saved. Energy saving is the amount of energy that is the difference between energy potentially used by an object, technical device or installations in a given period before one or more projects are implemented to improve energy efficiency, and energy consumed by this facility, technical device or installation in the same period, after completing these projects and taking into account normalized conditions affecting energy consumption [17]. A venture aimed at improving energy efficiency is an action consisting in introducing changes or improvements in a facility, technical device or installation, as a result of which energy savings [18] are obtained.

Improving energy efficiency can serve the following types of projects:

(1) insulation of industrial installations
(2) reconstruction or renovation of buildings
(3) modernization:
   a. electrical and electronic devices (computers, monitors, etc.)
   b. lighting
   c. devices and installations used in industrial processes
   d. local heating networks and local heat sources
(4) energy recovery in industrial processes
(5) reduction:
   a. reactive power flows
   b. network losses in linear sequences
   c. losses in transformers
(6) using energy for heating or cooling produced in own or connected to the network of renewable energy sources.

3   Use of Energy in SME Production Enterprises in Poland

In the years 2014–2016, a series of audits related to the implementation of the ISO 14001 system were carried out in 66 production companies belonging to the Polish SME sector. During the audits, group of experts examined every company in terms of process and amount of power consumption. Conducted surveys aimed at gathering information about places and processes that consume energy, energy sources and plans for work reorganization of and investments aimed at reducing energy consumption. In the vast majority of these enterprises there was a relatively low direct impact on the environment limited to the production of small amounts of hazardous waste and emissions from fuel combustion (mainly in cars and other vehicles). By far the largest environmental burden for SMEs in the SME sector is demonstrated by indirect aspects - the use of energy in both primary and secondary forms. This relationship, together with the fact that energy in various forms is one of the basic operating costs of any organization, means that energy management should become one of the basic elements of enterprise management of SMEs [7].

Nearly 95% of surveyed enterprises declare that energy consumption issues are taken into account in the company. In the vast majority of enterprises, this is related to the costs of energy use. In part of the enterprises (60%) attention was also focused on the aspects linked to environmental pollution, made during the production of energy. Such considerable attention devoted to SMEs in energy consumption issues makes it necessary to analyze the main energy consumption points in the functioning of the company and to consider the possibility of lowering or eliminating those points. In order to conduct such an analysis, it is necessary to distinguish sources of energy supplied (primary and secondary energy sources) and diagnose processes in which it is used.

Primary energy is used mostly for the purposes of main processes (production). In the analyzed group of the SME sector it was used only in 27% of enterprises. This energy was supplied in 95% of cases in a gaseous form (acetylene, natural gas or LPG), and in the case of the remaining 5% in the form of firewood. The production gas was mainly used to supply production furnaces used for hot metal processing. This mainly applies to processes such as forging, rolling or gas welding. It is worth mentioning that around 27% of SMEs in Poland deal with metalworking, which makes the use of primary energy in this sector widespread. In other situations, the production gas was used to heat drying chambers used for curing coatings applied to metal products [19]. Firewood in production processes was mainly used for heating wood driers and varnished wooden products. It should be noted that the choice of raw energy material in form of wood was dictated mostly by the possibility of using in these plants by-products of its own production, mainly in the form of wood shavings and dust and used for heating processes, including social and living purposes.

Secondary energy in the form of electricity was used in all enterprises of the studied group, in 63% of cases it was the only source of energy in manufacturing processes. Electricity in production is used in particular to power production machines and devices, such as milling lathes, injection molding machines, etc. In terms of energy use, do not forget about the use of hand tools (drills, grinders) that are often neglected when creating specifications for existing production equipment.

In the surveyed group of companies, almost 23% declare the exchange of their production equipment with newer ones with greater energy efficiency. 18% of companies have machinery and equipment younger than 3 years, so their replacement would not be associated with an increase in the efficiency of energy use. 59% of surveyed enterprises do not plan to replace the machine park within the next 3 years.

One of the basic purposes of using energy for social and living purposes is heating the interior of buildings and utility water. The basic parameter affecting the size of this consumption, apart from the size of the heated rooms, is the quality and condition of the building insulation. In the surveyed group of enterprises, 70% do not anticipate thermo-modernization of buildings within the next 3 years. Less than 15% do not plan any investments related to the insulation of buildings due to the fact that the buildings they occupy are new or have been renovated afterwards, including thermo-modernization. 15% of companies declare that work related to the thermal modernization of buildings or the construction of new facilities with better insulation parameters will take place within 3 years from the end of the project.

Due to the fact that the vast majority of enterprises in the surveyed group benefited from the assistance of external suppliers of secondary energy (electricity and heat), the investment in own renewable energy sources was perceived as ineffective in terms of finance. On the day of the survey, only 15% of the surveyed companies used renewable energy, mainly in the form of recuperation and heating solar panels. Among other companies only 3% declare their willingness to purchase renewable energy sources within the next 3 years.

The energy is used in production on the 3 main objectives: transportation, infrastructure maintenance and manufacturing processes. Figure 1. presents the share of particular types of primary energy used in processes related to the transport of people, raw materials and manufactured goods [20]. The diagram shown at Fig. 2 presents

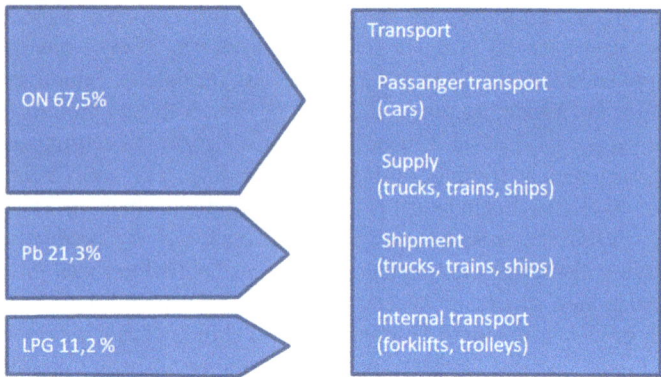

Fig. 1. Share of energy sources in transport

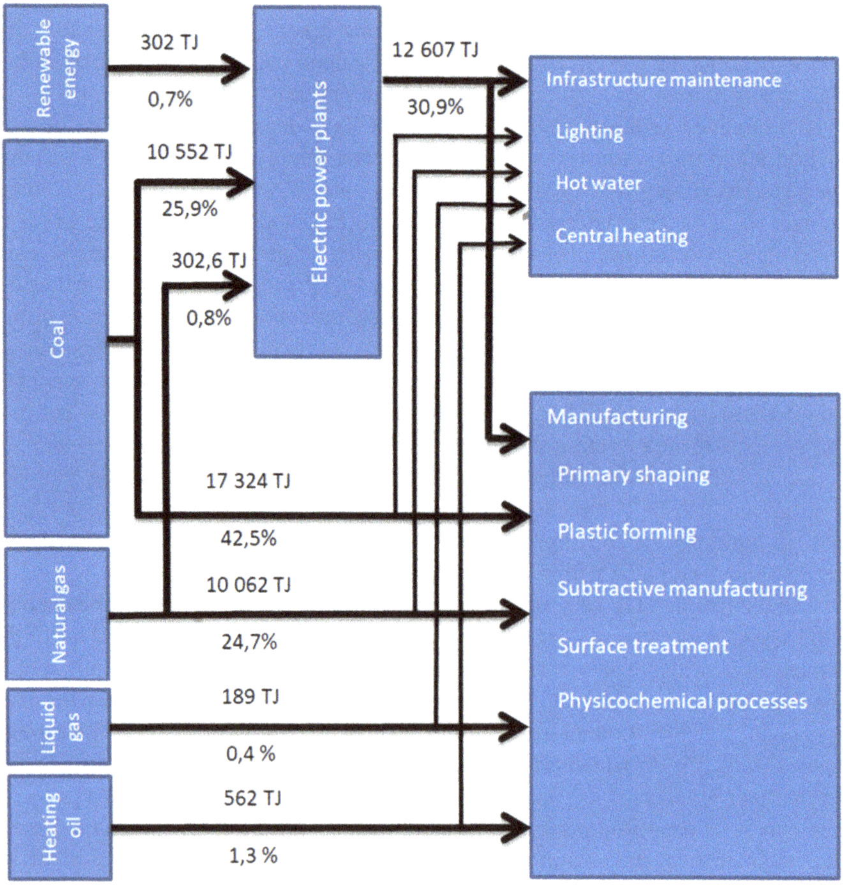

Fig. 2. Sources of energy used in industry in Poland

sources of primary and secondary energy, amount of energy consumed in and the percentage share of a given energy source in the final consumption in the industrial sector in Poland [20].

4 Conclusion

The increasing pressure on the aspect related to the use of energy in enterprises is confirmed by as many as 95% of the surveyed companies in the SME sector in Poland. Entrepreneurs pay attention to this aspect mainly due to the high share in the costs of maintaining the company (up to 30% of the company upkeep) and the significant impact of energy production on the natural environment (practically 68% of energy in Poland comes from coal combustion). The increase in interest in energy consumption controls results in increasing demand for implementation and maintenance of the Energy Management System compliant with the international ISO 50001 standard.

Acknowledgments. The presented results of the research, carried out under the theme No. 02/23/DSPB/7716, was funded with a grant to science granted by the Ministry of Science and Higher Education.

References

1. Antosz, K., Chandima Ratnayake, R.M.: Machinery classification and prioritization: empirical models and AHP based approach for effective preventive maintenance. In: 2016 IEEE International Conference on Industrial Engineering and Engineering Management, pp. 1380–1386 (2016)
2. Antosz, K., Stadnicka, D.: Lean philosophy implementation in SMEs - Study Results, 7th International Conference On Engineering, Project, and Production Management, Procedia Engineering, vol. 182, pp. 25 32 (2017)
3. Kujawińska, A., Diering, M., Rogalewicz, M., Żywicki, K., Hetman, Ł.: Soft modelling-based methodology of raw material waste estimation. In: Burduk, A., Mazurkiewicz, D. (eds.) Intelligent Systems in Production Engineering and Maintenance, ISPEM 2017. Advances in Intelligent Systems and Computing, vol. 637, pp. 407–417. Springer (2018)
4. Rewers, P., Trojanowska, J., Diakun, J., Rocha, A., Reis, L.P.: A study of priority rules for a levelled production plan. In: Hamrol, A., Ciszak, O., Legutko, S., Jurczyk, M. (eds.) Advances in Manufacturing. Lecture Notes in Mechanical Engineering, pp. 111–120. Springer (2018)
5. Trojanowska J., Kolinski A., Varela M.L.R., Machado J.: The use of theory of constraints to improve production efficiency–industrial practice and research results. DEStech Transactions on Engineering and Technology Research, (ICPR 2017), pp. 537–542 (2017)
6. Trojanowska, J., Kolinski, A., Galusik, D., Varela, M.L.R., Machado, J.: A methodology of improvement of manufacturing productivity through increasing operational efficiency of the production process. In: Hamrol, A., Ciszak, O., Legutko, S., Jurczyk, M. (eds.) Advances in Manufacturing. Lecture Notes in Mechanical Engineering, pp. 23–32. Springer (2018)
7. McKanne, A.: Thinking Globally: How ISO 50001 – Energy Management can make industrial energy efficiency standard practice. Ernest Orlando, Lawrence (2009)

8. Giacone, E., Mancò, S.: Energy efficiency measurement in industrial processes. Energy **38** (1), 331–345 (2012)
9. Lindgreen, A., Swaen, V.: Corporate social responsibility. Int. J. Manag. Rev. **12**, 1–7 (2010)
10. Finnerty, N., Sterling, R., Coakley, D., Contreras, S., Coffey, R., Keane, M.M.: Development of a global energy management system for non-energy intensive multi-site industrial organisations: a methodology. Energy **136**, 16–31 (2017)
11. Limiting Global Climate Change to 2 degrees Celsius – The way ahead for 2020 and beyond, Communication by the Commission to the European Council. http://eur-lex.europa.eu/LexUriServ/LexUriServ.do?uri=COM:2007:0002:FIN:EN:PDF. Accessed 07 Feb 2018
12. EN ISO 50001, Energy management systems – Requirements with guidance for use, ISO, Geneva (2011)
13. Paska, J., Biczel, P., Kłos, M.: Hybrid power systems – an effective way of utilising primary energy sources. Renew. Energy **34**(11), 2414–2421 (2009)
14. The Energy Efficiency Act, Dz. U. 2011 r. No 94, poz. 551 (in Polish)
15. Capturing the Multiple Benefits of Energy Efficiency. http://www.iea.org/publications/freepublications/publication/Multiple_Benefits_of_Energy_Efficiency.pdf. Accessed 07 Feb 2018
16. Regulation of the Minister of Economy of 10 August 2012 on the detailed scope and method of preparation of the energy efficiency audit, model of the energy efficiency audit card and methods for calculating energy savings. Dz.U. 2012, poz 962 (in Polish)
17. Tanaka, K.: Assessment of energy efficiency performance measures in industry and their application for policy. Energy Policy **36**(8), 2887–2902 (2008)
18. Kanneganti, H., Gopalakrishnan, B., Crowe, E., Al-Shebeeb, O., Yelamanchi, T., Nimbarte, A., Currie, K., Abolhassani, A.: Specification of energy assessment methodologies to satisfy ISO 50001 energy management standard. Sustain. Energy Technol. Assessments **23**, 121–135 (2017)
19. Osiński, F., Grudzień, Ł., Hamrol, A.: Environmental awareness of the SME sector based on the implementation of the project "Implementation of environmental management systems". Innovation in management and production engineering. PTZP. Opole (2016)
20. Consumption of fuels and energy carriers. Central Statistical Office (GUS), Warsaw, December 2017

Autonomous Production Control:
A Literature Review

L. Martins[1](✉) ⓘ, Nuno O. Fernandes[2] ⓘ, and M. L. R. Varela[1] ⓘ

[1] Department of Production and Systems,
University of Minho, Guimarães, Portugal
luis_miguel-17@hotmail.com, leonilde@dps.uminho.pt
[2] Instituto Politécnico de Castelo Branco, Castelo Branco, Portugal
nogf@ipcb.pt

Abstract. Autonomous production control (APC) aims at improving production systems performance through fast and flexible reaction to changes in dynamic production environments. APC shifts the power of decision from a central planning unit, towards single intelligent and distributed logistic objects that can cope with the rising complexity of today's manufacturing systems. The purpose of this research is to analyse the current state of art on APC in discrete manufacturing industries, throughout a systematic literature review. The study's objectives were to: (1) identify specialists in APC; and (2) identify theoretical developments and practical implementations. The reviewed was obtained by searching the Scopus database - a total of 49 papers have been analysed. The findings revealed that there is not much work carried out on APC until now, and that most of the contributions are theoretical. It is hoped that this literature review will contribute to enhance research in this science field.

Keywords: Autonomous Production Control · Literature review

1 Introduction

The future of manufacturing industries depends largely on their capability to respond to the customers' expectations while maintaining a competitive advantage on their markets. Manufactures must produce a growing number of customized products [1, 2], with reduced life cycles, short delivery times and high adherence to the order' due dates.

Industry 4.0 allows to respond to these needs through autonomous Production Planning and Control (PPC). Traditional Production Planning and Control (PPC) systems are characterized by a central planning and controlling of the factory. However, the dynamics imposed by mass customization makes these systems less useful in the current days. This is due to the lack of flexibility to adapt to disturbances in the production environment, resulting from e.g. rush orders, machine breakdowns, etc. [3]. The development of Cyber-Physical System (CPS), obtained by advances at wireless communication, sensors and computing power, and the growth of Internet of Thing (IoT), in the last decade [4, 5], allows for new approaches to production control. One of these approaches is Autonomous Production Control (APC).

© Springer International Publishing AG, part of Springer Nature 2019
J. Machado et al. (Eds.): HELIX 2018, LNEE 505, pp. 425–431, 2019.
https://doi.org/10.1007/978-3-319-91334-6_58

Autonomous production control (APC) aims at improving production systems performance through fast and flexible reaction to changes in dynamic production environments. APC can cope with the rising complexity of today's manufacturing systems due to the decentralized decision-making. APC enables each logistics object, e.g., resource, pallet or order, to make decisions on their own, concerning the way jobs (or orders) will be produced, sharing information between them, with no human interference [6]. This permits a holistic vision of real-time production towards making the best production control decisions.

This paper makes a systematic literature review on the theme of APC for discrete manufacturing industries. The focus is on following production control functions: (1) job release, which determines the jobs to be released to the shop floor and that decide when they should be released; (2) detailed scheduling, that allocates jobs to resources and sets the jobs priorities on the shop floor

The remainder of the paper is organised as follows. The next section describes the research methodology used in this study, in Sect. 3 the findings are presented and discussed and finally Sect. 4 presents the conclusions and provides directions for future research work

2   Methodology

According to [7] a systematic literature review typically has the following nine purposes/benefits: (1) reduces large amounts of information; (2) integrates pertinent pieces of information for decision-making, research and policy; (3) it is a scientific technique typically less costly than new research, especially if it is updated permanently; (4) facilitates the generalizability of findings; (5) permits a systematic valuation of relationships among variables; (6) helps to explain and discover inconsistent data or contradictory findings in a specific field; (7) empowers the statistical analyses in quantitative results; (8) increases the precision for the estimation of statistical risks; (9) improves the accuracy and allows verification due to the systematic methodology.

In this paper, a systematic literature review is accomplished using the main steps referred by [8, 9] namely: define research question, determine required characteristics, retrieve sample of potentially relevant literature, select pertinent literature, synthesize literature and report results.

2.1   Planning the Review

To initiate this process the topic has been disputed in a brainstorming session with the authors to understand what is important to study. Then, the research questions were formulated. Due to the advent of Industry 4.0 and the increasing importance of Autonomous Production Control (APC) for manufacturing [6], it becomes relevant to (1) find out which are the specialist in the area of APC; and (2) known how is APC research evolving - throughout theoretical and empirical research work? This research questions are the basis of this work. The next step was to select the criteria of inclusion or exclusion of papers. The approach used was to include all the documentation concerning autonomous control in production or manufacturing, involving production control functions such as sequencing, releasing, dispatching, between others. The content of the papers may be a framework, an application, a technique, a new method, etc. to perform typical tasks of production control in the context of autonomous decision-making in discrete manufacturing industries.

2.2 Conducting the Review

The review was conducted in two searches. To keep results to a manageable number, the search was restricted to the title, abstract, and keywords of papers. The first search resulted in 2176 documents using "Autonomous Production Control". This search resulted in a set of documents out of the scope of our study (see next section). In a second search, this was further reduced to 97 documents by excluding apparently unrelated articles. These 97 papers are the result of a search conducted in March 2018 in one of the most popular and largest coverage database of scientific documents – Scopus, using the following script:

```
TITLE-ABS-KEY (Autonomous Control) AND TITLE-ABS-KEY
(production OR shop floor OR manufacturing) AND ALL (se-
quencing OR dispatching OR release OR capacity OR alloca-
tion OR scheduling OR planning OR programming)
```

After the abstract reading of these papers, only 49 were relevant to the scope of our study. The next section presents and discusses the findings obtained.

3 Findings and Discussion

In the past decades, there was a growing interest by the scientific community in Autonomous Production Control (APC). This can be noticed by observing Fig. 1 that resulted from the first search. Over the past decade the number of papers published in APC more than duplicated, achieving 174 in 2017 from a total of 2167 published in the period between 1960 and 2017.

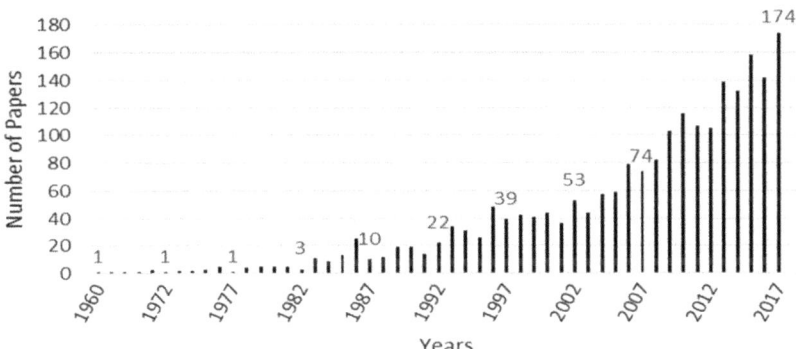

Fig. 1. Distribution of papers by year of publication.

This first search resulted in documents about areas such as enabling technologies for APC (e.g., sensors, RFID) and in process industries, such as oil, fluid and gas production, which is out of the scope of this study. VOSviewer software [10] was used to give a general view of the main research fields in APC Fig. 2.

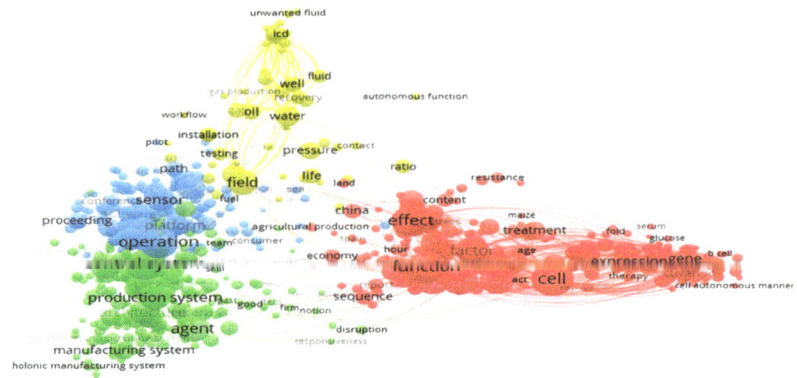

Fig. 2. Research fields in APC resulting from the first search.

Concerning the relevant papers, that resulted from reading all abstract, Figs. 3 and 4 makes a characterization concerning the journals that have published in APC and region (country) the authors of the papers are from, respectively.

Fig. 3. Distribution of publications by: Journal

Next, we present and discuss the findings associated with each one of the research questions (RQ).

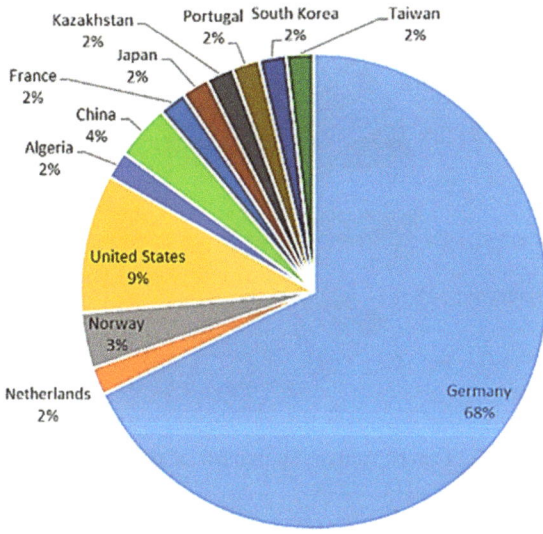

Fig. 4. Distribution of publications by: region

3.1   RQ1: Who Are the Specialists in Autonomous Production Control?

Figure 5 shows the top 10 authors in number of publications in APC from a universe of 77. By a quick analysis to the Fig. 5 an eye-catching author appears: Bernd Scholz-Reiter. This author appears in 23 of the 49 selected papers, i.e., in 47% of the papers. Bernd Scholz-Reiter, start writing about APC in 1999, about 20 years ago, with [11] referring that traditional MRP systems don't have the flexibility to react effectively to disturbances in the production environment. Bernd Scholz-Reiter was head of the Fraunhofer Application Center for Logistics System Planning and Information Systems, which he founded, and headed the Bremen Institute for Production and Logistics (BIBA).

It is important to refer that a total 77 co-authors are responsible for these 49 papers. This number indicates a relatively small worldwide scientific community working in APC. Given the novelty of the field and the interest Industry 4.0 has aroused in the scientific community, it is expected that in near future these community can increase significantly.

3.2   RQ2: How Is APC Research Evolving - Throughout Theoretical or Empirical Research Work?

In this case it's necessary to explain the differences between these research approaches. Theoretical research presents new frameworks, new methods, etc. in hypothetical contexts. The researcher does not use empirical data to build the theory but uses hypothetical examples. Neither does the researcher analyses concrete and specific works. Empirical research uses empirical evidence. Empirical research starts from specific concrete examples or observations to create a model and subsequently, a theory.

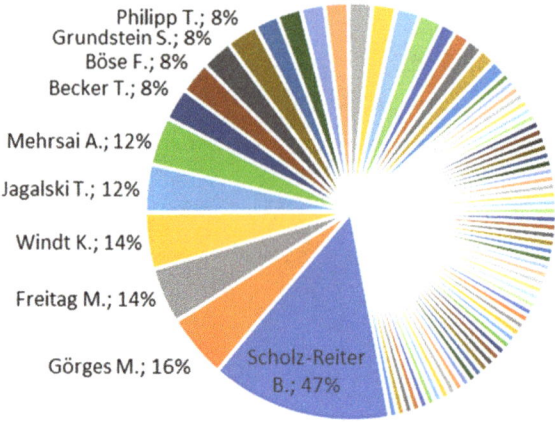

Philipp T.; 8%
Grundstein S.; 8%
Böse F.; 8%
Becker T.; 8%
Mehrsai A.; 12%
Jagalski T.; 12%
Windt K.; 14%
Freitag M.; 14%
Görges M.; 16%
Scholz-Reiter B.; 47%

Fig. 5. Top 10 Authors in number of publications.

This research question helps to understand which path APC research work is following. Most of the publications, 44, covers theoretical issues. Through Table 1 we can realize a focus of the authors in one kind of paper for the development of new methods for APC. Table 1 categorizes papers according to: (1) theoretical vs. empirical; and (2) the proposal (new method, framework, methods comparison, other). Theoretical represent 90% of the papers, new method represents 49%, framework 21%, comparison between methods 18% and other areas researched 12%. This will help future research to identify theoretical developments, and practical issues that may lead to new theory build.

Table 1. Classification of the published papers.

Research type	New method	Framework	Methods comparison	Others
Theoretical	43%	19%	16%	12%
Empirical	6%	2%	2%	

4   Conclusions

The main purpose of this paper was to carry out a literature review to better understand the current state of the research work in Autonomous Production Control (APC). In this context, two research questions emerged, namely: (1) who are the specialists in Autonomous Production Control? (2) how is this research evolving - throughout theoretical or empirical research work? The distribution of the number of publications in time, by editorial publication and by geographical region was also considered in the study. The main findings are summarized in the following. Research work in APC is still in the beginning, with most of the contributions being theoretical and concentrated in some geographical regions, such as Germany and USA (whereas in regions with developing countries few studies are found), with a huge contribute of one author -

Bernd Scholz-Reiter. Future research work should extend the research questions into more detail, to better understand the contribution of APC to the company's competitiveness.

References

1. Hu, S.J.: Evolving paradigms of manufacturing: From mass production to mass customization and personalization. Procedia CIRP **7**, 3–8 (2013)
2. Niehues, M., Blum, M., Teschemacher, U., Reinhart, G.: Adaptive job shop control based on permanent order sequencing: balancing between knowledge-based control and complete rescheduling. Prod. Eng., 1–7 (2017)
3. Scholz-Reiter, B., Dashkowskiy, S., Görges, M., Jagalski, T., Naujok, L.: Autonomous decision policies for networks of production systems. In: Decision Policies for Production Networks, pp. 235–263 (2012)
4. Roblek, V., Meško, M., Krapež, A.: A complex view of industry 4.0. SAGE Open **6**(2), 215824401665398 (2016)
5. Khan, A., Turowski, K.: A perspective on industry 4.0: from challenges to opportunities in production systems. In: Proceedings of the International Conference on Internet of Things and Big Data, no. IoTBD, pp. 441–448 (2016)
6. Taphorn, C.: Factors for a decentralized production and sequence planning from the perspective of products and resources. In: FAIM 2014 – Proceedings of the 24th International Conference on Flexible Automation and Intelligent Manufacturing. Advanced Manufacturing Enterprise Transforming, vol. 9, pp. 1057–1063 (2014)
7. Mulrow, C.D.: Systematic reviews: rationale for systematic reviews. BMJ **309**(6954), 597 (1994)
8. Durach, C.F., Kembro, J., Wieland, A.: A new paradigm for systematic literature reviews in supply chain management. J. Supply Chain Manag. **53**(4), 67–85 (2017)
9. Thomé, A.M.T., Scavarda, L.F., Scavarda, A.J.: Conducting systematic literature review in operations management. Prod. Plan. Control **27**(5), 408–420 (2016)
10. Wong, D.: VOSviewer. Tech. Serv. Q. **35**(2), 219–220 (2018)
11. Scholz-Reiter, B., Nathansen, K.: Non linear dynamics in production systems. ZWF Zeitschrift fuer Wirtschaftlichen Fabrikbetr. **94**(12), 746–753 (1999)

New Trends in Mechanical Systems Development

Experimental Shear Fracture Characterization of Adhesively-Bonded Joints

Alberto J. S. Leal[1] and Raul D. S. G. Campilho[1,2(✉)]

[1] Departamento de Engenharia Mecânica,
Instituto Superior de Engenharia do Porto, Instituto Politécnico do Porto,
Rua Dr. António Bernardino de Almeida, 431, 4200-072 Porto, Portugal
raulcampilho@gmail.com
[2] INEGI – Pólo FEUP, Rua Dr. Roberto Frias, 400, 4200-465 Porto, Portugal

Abstract. Adhesive bonding is widespread in several fields, ranging from construction, packaging, automotive industry, aeronautics and medical applications. As a result, it becomes increasingly necessary to reliably estimate their mechanical behaviour. In the context of Cohesive Zone Modelling (CZM), the shear fracture toughness (G_{IIC}) is one of the most influential parameters for the strength prediction. This work aims to make an evaluation of the End-Notched Flexure (ENF) and Four-Point End-Notched Flexure (4ENF) tests to determine G_{IIC} of bonded joints. Three adhesives were tested: Araldite® AV138, Araldite® 2015 and SikaForce® 7752. It was concluded that the G_{IIC} values for each adhesive obtained by the ENF and 4ENF tests are in good agreement.

Keywords: Adhesive joint · Fracture toughness · End-Notched Flexure
Four-Point End-Notched Flexure

1 Introduction

In modern structures, the use of adhesive bonding joints has increased when comparing with other traditional methods. As a result, and to increase the confidence in this joining method, it is important to accurately predict the mechanical strength of bonded structures. In the fracture mechanics approach, an energetic analysis is often applied, where the main parameter to predict damage evolution is the critical strain energy release rate (G_C). An adhesive joint working under real conditions is usually subjected to combined efforts, tensile and shear (mixed-mode). G_C can also be divided into tensile (G_{IC}) and shear (G_{IIC}) fracture toughness. In mode I, the Double-Cantilever Beam (DCB) and Tapered Double-Cantilever Beam (TDCB) tests are widely used [1]. In mode II, several authors have studied the G_{IIC} estimation, although much of the studies uses the ENF test [2]. Despite its simplicity, this test tends to promote unstable crack propagation, which leads to measuring errors during crack propagation. Nonetheless, there are other methods for mode II testing, such as the 4ENF test, whose loading configuration leads to stable crack propagation [3]. In mode II loading by the ENF test, there are several data reduction schemes to estimate G_{IIC}, which can be separated in methods that require crack length (a) measurement, equivalent crack length methods and J-integral techniques. Between the methods requiring measurement

© Springer International Publishing AG, part of Springer Nature 2019
J. Machado et al. (Eds.): HELIX 2018, LNEE 505, pp. 435–441, 2019.
https://doi.org/10.1007/978-3-319-91334-6_59

of *a* for the ENF test, the Compliance Calibration Method (CCM), Direct Beam Theory (DBT) and Corrected Beam Theory (CBT) are often used. The Compliance-Based Beam Method (CBBM) method is well established within the techniques that do not require this parameter. On the other hand, for the 4ENF test the most commonly used data reduction scheme is the CCM, which depends on the measurement of *a*, and the CBT and Effective Crack Method (ECM), whose formulation does not require this value. Contrary to the ENF test, for the 4ENF test there is still no literature for the application of the *J*-integral.

This work aims to make an evaluation of the ENF and 4ENF tests to determine G_{IIC} of bonded joints. Three adhesives were tested: the brittle Araldite® AV138, the moderately ductile Araldite® 2015 and the highly ductile SikaForce® 7752.

2   Experimental Details

2.1   Materials

The following adhesives were chosen (with increasing degree of ductility): epoxy Araldite® AV138, epoxy Araldite® 2015 and polyurethane Sikaforce® 7752. Regarding the adherends, a high tensile yield stress (σ_y) aluminium alloy was used (AW6082 T651). This alloy presents the following mechanical characteristics [4]: Young's modulus (*E*) of 70.07 ± 0.83 GPa, σ_y of 261.67 ± 7.65 MPa, tensile failure strength (σ_f) of 324.00 ± 0.16 MPa and tensile failure strain (ε_f) of 21.70 ± 4.24%. Several studies [4–6] were conducted to characterise the chosen adhesives. The tensile mechanical properties (*E*, σ_y, σ_f and ε_f) were estimated from bulk tensile tests. The shear mechanical properties were calculated from Thick-Adherend Shear Tests (TAST). DCB and ENF tests enabled to obtain, respectively, G_{IC} and G_{IIC} using conventional data reduction schemes. Table 1 shows the most relevant mechanical and fracture data.

2.2   Geometry and Testing

Figure 1 (a) depicts the geometry of the ENF tests: mid-span *L* = 100 mm, initial crack length $a_0 \approx 60$ mm, adherends' thickness *h* = 3 mm, width *B* = 25 mm and adhesive thickness $t_A = 0.2$ mm. Figure 1(b) shows the geometry of the 4ENF tests: *L* = 130 mm, loading span $L_i = 130$ mm, $a_0 \approx 60$ mm, *h* = 3 mm, *B* = 15 mm and $t_A = 0.2$ mm.

Seven specimens were fabricated for each adhesive of the ENF and 4ENF tests. All tests were carried at room temperature in a Shimadzu AG-X 100 electro-mechanical testing machine. The load (*P*) measurement was performed by a load cell of 100 kN. During the tests, and to ensure *a* measurement, several images were acquired using an 18 MPixel digital camera with no zoom and fixed focal distance of 100 mm. The resolution of the optical method was 11 μm. The crack tip position was inferred from the last point of failure of the white brittle paint applied to the specimens' side.

Table 1. Properties of the Araldite® AV138, Araldite® 2015 and Sikaforce® 7752 [4–6].

Property	AV138	2015	7752
Young's modulus, E [GPa]	4.89 ± 0.81	1.85 ± 0.21	0.49 ± 0.09
Poisson's ratio, v	0.35[a]	0.33[a]	0.30[a]
Tensile yield stress, σ_y [MPa]	36.49 ± 2.47	12.63 ± 0.61	3.24 ± 0.48
Tensile strength, σ_f [MPa]	39.45 ± 3.18	21.63 ± 1.61	11.48 ± 0.25
Tensile failure strain, ε_f [%]	1.21 ± 0.10	4.77 ± 0.15	19.18 ± 1.40
Shear modulus, G [GPa]	1.56 ± 0.01	0.56 ± 0.21	0.19 ± 0.01
Shear yield stress, τ_y [MPa]	25.1 ± 0.33	14.6 ± 1.3	5.16 ± 1.14
Shear strength, τ_f [MPa]	30.2 ± 0.40	17.9 ± 1.8	10.17 ± 0.64
Shear failure strain, γ_f [%]	7.8 ± 0.7	43.9 ± 3.4	54.82 ± 6.38
Toughness in tension, G_{IC} [N/mm]	0.20[b]	0.43 ± 0.02	2.36 ± 0.17
Toughness in shear, G_{IIC} [N/mm]	0.38[b]	4.70 ± 0.34	5.41 ± 0.47

[a]manufacturer's data
[b]estimated in reference [4]

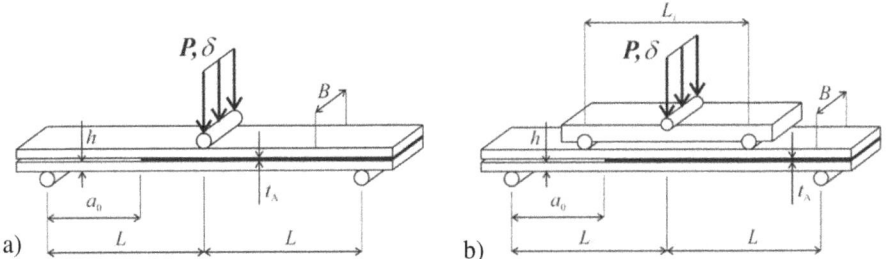

Fig. 1. Geometry and dimensions of the ENF (a) and 4ENF joints (b).

2.3 Data Reduction Methods

ENF Test. The CCM, DBT and CBT rely on the measurement of a and are based on the beam theory. The CCM applies the Irwin-Kies expression to the data and is described in reference [7]. The DBT does not include any a corrections to account for shear deformation (oppositely to the CBT), and is presented in reference [8]. The CBT was formulated by Wang and Williams [9]. The CBBM provides G_{IIC} only from C measured during the tests and is described in Azevedo et al. [2]. This data reduction methods uses an equivalent crack length (a_{eq}), which accounts for Fracture Process Zone (FPZ) that generates around the crack tip due to the materials' plasticity.

4ENF Test. The 4ENF is a modified version of the ENF, but it is not standardized. Mainly two data reduction methods are available for G_{IIC} estimation, namely the CCM, requiring the measurement of a, and the CBT, not requiring this parameter. The CCM is described in the work of Schuecker and Davidson [10]. According to Reeder et al. [11], this is the most widespread data reduction method applied to the 4ENF test.

The CBT expression including friction correction was derived from the work of Wang et al. [3]. Other authors [12] used the CBT without friction correction.

3   Results

3.1 G_{IIC} Estimation

The $P-\delta$ response showed that the Araldite® AV138 failed abruptly, whereas the other adhesives revealed smooth crack propagation. The CCM, apart from requiring measurement of a, involves estimation dC/da, obtained by differentiation of $C = f(a)$. Figure 2(a) gives an example, for the Araldite® AV138, of R-curves by the different methods.

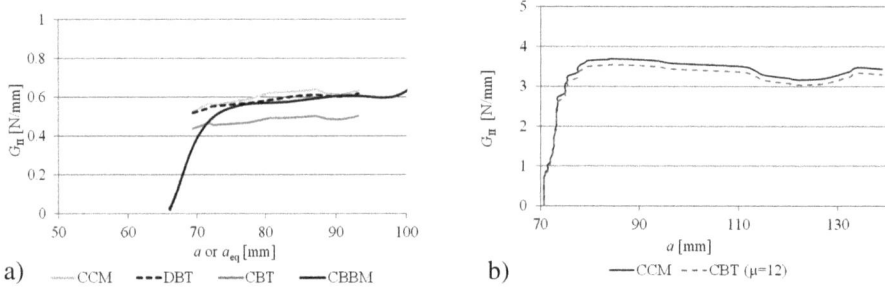

Fig. 2. Example of R-curves for an ENF specimen bonded with the Araldite® AV138 (a) and for a 4ENF specimen bonded with the Araldite® 2015 (b).

For the Sikaforce® 7752, only the CBBM was evaluated. G_{IIC} was considered as the average G_{II} during stable crack growth [13]. In general, the correspondence between the CCM, DBT and CBBM was good, while the CBT resulted in smaller G_{II}. This deviation was previously discussed [2]. Table 2 reports all G_{IIC} values (in N/mm) for the ENF test and the three adhesives. The tests with * were not considered in the analysis because of complications in the calculation of dC/da. The G_{IIC} results showed a good repeatability for each method/adhesive set. Between methods of the same adhesive, the correspondence was also acceptable, except for the CBT, which underestimates the CBBM by 17.7% for the Araldite® AV138 and 17.0% for the Araldite® 2015. The deviation for each adhesive/method is typically under 10% of the respective average G_{IIC}.

For the 4ENF specimens, the $P-\delta$ curves showed that crack growth for 4ENF tests takes place with a constant load. The CCM technique, equally to the ENF test, requires the calculation of dC/da from the $C = f(a)$ curve during crack propagation. Figure 2(b) shows the CCM and CBT R-curves for a single specimen bonded with the Araldite® 2015. A friction coefficient (μ) correction had to be applied to the CBT curves, because of the significant friction effects between cracked faces [10].

Table 2. Values of G_{IIC} [N/mm] for the three adhesives obtained by all methods (ENF test).

Adhesive method	Araldite® AV138				Araldite® 2015				Sikaforce® 7752
	CCM	DBT	CBT	CBBM	CCM	DBT	CBT	CBBM	CBBM
1	0.469	0.566	0.440	0.572	3.029	3.083	2.644	3.420	5.825
2	*	0.709	0.566	0.712	3.675	2.401	2.177	2.545	5.877
3	*	0.650	0.608	0.724	3.214	2.916	2.544	2.943	5.474
4	*	0.578	0.519	0.594	2.812	2.741	2.476	2.801	4.813
5	0.568	0.579	0.487	0.562	3.357	3.088	2.644	3.136	5.676
6	0.605	0.581	0.478	0.576	2.696	2.831	2.624	2.901	5.648
7	0.603	0.583	0.481	0.585	3.008	2.952	2.512	3.025	5.619
Average	0.561	0.606	0.511	0.618	3.113	2.859	2.517	2.967	5.562
deviation	0.064	0.053	0.058	0.069	0.334	0.238	0.164	0.273	0.356

*Polynomial fitting difficulties

Table 3 shows all G_{IIC} (in N/mm) and standard deviation for the 4ENF test. In all cases, the deviation was below 10% of the respective average G_{IIC}. The comparison between the CCM and CBT ($\mu = 0$) clearly shows that the CBT without correction overshoots the CCM by a significant difference. In average, G_{IIC} for the CBT ($\mu = 0$) is higher than the CCM by 189.0% (Araldite® AV138), 162.3% (Araldite® 2015) and 70.2% (Sikaforce® 7752). It was found by visual inspection during the 4ENF tests that significant friction occurs between the cracked faces of the adhesive layer. This friction results from the combination of the compression between the adherends and the high roughness of the cracked faces. The μ values specified in Table 3 were found by an iterative fitting process to make the CBT R-curves match the CCM curves. These values are not reasonable and make the CBT method not feasible to obtain G_{IIC}.

Table 3. Values of G_{IIC} [N/mm] for the three adhesives obtained by all methods (4ENF test).

Adhesive specimen	Araldite® AV138			Araldite® 2015			Sikaforce® 7752		
	CCM	CBT ($\mu = 0$)	CBT ($\mu = 13,5$)	CCM	CBT ($\mu = 0$)	CBT ($\mu = 12$)	CCM	CBT ($\mu = 0$)	CBT ($\mu = 9$)
1	0.700	2.020	0.691	3.364	8.117	3.230	-	-	-
2	0.669	2.139	0.731	3.212	6.810	2.710	5.276	9.629	5.472
3	0.699	2.261	0.773	2.781	8.302	3.303	5.425	9.122	5.184
4	0.701	2.213	0.756	3.053	7.873	3.132	-	-	-
5	0.674	1.926	0.658	3.674	8.940	3.557	5.293	9.632	5.474
6	0.741	1.881	0.643	-	-	-	5.555	9.366	5.322
7	0.760	1.846	0.631	3.215	8.529	3.393	5.318	7.983	4.536
Average	0.706	2.041	0.698	3.087	8.095	3.221	5.373	9.146	5.198
deviation	0.033	0.166	0.057	0.220	0.727	0.289	0.117	0.684	0.389

- discarded tests

4 Conclusions

This work aimed to perform an evaluation of the ENF and 4ENF tests for G_{IIC} measurement in adhesive joints. For both test configurations, data reduction methods were evaluated that to not need measuring a. The methods for the ENF test agreed, except for G_{IIC} under estimations by the CBT. The CCM data for the 4ENF test showed reliable results compared to the ENF values, although being dependent on estimating dC/da. The CBT, applied to the 4ENF test, is highly dependent on μ correction, and it only provides close values to the other methods with very high μ values. It is concluded that the ENF is efficient in estimating G_{IIC}, it has a simple setup and is not affected by friction effects, although it can be affected by unstable crack growth. On the other hand, crack growth for the 4ENF test is stable, but it is seldom addressed in the literature. Moreover, only one method (CCM) is capable of producing good results (requiring a measurement) while the CBT is affected by friction effects, and the test setup is more complicated. As a result, the ENF test is recommended over the 4ENF test for G_{IIC} calculation of adhesives.

References

1. Blackman, B.R.K., Kinloch, A.J., Paraschi, M., Teo, W.S.: Measuring the mode I adhesive fracture energy, GIC, of structural adhesive joints: the results of an international round-robin. Int. J. Adhes. Adhes. **23**(4), 293–305 (2003)
2. Azevedo, J.C.S., Campilho, R.D.S.G., da Silva, F.J.G., Faneco, T.M.S., Lopes, R.M.: Cohesive law estimation of adhesive joints in mode II condition. Theor. Appl. Fract. Mec. **80**, 143–154 (2015)
3. Wang, W.-X., Nakata, M., Takao, Y., Matsubara, T.: Experimental investigation on test methods for mode II interlaminar fracture testing of carbon fiber reinforced composites. Composites Part A: Appl. Sci. Manufact. **40**(9), 1447–1455 (2009)
4. Campilho, R.D.S.G., Banea, M.D., Pinto, A.M.G., da Silva, L.F.M., de Jesus, A.M.P.: Strength prediction of single- and double-lap joints by standard and extended finite element modelling. Int. J. Adhes. Adhes. **31**(5), 363–372 (2011)
5. Campilho, R.D.S.G., Banea, M.D., Neto, J.A.B.P., da Silva, L.F.M.: Modelling adhesive joints with cohesive zone models: effect of the cohesive law shape of the adhesive layer. Int. J. Adhes. Adhes. **44**, 48–56 (2013)
6. Faneco, T., Campilho, R., Silva, F., Lopes, R.: Strength and fracture characterization of a novel polyurethane adhesive for the automotive industry. J. Test. Eval. **45**(2), 398–407 (2017)
7. Compston, P., Jar, P., Burchill, P., Takahashi, K.: The effect of matrix toughness and loading rate on the mode-II interlaminar fracture toughness of glass-fibre/vinyl-ester composites. Compos. Sci. Technol. **61**, 321–333 (2001)
8. Elmarakbi, A.: Advanced composite materials for automotive applications: Structural integrity and crashworthiness. Wiley, Hoboken (2014)
9. Wang, Y., Williams, J.: Corrections for mode II fracture toughness specimens of composite materials. Compos. Sci. Technol. **43**, 251–256 (1992)
10. Schuecker, C., Davidson, B.D.: Evaluation of the accuracy of the four-point bend end-notched flexure test for mode II delamination toughness determination. Compos. Sciites. Technol. **60**(11), 2137–2146 (2000)

11. Reeder, J.R., Demarco, K., Whitley, K.S.: The use of doubler reinforcement in delamination toughness testing. Compos. Part A: Appl. Sci. Manufact. **35**(11), 1337–1344 (2004)
12. Sun, X., Davidson, B.D.: Numerical evaluation of the effects of friction and geometric nonlinearities on the energy release rate in three- and four-point bend end-notched flexure tests. Eng. Fract. Mech. **73**(10), 1343–1361 (2006)
13. Ameli, A., Papini, M., Schroeder, J.A., Spelt, J.K.: Fracture R-curve characterization of toughened epoxy adhesives. Eng. Fract. Mech. **77**(3), 521–534 (2010)

Modal Analysis and Crack Detection Through Harmonic Analysis with Numerical Simulations of an Thermoplastic Injection Mold

Cedrico R. Oliveira[✉], N. Peixinho, and J. Meireles

MEtRiCs - Mechanical Engineering and Resource Sustainability Center,
University of Minho, Guimarães, Portugal
cedric.resende.oliveira@gmail.com

Abstract. The use of injection molds is very common nowadays but the lack of tools or mechanisms to detect any structural changes during its life-time is a problem. Since this is a dynamic system the measurement of the system response could be a tool to create a non-intrusive mechanism to detect any structural problems. In this paper a modal analysis is performed of an injection mold to understand the order of its natural frequencies then an harmonic analysis to check the availability of using accelerometers to detect cracks. By the results obtained by this work the use of accelerometers need to be tested in a live trial to verify if they are susceptive to external factors in order to interfere with the measurements and make impossible to detect the cracks initiations.

Keywords: Thermoplastic injection mold · Modal analysis · Crack detection Harmonic analysis

1 Introduction

The injection mold is a precision tool and one of the most common process to fabricate thermoplastic parts [1]. The modal analysis studies for these components isn't very common, this is a system that works in low frequencies and has a high stiffness and masses so the resonance frequencies usually are very high. But this system can be excited since the injection mold works inside an injection machine and this machine can induce vibrations to the injection mold, to have an idea how an injection machine behaves in this field the study made by Lee et al. [2] where they conclude that most components could be excited with frequencies lower than 30 Hz.

Another area where this studies could be applied is for the crack detection, because the crack initiation should change the system response and it should be detectable with the proper sensors. The use of sensor to detect cracks with the changes of the system response is used for other systems, for instance the works presented by Gillich et al. [3] and Her and Lin [4] where with the use of modal analysis or ultrasonic frequencies they could detect changes in the system response to detect the crack or to measure it's depth.

To study this phenomenon's in this paper will be presented a modal analysis of an injection mold to check its modal shapes and natural frequencies, than with the same

© Springer International Publishing AG, part of Springer Nature 2019
J. Machado et al. (Eds.): HELIX 2018, LNEE 505, pp. 442–449, 2019.
https://doi.org/10.1007/978-3-319-91334-6_60

model under the same conditions a harmonic analysis will be presented to understand how the crack can influence the system response and take some conclusions of how it can be measured in a live trial.

2   Model to Be Studied

To perform the Modal analysis of the injection mold a simple model was used to reduce the computation time and since the goal is to check the influence of each component on each other. So to perform the study the injection tool was designed as presented in the Fig. 1.

Fig. 1. Model of the injection tool used for the study.

The physical properties and considered materials are described in the Table 1. For the bolts will be used structural steel to characterize its properties, these are connecting elements and they are not the object of the study.

Table 1. Mold components physical properties.

Components	Material	Volume [mm^3]	Mass [kg]
Injection machine plate	Cast iron	3.86E+06	27.763
Support plate core side	Structural steel	3.13E+05	2.4556
Support plate cavity side	Structural steel	3.13E+05	2.4588
Rail (data for 1 rail)	Structural steel	1.02E+05	0.80292
Cavity	Bohler-Uddeholm p20	4.45E+05	3.4757
Core	Bohler-Uddeholm p20	7.90E+05	6.1713

The simulations will be performed with the software ANSYS and for model interactions the following connections will be used: for the bolt thread and bolt head connection will be performed with the **bonded** contact; the bolt pretention won't be considered; all the other surfaces will be connected with the **frictional** contact using a general friction coefficient of 0,2.

3   Modal Analysis Results and Discussion

In the modal analysis made in the injection mold, the first 6 modal shapes and frequencies will be extracted for each part of the tool and with it is pretended to understand how each part interacts with each other and how it affects the modal shapes.

For each part were applied a rigid constraint in the regions presented on Fig. 2 which will block any movements for those surfaces.

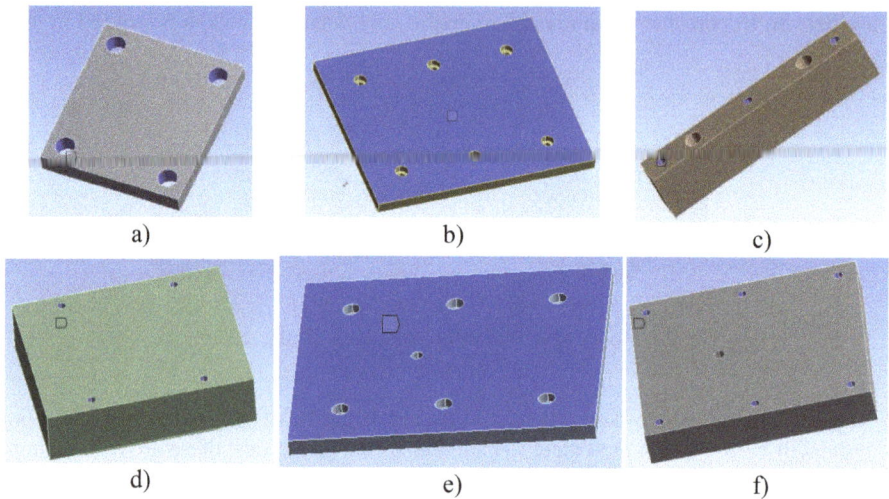

Fig. 2. Constraints for each part are represented in purple where (a) injection machine plate, (b) support plate core side, (c) rail, (d) core, (e) support plate cavity side and (f) cavity.

This part of the study is meant just to retrieve de modal shapes and frequencies to understand the influence of each part in the mold modal shapes. In the Table 2 will be presented the first ten resonance frequencies for each part.

Table 2. Resonance frequencies in Hz for each part for the first 10 eigenvalues.

Mode	Injection machine plate	Support plate core side	Rail	Core	Support plate cavity side	Cavity	Injection mold
1	1.96E+03	7.57E+04	7.78E+03	6.78E+03	7.48E+04	5.84E+03	**3.54E+3**
2	**2.79E+03**	7.58E+04	9.65E+03	6.86E+03	7.48E+04	6.79E+03	**3.58E+3**
3	**3.18E+03**	7.58E+04	1.39E+04	9.06E+03	7.49E+04	1.00E+04	**3.82E+3**
4	**3.32E+03**	7.58E+04	1.47E+04	1.02E+04	7.49E+04	1.07E+04	**3.89E+3**
5	**4.06E+03**	7.65E+04	1.58E+04	1.03E+04	7.56E+04	1.22E+04	**4.39E+3**
6	**4.34E+03**	7.67E+04	1.60E+04	1.21E+04	7.58E+04	1.27E+04	4.81E+3

As it will be presented in Table 3 the element that will have the biggest influence in the injection mold modal shapes is the injection machine plate, as expected due to its geometry, mass and constrains which has most of its surfaces unconstrained, not like the other elements that have most of its surfaces constrained by other elements. Also its flat geometry also increases its capability to reach very easily resonance modes.

Table 3. Mode shapes of the injection machine plate and injection mold.

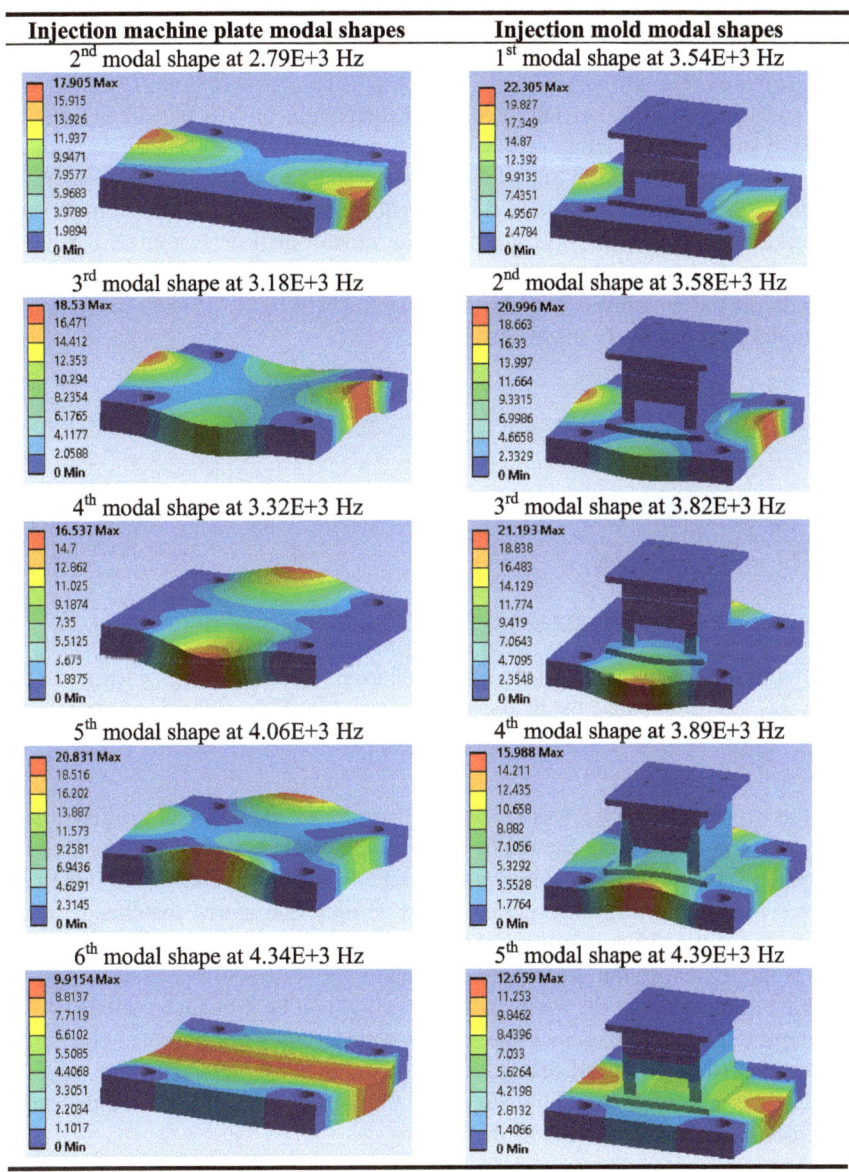

Taking under account the results presented before is easily understandable that the injection machine plate has the biggest part in the modal shapes of the mold, the natural frequencies doesn't match due to the other elements which in most of the natural frequencies makes them higher as expected, since in this case the system damping is exclusively made by the system mass. It can be also observed that the fixed side of the mold, i.e. the side of the cavity doesn't present any influence on the modal shapes of the injection mold, this is caused by the stiffness of this part of the mold [5, 6].

4   Crack Detection Through Harmonic Analysis Results and Discussion

The spectral analysis is already used for crack detection, on this part of the work is pretended to study the feasibility of this method in injection molds, due to its big mass and this being a system with a high stiffness it's pretend to understand how the crack will influence the spectral results and if it's possible to detect a small crack in the tool.

To make this analysis the tool used for the modal analysis was used but here a 3 MPa load with a duration of 1 Hz were placed at the region which should be in contact with the molten plastic (Fig. 3), the constrains will be the same as the ones used for the modal analysis study made above.

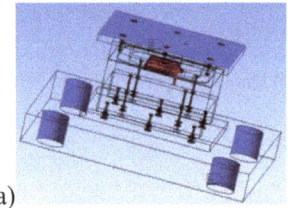
a)

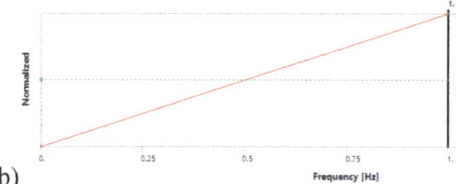
b)

Fig. 3. Constraints and loads applied in the injection mold (a) In red the region where the load of 3 MPa was placed, in purple the rigid constrains placed in the injection mold and injection machine plate; (b) load incrementation during one Hz.

For this concept a crack was modeled in the core of the mold and different simulations were made changing the depth of the crack but keeping its length. This is a simplification since the cracks change in both dimensions but to understand if they can be detected in a body with this stiffness it will be only considered the change in one dimension due to the difficulty to model this phenomenon. The crack with 0.11 mm in length was modeled with the shape presented in the Fig. 4, this simple crack shape was made taking under account the mesh quality which would be difficult to achieve with a more complex shape and the crack meshing is outside this work purpose. Having the parameters described above all defined in the model it will be extracted the values of the acceleration amplitude through the range of 0 to 10000 Hz and extracted the curves with a probe placed at the corner of the core.

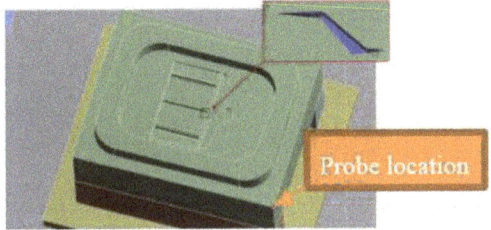

Fig. 4. Crack representation and location in the injection mold core.

By the results obtained from numerical simulations which are presented in the Figs. 5, 6 and 7 it's possible to check how the amplitude changes with the crack, where it is possible to understand how the amplitude changes when the crack appears, for all crack lengths the amplitude values are similar up to the 8000 Hz but they show different values when compared with the spectrum curve for the tool without any crack.

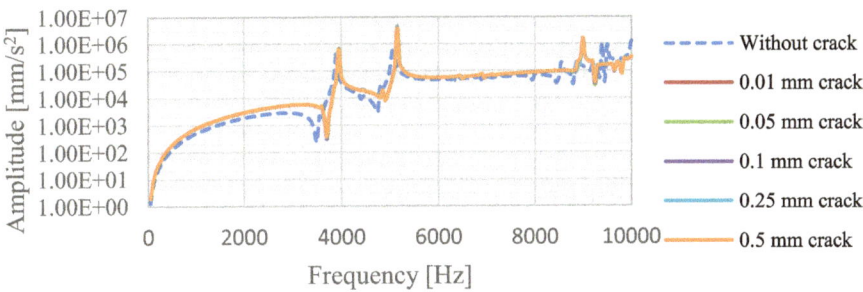

Fig. 5. Amplitude versus frequency at the probe location at the x direction.

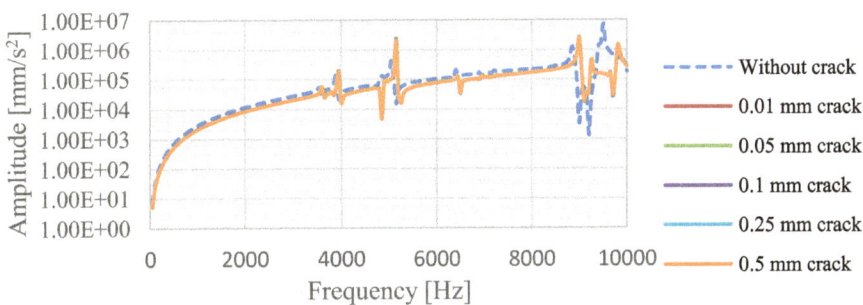

Fig. 6. Amplitude versus frequency at the probe location at the y direction.

Another problem to measure this results in a live trial is the low difference of results during the first 3000 Hz, because as it was observed before this is a system with a high stiffness and the mass concentration helps damping the vibrations caused by the

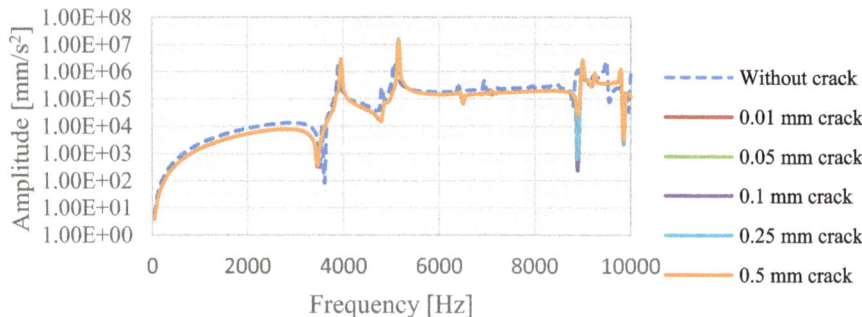

Fig. 7. Amplitude versus frequency at the probe location at the z direction.

external factors. Taking this under account the inputs used for this work can't be forgotten that this is made under an ideal environment, so the noise (all environmental factors which are not taken under account on the numerical simulation) isn't taken under account and with the low differences presented in the numerical results for lower frequencies might hinder the achievement of the pretended.

It can be observed that the axis where the differences are higher is the z axis, so in a live trial, it could be used only one accelerometer placed in order to measure de accelerations by this axis.

5   Conclusions and Future Researches

From this work is possible to conclude that this system due to its mass and shape presents high values for its natural frequencies, so there shouldn't be a problem of amplifying the amplitude of the vibrations when the mold is working in the injection machine since it should work with lower frequencies. It's also visible the lower influence of the elements from the core side into the cavity side, mostly due to the connection between those parts which allows the dissipation of energy.

For the study of the response of the system when a crack appears it was observed that the differences are very small up to 3000 Hz and this is the region which is more interesting to measure since this is a system that should work with low frequencies. And with this small differences the risk of disregard is high due to the noise cause by other elements which don't belong to the system under study.

For future researches and having in mind the use of non-intrusive systems to detect the start of cracks or their evolution, is proposed to conduct live trials with accelerometers to check if the interference of outside agents could corrupt the data. It's also proposed with the same objective the use of acoustic sensors to detect any structural changes in the injection mold system, if this sensor could detect mechanical deformations before the beginning of the mechanical failure it could be possible to create tools to alert before these failures happen.

References

1. Ferreira, I., Cabral, J.A., Saraiva, P., Oliveira, M.C.: A multidisciplinary framework to support the design of injection mold tools. Struct. Multidiscip. Optim. **49**(3), 501–521 (2014)
2. Lee, J., Park, W., Kim, J.: Structural dynamic analysis of the electric injection molding machine
3. Gillich, G.-R., Maia, N.M.M., Mituletu, I.-C., Tufoi, M., Iancu, V., Korka, Z.: A new approach for severity estimation of transversal cracks in multi-layered beams. Lat. Am. J. Solids Struct. **13**(8), 1526–1544 (2016)
4. Her, S.-C., Lin, S.-T.: Non-destructive evaluation of depth of surface cracks using ultrasonic frequency analysis. Sensors **14**(9), 17146–17158 (2014). (Basel)
5. "ANSYS Product Help." https://ansyshelp.ansys.com/. Accessed 15 Feb 2018
6. He, J., Fu, Z.-F.: Modal analysis. Butterworth-Heinemann (2001)

Evaluation of Mixed-Mode Fracture in Adhesively-Bonded Joints

Filipe A. A. Nunes[1] and Raul D. S. G. Campilho[1,2(✉)]

[1] Departamento de Engenharia Mecânica, Instituto Superior de Engenharia do Porto, Instituto Politécnico do Porto, Rua Dr. António Bernardino de Almeida, 431, 4200-072 Porto, Portugal
raulcampilho@gmail.com
[2] INEGI – Pólo FEUP, Rua Dr. Roberto Frias, 400, 4200-465 Porto, Portugal

Abstract. To enable the widespread use of adhesive joints in industry application, it is fundamental to be able to predict their strength. Advanced predictive techniques such as cohesive zone models (CZM) require the estimation of both mechanical and fracture properties of adhesives. In this work, the Asymmetric Tapered Double-Cantilever Beam (ATDCB) mixed-mode test is studied to estimate the fracture envelope of three adhesives. The *R*-curves were calculated and the fracture envelopes of each adhesive were built, which enabled calculating the most suitable propagation criterion in mixed-mode to apply in numerical simulations. With this work, the mixed-mode propagation criterion of three adhesives was proposed for strength prediction of adhesive joints.

Keywords: Adhesive joint · Experimental testing · Fracture toughness
Asymmetric Tapered Double-Cantilever Beam

1 Introduction

Due to the increasing use of adhesive joints in structural applications it is very important to know and be able to predict the strength and crack growth resistance of an adhesive layer, which is measured by the fracture toughness (G_C). For this purpose, several fracture characterization tests were developed for the various fracture modes. Mixed-mode loading consists of the combination of at least two of the pure-mode modes. In mode I, the joint is subjected to loads that force its opening and respective crack propagation perpendicularly to the plane of loading. The most used and well accepted test in this loading mode is the Double-Cantilever Beam (DCB). Another possibility is the Tapered Double-Cantilever Beam (TDCB) test. In mode II, the joint is subjected to a shear loading. Several tests have been developed, including the End-Notched Flexure (ENF) and End-Loaded Split (ELS) tests, which are the most popular [1]. The mixed-mode loading consists of a combination of mode I and mode II efforts. There are several tests developed to test adhesive joints in mixed-mode. The Mixed-Mode Bending (MMB) test is a combination between the DCB and ENF tests [2]. Its major advantage over the rest of the available mixed-mode tests is to allow the mixed-mode ratio to be varied almost without limitation. The Asymmetric Double-Cantilever Beam (ADCB) is an asymmetric variation of the DCB test, whose

© Springer International Publishing AG, part of Springer Nature 2019
J. Machado et al. (Eds.): HELIX 2018, LNEE 505, pp. 450–455, 2019.
https://doi.org/10.1007/978-3-319-91334-6_61

asymmetry that causes the II mode loading in addition to mode I is induced by using adherends with different thickness. The Single-Leg Bending (SLB) test is a more limited test than the MMB but, on the other hand, does not require specialized equipment and the experimental procedure is easier [3, 4]. The ATDCB test is a relatively recent test [5], and it consists of a variation of the TDCB test in which only one of the adherends has the same shape as the TDCB adherends, while the other is a DCB adherend. In this test, the mixed-mode ratio slightly varies during crack growth [5], with the average of the mixed-mode ratio being 24° [5]. It has as its main drawbacks the more complex geometry of the TDCB adherend and its limited range of mixed-mode ratio variation. Park and Dillard [6] conducted ATDCB tests of bonded joints using 6061-T6 aluminium adherends and a Betamate® Lesa adhesive. A G_C comparison was made using linear beam theories and the Finite Element Method (FEM). The results showed that FEM predicted a slightly higher G_C than the corrected beam theory because of the high adherend ratios of thickness to length.

In this work, the ATDCB mixed-mode test is studied to estimate the fracture envelope of three adhesives. The experimental R-curves were calculated and the fracture envelopes of each adhesive were built, which enabled calculating the most suitable propagation criterion in mixed-mode to be applied in numerical simulations.

2   Experimental Work

2.1   Adherends and Adhesives

This work evaluated three adhesives with different strength and ductility: the brittle epoxy Araldite® AV138, the ductile epoxy Araldite® 2015 and the structural polyurethane SikaForce® 7752, which combines high ductility with moderate strength. The tensile and shear mechanical properties of these adhesives are presented in Table 1.

Table 1. Properties of the Araldite® AV138, Araldite® 2015 and Sikaforce® 7752 [7–9].

Property	AV138	2015	7752
Young's modulus, E [GPa]	4.89 ± 0.81	1.85 ± 0.21	0.49 ± 0.09
Poisson's ratio, v	0.35 [a]	0.33 [a]	0.30 [a]
Tensile yield stress, σ_y [MPa]	36.49 ± 2.47	12.63 ± 0.61	3.24 ± 0.48
Tensile strength, σ_f [MPa]	39.45 ± 3.18	21.63 ± 1.61	11.48 ± 0.25
Tensile failure strain, ε_f [%]	1.21 ± 0.10	4.77 ± 0.15	19.18 ± 1.40
Shear modulus, G [GPa]	1.56 ± 0.01	0.56 ± 0.21	0.19 ± 0.01
Shear yield stress, τ_y [MPa]	25.1 ± 0.33	14.6 ± 1.3	5.16 ± 1.14
Shear strength, τ_f [MPa]	30.2 ± 0.40	17.9 ± 1.8	10.17 ± 0.64
Shear failure strain, γ_f [%]	7.8 ± 0.7	43.9 ± 3.4	54.82 ± 6.38
Toughness in tension, G_{IC} [N/mm]	0.20 [b]	0.43 ± 0.02	2.36 ± 0.17
Toughness in shear, G_{IIC} [N/mm]	0.38 [b]	4.70 ± 0.34	5.41 ± 0.47

[a] manufacturer's data
[b] estimated in reference [8]

The adhesives were previously characterized regarding the mechanical and toughness properties [10–12]. To obtain the Young's modulus (E), tensile yield stress (σ_y), tensile strength (σ_f) and tensile failure strain (ε_f) bulk specimens with dogbone shape for tensile testing were tested. Shear characterization of the adhesives was performed by Thick Adherend Shear Tests (TAST) in previous works [10, 11]. The DCB test was selected to obtain the tensile toughness (G_{IC}) and the ENF test was used for the shear toughness (G_{IIC}). The adherends were fabricated in C45E steel. Its mechanical properties were previously characterized [13] according to the ASTM-E8 M-04 standard. However, in this work, only E is relevant due to the lack of plasticity.

2.2 Specimens' Geometry and Testing

The geometry of an ATDCB specimen results from the combination of a DCB and a TDCB adherend, as shown in Fig. 1.

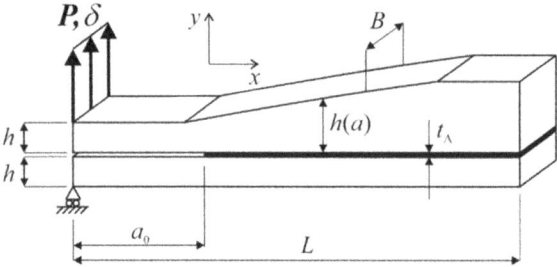

Fig. 1. Geometry and dimensions of the ATDCB specimens.

Both adherents have a length of $L = 241.3$ mm and a width of $B = 25.4$ mm. While the DCB adherend has a constant thickness of $h = 12.7$ mm, the TDCB adherend has the same thickness at one end, but h then varies as a function of the crack length $h = h(a)$. However, this variation should follow the geometric factor m, which is given by the ISO 25217 standard. Both adherends are bonded together by a thin layer of the adhesive with $t_A = 1$ mm. The initial crack length a_0 was 50 mm. All the tests, seven for each adhesive, were performed in a Shimadzu AG-X 100 testing machine with a load cell of 100 kN. The equipment recorded the load (P) and displacement (δ) values every 0.2 s, providing the P-δ curve. Crack propagation was recorded every 5 s using a Canon® EOS 700 D digital camera with a resolution of 20.2 MPixel.

2.3 Fracture Toughness Estimation

The available data reduction method to estimate G_C by the ATDCB test is based on Linear Elastic Fracture Mechanics (LEFM). G_C is obtained recurring to the Euler Bernoulli beam theory considering an equivalent system [5], which separated the ATDCB specimen into DCB and TDCB components. G_C for the ATDCB specimen is given by half of the components of these two sub-systems. This is reasonable if the adhesive layer remains perpendicular to the loading direction during the test [6]. Full details of this model are given in the work of Da Silva et al. [5].

3   Results

3.1   Mixed-Mode Fracture Energy

The P-δ curve analysis enabled plotting the R-curves. Following, the fracture envelopes allow evaluating the most suitable mixed-mode criterion for the mixed-mode simulation of the adhesive joints. Mixed-mode criteria enable combining the tensile and shear pure modes and, for each adhesive, a mixed-mode criterion of the type

$$\left(\frac{G_{\mathrm{I}}}{G_{\mathrm{IC}}}\right)^{\alpha} + \left(\frac{G_{\mathrm{II}}}{G_{\mathrm{IIC}}}\right)^{\beta} = 1 \tag{1}$$

is chosen for each adhesive. In this work, $\alpha = \beta$ will be considered.

Figure 2 shows one representative example of R-curves for each tested adhesive. All R-curves recognizably present crack growth at constant G_{I} or G_{II}. For all adhesives, the values of $G = G_{\mathrm{I}} + G_{\mathrm{II}}$, G_{I} and G_{II} obtained in the tests were very consistent between specimens. For the set of specimens bonded with the Araldite® AV138, the percentile standard deviation of the measured G, G_{I} and G_{II} was under 3% with respect to the respective average values. A slightly higher scatter was found for the joints bonded with the Araldite® 2015 (approximately 9%). The test results for the joints bonded with the SikaForce® 7752 adhesive were somewhat dispersed.

3.2   Mixed-Mode Fracture Energy

Figure 3 presents the fracture envelopes and the positioning of the ATDCB points. The fracture envelopes were built using pure-mode data applied at the horizontal and vertical axes, and four exponents α in the criterion of expression (1): 1/2, 1, 3/2 and 2.

A relatively low dispersion was found for the $G_{\mathrm{I}}/G_{\mathrm{II}}$ data points of the Araldite® AV138, which allows to position all the specimens in a small area of the fracture envelope, quite close to the $\alpha = 1/2$ criterion and, thus, to easily identify the 1/2 criterion as the most suitable mixed-mode criterion. The results for the Araldite® 2015 equally show a small dispersion of data points, which allows to easily locate the optimal propagation criterion for this adhesive ($\alpha = 1/2$). The fracture envelope for the SikaForce® 7752 reveals a small dispersion of data points and a behavior between the $\alpha = 3/2$ and $\alpha = 2$ criteria. Nonetheless, the $\alpha = 2$ criterion generally suits better and is thus selected for this adhesive.

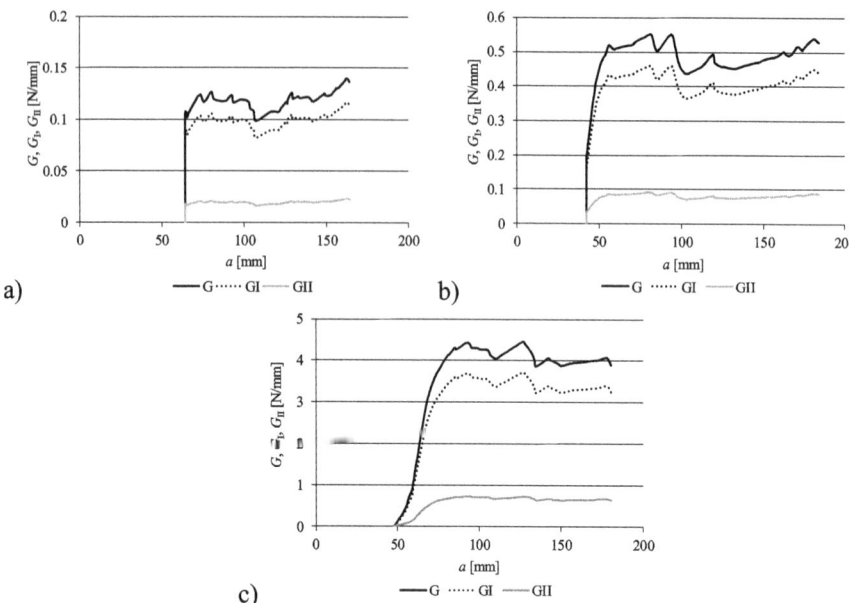

Fig. 2. Example of *R*-curves (*G*, *G*_I and *G*_{II}) for an ATDCB specimen with the adhesive Araldite® AV138 (a), Araldite® 2015 (b) and Sikaforce® 7752 (c).

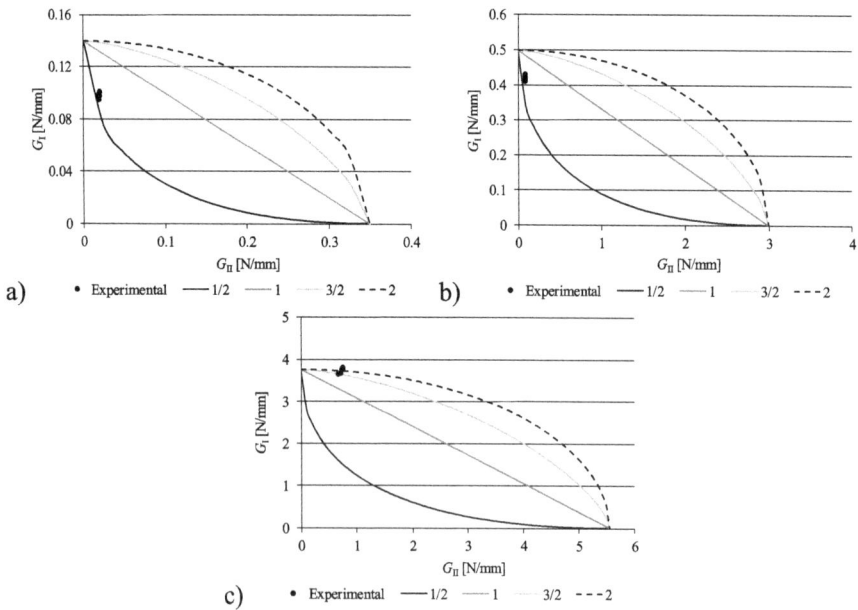

Fig. 3. Experimental G_I/G_{II} data plotted against the ideal fracture envelopes for the adhesive Araldite® AV138 (a), Araldite® 2015 (b) and Sikaforce® 7752 (c).

4 Conclusions

The main objective of this work was the estimation and numerical validation of the fracture envelope of three adhesives, using the ATDCB test. In the experiments, the tests revealed good repeatability. The obtained R-curves showed identifiable steady-state G_I and G_{II} values, used to construct the fracture envelopes. The low dispersion of G_I and G_{II} values allowed to position the points in the fracture envelope and to identify the mixed-mode propagation criterion suitable for each adhesive. The Araldite® AV138 and Araldite® 2015 were best fitted with the $\alpha = 1/2$ criterion, whilst the Sikaforce® 7752 could be best represent with the $\alpha = 2$ criterion. This information can now be used for mixed-mode strength prediction of bonded joints with these adhesives in techniques such as CZM that account for the mixed-mode behavior of materials.

References

1. da Silva, L.F., Öchsner, A., Adams, R.D.: Handbook of adhesion technology. Springer, Heidelberg (2011)
2. da Silva, L.F.M., Dilard, D.A., Blackman, B., Adams, R.D.: Testing Adhesive Joints. Wiley, Weinheim (2012)
3. Chaves, F.J., Da Silva, L., De Moura, M., Dillard, D., Esteves, V.: Fracture mechanics tests in adhesively bonded joints: a literature review. J. Adhes. **90**(12), 955–992 (2014)
4. Fernández, M.V., de Moura, M.F.S.F., da Silva, L.F.M., Marques, A.T.: Mixed-mode I + II fatigue/fracture characterization of composite bonded joints using the Single-Leg Bending test. Compos. A Appl. Sci. Manuf. **44**, 63–69 (2013)
5. Da Silva, L., Esteves, V., Chaves, F.: Fracture toughness of a structural adhesive under mixed mode loadings. Materialwiss. Werkstofftech. **42**(5), 460–470 (2011)
6. Park, S., Dillard, D.A.: Development of a simple mixed-mode fracture test and the resulting fracture energy envelope for an adhesive bond. Int. J. Fract. **148**(3), 261–271 (2007)
7. Campilho, R.D.S.G., Banea, M.D., Neto, J.A.B.P., da Silva, L.F.M.: Modelling adhesive joints with cohesive zone models: effect of the cohesive law shape of the adhesive layer. Int. J. Adhes. Adhes. **44**, 48–56 (2013)
8. Campilho, R.D.S.G., Banea, M.D., Pinto, A.M.G., da Silva, L.F.M., de Jesus, A.M.P.: Strength prediction of single- and double-lap joints by standard and extended finite element modelling. Int. J. Adhes. Adhes. **31**(5), 363–372 (2011)
9. Faneco, T., Campilho, R., Silva, F., Lopes, R.: Strength and fracture characterization of a novel polyurethane adhesive for the automotive industry. J. Test. Eval. **45**(2), 398–407 (2017)
10. Campilho, R.D., Banea, M.D., Neto, J., da Silva, L.F.: Modelling adhesive joints with cohesive zone models: effect of the cohesive law shape of the adhesive layer. Int. J. Adhes. Adhes. **44**, 48–56 (2013)
11. Faneco, T., Campilho, R., Silva, F., Lopes, R.: Strength and fracture characterization of a novel polyurethane adhesive for the automotive industry. J. Test. Eval. **45**(2), 398–407 (2016)
12. Campilho, R., Moura, D., Gonçalves, D.J., da Silva, J., Banea, M.D., da Silva, L.F.: Fracture toughness determination of adhesive and co-cured joints in natural fibre composites. Compos. B Eng. **50**, 120–126 (2013)
13. Campilho, R.D.S.G., Pinto, A.M.G., Banea, M.D., da Silva, L.F.M.: Optimization study of hybrid spot-welded/bonded single-lap joints. Int. J. Adhes. Adhes. **37**, 86–95 (2012)

Numerical Analysis of the Stress Distribution in Fillet Welds and Double Fillet Welds

Francisco Neves[1(✉)], Tiago Soares[2], Marques Pinho[1], and José Meireles[1]

[1] Department of Mechanical Engineering,
University of Minho, Guimarães, Portugal
`a68601@alunos.uminho.pt`,
`{acmpinho,meireles}@dem.uminho.pt`
[2] Project Department, Bysteel, Braga, Portugal
`Tiago.soares@bysteel.pt`

Abstract. Welding processes are extremely used in the metalworking industry all over the world. Sometimes the engineer has difficulties in finding solutions for the analytical calculation welded joints with eccentricity. In addition, calculations and analytical principles generally do not coincide with numerical results. In this work, analytically calculated stresses, are compared with numerical stresses for a double-fillet weld and the reasons for not having a complete agreement of results are clarified. In addition, an analytical method for fillet welds is compared, with numerical results.

Keywords: Eccentric joints · Fillet weld · Eurocode · Numerical simulation

1 Introduction

Welded joints have smaller dimensions than the parts they are connecting and are usually located in transition zones, i.e. the weld itself is a stress concentration.

When someone tries to understand what happens in the distribution of the stresses, it is observed that these are not constant. For example, in [1] an analytical analysis is made, and it has been found that the stresses that cause the weld failure do not occur in the plane of the throat, but closer to the base plate. Through finite element analysis [2], the results of the stress distribution complexity were obtained, showing the stress distribution on the faces of the geometry of the weld that were not constant.

In the cases verified in academic literature, there was no study directly related to the eccentricity provoked by the situation under analysis in this work. In [3] no results were obtained that could be justified to use in the project, because the produced results are not significantly close to those observed in simpler methods. On the other hand, in Eurocode 3 part 1: 8, [4], two calculation methods for weld joints are presented. In Sect. 2 the directional method will be explained.

© Springer International Publishing AG, part of Springer Nature 2019
J. Machado et al. (Eds.): HELIX 2018, LNEE 505, pp. 456–463, 2019.
https://doi.org/10.1007/978-3-319-91334-6_62

2   Analytical Methods for Calculation of Fillet Welds

2.1   Non-eccentric Fillet Welds

Initially, the directional stresses on the resistance plane of the weld, shown in Fig. 1, are calculated.

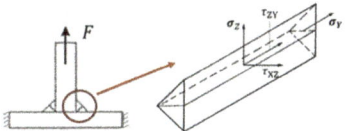

Fig. 1. Theoretical model of the weld and the directional stresses (adapted from [4]).

The directional stresses are constant for the entire resistance plane, and the σ_Y stress should not be considered. The engineer must put the load, F, on the resistance plane and calculate the directional stresses and, later, the von Mises stresses, Eq. 1.

$$\sigma_{von\ Mises} = \sqrt{\sigma_Z{}^2 + 3(\tau_{XZ}{}^2 + \tau_{ZY}{}^2)} \tag{1}$$

For the case shown in Fig. 1, there is a normal stress, σ_Z, and a shear stress, τ_{XZ}. Both are equal and calculated by Eq. 2,

$$\sigma_Z = \tau_{XZ} = \frac{F/2\cos 45°}{a_{eff} L_{eff}} \tag{2}$$

where, a_{eff} and L_{eff} are, respectively, the throat and length of the weld.

Then, the sizing criteria for normal and von Mises stresses are verified, with Eq. 3,

$$\sigma_z \leq \frac{0,9 f_u}{\gamma_{M2}}, \quad \sigma_{von\ Mises} \leq \frac{f_u}{\beta_w \gamma_{M2}} \tag{3}$$

where γ_{M2}, β_w and f_u, are, respectively, safety coefficients, material coefficients and the tensile stress of the weakest element of the bond [4].

2.2   Eccentric Fillet Welds

When there is only one fillet weld, the applied force is eccentric from the weld axis. This creates a moment at the root that must be considered, Fig. 2.

In addition to what was mentioned in the previous paragraph, the Eurocode, [4], does not mention anything else, leaving to the engineer the freedom to consider the best approach. In this way, a possible method is the application of the moment, M_e, on the plane of the weld and calculate the stresses with the Eq. 4.

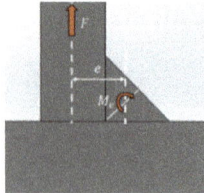

Fig. 2. Creation of the moment due to the eccentricity, "M_e".

$$\sigma_{M_e} = \frac{M_e y}{I_c}, where\ y = \frac{a_{eff}}{2} \tag{4}$$

In this case, the strength of the weld is given by the moment of inertia of the resistance plane on its central axis, Eq. 5.

$$I_c = \frac{a_{eff}{}^3 L_{eff}}{12} \tag{5}$$

The stresses resulting from Eq. 4 are normal stresses and positive at the root of the weld and negative at the outside face, which are added to those previously calculated. The same sizing criteria referred in Eq. 3 are used. This result is 2.5 times higher than the previous result [5].

3   Numerical vs Analytical Stresses

3.1   Non-eccentric Fillet Welds

The finite element analysis of the double fillet weld was made in SolidWorks, with tetrahedral elements. This type of welding does not introduce any eccentricity into the plates being jointed at any of the welds, so it is a case verified by the Eurocode. The analyzed model is shown in Fig. 3, with the proper dimensions, having 80 mm of length.

This model is made of four bodies, two plates and two weld beads. The material used was a S355 steel with a 210 GPa Young's Modulus, 0,3 Poisson's ratio and a

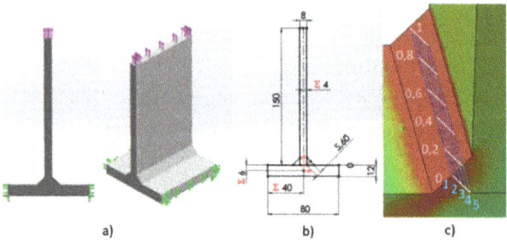

Fig. 3. (a) SolidWorks model, (b) dimensions of the model and (c) stresses lines studied.

355 MPa yield strength [6, 7], modeled as an elastic isotropic material. The welds are rigidly connected to the plates by bonded contacts. The plates are not connected to each other and have zero clearance. It was verified that the stress concentration will increase if some clearance between plates were considered [5]. Thus, the introduction of this variable was not considered relevant. The base plate is fixed at its ends.

For a load of 72 kN, the analytical results have the values represented in Table 1.

The values for σ_{Z^*} and $\sigma_{\text{von Mises}^*}$ (see first row of Table 1), are obtained by placing Eq. 3 in order to the tensile stress, f_u, naming them "penalized normal" and "von

Table 1. Analytical results for a load of 72 kN.

σ_Z	σ_{Z^*}	σ_Y	τ_{YZ}	τ_{XZ}	$\sigma_{\text{von Mises}}$	$\sigma_{\text{von Mises}^*}$
57	79	0	0	57	114	129

Mises" stresses. These stresses are constant along the resistant plane.

In numerical terms, the distribution of von Mises stresses along the weld is given by the plot of Fig. 4, also representing the analytical values of the Table 1. The stresses were read on five equidistant lines along the length of the weld, as well as six other lines at 90° from those, representing the welding throat (see Fig. 3). These lines are on the theoretical resistance plane.

It is possible to observe that the stresses are not constant along the whole length of the weld. The stress values diminish from the outer face (line 1) to the root (line 5) of

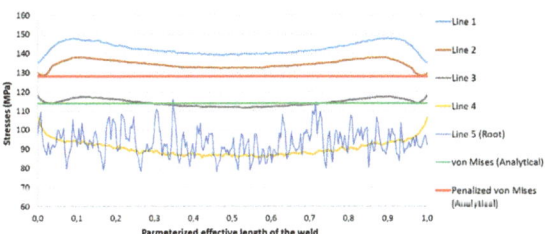

Fig. 4. Von Mises stresses along the length of the weld.

the weld because the outer face it is pulled more severely by the flexing of the base plate. In the root of the weld, there is a more disordered variation of the stresses, due to the amount of differences of boundary conditions in this model. This variation was verified in all simulations. Consequently, there are approximately linear variations along the weld throat, except for the weld root. By observing von Mises analytical stress, a result almost equal to the numerical value is obtained for the central axis of the weld (line 3). This may happen because the weld rotates on this axis. Although there is no moment due to some eccentricity, there is flexion due to the flexibility of the base plate. In turn, the penalized stresses are higher than line 3, but lower than line 4.

In Fig. 5, the stresses for the weld throat at each line along the length of the weld (see Fig. 3) are shown.

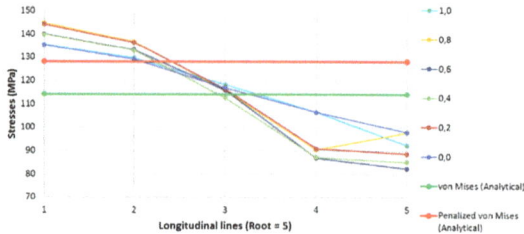

Fig. 5. Von Mises stresses at the throat along weld length.

From Fig. 5 it is noted that the stresses are not constant along the weld throat. These are only close to the non-penalized analytical value ($\sigma_{von\ Mises}$) on the line 3 values, as well as the penalized stress ($\sigma_{von\ Mises^*}$) is between line 2 and 3 values (see "Longitudinal lines"-axis). This can happen because the weld rotates on an axis comprised between these two locations. The values of the longitudinal line 5 show the already mentioned random variation.

Considering that the analytical method discriminates the stresses in its directions (normal Y and Z, and shear XZ), numerical directional stresses were also analyzed, and Fig. 6 shows the results for longitudinal line 3.

According to Fig. 6, none of the stresses is constant along the entire weld, especially at its ends. The results that are not predictable analytically are the stresses in the

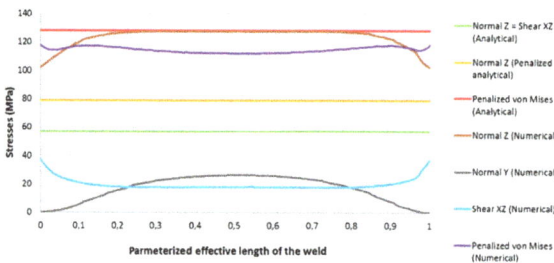

Fig. 6. Directional and von Mises numerical stresses along the weld bead (Line 3).

longitudinal direction of the weld. These stresses have a quadratic variation with the maximum value at half-length of the weld. Likewise, the longitudinal shear stresses are not predicted in the analytical method, and these have a negligent value relative to the other stresses. The normal Z stresses are the ones with the highest value and most contribute to the von Mises stresses, since the normal Y stresses reduce the values of the von Mises stresses. On the other hand, at the length ends, the XZ shear stress causes an increase of the von Mises stress. Since both analytical directional stresses are equal, they appear to be overlapping, and the normal Z analytic stress is twice as small as the numerical one, whereas the analytical shear stress XZ section is three times lower than the numerical one. Another interesting aspect is that the von Mises penalized analytical stresses are equal to the normal Z numerical stresses.

When the principal stresses are observed, the normal stresses Z and Y are practically coincident with the principal stresses, so it means that the plane of resistance of the weld is almost coincident with the plane of the principal stresses, making the shear stresses less relevant than the normal ones. On the other hand, the justification for the existence of normal stresses Y, and for the stresses variations along the weld, is only found if the Poisson ratio is null, as it can be seen in Fig. 7.

As observed in Fig. 7, the values are practically constant along the length of the weld, and, therefore, along the throat, apart from the weld root, for the same reason

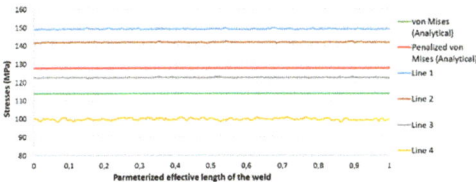

Fig. 7. Von Mises stresses variations along the weld with a null Poisson ratio.

previously mentioned, so it was not represented. In comparative numerical/analytical terms, the stresses of line 3 are closer to the von Mises stresses than to the non-penalized stresses, contrary to what has been observed previously. In directional terms, normal stresses Y, Z and shear XZ have become constant with values of 124 MPa and 19.5 MPa, in line 3.

3.2   Eccentric Welds

The numerical model studied for this part of the work is the same as the one in Fig. 3, except the double fillet weld, is used instead of the single fillet weld. The stress results obtained are showed in the Fig. 8, where the stress plots are limited to 355 MPa. In Fig. 9 the same plot is placed as iso clipping, allowing to see the zones where the stresses are equal to and greater than 355 MPa.

The numerically obtained stresses have higher values than 925 MPa, but only in some nodes of the transition elements between the web/weld/base. Throughout the remaining weld, the stresses are not higher than 500 MPa.

Another aspect that was firstly observed is that a joint in these loading conditions

Fig. 8. Results of von Mises stresses.

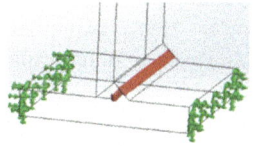

Fig. 9. Iso clipping plot of the von Mises stresses.

will hardly be totally free at the place of application of the load. In this way, simulations were carried out where the web plate was laterally restricted, obtaining the result in Figs. 10 and 11.

It is observed that the stress values decrease with this restriction.

Figure 12 shows the comparative plot between the analytical and numerical stresses

Fig. 10. Von   Mises stress results

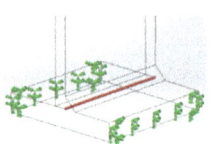

Fig. 11. Iso clipping plot of the von Mises stresses.

along the throat at the half-length of the weld, with the web plate free or restricted.

With this plot, it is notable that the analytical method is very conservative, being only close to the numerical results in the middle of the weld for the free web, as observed in Sect. 3.1.

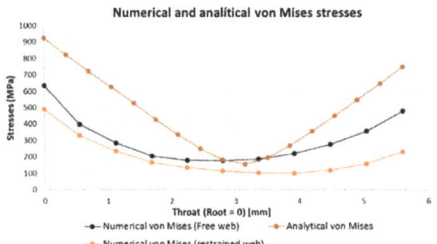

Fig. 12. Plot of numerical and analytical stresses for two boundary conditions.

4   Conclusions

It has been found that the numerical von Mises stresses in a weld bead are not constant in the theoretical resistance plane. Additionally, this is practically coincident with the plane of principal stresses. In turn, the directional stresses are not close to the analytical ones, with a higher proportion of normal stresses than shear stresses.

Also, it has been observed numerically, that analytically neglected longitudinal stresses to the axis of the weld, exist, which diminish von Mises equivalent stresses. Both these and the variations of the von Mises stresses along the weld are due to the "effect" of the Poisson ratio.

With the study of the fillet weld, it was detected that the analytical method that comes from an interpretation of the Eurocode is very penalizing when compared with the numerical results. On the other hand, in the numerical analysis, it was seen that the web restriction causes an increase in the weld strength.

References

1. Kato, B., Morita, K.: Strength of transverse fillet welded joints. Weld. Res. Suppl. **53**, 59–65 (1974)
2. Machado, I.G.: Dimensionamento de Juntas Soldadas de Filete: Uma Revisão Crítica. Soldagem e Inspeção **16**(2), 189–201 (2011)
3. Hicks, J.: Welded Design – Theory and Practice, 1st edn. Woodhead Publishing Limited, Cambridge (2000)
4. Eurocódigo 3 – Projeto de estruturas de aço. Parte 1-8: Projeto de ligações, NP EN 1993-1-8 (2010)
5. Neves, F.: Análise paramétrica de tipologias de soldadura – Ensaios e dimensionamento numérico, Dissertação de Mestrado Integrado em Engenharia Mecânica, Universidade do Minho, Guimarães (2017)
6. BS EN 10025-2:2004 - Hot rolled products of structural steels. Part 2: Technical delivery conditions for non-alloy structural steels
7. Eurocódigo 3 – Projeto de estruturas de aço. Parte 1-1: Regras gerais e regras para edifícios, NP EN 1993-1-1 (2010)

Inchworm Locomotion of an External Pipe Inspection and Monitoring Robot

Constantin Niţu, Bogdan Grămescu$^{(\boxtimes)}$, Ahmed Sachit Hashim,
and Mihai Avram

Department of Mechatronics and Precision Mechanics,
University "POLITEHNICA" of Bucharest, Bucharest, Romania
{constantin.nitu,bogdan.gramescu}@upb.ro,
Ahmedhashim774@yahoo.com, mavram02@yahoo.com

Abstract. The paper presents the external pipe locomotion of a robot for inspection and monitoring of the pipelines for use in oil and gas industry. During regular motion, the robot resembles to a crank-slider mechanism with equal lengths of the crank and rod, while for stepping over flanges it becomes an open chain with double actuated leverage. A bond graph model and simulation results by use of 20-sim environment are also presented.

Keywords: Mobile robot · Inchworm locomotion · External pipe inspection
Monitoring

1 Introduction

The importance of oil and gas industry is obvious at least in the next few decades for the world economy. The climate conditions of the areas where these resources are mainly exploited, and the personnel costs are reasons for use of robotics and automation equipment within this industry [1]. As concerns robotics, it is expected to help for inspection and detection of pipelines defects like cracks, corrosion, deposition and leakage, or for maintenance operations, especially cleaning. Performing all these tasks requires movement along the pipe, which most of the known robots do inside the pipe [2–4]. The inner surface of the pipe is used for locomotion, by help of wheels, inchworm or screw techniques. The main advantage of these developments is the adjustment to variable pipe diameters and detection/cleaning of the internal depositions, while the main disadvantage is the need for pipe dismantling in order to perform the robot tasks. An alternative is to use the outer surface of the pipe for locomotion and inspection tasks [5, 6], for which the inchworm technique or actuated wheels are regular solutions.

2 Structure and Kinematics of the Robot

The proposed robot is meant to travel along the pipelines, carrying sensorial equipment and communication devices, in order to perform the inspection of the pipes. The most important function of the robot structure is the locomotion, which is, in this case, of inchworm type (see Fig. 1).

© Springer International Publishing AG, part of Springer Nature 2019
J. Machado et al. (Eds.): HELIX 2018, LNEE 505, pp. 464–470, 2019.
https://doi.org/10.1007/978-3-319-91334-6_63

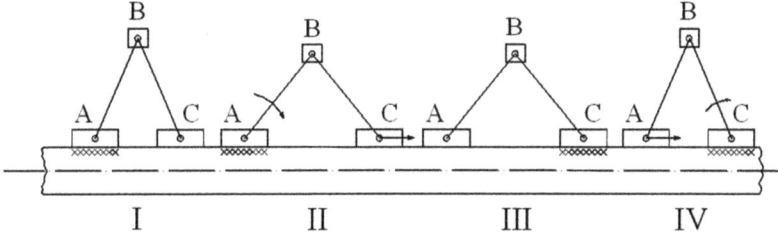

Fig. 1. Inchworm locomotion of the robot along a pipe

For performing one step, the robot is working like a crank-slider mechanism with equal length of the crank and rod respectively. The bodies A and C are equipped with clamping mechanisms (see Fig. 2a), which are alternately activated. In Fig. 1, the stage I shows the clamping phase of A and the release one of B, with the minimum distance AC. In the stage II, AB becomes crank and C slider, which translates to the maximum distance AC. During the stage III, the clamp A is released, while C is activated. Finally, during stage IV, BC becomes crank and A slider, for moving to minimum AC distance. This motion strategy requires from joints A and C to be both free and actuated, depending on the motion stage.

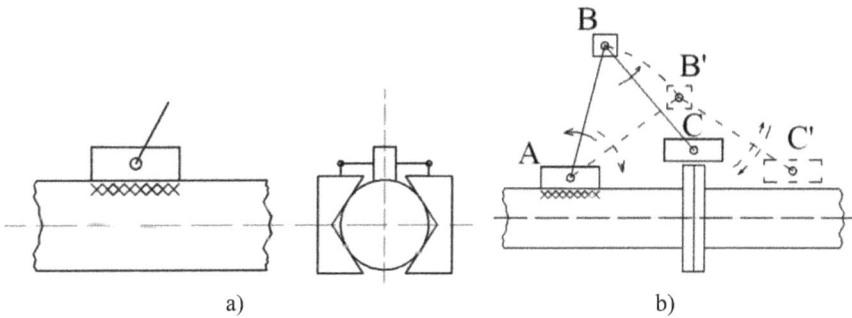

Fig. 2. (a) Clamping symbol and mechanism; (b) Robot jump over a flange obstacle

Another task which could be accomplished by the robot during its travel, along a pipe, is to step over a larger diameter, encountered where the pipes sections are connected by flanges (see Fig. 2b). This action requires, additionally, an active joint B, provided with an actuator, while it is also beneficially for moving at turns down and up (see Fig. 3a). For turning left and right, the body B of the robot has to be provided with a joint having rotation in a plane perpendicular to the crank-slider mechanism (see Fig. 3b). So, Figs. 1, 2 and 3 define the kinematic capabilities of the proposed robot, which should be able to dynamically accomplish them.

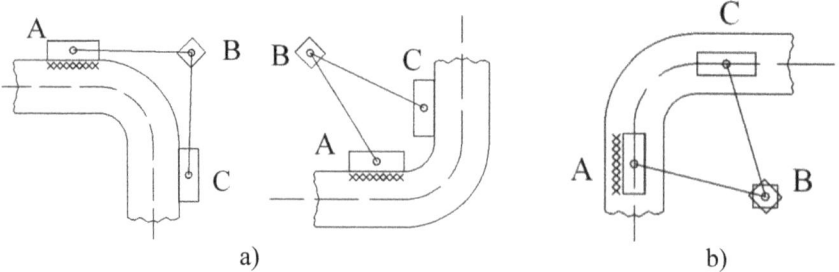

Fig. 3. (a) Robot at turns down and up; (b) Robot at turns right and left

3 Modeling and Simulation of the Robot Behavior

A dynamic model of a mechatronic system, for which a mobile robot is a key representative, can be developed by help of proprietary or open source software handling
block diagrams, iconic ones and bond graphs. Block diagrams require the mathematical
model of the system, describing the physical behavior by time depending equations or
by networks of impedances. They are especially useful when the model includes a
control system. Both iconic modeling and bond graphs networks are based on power
ports, which can overtake the entire physical phenomena. Examples of available
software, which can handle both block diagrams, iconic diagrams and bond graphs
networks are 20-sim, Dymola, Open Modelica, a.o.. Anyway, their libraries do not
satisfy all the requirements of a desired particular model, as happens with the
crank-slider mechanism from the iconic library of 20-sim, which ignores the inertia of
its all components. As usual with mechanisms, an equivalent inertia will depend on the
input position, causing a nonlinear behavior of the mechanical subsystem.

In order to model the robot locomotion, the crank-slider mechanism is considered
for transition from state III to IV, as in Fig. 1, due to the torques generated by the
gravity forces of the mechanism components, which are opposite to the motor torque.
This is depicted in Fig. 4a, where: G, G' - mass centers of the crank and rod; m - mass
of each arm; J - inertia of each arm; M - mass of the body attached to point B; m_s - mass
of the slider (point A); l, l_G - dimensions; θ - position angle between each arm and x
axis; Mg, mg - gravity forces; T_m - motor toque. According to Fig. 4:

$$AC = 2l\cos\theta \tag{1}$$

So, the robot step for an entire locomotion cycle (I–IV) is:

$$s = AC_{max} - AC_{min} = 2l(\cos\theta_{min} - \cos\theta_{max}) \tag{2}$$

For a bond graph representation of the robot, its kinematics and dynamics are
analyzed from Fig. 4. The dynamic model of the mechanism requires isolation of the

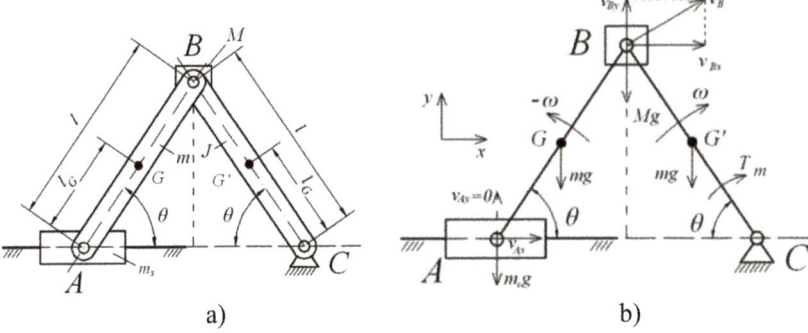

Fig. 4. Crank-slider mechanism in transition steps III to IV: (a) dimensions; (b) forces and velocities

two arms. The crank (arm BC) has a rotation movement around the fixed-point C, which generates a linear velocity of the point B, with its orthogonal components:

$$v_{Bx} = l\omega\cos\theta; v_{By} = l\omega\sin\theta \tag{3}$$

The motor torque is diminished by the moment of the gravity forces acting on the crank:

$$T_G = (ml_g + Ml)g\cos\theta \tag{4}$$

The Eqs. (3), (4) and the physical parameters of the crank arm are used for the bond graphs diagram presented in Fig. 5.

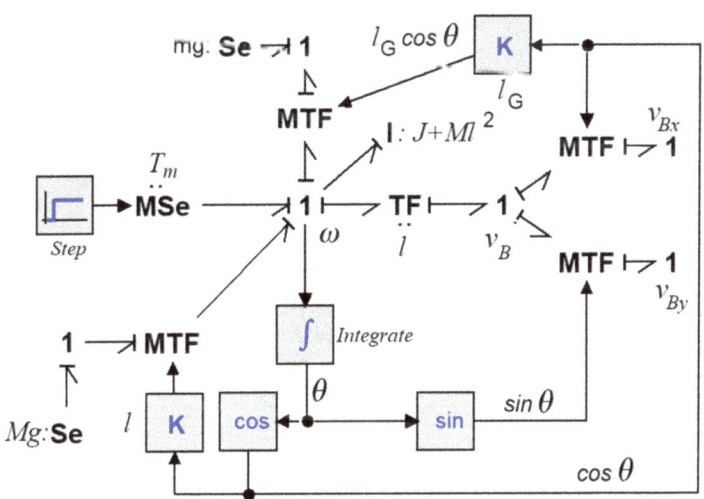

Fig. 5. Bond graph model of the crank in 20 sim

The rod, connected to the crank at point B, as shown in Fig. 4b, has a planar motion consisting of mass center translation with linear speed, v_G and rotation around G, with angular speed, $\omega_{rod} = -\omega$. So, the orthogonal components of the rod ends velocities are expressed as:

$$v_{Ax} = v_{Gx} + l_G\omega \cdot sin\theta \tag{5}$$

$$v_{Ay} = v_{Gy} - l_G\omega \cdot cos\theta = 0 \tag{6}$$

$$v_{Bx} = v_{Gx} - (l - l_G)\omega \cdot sin\theta \tag{7}$$

$$v_{By} = v_{Gy} + (l - l_G)\omega \cdot cos\theta \tag{8}$$

The bond graphs diagram of the rod and slider was built by use of the Eqs. (5)–(8) and the physical parameters of the rod arm and it is presented in Fig. 6. As known, each 1-junction is assigned to absolute velocities, while the four 0-junctions, connected to MTFs represent the summations defined by the Eqs. (5)–(8). Two more 0-junctions and small value compliances, C, were added for keeping the integral causality of the inertances m and m_s. There are two effort sources, Se, for taking into account the friction force acting upon the slider and the gravity force of the rod BC, which develops the resistant torque $mgl_G.cos\theta$. A flow source, Sf, is necessary to force a null component of the point A velocity along the y axis.

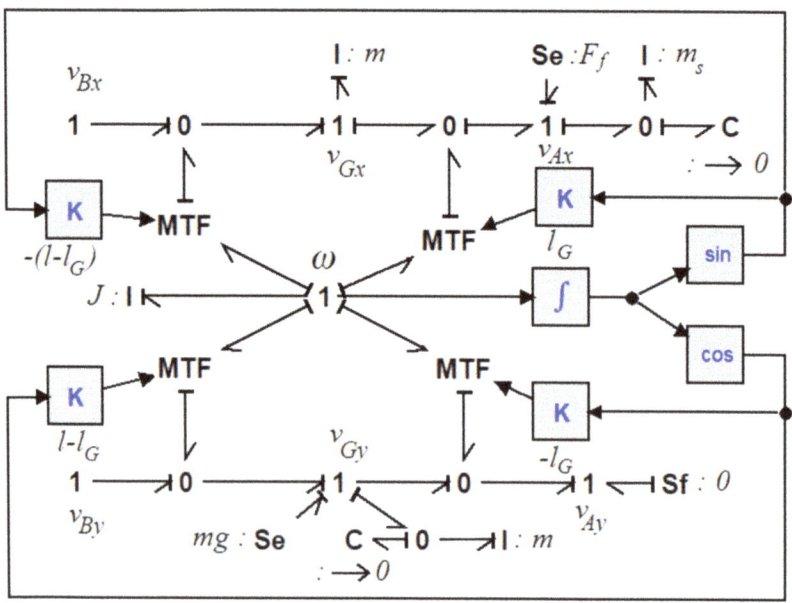

Fig. 6. Bond graph model of the rod and slider in 20 sim

The diagrams in Figs. 5 and 6 must be connected and embedded into a sub model with an input/output power ports for further use. As it can be noticed, the 1-junctions defining v_{Bx} and v_{By} in Figs. 5 and 6 need to be connected by 0-junctions and small value compliances, in order to satisfy the causality rules for each junction.

For simulation, the parameters were established starting from the assumed pipe diameter (50 mm) and the design of a small demonstrator model. The arms lengths of the crank-slider mechanism were chosen as l = 0.25 m, in order to limit the values of the gravity forces moments, which can oppose to the motor torque in transition from stage III to IV and help it I during transition from stage I to II. In the same time, a shorter arm leads to a smaller own inertia, of while a larger value helps for achieving a larger step. The masses of the bodies from A, B, C were evaluated as m_s = 0.6 kg and M = 0.5 kg, while the arms mass was taken as m = 0.4 kg. Anyway, with these values, there are large variations of the gravity force moments within the range 0.8 – 2.3 Nm and of the equivalent inertia of the mechanism $(0.8 - 1.8 \times 10^{-3} \text{ kgm}^2)$, depending on the position θ. In these circumstances, a large ratio of the gearbox is necessary, and an adequate DC motor should be used. Several simulations were performed in order to decide that Faulhaber 2237012CXR motor and 22EKV planetary gearhead, with a reduction ratio 809:1, are the appropriate actuation devices.

It was imposed for simulation an angular motion of the arms within the range *30°–75°*, which geometrically corresponds to a *0.44 m* stroke of the slider. Even a non-avoidable stationary error is encountered, a proportional control was chosen for simulation, to demonstrate the locomotion robot behavior. The simulation model, with dc motor, gear, crank-slider sub model and proportional control is presented in Fig. 7.

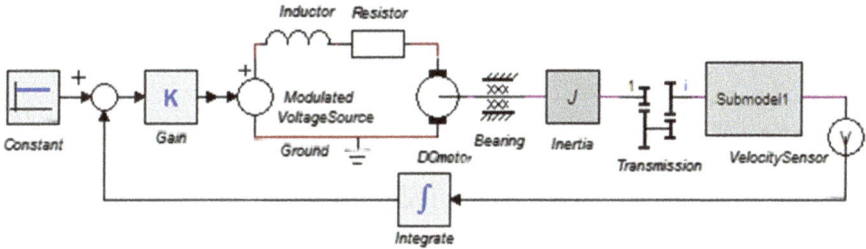

Fig. 7. Robot model in 20 sim with proportional control (iconic diagrams)

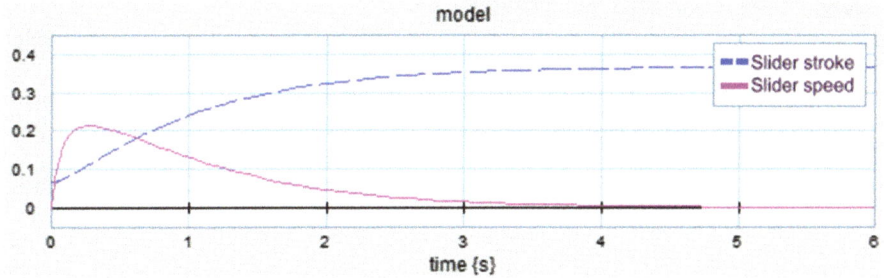

Fig. 8. Simulated slider stroke and speed

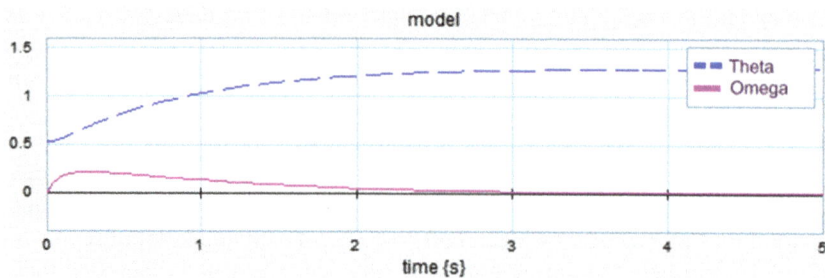

Fig. 9. Angular position and speed of the crank

The reference value (Constant) is the desired stroke of the slider A, geometrically calculated as 0.44 m, which is not reached (see Fig. 8). Similar results are for the crank angular position and speed. (see Fig. 9).

4   Conclusion and Future Work

The most important purposes of the simulation were both to verify the model and the robot capability to perform the proposed locomotion, when estimated values of the geometrical and physical parameters were assigned, based on a preliminary list of the necessary robot equipment (power source, motors, gearboxes, control electronics, sensors, wires, communication devices, etc.). The simulation has proved the robot locomotion capability, while control improvement, mechanical design and modeling of performing the other tasks are future objectives.

References

1. Shukla, A., Karki, H.: Application of robotics in onshore oil and gas industry-a review part I. Robot. Auton. Syst. **75**, 490–507 (2016)
2. Sharma, S.L., Qavi, A., Kumari, K.: Oil pipelines/water pipeline crawling robot for leakage detection/cleaning of pipes. Glob. J. Res. Eng. H Robot. Nano-Tech **14**(1), 31–37 (2014)
3. Schempf, H., Mutschler, E., Gavaert, A., Skoptsov, G., Crowley, W.: Visual and nondestructive evaluation inspection of live gas mains using the Explorer™ family of pipe robots. J. Field Robot. **27**(3), 217–249 (2010)
4. Kotawad, A., Lad, K., Jadhav, S., Mandlik, R.: Identify the deterioration in pipe by using wheel operated robot. Int. J. Res. Appl. Sci. Eng. Technol. **4**(II), 278–281 (2016)
5. Chatzakos, P., Markopoulos, Y.P., Hrissagis, K., Khalid, A.: On the development of a modular external-pipe crawling omni-directional mobile robot. Ind. Robot: Int. J. **33**(4), 291–297 (2006)
6. Singh, P., Ananthasuresh, G.K.: A compact and compliant external pipe-crawling robot. IEEE Trans. Robot. **29**(1), 251–260 (2013)

Protection and Commercialization of Patents in Portuguese Universities: Motivations and Perception of Obstacles by Inventors

Liliana Alves$^{(\boxtimes)}$ and Ana Dias Daniel

Department of Economics, Management, Industrial Engineering and Tourism,
University of Aveiro, Aveiro, Portugal
lalves@ua.pt

Abstract. This research aims to understand the motivations of Portuguese university researchers to patent, as well as main obstacles found in this process of patent protection and commercialization. We interviewed 19 academic researchers from five Portuguese public universities. The researchers interviewed are inventors in university's patents, currently granted and in force, corresponding to a total of 63 patents in force, being 13 of those commercialized. As main motivations researchers have highlighted curricular valorization, followed by the universities' incentives to patent and the commercialization potential of technologies. In the case of main obstacles, the inventors have stressed the level of development of the patent technology, the university administrative procedures and the valorization of patents by industry. More specifically, the need for further R&D before technology reach the market, the bureaucracy of internal university processes, inflexibility of university representatives in negotiations with industry, little appreciation of R&D by industry and lack of financial capacity to implement the patent technology are the most frequently cited obstacles.

Keywords: University · Patent · Motivations · Obstacles
University-industry collaboration

1 Introduction

In Portugal, as well as in other countries, public universities are more and more concerned in gathering additional funding, in addition to government subsidies. This additional funding stream aims to minimize the impact of financial cuts on teaching and research activity. Funded research projects, consortia with industry, services to industry are some of the ways to achieve additional funds and resources [1]. In addition, universities have developed a culture of encouragement of researchers to patent the results of their investigations in order to promote financial return on inventions. The legal protection of technical knowledge with potential industrial application could, in turn, boost the valuation and transfer of knowledge to industry, bringing academia and industry closer together and, consequently, generate profits for both parties [2, 3]. In Portugal, the number of patent applications by universities increased strongly since year 2000. According to Cartaxo and Godinho [1], the creation of GAPIs (Offices for

© Springer International Publishing AG, part of Springer Nature 2019
J. Machado et al. (Eds.): HELIX 2018, LNEE 505, pp. 471–477, 2019.
https://doi.org/10.1007/978-3-319-91334-6_64

the Promotion of Industrial Property) in 2000, and OTICs (Technology and Knowledge Transfer Offices) in 2006 were a key factor for the increase number of patent applications since the beginning of the millennium. The type of services provided by these Technology Transfer Offices (TTO) throughout the process has highlighted the relevance of those offices in the process of patent application [1, 4]. Almost 20 years from the creation of the first GAPIs in Portugal, we sought to understand the main motivations of inventors to patent, especially in the case of researchers from academia, as well as the role of university-industry collaboration on the promotion of patents. Thus, we conducted 19 semi-structured interviews with several researchers, which are inventors in patents currently granted and in force, in five Portuguese universities. The theoretical basis, methodology and results of our study are presented in next sections.

2 Literature Review

In the current era of the knowledge-based economy, universities are becoming important agents for building innovative environments [5]. Etzkowitz and Leydesdorff [6] considered as characteristics of current firm environment its accelerated mutability and the increasing need to create value through innovation. Building competitive advantage through innovation has given rise to the demand of companies for knowledge-generating partnerships, among those university stands out. The university-industry approach is slow and distinct from country to country [7]. In the 1980s, the Bayh-Dole Act was an important step in the university-industry approach in the US [8]. Bayh-Dole instituted a uniform patent policy, removed many restrictions on licensing, and allowed universities to own patents arising from federal research grants. The framers of this legislation asserted that university ownership and management of intellectual property would accelerate the commercialization of new technologies and promote economic development and entrepreneurial activity [4]. The University-Industry Technology Transfer (UITT) policies have led to the creation of Technology Transfer Offices (TTOs) as the university's internal body to support researchers and companies throughout the process of patent protection and commercialization [4].

Siegel et al. [9] presented a primary motive and a secondary motive for patent protection by university scientist: the recognition within the scientific community – publications, grants (especially if untenured) as primary motive, and the financial gain and desire to secure additional research funding (mainly for graduate students and lab equipment) as second motive. Baldini et al. [10] conducted a study on the motivations of Italian university researchers to patent inventions. In their study, they surveyed several university academic inventors involved in patenting. The authors noted that the prestige/visibility/reputation, and the "look for new stimuli for their research" were the main motivations for academics to patent an invention. On the other hand, Lam [11] analyzed the UK academic researches to engage in research commercialization. Through three concepts – 'gold' (financial rewards), 'ribbon' (reputational/career rewards) and 'puzzle' (intrinsic satisfaction) - concluded that many do it for reputational and intrinsic reasons, and that financial rewards play a relatively small part.

Many authors concluded that there is ample evidence of increased academic engagement in commercial activities, such as patenting and spin-off company formation [11–13]. However, the implementation of a Bayh–Dole-like legislation outside the US is still a great challenge [14]. Geuna and Nesta [15] analyzed university patenting and its effects on academic research, and observed that European university patenting is growing, although there are heterogeneous scenarios across countries and disciplines. The authors found some evidence that university licensing is not profitable for most universities, although some do succeed in attracting substantial additional revenues. Scarce knowledge of university-level patent regulations, "open science" mentality of university, too heavy teaching duties, too heavy administrative duties, scarce possibility for commercial exploitation/industrialization, difficulties in evaluating the commercial potential, scarce interest from industry toward academic research, excessive bureaucracy and rigidity of university administrators, insufficient reward for researchers, lack of support in the patenting activity, lack of a TTO, and lack of funds to cover patenting costs are some relevant obstacles identified in several studies [9, 10, 14].

3 Methodology

Although there are studies on the motivations and obstacles experienced by academic researchers in the process of patent protection and commercialization, the Portuguese case is still scarcely studied. Thus, the research question of our case study was: How do Portuguese university's researchers which patent their inventions perceive the process of patent protection and its subsequent commercialization?

Given the exploratory nature of the research, we opted to carry out semi-structured interviews with researchers from Portuguese public universities with a relevant track record in patent protection. Thus, the selected universities were the Portuguese public universities with the best rankings in QS World University Rankings 2018: (*sort order*) University of Porto, University of Lisbon, Universidade NOVA de Lisboa, University of Coimbra, University of Aveiro and University of Minho. We then searched the databases of the National Institute of Industrial Property (INPI) and the European Patent Office (EPO) and we identified the main inventors of university patents currently granted and in force, as well as, the three inventors by each university that have a greater number of patents currently granted and in force (Top Inventors). We contacted the inventors and requested a face-to-face or Skype interview. The interviewees considered three key moments: research that led to the patent, motivation to protect the invention and commercialization of the patent. Throughout the interview we asked to inventor to recall the processes leading to the patent application and the commercialization (if applicable) of patents currently granted and in force. All the interviews were recorded and later transcribed. The data were then cataloged by topic and counted the frequencies of each type of motivation and obstacle mentioned by the inventors [9]. In the following section the results of the investigation are presented.

4   Results

Between October 2017 and January 2018, we conducted 19 interviews with inventors from five universities: (alphabetical order) Universidade NOVA de Lisboa, University of Aveiro, University of Lisbon, University of Minho and University of Porto. We received no positive response from the University of Coimbra. Among those interviewed, 6 were top inventors at their university, according to the criterion indicated above. 4 inventors were interviewed at NOVA, at the University of Aveiro and at the University of Minho; 5 at the University of Porto; and 2 from the University of Lisbon. In total, 63 cases of patent applications currently granted and in force were considered.

Concerning the inventors' motivations to protect their invention through patent, it was asked the interviewees: "What was the motivation to legally protect the invention (objective of future commercialization, curricular valorization, existence of financial incentives, imposition of the organization/institution policy, prospect of financial gains obtained by the commercialization of the invention)?" The generic options presented provided a starting point for interviewer to explore researchers' motivations.

The analysis of the answers enabled the identification of three main motivations: curricular valorization, university incentive and the potential of commercialization. The curricular valorization motivation was associated with 43 of the 63 patents analyzed and mentioned by most top inventors. This motivation should be carefully considered. Several inventors mentioned that having patents is undervalued in academic assessment of curriculum vitae when compared to papers published in high scientific impact journals. Thus, inventors stated that curricular valorization is indirect: more possibility to get funded projects and partnerships with industry, and valorization of their research group.

In turn, the university's incentive along with the incentive to the research group was referred as the motivation for filing 32 patent applications. Regarding the university's incentive, the inventors of different universities refer to a period in time which universities had financial incentives to protect patents. This financial incentive was a strong motive for the inventors to patent, alongside other motivations, such as curricular valorization, personal satisfaction or objective of commercialization in the future.

The 'objective of commercialization in the future' was referred as a motivation for initiating 35 patent application processes. This motivation was associated with other motivations - curricular valorization and university incentive - and it was a consequence of the researcher's perception of an opportunity/need in the market, and those were willing to license the patent. It should be noted that the time to market differs from process to process. However, in many cases the objective had a long-term perspective since more R&D would be required, and the need to establish partnerships with industry.

In turn, throughout the interview we raised issues related to the obstacles in the patent application process and subsequent commercialization. According to the answers obtained during the interviews with the inventors, we found out that among the 63 patent processes analyzed, only 20% (13 patents) are being commercialized. The analysis of the inventors' answers makes clear that the perception of obstacles throughout the process of protection and commercialization (or attempted commercialization) is related to three fundamental questions: the level of development of patent technology, university procedures and valorization of the patent by industry. Regarding the level of development of the technology, several inventors consider that the patent is not being commercialized because more research is needed to enable its transfer to the market. This argument was especially relevant for inventions with potential pharmaceutical application, where it was often pointed out that one of the major obstacles was the need for high financial investments to carry out the various mandatory tests for drug approval.

On the other hand, the role of the university in the process of protection and technology transfer was debated by all inventors. Several acknowledge a major change in university culture since the year 2000. European and National government's incentive through the opening of calls for funded projects with industry, funding of patent protection, creation of offices to support the protection and transfer of technology, as well as encouraging speech for filing of patents were mentioned by inventors as indicatives of a change in university culture. However, the inventors refer important issues that they feel should be rethought by universities. Firstly, inventors feel the need for more support in the writing of a patent, since the requirement to writing a patent are very different those of a scientific article. Secondly, as stated by inventors from different universities, the university's policy is to pay only the costs of national patent protection. Thus, protection at European or international level is dependent on the existence of an industrial partner interested in the commercialization of the patent. Third, the inventors consider that there is excessive bureaucracy and rigidity in internal university processes, which hinders the fluid interaction with companies. Finally, in the fourth place, the process of negotiating the transfer of intellectual property rights between the university and the industry is considered difficult. The incapacity to reach an agreement about financial return of the patent was the reason for the non-commercialization of several patents in different universities. As mentioned at the beginning of this section, the valorization of the patent by industry is other key obstacle throughout the university patent protection and commercialization process. On the one hand, some inventors consider that the Portuguese industry does not value university patents because it does not have sufficient financial and/or R&D capacity to apply the patent. On the other hand, the inventors refer to companies' preference, in some cases, for industrial secret. A summary of motivations and obstacles is presented in Table 1.

Table 1. Motivations and obstacles: summary table

Motivations	Frequency	Obstacles	Frequency
Curricular valorization	43	University procedures	40
Commercialization	35	Funds to international patents	16
University incentive	32	Patent negotiation	10
		University bureaucracy	8
		Patent writing	6
		Valorization by industry	24
		Development of technology	16

5 Conclusion and Practical Implications

The interviews carried out with the 19 inventors of 5 Portuguese public universities allowed a better understanding of the Portuguese motivations and obstacles in patent protection process, as well as the scenario of university-industry interaction to transfer university patents. We observed that the main motivations to patent were the curricular valorization, the university's incentive and the commercialization of technology. Similar to the results of Siegel et al. [9], Baldini et al. [10] and Lam [11], the prestige/visibility/reputation is still a strong motive for patenting. Although some Portuguese inventors have stated that the patent is not yet highly valued in the academic evaluation of the curriculum, they have recognized that there are advantages associated with being inventor of a patent. Throughout the process of patent protection and commercialization, the inventors have identified obstacles primarily related to the level of development of patent technology, university procedures, and valorization of the patent by industry. These obstacles are similar to the obstacles identified in some studies in Europe and in USA [9, 10, 14, 15]. Nevertheless, we highlight the predominance of responses related to the bureaucratization of internal university processes and the rigidity of the university in negotiating the patent with industry. These obstacles, associated with the others identified, may partially explain the low number of university patents being commercialized.

Governmental and university policies to enhance university-industry collaboration, namely technology transfer, seem to have impact in increasing number of universities patent applications. However, our results suggest that most of the patents are not being commercialized due to a lack of coordination between university-industry relations. It is important to look deeply into the strategy of Portuguese universities for this issue. The relationship between personal incentives versus number and quality of patents could be explored at policy level.

Acknowledgements. We are grateful to all inventors who participated in this study and shared their deep personal experience in patent protection and commercialization. We are grateful to each shared experience which provided important value for this study.

References

1. Cartaxo, R.M., Godinho, M.M.: How institutional nature and available resources determine the performance of technology transfer offices. Ind. Innov. **24**(7), 713–734 (2017)
2. Arqué-Castells, P., et al.: Royalty sharing, effort and invention in universities: evidence from Portugal and Spain. Res. Policy **45**, 1858–1872 (2016)
3. Marques, J.P.C., Caraça, J.M.G., Diz, H.: How can university-industry-government interactions change the innovation scenario in Portugal? - the case of the University of Coimbra. Technovation **26**, 534–542 (2006)
4. Siegel, D.S., et al.: Toward a model of the effective transfer of scientific knowledge from academicians to practitioners: qualitative evidence from the commercialization of university technologies. J. Eng. Technol. Manag. **21**, 115–142 (2004)
5. Perkmann, M., et al.: Academic engagement and commercialisation: a review of the literature on university–industry relations. Res. Policy **42**(2), 423–442 (2013)

6. Etzkowitz, H., Leydesdorff, L.: The future location of research: a triple helix of university–industry–government relations II. EAAST Rev. **15**(4), 20–25 (1996)
7. Kalar, B., Antoncic, B.: The entrepreneurial university, academic activities and technology and knowledge transfer in four European countries. Technovation **36–37**, 1–11 (2015)
8. Thursby, J.G., Kemp, S.: Growth and productive efficiency of university intellectual property licensing. Res. Policy **31**(1), 109–124 (2002)
9. Siegel, D.S., et al.: Commercial knowledge transfers from universities to firms: improving the effectiveness of university-industry collaboration. J. High Technol. Manag. Res. **14**(1), 111–133 (2003)
10. Baldini, N., Grimaldi, R., Sobrero, M.: To patent or not to patent? a survey of Italian inventors on motivations, incentives, and obstacles to university patenting. Scientometrics **70** (2), 333–354 (2007)
11. Lam, A.: What motivates academic scientists to engage in research commercialization: 'Gold', 'ribbon' or 'puzzle'? Res. Policy **40**, 1354–1368 (2011)
12. Siegel, D.S., Wright, M., Lockett, A.: The rise of entrepreneurial activity at universities: organizational and societal implications. Ind. Corp. Change **16**(4), 489–504 (2007)
13. D'Este, P., Patel, P.: University–industry linkages in the UK: what are the factors underlying the variety of interactions with industry? Res. Policy **36**(9), 1295–1313 (2007)
14. Baldini, N.: Implementing Bayh–Dole-like laws: faculty problems and their impact on university patenting activity. Res. Policy **38**, 1217–1224 (2009)
15. Geuna, A., Nesta, L.J.J.: University patenting and its effects on academic research: the emerging European evidence. Res. Policy **35**, 790–807 (2006)

Main Frame Structural Optimization for Power Transformers in Short Circuit

C. Linhares[1(⊠)], V. H. Carneiro[2], J. Meireles[2], A. C. M. Pinho[2], and H. Mendes[1]

[1] Power Transformer R&D, EFACEC Energia S.A., São Mamede de Infesta, Portugal
cassiano.linhares@efacec.com
[2] MEtRiCs - Mechanical Engineering and Resource Sustainability Center, University of Minho, Guimarães, Portugal

Abstract. Electric power market is a contemporary challenge, due to the permanent increase in energy distribution demands, suppliers of power transformers have to constantly improve their engineering solutions for the design of equipment that are able to adapt to this permanent evolution. A fundamental challenge is the research of structural components embedded in power transformers tank that are able to withstand loading due to regular production loading, as vacuum, or during service like short-circuit events. This paper presents the results of experimental testing on different main frame geometries of power transformers shell type to predict structural behavior due to short-circuit electromagnetic force The resulting data allows the determination of novel structural lay-outs that are optimized to lower deformation due to these specific loading regimes, are characterized by a reduced use of material and lowering of the overall structural weight.

Keywords: Power transformer · Main frame · Short-circuit
Structural analysis

1 Introduction

The structural project of a transformer is even nowadays a complex, rigorous and empirical process [1], based on the classic law of electromagnetic induction first introduced by Faraday and then generalized by Maxwell [2]. This can be justified by the safety, economic and quality that the energy distribution process has to fulfill [3] and the high values of these equipment's. However, the increase of worldwide competitiveness between power transformers suppliers demand a constant optimization in the equipment of energy transmission industry [4].

One way of accomplishing this objective is the weight reduction of these components. This will allow the decrease of transportation and assembly related costs, a reasonable use of natural resources and a more economical fabrication. There are different strategies to execute the optimization of tank of a power transformer, e.g. the adoption of new materials and the use of more efficient structural geometries [5], optimizing structural components as the main frame used to clamp the windings on

© Springer International Publishing AG, part of Springer Nature 2019
J. Machado et al. (Eds.): HELIX 2018, LNEE 505, pp. 478–486, 2019.
https://doi.org/10.1007/978-3-319-91334-6_65

power transformers shell type. Another route is to find more accurate ways to perform the dimensioning of these structures and reduce or eliminate the empirical factor of the current project based on a consolidated knowledge. Since all changes are correlated, this is an iterative process and these developments must also generate a relative advantage to the final product, when compared to competitors. However, the resulting changes must not interfere with product duties. It must be able to resist the functional and short-circuit loading throughout its life-time.

In normal operation, the transformer electromagnetic loads are relatively small [6]. However, short-circuits generate high electro-dynamic loads [7] which are one of the main reasons of transformer malfunction [8] and may damage or even destroy the transformer structure [9, 10]. Thus, this is one of the main design challenges, an important transformer feature [11] and the main project criteria required by manufacturers [12].

This study shows the design of a novel generation of main frame that is able to better withstand short-circuit loading and reduce the overall structural weight. Experimental procedures were performed in different main frame prototypes to represent a short-circuit event. The values of deformation were monitored to represent approximate scenarios and to obtain three key results: (i) lower the overall structural deformations (ii) decrease the overall weight; (iii) select the most efficient structural layout.

2 Methodology

2.1 Main Frame Prototypes

The first objective of this study is to estimate the deformations in three main frame prototypes when subjected to a short-circuit event. These prototypes consist of four box beams welded, with interior reinforcements (Fig. 1), using different thicknesses of plates. Two beams parallels to the windings where the short-circuit force is applied, and another two perpendicular to windings. The difference between prototypes is their parallel beams. Prototype 1 (Fig. 1(b)) is a standard design and is used to compare the performance of the designs used in prototypes 2 (Figs. 1(c)) and 3 (Fig. 1(d)) with the same width and different height. Prototypes 2 and 3 are designed to obtain a superior moment of inertia and reduce the weight of the main frame. While in prototype 1, the used sheet thickness is 16 [mm], the sheet thickness in prototype 2 and 3 is 12 [mm]. Additionally, prototype 3 is was manufactured by increasing the lateral walls height.

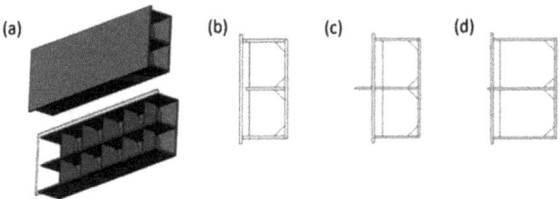

Fig. 1. Main frame (a) internal reinforcement: (b) prototype 1, (c) 2 and (d) 3.

2.2   Base Material Properties

To eliminate material induced differentiation, all three prototypes were built using sheets of S275JR structural steel. This material was produced according to EN10025, in the form of sheet and assembled by welding. Specimens were manufacture and tested according to ISO 6892-1 – specimen Type D, recurring to a universal testing machine. The instantaneous loads and axial/transverse deformations were monitored to determine the stress-strain behavior and Poisson's ratio of five equivalent specimens. Furthermore, samples were tested according to the parallel and perpendicular direction of the sheet lamination.

2.3   Instrumentation Design by FEA

An estimation of the areas of the prototypes with higher and lower deformations was performed by an initial structural simulation using FEA. Two rigid plates were assembled in the internal faces of the main frame and a displacement of 10 [mm] was individually imposed in the perpendicular thru-face direction (Fig. 2). The input parameters of the simulation may be observed in Table 1. Deformation results were used to select the point where strain-gauges are placed in the experimental testing.

Fig. 2. Main frame loading

2.4   Prototype Testing

The experimental testing of the prototypes was performed by the application of a quasi-static distributed load on the parallel beam of the prototypes (Fig. 3(a)). The load was transmitted by five parallel equally distant hydraulic cylinders (Fig. 3(b)) that generated an approximately distributed load, while its orientation was consistent with the one that happens in a short-circuit event. Loads from 25 to 300 [tonf] were applied to each prototype upper beam, according to a protocol (Fig. 3(c)): initial pre-load of 5 [tonf] to accommodate the hydraulic cylinders on to the main frame walls; loading returned to zero, and afterwards, increments of 25 [tonf] up to 300 [tonf] were added as the imposed deformations were monitored in the critical locations defined by the preliminary FEA.

Table 1. FEA input details.

Material	S275 – Isotropic strain hardening	
Contacts	Type	Frame-plate: bonded
		Plate-plate: frictionless
	Formulation	Pure penalty
Meshing	Element	SOLID187
	Description	Quadratic-high order

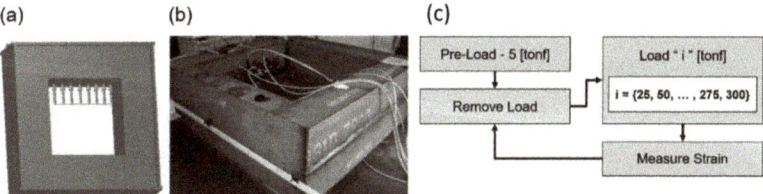

Fig. 3. Upper beam (a) loading scheme, (b) actual testing apparatus and (c) testing method.

3   Results and Discussion

3.1   Base Material Properties and Instrumentation Design by FEA

Tensile testing on the base material allowed the determination of the stress-strain behavior (Fig. 4(a)) and the Poisson's ratio. These basic elastic and plastic characteristics were used as input in the performed FEA and the resultant data was used to determine the strain and critical points according to each axis (Figs. 4(b) to (d)). Based on the referred numerical results there were selected eight point for strain-gauge measurement. These sensors were placed according to Fig. 5 and described in Table 2.

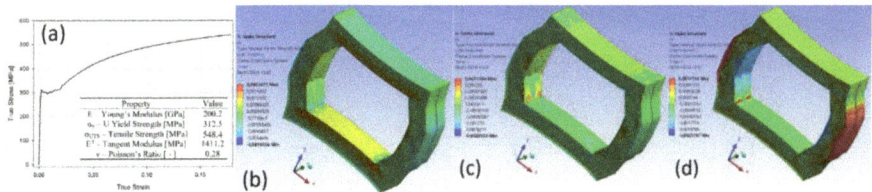

Fig. 4. Material and FEA results: (a) s275 true stress-strain plot; and numerical strains in the (b) XX-axis, (c) YY-axis and (d) ZZ-axis.

Based on previous FEA of main frame, the sensors were placed according to Fig. 5 and described in Table 2.

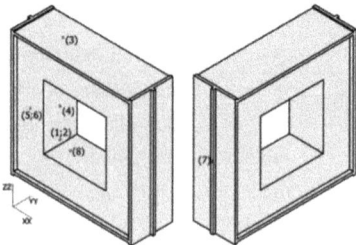

Fig. 5. Strain-gauge locations for experimental testing

Table 2. Description of the defined strain-gauges

Gauge	Description	Direction
1	Rosette	ZZ axis
2		YY axis
3	Simple	YY axis
4	Simple	ZZ axis
5	Rosette	ZZ axis
6		XX axis
7	Simple	ZZ axis
8	Simple	XX axis

3.2   Prototype Testing

The first rosette was applied in the interior face of the upper beam. It was composed by strain-gauges number 1 and 2, which measured the deformation in the ZZ and YY axis. The measurements for the corresponding applied load are shown in Fig. 6.

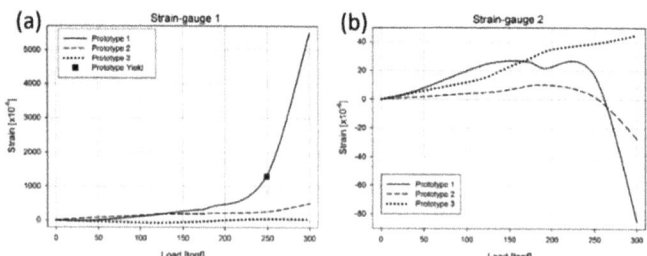

Fig. 6. Experimental strain values in strain-gauges (a) 1 and (b) 2.

Figure 6(a) shows that for prototype 1 the strain values peaked and the deformations in the ZZ axis reached far more significant values in this prototype. It is shown that only prototype 1 presented plastic deformation in this location, where yield extension is $\varepsilon_{yield} \approx 1300^{-6}$. In the data collected by strain-gauge 2 (Fig. 6(b)) may be observed that the more efficient design for deformations in this point is prototype 2. Even though prototype 1 has a lower performance than the other designs, the deformations are relatively low in all prototypes.

Strain-gauge 3 was positioned on the exterior lateral face of the perpendicular beam. The deformations monitored can be observed in Fig. 7(a). As the main reinforcements and differences between the prototypes are localized in their parallel beams, it would be expected that the values of deformation in the lateral faces to be similar. However, it can be concluded by the data gathered by strain-gauge 3 that the improved geometries reduced the deformation values. Prototypes 2 and 3 show similar values of deformations, while prototype 1 presented significant and plastic deformation. Strain-gauge 4 was placed in the center of the interior plate of the upper beam. The deformations that resulted from the applied loads are shown in Fig. 7(b).

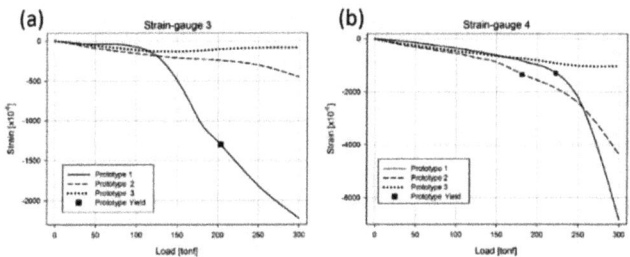

Fig. 7. Experimental strain values in strain-gauges (a) 3 and (b) 4.

Prototype 1 reveals higher deformations then the other prototypes. However, for the first time, there is a significant difference between prototypes 2 and 3. It is shown that the structural lay-out of prototype 3 is able to remain in the elastic domain, while prototypes 1 and 2 display permanent deformation.

The second rosette was placed in the lateral face of the parallel beam and is composed by strain-gauges 5 and 6, which monitors the deformations in the ZZ and XX axis. In Fig. 8 are expressed the deformations imposed by the applied loads.

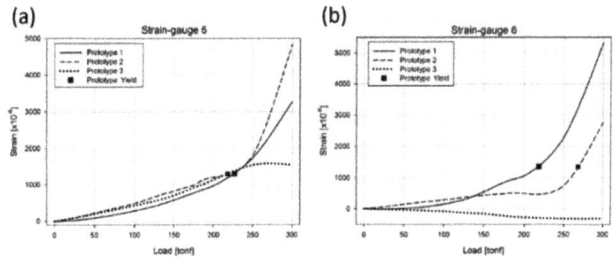

Fig. 8. Experimental strain values in strain-gauges (a) 5 and (b) 6.

In the data collected by strain-gauge 5, all prototypes presented permanent deformation, while in strain-gauge 6 only prototypes 1 and 2 present permanent
deformation.

Strain-gauge 7 is placed on top of the parallel beam and shows the deformations
that occur in the superior reinforcement (Fig. 9(a)). Prototype 2 presented an inferior
performance than the other prototypes and plastic strain is reached by prototypes 1 and
2. Strain-gauge 8 is placed in the interior face of the perpendicular beam (Fig. 9(b)) and
shows that the performance of prototypes 2 and 3 are similar. However, at higher
loading prototypes 1 and 2 enter the plastic domain.

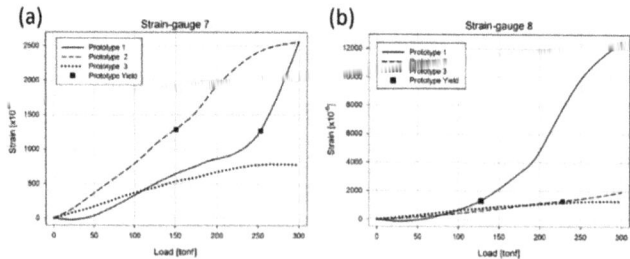

Fig. 9. Experimental strain values in strain-gauges (a) 7 and (b) 8.

Another fundamental objective is the overall weight reduction, being the specific
weight of the presented prototypes shown in Table 3. It may be seen that the prototype
2 is the one that carried the most elevated weight reduction.

Table 3. Prototype mass comparison.

Prototype	Mass [kg]	Overall weight reduction [%]
1	1814	0
2	1603	11.63
3	1684	7.17

It is clear that the values of deformation make prototype 1 the design with lower
performance. In the localizations of strain-gauges 1, 3, 8 in some measurements, the
observed deformation is approximately the same for prototypes 2 and 3. However,
prototype 2 display more places with plastic strain, relatively to prototype 3. It is clear
that the best structural performance is shown by prototype 3, even though it is not able
to display the most elevated weight reduction.

4 Conclusions

An experimental procedure, applying a distributed load in the parallel beam of three prototypes of power transformer main frames has been performed. It was shown that the prototype with the best design is the one with the lateral walls with most elevated height and 7.17% of the material weight removed (prototype 3). Further lowering of structural weight is possible, however, the values of plastic strain are increased. Due to these results a novel generation of lighter main frame was designed for short-circuit loads.

Acknowledgment. This work is supported by the project with acronym iCubas5D, POCI-0247-FEDER-017584, co-funded by the European Regional Development Fund (ERDF) through COMPETE2020 - Programa Operacional Competitividade e Internacionalização (POCI) under Programme "Portugal 2020".

Cofinanciado por:

References

1. Hernandez, C., Arjona, M.A.: An intelligent assistant for designing distribution transformers. Expert Syst. Appl. **34**, 1931–1937 (2016)
2. Funaro, D.: Electromagnetism and Structure of Matter. World Scientific, New Jersey (1999)
3. Morar, D., Guzun, B., Rodean, I.: Operational reliability in transmission power grid. J. Energy Power Eng. **6**, 150–154 (2012)
4. Almeida, E.E., Mendes, H.G., Pinho, A.M.: Experimental validation of a core type power transformer. In: ASME Mechanical Engineering Congress and Exposition, vol. 11, pp. 129–136. ASME, United States of America (2009)
5. Tsili, M.A., Kladas, A.G., Georgilakis, P.S.: Computer aided analysis and design of power transformers. Comput. Indus. **59**, 338–350 (2008)
6. Rosentino, A.J.P., Saraiva, E., Delaiva, A.C., Guimarães, R., Lynce, M., Oliveira, J.C., Fernandes Jr., D., Neves, W.: Modelling and analysis of electromechanical stress in transformers caused by short-circuits. Renew. Energy Power Qual. J. **9**, 432 (2011)
7. Wang, Y., Pan, J., Jin, M.: Finite element modeling of the vibration of a power transformer. In: Proceedings of ACOUSTICS, paper no. 34, Gold Coast, Australia (2011)
8. Bartley, W.H.: Analysis of transformer failures. In: International Association of Engineering Insurers - 36th Annual Conference of IMIA, Stockholm, Sweden (2003)
9. Linhares, C.C., Carneiro, V.H., Pinho, A.C.M., Mendes, H.G.: Elasto-plastic analysis of an active part support on shell-type transformers during a short-circuit. In: proceedings of 13th Spanish-Portuguese Conference on Electrical Engineering, paper no. 198, Valencia, Spain (2013)
10. Gilany, M., Al-Hasawi, W.: Reducing the short circuit levels in Kuwait transmission network (a case study). J. Energy Power Eng. 4, 45–51 (2010)

11. Krause, C.: Short-circuit resistant power transformers – prerequisites for reliable supply of electrical energy. In: Cigré 6th Southern Africa Regional Conference, Cape Town, South Africa (2009)
12. Faiz, J., Ebraimi, B.M.: Computation of static and dynamic axial and radial forces on power transformer windings due to inrush and short-circuit currents. In: IEEE Jordan Conference on Applied Electrical Engineering and Computing Technologies, Amman, Jordan (2011)

Alignment Issues for any Sample Direction BTF Measurement

Jan Hošek[1(✉)], Vlastimil Havran[2], Jiří Čáp[1], and Šárka Němcová[1]

[1] Faculty of Mechanical Engineering, Czech Technical University, Technická 4,
16607 Praha 6, Czech Republic
{Jan.Hosek,Jiri.Cap,Sarka.Nemcova}@fs.cvut.cz
[2] Faculty of Electrical Engineering, Czech Technical University, Karlovo nám.
13, 12135 Praha 2, Czech Republic
havran@fel.cvut.cz

Abstract. Measurement of the bidirectional texture function (BTF) of the surfaces with portable instruments allowing on-site measurement outside the laboratory conditions is a newly emerging technique enabling data acquisition of the real objects without the necessity of their extraction from the environment. A practical issue of the sample measurement is the instrument alignment regarding the measured surface. The aim is to measure a sample surface under any orientation in space and to align the measurement instrument to the surface so to measure surface reflectance of the sample surface. This paper describes the design of two setups allowing adjustment of a portable BTF measurement instrument regarding the sample surface at any position. The final instrument alignment is performed with a help of feedback information provided by a simple laser autocollimator.

Keywords: System alignment · BTF measurement · Feedback
Autocollimator · Kinematic constraint

1 Introduction

The appearance of real surfaces depends on the kind and direction of illumination and the direction of observation, and therefore the actual surface perception depends on the conditions of its observation. This is a problem in virtual representation of the surface appearance in digital visualization, where a faithful representation of the surface need to be represented by a set of thousands of images taken under various combinations of illumination and observation directions. Surface reflectance of a single point was formalized by Nicodemus et al. [1] as bidirectional reflectance distribution function (BRDF). This is a four-dimensional function, where two values specify the direction of incident light and two the direction of outgoing light. Spatially varying description of surface reflectance, a seven-dimensional function, is called bidirectional texture function (BTF) and was introduced by Dana et al. [2].

It is technically difficult to realize an instrument capable of acquiring thousands of images with different illumination and observation directions of the sample surface within reasonable time, so there are only a few realized BTF setups. All of them are

© Springer International Publishing AG, part of Springer Nature 2019
J. Machado et al. (Eds.): HELIX 2018, LNEE 505, pp. 487–493, 2019.
https://doi.org/10.1007/978-3-319-91334-6_66

experimental and the majority of them working in laboratory conditions. Portable BTF measurement setups are rare, but it can be found a few that could be realized as portable ones [3–8]. Recently, Havran et al. [9] and Hošek et al. [10] developed BTF portable measurement instruments allowing measurement of real surfaces in the nature on-site directly. Necessary and practical issue of such approach is the ability to set the measurement system regarding the naturally placed measured surface with the required precision. This paper shortly introduces the principle of the proposed measurement instrument in the next chapter and required alignment conditions. In the next chapters there are presented two setups allowing to set the instrument to measure the sample surface at any inclination position and the laser autocollimator principle used for alignment feedback.

2   Portable Instrument for BTF On-Site Measurement

The measurement system was designed as a compact and portable system which tackles dimension limitation and acquisition speed enhancement by a combination of the parallelization of the illuminators and cameras with their mutual motions regarding the sample surface. This approach even more allows for unlimited number of possible combinations of illumination and acquisition directions, as cameras can move to any angular position with two degrees of freedom as it is shown in Fig. 1.

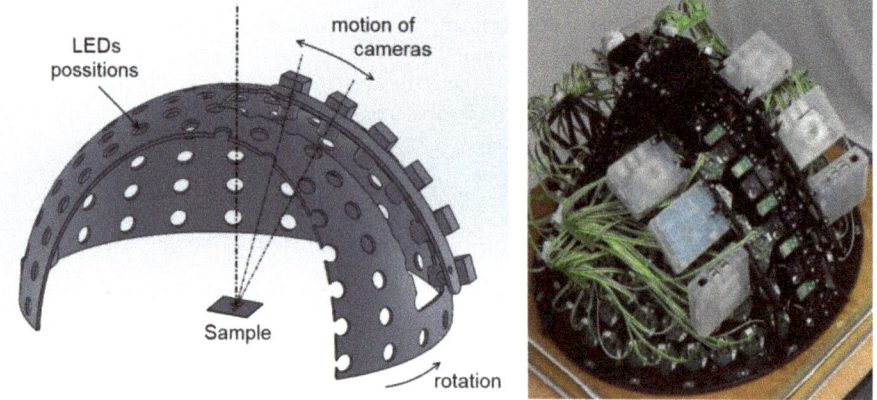

Fig. 1. Left - conceptual design of the BTF measurements instrument. Right - realization.

The system is equipped with totally 139 LEDs illuminating the sample surface during a measurement sequence. Surface appearance data are acquired with 6 cameras simultaneously. Due to practical reasons the single sample measured data are limited to 16,680 high-dynamic-range images taken at 20 different rotational positions set with Servomotor TGN2 provided with 20-bit multi-turn absolute encoder and geared with harmonic drive CSG-20-80-2UH. Complete set of images was taken within 1020 s. More details regarding the system design can be found in [10].

The data acquisition and processing algorithm, described in detail in [9], utilizes a specially designed marker sticker with a circular hole imaged by cameras together with the measured surface sample. The acquired image data are first transferred from the device to the ordinary computer and than processed with a multi-step algorithm that rectifies and aligns the images with subpixel precision. We evaluated all possible uncertainties which can affect the final BTF data uncertainty. Among geometrical parameters we assessed optic's depth of field 51 mm, which limits the maximum zenith angle to 75°. Next the uncertainty of the sample illumination beam angle, which is $\pm 0.59°$, is mainly caused by high fabrication tolerance of the LED chip inclination. The maximum camera total angular uncertainty was calculated as $\pm 0.167°$ in the meridional direction and $\pm 0.184°$ in the zonal direction. Those uncertainties are valid under the assumption of maximum misalignment of the measurement instrument regarding the sample ± 0.5 mm and the rotary axis perpendicular towards the sample inclination $\pm 0.1°$. Mentioned conditions bring tight demands on alignment capabilities of the core of the instrument a 580 mm wide cylinder like spinning structure of total mass of 11.2 kg.

3 Adjustment External Holder for the BTF Measurement Instrument

Majority of reported BTF measurement instruments was designed as laboratory instruments intended for measurement of a horizontal plane sample (on the table/floor) or a vertical plane sample (e.g. detail of a wall). The aim of the external adjustment frame for the portable instrument and on-site measurement is to achieve sample surface measurement under any inclination angle and to keep the position during whole measurement time. The instrument still needs to be light enough to be portable.

We analyzed possibilities of available portable holder capable of adjustment, fixing and holding the set position under any sample normal direction, but the solutions split into two different ways. There exists light and stiff enough tripods provided with an angular adjustment mechanism. We found it suitable for measurement angles from vertical sample plane - wall up to the top horizontal plane - ceiling. In the case of the sample surface axis in the lower quadrant less than 20° from the horizontal plane the tripod solution is rather impractical. We solved the lower quadrant measurement angles with specially designed frame holder allowing to set and keep any bottom measurement axis angles.

3.1 Tripod Based External Holder

We found there exist light stiff enough tripods provided with an angular adjustment mechanism, but without the possibility of precise linear motions. Another limitation of the tripod solution emerged from the instrument mass and dimensions. Due to the high position of the center mass point of gravity of the instrument using the tripod's angular adjustment mechanism would lead to a misbalancing of the system. For this reason we provided a heavy-duty tripod Manfroto with an XY stage of 50 mm motion range and a specially designed alt-azimuth mount provided with a third linear motion. This mount allows us to place the instrument's center mass of gravity in the cross section of the vertical and horizontal axes of angular motions. The vertical rotational axis was mounted onto the XY stage. The horizontal axis was placed at the end of arms of the U-shaped frame. As the instrument needs to be adjusted with respect to the sample surface in all six degrees of freedom, we placed the horizontal axis joints on linearly adjustable slides. The tripod based alignment system is shown in Fig. 2.

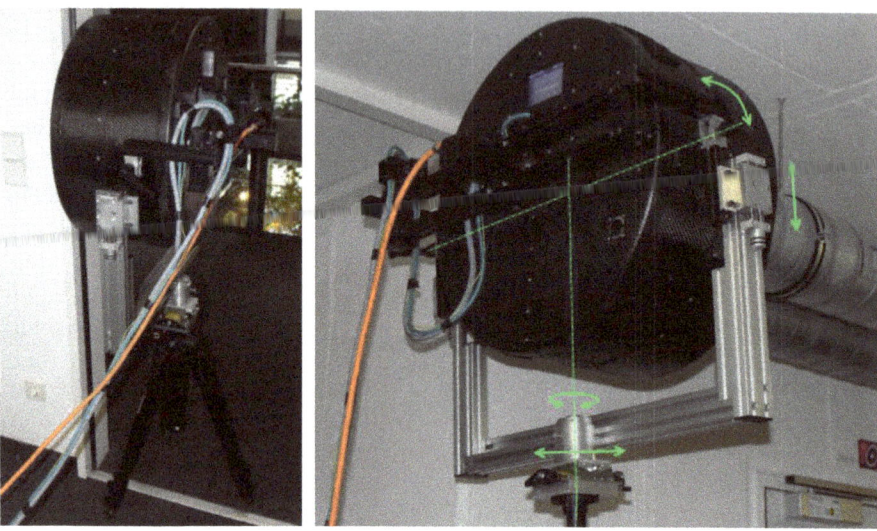

Fig. 2. Left - the tripod-based alignment system. Right - marked alignment axes and motions.

3.2 Table Top Frame Based External Holder

The table top frame based external holder was designed to cover the sample surface axis in lower quadrant directions. The measurement instrument can be disengaged from the tripod U-frame in vertical axis supports and replaced to the same support's mount on the table top holder. It keeps a fixed position of the instrument's center mass of gravity regarding the moving axis. The angle can be set from the bottom perpendicular view up to the horizontal view angle. The sliding motions are distributed over the holder structure. Horizontal y axis motion is realized by moving the sliders holding the horizontal axis supports. Vertical z motion is realized by leveling three set screws in the frame's legs. The legs with screws were designed telescopic to compensate ground irregularities larger than motion range of the set screws. The x motion is performed with a precise linear guide. To keep the exact position during alignment the tips of the leg's set screws are kinematically constrained to the ground. The set screw engaged to the linear guide is permanently fixed allowing the angular inclination only. Rest two leg's screws are provided with ball transfer units. The ball unit on the leg in the direction of linear slide motion is engaged in a V-groove in the direction of the motion. This removes another two degrees of motion of the holder. The last ball unit in the third leg is freely sliding over a flat metal pad. This design allows for an exact position at any angular position θ of the instrument where the final position of the instrument regarding its rotary axis in the yz plane are given by the coordinate transfer relations:

$$\begin{bmatrix} x' \\ y' \\ z' \end{bmatrix} = \begin{pmatrix} 1 & 0 & 0 \\ 0 & \cos\theta & -\sin\theta \\ 0 & \sin\theta & \cos\theta \end{pmatrix} \begin{bmatrix} x \\ y \\ z \end{bmatrix} \quad (1)$$

The table top frame-based external holder 3D design and its final realization is shown in Fig. 3.

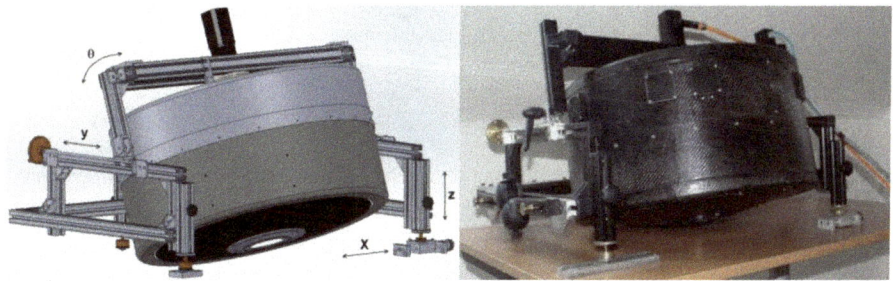

Fig. 3. Left - 3D design of table top frame based alignment system. Right - realization.

3.3   The Laser Autocollimator Alignment Feedback

Presented alignment holders were designed so to set the instrument position and angles regarding the measured sample surface with high enough resolution and stability. The alignment motions are controlled with feedback information provided by the instrument. When a coarse sample position is set, checked with the camera image, the rotation given by the servomotor with a gear box axis needs to be aligned perpendicularly to the sample with a help of in-built laser autocollimator set to the instrument rotary axis.

The laser autocollimator consists of an adjustable 1 mW laser module and lateral displacement beamsplitter made of a beamsplitter cube and 90° mirror prism. The laser beam is split into a reference beam making fixed laser spot on a diffuser attached to the beamsplitter prism and a measurement beam aligned in the instrument rotary axis. The sample surface temporarily provided with a thin mirror reflects the beam back through the prisms to the diffuser. The angular misalignment is indicated as a distance between the reflected and the reference laser spots imaged with 5 Mpixel camera. On 25 mm wide diffuser and the light travel distance 346 mm between the mirror and the diffuser, we can detect angle changes in the range ±2.65° with theoretical resolution 0.0027°. This feedback information helps to set the instrument rotary axis perpendicular towards the sample with accuracy ±0.1°.

4   Results

We checked the functionality of both external alignment frames. They proved the capability to set the measurement instrument to a measured sample with any inclination angle. We took measurements of many samples and the overall geometrical repeatability was ± 1.2 px/0.011°. The BTF data sets were successfully measured, processed and applied for rendering on 3D virtual objects, with an example shown in Fig. 4.

Fig. 4. Stone sample surface and its rendering on a sphere illuminated by an environment map approximated by 256 directional light sources and a 3D objects coated with the measured material.

The image on the left is one of 16,680 taken images. The next three images show the application of the BTF to surface of three different 3D object models (sphere, dragon, ashtray) illuminated by environment map. The last image shows the different appearance of the image on the 3D model of ashtray, because of the change of illumination. The 3D object of ashtray is illuminated by a single point light. The software for processing the acquired images is custom, written in ANSI C++. The taken images have to be rectified and processed to a set of square images as they were viewed from a normal direction by orthographic camera. As the size of the whole image dataset for one material is big (about 10 GBytes after processing), they are optionally compressed by lossy compression method such as principal component analysis, the details in [7]. During the rendering in a custom application the data are retrieved from the processed images, possibly after decompression and applied to the surface. The BTF data for incident light direction and outgoing view direction that were not acquired during the measurement are computed via interpolation or extrapolation (for zenith angle larger than 75°). The details of the whole processing pipeline are in [9, 10].

5   Conclusions

We presented a portable instrument for BTF measurement and solutions allowing alignment of the measurement system against the sample surface under any inclination angle. The task was split into two solutions each for lower and upper quadrant angles. Both solutions have ability to perform measurement in the second quadrant limited up to 20° angle. We presented a simple laser autocollimator system that enables to align the instrument rotary axis to the sample surface normal within angle range better than ± 0.1°. The functionality and rigidity was confirmed by a set of *in situ* measurements of the sample surfaces under different inclination angles and final BTF data processing and rendering.

Acknowledgements. This work was supported by the Czech Science Foundation of the Czech Republic, under research project number GA14-19213S and SGS17/176/OHK2/3T/12.

References

1. Nicodemus, F.E., Richmond, J.C., Hsia, J.J., Ginsberg, I.W., Limperis, T.: Geometric Considerations and Nomenclature for Reflectance. National Bureau of Standards (US), Jones and Bartlett Publishers, Burlington (1977). Monograph 161
2. Dana, K., Van-Ginneken, B., Nayar, S., Koenderink, J.: Reflectance and texture of real world surfaces. ACM Trans. Graph. **18**, 1–34 (1999)
3. Han, J.Y., Perlin, K.: Measuring bidirectional texture reflectance with a kaleidoscope. ACM Trans. Graph. **22**(3), 741–748 (2003)
4. Müller, G., Meseth, J., Sattler, M., Sarlette, R., Klein, R.: Acquisition, synthesis, and rendering of bidirectional texture functions. Comput. Graph. Forum **24**(1), 83–109 (2005)
5. Holroyd, M., Lawrence, J., Zickler, T.: A coaxial optical scanner for synchronous acquisition of 3D geometry and surface reflectance. ACM Trans. Graph. **29**(4) (2010). Article 99
6. Mukaigawa, Y., Tagawa, S., Kim, J., Raskar, R., Matsushita, Y., Yagi, Y.: Hemispherical Confocal Imaging Using Turtleback Reflector, pp. 336–349. Springer, Heidelberg (2011)
7. Schwartz, C., Sarlette, R., Weinmann, M., Klein, R.: DOME II: a parallelized BTF acquisition system. In Proceedings of the Eurographics 2013 Workshop on Material Appearance Modeling: Issues and Acquisition (MAM 2013). Eurographics Association, Aire-la-Ville, Switzerland, pp. 25–31 (2013)
8. Filip, J., Vávra, R., Krupička, M.: Rapid material appearance acquisition using consumer hardware. Sensors **14**(10), 19785–19805 (2014)
9. Havran, V., Hošek, J., Němcová, Š., Čáp, J., Bittner, J.: Lightdrum – portable light stage for accurate BTF measurement on site. Sensors **17**(3), 7373–7384 (2017)
10. Hošek, J., Havran, V., Němcová, Š., Bittner, J., Čáp, J.: Optomechanical design of a portable compact bidirectional texture function measurement instrument. Appl. Opt. **56**(4), 1183–1193 (2017)

Fatigue Analysis of T-filleted Welded Joints in Press Cutting Machines

L. Marques[1], V. H. Carneiro[2(✉)], and J. Meireles[2]

[1] Department of Mechanical Engineering,
University of Minho, Guimarães, Portugal
[2] MEtRiCs - Mechanical Engineering and Resource Sustainability Center,
University of Minho, Guimarães, Portugal
vitorhcarneiro@hotmail.com

Abstract. Welded Joints in components or structures submitted to cyclic loading can fail by crack initiation and progression due to fatigue. Thus, they are critical features in the overall manufacturing of press cutting machines and display an important part of production costs. In this study, the decrease of throat thickness and weld penetration are analyzed in T-filleted joints. Overall, it is determined that throat reduction only represents a real challenge when weld penetration is fairly reduced. In order to obtain infinite fatigue life, a minimum 27 [%] weld penetration must be assured for a 6 [mm] throat, while in 8 [mm] throat designs only 15 [%] is required.

Keywords: Welding · T-joint · Penetration · Throat reduction
Fatigue

1 Introduction

Welded joints in components or structures submitted to cyclic loading can fail by crack initiation and progression due to fatigue [1]. The fatigue strength of a welded joint can be affected by a large number of parameters, such as global and local geometry, material, loading type, structural discontinuities, stress distribution, cyclic stress, mean and residual stresses, etc. [2].

Given these challenges, researchers have focused their efforts in the developing new and better methods and procedures for accurate determination of stress concentrations and fatigue life of welded components, e.g. "Eurocode 3" code [3] and "IIW recommendations for fatigue design of welded joints and components".

Recently, there has been significant research concerning the life prediction of welded structures based on FEA, although this procedure can be done in several different ways with quite different level of accuracy [4].

In this study the life prediction of T-fillet welded joint of press cutting machines is studied by variation of weld throat thickness and weld penetration employing FEA routines. The objective of this research is the analysis of the possible industrial implementation of throat and weld penetration reduction.

© Springer International Publishing AG, part of Springer Nature 2019
J. Machado et al. (Eds.): HELIX 2018, LNEE 505, pp. 494–500, 2019.
https://doi.org/10.1007/978-3-319-91334-6_67

2 Methodology

Effective notch stress is the total stress at the root of a notch, obtained assuming linear-elastic material behavior. To take account of the variation of the weld shape parameters, as well as of the non-linear material behavior at the notch root, the actual weld contours were replaced by an effective one. Given the purpose of this work, a sub-model with 400 [mm] of thickness, according to the standard norm AWS D1.1:2000 (p. 230, Table 5.4) [5] and the dimensions in Fig. 1 was studied by FEA.

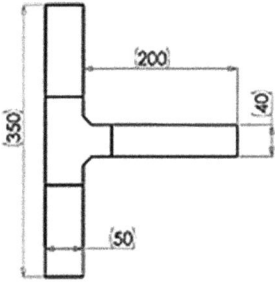

Fig. 1. Model of T-filleted joint (dimensions in [mm]).

According to the purpose of this study, the referred model was designed to assume weld throat thicknesses of 8 [mm] and 6 [mm]. Additionally, weld penetration was also changed. While an initial full penetration state was analyzed (Fig. 2(a)), intermediate states with 50 [%], 25 [%] (Fig. 2(b)), 12.5 [%] and without penetration (0 [%] – Fig. 2 (c)) were also considered.

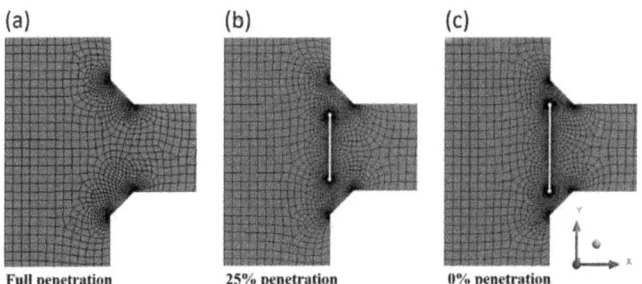

Fig. 2. Examples of models with (a) full, (b) 25[%] and without penetration (8 [mm] throat).

2.1 Material Characterization

To eliminate material induced differentiation, all three prototypes were built using S275 structural steel. This material was produced according to EN10025, in the form of sheet metal and assembled by welding. Specimens were manufacture and tested according to

ISO 6892-1 – specimen Type D, recurring to a universal testing machine in displacement control (strain rate of 0.02 [/s]). The instantaneous loads and axial/transverse deformations were monitored to determine the stress-strain behavior and Poisson's ratio of five equivalent specimens. Data concerning the behavior of this material in fatigue regime was obtained in [6] by the definition of its S-N curve.

2.2   FEA Inputs – Structural and Fatigue Simulation

Static structural finite element analysis was performed in the developed models recurring to ANSYS 17.0 and using the tested material properties to predict the values of maximum Von Mises Stress (σ) and number of cycles (N) before failure by fatigue (using the classic Goodman formulation). According to Fig. 3, a 2D model of the T-filleted joint was fixed (u = 0; v = 0) in the top and bottom of the vertical plate, while an initial static load (W) of 27 [kN] in the YY-axis was applied in the horizontal plate. The adopted 2D approach is justified by the overall classic Plane Strain configuration. Meshing (SHELL181 rectangular 4-noded elements) with refinement in the predicted critical areas was performed to permit a smooth response surface and reduce meshing response errors.

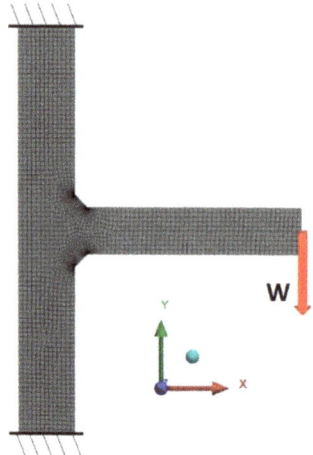

Fig. 3. Representation of FEA boundary conditions and meshing.

3   Results and Discussion

3.1   S275 – Material Characterization

Results concerning the stress-strain behavior and the determined properties of the S275 steel may be observed in Fig. 4(a). In Fig. 4(b) is plotted the S-N curve of the adopted steel. The data in Fig. 4 was used as input parameters in the subsequent FEA routines.

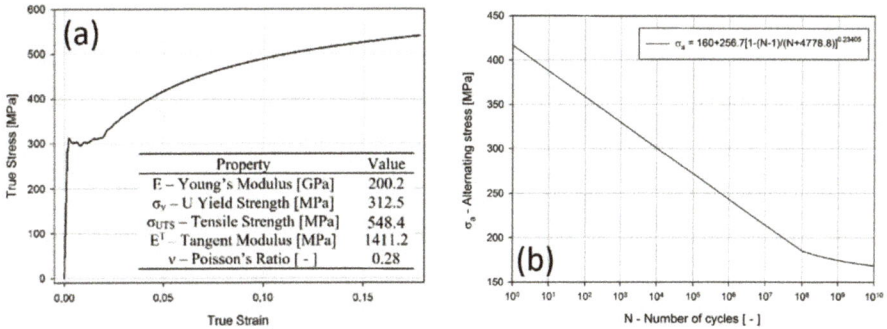

Fig. 4. S275 mechanical properties: (a) Stress-strain and (b) S-N curves.

3.2 FEA Fatigue Comparison

Interpreting the numerical results, the critical zones (areas with higher Von Mises stress values) were determined for all adopted designs. According to the static structural analysis, a primordial parameter to determine the critical zones is defined by the presence of penetration. In samples with full penetration (Fig. 5(a)) the imposed bending moment imposes high stress values in the lower fillet of the T-joint, while samples without full penetration (Fig. 5(b)) shown maximum Von Mises stress in the upper fillet. Additionally, in the latter, high values of stress are also present in the weld root.

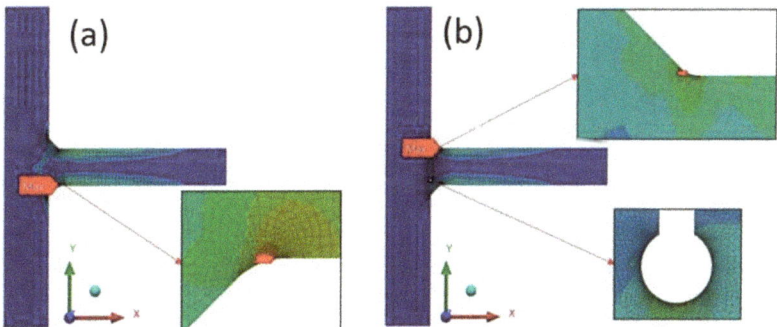

Fig. 5. Graphical representation of Von Mises stress distribution in (a) full and (b) partial penetration samples.

Relatively to these maximum stress areas, the determined maximum values of Von Mises stress according to throat thickness and weld penetration may be observed in Fig. 6. It is determined that a reduction in penetration implies an increase in the overall Von Mises stress values according to an exponential decay function, represented by the calculated non-linear regressions. It may be observed that a reduction of throat thickness form 8 [mm] to 6 [mm] generates an increase in stress. However, this

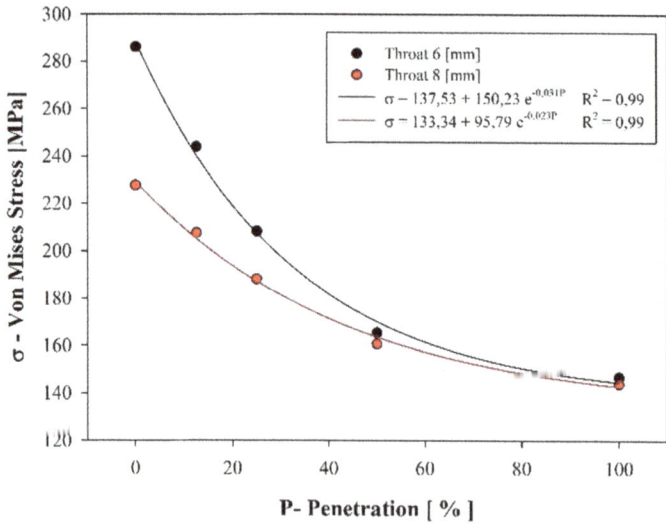

Fig. 6. Maximum Von Mises stress by the change in throat thickness and penetration.

differences are much more significant when the penetration is reduced. In full penetration samples this difference is negligible (~ 3 [%] difference).

Interpreting the maximum stress values that were determined by FEA, the representation of the stress amplitude (σ_a) for the simulated conditions were plotted in Fig. 7. These defined stress cycles impose different fatigue regimes, whose maximum number of cycles (N) were also determined by FEA and are presented in Fig. 8.

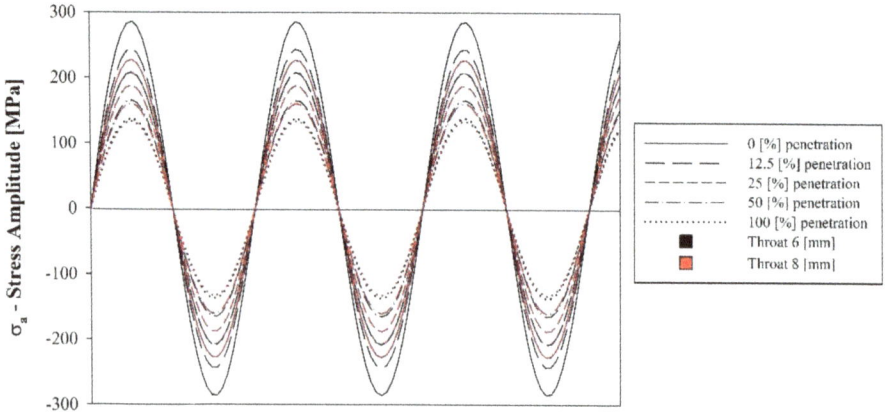

Fig. 7. Representation of numerical stress amplitude for the simulated models.

According to Fig. 8, it may be observed that both throat thickness and weld penetration are primordial in the fatigue response of the T-joint. It may be observed that a

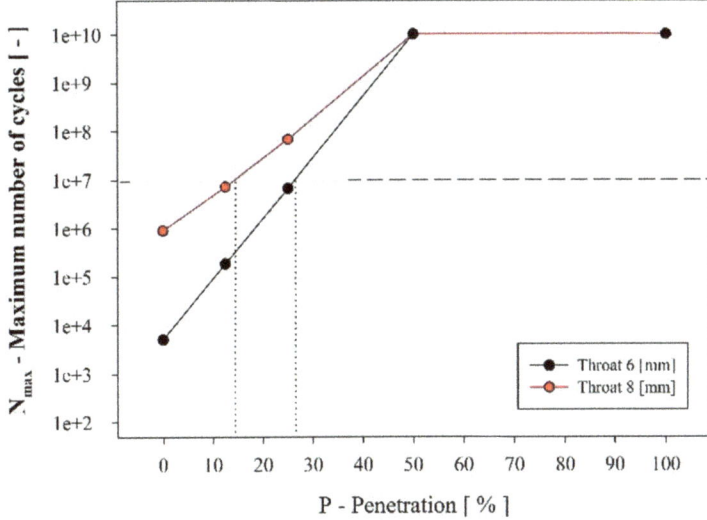

Fig. 8. Fatigue analysis by the determination of the maximum number of cycles by throat and penetration reduction.

throat reduction implies a fair lowering in the maximum number of cycles and, thus, a premature failure. Generally, to obtain an infinite life in the applied conditions a minimum of 27 [%] and 15 [%] penetration, respectively, for 6 [mm] and 8 [mm] throat thicknesses.

4   Conclusions

The role of throat thickness and weld penetration in the structural and fatigue behavior of a T-filleted joint subjected to a cantilever-like loading is analyzed in this study by FEA. The input conditions, namely, the used material, loading and boundary conditions correspond to a representation of actual applications in the manufacturing of press cutting machines. After such study, the following conclusions were drawn:

- T-joints with full weld penetration display the maximum Von Mises stress in the lower weld fillet. However, when penetration is not full, this is observed in the upper weld fillet.
- The maximum Von Mises stress value may be related to the penetration thru an exponential decay function.
- The role of throat thickness reduction form 6 [mm] to 8 [mm] is only significant as penetration decreases.
- To obtain infinite fatigue life in this particular application a minimum penetration of 27 [%] and 15 [%] must be assured for throat thicknesses of 6 [mm] and 8 [mm].
- A future economic study, concerning the costs of throat reduction by increasing penetration may be interesting to determine the economic viability of this procedure and its impact in the overall welding costs in T-fillet joints.

References

1. Branco, C.M., Ferreira, J., Domingos, J., Silva, A.: Projecto de Órgãos de Máquinas. Fundação Calouste Gulbenkian, Lisbon (2005)
2. Eurocode 3: Design of steel structures – Part1–9: General – Fatigue strength (2005)
3. Hobbacher, A.: Recommendations for Fatigue Design of Welded Joints and Components. International Institute of Welding, Paris (2008)
4. Baumgartner, J.: Review and considerations on the fatigue assessment of welded joints using reference radii. Int. J. Fatigue **101**, 459–468 (2017)
5. Structural welding code – Steel, American Welding Society, Miami, USA (2000)
6. Boyer, H.E.: Atlas of Fatigue Curves. ASM International, Ohio (1985)

Vierendeel Bending of Beams with Web Openings – Automatic Calculation

Francisco Silva[1]([✉]), João Marques[2], José Meireles[1],
and Nuno Peixinho[1]

[1] Department of Mechanical Engineering,
University of Minho, Guimaraes, Portugal
a70253@alunos.uminho.pt,
{meireles, peixinho}@dem.uminho.pt
[2] Design Department, bysteel, Braga, Portugal

Abstract. This paper addresses the design of beams with web openings. The Vierendeel bending design procedures are described, and the related numerical equations are given. An automatic calculation model, for sections design is briefly described, and an input and output flowchart is also presented.

Keywords: Steel · Beam with web openings · *Vierendeel* bending

1 Introduction

With the increasing application of steel beams with web openings in structures, and bearing in mind that, today, the EN 1993-1-13 Eurocode 3 - Design of steel structures - Part 1–13: Rules for beams with large web openings standard is being drafted, this paper pretends to address the section design of these beams.

According to the British guide SCI-P355: Design of Composite Beams with Large Web Openings (Lawson and Hiks) {adapted from [1] 2011}, the Vierendeel bending mechanism is regarded as one important phenomenon to be analysed in this type of elements. Having as guidance, the Portuguese version of the European standard for steel structures, NP EN 1993-1-1 {adapted from [2]} and the British guide, a section design automatic calculation model was developed. Scientific publications {adapted from [3] 2012}, were also consulted during the development and calibration of the automatic calculation model. On the other hand, a scientific essay {adapted from [4]} was helpful in theoretical terms.

2 Vierendeel Bending Design

In a beam with a web opening, the cross-section on the opening zone is composed by two "T" sections (top and bottom), and the shear transfer occurs by Vierendeel bending of the T-sections at the four corners of the opening. This phenomenon is exemplified in Fig. 1.

© Springer International Publishing AG, part of Springer Nature 2019
J. Machado et al. (Eds.): HELIX 2018, LNEE 505, pp. 501–507, 2019.
https://doi.org/10.1007/978-3-319-91334-6_68

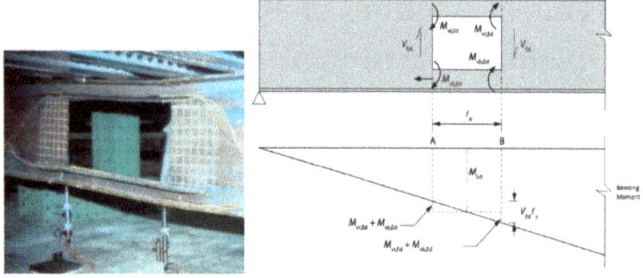

Fig. 1. Balance of forces in an aperture {adapted from Lawson and Hicks, 2011}

The Vierendeel bending resistance is oriented by the web classification of the T's, and in case of class 3 or 4 webs, only the elastic resistance can be used. However, if the sections meet the limits for class 2, the Vierendeel resistance can be determined based on the plastic resistance of the T's. Thus, the plastic bending resistance the T in the absence of axial force (and in the absence of high shear), is given by the following expression, assuming that the plastic neutral axis (PNA) is in the flange of the T, (Eqs. 1 and 2), {adapted from [5]}:

$$M_{pl,Rd} = \frac{A_{w,T} f_y}{\gamma_{M0}} \left(\frac{h_w}{2} + t_f - z_{pl} \right) + \frac{A_f f_y}{\gamma_{M0}} \left(\frac{t_f}{2} - z_{pl} + \frac{z_{pl}^2}{t_f} \right) \tag{1}$$

$$z_{pl} = \frac{A_f + A_{w,T}}{2bf} \tag{2}$$

To determine if the PNA is located in the flange or in the web of the T, an automatic calculation model has been developed that assumes the position of the PNA as x, and with a condition solver, determinates the actual PNA position. Therefore, if the PNA is located in the flange (Eq. 3):

$$M_{pl,Rd} = \frac{\left(\frac{x * b_f * x}{2} \right) + (t_f - x) * b_f * \left(\frac{t_f - x}{2} \right) + h_w * t_w * \left(t_f - x + \frac{h_w}{2} \right)}{f_y} \tag{3}$$

If it is in the web (Eq. 4):

$$M_{pl,Rd} = \frac{A_f \left(x - \frac{t_f}{2} \right) + (x - t_f) * t_w * \left(\frac{x - t_f}{2} \right) - (h - x) * t_w}{f_y} \tag{4}$$

where:

Mpl, Rd - plastic bending resistance (of the T);
Af - cross-sectional of the flange;
tf - flange thickness;
tw - web thickness;

Aw, T - cross-section area of web;

hw - clear web depth of beam between flanges;

x - solver result;

zpl – depth of plastic neutral axis of T from outer face of flange;

The plastic bending resistance shall be reduced if there is co-existent axial force. In the presence of axial force, the value of the plastic resistant bending moment shall be reduced according to the following equation given in ENV 1993-1-1:

$$M_{bT,N,Rd} = M_{pl,N,Rd} = M_{pl,Rd}\left(1 - \left(\frac{N_{Ed}}{N_{pl,Rd}}\right)^2\right) \tag{5}$$

onde:

MbT, N, Rd - bending resistance of the bottom T, reduced for axial tension and shear;

Mpl, Rd - plastic bending resistance (of T section);

NEd - actuant strain of T;

Npl, Rd - resistance strain plastic of T;

Thus, the total Vierendeel bending resistance at the four corners should not be less than the design value of the difference in bending moment from one side of the opening to the other, due to the shear force, as defined by the following (Eq. 6): (Fig. 2)

$$2M_{bT,NV,Rd} + 2M_{tT,NV,Rd} \geq V_{Ed}l_e \tag{6}$$

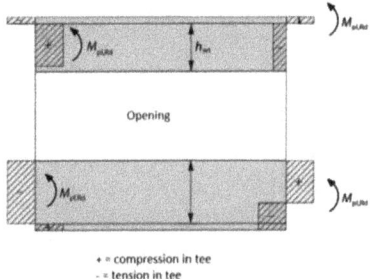

Fig. 2. Tensions due to the plastic bending moment in an aperture. {adapted from Lawson and Hicks, 2011}

2.1   Automatic Calculation Model

In order to parameterize the calculation of the alveolar beams sections, an automatic calculation model has been designed in a spreadsheet environment. This model allows the dimensioning of simply supported alveolar beams in metallic working. An exhaustive library, according to ArcelorMittal's catalog {adapted from [6]}, containing the commercially available sections on the market has been included on the spreadsheet

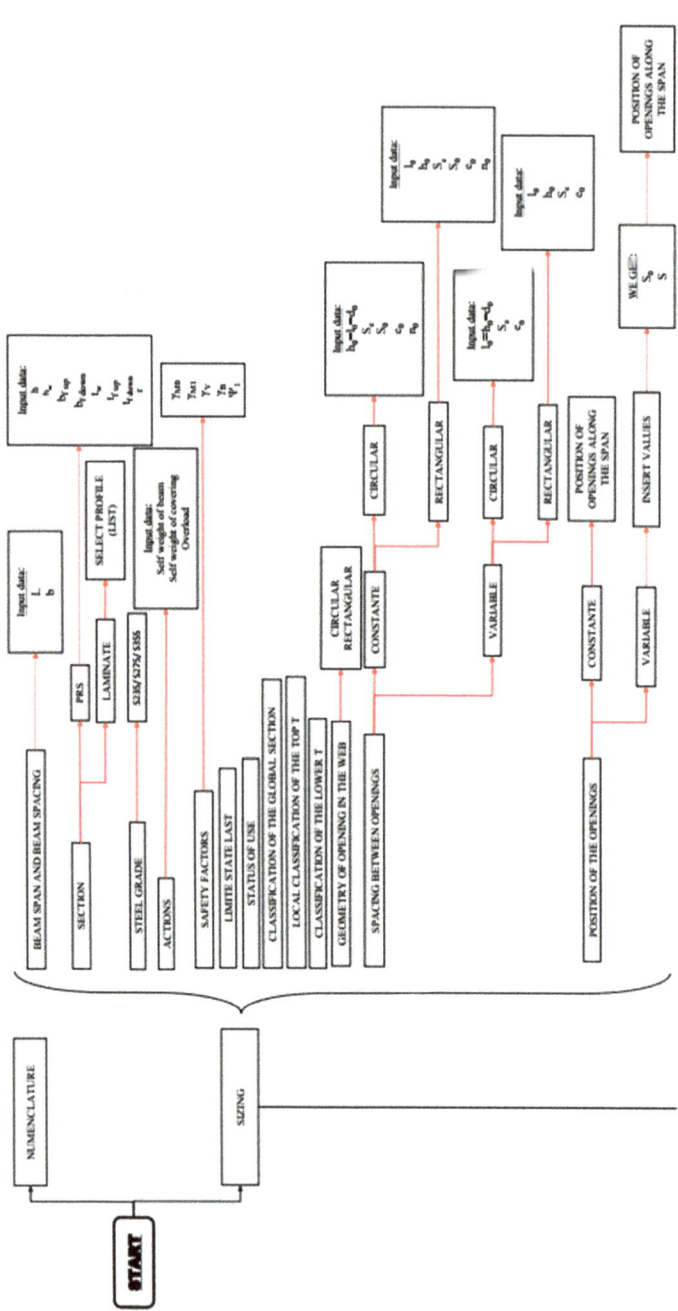

Fig. 3. Flowchart of the automatic calculation model of alveolar beams

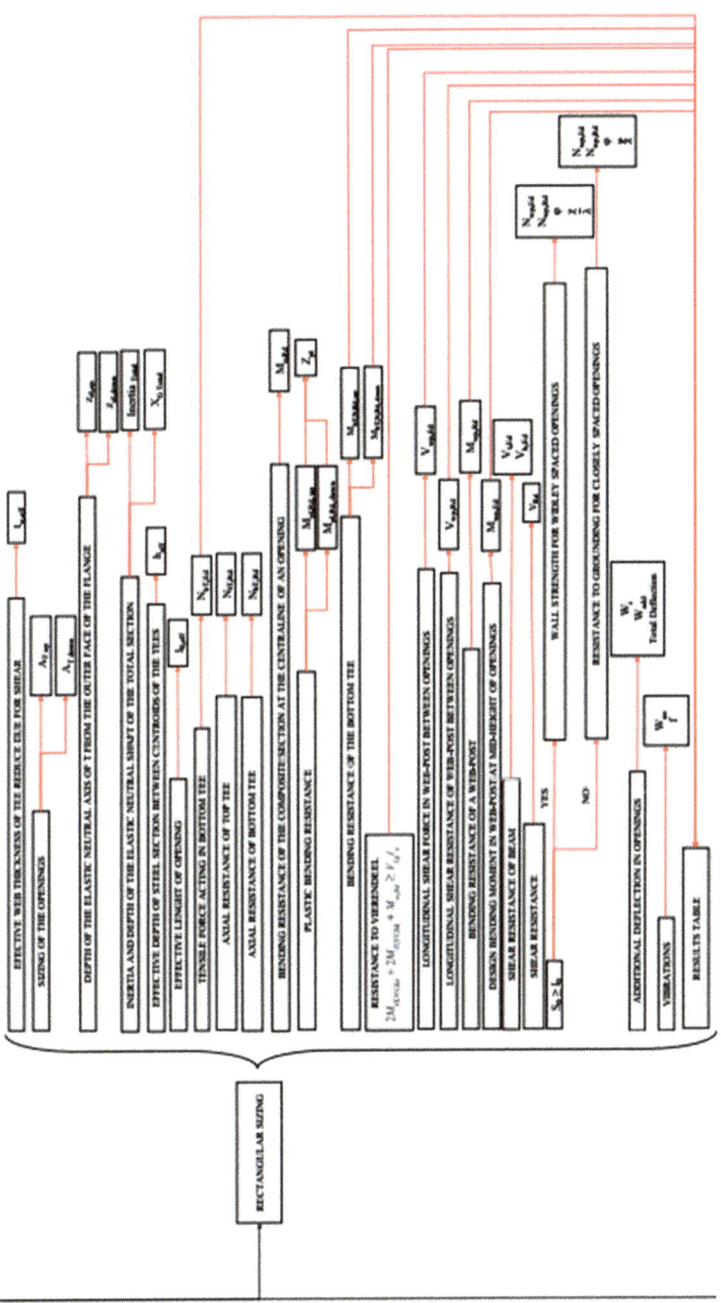

Fig. 3. (*continued*)

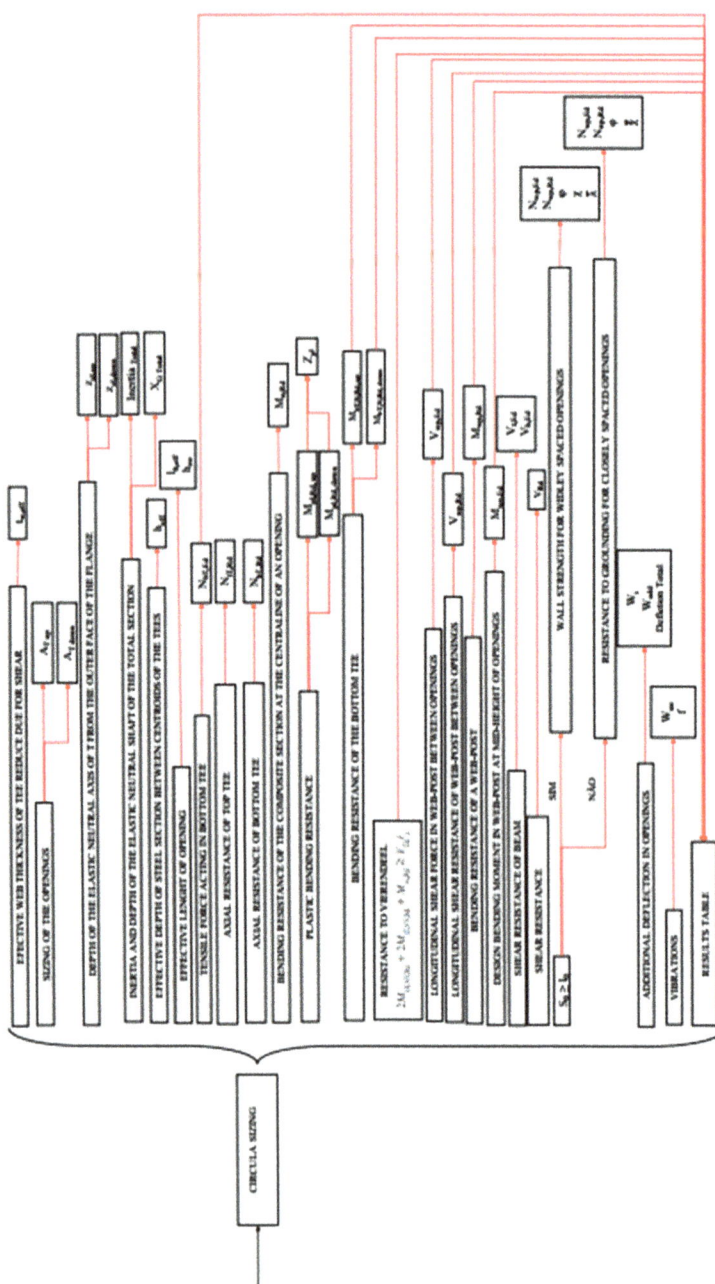

Fig. 3. (*continued*)

environment. This environment allows the customization of asymmetric sections with different flanges' width and thickness. For the openings, circular or rectangular geometry with variable or constant spacing can be chosen. Nonetheless, for practical purposes, the number of apertures on the beams has been restricted to 20. This model also allows for the possibility to insert eccentricity in the aperture relative to the web axis and to include beams under distributed loads. Figure 3 presents a flowchart containing the organized structure of the model's functionalities.

3 Conclusion

It was analysed the calculation approach recommended in the British sizing guide SCI-P355: Design of composite beam with large web openings (Lawson and Hiks 2011), namely the Vierendeel flexing mechanism. An automatic calculation model was also drawn up and briefly described, which aims to provide an expeditious approach to the calculation of structures whose size is not, as of the date of execution of this publication, set out in legal documentation.

References

1. Lawson, R.M., Hicks, S.J., SCI-P355: Design of composite beams with large web openings: in accordance with Eurocodes and the UK National Annexes, Steel Constructional Steel Research (2011)
2. NP 1993-1-1: Eurocódigo 3- Projeto de estruturas de aço, Parte 1-1: Regras gerais e regras edifícios (2010)
3. Erdal, F., Saka, M.P.: Ultimate load carrying capacity of optimally designed steel cellular beams. J. Constr. Steel Res. **80**, 355–368 (2012)
4. Alminhana, G.W.: Vigas metálicas alveolares: Análise comparativa com vigas de alma cheia, p. 143. Universidade Federal do rio grande do Sul (2014)
5. Silva, F.: Análise paramétrica de vigas alveolares - Ensaios e Dimensionamento Numérico. Dissertação de Mestrado Integrado em Engenharia Mecânica, Universidade do Minho (2017)
6. ArcelorMittal Europe: Profilés et Aciers Marchands Sections and Merchant Bars Profil-und Stabstahl (2017)

Mechanical Design in Industry 4.0: Development of a Handling System Using a Modular Approach

Joel Galvão[1(✉)], João Sousa[1], José Machado[1], João Mendonça[1], Toni Machado[2], and Pedro Vaz Silva[2]

[1] MEtRICs Research Centre, University of Minho, Guimarães, Portugal
jmrgalvao@hotmail.com
[2] Bosch Car Multimedia Portugal S.A, Braga, Portugal

Abstract. In the nowadays-global market, it is necessary to consider the new paradigm of industry 4.0 that presents a lot of features in the development completive solution. Several approaches are being used to considering the problem. In this paper a modular approach for developing a handling system to use in the shop floor production line to move raw materials using autonomous vehicles is presented is present. In addition, the final solution for the module to transfer returnable packages is discussed.

Keywords: Industry 4.0 · Modular Mechanical Design · Mechanical design

1 Introduction

Industry 4.0 tries to create smart features [1] where products are equipped with embedded systems, sensors and actuators as well as the ability to communicate with each other, that is called Cyber-Physical Systems (CPS) [2].

The possibility of CPS to revolutionize a variety of aspect in your daily life is enormous. Concepts such as autonomous guide vehicles [3], intelligent buildings [4], smart electric grid [5], smart factories [6], and implanted medical devices [7] are just some of the practical examples that have already emerged.

This new paradigm [8–10], brings to the table social, economic, and political changes in particular; short development periods, individualization on demand, flexibility, decentralization and resource efficiency. This leads to exceptional technology-push in areas like mechanization and automation; digitalization and networking; and miniaturization. The development team of Ang [11] proposes a 5 level architecture for the application of industry 4.0 paradigm to the production line.

In fact, with the expanding of the industry 4.0 [12], the need for combining different areas of knowledge is imperative, and there is still is a gap in what really is Industry 4.0 is and what are its design steps.

The necessity to incorporate different areas of knowledge and a lot of components with different requirements, brings to the mechanic design, new features to consider, as discussed in [13–15], One possible technique, to respond to high-performance specifications of Industry 4.0, is a mechatronics approach [16, 17], where a coordinated and

© Springer International Publishing AG, part of Springer Nature 2019
J. Machado et al. (Eds.): HELIX 2018, LNEE 505, pp. 508–514, 2019.
https://doi.org/10.1007/978-3-319-91334-6_69

simultaneous philosophy between mechanics, electronics, and intelligent control in the manufacturing of equipment is applied [18]. The systems developed with this technique are considerably complex. This is due to a lot of components when compared to a usual mechanic system. This problem was considered in the preliminary stages of the project due to the interaction between different component's impact by the final system behavior [19].

Other possible methodologies consider a modular approach to solve problems individually and then join together each different module in a global solution [20–23]. This approach has several advantages [23], as discussed below:

- Modules can be manufactured simultaneously and separately using different manufacturing tactics;
- Assembly and disassembly for maintenance are eased;
- Several alternatives for the same product can be executed modifying only one of the modules;
- Manufacturing difficulties are reduced because a simpler module may be easier to obtain than a more complex module;
- End of life of the product facilitates reuse and recycling.

On the other hand, there were also some challenges in the mechanic design point of view due to the necessary to implement this approach from the beginning of the development of a new equipment, namely it was essential to be considered from the very early stage when solutions are being developed. The modules necessity to be well designed in order to that the obtained final product had the same quality standards of a one-piece machine [22]. In fact, the design using a modular approach presents the major challenge of management of a large volume of information like the materials [24], rigor dimensions of each module [22], and an interconnection between the different modules [25].

This scientific research is part of a partnership between the Bosch BrgP and the University of Minho called Innovative car HMI [26] with the focus on software and embedded applications for intelligent vehicles. The research team was focused on developing innovate cyber-physic solution to use in the industrial shop floor.

This paper is organized with a Sect. 2 where the design intent is described and its requirements. In Sect. 3, conclusions and future work are presented.

2 The Project

The main objective of this project is to develop a system for the internal movement of raw materials using autonomous guided vehicles (AGVs) adaptable to different environments. The objective of increasing efficiency and lower costs of this operation, according to the model Just in Time [27].

2.1 The Handling System

One of the requirements is to develop a handling system capable of autonomously transfer Returnable Packages (RPs) from the trolleys to the dynamic racks. The system comprises the following sub-systems: reception station (Fig. 1c), delivery station (Fig. 1a) dynamic racks (Fig. 1e), the autonomous guide vehicle (Fig. 1b) and the trolleys (Fig. 1d).

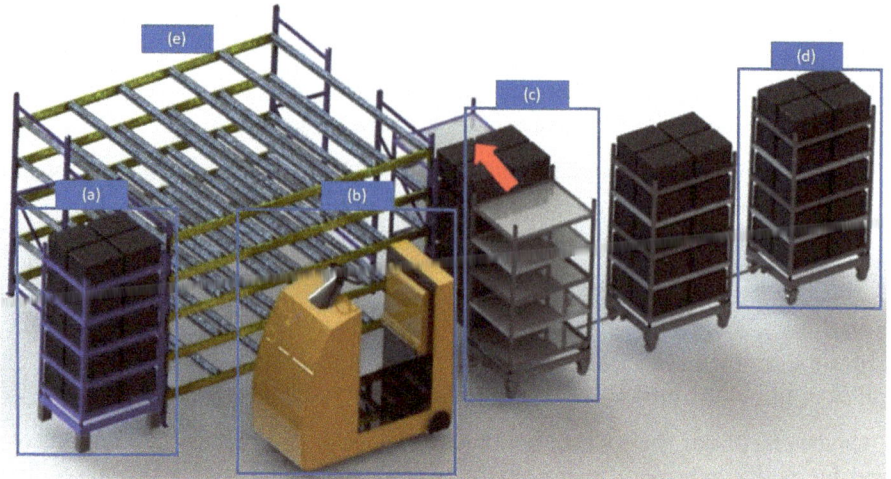

Fig. 1. Trolley transferring all RPs to the reception station

In Fig. 1 a global view of the project is represented. It is possible to see each main physical part. For each logistical cycle, the autonomous vehicle and the trolleys filled with RPs to deliver them to the reception station (Fig. 1c). Then it drives to the delivery station (Fig. 1a) to receive the empty RPs, always without any human interference. This lead to the following modules:

- Autonomous guide vehicle;
- Trolleys main body;
- Returnable packages translation mechanism;
- Returnable packages lateral stopping mechanism;
- Energizing system;
- Dynamic racks;
- Reception station;
- Delivery station;
- Coupling mechanism between trolleys;

This paper discusses the returnable packages translation mechanism, as represented in Fig. 1. The purpose of this mechanism is to transfer returnable packages from the reception station. In the case, of the RPs translation mechanism is able to:

- Easy access, no obstruction of the front of the dynamic rack, because manual workers need to be able to reach it to unload returnable packages manually if required;

- Designed to operate with the following returnable packages: RK22, RK17, RK22P, RK22G e RK12PP;
- Automatically remove all RPs from each level;
- Provide constant linear velocity for extracting the RPs of 0,018 m/s.

2.2   Developed Solution

For this work to develop the CAD models, CAD software SOLIDWORKS [28] version 2016b was used.

The electric motor (fixed under the lower level) transmits rotational motional to the lateral sprockets (orange components in Fig. 2) using chain rollers (not represented). The rotating sprockets are connected to the pulleys by a common shaft leading to the rotation of the timing belts.

Fig. 2. Handling mechanism with returnable packages translation mechanism

A steel rod is connected to each pair of timing belts (a pair for each level) by an insert connected to the timing belt by heat treatment. The rotation of each pair of timing belts will then lead to the translation of the returnable packages by pushing them with the steel rod (Fig. 3).

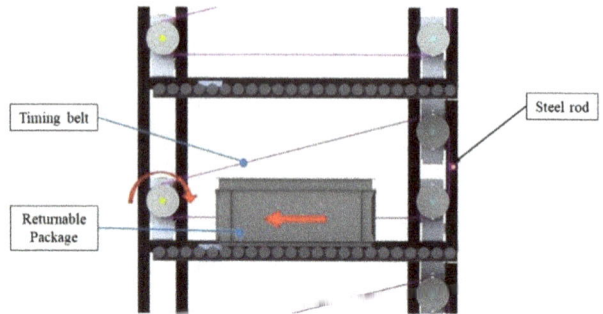

Fig. 3. Lateral view of the returnable packages translation mechanism

3   Conclusions

In the nowadays-global market, it is necessary to consider a multidisciplinary approach to respond to highly challenging demands. In this paper, a modular approach to solve individually all the necessities of the project is used. This approach has several advantages as, simple and faster to manufacture, easier to assemble and disassemble, easier to reuse and recycle but was also a lot of challenges as implement this approach from the beginning of development and a great deal of information to be managed.

A handling system to be used in the transport of raw materials according to a methodology of Just in Time with autonomous guide vehicles is presented. The solution of the adopted model has three sub-modules: an electric motor with gearbox and transmission chain rollers, sprockets for power transmission from the motor to each different level and timing belts and pulleys for linear displacement of a steel rod. The translation mechanism is responsible for removing all RPs from each level.

The modular approach was primordial to achieve this solution and the communication between different areas of knowledge was a differentiating factor for the project's success.

Future work is related to dynamic simulation using Matlab's Simscape capabilities [29] to determinate the best coupling mechanism between the trolleys and the autonomous guide vehicle using the technique proposed in [30].

Acknowledgment. This work is supported by: European Structural and Investment Funds in the FEDER component, through the Operational Competitiveness and Internationalization Programme (COMPETE 2020) [Project n° 002814; Funding Reference: POCI-01-0247-FEDER-002814].

References

1. Anderl, R., Picard, A., Albrecht, K.: Smart product engineering, pp. 10–11 (2013). https://doi.org/10.1007/978-3-642-30817-8
2. Monostori, L.: Cyber-physical production systems: roots, expectations and R&D challenges. Procedia CIRP **17**, 9–13 (2014). https://doi.org/10.1016/j.procir.2014.03.115
3. Drath, R., Horch, A.: Industrie 4.0: hit or hype? [Industry Forum]. IEEE Ind. Electron. Mag. **8**, 56–58 (2014). https://doi.org/10.1109/MIE.2014.2312079
4. Rodriguez-Andina, J.J., Rieger, C.: Intelligent Buildings of the Future, pp. 32–49 (2016)
5. Lom, M.: Industry 4. 0 as a Part of Smart Cities, pp. 2–7 (2016)
6. Shrouf, F., Ordieres, J., Miragliotta, G.: Smart factories in industry 4.0: a review of the concept and of energy management approached in production based on the Internet of Things paradigm. In: International Conference on Industrial Engineering and Engineering Management, January 2015, pp. 697–701 (2014). https://doi.org/10.1109/ieem.2014.7058728
7. Hu, F.: Cyber-Physical Systems: Integrated Computing and Engineering Design, 1st edn. CRC Press Inc., Boca Raton (2013)
8. Müller, J.M., Maier, L., Veile, J., Voigt, K.-I.: Cooperation strategies among SMEs for implementing industry 4.0 (2017)
9. Nowotarski, P., Paslawski, J.: Industry 4.0 concept introduction into construction SMEs. IOP Conf. Ser. Mater. Sci. Eng. **245**, 52043 (2017). https://doi.org/10.1088/1757-899X/245/5/052043
10. Lasi, H., Fettke, P., Kemper, H.G., et al.: Industry 4.0. Bus. Inf. Syst. Eng. **6**, 239–242 (2014). https://doi.org/10.1007/s12599-014-0334-4
11. Lee, J., Bagheri, B., Kao, H.A.: A cyber-physical systems architecture for industry 4.0-based manufacturing systems. Manuf. Lett. **3**, 18–23 (2015). https://doi.org/10.1016/j.mfglet.2014.12.001
12. Hermann, M., Pentek, T., Otto, B.: Design principles for industrie 4.0 scenarios. In: Proceedings Annual Hawaii International Conference on System Sciences, March 2016, pp. 3928–3937 (2016). https://doi.org/10.1109/hicss.2016.488
13. Ang, J.H., Goh, C., Saldivar, A.A.F., Li, Y.: Energy-efficient through-life smart design, manufacturing and operation of ships in an industry 4 0 environment. Energies **10**, 1–13 (2017). https://doi.org/10.3390/en10050610
14. Vogel-Heuser, B., Hess, D.: Guest editorial industry 4.0-prerequisites and visions. IEEE Trans. Autom. Sci. Eng. **13**, 411–413 (2016). https://doi.org/10.1109/TASE.2016.2523639
15. Damrath, F., Strahilov, A., Bär, T., Vielhaber, M.: Method for energy-efficient assembly system design within physics-based virtual engineering in the automotive industry. Procedia CIRP **41**, 307–312 (2016). https://doi.org/10.1016/j.procir.2015.10.004
16. Rivera, C.A., Poza, J., Ugalde, G., Almandoz, G.: A knowledge based system architecture to manage and automate the electrical machine design process. In: 2017 IEEE International Work Electron Control Measurement Signals their Application to Mechatronics, pp. 1–6 (2017). https://doi.org/10.1109/ecmsm.2017.7945875
17. Eigner, M., Dickopf, T., Apostolov, H.: The evolution of the V-Model: from VDI 2206 to a system engineering based approach for developing cybertronic systems. In: Ríos, J., Bernard, A., Bouras, A., Foufou, S. (eds.) Product Lifecycle Management and the Industry of the Future: 14th IFIP WG 5.1 International Conference, PLM 2017, Seville, Spain, 10–12 July 2017, Revised Selected Papers, pp. 382–393. Springer International Publishing, Cham (2017)

18. Bolton, W.: Introdução à mecatrônica. In: Mecatrônica: uma abordagem multidisciplinar, Bookman, pp. 11–13 (2010)
19. Gausemeier, J., Moehringer, S.: VDI 2206-a new guideline for the design of mechatronic systems. In: IFAC Conference on Mechatronic Systems, Berkeley, USA (2002)
20. Gu, P., Slevinsky, M.: Mechanical bus for modular product design. CIRP Ann. – Manuf. Technol. **52**, 113–116 (2003). https://doi.org/10.1016/S0007-8506(07)60544-7
21. Ristevski, S., Cakmakci, M.: Mechanical design and position control of a modular mechatronic device (MechaCell). In: IEEE/ASME International Conference on Advanced Robotics and Mechatronics, AIM, August 2015, pp. 725–730 (2015). https://doi.org/10.1109/aim.2015.7222623
22. Stone, R.B., McAdams, D.A., Kayyalethekkel, V.J.: A product architecture-based conceptual DFA technique. Des. Stud. **25**, 301–325 (2004). https://doi.org/10.1016/j.destud.2003.09.001
23. Kerbrat, A.O., Mognol, P., Hascoët, J.: A new DFM approach to combine machining and additive manufacturing Comput. Ind. **33**, (2011)
24. Mustata, M.Y.: Design of (PEM) fuel cell for optimized manufacturing and performance. In: 2016 International Symposium on Small-Scale Intelligent Manufacturing Systems, pp. 65–70 (2016). https://doi.org/10.1109/sims.2016.7802901
25. Sundin, E., Björkman, M.: Development of a design for manufacturing and assembly (DFM/A) methodology concerning products and components made in composites of carbon fiber reinforced plastics (CFRP) used in the aerospace industry. In: Swedish Production Symposium (SPS-16) (2016)
26. Programa INNOVATIVE CAR HMI. https://www.eng.uminho.pt/pt/investigareinovar/projetoscomempresas/Paginas/programainnovativecarhmi.aspx. Accessed 26 Dec 2017
27. Ghinato, P.: Sistema Toyota de produção: mais do que simplesmente Just-in-Time. Production **5**, 169–189 (1995). https://doi.org/10.1590/S0103-65131995000200004
28. SOLIDWORKS Products - Google AdWords—SOLIDWORKS. http://www.solidworks.com/sw/products/solidworks-products-adwords.htm?mktid=7771&gclid=EAIaIQobChMIrczOqOnw2QIVTrcbCh1JhgieEAAYASAAEgKa4vD_BwE. Accessed 16 Mar 2018
29. Features - Simscape - MATLAB & Simulink. https://www.mathworks.com/products/simscape/features.html. Accessed 29 Dec 2017
30. Mariappan, S.M., Veerabathiran, A.: Modelling and simulation of multi spindle drilling redundant SCARA robot using SolidWorks and MATLAB/SimMechanics. Rev. Fac. Ing. **2016**, 63–72 (2016). https://doi.org/10.17533/udea.redin.n81a06

Modeling and Analysis of Low Velocity Impact on Composite Plate with Different Ply Orientations

Kasam Santosh Kumar, Shreekant Patil,
and D. Mallikarjuna Reddy[(✉)]

Mechanical Engineering, VIT University, Vellore 632014, India
dmreddy@vit.ac.in

Abstract. There has been a predominant growth in the application of composite structures in the engineering fields, particularly in automobile and aerospace industries. This paper is concentrated on the finite element analysis of the low-velocity impact on composite laminates of unidirectional Glass/Epoxy and Carbon/Epoxy material with different ply orientations. The modeling, meshing and simulations are performed using the Finite Element Package-Abaqus/ Explicit. Parameters like deflection and Contact force are studied under fixed boundary condition by varying the ply orientation in the models and maintained a constant angle between the consecutive plies in the plate. A good agreement is seen between the results of Glass/Epoxy and Carbon/Epoxy laminates and discussed.

Keywords: Low velocity impact · Abaqus/Explicit · Consecutive plies
Composite

1 Introduction

Now a day composite structures are used in many applications like aerospace, marine etc., because if its special properties like high strength to weight ratio and lightweight. Low impact velocity causes matrix cracking, delamination, fiber breakage and penetration. Impact on a composite structure damages the structures and reduces the strength and stability and damages are very difficult to detect by naked eye. Therefore, it is very important to predict the failure of the composite structure and also very difficult to understand the behavior of composite structures [1–3]. In order to improve the composite structure, it is important to understand the dynamic behavior and damage propagation under impact loads [4]. From the experiments, low impact velocity on composite structures attains less reduction in their properties like tensile, compression and shear strengths by Sevkat et al. [5] and damages are mainly due to matrix failure and delamination, not by fiber failure by Mikkor et al. [6]. In the cases of metals, impact energy absorbed by elastic region and plastic region but it is very difficult to predict on composite materials because of its complex nature.

In engineering design in order to predict the static and dynamic responses by using FE packages. Abaqus/Explicit has been used to simulate the low-velocity impact on

© Springer International Publishing AG, part of Springer Nature 2019
J. Machado et al. (Eds.): HELIX 2018, LNEE 505, pp. 515–521, 2019.
https://doi.org/10.1007/978-3-319-91334-6_70

composite structures. Study of impact responses, failure and damage progression in composite material has become the most interesting research topic in last two decade [7, 8] and many researchers have reviewed the papers on these topics [9, 10]. Impact behavior of composite structure was studied by both experimental and numerical [11–13], in order to predict the behavior of the composite structure. Some are proposed the analytical formulation [14], analytical models are only applicable for simplified models. In order to overcome experimental approach and analytical method, an efficient numerical analysis is developed. Many researchers have been done work related to low impact velocity on composites by studying the effects of experimental and numerically composite models [15, 16], impact parameters, impact characterization and damage characterization on the impact response and studied the effects on impactor mass, varying impact energy and velocity and damage of laminated composites, experimental and numerically.

In this particular study, the low impact velocity is carried out on composite plate by considering consecutive ply orientation for laminates for the first time. The simulation is carried out by considering a fixed boundary condition. The effect of impactor mass, velocity and its boundary condition, maximum contact force and stress developed in the unidirectional Glass/Epoxy and Carbon/Epoxy plate are investigated and discussed.

2 Formulation of Impact Analysis

2.1 Impact Problem Formulation

In the particular study, the composite model has normally impacted by intender means neglecting friction force with simply supported boundary condition. Local damage developed by rigid spherical indenter, it can be given by governing Eq. 1.

$$M \ddot{u}(t) + K u(t) = F(t) \tag{1}$$

Where M and K are mass and stiffness matrices, u is nodal coordinate vector and $F(t)$ is force vector due to impact.

2.2 Failure Criterion

The popular Hashin's damage criterion is considered for the composite laminate models to predict the progressive damages. Johnson et al. [17] and Fan et al. [18] have also used the Hashin's damage criteria to predict the damage in composite plates under impact, damage of structures are discussed [19]. The damage on the composite plate is predicted by following expressions.

Fiber tension:

$$\left(\frac{\sigma_{11}}{X_T}\right)^2 + r\left(\frac{\sigma_{12}}{S_L}\right)^2 = 1, 0 < r > 1 \tag{2}$$

Matrix tension:

$$\left(\frac{\sigma_{21}}{Y_T}\right)^2 + \left(\frac{\sigma_{22}}{S_L}\right)^2 = 1 \tag{3}$$

Fiber compression:

$$\left(\frac{\sigma_{11}}{X_C}\right)^2 = 1 \tag{4}$$

Matrix compression:

$$\left(\frac{\sigma_{22}}{2S_T}\right)^2 + \left[\left(\frac{Y_C}{2S_T}\right)^2 - 1\right]\left(\frac{\sigma_{22}}{Y_C}\right)^2 + \left(\frac{\sigma_{11}}{S_L}\right)^2 = 1 \tag{5}$$

Where, X_T is the tensile strength and X_c is compressive strength in the fiber direction, Y_T and Y_c tensile and compressive in matrix direction and S_T and S_c are longitudinal and transverse shear strength and r is the coefficient of contribution shear stress to tensile.

3 Material Properties

The laminate is a 300 mm × 300 mm square plate of 2 mm thickness. The Indenter is a spherical analytical rigid body of the 8 mm radius and 3 kg mass with an initial velocity of 3 m/s (impact energy = 13.5 J). Two unidirectional materials Glass/Epoxy and Carbon/Epoxy are considered for similar laminate models to compare the results and their material properties are given in Table 1.

Table 1. Composite laminate's material properties [20, 21].

Properties	Glass/Epoxy	Carbon/Epoxy
Density	1230 kg/m^3	1230 kg/m^3
Elastic properties	$E_1 = 44.87$ GPa; $E_2 = E_3 = 12.13$ GPa; $G_{12} = G_{13} = G_{23} = 3.38$ GPa; $\upsilon_{12} = \upsilon_{13} = 0.30$; $\upsilon_{23} = 0.5$;	$E_1 = 152$ GPa; $E_2 = E_3 = 8.71$ GPa; $G_{12} = G_{13} = G_{23} = 3.35$ GPa; $\upsilon_{12} = \upsilon_{13} = \upsilon_3 = 0.3$;
Strengths [MPa]	$X_t = 1006.30$; $X_c = 487.00$; $Y_t = 45.95$; $Y_c = 131.90$; $S_{12} = S_{13} = S_{23} = 49.51$;	$X_t = 1930$; $X_c = 962$; $Y_t = 41.4$; $Y_c = 276$; $S_{12} = S_{13} = S_{23} = 82.1$;

The composite models consists of 8 plies and their ply orientation is calculated by,

$$Ply\, orientation = (n-1)\theta \tag{6}$$

Where θ = [0, 18, 36, 54, 72, 90] and n = 1 to 8 and θ is constant angle between the consecutive layer of the laminate, it is unique in range of $0 < \theta > 90$ for each laminate model considered and given in Table 2.

Table 2. Geometry configuration for composite laminate

Laminates	Orientation [degree]
1	0/0/0/0/0/0/0/0
2	0/18/36/54/72/90/18/36
3	0/36/72/18/54/90/36/ 72
4	0/54/18/72/36/90/54/18
5	0/72/54/36/18/90/72/54
6	0/90/0/90/0/90/0/90

4   Finite Element Modeling and Simulation

The simulations are carried out in Abaqus/Explicit. The composite laminate and the indenter are modeled in Abaqus 6.14. The laminate is a deformable shell of extrusion type whereas the indenter is an analytical rigid body of revolution type. The interaction is defined as a surface to surface contact where the first surface is surface of the topmost layer and the second surface is the outer surface of the spherical indenter. It is mechanically constrained by kinematic contact method and governed by the finite sliding formulation.

The edges of the laminate are constrained by fixed boundary condition; the indenter is constrained in all the degrees of freedom except the direction of velocity. The laminate has meshed fine in the impact region and course as it progresses towards edges using the surface partition and edge seeding tools considering C3D8R elements as shown in Fig. 1.

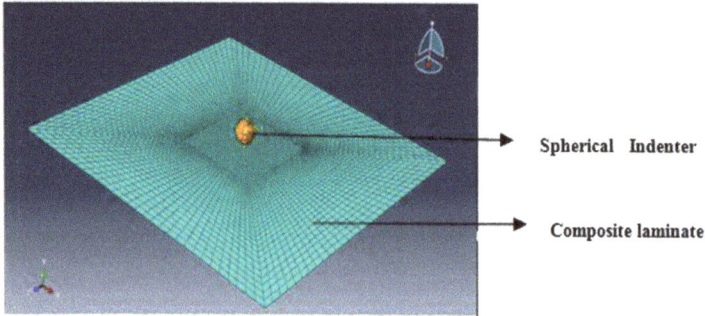

Fig. 1. Composite laminate's mesh for simulation

5   Results and Discussion

The results of maximum deflection and contact force are developed under the impact of the models mentioned in Table 1, analyzed by Finite Element Method. From Figs. 2 and 3 the relation between the angle between consecutive plies to deflection and contact force are observed from the results of the carbon fiber and glass fiber composite laminates.

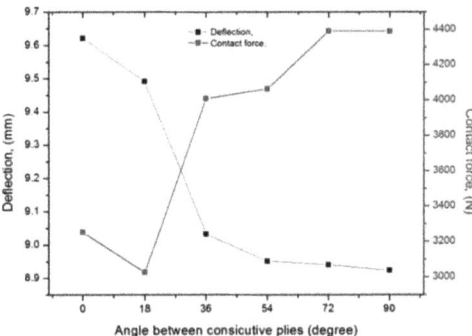

Fig. 2. Deflection and contact force vs. angle between consecutive plies of carbon fiber composite laminates.

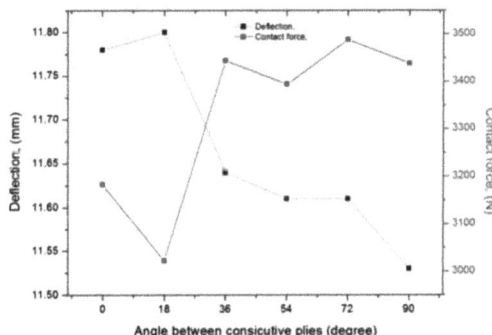

Fig. 3. Deflection and contact force vs. angle between consecutive plies of glass fiber composite laminates.

In case of laminate with consecutive ply angle of 90°, least deflection was observed and the laminate with a consecutive ply angle of 18° has the maximum capacity to absorb the impact energy. Both the carbon fiber and glass fiber composite laminates exhibit similar behavior in terms of deformation under impact observed from models of various constant ply angles and carbon fiber is having high energy absorbing capacity compare to glass fiber.

6 Conclusions

Abaqus/Explicit were efficiently used to model the composite laminate and the indenter. The numerical model was simulated and measured the material data in order to predict the impact behavior for the consecutive staking sequences, deformation and the contact force developed in various laminates under normal impact for considering unidirectional Glass/Epoxy and Carbon/Epoxy material with different ply orientations. The relation between the consecutive ply angle and the maximum deflection is studied. A good agreement is seen between the results of Glass/Epoxy and Carbon/Epoxy laminates and discussed. The parametric study with varying impact energy and boundary conditions is helpful to understand the response of composite laminate under low-velocity impact.

Further, similar studies can be carried out on cylindrical shells and beams by varying the impact energy and limpactor mass under different boundary conditions.

Acknowledgements. The authors are thankful for the support from Science and Engineering Research Board (File Number: ECR/2017/000512) and Department of Science and Technology, Government of India.

References

1. Cantwell, W.J., Morton, J.: The impact resistance of composite materials—a review. Composites **22**, 347–362 (1978)
2. Abrate, S.: Impact on laminated composite materials: recent advances. Appl. Mech. Rev. **47**, 517 (1994)
3. Richardson, M.W., Wisheart, M.J.: Review of low-velocity impact properties of composite materials. Composites **27**, 1123 (1994)
4. Mili, F., Necib, B.: Impact behavior of cross-ply laminated composite plates under low velocities. Compos. Struct. **51**, 237–244 (2001)
5. Sevkat, S., Liaw, B., Delale, F., Basavaraju, B.R.: Drop weight impact of plain woven hybrid glass graphite/toughened epoxy composites. Compos. Part A **40**(8), 1090–1110 (2009)
6. Mikkor, K., Thomson, S., Herszberg, I., Weller, T., Mouritz, A.P.: Finite element modeling of impact on preloaded composite panels. Compos. Structure. **75**(1–4), 501–513 (2006)
7. Christoforou, A.P., Yigit, A.S., Majeed, M.: Low-velocity impact response of structures with local plastic deformation: characterization and scaling. ASME J. Comput. Nonlinear Dyn. **8** (1), 011012 (2012)
8. Lopes, C.S., Camanho, P., Gurdal, Z., Maimi, P., Gonzalez, E.V.: Low velocity impact damage on dispersed stacking sequence laminates. Part II: numerical simulations. Compos. Sci. Technol. **69**(7–8), 937–947 (2009)
9. Abrate, S.: Impact on laminated composite materials: recent advances. Mech. Rev. **47**, 517 (1994)
10. Richardson, M.O.W., Wisheart, M.J.: Review of low-velocity impact properties of composite materials. Composites A **27**, 1123 (1996)
11. Anderson, T., Madenci, E.: Experimental investigation of low-velocity impact characteristics of sandwich composites. Compos. Struct. **50**, 239 (2000)

12. Tita, V., Carvalho, J., Vandepitte, D.: Failure analysis of low velocity impact on thin composite laminates: experimental and numerical approaches. Compos. Struct. **83**, 413 (2008)
13. Wang, J., Waas, A.M., Wang, H.: Experimental and numerical study on the low velocity impact behavior of foam-core sandwich panels. Compos. Struct. **96**, 298 (2013)
14. Abrate, S.: Modeling of impacts on composite structures. Compos. Struct. **51**, 129 (2001)
15. Salvetti, M., Gilioli, A., Sbarufatti, C., Manes, A., Giglio, M.: Analytical model of the dynamic behavior of CFRP plates subjected to low velocity impacts. Compos. A **142**, 47–55 (2018)
16. Jagtap, K.R., Ghorpade, S.Y., Lal, A., Singh, B.N.: Finite element simulation of low velocity impact damage in composite laminates. Mater. Today Proc. **4**, 2464–2469 (2017)
17. Johnson, H.E., Louca, L.A., Mouring, S., Fallah, A.S.: Modeling impact damage in marine composite panels. Int. J. Impact Eng. **36**(1), 25–39 (2009)
18. Fan, J., Guan, Z.W., Cantwell, W.J.: Numerical modeling of perforation failure in fibre metal laminates subjected to low velocity impact loading compos. Structure **93**(9), 2430–2436 (2011)
19. Mallikarjuna Reddy, D., Jayaprakash, G., Swarnamani, S.: Detection and localization of damage in bridge model by spatial based wavelet approach. Int. J. Model. **3**, 156–161 (2008)
20. Perillo, G., Jorgensen, J.K.: Numerical/Experimental study of the impact and compression after on GFRP composite for wind/marine applications. Presidia Eng. **167** (2016)
21. Liu, H., Falzon, B.G., Tan, W.: Predicting the Compression-After-Impact (CAI) strength of damage-tolerant hybrid unidirectional/woven carbon-fiber reinforced composite laminates, composites. Compos. A Appl. Sci. Manuf. **105**, 189–202 (2018)

Risk Response Strategies for Collaborative University-Industry R&D Funded Programs

Gabriela Fernandes[1(✉)], Ana R. Martins[1], Eduardo B. Pinto[2],
Madalena Araújo[1], and Ricardo J. Machado[1,2]

[1] Centro ALGORITMI, University of Minho,
Campus de Azurém, Guimarães, Portugal
g.fernandes@dps.uminho.pt
[2] CCG/ZGDV Institute, Campus de Azurém, Guimarães, Portugal

Abstract. Universities are centers of knowledge in our societies and their role when it comes to innovation has become more important over the years. Companies have several reasons to engage in research collaborations with universities, namely to gain access to innovative technologies. University-Industry R&D collaborations are expected to play an important role in regional economies, and to fulfill the industry's demand for innovative products, technologies and processes. However, the knowledge on what are the potential risks resulting from these collaborations and the risk response strategies to reduce the negative risk impacts and to enhance positive risk impacts is still limited. Thus, this paper aims to fill the gap in literature when it comes to risk identification and risk responses' planning, by identifying, based on a case study analysis, 19 potential risks and 53 potential risk response strategies.

Keywords: University-industry collaboration · Risk identification
Risk responses · Program and project management · R&D funded programs

1 Introduction

University-Industry (UI) Research and Development (R&D) collaborations have been increasing over time, especially in the last couple of decades, accompanying the globalization of economy and the increased complexity of industrial systems [1, 2]. These are encouraged by the government as a mean to increase national competitiveness and wealth creation [3]. Several reasons make companies want to engage in research collaboration with universities. Perkmann et al. [4] identified 4 main reasons: (1) many public programs for R&D funding request the involvement of universities; (2) companies need to have access to research and critical skills, which allow them to reach the very edge of technology and to push it further; (3) companies aim to improve their problem-solving competences and academic researchers are hired to resolve problems; and (4) these collaborations result in several other benefits, e.g., capturing talented collaborators and increasing the enterprise's reputation and visibility. Tartari and Breschi [5] identified that the main motivations for researchers to engage with industry is the access to equipment and additional research resources.

© Springer International Publishing AG, part of Springer Nature 2019
J. Machado et al. (Eds.): HELIX 2018, LNEE 505, pp. 522–529, 2019.
https://doi.org/10.1007/978-3-319-91334-6_71

While the literature provides some guidance on how to manage programs and projects [6] and on Risk Management (RM) [7, 8], the specific context of UI collaboration, with multiple stakeholders and a complex governance model, demands a strong research effort to produce effective guidelines. There are additional challenges to be addressed with a UI consortium structure [9], since PM is highly dependent on the organizational context [10]. UI collaborative research projects face various challenges, since they are generally associated with high uncertainty and risks, significant pressure in terms of creativity and innovativeness, individually-oriented employees, and project members which are settled at different locations [1, 11]. The cultural gap also presents a threat that can cause conflicts over ownership of Intellectual Property, interfere in academic freedom to publish, and create different priorities, time horizons and areas of interest [12].

This paper aims to present and discuss the results of an exploratory study applied to a case study R&D program that covers several projects, and that has joined a company and a university in a research and technological development initiative targeting critical R&D regarding the development and production cycle of advanced multimedia systems for the automobile industry, in order to answer two research questions: (RQ1) What are the potential risks in UI R&D collaborative funded programs? (RQ2) What are the key potential risk response strategies to manage the identified risks?

This paper follows a commonly used structure. The second section discusses RM concepts. The third describes the research methodology applied in this study. The fourth section presents the main findings that emerged from this study. Finally, the conclusions and suggestions for future work are discussed.

2   Risk Management

The Project Management Institute (PMI) defines risk as "an uncertain event or condition that, if it occurs, has a positive or a negative effect on a project's objectives" [13]. The definition of risk includes both uncertain events which could impact the project negatively (threats), as well as those which may cause positive effects on the project's objectives (opportunities) [8]. However, this study focuses only on the negative risks.

The researchers focused on two RM processes in this study: (a) Identify Risks and (b) Plan Risks Responses. The (a) process aims to determine which risks may affect the project, as well as document their characteristics [13]. Key stakeholders should participate in risk identification activities to define responsibilities over the risks and their planned responses. During the project life cycle, some risks may evolve, and other may arise, so it is necessary to meet with key stakeholders on a regular basis [13]. The (b) process goal is to develop strategies to reduce negative impacts and enhance positive impacts on project objectives. It addresses the risks according to their exposure, adding activities and resources to the budget and adjusting the project schedule [13].

3   Research Methodology

An exploratory research was carried out in one UI R&D collaborative program case study, aiming to learn from the experience of program and project managers and other program stakeholders. The research methods applied to the case study were document analysis, participant observation and unstructured focus groups.

The analysis of several documents was conducted to better understand the case study context and to identify risks, namely the established governance model, the management register that holds the identified risks and issues, among other aspects of the program, as well as several documents that supported the management of the program and its projects.

Participant observation plays an important role in the context of this research. It is a complex research method because it often requires the researcher to assume different roles and use several techniques to collect data, without forgetting her/his primary role [14]. The observer enters the social world of those to be observed and attempts to participate in their activities by becoming a member of their workgroup, organization or community [15]. Different stakeholders were observed during regular meetings. It was hence possible to understand the organizational context and identify its risks.

The unstructured focus groups consisted in gathering a group of experts to collectively identify the risks of this program typology and collect their opinions on the risk responses that should be performed [16]. Two sessions were conducted without a strict structure, allowing free-flowing discussions, with the researcher's moderation. The first focused on risks identification and the later on risk responses. The preparation and the conduction of the focus group sessions are similar to those of interviews, as they involve formulating questions in advance and providing feedback [17]. The first session was prepared taking in consideration the results obtained from documents analysis and participant observation, while the preparation of the second session took in account the results obtain from the first session. Focus groups' advantage over interviews and surveys is the ease of discussion and participation, enriching the information collected [18]. Both sessions had the participation of 8 experts – the Program Manager, 4 Program and Project Management Officers, 2 Program and Project Management Communication members and 1 Project Team member.

4   Results

Several potential risks were identified as a result of this case study (Table 1).

Risk categorization is a good RM practice [13]. Thus, the researchers tried to assign each risk to one of the 3 categories presented by Krane et al. [19]: operational, short-term strategic, or long-term strategic risks. However, the researchers agreed that this categorization was not adequate to risks related to UI collaboration, despite suiting well megaprojects' risks [19]. The researchers decided to associate risks with the phases when they can occur, and followed the Program and Project Management (PgPM) approach for collaborative UI R&D funded programs created by Fernandes et al. [20]. The PgPM approach is easily comprehensible and applicable, and is proven to deliver successful results, which makes it a suitable approach for the risk

Table 1. Potential risks identified and potential proposed response strategies.

Phase	Risk description	Proposed risk response
A, B, C, D	R1. Inadequate program stakeholders' engagement	(1) Create a sense of belonging, and define the program's vision, mission and values; (2) Communicate the benefits arising from participation in UI R&D collaborations for the career development; (3) Assign a maximum of 3 projects to each project leader, to avoid work overload; (4) Demonstrate the importance of the program expected benefits; (5) Promote moments of sharing program results among stakeholders
	R2. Program governance mechanisms are not fully implemented	(1) Disclose the governance model among all program stakeholders; (2) Formally create all the governance mechanisms; (3) Demonstrate the value of a fully implemented governance mode
	R3. Disturbances in information flows and communication between stakeholders	(1) Establish different communication channels for different stakeholders; (2) Develop team building activities; (3) reinforce the supporting role of program managers requiring the teams to provide them with reliable information
	R4. Non-exploitation of the generated knowledge	(1) Develop talent management policies to keep the key human resources (HR); (2) Collect, analyze, archive and disclose the lessons learned, risks and issues for future use in new R&D collaborative programs
	R5. Strategic misalignment	(1) Frequent meetings between stakeholders to analyze the strategic alignment; (2) Develop a contribution matrix for each project and frequently register the cumulative percentage of agreement with the desired outputs; (3) Clarify the hierarchical structure of program management decision making to all stakeholders
	R6. Significant changes in the project/program environment	(1) Frequent monitorization of the external environment, economic, technologic, etc.
A	R7. The effective start date differs from the official kickoff date	(1) Prepare the consortium for self-funding until the funding contract is signed; (2) Prepare the funding application 12 to 18 months prior to the planned start date in the funding application

(*continued*)

Table 1. (*continued*)

Phase	Risk description	Proposed risk response
C	R8. Delays in program financial execution	(1) Plan the investments in equipment and materials attending to the organizational financial restrictions; (2) Monitor the procurement processes to initiate them as soon as possible
	R9. Conflict in the attribution of the authorship of Intellectual Property	(1) Establish explicit pre-agreements on Intellectual Property rights, identifying the rightful authors; (2) Involve all stakeholders to ensure that there is an agreement
	R10. Delays in HR recruitment	(1) Improve the visibility of available research grants; (2) Improve employment contract conditions
	R11. Conflicting objectives between projects	(1) Promote workshops among projects in the same area of knowledge; (2) Create a subprogram level of management
D	R12. Failure to comply with the contract's clauses	(1) Implement a contract clauses' monitoring system
E, F, G, H	R13. Lack of project sponsorship	(1) Nominate a project sponsor; (2) Clearly communicate the program and projects benefits; (3) Link the program and project objectives to the partner organization strategic objectives
E, G	R14. Impossibility to achieve results according to the industry guidelines	(1) Propose the participation of the project members in the definition of the industry guidelines in development; (2) Report to the program coordination to develop the inclusion mechanisms to meet project needs
G	R15. Non-innovative project results	(1) Operationalize a program innovation management team; (2) Consult the program innovation management team on a regular basis on the technology roadmaps, to adapt the final product to the market's needs; (3) Hire senior research fellows; (4) Request project sponsorship; (5) Require scope changes when necessary
	R16. Key HR leave the project during its life cycle	(1) Create employment contracts for key researchers to fill the precariousness of research fellowship contracts; (2) Create a professional career perspective at industry and/or continue their research work on new projects at university; (3) Promote collaboration among team members;

(*continued*)

Table 1. (*continued*)

Phase	Risk description	Proposed risk response
		(4) Provide better working conditions; (5) Integrate research fellows in university degree cycles; (6) Provide technical training; (7) In the case of early departure, provide a period for the transmission of knowledge to the new resource; (8) Archive the developed knowledge
	R17. Failure to adopt PM practices in projects	(1) Agree previously standardized practices with stakeholders, to adapt them to the needs of stakeholders, and involving them in the decision-making process; (2) Promote workshops about PM
	R18. Misalignment between the plan and the project execution	(1) Identify the project leaders as soon as possible, ideally during the funding application development; (2) Guarantee the commitment of the project responsible with the set of deliverables planned in the funding application; (3) Identify the necessary changes to the project as early as possible, minimizing its impact in the project
G, H	R19. Failure to meet the project's requirements	(1) Develop an explicit and detailed research plan; (2) Request the project scope's change to the funding entity

categorization. PgPM distinguishes between programs and projects. In programs covering a group of related projects, their management must be coordinated, and synergies must be created, so that projects can generate greater benefits than they would if managed individually [21, 22]. Nevertheless, the management of a program encompassing several projects demands the management of them all. Thus, PgPM establishes a PM layer bellow the program management layer [20].

The life-cycle of the program management layer is divided into 4 phases: Program Preparation (A), Program Initiation (B), Program Benefits Delivery (C), and Program Closure (D). The lifecycle of PM layer is divided into 4 phases as well: Project Initiation (E), Project Initial Planning (F); Project Execution, Monitoring and Controlling, and Replanning (G), and Project Closure (H) [20].

After the identification of potential risks, the researchers performed a root cause analysis to plan the risk response strategies. Each strategy falls into one of these 4 types: (1) Take actions required to avoid the risk, (2) Transfer the risk to a third party, (3) Mitigate the risk to decrease its probability and/or impact, (4) Accept the risk and take no action unless it occurs [8]. The planned responses are presented in Table 1.

5   Discussion and Conclusions

Identifying risks in UI collaborative R&D funded programs and planning its responses demands a strong effort to produce effective strategies to reduce the risks' negative impacts. Therefore, this case study gives answer contributions to two relevant questions: What are the potential risks of UI R&D collaborative funded programs? What are the key potential risk response strategies to manage the identified risks?

A total of 19 potential negative risks were identified (see Table 1), such as "Inadequate program stakeholders' engagement", and "Lack of project sponsorship"; each one of them was assigned to PgPM [20] phases. Program Benefits Delivery and Project Execution, Monitoring and Controlling, and Replanning phases present a larger number of potential risks when compared to other phases, which was expected, as they have longer duration than the others and require higher effort.

The researchers planned a total of 53 recommended potential response strategies to reduce or eliminate risk impacts. For each risk, at least one response strategy is identified (see Table 1). As an example, for the potential risk: "Lack of project sponsorship", three risk responses were identified: "Nominate a project sponsor", "Clearly communicate the program and projects benefits", and "Link the program and project objectives to the organization strategic objectives". Although researchers have only focused on 2 RM processes, Identify Risks and Plan Risk Responses, the other processes are also relevant and will be studied in future work, in order to create a full RM methodology for UI R&D collaborations, such as the one proposed by Peixoto et al. [23] for an electric energy organization. As future work, the researchers also aim to perform a qualitative risk assessment to identify critical risks.

Acknowledgments. This work was supported by the FCT - Fundação para a Ciência e a Tecnologia (SFRH/BPD/111033/2015); and by the Portugal Incentive System for Research and Technological Development. Project in co-promotion nº 36265/2013 (Project HMIExcel - 2013-2015).

References

1. vom Brocke, J., Lippe, S.: Managing collaborative research projects: a synthesis of project management literature and directives for future research. Int. J. Proj. Manag. **33**(5), 1022–1039 (2015)
2. Hanel, P., St-Pierre, M.: Industry – university collaboration by Canadian manufacturing firms. J. Technol. Transf. **31**, 485–499 (2006)
3. Barnes, T., Pashby, I., Gibbons, A.: Effective university - Industry interaction: a multi-case evaluation of collaborative R&D projects. Eur. Manag. J. **20**(3), 272–285 (2002)
4. Perkmann, M., Neely, A., Walsh, K.: How should firms evaluate success in university-industry alliances? A performance measurement system. R&D Manag. **41**(2), 202–216 (2011)
5. Tartari, V., Breschi, S.: Set them free: Scientists' evaluations of the benefits and costs of university-industry research collaboration. Ind. Corp. Chang. **21**(5), 1117–1147 (2012)

6. May, C.C.M., Hwa, Y.E., Spowage, A.: Developing and evaluating a project management methodology (PMM) for university-industry collaborative projects. Prod. Manag. Dev. **9**(2), 121–135 (2011)
7. Aven, T.: Risk assessment and risk management: review of recent advances on their foundation. Eur. J. Oper. Res. **253**(1), 1–13 (2016)
8. Project Management Institute: Practice standard for project risk management, 4th ed. Project Management Institute (2009)
9. Peterson, S.: Consortia partnerships: linking industry and academia. Comput. Ind. Eng. **29**(1–4), 355–359 (1995)
10. Besner, C., Hobbs, B.: Contextualized project management practice: a cluster analysis of practices and best practices. Proj. Manag. J. **44**(1), 17–34 (2012)
11. König, B., Diehl, K., Tscherning, K., Helming, K.: A framework for structuring interdisciplinary research management. Res. Policy **42**(1), 261–272 (2013)
12. Barnes, T.A., Pashby, I.R., Gibbons, A.M.: Managing collaborative R&D projects development of a practical management tool. Int. J. Proj. Manag. **24**(5), 395–404 (2006)
13. Project Management Institute: Project Management Body of Knowledge: A Guide to the Project Management Body of Knowledge, 6th ed. Project Management Institute (2017)
14. Baker, L.: Observation: a complex research method. Libr. Trends **55**(1), 171–189 (2006)
15. Saunders, M., Thornhill, A., Lewis, P.: Research Methods for Business Students, 7th edn. Pearson Education Limited, Edinburgh (2016)
16. Krueger, R.A., Casey, M.A.: Focus Groups: A Practical Guide for Applied Research. SAGE Publication, Thousand Oaks (2015)
17. Langford, J., McDonagh, D.: Focus Groups: Supporting Effective Product Development. London (2003)
18. Fernandes, J., Machado, R.J.: Lecture Notes in Management and Industrial Engineering: Requirements in Engineering Projects. Springer (2015)
19. Krane, H.P., Rolstadås, A., Olsson, N.O.E.: Categorizing risks in seven large projects—which risks do the projects focus on? Proj. Manag. J. **41**(1), 81–86 (2010)
20. Fernandes, G., Pinto, E.B., Machado, R.J., Araújo, M., Pontes, A.: A program and project management approach for collaborative university-industry R&D funded contracts. Procedia Comput. Sci. **64**(September), 1065–1074 (2015)
21. Project Management Institute: Program Management Standard, 3rd edn. Project Management Institute Inc, Pennsylvania (2013)
22. Office of Government Commerce: Managing Successful Programmes, 4th edn. Office of Government Commerce, London (2011)
23. Peixoto, J., Tereso, A., Fernandes, G., Almeida, R.: Project risk management methodology: a case study of an electric energy organization. Procedia Technol. **16**, 1096–1105 (2014)

Innovative Solutions for Pneumatic Positioning Systems

Lucian Bogatu, Ciprian Rizescu, Bogdan Grămescu$^{(\boxtimes)}$,
and Dana Rizescu

Department of Mechatronics and Precision Mechanics,
University "POLITEHNICA" of Bucharest, Bucharest, Romania
{lucian.bogatu,ciprian.rizescu,bogdan.gramescu,
dana.rizescu}@upb.ro

Abstract. A pneumatic system represents a safe and "clean" actuation mechanism. The paper presents several innovative solutions for such kind of systems. The aim is to increase the precision of positioning and especially its preservation, as it is known that, in the case of pneumatic positioning systems, these problems have not yet been solved. The proposed units provide an accurate positioning only on certain work areas, set by manual adjustment. At this moment, the presented solutions are in the prototype stage in the tests.

Keywords: Pneumatics · Positioning systems · Modular design

1 Introduction

The traditional actuating systems, depending on the used energy, may be: electrical, hydraulic, pneumatic or mixed. The optimal solution from the operational and economical point of view shall be established for each specific application by an analysis of the system [1], based on objective criteria that highlight the advantages and disadvantages which these systems shall submit them for the process to be analyzed. In most cases the actuating systems referred to not only that they are mutually exclusive, but on the contrary, shall be completed in a harmonized manner, resulting in superior performance. Pneumatic systems, as well as the electrical ones, represents a "clean" and reliable actuating solution. At the same time these systems, as well as the hydraulic ones, have a simple structure, driven load can be coupled directly with the output element of the motor. On the other hand, in relation with the other systems, the pneumatic have the best specific economic indicators (weight/unit of energy, volume/unit of energy, price/unit of energy). The modern pneumatic actuating systems [2–6] have in their structure in addition to the automation classics and proportionate pneumatic equipment (compressor, equipment that regulates and controls the flow rate and pressure, pneumatic motors), several sensors and transducers for mechanical parameters and an electronic unit command.

Practically, the implementation of a computing unit drive to intelligent systems [7], which carries drive function with an imposed precision, in the conditions under which during operation appear only perturbations. Information on the time variations of the

© Springer International Publishing AG, part of Springer Nature 2019
J. Machado et al. (Eds.): HELIX 2018, LNEE 505, pp. 530–536, 2019.
https://doi.org/10.1007/978-3-319-91334-6_72

parameters defining as regards the evolution of the system are sensed by the sensors and transmitted to the control unit. These signals are processed and analyzed and on this basis, the control unit shall adopt, in real time a decision, several possible solutions, so that the system perform the function under the pre-set conditions.

Often, in the processes of automation the actuation pneumatic systems are used for positioning of the handled load with a required precision [8, 9]. Increasing the accuracy of the positioning and especially the preservation in time remain unresolved issues yet. In addition, during the movement to the target position and during the periods for parking in this position, there are several perturbation phenomena which are difficult to control, and which affects the dynamics of the movement and the conservation of position. To solve these aspects are necessary extensive theoretical and applicative research. It is followed by this limitation of the negative effects due to the physical properties of the environment refrigerant: the viscosity of the low and high compressibility. These properties are the main obstacles to the establishment of systems with precise pneumatic actuator.

In practice it may encounter one of the following situations:

- handled load must be positioned precisely in two points on the working stroke; these may be ends of the stroke or points on the working stroke delimited by means of the mechanical stops; they are positioned manually and then locked in the desired positions; such application does not develop special problems [10, 11];
- handled load must be positioned at several points on the working stroke; for such applications there are two possibilities to resolve, namely: use of intermediate stops located in the points of interest; the stops are positioned manually and then locked in the desired positions; such an application does not develop special problems; use certain special pneumatic motors [7];
- handled the load must be positioned with accuracy within a certain area of work; now, there is a solution dedicated for such applications.

In this work the authors propose to find a solution for this type of applications.

2 The Description of the Proposed Solution

The basic idea (Fig. 1) consists in a mechanical positioning of a support O close to the work area; on this support is fixed the stop positioning unit UPO (stroke of which is correlated with the working area). The drive into motion by the translation of the load M is done by means of the pneumatic actuator, system that incorporates the following equipment: pneumatic linear motor MPL; pneumatic proportional valves with priority position, electrically controlled DP5/3, DP3/2 and DP2/2; selective unique sense, controlled system, unlockable S1 and S2; the anti-return valve single unit; braking restrictor D_f.

The operation of the pneumatic actuating system is the following:

- In the first stage the electromagnets EM2 and EM4 shall be powered; in these conditions the main valve DP5/3 will supply the room from the right-hand side of the motor MPL (via S2) and put it in connection with the atmosphere of the room

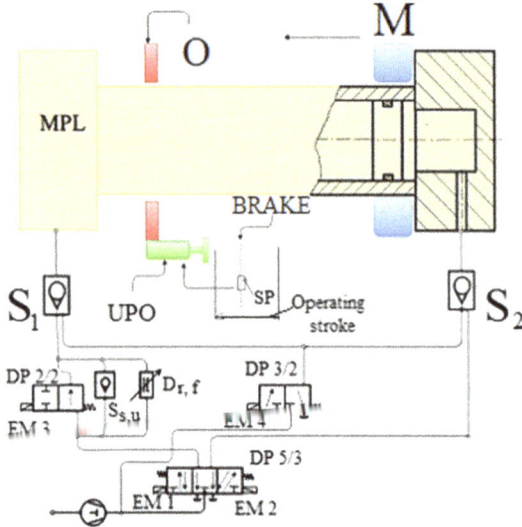

Fig. 1. The schematic diagram of the positioning system

from the left through the valve S1 (unlocked because of the existence of the electromagnet supply EM4) and by the valve DP5/3);

- When the load fall within the range of the proximity sensor SP it will generate a signal which will start the power supply of the electromagnet EM3 and cancellation of the electromagnet EM2 supply; in these conditions: valve DP2/2 will toggle, cutting off the connection between the two holes; from this moment the air will be expelled from the active room via the restrictor of braking D_{rf}; flow section rate through the throttle valve is previously adjusted to a value less than the nominal section, which give rise to a reduction of the load movement speed; in this situation the impact of the mobile assembly with the stop, in programmed position, will be carried out with less intensity; valve 5/3 DP will go into the priority position, in which case the two active rooms of the motor are connected with the atmosphere; the movement of the mobile assembly will continue due to inertia.

To return the load in the zero position will supply the electromagnets EM1 and EM4.

Regarding the positioning unit, which has a mechanical stop, this can be standardized (made available by the companies producing) or may be specifically designed for this purpose. An example of a standardized unit could be a unit manufactured by SMC. This construction consists in an electric motor and a mechanism for rotation-translation conversion: screw and nut mechanism. The movement is transmitted from the electric motor shaft to the screw drive shaft by a mechanism with the toothed belt. The motor can be placed with the longitudinal axis parallel to the axis of the bolt and it can be a stepper motor or a servomotor. The nut translation is transmitted outside through an element connected with the nut. The repeatability error of these units is ±0.02 mm. Maximum stroke can be chosen between 30 … 300 mm.

Table 1. The schematic diagrams for the analized variants of positioning systems

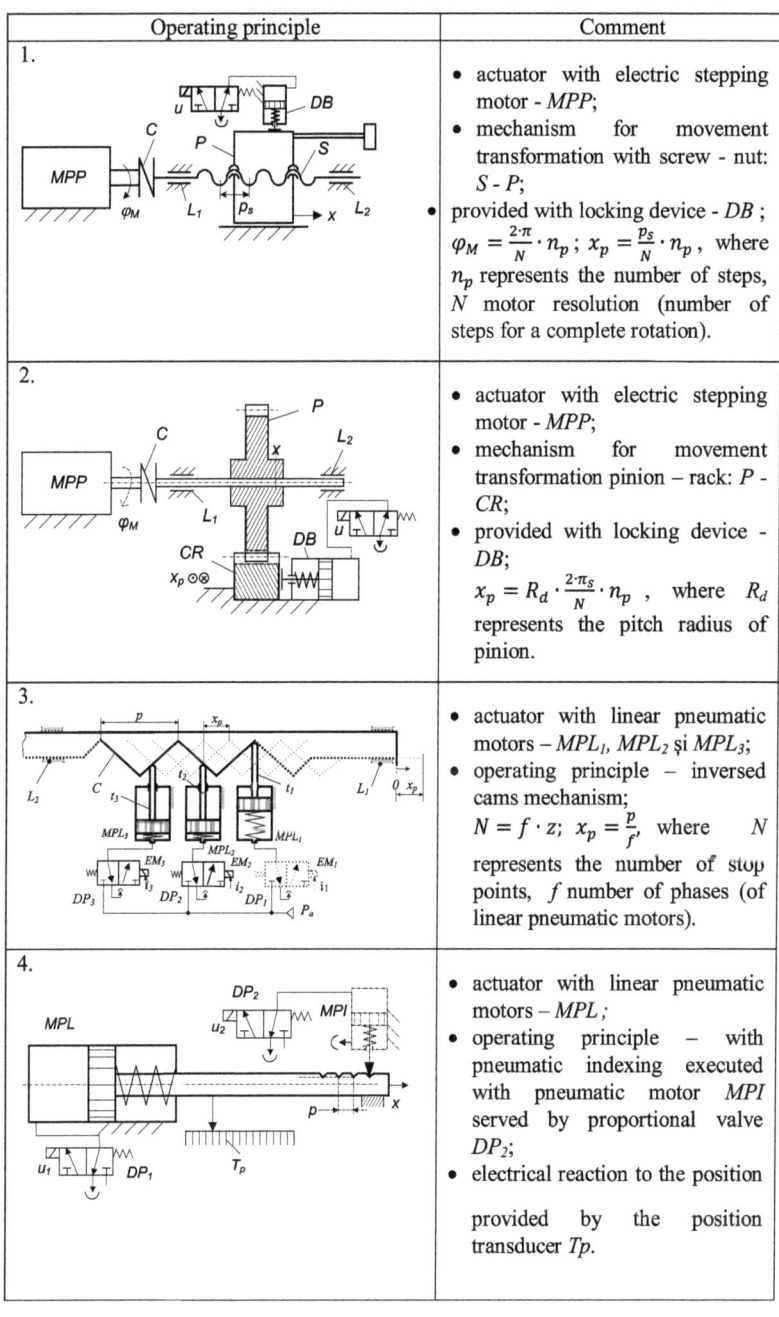

Operating principle	Comment
1.	• actuator with electric stepping motor - *MPP*; • mechanism for movement transformation with screw - nut: *S - P*; • provided with locking device - *DB* ; $\varphi_M = \frac{2 \cdot \pi}{N} \cdot n_p$; $x_p = \frac{p_s}{N} \cdot n_p$, where n_p represents the number of steps, N motor resolution (number of steps for a complete rotation).
2.	• actuator with electric stepping motor - *MPP*; • mechanism for movement transformation pinion – rack: *P - CR*; • provided with locking device - *DB*; $x_p = R_d \cdot \frac{2 \cdot \pi_s}{N} \cdot n_p$, where R_d represents the pitch radius of pinion.
3.	• actuator with linear pneumatic motors – *MPL₁*, *MPL₂* şi *MPL₃*; • operating principle – inversed cams mechanism; $N = f \cdot z$; $x_p = \frac{p}{f}$, where N represents the number of stop points, f number of phases (of linear pneumatic motors).
4.	• actuator with linear pneumatic motors – *MPL ;* • operating principle – with pneumatic indexing executed with pneumatic motor *MPI* served by proportional valve *DP₂*; • electrical reaction to the position provided by the position transducer *Tp*.

The authors have proposed to design the stopper positioning unit. For this purpose, the possibilities for such units were analyzed; Table 1 presents in principle several possible variants.

3   Constructive Solutions

By analyzing the principle sketches presented in the previous paragraph one can identify the structure of the positioning units which will be designed and manufactured. This structure (Fig. 2) contains the following blocks: MR – rotary motor, which may be

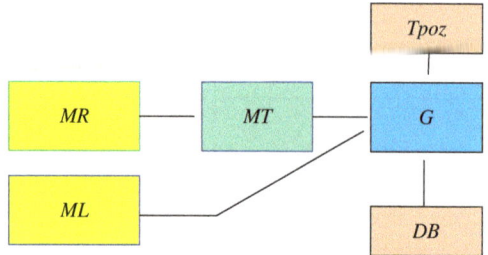

Fig. 2. The block diagram of the positioning unit

a stepper motor or a servomotor (Table 1, positions 1 and 2); MT - mechanism for movement conversion with screw – nut mechanism (Table 1, position 1) or pinion - rack mechanism (Table 1, position 2); ML - linear motor, which can be incremental motor (Table 1, position 3), with the indexing (Table 1, position 4); G - guide; DB - locking device; Tpoz - the position transducer.

The guide represents the core element of the proposed constructions, which will be a multipurpose one, that will be used for all analyzed variants. Figure 3a presents the proposed solution. Elements of this sub-assembly are: slider 1, double roller guide 2, and the longitudinal rods guide 3, the body in the U shape, 4, and top cover 5. The slider 1 is mounted to the lower part of a rack "C", and perpendicular to the longitudinal axis, it has processed a number of equidistant calibrated holes g. The slider has an axial hole "O". On the right end of the slide is a hydraulic shock absorber A. The Fig. 3b, c, d, e presents the 3D models for the four principle variants of units presented in Table 1. Each variant uses the same proposed sub-assembly Fig. 3a.

The solution presented in Fig. 3d shows the principle diagram from Table 1, line 1. The driver motor 1 is an electric stepper motor or a servomotor. Rotation conversion of the motor shaft into translation motion of the slider 4 is made through a screw-nut mechanism [12]. The nut of this mechanism is mounted pressed in the axial hole of the slider. Engaging the motor shaft with the drive shaft 2 is carried out by means of an elastic coupling. Motor mounting guide G shall be by means of part 3.

For solution shown in Fig. 3b, which reflects the principle diagram from Table 1, line 2, the transformation of motor drive shaft rotation - shaft 1 into translation, shall be achieved by a pinion - rack mechanism. The pinion (not represented) shall be fixed on

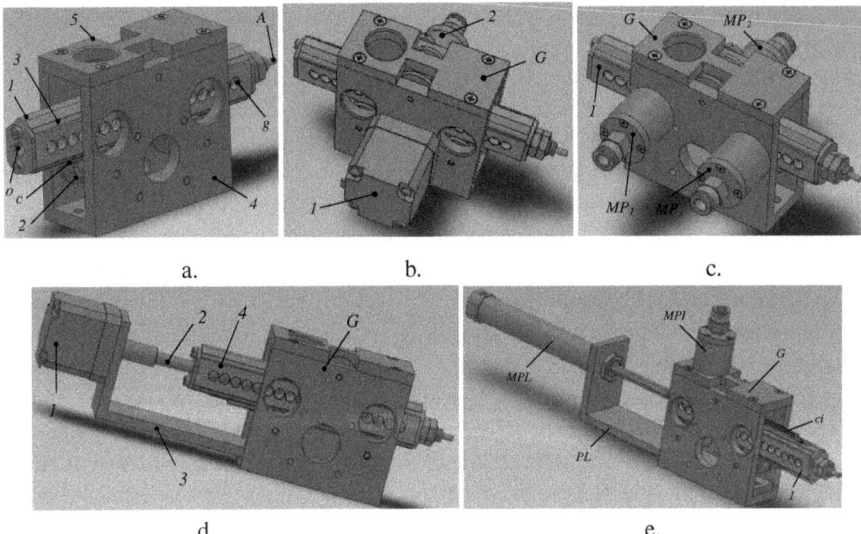

Fig. 3. The design of the proposed solution (a) used in 4 constructive variants (b, c, d, e)

the motor shaft and the rack c (Fig. 3a) is mounted, as it has already shown on the slider 1 of the guide G. For locking the slider in the programmed positions, using the pneumatic tool for locking 2, secured with screws on the fixed part of the guide, G.

The solution presented in Table 1, line 3, rack C is replaced by the slider 1 (Fig. 3c). In along which are processed several equidistant holes [2] what they have in the area of contact with the followers of a certain geometry. The followers, three pieces, are processed on the linear pneumatic motor rods MP1, MP2 and MP3 (Fig. 3c). These motors are fixed to the body in the form of the U of the guide (Fig. 3a) in very well-defined positions [2]. Permanently one of the followers is indexed (axis coincides with the axis of a hole) while the other two are offset with a step in one direction or another to the adjacent holes axis.

For solution 4, of Table 1, the slider drive will be made by means of a standardized linear pneumatic motor, MPL, whose rod will be engaged with the slider. The indexation will use the rack but secured with screws along the slider 1 and that indexer will use a linear pneumatic motor, MPI (Fig. 3e).

4   Conclusions

High compressibility and the extremes reduced viscosity, which characterizes the compressed air, limits the position accuracy of the pneumatic actuation units. It is known, that stopping the mobile assembly of a motor system may be carried out with precision on the ends of the strokes, or on the mechanical stops fitted manually on the working stroke.

The authors propose an innovative solution for a precise positioning pneumatic system to which the handled load can be positioned with accuracy within a certain area of work. The solution involves changing the controlled of (stopper) position locker/lockers which delimit the working stroke. For a precise positioning of the stopper there were conceived and designed four solutions of the positioning units: two electrically actuated and other two - pneumatically actuated. Characteristic of these solutions is the fact that all have in their structure a basic structure, called "guide". Integrating such a unit in the structure of a linear pneumatic actuated system composed of standardized equipment to achieve a precise positioning unit, which can position the driven load in a certain area of work. Changing this area require a change in the position of the support on which fits the positioning, through a manual adjustment.

The main advantage of the proposed solution consists in the fact that the entire mobile positioning stop unit has an equivalent inertia mass much less than the mobile assembly (piston rod, load), which leads to a top dynamic behavior and promote greater positioning accuracy. The proposed unit may replace "pneumatic axes" in many applications - positioning units that can position the load in any point of the stroke. They have in their structure a position transducer and proportional pneumatic equipment; for this reason, they are expensive and in many applications it is not economically justified to use them. The advantages of the proposed systems are obvious: they can develop higher driving forces, they are robust, reliable and can assume the positioning precision compared to the electric drive systems.

References

1. Avram, M.: Acţionări hidraulice şi pneumatice – Echipamente şi sisteme clasice şi mecatronice. Editura Universitară, Bucureşti (2005)
2. Avram, M., Constantin, V., Nitu, C., Bucsan, C.: A method for improving the positioning accuracy of linear pneumatic actuators. Control Eng. Appl. Inform. **19**(1), 85–93 (2017)
3. Avram, M., Bucşan, C., Constantin, V., Negrila, C.: Indexing module for accurate pneumatic actuating. Rom. Rev. Precis. Mech. Opt. Mecatronics (44), 68–72 (2013). ISSN 1584-5982
4. Avram, M., Bucsan, C.: On improving the performances of pneumatic positioning systems. Rom. Rev. Precis. Mech. Opt. Mechatronics (41), 132–135 (2012). ISSN 1584-5982
5. Avram, M., Duminică, D., Gheorghe, V.: Experimental research of a pneumatic unit in mechatronic concept. Rom. Rev. Precis. Mech. Opt. Mechatronics (34), 41–44 (2008). ISSN 1584-5982
6. Sorli, M., Figliolini, G., Pastorelli, S., Rea, P.: Experimental identification and validation of a pneumatic positioning servo-system. In: Bath Workshop on Power Transmission and Motion Control, PTMC 2005, pp. 365–378 (2005)
7. Avram, M., Bucşan, C.: Sisteme de acţionare pneumatice inteligente. Editura Politehnica PRESS, Bucureşti (2014). ISBN 978-606-515-557-2
8. Belforte, G.: New developments and new trends in pneumatics. In: 6th Triennal International Symposium on Fluid Control Measurement and Visualization: FLUCOME 2000, 13–17 August 2000 (2000)
9. Ferraresi, C., Quaglia, G.: High accuracy pneumatic positioners made up of two cooperating actuators. In: 1999 Proceedings of the Forth JHPS International Symposium on Fluid Power, Tokyo, pp. 207–212 (1999). ISBN 4-931070-04-3
10. Constantin, V., Belforte, G., Donţu, O., Avram, M.: Pneutronic positioning system, U.P.B. Sci. Bull. Ser. D **76**(4), 57–68 (2014). ISSN 1454-2358
11. Avram, M., Constantin, V., Bucşan, C., Besnea, D., Spânu, A.R.: Hardware structures for high precision pneutronic systems. Appl. Mech. Mater. **658**, 541–546 (2014). ISSN 1662-7482
12. Mauro, S., Pastorelli, S., Johnston, E.: Influence of controller parameters on the life of ball screw feed drives. Adv. Mech. Eng. **7**(8), 1–11 (2015)

Development of a Hydraulic Robotic Arm – Determination of the Kinematic Parameters

Emanuel Moutinho Cesconeto$^{(\boxtimes)}$ ⓘ and Eduardo André Perondi

Mechanical Engineering Department, Federal University of Rio Grande do Sul,
Porto Alegre 90050-170, Brazil
emanuelemc@hotmail.com, perondi@mecanica.ufrgs.br

Abstract. This paper proposes a new systematic strategy for the determination of the kinematic parameters of a planar manipulator arm with two links and two linear actuators. This kind of mechanism is used in various types of equipment, such as cranes, excavators, etc., and usually consists of the second and third links and joints. The proposed strategy aims to calculate the optimum length of the links and the angular limits of the joints that result in an arm with the best possible kinematic performance characteristics within a desired design-defined workspace, while considering the possible costs involved. It is versatile and easily expandable, and can thus be used in diverse cases. To verify the strategy a case study is analyzed, and the result is an arm with ideal parameters.

Keywords: Robot design · Kinematic parameters · Performance optimization

1 Introduction

Kinematics synthesis is an important preliminary stage of the development of any robotic manipulator. The arm of a robot is arguably its most critical part, because it must support a great load without being too heavy, can cause positioning errors due to flexibility and geometric inaccuracies, and its kinematic parameters define most of the workspace. It can also influence the accuracy and repeatability of the robot by mechanically amplifying or attenuating the positioning error of the actuators. As such, it is very important to develop a method to properly design its kinematic parameters. However, this stage is usually not deeply addressed in regular research papers. Therefore, this paper proposes a new approach for the definition of the optimal kinematic parameters of a type of planar robot arm, more specifically one with hydraulic linear actuators that are directly coupled to the links of the robot, that is, without the use of cables or other mechanisms. The application of this method, along with others that can define the other parameters necessary for the construction of a robotic arm, can drastically reduce the guesswork involved its design because the algorithms decide on the best shape for the arm, based on documented criteria and a proper analysis of the tasks the robot must complete. As the process is deterministic it is also fully replicable.

In this work "arm" refers to the typical second and third links of an anthropomorphic robot, along with its second and third joints. It is moved by two linear hydraulic actuators that are directly connected to the links, as shown in Fig. 1.

© Springer International Publishing AG, part of Springer Nature 2019
J. Machado et al. (Eds.): HELIX 2018, LNEE 505, pp. 537–544, 2019.
https://doi.org/10.1007/978-3-319-91334-6_73

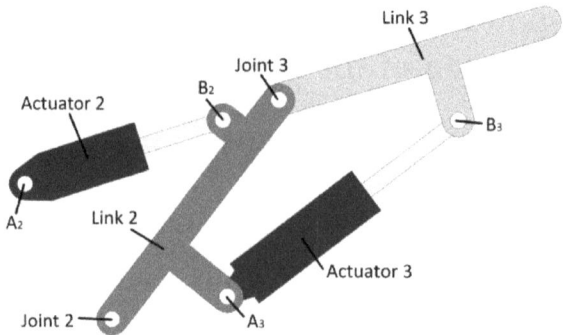

Fig. 1. Sketch showing the links, joints and actuators of a generic arm, as well as the coupling points (A_2, B_2, A_3 and B_3) of the actuators.

The kinematic parameters refer to the link lengths, a_1 and a_2, and the angular limits of the joints, γ_{1min}, γ_{1max}, γ_{2min} and γ_{2max}. They influence the kinematic properties of the arm, namely the workspace and the mechanical transmission ratio between the movement of the actuators and the movement of the end of the arm.

The mechanical transmission ratio is the relation between the speed of the actuators and the speed of the end of the arm, and is thus equivalent to the Jacobian matrix. It is important because it can amplify or attenuate the position errors of the actuators. The Jacobian matrix for a generic robotic arm can be found in [1], but for arms with linear actuators a modified matrix must be used. This can be found in [2], where the inverse kinematics of the arm are also shown.

2   Performance Indexes and the Global Performance Index

The performance indexes are indicators useful in the analysis of the controllability of a robotic arm. They can express how easy it is to control the arm, how close it is to a singularity, and the maximum speed it can have for a given actuator speed. They are calculated based on the Jacobian matrix, and thus are position-dependent. Four indexes defined in the literature were chosen to analyze a proposed design of the arm: the manipulability, the velocity minimum, the velocity isotropy and the dexterity. The definition of the first three can be found in [3], while the last one is presented in [4].

The manipulability is a measure of the overall capability of the arm to move. The velocity minimum indicates the minimum velocity transmission ratio, and thus the best accuracy. The velocity isotropy is the ratio between the transmission ratios of the actuators, which should be as close to 1 as possible to equalize the accuracy in both directions, making the arm easier to control. The dexterity index expresses the transmission accuracy, and the greater it is the greater is the accuracy of the solution of the inverse kinematics, which makes the control easier.

When designing an arm it is desirable to achieve the best possible performance throughout the whole workspace, and thus we need global values rather than local ones. [5] proposes a method for evaluating a single value that expresses the performance of the arm, the Global Performance Index, or GPI, composed by combining global values calculated for each of the four performance indexes.

According to [5], to calculate a global value for a performance index four statistical values are calculated over the desired workspace. These are the average, the deviation, the skew and the kurtosis. A vector α with 4 elements of arbitrary lengths is then defined, so that the global value for an index is the sum of α_1 times the average, α_2 times the deviation, α_3 times the skew and α_4 times the kurtosis. After the four global index values are calculated they are combined by using another set of arbitrary relative weights, β, resulting in the GPI, which is the sum of β_1 times the manipulability, β_2 times the velocity minimum, β_3 times the velocity isotropy and β_4 times the dexterity index. Since these relative weights are arbitrary they can be changed to suit the use case, increasing the versatility of the strategy.

3   Methodology

Since the kinematic parameters define the work area of the arm and influence the overall performance of the robot, and both of these are design variables, it is necessary to do the reverse operation and define the parameters as a function of the desired work area and performance. A Matlab® program that implements this methodology was developed, and Fig. 2 shows a flowchart of its simplified operation.

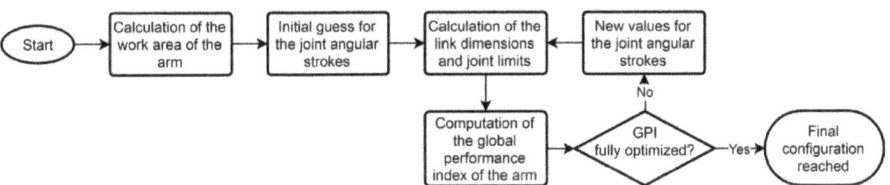

Fig. 2. Flowchart showing the simplified methodology for the definition of the kinematic parameters.

3.1   Calculation of the Work Area of the Arm

For the type of robot studied a bigger workspace results in lower accuracy, because the mechanical transmission ratio must be higher. As we prioritize accuracy inside the space, the tasks that the robot must accomplish must be analyzed and the required work volume defined, so that the robot can be developed to reach all points inside it but without its workspace being bigger than needed. It is usually easier to define a prismatic work volume, with width V_x, length V_y, and height V_z, while the workspace of the robot is much more complex, so a series of equations were developed to define the required kinematic parameters as a function of this defined prismatic region.

The first step is to transform the 3D work volume into a 2D work area, since the arm to be designed is only capable of moving in a 2D space. This area has a width, V_x', and a height, V_z', and the distance between this area and the first joint of the arm is *Lmin*. This area can be seen in Fig. 3.

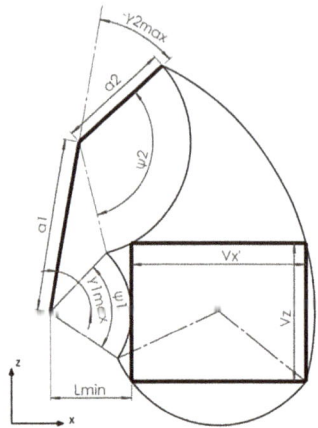

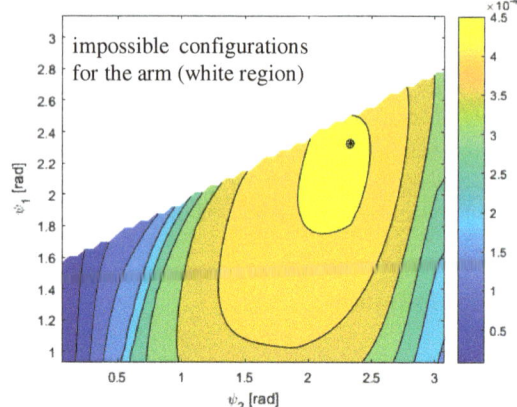

Fig. 3. Relation between the workspace of the arm and the desired rectangular work area.

Fig. 4. *GPI*$_r$ as a function of the joint angular strokes ψ_1 and *psi*$_2$

In the complete robot the arm can rotate around the vertical axis due to the joint of the base, and when it does so the work area is extruded into the work volume, which is a section of a toroid. This toroid must include the previously defined prismatic work volume, and to do so the following equations must be satisfied:

$$Vx' = \sqrt{(Lmin + Vx)^2 + \left(\frac{Vy}{2}\right)^2} - Lmin, \, V_z' = V_z \tag{1}$$

3.2 Calculation of the Link Dimensions and Joint Limits

The next step is to define relations between the work area and the workspace of the arm. Figure 3 shows these relations, which define four of the kinematic parameters, plus an x distance, *Lmin*, between the first joint of the arm and the work area. The position of this first joint must also be set so that it is vertically centered with respect to the area. Two of the six kinematic parameters remain arbitrary and must be defined later. Through a simple variable change, the arbitrary parameters chosen are the angular strokes of the first and second joints of the arm, ψ_1 and ψ_2 respectively. All the remaining parameters can then be expressed as a function of these angular strokes and of the work volume lengths, as given by Eq. (2) through (9).

$$Lmin = V_z / \left[2\tan\left(\frac{\psi_1}{2}\right)\right], \quad Lmax^2 = \left(\frac{V_z}{2}\right)^2 + \left(V_x' + Lmin\right)^2 \tag{2}$$

$$2a_2{}^2 = \left(Lmin^2 - Lmax^2\right)/\sin\left(\frac{\psi_2}{2}\right)^2 + C - \left[(Vx' + Lmin)\cotg\left(\frac{\psi_2}{2}\right) + \frac{Vz}{2}\right]D \tag{3}$$

$$C = \left[(Vx' + Lmin)\cotg\left(\frac{\psi_2}{2}\right) + \frac{Vz}{2}\right]^2 + Vx'\left[(Vx' + Lmin) - \frac{Vz}{2}\cotg\left(\frac{\psi_2}{2}\right)\right] \tag{4}$$

$$D = \cotg\left(\frac{\psi_2}{2}\right)\sqrt{Lmin^2 + Vz\,Lmin\,\mathrm{tg}\left(\frac{\psi_2}{2}\right) - \frac{Vz^2}{4}} \tag{5}$$

$$a_1{}^2 = \left(Lmin + \frac{Vx'}{2}\right)^2 + \left[\sqrt{a_2{}^2 - \left(\frac{Vx'}{2}\right)^2} - \frac{Vz}{2}\right]^2 \tag{6}$$

$$\gamma_{1min} = \mathrm{atan}\left[\left(\sqrt{a_2{}^2 - \left(\frac{Vx'}{2}\right)^2} - \frac{Vz}{2}\right) \Big/ \left(Lmin + \frac{Vx'}{2}\right)\right] \tag{7}$$

$$\gamma_{2min} = \mathrm{acos}\left[\left(a_1{}^2 + a_2{}^2 - Lmin^2\right)/(2a_1 a_2)\right] - \pi \tag{8}$$

$$\gamma_{1max} = \gamma_{1min} + \psi_1, \quad \gamma_{2max} = \gamma_{2min} + \psi_2 \tag{9}$$

3.3   Computation and Optimization of the GPI of the Arm

The methodology of this optimization is presented by [5], though here the joint angular strokes ψ_1 and ψ_2 are used as the design variables rather than the link lengths. This makes it easier to define boundaries and compare different configurations of the arms, as all of them have the same desired workspace. As the computation of the GPI is similar it will not be depicted here. Matlab® functions *quad2d* and *fmincon* were used to perform the required numerical integrations and to execute the optimization, respectively.

The GPI, as proposed by [5], does not take into consideration the monetary cost involved in the construction of the arm. Namely, a bigger arm is more expansive to build and operate than a smaller one. Due to this, it is proposed that the final GPI be divided by a factor, as shown in Eq. (10), where p is a weighting factor that relates the costs associated with a longer arm and the benefits of a better performing arm.

$$GPI_r = GPI/(a_1 + a_2)^p \tag{10}$$

4 Results

As an example, a case study with the parameters shown in Table 1 was considered. This case corresponds to a generic arm, which could be used for manipulating heavy objects between assembly lines. This resulted in arm with the kinematic parameters shown in Table 2, and the individual indexes that make up the final GPI_r value are shown in Table 3. The optimization procedure took 78.11 s to run on a PC with an Intel® Core2Duo™ E8400 3 GHz processor and 4 GB of RAM, using Matlab® version 2012b. Figure 4 shows a graph of the GPI_r of the arm as a function of the angular strokes.

Table 1. Desired parameters for the example case

Work volume	$V_x = 1000$ mm, $V_y = 2000$ mm, $V_z = 1000$ mm
Statistical weights	$\alpha = \begin{bmatrix} 1 & -1\zeta_{avg} & -0.1\zeta_{avg} & -0.1\zeta_{avg} \end{bmatrix}$
Performance index weights	$\beta = \begin{bmatrix} -0.1 & -0.1 & 1 & 1 \end{bmatrix}$
Cost weighting factor	$p = 1$

Table 2. Optimized kinematic parameters and arm characteristics

Link lengths	$a_1 = 894.58$ mm, $a_2 = 827.62$ mm
Joint angular limits	$\gamma_{1min} = -0.0305$ rad, $\gamma_{1max} = 2.2994$ rad, $\gamma_{2min} = -2.9038$ rad, $\gamma_{2max} = -0.5775$ rad
Distance from work area	$Lmin = 214.84$ mm
Angular strokes	$\psi_1 = 2.3299$ rad, $\psi_2 = 2.3263$ rad
Final GPI_r	$GPI_r = 4.610 \cdot 10^{-4}$

Table 3. Table showing the individual performance index values that compose the final GPI.

Index	Average	Deviation	Skew	Kurtosis
M_r	$3.8889 \cdot 10^{-6}$	$3.2486 \cdot 10^{-1}$	1.3528	$8.4807 \cdot 10^{-1}$
$\sigma_{r,min}$	$1.2727 \cdot 10^{-3}$	$1.6749 \cdot 10^{-1}$	$4.2693 \cdot 10^{-2}$	$8.7441 \cdot 10^{-1}$
μ_{iso}	$9.0148 \cdot 10^{-1}$	$9.0684 \cdot 10^{-2}$	$-9.239 \cdot 10^{-1}$	$-7.194 \cdot 10^{-9}$
k_j	$2.4263 \cdot 10^{-1}$	$6.8327 \cdot 10^{-1}$	$5.1488 \cdot 10^{-1}$	$-2.790 \cdot 10^{-1}$

The coupling points of the actuators are determined by another method, outside the scope of this paper, but are required to calculate the Jacobian matrix. They can be expressed as a vector $[X \quad Y]^T$, in the local reference frame of the link they belong to, as in the D-H convention. The values used for the final arm are $A_2 = [-227.68 \; 141.10]^T$ mm, $B_2 = [894.58 \quad 0]^T$ mm, $A_3 = [0 \quad 0]^T$ mm, and $B_3 = [278.46 \quad 12.81]^T$ mm.

Table 4. Non-optimized kinematic parameters and arm characteristics

Link lengths	$a_1 = 1000.0\,\text{mm}, a_2 = 1000.0\,\text{mm}$
Joint angular limits	$\gamma_{1min} = -0.2448\,\text{rad}, \gamma_{1max} = 2.3051\,\text{rad},$ $\gamma_{2min} = -2.8403\,\text{rad}, \gamma_{2max} = -1.0813\,\text{rad}$
Distance from work area	$Lmin = 300.16\,\text{mm}$
Angular strokes	$\psi_1 = 2.0603\,\text{rad}, \psi_2 = 1.7590\,\text{rad}$
Final GPI_r	$GPI_r = 4.391 \times 10^{-4}$

Table 5. Table showing the individual performance index values that compose the final GPI of the non-optimized arm.

Index	Average	Deviation	Skew	Kurtosis
M_r	$2.2346 \cdot 10^{-6}$	$3.4537 \cdot 10^{-1}$	$9.1820 \cdot 10^{-1}$	$2.1064 \cdot 10^{-1}$
$\sigma_{r,min}$	$9.9052 \cdot 10^{-4}$	$1.1761 \cdot 10^{-1}$	-1.5502	1.4575
μ_{iso}	$9.2012 \cdot 10^{-1}$	$6.1377 \cdot 10^{-2}$	$-7.931 \cdot 10^{-1}$	$-1.335 \cdot 10^{-1}$
k_j	$2.4823 \cdot 10^{-1}$	$5.3322 \cdot 10^{-1}$	$2.4447 \cdot 10^{-1}$	$-3.974 \cdot 10^{-1}$

For comparison, a non-optimized arm is also shown below. This arm possesses links with the same length, and the characteristics shown in Table 4. In this case, the coupling points used are $A_2 = [-290.64 \quad 145.43]^T\text{mm}$, $B_2 = [1000 \quad 0]^T\text{mm}$, $A_3 = [0 \quad 0]^T\text{mm}$, and $B_3 = [364.34 \quad 56.10]^T\text{mm}$. Table 5 shows the individual values that compose the GPI_r value.

5 Conclusion

A method was proposed to calculate the kinematic parameters of a hydraulic arm with linear directly coupled actuators. This method extends the method proposed by [5] by optimizing the work area of the arm, which is important for the type of arms considered, and by taking into consideration the costs associated with bigger arms. A program was developed to implement the method, and its results show that it can be used to design an arm with optimal performance inside the desired workspace. By comparing the results with the non-optimized design we can observe a 4.98% improvement in the GPI_r, but the main benefit of using the proposed strategy is to automate the process of defining the arm's parameters. When combined with other such strategies it allows the designer to quickly develop a new robotic arm for a specific task or set of tasks, based on a documented procedure.

Acknowledgments. The research described in this paper was financially supported by the National Council of Scientific and Technological Development CNPq-Brazil (grant 148887/2016-3) from the Ministry of Science, Technology and Innovation.

References

1. Siciliano, B., Sciavicco, L., Villani, L., Oriolo, G.: Robotics Modelling. Planing and Control. Springer, London (2009). https://doi.org/10.1007/978-1-84628-642-1
2. Valdieiro, A.: Controle de robôs hidráulicos com compensação de atrito. Doctorate thesis, Universidade Federal de Santa Catarina, Florianópolis (2005)
3. Kim, J.O., Khosla, P.: Dexterity measures for design and control of manipulators. In: IEEE/RJS International Workshop on Intelligent Robots and Systems, Osaka, Japan (1991)
4. Gosselin, C.: Dexterity indices for planar and spatial robotic manipulators. In: IEEE International Conference on Robotics and Automation, Cincinnati, OH, USA (1990)
5. Zhang, P., Yao, Z., Du, Z.: Global performance index system for kinematic optimization of robotic mechanism. J. Mech. Des. (2013). https://doi.org/10.1115/1.4026031

Advanced Materials and Innovative Applications

Theoretical and Experimental Considerations on Acoustic Attenuation of Plenum Boxes Lined with Different Materials

Carmen Bujoreanu[1]([⊠]), Marcelin Benchea[1], Gelu Ianuş[1], and José Machado[2]

[1] Technical University "Gheorghe Asachi" Iaşi, Iaşi, Romania
carmen.bujoreanu@gmail.com, marcelin_ben@yahoo.com,
gianus2002@yahoo.com
[2] University of Minho, Guimarães, Portugal
jmachado@dem.uminho.pt

Abstract. Plenum boxes are used for supply and exhaust of air through diffusers and grilles from heating and ventilating air conditioning (HVAC) systems. The plenum boxes ensure a continuous flow and moreover, they are usually lined with different materials in order to realize an acoustic insulation. These silencers/attenuators are of great importance in realizing a comfortable acoustical environment. The paper presents a theoretical and experimental investigation on some commercial plenum boxes with various sizes and lined with different sound absorbing materials. The sound transmission loss TL is the performance criterion for the silencers acoustical behaviour. The test procedure agrees the requirements of ISO 7235:2010 and the measurements were made in an anechoic room. Third-octave band analysis of experimental data is compared with theoretical ones. The results evidence the importance of silencer design which must take account of its geometry and the proper sound absorbing material as a function of frequency in order to optimize its acoustic efficiency.

Keywords: Plenum box · Transmission loss · Third band octave analysis

1 Introduction

Heating, ventilating and air-conditioning (HVAC) systems represent an important noise source negatively affecting the acoustical environment of buildings and rooms.

Consequently, the use of a silencer is prompted by the need to reduce the generated noise in the systems elements [1]. Plenum boxes are dissipative silencers most widely used to attenuate broadband noise in ducts with a minimum of pressure drop across the silencer. They are available in various shapes and different sizes in accord to the duct design. Usually, a plenum box consists of a sheet metal casing with length commonly ranging from 1–3 m. They are commonly lined with different materials in order to realize an acoustic insulation. There are also other silencer configurations including a centrally located bar or several splitters mounted horizontally or vertically inside the box [2].

© Springer International Publishing AG, part of Springer Nature 2019
J. Machado et al. (Eds.): HELIX 2018, LNEE 505, pp. 547–553, 2019.
https://doi.org/10.1007/978-3-319-91334-6_74

The European Standards EN ISO 7235:2010 (EN ISO 2010a) and EN ISO 11691:2009 (EN ISO 2009b) and the American Standard ASTM E477-06a (ASTM 2006) set the experimental measurement procedures in order to determine the sound attenuation of ducted silencers and plenum boxes, for frequency bands of interest.

From the theoretical point of view, analytical models [3, 4] and also, several numerical methods [5, 6] are developed in order to calculate in-duct sound attenuation. The analytical approaches are appropriate for simple geometries, while numerical methods need to be used for silencers with complex configurations or inhomogeneous absorbing material properties. The absorbing material has a complex structure and its acoustic behaviour is usually obtained by experiments.

Our paper presents a theoretical approach and an experimental investigation of acoustic performance of plenum boxes of various sizes and different lining material (glass fiber, mineral wool, nylon wool). The transmission loss of the silencer, as a performance criterion for noise, is discussed. The experimental data acquired from the tested boxes are compared in order to establish their efficiency. Transmission loss measurements are made using the test facility that agrees the requirements of ISO 7235:2010 [7]. The sound source equipment consists in a centrifugal fan of an air handling unit from a HVAC system, a white noise generator, an amplifier and broadband loudspeakers. The acoustical power contribution of the source follows the variable airflow provided by the centrifugal fan of the air handling unit. Fan power can vary between 1 V (corresponding to minimum airflow) and 10 V (corresponding to maximum airflow). This is resulting in changes in sound attenuation. The third-octave band analysis of generated noise is performed. The results highlight the influence of the silencers geometry, sound absorbing lining material and variable airflow on the sound attenuation level and lead to a better silencer design.

2 Theoretical Background

The basic components of a silencer system are the noise source, plenum box, connecting pipes and environment. The literature agrees a corresponding electrical analogy for the acoustical system in order to use more easily the sound transmission lines [8].

This model assimilates the electric voltage to the sound pressure p and the electric current intensity to the mass velocity $\rho_0 S u$ (including the air density ρ_0, the duct cross-sectional area S and the air velocity u at the interface of the silencer system components). The sound source has an acoustic pressure p_S and internal impedance Z_S and the system end also has an impedance Z_T (representing the ratio between sound pressure and mass velocity for the specific location). The transfer matrices T and D (also known as four-pole parameter representation) link the elements of the silencer system through the above state variables and their analytical expressions are settled using the standing wave theory [9]. These considerations lead to the following relations describing the situation from Fig. 1.

$$p_S = p_3 + \rho_0 Z_S S_3 u_3$$

$$\begin{pmatrix} p_3 \\ \rho_0 S_3 u_3 \end{pmatrix} = \begin{pmatrix} T_{11} & T_{12} \\ T_{21} & T_{22} \end{pmatrix} \begin{pmatrix} p_2 \\ \rho_0 S_2 u_2 \end{pmatrix}$$

$$\begin{pmatrix} p_2 \\ \rho_0 S_2 u_2 \end{pmatrix} = \begin{pmatrix} D_{11} & D_{12} \\ D_{21} & D_{22} \end{pmatrix} \begin{pmatrix} p_1 \\ \rho_0 S_1 u_1 \end{pmatrix}$$

$$p_1 = \rho_0 Z_T S_1 u_1$$

(1)

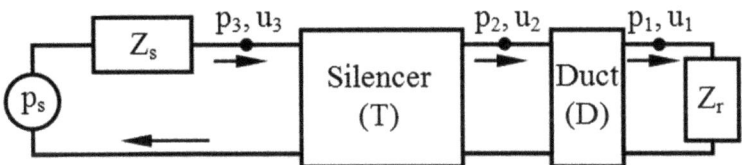

Fig. 1. Plenum box system representation using the analogy with an electric system

The resulting values for sound pressure p_i and mass velocity $\rho_0 S_i u_i$ ($i = 1, 2, 3$) are then used in acoustic performances evaluation of the silencer.

The acoustic performance criteria refer to the insertion loss (IL), noise reduction (NR) and transmission loss (TL) [8]. Our study evaluates the plenum acoustic performances through the transmission loss criterion (TL), representing the attenuation in the level of sound power behind and after the silencer insertion [7]. Transmission loss TL and insertion loss IL of a silencer are quantitative identical when both the noise source and the silencer termination are anechoic ($Z_S = Z_T = c/S$) as depicted in Fig. 2.

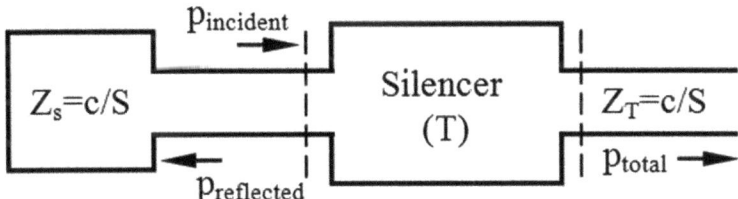

Fig. 2. Transmission lines in a silencer with an anechoic termination

According to Fig. 2, in terms of the corresponding transfer matrices, Eq. (1) becomes:

$$TL = 20 \log \left| \frac{T_{11} + (S/c)T_{12} + (c/S)T_{21} + T_{22}}{2} \right|$$

(2)

where c is the sound speed.

Applied to our plenum boxes experimentally tested, the sound attenuation considers their size and the connecting ducts size, the inlet – outlet angle of the plenum and the absorption behavior of the lining material (Fig. 3). Therefore, it can be written the following relation [9]:

$$TL = -10\log\left[S_{inlet}\left(\frac{Q\cos\theta}{4\pi r^2} + \frac{1}{R_C}\right)\right] \tag{3}$$

where: S_{inlet} is the area of inlet or outlet duct section (in our study they are equal), Q is the directivity factor (in our study $Q = 1$, as the measurements are realized in an anechoic room), θ represents the orientation between silencer inlet and outlet (we have tested straight silencers, therefore $\cos\theta = 1$), r is the distance between the inlet and outlet center of the plenum, $R_C = \left(S_{total} \cdot \alpha_{average}\right)/\left(1 - \alpha_{average}\right)$ with S_{total} representing total plenum surface and $\alpha_{average}$ the average absorption coefficient of the lining material [9].

Fig. 3. Plenum box geometry

3 Experimental Methodology

The experimental investigation on acoustic behavior of plenum boxes have been realized in an anechoic room, essentially in free field conditions, according to ISO 7235: 2010. We have tested three types of plenum boxes with different sizes and lining materials for different airflows variations of the centrifugal fan (from 1 V to 10 V with a ratio of 1 V). The silencers characteristics are depicted in Table 1.

Table 1. Tested silencers characteristics

Inlet/outlet [mm]	Width/height/length [mm]	Lining material
Φ125/Φ125	170/170/600	Green/ blue/ black
Φ200/Φ200	250/250/900	Green/ blue/ black
Φ315/Φ315	400/400/1250	Green/ blue/ black

The lining materials have the same thickness and porosity for all the tested silencers.

The green, blue, black notations correspond to three different material types based on glass, mineral and nylon fibers respectively. The average absorption coefficients of each material type are 0.65, 0.7, and 0.55, respectively, as they are indicated by the manufacturer.

The experimental setup [10] for our measurements consists of: centrifugal fan providing a variable airflow (from 1 V to 10 V with a ratio of 1 V), device for measuring the flow rate, device for measuring the pressure drop, special sound-source equipment (consists of a white noise generator B&K, an amplifier and a broadband loudspeaker), the test object (the dissipative silencers), the transition elements on either side of the test object, special receiving-sound equipment (soundmeter B&K connected with the microphone and NIDAQ board, laptop with LabVIEW soft compatible with National Instruments DAQPad for data processing).

Third-octave analysis with LabVIEW soft give us the sound pressures level dB(A) for nominal frequencies of 63, 125, 250, 500, 1000, 2000, 4000, 8000 Hz and total band power (within the specified frequency band). The attenuation expresses in dB represents the difference between the measured noise values of system without and with the tested silencer, for every frequency band. The experimental setup and the test procedure are detailed in [10].

4 Results and Discussion

The fan maximum airflow (corresponding to 10 V, according to the manufacturer) also implies an increase in the noise generated by the HVAC system, and adequate sound attenuation must be ensured, for the acoustic comfort of the beneficiaries. As consequence, the results and discussions are focused on this situation that is the most disturbing.

Figure 4 compares the attenuation (TL) experimental data processed through LabVIEW third-octave analysis, for the same size silencers, but different linings.

Therefore, the best attenuation for 125 silencer is provided up to 500 Hz in the 1/3 octave band center frequency by the black lining, and over 500 Hz by the blue lining (Fig. 4a). The best attenuation for 200 silencer is realized up to 1500 Hz in the 1/3 octave band center frequency by the blue lining, between 1500–3000 Hz by the black lining and over 3000 Hz there are no significant differences between the three linings (Fig. 4b).

The best attenuation for 315 silencer is provided up to 500 Hz in the 1/3 octave band center frequency by the black lining, between 500–1500 Hz by the blue lining and over 1500 Hz by the black lining (Fig. 4c).

Figure 5 compares the attenuation experimental data for the three silencers with the same linings, but different sizes. The green and blue linings of 200 silencer assure the best attenuation. The black lining of the 200 silencer provides a better attenuation up to 2500 Hz in the 1/3 octave band center frequency. A good attenuation of black lining is also realized between 2500–4500 Hz by the 200 silencer, by the 315 silencer and over 4500 Hz by the 200 silencer.

Table 2 presents comparative values of transmission loss TL obtained both by experimental procedure (according to ISO 7235: 2010 standard) and theoretical evaluation (according to Eq. 3) of the samples acoustic performances.

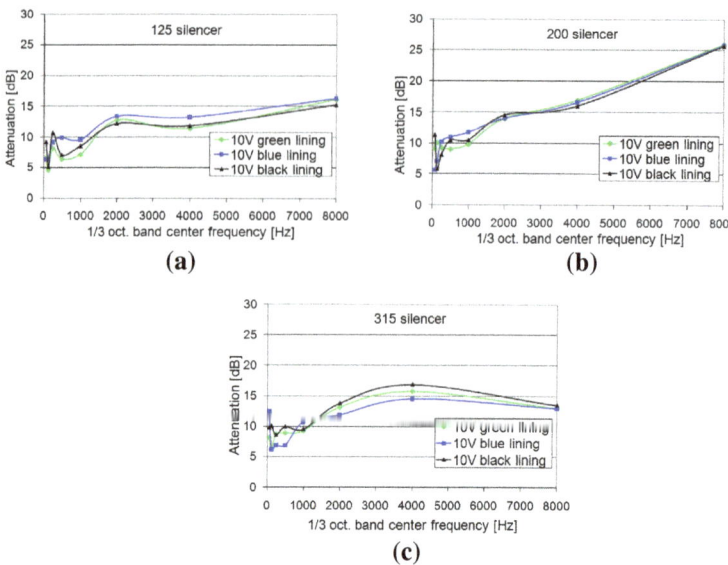

Fig. 4. Attenuation values for three silencers with the same size and different linings

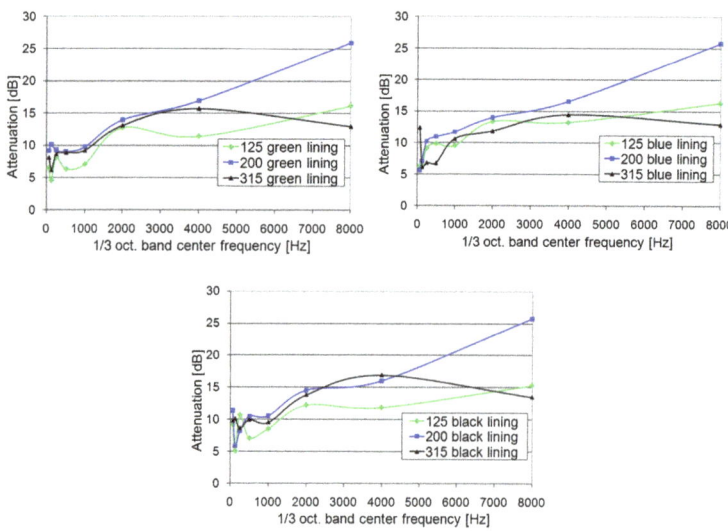

Fig. 5. Attenuation values for three silencers with different sizes and the same linings

The TL experimental values indicate that the 200 silencer blue lining can be preferred from acoustic behaviour point of view. However, we can notice that the sound absorption coefficient varies with the frequency and its average value used in calculations is not a consistent approach.

Table 2. Experimental and theoretical evaluation of samples transmission loss TL.

Inlet/outlet		$\Phi125/\Phi125$	$\Phi200/\Phi200$	$\Phi315/\Phi315$
TL green	Experimental	9.096	13.01	10.365
	Theoretical	11.553	10.804	10.333
TL blue	Experimental	10.497	13.743	10.286
	Theoretical	12.421	11.692	11.213
TL black	Experimental	9.992	12.828	11.538
	Theoretical	9.905	9.129	8.669

5 Conclusions

The paper deals with a theoretical and experimental investigation on sound attenuation performances of some commercial plenum types used for HVAC systems. The results, presented for the most demanding situation in the exploitation, highlight the important influence in attenuation of the absorptive material lining. The silencer lining influences the sound transmission loss more than the geometry in certain situations. An average sound absorption coefficient and an average attenuation are not conclusive for assessing the acoustic behaviour of a sound absorber, as these parameters vary with frequency. The experimental procedures in rating the HVAC systems noise lead to greater confidence. The inlet-outlet size of the silencer, the geometrical dimensions and the acoustical properties of the lining are important in its acoustic design.

References

1. Borelli, D., Schenone, C., Pittaluga, I.: Theoretical and numerical modelling of a parallel-baffle rectangular duct. In: Proceedings of Meetings on Acoustics, vol. 14, p. 040004 (2012)
2. ASHRAE – Handbook-Sound and Vibration Control, chap. 48, p. 279, USA (2011)
3. Peat, K.S.: A transfer matrix for an absorption silencer element. J. Sound Vib. **146**, 353–360 (1991)
4. Munjal, M.L., Beheera, B.K., Thawanib, P.T.: Transfer matrix model for the reverse-flow, three-duct, open end perforated element muffler. Appl. Acoust. **54**, 229–238 (1998)
5. Wu, T.W., Cheng, C.Y.R., Zhang, P.: A direct mixed-body boundary element method for packed silencers. J. Acoust. Soc. Am. **111**, 2566–2572 (2002)
6. Wu, T.W., Cheng, C.Y.R., Tao, Z.: Boundary element analysis of packed silencers with protective cloth and embedded thin surfaces. J. Sound Vib. **261**, 1–15 (2003)
7. EN ISO 7235: Acoustics-Measurement procedures for ducted silencers-Insertion loss, flow noise and total pressure (2010)
8. Vér, I.L., Beranek, L.L.: Noise and Vibrations Control Engineering: Principles and Applications, p. 501. Wiley, Hoboken (2006)
9. Munjal, M.L.: Acoustics of Ducts and Mufflers, p. 416. Wiley, Hoboken (2014)
10. Bujoreanu, C., Benchea, M.: Experimental investigation of noise characteristics for HVAC silencers. In: MATEC Web of Conferences, vol. 112, p. 07001 (2017)

Influence of Glass Fibre Fabrics/Epoxy Hybrid Adhesive Layer on Mechanical Properties of Adhesive Bond

Miroslav Müller[(⊠)] and Petr Valášek

Faculty of Engineering, Department of Material Science and Manufacturing Technology, Czech University of Life Sciences Prague, Kamýcká 129, Prague, Czech Republic
muller@tf.czu.cz

Abstract. An experimental program describes mechanical properties of hybrid adhesive bonds created with a two-component structural adhesive and a reinforcing glass fibre fabric (GF) tested at 20, 40 and 60 °C. The reinforcing fabric was of a type E glass fibres which was implemented into the adhesive bonds without a treatment and with a plasma treatment with a power of 175 W and 350 W. The adhesive bonds without the reinforcing GF showed at the temperature 20 °C a shear tensile strength 11.51 ± 0.42 MPa. This shear tensile strength was considerably decreased with increasing temperature. The hybrid adhesive bonds with the reinforcing GF of weights in grams 110 and 160 g/m^2 showed a significant increase of the shear tensile strength, namely up of ca. 63%. An influence of the plasma treatment was not significant. The significant increase of the shear tensile strength was reached by adding the reinforcing fabric. Higher values of the shear tensile strength were reached at the hybrid adhesive bonds reinforced with the GF of the weight in grams 160 g/m^2.

Keywords: Plasma treatment · Shear tensile strength · Strain at break

1 Introduction

An adhesive bonding technology is prospective method of diverse materials connecting which is used in various production branches. It is necessary to watch a safety and a design of a production process, namely in an automotive industry [1]. The adhesive bonding technology is used for connecting of e.g. composite materials which are difficult to connect by other method. They are e.g. natural and synthetic fibre composite materials [2–4]. Many research studies deal with a research on mechanical properties of both composite materials themselves and their connecting and mutual adhesive properties [2–7]. A different deformation of the adhesive layer occurs at the adhesive bonds where uniform layer of the adhesive is not secured [8, 9]. It comes to a failure of adhesive and cohesive bonds [8, 9]. This state decreases an adhesive bond strength [8, 9]. This negative influence can be eliminated by adding distance means into the layer of the adhesive which secure uniform layer of the adhesive [6, 10]. Many methods are used in the technical practice. However, it is necessary to search for ways which secure not only uniform layer of the adhesive but also do not decrease a tact of a production in

© Springer International Publishing AG, part of Springer Nature 2019
J. Machado et al. (Eds.): HELIX 2018, LNEE 505, pp. 554–560, 2019.
https://doi.org/10.1007/978-3-319-91334-6_75

the same time and ideally increase the adhesive bond strength and reduce a unit price per one bond at the same time. Researches focused on the increase of the adhesive bond strength showed adding a reinforcing synthetic or biological fabric into the layer of the adhesive as one of possibilities [6, 11]. The reinforcing fabric improves a uniformity of the adhesive layer thickness and minimizes bad integrity inside the layer of the adhesive at the same time [10, 12]. The research with the fabric in the form of glass fibres showed a positive influence on the adhesive bond strength increase [13, 14]. Mechanical properties of fibre composite materials depend on a composition of particular layers, their orientation, a specific weight and a chemical treatment influencing a wettability [15, 16]. A construction of agricultural machines where it is necessary to increase a rigidity of the construction owing to high shocks at a drive on unpaved roads and fields is an example of an application area of hybrid adhesive bonds with the reinforcing fabric. A plasma treatment is one of possibilities of the adhesive bonded surface activation [2]. Generally, plasma treatments have the ability to change the surface properties through the formation of free radicals, ions and electrons in the plasma stream [17]. During the plasma treatment, the surface of the substrate is bombarded with high–energy particles [17]. Surface properties such as the wettability, the surface chemistry and the surface roughness of the substrate can be altered as a result, without the need for any hazardous chemicals or solvents [17]. The use of the plasma technology is among the most efficient and economical methods of modifying the surface properties of polymers and fibres without affecting the internal structure [17]. The plasma modifies the surface of microfibres by removing weakly attached surface layers and forming new functional groups on the surface [17].

The aim of the experiment was to describe an influence of the reinforcing GF (110 and 160 g/m^2) on the shear tensile strength and a deformation of the adhesive bond exposed to increased temperatures acting, i.e. 20, 40 and 60 °C. Further the influence of the plasma treatment of the GF on the strength and the deformation of the adhesive bond and the influence of adhesive bonds between the adhesive layer and the reinforcing fabric were investigated. The research on this hybrid adhesive bond is essential owing to a combination of three different layers, i.e. the layer of the adhesive bonded material, the layer of the adhesive and the layer of the reinforcing fabric inserted into the layer of the adhesive.

2 Materials and Methods

Adhesive bonds were prepared in accordance with requirements of the standard CSN EN 1465. Adherents from structural carbon steel S235J0 (100 × 25 × 1.5 mm, an overlapping length of the adhesive bond was 12.5 ± 0.25) were used for the research. A structural two-component epoxy adhesive of a significant Czech producer (CHS Epoxy 1200, marked R – resin) was used for the research. Adhesive bonds with the use of the resin without the reinforcing fabric were the comparing standard. The reinforcing GF was inserted into the layer of the adhesive. The fabric was composed of the type E glass fibres in a plain weave. Weights in grams of fabric in the extent of 110 and 160 g/m^2 (fabrics without the treatment) were used for an optimization of the composite bonds properties. GF belts of the weight in grams 110 g/m^2 and 160 g/m^2

were plasma-irradiated before the adhesive bonding process, subsequently they were cut to the dimension 30×20 mm and implemented into the layer of the adhesive bond. A time delay 48 h was between the plasma treatment and the adhesive bonding. The plasma was generated by the plasma generator KPR 200 mm while supplying the reaction gas (oxygen) and maintaining the reactor's pressure at 0.1 Torr with the use of a vacuum pump. It was the plasma reactor based on Diffuse Coplanar surface Barrier Discharge. To determine the properties that depend on the discharge power and the treatment time, the plasma treatment was conducted in the power range 175 and 350 W for 90 s.

A reduced designation of variants of the experiment is used in following texts for a clarity of figures and a table: R means the adhesive bond without the reinforcement, hybrid adhesive bonds are marked by a record e.g. R/GF110Plasma175 W - GF means the GF reinforcing the adhesive bond and the number means the weight in grams of the GF; in case the fabric is treated by the plasma there is also the number indicating the plasma treatment power range in Watts at the variant.

The adhesive bonded surface of the structural carbon steel S235J0 was mechanically treated – grit blasted by Garnet MESH 80 and chemically treated – cleaned in the acetone bath. Roughness parameters were measured with a portable profilemeter Mitutoyo Surftest 301. A limit wavelength of the cut-off was set as 0.8 mm. The surface roughness was Ra = 1.75 ± 0.23 μm, Rz 11.62 ± 1.37 μm.

A principle consisted in putting the adhesive layer on the first adhesive bonded part, subsequently the fabric was applied and the second adhesive bonded part on which the layer of the adhesive was also deposited was attached. Adhesive bonds were hardened for 72 ± 5 h at the temperature 22 ± 2 °C. A bond was fixed by a weight of 750 g.

The tensile strength test was performed using the universal tensile strength testing machine LABTest 5.50ST (a sensing unit AST type KAF 50 kN, an evaluating software Test&Motion) with temperature chamber TJR 70-200. The destructive testing was performed at the laboratory temperature 22 ± 2 °C and at increased temperatures, i.e. 40 ± 2, 60 ± 2 °C. These temperatures are stated as preferably used temperatures in the standard CSN EN ISO 9142. The temperature range 20 to 60 °C corresponds to the most often environment temperatures which exist during the use of the agricultural machines. The thermal degradation of the adhesive bonds at a simultaneous adhesive bond loading belongs among often failure causes of adhesive and cohesive bindings inside the adhesive bond. Subsequently, the test sample was placed into a self-locking jaws and tempered at the chosen temperature without any other loading for 10 min so that the entire sample was sufficiently heated, then the sample was loaded by a speed 2 mm/min. The failure type was determined at the adhesive bonds according to ISO 10365. Fracture surfaces were examined with SEM (scanning electron microscopy) using a microscope MIRA 3 TESCAN at the accelerating voltage of the pack (HV) 5.0 kV. The samples were dusted with gold by means of the equipment Quorum Q150R ES. Measured values were processed by means of statistical analysis methods (ANOVA F-test, $H_0 = p > 0.05$).

3 Results and Discussion

The experiment results proved a positive influence of the reinforcing GF on the shear tensile strength of the adhesive bonds (Fig. 1). The considerable increase of the shear tensile strength of the adhesive bonds in the interval 20.3 to 63.0% occurred at the temperatures 20 and 40 °C at the adhesive bonds reinforced with the GF 110 and 160 g/m². The shear tensile strength of the adhesive bonds reinforced with the GF 160 g/m² was increased of 16.2% at the increased temperature 60 °C. The shear tensile strength of the adhesive bonds reinforced with the GF 110 g/m² did not considerably changed at the increased temperature 60 °C. The shear tensile strength was not unambiguous by comparing the adhesive bonds reinforced with the GF 110 g/m² without the treatment and with the plasma treatment of a power 175 W and 350 W before the adhesive bonding. A mild fall in the interval from 4 to 5.2% occurred at the temperatures 20 and 60 °C. The increase of the shear tensile strength occurred only at the temperature 40 °C. The shear tensile strength was not unambiguous by comparing the adhesive bonds reinforced with the GF 160 g/m² without the treatment and with the plasma treatment of a power 175 W and 350 W before the adhesive bonding. The results also did not show the influence of the temperature. A difference in the shear tensile strength of the adhesive bonds at the compared sets exposed to the temperatures 20, 40 and 60 °C in the significance level 0.05 among single tested variants stated in Fig. 1, i.e. the comparison of the adhesive bond without the reinforcing fabric (R) and with the use of the reinforcing GF 110 g/m² was proved by the statistical testing. The tested parameter p ranged in the interval 0.0000 to 0.0091.

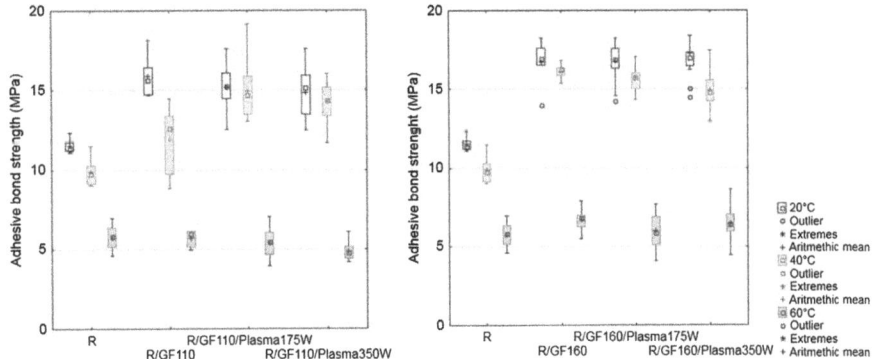

Fig. 1. Shear tensile strength - adhesive bonds reinforced with GF 110 g/m² (left) and 160 g/m² (right)

Analogous results were proved at the testing of the adhesive bonds reinforced with the GF 160 g/m² at the temperature 20 and 40 °C, i.e. the difference in the shear tensile strength of the adhesive bonds was proved (Fig. 1). The tested parameter p was 0.0000. The statistical difference was not proved at the tested temperature 60 °C, i.e. the hypothesis H_0 was certified. The statistical testing proved that there is no difference in the shear tensile strength, i.e. parameter p = 0.2457.

The adhesive layer thickness determined from the adhesive bond cut by means of an optical analysis was 180 ± 59 μm (a dispersion ca. 33%) at the adhesive bonds non-reinforced with the GF and 256 ± 32 μm (the dispersion ca. 12.5%) at the adhesive bonds reinforced with the GF. It is obvious from above stated results that adding the reinforcing fabric increases a homogeneity of the adhesive layer.

The difference in the shear tensile strength in the significance level 0.05 at the GF of the weight in grams 110 g/m² at the temperature 40 °C (p = 0.0011), 60 °C (p = 0.0102) and of the weight in grams 160 g/m² at the temperature 40 °C (p = 0.0191) was proved by the statistical testing evaluating the adhesive bonds reinforced with the GF with the plasma treatment and without it. No difference in the shear tensile strength in the significance level 0.05 at the GF of the weight in grams 110 g/m² at the temperature 20 °C (p = 0.3139) and of the weight in grams 160 g/m² at the temperature 20 °C (p = 0.6259), 60 °C (p = 0.2525) was proved by the statistical testing evaluating the adhesive bonds reinforced with the GF with the plasma treatment and without it.

The adhesive bonds having only the layer of the adhesive were distinguished for an adhesive fracture surface type. A fracture surface was changed to an adhesive-cohesive type by adding the reinforcing GF (Fig. 2).

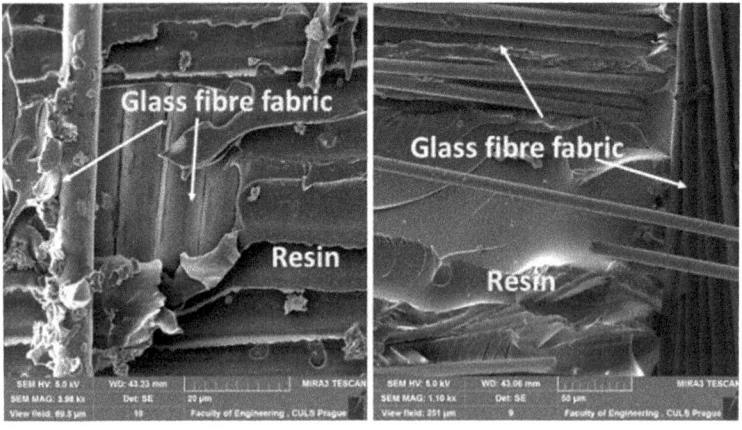

Fig. 2. Fracture surfaces of adhesive bonds reinforced with GF 110 g/m² (left) and 160 g/m² plasma treatment (right)

The increased temperature 60 °C negative influenced the shear tensile strength of the adhesive bonds at both tested variants. Messler came to similar conclusions about the thermal degradation [10]. The negative effect of the increased temperature is more significant at the simultaneous loading of the adhesive bond. This assumption was certified. The negative influence of the increased temperature is characterized namely by a rise of soft cracks which gradually grow and this their growth, namely in a presence of liquids, can lead up to cracking of products, mainly in the most loaded parts, e.g. in bends [10]. Hu et al. e.g. state that a critical fall of the strength of all adhesives is usually in the interval

Table 1. Strain at break of adhesive bonds (*R means adhesive bond without reinforcing GF, **hybrid adhesive bonds reinforcing with GF of relevant weight in grams, *** hybrid adhesive bonds with reinforcing GF of relevant weight in grams plasma treated by given power range)

Variant of experiment	20 °C	40 °C	60 °C
R*	2.46 ± 0.11	1.59 ± 0.11	2.28 ± 0.19
R/GF110**	2.83 ± 0.28	2.65 ± 0.22	2.11 ± 0.18
R/GF160**	2.78 ± 0.11	3.32 ± 0.28	2.82 ± 0.30
R/GF110Plasma175 W***	2.73 ± 0.22	3.33 ± 0.28	2.29 ± 0.24
R/GF110Plasma350 W***	2.61 ± 0.27	3.00 ± 0.32	2.58 ± 0.30
R/GF160Plasma175 W***	3.11 ± 0.23	3.82 ± 0.35	2.40 ± 0.20
R/GF160Plasma350 W***	2.96 ± 0.17	3.64 ± 0.25	2.44 ± 0.26

of the temperatures 60 to 100 °C [18]. The results confirmed the assumption that increasing temperature decreased the strength of the bond [9]. There was a significant difference in the shear tensile strength at 40 and 60 °C. A thermal degradation occurs in a polymer owing to the temperature. A reliability of the adhesive bonds depends also on their resistance to different temperatures acting. That means their ability to keep their properties at long-term acting of increased or decreased temperature is essential for the practical application. The results proved a positive influence of the reinforcing GF on a strain at break of the adhesive bonds (Table 1), mainly at 40 °C.

4 Conclusion

The results confirmed the positive influence of the reinforcing GF on the shear tensile strength and the strain at break of the adhesive bonds. The hybrid adhesive bonds with the reinforcing GF of the weight 110 g/m^2 showed considerable increase of the shear tensile strength at the loading temperature 20 °C of ca. 37.7% and up of ca. 50.0% at the reinforcing GF of the weight in grams 160 g/m^2. The significant fall of the shear tensile strength of the adhesive bonds without the reinforcing GF occurred owing to the increased temperature 40 and 60 °C. The adhesive bonds without the reinforcing GF showed the shear tensile strength 11.51 ± 0.42 MPa at the temperature 20 °C. The shear tensile strength was reduced of 14% at the adhesive bonds exposed to the temperature 40 °C and up of 49.9% to the temperature 60 °C. The shear tensile strength was not reduced by adding the reinforcing GF at the hybrid adhesive bonds even at the temperature 40 °C, i.e. the shear tensile strength was increased up of 63% compared to the adhesive bonds without the reinforcing fabric. The influence of the plasma treatment was not significant. The significant increase of the shear tensile strength was reached at the hybrid adhesive bonds reinforced with the GF of the weight in grams 160 g/m^2. The increased temperature 60 °C had the negative influence on the shear tensile strength of the adhesive bonds in all tested variants. Analogous trend as at the shear tensile strength was also determined at the strain at break. The experiment results proved the positive influence of the reinforcing GF on the strain at break of the adhesive bonds. The considerable increase of the strain at break occurred at the adhesive bonds reinforced with the GF at the temperature 40 °C.

Acknowledgements. This paper has been supported by the grant IGA TF CZU 2017.

References

1. Machado, J., Seabra, E., Campos, J., Soares, F., Leao, C.: Safe controllers design for industrial automation systems. Comput. Ind. Eng. **60**, 635–653 (2011)
2. Valášek, P., Müller, M., Šleger, V.: Influence of plasma treatment on mechanical properties of cellulose-based fibres and their interfacial interaction in composite systems. BioResources **12**(3), 5449–5461 (2017)
3. Ruggiero, A., Valášek, P., Müller, M.: Exploitation of waste date seeds of Phoenix dactylifera in form of polymeric particle biocomposite: investigation on adhesion, cohesion and wear. Compos. B Eng. **104**(1), 9–16 (2016)
4. Valášek, P., Ruggiero, A., Müller, M.: Experimental description of strength and tribological characteristic of EFB oil palm fibres/epoxy composites with technologically undemanding preparation. Compos. B Eng. **122**, 79–88 (2017)
5. Mizera, Č., Herák, D., Hrabě, P., Müller, M., Kabutey, A.: Mechanical behavior of ensete ventricosum fiber under tension loading. J. Nat. Fibers **14**(2), 287–296 (2017)
6. Müller, M., Valášek, P., Rudawska, A.: Mechanical properties of adhesive bonds reinforced with biological fabric. J. Adhes. Sci. Technol. **31**(17), 1859–1871 (2017)
7. Müller, M., Valášek, P., Ruggiero, A.: Strength characteristics of untreated short-fibre composites from the plant ensete ventricosum. BioResources **12**(1), 255–269 (2017)
8. Müller, M., Herák, D.: Dimensioning of the bonded lap joint. Res. Agric. Eng. **56**(2), 59–68 (2010)
9. Grant, L.D.R., Adams, R.D., Lucas da Silva, F.M.: Experimental and numerical analysis of single-lap joints for the automotive industry. Int. J. Adhes. Adhes. **29**(4), 405–413 (2009)
10. Messler, R.W.: Joining of Materials and Structures from Pragmatic Process to Enabling Technology, p. 816. Elsevier, Burlington (2004)
11. Naito, K., Onta, M., Kogo, Y.: The effect of adhesive thickness on tensile and shear strength of polyimide adhesive. Int. J. Adhes. Adhes. **36**, 77–85 (2012)
12. Hafiz, T.A., Abdel Wahab, M.M., Crocombe, A.D., Smith, P.A.: Mixed-mode fracture of adhesively bonded metallic joints under quasi-static loading. Eng. Fract. Mech. **77**(17), 3434–3445 (2010)
13. Zavrtálek, J., Müller, M., Šleger, V.: Low-cyclic fatigue test of adhesive bond reinforced with glass fibre fabric. Agron. Res. **14**, 1138–1146 (2016)
14. Zavrtálek, J., Müller, M.: Research on mechanical properties of adhesive bonds reinforced with fabric with glass fibres. Manuf. Technol. **16**(1), 299–304 (2016)
15. Karbhari, V.M., Abanilla, M.A.: Design factors, reliability, and durability prediction of wet layup carbon/epoxy used in external strengthening. Compos. B Eng. **38**(1), 10–23 (2007)
16. Maheri, M.R.: The effect of layup and boundary conditions on the modal damping of FRP composite panels. J. Compos. Mater. **45**(13), 1411–1422 (2010)
17. Boruvka, M., Ngaowthong, Ch., Cerman, J., Lenfeld, P., Brdlik, P.: The influence of surface modification using low-pressure plasma treatment on PE-LLD/α-cellulose. Compos. Prop. **16**(1), 29–34 (2016)
18. Hu, P., Han, X., Li, W.D., Li, L., Shao, Q.: Research on the static strength performace of adhesive single lap joints subjected to extreme temperature environment for automotive industry. Int. J. Adhes. Adhes. **41**, 119–126 (2013)

Technology of Preparing Coatings with Nanoparticles

Jan Novotný[✉], Irena Lysoňková, Jaroslava Svobodová, and Štefan Michna

Faculty of Mechanical Engineering, J. E. Purkyně University in Ústí nad Labem, Pasteurova 3334/7, 400 01 Ústí nad Labem, Czech Republic
{jan.novotny, irena.lysonkova, jaroslava.svobodova, stefan.michna}@ujep.cz

Abstract. This article is dealing with the formation of nanocomposite coatings. These coatings can be of wide use. His particular research has been carried out to extend the durability of the molds used to vulcanize rubber materials in the automotive industry. A base material in this case is an aluminum alloy, specifically Al-Si. This alloy is coated with a nanocomposite coating. In this case the PTFE coating, which can be used on its own, is enriched with nanoparticles based on titanium dioxide, tungsten carbide and aluminum oxide. This paper is describing the specific technology process, which is the coating, and subsequently the verification of the uniform coating by electron microscopy. The results obtained using electron microscopy is showing the confirmation of the equilibrium distribution and the content of each of the elements. Furthermore, the components of the nanocomposite are listed here. In the future, the application of these nanocomposite coatings can be extended to other industries.

Keywords: Nanoparticles · Coatings · Electron microscopy

1 Introduction

The area of nanoparticles and their use is a major subject of development today. Particles in scale "nano" may exhibit different properties relative to particles in the size "micro" and more. They are often changing their basic properties such as electrical conductivity, color, or strength. However, resizing to a nanometric area is not an easy process. This is a particle magnification, called "The Bottom Up" procedure, and particle reduction called "The Top Down" procedure. This research is dealing with The Top Down procedure. There are many ways how to achieve the particle reduction, for example a mechanical grinding, a laser ablation and a lithographic method. It was used by a mechanical grinding with a ball mill, as described in the next chapter.

After the preparation of the samples, they were coated using the above mentioned technological method. Afterwards the samples were verified using the electron micro-scope, meaning if the particles are in "nano" size, if the particles are attached to the coating, and if they are not lasting uniform distribution (not cluster of particles), which causes uniform properties of overall coating. In all three cases, these properties were positively tested [1].

© Springer International Publishing AG, part of Springer Nature 2019
J. Machado et al. (Eds.): HELIX 2018, LNEE 505, pp. 561–567, 2019.
https://doi.org/10.1007/978-3-319-91334-6_76

2 Preparing of Nanoparticles

As it was mentioned in the introduction, all the particles were prepared in a planetary ball mill, which has many advantages, such as financial variety, ecology, ecological disposal, relatively simple influencing of degradation of processes associated with high temperature, compatibility of immiscible mixtures. This type of mill can be not only used for mechanical grinding but also for mechanical alloying. Milling is to be held in a liquid atmosphere (colloid) or gaseous atmosphere (not affected – air, possibly with a particular gas). The base of the planetary ball is a container placed on a rotating support disk performing a rotation movement around its own axis. The planetary motion is characterized by the so called epicycloids. In the grinding containers there are grinding bodies which are usually in the uniform size. The balls are different sizes, but there are always the same one in the grinding cycle. The above-mentioned grinding bodies grind the above-described "Top down Method" and formed nanoparticles, [2] The milling process goes to some point a complicated process therefore it must be observed under certain conditions, for example it is necessary to avoid overheating and following thermal influence on the milled particles. The parameters that can affect the milling are following:

- The type of mill
- The grinding container
- The speed of milling
- The time of grinding
- Type, size and distribution of the grinding medium
- The weight ratio of the ball to the powder
- The degree of filling of the container
- The grinding atmosphere
- The motion force
- The milling temperature

The particles were grinded on a planetary ball mill PM 100. In this case, there was used a colloidal grinder in demineralized water. Each one of the powders was ground separately. A specific example of titanium dioxide for this experiment is given in Table 1. When grinding on ball mill, different sized particles are being formed, and it is necessary to separate the various sizes subsequently. This separation is carried out on an AS 200 sieving machine [3].

3 Technology of Coating

This chapter describes the entire coating process. A phase "I" is added to a particular type nanoparticles. It is also important to insert a suitable percentage of the particle type into the coating bath. These particles have to be stirred; otherwise they can settle down on the bottom of the coating container. Because of the size of the samples, there was used an electromagnetic stirrer in this particular case (Table 2).

Table 1. Parameters of particle preparation at ball mill.

Input sample	TiO2	Weight of balls	500 g
Total grinding time	150 h	Ball material	ZrO2
Grinding interval	10 min	Volume of balls	130 ml
Interval pause	50 min	Sample weight	30 g
Pure grinding time	25 h	Sample volume	7.7 ml
Revolutions	300 rpm	Input sample size	<20 μm
Container volume	250 ml	BPR	17:1
Container	ZrO2	Grinding media	Water
Balls diameter	3 mm	Vol. of grinding media	100 ml

Table 2. Coating conditions [4]

	Bath composition	Temperature	pH	Time	Note
Degreasing and depickling	15 g/l P1 + 3 g/l P2 + demineralized water	50–55 °C	11.0–12.5	2–3 min	The content of bath must be below 10 g/l in solution
Rinse	Demineralized water	20–30 °C		Briefly	
Coating phase I	40 g/l P3 + demineralized water	25–30 °C	4.8–5.2	1–2 min	The bath pH must not exceed 5.2
Rinse	Demineralized water	20–30 °C		Briefly	
Drying	Hot and drying kiln	110–115 °C		20–25 min	
Coating phase II	10–12 g/l P4 + demineralized water	60–65 °C	7.8–8.5	15–16 min	Material after phase II. Do not rinse the coating
Drying	Kiln	100 °C		30 min	Preheated oven

4 Sample Evaluation

4.1 Coefficient of Friction

Abrasion resistance measurements were performed on a tribolab according to ASTM G132-96. The first sample was coated with Teflon alone without the addition of nanoparticles. The measured value was 0.09794. A second specimen coated with titanium dioxide nanoparticles coated Teflon coating with a score of 0.5584. In this case, a five-fold increase of this coefficient was found.

4.2   Electron Microscopy

In this experiment, the samples were evaluated on a Zeiss SEM-FIB Cross Beam Auriga scanning electron microscope. Because it is important to link the coating and nanoparticles, we can see this connection in Fig. 1. There is captured a place that has been subjected to a more thorough examination. Figure 2 is showing only the image of the layout of all the elements contained within. The graphic representation and weight percentages are shown in Fig. 3. In Fig. 4 there are pictures of an each element. Aluminum, silicon and magnesium are captured from the base material. The fluorine, carbon and oxygen elements are contained in a Teflon coating. Oxygen and titanium exhibit titanium dioxide enrichment particles. In the following part of the experiment, we were using an electron microscope TESCAN Vega 3 with EDX analyzer BRUKER X-FLASH. In Fig. 5 there is shown the place of study. Figure 6 is showing a coating with an even distribution of Al_2O_3. The graphic representation of each element is in Fig. 7. Aluminum and silicon are contained in the base material. Carbon, zirconium, oxygen and fluoride are contained in the PTFE coating. Aluminum and oxygen are contained in alumina.

Fig. 1. Rated place of coating reached TiO_2

Fig. 2. Overall element map of coating TiO_2

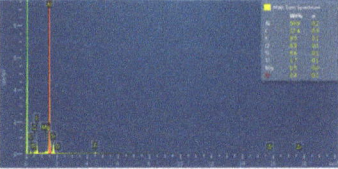

Fig. 3. Graphic and percentage representation of individual elements of coating reached TiO_2

In the following section we were using the same microscope. In this case, nanocomposite PTFE coating and carbide tungsten were evaluated. Figure 8 is showing a uniform distribution of the particles and the site under investigation. In Fig. 9 there is shown overall element map. The capture graph in Fig. 10 is indicating the individual

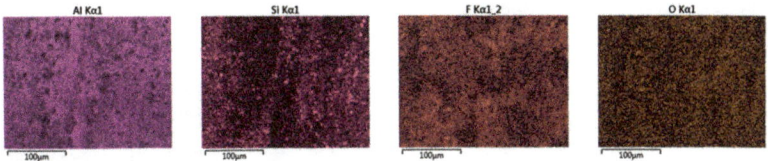

Fig. 4. Element map of each element of coating reached TiO$_2$

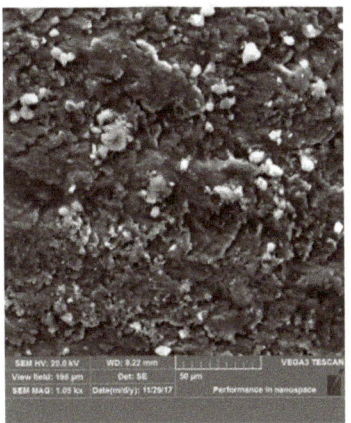

Fig. 5. Rated place of coating reached Al$_2$O$_3$

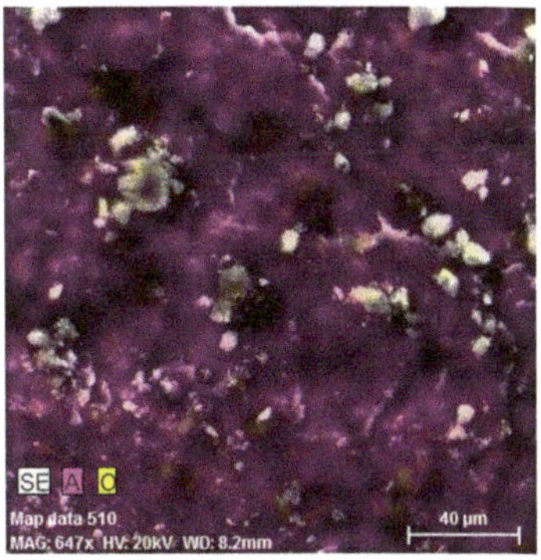

Fig. 6. Overall element map of coating reached Al$_2$O$_3$

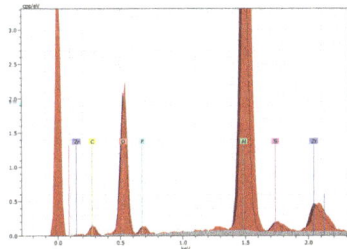

Fig. 7. Graphic representation of individual elements of coating reached Al$_2$O$_3$

samples. Aluminum is part of the base material. Zirconium, oxygen and fluorine are part of the PTFE coating, and tungsten and carbon are made of tungsten carbide.

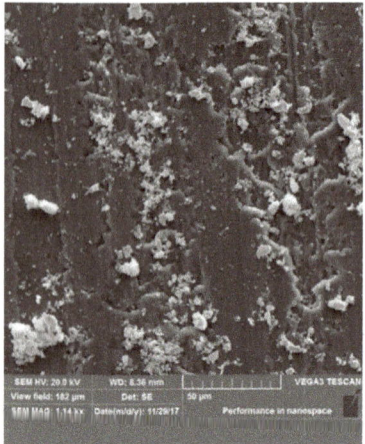

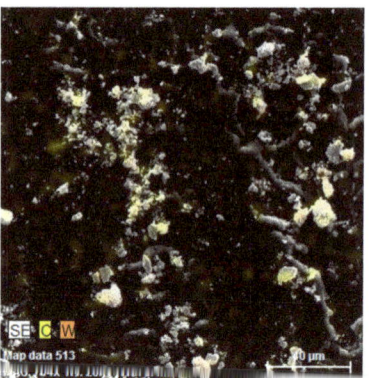

Fig. 8. Rated place of coating reached WC

Fig. 9. Overall element map of coating reached WC

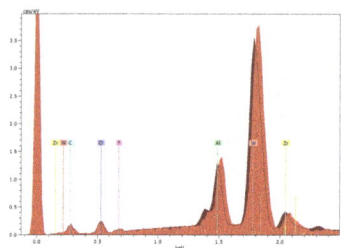

Fig. 10. Overall element map of coating reached WC

5 Conclusions

This article was dealing with the technology of forming a nanocomposite PTFE coating enriched with titanium dioxide particles. This coating is subsequently confirmed by electron microscopy. It has been confirmed that clumps are clearly clogged in the coating and that they are evenly distributed. The coating was thus formed according to requirements.

In addition, two other coatings were created similar to the technology, namely PTFE and Al_2O_3 and PTFE and WC. With these two coatings the procedure is very similar to small differences. As with the previous type of coating, evaluation was performed on a scanning electron microscope.

References

1. Svobodová, J., Kuśmierczak, S.: New trends in surface in pre-treatment. Toyotarity. Value Engineering of Production Processes. Borkowski, S., Stasiak-Betlejewska, R. Faculty of Logistics, University of Maribor: AWR Editor (2012). ISBN 978-961-6562-60-7
2. Žárská, S.: Výzkum možností aplikací vybraných mechanicky legovaných kovových prášků. UJEP, Ústí nad Labem (2017)
3. Jaskevič, M.: Příprava a analýza velmi malých částic vhodných k povlakování hliníkových slitin. UJEP, Ústí nad Labem (2017)
4. Utility Pattern: 13420/13:43885462 Zavedení technologie chemického povlakování Al forem za vzniku ochranných nanovrstev

Friction Properties of Polyoxymethylene (POM) Materials in Dry and Lubricated Conditions

Seyyed M. M. Sabet[1(✉)], Marco Dourado[2], António Pereira[1],
Ana Silva[1], Flávia Carvalho[1], Irene Brito[3], José Machado[1,4],
José Meireles[1,4], José Azevedo[2], Sergio Carvalho[1], José Gomes[1,5],
and Cristiano Abreu[1,6]

[1] Mechanical Engineering Department, University of Minho, Guimarães,
Portugal
{sabet, b7043, b7955, b8014, jmachado, meireles, sergioc,
jgomes}@dem.uminho.pt
[2] Bosch Car Multimedia SA, Max Grundig, 35 Lomar, Braga, Portugal
{marco.dourado, JoseAzevedo.Goncalves}@pt.bosch.com
[3] Center of Mathematics, University of Minho, Guimarães, Portugal
ireneb@math.uminho.pt
[4] MEtRICs - Mechanical Engineering & Resource Sustainability Center,
University of Minho, Guimarães, Portugal
[5] CMEMS - Center for Microelectromechanical Systems, University of Minho,
Guimarães, Portugal
[6] Physics Department, Porto Superior Engineering Institute, ISEP, Porto,
Portugal
csa@isep.ipp.pt

Abstract. This work is based on dry and lubricated friction properties of Polyoxymethylene (POM) plastics widely used in multimedia and electronic sector of automotive industries. The results show that lubrication does not seem to influence the static coefficient of friction; however, significantly influences the dynamic one and increases it by two-fold.

Keywords: Static and dynamic coefficient of friction
Polyoxymethylene (POM) · Steering angle sensors (SAS) · Tribometer

1 Introduction

Noise reduction plays an important role for every car manufacturer with an increasing trend to produce components with lower noise generation levels. Interactions of plastic components are often associated with friction and noise generation. As of 1929, the external noise produced by vehicle has been continuously regulated. The trend of the regulation was to reduce its maximum limits from 82 dB (A) in 1978 to 74 dB (A) in 1996. The European Union has set a limit of 68 dB (A) till 2026 [1].

As tendency of the automotive industry is to increase the number of electric cars, there is a growing demand for cars with quieter cabins where small components such as

© Springer International Publishing AG, part of Springer Nature 2019
J. Machado et al. (Eds.): HELIX 2018, LNEE 505, pp. 568–573, 2019.
https://doi.org/10.1007/978-3-319-91334-6_77

the steering angle sensors (SAS), contribute to the general acoustic behavior of the car cockpit [2].

The SAS is a sensor that measures the angle, position and rotational velocity of the steering wheel. The measurements from SAS sensor, along with the information of other sensors, are used to control various driving assistance systems, such as steering power and active steering control [3].

The SAS consists of 6 different parts (as shown in Fig. 1): a housing, a cover, a gear system containing two small gears and a larger one (hub) driven by the steering column and a PCB board, which measures the steering information while driving.

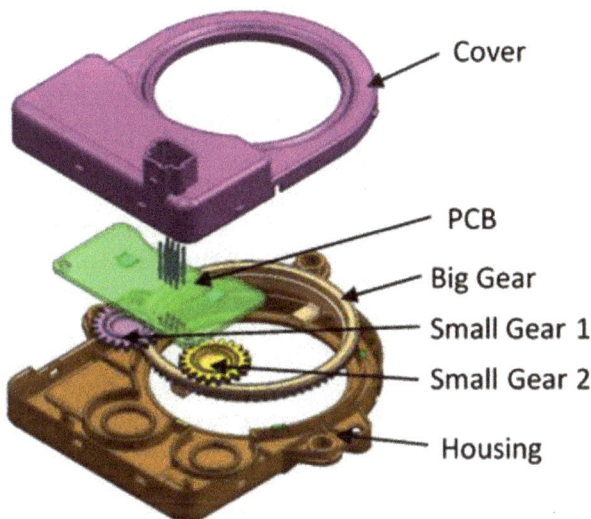

Fig. 1. The components of a SAS sensor.

Noise is transmitted in two ways: structure-borne or air-borne noise transmission [4]. Structure-borne noise transmission is caused by a structure whose vibration is transferred to the neighboring structures. Air-borne noise transmission is generated by a vibrating surface moving the air molecules. In case of SAS, noise is transmitted in both ways. The rotation of the gears and impact of the teeth promote the structural vibration of the components (structure-borne), which subsequently excites the air molecules in the sensor to vibrate causing air-borne noise [5].

The large gear (or hub) in Fig. 1 is the most important part in the SAS assembly as it is contact with all the other parts and is the main source of noise generation due to its sliding contact on housing and cover and impact with the smaller gears (1, 2).

This work is focused on the study of dry and lubricated friction properties of POM materials used in the steering angle sensors.

2 Methodology

Bosch Car Multimedia manufactures nearly 5 millions of SAS per year for major car manufactures and is committed to keep the noise levels within 32 to 36 dB (A). Housing, cover and hub of a large of number of SAS manufactured by Bosch are made of POM materials and due to the large contact between hub and housing/cover, the friction behavior of these materials their contribution to noise generation is of great importance.

POM, which is also known as acetal or polyacetal, is a thermoplastic with high stiffness, low friction, and excellent dimensional stability and is widely used in the automotive and electronics industries. Based on the literature [6], POM is the best commercial plastic for high friction applications.

The housing and cover of SAS are made of Ultraform S2320-003 POM supplied by BASF, while HUB is produced by Ultraform W2320-003 PRO TR UN POM (also supplied by BASF). These two POM grades are slightly different in chemical composition to avoid high friction due to self-adhesion.

The two polymers were produced by injection molding without any abrasive pre-treatment. The S2320-003 POM (housing/cover) specimen was characterized by a flat surface while the W2320-003 PRO TR UN POM (hub) sample corresponded to a gear with 65 mm in diameter.

A specially designed sample holder was used to hold the hub, allowing it to rotate in contact with the housing surface in order to measure the static and kinetic friction of the assembly. Figure 2 shows the exploded and assembly views of the sample holder, respectively. Figure 3 shows a general perspective of the sample holder mounted on a Bruker UMT-2 tribometer.

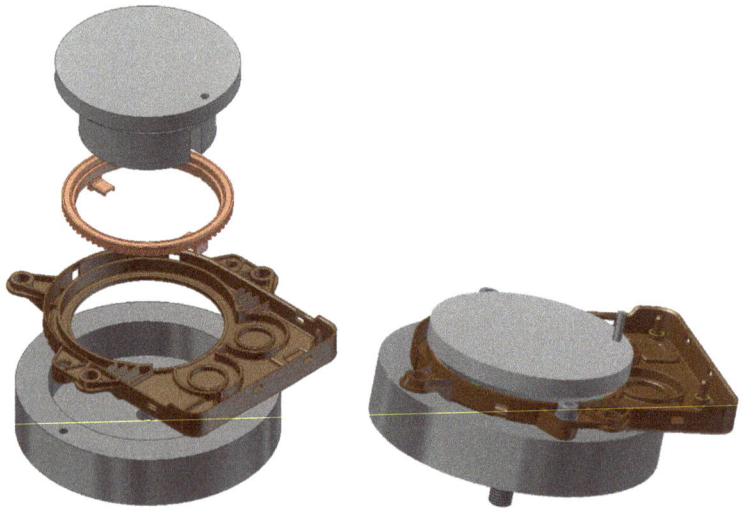

Fig. 2. CAD drawing of custom designed sample holder: Exploded view (left), and assembly view (right).

Fig. 3. The sample holder mounted in Bruker UMT-2 tribometer.

The sliding contact between the specimens was lubricated by lubricant grease, ISOFLEX TOPAS L 32. According to the supplier, the general purpose of this grease is noise reduction for plastic gears.

Throughout the tribological tests, the applied normal load and angular velocity were kept constant at 6 N and 60 rpm, respectively.

3 Experimental Results

The experiments were run with 3 sensors, whereby each sensor was tested three times, in clockwise and counter clockwise rotation accounting for a total number of 12 tests. The static and dynamic friction coefficients were obtained for the hub and for the gears in dry and lubricated conditions.

The acquired friction data was analyzed statistically and point and interval estimates were derived for static and dynamic friction coefficients for dry and lubricated conditions. Tables 1 and 2 contain the sample mean ($\bar{x}$), the standard deviation (s) and

95% confidence intervals (CI) for the mean of the friction coefficients, in dry and lubricated conditions, respectively.

Table 1. Statistical results in dry conditions.

Quantity	Static	Dynamic
$\bar{x}$	0.314	0.040
S	0.079	0.012
CI for μ	[0.261, 0.367]	[0.032, 0.048]

The results show that in dry sliding conditions, the estimated static friction coefficient of the two POM materials sliding against each other is approximately 0.314, with an estimated standard deviation of 0.079, while the dynamic friction is approximately 0.040, with a standard deviation of 0.012 (see Table 1). Lubricated sliding was characterized by friction coefficient values of 0.281 and 0.139 for static and kinetic regimes, respectively. The corresponding standard deviations were 0.102 and 0.098, respectively (see Table 2).

Table 2. Statistical results in lubricated conditions.

Quantity	Static	Dynamic
$\bar{x}$	0.280	0.139
S	0.102	0.098
CI for μ	[0.212, 0.349]	[0.074, 0.204]

As it can be seen, it seems that the lubricant used in this study has decreased the static coefficient of friction, however a hypotheses test with 5% significance level reveals that statistically there is no significant difference between the static coefficient of friction in dry and lubricated conditions. Concerning the dynamic properties, unlike expected [7, 8], the lubricant does not improve the static dynamic response of the tribosystem, on the contrary, it leads to a twofold increase in the dynamic friction coefficient, which is statistically significant. This may be explained as the result of the high viscosity of the grease, which in turn leads to high friction among layers of grease, which modifies the system's energy dissipation under sliding. In other words, the contact of smooth hard polymers with negligible adhesion and low deformation leads to low dynamic friction levels. The results obtained in this study are in agreement with those reported by [9–11] for POM materials.

4 Conclusions

The results showed that in dry conditions the static coefficient of friction is approximately 0.314, while the dynamic one is 0.040. It was also shown that the lubricant used in this work does not influence significantly the static friction, but increases by twofold the dynamic friction coefficient. The coefficients of friction were 0.280 and 0.139 for static and dynamic, respectively in lubricated conditions.

Acknowledgements. This work is supported by: European Structural and Investment Funds in the FEDER component, through the Operational Competitiveness and Internationalization Programme (COMPETE 2020) [Project n° 002797; Funding Reference: POCI-01-0247-FEDER-002797].

References

1. VCA Homepage. http://www.dft.gov.uk/vca/fcb/cars-andnoise. Accessed 15 Feb 2018
2. European Environment Agency Homepage. http://www.eea.europa.eu/articles/electric-vehicles-moving-towards-a. Accessed 10 Feb 2018
3. Hunter Engineering Company Homepage. https://www.pro-align.co.uk/wp-content/uploads/2016/05/6158T-Steering-Angle-06-13.pdf. Accessed 10 Feb 2018
4. Bolton, J.S., Bono, R.W.: Airborne sound transmission into vehicle cabins through door like structures. In: Proceedings of 13th International Modal Analysis Conference-Going Beyound Modal Analysis (1995)
5. Flores, P., Gomes, J.: Ginemática e Dinâmica de Engrenagens: Aspetos Gerais sobre Engrenagens (in Portuguese). Univesidade do Minho, Guimarães (2014)
6. Craftech Homepage. http://www.craftechind.com/the-best-plastic-materials-for-high-friction-applications. Accessed 10 Feb 2018
7. Hutchings, I.M., Shipway, P.: Tribology: Friction and Wear of Engineering Materials, 2nd edn. Elsevier, London (1992)
8. Zsidai, L., De Baets, P., Samyn, P., Kalacska, G., Van Peteghem, A.P., Van Parys, F.: The tribological behaviour of engineering plastics during sliding friction investigated with small-scale specimens. Wear **253**, 673–688 (2002)
9. Chaudri, A.M., Suvanto, M., Pakkanen, T.T.: Non-lubricated friction of polybutylene terephthalate (PBT) sliding against polyoxymethylene (POM). Wear **342–343**, 189–197 (2015)
10. Mergler, Y.J., Schaake, R.P., Huis, A.J.: Material transfer of POM in sliding contact. Wear **256**, 294–301 (2003)
11. Dourado, M., Sabet, S.M.M., Pereira, A., Figueiredo, L., Brito, I., Machado, J., Meireles, J., Pinto, F., Carvalho, S., Gomes, J.R., Abreu, C.S.: Dry and lubricated friction properties of polybutylene terephthalate (PBT) against polyamide 12 (PA12). In: IBERTRIB 2017 – IX Iberian Conference on Tribology, Guimarães, Portugal (2017)

Tensile Characteristics of Epoxy/Jute Biocomposites Prepared by Vacuum Infusion

Petr Valášek[✉] and Miroslav Müller

Faculty of Engineering, Department of Material Science and Manufacturing Technology, Czech University of Life Sciences Prague, Kamýcká 129, Prague, Czech Republic
valasekp@tf.czu.cz

Abstract. Developmental trends in composite systems include substitution of synthetic fillers by renewable sources. Prospective renewable sources usable in the field of composite systems can be denoted natural plant fibers. The paper describes a composite system prepared by a vacuum infusion. As a matrix was used an epoxy resin which was filled with arranged and randomly oriented short jute fibers in the form of yarn residues of a length 42.63 ± 16.52 mm. The tensile strength and strain at break of these composites are described experimentally without surface treatment of the fibers. Electron microscopy was used to describe the interfacial interaction. The short fibers, which were oriented in the direction of the load force, increased the tensile strength by 11 MPa versus the non-filled resin, the non-oriented fibers reduced the tensile strength of 7 MPa.

Keywords: Corchorus · Tensile strength · Cellulose-based fibres

1 Introduction

Composites are perspective materials that combine the properties of their phases [1, 2]. As a biocomposite can be considered a material that uses the properties of its sub-components, of which at least one component is from natural origin - a plant or animal. In the industrial sector, these materials are of great application, for example in the automotive industry. Very promising in this area are in particular natural fibers, which are used in conjunction with a thermoplastic or reactoplastic matrix - for example, coconut, jute, banana fibers [3–6].

Composites with natural fibers are mainly used due to their cost, they have good mechanical properties and are a renewable source. Natural fibers also have some disadvantages, among which is nature of fibers, which is reflected, for example, in the possible variance of the measured values [7, 8]. The use of bio-composites with a filler in the form of a yarn is wide [9].

Secondary products can be considered as very interesting materials from an environmental and ecological point of view. For example, during the processing of fruit palms, EFB (empty fruit bunch) fibers are produced, the meaningful material use of which can be an alternative to their combustion [4, 7, 8, 10].

© Springer International Publishing AG, part of Springer Nature 2019
J. Machado et al. (Eds.): HELIX 2018, LNEE 505, pp. 574–580, 2019.
https://doi.org/10.1007/978-3-319-91334-6_78

During processing of jute fibers on different mats, for example, yarn cuttings of these fibers can be produced, which can be referred to as a secondary product.

According to conclusions of Gopinath et al. [11], the jute fibers can be used to fill technical polymers - the inclusion of 5–6 mm long jute fibers treated with alkali - aqueous NaOH resulted in an 18.67% increase in tensile strength of the polyester matrix and increase of 16.67% at epoxy matrix. Alkali treatment led according to Vilaseca et al. [12] to increase in the composite tensile strength from 22.2 MPa to 26.3 MPa, the same time an increase in tensile strength was observed with the increasing proportion of fibers in the matrix. Jute fibers can be used in composite systems with other natural or synthetic fibers [13].

The aim of the paper is to experimentally describe the tensile strength and hardness of the composites formed by reaction resin and jute fibers in the form of yarn residues. The test bodies were cut by a water jet from the plates prepared by a vacuum infusion. The described composite system uses yarn residues as a filler, thus it is the use of secondary products, for this reason, the fibers have not been modified - this is a description of the possibility of using the secondary product without increasing the complexity of the technological process of preparation.

2   Materials and Methods

Jute fibers are obtained from the Corchorus plant of the family Malvaceae. The fibers were used in the form of yarn residues (twisted spinning fibers) that had an average length of 42.63 ± 16.52 mm. Jute fibers are considered as textile fibers and are widely used. Jute is mainly grown in tropical and subtropical climates. After harvest, the stems are soaked and the woods are removed via the breakers. In the case of subsequent yarn processing, the cuts are produced in many cases, and these cuttings have been used to prepare composite systems. The basic characteristics of jute fibers are shown in Table 1. Fibers grown in China were used in the experiment.

Table 1. Juta fibres: basic properties [11]

Juta fibres properties	Range
Cellulose content (%)	50–57
Lignin content (%)	8–10
Elongation at break (%)	1.8
Tensile strength [MPa]	400–800
Young's modulus [GPa]	30

The two-component epoxy resin (based on bisphenol A and F) with a viscosity suitable for vacuum infusion technology was used as a matrix.

The fibers were placed in a mold space made of plastic, the inside of the mold was treated with a liquid separator to remove the composite plate from the mold after curing. The board always had the following dimensions: 200 × 300 × 4 mm. For the vacuum infusion process, the following parameters were used: 55 l/min, the absolute pressure of 100 mbar abs was used for the experiment. The resin was mixed with the hardener and the mold was swirled with this mixture. After removing the plate, the test specimens were cut to correspond with the standard CSN EN ISO 3167 (AWJ cutting - speed of the cut - 100 mm · min^{-1}, diameter of the nozzle - 0.8 mm, working pressure - 380 MPa). Composite plates were prepared with 20 vol.% of fibers (see Fig. 1) with three orientations of the fibers: with fibers oriented in the direction of the loading force (labeled 0°) with fibers oriented perpendicular to the loading force (labeled 90°) and with non-oriented fibers.

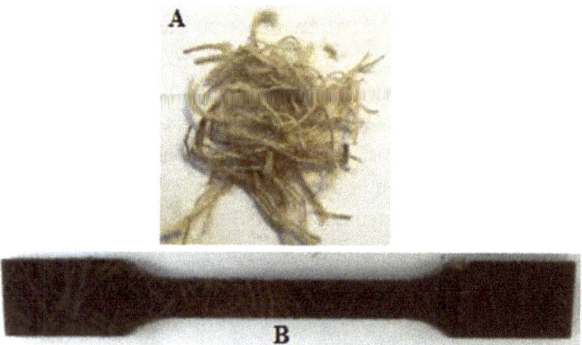

Fig. 1. Fibers in the form of yarn waste (a), Test sample: non-oriented fibers (b)

Experimental tests – tensile strength were carried out in accordance with the standard CSN EN ISO 527 and were carried out on universal testing machine, where the speed of crosshead movement corresponded to 6 mm · min^{-1}. The image analysis of the fibres, fracture surfaces and the interfacial interaction was performed on the electron microscope Tescan Mira 3 GXM (the samples were gilded). The experimentally measured data were evaluated by means of the analysis ANOVA.

3   Results and Discussion

The diameter of yarn residue was evaluated by stereoscopic microscopy - the diameter corresponded to 3.21 ± 0.12 mm (the diameter of the fibers from which the yarn is formed corresponds to 93 ± 36 µm). In terms of the ratio between fiber length and diameter, the short-fiber composite can be considered as having a ratio of less than one hundred. In this case, the ratio was 13.28. Structure of fibres is clear from electron microscopy Fig. 2.

The structure of the surface of the fibers was evaluated on an electron microscope. In Fig. 3 is a surface of single fiber from which the yarn is made.

The tension, in short-fiber composites, is transmitted via the matrix through the fiber ends and the cylindrical surface of the fibers, that is through the adhesion. For short-fiber composites, the influence of the ends cannot be neglected and the properties of the composite are a function of fiber length.

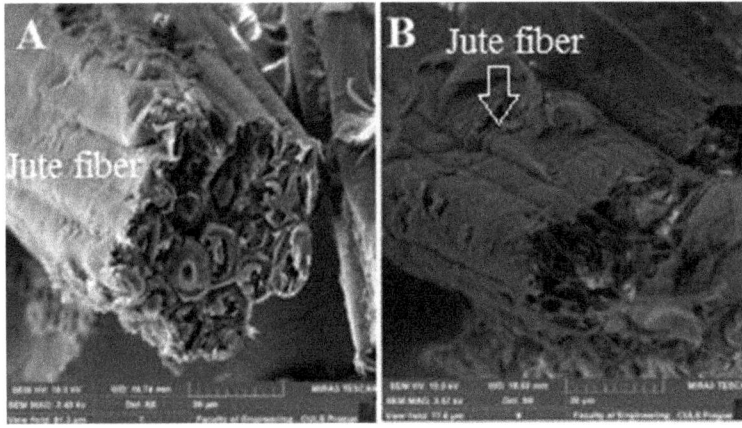

Fig. 2. SEM analysis – microstructure of fibers from which the yarn is formed: Mag. 3.40 kx (a), Mag.3.57 kx (b)

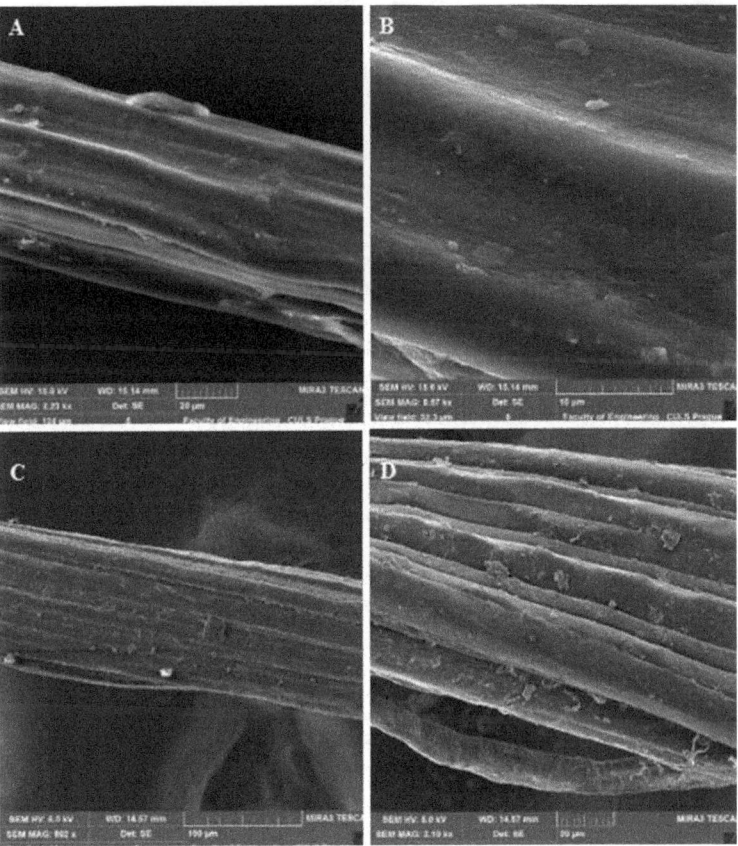

Fig. 3. SEM analysis –surface of fibers from which the yarn is formed: Mag. 2.23 kx (a), Mag. 8.57 kx (b), Mag. 882 kx (c), Mag. 2.10 kx (d)

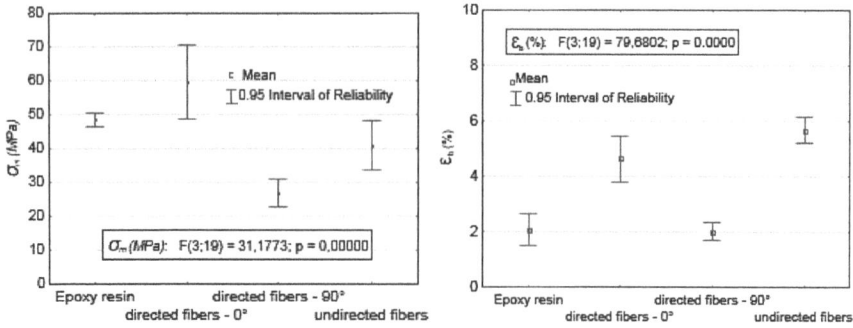

Fig. 4. Tensile strength σ_m (left) and Strain at break ε_b (right)

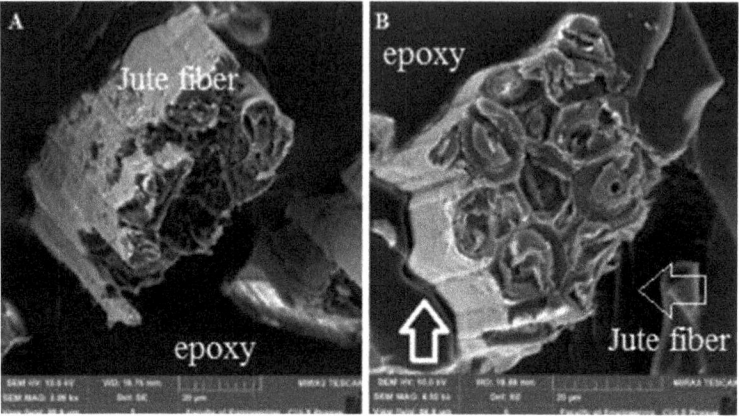

Fig. 5. SEM analysis - interfacial interaction juta/epoxy: Mag. 3.08 kx (a), Mag. 4.02 kx (b)

The epoxy resin without a filler had a tensile strength of 48.63 ± 1.79 MPa. The presence of jute fibers in the form of yarn residues oriented in the direction of the loading force resulted in an increase of the strength by 11.02 MPa, while the variation coefficient increased from 3.7% to 13.2% - this increase is related to increased variance of the measured values. The presence of un-oriented fibers led to a drop of 7.0 MPa, and the fibers oriented perpendicular to the direction of the load force reduced the strength of the composite system by 22.2 MPa (see Fig. 4). Interfacial interaction was evaluated by electron microscopy, see Fig. 5.

While long-fiber composites can be considered as an alternative to structural materials, short-fiber composites are more perceived as a reinforced alternative to plastics. Jute composite with randomly oriented fibers can be defined as isotropic in contrast to oriented systems. Short-fiber composites with random-oriented fibers are suitable for applications where the exact direction of loading is not predetermined - they have roughly the same properties in all directions. For applications where the direction of loading is known in advance, it is inappropriate to use the composites with fibers in the transversal direction to the loading force, because the cracking initiation occurs in layers oriented in the transverse direction to the applied force. This basic assumption was verified by the experiment and the strength limits for the jute/epoxy composites were defined where the jute fibers were in the form of yarn waste from the processing.

4 Conclusion

The experimental program describes the tensile properties of the epoxy resin filled with jute fibers in the form of yarn residues from its processing. The experiment describes the formation of a composite system without surface treatment of fibers, although it is apparent from the search [11, 13] that the treatment of jute fibers, for example by alkali, leads to an increase in interfacial interaction with polymeric matrices. The orientation of the short fibers in the direction of the loading force increased the tensile strength by 22.7%, which confirms the conclusions of Gopinath et al. [11], who achieved a roughly 19% increase in tensile strength in the polyester matrix. On the contrary, the presence of short fibers, which were 90° oriented to the loading force, led to a 45.7% decrease in the strength on the test bodies (standard CSN EN ISO 3167).

Acknowledgement. This paper has been made with the assistance of the grant IGA TF CZU 2017:31140/1312/3113.

References

1. Valášek, P., Müller, M.: Composite based on hard-cast irons utilized on functional parts of tools in agrocomplex. Scientia Agriculturae Bohemica **3**, 172–177 (2013)
2. Valášek, P.: Mechanical properties of epoxy resins filled with waste rubber powder. Manufact. Technol. **14**(4), 632–637 (2014)
3. Fowler, P.A., Hughes, J.M., Elias, R.M.: Biocomposites: technology, environmental credentials and market forces. J. Sci. Food Agric. **86**(12), 1781–1789 (2006)
4. Valášek, P., Müller, M., Šleger, V.: Influence of plasma treatment on mechanical properties of cellulose-based fibres and their interfacial interaction in composite systems. BioResources **12**(3), 5449–5461 (2017)
5. Kumar, N., Das, D.: Fibrous biocomposites from nettle (Girardinia diversifolia) and poly (lactic acid) fibers for automotive dashboard panel application. Compos. Part B: Eng. **130**(1), 54–63 (2017)
6. Ruggiero, A., Valášek, P., Müller, M.: Exploitation of waste date seeds of Phoenix dactylifera in form of polymeric particle biocomposite: investigation on adhesion, cohesion and wear. Compos. Part B: Eng. **104**(1), 9–16 (2016)
7. Valášek, P., Ruggiero, A., Müller, M.: Experimental description of strength and tribological characteristic of EFB oil palm fibres/epoxy composites with technologically undemanding preparation. Compos. Part B: Eng. **122**, 79–88 (2017)
8. Jawaid, M., Abdul Khalil, H.P.S.: Cellulosic/synthetic fibre reinforced polymer hybrid composites: a review. Carbohyd. Polym. **86**(1), 1–18 (2011)

9. Gonçalves, N., Carvalho, V., Soares, F., Vasconcelos, R., Belsley, M., Machado, J.: Yarn features extraction using computer vision – a study with cotton and polyester yarns. Measur. J. **68**, 1–15 (2015)
10. Shinoj, S., Visvanathanb, R., Panigrahi, S., Kochubabua, M.: Oil palm fiber (OPF) and its composites, a review. Ind. Crops Prod. **33**(7), 22 (2011)
11. Gopinath, A., Kumar, M.S., Elayaperumal, A.: Experimental investigations on mechanical properties of jute fiber reinforced composites with polyester and epoxy resin matrices. In: 12th Global Congress on Manufacturing and Management, GCMM (2014)
12. Vilaseca, F., Mendez, J.A., Pèlach, A., Llop, M., Cañigueral, N., Gironès, J., Turon, X., Mutje, P.: Composite materials derived from biodegradable starch polymer and jute strands. Process Biochem. **42**(3), 329–334 (2007)
13. Ramesh, M., Palanikumar, K., Hemachandra Reddy, K.: Mechanical property evaluation of sisal–jute–glass fiber reinforced polyester composites. Compos.: Part B **48**, 1–9 (2013)

Influence of Ca, Sb and Heat Treatment on AlSi9CuMnNi Alloy in Frame of Their Structure

Natasa Naprstkova$^{(\boxtimes)}$, Pavel Kraus, and Jan Novotny

Faculty of Mechanical Engineering, Jan Evangelista Purkyne University in Usti nad Labem, Pasteurova 7, 400 96 Usti nad Labem, Czech Republic
natasa.naprstkova@ujep.cz

Abstract. Treatment of metal alloys is one of the ways to affect the quality and properties of the material at the Faculty of Mechanical Engineering of the Jan Evangelista Purkyně University in Ústí nad Labem, one part of the research consists of searching for the influence of various modifying and inoculating elements on selected aluminum alloys. One of the alloys undergoing the present research is the hardening aluminum alloy AlSi9CuMnNi. The article describes one from these experiments. AlSi9CuMnNi alloy was modified with various amounts of calcium (0.05, 0.1 and 0.15 wt% Ca) and 0.2 wt% Sb. The alloy without modification and with 0.2 wt% Sb was heat-treated, too. In the experiment were made three castings for each type of alloy. The final structure of each casting was analyzed.

Keywords: Alloy · Modification · Heat treatment

1 Introduction

For various reasons, the alloy are adjusting (alloycd, inoculated, modified or heat-treated), which of course also applies to aluminum alloys. These reasons are e.g. to improve the physical and technological properties. At the FME in this context implement various experiments just with aluminum alloys. One of them is research the effect of different amounts of selected elements to certain aluminum alloys and using of heat treatment, if it is possible. In this case investigated elements were Ca and Sb [1–3]. These elements in the role of modifiers can be considered relatively less common in the present area, Na and Sr are more commonly used. The modifying effect of Ca and Sb is different from the "classical" modifiers (Na, Sr) both in terms of the resulting Si particle morphology and the modification mechanism. The effect of Na and Sr on the Si morphology is such that small spherical or slightly elongated rounded Si particles are visible in the cross section whereas Ca and Sb remain thicker slightly rounded Si needles. In the case of Ca and Sb, it is more likely to be a partial modification or another "different" modification compared to Na and Sr [4–6]. The modification by the Ca and Sb elements is actually only a secondary product, because primarily these elements are added to Al-Si alloys for another reason (Ca mainly for fluidity improving, lowering the burn through of Al, etc., Sb for increasing the ductility).

© Springer International Publishing AG, part of Springer Nature 2019
J. Machado et al. (Eds.): HELIX 2018, LNEE 505, pp. 581–587, 2019.
https://doi.org/10.1007/978-3-319-91334-6_79

Ca but also has negative effects, e.g. deterioration of the homogeneity of the resulting structure and is therefore considered as the harmful element. Sb is often used as a minor component of various alloys. They serve to improve the mechanical properties and increased resistance to chemical influences [7–9].

Some types of aluminium alloys can be heat-treated. Heat treatment of aluminium alloys can be in the sense for CSN 42 0056 defined as the process by which a product or product parts in solid state is subjected to one or more cycles of annealing to achieve the desired structure or substructure and properties [10–12].

The article describes the one part of the experiments that are carried out at FME JEPU in Usti nad Labem in presented area based on the requirements of companies and deals with the analysis of structures resulting in the production of experimental samples [5–7, 12–14].

2   Experiment

For the experiment were made castings from alloy AlSi9CuMnNi for each type of alloy. AlSi9CuMnNi belongs to the Al - Si - Cu group, which can be heat treated. In these alloys, the mechanical properties after the heat treatment vary greatly depending on the copper content. These alloys also have good foundry and mechanical properties, but the main disadvantage is low corrosion resistance. They are used for thin-walled castings in engine manufacturing, such as blocks, crankcase, air-cooled cylinder heads or carburettor parts. These alloys are heat-treatable (hardenable) and, in this state, they can achieve maximum strength properties [1, 5, 6].

The alloy AlSi9CuMnNi was made at FME from individual components. Four casting groups were made, which alloyed by varied amounts of calcium (0, 0.05, 0.1 and 0.15 wt%) and 0.2 wt% Sb and unmodified castings and casting modified by 0.2 wt% Sb were heat-treated, too. Castings were carried out gravitationally into a suitable metal mould of the required dimensions and the equipment was used which is available at FME. All the mould dimensions have to comply with requirements for the size of the casting, order to be able subsequently carry out other planned experiments. Besides modification was also the part of the castings heat-treated. As a kind of heat treatment was carried out solution annealing leading to an increase in material hardness and uniformity of the chemical composition. In the first part, the hardening of the material supercritical cooling, and in the second part of the artificial aging. The entire course of the heat treatment process (Fig. 1) took approximately 16 h [1, 12, 14].

3   Microscopical Structure Analyses

For checking the correct chemical composition of the samples was carried out spectrographic analysis. The composition of the samples was examined equipment INNOV Delta X, several randomly selected samples were analyzed for comparison by using equipment Q4 Tasman. For the microscopic observation of the structure metallographic samples were made and the confocal laser scanning microscope Lext OLS 3100 was used for microstructure analysis [15–17].

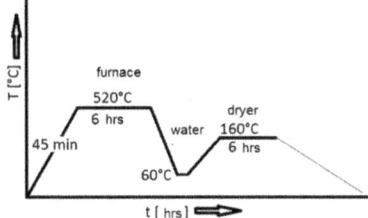

Fig. 1. The procedure of heat treatment

Figure 2 shows the structure of castings without modification. From these images it is evident that the structure is not affected, as was expected. There were visible large silicon plates (shown as needles), which are sharply bounded. Silicon needles were relatively large and irregularly arranged. From these images it is evident that the structures are not affected. The structure of all samples without modification and without heat treatment was very similar. Silicon needles were relatively large and irregularly arranged. Generally, it can be observed in castings occasional occurrence of porosity, too. It was a consequence of casting technology. All samples were prepared by gravitational casting.

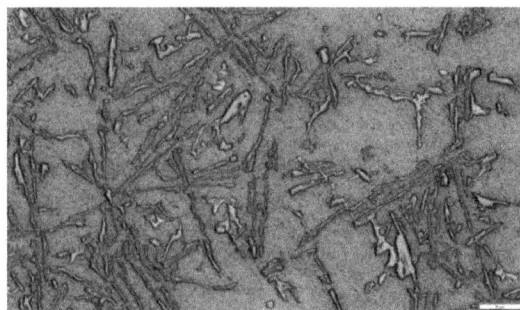

Fig. 2. Example of the unmodified casting structure (0 wt% Ca, 0 wt% Sb), mag. 100x

Further samples were analyzed without modification, but after heat treatment. The example of microstructure of these samples is shown in Fig. 3. On the microstructure were evident more uniform distribution needles of silicon compared to the unmodified and untreated samples.

Figure 4 shows the structure of castings after modification by 0.05 wt% Ca. The modifying effect of Ca was visible. From these images it is evident that the structure is affected, as was expected. There are visible smaller silicon plates (shown as needles) compared to an unmodified alloy. Silicon needles are smaller and irregularly arranged. It was possible to observe a directed dendritic structure in the direction of heat dissipation. The modified eutectic silicon particles on the cut show a slight rounding.

Figure 5 shows the structure of castings after modification by 0.1 wt% Ca. The modifying effect of Ca was visible. From these images it is evident that the structure is

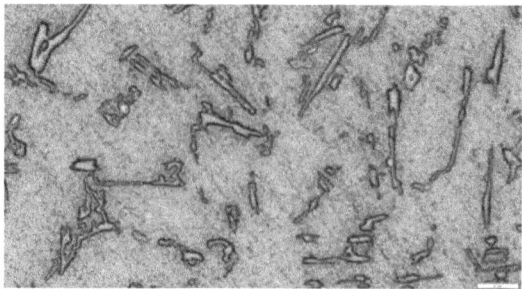

Fig. 3. Example of the unmodif. casting structure (0 wt% Ca, 0 wt% Sb) after HT, mag. 100x

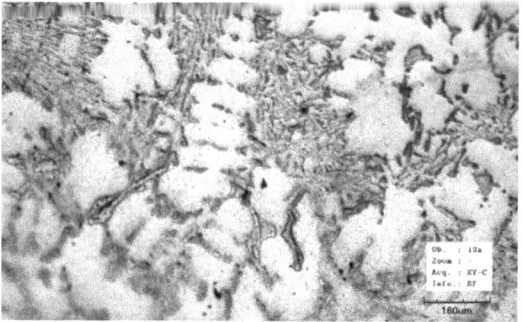

Fig. 4. Example of modified casting structure (0.05 wt% Ca), mag. 100x

affected, as was expected. There are visible smaller silicon plates (shown as needles) compared to an unmodified alloy. Needles hade similar size, like for alloy modified by 0.05 wt% Ca. The modified eutectic silicon particles on the cut show a slight rounding. The effect of the modifying effect was very similar for both alloys. There was visible a large porosity, too. This was true for all castings.

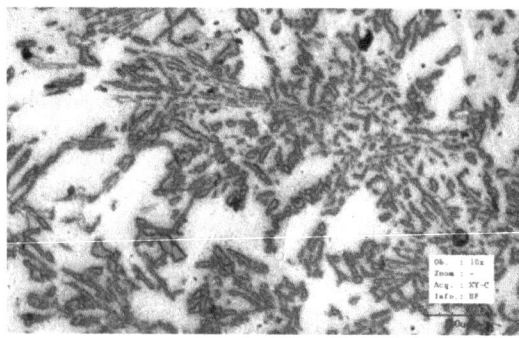

Fig. 5. Example of modified cast. struct. (0.1 wt% Ca), mag. 100x

Figure 6 shows the structure of alloy modified by 0.15 wt% Ca. From this image it is evident that the structure is not so affected. There are visible silicon plates (shown as needles) in similar size compared to an unmodified alloy. Silicon plates are thicker than lower Ca. The structure of this alloy already shows signs of overmodifying. This is no longer a desirable state. There was visible porosity, too.

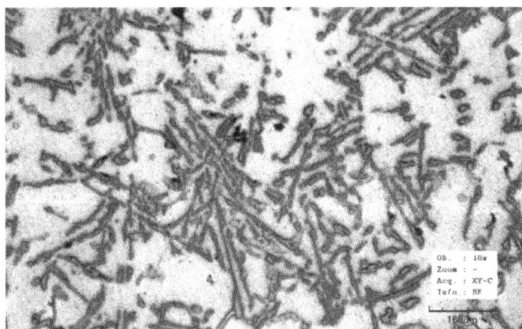

Fig. 6. Example of modified casting structure (0.15 wt% Ca), mag. 100x

Other evaluated samples were modified by antimony. The example of structures shows in Fig. 7. From this figure it is evident that modification to the microstructure was significantly affected and there is therefore seen modifying effect of antimony. The microstructure consists of eutectic structure (composed of fine platelets eutectic silicon in solid solution α) and α phase. Silicon had the shape of finer needles, which were on the edges slightly rounded.

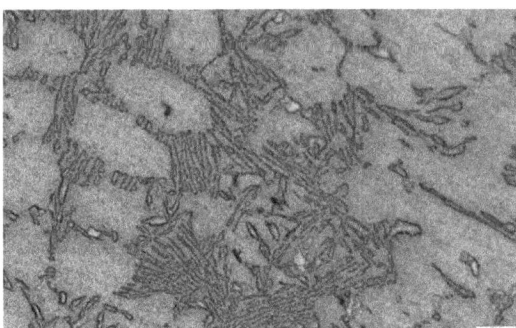

Fig. 7. Example of structure modified casting (0.2 wt% Sb), mag. 100x

On the samples which were modified at the same time antimony and undergo a process of heat treatment was apparent modifying effect of antimony (eutectics formed by fine eutectic silicon in solid solution α), and also the effect of heat treatment on more uniform distribution of structure. The last samples were evaluated, which have been modified and have undergone thermal treatment. On Fig. 8 is example of obtained structure. This structure apparent modifying effect of antimony (eutectic formed by fine eutectic silicon in solid solution α) and also effect of heat treatment on the more uniform distribution of structure. Silicon particles are finer and have rounded edges.

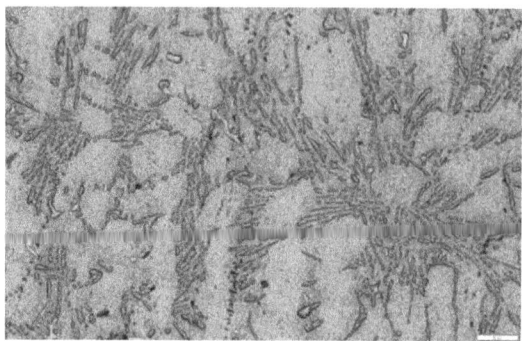

Fig. 8. Example of structure modified casting (0.2 wt% Sb) after HT, mag. 100x

4 Conclusion

In the context of the presented experiment the alloy AlSi9CuMnNi was made at FME using pure metals and were prepared samples with defined amount of Ca and Sb and some types of presented alloy were heat-treated. Castings were analyzed in terms of their microstructure.

Based on the observations, it was possible to state that in the structure of eutectic unmodified silumin (alloy AlSi9CuMnNi) for 0 wt% Ca content, were evident coarse Si needles (hexagonal plates in 3Dd). The effect of calcium modification was apparent to a maximum of 0.1 wt% Ca. Then, with a Ca content of 0.15%, the modifying effect was modest and the rounded Si particles were visible in the structure, but already very coarse and there it began production of the intermetallic phase $AlCa_4$. The best modifying effect was for 0.05 wt% Ca, where the fine Si particles are well rounded, but in 0.1% Ca the Si particles are larger even though rounded, compared with content of 0.05 wt% Ca. The effect of Sb on the modification is different from Ca.

Sb had not so effect on the particle Si size, but more on the rounding of the needles with a Sb content of 0.2% (its size compared to the 0% Sb structure decreased only slightly). After the heat treatment, the modified alloy structure was more uniform. The influence of heat treatment on the structure of the alloy at 0.2 wt% Sb content did not occur. It was also possible to observe interdentritic porosity and gasification of the castings were evident on samples modified with some Ca, which could be expected with respect to the amount of Ca in the alloy and the casting process.

On the present samples, selected properties of the individual types of AlSi9-CuMnNi alloy were further evaluated, too. The presented results are part of a larger research, which is realized at FME JEPU in Ústí nad Labem in cooperation with the industrial sphere.

Acknowledgement. Authors are grateful for the support of grant No. CZ.1.05/4.1.00/11.0260 EDIMARE and SGS FME JEPU in Ústí nad Labem for 2018.

References

1. Michna, S., Lukac, I., Ocenasek, V., Koreny, R., Drapala, J., Schneider, H., Miskufovaá, A., et al.: Encyklopedie hliniku, 1st edn. Adin, Presov (2005)
2. Tillova, E., Farkasova, M., Chalupova, M.: The role of antimony in modifying of Al-Si-Cu cast alloy. Manuf. Technol. **13**(1), 109–114 (2013)
3. Formanek, J., Kucerova, L., Tochylin, M.: The hazards of defects and poor quality of aluminium materials. In: Proceedings of the 24th International Conference on Metallurgy and Materials, METAL 2015, pp. 1552–1557 (2015)
4. Naprstkova, N., Kusmierczak, S.: Analysis of decrease machinability possible causes for claimed alloy. Adv. Sci. Technol.-Res. J. **10**(31), 94–101 (2016)
5. Cais, J., Kraus, P., Lysonkova, I.: Influence of the homogenization temperature on the microstructure and properties of AlSi10CuNiMgMn alloy. Adv. Sci. Technol.-Res. J. **11**(1), 104–110 (2017)
6. Michna, S., Naprstkova, N.: The mechanical properties optimizing of Al - Si alloys precipitation hardening and the effect on the character of the chip. Acta Metallurgica Slovaca **18**(3), 193–199 (2011)
7. Naprstkova, N., Kalincova, D.: Influence of additional chemical components on machining properties of selected aluminium-silicon alloy. In: Malinovska, L., Osadcuks, V. (eds.) International Scientific Conference Engineering for Rural Development, vol. 14, pp. 766–771 (2015)
8. Strihavkova, E., Weiss, V., Michna, S.: Study of the structure and fluidity of alloy of the Al-Si-Mg system with a different calcium content. Metallurgist **56**(9–10), 708–713 (2013)
9. Strihavkova, E.: Analysis of new type Al-Si-Mg alloy structures with different contents of Ca due to chemical properties. In: Malinovska, L., Osadcuks, V. (eds.) International Scientific Conference Engineering for Rural Development, vol. 16, pp. 521–527 (2017)
10. Kučerová, L.: The effect of two-step heat treatment parameters on microstructure and mechanical properties of 42SiMn steel. Metals **7**(12), 573 (2017)
11. Michna, S., Naprstkova, N., Lukac, I.: Mechanical properties optimization of AlSi12-CuMgNi alloy by heat treatment. Metallofiz. Noveishie Tekhnol. **33**(11), 1559–1568 (2011)
12. Vajsova, V., Michna, S.: Optimization of AlZn5.5Mg2.5Cu1.5 alloy homogenizing annealing. Metallofiz. Noveishie Tekhnol. **32**(7), 949–958 (2010)
13. Kusmierczak, S.: Udage of technical equipment in teaching technical subjects. In: Malinovska, L., Osadcuks, V. (eds.) International Scientific Conference Engineering for Rural Development, vol. 14, pp. 748–752 (2015)
14. Weiss, V., Strihavkova, E.: Influence of the homogenization annealing on microstructure and mechanical properties of ALZn5,5Mg2,5Cu1,5 alloy. Manuf. Technol. **12**(13), 297–302 (2012)
15. Lipiński, T.: Microstructure and mechanical properties of the AlSi13Mg1CuNi alloy with ecological modifier. Manuf. Technol. **11**(11), 40–44 (2011)
16. Michalcova, A., Vojtěch, D.: Structure of rapidly solidified aluminium alloys. Manuf. Technol. **12**(13), 166–169 (2012)
17. Svobodová, J., Cais, J., Michna, Š., Brůha, M.: Research of corrosion properties of Al-Si alloys antimony alloyed. Manuf. Technol. **13**(3), 404–409 (2013)

A Thermomechanical Study of POM-Based Polymers

Ana Figueiredo[1,2], Seyyed M. M. Sabet[2(✉)], Pedro Bernardo[1], and José Meireles[2]

[1] Bosch Car Multimedia SA, Max Grundig, 35 Lomar, Braga, Portugal
{ana.figueiredo, pedro.bernardo}@pt.bosch.com
[2] Mechanical Engineering Department, University of Minho, Guimarães, Portugal
{sabet, meireles}@dem.uminho.pt

Abstract. This work is focused on the study of thermomechanical behavior of a Polyoxymethylene-based gear at high temperatures using ANSYS16. The deformation of the gear teeth due to thermal loads was also studied through coupled thermal-structural analyses. The displacement results are obtained by numerical and experimental studies and compared with each other. It was concluded that the experimental and numerical results are in agreement.

Keywords: Thermomechanical properties · Displacement · Heat transfer FEM

1 Introduction

The use of plastics gears has been consistently increasing in recent years, due to their advantages over metals, such as low density, reduced friction and shock and vibration absorption properties [1]. However, the thermal stability of plastics is generally lower than metals [2]. Plastics can be used with minimal or no lubrication, due to their self-lubricating properties and low audible noise [3]. On the other hand, plastics exhibit reduced load-carrying capacity, due to their low yield and ultimate tensile strength values [4]. Due to large coefficients of thermal expansion and moisture absorption, plastic gears suffer dimensional instability [5]. Due to these properties, the thermo-mechanical properties of plastics at elevated temperatures has been the subject of various studies [6, 7]. The deformation of plastic gear teeth has also been reported using experimental and numerical approaches [8].

POM or acetal is a plastic dominantly used for precision parts that require high stiffness, low friction, and excellent dimensional stability in several industries mainly electronics, namely as Bosch.

In this work, the deformation of POM-based gears and their teeth are studied by experimental and numerical methods in order to obtain an understanding of the behavior of these gears in actual working conditions at various temperatures. The numerical results are validated by experimental ones. Initially the geometry will be presented, and then the meshing, boundary conditions and the studies will be defined. Finally the conclusions drawn from each set of tests are to be compared and analyzed.

© Springer International Publishing AG, part of Springer Nature 2019
J. Machado et al. (Eds.): HELIX 2018, LNEE 505, pp. 588–594, 2019.
https://doi.org/10.1007/978-3-319-91334-6_80

2 Numerical Methodology

The finite volume method (FVM) and thermal module of ANSYS (FLUENT) were used for the numerical studies of this work to simulate the fluid-structures interactions. As claimed by Fallah et al. [9] the accuracy of the finite volume solution is comparable to the finite element method. This means that the FVM method can be considered as an equally appropriate candidate as the Finite Element (FE) involving problems related to deformations and stresses. Hence, this provides the opportunity to solve multiphysics phenomena (i.e., fluid-structure interaction) using a single modeling formulation.

The governing equations of the FLUENT solver can be found in [9, 10]. The solver employed to process the equations of energy and momentum is the second order upwind which may allow a slower convergence as it uses more data points to define the approximation of spatial derivative. On the other hand, it demonstrates a second order accuracy, essential with tetrahedral meshes [9].

3 Problem Statement

The gear in this study has a complex geometry, shown in Fig. 1, and is used for head-up displays where high temperatures could be reached. This induced heat can cause displacements and, consequently, threaten the operation of the system. This gear is made of POM ZLV40 [11], has 95 teeth is 76 mm in diameter and 0.95 mm in thickness. The pressure angle of the teeth is 15°.

Fig. 1. The geometry of the gear: front (left) and rear (right).

Table 1 lists the properties of POM ZLV40 based on the supplier [11]. The density, thermal conductivity and the specific heat are used in the simulations while the rest in structural simulations. The viscosity of the air is 1.79×10^{-5} kg/ms. The density of POM ZLV40 at various temperatures was adopted from the supplier [12].

Table 1. Materials properties.

Material	Density (kg/m^3)	Conductivity (W/m.k)	Specific heat (J/kg.K)	Tensile modulus (MPa)	Poisson ratio	Thermal expansion (K^{-1})	Yield strength (MPa)
POM	1410	0.23	1465	2700	0.35	1×10^{-4}	61
Air	–	0.02	1006.43	–	–	–	–

4   CFD Analysis

Figure 2 shows the geometry of the gear (surrounded by air) and the unstructured mesh of the assembly. The dimensions of the fluid volume are 85 mm × 85 mm × 40 mm^3. It is notable to mention that both parts are joined together as one body. This allows to correctly evaluate the heat transfer in the simulation.

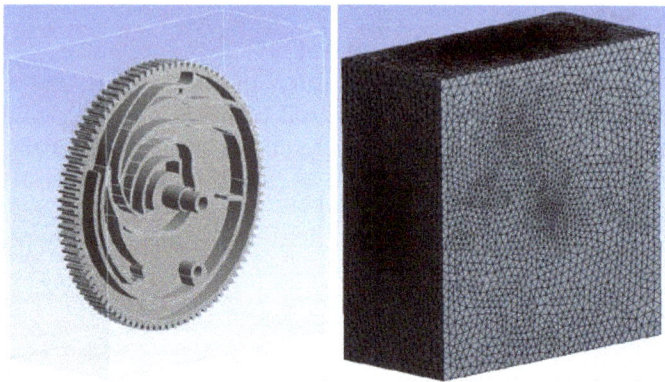

Fig. 2. The geometry of the gear with enclosure and respective mesh.

The constitutive element is SOLID72 which is a tetrahedron defined by 4 nodes, 1 per each vertex. The minimum element size was 0.2 mm and the maximum 3 mm. Ideally, the minimum length would have to be smaller due to the difficulty in meshing the transition air-component. The number of nodes is 2,886,960 and that of cells is 17,204,221. The average value for element quality of these cells (varying between 0 and 1) is 0.85. The skewness parameter determines how close to ideal a face or cell is and the perfect value is 0. This value was set as 0.21.

After some trial simulations, a number of zones were identified as potential causes of error due to geometric discontinuities. Instead of reducing the minimum element size, it was decided to employ local refinement on the identified zones, therefore, 2 spheres of 7.5 mm in diameter were placed as shown in Fig. 3 (left). The elements in the two spheres are only 0.18 mm in length, which can be seen in Fig. 3 (right).

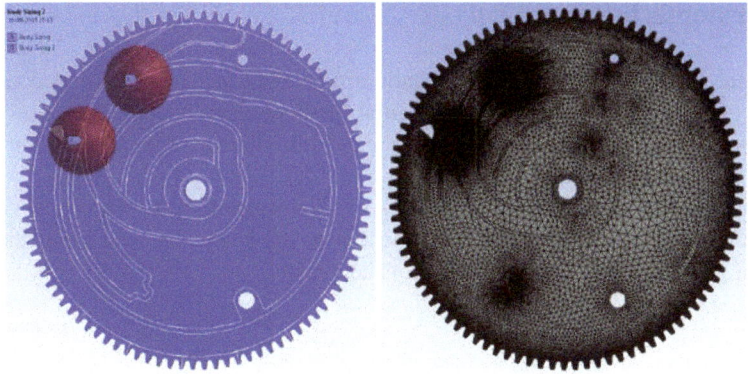

Fig. 3. The spheres created for the numerical studies.

In order to define the boundary conditions, a steady-state and gravitational acceleration of -9.8 m/s^2 was considered. The energy equations were adopted as a laminar regime. The inlet and outlet were considered on the front and rear sides of the gear, respectively (Fig. 1). This allows ensuring that the air follows that trail and goes through the body as intended. The total fluid volume is at 90 °C, which is the maximum working temperature of the gear. This means that the air enters and exits the gear at the same temperature.

An outflow boundary condition was defined for the outlet, which is used to model flows where velocity and pressure are not known prior to solving the problem. Finally, the convection coefficient at 90 °C was set as 8 W/m^{2}K. Figure 4 shows the heating process of the gear at two different phases. The pressure-based solver was used for incompressible and semi-compressible flows, which can be applied to a wide range of flows. A segregated algorithm is used in which the governing equations are solved sequentially employing the pressure velocity algorithm SIMPLE, which improves the convergence speed.

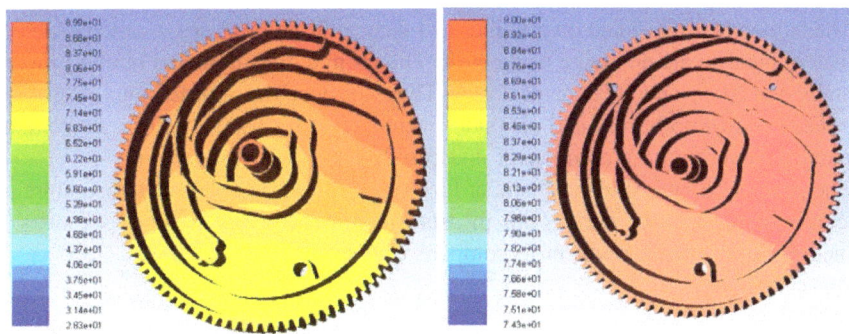

Fig. 4. The heating stages of the model.

5 Thermomechanical Analysis

The thermal results were imported to the static structural module as a boundary condition. The axis of the gear was fixed and the initial temperature was also defined as 20°C. The mechanical study was completed with the preconditioned conjugate gradient solver using a convergence tolerance of 1×10^{-8}.

The simulation results are shown in Fig. 5. The maximum displacement value is nearly 0.27 mm, which is only present on the teeth. It can be concluded that displacement of the teeth is higher the rest of the gear.

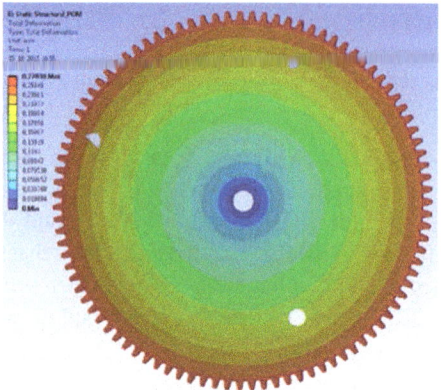

Fig. 5. The displacement results.

6 Experimental Analysis

The displacement of the POM-based gear was also studied experimentally from 20 °C to 90 °C with a COMET 6 16 M 3D-sensor, supplied by Steinbichler [13] and whose precision for part size of up to 150 mm is 30 μm. COMET 6 can locally adapt the light quantity projected onto the object surface (Fig. 6).

The component was fixed on a holder and then subjected to laser in order to detect the material and create an image of the gear. The room temperature and humidity were maintained at 20 °C and 60%, respectively. Then, the gear was inserted into a chamber and rotated so that both front and rear sides would reach 90 °C. This allowed avoiding the manual rotation of the component that could cause positioning errors.

The gear was kept inside the chamber for one hour to ensure that the temperature of the component is stable during the measurements. The software first generates a scanned model of the component to compare with the CAD geometry.

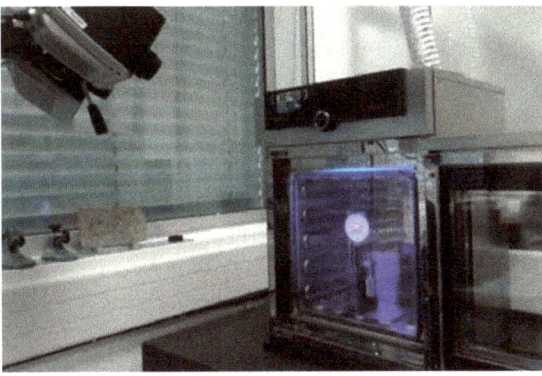

Fig. 6. The experimental setup.

7   Results and Discussions

The displacement measurements were performed every other 5 teeth, which accounts for a total number of 19 points. Table 2 compares the maximum deformation of the teeth obtained by experimental and numerical studies.

Table 2. The comparison of numerical and experimental results.

Angle (°)	Experimental (mm)	Numerical (mm)	Difference (%)
18.95	0.220	0.277	23
56.84	0.234	0.277	17
94.74	0.243	0.274	12
132.63	0.252	0.274	8
170.53	0.253	0.275	8
208.42	0.262	0.275	5
246.32	0.257	0.275	7
284.21	0.241	0.275	13
322.11	0.229	0.276	19
360.00	0.230	0.277	18

It is notable that the radial displacements in numerical analysis are higher than those of the experimental ones. It can also be seen that the highest difference between the experimental and numerical results is 23%, while the average of the difference at various angles is about 12.8%. Interestingly, a symmetry can be observed among the values, as they decrease form a maximum of 23% to a min of 4% (for 227.37°) and then increase again until 360°. It can be concluded that the gear does not increase in diameter in the same way as in other directions. There is a good agreement with these results and those reported in similar studies [14]. The difference among the experimental and numerical results could be due to: the correlation error of the software in every simulation, the material properties, and the local temperature changes. It is very difficult to reproduce the experiment in the exact same way by the numerical model.

8 Conclusions

In this work, the thermomechanical behavior of a POM-based gear was studied at 90 °C using numerical and experimental approaches. The results showed a good agreement among the numerical and experimental result and the maximum difference of 23% and an average of 12.8% was reported. The 23% difference is possibly due to the combination of geometry and process (equipment) tolerances. It would be interesting to repeat both approaches at various temperatures in order to obtain a graph of temperature vs. the expansion rate.

References

1. Geethamma, V.G., Asaletha, R., Kalarikkal, N., Thomas, S.: Vibration and sound damping in polymers. J. Sci. Educ. **19**(19), 821–833 (2014)
2. Luo, R.K., Wu, X.P., Mortel, W.J.: Rubber unloading-behaviour evaluation using product-orientated specimen based on a resilience test. Polym. Test. **34**, 49–57 (2014)
3. Pogosian, A.K., Hovhannisyan, K.V., Isajanyan, A.R.: Friction transfer and self-lubrication of polymers. J. Frict. Wear **31**(1), 81–88 (2010)
4. Umanskii, E.S., Shidlovskii, N.S., Kryuchkov, V.V., Grishko, S.V., Stezhko, L.L.: Load-carrying capacity of polymer tape rolls under impact loads. Strength Mater. **22**(4), 611–617 (1990)
5. Malkin, A.Y.: Surface instabilities. Colloid J. **70**(6), 673–689 (2008)
6. Düzcükoglu, H.: Study on development of polyamide gears for improvement of load-carrying capacity. Tribol. Int. **42**, 1146–1153 (2009)
7. Mao, K., Li, W., Hooke, C.J., Walton, D.: Polymer gear surface thermal wear and its performance prediction. Tribol. Int. **43**, 433–439 (2010)
8. Kadashevich, I., Beutner, M., Karpuschewski, B., Halle, T.: A novel simulation approach to determine thermally induced geometric deviations in dry gear hobbing. Procedia CIRP **31**, 483–488 (2015)
9. Fallah, N., Bailey, C., Cross, M., Taylor, G.: Comparison of finite element and finite volume methods. Appl. Math. Model. **24**, 439–455 (2000)
10. Moukalled, F., Mangani, L., Darwish, M.: The Finite Volume Method in Computational Fluid Dynamics. Springer, New York (2015)
11. Material Data Center Homepage. http://www.materialdatacenter.com/ms/en/tradenames/Tenac/Asahi+Kasei/TENAC%E2%84%A2-C+ZLV40/6bbb0224/212/. Accessed 15 Feb 2018
12. AKChem Homepage. http://www.akchem.com/en/enpla/products/tenac/grade.html. Accessed 15 Feb 2018
13. Steinbichler Homepage. http://www.steinbichler.com/. Accessed 15 Feb 2018
14. Li, Y., Liu, W.Y., Frimpong, S.: Effect of ambient temperature on stress, deformation and temperature of dump truck tire. Eng. Fail. Anal. **23**, 55–62 (2012)

Integrating Technologies into Fashion Products: Future Challenges

Nelson Oliveira(iD) and Joana Cunha$^{(\boxtimes)}$(iD)

School of Engineering, University of Minho, Guimarães, Portugal
jcunha@det.uminho.pt

Abstract. The integration of technological elements into textile products and fashion accessories is clearly not a new phenomena. But after more than two decades since its boom, it is important to look at the progress made, the applicability, the market availability of the technologies, their consequent acceptance or refusal by consumers and the future prospects for these products. The present study presents a review on the subject, focusing on the relationship between science and design, in particular the fusion between technology and fashion.

The findings of the study indicate that the future market is amenable to this type of products. However, there is a need to focus on the creation of a new profile of fashion designer, with the skills to support innovative decision making in sourcing and selection of suitable materials, technologies and construction methods for the development of functional garments, which also meet the end user's demand, needs and cultural contexts. In this sense, the use of alternative design methodologies, in particular fiction design and working with transdisciplinary design teams can enhance the success in the creation of innovative fashion products.

Keywords: Interactive fashion · Smart wearable · Smart clothing

1 Introduction

The evolution of civilization and its consequent technological development have resulted in substantial changes to the improvement of people's living conditions. As a result of continued scientific research and the application and implementation of the knowledge acquired, the pace of change tends to be increasing and in ever shorter time periods.

The universe of textiles, and consequently fashion, is no exception. By taking a historical retrospective, and not going much further than the Industrial Revolution to the present day, changes and increases in materials, technologies, manufacturing processes and paradigms have been the subject of constant improvements, discoveries and innovations.

In particular the integration of technological elements into textile products and fashion accessories is not something new. In fact, since the beginning of the 20th century there have been products with wearable technology, but this type of development has been intensifying, and for more than twenty years the theme has been debated, analyzed and developed at the academic and industrial level [1].

© Springer International Publishing AG, part of Springer Nature 2019
J. Machado et al. (Eds.): HELIX 2018, LNEE 505, pp. 595–601, 2019.
https://doi.org/10.1007/978-3-319-91334-6_81

Thus, after more than two decades since the boom of this theme, it is important to look at the progress made, the applicability, the availability on the market of technologies, their consequent acceptance or refusal by consumers and the future prospects for these products.

The futuristic predictions, developed in particular throughout the twentieth century, of the current reality, often portrayed in films and works of science fiction literature, placed today's society at a very advanced level of technological development, where interfaces and technologies are a fundamental part of man's existence. In particular, clothing products and fashion accessories imagined in iconic films such as 2001: Kubrick's Space Odyssey and Ridley Scott's Blade Runner all have in common the integration of wearable technology elements that serve as the interface between the individual and the world, where the boundaries between body, clothing and physical space are blurred [?]

Thus, based on these futuristic visions of science fiction and in order to get a better understanding of the reality of the subject concerned, the pursued line of research is based on the relationship between science and design, in particular the fusion of technology and fashion.

Over the last decades, more and more researchers and creatives have been gaining interest in this field of design and engineering, fostering the constant expansion of technology applied to fashion. On the one hand, those interested in the futuristic and image aspects appear, on the other the approaches are aimed at taking advantage of the technical potential and the combination of hard technology with the soft textile [1].

2   Wearable Technology

There are several terminologies for designing the integration of technological interfaces into textile and clothing products, all very similar to each other, but two in particular seem to be the most consensual:

- Smart wearable: smart, portable wearable device, electronic used in the human body and offers some level of internal processing and performs some kind of integrated software. Most wearables have built-in sensors of various types as well as connectivity options (wired or wireless) on other smart devices such as smartphones [3].
- Smart clothing is a product that should be appealing and appropriate to the culture of the end user, with clothing functionality reinforced by the incorporation of technologies such as electronics and computing in the garment itself [4].

The beginning of the research and development of such products refers to the last decade of the twentieth century, first for military purposes then followed by medical and sports. According to Suh et al. [5], the research and development of smart apparel products arises from the cross between textile technology, human physiology and design research. The level of technological integration can be high or low, varying with the ability of the interfaces to be implanted (introduced in the users' body), coupled (integration into textile fibers and yarns, embedded or coupled), or simply transportable (carried by the user) [1].

Regarding the possible spectrum of application of wearable technology, researchers Bhat et al. [6], highlight the areas of fitness, medical, lifestyle, games and infotainment.

However, this classification is not universal, and other classifications by other theorists and researchers, such as that developed by Cranny-Francis and Hawkins [7], emerge. According to these, the universe of wearable technology, and its applications, is extremely vast, suggesting that there are seven main areas: security; medical; welfare; sport and fitness; lifestyle computing; communication; and glamor [7].

According to McGregor [8], president and CEO of the Broadcom electronics company, the main key attributes for usable technologies are: energy efficiency; small dimension, connection; safety; cost effective; battery life; discretion; appealing design; accurate sensors and accessibility. Similarly, connectivity is an essential factor for the development of a technological product nowadays. It is this feature that will enhance the integration of interfaces in fashion products, which can be realized by: Wi-Fi; Bluetooth, GPS or RFID, allowing to connect everything to the internet [8].

The beginning of scientific development in this area arises, as is often the case in almost all technological areas, in military research. According to Scataglini et al. [9], the first idea to put sensors incorporated into clothing was developed in the military field by Sundaresan Jayaraman and his research team from the Georgia Institute of Technology in the United States of America. This product, called Wearable Motherboard ™ or Smart Shirt, consisted of a wearable technology t-shirt and was considered Millennium's best invention in 1998, and TIME named it as one of the best inventions in 2001 [9, 10]. This first prototype consisted in the creation of a portable information infrastructure, being a precursor in the processing of personalized information, health and telemedicine, and even space exploration. Up until now, it had not been possible to create a personal information processor that was customizable, wearable and comfortable, nor was there any garment that could be used for non-invasive monitoring of the vital signs of humans on Earth or space, allowing to obtain information regarding the monitoring of health, safety of the physical environment; management of stress, strengthening of human skills [10].

This was the first step to the current applications of wearable technologies not only in smart clothing, but also integrated in small accessories such as bracelets, necklaces, rings, glasses, etc. [9].

In Europe we also see government investment in research projects of this type of wearable technology, with technical purposes. Namely ProeTEX, which developed textile systems with technological interfaces for rescue workers and firefighters, namely: a sweater with heart rate sensor, respiratory sensor and temperature sensor; a jacket with temperature sensor, accelerometers to detect activity, data processing unit (Personal Electronic Box), textile antennas for communication with the base, flexible batteries and audible or visual alarm; and a pair of boots with gas sensors. This allows continuous monitoring of body signals, breathing and heart rate; monitoring of activities; internal and external temperature verification; chemical detection; wireless communication between the garment and a base station; and energy supply [11].

The development of wearables capable of producing energy has also been the subject of scientific research. An example of this is the development of a silicone polymer capable of producing electrical energy. According to researcher and mechanical engineer Michael McAlpine of Princeton University, his research team was able to develop a flexible silicone square that has a tape of a piezoelectric material called PZT, the most efficient of this category. This type of material has special properties, through the mechanical movement is able to generate a voltage that can be used to produce electric current and vice versa. Flexibility thus emerges as a differentiating factor and a potential enhancer for this type of product, increasing the range of possible applications, although it is indicated that the biomedical application is the one with the greatest potential, in order to reduce the need for surgical interventions in implants that need batteries [12].

3   Fiction Design

In imagining the future Man tends to associate technological development as an essential and present element in all moments and situations of daily life. This tendency is well present in the most diverse theories and works of fiction-science developed throughout the XX and XXI centuries, and often these exercises of imagination and visionary vision act as leverage for the technological development.

In this sense, fiction design emerges as a design chain that aims to explore the imagination and seek to materialize these mental conceptions. Thus, this design results from the mixing of the unreal, the thoughtful and the introspective (artistic dimension) with the reality and hardness of the facts (engineering), maximizing creativity, as well as the development of materials, technologies and productive processes. The fiction design is a fusion of design, science and science fiction. It is an amalgam of practices that, together, shapes expectations about what each of these areas does on their own and fuses them into something new. It is a way of materializing ideas and speculations without the pragmatic reduction that often happens when dead weights are fixed to the imagination [13].

Like science fiction, fiction design creates imaginative dialectics about possible future worlds. Like some forms of science fiction, it seeks to speculate about the near future by extrapolating from the present. In this extrapolation, fiction design casts a critical eye on the forms and functions of the current objects and on the rituals of interaction with the individual. This allows us to speculate without the usual constraints imposed by reality, being an innovative way of thinking about design. Thus, as Bleecker [13] (p. 8) says, "fiction design works in the space between arrogance of scientific fact, and the seriously playful imaginary of science fiction, doing things that are both real and false, but aware of the irony present in the confusion".

This practice of design seeks to develop through the design and exploitation of ideas the materialization of the imagination, also using modeling, prototyping and storytelling. Throughout the process you can discover innovative solutions and develop new products that have never been thought of before. Ideas to develop can come from multiple sources: the designer's imagination, works of literature, film and art in general. Directors such as Kubrick, Hitchcock, Ridley Scott, among others, are often visionaries who inspire this type of design [13].

Thinking through fiction to understand the action of design is a way of extrapolating the design process beyond the routine and reducing notions of what design does.

This method of designing surpasses the barriers of conformism, leading to a reflection on the unexpected, the unconventional, the unruly and the unpublished [14].

Thus, what may initially be a "simple" science fiction film, can become an object of reflection and analysis, a real challenge for science and for design. From these artifacts it is possible to develop an exploratory methodology that addresses multiple approaches limited only by the imagination. As much as science dictates that it is not possible, the challenge is precisely to go beyond these conventions and dogmas to take the next step in innovation and development of new products [14].

According to Flanagan and Veja [2] visionary and sometimes utopian forecasts of products presented in films and other works of science fiction do not particularize materials, components or technical specificities, but indicate the purpose and function of the same, thus identifying a need or desire to be worked and, as such, leverage the development of technologies and interfaces applicable to products, actually managing to anticipate a future condition. According to the authors, futuristic products besides the aesthetic dimension combine empathy, data communication and reaction to stimuli [2].

Despite all the research and development of clothing products and accessories with technological elements there are still many gaps in the process, which result in the difficulty of products to reach success and being accepted by consumers.

4   The Need for a Different Approach

According to McCann et al. [4], there is a gap in the profile of designers who are dedicated to the development of this type of products. The design of smart, eye-catching clothing with truly embedded technologies requires the fusion of science and technology with fashion and design. The creation of smart clothing raises quite different and broader issues, which go beyond those of fashion design, namely the determination of the balance between a mixture of creativity and the more systematic and practical design process. According to the researchers, "… smart clothing with truly integrated technologies is emerging in areas such as sport and fitness, health and wellness, and products for the aging population. For these clothing sectors, an in-depth analysis of the end-user's needs is required before starting the development of the project, which can be better understood in the graphic, product or industrial design disciplines" (p. 13).

Thus, there is a gap in providing designers with a common language with other areas of knowledge linked to technology, to understand the mix of requirements as distinct as aesthetic, technical and cultural needs of the potential market for smart clothing. However, there is a growing demand for designers capable of effectively applying and developing smart textiles and wearable technologies in the development of functional gamut design products, tailoring products to the end-user's cultural needs and contexts. This requires an aesthetic awareness, general creative design skills and an adequate knowledge of clothing, modeling and technology applied to the textile [4].

The researchers also point out the need to develop a specific design methodology for this type of product, which starts with the identification of the needs of the final consumer; followed by textile development; garment development; the integration of intelligent and wearable technology; production; the distribution and launch of the product; and end-of-life [4].

According to some studies, although the research and development of wearable integrated devices has already reached the household goods market, its high cost is a factor that hinders its acceptance [3].

According to the Sutardja Center for Entrepreneurship & Technology, associated with the University of Berkeley, California, there are some barriers that hinder the acceptance of products with integrated interfaces and technologies, namely: cultural challenges; and technical challenges [15].

In turn, the value bestowed by the wearer on clothing comprises both functional and symbolic attributes. Functional value relates to protection and comfort, while the symbolic value with the impression that the user imparts to others by the outward appearance, as well as the emotional relationships with the objects. Thus, the challenges to foster the acceptance of this type of products are based on increasing the development of the functional and aesthetic aspect [15].

5   Future Challenges

For a product with integrated technology to become usable and accepted by consumers it must meet several requirements, namely, these should be more than just functional or innovative, they have to be appealing. Consumers have to want to use these products, a fact that until recently was not taken into account by the technology industry, not reflecting, or secondarily, the aesthetic value of the product. However, it is not possible to separate fashion from aesthetics. A trendy yet functional product should be concerned with aesthetic issues, becoming part of the user's physical, psychological and emotional identity.

In this new technological phase, the smart fashion designer should be seen as an element to be integrated into an R&D team, along with textile technologists, clothing engineers, electronics specialists, biologists, digital communication experts and computer scientists; and must be able to communicate with all, mastering the lexicon and cross-cutting technical specificities. The profile of the fashion designer should be specialized to meet the particularities of development of fashion products that contain integrated technological elements. This will be the key element for a better acceptance of this type of products by the market.

Smart clothing designers require guidance and skill development at usability, production, fashion, consumer culture, recycling, sustainability and end user needs. This "new" area of design is a very challenging topic as it crosses the boundaries of specialized knowledge inside and outside the space of art and design.

Acknowledgments. This work is financed by FEDER funds through the Competitivity Factors Operational Programme - COMPETE and by national funds through FCT – Foundation for Science and Technology within the scope of the project POCI-01-0145-FEDER-007136

References

1. Seymour, S.: Fashionable Technology: The Intersection of Design, Fashion, Science, and Technology. Springer, Vienna (2009)
2. Flanagan, P.J., Vega, K.F.C.: Future fashion – at the interface. In: Marcus, A. (ed.) Proceedings of the International Conference of Design, User Experience, and Usability DUXU 2013, Part I, pp. 48–57. Springer, Heidelberg (2013)
3. O'Donnell, B.: The Slow Build : Smart Wearables Forecast, 2014–2020 (2015). http://www. technalysisresearch.com/downloads/TECHnalysisResearchSmartWearablesForecast SummaryMay2015.pdf. Accessed 20 Feb 2017
4. McCann, J., Hurford, R., Martin, A.: A design process for the development of innovative smart clothing that addresses end-user needs from technical, functional, aesthetic and cultural view points. In: Ninth IEEE International Symposium on Wearable Computers (ISWC 2005) (2005). https://doi.org/10.1109/ISWC.2005.3
5. Suh, M., Carroll, K., Cassill, N.: Critical review on smart clothing product development. J. Text. Appar. Technol. Manag. **6**(4), 1–18 (2010)
6. Bhat, A., Badri, P., Reddi, U.S.: Wearable Devices : The Next Big Thing in CRM. New York, NY, USA (2014). https://www.cognizant.com/services-resources/. Accessed 10 Feb 2018
7. Cranny-Francis, A., Hawkins, C.: Wearable technology: towards function with style, vol. 7. Beecham Research, London (2014)
8. Mcgregor, S.: Wearable technologies, vol. 12. WSA (2013). https://bi.snu.ac.kr/Courses/ 4ai14s/140522.pdf. Accessed 20 Feb 2017
9. Scataglini, S., Andreoni, G., Gallant, J.: A review of smart clothing in military. In: Proceedings of the 2015 Workshop on Wearable Systems and Applications, pp. 53–54, NewYork (2015)
10. GaTech: Georgia Tech Wearable Motherboard (2003). http://www.smartshirt.gatech.edu/. Accessed 20 Feb 2017
11. Curone, D., Secco, E.L., Tognetti, A., Loriga, G., Dudnik, G., Risatti, M., Whyte, R., Bonfiglio, A., Magenes, G.: Smart garments for emergency operators: the ProeTEX project. IEEE Trans. Inf. Technol. Biomed. **14**(3), 694–701 (2010)
12. Qi, Y., Jafferis, N.T., Lyons, K., Lee, C.M., Ahmad, H., McAlpine, M.C.: Piezoelectric ribbons printed onto rubber for flexible energy conversion. Nano Lett. **10**(2), 524–528 (2010)
13. Bleecker, J.: Design Fiction: A Short Essay on Design, Science, Fact and Fiction. Near Future Laboratory, 49, March 2009
14. Bleecker, J.: Design fiction: from props to prototypes. In: Proceedings of the 6th Swiss Design Network Conference, pp. 58–67 (2010)
15. Hanuska, A., Chandramohan, B., Bellamy, L., Burke, P., Ramanathan, R., Balakrishnan, V.: Smart Clothing Market Analysis, California (2014)

Novel Naturals Colorants

Sandra Heffernan[(✉)]

Massey University, Wallace St, Wellington 6041, New Zealand
s.l.heffernan@massey.ac.nz

Abstract. The research developed a novel approach to dyeing and finishing a contemporary silk and wool textile length collection. The colorant is extracted from an invasive weed, clematis vitalba, a forest destroying invasive pest in New Zealand. Originally released as an ornamental garden plant, clematis vitalba escaped into the sub-alpine environment, preventing forest regeneration. Reducing the impact of dyeing and finishing is a key aim in this eco dye research. Kate Fletcher states the textile and the garment manufacturing sector is the largest user of water and scores worse than any other on the UK Environment Agency's pollution risk report [1]. Safe colour modifying mordants and sustainable degumming techniques enhanced the results and design aesthetic. The clematis vitalba dyed silk and wool fabric shows a high degree of light fastness. The design concept explored both a sense of place and a sense of time, referencing visual rhythms and forms of local landscapes.

Keywords: Degumming dye sustainable weed

1 Introduction

Traditional textile production consumes large amounts of water, energy and chemicals and maybe environmentally unsustainable. This is a highly competitive industry with small profit margins. The need for ethical and responsible production along the textile lifecycle is widely recognised. Consumers and retailers demand suppliers to adopt more eco-friendly processes. More than 700,000 tons of dyestuffs are applied to 40 million tons of fabric per year and evidence suggests the fashion and textile sector is environmentally damaging in a similar way to the chemical industry [1]. The textile industry struggles with a tainted image. Responsible practice is required in water and energy usage, chemical pollution and worker conditions. The use of chemicals is one of the main areas of concern to consumers and is now advocated through regulatory control of chemical usage are in Europe. REACH requires registration, evaluation, authorisation and restriction of chemicals. Other schemes include Blue Sign, Oeko-Tex Standard 100 and Zero Discharge of Hazardous Chemicals initiative. These are unmistakable, small signs that change is coming.

The objective of this research is to use alternative dye sources to help trigger new dye practices to reduce water use and toxicity. It explores the use of an invasive New Zealand weed, clematis vitalba, a lichen, as an alternative sustainable colorant for textiles. In New Zealand, clematis vitalba was originally released as an ornamental garden plant, but in the 1800s it escaped into the sub-alpine environment. It can cause

J. Machado et al. (Eds.): HELIX 2018, LNEE 505, pp. 602–607, 2019.
https://doi.org/10.1007/978-3-319-91334-6_82

significant damage in native forests by smothering large areas of trees. Commonly called Old Man's Beard, the peculiar greyish-yellow-tinged, hairy moss, clings to trees and old wood, but it contains anti fungal and anti-biotic properties and produces interesting color (Fig. 1). Since the 1990s several species were introduced for the biological control of clematis vitalba, including a leaf fungus, a leaf miner and sawfly [2].

Fig. 1. Clematis vitalba weed

A science and textile design approach identifies the impact of new technological methods to extract colorants from a forest destroying invasive pest clematis vitalba. Simple colour extraction within the dye process is coupled with select finishing processes. An historical approach to lichen collecting and dyeing in New Zealand and the Sub-Antarctic islands informs the contemporary dyeing. This research is inspired by images from the Museum of New Zealand's Nancy Couzin's lichen catalogue [3]. The lichen samples were collected by Scott's 1901-4 expeditions to the Sub Antarctic Campbell Islands, located at 78° south and 162° east. Couzins, based in Christchurch, documented the Campbell Island lichen samples and others collected from sub alpine forests in Canterbury.

Best eco-practice techniques are developed to offer value opportunities using invasive weeds to reduce the impact of dyeing. It demonstrates how an optimisation of traditional and improved natural dye process provides eco-textile dye practice solutions. The initial challenge was to establish the most effective means of extracting colorant. New processes, new concepts and raw materials highlight a more sustainable method for the colouration of textiles with enhanced colour fastness for a collection of textile lengths.

2 Experimental

Following a long, hot summer, during early winter, the clematis vitalba was gathered from northern Hokianga forests. The clematis vitalba was harvested from an area of both high rainful and sunshine hours in Northland, New Zealand, was soaked in water for 12 hour, then processed at 60C for one hour. The process involved heat extraction, filtration, dyeing and modifying to produce an extended colour palette. The solution was then filtered to separate the residue. Another solution was prepared without application of heat (Fig. 2).

Fig. 2. Clematis vitalba weed plant residue strained after dye use. Photo Sandra Heffernan.

Commercially produced wool and silk fabrics were scoured with an aqueous nonion surfactant solution at a temperature of 45C for 30 min, then rinsed and air dried at room temperature. The pH of the dyeing solution was adjusted to 4.0. Clematis vitalba is a substantive dye but colours may be changed by using a mordant. A search for different colour tones led to the use of colour modifying mordants, the chemicals which increase natural dye uptake by textile fibres. Colour modifying mordants were applied after dye application. Aqueous solutions of ferrous sulphate (iron) and potassium aluminium sulphate (alum) thickened with guar gum are used in varying ratios. Alum brightens the colour and is cheap easily available and a safe to use mordant. Iron darkens and blackens colours and is readily available. Colour fastness to light was determined using Microsal light fastness tester having a xenon vapour lamp [4]. The penetration of the dye was negotiated through pleating, shaping, resisting and layering The textile designs reference visual rhythms and forms of local landscapes (Fig. 3).

Fig. 3. Woven wool fabric resist dyed using *clematis vitalba*, and indigo. Photo Sandra Heffernan.

A finishing process to degum silk by selectively removing one of the two proteins, sericin, from the silk organza fiber to enhance the aesthetic was developed. Archroma's Hostapal MRN liquid, a nonionic wetting, washing and cleaning agent free of APEO (Alkylphenol ethoxylates), of silicone and of solvent was used. This process used in combination with resist techniques on silk and wool fabrics produced patterning of both visual and textural interest (Fig. 4).

Fig. 4. Pleated, degummed and indigo dyed silk organza work exploring the effectiveness of the combination of techniques. Photo Sandra Heffernan.

3   Results

The results include carefully calibrated recipes and processes for a range of dyed cloths. To avoid reducing the strength of the natural fibres, samples were dyed for 60 min at 80C. pH 4 was established as the optimum pH for dyeing wool. A reduced dye liquor of 8% ratio of increased contact. A selection of samples were treated with a range of mordants and dyed for 30 min at 80C. Colourfastness to light testing and wash fast testing was used to confirm the effectiveness of the dye process (Table 1).

Table 1. Effect of post mordants and lightfast-ness of *clematis vitalba* dye.

Title	Dyeing with *clematis vitalba*				
Mordant	g/l	Fibre colour wool	Fibre colour silk	Lightfast wool	Lightfast silk
Without mordant	8	golden peach	golden peach	5	4/5
AlK (SO4) 2	8	orange gold	orange gold	7	5
FeSO4	8	dark sienna	dark sienna	7	5

4   Discussion

Clematis vitalba dye showed good scope in the environmental natural dye process, producing good colour, from golden oranges to gold to deep sienna browns on both silk and wool. The colour range on silk is similar to colours achieved using natural osage dye. Different types of mordants yield different colours for the same natural fibres. The type of mordant influenced the color strength and hue. The mordanting process renders the dye more bonded and more aggregated on the fibre, thereby the surface area of the dye accessible to light is reduced [5]. Wool fabrics dyed with clematis vitalba showed a higher colour strength than the silk fabrics.

5   Conclusion

Dyeing using clematis vitalba is novel, recognising responsibility to environmental concerns and helps to make a textile product more "green" and sustainable. The dye process did not use any salt or toxic chemicals and used a mimimal amount of water. It explored key areas of environmental impact knowledge and provides an insight into future action for creative change in textile dyeing and finishing. It found a use for a pest plant which has a positive effect on bacterial growth, something that could be further exploited.

Dye from the tree smothering invasive pest, clematis vitalba, produced innovative dye solutions for protein fibre-based materials in a fashion focused contemporary textile design collection. It reveals memory of pleat, resist and clamp process akin to an abstracted topographical map of the Hokianga region.

Future research could involve the use of solar energy to the impact minimise energy used during the dye process or to assess the impact of time and sun on the fixation (Fig. 5).

Fig. 5. Detail of wool fabric resist dyed using *clematis vitalba*. Photo Sandra Heffernan.

References

1. Fletcher, K.: Sustainable Fashion and Textiles: Design Journeys. Routledge, Abingdon (2014)
2. Sustainable solutions: Textile Month International, no. 4, pp. 23–25 (2015)
3. Nature: Pests and threats. Department of Conservation. http://www.doc.govt.nz/nature/pests-and-threats/, Accessed 21 Dec 2017
4. Couzin, N.: 1901-4, Lichen catalogue. Te Papa Tongarewa Museum of New Zealand, Wellington, New Zealand (1910)
5. Mansour, H., Heffernan, S.: Environmental aspects on dyeing silk fabric with sticta coronata lichen using ultrasonic energy and mild mordants. Clean Technol. Environ. Policy **13**, 207–213 (2011)

Investigation of Copper and Zinc Contamination on the Work Piece Surface with WEDM

V. K. Manupati[1], G. Rajyalakshmi[1], M. L. R. Varela[2]([⊠]),
J. Machado[3], and G. D. Putnik[2]

[1] Department of Manufacturing, School of Mechanical Engineering,
VIT, Vellore, India
manupativijay@gmail.com, rajyalakshmi@vit.ac.in
[2] Department of Production and Systems, University of Minho,
Guimarães, Portugal
{leonilde,putnikgd}@dps.uminho.pt
[3] Department of Mechanical Engineering, School of Engineering,
University of Minho, Guimarães, Portugal
jmachado@dem.uminho.pt

Abstract. This work presents an investigation on Wire Electric Discharge Machining (WEDM) of Pure Titanium (Grade-2). Pure Titanium and its alloys are the most suitable metallic materials for biomedical applications. In health care, they are used for implant devices like knee and hip implants, fixing materials for bones like nails and screws, and in the oral and maxillofacial surgery. Properties like good biocompatibility, high corrosion resistance, and high fracture strength make titanium parts the perfect solution for bio-implants. The more frequent use of the titanium necessitates an economical machining process. Conventional processes like milling, grinding or drilling have their limits because of the material's mechanical properties which cause high tool wear. WEDM proves to be an alternative especially for manufacturing complex parts as there is no actual contact with tool and the work piece. Within this study, the capabilities of Wire-EDM for the manufacturing of clean surfaces with distinct Surface Roughness (SR) and Machining Speed (MS) are conducted through optimization of process parameters. Taguchi technique is used for optimization and the experiments are conducted according to Taguchi Design of Experiments (DOE). Different wire electrode materials have been used considering a possible copper and zinc contamination on the work piece surface.

Keywords: WEDM · Titanium grade 2 · Taguchi Design · Optimization

1 Introduction

Increased use of titanium and its alloys as biomaterials stems from their lower modulus, superior biocompatibility, and better corrosion resistance when compared to more conventional stainless and cobalt-based alloys. As a hard tissue replacement, the low elastic modulus of titanium and its alloys is generally viewed as a biomechanical

© Springer International Publishing AG, part of Springer Nature 2019
J. Machado et al. (Eds.): HELIX 2018, LNEE 505, pp. 608–615, 2019.
https://doi.org/10.1007/978-3-319-91334-6_83

advantage because the less elastic modulus can result in lower stress shielding. Hence, for the present research pure Titanium has been chosen as the test material.

Titanium and its alloys are often used in the medical sector. It is used for joint substitutions like knee and hip implants, fixing materials for bones like nails and screws, and in the oral and maxillofacial surgery because of the materials' characteristics like good biocompatibility and bio adhesion in addition to corrosion resistance and specific mechanical properties. However, problems arise during manufacturing of titanium implants with conventional processes like milling, grinding or drilling obtain their limits because of the materials mechanical properties which cause e.g. high tool wear.

Electro Discharge Machining (EDM) represents an alternative especially for manufacturing of filigree parts. Furthermore, it is applicable for flexible manufacturing process that allows to machine implants which can be adapted to specific patients' demands. In terms of a single-unit production, it is hereby an economic process. Further differences between EDM and conventional manufacturing processes can be pointed out in terms of surface integrity. Through machining by cutting processes smearing and plastic deformation (e.g. micro ploughing) might lead to small cavities. These negative surface conditions do not occur after machining by EDM because of a nearly force free process. This leads to a surface which can be very well sterilised. Especially the Wire-EDM process leaves residues like copper and zinc on surfaces. These residues substantially affect the biocompatibility. One way to decrease or eliminate this problem is to use non-common wire electrodes. These are electrodes made of materials which themselves feature a high biocompatibility. Another option could be to use Trilayered wires but these are very expensive to produce or even non-producible. Figure 1 shows the schematic diagram of the Wire Electric Discharge Machining (WEDM) process.

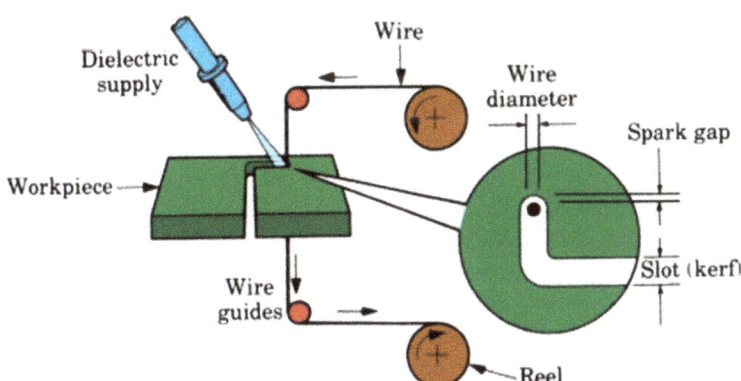

Fig. 1. Wire EDM process diagram Tsai et al., (2003)

Advantages of WEDM Process can be summarized through: as continuously travelling wire is used as the negative electrode, so electrode fabrication is not required as in EDM; there is no direct contact between the work piece and the wire, eliminating the mechanical stresses during machining; WEDM process can be applied to all electrically conducting metals and alloys irrespective of their melting points, hardness, toughness or brittleness; and users can run their work pieces over night or over the weekend unattended.

Disadvantages of WEDM Process, can be summarizes through: high capital cost that is required for WEDM process; there is a problem regarding the formation of recast layer; WEDM process exhibits very slow cutting rate; and it is not applicable to very large work piece.

This paper focuses on a basic study on the capabilities of Wire-EDM for the manufacturing of clean surfaces with distinct roughness and a high surface integrity. Common wire materials will be analysed regarding a possible copper and zinc contamination of the work piece surface due to re-solidification of small particles originating from the wire. Furthermore, the process itself will be optimised by parametric optimization of various process parameters to obtain distinct surface roughness and better machining speed of WEDM.

2 Literature Review

Electric Discharge Machining amongst the many non-conventional machining methods has numerous applications in the current industrial processes. The EDM process converts the electrical energy generated in a channel between a cathode and an anode to thermal energy which creates sufficient high temperature for the erosion of work piece material [1]. This distinct feature of thermo-electric conversion of EDM makes it suitable for manufacturing of complex parts particularly for aerospace and automotive industry, and surgical instruments [2]. In EDM process, the desired shape of the object is obtained by electrical discharges i.e., sparks. A localised spark is produced between the work-piece and the electrode in the presence of a dielectric fluid. Here, generally deionized water acts as a dielectric fluid which filters and directs the sparks. The water also acts as a coolant and also helps in flushing away the residual metal particles [3]. In Wire EDM, a moving metallic wire is used as the electrode to cut a programmed contour. The wire movement is controlled numerically which provides accuracy for machining of complex two and three dimensional work piece configurations. Wire EDM is a non-contact type machining method which provides an economical and efficient way for cutting Metal Matrix Composite materials. The work material Hardness does not possess any disadvantage in cutting as there is no relative contact between the wire and the work material [4]. Instead of cutting the material, the WEDM produces high temperature which melts or vaporizes the material, leaving very little debris providing accurate results. A taut thin wire discharges the electrified current which acts as a cathode guided alongside the desired cutting path. The work piece undergoing machining is usually fixed on a table. High accuracy instrumentation comprising of CNC systems is used to control the movements of work piece which reduces the positioning errors to a large extent [5].

Some of the common wire electrode is made up of copper because of its higher electrical conductivity but it wears rapidly and its tension ability is poor [6].

From literature, it has been found that many researchers have focused on the developments of EDM, WEDM, and micro-WEDM. Hence, in this study, an attempt has been made to determine the effect of process parameters and wire material on the responses like MRR and surface roughness. Taguchi's orthogonal array has been used for conducting the experiments and presented the results of wire materials influence on performance of WEDM process for Titanium grade 2.

3 Design of Experiments

Design of experiments (DOE) or experimental design is the design of any information gathering exercises where variation is present, whether under the full control of the experimenter or not. Classical experimental design often requires a large number of experiments to be carried out when number of machining parameters increase.

Genichi Taguchi developed this statistical method for improving the quality of the products and manufacturing goods which finds many applications in the field of engineering [7].

This method provides an efficient, systematic and simplistic approach for optimization of the design for better performance, higher quality and cost cutting. This method is more effective when used with the discrete and qualitative design parameters.

The optimization process in a Taguchi method is carried out in 3 steps: System design, parameter design, and tolerance design. The main thrust of this method lies in the usage of a method for product and process design i.e. Parameter design, which focuses on deciding the parameters for developing the best of quality characteristics with minimal variations. The S/N ratio is used to measure the quality characteristics deviating from the desired values. The analysis includes 3 categories of quality characteristics: (i) Lower-the-better, (ii) Higher-the-better, and (iii) Nominal-the-better. The greatest S/N ratio value should be selected for best quality characteristics. In this study, 4 machining parameters were used as control factors and each parameter was designed to

Table 1. Taguchi's L9 orthogonal array.

Exp. no.	T ON	T OFF	IP	Wire feed
1	1	1	1	1
2	1	2	2	2
3	1	3	3	3
4	2	1	2	3
5	2	2	3	1
6	2	3	1	2
7	3	1	3	2
8	3	2	1	3
9	3	3	2	1

Table 2. Process parameters for WEDM and their considered levels.

Symbol	Control factor	Unit	Level 1	Level 2	Level 3
A	Pulse on time (T ON)	µs	50	70	90
B	Pulse off time (T OFF)	µs	4	6	9
C	Current (IP)	A	3	4	5
D	Wire feed	mm/min	50	70	90

Table 3. Properties of different types of wire used.

	Material	Coating	Tensile	Elongation	Conductivity	Diameter
Coated wire	Cu Zn 36	Zn	900 N/mm^2	1.50%	22% IACS	0.25 mm
Annealed wire	Cu Zn 37	Zn	900 N/mm^2	2%	22% IACS	0.25 mm

Table 4. Experimental readings of SR, MS and S/N ratio values for machining speed and SR.

Wire	Exp. no.	T ON	T OFF	IP	Wire feed	MS (mm/min)	SR (μm) R_a	S/N ratio MS	S/N ratio SR
Coated wire	1	1	1	1	1	1.24	11	1.868433703	−20.8278537
	2	1	2	2	2	1.18	8.511	1.437640146	−8.59961181
	3	1	3	3	3	0.27	9.2376	−11.37272172	9.31118300
	4	2	1	2	3	0.5	10.0069	−6.020599913	−20.0059912
	5	2	2	3	1	0.498	8.895	−6.055413145	−8.98291905
	6	2	3	1	2	0.59	10.659	−4.582959767	−0.55432924
	7	3	1	3	2	0.85	7.5138	−1.411621486	−7.51719262
	8	3	2	1	3	0.78	8.3414	−2.158107946	−8.42477895
	9	3	3	2	1	0.28	7.3834	−11.05683937	−7.36512794
Annealed wire	10	1	1	1	1	1.112	11.5695	0.922095745	−1.26629181
	11	1	2	2	2	1.37	9.2105	2.734411343	−9.28566414
	12	1	3	3	3	0.608	11.3269	−4.321928415	−1.08222133
	13	2	1	2	3	0.392	8.5086	−8.13427866	−8.59716215
	14	2	2	3	1	0.18	9.176	−14.8945499	−19.2530681
	15	2	3	1	2	0.57	8.6392	−4.882502887	−8.72947056
	16	3	1	3	2	0.82	8.2458	−1.723722952	−8.32465594
	17	3	2	1	3	0.435	7.8961	−7.230214861	−17.9482528
	18	3	3	2	1	0.16	8.4904	−15.91760035	−8.57856302

have 3 levels. According to the Taguchi quality design concept, a L9 orthogonal arrays table with 9 rows (corresponding to the number of experiments) was chosen for the experiments. Table 1 shows the distribution of levels of different parameters in a Taguchi Orthogonal Array. In Tables 2 and 3 the properties of the wires and process parameters of the WEDM is shown. The analysis using Taguchi is portrayed in Table 4.

4   Method of Analysis

A single response optimization is carried out to investigate the effects of machining parameters on Machining Speed (MS) and Surface Roughness (SR). According to the Taguchi method, S/N ratios in Eqs. 1 and 2 were calculated for each experiment. The objective of optimization is to maximize the MS and minimize the SR. The response table for S/N ratios of MS is calculated considering the fact that MS is a larger-the-better performance characteristic; the maximization of the quality characteristic of interest is sought and is expressed as:

$$S/NRatio \ = \ -10\log\left(\frac{1}{n}\right)\sum_{i=1}^{n}\frac{1}{y^2} \qquad (1)$$

yij = observed response value i = 1, 2... ...n; j = 1, 2 ... k, n = number of replications.

The surface roughness is the lesser-the-better performance characteristic and the S/N ratio for SR is calculated by:

$$S/NRatio \ = \ -10\log\left(\frac{1}{n}\right)\sum_{i=1}^{n}y_{ij}^2 \qquad (2)$$

SEM and EDX analysis are performed on the samples obtained from single objective optimization of Surface Roughness of both Coated and Annealed wire. The objective of the analysis is to check the contamination levels of copper and other materials in the sample after wire electric discharge machining. Figure 2, shows the SEM results and composition of different materials that are present on the surface of the machined surface which are cut through coated wire for three spectrums shown in Table 5.

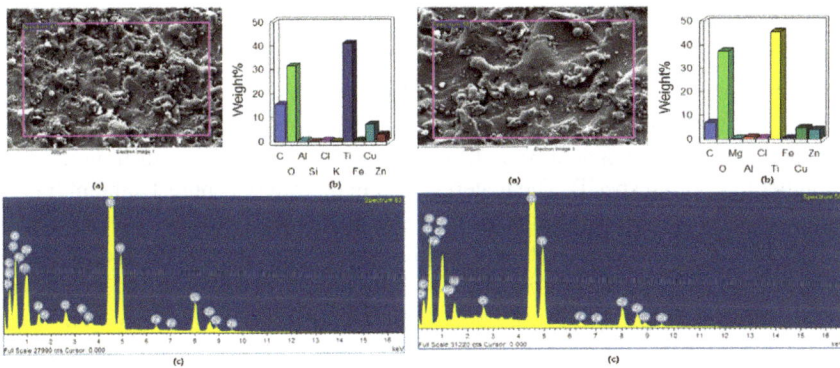

Fig. 2. (a) SEM micrograph for spectrum 1, 2 and 3 (b) composition graph for spectrum 1, 2 and 3, (c) EDX analysis for spectrum 1, 2 and 3, for coated wire, respectively.

According to SEM images of sample of coated wire, the fibrous network is having maximum Cu content as 11.0%, and in the globular phases, the Cu content is reduced to 9.58%.

Figure 2 shows the SEM results and composition of different materials that are present on the surface of the machined surface which are cut through annealed wire for three spectrums shown in Table 5 respectively. According to SEM images of sample of annealed wire, the surface appears as beach sand shaped with globular phases over it. There are some cracks visible to across the whole surface. At various points, EDAX analysis were performed. In the globular phases, more amount of Cu (4.72%) is seen. In the remaining matrix, the Cu content is reduced, which tells that there has been

Table 5. Chemical composition of sample for all spectrums.

Element	Spectrum 1		Spectrum 2		Spectrum 3	
	Weight (%)	Atomic (%)	Weight (%)	Atomic (%)	Weight (%)	Atomic (%)
C	10.7	22.84	11.51	23.9	15.51	29.84
O	28.99	46.43	30.25	47.14	31.56	45.59
Al	0.54	0.51	0.51	0.47	0.46	0.39
Si	0	0	0	0	0.09	0.08
S	0.12	0.09	0	0	0	0
Cl	0.77	0.56	0.58	0.4	0.45	0.3
K	0.31	0.21	0.22	0.14	0.15	0.09
Ca	0.12	0.08	0.13	0.08	0	0
Ti	43.6	23.32	43.68	22.73	41.16	19.86
Fe	0.32	0.15	0.3	0.13	0.49	0.2
Cu	11	4.44	9.58	3.76	7.09	2.58
Zn	3.53	1.38	3.24	1.24	3.05	1.08

Element	Spectrum 1		Spectrum 2		Spectrum 3	
	Weight (%)	Atomic (%)	Weight (%)	Atomic (%)	Weight (%)	Atomic (%)
C	6.83	14.08	6.49	14.07	6.47	14.32
O	37.36	57.81	32.81	53.39	30.93	51.41
Mg	0.08	0.08	0	0	0	0
Al	0.76	0.7	0.77	0.74	0.79	0.77
Cl	0.44	0.3	0.33	0.24	0.29	0.21
Ti	45.56	23.55	53.64	29.15	55.24	30.67
Fe	0.3	0.13	0.15	0.07	0.18	0.09
Cu	4.72	1.84	2.51	1.03	3.11	1.3
Zn	3.96	1.5	3.3	1.32	3.01	1.22

micro-segregation of Cu, with Cu accumulating in the globular phases. In the matrix dense fibrous network of the alloy is seen Fig. 2. Moreover, we can realise that the Cu content is reduced (3.11%), along with some Cu less phases.

5   Conclusions

The present work focuses on optimization of parameters of wire electric discharge machining and their effects of different wires to improve the surface quality (reduce contamination), surface characteristics (distinct surface roughness), and time of operation (higher machining speed). The material used in this study is pure Titanium (Grade 2) owing to its better biocompatibility. Various parameters affecting the machining conditions are considered for single and multi-response optimization. The conclusions obtained are as follows. From single response optimization, it is observed that machining through annealed wire provides higher machining speed. However, it also results in higher surface roughness. The SEM and EDAX results show that the contamination levels in annealed wire, especially of copper, is significantly lower than the Zinc coated wire. This is due to better diffusion of zinc and copper in the annealed wire which avoids spark erosion from the wire to the material. Overall, it is recommended to use annealed wire during Wire Electric Discharge Machining for biological applications as it provides better results in terms of both machining speed and biocompatibility.

Future work can be performed by studying effects of other types of wire like molybdenum wire, brass wire and composite wires for obtaining better results. Different biocompatible materials like titanium alloys, stainless steel, magnesium alloys can also can be studied in terms of better machinability and surface characteristics.

Acknowledgements. This work is supported by Fundação para a Ciência e Tecnologia with ref. PEst2015-2020.

References

1. Tsai, H.C., Yan, B.H., Huang, F.Y.: EDM performance of Cr/Cu-based composite electrodes. Int. J. Mach. Tools Manuf **43**(3), 245–252 (2003)
2. Abbas, N.M., Solomon, D.G., Bahari, M.F.: A review on current research trends in electrical discharge machining (EDM). Int. J. Mach. Tools Manuf. **47**(7), 1214–1228 (2007)
3. Spedding, T.A., Wang, Z.Q.: Study on modeling of wire EDM process. J. Mater. Process. Technol. **69**(1), 18–28 (1997)
4. Ramakrishnan, R., Karunamoorthy, L.: Multi response optimization of wire EDM operations using robust design of experiments. Int. J. Adv. Manuf. Technol. **29**(1–2), 105–112 (2006)
5. Kumar, A., Kumar, V., Kumar, J.: Multi-response optimization of process parameters based on response surface methodology for pure titanium using WEDM process. Int. J. Adv. Manuf. Technol. **68**(9–12), 2645–2668 (2013)
6. Kanlayasiri, K., Boonmung, S.: Effects of wire-EDM machining variables on surface roughness of newly developed DC 53 die steel: design of experiments and regression model. J. Mater. Process. Technol. **192**, 459–464 (2007)
7. Rosa, J.L., Alain, R., Silva, M.B., Baldan, C.A., Peres, M.P.: Electrodeposition of copper on titanium wires: taguchi experimental design approach. J. Mater. Process. Technol. **209**(3), 1181–1188 (2009)

Waste to Energy and Sustainable Environment

Characterization of Municipal, Construction and Demolition Wastes for Energy Production Through Gasification - A Case Study for a Portuguese Waste Management Company

Octávio Alves[1,2]($\boxtimes$) ⓘ, Jeysa Passos[2] ⓘ, Paulo Brito[2] ⓘ,
Margarida Gonçalves[1,2] ⓘ, and Eliseu Monteiro[2] ⓘ

[1] MEtRICs - Mechanical Engineering and Resource Sustainability Center,
Department of Science and Technology of Biomass, Faculty of Science
and Technology, Universidade NOVA de Lisboa, Lisbon, Portugal
o.alves@campus.fct.unl.pt
[2] VALORIZA - Research Center for Endogenous Resource Valorisation,
Polytechnic Institute of Portalegre, Portalegre, Portugal

Abstract. Gasification of wastes is considered a promising alternative for energy generation due to its lower environmental impacts when compared with conventional landfilling and incineration. Valorisation of such wastes improves sustainability of resource management and of energy production. However, an appropriate characterisation of wastes in terms of physical and chemical properties is essential for the prediction of their behaviour during gasification, allowing to identify possible problems for the environment and installed equipment and also to define which materials present a greater energy potential. This study aimed to characterise 10 different fractions from municipal, construction and demolition wastes received in different fluxes by a Portuguese waste management company. These fractions included wood (44.83 wt%), plastic (22.15 wt%), paper/card (0.04 wt%), mixtures of paper and plastic (14.67 wt%) and sewage sludge (18.31 wt%). For this purpose, determination of density, proximate and ultimate analysis, higher heating value (HHV), thermogravimetric profiles and inorganic composition of ashes were performed for each fraction. Analysis revealed that plastics and their mixtures with paper/card possess the highest HHV's (25–45 MJ/kg db), thus exhibiting a greater capacity for energy production. High levels of ashes found in dried sewage sludge (50 wt % db) indicate that a lot of by-product will be generated after gasification, possibly increasing the treatment costs. A gasification unit operating at 50 kg/h and admitting a mixture of all these wastes would generate 109.7 kW of total power, having capacity to receive more waste fluxes along the year.

Keywords: Municipal solid waste · Construction and demolition waste
Property analysis · Gasification

© Springer International Publishing AG, part of Springer Nature 2019
J. Machado et al. (Eds.): HELIX 2018, LNEE 505, pp. 619–625, 2019.
https://doi.org/10.1007/978-3-319-91334-6_84

1   Introduction

Traditional methods for the treatment and disposal of wastes like landfilling, inciner-ation or application in agriculture raise several health and environmental concerns. Recent European directives like 2008/98/EC and the future RED II for 2021–2030 have encouraged the adoption of new waste-to-energy solutions using advanced technolo-gies [1]. In this context, gasification may be considered a valid alternative. It consists in a thermochemical conversion of wastes at c.a. 800 °C with limited amounts of oxygen to produce a combustible gas (syngas) that can be used for energy production. The process has lower toxic gas emissions than incineration and may reduce the depen-dence on fossil fuels [2]. The implementation of a gasification unit implies a good knowledge about the properties of wastes in order to estimate the potential energy that will be generated and to predict possible issues for the equipment and health: for example, chlorine and alkali metals are known to cause corrosion and obstruction problems [3, 4]. Knowledge of these properties is required for an adequate choice of pre-treatments and components for the facility. Municipal solid (MSW) and con-struction and demolition wastes (CDW) are heterogeneous residual fluxes containing organic fractions like plastics, wood and paper/card, all of them viable for gasification due to the good calorific values. A few works may be found in the literature that focused on the characterization and classification of the different materials present in wastes [1, 5]. The present study aimed to evaluate the physical and chemical properties of 10 distinct and separated fractions present in MSW and CDW received by a Por-tuguese waste management company, in order to evaluate the contribution of each fraction to the overall gasification process.

2   Materials and Methods

Waste samples were collected in a Portuguese waste management company and are identified in Table 1, along with their corresponding code number in the European list of wastes (ELW) and with the waste amounts received and processed during 2015.

Table 1. Identification and received amounts of wastes by the company (year 2015).

Sample	ELW number	Brief description	Amount (kg)
R1	170201	Wood (CDW)	56 910
R2	200138	Wood (MSW)	4 280
R3	200101	Paper/card (MSW)	50
R4	150102	Plastic packages	13 840
R5	150105	Composite packages (99.9% paper + 0.1% plastic)	7 850
R6	150106	Mixture of packages (65% paper/card + 35% plastic)	12 180
R7	170203	Plastics (CDW)	4 400
R8	200139	Plastics (MSW)	7 820
R9	170604	Polymeric insulations (CDW)	4 180
L1	200306	Sludges from sewage cleaning	25 000

Since samples R4 and R8 were similar in composition, analysis of R8 was excluded and considered equivalent to R4. The solid wastes (i.e. all but the sludges) were crushed, milled and sieved in particle sizes of c.a. 12 mm, <0.425 mm and <0.250 mm. The first-class size was used to determine bulk density (according to EN 15103), the second for moisture content (ASTM E949-88) and the third for a thermogravimetric analysis under a N_2 atmosphere (using a PerkinElmer STA 6000). Samples R4 and R6 were highly malleable and thus were not possible to be milled in small grains; instead, they were cut with scissors until the minimum size that was possible to reach. Sludge was analyzed as received for the determination of moisture content and thermogravimetric profile, however a different method was employed to determine density (2710F from Standard Methods). Then the sludge was completely dried at 105 °C and milled to produce grains of <0.425 mm. Posteriorly, all samples with <0.425 mm were dried at 105 °C in order to determine their higher heating value (HHV, with a IKA C200) and their ultimate analysis (in a ThermoFisher Scientific Flash 2000 CHNS-O). Contents of volatile matter, ash and fixed carbon were measured following ASTM E897-88, ASTM E830-87 and calculated by difference, respectively. Inorganic composition of ashes was analysed by X-ray fluorescence using a Thermo Scientific Niton XL 3T Gold++.

3 Results and Discussion

Analysis of the received waste fluxes reported in Table 1 indicated that woods from RCD (R1) and sludges (L1) were the most representative, with weight fractions >18 wt % ar. Paper/card waste was almost negligible (0.04 wt% ar). Table 2 shows the results obtained for density, HHV and proximate analysis of each sample.

Table 2. Density, HHV and proximate analysis of sample wastes.

Sample	Density (kg/m³ ar)	HHV (MJ/kg db)	Moisture (wt% ar)	Volatiles (wt% db)	Fixed carbon (wt% db)	Ash (wt% db)
R1	253	18.9	10.9	88.0	9.3	2.7
R2	286	19.4	12.7	89.3	9.4	1.3
R3	37	14.9	6.2	82.0	0.0	18.3
R4/R8	26	44.9	0.6	98.0	0.0	2.5
R5	25	17.0	6.1	94.5	0.0	6.1
R6	43	26.5	4.3	85.7	5.2	9.1
R7	543	25.0	0.4	81.8	7.2	10.9
R9	19	30.3	1.0	93.6	4.4	1.9
L1	1070	24.6	89.2	47.9	2.3	49.8

HHV was greater for plastic materials (R4, R8, R7 and R9) and for the mixture of plastics + paper/card (R6), with values ranging from 25–45 MJ/kg db, indicating that these wastes possess high energy levels available for gasification. However, they also present the lowest densities (≤ 26 kg/m3), and therefore rise costs of transportation from collection points to the gasification plant. Waste R7 (CDW) had the most balanced results among HHV's and densities (25 MJ/kg db and 543 kg/m3, respectively), thus being a material that may be directly used while others should be mixed to dilute undesirable characteristics. Considering that gasification generally admits materials with moisture contents less than 15 wt%, all wastes are appropriate for the process with the exception of L1 (89 wt% ar) that must be submitted to an intensive drying pre-treatment. There is no apparent correspondence between HHV and volatile matter: for example, plastics R4, R8 and R9 possess high values in both parameters (>30 MJ/kg db and >93 wt% db, respectively), but on the contrary the composite waste R5 has a lower HHV and a high volatile matter content (17 MJ/kg db and 95 wt% db).

Ash contents were significantly larger in the case of L1 (sludge), achieving almost 50 wt% db. This may be a problem because a lot of inorganic by-product is generated during gasification what imposes the need of convenient post-treatment or disposal methods, thus increasing operational costs. Therefore, a definition of valorisation pathways for these ashes is relevant, namely, it may be opportune to study their application in remediation of effluents or their use in agriculture.

The mixture of paper/card with plastics (R6) seems to be advantageous for gasification because it not only reduced the ash content (18 to 9 wt% db) but also increased the HHV (15 to 27 MJ/kg db), when compared with paper/card alone (R3).

Table 3 shows the measured results for ultimate analysis, H/C and O/C ratios.

Table 3. Ultimate analysis and atomic ratios O/C and H/C of sample wastes.

Sample	Element (wt% db)					Atomic ratio (db)	
	N	C	H	S	O	O/C	H/C
R1	2.0	45.8	5.8	0.0	43.7	0.95	0.13
R2	0.1	47.8	6.0	0.0	44.8	0.94	0.13
R3	0.3	38.6	5.2	0.1	37.5	0.97	0.13
R4/R8	0.0	83.9	14.0	0.0	0.0	0.00	0.17
R5	0.2	40.2	5.4	0.1	48.0	1.19	0.13
R6	0.0	41.8	5.9	0.0	43.2	1.03	0.14
R7	0.3	50.9	6.6	0.1	31.2	0.61	0.13
R9	3.3	70.9	7.4	0.0	16.5	0.23	0.10
L1	0.1	33.4	4.6	0.4	11.7	0.35	0.14

Typically, lower O/C and H/C ratios are associated with solid fossil fuels like vegetable coal, meaning that materials with lower ratios present greater HHV's [6]. As seen in Table 3, all samples have similar H/C ratios but show some variations in O/C ratios, so this last parameter apparently has a higher impact on HHV. In fact, plastic samples R4, R8 and R9 possess the lowest O/C ratios (<0.3) which justify the highest HHV's determined in Table 2. On the other hand, wood samples R1 and R2, paper/card R3 and composite R5 showed higher O/C ratios (>0.9) that decreased their

HHV's. Mixture R6 seems to escape to this rule, possibly because problems of homogenization of its constituents occurred during the determination of ultimate analysis and which were associated to the difficulty of obtaining fine grains. The corresponding O/C ratio (1.03) did not agree well with the literature (0.7 [7]). Contents of S were relatively low for the generality of wastes (≤ 0.4 wt% db) and therefore formation of toxic compounds SO_2 and H_2S during gasification would be negligible. However, wastes R1 and R9 possess high quantities of N (≥ 2 wt% db) that promotes the formation of NO_x gases, so a careful regulation of the O_2 injected in the gasifier must be done. Table 4 presents the composition of ashes of all wastes in terms of the relevant inorganic oxides.

Table 4. Inorganic composition of ashes derived from waste samples.

Sample	Oxide (wt%)									
	Al_2O_3	CaO	CuO	Fe_2O_3	K_2O	P_2O_5	SO_3	SiO_2	TiO_2	ZnO
R1	0.0	26.5	0.9	5.5	4.1	1.3	3.2	8.1	1.9	0.4
R2	0.0	30.6	0.3	3.0	7.4	2.3	2.3	4.0	0.6	0.2
R3	3.3	40.7	0.1	1.6	0.1	0.0	1.6	9.6	0.4	0.1
R4/R8	0.0	29.1	0.2	3.9	1.6	0.0	2.5	9.9	4.1	0.4
R5	0.0	34.7	0.2	3.1	1.1	0.0	3.7	7.9	0.8	0.1
R6	2.1	35.4	0.1	1.5	0.3	0.0	1.0	8.9	0.6	0.1
R7	0.0	19.8	0.0	1.2	0.1	0.5	4.4	3.8	6.7	0.0
R9	0.0	16.7	0.3	9.7	0.6	2.7	0.7	39.4	0.9	1.6
L1	7.6	9.0	0.0	6.2	2.5	1.8	4.1	35.7	0.7	0.2

Generally, CaO and SiO_2 were the most representative oxides with mass fractions that achieved around 40 wt%. Compositions were similar for samples of woods (R1 and R2) and those based on paper/card (R3 and R5), but they differed significantly in the case of polymers (R4, R7, R8 and R9). The relatively high concentrations of CaO (26–31 wt%), P_2O_5 (1–3 wt%), Fe_2O_3 (3–6 wt%) and K_2O (4–8 wt%) found in wood ash make them viable for fertilization or pH correction in agriculture, but it is necessary to ensure that heavy metals remain under the legal limits by applying appropriate decontamination post-treatments. Figure 1 presents the thermogravimetric profiles and derivative curves obtained for four different samples in terms of structure and composition (R1, R6, R7 and L1). The first concavity in the derivative curve corresponds to the release of moisture, notably higher for sample L1 and almost absent for sample R7. The weight loss between 300–400 °C is generally high for lignocellulosic materials due to the degradation of hemicellulose and cellulose fibers, a behavior observed for R1 and R6 samples. On the other hand, R7 showed a wider range of decomposition temperatures (300–500 °C), corresponding to the pyrolytic decomposition between 300 °C and 400 °C followed by the char volatilization and oxidation in the range from 400 °C to 500 °C, a behavior frequently observed for polymeric materials [8, 9].

Regarding R6 (mixture of paper/card and plastic), two pronounceable concavities were found: the first reports to the decomposition of the lignocellulosic structures of

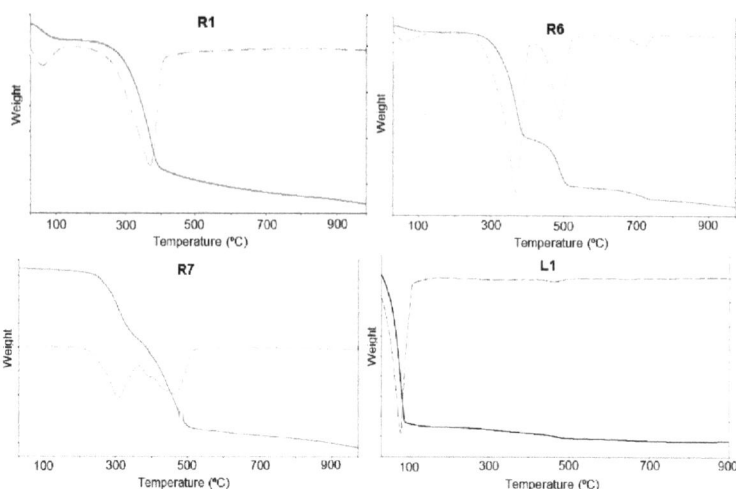

Fig. 1. Thermogravimetric profiles for samples R1, R6, R7 and L1 (continuous line represents the loss of mass while the dashed one is the derivative).

paper/card and the second to plastics. According to these profiles, if a pre-treatment of carbonization for enhancing waste properties is to be considered (particularly in the case of polymers), then it must be performed at temperatures ≤ 300 °C to avoid a significant loss of mass. A fuel mixture containing all these wastes in the proportions mentioned in Table 1 would have an average HHV of 25.2 MJ/kg db and an average ash content of 12 wt% db. For a gasification unit operating continuously at 50 kg/h and producing 0.916 Nm^3 of syngas per kg of compound with a lower heating value of 8.62 MJ/Nm^3, then it would generate 16.5 ton of ash and 109.7 kW of total power. Operation time would be short (114 days/year), meaning that the unit would be able to receive higher quantities of waste therefore expanding its activity.

4 Conclusions

A characterization of the properties of 10 distinct wastes received by a Portuguese waste management company was carried out in this study in order to evaluate their potential use for energy generation through gasification. It was concluded that plastic wastes and their mixtures with paper/card showed the highest HHV's, thus exhibiting good capacity for energy conversion. High ash contents found in dried sewage sludge may promote the formation of large quantities of inorganic by-product that has to be treated, increasing the overall costs. A possible pre-treatment of carbonization of all wastes must be executed at no more than 300 °C in order to avoid an appreciable loss of mass. A gasification unit projected to operate at 50 kg/h and admitting a mixture of these wastes would produce 109.7 kW of total power. The total amount of wastes is insufficient to keep the unit working continuously all the year, and therefore it is capable to receive more waste fluxes eventually from other sources. Determination of chlorine contents of all wastes (especially in the case of plastics) is relevant and suggested for future works with the aim of evaluating possible problems for equipment and to define appropriate decontamination treatments.

Acknowledgements. Authors acknowledge the financial support received by the Foundation for Science and Technology from the Portuguese Ministry of Science, Technology and Higher Education (grant no. SFRH/BD/111956/2015), and project POCI-01-0145-FEDER-024020 (RDFGAS - Aproveitamento energético dos combustíveis derivados de resíduos e lamas secas), co-financed by COMPETE 2020 - Programa Operacional Competitividade e Internacionalização, Portugal 2020 and União Europeia through FEDER. An acknowledgement is sent to the company Pragosa Ambiente, S. A. for providing the waste samples, and also to Luís Calado, Bruno Garcia, Miltiadis Samanis and Paula Rodrigues.

References

1. Zhou, H., Meng, A., Long, Y., Li, Q., Zhang, Y.: Classification and comparison of municipal solid waste based on thermochemical characteristics. JAPCA J. Air Waste Manag. Assoc. **64** (5), 597–616 (2014)
2. Di Fraia, S., Massaroutti, N., Vanoli, L., Costa, M.: Thermo-economic analysis of a novel cogeneration system for sewage sludge treatment. Energy **115**, 1560–1571 (2016)
3. Silva, R., Fragoso, R., Sanches, C., Costa, M., Martins-Dias, S.: Which chlorine ions are currently being quantified as total chlorine on solid alternative fuels? Fuel Process. Technol. **128**, 61–67 (2014)
4. Asadullah, M.: Biomass gasification gas cleaning for downstream applications: a comparative critical review. Renew. Sustain. Energy Rev. **40**, 118–132 (2014)
5. Zhou, H., Meng, A., Long, Y., Li, Q., Zhang, Y.: An overview of characteristics of municipal solid waste fuel in China: physical, chemical composition and heating value. Renew. Sustain. Energy Rev. **36**, 107–122 (2014)
6. Prins, M., Ptasinski, K., Janssen, F.: More efficient biomass gasification via torrefaction. Energy **31**, 3458–3470 (2006)
7. Kobayashi, J., Kawamoto, K., Fukushima, R., Tanaka, S.: Woody biomass and RPF gasification using reforming catalyst and calcium oxide. Chemosphere **83**, 1273–1278 (2011)
8. Xu, J., Liu, C., Qu, H., Ma, H., Jiao, Y., Xie, J.: Investigation on the thermal degradation of flexible poly(vinyl chloride) filled with ferrites as flame retardant and smoke suppressant using TGA-FTIR and TGA-MS. Polym. Degrad. Stab. **98**, 1506–1514 (2013)
9. Ali, M., Qureshi, M.: Transportation fuels from catalytic co-pyrolysis of plastic wastes with petroleum residues: evaluation of catalysts by thermogravimetric analysis. Pet. Sci. Technol. **31**(16), 1665–1673 (2013)

Spatial Multicriteria GIS-Based Analysis to Anaerobic Biogas Plant Location for Dairy Waste and Wastewater Treatment and Energy Recovery (Barcelos, NW Portugal)

Cristóvão Rodrigues[1], Ana Cristina Rodrigues[1,2], Cândida Vilarinho[3],
Madalena Alves[1], and Joaquim Mamede Alonso[2(✉)]

[1] CEB – Centre of Biological Engineering, University of Minho, Braga, Portugal
[2] ProMetheus; Materials, Energy and Environment Research Unit – Instituto
Politécnico de, Viana do Castelo, Portugal
malonso@esa.ipvc.pt
[3] Metrics – Mechanical Engineering and Resource Sustainability Centre,
University of Minho, Guimarães, Portugal

Abstract. Intensification, concentration and specialization of dairy cow farms originate many activity by-products, such as wastewater and slurry which implies treatment costs, reduce environmental quality and promote social conflicts between the urban and rural population. These (by)products present a high potential for energy recovery from biogas produced in anaerobic digestion processes, thus contributing to sustainable development. This study is focused on the development of a GIS-based spatial support decision system that supports Multicriteria Analysis and Weighted Hierarchical Analysis (AHP) models aiming to identify site locations with appropriate conditions for the implementation of biogas production units, using waste and wastewater produced by the dairy cow farms in Barcelos municipality (NW Portugal). This spatially explicit model considers environmental, social and economic factors, as well as, legal location constraints tested with a consistent sensitivity analysis of the modelling processes and results. The results indicate sites with appropriate conditions for the location of biogas production units, considering the minimal distance to the dairy cow farms, electric and road networks, in higher areas with forest land cover and maximizing the distance from the urban spaces, as well as, from the river/ground water surfaces and agricultural/ecological reserves. In the next studies it´s relevant to explore biogas units dimensioning concerns, explore other complementary biomass sources in anaerobic (co)digestion solutions and expand collected spatial data to spatial decision support system during future biogas units operation phase.

Keywords: Bioenergy · Bioeconomy · Spatial location · Energy system

1 Introduction

Land suitability, scientific and technological development as well as agribusiness markets demand the intensification, concentration and specialization of vegetal and animal production activities at farm scale, particularly in dairy cow farming. These

© Springer International Publishing AG, part of Springer Nature 2019
J. Machado et al. (Eds.): HELIX 2018, LNEE 505, pp. 626–632, 2019.
https://doi.org/10.1007/978-3-319-91334-6_85

activities originate waste and wastewater with potential impacts on natural resources and environmental conditions, social pressures and conflicts, as well as increase legal, political and technical requirements in order to design and implement socio-economics and organizational sustainable solutions. Among the different waste and wastewater treatment and energy recovery systems, the interest and potential of anaerobic (co) digestion units for biogas production is highlighted in the context of biorrefineries and biobased economy. Circular and blue economy promote the integration of waste energy recovery systems in innovative periurban land use planning, local governance models and renewable energy policy.

Biological process of anaerobic digestion allow biogas production from organic waste leading to negative environmental impacts reduction, namely impacts associated with odours and greenhouse gas emissions as well as, the production of soil fertilizers with potential to be used in organic agriculture. The improvements in the physical, chemical and technological processes of biogas production require the assessment of the spatial-temporal patterns of waste and wastewater production, as well as the use of mass and energy balances at farm and local level aiming to identify and manage treatment and energy recovery processes and systems, which are related to biogas production unit location, dimensioning and supporting logistic associated/adequate solution in operational context.

The scientific and technological advances concerning the operation of anaerobic digestion plants and the optimization of the biogas production process should provide data to improve biogas plant location decision models at local or regional level. Environmental and energy policies with local socio-ecological systems analysis/ modelling require integrated solutions to locate, size and optimize organization and operation of anaerobic digestion plants for biogas production.

Spatial data on environmental conditions (climate, geology, soil, orography, hydrology, species distribution) and socio-ecological systems, including population, demography and economy data namely road, transport, wastewater and associated energy infrastructure, as well as biomass production and availability patterns are central to spatially explicit models development for biogas plant location. The complex and evolutionary nature of location analysis indicate the GIS based multicriteria spatial models that consider alternative and conflicting environmental (protection), social (responsibility and safety) and economic (viability) factors, as well as multiple objectives between sites, moments, operators and other stakeholders.

2 Methodology

A multidisciplinary, system and spatial approach aims to integrate different, complex and conflictual environmental, social and economic criteria considering natural resources, legal framework, social requirements and biogas unit locations costs. Spatial multicriteria decision-making development model include different (sub)tasks related to: (1) *scope and problem definition* using available data and interviews with farmers, researchers, political decision makers and other dairy stakeholders in Barcelos Municipality area, NW Portugal); (2) *introduction and definition of informative multicriteria/factors based on extended bibliographical review/research* and thematic and spatial knowledge experts resulting in environmental (5), social (7) and economic (5) criteria (see Table 1); (3) *reference and thematic spatial data collected was submitted to spatial and format transformations* (data collection); primary data related to dairy cow farms produced by local farmers cooperative, after georeference, data was processed using production technical parameters and spatial interpolations techniques; the multidisciplinary nature of spatial datasets implied data parametrization using amplitudes and histograms values variability analysis; all factors are reclassified in categorical scale from 0 (without conditions to locate biogas plant), 1 (minimum conditions) to 5 (optimum conditions) using spatial modelers and energy planning expert knowledge; (4) *spatial modelling analysis process* includes distance analysis, cluster and raster overlay operations; (5) *analytic hierarchy process (AHP)* used in locations sites GIS-based models compares relative importance between two/pair of factors (see Table 1) using a comparison categorical scale that's permits define environmental, social and economic factors weight in biogas plant location modelling;

Table 1. Environmental, social and economic criteria selected to support GIS-based biogas site location model.

Type	Name	Bibliographical References	AHP weight
Economic	Dairy cow farm (distance)	[16, 17, 27]	0.06
	Main grid road (distance)	[7, 8, 10–12, 14, 16, 18, 20–24, 27]	0.04
	Secondary grid road (distance)	[7, 8, 10–12, 14, 16, 18, 20–24, 27]	0.07
	National electric grid (distance)	[12, 14, 17, 22, 23, 25, 27]	0.06
	National grid of electricity injection points (distance)	[12, 14, 17, 22, 23, 25, 27]	0.06
Social	Land cover and land use (classes)	[8, 11, 17, 20, 23–25]	0.05
	Forest areas (distance)	[18]	0.05
	Build areas (distance)	[4]	0.15
	Urban areas (distance)	[4, 8, 11, 12, 14, 16, 18, 20, 22–24];	0.09
	Industrial areas (distance)	[22]	0.03
Environmental	National ecological reserve area (classes)	[22]	0.05
	National Agriculture Reserve Area (classes)	[4, 18]	0.06
	Main and secondary rivers (distance)	[4, 8, 22, 24, 25]	0.02
	rivers (distance)	[4, 14, 17, 18, 20, 25];	0.13
	Altitude (classes)	[8, 23]	0.02
	Slopes (classes)	[4, 11, 12, 16–18, 20, 22–25];	0.02
	Sun aspects (classes)	[4, 23]	0.03

(6) the first results *indicate the suitable areas to locate a biogas plant*; a sensitive analysis of spatial model was implemented by modifying in each test ±20% the weight in environmental, social and economic factors in final results (7) *permitting analyze the adequate areas, develop a critical methodological analysis and future improvements.*

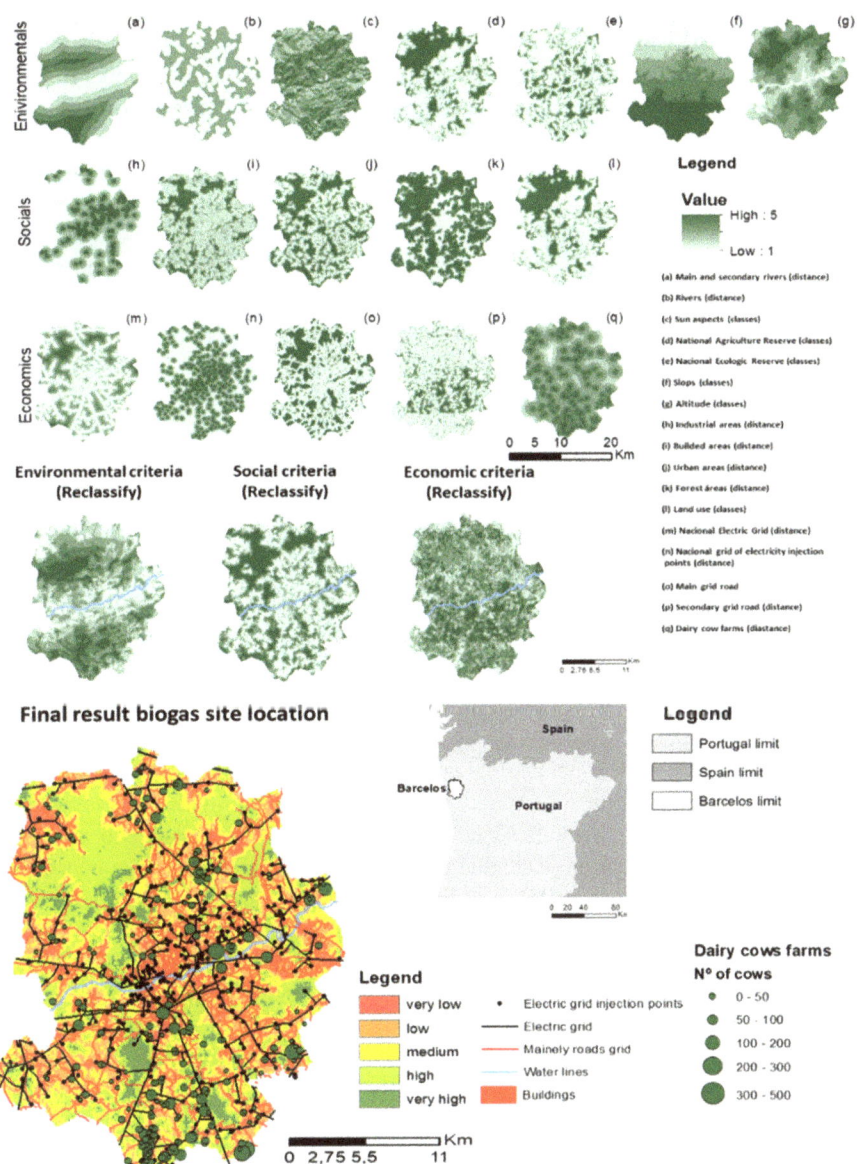

Fig. 1. Environmental, social and economic multicriteira maps and final result in biogas unit site location related to dairy cow farms (location and dimension) and road/electric infrastructures.

3 Results

Barcelos Municipality (379 km² and 120 391 inhabitants; 2011) is located in the lower third and is divided by Cávado river with granite soils origin, light texture and good physical structure resulting in high agro-forestry suitability. This area presents a dynamic economy, demography and traditional agricultural matrix. In the last decades the mini or micro property structure led to the development of local dairy cow sector. However, the increasing pressures from public environmental policies and public health, as well as the evolution of markets resulted in a decrease in the number of farms, increased size and productive concentration in larger farms. At the same time, the production intensification increased conflicts with urban activities and inhabitants, and associated organizational, spatial and sectorial planning solutions, licensing and waste and wastewater treatment options.

The weighted sum of the environmental, social and economic criteria (AHP) shows that: (i) the distance to the main and secondary rivers as well as the limitations in deep alluvial soils at lower altitudes and in agricultural and ecological reserves determine the possibility of preferential biogas units location in mid-altitude spaces; (ii) there is a strong complementarity between the most densely populated areas (urban areas) and higher energy consumption (industrial zones), forest areas and other classes of land cover/land use; (iii) the number and proximity to (the points of injection) in the electricity network, the road network and the distributed farms result in suitable areas throughout the territory. The final results validated by the sensitivity analysis, indicate considerable areas with high/very high suitability to biogas units location. More restrictive criteria or final results reclassification can help in finding the locations with optimal considered conditions (Fig. 1).

4 Conclusions

Sustainable development requires governance models and organizational solutions located in adequate sites with technological and functional operations. In planning and project phases the development of spatial multicriteria decision models in biogas plant location considers environmental, social and economic factors related to anaerobic processes of biogas production, technical parameters and operational costs concerning the technical and economic feasibility of the biogas production unit. The results show the model interest/capacity in finding suitable sites for installation of distribute systems of biogas units in higher elevations, forest areas, near the farms, road network, electric injection points but far away from the water resources, and outside of the agricultural and ecological reserves.

In next phases/studies, it's important to consider the design and define the number of biogas units associated with multiple objectives, namely considering elements of spatial econometrics in resources allocation, investment or operating costs. At the same time, other local sources of biomass should be identified to improve anaerobic (co) digestion processes, to calculate local biomass balances, to improve functional integration between distributed biogas plants, energy distribution systems and local energy consumption patterns. These data and knowledge are critical to improve spatial decision support systems in location and dimensioning phases as well as, to support the management and operation of future biogas production systems/local units.

References

1. Alçada-Almeida, L., Coutinho-Rodrigues, J., Current, J.: A multiobjective modeling approach to locating incinerators. Soc.-Econ. Plan. Sci. **43**(2), 111–120 (2009)
2. Municipio de Barcelos: Plano Director Municipal do Concelho de Barcelos (2015)
3. Brito, L.M., Alonso, J.M., Rey-Graña, J.: Gestão de efluentes nas explorações leiteiras do Entre Douro e Minho. Revista de Ciências Agrárias **36**(2), 80–93 (2011)
4. Cabral, A.V.: Análise multicritério em sistemas de informação geográfica para a localização de aterros sanitários: O caso da região Sul da Ilha de S. Tiago, Cabo Verde. Diss. Mestrado em Gestão do Território, FCSH-Universidade Nova de Lisboa, 113 p. (2012)
5. Censos: Instituto Nacional de Estatística (2001)
6. Censos: Instituto Nacional de Estatística (2011)
7. Comber, A., Dickie, J., Jarvis, C., Phillips, M., Kevin Tansey, K.: Locating bioenergy facilities using a modified GIS-based location–allocation-algorithm: considering the spatial distribution of resource supply. Appl. Energy **154**, 309–316 (2015)
8. Estratégia Nacional para os Efluentes Agro-pecuários e Agro-industriais: Ministério do Ambiente, do Ordenamento do Território e do Desenvolvimento Regional (2007)
9. Ferretti, V., Pomarico, S.: Integrated sustainability assessments: a spatial multicriteria evaluation for siting a waste incinerator plant in the province of Torino (Italy). Environ. Dev. Sustain. **14**(5), 843–867 (2012)
10. Furtado, M.: Gestão de chorumes bovinos, fluxos e disponibilização de nutrientes em explorações leiteiras do concelho de Barcelos. ESA - Instituto Politécnico de Coimbra (2012)
11. Gonçalves, C.S.: Avaliação do Potencial de Geração de Biogás a partir de Resíduos de Boviniculturas na Área Metropolitana do Porto. Diss. Mestrado em Eng. Ambiente, FEUP-Universidade do Porto, 135 p. (2010)
12. Höhn, J., Lehtonen, E., Rasi, S., Rintala, J.: A Geographical Information System (GIS) based methodology for determination of potential biomasses and sites for biogas plants in Southern Finland. Appl. Energy **113**, 1–10 (2014)
13. Latinopoulos, D., Kechagia, K.: A GIS-based multi-criteria evaluation for wind farm site selection. A regional scale application in Greece. Renew. Energy **78**, 550–560 (2015)
14. Ma, J., Scotta, N., DeGloria, S., Lembo, A.: Siting analysis of farm-based centralized anaerobic digester systems for distributed generation using GIS. Biomass Bioenergy **28**(6), 591–600 (2005)
15. Marchettini, N., Ridolfi, R., Rustici, M.: An environmental analysis for comparing waste management options and strategies. Waste Manag **27**(4), 562–571 (2007)
16. Noorollahi, Y., Yousefi, H., Mohammadi, M.: Multi-criteria decision support system for wind farm site selection using GIS. Sustain. Energy Technol. Assess. **13**, 38–50 (2016)
17. Panichelli, L., Gnansounou, E.: GIS-based approach for defining bioenergy facilities location: a case study in Northern Spain based on marginal delivery costs and resources competition between facilities. Biomass Bioenergy **32**(4), 289–300 (2008)
18. Peng, L., Chen, W., Li, M., Bai, Y., Pan, Y.: GIS-based study of the spatial distribution suitability of livestock and poultry farming: the case of Putian, Fujian, China. Comput. Electron. Agric. **108**, 183–190 (2014)

19. Sánchez-Lozano, J.M., Teruel-Solano, J., Soto-Elvira, P.L., García-Cascales, M.S.: Geographical Information Systems (GIS) and Multi-Criteria Decision Making (MCDM) methods for the evaluation of solar farms locations: case study in south-eastern Spain. Renew. Sustain. Energy Rev. **24**, 544–556 (2013)

20. Shah, S.A., Wani, M.A.: Geospatial based approach for enhancing environment sustainability of Srinagar city - a study on solid waste disposal. Int. J. u-, e-Service, Sci. Technol. **8** (3), 289–302 (2014)

21. Kientga, S.: Contribution du SIG à l'analyse des liens déchets-santé en milieu urbain dans les pays en développement. Cas de deux secteurs de la ville de Ouagadougou Burkina Faso. École Polytechnique Fédérale de Laussanne (2008)

22. Spigolon, L.M.G., Souza, N.C., Larocca, A.P.C., Giannotti, M.A., Russo, M.A.T., Alonso, J. M.: Seleção de áreas adequadas para a instalação de aterro sanitário utilizando SIG e analise multicritério - estudo de caso: UGRHI 5 (Piracica-ba/Capivari/Jundiaí). Anais XVII Simpósio Brasileiro de Sensoriamento Remoto   SDSR, João Pessoa-PB, Brasil, pp. 1983–1990 (2015)

23. Suárez-Vega, R., Santos-Peñate, D.R., Dorta-González, P., Rodríguez-Díaz, M.: A multi-criteria GIS based procedure to solve a network competitive location problem. Appl. Geogr. **31**(1), 282–291 (2011)

24. Sultana, A., Kumar, A.: Optimal siting and size of bioenergy facilities using geographic information system. Appl. Energy **94**, 192–201 (2012)

25. Tavares, G., Zsigraiová, Z., Semião, V.: Multi-criteria GIS-based siting of an incineration plant for municipal solid waste. Waste Manag **31**, 1960–1972 (2011)

26. Tavares, D.: Modelização cartográfica para a localização ótima de um aterro sanitário na Ilha de S. Tiago. Relatório de projecto apresentado em cumprimento parcial dos requisitos para obtenção do grau de Mestre em Ciência e Sistemas de Informação Geográfica pela Universidade Nova de Lisboa e Universidade de Cabo Verde, 124 p. (2012)

27. Thompson, E., Wang, Q., Li, M.: Anaerobic digester systems (ADS) for multiple dairy farms: a GIS analysis for optimal site selection. Energy Policy **61**, 114–124 (2013)

28. Ward, A., Hobbs, P., Holliman, P., Jones, D.: Optimisation of the anaerobic digestion of agricultural resources. Bioresour. Technol. **99**(17), 7928–7940 (2008)

29. Zubaryeva, A., Zaccarelli, N., Giudice, C., Zurlini, G.: Spatially explicit assessment of local biomass availability for distributed biogas production via anaerobic co-digestion – mediterranean case study. Renew. Energy **39**(1), 261–270 (2012)

30. Copeland, C.: Animal Waste and Hazardous Substances: Current Laws and Legislative Issues. Congressional Research Service, Washington, D.C. (2010)

31. Correia, R.B.: Modelação cartográfica em ambiente SIG de suscetibilidade à erosão hídrica dos solos, caso da bacia da Ribeira dos Picos, Santiago (Cabo Verde), p. 162. Universidade de Coimbra, Tese de Mestrado (2007)

32. Figueiredo, R.F.: Modelação cartográfica em ambiente SIG para apoio à decisão: aplicação ao estudo da afetação potencial de usos do solo no sector Norte do Maciço Marginal de Coimbra. Diss. de Mestrado. Univ. Coimbra. 204 p. (2001)

33. Rodrigues, A., Guimarães, J., Oliveira, C.: Rentabilidade das explorações leiteiras em Portugal - dados técnicos e económicos. In: Livro de Resumos, V Jornadas de Bovinicultura, 30–31 Março, pp. 109–129. IAAS-UTAD, Vila Real (2012)

34. Saaty, T.: Método de Análise Hierárquica. McGraw-Hill, Makron, São Paulo (1991). 367 p.

A Brief Assessment on the Application of Torrefaction and Carbonization for Refuse Derived Fuel Upgrading

Catarina Nobre[1](✉) [iD], Margarida Gonçalves[1] [iD],
and Cândida Vilarinho[2] [iD]

[1] MEtRICs, Departamento de Ciências e Tecnologia da Biomassa,
Faculdade de Ciências e Tecnologia, Universidade Nova de Lisboa,
2829-516 Caparica, Portugal
cp.nobre@campus.fct.unl.pt
[2] MEtRICs, Departamento de Engenharia Mecânica, Escola de Engenharia,
Universidade do Minho, 4804-533 Guimarães, Portugal

Abstract. Refuse derived fuel (RDF) represents a very robust and endogenous resource that has the potential to minimize landfilling of solid waste, reduce greenhouse gas emissions through its biogenic component and contribute to national energy provision whilst diversifying solid fuel supplies. This waste derived fuel can be produced from different waste streams after significant mechanical and biological processing. In spite of their processing, these waste derived fuels still have a high degree of heterogeneity, presenting variable fuel properties and some negative characteristics such as high moisture, ash and chlorine contents. Although RDF is used for energy generation in some high energy demanding industrial applications or dedicated energy recovery facilities, its physical-chemical characteristics can result in significant technical and environmental problems that may benefit from an upgrading treatment. Torrefaction and carbonization are thermal treatments that have the potential to upgrade RDF, producing a waste derived char with reduced moisture and chlorine contents, more homogeneous and friable, which are characteristics of great importance for feeding systems in gasification and combustion facilities. Using waste derived chars could result in major environmental and waste management advantages, with potential to help with the waste management crisis, reducing waste volume and corresponding to the present European guidelines for energy recovery from wastes, fitting perfectly in the concept of circular economy.

Keywords: RDF · Energy recovery · Torrefaction · Carbonization
Waste derived char

1 Introduction

Waste generation has become one of the most pressing issues of our times. Resource consumption increases as the population and its demands grow, yielding unsustainable amounts of waste as significant as 1.3 billion tons per year of municipal solid waste

© Springer International Publishing AG, part of Springer Nature 2019
J. Machado et al. (Eds.): HELIX 2018, LNEE 505, pp. 633–640, 2019.
https://doi.org/10.1007/978-3-319-91334-6_86

(MSW), worldwide. This value tends to increase with projected values of 2.2 billion tons by the year 2025, and 9.5 billion by 2050 [1–3].

Waste-to-Energy (WtE) conversion technologies are a very promising way of using MSW to produce renewable energy, namely because MSW represents a very abundant and endogenous feedstock, granting energy provision and a reduced environmental impact. Waste incineration is the most applied thermal disposal technique with energy recovery, but it still has major drawbacks mostly related to harmful emissions, low energy efficiencies and poor quality of the used fuels [4, 5].

Currently waste management strategies involved recycling of a significant part of the wastes produced from various sources but there is still an organic non-recyclable fraction, composed mainly of textiles, paper, cardboard, plastics and non-recyclable rubber, which is converted to refuse derived fuel (RDF) or landfilled. Some years ago, landfilling was the only final destination for these wastes, however, nowadays, this fraction is mostly directed to RDF production, for use as fuel in cement kilns. RDF production follows the principles of European energy and environmental policies, which encourage use and demand of renewable and sustainable energy sources whilst reducing greenhouse gases emissions.

2   RDF Main Characteristics and Applications

In 1973, Dr. Jerome Collins named the secondary fuel produced from processed MSW as Refuse Derived Fuel [6]. As such, RDF is the broad designation for fuels obtained from waste not obeying specific technical characteristics and this designation is generally associated with low quality fuels [7]. To provide RDF of a classification based on specific technical criteria, the European Commission has drawn up a set of standards for this fuel and named it as Solid Recovered Fuel (SRF) [8].

RDF can be defined as an alternative fuel that arises from the treatment of several waste streams such as MSW, regular industrial waste (RIW), construction and demolition waste (CDW) or sludge produced from potable water supply and sanitation in treatment plants as well as other non-hazardous waste flows [9]. It results from operations such as dehydration, separation and shredding, and includes the removal of the non-combustible materials such as metals and minerals that are directed to specific valorization pathways. Figure 1 represents two RDF samples produced from MSW and RIW.

These RDF/SRF production operations usually take place at mechanical and biological treatment plants (MBT) mainly for MSW processing, or mechanical treatment facilities (MT) for RIW or CDW processing [10]. As such, when compared to MSW, RDF has better fuel properties including a higher homogeneity and an improved calorific value.

RDF is mainly used for energy recovery purposes through thermochemical conversion in WtE plants mostly through incineration or co-incineration. Gasification is another WtE technology with several benefits when compared to combustion, mainly because syngas is easier to handle, has lower combustion associated emissions than the original RDF and offers a wide range of outcomes, such as direct burning, advanced liquid fuels production or chemical synthesis [11].

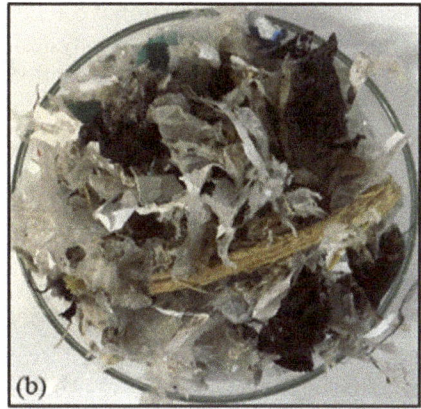

Fig. 1. Two RDF samples: (a) RDF from MSW, (b) RDF from RIW.

RDF characteristics vary greatly depending on their source and production line scheme, making it very important to evaluate its physical and chemical properties and the performance of the thermal treatment that this fuel will undergo [12]. The distribution of proximate analysis parameters will influence the performance of the burning bed and the heat distribution along the conversion system. Table 1 presents typical values for RDF fuel characterization.

Table 1. Proximate analysis results for different samples of RDF from different sources *(adapted from [12–15])*.

Sample	RDF	RDF	RDF	RDF	RDF	RDF	SRF
Waste stream	-	MSW	ICW	MSW	MSW	MSW	MSW
Moisture (%wt)	8.33	0.99	1.30	3.70	1.70	19.70	28.11
Ash (%wt)	0.37	10.04	16.10	18.90	17.70	20.40	9.65
Volatile matter (%wt)	71.11	79.10	71.90	67.60	73.60	49.10	55.99
Fixed carbon (%wt)	20.19	9.60	10.70	9.80	7.00	10.80	6.36
HHV (MJ/Kg)	14.04	23.90	22.35	22.30	24.60	13.90	14.78

SRF/RDF produced from MSW generally presents larger ranges of proximate composition and calorific value due to its highly heterogeneous composition, affected by source and seasonal variability. The direct use of these wastes in energy recovery systems can lead to operational problems or toxic emissions, namely corrosion due to the formation of HCl and production of highly hazardous products such as dibenzo-dioxins (PCDD) and dibenzofurans (PCDF) [16]. Also, high chlorine combined with a high ash content can lead to slagging and fouling phenomena in boilers and gasifiers [17]. On the other hand, the presence of polymeric fractions can create problems of clogging in the feeding systems and char accumulation in the fluidized beds.

Therefore, the use of RDF for combustion or gasification units currently in use, requires a characterization of its physical and chemical properties as well as their thermal behavior. Moreover, RDF negative characteristics could be minimized by fuel pre-treatment technologies, such as torrefaction or carbonization. These thermal conversion processes have proven to improve biomass and waste fuel properties, thereby reducing operational problems and environmental impacts associated with the application of these feedstocks in WtE installations [10, 11, 18].

3   Torrefaction and Carbonization for RDF Upgrading

Torrefaction corresponds to a thermal treatment that takes place at operating temperatures of typically 200 to 300 °C in the absence of oxygen [11]. This process can greatly improve the physical, chemical and energy characteristics of different raw materials, contributing to their homogenization, increasing their grindability, density, hydrophobic character and gross calorific value while significantly reducing its volume [18–20]. Figure 2 shows the significant increase in homogeneity and volume reduction after submitting an industrial RDF to torrefaction at 300 °C for 30 min.

Fig. 2. Impact of torrefaction in homogeneity and volume reduction. (a) RDF as received, (b) RDF char.

Chosen torrefaction conditions (temperature and residence time) are pivotal for the final product. Typically, temperatures above 300 °C yield chars with substantial ash content, they favor volatilization of a larger proportion of carbon and the formation of PCDDs and PCDFs [19]. At low temperatures (around 200 °C) there are no significant physical changes of the raw material regarding density or homogeneity. But between 200–300 °C, water is volatilized and functional groups at the surface of the raw material are eliminated, causing permanent changes in the structure and properties of the torrefied materials. The decrease in the O/C ratio, increase in the hydrophobic character, enhanced heating value and increase of density are relevant modifications for the storage and final use of these fuels.

Waste derived char is being studied as an alternative fuel mainly because of its higher quality and better fit as a substitute in existing coal-fired power plants, presenting itself as an interesting alternative to other carbonaceous materials and fossil fuels [21, 22].

Research on RDF torrefaction is mainly focused on RDF char characterization, process emissions or fuel upgrading for application in advanced thermal conversion processes, such as gasification.

Białowiec et al. [10] tested RDF torrefaction at different temperatures and found that the process could reduce moisture content about 21.5%, consequently increasing the lower heating value (LHV) by 22.5%. The authors concluded that this method could increase the attractiveness of RDF as a fuel while homogenizing RDF characteristics for market purposes.

Torrefaction as a way to improve SRF properties in order to optimize its use in gasification was studied by Recari et al. [11]. The authors applied torrefaction temperatures between 290–320 °C and verified that torrefied SRF showed improved gasification parameters, such as lower tar, higher carbon conversion and higher H_2/CO ratio. Moreover, torrefaction reduced the chlorine content in SRF leading to lower HCl concentrations in the producer gas.

Evaluation of toxic emissions is a major concern in the evaluation of waste derived fuels. As such, Edo et al. [16] assessed the emissions of the torrefaction process in different wastes, including RDF. This work showed that the chars had reduced chlorine concentrations, which implied a reduction in the potential for PCDD and PCDF formation, and this was experimentally demonstrated by combustion of the torrefied RDF and analysis of the effluent gases. This study reinforces the concept that applying the torrefaction process to waste fuels can significantly reduce the environmental impacts of their energetic valorization.

Carbonization is another a thermal process used to upgrade solid fuels, that also occurs in an inert atmosphere but at a higher temperature range than torrefaction (300–500 °C), yielding biomass or waste derived char [23]. Although energy valorization is a major application of the waste derived char, carbonization can also be seen as a pre-treatment before landfilling since the carbonized RDF has substantial advantages relatively to its density, hydrophobic nature and homogeneity when compared with raw RDF [24]. Some studies also encompass the use of waste derived char as an adsorbent and precursor for activated carbon [25–27].

The properties of RDF char obtained through carbonization were investigated by Haykiri-Acma et al. [28]. Carbonization was carried out in a temperature range of 400–900 °C and the increase in carbonization temperatures not only led to a decrease in volatile content, but also to a loss in calorific value. The authors found that the ash content increased very significantly with temperature, exceeding the increase in fixed carbon, and therefore negatively impacting the HHV. These results indicate that RDF carbonization should not be done in temperatures above 400 °C. Up to this temperature around half of the oxygen content can be removed, there is a significant increase in fixed carbon and the calorific value of the RDF chars reach their maximum value.

Both torrefaction and carbonization carried out in optimal conditions represent a way to produce upgraded RDF chars that have a significant heating value, lower moisture content and high density. These new characteristics allow this char to be used not only as a solid biofuel but as carbon black or as an activated carbon precursor. If the energetic or material valorization pathways are not available for this material, it will also present advantages when landfilled mainly due to its lower volume when compared to the untreated RDF. The choice of the temperature operating range will mainly depend on the RDF initial composition, being that torrefaction can be more adequate for RDF with high contents of lignocellulosic materials while carbonization leads to better results for RDF with a large polymeric fraction.

4 Concluding Remarks

The production of waste derived fuels in Europe has been quickly rising, bringing assessable economic and ecologic outcomes such as a significant reduction in CO_2 emissions and fossil fuel savings. Nevertheless, the thermal applications using SRF/RDF are conditioned by physical-chemical properties of these wastes, namely their heterogeneity and high moisture, chlorine and ash contents. Torrefaction and carbonization have proven to be adequate processes to upgrade solid fuels such as biomass, and are currently being thoroughly studied for their application in waste derived fuels. These pre-treatments yield chars with increased homogeneity, lower chlorine content and increased heating value. These improved characteristics could eliminate some of the operational and environmental problems associated with the use of raw RDF.

Improving energy recovery whilst adding material value to RDF is also an expanding area, where carbonization of the waste derived fuel yielding a char, plays an important role in transforming the fuel into a more homogeneous and appealing product. Using carbonized RDF to extract valuable compounds, heavy metals or dyes, represents a whole new perspective for these waste derived fuels, and also a way to reclaim landfill space.

References

1. Bolyard, S.C., Reinhart, D.R.: Application of landfill treatment approaches for stabilization of municipal solid waste. Waste Manag. **55**, 22–30 (2016)
2. Kawai, K., Tasaki, T.: Revisiting estimates of municipal solid waste generation per capita and their reliability. J. Mater. Cycles Waste Manag. **18**, 1–13 (2015)
3. Beyene, H.D., Werkneh, A.A., Ambaye, T.G.: Current updates on waste to energy (WtE) technologies: a review. Renew. Energy Focus **24**, 1–11 (2018)

4. Malkow, T.: Novel and innovative pyrolysis and gasification technologies for energy efficient and environmentally sound MSW disposal. Waste Manag. **24**, 53–79 (2004)
5. Násner, A.M.L., Lora, E.E.S., Palacio, J.C.E., et al.: Refuse Derived Fuel (RDF) production and gasification in a pilot plant integrated with an Otto cycle ICE through Aspen plus™ modelling: thermodynamic and economic viability. Waste Manag. **69**, 187–201 (2017)
6. Casado, R.R., Rivera, J.A., García, E.B., et al.: Classification and characterisation of SRF produced from different flows of processed MSW in the Navarra region and its co-combustion performance with olive tree pruning residues. Waste Manag. **47**, 206–216 (2016)
7. Nasrullah, M., Vainikka, P., Hannula, J., Hurme, M.: Elemental balance of SRF production process: solid recovered fuel produced from commercial and industrial waste. Fuel **145**, 1–11 (2015)
8. European Committee for Standardisation. CEN/TC343/WG 2, Solid recovered fuels – specifications and classes. Draft European Standard (2005)
9. Rada, E.C., Andreottola, G.: RDF/SRF: which perspective for its future in the EU. Waste Manag. **32**, 1059–1060 (2012)
10. Białowiec, A., Pulka, J., Paweł, S., et al.: The RDF/SRF torrefaction: an effect of temperature on characterization of the product – carbonized refuse derived fuel. Waste Manag. **70**, 91–100 (2017)
11. Recari, J., Berrueco, C., Puy, N., et al.: Torrefaction of a solid recovered fuel (SRF) to improve the fuel properties for gasification processes. Appl. Energy **203**, 177–188 (2017)
12. Hernanéz-Atonal, F.D., Ryu, C., Sharifi, V.N., Swithenbank, J.: Combustion of refuse-derived fuel in a fluidised bed. Chem. Eng. Sci. **62**, 627–635 (2007)
13. Duan, F., Liu, J., Chyang, C.S., et al.: Combustion behavior and pollutant emission characteristics of RDF (refuse derived fuel) and sawdust in a vortexing fluidized bed combustor. Energy **57**, 421–426 (2013)
14. Grammelis, P., Basinas, P., Malliopoulou, A., Sakellaropoulos, G.: Pyrolysis kinetics and combustion characteristics of waste recovered fuels. Fuel **88**, 195–205 (2009)
15. Agraniotis, M., Nikolopoulos, N., Nikolopoulos, A., et al.: Numerical investigation of solid recovered fuels' co-firing with brown coal in large scale boilers – evaluation of different co-combustion modes. Fuel **89**, 3693–3709 (2010)
16. Edo, M., Skoglund, N., Gao, Q., et al.: Fate of metals and emissions of organic pollutants from torrefaction of waste wood, MSW, and RDF. Waste Manag. **68**, 646–652 (2017)
17. Silva, R.B., Fragoso, R., Sanches, C., et al.: Which chlorine ions are currently being quantified as total chlorine on solid alternative fuels? Fuel Process. Technol. **128**, 61–67 (2014)
18. Yuan, H., Wang, Y., Kobayashi, N., et al.: Study of fuel properties of torrefied municipal solid waste. Energy Fuels **29**(8), 4976–4980 (2015)
19. Chew, J.J., Doshi, V.: Recent advances in biomass pretreatment – torrefaction fundamentals and technology. Renew. Sustain. Energy Rev. **15**, 4212–4222 (2011)
20. Verhoeff, F., Adell i Arnuelos, A., Boersma, A.R., et al.: Torrefaction technology for the production of solid bioenergy carriers from biomass and waste. TorTech Project Report (2011). https://www.ecn.nl/docs/library/report/2011/e11039.pdf
21. Vassilev, S.V., Braekman-Danheux, C., Laurent, P.: Characterization of refuse-derived char from municipal solid waste 1. Phase-mineral and chemical composition. Fuel Process. Technol. **59**(2–3), 95–134 (1999)
22. Hwang, I.H., Kawamoto, K.: Survey of carbonization facilities for municipal solid waste treatment in Japan. Waste Manag. **30**, 1423–1429 (2010)
23. Qi, J., Zhao, J., Xu, Y., et al.: Segmented heating carbonization of biomass: yields, property and estimation of heating value of chars. Energy **144**, 301–311 (2018)

24. Hwang, I.H., Matsuto, T., Tanaka, N., et al.: Characterization of char derived from various types of solid wastes from the standpoint of fuel recovery and pretreatment before landfilling. Waste Manag. **27**, 1155–1166 (2007)
25. Nakagawa, K., Tamon, H., Suzuki, T., Nagano, S.: Preparation and characterization of activated carbons from refuse derived fuel (RDF). J. Porous Mater. **9**, 25–33 (2002)
26. Wu, F.C., Wu, P.H., Tseng, R.L., Juang, R.S.: Use of refuse-derived fuel waste for the adsorption of 4-chlorophenol and dyes from aqueous solution: Equilibrium and kinetics. J. Taiwan Inst. Chem. Eng. **45**, 2628–2639 (2014)
27. Buah, W.K., Williams, P.T.: Activated carbons prepared from refuse derived fuel and their gold adsorption characteristics. Environ. Technol. **31**, 125–137 (2010)
28. Haykiri-Acma, H., Kurt, G., Yaman, S.: Properties of biochars obtained from RDF by carbonization: Influences of devolatilization severity. Waste Biomass Valor. **8**(3), 539–547 (2017)

Waste-to-Energy Technologies Applied for Refuse Derived Fuel (RDF) Valorisation

André Ribeiro[1]([✉]), Margarida Soares[1], Carlos Castro[1], André Mota[1],
Jorge Araújo[1], Cândida Vilarinho[2], and Joana Carvalho[1,2]

[1] CVR – Centro para a Valorização de Resíduos, Guimarães, Portugal
aribeiro@cvresiduos.pt
[2] Departamento de Engenharia Mecânica,
University of Minho, Guimarães, Portugal

Abstract. Refuse Derived Fuel (RDF) is a solid fuel made after basic processing steps or techniques that increase the calorific value of municipal solid waste (MSW), commercial or industrial waste materials. Therefore, energy production from RDF can provide economic and environmental benefits as it reduces the amount of wastes sent to landfill and allows the energy recovery from a renewable source.

In this work, it was studied the application of waste-to-energy technologies to RDF valorisation, namely pyrolysis and gasification. This study intended to evaluate the effect of temperature and different technologies (gasification and pyrolysis) in gas production, gas composition and mass conversion of RDF. Experiments of RDF gasification and pyrolysis were performed in a laboratory scale fixed bed gasifier, under different conditions.

The effect of reaction temperature was studied at 450, 600 and 750 °C in pyrolysis experiments and at 750 °C and 850 °C in gasification. Results showed that, for the same operational conditions, pyrolysis was more efficient at 750 °C. At this temperature, it was obtained a syngas of 11.2 MJ/m^3 and a specific gas production of 0.43 m^3 syngas/kg RDF. Results also proved that, for the same operational conditions, the rise of temperature improved gas production ratio (Nm^3/kg RDF), gas low heating value (LHV) and mass conversion. Regarding to gas production ratio the utilization of air at equivalence ratio (ER) of 0.6 induced the formation of 1.5 m^3 gas/kg RDF. Differently, steam gasification only allowed the production of 0.5 m^3 gas/kg RDF. Mass conversion and carbon conversion achieved almost 100% in air gasification at highest molar ratio.

Keywords: Refuse Derived Fuel (RDF) · Waste-to-energy · Gasification
Pyrolysis

1 Introduction

SRF or Solid Recovered Fuel is defined by EN 15359/2011 as solid fuel made from non-hazardous waste used for energy recovery in incineration and co-incineration plants that can be production by specific waste, municipal solid waste, industrial waste, commercial waste, construction and demolition waste and sewage sludge [1]. Nevertheless, RDF production is most related to municipal solid wastes (MSW) treatment [2].

© Springer International Publishing AG, part of Springer Nature 2019
J. Machado et al. (Eds.): HELIX 2018, LNEE 505, pp. 641–647, 2019.
https://doi.org/10.1007/978-3-319-91334-6_87

According to Edo-Alcón et al. [3], in EU the amount of RDF produced from MSW its about 12 million tonnes per year [3]. Therefore, the utilization of RDF as a robust endogenous resource can respond to a range of problems associated with waste management, making it possible to: minimize the deposition of MWS in landfill; contribute to national self-sufficiency in energy production; contribute to the reduction of greenhouse gases through the reduction of CH_4 emission from landfill; diversify the sources of fuels [4].

Concerning to waste-to-energy several technologies presenting an alternative to conventional combustion and incineration processes are currently being developed. These technologies include mainly the wet oxidation process, pyrolysis and the gasification process [5]. Among them, gasification and pyrolysis could play an import role in greenhouse gas (GHG) emissions cutting due to its pollution minimization effects, higher overall efficiency and ability to produce a syngas with multiple exploration methods [6].

In general, pyrolysis represents a process of thermal degradation of the waste in the total absence of air that produces recyclable products, including char, oil/wax and combustible gases. Pyrolysis has been used to produce charcoal from biomass for thousands of years. According to Velghe et al. [7], pyrolysis temperature varies from 300 to 900 °C, but the typical running temperature is around 500–550 °C with liquid products as major portion of products. At temperatures higher than 700 °C, syngas is the vital product. Most of the research studies paid more attention to the liquid and syngas products than the char due to the fact that the oil and syngas are more valuable. Additionally, the yields and composition of pyrolysis oil and syngas are mainly changing with temperature [7].

On the other hand, conventional gasification is a thermal process, which converts carbonaceous materials, such as RDF, into syngas using a gasifying agent or agents as steam, carbon dioxide or air [8]. These gasifying agents reacts with chars, tars, and gases to produce syngas [9]. The main syngas products (carbon monoxide (CO) and hydrogen (H_2)) are used to produce heat and electricity or are converted into liquid hydrocarbons [9]. Syngas composition also presents methane (CH_4) and carbon dioxide (CO_2) that are produced and play an import role in gasification reactions. Gasification occurs through a sequence of complex thermochemical reactions that include: drying zone, pyrolysis or thermal decomposition zone, partial oxidation or combustion zone and reduction zone (utilization of gasifying agents) [9].

Therefore, RDF valorisation has a twofold aim. The first is environmental, and it is related to the constitution of an alternative of disposal this material in landfill; the second is to provide an alternative energy source that can be used for electric or heat production [10].

2 Materials and Methods

2.1 Refused Derived Fuel (RDF) Characterization

Refused Derived Fuel (RDF) used in the present paper was collected from a local company. According to the company, this RDF is a mixture of MSW from mechanical and biological treatment (MBT) plants and non-hazard industrial wastes. Physical-Chemical characterization of RDF was performed according to EN 15359:2011 (Solid recovered fuels—Specifications and classes).

2.2 Pyrolysis and Gasification of Refuse Derived Fuel (RDF)

Pyrolysis and gasification experiments were performed for approximately 120 min in a laboratory scale fixed bed reactor with 36 L of total volume using an initial feedstock of 450 g of RDF. This reactor consisted of a vertical stainless steel tube with an inner diameter of 15 cm.

The effect of temperature in pyrolysis and gasification was evaluated in syngas composition, syngas heating value (kJ/m3 N), specific syngas flow rate (m3 N/kg RDF) and carbon conversion efficiency. The effect of reaction temperature in pyrolysis was studied at 450, 600 and 750 °C. The effect of reaction temperature in gasification was studied at 750 °C and 850 °C with feeding molar ratios of 1.0 (S/B) with steam and 0.6 (ER) with air.

Lower Heating Value of syngas produced was calculated according to Eq. 1. Molar percentages of syngas components were presented in this study are in N2 basis free, in other words, without nitrogen. Carbon conversion efficiency (CCE) (%) was calculated using Eq. 2.

$$LHV = 1/1000 * (CO * 126.36 + H_2 * 107.98 + CH_4 * 358.18) \tag{1}$$

where, CO, H_2, CH_4, are the molar percentages of components of syngas.

$$CCE = ((12Y * (CO + CO_2 + CH_4))/(22.4 * C\%)) * 100 \tag{2}$$

where, Y is the dry gas yield (Nm^3/kg), $C\%$ is the mass percentage of carbon in ultimate analysis of biomass feedstock, and the other elements are the molar percentage of syngas components.

3 Results and Discussion

3.1 Refused Derived Fuel (RDF) Characterization

Refuse Derived Fuel (RDF) samples were characterized in order to evaluate their energetic potential and according to EN 15359:2011. Table 1 shows the chemical and physical characterization of RDF.

Results demonstrated that RDF presents high potential for energetic valorisation because has low content of water (5.7%), high volatile content (77.2%) and high calorific value (24330 J/g). It is also possible to notice that carbon is the major component of this RDF (56.2%). However, results also reveal that RDF has high concentrations of sulphur (2150 mg/kg) and chlorine (9120 mg/kg) which could represent operational problems on a large-scale exploration.

These results are quite similar to other authors, e.g., Genon and Brizio [11], for instance, characterized RDF samples in a life cycle assessment study. These authors

Table 1. Chemical-physical characterization of Refuse Derived Fuel (RDF)

Parameters	Unit	Results
Ash content	%	15.00
Moisture	%	5.70
Low heating value	[J/g]	24330.00
Bulk density	[g/cm^3]	0.10
Biomass content	%	39.70
Volatile matter	%	77.20
Elemental analysis		
Carbon	%	56.20
Nitrogen	%	7.14
Hydrogen	%	0.91

Metals	Results [mg/kg]	Metals	Results [mg/kg]
Antimony	54.00	Nickel	15.00
Arsenio	<0.80	Thallium	<0.20
Cadmium	0.30	Vanadium	5.00
Lead	42.00	Metallic aluminium	0.25
Cobalt	2.00	Chlorine (Cl)	9120.00
Copper	2970.00	Sulphur (S)	2150.00
Chromium	520.00	Bromine (Br)	513.00
Manganese	52.00	Fluorine (F)	95.00
Mercury	<0.05		

verified that RDF also had high calorific value (20000 J/g), high carbon content (53%) and high sulphur content (5%) [11]. Dalai et al. [12] also observed that RDF from municipal wastes had high volatile content (84.7%), high carbon content (46.7%) and low moisture content (3.2%). In this case, authors only detected 0.5% of sulphur in RDF which could be related to the sample origin [12].

3.2   Effect of Temperature in RDF Pyrolysis

Temperature is one of the most relevant parameters in waste-to-energy processes, since it modifies syngas composition and calorific value. In order to evaluate the effect of temperature in RDF conversion, gasification experiments were carried out at 450, 600 and 750 °C. Table 2 show the effect of this parameter in syngas production, syngas composition, carbon conversion efficiency (CCE), syngas heating value and specific syngas flow rate, respectively.

Results expressed in Table 2 shows a trend in syngas production with increasing temperature. In fact, syngas gas production increased from 24 L at 450 °C to approximately 120 L of syngas at 750 °C.

These results represented also an increase of specific syngas flow rate from 0.104 to 0.314 m^3 syngas/kg RDF at 450 °C to 750 °C, respectively. Regarding to syngas gas

Table 2. Results of RDF pyrolysis and gasification at different temperatures, gasifying agents and molar ratios (ER of 0.6 and S/B of 1.0)

Temperature (°C)	Pyrolysis/Gasifying agent	CCE (%)	LHV (MJ/m^3)	Syngas flow rate (m^3/kgRDF)	Syngas composition (%)			
					H$_2$	CO	CO$_2$	CH$_4$
450	Pyrolysis	15.0	2.70	0.10	2.30	4.80	25.1	4.30
600	Pyrolysis	16.0	13.5	0.23	8.10	9.50	8.90	13.5
750	Pyrolysis	25.6	24.3	0.31	5.40	5.80	11.0	24.3
750	Air	86.6	8.70	1.38	22.3	32.0	39.8	5.90
850	Air	97.5	8.00	1.48	30.8	26.5	39.2	3.50
750	Steam	24.7	9.10	0.52	50.7	11.2	32.4	5.70
850	Steam	36.0	7.10	0.59	35.7	14.2	46.5	3.60

composition, Table 2 show the influence of temperature in this parameter. Results plotted in Table 2 reveal also that the increasing of temperature promoted the formation of a methane rich gas. At 750 °C methane yield increased to 24.3% from 4.3% at 450 °C, mostly due to the cracking of larger molecules presented in oil products. In contrast, carbon dioxide yield decreased from 25.1% at 450 °C to 11% at 750 °C. The increasing of methane yield at 750 °C resulted also in the increasing of calorific value of the syngas produced. At 750 °C methane yield increased to 11.2 MJ/m^3 from 2.7 MJ/m^3 at 450 °C (Table 2). Wang et al. (2014) studied the pyrolysis of RDF from MSW at 500, 700 and 900 °C. Authors also reported an increasing of specific syngas flow rate and gas LHV with temperature increasing. The authors obtained an increase of 0.13 to 0.43 m^3 syngas/kg RDF at 500 to 900 °C, respectively. The higher heating value (HHV) of the pyrolysis gas from this RDF were 10.4, 17.0 and 19.1 MJ/kg for temperatures of 500, 700 and 900 °C, respectively [13].

Regarding to carbon conversion efficiency (CCE), results represented in Table 2 show that temperature also increased the carbon conversion from 12% at 450 °C to approximately 50% at 750 °C. Buah et al. [14] also found that carbon conversion in pyrolysis increases with temperature increasing. These authors reported that at 400, 500, 600 and 700 °C, the calorific values of the chars were 20.4, 16.7, 16.4 and 11.2 MJ/kg, respectively, with a yield decrease from 50% to approximately 31% [14].

3.3   Effect of Temperature in RDF Gasification

In order to evaluate the effect of temperature in RDF conversion, gasification experiments were carry out with air and steam at 750 and 850 °C. Table 2 show the effect of this parameter in syngas production. In both cases, the increase of temperature also improved slightly syngas production. However, it is possible to observe that LHV of syngas was reduced by the increase of temperature in both cases. In fact, when temperature was increased from 750 °C to 850 °C, LHV was reduced from 8.7 to 8.0 MJ/m^3. Nevertheless, syngas gas composition (Table 2) is richer in H$_2$ in air gasification at 850 °C. These results are consistent with the current literature. Barba et al. [15] studied the effect of temperature in gas composition and LHV from RDF gasification with air. These authors also observed that syngas LHV decreased with temperature increasing. Results showed that from 800 °C to 900 °C, LHV was reduced from 7.01 to 5.79 MJ/m3 mostly due to CH_4 and CO decreasing. These authors also observed that at 1000 °C, syngas LHV increased to 7.23 MJ/m^3 due to H_2 increasing, which represents almost 50% of total syngas composition [15].

Instead, steam gasification registered better results at 750 °C, because when temperature was increased from 750 °C to 850 °C, H_2 production was reduced from 51 to 36%. Dalai et al. [12] reported similar results in RDF gasification with steam. According to these authors, H_2 production was reduced by increasing temperature although CO production was enhanced [16]. Regarding to carbon conversion efficiency (CCE), results in Table 2 show that in both cases was registered a slightly increase of carbon conversion at higher temperatures. When temperature was increased from 750 °C to 850 °C, CCE was also increased from 87% to 98% with air and 25% to 36 with steam. Differences between CCE results in both gasification experiments are related to the occurrence of more oxidation reaction in air gasification [17]. In Table 2 all the results of RDF valorisation trough pyrolysis and gasification are presented.

4 Conclusions

The utilization of Refuse derived fuels or RDF as an alternative and renewable solid fuel can respond to a range of problems a robust endogenous resource can respond to a range of problems related to waste management.

This study intended to evaluate the effect of temperature in syngas production, composition, heating value and carbon conversion efficiency. Results proved that pyrolysis is more efficient at 750 °C. At this temperature, it was obtained a syngas calorific value of 11.2 MJ/m^3 and a specific gas production of 0.43 m^3 syngas/kg RDF. Results also proved that the increasing of temperature in pyrolysis reaction enhanced methane production and decreased carbon dioxide content. Concerning to gasification, results proved that steam gasification is more efficient at 750 °C. Instead, air gasification proved to be more efficiency at 850 °C. Comparatively, air gasification produced more syngas flow rate that steam gasification. However, due to nitrogen dilution effect and more oxidation reactions, syngas produced in air gasification has less calorific or heating value than steam gasification syngas. In contrast, gasification with air allowed to obtain more carbon conversion than steam gasification. Although purely suggestive, these results represent very promising data for RDF gasification in pilot or industrial scale. Furthermore, more studies need to be conducted in order to understand the effect of these parameters in chair and tar composition.

References

1. EN 15359:2011. Solid recovered fuels. Specifications and classes
2. Gendebien, A., Leavens, A., Blackmore, K., Godley, A., Lewin, K., Whiting, K.J., Davis, R.: Refuse derived fuel, current practice and perspectives – Final report. European Commission – Directorate General Environment (2003)
3. Edo-Alcón, N., Gallardo, A., Colomer-Mendoza, F.: Characterization of SRF from MBT plants: influence of the input waste and of the processing technologies. Fuel Process. Technol. **153**, 19–27 (2006)
4. Heidenreich, S., Foscolo, P.: New concepts in biomass gasification. Progress Energy Combust. Sci. **46**, 72–95 (2015)
5. Strezov, V., Patterson, M., Zymla, V., Fisher, K., Evans, T.J., Nelson, P.F.: Fundamental aspects of biomass carbonization. J. Anal. Appl. Pyrol. **79**, 91–100 (2007)
6. Ionescu, G., Rada, E., Ragazzi, M., Marculesc, C., Badea, A., Apostol, T.: Integrated municipal solid waste scenario model using advanced pretreatment and waste to energy processes. Energy Conv. Manag. **76**, 1083–1092 (2013)
7. Velghe, I., Carleer, R., Yperman, J., Schreurs, S.: Study of the pyrolysis of munici-pal solid waste for the production of valuable products. J. Anal. Appl. Pyrol. **92**, 366–375 (2011)
8. Young, G.: Municipal Solid Waste to Energy Conversion Processes: Economic, Technical, and Renewable, 396 p. Wiley, Hoboken (2010)
9. Mahinpey, N., Gomez, A.: Review of gasification fundamentals and new findings: reactors, feedstock, and kinetic studies. Chem. Eng. Sci. **148**, 14–31 (2016)
10. Sansaniwala, S.K., Pala, K., Rosenb, M.A., Tyagia, S.K.: Recent advances in the development of biomass gasification technology: a comprehensive review. Renew. Sustain. Energy Rev. **72**, 363–384 (2017)
11. Genon, G., Brizio, E.: Perspectives and limits for cement kilns as a destination for RDF. Waste Manag. **28**, 2375–2385 (2008)
12. Dalai, A., Batta, N., Eswaramoorthi, I., Schoenau, G.: Gasification of refuse derived fuel in a fixed bed reactor for syngas production. Waste Manag. **29**, 252–258 (2009)
13. Wang, N., Chen., D.Z., He, P.J.: Reforming of MSW pyrolyis volatile on their char and the change of syngas. In: Proceedings of the 5th International Symposium on Energy from Biomass and Waste, Venice, 17–20 November 2014
14. Buah, W.K., Cunliffe, A.M., Williams, P.T.: Characterization of products from the pyrolysis of municipal solid waste. Process Saf. Environ. **85**, 450–457 (2007)
15. Barba, D., Capocelli, M., Cornacchia, G., Matera, D.: Theoretical and experimental procedure for scaling-up RDF gasifiers: the Gibbs gradient method. Fuel **179**, 60–70 (2016)
16. Couto, N., Silva, V., Rouboa, A.: Thermodynamic evaluation of Portuguese municipal solid waste gasification. J. Clean. Prod. **139**, 622–635 (2016)
17. Sikarwar, V., Zhao, M., Clough, P., Yao, J., Zhong, X., Memon, M., Shah, N., Anthony, E., Fennell, P.: An overview of advances in biomass gasification. Energy Environ. Sci. **9**, 2939 (2016)

Composition of Producer Gas Obtained by Gasification of Pellet Mixtures Produced with Residual Lignocellulosic Biomass, Cork Wastes, Polymers and Polymer Derived Chars

Andrei Longo[1](✉)[iD], Margarida Gonçalves[1,2][iD], Catarina Nobre[1][iD],
Octávio Alves[1,2][iD], Luís Calado[2], and Paulo Brito[2][iD]

[1] MEtRICs, Departamento de Ciências e Tecnologia da Biomassa, Faculdade de
Ciências e Tecnologia, Universidade Nova de Lisboa, 2829-516 Caparica,
Portugal
mmpg@fct.unl.pt
[2] VALORIZA, Instituto Politécnico de Portalegre, Portalegre, Portugal

Abstract. In this work, the gasification of pellets produced from residual pine biomass and their mixtures with cork wastes, polymeric wastes and polymer derived char pellets was studied. The gasification tests were performed in a fixed bed downdraft gasifier at the temperatures of 700 °C and 800 °C. The influence of pellet combinations and gasification temperature in the composition and high heating value of the product gas was evaluated. The results were compared with commercially available pine pellets. At 800 °C, all the tested pellet mixtures exhibited higher CO and H_2 concentrations than the commercial pellets, with 9.3% mol CO and 6.2% mol H_2 for the mixture with cork wastes pellets, for example. The high heating value of the product gas for the different pellet mixtures presented values between 1.04 and 3.84 MJ/m^3, for different gasification conditions. Residual lignocellulosic biomass and mixed wastes have the potential to be used as sustainable raw materials for energy production by gasification. Gasification operational parameters, such as temperature and equivalence ratio are decisive factors in product gas output, that need to be optimized in order to efficiently take advantage of the energy potential from these raw materials. Furthermore, this approach can contribute to the coupling of waste management and energy production through endogenous resources, reducing the deposition of these waste materials in landfills.

Keywords: Pellets · Gasification · Biomass wastes · Cork · Polymeric wastes

1   Introduction

Gasification is the thermochemical conversion of carbon-based feedstock into a combustible gas through the controlled supply of a gasification agent changing the chemical structure of the fuel particles due to the high temperatures (>700 °C) reached in the process. This technique produces syngas, a mixture of CO, H_2, CO_2 and minor amounts

© Springer International Publishing AG, part of Springer Nature 2019
J. Machado et al. (Eds.): HELIX 2018, LNEE 505, pp. 648–654, 2019.
https://doi.org/10.1007/978-3-319-91334-6_88

of other combustible gases that may later be used as a feedstock for the chemical industry, or as a fuel for efficient production of electricity and/or heat [1].

Several parameters, such as biomass composition, reactor type and temperature, may influence the quality of product gas. Co-gasification of biomass with wastes can overcome several problems related to tar formation and can also enhance product quality. Gasification of biomass with different wastes blending ratio may elucidate some mechanisms involved in their interaction [1].

Although some waste streams can be submitted to gasification without any pre-treatment, such as paper wastes and waste from agroindustry, usually most of the wastes benefit from pre-treatment processes such as pelletization or torrefaction to improve their fuel quality [2]. Pelletization is the mechanical process that compacts biomass or wastes into uniformly sized particles with pellet or briquette shape, raising its density and lowering the moisture content. These processed forms of waste result in higher calorific value, more homogeneous physical and chemical compositions, lower pollutant emissions, lower ash content, reduced excess air requirements during combustion and finally easier storage, handling and transportation giving access to a more affordable subsequent thermal conversion [1].

Torrefaction has been used as pretreatment of biomass or waste prior to the application of thermochemical treatment methods such as pyrolysis or gasification, improving the physical and chemical characteristics of the material and increasing process efficiency [3]. Compared with the original material, torrefaction reduces significantly the moisture and volatiles content while increasing the specific heat of the fuel. The properties of the material produced after torrefaction depend on the nature of the biomass or waste used, as well as the torrefaction temperature and residence time [4].

As expected, different gasification fuels feature different syngas compositions. Hence, when mixing two (or more) feedstocks in different proportions, distinct characteristics and distribution of the components are expected in the producer gas. As already stated, blending biomass and wastes seems to afford higher yields of volatile products thus, the energy content of the produced syngas is expected to be higher than for biomass gasification alone [1]. Co-gasification of biomass and plastic wastes may be a promise alternative to valorize mixed wastes and obtain a product gas with more quality [5].

This work aims to evaluate the composition and HHV of product gas, during gasification of pure biomass and biomass with mixed wastes. Synergetic effect of biomass and wastes co-gasification may address some problems regarding product gas composition and lower tar formation.

2   Materials and Methods

2.1   Raw Materials and Pellet Production

In this work, biomass and waste pellets were produced from pine wastes, cork wastes, polymer wastes and polymer derived char and were used in the gasification tests. Pine biomass wastes (PBW) were supplied by Nova Lenha, Lda. Cork wastes (CW) are a mixture of cork contaminated with polymers and were supplied by Prélis Carâmica, Lda. The polymer wastes (PW) were obtained from Cimpor - Cimentos de Portugal, S. A and were also used to produce the polymer derived char sample (PWchar), in a muffle furnace (Stuart Scientific Furnance - 50 W) at 300 °C for 30 min. The raw materials used in pellet production are represented in Fig. 1.

Fig. 1. Raw materials studied in this work: (a) Pine biomass wastes (PBW), (b) Cork wastes (CW), (c) Polymeric wastes (PW) and (d) Polymer derived char (PWchar)

Each raw material was pelletized in a Kahl 14–175 (3 kW) pelletizer with 8 mm diameter holes and commercially available pine pellets (CPP) were used as reference. Prior to analytical determinations, pellet samples were grinded and sieved to a particle diameter of 500 μm. Total moisture content was determined according to ISO 18134-1:2015. Ash content was evaluated gravimetrically after incineration of the samples at 550 °C according to the standard ISO 18122:2015. Volatile matter content of each sample was determined through the method described in ISO 18123:2015. Fixed carbon was calculated by difference.

2.2   Gasification Tests

Gasification tests were conducted at 700 °C and 800 °C for a period of about 2 h, using a PP20 Power Pellet fixed bed downdraft gasifier, with a fuel consumption rate of 4.8–19.2 kg/h. Commercial pine pellets and pine wastes pellets were gasified alone. Cork wastes, polymer wastes and polymer derived char pellets were co-gasified along with pine wastes pellets corresponding to a mass proportion of 15%. Experimental setup is described in Table 1.

During each run, producer gas was collected in 3 Tedlar bags, and major gas components (CO, H_2, CH_4, CO_2, $N_2 + O_2$) were analyzed by gas chromatography (Varian 450-GC). Air flow varied from 0.3 to 0.4 dm^3/s and the producer gas flow from 0.56 to 0.61 dm^3/s.

Table 1. Experimental setup of the gasification tests

Test number	Sample	T (°C)	Test time (h)	Fuel consumption (kg)
1.1	CPP	700	2 h 45 min	15
1.2		800	2 h 45 min	
2.1	PBW	700	2 h 22 min	12
2.2		800	1 h 42 min	
3.1	85% PBW + 15% CW	700	2 h 05 min	10.2
3.2		800	3 h 10 min	
4.1	85% PBW + 15% PW	700	2 h 49 min	6.5
4.2		800	2 h 00 min	
5.1	85% PBW + 15% PWchar	700	1 h 40 min	9.5
5.2		800	2 h 00 min	

3 Results and Discussion

Proximate composition of the pellet samples used in the gasification experiments is shown in Table 2. Moisture content is lower on the samples with more polymeric wastes but, on the other hand, ash is higher, due to the presence of mineral additives that are commonly present in these polymeric wastes.

Table 2. Proximate analysis results for the pellet samples used in this work

Sample	CPP	PBW	CW	PW	PWchar
Proximate analysis (%, dry basis)					
Moisture[a]	7.1 ± 0.2	8.4 ± 0.7	4.7 ± 0.0	3.1 ± 0.0	3.2 ± 0.1
Ash	0.7 ± 0.1	7.3 ± 1.2	18.5 ± 0.1	12.8 ± 1.2	17.2 ± 0.2
Volatile matter	84.6 ± 0.8	77.3 ± 0.6	69.0 ± 1.1	82.7 ± 1.3	70.5 ± 0.8
Fixed carbon	14.8 ± 0.7	15.4 ± 1.7	12.5 ± 1.1	4.5 ± 1.1	12.3 ± 0.7

[a]*Moisture is presented in an "as received" basis*

Since ash was determined at 550 °C there could be an overestimation of the ash content for the polymer containing wastes that may require higher calcination temperatures. On the other hand, the determination of ash from biomass materials should not be done at temperatures higher than 550 °C to prevent the loss of some mineral components that occurs at higher temperatures.

Mixing biomass and polymer pellets can reduce the incidence of fouling and slagging problems associated with high ash contents enabling the recovery of the carbon fraction from these wastes.

Regarding the composition of the product gas it can be observed that in the experiments performed at 700 °C, higher amounts of combustible gases, such as CO, H_2 and CH_4, were obtained for pure pine and pine wastes samples (Fig. 2).

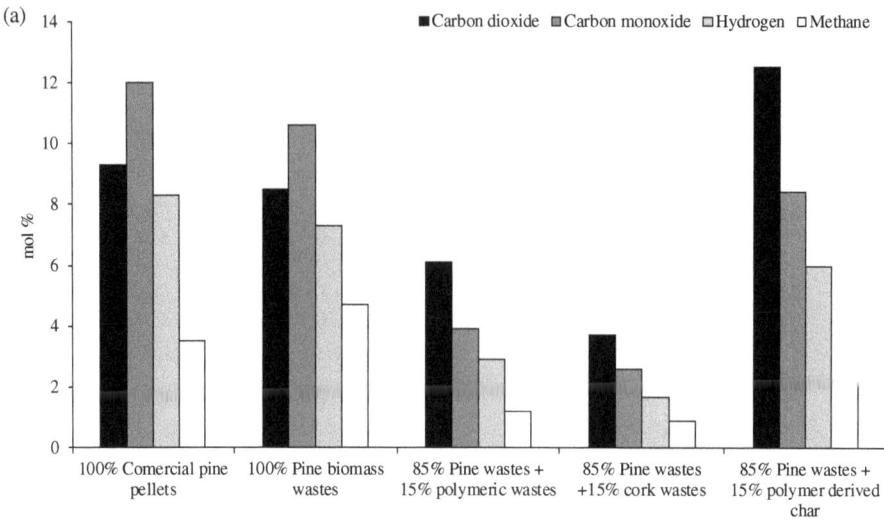

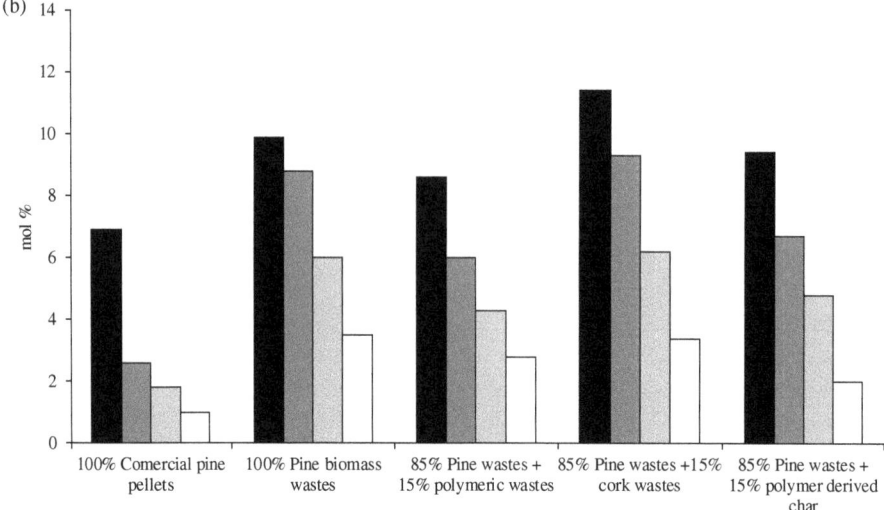

Fig. 2. Product gas composition for the pellet samples at different gasification temperatures. (a) T = 700 °C and (b) T = 800 °C.

The fuel mixtures with 15% polymeric wastes or 15% cork wastes showed a much lower gasification efficiency at this temperature, what is coherent with thermogravimetric data of polymer materials that generally present a wider decomposition temperature range than lignocellulosic materials [6]. On the other hand, the fuel mixture with 15% polymer derived char presented higher concentrations of CO, H_2 and CH_4, than the mixture with the raw polymeric wastes, indicating that the torrefaction process improved the availability of the polymeric material to undergo oxidative decomposition.

Nevertheless, the CO_2 concentration of this fuel mix is the highest, maybe due to the complete oxidation of tar molecules present in the char pores.

At 800 °C, the gasification efficiency is improved for all polymer containing fuel mixtures and for the waste pine biomass pellets, indicating that this temperature is more adequate for the pyrolytic decomposition and oxidative gasification of more stable materials such as polymers or lignin (Fig. 2).

The HHV of the product gas (Fig. 3) was evaluated taking into account its composition and the results obtained indicate that:

- Pine biomass wastes yielded a product gas with higher HHV than commercial pine pellets at both temperatures tested, probably due to the presence of lignin-rich fractions such as pine bark;
- the gasification of polymer containing wastes was more effective at 800 °C, as a consequence of the chemical stability of those materials, yielding product gases comparable to the pine biomass wastes.

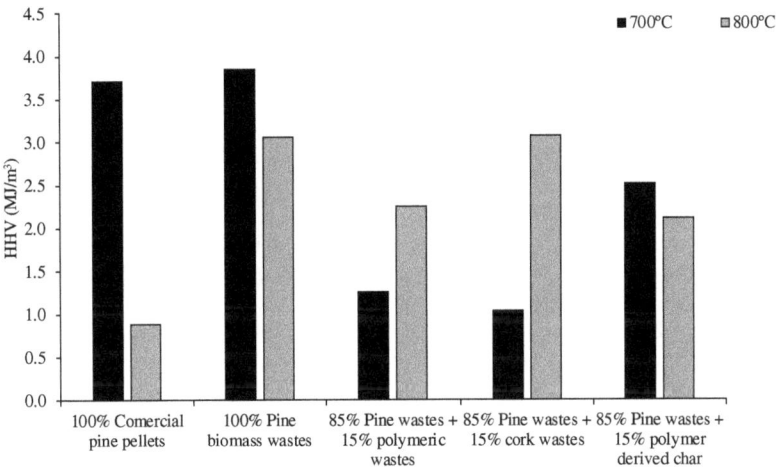

Fig. 3. Product gas heating value (MJ/m^3) for the fuel mixtures at different gasification temperatures

It should be noticed that the product gas obtained with the polymer derived char mix, at 700 °C presented a HHV higher than the gas obtained with the same material at 800 °C.

The observed lower calorific value of the gases obtained from the fuel mixtures with higher concentrations of polymers may also be explained by a loss of the carbon content of the original fuels, in the form of tars and chars, therefore decreasing the efficiency of conversion in combustible gases [7].

4 Conclusions

The use of alternative fuels in gasification allows the preservation of biomass resources and reduces the environmental impact of waste management systems.

The gasification of lignin-rich biomass wastes yielded a product gas with higher calorific value than commercial pine biomass pellets. The incorporation of small amounts of plastic RDF reduces HHV of the product gas and can create some problems on the feeding systems. The use of higher gasification temperatures and the introduction of a torrefaction pre-treatment may improve the overall efficiency of the process regarding the incorporation polymeric wastes. The optimization of the conversion conditions for polymeric wastes is essential to reduce the deposition of such wastes in landfills and promote their energetic valorization through advanced techniques.

References

1. Ramos, A., Monteiro, E., Silva, V., Rouboa, A.: Co-gasification and recent developments on waste-to-energy conversion: a review. Renew. Sustain. Energy Rev. **81**, 380–398 (2018)
2. Belgiorno, V., Feo, G., Rocca, C.N., Napoli, R.M.A.: Energy from gasification of solid wastes. Waste Manag. **23**, 1–15 (2003)
3. Chew, J.J., Doshi, V.: Recent advances in biomass pretreatment – torrefaction fundamentals and technology. Renew. Sustain. Energy Rev. **15**, 4212–4222 (2011)
4. Prins, M.J., Ptasinski, K.J., Janssen, F.J.J.G.: More efficient biomass gasification via torrefaction. Energy **31**, 3458–3470 (2006)
5. Alipour, R., Osmieri, L., Specchia, S., et al.: H₂-rich syngas production through mixed residual biomass and HDPE waste via integrated catalytic gasification and tar cracking plus bio-char upgrading. Chem. Eng. J. **308**, 578–587 (2017)
6. Haykiri-Acma, H., Kurt, G., Yaman, S.: Properties of biochars obtained from RDF by carbonization: influences of devolatilization severity. Waste Biomass Valorization **8**(3), 539–547 (2016)
7. Haykiri-Acma, H., Lopez, G., Artetxe, M., Amutio, M., Bilbao, J.: Recent advances in the gasification of waste plastics. a critical overview. Renew. Sustain. Energy Rev. **82**, 576–596 (2018)

Production of High Calorific Value Biochars by Low Temperature Pyrolysis of Lipid Wastes and Lignocellulosic Biomass

Luís Durão[1]([⊠])(iD), Margarida Gonçalves[1](iD), Catarina Nobre[1](iD), Octávio Alves[1,2](iD), Paulo Brito[2](iD), and Benilde Mendes[1](iD)

[1] MEtRICs, Departamento de Ciências E Tecnologia Da Biomassa, Faculdade de Ciências E Tecnologia Da, Universidade Nova de Lisboa, 2829-516 Caparica, Portugal
mmpg@fct.unl.pt

[2] VALORIZA, Instituto Politécnico de Portalegre, Portalegre, Portugal

Abstract. Low quality waste oils and fats, with elevated levels of acidity, water and other contaminants are not appropriate for biodiesel production or other material recycling processes. Nevertheless, they can be converted to energy-dense bio-oils by pyrolysis, in an oxygen deficient atmosphere. These bio-oils may be used in combustion or upgraded to yield liquid biofuels for internal combustion engines although their chemical stability still creates some limitations during storage. This work aimed at the evaluation of biochar production by pyrolysis of mixtures of lipid wastes and pine sawdust. For this purpose, thermal conversion was performed with initial vacuum, at temperatures between 573.15 and 673.15 K and a reaction time of 60 min. Biomass incorporation varied between 0 and 38% (w/w). Biochar formation was favored by biomass incorporation and by the increase in temperature. The biochar yield varied from 1–28% and its high calorific value ranged between 24.3 to 36.6 MJ/kg. For incorporations of pine sawdust higher than 23% the carbon content and the calorific value of the biochar decreased due to a higher oxygen content in the raw materials. Highly calorific bio-oils were also produced in these conditions with yields from 24 to 83% (w/w), and high calorific values from 37.8 to 43.4 MJ/kg. These bio-oils have a carbon content higher than 75% and contain high molecular weight components. As such, they could be used in direct combustion in boilers, or as pellet additives. This approach contributes to the implementation of the renewed hierarchy for wastes as defined in the Directive 2008/98/EC namely by identifying alternatives to the deposition of the used cooking oils in landfills.

Keywords: Used cooking oil · Pine sawdust · Biochar

1 Introduction

The negative environmental impacts of used cooking oils (UCO) are associated with inadequate practices of their treatment and disposal. An important part of the organic load present in urban wastewater (UWW) is lipid waste, responsible for 30 to 40% of

© Springer International Publishing AG, part of Springer Nature 2019
J. Machado et al. (Eds.): HELIX 2018, LNEE 505, pp. 655–661, 2019.
https://doi.org/10.1007/978-3-319-91334-6_89

the chemical oxygen demand in wastewater treatment plants (WWTP) [1]. The European Directive 2008/98/EC [2] establishes a renewed hierarchy for waste, prioritizing prevention, preparation for reuse, recycling and other forms of recovery. It is estimated that more than 1.1 million tons of UCO were processed in Europe, in 2013. A substantial part of this waste, around 700,000 ton, comes from within the EU, while the remaining (about 250,000 ton) was imported mostly from the USA [3].

Usually, collected used cooking oils are subjected to some preliminary treatment operations, including filtration and decantation. The treated oils are then directed to recovery units for refinement, yielding a recycled oil with good quality. These stages of treatment originate lipid fractions that are retained in filters and decanters and have a high acid content and some emulsified water, so they do not have sufficient quality to produce biodiesel by transesterification. These low-quality lipid wastes may be converted to free fatty acids or simply deposited in landfills, both management solutions with significant environmental impact [4, 5].

Pyrolysis is a thermal conversion process that occurs in an inert atmosphere, with subsequent rearrangement of the fragmented molecules, originating three products: gases, liquids and solids [6]. By varying the main process parameters (temperature, heating rate, residence time, catalyst and atmosphere) it is possible to influence the distribution and characteristics of the pyrolysis products [7, 8]. Most pyrolysis tests are performed under inert atmospheres, typically N_2 or He [9].

Vegetable oils and animal fats of different sources have been used for the production of pyrolysis bio-oils that can be used in boilers as bio-liquid fuels or distilled to yield liquid biofuels with sufficient quality to be used in internal combustion engines [10, 11]. On the other hand, lignocellulosic wastes from the wood industry have a high organic matter being composed essentially from cellulose, hemicellulose and lignin, yielding pyrolysis oils with much higher contents of oxygenated and aromatic compounds and therefore higher chemical instability and lower calorific value than the bio-oils produced from lipid materials [12].

Zhang et al. [9] studied the fast pyrolysis of corn cob under different modified atmospheres of N_2, CO_2, CO, CH_4 and H_2. The formed products were bio-oils and biochars with a large number of aromatic and oxygenated functional groups. The yields of liquid products varied from 49.6 to 58.7%, and their high heating values varied from 17.2 to 24.4 MJ/kg which are comparable to the HHV of various biomass-based solid fuels or biochars [12, 13].

Chen et al. [14] studied the co-pyrolysis of corn cob and UCO, at temperatures from 500 to 600 °C and concluded that the bio-oil and biochar yields are strongly dependent on the reaction temperature and on the mass ratio between corn cob and UCO. The increase in corn cob biomass favors the production of biochars while the increase in UCO mass promotes the formation of bio-oils.

In this work, the co-pyrolysis of UCO and pine sawdust, at temperatures of 300 and 350 °C and under initial vacuum was studied. The purpose of the work is to find alternative valorization pathways for these two waste materials, namely the production of high calorific value biochars and bio-oils that could be used as additives for solid biofuels.

2 Materials and Methods

2.1 Characterization of Raw Materials and Pyrolysis Products

Used cooking oil (UCO) was supplied by EcoMovimento and corresponds to a blend of used cooking oil from different sources (vegetable and animal). Pine sawdust (PS) was supplied by CMC Biomassa, Lda. The pine sawdust was sieved to a particle size of less than 500 μm before any determination. The elemental composition of the raw materials and of liquid and solid products (biochars and bio-oils) was determined in an elemental analyzer (Thermo Finnigan Flash EA 112 CHNS), and their high heating value (HHV) was measured by calorimetry (model C200 calorimeter, IKA). The density of the used cooking oil was determined gravimetrically and its acid value was determined according to the standard method EN 14104.

2.2 Pyrolysis Tests

Pine sawdust was added to the used cooking oil at fortification levels of 0, 9, 23 and 38% wt and a sample of 100 g of each mixture was subject to pyrolysis at the temperatures of 300 and 350 °C, during 60 min, under initial vacuum. The pyrolysis tests were performed in a pilot scale laboratory installation (Parr Instruments), equipped with a stainless-steel autoclave (1L), with internal stirring. Product yield was determined gravimetrically for gases and solids and liquid yield was determined by difference.

3 Results and Discussion

Table 1 shows the results for the characterization of the raw materials used in the pyrolysis tests.

Table 1. Properties of used cooking oil (UCO) and pine sawdust (PS).

Parameter	UCO	PS
Elemental composition (%wt, dry basis)		
C	62.6	44.9
H	11.3	6.0
N	0.2	2.2
S	0.0	0.2
O	25.9	46.7
HHV (MJ/kg)	38.9	18.0
Density (g/cm^3)	0.9	–
Acid value (mg KOH/g)	15.8	–

The UCO is a low quality waste oil with high acidity that is not adequate for the production of biodiesel by transesterification, but it is a material with high carbon content (62.6%) and high calorific value (38.9 MJ/kg) clearly appropriate for energetic valorization. The co-pyrolysis of this lipid waste and of its mixtures with pine biomass, at relatively low temperatures (300 and 350 °C), and long residence time (60 min) yielded mixtures of dense bio-oils and biochars (Table 2).

Table 2. Gaseous, liquid and solid product yields resulting from pyrolysis of UCO with different rates of incorporation of PS at 300 and 350°C for 60 min under initial vacuum.

PS concentration (%wt)	Reaction temperature (°C)	Pyrolysis yield (%wt)			
		Gases	Aqueous phase	Bio-oils	Solids
0	300	9.04	7.58	82.55	0.83
0	350	11.98	10.46	76.53	1.03
9	300	10.13	16.75	67.98	5.14
9	350	12.75	17.67	63.84	5.74
23	300	23.19	19.44	36.94	20.43
23	350	25.68	21.30	31.48	21.54
38	300	25.18	22.42	24.27	28.13
38	350	26.09	24.57	23.84	25.50

The increase in the concentration of pine sawdust caused an increase in the yield of biochar and a decrease in the yield of bio-oil because the pyrolysis of lignocellulosic biomass originates primary products with oxygenated functions that tend to condense forming larger carbonaceous structures [12].

The formation of an aqueous phase large enough to spontaneously separate from the organic fraction (bio-oil) during the cooling phase, is mainly due to dehydration and deoxygenation of the pine sawdust, but some water emulsified in the waste cooking oil and resulting from its deoxygenation also contribute to this water phase as it can be seen from the experiments performed without incorporation of pine sawdust.

Increasing the temperature to 350 °C favored the formation of gas products and of aqueous phase because it facilitates bond cleavage but it also increased the extent of some recombination reactions, therefore promoting the formation of solid products [12, 13, 15]. The higher percentage of pine sawdust also led to the formation of more gaseous products causing the pressure to increase since the experiments were performed in a batch reactor. Higher pressures favor recombination reactions between primary pyrolysis products thus increasing the yields of biochar [15].

Pine sawdust incorporations improved the carbon content and the HHV of the correspondent biochar products when compared with the biochars obtained from UCO alone, as seen in Table 3. The biochars produced with 23% PS at 350 °C and 38% PS at 300 °C had the highest calorific values (35.5 and 36.6 MJ/kg, respectively). The HHV determined for these biochars are in the same range as bituminous coal (33 M/kg) [15]. The positive correlation between the incorporation of PS and the carbon content of the biochars reaching values higher than the raw pine sawdust or the biochar produced from 100% UCO, is an evidence that the carbonaceous structure of the pine biomass suffers deoxygenation and rearrangement, even at these moderate temperatures.

Table 3. High heating value (HHV) and elemental composition of the biochar obtained in UCO pyrolysis tests with PS incorporation, at 300 and 350°C.

PS concentration (%wt)	Reaction temperature (°C)	HHV (MJ/kg)	Elemental composition (%wt)				
			N	C	H	S	O
0	300	24.3	1.3	50.0	5.5	0.2	43.0
0	350	29.9	1.0	65.1	5.9	0.1	27.9
9	300	33.3	1.1	71.7	8.3	0.1	18.8
9	350	35.5	0.9	68.3	5.2	0.1	25.5
23	300	34.0	1.8	74.8	7.3	0.1	16.0
23	350	36.6	1.6	74.0	5.0	0.1	19.3
38	300	34.5	1.4	66.3	5.8	0.1	26.4
38	350	32.8	2.9	68.7	4.6	0.1	23.7

Carbon contents of the produced biochars are similar to those found in lignite or bituminous coal, and higher than peat [16]. Table 4 shows the density, HHV and elemental composition results of the bio-oils produced in each pyrolysis test. The bio-oils presented carbon contents between 75.6 and 83.3% and calorific values between 37.8 and 43.4 MJ/kg, which are comparable to ethanol or biodiesel [17]. These characteristics give them excellent fuel properties to be used in their crude form, as liquid fuels for boilers or as solid biofuel additives. Elemental analysis results show that the deoxygenation reactions of the bio-oil were more effective at 350 °C than at 300 °C, a behavior that was not observed for the biochars. Deoxygenation was further enhanced as the incorporation of sawdust into UCO, reaching a minimum value of 3.4% for the sample with 23% PS incorporation pyrolyzed at 350 °C. These results point out to a predominant retention of the oxygenated functions in the aqueous phase and in the solid products, leaving higher concentrations of non-polar components in the liquid and gas phases.

Table 4. Properties of the organic fraction of pyrolysis liquids.

PS concentration (%w/w)	Reaction temperature (°C)	Density (g/cm³)	HHV (MJ/kg)	Elemental composition (%wt)				
				N	C	H	S	O
0	300	0.769	37.9	0.3	75.9	11.7	0.0	12.1
0	350	0.801	39.5	0.4	77.5	12.0	0.0	10.1
9	300	0.767	37.8	0.5	75.6	11.9	0.0	12.0
9	350	0.819	40.4	0.8	79.9	12.1	0.0	7.2
23	300	0.766	37.8	1.1	77.5	12.1	0.0	9.3
23	350	0.864	43.4	1.1	83.3	12.2	0.0	3.4

The O/C and H/C atomic ratios of the raw materials and of the bio-oils and biochars obtained at 350 °C are presented in Fig. 1, to illustrate the composition transformations that occurred during pyrolysis.

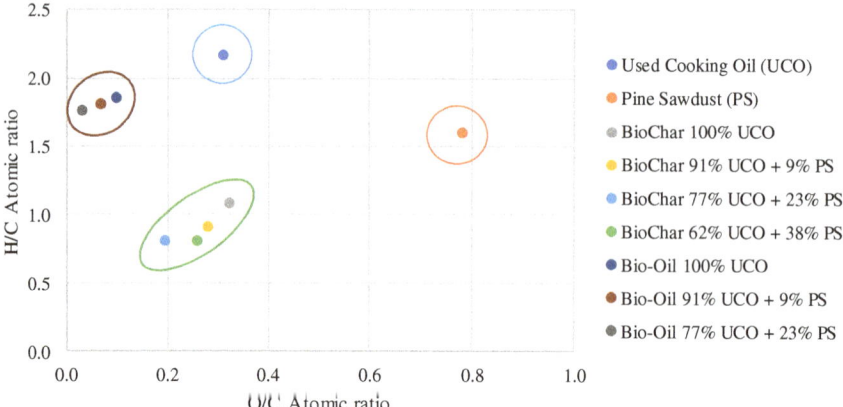

Fig. 1. van Krevelen diagram of raw materials, biochars and bio-oils from the pyrolysis tests conducted at 350 °C.

The diagram shows that the bio-oils are practically deoxygenated and still preserve a high hydrogen content. The biochars have an O/C ratio similar to the UCO but a lower H/C ratio, indicating a large recombination of the carbonaceous structure. The retention of oxygenated functions in the aqueous phase is demonstrated by the low O/C ratios of the biochars, even when the incorporation of pine sawdust reached values of 38%.

4   Conclusions

Low-temperature pyrolysis proved to be an effective method to obtain carbon-rich biooils and biochars with high calorific value. Combining biomass and lipid wastes allowed the formation of two different fuels: the biochars whose yield and calorific value are improved by incorporation of pine biomass and the bio-oils whose composition and properties are strongly influenced by the presence of the UCO.

The elimination of emulsified and constitution water in the form of an aqueous phase that also retains polar pyrolysis products such as phenolic compounds prevents the dissolution of those components in the bio-oil thus contributing to its upgrading.

Low temperature pyrolysis permits the thermal conversion of low quality lipid wastes in good quality fuels, under mild operational conditions, therefore contributing to a reduction of their deposition in landfills.

References

1. Chipasa, K.B., Medrzycka, K.: Behavior of lipids in biological wastewater treatment processes. J. Ind. Microbiol. Biotechno. **33**, 635–645 (2006)
2. Directive 2008/98/EC: Directive 2008/98/EC of the European parliament and of the council of 19 November 2008 on waste. Off. J. Eur. Union L 312/4, 3–30 (2008)

3. Harrison, P.: WASTED - Europe's untapped resource: an assessment of advanced biofuels from wastes and residues (2015)
4. Doğan, T.H.: The testing of the effects of cooking conditions on the quality of biodiesel produced from waste cooking oils. Renew. Energy **94**, 466–473 (2016)
5. Ibrahim, H.G.: Recycling of waste cooking oils (WCO) to biodiesel production. J. Multidiscip. Eng. Sci. Technol. **2**(4), 1 (2015)
6. Bridgwater, A.V.: Review of fast pyrolysis of biomass and product upgrading. Biomass Bioenergy **38**, 68–94 (2012)
7. Li, S., Xu, S., Liu, S., Yang, C., Lu, Q.: Fast pyrolysis of biomass in free-fall reactor for hydrogen-rich gas. Fuel Process. Technol. **85**(8–10), 1201–1211 (2004)
8. Bridgwater, A.V., Meier, D., Radlein, D.: An overview of fast pyrolysis of biomass. Org. Geochem. **30**(12), 1479–1493 (1999)
9. Zhang, H., Xiao, R., Wang, D., He, G., Shao, S., Zhang, J., Zhong, Z.: Biomass fast pyrolysis in a fluidized bed reactor under N2, CO2, CO, CH4 and H2 atmospheres. Bioresour. Technol. **102**(5), 4258–4264 (2011)
10. Lima, D.G., Soares, V.C.D., Ribeiro, E.B., et al.: Diesel-like fuel obtained by pyrolysis of vegetable oils. J. Anal. Appl. Pyrol. **71**(2), 987–996 (2004)
11. Melo-Espinosa, E.A., Piloto-Rodríguez, R., Goyos-Pérez, L., Sierens, R., Verhelst, S.: Emulsification of animal fats and vegetable oils for their use as a diesel engine fuel: an overview. Renew. Sustain. Energy Rev. **47**, 623–633 (2015)
12. Liu, C., Wang, H., Karim, A.M., Sun, J., Wang, Y.: Catalytic fast pyrolysis of lignocellulosic biomass. Chem. Soc. Rev. **43**(22), 7594–7623 (2014)
13. Fernandez, A., Mazza, G., Rodriguez, R.: Thermal decomposition under oxidative atmosphere of lignocellulosic wastes: Different kinetic methods application. J. Environ. Chem. Eng. **6**(1), 404–415 (2018)
14. Chen, G., Liu, C., Ma, W., Zhang, X., Li, Y., Yan, B., Zhou, W.: Co-pyrolysis of corn cob and waste cooking oil in a fixed bed. Bioresour. Technol. **166**, 500–507 (2014)
15. Mánya, J.: Pyrolysis for biochar purposes: a review to establish current knowledge gaps and research needs. Environ. Sci. Technol. **46**(15), 7939–7954 (2012)
16. Ahmad, M., Subawi, H.: New Van Krevelen diagram and its correlation with the heating value of biomass. Res. J. Agric. Environ. Manag. **2**(10), 295–301 (2013)
17. Agarwal, A.K.: Biofuels (alcohols and biodiesel) applications as fuels for internal combustion engines. Prog. Energy Combust. Sci. **33**(3), 233–271 (2007)

Assessment of Municipal Solid Wastes Gasification Through CFD Simulation

Eliseu Monteiro[1,2(✉)], Nuno Couto[2], Valter Silva[1], and Abel Rouboa[2]

[1] VALORIZA - Research Center for Endogenous Resource Valorisation,
Polytechnic Institute of Portalegre, Portalegre, Portugal
`eliseu@estgp.pt`
[2] LAETA-INEGI/Faculty of Engineering, University of Porto, Porto, Portugal

Abstract A two-dimensional CFD model for MSW gasification has been used to predict and analyze the viability of the hydrogen generation from MSW gasification. The model is based in an Eulerian-Eulerian approach to describe the transport of mass, momentum and energy for the solid and gas phases. The model is applied to a fluidized bed gasifier to full predict and analyze the viability of the hydrogen generation from MSW gasification taking into account the equivalence ratio and steam-to-waste ratio. Conclusion could be drawn that the increase of equivalence ratio has a negative effect on hydrogen production because the oxidation reactions are favored. The introduction of steam to MSW gasification is favorable for improving hydrogen yield, because it increases the partial pressure of steam inside the reactor which favors the gas-phase reactions.

Keywords: Gasification · Municipal solid wastes · CFD

1 Introduction

Hydrogen as a clean energy carrier is expected to satisfy a considerable portion of the world's future energy needs [1]. It can be used in internal combustion engines as well as fuel cells with less pollution on the environment, since the combustion with oxygen produces water as its only product [2]. Moreover, it has the highest energy content in comparison to other common fuels.

A problem that rises is that fossil fuels make up by far the largest contemporary source of hydrogen [3]. Water electrolysis is reported to be energy intensive resulting in high hydrogen production cost [4]. In contrast, biomass as carbon-neutral energy resource has other advantageous features like the local availability, large commercialization potential, make the route feasible for sustainable hydrogen production [5].

Biagini et al. [6] conducted an experimental study to evaluate the performance of the different thermo-chemical technologies for hydrogen production from biomass. They concluded that gasification of biomass is the most efficient and economical route for hydrogen production.

Among the biomasses sources, the municipal solid wastes (MSW) are the largest volume of residues produced worldwide. The disposal of MSW has become a critical and costly problem. The traditional landfilling method requires large amounts of land and contaminates air, water and soil [7]. Furthermore, incineration has drawbacks as

© Springer International Publishing AG, part of Springer Nature 2019
J. Machado et al. (Eds.): HELIX 2018, LNEE 505, pp. 662–667, 2019.
https://doi.org/10.1007/978-3-319-91334-6_90

well particularly harmful emissions of acidic gases, dioxin and leachable toxic heavy metals [8]. Gasification is a high-temperature partial oxidation process in which a solid carbonaceous feedstock is converted into a gaseous mixture (H_2, CO, CO_2, CH_4, light hydrocarbons, tar, char, ash and minor contaminates) using some oxidizing agent [9].

Many studies have been performed to increase the hydrogen production yield from biomass gasification. Onel et al. [10] presents a gasifier model using MSW towards the production of liquid fuels. Using a nonlinear parameter estimation approach, the unknown gasification parameters are obtained to match the experimental gasification results. The results suggest that a generic MSW gasifier model can be obtained with an average uncertainty of 8.75%. Couto et al. [11] developed a two-dimensional CFD model for MSW gasification based on an Eulerian-Eulerian approach to describe the transport of mass, momentum and energy for the solid and gas phases. This model is validated using experimental data from the literature. The numerical results obtained are in good agreement with the reported experimental results.

In the present work, a two dimensional CFD model for MSW gasification has been used to predict and analyze the viability of the hydrogen generation from MSW gasification taking into account the equivalence ratio and steam-to-waste ratio.

2   Materials and Methods

MSW was simulated according to the average proportion of organic components (dry basis) in actual MSW of Portugal and used as feedstock for the simulations, as shown in Table 1.

Table 1. Chemical composition of the MSW [8]

Category	% weight	Chemical formula
Cellulose	49.34%	$C_6H_{10}O_5$
Hemicelulose	13.72%	$C_5H_8O_4$
Lignin	22.16%	$C_{20}H_{22}O_{10}$
Polyethylene	11.14%	$(C_2H_4)_n$
Polyethylene terephthalate	2.05%	$(C_{10}H_8O)_n$
Polypropylene	0.82%	$(C_3H_6)_n$
Polystyrene	0.77%	$(C_8H_8)_n$

It is assumed a MSW pre-treatment that gives rise to a refuse derived fuel which contains cellulosic and plastics only [8]. Cellulosic materials are mainly composed of cellulose, hemicelluloses, and lignin. Plastic residues are composed of polyethylene, polystyrene and polypropylene.

A global chemical formula is obtained by dividing the values found in the ultimate analysis of each chemical element (C, H, O) by the value of the reference element carbon (C). This MSW global chemical formula was obtained based on its chemical characterization for the Lipor MSW as shown in Table 1.

3   Mathematical Model

The gasification was modeled using Fluent data base for a two dimensional model and multi-phase (gas and solid) model. The solid phase was treated as an Eulerian granular model while the gas phase is considered as continuum. The main interaction between the phases is also modeled, heat exchange by convection, mass (the heterogeneous chemical reactions), and momentum (the drag in gas and solid phase). Numerical procedure can be seen in Ref. [11]

4   Model Validation

In order to validate the model, the results obtained numerically were compared with the experimental results of Xiao et al. [12]. The operational gasification conditions are defined in Table 2.

Table 2. Experimental operating conditions used for model validation [12]

Temperature (°C)	720	705	687	691
MSW (kg/h)	2.3	3	4	6
Air flow (Nm^3/h)	6	6	6	6
Equivalence ratio	0.5	0.38	0.29	0.19

Relative errors between numerical and experimental results are depicted in Table 3.

Table 3. Relative error of the main syngas species for various operational conditions (%)

Temperature (°C)	H$_2$	CO	CO$_2$	CH$_4$	N$_2$
720	−27.5	−1.69	−13.37	1.92	3.54
705	7.86	3.6	3.95	11.84	6.88
687	6.9	6.92	8.97	16.52	2.51
691	−7.41	2.94	4.17	1.75	4.18

The model estimates reasonably well the main species of the syngas in a large range of operating conditions. The most important errors are found for species at minor molar fractions.

Despite these numerical results being within the range of values of the experimental results, there are still a few deviations which are justified by the simplifications performed in the model, namely: the kinetic constants were taken from the literature and can differ greatly from source to source and the lack of data on the characteristics of waste has led to the use of constants of other biomasses which may differ from the actual one.

5   Results and Discussion

The performance of the MSW gasification in fluidized bed gasifier was evaluated by investigating the effects of various operating parameters (equivalence ratio (ER) and steam to waste ratio (SWR)) on the amount of hydrogen, dry gas yield, gas composition, and tar amount.

5.1   Influence of the Equivalence Ratio

Equivalence ratio is one of the most significant parameters, which have effect on the gasification process including syngas composition. ER is the ratio of the actual air/fuel ratio to the stoichiometric air/fuel ratio. ER is unitary for the stoichiometric combustion and typically range from 0.2 to 0.4 for biomass gasification. The model predictions about the influence of ER on hydrogen production are shown in Fig. 1.

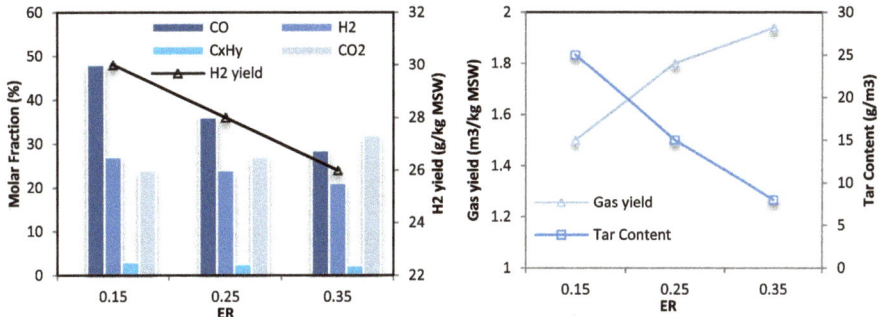

Fig. 1. Influence of ER on syngas molar fraction and hydrogen yield (left) and tar and gas yield (right). Dry and N_2-free basis. (Operating conditions: Temperature –700°C; MSW admission – 25 kg/h).

It can be observed in Fig. 1 that when ER rose, the CO_2 content increased, while CO was reduced, as oxidation reactions were favored. A decrease in H_2 and C_nH_m was also observed, because the oxidation reactions were favored when the reaction medium had higher contents of oxygen.

The influence of ER on tar content and gas yield is shown in the Fig. 1. A decrease in tar release of around 68% with the rise of ER to 0.35 is observed. As air feed rate increase with ER, more N_2 is added into the gasifier and though the output gas rate is higher, but containing more N_2. The gas yield increased with the rise of ER, but gas HHV will be lower, not only due to the diluting effect of N_2, but also because H_2 and hydrocarbons contents are lower.

5.2   Influence of the Steam to Waste Ratio

As the gasification medium is a very important parameter in governing the gas yield and composition, the effect of using steam instead of air was also studied. The steam to waste ratio is defined as the steam mass flow rate divided by the waste mass flow rate in dry basis. The use of steam as a gasifying agent increases the partial pressure of H_2O inside the gasification reactor which favors the water gas, water gas shift and steam reforming reactions, leading to increased H_2 production.

The SWR was varied over a range of values from 0 to 1.5 by holding the other variables constant. Figure 2 shows the syngas molar fractions as a function of the SWR.

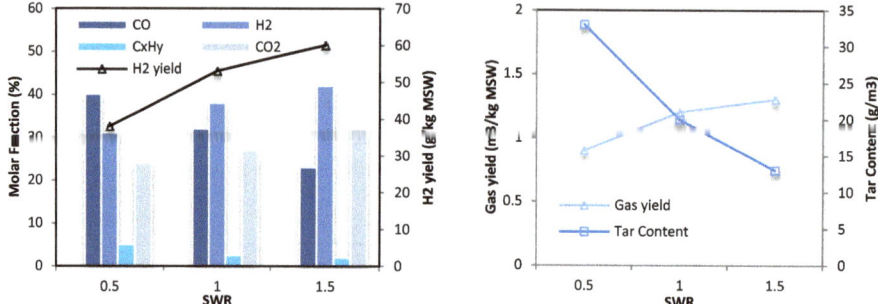

Fig. 2. Influence of SWR on syngas molar fraction and hydrogen yield (left) and tars and gas yield (right). Dry and N_2-free basis. (Operating conditions: Temperature –700°C; MSW admission – 25 kg/h).

The presence of steam in gas-phase reactions results in the decomposition of hydrocarbons C_nH_m and increasing contents of H_2 and CO_2 as reaction products.

The use of steam might also have favored water gas shift reaction, which would also explain the increase in CO_2 and in H_2 and the decrease in CO observed in Fig. 2. Moreover, it can be found from Fig. 2 that with the introduction of steam, tar content remarkably decreases, which is attributed to steam reforming of the tar with an increased partial pressure of steam. This promotion of the steam reforming reactions led to a rapid increase of dry gas yield as shown in Fig. 2, which agrees with the reduction of tar.

6 Conclusion

A two dimensional CFD model for MSW gasification has been used to predict and analyze the viability of the hydrogen generation from MSW gasification.

From the simulated results it is possible to verify that the increase of ER has a negative effect on H_2 production because the oxidation reactions are favored when the reaction medium had higher contents of oxygen. On the other hand, the increase of ER has a positive effect on the reduction of tar content with increased gas yield.

The use of steam as a gasifying agent in gas-phase reactions results in the decomposition of hydrocarbons and increasing contents of H_2. The introduction of steam also leads to more tar participating in steam reforming, which led to a rapid increase of gas yield and tar reduction because of higher conversion efficiency.

References

1. Balat, H., Kirtay, E.: Hydrogen from biomass – present scenario and future prospects. Int. J. Hydrogen Energy **35**, 7416–7426 (2010)
2. Tanksale, A., Beltramini, J.N., Lu, G.Q.M.: A review of catalytic hydrogen production processes from biomass. Renew. Sustain. Energy Rev. **14**, 166–182 (2010)
3. Koroneos, C., Dompros, A., Roumbas, G.: Hydrogen production via biomass gasification – a life cycle assessment approach. Chem. Eng. Process. **47**, 1261–1268 (2008)
4. Bartels, J.R., Pate, M.B., Olson, N.K.: An economic survey of hydrogen production from conventional and alternative energy sources. Int. J. Hydrogen Energy **35**, 8371–8384 (2010)
5. Ni, M., Leung, D.Y.C., Leung, M.K.H., Sumathy, K.: An overview of hydrogen production from biomass. Fuel Process. Technol. **87**, 461–472 (2006)
6. Biagini, E., Masoni, L., Tognotti, L.: Comparative study of thermochemical processes for hydrogen production from biomass fuels. Bioresour. Technol. **101**, 6381–6388 (2010)
7. Porteous, A.: Energy from waste incineration – a state of the art emissions review with emphasis on public acceptability. Appl. Energy **70**, 157–167 (2001)
8. Teixeira, S., Monteiro, E., Silva, V., Rouboa, A.: Prospective application of municipal solid wastes for energy production in Portugal. Energy Policy **71**, 159–168 (2014)
9. Couto, N., Rouboa, A., Silva, V., Monteiro, E., Bouziane, K.: Influence of the biomass gasification processes on the final composition of syngas. Energy Procedia **36**, 596–606 (2013)
10. Onel, O., Niziolek, A.M., Hasan, M.M.F., Floudas, C.A.: Municipal solid waste to liquid transportation fuels e part I: mathematical modeling of a municipal solid waste gasifier. Comput. Chem. Eng. **71**, 636–647 (2014)
11. Couto, N., Silva, V., Monteiro, E., Teixeira, S., Chacartegui, R., Bouziane, K., Brito, P.S.D., Rouboa, A.: Numerical and experimental analysis of municipal solid wastes gasification process. Appl. Thermal Eng. **78**, 185–195 (2015)
12. Xiao, G., Jin, B.-S, Zhong, Z.-P., Chi, Y., Ni, M.-J., Cen, K., Xiao, R., Huang, Y.-J., Huang, H.: Experimental study on MSW gasification and melting technology. J. Environ. Sci. **19**, 1398–1403 (2007)

Strategies for the Design of Domestic Pellet Boilers

Pedro Ribeiro, Candida Vilarinho, Manuel Ferreira, Eurico Seabra,
Lelis Fraga, and José Teixeira[✉]

Universidade do Minho, 4800-058 Guimarães, Portugal
jt@dem.uminho.pt

Abstract. Domestic and industrial heating are amongst the most attractive applications for biomass as it combines high efficiency and ease-of-use. Boilers and furnaces in this range are usually of simple design and control algorithm while expected to operate over a wide range of thermal loads, typically from up 30% of the nominal load. Such constrainments often limit the overall efficiency of biomass use. In addition, from the manufacturing point of view, a product with flexibility of operation would provide a quasi-universal solution for a wide range of applications.

The present paper investigates the influence of the fuel grate dimensions and the heat exchanger design on the thermal efficiency of a 25 kW pellet boiler. The prototype has a controllable independent supply of primary and secondary air and the temperature is monitored at critical locations. The continuous analysis of the flue gases for CO_2, NOx, O_2 and CO enabled the definition of an operation envelope that resulted in CO emissions as low as 10 ppm, using standard A1 grade pellets as fuel.

Keywords: Combustion · Biomass · Emissions

1 Introduction

The design of a pellet boiler that can be thermally efficient and with reduced emissions is directly associated with three components: the fuel grate, the combustion chamber and the heat exchanger.

The air supply is of paramount relevance for both total the amount and location of the supply ports and it will be balanced between the low emissions (particularly CO) and high thermal efficiency. On one hand low CO would require an air flow rate above the stoichiometric conditions; on the other hand, the excess air (EA) should be as low as possible in order minimize the thermal losses by the stack. Low flame temperature may also contribute to incomplete combustion which would contribute to higher CO and lower efficiency. In addition, because the immediate route is the preferred mechanism for NOx formation in pellet boilers, and it is promoted by the presence of oxygen, the EA will contribute directly to the NOx emissions.

For small scale pellet boilers for domestic applications, the most effective and easily implemented technique for NOx control is by air staging. This technique is characterized by the creation of an intermediate reduction zone by separate the air

© Springer International Publishing AG, part of Springer Nature 2019
J. Machado et al. (Eds.): HELIX 2018, LNEE 505, pp. 668–674, 2019.
https://doi.org/10.1007/978-3-319-91334-6_91

supply through two zones: *(i)* Primary Air (PA) introduced through the fuel bed to promote its devolatilization; and *(ii)* Secondary Air (SA), supplied downstream of PA, to complete the fuel oxidation.

In order to be effective, the following conditions should be observed: (i) in order to limit the oxygen availability in the fuel bed, the PA should be in the range $0.7 < \lambda < 0.8$; (ii) in the reduction zone the temperature should be in the region of 1,100–1,200 °C and the residence time within the 0.3 to 0.5 s for ensuring complete combustion [1, 2]. As such, the flow interaction between the volatiles and the SA, is crucial to the design of the boiler.

The size for a gravity fed (the pellets are introduced from above) fixed grate boiler are directly related with the power rate of the device. Ronnback et al. [3] investigated the ignition rate in pellet fuel beds as a function of the PA supply rate. The ignition rate determines the power level of the boiler as it correlates with the fuel consumption. The results show that the ignition rate is a function of the PA flow rate and reaches a maximum at approximately at PA = 0.3 kg_{air}/m^2s, leveling off beyond this point. The maximum ignition rate is approximately 0.07 kg_{fu}/m^2s [3]. This behavior suggests that the thermal load can be controlled by adjusting the PA and is limited to a maxim value of 0.07 kg_{fu}/m^2s.

One of the certification requirements for pellet boilers is that they must operate at a power rate down to 30% of the nominal level. In such conditions, when the burner is designed to operate at the nominal level, one may find that the ignition rate can be higher than the fuel supply rate. This will lead to a decrease of the fuel on the bed and eventually to the flame extinction. Furthermore, in order to keep the CO emissions low, pellet boilers are usually operated at low power levels with higher EA than that expected at full load. This will make the problem worse. In such scenario, one technique would be to adjust the PA supply rate through the grate by combining a reduction of the exhaust blower velocity with throttling the PA supply.

The heat exchanger, that transfers the thermal energy to water, should be of simple design and cost effective. Its efficiency is related with the exhaust temperature of the flue gases which should be above the dew point to prevent condensation in the stack.

The present paper addresses these issues regarding the boiler design. The experimental work was carried out on a purpose built prototype whose main components can be tailored for evaluating the influence of the grate design, the heat exchanger and the combustion chamber amongst others.

2 Experimental Details

The experiments were carried out using 100% pine sawdust pellets, 6 mm in diameter. These are primarily used for domestic applications as they meet the EN Plus standards for Class A [2].

Table 1 summarizes the physical characteristics of the pellets.

The heat value was measured by a calorimeter Leco AC500, according to CEN/TS 14918 standard. The elemental analysis was performed according to standard CEN/TS 15104 using a LECO TruSpec Series. The composition in C, S and H was measured by

Table 1. Physical characterization of pellets.

Parameters	
Moisture content (%)	5.1
Bulk density (kg/m³)	624
Ash content (%)	0.9
Low heat value (kJ/kg)	17,500
C (%)	51.1
N (%)	0.1
O (%)	42.5
H (%)	6.3
S (%)	0.73

infrared absorption and that of N by thermal conductivity. Other elements were measured by an X-ray fluorescence spectrometry technique.

The prototype boiler is a mono-block type, which means that it includes the burner, combustion chamber and heat exchanger integrated into a single structure. The heat exchanger has two passes; the first consists of fourteen tubes with 370 mm length and the second, downwards, is made up of six with 900 mm length, all of them with inside diameter of 42 mm. The flue gases exhaust is made up by a fan to extract the flow of the gases from inside the boiler. The entire apparatus operates in draft, below atmospheric pressure.

The cooling system includes a cross-flow air/water heat exchanger that incorporates a fan for air circulation. Water circulation is provided by a water pump which is rated at 500 L/hr. A calibrated rotameter for water flow rate measurement is also included in the cooling loop.

The prototype burner has been described in detail in other publication [4]. It includes independent primary (PA) and secondary (SA) air inlets. Primary air is injected under the grate and the secondary air is injected above the grate through 25, 5 mm diameter orifices uniformly distributed along each one of the burner walls, 70 mm above the grate. Each air inlet includes a valve for flow rate control. The burner construction allows for an easily change the grate design for testing purposes. The fuel is supplied by gravity from an external conveyer to the boiler by a helical screw. A digital scale under the pellet storage tank directly measured the fuel consumption. This has an accuracy of 0.05 kg.

The gas sample was extracted from the final section of the stack by means of a vacuum pump. Upstream of the vacuum pump the sample was filtered and cooled in order to remove particles and moisture. A Multi Gas Analyzer, model SIGNAL 9000MGA, was used for continuous monitoring the flue gas emissions: O_2, CO_2 and CO which are monitored through a NI PXI-1052 data acquisition system. The details of the assembly and procedure are described elsewhere [4].

3   Results and Discussion

One of the crucial details in designing a burner is to size the fuel grate for the desired range of thermal loads. For this paper were tested three different dimension grates: *(i)* 135 × 135 mm grate; *(ii)* 135 × 115 mm grate; *(iii)* 135 × 95 mm grate. Because the designed burner has independent PA and SA supply control that allows for slowing down the burning process by throttling down the PA supply. In this way, the burner should be designed in order to ensure that the fuel supply does not exceed the maximum ignition rate for wood pellets on a fixed grate. Ronnback et al. or plotted the ignition rate as a function of the superficial air (PA) velocity through the grate as shown in Fig. 1 by the open circles. For this purpose, it was manufactured a 135 × 135 mm grate, which corresponds to a fuel supply for the nominal load (25 kW) of less than 0.08 k_{gfu}/m^2s, as represented by the red dot in Fig. 1.

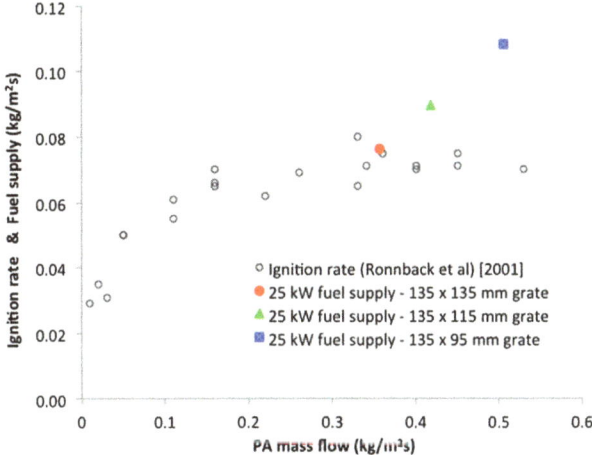

Fig. 1. Specific ignition rate and fuel supply for three different dimension grates in function of the PA mass flow [3].

The data suggests that there is a limit for the devolatization rate as a function of the superficial velocity. On other words, for higher power outputs, the size of the grate has to be increased. Smaller grates (blue and green symbols in Fig. 1) would be unbalanced has the fuel supply rate is greater than the combustion rate. Three fuel grates were tested. For the larger ones it was observed that the burner was capable to operate at any load without being overloaded of pellets.

However, especially at partial load, it was noted that the fed pellets tended to accumulate on preferential regions of the grate, aligned with the feeding direction and a portion of its surface would be free of fuel. This phenomenon leads to a misallocated PA supply, where most of the air would flow through the grate area that had less or none pellets.

On these circumstances, it was observed an effective lack of air on the fuel bed causing incomplete combustion and high CO emissions (650 ppm average at 13% O_2 for the largest grate; 350 ppp for the middle size) and, because of the fact that the majority of the oxygen passed through the grate without being reduced and the combustion was characterized by a high level of EA.

A further reduction of the grate size was achieved by the construction of an even smaller and rectangular grate, with a cross section of 135 × 95 mm. With this grate it was achieved an even fuel bed characterized by a very good CO emission level (10 ppm @ 13% O_2), as depicted in Fig. 2. The PA and fuel were well mixed, allowing the boiler to operate at a low EA, improving the combustion efficiency. With this particular grate, the problem of burner overload became important at full load (25 kW). In such conditions it was observed occasionally situations of fuel overload that forced to interrupt the feeding cycle once at approximately every hour of operation. The residence time of the gases at high temperatures is an essential condition for a complete burnout of the particles and thus to enhance a low level of CO emissions. So, in addition to the volume, the height of the combustion chamber is the most important dimension.

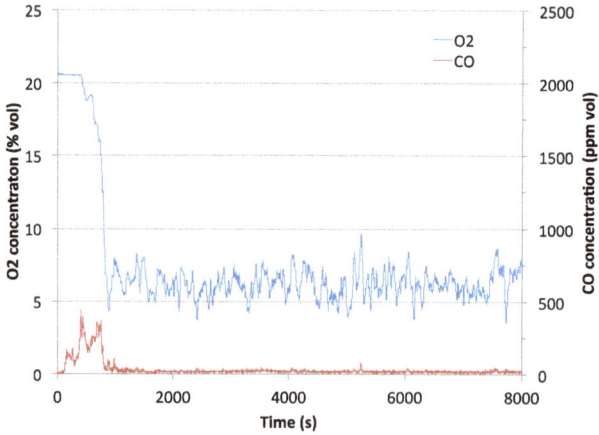

Fig. 2. O_2 and CO concentration on flue gases for test at 20 kW with 135 × 95 mm grate.

Previous experiments on early designs for the boiler had revealed that at full load, part of the flame reach the heat exchanger entrance, forcing the gases to cool down before complete combustion. This had been taken into account on the design of the present prototype. The combustion chamber has 253 × 320 mm of section area and 530 mm high.

Because the combustion chamber has wet walls the gases tend to suffer a severe cooling before entering in the heat exchanger. This may lead to an incomplete combustion of some gaseous particles, especially at high loads. To prevent this, a thermal insulation, out by 25 mm fire clay boards, was tested. The results show that the CO emissions can be reduced by a factor of 3–4.

Because the space heating domestic pellet boilers need to modulate the power to at least 30% of the nominal load [6, 7], one of the biggest challenges in the design of an efficient pellet boiler is the heat exchanger geometry. The efficiency of the heat exchanger is directly related to the exhaust temperature. Due to the fact that the exhaust temperature increases with the load, the efficiency of the heat exchanger decreases. So, typically, a pellet boiler has the maximum thermal efficiency at low loads and minimum thermal efficiency at nominal loads. In fact, there exists a minimum acceptable temperature for the flue gases that limits the heat exchanger efficiency: if the gases cool down to the dew point, condensation of gases and soot on the chimney and corrosion may occur [8].

Two experiments were made: *(i)* with 4 tubes of the 1st pass intentionally blocked; *(ii)* with 4 tubes of the 1st pass and one tube of the 2nd pass intentionally blocked. The results are depicted in Fig. 3 and suggest that it is not possible to reasonably control the exhaust temperature of a fume type domestic boiler by the disabling some of the heat exchanger tubes.

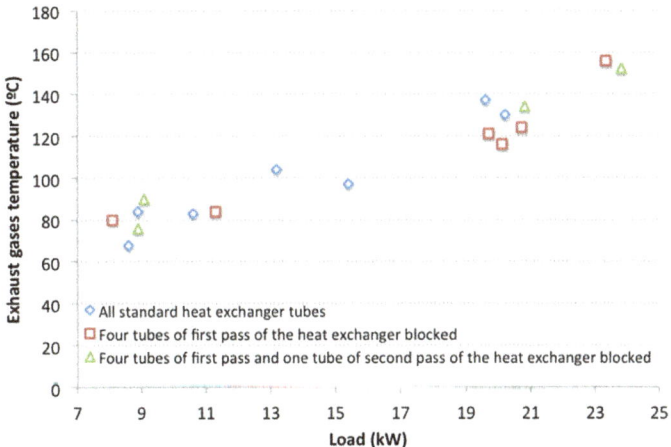

Fig. 3. Average exhaust temperature versus load for different heat exchanger configurations.

When the flow pass is blocked on some tubes, the velocity of the gases through the heat exchanger increases, due to the reduction of the section area, increasing the convective heat transfer coefficient.

4 Conclusions

The present work discusses and evaluates various alternatives in the design of small scale pellet burners. These aim to improve the overall efficiency while keeping the emissions (CO) at a very low level.

The design of the fuel grate (mostly the cross section area) has to be able to devolatize the fuel at rate equal to the fuel supply. Although this is easy to attend for a fixed load, it is troublesome to provide stable conditions for a wide range of operating conditions. The most efficient and cost effective alternative is to size the grate for a medium thermal load and introduce routines for periodically cleaning the grate in order to prevent fuel accumulation.

The application of a refractive lining on the combustion chamber prevents the flame temperature to drop and enabling the oxidation reaction to be completed inside the combustion chamber. This will contribute to the reduction of the emissions.

The correct adjusting of the surface area to the thermal load is difficult to implement in a simple manner such as reducing the number of tubes. In fact, for a fixed flow rate of gases, the heat exchanged only depends upon the length of the tubes and not on their diameter. Nonetheless, the problem could be circumvented by operating the boiler at the full thermal load during the start-up period which is the most likely to promote vapor condensation in the stack.

Acknowledgments. This work was financed by National Funds-Portuguese Foundation for Science and Technology, under Strategic Project and PEst-C/EME/UI4077/2011 and the private company AMARO.

References

1. Hao, L., Chaney, L., Li, L., Sun, C.: Control of NOx emissions of a domestic/small-scale biomass pellet boiler by air staging. Fuel **103**, 792–798 (2013)
2. Nussbaumer, T.: Primary and secondary measures for the reduction of nitric oxide emissions from biomass combustion. In: Developments in Thermochemical Biomass Conversion. Blackie Academic (1997)
3. Ronnback, M., Axell, M., Gustavsson, L.: Combustion processes in a biomass fuel bed – experimental results. In: Progress in Thermochemical Biomass Conversion, pp. 743–757 (2001)
4. Ribeiro, P., Teixeira, J., Ferreira, M.: Ash sintering in a biomass pellet boiler. In: ASME Conference Proceedings IMECE 2013, San Diego, CA, paper IMECE2013-65246, Volume 8A: Heat Transfer and Thermal Engineering (2013)
5. Bostrom, D., Skoglund, N., Grimm, A., Boman, C., Ohman, M., Brostr, M., Backman, R.: Ash transformation chemistry during combustion of biomass. Energy Fuels **26**, 85–93 (2012)
6. Obernberger, I., Thek, G.: The Pellet Handbook: The Production and Thermal Utilisation of Pellets. Earthscan, London (2010)
7. European Standard EN 14785: Residential Space Heating Appliances Fired by Wood Pellets-Requirements and Test Methods, European Committee for Standardization (2003)
8. Hansen, M., Jein, A., Hayes, S., Bateman, P.: English Handbook for Wood Pellet Combustion, Pelletsatlas (2003)

Experimental Biogas Production and Biomethane Potential of Swine Wastewater Among Different Production Stages

G. Lourinho[⊠], P. S. D. Brito, and L. F. T. G. Rodrigues

VALORIZA - Research Center for Endogenous Resource Valorization,
Campus Politécnico, 10, 7300-555 Portalegre, Portugal
glourinho@gmail.com

Abstract. Effluent streams originating from swine production can be a major cause of point-source pollution and very dangerous for the surrounding environment. In this work, biogas and biomethane potential of swine-derived effluents were assessed under mesophilic conditions (37 °C) among different production stages (Farrowing, Gestation, Weaners, and Fattening) in order to investigate the suitability of anaerobic digestion as a treatment technology and energy recovery tool. Anaerobic biodegradability of the different wastewaters were also evaluated in order to better understand the degradation patterns of each waste stream. The specific methane yields for each substrate wastewater were successfully determined gravimetrically and ranged from 293.9 ± 31.6 mL CH4.gVS added^{-1} for Gestation, 248.0 ± 246.6 mLCH4.gVS added^{-1} for Fattening, and 172.4 ± 1.62 mLCH4.gVS added^{-1} for Weaners. Farrowing wastewater presented no detectable biomethane production in the studied conditions probably due to very low organics content. Anaerobic biodegradability results showed that wastewaters from Gestation (60.28%) had a higher biodegradability index when compared with Fattening (37.96%) and Weaners (45.97%) stages. The results obtained in this work should be encouraging for on-farm energy recovery and anaerobic digestion technology viewed as an important contributor to alleviate the increasing energy demand within the swine industry. However, as indicated by biodegradability data, biological treatment does not comprise a complete solution and should be complemented with other treatment methods.

Keywords: Swine wastewater · Anaerobic digestion · Biogas production Biodegradability

1 Introduction

Swine wastes are a complex biogenic organic-inorganic waste generated by the natural process of animal food digestion and can be defined as a mixture of animal excrements (45% solid and 55% liquid) [1], and process waters used for sanitary purposes in swine farms [2, 3]. As the main by-product of swine farms, the liquid fraction of swine manure designated as swine wastewater (SW), is typically generated in large quantities

© Springer International Publishing AG, part of Springer Nature 2019
J. Machado et al. (Eds.): HELIX 2018, LNEE 505, pp. 675–681, 2019.
https://doi.org/10.1007/978-3-319-91334-6_92

in non-bedding intensive animal farming, being the subject of great environmental concern in the modern world as a major cause of point-source pollution from wastewater discharge or overfertilization of soils [1, 3, 4]. At present, the main challenge in swine waste management in large farms is thus effluent processing. As a consequence, extensive research has been carried out on the use of novel technologies to adequately handle effluent generation at a farm level. There are three primary approaches for swine waste remediation at farm scale: solid/liquid separation methods, solid fraction treatments, and liquid fraction treatments. From those mentioned, the most common process to improve the characteristics of SW is anaerobic digestion (AD). AD is the biotechnological application of methanogenesis, a process which uses microorganisms to achieve partial degradation of organic matter in the absence of oxygen. In AD, complex microbial communities mediate the conversion process in a series of consecutive steps (hydrolysis, acidogenesis, acetogenesis, and methanogenesis) and yield a mixture of gaseous products comprising mainly CH_4 and CO_2 called biogas. The presence of CH_4 in high quantities (typically 60–65%) in biogas has established its use as a fuel within many applications [5, 6].

Some important studies have been published dealing with the digestion characteristics of liquid swine wastes in many farms. Nevertheless, there are few studies which portrait a complete picture of animal growth cycle as these studies often do not include the variability between the various growth stages in which large farms usually operate. This paper is thus focused on the characterization and evaluation of biogas and biomethane potential of swine-derived effluents from different production stages. The study specifically assesses the degradation patterns of the swine waste streams produced in a large scale farm using BMP assays. The other aim of this work is to demonstrate the inadequacy of AD as a full-scale treatment technology of especially dirty effluents such as swine wastewater while simultaneously providing valuable insights for a future *in situ* implementation of the technology, in this way not devaluing its importance as an energy recovery tool for agro-industries.

2   Materials and Methods

2.1   Farm Description, Substrates, and Inoculum

The swine wastewater used in the study was obtained from a large-scale pig farm operating in a closed cycle, i.e., pigs are raised in separated breeding units, located in Alentejo, Portugal. Swine wastewater samples were collected from drainpipes that carried wastewater from slatted floors inside buildings to external tanks before any treatment was applied and were characterized based on selected physicochemical and biochemical parameters relevant for the digestion process according to standard methods [7]. The main characteristics of the wastewater collected in the different production stages were as follows (Gestation, Farrowing, Weaners, and Fattening, respectively): pH 7.8 ± 0.01, 7.3 ± 0.01, 8.15 ± 0.01, 8.05 ± 0.06; 4651 ± 101 mg.L^{-1} TS, 752 ± 38 mg.L^{-1} TS, 6158 ± 8 mg.L^{-1} TS, 12088 ± 630 mg.L^{-1} TS; 1820 ± 167 mg.L^{-1} TVS, 398 ± 15, 2348 ± 22 TVS, 5220 ± 793 TVS. The inoculum was fresh cow manure obtained from a local farm in Portalegre, Portugal, with the following characteristics: pH 7.3,

20.44 ± 0.45%TS and 60.96 ± 1.39% VS (%TS). Each sample was classified according to the different growth stages and stored at −4 °C until usage in the present work.

2.2 Biogas and Biomethane Assay Experimental Procedure

The experimental procedure followed in this study was based on the principles described by [8] and later revised by other researchers [9, 10]. Specifically, BMP assays were carried out in serum bottles (Schott Duran, Germany) of 1000 mL, with a working liquid volume of 600 mL and a headspace volume of 400 mL. Bottles contained appropriate quantities of inoculum/substrate in a ratio of 3.0 g VS inoculum/g VS substrate. The bottles were flushed with nitrogen gas, sealed with 5 mm thick silicone discs (Schott Duran, Germany) and closed by a plastic screw cap (Schott Duran, Germany). Each bottle was then placed in an incubator at a constant mesophilic temperature (37 °C). Biogas production was measured indirectly by mass loss (gravimetric method) as detailed by [11]. The mass of each reactor was determined to the nearest 10 mg before and after any period of biogas production, during which time biogas was removed/collected in gas bags by puncturing with a hypodermic needle until atmospheric pressure/equilibrium was reached. The collected biogas was immediately characterized by a portable gas analyzer (GasData GFM406) which allows measuring the volume percentages of CO_2, CH_4, O_2, H_2S, CO in the mixture. Mixing was performed manually during the incubation period at regular times. All experimental assays were performed in duplicate and control/blank bottles with only inoculum/water were included in order to correct the obtained methane production. All values are expressed at standard temperature and pressure.

2.3 Theoretical BMP, and Methane-Based Biodegradability, and Data Analysis

The amount of biogas produced from the degradation of a specific sample, in the case SW from different production stages, was theoretically estimated using its elemental composition (see [12] for more details) and COD characterization (see [13, 14]. Statistical analysis of the experimental results with respect to samples characterization and biogas and biomethane potential was carried out by means of R software [15]. A one-way ANOVA test was implemented to evaluate the different growth stages, after which post hoc multiple comparisons were carried out by means of the Tukey HSD test at the 95% confidence level. All values correspond to the mean of two independent replicates (n = 2) ± standard deviation (SD).

3 Results and Discussion

3.1 Experimental Biogas and Biomethane Production

Results concerning the cumulative biogas production (Fig. 1), daily biogas production, and methane content in the biogas were obtained during the digestion of SW from different production stages under mesophilic conditions. Biogas production started immediately on the first 5 days of digestion in all reactors, but those containing weaners wastewater. Biogas production was very low in these digesters until day 9 when the gas production began to sustainably increase until day 14. Thereafter, biogas production remained relatively constant until day 21 before slowly decreasing and ceasing production in day 54. In the reactors containing gestation and fattening wastewaters, biogas production rapidly increased until days 14 and 9, respectively, before continuously decrease until the end of the experiment. Peak values for the daily biogas production rate were calculated to be 25.0 ± 0.78, 13.8 ± 0.74, and 17.9 ± 1.49 mL.gVS added^{-1} after 14, 14, and 9 days for swine wastewater from gestation, weaners, and fattening stages, respectively. Farrowing wastewater showed no detectable biogas production in the studied conditions due to their very low organics content as measured by VS and COD; in reality, however, some biogas was likely produced during the experiment but in such a low amount that it was not measurable gravimetrically. Gestation wastewater presented the largest average biogas potential after 54 days (461.3 ± 33.7 mL.gVS added^{-1}) followed by weaners (245.8 ± 3.16 mL.gVS added^{-1}) wastewater. Fattening wastewater, in turn, presented a biogas potential of 348.6 ± 342.54 mL.gVS added^{-1}) after 61 days. ANOVA on the cumulative biogas production showed no statistically significant differences between different stages. As shown in Fig. 1, most of the final biogas yields were obtained in the first 28–36 days of digestion for each production stage, specifically 89.8%, 90.1%, and 95.8% for fattening, gestation and weaners wastewater, after 36, 28, and 36 days, respectively. These values may be used as suitable hydraulic retention times in the treatment of each wastewater in a continuous system [16]. The methane content in the biogas produced presented a similar trend in all digesters. Fattening wastewater presented the higher methane contents with values increasing from $28.85\% \pm 7.00$ to $60.2\% \pm 8.34$ between day 5 and day 21. With respect to gestation wastewater, the methane content increased from $8.70\% \pm 3.54$ in day 5 to a maximum value of $47.1\% \pm$ after about 36 days. In the digesters containing weaners wastewater, a peak value of $45.45\% \pm 0.07$ methane was achieved in day 36, increasing from $4.6\% \pm 0.00$ in day 5 of digestion. The specific methane yield for each substrate wastewater was 248.0 ± 246.6 mLCH$_4$.gVS added^{-1}, 293.9 ± 31.6 mLCH$_4$.gVS added^{-1}, and 172.4 ± 1.62 mLCH$_4$.gVS added^{-1} for fattening, gestation, and weaners wastewater. From these results, it is apparent that the digesters containing fattening wastewater showed a very high variability between duplicates. Despite similar sampling procedures during collection, issues may have occurred during sample management prior to storage resulting in uneven samples for digestion. In fact, according to previous research, representative and uniform samples cannot be obtained without using proper agitation methods during sample collection and storage [17]. Nevertheless, when consistently comparing these results with typical methane values obtained by other researchers, similarities can be found suggesting that the overall yield obtained for fattening wastewater is reasonable [18, 19].

3.2   Theoretical BMP and Methane-Based Biodegradability

In order to evaluate the biodegradability of the tested wastewaters, the specific methane yield obtained in the BMP assay was compared with the theoretical methane yield estimated using its elemental composition and organics content measured via COD.

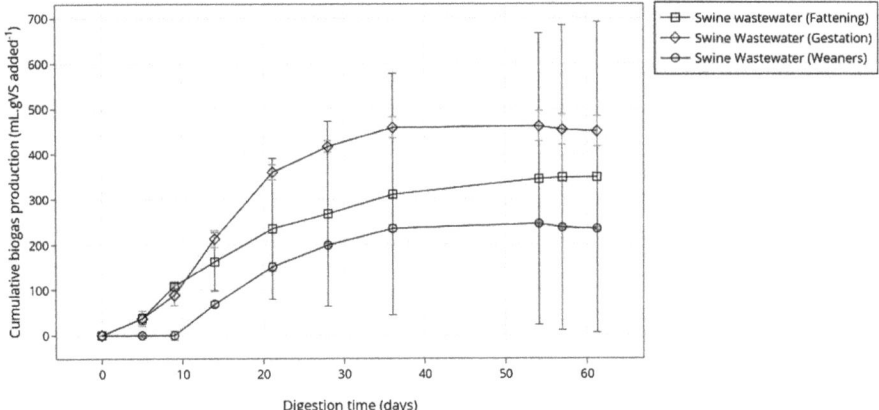

Fig. 1. Cumulative biogas production (ml.gVS added^{-1}) of fattening, gestation and weaners wastewater.

Theoretical methane potential calculation results showed that wastewaters from the fattening stage (BMP$_{th}$ = 653.46, BMP$_{thCOD}$ = 991.47) had a higher theoretical potential than gestation (BMP$_{th}$ = 487.65, BMP$_{thCOD}$ = 471.69) or weaners (BMP$_{th}$ = 374.98, BMP$_{thCOD}$ = 618.84) wastewater. In general, the theoretical results obtained from elemental composition presented a better agreement with the experimental results. According to this methodology, the biodegradability of gestation, fattening, and weaners wastewater was 61.45%, 39.57%, and 46.49%. The results are in the range of those obtained by other authors [19] who experimented with similar swine wastewater streams. Despite having the higher theoretical potential, experimental values for fattening wastewater were lower when compared to gestation wastewater due to lower biodegradability, indicating that gestation wastewater is a more desirable substrate for anaerobic digestion. Also, the superior theoretical potentials in terms of COD for fattening and weaners wastewater suggest the presence of recalcitrant organic compounds not easily treated via biological methods. This is likely the case for fattening wastewaters as BOD$_5$/COD ratio, which may be conceived as an acceptable index of biological treatability, is relatively low (0.15) [20]. For weaners wastewater, however, some level of process inhibition may have occurred probably due to nonoptimized conditions.

4   Conclusion

In this study, swine wastewaters from a farrow to finish farm were anaerobically digested under mesophilic conditions. In general, the characteristics of gestation, weaners, fattening, and farrowing wastewaters were significantly different due to different feed strategy and nutrient digestibility at different growth stages. Fattening wastewater presented higher solids and organics contents, while farrowing effluent streams were considered low strength due to mild physicochemical characteristics. The specific methane yields for each substrate wastewater was successfully determined gravimetrically and ranged from 293.9 mLCH$_4$.gVS added^{-1} for gestation and 172.4 mLCH$_4$.gVS added^{-1} for weaners. Farrowing wastewater presented no detectable biogas production in the studied conditions probably due to very low organics content. These values are encouraging for on-farm energy recovery and may provide an important contributor to alleviate the increasing energy demand within the industry. With this in mind, experimental data suggest an HRT between 28–36 days for an effective treatment under continuous operations. However, as indicated by the partial biodegradability of the waste streams, AD should not be viewed as a complete solution and should be integrated with other treatment methods, either as pre-treatment or post-treatment.

Acknowledgments. G. Lourinho would like to thank FCT- Fundação para a Ciência e Tecnologia for financial support regarding the grant SFRH/BDE/111878/2015.

References

1. Villamar, C.A., Canuta, T., Belmonte, M., Vidal, G.: Characterization of swine wastewater by toxicity identification evaluation methodology (TIE). Water. Air. Soil Pollut. **223**, 363–369 (2012)
2. Vassilev, S.V., Baxter, D., Andersen, L.K., Vassileva, C.G.: An overview of the chemical composition of biomass. Fuel **89**, 913–933 (2010)
3. Marszałek, M., Kowalski, Z., Makara, A.: Physicochemical and microbiological characteristics of pig slurry. Tech. Trans. Chem., 81–91 (2014)
4. Kowalski, Z., Makara, A., Fijorek, K.: Changes in the properties of pig manure slurry. Acta Biochim. Pol. **60**, 845–850 (2013)
5. Mata-Alvarez, J., Dosta, J., Romero-Güiza, M.S., Fonoll, X., Peces, M., Astals, S.: A critical review on anaerobic co-digestion achievements between 2010 and 2013. Renew. Sustain. Energy Rev. **36**, 412–427 (2014)
6. O'Flaherty, V., Collins, G., Mahony, T.: Anaerobic digestion of agricultural residues. In: Environmental Microbiology, pp. 259–279. Wiley, Hoboken (2010)
7. APHA/AWWA/WEF: Standard Methods for the Examination of Water and Wastewater. Stand. Methods. 541 (2012)
8. Owen, W.F., Stuckey, D.C., Healy, J.B., Young, L.Y., McCarty, P.L.: Bioassay for monitoring biochemical methane potential and anaerobic toxicity. Water Res. **13**, 485–492 (1979)
9. Angelidaki, I., Alves, M., Bolzonella, D., Borzacconi, L., Campos, J.L., Guwy, A.J., Kalyuzhnyi, S., Jenicek, P., Van Lier, J.B.: Defining the biomethane potential (BMP) of solid organic wastes and energy crops: a proposed protocol for batch assays. Water Sci. Technol. **59**, 927–934 (2009)
10. Holliger, C., Alves, M., Andrade, D., Angelidaki, I., Astals, S., Baier, U., Bougrier, C., Buffiere, P., Carballa, M., de Wilde, V., Ebertseder, F., Fernandez, B., Ficara, E., Fotidis, I., Frigon, J.-C., de Laclos, H.F., Ghasimi, D.S.M., Hack, G., Hartel, M., Heerenklage, J., Horvath, I.S., Jenicek, P., Koch, K., Krautwald, J., Lizasoain, J., Liu, J., Mosberger, L., Nistor, M., Oechsner, H., Oliveira, J.V., Paterson, M., Pauss, A., Pommier, S., Porqueddu, I., Raposo, F., Ribeiro, T., Rusch Pfund, F., Stromberg, S., Torrijos, M., van Eekert, M., van Lier, J., Wedwitschka, H., Wierinck, I.: Towards a standardization of biomethane potential tests. Water Sci. Technol., 1–9 (2016)

11. Hafner, S., Rennuit, C., Triolo, J.M., Richards, B.K.: Validation of a simple gravimetric method for measuring biogas production in laboratory experiments. Biomass Bioenergy **83**, 297–301 (2015)
12. Symons, G.E., Buswell, A.M.: The methane fermentation of carbohydrates. J. Am. Chem. Soc. **55**, 2028–2036 (1933)
13. Jingura, R.M., Kamusoko, R.: Methods for determination of biomethane potential of feedstocks: a review. Biofuel Res. J. **4**, 573–586 (2017)
14. Nielfa, A., Cano, R., Fdz-Polanco, M.: Theoretical methane production generated by the co-digestion of organic fraction municipal solid waste and biological sludge. Biotechnol. Reports. **5**, 14–21 (2015)
15. R Development Core Team: A Language and Environment for Statistical Computing (2011)
16. Li, Y., Zhang, R., Liu, X., Chen, C., Xiao, X., Feng, L., He, Y., Liu, G.: Evaluating methane production from anaerobic mono- and co-digestion of kitchen waste, corn stover, and chicken manure. Energy Fuels **27**, 2085–2091 (2013)
17. Zhu, J., Ndegwa, P.M., Zhang, Z.: Manure sampling procedures and nutrient estimation by the hydrometer method for gestation pigs. Bioresour. Technol. **92**, 243–250 (2004)
18. Guo, J., Clemens, J., Li, X., Xu, P., Dong, R.: Performance evaluation of a Chinese medium-sized agricultural biogas plant at ambient temperature. Eng. Life Sci. **12**, 336–342 (2012)
19. Zhang, W., Lang, Q., Wu, S., Li, W., Bah, H., Dong, R.: Anaerobic digestion characteristics of pig manures depending on various growth stages and initial substrate concentrations in a scaled pig farm in Southern China. Bioresour. Technol. **156**, 63–69 (2014)
20. Orhon, D., Ateş, E., Sözen, S., Çokgör, E.U.: Characterization and COD fractionation of domestic wastewaters. Environ. Pollut. **95**, 191–204 (1997)

Solutions Notes on Clean Textile Waste

Carina Jordão[1] , Regis Puppim[1,2(✉)] ,
and Ana Cristina Broega[1]

[1] Textile Engineering Department, Centre for Textile Science and Technology,
University of Minho, Azurém Campus, Guimarães, Portugal
carijordao@hotmail.com, regispuppim@gmail.com,
cbroega@det.uminho.pt
[2] Federal Institute of Goiás,
Aparecida de Goiânia Campus, Aparecida de Goiânia, Brazil

Abstract. In the contemporary moment, the textile and clothing industries are practicing and passing through a transition period into their management systems (specially on waste) due to consumer change in social behavior, which can, potentially, ensure the sustenance and well-being for the generations to come. Into this context, the present research aims to analyze sustainable strategies that are being implemented by the textile and clothing industry, searching for solutions for the management of clean solid waste, concerns with processed and non-processed textile fibers waste. In addition, it is mapped and pointed some gaps for the development of further and future researches. In order to achieve the proposed goal, it was searched a rigorous selection of projects and studies that expressed innovative solutions, concerning practices and theories application about the possibilities in the management of the clean textile waste. The results are optimistic, as they already point to some real action in terms of companies aware of the need for sustainable practices in their production.

Keywords: Textile sustainability · Clean textile waste · Waste management

1 Introduction

Sustainability has been an emerging issue on the strategic decisions of different industrial sectors in response to the demands of a more conscientious consumer, as well as more stringent environmental policies. In this way, industries are searching for innovative solutions to decrease generated impacts by waste, from their production processes and that can cause damage on the environmental, social and economic areas.

In the textile and clothing sectors, the themes related to the management of its products are more highlighted, due to its relevance in the world economy and the extension of its productive chair. This industry constantly develops products that represent worldwide consumption of about 89.1 million tons of fibers (based on 2013), within Asia accounting for 73% of the world production [1]. Regarding on the disposal of their waste, the data is alarming: Europeans discard 5.8 million tons of textile waste each year, 75% of this is sent to landfills or to incineration plants, and only 25% to recycling [2]. In Brazil, the production is estimated on 175,000 tons of solid waste from the clothing industry each year, which 90% are incorrectly disposed [3]. It is important

© Springer International Publishing AG, part of Springer Nature 2019
J. Machado et al. (Eds.): HELIX 2018, LNEE 505, pp. 682–689, 2019.
https://doi.org/10.1007/978-3-319-91334-6_93

to express that it is concerned on registered codes 040221 and 040222 by the European List of Waste (LER), specifying processed and non-processed textile fibers waste.

Into the presented scenario, it is impelled to understand the changes that are being implemented by the textile and clothing sectors searching for new waste management systems, which can add value the concept of sustainability, in an attempting to minimize ecological and social crises, aiming the maintenance and perpetuation of the resources for future generations [4].

This article aims to analyze the main tools and methodologies that are being used to implement sustainable actions for the reuse of clean textile waste, as well as to evaluate the sustainability dimensions that are being explored. For this, a systematic research has been conducted, on scientific papers that express researches and investigations on the possibilities of recycling for textile waste. The main point for the research was to find proposals with innovative solutions, which seek, mostly, to reach the three dimensions of sustainability: environmental, social and economic.

2 Sustainability on Textile and Clothing Industries

The concept of sustainability rises visibility from the mid-twentieth century as a reaction to the possible risks on the current operating industrial system, initially focusing on environmental impacts, expanding its debate to the dimensions of social, cultural, economic, territorial and political problems [5].

In this sense, the theory of the three pillars - or dimensions - of sustainability, also known as the triple bottom line (profit-planet-people), is still a relevant and current analysis proposing the creation of multidimensional values to a more sustainable capitalism, in the face of a global cultural revolution, in order to achieve economic prosperity, environmental protection and quality and social justice [6].

The segment of the textile and clothing industry also try to engage on this trend, due to the relevance of this sector, the environmental degradation caused by its production processes, as well as the extension and diversification of its manufacturing chain, with several responsible parties involved.

Fashion appears as an influencer of several stages on the textile chain and needs to be stimulated to a sustainable perspective, especially regarding the role of designers in the processes of product development. Thus, it is concerning around fashion products, planned for short life cycles, stimulating exacerbated consumption and disposal. In contrast to the flow of this current, it is proposed fashion with sustainability, which may be for designing products with low environmental impact and high social quality, analyzing the criteria, methods and investments of Life Cycle Design [7].

It is understandable the Life Cycle of a garment, following 5 stages with different social and environmental impacts: (i) design, (ii) production, (iii) distribution, (iv) use and (v) end of life, in which we can find indications of actions, such as recycling of fabrics, design by reuse, remanufacturing of existing materials, fashion from codesign, upcycling, known as a concept that involves the reuse and/or reuse of materials or products wasteful or discarded, transforming them into other products or materials, with greater value, quality and adoption of Closed Production Life Cycles [8].

After the construction of this preliminary framework, it is seen that there is an awareness of the sector for sustainability practices, in response to the new demands of the political, economic, environmental and social scenarios. In this sense, Fashion can be used as a cultural influencer, being a valuable tool for sustainability, contributing to the dissemination of an awareness of the impacts of actual consumption and the active role that we must play as an integral part of the environment.

3   Data Analysis

From the systematic research on scientific articles, from database as Google Scholar, Web of Science, Scopus, and books of proceedings from congresses and conferences of the textile specialty (Autex, Colóquio de Moda and Cimode), in the period from 2012 to 2017. The keywords used in the research were: textile/garment/clothing sustainability, fashion and sustainability, clean textile/clothing waste. It was observed that there is a great amount of publications on the issues related to new sustainable business models in fashion. However, specifically to the perspective of reuse of waste of this industry (recycling), there are few publications. It was selected 10 articles, which present innovative sustainable solutions for the management of clean waste.

As observed in the presented scenario, the great extension of the textile chain generates large amounts and different types of waste, which causes a series of environmental, social and economic problems. Studies in Japan show that the recycling rate for textile waste is about 20% and that there is a deficit use of an effective recycling system for textile waste. This is due, among other factors, to the composition of intimately mixing various materials such as cotton, wool, rayon, polyester, polyamides among others, which makes it difficult to separate the residues of textile materials into specific types of materials. The authors propose a solution to this problem, by the creation of a color recycling system, which consists of an additional cycle that interacts with the conventional system of textiles. The residues are grouped in color and recycled, and can be, later, converted into felt, or wire, or milled into pigments. The products created by this system can be recycled again and circulated through the system [9].

The Circular Economy is also presented as a tool that allows the transition of systems to sustainability and a survey carried out in the Dutch textile industry, which is seen as a pioneer in the transition of flows to circular materials, analyzes cases of collaborations between companies involved, proposing evaluating various types of strategies and interactions. One of the companies is the House of Denim, which aims to establish an industrial standard with the use of recycled materials after consumption in new fabrics [10].

It can be observed that the Brazilian scenario is highlighted by the high quantity of textile waste production, as well as its incorrect destination, but it appears with visibility by the creative quality of the waste reuse strategies. Two studies analyzed and presented at the Autex Conference, of great relevance to the textile environment, present Brazilian cases related to waste reduction, reuse and recycling of textiles. One of them, [11] studies initiatives in São Paulo, Minas Gerais and Rio de Janeiro, focusing on solidarity economy. Actions such as the COMAS[1] project that reuse men's shirts for new fashion products and ECOMODA[2], which also works with the interference of design on waste for the creation of products with environmental and social value, are presented. The other research [12] also makes a Brazilian case analysis highlighting the São Paulo Reuse Fabric Bank as an action of creative economy in the sustainable dimensions.

The use of Reuse, Recycling and Redesign, proposed by Brown [13], the methodology are reported on the article by Puppim [14], which presents experiences of fashionable sustainability with the development of garments and collections from textile waste, in a project developed on the downtown of Belo Horizonte city, with local labors, initially the goal is training and qualifying these people and later including them in the process of making the proposed collection. This project was developed by the designer in association with the local/state government.

Another study, considers the creation of new materials as a strategy to reduce the environmental impacts of textile waste, discussing sustainability approaches support a solid waste R&D business model. The theoretical foundation includes approaches to sustainability with a focus on design (Design for Sustainability - DfS, Life Cycle Design), chemistry (Green Chemistry) and engineering (Green Engineering, Industrial Ecology, Cleaner Production) [15].

Wider solutions were observed into the whole process with the creation of a tool that assists the consumer in the evaluation and decision to purchase sustainable brands. Thus, an eco-label was created with seven criteria: certification of raw materials, certification of textile processing, generation of waste in the manufacture of clothes, CO2 emission in product life, durability, maintenance in use and destination after use [16].

An case study has been also taken in consideration, presenting Vuelo, a Brazilian brand that develop clothing collections using discarded umbrellas textiles materials, within textile waste and rubber waste (from tires). The main difference on their product is, besides the very well-studied design, the production proposal: there are not seasonal collection, but a restricted production and sell by demand. I.e., the brand start a punctual production based on the amount of products demanded by a period of time [17].

On the perspective of sustainable technical construction, another research investigated the use of waste from the textile industry in construction applications. Fibers with 30% wool and 70% acrylic were used as a reinforcement fiber in the construction and experimental results already indicate that the composite material may be of interest from the technological, sustainable and economic point of view [18]. In this way, research is also worthy of attention [19], which aims to systematize the waste of urban textiles in the Netherlands and reuses them as a reinforcement with biological biodegradable resin.

The Table 1 summarizes the tools that are being used for the implementation of waste reuse strategies and the dimensions of sustainability that are being explored in these actions. In this sense, it is possible to visualize a trend, but only the initial perception that the methodologies/tools of the Product Life Cycle and the Circular

---

[1] https://comas.com.br/.

[2] https://www.facebook.com/pg/Eco-Moda-372274436283765/about/?ref=page_internal.

Table 1. Table solutions, tools and dimensions of sustainability

Article title	Solution presented	Tool/ Methodology	Dimension* Env/Soc/Econ
Waste: Designing New Materials as Sustainability Strategy [a] [15]	Design intervention for the development of new materials from waste	Design for Sustainability (DfS), Life Cycle Design (LCD), Green Chemistry and Life Cycle Analysis	X X X
Contribution to an efficient transmission of information to the textile fashion consumer and the influence in sustainable attitudes [16]	Eco-label: environmental certifications	Labelling/Life Cycle Design	X X X
Render reinforced with textile threads [18]	Use of waste as a reinforcing fibber in civil engineering	Technical Textile	X
Study on Recycling System of Waste Textiles based on Colour [9]	Recycling system that includes an additional cycle separating wastes by colour	"Colour Recycle System"	X
Experiences of Sustainability in Fashion[b] [14]	Development of clothing collections from textile waste	Reuse, Recycling and Redesign	X X
Institutional incentives in circular economy transition: The case of material use in the Dutch textile industry [10]	Interaction of companies through the circular economy with distinct strategies of operation	Circular Economy	X X X
Strategic design and business models for sustainable fashion: the case study VUELO[c] [18]	Development of non-seasonal clothing and accessories collections from textile and non-textile waste	Consumption by demand/ Recycling Textile	X X X
Towards reinforcement solutions for urban fibre/fabric waste using bio-based biodegradable resins [19]	Textile wastes used as backing materials	Technical Textiles - Bio-Degradable Bins	X

(*continued*)

Table 1. (*continued*)

Article title	Solution presented	Tool/ Methodology	Dimension* Env/Soc/Econ
Social and economic importance of textile reuse and recycling in Brazil. [11]	Study of Brazilian initiatives that use the intervention of fashion design to add value to the products	Creative and Solidarity Economy	X X X
Textile sustainability: reuse of clean waste from the textile and apparel industry [12]	Case study Brazilian initiative – Banco de Reuso de Tecidos	SDO (Sustainability Design Orienting Toolkit) Circular/ Creative Economy	X X X

[a] Freely authors translation: *Resíduos Têxteis: Design de Novos Materiais como estratégia de sustentabilidade Textile.*

[b] Freely authors translation: *Experiências de Sustentabilidade na Moda.*

[c] Freely authors translation: *Design estratégico e modelos de negócio para moda sustentável: o estudo do caso VUELO.*

[*] *Where "Env" is for Environmental, "Soc" is for Social and "Econ" is for Economical dimensions.*

Economy are more approach. Regarding the sustainability dimensions, one can more clearly perceive that sustainable strategies include mostly the environmental and economic dimension of sustainability and do not involve all the stakeholders of the productive processes.

4   Conclusions

The present research has shown that there is a large volume of studies on the importance of sustainable awareness for the management of textile and clothing industries worldwide, some in countries where industry is on development to the conjunction of Fashion and Sustainability (such as Brazil and Portugal), and some in countries where the segment is consolidated (such as Germany and Netherlands).

The issue of textile waste is also faced as a serious problem that challenges various sectors of society. Government policies are being implemented and improved in order to charge industries practices, for their responsibility on waste generation and new consumers with a more sustainable awareness, changing their consumption patterns, in order to achieve positive impacts, worried about environmental resilience. This scenario, even in ongoing process, has forced companies to rethink their management policies and, in particular, the textile sector, which was presented in this article, is moving towards in adaptation to this new panorama.

The perspective mapped is fertile for the development of research, regarding that companies are already aware of the need for sustainable practices in their production processes. However, it is also seen that there is still a great gap of studies and researches that deals specifically with the problem of clean textile waste in the industry and the dimensions of sustainability are not yet worked in an equitable way, the environmental issue being addressed as a priority.

In this way, the analyzed data shows that much is being revealed and discussed about the impacts of this industry, but that there are still many spaces to be explored by future and further researches in order to contribute with investigations that may indicate for the development of waste management systems, guided by the Circular Economy, that operate in all dimensions of sustainability in order to integrate the needs of the different stakeholders of this productive chain. These new systems will have the primary objective of valorizing textile waste to minimize socio-economic, cultural and environmental problems on the way that shall develop a more fair and egalitarian society.

Acknowledgements "This work is supported by FEDER funds through the Competitiveness Factors Operational Programme - COMPETE and by national funds through FCT – Foundation for Science and Technology within the scope of the project POCI-01-0145-FEDER-007136".

References

1. Abit (Associação Brasileira da Indústria Têxtil). Cenários-Desafios e Perspectivas: Agenda de Competitividade da Ind. Têxtil e de Confec. Brasileira 2015-18. Abit, S. Paulo (2015)
2. Eurostat – Statistic Explained. Homepage. http://ec.europa.eu/eurostat/statistics-explained/index.php/Environment. Accessed 20 Dec 2017
3. Abit. Têxtil e Confecção: inovar, Desenvolver e Sustentar. Brasília: CNI/ABIT. Confederação Nacional da Indústria.Associação brasileira da indústria Têxtil e de Confecção
4. Vezolli, C.: Design de Sistemas para a Sustentabilidade. Edufba, Salvador (2010)
5. Sachs, I.: Desenvolvimento includente, sustentável e sustentado. SP, Garamond (2008)
6. Elkington, J.: Canibais com Garfo e Faca. (Edição Histórica). M. Books, S. Paulo (2012)
7. Vezolli, C.: Cenário do Design para uma moda sustentável, (trad. Castilio, K.). In: Pires, D. (Coord.). Design de moda: olhares diversos, pp. 197-205. Estação das Letras, SP (2009)
8. Gwilt, A.: Moda Sustentável: um guia prático (trad. Márcia Longarço). GG, SP (2014)
9. Uchimaru, M., Kimura, T., Sato, T.: Study on recycling system of waste textiles based on colour. J. Text. Eng. **59**(6), 159–164 (2013). http://jlc.jst.go.jp/DN/JST.JSTAGE/jte/59.159?lang=en&from=CrossRef&type=abst
10. Fischer, A., Pascucci, S.: Institutional incentives in circular economy transition: The case of material use in the Dutch textile industry. J. Clean. Prod. **155**, 17–32 (2017). https://doi.org/10.1016/J.JCLEPRO.2016.12.038
11. Baruque, R., et al. Social and economic importance of textile reuse and recycling in Brazil. IOPConf. Ser.: Mater. Sci. Eng. **254**, 192003 (2017). https://doi.org/10.1088/1757-899x/254/19/1920
12. Broega, A.C., Jordão, C., Martins, S.B.: Textile sustainability: reuse of clean waste from the textile and apparel industry. IOP Conf. Ser.: Mater. Sci. Eng. **254**, 192006 (2017). https://doi.org/10.1088/1757-899x/254/19/192006
13. Brown, S.: Eco Fashion. Laurence King Publishing Ltd., London (2010)

14. Puppim, R.: Experiências de Sustentabilidade na Moda 8° Coloquio de Moda, Rio de Janeiro/RJ - Brazil (2012)
15. Sampaio, C., et al.: Resíduos têxteis: design de novos materiais como estratégia de sustentabilidade. 13° Coloquio de Moda, Bauru/SP- Brazil. (2017)
16. Carneiro, N., Refosco, E., Soares, G.: Contribution to an efficient transmission of information to the textile fashion consumer and the influence of sustainable attitudes, TIWC, PL (2016)
17. Freire, K.M., Araújo, R.Z.: Design Estratégico e Modelos de Negócio para moda sustentável: o estudo de caso Vuelo. 12° Coloquio de Moda, J. Pessoa/PB- Brazil (2016)
18. Pinto, J., Peixoto, A., et al.: Render reinforced with textile threads. Constr. Build. Mater. **40**, 26–32 (2013). https://doi.org/10.1016/j.conbuildmat.2012.09.099
19. Agrawal, P., et al.: Towards reinforcement solutions for urban fibre/fabric waste using bio-based biodegradable resins. IOP Conf. Ser.: Mater. Sci. Eng. **254**, 192001 (2017). https://doi.org/10.1088/1757-899x/254/19/192001

Use of Industrial Waste as a Substitute for Conventional Aggregates in Asphalt Pavements: A Review

Lucas Pereira do Nascimento[1]([✉]) (iD), Joel R. M. Oliveira[1] (iD),
and Cândida Vilarinho[2] (iD)

[1] CTAC, University of Minho, 4800-058 Guimaraes, Portugal
lucas.nascimento92@hotmail.com
[2] MEtRiCS, University of Minho, 4800-058 Guimaraes, Portugal

Abstract. The need to save natural resources has paradigmatic attitude changes, such as the need to value waste materials in order to reduce landfill and minimize the use of raw materials. The objective of this work is to show the contribution of existing literature about the incorporation of industrial wastes in asphalt pavements. Thus, some of the main industrial wastes that can be incorporated in asphalt mixtures as aggregate substitute are steel slag, foundry sand and fly ash. This wastes are produced on a large scale on the planet, and the literature shows that, when treated for use in road pavements, their mechanical properties are similar or superior to conventional aggregates. Thus, according to the studies presented here, the incorporation rates vary from 5 to 15% for foundry sand, 70% for steel slag and 30% for fly ash. When incorporated in asphalt pavements, the dangerous elements may be encapsulated in the mixture, reducing their environmental toxicity.

Keywords: Industrial wastes · Asphalt pavement · Waste valorization

1 Introduction

With the advent of the industrial revolution and large-scale production, the need for natural resources has increased exponentially, causing innumerable environmental impacts. However, in the middle of the 20th century, humanity began to change the paradigms regarding the environment. Among the innumerable industries of the present time, the construction industry can be highlighted as that with the largest consumption of raw materials. At the global level, the construction industry consumes more raw material (approx.. 3000 Mt/year) than any other economic activity [1].

Within the construction industry, the road construction sector causes numerous impacts on the environment, such as deforestation, loss of biological diversity, alteration of the natural drainage system and soil degradation [2].

With the evolution of technology and environmental concern, there are now several ways to minimize environmental impacts. One of them is to recover waste with the potential to replace the conventional aggregates used in road pavements.

© Springer International Publishing AG, part of Springer Nature 2019
J. Machado et al. (Eds.): HELIX 2018, LNEE 505, pp. 690–696, 2019.
https://doi.org/10.1007/978-3-319-91334-6_94

In this context, it was realized that resources are finite, having the need or obligation to reduce the consumption of raw materials and start using the waste as a resource with the potential to take advantage of new industrial matrices. Thus, different legislation and recommendations have been published worldwide regarding the waste and resource management. An example of the paradigm change can be seen in Fig. 1.

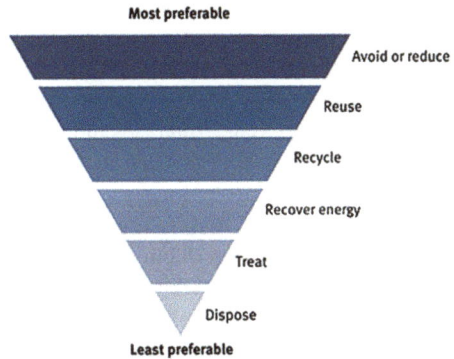

Fig. 1. The waste and resource management hierarchy [3]

In Portugal, Decree-Law no. 73/2011 (Article 7) defines targets for the recycling and reuse of waste in order to meet the targets for 2020 defined by the European Union. One of the targets is the use of at least 5% of waste incorporated in works of a civic nature. Thus, Decree-Law no. 18/2008, established the obligation to use at least 5% of recycled material in public works, in order to reduce the extraction of natural resources and the disposal of waste materials.

The purpose of this work is show the potential that some wastes have to replace natural aggregates used in road pavements without compromising its properties.

2 Asphalt Pavements Composition

A road pavement is a set of layers placed over the natural soil, whose main purpose is to resist the traffic and climate actions, providing the users with stability, comfort and safety [4–6]. Asphalt mixtures used in pavements comprise natural aggregates (typically around 95%) and bitumen (5%) [7]. Bitumen is a compound derived from petroleum [5, 6]. It essentially comprises hydrocarbons and heteroatoms, containing a small amount of elements such as Nitrogen, Sulfur and Oxygen atoms, forming the asphaltenes and maltenes, which are responsible for its viscoelastic behavior [8].

Currently, the main criterion used in the design of road pavements is no longer just the lowest cost, but the long-term impacts (economic, environmental or social) of this type of investments starts to be taken into account [9]. One of the possible ways to minimize their impacts is through the use of sustainable and recycled materials.

3   Recycled Aggregates for Road Pavements

In order to make the pavement more ecological and economical, one of the alternatives is to promote the use of wastes as partial or total substitutes for some materials, namely the aggregates [10]. In such scenarios, ecological advantages can be obtained regarding the environmental protection of new resources, thus reducing mining [6].

What is important in replacing traditional materials with waste (totally or partially) is to ensure that the products' final properties meet the specifications necessary for their application, without increasing their environmental impacts and if possible with economic advantages. Thus, this paper shows some examples found in the literature of wastes used in pavements, such as steel slag, foundry sands, fly ashes or CDW.

3.1   Steel Slag Aggregates

Steel slag comes from the industrial steel production process. Slags are the largest waste generated in this type of process (more than 60.0% of waste generation). In Portugal, there are two steel mills that use the electric arc furnace, Maia and Seixal steelworks, where in 2007 steel production was estimated at 1×10^6 tons in Seixal and 900×10^3 tons in Maia [11].

Slag may also be used as aggregates for the road construction industry. From a technical point of view, the mechanical properties of this waste may be equivalent to or greater than the materials commonly used in paving.

The benefits of slag incorporation are shown by several authors, where the incorporation of 40–70% of this material often ends up being superior in mechanical properties than natural aggregates and a reduction of environmental impacts with less greenhouses gases, and human toxicity [12–14]. The innate properties of iron and steel slag create it an ideal material for surface course mixtures [15]. The slag of electrical arc furnace has a large variation in its composition, being possible to obtain slags with 30% to 60% Calcium oxide (CaO), 0% to 35% iron oxide (Fe_2O_3) and 15% to 30% silica (SiO_2) in the same production day [16]. Thus, scanning electron microscopy (SEM) images (Fig. 3) and mineralogical analyzes are fundamental to understand its behavior [17].

3.2   Foundry Sand Aggregates

Foundry industry generates an extensive amount of wasted sand, known as foundry sand (sometimes referred to as green sands). This waste can be used in pavement layers, such as granular sub-base and base, or in bound layers like the surface course [18]. It was estimated that some 50,000 tons of foundry sands are produced annually in Portugal, and that these may replace a small part of the smaller aggregates in asphalt pavements, depending on their gradation [19].

The foundry sand (Fig. 2) has been used as fine aggregates in hot mix asphalts, to replace the thin part of the conventional aggregates. Results of previous works showed that replacing the natural aggregate by foundry sand in the proportion of 15% resulted in asphalt mixtures with a satisfactory mechanical performance [20].

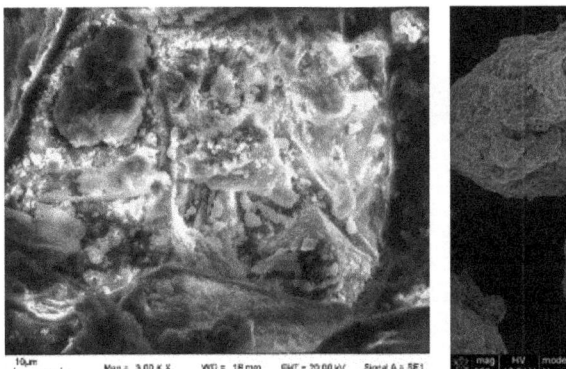

Fig. 2. Morphological analysis of steel slag (left) [17] and foundry sand (right) [21] by SEM

Another work showed a successful incorporation of 5% foundry sand in an asphalt concrete mixture (AC 20), partially substituting the natural aggregate, with a mechanical performance similar to that of a conventional mixture. In that study, the importance of performing an assessment of the leaching potential of asphalt mixtures with this type of waste incorporation was also highlighted, due to the composition of foundry sands (Table 1), which may contain hazardous elements (e.g. heavy metals) that may contaminate the soil and the underground water [21].

Table 1. Chemical composition of the foundry sand by XRF (wt%) [21]

SiO2	Al2O3	CaO	Fe2O3	K2O	MgO	Na2O
71,4	17,0	2,37	2,36	1,00	2,20	1,99

3.3 Fly Ash Aggregates

Estradas de Portugal [22] defined fly ash as "a fine powder consisting specifically of spherical and vitreous particles resulting from the burning of pulverized coal, with pozzolanic characteristics and consisting essentially of SiO_2 and Al_2O_3, because they are very small and light particles that are filtered by electrostatic precipitation or by filters that are used in thermoelectric plants to control emissions of pollutant particles. To be collected, the particles are electrostatically charged and then attracted by mechanisms that neutralize and collect them" (Fig. 3).

Another way of obtaining this waste is through waste incineration plants, creating one such type of material, which according to the legislation are considered hazardous waste [23].

One possibility of valuing and treating this waste is the incorporation in asphalt pavements. Studies show that the incorporation of fly ash in asphalt pavements results in a reduction of heavy metal leachates and of other environmental impacts, while their mechanical properties resemble or surpass those of the natural aggregates [24]. With the incorporation of fly ash from agricultural waste in building materials, it has been

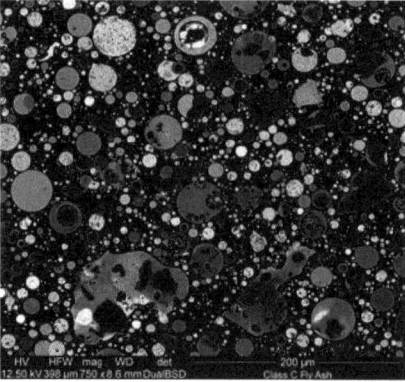

Fig. 3. Fly ash in scanning electron microscopy [26]

found that the values of chromium, sulfates and chlorides are reduced to non-hazardous levels [25].

Studies of the use of fly ash in the recycling of asphalt pavements (incorporating up to 30% ash) or as filler for asphalt mixtures have been carried out and the results show that the mixtures fulfilled the specification requirements to be used on road pavements [27, 28].

3.4   Other Wastes That Can Be Used as Aggregates for Road Pavements

Currently, construction and demolition waste (CDW) constitute a major environmental liability due to the volume of waste generated. By being 80% inorganic it contributes to rapidly filling the landfills. Another waste with a similar problem is the plastic waste. Thus, the valorization of these wastes as aggregates for construction would constitute great environmental and economic benefits.

Several studies show the feasibility of the use of these materials for this purpose. Costa et al. [29] showed that it is possible to partially incorporate plastic waste in an asphalt mixture (5.5% by volume). Ahmed et al. [30] have successfully used crumb rubber as a sustainable aggregate in a chip seal pavement.

One CDW widely used in road construction is the reclaimed asphalt pavement material itself. Previous studies [31, 32] show that the incorporation of 50% reclaimed asphalt material improves mechanical properties and may decrease energy consumption and emission of greenhouse gases.

4   Conclusions

Based on what the investigation carried in the present work, it is possible to confirm that there are several industrial wastes with added value, since they are materials with great potential to be used as partial substitute of natural aggregates, thus avoiding/ reducing the deposition of these wastes in landfills, as well as minimizing the use of scarce natural resources, reducing environmental impacts and promoting more sustainable construction practices.

Within several studies performed in the past regarding the incorporation of industrial wastes in road pavements, three wastes have stood out as those with most promising potential to substitute natural aggregates. These are the steel slag, the foundry sand and the fly ash. Nevertheless, previous studies also highlight the need to perform environmental assessments of the final products, namely by carrying out leaching tests, to confirm that the incorporation of these wastes do not present any undesirable impact in the surrounding natural habitats.

References

1. Torgal, F.P., Jalali, S.: A Sustentabilidade dos Materiais de Construção. Universidade do Minho, TecMinho, Portugal (2010)
2. Panazzolo, A., Aurélio, S., Frantz, L., Feiten, F., Cotrim, L.: O Papel da Gestão Ambiental na Mitigação de Impactos na Construção de Rodovias. In: 13º Congr Bras de Geol de Eng e Ambiental, São Paulo (2011)
3. Queensland Government: The waste and resource management hierarchy. The State of Queensland (2017). https://www.ehp.qld.gov.au/waste/qld-waste-strategy.html. Accessed 10 Feb 2018
4. Branco, F., Pereira, P., Santos, L.P.: Pavimentos Rodoviários. Alamedina, Portugal (2008)
5. Bernucci, L., Motta, L., Ceratti, J., Soares, J.: Pavimentação asfáltica: formação básica para engenheiros. ABEDA Rio de Janeiro, Petrobrás, Brasil (2006)
6. Azevedo, M.C.: Directivas para a concepção de pavimentos – Critérios de dimensionamento. INIR, Lisboa (2012)
7. Araújo, J.P., Oliveira, J., Silva, H.M.R.D.: Avaliação da influência da camada de desgaste na sustentabilidade dos pavimentos rodoviários. In: 7º CRP, Lisboa (2013)
8. Read, J., Whiteoak, D.: The Shell Bitumen Handbook. Thomas Telford, London (2003)
9. Araújo, J.P., Palha, C., Oliveira, J., Silva, H., Pereira, P.: Desenvolvimento de uma metodologia de avaliação da resistência ao rolamento de diferentes misturas betuminosas. In: XVIII Cong Ibero Latinoamericano del Asfalto (2015)
10. Souza, D.M., Teixeira, R.F., Ostermann, O.P.: Assessing biodiversity loss due to land use with Life Cycle Assessment: are we there yet? Glob. Chang. Biol. 21, 32–47 (2015)
11. Cavalheiro, A.: A evolução do processo siderúrgico em Portugal e a valorização das escórias. LNEC, Lisboa (2007)
12. Mladenovič, A., Turk, J., Kovač, J., Mauko, A., Cotič, Z.: Environmental evaluation of two scenarios for the selection of materials for asphalt wearing courses. J. Clean. Prod. 87, 683–691 (2015)
13. Andrade, R.M.A.: Caracterização Laboratorial de Misturas Betuminosas com Incorporação Agregados Siderúgicos Inerte para Construção. MSc thesis, ISEL (2015)
14. Chen, J.-S., Wei, S.-H.: Engineering properties and performance of asphalt mixtures incorporating steel slag. Const. Build. Mat. 128, 148–153 (2016)
15. Euroslag: Asphalt (2018). http://www.euroslag.com/applications/aggregates/asphalt/. Accessed 12 Feb 2018
16. Ferreira, S.R.: Comportamento mecânico e ambiental de materiais granulares: aplicação às escórias de aciaria nacionais. Ph.D. thesis, Universidade do Minho (2010)

17. Ahmedzade, P., Sengoz, B.: Evaluation of steel slag coarse aggregate in hot mix asphalt concrete. J. Hazard. Mater. **165**, 300–305 (2009)
18. Yazoghli-Marzouk, O., Vulcano-greullet, N., Cantegrit, L., Friteyre, L., Jullien, A.: Recycling foundry sand in road construction–field assessment. Const. Build. Mat. **61**, 69–78 (2014)
19. Castro, F., Vilarinho, C., Soares, D.: Gestão de resíduos industriais por incorporação em materiais para construção civil. In: VIII Congresso Iberoamericano de Metalurgia y Materiales, Quito (2004)
20. Bradshaw, S., Benson, C.: Using foundry sand in green infrastructure construction. In: Green Streets and Highways, pp. 280–298. Recycled Materials Resource Center (2010)
21. Nascimento, L., Oliveira, J., Vilarinho, C.: Analysis of foundry sand for incorporation in asphalt mixtures. In: 4th International Conference WASTES: Solutions, Treatments and Opportunities, Porto (2017)
22. EP: Caderno de encargos tipo obra 14.03-Pavimentação: Características dos materiais. Estradas de Portugal, S.A., Almada (2014)
23. LER: Lista de Resíduos da Diretiva 2008/98/CE. EU Off. J. Brussels (2014)
24. Dhir, R.K., Brito, J., Lynn, C.J., Silva, R.V.: Geotechnics and road pavements. In: Sustainable Construction Materials, pp. 197–237. Woodhead Publishing (2018)
25. de la Grée, G.D., Florea, M., Keulen, A., Brouwers, H.: Contaminated biomass fly ashes–Characterization and treatment optimization for reuse as building materials. Waste Manag. **49**, 96–109 (2016)
26. Kapinos, M.: Substitutes for Portland cement in concrete - fly ash and coal slag. Nohtern-Constructions, INC (2014). https://northernconcreteinc.com/substitutes-for-portland-cement-in-concrete-fly-ash-and-coal-slag/. Accessed 28 Feb 2018
27. Hoy, M., Horpibulsuk, S., Arulrajah, A.: Strength development of recycled asphalt pavement – fly ash geopolymer as a road construction material. Const. Build. Mat. **117**, 209–219 (2016)
28. Kar, D., Panda, M., Giri, J.: Influence of Fly Ash as a Filler in Bituminous Mixes. Department of Civil Engineering, NIT Rourkela, Odisha, India (2014)
29. Costa, L., Silva, H., Oliveira, J., Fernandes, S., Freitas, E., Hilliou, L.: Plastic waste use as aggregate and binder modifier in open-graded asphalts. In: 3rd International Conference WASTES: Solutions, Treatments and Opportunities, Viana do Castelo (2015)
30. Gheni, A., ElGawady, M.: Crumb rubber as a sustainable aggregate in chip seal pavement. In: Congrès International de Géotechnique–Ouvrages–Structures, Ho Chi Minh, Vietnam (2017)
31. Abreu, L., Oliveira, J., Silva, H., Fonseca, P.: Recycled asphalt mixtures produced with high percentage of different waste materials. Const. Build. Mat. **84**, 230–238 (2015)
32. Oliveira, J., Silva, H., Abreu, L., Gonzalez-Leon, J.: The role of a surfactant based additive on the production of recycled warm mix asphalts–less is more. Const. Build. Mat. **35**, 693–700 (2012)

Thermochemical Conversion of Waste Tires for Energy Recovery

L. Calado[(⊠)], B. Garcia, P. Brito, R. Panizio, and G. Lourinho

VALORIZA - Research Center for Endogenous Resource Valorisation,
Polytechnic Institute of Portalegre, Portalegre, Portugal
luis.calado@ipportalegre.pt

Abstract. The present work studies the possibility of energy recovery by thermal conversion of waste tires, a potential feedstock for combustion and gasification processes. Considering the difficulties of using these residues in isolation, cocombustion and co-gasification tests with acacia and *miscanthus* biomass were carried out in order to assess the characteristics of these residues as a fuel for thermochemical processes. Co-gasification tests were run in a fixed bed reactor at temperatures of about 800 °C. The results obtained demonstrate the viability of the technology, with the ideal conditions for the production of syngas with higher LHV (3.64 MJ/Nm^3) occurring in mixtures with 20% of waste tires. As for co-combustion, tests were performed in a pyro-tubular multi-fuel boiler with temperatures ranging from 400 °C to 500 °C. With respect to gaseous emissions (NO_x, NO, and SO_2), it was verified that increasing the percentage of tires in the mixture resulted in increased emissions of pollutant gases. These results present a problem since the obtained values are higher than those allowed by the Portuguese law. It was also concluded that gasification is a perfectly adjusted technology for the valorization of waste tires, being able to transform them into a fuel for energy recovery derived from residues with no other use.

1 Introduction

There are three main types of waste conversion process into energy and fuel: thermochemical, biochemical and chemical. These operations are generally referred to as 'Waste to Energy'. Gasification is a thermo-chemical process which is limited between combustion and pyrolysis. In general, gasification occurs when small amounts of gasification agent (air/steam/oxygen) are introduced into the reactor to allow some of the organic material to be partially "burned" in order to produce carbon monoxide and energy [1]. Combustion is another thermo-chemical technology for waste/biomass valorization which consists to burn them for power generation. In such cases, this type of material is burned in boilers for high-pressure steam production, which is introduced into a steam turbine coupled to a generator for electricity generation. The combustion process could be accomplished in fixed or fluidized-bed conditions. Successful waste tire combination experiments in fluidized-bed reactors were reported, being the most effective process for a lot of residual wastes management [2, 3]. However, this is relatively expensive due to high operating costs and considerable feedstock

© Springer International Publishing AG, part of Springer Nature 2019
J. Machado et al. (Eds.): HELIX 2018, LNEE 505, pp. 697–704, 2019.
https://doi.org/10.1007/978-3-319-91334-6_95

preparation. Tires are a solid waste that has become a major environmental challenge worldwide. The used tires have a high heat output (29–39 MJ/kg) and are made up of about 90% of organic materials, so they are an excellent source of fuel. With a higher heat output than coal, it is suitable for use in cement plants, boilers and melting furnaces. Compared with the combustion of coal, the burning of tires has lower emissions [4]. The combustion process of used tires, in particular their gaseous emissions, is influenced mainly by the combustion temperature, the excess of oxygen inside the reactor and the grain size of the material. With the proximate analysis, it was determined that there is a first phase of intense volatile combustion, followed by a less intense combustion phase with the simultaneous burning of volatiles and char, the burning times are considerably shorter than the coal [5].

The type of reactor used in the combustion process influences to a large extent the toxic gas emission. Several works present co combustion of used tires with coal as an effective way of energy recovery of used tires, where the emissions are effective reduce [5]. Other studies have proposed the co-combustion of coal and pulverized tires to reduce CO_2 emissions, thereby reducing emissions and solving the problem of tires as waste [4, 6]. Therefore, this work studied the possibility to recovering as fuels from residues, namely used tire, based on thermochemical processes of combustion and gasification.

2   Materials and Methods

2.1   Biomass Characteristics

Two different types of biomass with similar chemical characteristics, acacia and miscanthus, were selected to be inserted as feedstock for the test in co-gasification and cocombustion with used tires, respectively. In co-gasification, the mix between acacia and used tires, were tested in order to investigate the synthesis gas composition and it heating value in temperature between 790 and 800 °C. In co-combustion the tests were performed with a mix of miscanthus and used tires, to observe the gaseous emissions at different powers (100 and 157 kW) and temperatures (480 and 526 °C). For the used tires gasification tests, mechanical screening was necessary to separate the fractions between 1 cm and 4 cm in diameter, so that it could be processed in the Ibert reactor. After sorting by size, a manual sorting was carried out in order to avoid that pieces of tire still with wires were introduced in the reactor, since such materials could damage that system. To perform the tests, it was necessary to grind the tires with a 10 mm sieve to obtain more uniform and homogeneous tire particles to mix with the miscanthus. This biomass had to be milled in the hammer mill, but first had to remove the metallic remains because presents a risk of fire or even explosion. After milling, an additional classification was performed by separating the tire grains from the reinforcements of textile fibers that reinforced the tire.

2.2   Co-gasification

The gasification tests were carried out on an AllPowerLabs PP20 Power Pallets – 20 kW gasifier, a common downdraft reactor that combines an electric power generator and an electronic control unit. The tests were carried out in co-gasification with acacia, biomass material that presents a good gasification facility and produces a low quantity of ashes, trying to reach, if possible, the gasification of 100% of the residue. Each test was started with a 100% acacia chip, standard biomass for the tests, with the reactor temperature rises to 800 °C, until all the biomass in the hooper was consumed. Then, chip was introduced with a 20% mixture with the biomass to be tested, performing the test continuously in co-gasification. The percentage of the biomass was increased successively until there is no gasification (flare visible in the flare) or problems of feeding the biomass mixture, which prevent the continuation of the test. Each test was repeated twice, each lasting 8 h. The gas samples were withdrawn from the biomass particle filter into suitable bags with the aid of a vacuum pump. One of the samples was collected when the gasification process was stabilized (zero or near zero temperature variation) and another one at the end of the test, before the equipment was closed. The coals were trapped in the bottom of the reactor and in the cyclone filter, which was downstream of the reactor. The condensates were collected at the bottom of the bio-mass particulate filter.

2.3   Co-combustion

The combustion tests were performed on a pellet combustion unit D'Alessandro Ter-momeccanica S.R.L. CS SMALL 45. Pyro-tubular boiler, in steel, with door for internal inspection and cleaning, and burner with mechanical feed engine. Co-combustion tests were performed with Miscanthus and used tires. The process started with 100% pellets of Miscanthus, standard biomass for the boiler tests, raising the boiler temperature to the maximum that the equipment allows, a temperature near 500 °C. At this temperature it was intended to produce a hot water at about 60 °C and a flow rate of 11 L per minute. These conditions were maintained until all the biomass in the hooper was consumed. The residue was then introduced, previously prepared with mixtures. The percentage of the residue in the mixture was increased until it was technically possible to carry out the test.

During the tests, the boiler temperature, air temperature, water temperature at the inlet and outlet of the boiler, and water flow were controlled. Therefore, the "power" produced by the boiler was calculated ($Q = 100$ kWh). The exhaust gas analysis was determined on the basis of a flue gas analyzer connected to the boiler stack. The analyzer comes equipped with 6 non-dispersive infrared (NDIR) sensors and 3 elec-trochemical sensors. A Madur Photon II gas analyzer was used to identify and qualify combustion exhaust gases, O_2, CO, CO_2, NO_2, SO_2, H_2S, as well as combustion parameters of pressure and temperature.

2.4   Biomass and Char Analysis

The biomass tested and chars produced in the gasification and combustion tests, were analyzed in terms of mass and total volume, LHV, elemental composition (ultimate analysis), inorganic fraction composition (for chars), thermogravimetric profile (proximate analysis).

3   Results and Discussion

3.1   Biomass Characteristics

The average proximate, ultimate, and low heat value of the inserted fuels are illustrated in Table 1. Tires had such high values that could be a risk for the equipment, which was not prepared to receive this type of fuels. Biomass energy production equipment normally has a recommended limit of 18 MJ/kg for the calorific value of the fuel.

Table 1. Biomass analysis.

Analysis	Parameters	Units	Biomass		
			Acácia	Miscanthus	Used Tires
	Moisture	(%)	14.2	6.3	0.8
	Volatile Matter	(%)	49.7	64.7	64.5
Proximate	Fixed Carbon	(%)	32.1	26.0	29.6
	Ashes	(%)	4.0	3.0	5.1
	Nitrogen	(%)	0.3	0.4	0.0
	Carbon	(%)	44.1	44.5	75.5
Ultimate	Hydrogen	(%)	5.6	6.2	0.7
	Sulphur	(%)	0.0	0.0	5.6
	Oxygen	(%)	45.9	47.4	11.8
LHV		(MJ/kg)	17.0	18.1	38.6

The ultimate analysis also reveals the presence of a high sulfur content in the tires composition which is the main source of SO_2 emissions into the atmosphere.

3.2   Co-gasification Tests

The Table 2 shows the results of the co-gasification process of used tires. In the process of gasification of this residue the fuel gas that was produced in greater quantity, was the carbon monoxide, with percentages ranging from 1.9% to 11.4%. The molecular hydrogen produced, ranged from 0.8% to 6.9% and methane from 2.3% to 4.3%, the higher values obtained happened when additions in the mix were below 40%. The biomass intake was reduced from about 7 kg/h with 0% tires to near 1.8 kg/h with 40%

tires. Above 60% could no longer carry out the process. As for the calorific value of the gas, it was verified that it assumes higher values in co-gasification with small amounts of mixture. Considering the high percentage of carbon in the used tires, this can lead to conditions for the production of large quantities of ash, rich in coal, in the gasification process [7].

Table 2. Co-gasification analysis

Parameter	Unity	0% Tire				20% Tire				40% Tire				60% Tire			
Acácia	% (m/m)	100,0				80,0				60,0				40,0			
Tires	% (m/m)	0,0				20,0				40,0				60,0			
T Oxidation	°C	791,0	795,0			803,0	802,0			805,0	790,0			790,0	788,0		
T Reduction	°C	394,0	405,0			571,0	574,0			455,0	516,0			507,0	507,0		
Inlet Biomass	kg/h	7,1				5,0				1,8				2,8			
Tars	l/h	0,0															
Chars	kg/h	0,2															
Syngas CO₂		7,5	7,6	5,7	6,1	10,0	9,7	8,1	5,7	10,1	9,6	12,2	11,8	9,7	9,4	9,9	9,4
C₂H₄		0,1	0,1	0,0	0,1	0,4	0,4	0,2	0,2	0,3	0,3	0,2	0,3	0,2	0,2	0,1	0,2
C₂H₆	% mol	0,0	0,1	0,0	0,0	0,0	0,0	0,0	0,0	0,0	0,0	0,0	0,0	0,0	0,0	0,0	0,0
N₂		59,7	60,3	67,9	67,9	54,3	55,7	61,2	75,3	56,3	58,0	56,1	58,1	60,3	61,1	58,6	61,1
CH₄		4,1	4,3	3,6	3,3	3,6	3,8	3,4	2,3	3,3	3,4	2,3	2,5	3,1	3,2	3,2	3,2
CO		4,3	4,8	1,9	1,8	10,0	10,6	5,9	4,1	8,5	8,9	10,7	11,4	5,5	6,0	4,9	6,0
H₂		2,6	2,3	0,8	0,9	6,8	6,0	3,6	2,4	5,6	5,6	6,9	6,2	3,8	3,7	3,3	3,7
HLV Syngas		2,3	2,5	1,6	1,5	3,5	3,6	2,5	1,7	3,0	3,1	3,0	3,1	2,3	2,4	2,2	2,4
HLV Max		2,5				3,6				3,1				2,4			
HLV Med	MJ/m³	2,0				2,8				3,1				2,3			
HLV Min			1,6				1,7				3,0				2,2		

The results show that the ideal conditions are precisely the low percentages of tires (less than 40%). This can be one of the justifications presented so, with increasing of mixing, the calorific value of the gas falls, and there is also a lot of material that does not burn in the reactor at the end of the test [7].

Co-gasification Chars and Ashes

Although it was a process in co-gasification with acacia, and as such the composition of the acacia char and ash had a great preponderance, it allows to verify the influence of the residues. The material resulting from the co-gasification of the waste seems to have little value for energy recovery. However, these types of coal-rich ash have good characteristics for the treatment of effluents.

Table 3. Char/Ashes analysis of acacia and tires

Sample Char/Ashes	LHV (MJ/kg)	proximate analysis			ultimate analysis				
		Moisture	Volatiles	Fixed Carbon	N	C	H	S	O
		2.8		78.9	2	53.01	0.4	0	45.35
		0.4		92.3	0	10.4	0.9	4.2	84.5

Acacia 13.37 18.3 1. Tires 4.43 7.4

The calorific value of the chars of acacia were higher than the calorific value of the chars resulting from the mixtures. Effectively it allows to conclude that for similar conditions of the gasification process, the addition of residues allows to increase the rate of gasification of the material. Analyzing the inorganic composition of the used tires co-gasification, there was a significant presence of the elements Ca, K, Fe, S. The amounts of Ca and K also suggest the use as agricultural fertilizer, as well as cement substitute. The ashes were also analyzed in the elemental analyzer with the values shown in Table 3. The percentage of carbon in the ash resulting from the co-gasification was influenced by the temperature at which the reduction process takes place, as well as the equivalence ratio. Higher equivalence ratios present higher oxidation potential implying a lower amount of carbon formed.

Co-combustion Tests

Table 4 represents the results of co-combustion of used tires with Miscanthus.

Table 4. Co-combustion analysis

Parameter	Unity	0% Tire	20% Tire	40% Tire	60% Tire
Acácia	% (m/m)	100	80	60	40
Tires	% (m/m)	0	20	40	60
T Boiler	°C	499	523	523	481
Power Boiler	kW	100	103.81	133.6	156.9
Inlet Biomass	kg/h		9.6		
Chars	kg/h		0.508		
O_2	%	5.9	7.92	6.93	6.1
CO	ppm	708.7	232.25	243.33	1238.71
NO	ppm	180.7	629.75	216.67	224.29
NO_2	ppm	1	0.50	1.00	9.86
CO_2	%	14.7	11.70	12.26	12.54
SO_2	ppm	11.3	170.25	309.33	463.71
CO	mg/Nm^3	1.1	1.33	1.26	1.26
No_x	mg/Nm^3	900.4	288.48	290.40	4862.87
NO	mg/Nm^3	373.5	425.08	449.57	495.92
SO_2	mg/Nm^3	241.2	277.05	258.80	301.5
H_2S	mg/Nm^3	33.3	496.23	990.00	1304.99

The biomass consumption was constant during the test period, with an increase in the power produced by the increase of the percentage of tires. With the 40% and 60% mixtures, it was difficult to achieve an automatic boiler feed, requiring constant operator intervention. The results allow verifying the inverse proportionality between CO_2 and O_2 with a point of inversion in the 20% of the mix of tires. In this case, the values of O2 and CO increased up to 20% of the tires with 288.48 mg/Nm^3 and 11% of oxygen.

Co-combustion Chars/Ashes Analysis

Table 5 shows the LHV values of the ashes resulting from the combustion tests. A significant presence of the elements Zn, Ca, K, Fe and S was observed in the inorganic analysis, with mass percentages of 6.4%, 3.2%, 1.9%, 0.7 and 0.9%, respectively.

Table 5. Char/Ashes analysis of miscanthus and tires

Sample Char/Ashes	LHV (MJ/kg)	proximate analysis			ultimate analysis				
		Moisture	Volatiles	Fixed Carbon	N	C	H	S	O
Miscanthus	15.6	2.1	14.88	83.02	1.2	31.5	4.6	0	62.6
Tires	19.1	1.3	7.6	91	0	44.7	0	5.8	49.5

As expected, the fixed carbon contents of the chars/ashes were those that exhibited the highest mass proportions, average percentage of 90.0%. The low moisture content suggests the possibility of direct combustion of the coal to obtain energy, without the application of a drying treatment and the consequent additional energy consumption as long as the material is well conditioned in storage.

4 Conclusion

Biomass gasification is a promising technology to displace use of fossil fuels and to reduce CO2 emission. Among other alternative energy conversion pathways, it has great potential because of its flexibility to use a wide range of feedstock, and to produce energy and a wide range of fuels and chemicals. Combustion tests indicate the need to implement mechanisms to mitigate these emissions, which can be economically costly. This work demonstrated the feasibility of transforming this type of waste into thermal energy through the production of a valuable gas rich in hydrogen and other products that are highly relevant in terms of heating power and interesting to the industry.

Acknowledgement. Authors would like to thank to the project POCI-01-0145-FEDER-024020 - RDFGAS (Energy utilization of fuels derived from waste and dry sludge) Co-financed by Compete 2020 - Competitiveness and Internationalization Operational Program, Portugal 2020 and European Union through the FEDER.

References

1. Antonopoulos, I.S., Karagiannidis, A., Gkouletsos, A., Perkoulidis, G.: Modelling of a downdraft gasifier fed by agricultural residues. Waste Manag. **32**(4), 710–718 (2012)
2. Teng, H., Chyang, C.S., Shang, S.H., Ho, J.A.: Characterization of waste tire incineration in a prototype vortexing fluidized bed combustor. J. Air Waste Manag. Assoc. **47**(1), 49–57 (1997)

3. Mastral, A.M., Callén, M.S., García, T.: Fluidized bed combustion (FBC) of fossil and nonfossil fuels. A comparative study. Energy Fuels **14**(2), 275–281 (2000)
4. Rowhani, A., Rainey, T.J.: Scrap tyre management pathways and their use as a fuel - a review. Energies **9**(11), 1–26 (2016)
5. Atal, A., Levendis, Y.A.: Comparison of the combustion behaviour of pulverized waste tyres and coal. Fuel **74**(11), 1570–1581 (1995)
6. Singh, S., Nimmo, W., Gibbs, B.M., Williams, P.T.: Waste tyre rubber as a secondary fuel for power plants. Fuel **88**(12), 2473–2480 (2009)
7. Straka, P., Bučko, Z.: Co-gasification of a lignite/waste-tyre mixture in a moving bed. Fuel Process. Technol. **90**(10), 1202–1206 (2009)

Hydrothermal Torrefaction of Mixtures of Biomass and Hydrocarbon-Rich Sludge in the Presence of Fossil Fuels

Ana Paula Oliveira[1] ⓘ, Margarida Gonçalves[1,2(✉)] ⓘ,
Luís Durão[1,2] ⓘ, and Cândida Vilarinho[2,3] ⓘ

[1] MEtRICs - Department of Sciences and Technology of Biomass,
Faculty of Sciences and Technology, New University of Lisbon,
Caparica, Portugal
mmpg@fct.unl.pt
[2] VALORIZA - Polytechnic Institute of Portalegre, Portalegre, Portugal
[3] MEtRICs - Department of Mechanical Engineering, Engineering School,
Minho University, Guimarães, Portugal

Abstract. A new process for the torrefaction of mixtures of biomass and oily sludges is proposed. A fossil fuel (gasoline or diesel) was added to mixtures of biomass and hydrocarbon-rich sludge (10:1), and this mixture was subject to distillation until all liquids are recovered. The fossil fuel partially dissolved the sludge components and promoted their evenly distribution over the biomass particles. During distillation, the fossil fuel and all the distillable components present in the biomass and sludge were collected and the mixture was subject to temperatures at which a considerable transformation of the non-distillable fraction occurred. The biomass lost water and suffered partial decomposition and rearrangement to yield biochars with HHV of 23.6 and 33.2 MJ/kg. The original biomass had a HHV of 17.9 MJ/kg, but the hydrothermal torrefaction process as well as the fortification with heavy hydrocarbons from the oily sludge significantly increased its calorific value. The distillable liquids were recovered in the form of two immiscible liquid phases: (a) an organic phase mainly composed by the added fossil fuel but also containing the nonpolar volatile components present in the sludge and the biomass, and (b) an aqueous phase that contained the free and emulsified water present in the biomass and in the sludge, but also the water and polar organic components released from the mixture during this thermal treatment. This process takes place at atmospheric pressure and temperatures lower than 300 °C and can be applied to different sludges with high carbon content to promote their energetic valorization.

Keywords: Hydrothermal torrefaction · Biomass · Oily sludges

1 Introduction

The hydrocarbons present in oily sludge of petrochemical origin are compounds of high calorific value and great ecotoxicity, so that their recovery and energy recovery is a desirable objective both from an economic and an environmental point of view.

© Springer International Publishing AG, part of Springer Nature 2019
J. Machado et al. (Eds.): HELIX 2018, LNEE 505, pp. 705–711, 2019.
https://doi.org/10.1007/978-3-319-91334-6_96

However, the presence of water, surfactants and mineral components contributes to the stability of the emulsions between the hydrocarbons and the remaining components of these slurries, reducing the efficiency of traditional emulsion separation techniques such as centrifugation or filtration.

Thus, some authors have sought to explore other unitary operations such as extraction with solvents, sometimes associated with some pre-treatments such as dehydration [1] or freeze-thaw cycles [2] intended to reduce the water present in the slurry by evaporation or decantation and thereby to increase the efficiency of the next extraction step.

Liang and co-authors studied the extraction of hydrocarbons from oily sludges using different hydrocarbon and alcohol type solvents; these authors concluded that the partition coefficient of the hydrocarbons between the sludge and the extraction solvent was higher for less polar solvents such as cyclohexane and hexanol and this coefficient shows a positive correlation with the mud solids content and with the extraction temperature [3]. These studies highlight some of the factors that difficult the recovery of hydrocarbons from these sludges, namely the presence of water in an emulsified form that limits the transfer of apolar compounds to the solvent, also apolar, as well as the diffusion of the solvent in the sludge matrix.

2   Materials and Methods

2.1   Raw Materials

The oily sludge (OS) sample was collected by Carmona, S.A., and the pine biomass was provided by CMC Biomassa Lda. Gasoline and diesel were obtained in local distributors of fossil fuels for internal combustion engines.

2.2   Methods

The pine biomass was mixed with the oily sludge in the proportion of 10 g of biomass to 1 g of sludge. Gasoline or diesel were added to a concentration of 50 g/100 mL of liquid fuel, and the mixture was heated progressively in a simple distillation apparatus equipped with a condenser unit and a gas collection adapter. Gases were condensed at 0 °C, and liquids were kept at that temperature during the process. The process is stopped when no more liquid is condensing at the exit of the distillation flask. The yields of liquid and solid products were determined gravimetrically and the gases by difference to the initial materials. Moisture content of sludge, pine biomass and solid products, was determined according to the standard BS EN 14474-2:2009 [4]. Volatile hydrocarbons, non-volatile hydrocarbons and solids were evaluated by the methods proposed by Taiwo and Otolorin [5]. The ash content was evaluated according to standard ASTM-D482-03 [6]. The elemental analysis (N, C, H and S) was performed using an Elemental Analyzer (Thermo Finnigan-EC Instruments, Flash template and CHNS 112 series). Oxygen was evaluated by difference in a dry ash free basis. The gas phase composition (H_2, N_2, O_2, CH_4, CO and CO_2) was determined using GC-TCD (Thermo Electron Corporation) and the high heating value was determined using a calorimeter IKA C200. The low heating value was calculated according to the standard ASTM E711 [7]. The composition of the liquid phases was studied by GC-MS (Focus GC-Polaris Q spectrometer). A hydrocarbon mixture from C7 to C30 was used for the determination of carbon number distribution and relative concentrations.

3 Results and Discussion

The characterization of the materials used in the hydrothermal torrefaction involved the determination of water content, volatile and non-volatile hydrocarbon content, solids content and ash content. The elemental composition and calorific values were also determined, and results are presented in Table 1.

Table 1. Properties of the raw materials used in the process

Properties	Pine biomass	Oily sludge (OS)	Biomass + OS (10:1)
Water,[1] wt% a.r.	10.7 ± 0.2	8.9 ± 1.0	9.2 ± 0.2
Volatile hydrocarbons, wt% ar	n.d	6.5 ± 2.2	1.1 ± 0.2
Non-volatile hydrocarbons, wt% ar.	73.8 ± 1.2	45.6 ± 0.3	74.9 ± 0.1
Solids, wt% ar	16.6 ± 1.3	39.0 ± 1.0	14.8 ± 0.1
Ash wt% ar.	3.7 ± 0.7	29.2 ± 0.1	5.9 ± 0.1
Ash wt% d.b.	4.1 ± 0.7	32.0 ± 0.1	6.4 ± 0.1
C (wt%, daf)	47.9	74.4	49.5
N (wt%, daf)	0.7	0.8	0.5
H (wt%, daf)	6.3	11.0	6.6
S (wt%, daf)	0.0	4.0	4.7
O^2 (wt%, daf)	45.1	9.8	38.7
HHV (MJ/kg) d.b.	17.93	23.92	18.31
LHV[3] (MJ/kg) d.b.	16.58	21.54	16.88

a.r.: as received; d.b.: dry basis; daf: dry ash free; (1) evaluated according to BS -EN; (2): by difference; (3) calculated.

The oily sludge presented a high content of volatile and non-volatile hydrocarbons (52,1wt%, a.r.) and a calorific value higher than the pine biomass, characteristics that are typical of this type of sludges [8, 9] and motivates the interest on their energetic valorization. Nevertheless, their high ash content may cause slagging and fouling problems in boilers, as well as loss of efficiency by heat loss in large amounts of residual ash [10]. Dispersion of these sludges over a lower ash fuel, such as pine biomass maybe a strategy to recover its carbon content while lowering the negative impact of its high ash content [11]. However, direct use of this sludge as a pellet additive is limited by its high viscosity [8], that hinders a homogeneous distribution over the biomass, and by the presence of hazardous volatile components (6.5% wt% a. r.), that increase occupational hazard concerns during pelletization. Furthermore, as it can be seen in Table 1, using a 10:1 biomass to sludge mixing ratio, there is a minor increase in the calorific value (from 16.58 to 16.88 MJ/kg), but at the expenses of a 2.2% increase of the ash content. Also, the ratio 10:1 of additive in pellets, even for an industrial use, is above the limits established by ISO 17225-2:2014.

In this work, it is proposed a new approach for the thermal treatment of biomass and oily sludge mixtures, in order to avoid undesirable gaseous emissions and obtain a solid fuel with upgraded characteristics. This process was designated by hydrothermal torrefaction, because it occurs in a temperature range typical of torrefaction [12] and involves the contact between water and the materials under treatment at temperatures higher than 100 °C [13]. The process was applied to a mixture of pine biomass, oily sludge and a fossil fuel (gasoline or diesel) and involved heating this mixture at temperatures that promote the distillation of all volatile and semi-volatile components, to obtain gases, an organic liquid and an aqueous phase leaving in the reactor a solid product (Fig. 1).

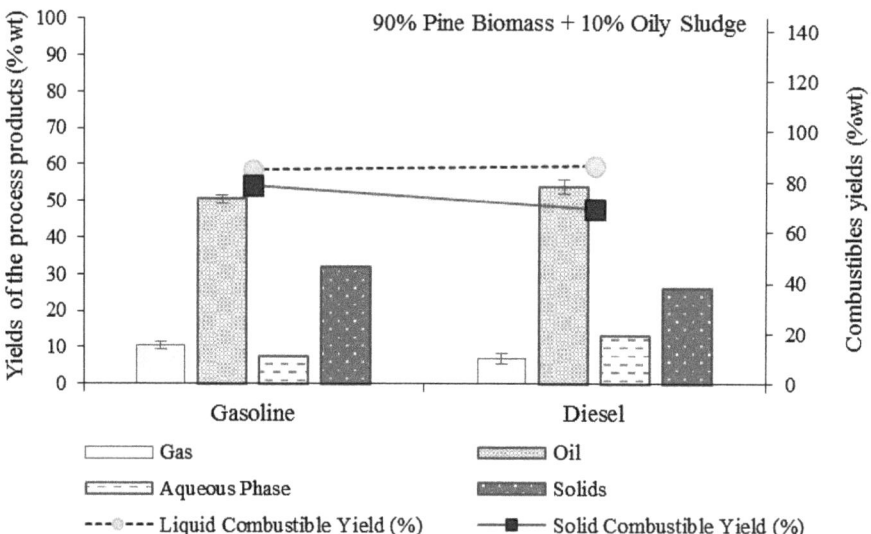

Fig. 1. Yields of the gas, liquid and solid products of the hydrothermal torrefaction and yield of liquid and solid fuel relatively to the added fossil fuel and pine biomass.

The process yielded 10.45 and 6.95% gas products that result from the volatilization of low boiling point components present in the raw materials but also from the formation of permanent gases during the partial decomposition of the raw materials. The organic liquids had mass yields of around 50% relatively to the initial mass of raw materials but that represented a liquid fuel yield higher than 84%, taking into account the added masses of gasoline or diesel. The remainder fossil fuel that is not recovered in the liquid form may be present in a nebulized form, in the gas products, or adsorbed in the pores of the solid products. The aqueous phase is formed by water released from the raw materials and other water-soluble components co-distilled with the fossil fuel.

The yield of the aqueous phase (7.3 and 13.2%), clearly exceeded the moisture content of the biomass: oily sludge mixture (9.2%), that after dilution with the fossil fuel should be around 3%. This result is an evidence of the release of water bound to the polymers of the lignocellulosic materials and eventually the formation of water by decomposition of hydroxyl groups from the biomass surface; the higher water phase yield obtained with diesel also indicates that exposure to higher temperatures in the distillation range of diesel probably led to the higher decomposition of the biomass. The presence of a water immiscible, high boiling point organic liquid, led to a progressive heating of the materials, reaching temperatures in the range of 200 to 300 °C, at which, thermal decomposition analogous to torrefaction occurs [12]. During this process, the water present in the biomass and in the sludge, is still interacting with the materials, promoting oxidation reactions, as it happens during hydrothermal liquefaction or carbonization [13]. The presence of the fossil fuel hinders the immediate release of the water vapor, therefore creating the conditions for the interaction between the water that is being released from the biomass and groups of the biomass surface, thus promoting their oxidative decomposition.

The composition of the gas phase (Fig. 2) shows that more permanent gases associated with the thermal decomposition of biomass (CO_2, CO, H_2, hydrocarbon gases) were produced during hydrothermal torrefaction with diesel, confirming the positive association between the higher distillation range and the higher extent of biomass decomposition. Residual air in the mixture, should be equivalent for the experiments with gasoline or with diesel, because it corresponds to air present in the apparatus that was displaced and diluted by the gases formed in the process.

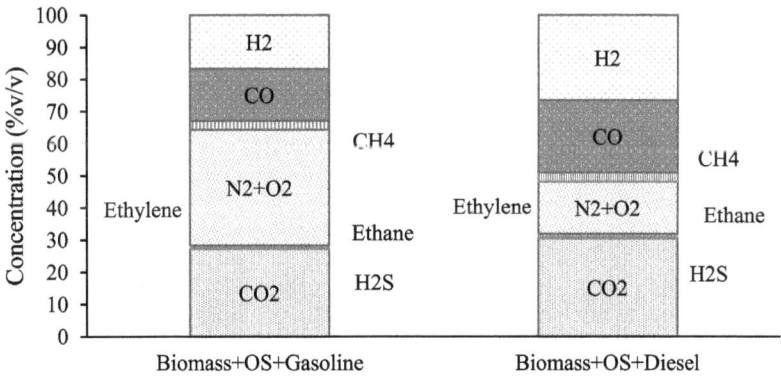

Fig. 2. Composition of the gas phase obtained by hydrothermal torrefaction in the presence of gasoline or diesel.

The organic liquid phases presented HHV of 41.8 and 44.7 MJ/kg, respectively for gasoline and for diesel experiments values close to the HHV of the original fuels [14]. This is in accordance with the composition of the organic liquids, as evaluated by GC-MS, and represented as the relative concentration of the components as a function of their carbon number (Fig. 3).

As expected, the composition of the organic liquid phase closely reflects the composition of the fuel used in the process, but the presence of the sludge and biomass,

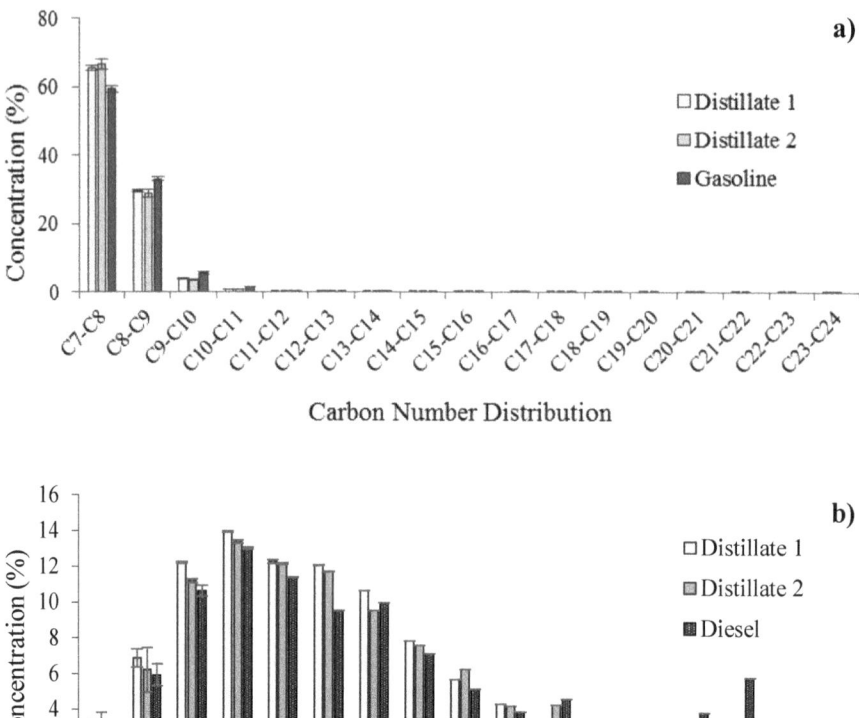

Fig. 3. Relative concentrations of the components of the distilled organic phases as a function of their carbon number distribution for: (a) gasoline experiments; (b) diesel experiments

contributed to a shift of the composition of the organic distillates towards lower carbon numbers, C7-C8 in the case of the experiments with gasoline and C7 to C17 for the experiments with diesel. The presence of organic compounds in the aqueous phase was also evaluated by GC-MS and the main functional groups found were organic acids and oxygenated aromatic hydrocarbons such as furans or phenols, components also found in pyroligneous acid that is the aqueous phase collected during production of charcoal [15]. The solid products (biochars) presented carbon contents of 47.2% and 60% and LHV of 22.6 and 31.9 MJ/kg for, respectively, the gasoline and diesel experiments, values well above the LHV of 16.58 MJ/kg measured for the raw pine biomass. The improved carbon content and calorific value of the char are probably due to the deoxygenation of the biomass fraction but also to the deposition of heavy hydrocarbons from the sludge in the porous structure of the biochar.

4 Conclusions

The hydrothermal torrefaction process allowed the upgrading of pine biomass fortified with oily sludge to yield a biochar with higher carbon content and higher calorific value than the original biomass. The hydrocarbons present in the sludge are distributed among the liquid and solid products and can therefore be energetically valorised. The use of a higher boiling point liquid such as diesel allows a higher degree of deoxygenation with advantaged for the fuel properties of the organic products. The process also allows the fractionation of liquid products among two separate immiscible phases that can be processed or valorised by appropriate methods.

References

1. Trowbridge, T.D., Holcombe, T.C., Trowbridge, T.D., et al.: Refinery sludge treatment/hazardous waste minimization via dehydration and solvent extraction. J. Air Waste Manage. Assoc. **45**(10), 782–788 (1995)
2. Hu, G., Li, J., Hou, H.: A combination of solvent extraction and freeze thaw for oil recovery from petroleum refinery wastewater treatment pond sludge. J. Hazard. Mater. **283**, 832–840 (2015)
3. Liang, J., Zhao, L., Du, N., et al.: Solid effect in solvent extraction treatment of pre-treated oily sludge. Sep. Purif. Technol. **130**, 28–33 (2014)
4. BS EN 14774-2:2009: "Solid biofuels – Methods for determination of moisture content-Part 3", ECS – European Committee for Standardization, Brussels (2009)
5. Taiwo, E.A., Otolorin, J.A.: Oil recovery from petroleum sludge by solvent extraction. Pet. Sci. Technol. **27**, 836–844 (2009)
6. ASTM-D482-03: Standard Test Method for Ash from Petroleum Products. Annual Book of ASTM Standards. Epub Ahead of Print (2003)
7. ASTM E711-87: Standard Test Method for Gross Calorific Value of Refuse-Derived Fuel by the Bomb Calorimeter. Annu B ASTM Stand. Epub Ahead of Print (2004)
8. Al-Futaisi, A., Jamrah, A., Yaghi, B., et al.: Assessment of alternative management techniques of tank bottom petroleum sludge in Oman. J. Hazard. Mater. **141**, 557–564 (2007)
9. Prithiraj, S., Kauchali, S.: Yields from pyrolysis of refinery residue using a batch process. South African J. Chem. Eng. **24**, 95–115 (2017)
10. Furimsky, E.: Gasification in petroleum refinery of 21st century. Oil Gas Sci. Technol. **54**, 597–618 (1999)
11. Deng, S., Wang, X., Tan, H., et al.: Thermogravimetric study on the co-combustion characteristics of oily sludge with plant biomass. Thermochim. Acta **633**, 69–76 (2016)
12. Mamvura, T.A., Pahla, G., Muzenda, E.: Torrefaction of waste biomass for application in energy production in South Africa. South African J. Chem. Eng. **25**, 1–12 (2018)
13. Dimitriadis, A., Bezergianni, S.: Hydrothermal liquefaction of various biomass and waste feedstocks for biocrude production: a state of the art review. Renew. Sustain. Energy Rev. **68**, 113–125 (2017)
14. Channiwala, S.A., Parikh, P.P.: A unified correlation for estimating HHV of solid, liquid and gaseous fuels. Fuel **81**, 1051–1063 (2002)
15. Mansur, D., Yoshikawa, T., Norinaga, K., et al.: Production of ketones from pyroligneous acid of woody biomass pyrolysis over an iron-oxide catalyst. Fuel **103**, 130–134 (2013)

Development of Bioplastic Film for Application in the Footwear Industry

Joana Carvalho[1,2(✉)], Margarida Soares[1], Carlos Castro[1],
André Mota[1], André Ribeiro[1], Jorge Araújo[1], and Cândida Vilarinho[2]

[1] CVR – Centro para a Valorização de Resíduos, Guimarães, Portugal
jcarvalho@cvresiduos.pt
[2] Departamento de Engenharia Mecânica,
Universidade do Minho, Guimarães, Portugal

Abstract. There is a growing interest in the use of biodegradable polymers that can help minimize the environmental impact of plastics. Biopolymers have been considered the most promising materials for this purpose. However, they generally have poorer mechanical properties. Starch is a low-cost polysaccharide derived from agricultural plants. To improve starch processing, the molecular order within the granules must be destroyed. This is generally achieved by heating the granular starch mixed with plasticizers. With this process, a conversion of the biopolymer's molecular structure into thermoplastic starch is obtained. In this way, much of the starch changes from a crystalline structure to an amorphous structure. Of the various plasticizers used, the most common are polyols, in which glycerol is included, allowing a good structuring, although it induces the phenomenon of recrystallization. Four bioplastics were developed, based on corn flour, which differed in thickness (0.25 mm and 0.45 mm, on average), with and without a natural pine resin. Regarding the tests carried out in bioplastics, it was concluded that the bioplastic with 0.25 mm and with resin is the one that presents a greater transparency and a greater tensile strength. In turn, the bioplastic with the highest elongation was the one presenting 0.45 with resin. It was also concluded that up to a certain thickness of bioplastic (0.34 mm), the resin adds a certain resistance, from which it withdraws. Through the FTIR analysis, it was confirmed that the resin provides transparency to the bioplastic and that it causes interference in the bonds between the starch and the glycerol.

Keywords: Bioplastics · Agrowastes · Starch · Cellulose · Shoe sector

1 Introduction

Plastics are synthetic polymeric compounds produced from petrochemicals, indispensable to human activities. The growth of the world population combined with the high consumerism, led to the generation of large quantities of plastic waste and, consequently, to a negative impact at the environmental level [1, 2]. As a result, there is a strong demand for low environmental impact materials, such as bioplastics [3, 4].

The development and improvement of conventional plastics results from decades of development, with diverse properties such as: tensile strength, durability, flexibility, low weight, resistance to microorganisms and weathering, low production costs [5, 6].

© Springer International Publishing AG, part of Springer Nature 2019
J. Machado et al. (Eds.): HELIX 2018, LNEE 505, pp. 712–718, 2019.
https://doi.org/10.1007/978-3-319-91334-6_97

In the footwear industry, plastics derived from non-renewable resources (hydrocarbons) are indispensable materials. However, efforts have been encouraged towards the development of biodegradable or biomass-based plastics. In this sense, bioplastics are presented as alternative to conventional plastics [2, 6, 7].

Bioplastics encompass a family of materials that differ from conventional plastics as they are wholly or partly derived from biomass, biodegradable components or both [8].

Biodegradability refers to the chemical process by which microorganisms convert materials into natural substances, such as water, carbon dioxide and biomass [9].

Biopolymers meet environmental concerns but exhibit some performance limitations (thermal resistance, mechanical and barrier properties) [2]. However, the demand for polymer materials from renewable sources that present technological, economic and environmental viability, makes bioplastics interesting candidates [9]. Among the most well-known and abundant renewable resources to produce bioplastics are starch, lignin and cellulose which, though not being plastic in their native form, can be modified to plastics by various approaches [2]. Considering the non-depletion of natural resources in the long term, the most environmentally acceptable strategy to produce biopolymers will be using biological waste [10]. In this context, a potential source of renewable resources to produce bioplastics are the waste generated by agroindustries (vegetable and cereal processing industries) [2, 3, 11–13].

1.1 Starch-Based Bioplastics

Due to the excellent biodegradability, low cost of production and obtaining from renewable resources, the starch is considered a viable alternative source to obtain biodegradable plastics [14].

Starch ($C_6H_{10}O_5$) is a fully biodegradable polysaccharide, biosynthesized by several plants, and one of the most abundant known renewable resources. In its granular form, the starch is composed by linear amylose and highly branched amylopectin. Thus, the starch can be considered as a crystalline material, having an overall crystallinity between 20 and 45%.

When mixed with water and subjected to heat and shear it undergoes a spontaneous restructuring. A homogeneous melt known as a thermoplastic starch is formed, having thermoplastic characteristics. The bioplastic is then obtained by breaking down the structure of the granular starch, when processed with low water content and with thermal and mechanical forces in the presence of plasticizers [14, 15]. The molecular order of the granules is destroyed, converting the molecular structure of the starch to thermoplastic, changing its crystalline structure to an amorphous structure [16].

The stability, processing and physical properties of the thermoplastic depend on the nature of the amorphous and crystalline zones in the structure of the granules. In addition to the water content, the transformation of the granular starch is influenced by process conditions such as temperature and plasticizer content. Glycerol is often used. Plasticizers are extremely important in that they act as lubricants, facilitating the mobility of polymer chains and retarding the retrogradation of bioplastics [15].

The starch-based bioplastic has several advantages, such as biodegradability, renewability and mechanical flexibility. However, the drawbacks are based on the retrogradation and the less satisfactory mechanical properties, particularly in humid environments.

A limitation of the starch bioplastics is therefore its hydrophilic characteristic. The bioplastic easily absorbs water from the environment, dilating. This situation results in loss in their mechanical properties [14].

To reduce the hydrophilicity of the bioplastics a process of modifying the starch can be carried out, reducing the hydrophilicity of the final product, lowering the gelatinization temperature, reducing the retrogradation and improve its flexibility [17, 18].

2 Experimental

Initially, the raw material was selected, based on its complexity of use, as well as the ease of obtaining. From this selection a procedure was studied for producing a bioplastic with suitable characteristics to apply in the footwear sector. Mechanical characteristics were based on a commercial bioplastic of starch with PLA [19] and on adherent film, regarding transparency.

2.1 Selection and Characterization of Agroindustry Wastes

The amount of starch in several samples was determined, to understand which would be the most interesting to be used in the subsequent production of the bioplastic. Among the most interesting were Corn Flour (32, 53%), Fried Wastes (52%) and Starch Residue (100%).

The starch residue was the most promising for bioplastic production. Thus, it was initially used for several tests. However, gelation did not occur in the same way, the texture and viscosity achieved were completely different from corn flour-based bioplastic. The potential justification for this, concerns the composition of the starch in amylose and amylopectin. The amylose content is relevant to the formation capacity of the starch films. However, the final characteristics of the films are strongly influenced by the interaction of amylopectin with the plasticizer [20]. According to Liu *et al.* (2006), the amylopectin content affects the enthalpy of gelatinization. The enthalpy of gelatinization of an amylopectin-rich starch (such as corn-amylose/amylopectin = 23/77) is larger than an amylose-rich starch [21].

Thus, it is assumed that the starch residue is not favorable for the formation of bioplastic, due mainly to lack of amylopectin. Accordingly, corn flour was used.

2.2 Modification of Extracted Starch

The bioplastic was produced through a thermomechanical procedure, mixing the corn flour with other ingredients (deionized water, glycerol and acetic acid) and molding in the glass.

Several tests were carried out to obtain the mixture with the most suitable texture and desired characteristics. Bioplastics with various compositions were produced, and modifications were made in the procedure, to obtain the most suitable preparation. Bioplastic films showed several air bubbles, impairing visual and mechanical characteristics of the final product, being necessary to use a vacuum pump, as well as a vacuum oven instead of a normal oven.

At the end, the blend was prepared with the following composition: 300 mL of water, 30 g of starch, 15 mL of 5% (v/v) acetic acid and 15 mL of glycerol.

Briefly, the mixture was heated with stirring to promote gelatinization (about 65 °C) until the desired viscosity (about 90 °C).

Subsequently, the vial was introduced into the ultrasound with connection to the vacuum pump, to remove the internal air bubbles.

The mixture was then removed and spread immediately on the hot glass. Bioplastics were made with two different thicknesses, 0.25 and 0.45 mm and dried at 70 °C on a desiccator under vacuum for 4 h. After this initial drying, the bioplastic was placed at room temperature for complete drying. Finally, it was removed from the glass.

Pine resin was added to the surface of the bioplastic to promote a potential improvement in the transparency and the mechanical properties of the material.

2.3   Chemical, Morphological and Mechanical Characterization of Bioplastic Films

To understand the applicability of the films, they were characterized in terms of turbidity, mechanical resistance and, chemically, through FTIR.

Results present in Table 1, related to turbidity showed that thicker bioplastic presented, as expected, greater turbidity and that the resin does indeed improve this parameter. Compared with the adhesive film, it is noticed that it is much more transparent than all the bioplastics. However, also much thinner (0 and 0.05 mm). In conclusion, although it was not possible to obtain a transparency similar to the reference value (3%), an interesting transparency was reached, which does not significantly affect the original color, especially with regard to bioplastics with resin.

Table 1. Mechanical properties and turbidity of the bioplastics

Film thickness (mm)	Tensile strength (MPa)	Elongation (%)	Turbidity (%)
0,45 w/resin	1,51 ± 0,18	57,48 ± 2,20	23,61 ± 0,20
0,45	1,70 ± 0,29	39,54 ± 5,39	52,67 ± 0,24
0,25 w/resin	1,77 ± 0,11	55,64 ± 6,80	16,40 ± 1,08
0,25	1,62 ± 0,056	54,63 ± 1,55	50,75 ± 3,24

Mechanical characteristics are dependent on the final use of the material. For shoes, it is of extreme importance to be resistant, having a high breaking stress, although it is also important to have a good elongation. Thus, by analyzing Table 1, commercial bioplastic (reference values) is considerably more resistant to breakage than the ones produced. This may be justified in part by the existence, even if minimized, of internal air bubbles. If there is a bubble, the breakdown voltage will be decreased, since the rupture will occur through that point of weakness. On the contrary, they have a much higher elongation. It is important to remark that the commercial bioplastic has a higher thickness (1–14 mm), which also influences the final results.

Additionally, in bioplastic with resin the rupture tension decreases with increasing thickness. This fact can be justified since the thicker the bioplastic, the harder the resin penetrates the structure, not conferring the desired resistance.

FTIR analysis was performed to qualitatively analyze the interactions between the plasticizer (glycerol) and the starch molecules and to verify how the resin interferes with the material. According to Zullo (2009), part of the peaks between 992 and 1200 cm^{-1} are associated to the interactions between starch molecules and the plasticizer, in this case glycerol, can be used to evaluate thermoplasticization. It also states that stronger interactions are associated with lower wave numbers, leading to more stable hydrogen bonds between starch and glycerol [16].

Analyzing Fig. 1, turbidity results are confirmed. In the Turbidity test, the bioplastic with resin presented greater transparency, and in FTIR it also shows a higher transmittance. Therefore, the resin gives greater transparency to the bioplastic.

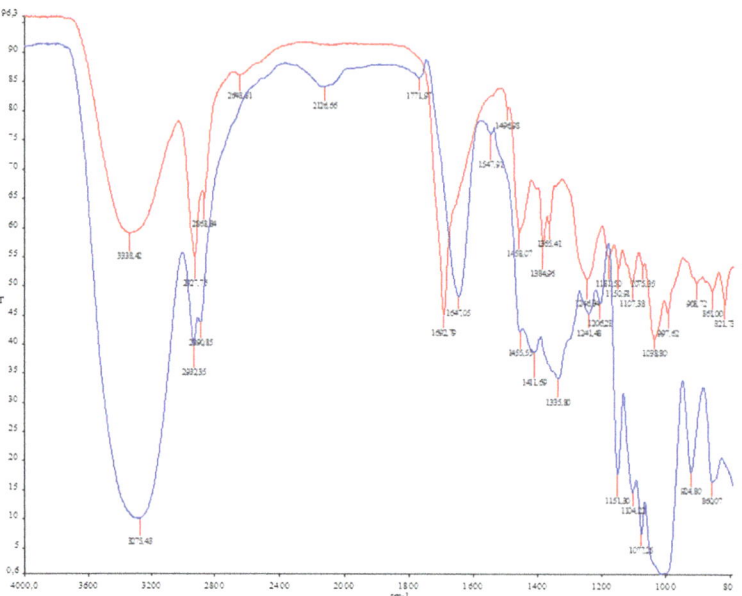

Fig. 1. FTIR analysis of the bioplastic with resin (red) and without resin (blue).

It is also concluded that the resin causes interference with respect to the interactions between the starch and plasticizer. It is not clear how this interaction occurs, because for the bioplastic with resin (red) there are peaks with higher wave numbers and peaks with less. There are also new peaks, such as 997.62 and 861.00 cm^{-1}.

3   Conclusions

Corn flour showed to be more suitable than the starch residue for bioplastic production, since it has amylopectin, unlike starch residue. The biggest challenge was to find a way to prevent internal air bubbles. A vacuum pump, ultrasound and vacuum oven were used to diminish to a large extent the existent bubbles.

Regarding transparency, turbidity was determined. It was concluded that the 0.25 mm bioplastic with resin showed the best result with a turbidity of $16.40 \pm 1.08\%$. It became clear that the resin confers turbidity reduction, improving the visual properties of the bioplastic. It was also visible that the thickness gives them less transparency, as expected.

Regarding the mechanical properties, the tensile strength and percentage elongation of the samples were determined using mechanical tests.

Regarding the tensile strength it was concluded that the values were very close to each other. However, the bioplastic that presented a higher breaking stress was 0.25 mm with resin. Considering a linear behavior, it was observed that, up to a certain thickness of bioplastic, resin adds a certain resistance.

With respect to elongation, it became clear that the increase in the thickness, in the bioplastic without resin, caused a decrease in the elongation, unlike what happened in the resin bioplastic. In this one, the resin was able to slightly increase this property.

Finally, FTIR analysis confirmed that resin interfered both in transparency and in the interactions between the starch and the plasticizer (glycerol).

An improvement in resin application would be necessary. The fact that it is applied to the surface may not be the best way to use it as it may not penetrate inside of the bioplastic, failing to provide the required strength. Incomplete penetration may be the justification for why it does not always provide a higher breaking-off stress. A possible change in the resin addition procedure could possibly improve the properties of the bioplastic. Immersing the bioplastic for some time in a dilute resin solution could be a potential bioplastic enhancement technique.

Future work includes a more in-depth practical study of other wastes that can generate bioplastics. A very promising case will be protein-rich residues.

References

1. Bassi, A.: 13 – Biotechnology for the Management of Plastic Wastes. Elsevier B.V. (2017)
2. Siracusa, V., Rocculi, P., Romani, S., Rosa, M.D.: Biodegradable polymers for food packaging: a review. Trends Food Sci. Technol. **19**(12), 634–643 (2008)
3. Ojeda, T.: Polymers and the environment. In: Polymer Science, p. 1. InTech (2013)
4. PlasticsEurope: Plastics – the Facts 2010. Plastics – the Facts 2010, p. zu finden unter (2010). www.plasticseurope.de/informations
5. Sivan, A.: New perspectives in plastic biodegradation. Curr. Opin. Biotechnol. **22**(3), 422–426 (2011)
6. Shah, A.A., Hasan, F., Hameed, A., Ahmed, S.: Biological degradation of plastics: a comprehensive review. Biotechnol. Adv. **26**(3), 246–265 (2008)

7. European Bioplastics: Bioplastics facts and figures. Institute for Bioplastics and Biocomposite, Nova-Institute, p. 6 (2014)
8. European Bioplastics: What are bioplastics?, p. 4 (2016)
9. European Bioplastics: Industrial Use of Agricultural Feedstock, p. 2 (2015)
10. Scott, G.: Green polymers. **68**, 1–7 (2000)
11. Zhang, Z., Wong, H.H., Albertson, P.L., Harrison, M.D., Doherty, W.O.S., Hara, I.M.O.: Bioresource technology effects of glycerol on enzymatic hydrolysis and ethanol production using sugarcane bagasse pretreated by acidified glycerol solution. Bioresour. Technol. **192**, 367–373 (2015)
12. Demelash, N., Bayu, W.: Current and residual effects of compost and inorganic fertilizer on wheat and soil chemical properties, pp. 357–367 (2014)
13. Tabone, M.D., Cregg, J.J., Beckman, E.J., Landis, A.E.: Sustainability metrics: life cycle assessment and green design in polymers. Environ. Sci. Technol. **44**(21), 8264–8269 (2010)
14. Andrade, C.T., Achete, C.A.: Polimerização Por Plasma, pp. 91–92 (2003)
15. Nafchi, A.M., Moradpour, M., Saeidi, M., Alias, A.K.: Thermoplastic starches: properties, challenges, and prospects. Starch/Staerke **65**(1–2), 61–72 (2013)
16. Zullo, R., Iannace, S.: The effects of different starch sources and plasticizers on film blowing of thermoplastic starch: correlation among process, elongational properties and macromolecular structure. Carbohydr. Polym. **77**(2), 376–383 (2009)
17. Da Róz, A.L., Carvalho, A.J.F., Gandini, A., Curvelo, A.A.S.: The effect of plasticizers on thermoplastic starch compositions obtained by melt processing. Carbohydr. Polym. **63**(3), 417–424 (2006)
18. Laycock, B.G., Halley, P.J.: Starch Applications: State of Market and New Trends. Elsevier B.V. (2014)
19. Gonzalez-Gutierrez, J., Partal, P., Garcia-Morales, M., Gallegos, C.: Development of highly-transparent protein/starch-based bioplastics. Bioresour. Technol. **101**(6), 2007–2013 (2010)
20. Mali, S., Grossmann, M.V.E., García, M.A., Martino, M.N., Zaritzky, N.E.: Effects of controlled storage on thermal, mechanical and barrier properties of plasticized films from different starch sources. J. Food Eng. **75**(4), 453–460 (2006)
21. Liu, H., Yu, L., Xie, F., Chen, L.: Gelatinization of cornstarch with different amylose/amylopectin content. Carbohydr. Polym. **65**(3), 357–363 (2006)

Production of Nanocellulose from Lignocellulosic Biomass Wastes: Prospects and Limitations

João R. A. Pires⬤, Victor Gomes Lauriano de Souza⬤,
and Ana Luisa Fernando$^{(\boxtimes)}$⬤

MEtRiCS, Departamento de Ciências e Tecnologia da Biomassa,
Faculdade de Ciências e Tecnologia, FCT, Universidade Nova de Lisboa,
Campus de Caparica, 2829-516 Caparica, Portugal
{jr.pires,v.souza}@campus.fct.unl.pt, ala@fct.unl.pt

Abstract. The search for renewable alternatives to petroleum products for industrial applications is increasing. Each year, many tons of inedible plant material is produced, much of which is landfilled. Cellulose, the most abundant biopolymer in nature, present in lignocellulosic biomass wastes can be extracted and converted to nanometric scale. Nanocellulose (NC) has unique characteristics for the development of new materials: abundance, renewability and biodegradability, mechanical properties and its nanometric dimensions open a wide range of possible properties and applications to be discovered. One of the most promising uses of NC is as reinforcement of mechanical properties in polymeric bionanocomposites. This review aims to give a recent view on this emerging nanomaterial, focusing on lignocellulosic biomass wastes extraction procedures, and its application in new technological developments. The challenges and future opportunities of bionanocomposites reinforced with NC will be discussed, as well as the remaining obstacles to its valorization and use.

Keywords: Bionanocomposites · Cellulose nanocrystals · Cellulose nanofibers
Lignocellulosic biomass · Nanocellulose

1 Introduction

Recently, the reuse of lignocellulosic biomass wastes has been explored as a green and viable alternative to the use of fossil resources [1]. This reuse has a dual purpose: to reduce the overload in landfills, once the annual production of this type of agricultural waste is tremendous (about 1.3×10^{10} metric tons) [2], and to reduce the dependency on fossil fuels, with all the environmental benefits associated with [3]. Agricultural wastes (straw, bark, shells, leaves, bagasse); forest residues (hardwood and softwood); energy crops; food wastes; and municipal and agro-industrial wastes are some lignocellulosic resources that can be reused [1].

Lignocellulosic fibers consist mainly of three biopolymers: cellulose (30–50% by weight); hemicellulose (19–45% by weight); lignin (15–35% by weight); and contribute to the hydrolytic stability and structural strength of the cell walls, as well as to

© Springer International Publishing AG, part of Springer Nature 2019
J. Machado et al. (Eds.): HELIX 2018, LNEE 505, pp. 719–725, 2019.
https://doi.org/10.1007/978-3-319-91334-6_98

prevent microbial degradation of the plant [4]. Cellulose is a polydisperse linear homopolymer with the formula $(C_6H_{10}O_5)_n$, a polysaccharide consisting of a linear chain of several hundred to many thousands of $\beta(1 \rightarrow 4)$ linked D-glucose units, including free hydroxyl groups (-OH) at the C-2, C-3, and C-6 atoms [5, 6]. Despite its simple chemical structure, depending on the source and the process by which it was obtained, the cellulose has very different supramolecular structures. These differences are mainly due to the biosynthesis conditions, specific enzymatic terminal complexes and chemical and physical configurations derived from the extraction procedures [7]. Cellulose is organized into microfibrils with amorphous regions interspersed with strongly hydrogen-bond crystalline domains of 2–20 nm in width and up to a few microns in length [8]. This biopolymer is widely used in traditional industries, such as fiber, paper, films or polymers, but new applications of materials have been explored, one of them being the production of cellulose at the nanoscale [9].

Nanotechnology refers to materials with measurements of about 100 nm in at least one dimension. In addition, nanometric materials have different physical, chemical or biological properties from those of bulk material [8]. Nanocellulose was extensively researched in the last decade due to their properties like low density, low cost, abundance, renewability, high mechanical properties, large surface area and aspect ratio, notable flexibility, specific barrier properties and low thermal expansion [10]. Even at low concentrations, the nanocellulose confers a greater rigidity to the polymer matrices, due to the fact of being able to form strong interconnected networks through hydrogen bonds and due to their great aspect ratio. For these reasons, nanocellulose can be consider as a potential reinforcing agent in polymer composites [5, 11]. Therefore, this work aims to review the possibilities of producing nanocellulose from lignocellulosic biomass wastes, and to discuss the challenges and future opportunities of bio-nanocomposites reinforced with NC.

2   From Lignocellulosic Biomass to Nanocellulose

The separation of cellulose from lignin and hemicellulose is not an easy procedure because the cell walls of the plants offer resistance to deconstruction (recalcitrance) [1]. The recalcitrance is due to the highly crystalline structure of cellulose which is embedded in a polymeric matrix of lignin and hemicellulose [12].

The biomass pretreatment has as main objective to overcome the recalcitrance, modifying the structure and size of the biomass, allowing opening of the material structure to facilitate access to the cellulose microstructure [6]. The fractionation of cellulose from the lignocellulosic biomass should be a cost-effective process, as such it must take into account certain requirements, such as avoiding structure rupture or loss of cellulose, hemicellulose and lignin, being economically profitable, not using large amounts of energy and avoiding the production of toxic and hazardous waste [1]. Several types of pretreatment can be used to open biomatrix structures: physical (milling and grinding); chemical (acid hydrolysis, alkaline hydrolysis, oxidation, use of organic solvents or ionic liquids); biological; and multiple or combined pretreatments (steam pretreatment/auto hydrolysis, hydrothermolysis and wet oxidation) [13]. The distinct lignocellulosic biomass pretreatment strategies can suffer variation in terms of pH, temperature, types of catalyst, and treatment time. These variations affect the severity of the pretreatment and the biomass composition during biomass degradation [1]. Chemical pretreatment is considered the most efficient in terms of processes and economic costs. Physical pretreatment, in addition to being less effective in the process of deconstructing biomass, also consume more energy, whereas enzymes for biological pretreatment is expensive and require time [1].

Nanocellulose can be defined in two principal specimens, cellulose nanocrystals (CNC)/nanowhiskers (CNW) or cellulose nanofibers (CNF) [9]. In addition to these two types, there are still three other types of nanocellulosic materials: microcrystalline cellulose (MCC), cellulose microfibrillated (CMF) and bacterial nanocellulose (BNC) [5]. The production of nanocellulose (NC) is generally performed in two steps. First, the raw material is pretreated to obtain "pure" cellulosic fibers. The second step (fibrillation) is when the fibers are transformed into CNCs or CNFs [6].

Cellulose nanocrystals are rod-shaped structures, with 3–20 nm wide and 50–500 nm in length, and are obtained through mechanical and chemical processes after the extraction of cellulose, in which acid hydrolysis is the most effective process [1, 6]. Usually, this process is performed in the presence of a mineral acid (sulfuric or hydrochloric) [14], but the use of other acids, like oxalic acid, can also be found in literature [15]. The crystalline and amorphous domains of the cellulose have different kinetics of hydrolysis leading to the removal of the amorphous part during this process, so the CNC exhibit high crystallinity (90% or higher) [11]. The hydrolysis is terminated by rapid dilution of the acid, followed by acid removal by centrifugation and/or dialysis [5]. The characteristics of nanocrystals from acid hydrolysis are highly dependent on various factors, such as origin of cellulose sources, types of acid, concentration of acid, reaction time, and temperature [1]. Mechanical processes, such as sonication, are usually applied following acid hydrolysis to avoid an aggregation of cellulose fibrils in order to obtain a homogeneous CNC suspension [5]. Conventional acid hydrolysis treatment entails environmental problems and is economically unfriendly since extra cost is required for effluent treatment. Environmentally, it is suggested that treatment by enzymatic hydrolysis is the most feasible, and in the future, when it is possible to overcome the economic and technical limitations, may be the most used method [1].

Cellulose nanofibers consist of a bunch of stretched cellulose chain molecules with long, flexible and entangled cellulose nanofibers of approximately 1–100 nm size. CNFs are preferably produced using intensive mechanical processes, with or without chemical and biological treatments. Mechanical approaches to diminish cellulosic fibers into nanofibers can be divided into refining and homogenizing, microfluidization, grinding, cryocrushing and high intensity ultrasonication [16]. Mechanical methods have as main problem the high energy consumption. Recently, research has focused on low cost methods for nanocellulose production, with high efficiency and limited environmental constraints [16]. Different pretreatments have been introduced before mechanical processes in order to reduce this energy consumption as well as to make the surface hydrophobic. Among the exploited pretreatments, TEMPO (2, 2, 6, 6-tetramethylpiperidine-1-oxyl) oxidation, acetylation, carbomethylation, alkaline pretreatment and enzymatic pretreatment are the most studied [8].

Among the pre-treatments, in the last years the literature has highlighted the potentialities of TEMPO oxidation. TEMPO oxidation is the transformation of cellulose hydroxyls groups to carboxyl moieties and actually, it has been proposed both to promote the nanofibrillation process, as an intermediate step to obtain grafting on cellulose chains, and as a chemical route to increase crystallinity and water dispersion of nanocellulose [7]. In comparison with alkaline or enzymatic pretreatments this type of oxidation offer clear advantages, such as, high reaction rate, high conversion ratio, high selectivity, partial decrease of molecular weight of polysaccharides during the process (if controlled) and low cost as co-oxidant [17].

3   Nanocellulose Application in Bionanocomposites: Opportunities and Challenges

Polymer nanocomposites (PNC) are mixtures of a certain polymer reinforced with small quantities of nanosized inorganic or organic fillers with particular size, geometry, and surface chemistry [18]. The nanofillers are incorporated with the purpose of increasing the mechanical and barrier properties of bionanocomposites [4]. Different production processes to incorporate cellulose nanoparticles into the polymeric bionanocomposites have been tested, with emphasis on the following methods: solvent casting, melt intercalation process and ring opening polymerization [9].

The most important effects of NCs on polymer matrices are on tensile properties. Both CNF and CNC have good reinforcement characteristics on polymer nanocomposites [11]. Generally, it is found that when inserted into the polymer matrix, the nanocellulose particles increase the modulus and strength, and decrease the gas permeability (especially to oxygen). This effect can be attributed to the strength and rigidity of the NCs as well as the strong intermolecular bonds of the nanocellulose [11, 19]. Additionally, nanocellulose has good barrier properties due to the dense structural network which is formed by nanofibers. In the case of CNCs, this density is attributed to the numerous small and uniform particles, whereas CNFs, a densification is reinforced by the inherent flexibility of the material [20]. In addition to the barrier and mechanical properties, other parameters are influenced by the incorporation of NCs, such as optical properties, electrical and thermal conductivity and water sensitivity [16]. Although water vapor permeability and water sensitivity are both related to material hydrophilicity, they are different properties. The permeability to water vapor is related to the transport of the water molecules through the bionanocomposite, whereas the water sensitivity refers to the degradation of the structure of the biofilm by water by swelling or dissolution [19]. Although the OH groups of the NCs make them hydrophilic, they have proved to be effective in increasing the water resistance of the matrix. This is due to the strong hydrogen bonds between the nanocellulose and the polymer, making the material more cohesive and more crystalline [19]. Unless the particle diameter is greater than 50 nm or if there is a significant agglomeration of the nanofibres, the nanocellulose does not affect the transparency of the bionanocomposites [11]. NCs have also been reported to improve thermal stability of biopolymers, whereas their effect on glass transition temperature (Tg) of nanocomposites have been disputed-sometimes increasing Tg, sometimes presenting negligible effects on it [19]. Despite the excellent properties presented by nanocellulose-reinforced composites, not everything is opti-mized. The properties of these new materials are very dependent on the possibility of strong interactions between the hydrogen bonds, further boosted by the size of the particles. The challenge at the moment to the researchers is to avoid particle agglom-eration, promoting the dispersion of cellulose nanoparticles and making the medium as homogeneous as possible throughout the processing, thus promoting favorable filler/matrix interactions and beneficial formation of a nanoparticle network [4].

Many opportunities and potential applications of nanocellulose have emerged recently. Among them, new functionalities have been added, including as hydrogels or drug delivery systems, such as antimicrobial agents, intelligent materials and as edible films for food packaging [4]. Lately, different biomasses were explored, as well as processes of extraction and production of nanocellulose. Faradilla et al. [21] charac-terized biofilms produced from banana pseudo-stem nanocellulose that was prepared by TEMPO-mediated oxidation showing the promising potential of this procedure. Pereira et al. [22] developed biodegradable films made of hemicellulose and reinforced with CNC extracted from wheat straw, showing that nanocellulose reinforced the films, improving mechanical and barrier properties. Another study was presented by Xu et al. [14] where chitosan (CS) biocomposites with nanocrystalline cellulose from rice straw residue was developed. The authors have evidenced the superior interfacial compati-bility of CS/CNC biocomposites with excellent tensile strength. Most of the papers reporting the development of nanocellulose-based composites for packaging present results on their barrier and mechanical properties, with some showing also results of antimicrobial and antioxidant properties, but few studies have evaluated the improvement of quality and shelf life of foodstuffs packaged with this kind of bio-nanocomposites. Also, migration, toxicity and ecotoxicity tests should be done, in order to avoid harmful effects on human health and on the environment [19].

4 Conclusion

Lignocellulosic biomass is the most abundant and biorenewable polymer on earth with great potential for sustainable nanocellulose production. Cellulose, has been considered a promising material for producing nanosize reinforcement for a broad range of applications such as papermarking and high-quality bionanocomposites as well as for health care products. The complex hierarchy structure of lignocellulose is the main obstacle for major components separation, so overcoming the recalcitrance of ligno-cellulosic biomass is a key step in separating the biopolymer.

Different approaches including mechanical treatments, pretreatments techniques and acid hydrolysis are being used for nanocellulose preparation. However, the current methods to convert cellulose to nanocellulose consume a lot of energy during and after the process and thus are deemed to be unprofitable and nonenvironmentally friendly. Yet, the excellent properties of nanocellulose materials and the fact that they can be considered green nanomaterials because they are carbon neutral, sustainable, biodegradable, recyclable and nontoxic, open excellent prospects for the increase of their use in novel biomaterials. Application of nanocellulose, even in very low amounts, leads to improvement of mechanical, thermal and barrier properties of polymer composites, which contributes to enhance the composites overall performance. The advantages that nanocomposites offer far outweigh the costs and concerns, and with time the technology will be further refined and more developed. Hence, the reinforcement of nanocellulose in polymer composites will resolve the various problems faced by the biopolymer industry at present. The safety issues of nanocellulose should, however, be monitored and controlled in order to confirm whether it has no harmful effects on human health and on the environment. In addition, several properties of these materials should be considered including the chemical composition, morphology, crystallinity, structure and thermal behavior.

References

1. Lee, H.V., Hamid, S.B.A., Zain, S.K.: Conversion of lignocellulosic biomass to nanocellulose: structure and chemical process. Sci. World J. **2014**, 1–20 (2014)
2. Deepa, B., Abraham, E., Cordeiro, N., Faria, M., Thomas, S., Pothan, L.A.: Utilization of various lignocellulosic biomass for the production of nanocellulose: a comparative study. Cellulose **22**, 1075–1090 (2015). https://doi.org/10.1007/s10570-015-0554-x
3. Souza, V.G.L., Fernando, A.L.: Nanoparticles in food packaging: biodegradability and potential migration to food—a review. Food Packag. Shelf Life **8**, 63–70 (2016). https://doi.org/10.1016/j.fpsl.2016.04.001
4. Mariano, M., El Kissi, N., Dufresne, A.: Cellulose nanocrystals and related nanocomposites: review of some properties and challenges. J. Polym. Sci. Part B Polym. Phys. **52**(12), 791–806 (2014). https://doi.org/10.1002/polb.23490
5. Miao, C., Hamad, W.Y.: Cellulose reinforced polymer composites and nanocomposites: a critical review. Cellulose **20**, 2221–2262 (2013). https://doi.org/10.1007/s10570-013-0007-3
6. Vilarinho, F., Sanches Silva, A., Vaz, M.F., Farinha, J.P.: Nanocellulose in green food packaging. Crit. Rev. Food Sci. Nutr. **0**(0), 1–12 (2017). https://doi.org/10.1080/10408398.2016.1270254
7. Li, F., Mascheroni, E., Piergiovanni, L.: The potential of NanoCellulose in the packaging field: a review. Packag. Technol. Sci. **28**(January), 475–508 (2015). https://doi.org/10.1002/pts.2121
8. Jonoobi, M., Oladi, R., Davoudpour, Y., Oksman, K., Dufresne, A., Hamzeh, Y., Davoodi, R.: Different preparation methods and properties of nanostructured cellulose from various natural resources and residues: a review. Cellulose **22**(2), 935–969 (2015). https://doi.org/10.1007/s10570-015-0551-0
9. Bharimalla, A.K., Deshmukh, S.P., Vigneshwaran, N., Patil, P.G., Prasad, V.: Nanocellulose based polymer composites for applications in food packaging: future prospects and challenges. Polym.-Plast. Technol. Eng. **56**(8), 1–71 (2017). https://doi.org/10.1080/03602559.2016.1233281
10. Liu, D., Chen, X., Yue, Y., Chen, M., Wu, Q.: Structure and rheology of nanocrystalline cellulose. Carbohyd. Polym. **84**(1), 316–322 (2011). https://doi.org/10.1016/j.carbpol.2010.11.039
11. Abdul Khalil, H.P.S., Saurabh, C.K., Adnan, A.S., Nurul Fazita, M.R., Syakir, M.I., Davoudpour, Y., Dungani, R.: A review on chitosan-cellulose blends and nanocellulose reinforced chitosan biocomposites: properties and their applications. Carbohyd. Polym. **150**, 216–226 (2016). https://doi.org/10.1016/j.carbpol.2016.05.028

12. Mosier, N., Wyman, C., Dale, B., Elander, R., Lee, Y.Y., Holtzapple, M., Ladisch, M.: Features of promising technologies for pretreatment of lignocellulosic biomass. Biores. Technol. **96**(6), 673–686 (2005). https://doi.org/10.1016/j.biortech.2004.06.025
13. Loow, Y.L., Wu, T.Y., Jahim, J.M., Mohammad, A.W., Teoh, W.H.: Typical conversion of lignocellulosic biomass into reducing sugars using dilute acid hydrolysis and alkaline pretreatment. Cellulose **23**(3), 1491–1520 (2016). https://doi.org/10.1007/s10570-016-0936-8
14. Xu, K., Liu, C., Kang, K., Zheng, Z., Wang, S., Tang, Z., Yang, W.: Isolation of nanocrystalline cellulose from rice straw and preparation of its biocomposites with chitosan: physicochemical characterization and evaluation of interfacial compatibility. Compos. Sci. Technol. **154**, 8–17 (2018). https://doi.org/10.1016/j.compscitech.2017.10.022
15. Poonguzhali, R., Basha, S.K., Kumari, V.S.: Synthesis and characterization of chitosan-PVP-nanocellulose composites for in-vitro wound dressing application. Int. J. Biol. Macromol. **105**, 111–120 (2017). https://doi.org/10.1016/j.ijbiomac.2017.07.006
16. Abdul Khalil, H.P.S., Davoudpour, Y., Islam, M.N., Mustapha, A., Sudesh, K., Dungani, R., Jawaid, M.: Production and modification of nanofibrillated cellulose using various mechanical processes: a review. Carbohyd. Polym. **99**, 649–665 (2014). https://doi.org/10.1016/j.carbpol.2013.08.069
17. Pierre, G., Punta, C., Delattre, C., Melone, L., Dubessay, P., Fiorati, A., Michaud, P.: TEMPO-mediated oxidation of polysaccharides: an ongoing story. Carbohyd. Polym. **165**, 71–85 (2017). https://doi.org/10.1016/j.carbpol.2017.02.028
18. Souza, V.G.L., Fernando, A.L., Pires, J.R.A., Rodrigues, P.F., Lopes, A.A.S., Fernandes, F. M.B.: Physical properties of chitosan films incorporated with natural antioxidants. Ind. Crops Prod. **107**, 565–572 (2017). https://doi.org/10.1016/j.indcrop.2017.04.056
19. Azeredo, H.M.C., Rosa, M.F., Mattoso, L.H.C.: Nanocellulose in bio-based food packaging applications. Ind. Crops Prod. **97**, 664–671 (2017). https://doi.org/10.1016/j.indcrop.2016.03.013
20. Ferrer, A., Pal, L., Hubbe, M.: Nanocellulose in packaging: advances in barrier layer technologies. Ind. Crops Prod. **95**, 574–582 (2017). https://doi.org/10.1016/j.indcrop.2016.11.012
21. Faradilla, R.H.F., Lee, G., Arns, J.Y., Roberts, J., Martens, P., Stenzel, M.H., Arcot, J.: Characteristics of a free-standing film from banana pseudostem nanocellulose generated from TEMPO-mediated oxidation. Carbohyd. Polym. **174**, 1156–1163 (2017). https://doi.org/10.1016/j.carbpol.2017.07.025
22. Pereira, P.H.F., Waldron, K.W., Wilson, D.R., Cunha, A.P., Brito, E.S.D., Rodrigues, T.H. S., Azeredo, H.M.C.: Wheat straw hemicelluloses added with cellulose nanocrystals and citric acid: effect on film physical properties. Carbohyd. Polym. **164**, 317–324 (2017). https://doi.org/10.1016/j.carbpol.2017.02.019

Low Cost System Based on Textile and Plastic Waste for Underground Irrigation in the Brazilian Semi-arid Region

Nicéa Ribeiro do Nascimento[1,3],
Luísa Rita Brites Sanches Salvado[1,3(✉)],
and Francisco Fechine Borges[2,3]

[1] Universidade da Beira Interior, Covilhã, Portugal
rita.salvado@ubi.pt
[2] Instituto Federal da Paraíba (PB), João Pessoa, Brazil
[3] Associação LETS (PB), João Pessoa, Brazil

Abstract. Irrigation is an ancient technique in order to guarantee the humidity required for a particular crop. Irrigation brings several benefits and some disadvantages if due care is not taken. It is of fundamental importance to choose the appropriate method of irrigation for each location and situation. Among the four major methods of irrigation - surface, sprinkler, localized and underground - the latter has promising characteristics for use in the semi-arid, such as reducing water consumption. On the other hand, the indiscriminate disposal of PET bottles and textile waste in the environment is a major problem in Brazil. This paper presents preliminary results of a low cost system based on reuse of textile waste and PET bottles for capillarity underground irrigation for family agriculture in the semiarid region, which offers the opportunity to better productive conditions to different regions in need, to contribute to the reuse of waste from the fast-fashion culture that dominates the fashion market today, as well as contribute to the empowerment of families who depend on family farming to survive. A pilot system was constructed at an experimental site in São Raimundo Nonato (PI), coordinates 8 ° 50'13.6 "S 42 ° 46'26.1" W, and has obtained very satisfactory qualitative results in the cultivation of manioc. The Capillary Underwater Irrigation System for Family Agriculture - SISCAFI, as the device was called, has the potential to impact the performance of perennial crops in the semiarid, since it combines complex concepts currently used only in expensive and sophisticated irrigation systems (capillarity, underground irrigation) with ease of construction, operation and maintenance, even in small semi-arid communities. In addition, its construction process is handcrafted, accessible to the population with low education. It reuses waste materials such as textile waste and PET bottles, contributing to the preservation of the environment and the generation of employment and income in the rural environment.

Keywords: Underground irrigation · Semiarid · Family farming
Fast-fashion

© Springer International Publishing AG, part of Springer Nature 2019
J. Machado et al. (Eds.): HELIX 2018, LNEE 505, pp. 726–731, 2019.
https://doi.org/10.1007/978-3-319-91334-6_99

1 Irrigation and the Water Problem in the Northeastern Semi-arid Region

Irrigation is an ancient technique of application of water to the soil, in addition to rainwater, in order to guarantee the humidity required for a particular crop. According to [1], irrigation has as its objective "to satisfy the water needs of crops, applying water evenly and efficiently, that is, that the largest amount of water applied is stored in the root zone and being available to the culture".

On the other hand, [2] of the Luiz de Queiroz Higher School of Agriculture in Piracicaba (SP), Brazil, extends the aforementioned classical concepts, including conservation and social aspects. In a conservationist view, irrigation is defined by this author as an "Artificial application of water to the soil through methods that best meet the conditions of the physical environment (crop water demand, topographic conditions of the land, soil water retention capacity...) and the desired objectives (maximize productivity, maximize profit...) with minimal environmental degradation". In a social view, irrigation is defined as "an agricultural practice capable of maximizing total benefits, including non-monetary benefits such as food security, job creation, improving the socio-economic conditions of rural communities, securing people in the countryside and protecting the quality of water".

As well, [2] classifies irrigation objectives according to two aspects:

- Financial, where "it is sought to maximize the benefit/cost ratio by increasing production, either in quantity or quality, or incorporating land that, without the use of irrigation, could not be cultivated"; and
- Social, where these aspects "are more relevant than financial (public projects of regional development): food security; fixation of man in the countryside, improvement of the socioeconomic conditions of rural communities".

Irrigation brings several benefits and some disadvantages if due care is not taken. Among the advantages, can be cited, based on the same author:

- Increasing crop productivity, profit and value of a rural property;
- The feasibility of staggering crops;
- The viability of two or more crops per year, in the same area;
- The introduction of nobler crops, with greater added value;
- Minimizing investment risk;
- The fixation of the man in the rural environment, by the increase of the demand of manpower.

Despite this, [2] defends that irrigation can also have some drawbacks depending on the situation:

- High implementation costs;
- High water consumption, in most cases, associated with high rates of waste;
- Lack of skilled labor for operation, which may render the system unfeasible;
- Environmental degradation and contamination of water resources;
- Salinization of soil and water.

Irrigation for food production is responsible for heavy waste of water. According to [3], "agriculture will remain the world's largest water consumer sector, which in many countries accounts for about 2/3 or more of the availability of rivers, lakes and aquifers". It is of fundamental importance to choose the appropriate method of irrigation for each location and situation. Consideration should be given to various aspects such as water availability, soil type, terrain relief, crop types, annual climate and rainfall, labor availability, associated technologies, among other factors.

Local and underground irrigation practices have attributes favorable to agriculture in semi-arid regions, including a significant reduction in the amount of water required and losses by evaporation. Some of these methods are already in use in the Northeast region of Brazil. Table 1 shows two examples of the use of this technology in this region.

Table 1. Examples of irrigation located in the semi-arid.

Local	Culture	Example	References
Lago de Sobradinho	Onion	"Through the Projeto Lago de Sobradinho, with the technical support of Embrapa Semiarido, the producer Neuwilton de Sousa implemented the systems, which resulted in a yield of almost 500% in four months of cultivation. The application of fertigation and drip irrigation systems in onion production saves about 50% of water, 80% of fertilizers and 30% of labor. In addition to reducing costs, there are improvements in quality and increased productivity"	[4]
Sobradinho	Melon and Watermelon	"Edvaldo Barbosa da Silva, with property in Novo São Gonsalo, in the rural area of Sobradinho, has been planting onion and melon for more than 20 years. After continuous failures with the cultures, was about to leave the work in the field and look for another way of life. In a second crop, after the start of the use of drip irrigation, the withdrawal jumped from 15 t/ha to almost 30 t/ha"	[5]

Several other examples of the use of localized irrigation with the simultaneous application of fertilizers, called fertigation, can be found in the semi-arid, as described by [6, 7].

2   A Possible Solution to the Problems of Textile and Plastic Waste

A large amount of textile waste is disposed annually in the environment. In Brazil alone, there are estimates of about 175 thousand tons per year, and more than 90% is wasted. With Law No. 12,305, which integrates the National Solid Waste Policy, and regulatory decree, the Brazilian government considers it a priority to reinforce the prevention of waste production and encourage its reuse and recycling.

One factor that draws attention is the use of textile waste coming from the fashion industry that became known as fast-fashion. In the 1970s the so-called Oil Crisis took place, as a result of an embargo by the OPEC (Organization of Petroleum Exporting Countries) and Persian Gulf member states on the distribution of oil to the United States and European countries. Because of the difficulties presented by this ban on the marketing of petroleum, the textile companies present in these countries have created a strategy to dispose of production in order to overcome the difficulties of the crisis, which was later called fast-fashion. This consumer behavior in fashion was purposely planned by the market, more specifically, by the fast fashion industry as shown by [8]: "In fast-fashion the company is structured to collect data and information, interpreted as elements of trend-fashion and verify their own hypotheses with the monitoring of consumers (and not only sales). The company in this model does not renounce its own "productive" role, but relativizes it in relation to a process in which a part of the product is made and built by the consumer".

The term fast-fashion, however, was only really coined in the mid-1990s, a way the media found to express the rapidly changing fashion market by major, trend-setting companies. The companies that work in this model use the notions of renowned brands in the fashion market to manufacture similar models, but of a much lower quality than would theoretically guarantee the consumption of such pieces, as [9] emphasize: "Fast-fashion describes the retail strategy of adapting merchandise assortments to current and emerging trends as quickly and efficiently as possible".

It should be noted, therefore, that despite encouraging the economy, the fast-fashion model is unsustainable, Fast-fashion clothing parts are used on average five times and generate 400% more carbon emissions than ordinary parts, which are used 50 times. And the production of clothes does not only pollute with carbon emission: to produce textile fibers it is necessary to deforest, to use fertilizers, pesticides and other forms of pollution. In addition, large-scale production of the fast fashion model encourages slave labor in Asian countries.

A similar situation occurs with the disposal of plastic bottles in the environment, especially PET (Polyethylene Terephthalate), generating a large problem. Although Brazil presents a PET recycling rate of 51% in 2015, approximately 537,000 tons of PET products were produced in 2015, that is, the other 263,000 tons were sent to landfills or were discarded directly in nature, impacting the entire environment. Moreover, most of the recycling companies are in the Southeast region.

In this context, the use of synthetic textile waste as well as discarded plastic materials has been proposed in an innovative way for the development of low-cost underground irrigation systems. This solution aims to favor family farming and reduce water consumption and environmental impacts in the northeastern semi-arid region and the world, reusing the discarded materials from major textile industries and PET bottles that would be discarded in the environment.

3   Objective and Methodology

The general objective of the project described in this work was to develop and test a low cost system for underground irrigation by capillarity, for use in family agriculture in the northeastern semi-arid region. The hypotheses of this study were:

- That the developed system, in fact, reduces the consumption of water for cassava and palm production, in a typical situation of cultivation in the semiarid;
- That the system be low cost and easy to build and operate;
- Reuse of waste from the fast-fashion textile industry;
- That there may be an awareness of the population regarding the reuse of textile waste in this process.

The system developed was called the Capillarity Underground Irrigation System for Family Agriculture - SISCAFI. It is composed of reused PET bottles, buried in inverted upright position (head-end), interconnected by irrigation hose with strips of synthetic fabric (leftovers and flaps) placed between the inside of the bottle and the soil to be irrigated, forming wicks. When the bottles are filled with water, the wick arries the liquid by capillary effect, irrigating the soil around the roots. Thus, plants are irrigated with an optimum amount of water, because of the significant reduction of loss by evaporation.

The methodological approach conceived in this work included planning, designing, confection and implantation of experimental irrigation units, technical-scientific monitoring of the crop and evaluation of the results achieved, besides the validation and improvement of the proposed solution. In order to conduct the technical-scientific studies and to validate the innovation, an experimental unit was built in São Raimundo Nonato (PI), in an experimental field on the banks of PI-140, coordinates 8 ° 50'13.6 "S 42 ° 46 '26.1 "W, where two typical northeastern crops are being cultivated: manioc (Manihot esculenta Crantz) and palm (Opuntia fícus-indica).

4   Results, Discussion and Conclusion

The system is stocked daily by the end bottles. Irrigation at root level is due to the principles of the communicating vessels and the capillarity of the wicks of synthetic fabric flaps. Capillarity has been sufficient to maintain the moisture required for plant growth, even in drought conditions.

The hypotheses were confirmed qualitatively for the pilot system: there is a significant reduction of water consumption in production, the system has low cost and is easy to build and operate. The synthetic fabric used, obtained from patchwork of a jacket company, also shows that such a system undoubtedly contributes to a new destination of waste from the textile industry and consequent reduction of the environmental impact caused by the indiscriminate disposal of this material.

SISCAFI can contribute to the region's water security by increasing food production for own consumption and marketing surpluses, job opportunities in the preparation of irrigation units and collecting recyclables, and, by extension, income for the rural families of the northeastern semi-arid region.

According to the empirical data observed it is possible to conclude that:

- Water savings of up to 60% compared to conventional irrigation systems;
- Provides water directly to the root zone;
- Use of the correct amount of water for a particular crop;
- Increases fertilizer efficiency;
- Allows safe, effective and efficient use of recycled water;
- Reduces the spread of weeds;
- Reduces vandalism and damage to the system due to its installation below the surface;
- The systems can be operated so that the surface remains dry while the roots are irrigated.

The experiments continue, aiming to quantify these benefits. The next steps of the research will include the assembly of seedbeds next to the SISCAFI, to compare the performance of the following seedbeds: (a) SISCAFI seedbed; (b) traditional irrigated land of the region; (c) land without irrigation, regarding the main parameters of the crop: reliability; volume of water used; growth of culture; presence of pests; need for labor and ease of operation.

SISCAFI is a simple irrigation system for family agriculture that has demonstrated its functionality under the conditions described in this paper. It is a SOCIAL TECHNOLOGY that has the potential to impact the performance of perennial crops in the semi-arid region, since it combines complex concepts currently used only in expensive and sophisticated irrigation systems (capillarity, underground irrigation) with ease of construction, operation and even in small semi-arid communities.

References

1. Carvalho, D.F.: Sistemas de Irrigação. In: Engenharia de água e Solo, pp. 1–67. UFRRJ, Rio de Janeiro (2014)
2. Frizzone, J.A.: Os Métodos de Irrigação, 1st edn. USP, São Paulo (2017)
3. Fao (Organização das Nações Unidas para a Alimentação e a Agricultura). http://www.fao.org.br. Accessed 12 Oct 2017
4. Embrapa.   https://www.embrapa.br/busca-de-noticias/-/noticia/1484364/cultivo-de-cebola-com-irrigacao-por-gotejamento-anima-produtores-do-nordeste. Accessed 12 Oct 2017
5. Ruralcentro/Embrapa. https://goo.gl/Yfqu2S. Accessed 12 Oct 2017
6. Coelho, E.F., Silva, T.S.M., Parizotto, I., Silva, A.J.P., Santos, D.B.: Sistemas de irrigação para agricultura familiar. In: Circular Técnica (106), pp. 1–7. EMBRAPA, Bahia (2012)
7. Pachico, I.W.L.: Avaliação de Sistemas de Irrigação Localizada de Baixo Custo Recomendado a Pequenos Agricultores da Região Semiárida. In: Master's Degree, pp. 1–67. UFERSA, Rio Grande do Norte (2014)
8. Cietta, E.: A revolução do fast-fashion: Estratégias e modelos organizativos para competir nas indústrias híbridas, 2nd edn. Estação das letras, São Paulo (2010)
9. Sull, D.; Turconi, S.: Fast-fashion lessons. J. Compilation Bus. Strategy Rev. Summer, 1–8 (2008)

Operational Research and Industrial Mathematics

A Hybrid Numerical Scheme
for Fractional-Order Systems

L. L. Ferrás[1,2], N. J. Ford[2], M. L. Morgado[3(✉)], and M. Rebelo[4]

[1] Institute for Polymers and Composites/I3N & CMAT-UM,
University of Minho, Guimarães, Portugal
[2] Department of Mathematics, University of Chester, Chester, UK
[3] CEMAT, Instituto Superior Técnico,
Universidade de Lisboa and Department of Mathematics,
University of Trás-os-Montes e Alto Douro, UTAD, Vila Real, Portugal
luisam@utad.pt
[4] Centro de Matemática e Aplicações (CMA) and Departamento
de Matemática, Faculdade de Ciências e Tecnologia,
Universidade NOVA de Lisboa, Caparica, Portugal

Abstract. In this work we present a hybrid numerical scheme for the solution of systems of fractional differential equations arising in several fields of engineering. The numerical scheme can deal with both smooth and non-smooth solutions, and, the idea behind the hybrid method is that of approximating the solution as a linear combination of non-polynomial functions in a region near the singularity, and by polynomials in the remaining domain. The numerical method is then used to study fractional RC electrical circuits.

Keywords: Fractional differential equations · Caputo derivative
Non-polynomial collocation method · Polynomial collocation method
Electrical circuits

1 Introduction

Fractional differential equations are becoming a hot topic in mathematics and engineering, and, in the last few decades we have witnessed a mass generalization of classical models to their fractional version. The reason is that several physical systems rely on *memory* [1,2] for their evolution, or, the fact that their rates of *evolution* need not to be classical derivatives of orders 1, 2, 3 ..., but instead, can be something in between. Therefore, fractional models allow a better modeling of physical phenomena by capturing information that is lost when we go from smaller scales to the usual continuum approach.

For example, fractional calculus plays an important role in control systems based on proportional-integral-derivative controllers, known as PID, and commonly used in industry, in instruments and laboratory equipment [1]. In these processes we arrive at systems of fractional differential equations that may need to be solved numerically.

© Springer International Publishing AG, part of Springer Nature 2019
J. Machado et al. (Eds.): HELIX 2018, LNEE 505, pp. 735–742, 2019.
https://doi.org/10.1007/978-3-319-91334-6_100

Other important examples of application of fractional calculus are electric circuits. As shown in the work of Gómez-Aguilar et al. [3] the use of fractional derivatives allows a much better fit to experimental data (see that of spectroscopy applied to an RC circuit in [3]). Since the analytical solution of fractional differential equations is only available for a small number of cases, and due to the fact that when using discrete techniques it takes a large amount of time to transform data from one domain to the other, in their work [3], Gómez-Aguilar et al. used the Numerical Laplace Transform to convert the time domain into the frequency domain (and vice-versa).

Another possibility is to discretize the integral representation of the problem directly. Therefore, in this work, we are concerned with the numerical solution of the linear systems of ordinary differential equations:

$$D^\alpha \mathbf{y}(t) = A\mathbf{y}(t) + \mathbf{F}(t), \quad t \in (0, T] \tag{1}$$

$$\mathbf{y}(0) = \mathbf{y_0}, \tag{2}$$

where A is a constant matrix $A = [a_{ij}]_{i,j=1,\dots,n}$, $\mathbf{y} = [y_1 \ y_2 \dots y_n]^T$, $\mathbf{F}(t) = [f_1(t) \ f_2(t) \dots f_n(t)]^T$ and $\mathbf{y_0} = [y_{01} \ y_{02} \dots y_{0n}]^T$, where $y_{0i} = y_i(0)$, $i = 1, \dots, n$. The order of the fractional derivative satisfies $0 < \alpha < 1$, and the fractional derivative is given in the Caputo sense, that is [2]:

$$\frac{d^\alpha u(t)}{dt^\alpha} = \frac{1}{\Gamma(1-\alpha)} \int_0^t (t-s)^{-\alpha} u'(s) \, ds.$$

In this work we present a hybrid numerical scheme for the solution of systems of fractional differential equations (FDE) arising in different fields of engineering. The numerical scheme can deal with both smooth and non-smooth solutions, and, the idea behind the hybrid method is that of approximating the solution as a linear combination of non-polynomial functions in a region near the singularity [4], and by polynomials in the remaining domain.

The work is organized as follows. In Sect. 2 we describe the numerical method, in Sect. 3 we present a verification of the method against an analytical solution and we also study a fractional RC electric circuit. The paper ends with some conclusions.

2   Numerical Method

Before presenting the numerical method, we will first present some preliminary results on the existence and uniqueness (see [2]) of solutions for system (1)–(2).

Lemma 1. *Assume that the solution $\mathbf{y}$ of (1)–(2) exists and is unique on $[0, T]$, for a certain $T > 0$. If $\alpha = \frac{p}{q}$, where $p \geq 1$ and $q \geq 2$ are two relatively prime integers and if each right-hand side function F_i can be written in the form $F_i(t, \mathbf{y}(t)) = \overline{F}_i(t^{1/q}, \mathbf{y}(t))$, $i = 1, \dots, n - 1$, where each $\overline{F}_i$ is analytic in a neighborhood of $(0, y_i(0))$, then the components of the unique solution of* the problem (1)–(2), $\mathbf{y}(t) = (y_1(t), \ldots, y_{n-1}(t))$, can be represented in terms of powers of $t^{1/q}$:

$$y_i(t) = \sum_{k=0}^{\infty} a_k^i t^{k/q}, t \in [0, r), \tag{3}$$

where $r < T$ and a_k^i are constants.

From the above Lemma it follows that for a fixed $m \in \mathbb{N}$ each component of the solution of (1)–(2) can be written as a sum $y_i(t) = y_i^1(t) + y_i^2(t)$, $i = 1, \ldots, n$, where, $y_i^1 \in C^m([0, T])$, $i = 1, \ldots, n$, and y_i^2 is the non-smooth part of y_i.

Note that if the order of the fractional derivative α is not of the form $\alpha = \frac{p}{q}$, with $p \geq 1$ and $q \geq 2$ two relatively prime integers, we can always replace α with the nearest rational number of this form, because rational numbers are dense in the reals, and, it has been proved in [6], that the solution of (1)–(2) depends continuously on the order of the derivative.

In order to approximate the solution of (1)–(2) we consider a nonuniform mesh on $[0, T]$, as in [7]. Given $N \in \mathbb{N}$, let i_0 be an integer such that $\left(\frac{N}{i_0}\right)^{\frac{m}{\alpha}} \leq N$ and $\left(\frac{N}{i_0-1}\right)^{\frac{m}{\alpha}} > N$ and let $N' = N - i_0 + 1$. The partition on $[0, T]$ is defined through the meshpoints:

$$t_0 = 0, \ t_i = \left(\frac{i_0 + i - 1}{N}\right)^{\frac{m}{\alpha}} T, \quad i = 1, 2, \ldots, N' - 1, \tag{4}$$

and the N' subintervals:

$$\sigma_0 = [0, t_1], \quad \sigma_i = (t_i, t_{i+1}], \ i = 1, 2, \ldots, N' - 1, \tag{5}$$

with lengths $\tau_i = t_{i+1} - t_i$, $i = 0, 1, \ldots, N' - 1$. Define also $\tau = \max \tau_i$ with $i = 0, 1, \ldots, N' - 1$.

Consider the space

$$\mathcal{V}_m^\alpha = span\left\{t^{i+j\alpha}, \ i, j \in \mathbb{N}_0, \ i + j\alpha < m\right\} = span\{t^{\nu_k}, \ k = 1, \ldots, \ell\}, \quad \ell = \#\mathcal{V}_m^\alpha.$$

Taking Lemma 1 into account, if near the origin we approximate the solution of (1)–(2) with a function spanned by elements of space $\mathcal{V}_m^\alpha$, then it will reflect the potential non-smooth properties of the solution near the singularity. Therefore, we define the space

$$S_\tau^m([0, T]) = \left\{u \in C([0, T]) : \ u\big|_{\sigma_0} \in \mathcal{V}_m^\alpha\big|_{\sigma_0}, \ u\big|_{\sigma_l} \in \mathcal{P}_{m-1}\big|_{\sigma_l}, \ l = 1, \ldots, N' - 1\right\},$$

where $\mathcal{P}_{m-1}$ is the space of polynomials of degree less than or equal to $m - 1$ and σ_i, $i = 0, 1, \ldots, N' - 1$, are defined by (5).

On the first interval of the partition, σ_0, we define ℓ collocation points $t_{0j} = c_j \tau_0$, $j = 1, \ldots, \ell$, $c_j \in [0, 1]$, and on the remaining intervals $\sigma_l, l = 1, \ldots, N'-1$, we consider m collocation points $t_{lj} = t_l + c_j \tau_l$, $j = 1, \ldots, m$, $c_j \in [0, 1]$. Noting that each equation of system (1) can be written as [2],

$$y_i(t) = y_{0i} + \frac{1}{\Gamma(\alpha)} \int_0^t (t - s)^{\alpha-1} \left(\sum_{k=1}^n a_{ik} y_k(s) + f_i(s) \right) ds, i = 1, 2, \ldots, n \quad (6)$$

we will then seek for a function $\mathbf{v} = [v_1 \, v_2 \ldots v_n]^T$ such that $v_i \in S_T^m ([0, T])$, $i = 1, 2, \ldots, n$, that satisfies

$$v_i(t_{0j}) = y_{0i} + \frac{1}{\Gamma(\alpha)} \int_0^{t_{0j}} (t_{0j} - s)^{\alpha-1} \left(\sum_{k=1}^n a_{ik} v_k(s) + f_i(s) \right) ds, \quad (7)$$

where $j = 1, \ldots, \ell$ if $p = 0$ and $j = 1, \ldots, m$ for $p = 1, \ldots, N' - 1$. In order to obtain approximations for each $v_i(t_{0j})$, $i = 1, \ldots, n$, $j = 1, \ldots, \ell$, we define the Lagrange functions, $\mathcal{L}_{0j}|_{\sigma_0} \in V_m^\alpha|_{\sigma_0}$, $j = 1, \ldots, \ell$, such that $\mathcal{L}_{0j}(t_{0k}) = \delta_{jk}$, $k = 1, \ldots, \ell$. Then, we can write

$$\mathcal{L}_{0j}(t) = \sum_{i=1}^\ell \beta_{ji} t^{\nu_i},$$

where, for each $j = 1, \ldots, \ell$, the coefficients β_{ji} may be obtained by solving the linear system $\mathcal{L}_{0j}(t_{0k}) = \delta_{jk}$, $k = 1, \ldots, \ell$.

Hence, for $t \in \sigma_0$, we use the following representation for $v_i \in V_m^\alpha|_{\sigma_0}$, $i = 1, \ldots, n$:

$$y_i(t) \approx v_i(t) = \sum_{k=1}^\ell v_i(t_{0k}) \mathcal{L}_{0k}(t).$$

The values $v_i(t_{0k})$, $k = 1, \ldots, \ell$ are obtained by imposing that the functions $v_i(t)$ satisfy the integral Eq. (7), for $p = 0$, at $t = t_{0k}$, $k = 1, \ldots, \ell$:

$$v_i(t_{0j}) = y_{0i} + \frac{1}{\Gamma(\alpha)} \int_0^{t_{0j}} (t_{0j} - s)^{\alpha-1} \left(\sum_{k=1}^n a_{ik} \sum_{p=1}^\ell v_k(t_{0p}) \mathcal{L}_{0p}(s) + f_i(s) \right) ds,$$
$$j = 1, \ldots, \ell. \quad (8)$$

$$y_i(t_{lk}) = y_{0i} + \frac{1}{\Gamma(\alpha)} \int_0^{t_1} (t_{lk} - s)^{\alpha-1} \left(\sum_{w=1}^n a_{iw} \sum_{p=1}^\ell y_w(t_{0p}) \mathcal{L}_{0p}(s) + f_i(s) \right) ds$$
$$+ \frac{1}{\Gamma(\alpha)} \sum_{j=1}^{l-1} \int_{t_j}^{t_{j+1}} (t_{lk} - s)^{\alpha-1} \left(\sum_{p=1}^n a_{ip} y_p(s) + f_i(s) \right) ds$$
$$+ \frac{1}{\Gamma(\alpha)} \int_{t_l}^{t_{lk}} (t_{lk} - s)^{\alpha-1} \left(\sum_{p=1}^n a_{ip} y_p(s) + f_i(s) \right) ds,$$
$$i = 1, \ldots, n, \; k = 1, \ldots, m.$$

On each subinterval σ_j, $j = 1, \ldots, N' - 1$, each y_i will be approximated by $v_i \in \mathcal{P}_{m-1}$, that is given by:

$$y_i(t) \approx v_i(t) = \sum_{\gamma=1}^{m} L_{j\gamma}(t) v_i(t_{j\gamma}), \quad t \in \sigma_j,$$

where $L_{j\gamma}$, $j = 1, \ldots, N' - 1$, $\gamma = 1, \ldots, m$, are the Lagrange polynomials associated with the collocations points $t_{j\gamma} = t_j + \tau_j c_\gamma$.

Substituting in (7), for $p = 1, 2, \ldots, N' - 1$, we obtain the following system of equations, from which the values $v_i(t_{lk})$, $i = 1, \ldots, n$, can be obtained:

$$
\begin{aligned}
v_i(t_{lk}) = y_{0i} &+ \frac{1}{\Gamma(\alpha)} \int_0^{t_1} (t_{lk} - s)^{\alpha-1} \left(\sum_{w=1}^{n} a_{iw} \sum_{p=1}^{\ell} v_w(t_{0p}) \mathcal{L}_{0p}(s) \right) ds \\
&+ \frac{1}{\Gamma(\alpha)} \sum_{j=1}^{l-1} \int_{t_j}^{t_{j+1}} (t_{lk} - s)^{\alpha-1} \left(\sum_{p=1}^{n} \sum_{\gamma=1}^{m} a_{ip} L_{j\gamma}(s) v_p(t_{j\gamma}) \right) ds \\
&+ \frac{1}{\Gamma(\alpha)} \int_{t_l}^{t_{lk}} (t_{lk} - s)^{\alpha-1} \left(\sum_{p=1}^{n} \sum_{\gamma=1}^{m} a_{ip} L_{l\gamma}^{(k)}(s) v_p(t_l + \tau c_k c_\gamma) \right) ds \\
&+ \frac{1}{\Gamma(\alpha)} \int_0^{t_{lk}} (t_{lk} - s)^{\alpha-1} f_i(s) ds, \\
& l = 1, \ldots, N' - 1, \quad k = 1, \ldots, m,
\end{aligned}
\tag{9}
$$

where $L_{l\gamma}^{(k)}$, $l = 1, \ldots, N' - 1$, $\gamma = 1, \ldots, m$, are the Lagrange polynomials associated with the points $t_l + \tau_l c_\gamma c_k$.

After solving (8) and (9), the approximate solution of system (1)–(2), $\mathbf{v} = [v_i]_{i=1}^{n}$ is given by:

$$
v_i(t) = \begin{cases}
\displaystyle\sum_{k=1}^{\ell} v_i(t_{0k}) \mathcal{L}_{0k}(t), & t \in \sigma_0 \\[2ex]
\displaystyle\sum_{k=1}^{m} v_i(t_{jk}) L_{jk}(t), & t \in \sigma_j, \ j = 1, \ldots, N' - 1
\end{cases}
\tag{10}
$$

Note that it was proved in the another work by the research group [5] that the convergence order of the method is m.

3 Numerical Results

In order to validate and illustrate the feasibility of the method, one example for which the analytical solution is known, and a case study involving fractional RC electric circuits are now presented. The numerical error is measured by determining the maximum error at the mesh points t_j:

$$\varepsilon_\tau = \max_{i=0,\ldots,N'-1, \ j=1,\ldots,l_i} |y(t_{ij}) - v(t_{ij})|, \tag{11}$$

where $l_0 = \ell$ and $l_i = m$ for $i = 1, \ldots, N' - 1$, and, v is the numerical solution.

3.1 Validation

For validation purposes we have considered the following system:

$$\begin{cases} D^\alpha y_1(t) = y_2(t) \\ D^\alpha y_2(t) = -y_1(t) - y_2(t) + t^{\alpha+1} + \frac{\pi \csc(\pi\alpha)t^{1-\alpha}}{\Gamma(-\alpha-1)\Gamma(2-\alpha)} + \frac{\pi t \csc(\pi\alpha)}{\Gamma(-\alpha-1)}, \end{cases} \quad (12)$$

together with the conditions $y_1(0) = 0$, $y_2(0) = 0$. The analytical solution is given by $y_1(t) = t^{1+\alpha}$ and $y_2(t) = \pi\alpha(\alpha+1)t \csc(\pi\alpha)/\Gamma(1-\alpha)$ with $\alpha \in [0,1]$.

For the numerical solution we consider a high order approximation with $m = 4$. We consider Eq. (12) with several values of alpha: $\alpha = 1/4, 1/2, 2/3$.

The numerical results listed in Table 1 suggest an experimental order of convergence of 4, as expected. Note that with standard finite difference schemes it would be impossible to attain such convergence rates. It should also be highlighted that fact that the new method is up to 338 times faster than the method in [4].

Table 1. Error and convergence order (p) obtained for the numerical solution of Eq. (12). SU - simulation time of the non-polynomial method [4] (the solution is a linear combination of non-polynomial functions in the all domain) divided by the simulation time of the hybrid method.

$\alpha = 1/4$				$\alpha = 1/2$				$\alpha = 2/3$			
N	N'	ε_τ	p	N'	ε_τ	p	SU	N'	ε_τ	p	SU
64	15	$2.11 \cdot 10^{-6}$	-	26	$1.94 \cdot 10^{-7}$	-	56	32	$5.53 \cdot 10^{-8}$	-	142
128	34	$1.45 \cdot 10^{-7}$	3.86	59	$1.27 \cdot 10^{-8}$	3.93	50	71	$3.58 \cdot 10^{-9}$	3.95	162
256	75	$9.43 \cdot 10^{-9}$	3.94	128	$8.13 \cdot 10^{-10}$	3.97	52	155	$2.28 \cdot 10^{-10}$	3.97	170
512	166	$6.00 \cdot 10^{-10}$	3.97	278	$5.14 \cdot 10^{-11}$	3.98	59	331	$1.44 \cdot 10^{-11}$	3.99	338

3.2 Case Study: Fractional RC Electrical Circuits

By following the work of Gómez-Aguilar et al. [3] the fractional differential equation governing a fractional RC electrical circuit (see Fig. 1(a)) is given by (for more details see [3]),

$$\frac{d^\alpha q(t)}{dt^\alpha} + \alpha^{1-\alpha}(RC)^{-\alpha}q(t) = C\alpha^{1-\alpha}(RC)^{-\alpha}E(t), \quad \alpha \in [0,1]. \quad (13)$$

We have solved the governing equation using the hybrid method (the parameters are shown in Fig. 1).

Note that we have considered the relationship given by Eq. (12) of [3], stating that $\sigma_R = \alpha RC$.

We compared our results with the results obtained with the numerical Laplace transform technique, where an excellent agreement is observed. Note that we use $\alpha = 0.96$, 0.98 which are close to 1 (the classical case), but, huge differences between the two values were observed in [3].

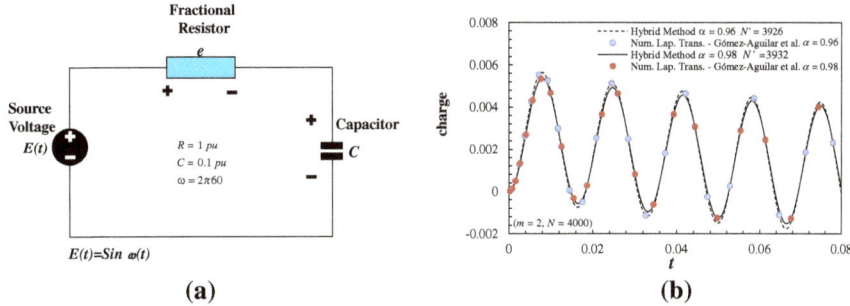

Fig. 1. (a) Schematic of the RC circuit. (b) Charge vs. time. Comparison with the results obtained by Gómez-Aguilar et al. [3] (the parameters used in the numerical solution of Eq. (13) are given in the figure). The simulations were performed with Mathematica 11.0 from Wolfram Research, Inc.

4   Conclusions

A hybrid numerical method was developed for the solution of systems of fractional differential equations. The method can deal with both regular and singular solutions extending in this way the results in [4] where only one equation was considered and where the non-polynomial approximation was used along all the time interval. The numerical results suggest that the computational effort of the new method is significantly lower and that it provides an optimal convergence order m that is independent of α. The hybrid method is also used to study fractional RC electrical circuits, as an alternative to the Numerical Laplace Transform.

Acknowledgments. The first, third and fourth authors would like to thank the funding by FCT-Portuguese Foundation for Science and Technology through scholarship and projects: SFRH/BPD/100353/2014 and UID/Multi/04621/2013, UID/MAT/00297/2013 (Centro de Matemática e Aplicações), respectively.

References

1. Podlubny, I.: Fractional Differential Equations: An Introduction to Fractional Derivatives, Fractional Differential Equations, to Methods of Their Solution and Some of Their Applications. Mathematics in Science and Engineering (1999)
2. Diethelm, K.: The Analysis of Fractional Differential Equations: An Application-Oriented Exposition Using Differential Operators of Caputo Type. Springer, Heidelberg (2010)
3. Gómez-Aguilar, J.F., Rosales-García, J., Razo-Hernández, J.R., Guía-Calderón, M.: Fractional RC and LC electrical circuits. Ing. Inv. y Tec. **15**, 311–319 (2014)
4. Ford, N.J., Morgado, M.L., Rebelo, M.: Non-polynomial collocation approximation of solutions to fractional differential equations. Fract. Calc. Appl. Anal. **16**, 874–891 (2013)

5. Ferrás, L.L., Ford, N.J., Morgado, M.L., Rebelo, M.: High-orders Methods for Systems of Fractional Ordinary Differential Equations and their Application to Time-Fractional Diffusion Equations (2018, to be submitted)
6. Diethelm, K., Ford, N.J.: Analysis of fractional differential equations. J. Math. Anal. Appl. **265**, 229–248 (2002)
7. Cao, Y., Herdman, T., Xu, Y.: A hybrid collocation method for volterra integral equations with weakly singular kernels. SIAM J. Numer. Anal. **41**, 364–381 (2003)

Integer Programming Model
for Ship Loading Management
A Case Study from Cement Industry

João Fonseca$^{(\boxtimes)}$ (iD), Ricardo Alves$^{(\boxtimes)}$ (iD), Ana Regina Macedo$^{(\boxtimes)}$ (iD),
José António Oliveira (iD), Guilherme Pereira (iD), and Maria Sameiro Carvalho (iD)

ALGORITMI Research Center, University of Minho, Braga, Portugal
jmsf.14@gmail.com, ralves_12_@hotmail.com, anareginamacedo@hotmail.com

Abstract. Industries only survive in the modern business world if they
are prepared to improve processes whenever it is necessary. Therefore,
companies need to adapt and/or modify processes to improve the level
of service and reduce needless costs. In this article the logistics problem
of planning a ship load operation of a Portuguese cement company is
addressed. The objective is to determine the best way to transport the
bagged cement from the warehouse to a ship reducing costs and optimizing forklifts operations. A mathematical programming model is proposed
showing this methodology as a powerful tool to provide effective support
to a more intelligent decision process.

Keywords: Operations research · Storage yard management
Logistics optimization · Cement industry · Repositioning model

1 Introduction

With the integration of technology in industrial systems, there is an increasing
need, for companies, to stay updated and alert or else they will not survive
in today's current competitive world.

The Cement Industry (CI) is the focus in this paper and in fact the Supply
Chain Management (SCM) in the CI is a topic with limited research. Studies
reflect that the three major concerns of the CI are the manufacturing process,
the cement material management and sustainability [1]. While topics such as
distribution and transportation were also studied [1], there is not much information about it. This work aims to contribute to these two main topics, with
the optimization of the transportation of the products from the warehouses to
the ship. More specifically, the logistics operations in the quay cranes area, that
includes the quay cranes and the storage yard [2].

The quay crane's area, is a critical area to any industry because it is the
vital link between the company and the international market. This specific area
of any industry's plant or container terminal should be efficient otherwise this

© Springer International Publishing AG, part of Springer Nature 2019
J. Machado et al. (Eds.): HELIX 2018, LNEE 505, pp. 743–749, 2019.
https://doi.org/10.1007/978-3-319-91334-6_101

link will be at risk [3]. That is, the satisfaction of the final client can be damaged, if anything goes wrong in the process to deliver the goods in perfect conditions and in time.

Cement is a cheap but heavy product and therefore transportation costs are higher compared to other variable costs in the industry [4]. This means that the operation of transportation has a large impact on a cement company and an effective planning and a good use of transportation resources are extremely important in this type of industry.

This paper describes a new approach to organize and manage the transport operations between the cement production area and the quay crane area, which is a crucial node in the Cement Industry's supply chain.

In the next sections of the paper a detailed explanation of the problem scenario and its particularities are going to be approached. The main goal of this study is to transform the current situation into a new optimized one, in order to reduce costs and improve the workflow.

2   Problem Description

The study involved a CI plant in Portugal. This company has a plant with a manufacture center and a warehouse near a quay. In this plant there was no coordination or planning for the process of transportation of the packets of bagged cement between the warehouse and the quay cranes that directly supplied the ship. This operation brings big issues to the company, because if anything goes wrong it would put at risk the loading process of the ship, and cause delays that would lead to higher costs. These delays often occur if the cranes' rate is not respected and the crane must stop working. The cranes' rate is the maximum number of packets that a crane can load the ship, in each period.

The bags of cement are usually grouped in large packets of about 50 small cement bags. Each of these large packets has about 2000 kg of cement. Each order usually asks for 5 to 6 thousand packets. The packets of cement are stored outside in the storage yard, where they will be waiting to be picked up and transferred close to the quay cranes.

2.1   Current Situation

The existing process starts with a randomly distribution of the packets of cement in the storage yard. Then, each distributed zone is transported to the quay crane, using the forklifts. After ending the transportation of the first zone, they go to the next adjacent zone, and transport the packets that are placed in there. They do this recurrently, until they finished loading the ship. This process takes, in average, 48 working hours, divided in shifts, with three forklifts to do the work.

This first approach brings benefits and disadvantages. This way of transport makes it more unlikely to damage the packets of cement because in this case the products are moved two times at most. On the other hand, this method of transportation withdraws part of the storage yard's area and implies, in most cases, that the forklifts travel some great distances. These unnecessary distances that the forklifts must travel can lead to delays. In return, these delays can interfere with the amount of time the ship has to stay in the quay or even interfere with the crane's rate. These two consequences of the delays are of great importance to this problem because the company would have to deal with high costs associated to time losses.

2.2 The Repositioning Idea

To improve the management of the storage yard's area and with the goal of optimizing the transportation method stated so far, a different method was studied - the repositioning method. With this approach it is intended to add some agility to the process in question and search a decrease of its working time.

After a first look over the available area for storing the packets, three distinct areas stood out, due to their distance to the crane. One, at the center, quite close to the quay cranes, and the other two - one in each side, left and right - that were further away from the crane. Being aware of the large distances between each area and the crane, as well as considering the forklifts' velocity, were the main reasons to look over the advantages of products' repositioning. In fact, it is much easier to keep up with the crane's rate if the products are closer [5]. The total yard storage considered consists of around $7000\,\mathrm{m}^2$, where thousands of packets of cement, with roughly $1.5\,\mathrm{m}^2$ each, are to be distributed.

Aiming to improve the workflow and reduce the disorder inside the storage yard, aisles in the middle of each area were included. Although losing inventory space, these empty areas of about $15\,\mathrm{m}$ of width, allow an easy and correct manoeuvring of the forklifts.

The central areas, designated C1 and C2, since they are closer to the crane they are the only ones that feed it directly. On the other hand, and regarding the method of repositioning, the cement stored in the side areas, the right ones (designated R1, R2 and R3) and the left ones (designated L1 and L2), of the storage yard, is only transported to the central areas (not to the crane). This will work as a compensation/repositioning of the cement absent in the center due to its transport to the crane and then to the ship.

3   Problem Formulation

The case presented in the previous section was studied in a real situation and because of that, there are many constraints and singularities that were considered. First, it was important to organize the data relative to the dimensions of each storage area.

The distances between the central areas and the crane were measured - about $20\,\mathrm{m}$, between the center and the left areas the distance varied from 82.5 to $137.5\,\mathrm{m}$, and between the center and the right areas it varied between 40 to $110\,\mathrm{m}$.

Afterwards, it was necessary to study the average velocity of the forklifts. The forklifts can reach different velocities, depending on the distance they travel.

Thus, the greater the distance, the higher the average velocity they reach. In this sense, it was considered an average velocity between the storage areas, being this average velocity between 7 km/h and 15 km/h.

Once these calculations were concluded, it was possible to determine the total time associated to each movement cycle. Each forklift can transport two cement packets each time. Thus, four stages compose each forklift cycle: (1) the time of loading the cement packets, (2) the time travelling between a storage area and the destination, (3) the unloading of the cement packets and (4) the time travelling back to the origin point.

The corresponding values of each cycle remained in few minutes. Then, with each cycle time, and considering that only one forklift is transporting between two storage areas, it was possible to calculate the maximum number of packets of cement that can be transported in one hour. This value is higher between the central area and the crane - 90 packets in one hour per forklift. On the other hand, it reaches a minimum value between L1 and C1, corresponding to 58 packets per hour per forklift.

4   Optimization Model

As stated before, the optimization model aims to improve the process of transporting the packets to the ship. In the next sections, the objective function, the decision variables, and the associated constraints are going to be presented.

The *Excel* contains a solver that can handle this type of problems. It allows to build an integer linear programming model through tables and cells, keeping it simple and organized. However, this solver is subject to a limitation on the number of constraints and variables.

Therefore, a different tool was used to overcome these limitations. The *Front-Line* Solver is an *Excel* extension that uses *Gurobi Optimization* and allows to use unlimited variables and constraints. Besides that, this solver is open source and allows fast computing. With this powerful tool, and with the data calculated so far, the optimization model was implemented.

$$\min \sum_{j=1}^{3} d_j \tag{1}$$

$$d_j = |h_j - h_{\mathrm{med}}| \tag{2}$$

$$f_{ij} \leq 2 \tag{3}$$

$$\sum_{i=1}^{7} f_{ij} \leq 5, \qquad \forall j \tag{4}$$

$$0.9 \cdot t_i \leq \sum_{j=1}^{24} p_{ij} \leq 1.1 \cdot t_i, \qquad \forall i \tag{5}$$

$$\sum_{j=1}^{3} c_j \leq 4 \tag{6}$$

$$\sum_{j=1}^{3} u_{ij} = 1, \qquad \forall i \tag{7}$$

$$0 \leq n_{ij} \leq 900 \tag{8}$$

In (1), h_j refers to the no. of working hours at shift j, the h_{med} is the average working hours per shift, the d_j the deviation from average working hours per shift, on shift j. This objective function tries to create an organized work environment, minimizing the deviation from an average number of hours per shift. These hours comprise the total work done by all forklifts at that shift. This aims that, in a perfect scenario, all the work shifts use, approximately, the same amount of time.

Aiming to limit the number of working hours and forklifts per hour and per storage area, constraints (3) and (4) were designed, where: f_{ij} is the no. of forklifts operating in area i, at hour j.

The number of packets transported, at each hour, between each pair of storage areas, is of great importance too. As stated in (5), where the p_{ij} equals the no. of packets transported from area i, at hour j and t_i is the no. of packets to be transported, from area i, the total number of packets, transported in each storage area, must be equal to the initial number of packets. However, equality constraints are too restrictive to this type of models - ending with exaggerated solving time, higher computer capacity required and, frequently, difficulty in finding feasible solutions. Thus a 10% deviation in the number of packets transported from each area was allowed, promoting the search for a feasible solution.

Aiming to minimize the number of hours to load the ship, it was verified that, part of the time, the number of packets arriving to the cranes' area was too big for one crane only. It is assumed that a crane can only handle 180 packets per hour and, allowing more packets in that area would only cause congestion. An optimization model does not have in consideration the workflow that a real case scenario must have. Thus, to overcome the dispersed usage of the second crane, it was necessary to implement the constraint presented in (6), where c_j is the no. of cranes used, at shift j.

A requirement of the problem is that the crane must not pause during working hours. At the center, the storage areas C1 and C2 are directly related to the crane. Thus, they must be in constant flow, to keep up with its cycle. On the other hand, forklifts operating in L1, L2 and R1, R2, R3, are only moving packets to the central areas, making sure all material is transported and that C1 and C2, respectively, do not be running on empty, with no packets. In (7) it is imposed that, once the movement of packets starts, from one of the sides, the job of transporting the packets from that area must be completed, during only one shift. In this restriction u_{ij} is 1 if there is transport, from area i, at shift j.

The number of packets in the center (C1 and C2) changes continuously - decreasing whenever they are transported from the central areas to the crane (or near) and increasing whenever the packets are transported from the sides (L1, L2, R1, R2 and R3) to the center. The constraint in (8), where n_{ij} is the no. of packets, in area i, at hour j, implies that the number of packets in each storage area must not be less than zero, nor greater than a specific upper limit.

5   Results

Applying the model introduced in the previous section, it was possible to determine the optimal solution, considering all the constraints. Several possible scenarios were tested, varying the number of forklifts and total hours utilized to complete the job. Although these instances lead to several solutions to the problem, in this section, only the most viable results are shown. During the results analysis, it is important to have in mind that, at present, the company is able to load the ship in about 48 working hours and with 3 forklifts.

In this instance, the number of available forklifts was increased to 5, and the labour spectrum was decreased to only 24 h, divided in 3 shifts of 8 h. Although using, at maximum, 2 more forklifts, it is verified that the ship loading period can be reduced to half.

Besides all the advantages in the working time, there are improvements in the utilization of resources. In fact, with this solution, it is possible for the ship to stay much less time in the quay, which can be of great appeal for the client and the company itself. Also, keeping the cranes' cycle of 180 packets per hour is not an easy task, requiring a constant transportation of material to its area. By increasing the number of forklifts operating simultaneously, it is possible to keep some of them closer to the cranes' area, maintaining this fast cycle, imposed to the system. The results of the computational solution are detailed below.

The total number of machine hours, given by the model, was 38 per shift, in every shift. Achieving a null deviation to the mean value, a perfect score for the objective function was reached. To complete the job, it was necessary a total of 114 h.

The total number of forklifts, working at the same time, does not exceed 5 and in each area, it does not exceed 2. As expected, there were some lot more operating hours in the center areas than in the sides, to keep up with the cranes' fast loading rate of 180 packets per hour. Therefore, while at lateral areas only 1 shift was used in each one as imposed by (7), at the center, forklifts were used in all the shifts.

With this configuration of working hours, it was possible to transport 5040 packets to the ship. As shown, this is an acceptable number, since in all areas the lower and upper limit from (5) were respected and the goal of 10 ton of cement packets was reached. Also, the minimum and maximum of packets allowed at each instant and area, imposed by (8), were respected.

In some periods, the number of packets arriving at the cranes area was more than it could handle. This could cause an accumulation of material and, that way, an extra crane was needed to load the excess. As forced by (6) these periods were concentrated in only one shift, to reduce costs associated with this extra resource. Thus, at shift one and three only one crane was utilized, while at the second shift, an extra crane was added.

6   Conclusion and Future Work

In this paper, a new method of load repositioning was tested, to improve the operations of loading a ship with huge orders of cement packets, using forklifts. To achieve such goal, an optimization model was built. This model successfully reduced the makespan of the job to 50% and allowed a better utilization of the available resources. It is important to refer that, to the best of our knowledge, this is the first paper tackling the repositioning method, modelling it and showing promising results.

Load repositioning method was proven to improve the flow of movements and reduce the chaos inside the plant, as well as the total number of working hours. In fact, this solution may be very helpful not only to the company - who sees its operating costs reduced - but also to the client, able to get the orders with no delays.

Methods such as load repositioning are normally set aside. In this paper a model for dealing with the referred problem is presented and the advantages are discussed. Also, it is important to refer that despite the optimization model being quite simple, it could have a huge impact in a company costs structure and in its client service level.

References

1. Agudelo, I.: Supply chain management in the cement industry. Massachusetts Institute of Technology (2009)
2. Diabat, A., Theodorou, E.: An integrated quay crane assignment and scheduling problem using branch-and-price. Comput. Ind. Eng. **73**, 115–123 (2014)
3. Türkoğullari, Y.B., Taşkin, Z.C., Aras, N., Altinel, I.K.: Optimal berth allocation, time-variant quay crane assignment and scheduling with crane setups in container terminals. Eur. J. Oper. Res. **254**(3), 985–1001 (2016)
4. Christiansen, M., Fagerholt, K., Flatberg, T., Haugen, O., Kloster, O., Lund, E.H.: Maritime inventory routing with multiple products: a case study from the cement industry. Eur. J. Oper. Res. **208**(1), 86–9 (2011)
5. Carlo, H.J., Vis, I.F.A., Roodbergen, K.J.: Transport operations in container terminals: literature overview, trends, research directions and classification scheme. Eur. J. Oper. Res. **236**(1), 1–13 (2014)

Determinants of SMEs Financial Performance: Empirical Evidences from Wholesale of Electrical Household Appliances Sector

Rui Coelho[(⊠)], Fábio Duarte, and Ana Borges

Master of Business Decision Support Methods, CIICESI,
ESTG/P.PORTO – Center for Innovation and Research in Business Sciences
and Systems of Information, School of Technology and Management,
Polytechnic of Porto, Porto, Portugal
fruicoelho@hotmail.com

Abstract. The topic of performance is a recurrent theme in most branches of management, especially regarding concepts and levels of analysis. Nonetheless, only a small part of literature explicitly devotes attention to the causality effects between organizational behavior, business strategies or economic environment and (ex post) firm performance, such as firm size, firm age, social responsibility, ownership structure, market diversification, innovation policies, industry concentration or macroeconomic conditions. Overcoming this empirical gap, this study addresses the performance of SMEs operating in the wholesale of electrical household appliances sector. We also provide first-hand results regarding the relation between innovation, internationalization and Portuguese SME´s business performance. Our results show that family firms have a positive impact on the performance and total debt and crisis have a negative impact on the performance of companies operating in this sector.

Keywords: SMEs · Performance · NACE 46430 · Crisis · Innovation
Internationalization

1 Introduction

The topic of performance is a recurrent theme in most branches of management, especially regarding concepts and levels of analysis [1]. Across empirical literature, the analyses of firm performance determinants are mainly conducted using financial information which is motivated by the research on business default predictive models [2, 3]. Such as in the research on risk analysis and business probability of default, liquidity, long-term solvency, tangibility and leverage are the main financial ratios used by the literature to explain firm performance [1, 4–6]. Though many of these studies are successful in predicting bankruptcy outcomes, they often fall short on identifying and explaining the characteristics that can be used as determinants of firm performance [6]. Only a small part of literature explicitly devotes attention to the causality effects between organizational behavior, business strategies or economic environment and (ex post) firm performance, such as firm size, firm age, social responsibility, ownership structure, market diversification, innovation policies, industry concentration or

© Springer International Publishing AG, part of Springer Nature 2019
J. Machado et al. (Eds.): HELIX 2018, LNEE 505, pp. 750–756, 2019.
https://doi.org/10.1007/978-3-319-91334-6_102

macroeconomic conditions [7–9]. Furthermore, given the nature of this information, the majority of studies addresses the topic of performance determinants for firms listed in the stock market [10]. SMEs are prominent in the business sector in most developed countries. In Portugal the trade sector represents approximately 19% of companies [11] and around 34% of the total turnover generated [12]. This sector can influence the performance of other sectors by its financial decision making and actions thereof. Considering the importance of this sector this study investigates the determinants of business performance aiming to provide guidelines for both managers and policy makers in terms of the best practices that maximize their probability of success. Furthermore, based on the assumption that during a financial crisis and economic recession the domestic demand tends to fall dramatically, affecting traditional, younger and smaller firms, we expect that business performances may be affected by macroeconomic conditions. For this purpose, we explicitly control the financial crisis that affected Portugal in the last decade. In order to mitigate the impact of the crisis in the economy, the Portuguese Government encouraged SMEs to develop innovative and internationalization practices. Nonetheless, there is a lack of firms-level analysis linking innovation and internationalization in terms of (ex post) performance.

This paper offers a threefold contribution. Firstly, this is the first study addressing the performance of SMEs operating in the wholesale of electrical household appliances sector. Secondly, the study provides first-hand results regarding the relation between innovation, internationalization and Portuguese SME′s business performance. Finally, at the best of our knowledge, this is the first study in Portugal, that explicitly controls the effect of the international financial assistance program on this subject. Regarding the structure of the remainder of this study: Sect. 2 provides a brief literature review and research hypothesis; Sect. 3 presents the research methodology; Sect. 4 reports preliminary results and conclusions.

2 Literature Review

2.1 Performance

Performance is defined as the potential for future successful implementation of actions in order to reach the objectives and targets [4]. Although easily understandable concept, measuring the performance is a complex and frustrating challenge. Empirically, performance has been broadly measured through the return on equity (ROE), return on assets (ROA), Tobin's Q, share price and business growth. In this study we measure the performance through the ROA (i.e., the ratio between the earnings before interest and taxes (EBIT) and total assets) [5].

2.2 Determinants

Firm Characteristics. Firm size boosts scale economy, diversified strategies and entry barriers to potential competitors [13, 14]. In other hand, agency costs might arise from the firm′s growth. Owners of bigger firms tend to have a limited ability to control managers' actions. This limitation might favour a self-interest manager′s behaviour reflected in investment decisions that increase their own prestige but which can lead to unprofitable results [15]. The age of firm is important also to overcome that problems. According to [16] firm owners and managers tend to become more efficient in the selection of the investment opportunities as the age of firm increases. Based on this argument, firms in more advanced stages of their life-cycle are more able to obtain higher rates of financial performance. Furthermore, older firms are more likely to benefit from reputation, experience and learning, and to have more and easier access to resources to finance growth opportunities [5]. Based on the arguments above, we formulate the following hypotheses: **H1:** Age and performance are positively related. **H2:** Size and performance are positively related.

Ownership structure might influence firm´s management and its performance. Family firms with large undiversified ownership stakes are usually long-term investors with substantial wealth at risk and wishing to pass the firm on to their heirs [17]. However, parental altruism and family ownership may also have a drawback as they seem to have the potential to exacerbate debt agency conflicts and corresponding agency costs [18]. Hence, we formulate the following hypotheses: **H3a:** Family firms and performance are positively related. **H3b:** Family firms and performance are negatively related.

Financial Indicators. Higher liquidity increases the capacity of firms to respond to increased competition [15] and to make efficient use of the various investment opportunities that arise, contributing to increased profitability [19, 20]. Hence, especially for younger and smaller firms characterized as opaque firms higher liquidity mitigates the dependence of external finance to fund profitable investment opportunities with positive impact on profits. Nonetheless, according to [21–23] excessive liquidity increases the ability to finance projects. However, this negative relation between liquidity and profitability will be minimal in the case firm´s with concentrated ownership which tend to be the case in SMEs [24]. Based on the arguments above, we formulate the following hypothesis: **H4:** Liquidity and performance are positively related.

Trade-off theory combines the advantages and disadvantages of debt. On the one hand, the debt is beneficial because of its tax benefit; on the other hand, high debt levels can induce to bankruptcy costs. For higher levels of indebtedness, agency and bankruptcy costs become significant, and the tax benefits of debt are covered by bankruptcy costs [22]. Thus, according to the trade-off theory, there is an optimal debt ratio, reached at the point where the costs of failure equal the tax benefits of debt [5]. Based on the arguments above, we formulate the following hypothesis: **H5:** There is a non-linear relation between leverage and performance.

Tangible assets are generally accepted as collateral for loans and can reduce borrowing agency costs [23]. Firms with higher tangible assets can use debt more easily as creditors believe these firms can fulfil their obligations more easily [25]. However, a higher level of tangibility would imply a higher level of indebtedness on the part of the company [26]. In both cases, tangibility affects the ability of firms to fund growth opportunities with impact on their performances. Based on the arguments above, we formulate the following hypotheses: **H6a:** Tangibility and performance are negatively related. **H6b:** Tangibility and performance are positively related.

Business Strategy. Innovation has been considered as the key factor for the survival, growth and development of SMEs [3, 27]. Innovative products, services and practices are faced as counterbalance vulnerability in a globalized business environment and in an economy that is now knowledge based [28]. Manufacturing SMEs in particular must continuously improve their manufacturing processes in order to ensure long-term sustainability [29]. Due to globalization and export growth, a robust export performance is increasingly a critical factor for a firm general performance and survival. For firms, internationalization steps involve high-risk decisions that includes sunk costs, exposition to exchange rate movements with impact on estimated revenues, changes on external market conditions, local competition and cultural barriers [30]. On the other hand, moving to external markets, firms diversify the business activity creating new growth opportunities which might reduce cash-flow volatility and the dependence to the domestic demand, decreasing the expected costs of bankruptcy. Based on the arguments above, we formulate the following hypotheses: **H7:** Internationalization and performance are positively related. **H8:** Innovation and performance are positively related.

3   Methodology

3.1   Database and Sample

This study uses a sample with information between 2007–2016, collected from SABI database. The number of active SMEs operating in wholesale of electrical household appliances included in our sample raises from 499 in 2007 to 711 in 2016. Almost 57% of firms are typed as family firms and 47% does not reveal any international commercial practices. Only 2% of firms reports R&D expenditures. For a detailed descriptive analysis see Table 1.

Table 1. Descriptive statistics.

VA	Measure	Mean	Min.	Max.	Std. Dev.
ROA	%	−2.82	−7 916.87	549.23	1.24
AGE	Units	17.1	1	95	13.91
SIZE	Units	9.96	1	217	19.67
FAMILY	(0;1)	0.57	0	1	0.50
LIQ	%	62 690	−60	57 724 290	13765.47
TD	%	83.53	0	9 423.13	1.98
TANG	%	14.28	0	98.94	0.19
TRADE	(1; 2; 3;4)	2.16	1	4	1.28
R&D	(0;1)	0.02	0	1	0.13
CRISIS	(0;1;2)	1	0	2	0.80

3.2   Econometric Model

In order to preliminarily test the effect of the covariates on firm performance, measured by ROA, we made use of the random effect panel model, since FAMILY is a time-invariant covariate. This model assumes that there is no correlation between the explanatory variables and the non-observed individual specific effects, hence, the individual effects are treated as terms of perturbation in a random way. The model formulation, considering Y_{it} as the ROA for the i firm at moment t, is as follows:

$$Y_{it} = \beta_0 + \beta_1 AGE_{it} + \beta_2 SIZE_{it} + \beta_3 FAMILY_{it} + \beta_4 LIQ_{it} + \beta_5 TD_{it} + \beta_6 TANG_{it} \\ + \beta_7 TRADE_{it} + \beta_8 R\&D_{it} + \beta_9 CRISIS_{it} + \alpha_i + \varepsilon_{it} \tag{1}$$

where α_i represents the individual effects, which are assumed to be independent and identically distributed with $\alpha_i \sim N(0; \sigma^2)$, and ε_{it} are N i.i.d. realizations of N $(0; \tau^2)$, representing the measurement error (variability non-specified). AGE is the Log (Firm age in years); SIZE is the Log (Total Employees); FAMILY is a binary variable that assumes the value 1 if more than 50% of the firm is owned by a single owner and 0 if otherwise; LIQ is the ration between Current Assets and Short debt; TD is the ratio between Total Debt and Total Assets; TANG is the ratio between Tangible Assets and Total Assets; TRADE is a categorical variables that assumes the value 1 if the firm does not trade in the international market; 2 if only exports; 3 if only imports and 4 if exports and imports simultaneously; R&D is a binary variable that assumes the value 1 if the firm reports R&D expenditures and 0 if otherwise; CRISIS is a categorical variables that assumes the value 0 if outside of a crisis context, 1 if during the international financial crisis and 2 if during the Program of Financial and Economic Assistance to Portugal.

4   Results

Table 1 reports descriptive statistics. The correlation matrix, reported in the Table 2, shows that age, debt and crisis are negatively correlated with ROA and, on the other side, size, family, liquidity and trade are positively correlated with ROA. The tangibility and R&D do not present a significant correlation with ROA.

Table 2. Correlation matrix.

VARIABLE	Measure		1	2	3	4	5	6	7	8	9	10
ROA	%	1	1									
AGE	Units	2	-0.08***	1								
SIZE	Units	3	0.13***	0.39***	1							
FAMILY	(0;1)	4	0,04***	-0.10***	-0.08***	1						
LIQ	%	5	0.21***	0.23***	0.08***	-0.06***	1					
TD	%	6	-0.33***	-0.29***	-0.21***	0.07***	-0.70***	1				
TANG	%	7	0.02	0.10***	0.28***	0.01	-0.15***	-0.03**	1			
TRADE	(1; 2; 3;4)	8	0.13***	0,20***	0.44***	0.01	0.07***	-0.12***	0.1***	1		
R&D	(0;1)	9	0	0.03**	0.07***	-0.01	0.01	-0.03**	0.03**	0.06***	1	
CRISIS	(0;1;2)	10	-0.07***	0,01	-0.02	0	0.02	0	-0.02	0	0.02	1

Note: * p-value <0.10; ** p-value <0.05; *** p-value <0.01

Table 3 reports the random effect model. The results suggest high concentrated ownership increases the firm performance in line with H3a. In line with H5 we found that total debt and firm performance are negatively related. Furthermore, we also confirm that macroeconomic conditions affect firm performance. However, there was no impact of innovation or internationalization on performance.

Table 3. Random effect model.

Variable	Constant	AGE	SIZE	FAMILY FIRM	LIQ	TD	TANG	TRADE (if=2)	TRADE (if=3)	TRADE (if=4)	R&D	CRISIS (if=1)	CRISIS (if=2)	R-Squared
Estimate	0.4116^{***}	2.472×10^{-4}	-7.738×10^{-4}	0.0776^{*}	-1.898×10^{-7}	-0.5492^{***}	-0.0547	0.0623	-0.0052	0.0016	-0.0403	-0.0401^{**}	-0.0561^{***}	0.6839

Note: * p-value <0.10; ** p-value <0.05; *** p-value <0.01

5 Preliminary Conclusions

Preliminary results suggest that family firms tend to have higher performances comparing to others. Total debt and firm performance are negatively related. Furthermore, we also confirm that macroeconomic conditions affect firm performance. However, the results suggest that R&D and International commercial practices, age, size, liquidity and tangibility does not influence firm performance. Future research should test take on account different measures of performance in line with financial literature. Also, we aim to improve regressors estimation adjusting a dynamic panel model that includes, among the analyzed covariates, one lagged values of the response variable making use of the GMM estimation method. Furthermore, it will be helpful for researchers and business managers in this sector to test if the determinants of firm performance also predict probability of business default.

References

1. Pantea, M., Gligor, D., Anis, C.: Economic determinants of romanian firms' financial performance. Procedia – Soc. Behav. Sci. **124**, 272–281 (2014)
2. Altman, E.: Driving lean and green project outcomes using BIM: a qualitative comparative analysis. J. Finance (1968)
3. Altman, E.I., Sabato, G., Wilson, N.: The value of non-financial information in small-sized enterprise risk management. J. Credit Risk **6**(2), 1–33 (2010)
4. Abbas, A., Bashir, Z., Manzoor, S., Akram, M.N.: Determinants of firm's financial performance: an empirical study on textile sector of Pakistan. Bus. Econ. Res. **3**(2), 76 (2013)
5. Vieira, E.S.: Debt policy and firm performance of family firms: the impact of economic adversity. Int. J. Manag. Finance **13**(3), 267–286 (2017)
6. Delen, D., Kuzey, C., Uyar, A.: Measuring firm performance using financial ratios: a decision tree approach. Expert Syst. Appl. **40**(10), 3970–3983 (2013)
7. Nunes, P.M., Viveiros, A., Serrasqueiro, Z.: Are the determinants of young SME profitability different? empirical evidence using dynamic estimators. J. Bus. Econ. Manag. **13**(3), 443–470 (2012)

8. Hansen, G.S., Wernerfelt, B.: Determinants of firm performance: the relative importance of economic and organization factors. Strateg. Manag. J. **10**(5), 399 (1989)
9. Ipinnaiye, O., Dineen, D., Lenihan, H.: Drivers of SME performance : a holistic and multivariate approach. Small Bus. Econ. **48**, 883–911 (2017)
10. Che, A., Abdullah, M.: Institutional ownership and market-based performance indicators: utilizing generalized least square estimation technique. Procedia – Soc. Behav. Sci. **164**, 477–485 (2014)
11. PORDATA: Total empresas por sector de atividade económica (2017). Accessed 15 Oct 2017
12. PORDATA: Volume de negócios das empresas: total e por setor de atividade económica (2017). Accessed 15 Oct 2017
13. Winter, R.A.: The dynamics of competitive insurance markets. J. Financ. Intermediation **3**, 379–415 (1994)
14. Gschwandtner, A.: Profit persistence in the "very" long run. evidence from survivors and exiters. Appl. Econ. **37**(7), 793–806 (2005)
15. Goddard, J., Tavakoli, M., Wilson, J.O.S.: Determinants of profitability in European manufacturing and services: evidence from a dynamic panel model. Appl. Financ. Econ. **15**(18), 1269–1282 (2005)
16. Jovanovic, B.: selection and the evolution of industry. Econometrica J. Econometric Soc. **50**, 649–669 (1982)
17. Steijvers, T., Voordeckers, W.: Private family ownership and the agency costs of debt. Family Bus. Rev. **22**(4), 333–346 (2009)
18. Greenwood, R.: Commentary on: "Toward a theory of agency and altruism in family firms". J. Bus. Ventur. **18**(4), 491–494 (2003)
19. Deloof, M.: Does working capital management affect profitability of Belgium firms? Faculty of Applied Economics UFSIA - RUCA - University of Antwerp, pp. 1–15 (2000)
20. Honjo, Y., Harada, N.: SME policy, financial structure and firm growth: evidence from Japan. Small Bus. Econ. **27**(4–5), 289–300 (2006)
21. Fama, E.F., Jensen, M.C.: Separation of ownership and control. Value Creating Board Corp. Govern. Organ. Behav. **26**(2), 90–111 (2008)
22. Myers, S.C.: The determinants of corporate borrowing. Sloan School of Management Massachusetts Institute of Technology, pp. 1–33 (1976)
23. Rajan, R.G., Zingales, L.: What do we know about capital structure? some evidence from international data. J. Finance **50**(5), 1421–1460 (1995)
24. Ang, J.S.: Small business uniqueness and the theory of financial management. J. Small Bus. Finance **1**(1), 11–13 (1991)
25. Moreira, A.: Determinants of the capital structure of Portuguese firms with investments in Angola. S. Afr. J. Econ. Manag. Sci. 885-10055-1-Pb(2), 1–11 (2014)
26. Avelar, E., Cavalcanti, J., Pereira, H., Boina, T.: Determinantes da Estrutura de Capital: Um Estudo sobre Empresas Mineiras de Capital Fechado. Revista Evidenciação Contábil Finanças **5**(2), 23–39 (2017)
27. Raymond, L., St-Pierre, J.: R&D as a determinant of innovation in manufacturing SMEs: an attempt at empirical clarification. Technovation **30**(1), 48–56 (2010)
28. Hoffman, K., Parejo, M., Bessant, J., Perren, L.: Small firms, R&D, technology and innovation in the UK: a literature review. Technovation **18**(1), 39–55 (1998)
29. Lagacé, D., Bourgault, M.: Linking manufacturing improvement programs to the competitive priorities of Canadian SMEs. Technovation **23**(8), 705–715 (2003)
30. Pacheco, L.: Capital structure and internationalization: the case of Portuguese industrial SMEs. Res. Int. Bus. Finance **38**, 531–545 (2016)

Numerical Simulation of Inertial Energy Harvesters using Magnets

André Gonçalves[1,4], M. Luísa Morgado[1,2(✉)], L. Filipe Morgado[1,3], Nuno Silva[1], and Raul Morais[1,4]

[1] Universidade de Trás-os-Montes e Alto Douro, UTAD, Vila Real, Portugal
luisam@utad.pt
[2] Center for Computacional and Sthocastic Mathematics,
Instituto Superior Técnico, Lisboa, Portugal
[3] Instituto de Telecomunicações, Lisboa, Portugal
[4] INESC TEC - INESC Technology and Science, Porto, Portugal

Abstract. Vibrational energy harvesters for powering wearable electronics and other electrical energy demanding devices are among the most used approaches. Devices that use magnetic forces to maintain the central mass in magnetic levitation, aligned with several coils as the emf generating transducer mechanism, are becoming a suitable choice since they do not need the usual spring that typically degrades over time. Modeling such energy harvesters poses different challenges due to the difficulty of getting the nonlinear closed-form expression that would describe the resulting magnetic force of the entire system. In this paper, modeling of the magnetic forces resulting from the system magnets interaction is presented. Results give valuable data about how the best energy harvester should be designed taking into account resonance frequency related to system's mass and dimensions.

Keywords: Magnetic energy harvester · Magnetic levitation
Magnetic force · Electromagnetic force · Numerical simulation

1 Introduction

Nowadays, taking into account the current needs of the human population, we can easily conclude that, in general, we all depend on the use of electronic devices constantly, which have as main limitation the duration of the useful life of their batteries. One way to overcome this problem is to use inertial generators to load these devices.

Using the human body as a source of kinetic energy is a very valid option for the production of electric energy. Since the human body is a vibratory structure, it has a frequency of excitation that according to [1] is less than 10 Hz. If the resonant frequency of the device does not match the frequency of ambient vibration, which in this case study will be the human body, then the output power of the generator significantly decreases. The resonant frequency is defined by the

© Springer International Publishing AG, part of Springer Nature 2019
J. Machado et al. (Eds.): HELIX 2018, LNEE 505, pp. 757–763, 2019.
https://doi.org/10.1007/978-3-319-91334-6_103

inertial generator, which is designed to operate at a single frequency. According to [1] a high resonance frequency means a major limitation for energy harvesting.

The kinetic energy must be collected adaptively by employing mechanisms that can adjust or tune to the resonant frequency of a generator so that the resonance frequency coincides with the ambient vibration frequency at all times or bandwidth of the generator.

According to [2] the resonance frequency adjustment can be obtained by changing the mechanical characteristics of the structure or electric charge in the generator. In addition, the broadening of the generator bandwidth can be achieved by using, for example, structures each having a different resonant frequency, an amplitude limiter, coupled oscillators or non-linear springs.

In the last years there have been several researches in this area that show that this is a very useful and sustainable way of producing energy [3 7]. Although this process produces relatively small amounts of energy, it should be used even way, since it allows us to obtain energy in places where any other energy sources are available.

Conventional inertial generators consist of a suspended mass within a structure. The suspended mass is magnetized and surrounded by a coil, which with the constant movement of the mass leads to the production of electromagnetic energy. A schematic of this type of generators is given in Fig. 1.

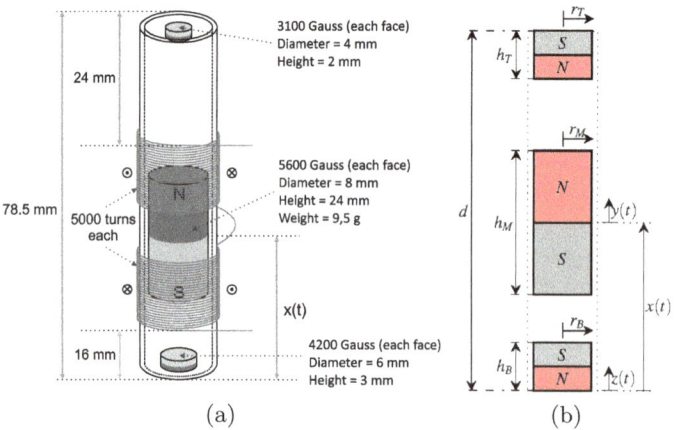

(a)   (b)

Fig. 1. Representation of the inertial generator in this case study: (a) 3D (b) Geometrical.

Initially in most case studies the mass was suspended due to the action of a spring. Here, we use a different device, in which the spring and the magnetized mass were replaced by magnets, as can be seen in the Fig. 1, where it is also presented the device considered in [8]. As it can be seen, the generator is composed by three magnets, one central and two fixed at both ends of the device, allowing in this way the central magnetized mass to oscillate within the structure according to the vibrations and movements it is subject to. The magnetized mass will oscillate preferentially in the area where the coils are placed, depending on the design of the device, generating a magnetic flux which in turn will generate an electromagnetic force.

The paper is organized as follows: In Sect. 2 we present the mathematical model we propose to use for this kind of devices. Here we follow the conclusions in [8], related to the expression for the resulting magnetic force. As it can be seen in that paper, where a closed expression form given in terms of complete elliptic integrals was provided, the use of linear functions to represent that force is not very accurate. Here, we use a nonlinear function to represent the magnetic force which is determined taking into account the magnets of the device, but with a simpler and easier to implement expression. In Sect. 3 we show through some numerical experiments how the solution of this model can be useful in the conception of these devices. We end with some conclusions.

2   Mathematical Model

In order to maximize the performance of the generator a mathematical model is used to simulate its functioning. Here, we follow the model used in [8] but the electromagnetic force is obtained by least squares fitting through experimental data, rather than by a closed expression for general magnets.

First two specific magnets were selected and the magnetic force between them, depending on the distance between the center of the two magnets, was measured. This experimental data is presented in Fig. 2. Using least squares approximation, the following function was obtained for the magnetic force between the two magnets: $G(x) = \left(1.75876 - 335.564x + 20317.3x^2\right)^{-1}$. The accuracy of this fitting can be observed in Fig. 2. Using this approximation for the magnetic force between two magnets, the resulting magnetic force, that is the force due to the interaction of the fixed top and bottom magnets with the

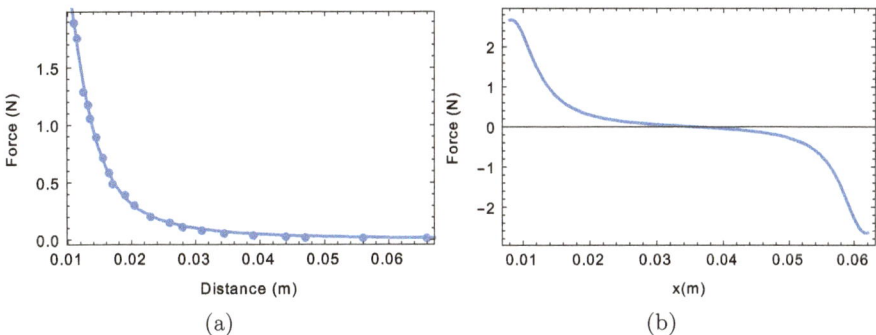

Fig. 2. (a) Magnetic force between two magnets: fitted function (continuous line) and experimental measurements (dots), (b) resulting magnetic force with $d = 0.075\,\mathrm{m}$.

central one was computed through $F_M(x) = G(x) - G(d - x)$, where d is the distance between the top and the bottom magnet (see Fig. 2).

The motion equation is, accordingly to the Newton's second law, given by

$$m\ddot{y}(t) = -c(\dot{y}(t) - \dot{z}(t)) + F_M(y(t) - z(t)) - mg, \tag{1}$$

where $c = c_m + c_e$ with c_m being the mechanical damping coefficient (see [8]), c_e the electrical damping coefficient (see [4]), m is the inertial mass, g is the acceleration of gravity, F_M is the resulting magnetic force acting on the inertial mass (the central magnet) and $\dot{y}$ and $\ddot{y}$ denote the first and second derivative of y, respectively, with respect to t. The external excitation is assumed to be harmonic and given, as usual, by

$$z(t) = Z_m \cos(\omega t), \tag{2}$$

for a specified amplitude Z_m and a frequency ω. Performing the variable substitution $x(t) = y(t) - z(t)$, (see Fig. 1(b)), Eqs. (1) and (2) may be written as

$$\ddot{x} + \frac{c}{m}\dot{x} - \frac{F_M(x)}{m} + g = Z_m \omega^2 \cos(\omega t). \tag{3}$$

The electromotive force, ϵ, is computed according to the Faraday's law

$$\epsilon(t) = -\frac{d\Phi}{dt}, \tag{4}$$

where Φ is the magnetic flux through the coils, and is given by $\Phi = \int_S \boldsymbol{B} \cdot \boldsymbol{ds}$, where S is the surface bounded by the coils and $\boldsymbol{B}$ is the magnetic field produced by the central magnet. Taking into account the geometry of the transducer here considered, it is sufficient to consider the axial component of the magnetic field, B_z, as long as S is perpendicular to the symmetry axes of the magnets, and therefore $\Phi = \int_S B_z ds$. Alternatively, the potential vector $\boldsymbol{A} = \boldsymbol{A_\varphi} + \boldsymbol{A_\rho} + \boldsymbol{A_z}$ (see [9], for example), can also be considered, which is such that $\boldsymbol{B} = \nabla \times \boldsymbol{A}$.

According to the symmetry of the device, only the azimutal component is needed, which in cylindrical coordinates is given by (see [10]):

$$A_\varphi(z, \rho) = \frac{J_M}{2}\sqrt{\frac{r_M}{\rho}}\left[\zeta k \left(\frac{k^2 + h^2 - h^2 k^2}{h^2 k^2} K(k^2) - \frac{1}{k^2}E(k^2)\right. \right. \tag{5}$$

$$\left. \left. + \frac{h^2 - 1}{h^2}\Pi(h^2|k^2)\right)\right]_{\zeta_-}^{\zeta_+},$$

where K, E and Π the complete elliptic integrals of first, second and third kind (see [8]), $h^2 = \frac{4r_M\rho}{(r_M + \rho)^2}$, $k^2 = \frac{4r_M\rho}{(r_M + \rho)^2 + \zeta^2}$, $\zeta_\pm = z \mp \frac{h_M}{2}$, and $[F(\zeta)]_{\zeta_-}^{\zeta_+} = F(\zeta_+) - F(\zeta_-)$.

The Stoke's theorem allows us to rewrite the magnetic flux in the form

$$\Phi = \int_\Gamma \boldsymbol{A} \cdot \boldsymbol{dl} = \int_\Gamma \rho A_\varphi \, d\varphi, \tag{6}$$

where Γ is the boundary of S, which in this case is a circumference. Hence, the flux through one turn of the coil (from the top, for example) may be given by $\Phi = 2\pi\rho_{tc}A_\varphi(z_{tc}, \rho_{tc})$, where ρ_{tc} is the radius of the turn of the top coil and z_{tc} is its position with respect to the central magnet, that is, $z_{tc} = z_{tc}(t) = x_{tc} - x(t)$, and $x(t)$ is the solution of (3) and (4).

Because each coil set may be viewed as a cylindrical ring with a certain height and thickness, the total flux may be approximated by

$$\Phi = 2\pi \left(\frac{n_t}{l_t t_t} \int_{\rho_{t_1}}^{\rho_{t_2}} \int_{z_{t_1}}^{z_{t_2}} \rho A_\varphi(z, \rho) dz d\rho - \frac{n_b}{l_b t_b} \int_{\rho_{t_1}}^{\rho_{t_2}} \int_{z_{b_1}}^{z_{b_2}} \rho A_\varphi(z, \rho) dz d\rho \right), \quad (7)$$

where n_t, l_t and t_t (n_b, l_b and t_b) are the total number of turns, the length and the thickness, respectively, of the top (bottom) coil, ρ_{t_1} and ρ_{t_2} (ρ_{b_1} and ρ_{b_2}) are the internal and external radii, respectively, of the top (bottom) coil and are such that $\rho_{t_2} - \rho_{t_1} = t_t$ ($\rho_{b_2} - \rho_{b_1} = t_b$). z_{t_1} and z_{t_2} (z_{b_1} and z_{b_2}) are the inferior and superior limits of the top (bottom) coil, respectively, and are such that $z_{t_2} - z_{t_1} = l_t$ ($z_{b_2} - z_{b_1} = l_b$). Here, the number of turns, the length, the thickness, the internal and external radii are considered to be equal on both coils.

Finally, the root mean square (rms) voltage and power are given in terms of the electromotive force:

$$V_{\text{rms}} = \sqrt{\frac{1}{T} \int_0^T \epsilon^2(t) \, dt}, \quad P_{\text{rms}} = \frac{V_{\text{rms}}^2}{R}$$

being T the length of a certain time interval and R the electrical resistance.

3   Numerical Simulations

In this section we illustrate the results of the mathematical model presented in the previous section. We start with some results in the case where the device is assumed to be vertically placed. In Fig. 3 we present the electromotive force and the root mean square voltage and power simulations for three frequencies and three different device sizes d. In Fig. 4 we present the electromotive force and the root mean square voltage and power simulations in the case where the generator is placed horizontally. By analyzing the graphs, it is possible to see that the generated electromotive force is greater with the increase of the size of the device or its frequency, verifying a relation between the factors in question. In conclusion, the larger the size of the device or its frequency, the greater the generated energy. Regarding the orientation (vertical/horizontal) it is not possible to draw final conclusions, since the obtained results do not show a clear tendency. Even though identical devices were simulated under equal conditions, varying only the orientation, we can say that for a horizontal position there is a smaller variation of the generated electromagnetic force with increasing frequency.

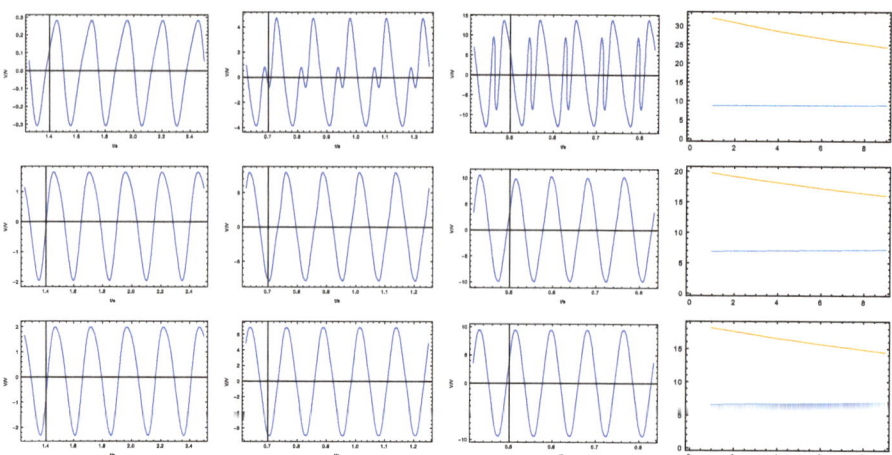

Fig. 3. Generator placed vertically. First row: $d = 5\,$cm, second row: $d = 8\,$cm, third row: $d = 11\,$cm. Three first columns: electromotive force (frequencies 4 Hz, 8 Hz and 12 Hz). Last column: root mean square voltage (V) and power (mW) curves (blue and yellow, respectively) as functions of the load resistance (from 0 to 800 Ω).

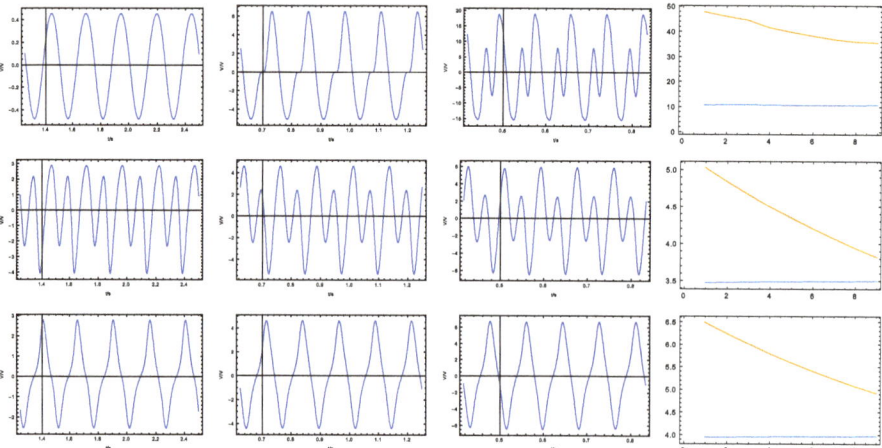

Fig. 4. Generator placed horizontally. First row: $d = 5\,$cm, second row: $d = 8\,$cm, third row: $d = 11\,$cm. Three first columns: electromotive force (frequencies 4 Hz, 8 Hz and 12 Hz). Last column: root mean square voltage (V) and power (mW) curves (blue and yellow, respectively) as functions of the load resistance (from 0 to 800 Ω).

4   Conclusions

In this paper a mathematical model is presented to simulate the performance of an inertial generator. Following the work initiated in [8], we use the same mechanical model but instead of a closed expression for the electromagnetic force, which is general for any kind of magnets, we have determined it by using least squares approximation to some experimental data. In this way, we have also avoided the use of the complete elliptic integrals a fact that relieves significantly the computational effort in the numerical simulations.

As it can be concluded from the results in the previous section, this model allows us to obtain the main guidelines in the building of such devices in view of their optimization.

Acknowledgements. First and fourth authors acknowledge funding from the ERDF – European Regional Development Fund through the Operational Programme for Competitiveness and Internationalisation - COMPETE 2020 Programme within project POCI-01-0145-FEDER-006961, and FCT – Fundação para a Ciência e a Tecnologia (Portuguese Foundation for Science and Technology) as part of project UID/EEA/50014/2013. Second author wish also to thank FCT trought projects UID/Multi/04621/2013.

References

1. Kazmierski, T.J., Beeby, S.: Energy Harvesting Systems, Modeling and Applications. Springer, Heidelberg (2011)
2. Zhu, D., Tudor, M.J., Beeby, S.: Strategies for increasing the operating frequency range of vibration energy harvesters. Meas. Sci. Technol. **21**, 022001 (2009)
3. Saha, C.R., O'Donnell, T., Wang, N., McCloskey, P.: Electromagnetic generator for harvesting energy from human motion. Sens. Actuators A Phys. **147**, 248–253 (2008)
4. Mann, B.P., Sims, N.D.: Energy harvesting from the nonlinear oscillations of magnetic levitation. J. Sound Vib. **319**, 515–530 (2009)
5. Morais, R., Silva, N., Santos, P., Frias, C., Ferreira, J., Ramos, A., Simões, J., Baptista, J., Reis, M.: Double permanent magnet vibration power generator for smart hip prosthesis. Sens. Actuators A Phys. **172**, 259–268 (2011)
6. Dallago, E., Marchesi, M., Venchi, G.: Analytical model of a vibrating electromagnetic harvester considering nonlinear effects. IEEE Trans. Power Electron. **25**, 1989–1997 (2010)
7. Yuen, S.C.l., Lee, J.M., Li, W.J., Leong, P.H.: An AA-sized vibration-based microgenerator for wireless sensors. IEEE Pervasive Comput. **6**, 64–72 (2007)
8. Morgado, M.L., Morgado, L.F., Silva, N., Morais, R.: Mathematical modelling of cylindrical electromagnetic vibration energy harvesters. Int. J. Comput. Math. **9**(1), 101–109 (2015)
9. Jackson, J.D.: Classical Electrodynamics. Wiley, Hoboken (1998)
10. Callaghan, E.E.: The magnetic field of a finite solenoid, Technical note D-465. Washington, USA, Nation Aeronautics and Space Administration (1960)

The Impact of Barrel Oil Price Variation in MIBEL Prices: A Longitudinal Approach

Ana Borges[(✉)] and Eliana Costa e Silva

School of Technology and Management, CIICESI Center
for Research and Innovation in Business Sciences and Information Systems,
Polytechnic of Porto, Porto, Portugal
{aib,eos}@estg.ipp.pt

Abstract. This paper seeks to analyse the impact of barrel oil price variation on the electricity prices. In particular, we analyse the impact of daily variation of barrel oil price on the short-term evolution of the Iberian Electricity Market (MIBEL) hourly prices, for biding optimization purpose, and also for scenario analysis in terms of pricing and strategy. The pursuit for accurate models to predict price movements can help to develop profit-maximizing trading strategies and optimal bidding techniques. In fact, important activities such as bidding strategies rely, nowadays, on price forecast information to improve decision making. For this purpose we make use of a mixed-effects longitudinal model to describe the variability of hourly prices from January 2015 to June 2016, in a total of 13 032 observations. We consider the value of barrel oil price, on the same day and on the previous day, as fixed effects, hour group as random effect and an autoregressive component of order 7 describing the within hour dependence. This analysis is an extension of [1], where the need of exogenous variables incorporation arose when analysing the longitudinal variation of electricity prices, with the purpose of enhance the longitudinal model performance. This study offers evidence that barrel oil price variation must be taken into account for electricity price forecasting, since it affects its variability.

Keywords: Longitudinal mixed-effects model · Electricity price
Oil price

1 Introduction

The complexity on electricity price modelling is due to its unusual characteristics: it is not storable, demand and supply must be constantly balanced, and demand is both volatile and inelastic [2]. A day-ahead market consists in a system where agents submit their bids and offers for the delivery of electricity for each hour of the next day before a certain market closing time [3]. Hourly prices for next day delivery are determined at the same time. In particular, the daily

© Springer International Publishing AG, part of Springer Nature 2019
J. Machado et al. (Eds.): HELIX 2018, LNEE 505, pp. 764–770, 2019.
https://doi.org/10.1007/978-3-319-91334-6_104

MIBEL electricity prices can be given as a strip of prices (one for each hour of the day), all simultaneously observed once at a given time of each day. Therefore, the daily market prices can be interpreted as a Longitudinal data. In fact, hourly prices within a day behave cross sectionally and hourly dynamics over days behave according to time-series properties [3]. [4] give an interest insight and justification on the application of a panel model to describe the dynamics in day-ahead hourly prices. The search for accurate models to predict price movements can help develop profit-maximizing trading strategies and optimal bidding techniques [5]. In fact, important activities such as bidding strategies rely, nowadays, on price forecast information to improve decision making. There is, in fact, an increased significant relevance of short term electricity forecasting in the energy price research for trading and bidding, since most of the electricity markets across the world are liberalized [5]. This work is a complement to a previous work [1], where the need of exogenous variables incorporation arose when analysing the longitudinal variation of MIBEL electricity prices, in order to enhance the longitudinal model performance. The necessity of studying the impact of oil prices in the variability of electricity prices is justifiable by the fact that, accordingly to The European Commission (EC)[1] information "fossil fuels such as oil, gas, and coal are non-renewable resources that account for around three quarters of the energy consumption in the EU." The electricity sector is by far the highest user of fossil fuels and the biggest CO_2 emitter [6]. In fact, the impact of barrel oil price is tested in several studies. Some of them conclude that the oil price has a significant impact on householding electricity prices [2] and even in electricity demand [7]. For example, [8] explore the relationships between electricity and oil, gas and coal future prices of Nordic countries, Continental Europe and the United Kingdom. While [9] study the relationships between electricity prices and energy prices for industrial and household sectors in 22 countries in Europe between 1996 and 2013. Through panel data analysis these authors found lower sensibility of electricity prices to fuel price changes for the short run. This is explained by the fact that the household final paid prices is only revised annually. Also in industries the authors found that short run price changes influence the long run contract values more than the short run one's. However, in our specific case, we are considering daily/hourly prices of electricity and therefore we expect that the daily variation of barrel oil prices will have an effect in the electricity price variation. As so, the decision of testing and introducing exogenous covariate such as barrel oil prices, in Euros, was introduced in the longitudinal model. In particular, we aim to assess how the value of barrel oil price, on the previous day, affects the electricity price.

2   Electricity Database

MIBEL, created in 2004, resulted from the cooperation between the Portuguese and Spanish Governments with the aim of promoting the integration of both

---

[1] http://ec.europa.eu/energy/en/topics/oil-gas-and-coal.

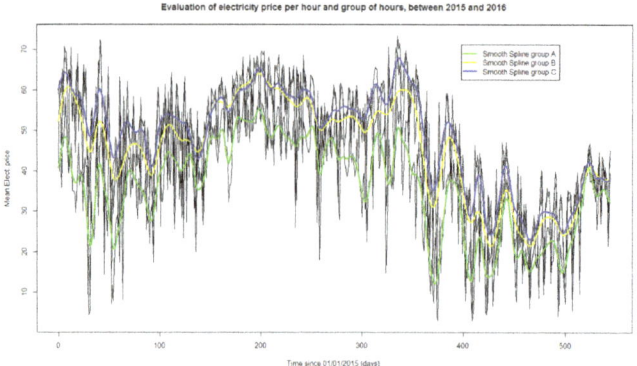

Fig. 1. Individual mean time progression for each group of hours (source: [1])

countries' electrical systems, involving the integration of their respective electric power systems and their previous electricity markets. MIBEL allows any
consumer in the Iberian region (mainland of Portugal and Spain) to purchase
electrical energy under a free competition regime from any producer or retailer
acting in that region [10]. Daily and intraday markets are organized in a daily
session, where next-day sale and electricity purchase transactions are carried
out, in six intraday sessions that consider energy offer and demand, which may
arise in the hours following the daily viability schedule fixed after the daily [10].
The daily market electricity prices can be given as a strip of prices (one for each
hour of the day), all simultaneously observed once at a given time of each day:
$Y_t = [y_{1t}, y_{2t}, \ldots, y_{nt}]$, where $n = 1, \ldots, 24$ (or 23 or 25) represent each hour of
the day, and $t = 1, 2, \ldots$ represents the day. It consists of disaggregated data,
i.e., hourly day prices, from January 1st 2015 to June 26th 2016, yielding a total
of 13 032 observations. Given that the present analysis precedes our work with
this same data exposed in [11], we have considered the prices on the logarithmic scale as done on the referred analysis. Since longitudinal models rely on
the assumption, among others, of independent subjects (in this particular case,
hours), we previously tested for correlation among the time series of electricity
prices for the 24 h of the day (for clarification see [1]). Which lead us to group
the 24 h in the following three main independent sets: (A) from the 1st until the
7th h; (B) from the 8th until the 18th h; and (C) from the 19th until the 24th h.
Hence, in the present analysis we consider these three groups, A, B and C as
independent subjects. The heterogeneity between the price dynamics in time for
each group is explicit in Fig. 1, that presents three smooth splines describing the
mean progression in time of electricity prices throughout the period analysed.
The mean progression in time for electricity prices for group A (green color),
that includes hours from midnight until 7, presents lower electricity price values,
comparing to the other two groups. While group C, representing the last hours of
the day, is the one that presents higher electricity price values on its progression
in time.

3 Methodology

Longitudinal data is usually characterized as response variables that are measured repeatedly through time for a group of individuals. They are useful since they can provide detailed representations of characteristics that are unique to each subject, thus accounting for a possible problem of heterogeneity. The main characteristic of longitudinal models is that it models both the dependence among the response on the explanatory variables and the autocorrelation among the responses. Ignoring correlation in longitudinal data could lead to incorrect inferences of the regression coefficients, inefficient estimates of the coefficients and, also, sub-optimal protection against biases causes by missing data [12]. Particularly, in the present study, we are dealing with hours of the day as subjects, and interested in modelling the progression in time of the electricity prices for those hours for each day throughout the year. We are in presence of a balanced longitudinal data, i.e., repeated measurements for each subject (hour), taking at the same moment (day). A mixed-effects model was adjusted to the data corresponding to the hourly prices from 01/01/2015 to 28/06/2016 divided in the three groups mentioned in the previous section. The analysis was performed using R Statistical Software (version 3.3.0) [13], in particular making use of the *nlme* package [14]. As explained in the previous section, there is, in our perspective, a difference on the dynamic of three distinct group of hours, A, B and C. As so, to account for the variability between the three groups of hours, we adjusted the linear mixed-effects model proposed by [15] where the n_i dimensional response vector y_{ij} for the i group, at a given time t_{ij} is given by:

$$y_{ij} = X_{ij}\boldsymbol{\beta} + Z_i\boldsymbol{b_i} + W_i(t_{ij}) + \epsilon_{ij}, \quad i = 1, ..., M, \tag{1}$$

where the p-dimensional vector of random effects b_i are M i.i.d realizations of $N(0, \Sigma^2)$, $W_i(t_{ij})$ is a continuous time Gaussian Process, representing the variability within subjects. Finally, ϵ_{ij} are N i.i.d. realizations of $N(0, \tau^2)$, representing the measurement error (variability non specified). The p dimensional vector of unknown population parameters β is associated with the known fixed-effect covariates matrix X_{ij} (of type $n_i \times p$), and Z_i (of type $n_i \times p$) is the random-effects covariates matrix. The random effects b_i and the within-group errors are assumed to be independent for different groups and to be independent of each other for the same group, i.e., the group of hours A, B and C are assumed independent as so are the days in each group of hours. For fixed-effects we considered the barrel oil price (on logarithmic scale) from the same day as also from the previous day.

The pattern of the empirical autocorrelation function [16] of the within-group residuals, (see [1]) suggests a strong correlation between two measurements in time of lag 7. Hence, we extend the basic linear mixed-effects model to take into account a serial correlation among observations in the same group of hours, modelling the dependence among the within-group errors by including an autoregressive component of order 7 in our model (1). As [17] explains, the general within-group correlation structure, for $i = 1, ..., M$ and $j, j' = 1, ..., n_i$ can be expressed as: $cor(W_i(t_{ij}), W_i(t_{ij'})) = h[d(p_{ij}, p_{ij'}), \rho]$, where ρ is a vector of correlation parameters and $h(.)$ is a correlation function, continuous in ρ, such that for two identical positions vectors $p_{ij} = p_{ij'}$ we have a correlation of $h(0, \rho) = 1$. The autoregressive model of order q, $AR(q)$ expresses the current observation as a linear function of previous observations as:

$$\epsilon_t = \phi_1 \epsilon_{t-1} + \cdots + \phi_q \epsilon_{t-q} + a_t \tag{2}$$

where $a_t - N(0, \theta^2)$ is a homoscedastic noise term, assumed independent of the previous observations. The coefficients ϕ_i, with $i = 1, ..q$, describe the dependence between the price on day t with previous days. Particularly, concerning the autoregressive model of order 7, AR(7), the correlation function can be defined recursively through the difference equation [16]:

$$h(k, \phi) = \phi_1 h(|k-1|, \phi) + ... + \phi_7 h(|k-7|, \phi), \quad k = 1, 2, ... \tag{3}$$

Note that in a previous analysis [11] the autoregressive coefficients of an adjusted VARX(7,0) explained the existence of dependence within the hourly prices, which corroborates our choice on the serial correlation structure. The estimates of the model were obtained by maximum likelihood methodology.

4   Main Results

Figure 2 presents the MIBEL daily prices of the 24 h from 01/01/2015 to 26/06/2016, along with the smooth spline (yellow color) describing the mean progression in time of daily electricity price, and the variation of the barrel oil price (blue line) for the same period.

Table 1. Estimated Parameters for MIBEL Electricity Prices Longitudinal Model

	Est	(p-value)
Intercept	3.179	(<0.001)
log(Oil price)	0.474	(0.0788)
log(Oil price −1)	−0.349	(0.1707)
σ^2	0.00181	
ϕ_1	0.5210	
ϕ_2	−0.0699	
ϕ_3	0.0644	
ϕ_4	0.0617	
ϕ_5	−0.0302	
ϕ_6	0.1034	
ϕ_7	0.2666	
τ^2	0.2095	
Log likelihood	−239.5538	
AIC	503.1075	

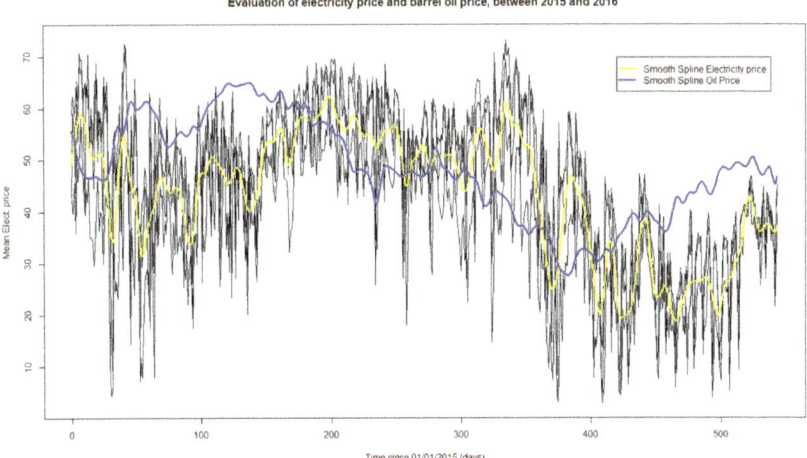

Fig. 2. MIBEL hourly prices from 01/01/2015 to 26/06/2016

Analysing the estimates and its respective p-values of the longitudinal model fitted (Table 1), contrary to what was expected, values of the barrel oil price on the previous day do not affect significantly the variation of electricity price. A significant effect was register for the barrel oil price of the current day, on the variation of electricity price.

For model diagnose we constructed the boxplot of residuals by group of hour (A, B and C) for the fitted model where we could observe that the residuals are centered at zero, but that the variability changes with group. However, the standardized residuals are small, suggesting that the model is successful in explaining the variation of electricity price. Result that was also obtained by inspection of the plot of the observed responses versus the within-group fitted values, where the fitted values are in close agreement with the observed electricity prices.

5   Conclusions and Further Work

This study has investigated the impact of barrel oil variation, together with other covariates, on MIBEL electricity prices using longitudinal model, aiming to contribute with possible factors associated with the complex dynamic of electricity prices. It has already been established that an adequate model to predict and describe price dynamic can lead to profit-maximizing trading strategies and optimal bidding techniques [1]. The mixed-effects longitudinal approach to model the MIBEL electricity prices allow us to infer about the affect of barrel oil price whilst accounting for the variability amongst the three group of hours. The model is able to explain the intra-day and intra-hour dynamics of the hourly prices, by the incorporation of an AutoRegressive component of order 7 describing the within hour dependence. Results suggest that price electricity variation may be explained by oil price variation, but not associated with barrel oil prices from the previous day. As future work we intend to extend the present study to other time periods, in order to assess if the same relation is observed comparing pre and post crisis eras.

References

1. Borges, A., Costa e Silva, E., Covas, R.: A longitudinal model for MIBEL energy prices. WSEAS Trans. Syst. Control **13**, 26–33 (2018)
2. da Silva, P.P., Cerqueira, P.: Drivers for household electricity prices in the EU: a system-GMM panel data approach (No. 2014-13). GEMF, Faculty of Economics, University of Coimbra (2014)
3. Weron, R.: Electricity price forecasting: a review of the state-of-the-art with a look into the future. Int. J. Forecast. **30**(4), 1030–1081 (2014)
4. Huisman, R., Huurman, C., Mahieu, R.: Hourly electricity prices in day-ahead markets. Energy Econ. **29**(2), 240–248 (2007)
5. dos Santos Coelho, L., Santos, A.A.: A RBF neural network model with GARCH errors: application to electricity price forecasting. Electr. Power Syst. Res. **81**(1), 74–83 (2011)
6. Kirat, D., Ahamada, I.: The impact of the European Union emission trading scheme on the electricity-generation sector. Energy Econ. **33**(5), 995–1003 (2011)
7. Akarsu, G.: Analyzing the impact of oil price volatility on electricity demand: the case of Turkey. Eurasian Econ. Rev. **7**(3), 371–388 (2017)
8. Frydenberg, S., Onochie, J.I., Westgaard, S., Midtsund, N., Ueland, H.: Long evidence from Nordic countries, Continental Europe and the United Kingdom. OPEC Energy Rev. **38**(2), 216–242 (2014)
9. Madaleno, M., Moutinho, V., Mota, J.: Time relationships among electricity and fossil fuel prices: industry and households in Europe. Int. J. Energy Econ. Policy **5**(2), 525–533 (2015)
10. Monteiro, C., Ramirez-Rosado, I.J., Fernandez-Jimenez, L.A., Conde, P.: Short-term price forecasting models based on artificial neural networks for intraday sessions in the iberian electricity market. Energies **9**(9), 721 (2016)
11. Costa e Silva, E., Borges, A., Teodoro, M.F., Andrade, M.A.P., Covas, R.: Time series data mining for energy prices forecasting: an application to real data. In: Madureira, A.M., et al. (eds.) Intelligent Systems Design and Applications, Advances in Intelligent Systems and Computing, vol. 557 (2017)
12. Diggle, P., Heagerty, P., Liang, K.Y., Zeger, S.: Analysis of Longitudinal Data, 2nd edn. Oxford University Press, Oxford (2002)
13. R Core Team.: R: A Language and Environment for Statistical Computing. R Foundation for Statistical Computing, Vienna, Austria (2016)
14. Pinheiro, J., Bates, D., DebRoy, S., Sarkar, D., R Core Team.: nlme: Linear and Nonlinear Mixed Effects Models. R package version 3.1-118 (2014). http://CRAN.R-project.org/package=nlme
15. Laird, N.M., Ware, J.H.: Random-effects models for longitudinal data. Biometrics **38**, 963–974 (1982)
16. Box, G.E.P., Jenkins, G.M., Reinsel, G.C.: Time Series Analysis: Forecasting and Control, 3rd edn. Holden-Day, San Francisco (1994)
17. Pinheiro, J.C., Bates, D.M.: Mixed-effects models in S and S-PLUS. Springer, New York (2000)

Policies and Economies Impact's on CO_2 Emissions in the World

José Abreu[1,2], Vítor Braga[1,2], and Aldina Correia[1,2(✉)]

[1] CIICESI — Center for Research and Innovation in Business Sciences
and Information Systems, P.Porto – Polytechnic of Porto, Felgueiras, Portugal
[2] ESTG – School of Technology and Management,
P.Porto – Polytechnic of Porto, Felgueiras, Portugal
{8170011,vbraga,aic}@estg.ipp.pt

Abstract. Protecting the environment is at the moment essential for ensuring sustainability for future generations. Climate change and carbon dioxide (CO_2) emissions are primarily responsible for the process and delivery of the *Global Warming*. This paper aims to study how the various policies and economies in the world influence CO_2 emissions. For this purpose, a research was done about which variables were considered in literature. It was concluded that there are references to various aspects such as: use of fossil fuels as a source of energy, renewable energies, energy efficiency, international technological cooperation, population information and several economic factors or development of the country. We use data for the year 2014 available from the World Bank on the various issues under consideration and over 64 countries. A multivariate linear regression was used to study the statistical significance of the variables considered in the CO_2 emissions of these countries.

Keywords: Carbon dioxide (CO_2) emissions
Policies and economies factors · International comparison

1 Introduction

Concerns about the well-being of the planet have been increasingly debated in major international conferences and its importance has gained prominence in international organizations, and is already a major goal of the United Nations (United Nations) for the millennium: Ensure environmental sustainability, [17].

In this paper, the aim is to explain CO_2 emissions by trying to understand the factors that influence them and, in particular, the impacts of the different policies of each country on global warming. This will be considered a large number of countries, trying to have different policies and different types of development, having in mind to have a representative of the world sample. Next, a review of the literature will be made to identify the considered explanatory variables of CO_2 emissions. Subsequently a multivariate statistical analysis will be performed, in particular a multivariate linear regression in order to study the significant effects

© Springer International Publishing AG, part of Springer Nature 2019
J. Machado et al. (Eds.): HELIX 2018, LNEE 505, pp. 771–778, 2019.
https://doi.org/10.1007/978-3-319-91334-6_105

of each one in the CO_2 emissions. The approximate variables will be taken from the World Bank database in *Sustainable Development Goals (SDGs)* [18], and will then be analysed. With this work we hope to understand and explain better the different policies of the countries in question concerning with sustainability, and to determine which ones have the greatest influence on CO_2 emissions.

2   Literature Review

Climate change is a topic that is causing more and more concern [3,7,8,12,14], especially since Al Gore, the former vice president of the United States, released the book and film "An Inconvenient Truth" which has considerably increased its impact. This served as a wake-up call for the entire planet and to identify a problem, which up to that time, many were still unaware or did not give importance.

Rapid economic growth has led to an increase in the consumption of fossil fuels and consequent release of CO_2 (Hopke [5]). The need for economic growth in countries often leads them to overlook the necessary aspects of sustainable development, taking a "Develop first and clean up later" vision (Fang [10]). The greatest force of climate change is the increase in concentrations of greenhouse gases, especially CO_2 (Hopke [5]). Song [13] identifies the key role of CO_2 in the carbon cycle of the earth, as well as the life cycle of many animals and plants, making it essential for life on earth. The CO_2 molecule is linear with a double bond between the carbon atoms and that of oxygen (O-C-O). CO_2 is a colorless, odourless gas that occurs in nature and serves as a carbon source for photosynthesis of plants and crops. High levels of CO_2 lead to an increase in the planet's temperature, which can lead to severe climate change and irreversible damage in different regions and sectors and to the rising sea level, precipitation and warming of the atmosphere, which consequently may lead to an increase in the occurrence of catastrophic atmospheric phenomena. In addition, its high concentration can lead to serious respiratory problems [1], as the *World Health Organization* has found, which estimates that 2.7 million deaths worldwide are related to the effects of air pollution (Zhang [11]). In order to control global warming, the international community has made efforts to mitigate CO_2 emissions, having reached, in December 2015, the first legal agreement - the Paris agreement (Zhao [19]). This agreement has as main objective to limit this heating below $2\,°C$ (Rogelj [9]). To this end, countries agreed to take various measures to limit their greenhouse gas emissions and improve their sustainability. As the problems of air pollution and global warming are caused mainly by the release of gases in the combustion of fossil fuels for energy production and use, these problems will only be solved with large-scale changes in the energy sector of all countries (Jacobson [6]). The reduction of primary energy consumption and the improvement in technological and technical production are presented as long-term measures that will, in a way, help reduce pollution (Fang [10]).

In addition, adopting alternative energy production processes and improving their performance is likely to lead to a reduction in energy losses. For this to take

Table 1. Variables identified in the literature and expected signal

Variables	Literature	Expected sign
Consumption of fossil fuels	(Mosler [4]) (Zhao [19])	+
Renewable energy	(Fang [10]) (Jacobson [6]) (Wakiyama [15]) (Zhao [19])	−
Energy efficiency	(Fang [10]) (Jacobson [6]) (Zhao [19]) (Zhang [20]) (Wakiyama [15])	−
Technological cooperation between countries	(Rogelj [9]) (Zhang [20])	−
Level of population information	(Zhang [20]) (Wakiyama [15]) (Schleich [8])	−
Economic factors/country development	(Hopke [5]) (Fang [10]) (Zhao [19])	+

place, it is necessary to accelerate the process of updating the equipment, namely through alternative processes, such as solar, wind, water and biomass (Song [13]). International cooperation in the development and diffusion of technologies can help to achieve the objectives that each country has conditioned, and may even help to improve the goals they have set themselves (Rogelj [9]), through the application of strategic policies to sub-sectors in order optimize the productive processes and to update their energy efficiency, increase fuel diversification and replace fossil fuels with clean or renewable energy (Zhao [19] and Mosler and Martens [4]). In China, during the first 13 years of the new millennium, a large part of the variables related to environmental innovations had a great impact on the reduction of CO_2 emissions, in particular energy efficiency, the number of patents issued and the level of information of the population, (Zhang in [20]). Zhang in [20], a study carried out in Japan, found that although the residential sector is not the most energy-consuming sector, only 14% have great potential for improvement and consequently reduce CO_2 emissions. To this end, the authors advise the implementation of reforms in the energy market, in order to encourage their inhabitants to reduce energy consumption (Wakiyama [15]). Anthropogenic CO_2 emissions are strongly linked to the growth of a country's Gross Domestic Product (GDP). In order to reverse this situation, it would be necessary to restructure the primary energy system of each country, which is quite complicated, requiring great financial capacity and time (Quéré [2]) According to Thomas ([16]) the response of species to recent and past climate change increases the possibility of these changes serving as a cause of extinction in the near future, which warns of an urgent need to reduce these emissions and limit global warming. Table 1 presents a summary of all the variables identified in the literature and their expected signal to explain CO_2 emissions.

3   Methodology

After the research in the literature and identified the variables presented in Table 1, which define the research hypotheses, all related to the effect of the indicated variables on the CO_2 emissions, it is time to collect the data. The data for the similar variables were obtained through the database of the world bank's website and were selected specifically according to the theoretical research carried out. Much of this data was taken from the database entitled Sustainable Development Goals (SDGs) [18]. We selected this database because it seemed to be the most comprehensive in view of the issue of this work, as it addressed the economic and environmental aspects, but also a list of the various countries worldwide.

Table 2. Proxies for the variables identified in the literature

Variables	Data
Consumption of fossil fuels	Electric production through oil, gas and coal source (% total produced) Energy consumption of fossil fuels (% of total produced)
Renewable energy	Renewable energy (% total consumed) Renewable energy (% total produced)
Energy efficiency	Renewable fuels and waste (% of total energy) Energy efficiency (% energy lost) Losses of transmission and distribution of energy (% of energy output)
Technological cooperation	Development of technologies related to the environment (% all technologies) Research and development expenditures (% of GDP)
Population information	Government expenditure on education (% of GDP) % Population with third cycle with more than 25 years (% total)
Economic factors/country development	Industry added value (% GDP) GDP per capita (US $) GDP growth (% per annum) Urban population (% of total)

Final data includes sixty-four countries. Countries with some available data in [18] but with too many missing data for the chosen variables were deleted. The choice of variables was made taking into account the similarity with the variables mentioned in the literature review. For each of the variables identified in the literature, at least two representative variables were chosen, according to Table 2. The year selected for the study was 2014 because it presented the highest number of values for the various variables considered. The quantitative methodology to be used will be a multivariate linear regression, with the objective of verifying how the variables considered (15 in total, concerning with the 6 concepts) explain the dependent variable CO_2 emissions, measured in tons per capita.

4   Results

In order to study if the parameters under analysis affects the CO_2 emissions, measured in tons per capita, a multivariate linear regression model was tested in agreement with the hypothesis formulated. Then the dependent variable is the CO_2 emissions, measured in tons per capita, and the variables presented in Table 2 are the independent variables. The CO_2 emissions is measured in tons per capita and have values between 0.31 and 19.53. The average value for the 64 countries is around 6 tons per capita with a standard deviation of 4.51 tons per capita. In order to exclude non-significant parameters, the stepwise method is considered. The model has an Adjusted R Square 0.619 . Then, we can say that the significant independent variables explain around 62% of the variance occurred on the dependent variable CO_2 emissions, measured in tons per capita. With these results, it was only possible to prove the significance of two independent variables, and the same ones had a positive influence on CO_2 emissions. They are the Electric Production through oil, gas and coal source (% Total Produced) and the GDP per capita (US $).

Considering the Unstandardised Coefficients we see that, if the other variables remain constant, the increase of 1 unit in the variable *GDP per capita* implies a grown up in the CO_2 emissions in 0.802 tons per capita. Considering the Unstandardised Coefficient o the second significant independent variable *Electricity production from oil, gas and coal sources (% of total)* we see that, if the other variables remain constant, the increase of 1 unit in the variable that variable implies a grown up in the CO_2 emissions in 0.610 tons per capita. Despite this, it is not possible to refute the other hypotheses although it has not been possible to prove that they have a significant effect on the dependent variable, it can be because it is very difficult to evaluate their impact on the emissions or its effect is very small. It may also have happened that the variable used in each hypothesis may not have been the most correct or vary from country to country. For example, in our country we take more environmental awareness at the beginning of the third cycle, because of that we have selected this variable as approximate to the information level of the population, but in another country this value may not be the most adequate.

Linear regression models require that residuals had a normal distribution with zero mean and constant variance, and that they must be independent. With the independence Durbin-Watson test close to the value 2, we can conclude that there is no evidence to consider that the residuals are correlated. In Fig. 1 we can

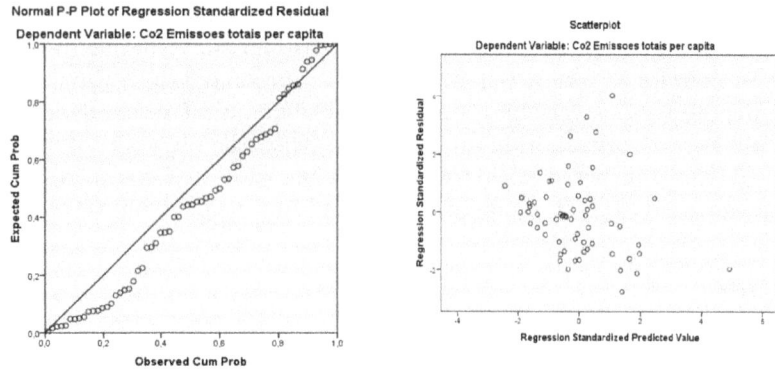

Fig. 1. Normal P-P Plot and Scatterplot

see that in Normal P-P Plot the points are close to the diagonal, so it is expected that the residuals have an approximately normal distribution. In addition, in the Scatterplot we can see that the dispersion of the residuals are around the mean value (zero), is more or less random and its variance appears to be relatively homogeneous.

5   Conclusions and Future Work

This work allowed to corroborate that the economic aspects of a country and the use of fossil fuels as sources of electric energy production have a great influence on the emissions of carbon dioxide per capita, explaining around 62% of the variance of our dependent variable. Moreover, by not corroborating the other hypotheses of investigation, it allows us to draw some conclusions regarding both their significance in relation to our variable and the limitations of the work. These limitations may be that the choice of approximate variables may not be the most indicated. Or that, in fact, it is quite difficult to reverse the situation we are in, and therefore none of the variables prove to be really effective in reducing CO_2 emissions.

For the future, it would be interesting to go deeper into the literature review, trying to identify other relevant variables or even to study other databases to not only explain the dependent variable, but also to find a variable that has an opposite effect on CO_2 emissions. Besides that, once the constant of the linear model obtained is not significant, a similar model without the constant must be considered. We still intend to classify or group countries, taking into account these variables, to detect the main differences between them and to suggest some indications for differentiated intervention.

References

1. Anderson, H.R.: Air pollution and mortality: a history. Atmos. Environ. **43**(1), 142–152 (2009)
2. Le Quéré, C., Raupach, M.R., Canadell, J.G., Marland, G., Bopp, L., Ciais, P., et al.: Trends in the sources and sinks of carbon dioxide. Nat. Geosci. **2**(12), 831 (2009)
3. Jacob, D.J., Winner, D.A.: Effect of climate change on air quality. Atmos. Environ. **43**(1), 51–63 (2009)
4. Mosler, H.J., Martens, T.: Designing environmental campaigns by using agent-based simulations: strategies for changing environmental attitudes. J. Environ. Manage. **88**(4), 805–816 (2008)
5. Hopke, P.K.: Contemporary threats and air pollution. Atmos. Environ. **43**(1), 87–93 (2009)
6. Jacobson, M.Z.: Review of solutions to global warming, air pollution, and energy security. Energy Environ. Sci. **2**(2), 148–173 (2009)
7. Gaffney, J.S., Marley, N.A.: The impacts of combustion emissions on air quality and climate-from coal to biofuels and beyond. Atmos. Environ. **43**(1), 23–36 (2009)
8. Schleich, J., Faure, C.: Explaining citizens' perceptions of international climate-policy relevance. Energy Policy **103**, 62–71 (2017)
9. Rogelj, J., Den Elzen, M., Höhne, N., Fransen, T., Fekete, H., Winkler, H., et al.: Paris Agreement climate proposals need a boost to keep warming well below 2 C. Nature **534**(7609), 631 (2016)
10. Fang, M., Chan, C.K., Yao, X.: Managing air quality in a rapidly developing nation: China. Atmos. Environ. **43**(1), 79–86 (2009)
11. Minsi Zhang, Y.S.: Economic assessment of the health effects related to particulate matter pollution in 111 Chinese cities by using economic burden of disease analysis. J. Environ. Manage. **88**, 947–954 (2008)
12. Cox, P.M., Betts, R.A., Jones, C.D., Spall, S.A., Totterdell, I.J.: Acceleration of global warming due to carbon-cycle feedbacks in a coupled climate model. Nature **408**(6809), 184 (2000)
13. Song, C.: Global challenges and strategies for control, conversion and utilization of CO_2 for sustainable development involving energy, catalysis, adsorption and chemical processing. Catal. Today **115**, 2–32 (2006)
14. Solomon, S., Plattner, G.K., Knutti, R., Friedlingstein, P.: Irreversible climate change due to carbon dioxide emissions. Proc. Nat. Acad. Sci. **106**(6), 1704–1709 (2009)
15. Wakiyama, T., Kuramochi, T.: Scenario analysis of energy saving and CO_2 emissions reduction potentials to ratchet up Japanese mitigation target in 2030 in the residential sector. Energy Policy **103**, 1–15 (2017)
16. Thomas, C.D., Cameron, A., Green, R.E., Bakkenes, M., Beaumont, L.J., Collingham, Y.C., Hughes, L., et al.: Extinction risk from climate change. Nature **427**(6970), 145 (2004)
17. United Nations: Transforming Our World: The 2030 Agenda for Sustainable Development (2015). Accessed https://sustainabledevelopment.un.org/post2015/transformingourworld

18. The World Bank: Sustainable Development Goals (SDGs). http://databank. worldbank.org/data/reports.aspx?source=sustainable-development-goals-(sdgs)
19. Zhao, X., Zhang, X., Shao, S.: Decoupling CO_2 emissions and industrial growth in China over 1993–2013: the role of investment. Energy Econ. **60**, 275–292 (2016)
20. Zhang, Y.J., Peng, Y.L., Ma, C.Q., Shen, B.: Can environmental innovation facilitate carbon emissions reduction? Evidence from China. Energy Policy **100**, 18–28 (2017)

Sector Analysis of Debt Maturity
in Portugal and Spain

Nélson Ferreira Alves[1], Isabel Cristina Lopes[1,2,3(✉)] (iD),
and Mário Queirós[1]

[1] ISCAP-P.Porto - Accounting and Business School, Polytechnic of Porto,
R. Jaime Lopes Amorim 4465-004, S. Mamede de Infesta, Portugal
nsfalves@hotmail.com, cristinalopes@iscap.ipp.pt,
mqueiros.iscap@gmail.com
[2] CEOS.PP – Centre for Organisational and Social Studies of P.Porto,
Porto, Portugal
[3] LEMA - Mathematical Engineering Lab, Porto, Portugal

Abstract. The main purpose of the present study is an empirical analysis of the
factors influencing debt maturity in the different sectors of activity, using 16 494
non-financial Portuguese and Spanish companies, both listed and unlisted, over
a time horizon from 2010 to 2015. Using the fixed effects model, we analyzed
the influence of the companies characteristics' explanatory variables, as well as
the influence of the variables of country characteristics. The results show that the
companies characteristics are the major drivers in the choice of debt maturity,
given the empirical evidence found in all sectors analyzed. Despite this disclosure, only the high rating variable shows a consistent and significant negative
influence in all sectors of activity. In the other hand, the characteristic of corporate indebtedness shows a positive and significant influence on debt maturity
in most sectors of activity, failing only in the transport sector. Regarding to the
countries characteristics, the results point to a weak evidence of the influence on
debt maturity, not being possible to find a characteristic that influences equally
all the sectors of activity.

Keywords: Capital structure · Debt maturity · Panel data

1 Introduction

The definition of the company's optimal capital structure plays an important role in the
company's financial management, where managers have to make at least two important
decisions: (i) choose between debt and/or equity and (ii) choose the type of capital
maturity, and companies should choose between short-term and long-term debt. That
is, it is the choice of the level of indebtedness and the maturity of the debt that best fit
the objectives of the company and that maximize its value. In the vast literature that
studies this subject, several authors consider that the choice of the optimum point of the
maturity of the debt is influenced by several factors. It was from the article by Modigliani and Miller [1], on the irrelevance of the capital structure, that several hypotheses
were formulated to determine the financing choices of companies. It was from [2] that
we considered taxation, bankruptcy costs, information asymmetries, agency costs, (in)

© Springer International Publishing AG, part of Springer Nature 2019
J. Machado et al. (Eds.): HELIX 2018, LNEE 505, pp. 779–785, 2019.
https://doi.org/10.1007/978-3-319-91334-6_106

flexibility of financing options, (in)efficiency of markets, among other factors that influence the choice of debt maturity. However, there are several authors who have made different critiques (and obtained different conclusions) to the study of Modigliani and Miller [1] and revealed that there are several factors that influence the markets, namely the choice of debt or capital as a source of financing. Since the literature presents us with different conclusions about the various factors that influence the financing choices and, consequently, what influences the maturity of the debt, this reason has increased the interest in wanting to study this subject. In this way, we try to answer the following research question:

Is the maturity of debt influenced by the characteristics of the country or the company itself?

In line with the literature, we identified and tested the influence of the variables related to the characteristics of the companies and the characteristics of the countries, by sector of activity, on the level of corporate debt maturity.

2   Sample Data Set

To carry out this study, we used a sample composed of only non-financial companies, both listed and not listed on the stock exchange, from two European countries (Portugal and Spain), whose data were extracted in March 2016 through the SABI platform, within a time horizon of 2010 to 2015. Information collected on countries has been obtained from a number of sources, such as Eurostat, the World Bank and Transparency International. The final sample consists of 4175 listed and unlisted companies in Portugal and 12319 listed and unlisted companies in Spain, making a total of 16494 companies. After determining the final sample, we decided to divide the sample by the sectors of activity in which the companies are inserted (Agriculture, Industry, Energy, Services, Construction, Commerce, Transportation, Real Estate). This division of the companies by sector of activity was carried out through the CAE rev. 3 (Classification of Economic Activities) of Portuguese and Spanish companies.

The explanatory variables and the expected relationship between the explained variable and the different explanatory variables chosen are presented below.

2.1   Explained Variable

The explained variable in this study is the Maturity of Debt (MAT). According to [3, 4], debt maturity (MAT) is defined as the ratio of long-term debt to total book value of the company. We consider long-term debt, all debt maturing over one year, as it is in most studies.

2.2   Explanatory Variables Related to the Company Characteristics

Size of the Company (DIME): We followed the authors Fan et al. [3] using the natural logarithm of total company assets as an indicator of the size of the company, thus expecting a positive relationship between debt maturity and company size.

Maturity of Assets (MATA): Based on Majumdar [5], asset maturity is defined by the ratio of the book value of net tangible assets to depreciation for the period. Following the theory that indicates that the greater the maturity of the assets, the greater the maturity of the debt, we expect to find a positive correlation between the variables.

Quality of the company (QUAL): Using the indicator of Kirch and Terra [4], we use the ratio of earnings before interest and taxes on total assets. In this context, according to Flannery's theory [6], the relationship between company quality and debt maturity is negative, since in equilibrium, higher quality companies emit more debt in the short term, while lower quality companies emit more debt in the long run. Thus, we expect in our study a negative relation between quality of the company and debt maturity.

Indebtedness (END): According to most authors, such as [5, 7], indebtedness is defined in our study as the ratio between total debt and asset. According to authors who argue that more indebted companies use more long-term debt, as well as issuing long-term debt reduces liquidity risk and bankruptcy risk, it is expected that we will find in our study a positive relationship between indebtedness and debt maturity.

Liquidity Risk (Rat.BX) and High Rating (Rat.AT): A company's credit risk rating is often flagged by assigning a rating. According to [4], we created two synthetic dummy variables based on the coverage ratio (the division between EBIT and Financial Losses), where: (i) the low rating (Rat.BX) takes the value 1 when the coverage ratio is less than 1, otherwise it is set to 0; (ii) the high rating (Rat.AT) takes the value 1 when the coverage ratio is in the upper quintile, otherwise it takes the value of 0. Thus, we expect to capture the relationship between debt maturity and credit risk predicted by Diamond [8], where companies with a very high rating and those with a very low rating often use short-term debt due to the low refinancing risk (companies with a high rating) and due to the lack of choice (companies with low rating).

Effective Tax Rate (IMP): The effective tax rate is defined, in our study, by the ratio between taxes on income for the period and earnings before taxes. In this context, we expect that there will be a negative sign in the relationship between the effective tax rate and the maturity of the debt, since when the effective tax rates are low, the tax benefits of the debt are reduced, which makes it issuance of long-term debt.

Interest Rate Volatility (VOLTJU): To measure interest rate volatility, we used the annual standard deviation of monthly interbank rates at 12 months, according to [9]. Following the model of Kim et al. [10], which suggests that corporate debt maturity increases with increasing interest rate volatility, we expect to find in our paper a positive relationship between interest rate volatility and debt maturity.

2.3 Explanatory Variables Related to the Country Characteristics

Macroeconomic Variables: Following Wang et al.[11] and Fan et al.[3], the macroeconomic variables are represented in the model by: (i) **Gross Domestic Product (GDP)** is measured by the gross domestic product growth rate; (ii) **Inflation Rate (TXINF)** is measured through the annual variation of the consumer price index;

(iii) **Volatility of the Inflation Rate (VOLTINF)** is measured by the standard deviation of the inflation rate of the previous five years. In this context, since macroeconomic conditions can affect debt because they represent the growth opportunities of a country's economy, given the ease of access to long-term financing when the economy is growing [12], we expect to find in our study a positive relationship between GDP growth rate and debt maturity. As for the inflation rate and the volatility of the inflation rate, we expect a negative relationship between these variables and the debt maturity, since significant increases in the rate of inflation decrease the real tax benefits, as well as the maturity of the debt [11], while the high volatility of the inflation rate manifests a greater uncertainty about the future prospects of inflation, causing creditors to grant more short-term debt [3].

Banking Sector Dimension (DIMSB): To measure the size of the banking sector, we follow the model proposed by Fan et al. [3], which analyzes the internal credit granted by the banking sector (as a percentage of GDP). Given that firms tend to finance themselves more frequently in short-term debt, in countries with larger banking sectors, because of bank bonds with reduced maturity [3], it is expected to find in our work a negative relationship between the size of the banking sector and the maturity of the debt.

Efficiency of the Legal System (EFJUR): To analyze the efficiency of the legal system we use, according to Fan et al. [3], the corruption perception index that is made available annually by Transparency International, ranging from 0 to 10, where 0 means low level of corruption or high transparency and 10 means high level of corruption or low transparency. In this context, we can expect a positive relationship between the efficiency of the legal system and debt maturity, since when a country's legal system is not very efficient (high corruption rates), companies tend to issue more short term debt [3].

Country (DuPais): It serves to capture the effects of country differences on corporate debt maturity, particularly with regard to the type of legal and financial system, as the differences between countries' legal and financial systems affect the use of finance for business growth. Thus, the dummy variable 1 corresponds to a Spanish company, while the dummy variable 0 corresponds to a Portuguese company.

3 Methodology

For the empirical analysis, we use the panel data methodology, which allows us to observe the behavior of companies over time, as well as offers more informative data, allows greater variability and it is adequate to examine the dynamics of change. The models generally used to address panel data are: (i) the *Ordinary Least Squares* (OLS) model (Eq. 1), whose technique proposes to find the coefficients that minimize the sum of the squares of the residuals, (ii) the *Fixed Effects model* (Eq. 2), where it is assumed that the heterogeneity between individuals remains constant over time; and (iii) the *Random Effects model* (Eq. 3), where it is assumed that there is no correlation between the explanatory variables and the individual specific effects not observed, that is, the individual effects are treated as terms of perturbation, specifying the individual effects, not in a fixed form, but in a random way.

$$\text{MAT}_{it} = \text{X}_{it}\,\beta + \varepsilon_{it} \tag{1}$$

$$\text{MAT}_{it} = \text{X}_{it}\,\beta + \alpha_i D_i + \varepsilon_{it} \tag{2}$$

$$\text{MAT}_{it} = \text{X}_{it}\,\beta + v_i + \varepsilon_{it} \tag{3}$$

In the equations, MAT_{it} is the explained variable observed for firm i at time t; X_{it} is the matrix of the explanatory variables observed over time t for each firm i; β is the matrix of the regression coefficients associated with each variable X, which may or may not include the constant term β_0; α_i are the individual effects for each company; D_i are the dummy variables that represent each company; v_i are random variables representing the individual effects, which are assumed to be independent and identically distributed $v_i \sim N(0, \sigma_v^2)$ and also independent of errors; ε_{it} represents the perturbation term which is assumed to be N replicates of independent and identically distributed variables $\varepsilon_{it} \sim N(0, \sigma_\varepsilon^2)$; $i = 1,\ldots, N$; $t = 1,\ldots, T$. It is also assumed in the fixed effects model that the explanatory variables X_{it} are independent of the errors ε_{it} for all i and all t, as it is assumed that X_{it} are independent of v_i and of ε_{it} [13].

Since these are the models to be used to analyze the factors influencing debt maturity, we then proceed to analyze the impact of the variables of the three models through the use of statistical software R with package *plm*. For the selection of the most appropriate model, we proceed to the elaboration of several tests, such as: (i) the Breusch Pagan Lagrange Multiplier (LM) test, to test for individual specific effects; and (ii) the Hausman test, to verify if the individual specific effects are correlated with the explanatory variables.

4 Analysis of Results

Our results in the LM test showed p-value values lower than 5% in all sectors of activity, with values very close to zero, which leads us to conclude that the most appropriate model is not an OLS but a panel data model. However, it remains to be seen whether a model of fixed effects or random effects is better.

As in the LM test, the results obtained by the Hausman test also presented a p-value of less than 5% in all sectors of activity, which led us to conclude that the best panel data model is the fixed effects model, since the null hypothesis is rejected. Thus, the fixed effect model is the most adequate model for the analysis of the determinants of debt maturity in the different sectors of activity.

Table 1 presents the regression coefficients of the debt maturity factors of listed and unlisted non-financial companies of Portugal and Spain divided by 8 activity sectors over a time horizon from 2010 to 2015. In parentheses, the expected signal of the relationship between the explained variable and the explanatory variables, as well as the robust standard errors of each explanatory variable are presented. Some of the coefficients of explanatory variables are followed by [***], [**], [*] and ['], which

Table 1. Results of the regressions with the fixed effects model

Explanatory variable	Expected signal	Agriculture	Commerce	Construction	Energy	Real Estate	Industry	Services	Transportation
DIME	(+)	0,1638**	−0,0139	0,0105	0,2202***	0,0645*	0,0692***	0,1271***	0,2334***
		(0,0509)	(0,0101)	(0,0164)	(0,0434)	(0,0261)	(0,0094)	(0,0113)	(0,0207)
MATA	(+)	−0,0001*	−0,00001**	0,0000005	0,00000006	0,00000007	0,0000006	0,0000002˙	−0,00000002
		(0,00007)	(0,000005)	(0,0000008)	(0,00000006)	(0,00000006)	(0,0000007)	(0,00000008)	(0,000001)
QUAL	(−)	−0,0758	0,0419**	0,0142*	0,0688	−0,0671**	−0,0232*	0,0153	0,0081
		(0,0658)	(0,0159)	(0,0061)	(0,0619)	(0,0237)	(0,0106)	(0,0117)	(0,0116)
END	(+)	0,1789***	0,1175***	0,0327***	0,1619***	0,0477**	0,0293***	0,0423***	−0,0091
		(0,0372)	(0,0077)	(0,0055)	(0,0390)	(0,0150)	(0,0031)	(0,0058)	(0,0057)
Rat.BX	(−)	−0,0264*	0,0072**	0,0097*	−0,0291˙	−0,0146*	0,0058*	0,0005	0,0174**
		(0,0112)	(0,0028)	(0,0046)	(0,0149)	(0,0063)	(0,0024)	(0,0036)	(0,0059)
Rat.AT	(−)	−0,0325*	−0,0278***	−0,0258***	−0,0779***	−0,0316***	−0,0314***	−0,0518***	−0,0619***
		(0,0137)	(0,0033)	(0,0057)	(0,0180)	(0,0080)	(0,0029)	(0,0047)	(0,0077)
IMP	(−)	−0,0048**	−0,00002	0,0003	0,0005	−0,00002	0,00003	0,0004	0,00003
		(0,0018)	(0,00004)	(0,0007)	(0,0010)	(0,0002)	(0,00009)	(0,0005)	(0,00002)
VOLTIU	(+)	9,0608	3,0300	−3,2284*	−1,9473	−0,1284	−0,0043	−3,3123˙	0,151
		(5,2251)	(1,2535)	(2,3737)	(6,2560)	(2,9929)	(1,1248)	(1,6963)	(2,9256)
GDP	(+)	0,0739	−0,2312**	−0,2934*	0,7889*	0,0734	0,0146	−0,0302	0,3031˙
		(0,3191)	(0,0731)	(0,1384)	(0,3923)	(0,1841)	(0,0667)	(0,1036)	(0,1705)
TXINF	(−)	−0,0404	0,1162	0,2641	1,163˙	0,0233	0,0387	0,0415	−0,1617
		(0,4853)	(0,1095)	(0,2081)	(0,5953)	(0,2786)	(0,1007)	(0,1570)	(0,2572)
VOLTINF	(−)	0,6746	−2,5391**	−3,9531*	0,1159**	0,3141	−0,7253	−2,9405*	1,9467
		(3,5894)	(0,8250)	(1,5726)	(4,4024)	(2,0631)	(0,7537)	(1,1637)	(1,9367)
DIMSB	(−)	−0,0005	0,0029	−0,0171	−0,1271**	0,0282	0,009	0,0158	0,1016***
		(0,0334)	(0,0079)	(0,0155)	(0,0400)	(0,0191)	(0,0071)	(0,0107)	(0,0183)
EFJUR	(+)	−0,0075	0,0195**	0,0341*	0,0551	0,0189	0,0071	0,022*	−0,0047
		(0,0325)	(0,0080)	(0,0151)	(0,0390)	(0,0186)	(0,0071)	(0,0106)	(0,0185)
COUNTRY	(+)/(−)	−0,7115***	0,1656**	0,0213	−0,0933	0,0411	−0,2451***	−0,0796	−0,6036***
		(0,1224)	(0,0631)	(0,0922)	(0,1202)	(0,1217)	(0,0718)	(0,0904)	(0,0782)
R^2		0,8013	0,8204	0,8288	0,7014	0,7347	0,8015	0,8161	0,8725
F Statistic		19,4	22,74	24,04	11,31	13,71	20,12	22,1	33,62

indicates that they are statistically significant at a level of 0.1%, 1%, 5% and 10%, respectively. The model does not show multicollinearity problems, as most of the regressors have a VIF value close to 1 and none greater than 7.

Given the obtained coefficients in the fixed effects model, we can observe that most of the characteristics of the companies significantly influence the choice of corporate debt maturity in the different sectors of activity, which contrasts with the weak significance of the characteristics of the countries towards debt maturity. The sectors of agriculture and trade are the ones that have the best results in terms of the influence of the characteristics of companies in the choice of debt maturity, while the trade and energy sectors are the ones that show the best results in terms of the influence of the characteristics of the countries. This shows that, in general, the commerce sector is the only one that presents the best empirical results to explain the maturity of the debt.

5 Conclusions

The results obtained by the chosen model allowed us to conclude that, in all sectors of activity analyzed, what influences more the choice of debt maturity are the characteristics of the companies, with the variables related to liquidity risk theory contributing the most, especially the high rating variable. Contrary to the characteristics of the companies, we have not found enough empirical evidence to conclude that the characteristics of the countries significantly influence the choice of debt maturity in all sectors of activity. In summary, our study allowed us to determine that the characteristics of the companies, as well as the characteristics of the countries, do not influence the choices of the level of debt maturity equally for all the different sectors of activity, apart from some exceptions. Thus, only the high rating binary variable evidences an influence on corporate debt maturity, in the same way in all sectors analyzed, as well as the indebtedness variable, although it fails in a single sector. As future work, regression models considering dummy variables for listed and unlisted companies, and their size can be considered.

References

1. Modigliani, F., Miller, M.: The cost of capital, corporation finance and the theory of investment. Am. Econ. Rev. **48**(3), 261–297 (1958)
2. Modigliani, F., Miller, M.: Corporate income taxes and the cost of capital: a correction. Am. Econ. Rev. **53**(3), 433–443 (1963)
3. Fan, J.P.H., Titman, S., Twite, G.: An international comparison of capital structure and debt maturity choices. J. Finan. Quantit. Anal. **47**(1), 23–56 (2012)
4. Kirch, G., Terra, P.R.S.: Determinants of corporate debt maturity in South America: do institutional quality and financial development matter? J. Corp. Finance **18**(4), 980–993 (2012)
5. Majumdar, R.: The determinants of corporate debt maturity: a study of Indian firms. J. Appl. Finance **16**(2), 70–80 (2010)
6. Flannery, M.J.: Asymmetric information and risky debt maturity choice. J. Finance **41**(1), 19–37 (1986)
7. García-Teruel, P.J., Martínez-Solano, P.: Ownership structure and debt maturity: new evidence from Spain. Rev. Quant. Financ. Acc. **35**(4), 473–491 (2010)
8. Diamond, D.W.: Debt maturity structure and liquidity risk. Q. J. Econ. **106**(3), 709–737 (1991)
9. López-Garcia, J., Mestre-Barberá, R.: Tax effect on Spanish SME optimum debt maturity structure. J. Bus. Res. **64**(6), 649–655 (2011)
10. Kim, C.S., Mauer, D.C., Stohs, M.H.: Corporate debt maturity policy and investor tax-timing options: theory and evidence. Financ. Manage. **24**(1), 33–45 (1995)
11. Wang, Y., Sun, Y., Lv, Q.: Empirical study on the debt maturity structure based on the macroeconomic variables. Int. J. Bus. Manag. **5**(12), 135–140 (2010)
12. Jõeveer, K.: Firm, country and macroeconomic determinants of capital structure: evidence from transition economies. J. Comp. Econ. **41**, 294–308 (2013)
13. Baltagi, B.H.: Econometric Analysis of Panel Data, 3rd edn. Wiley, West Sussex (2005)

Optimization of Purchasing Management of a Portuguese Company in the Retail Sector

Ana Teixeira$^{(\boxtimes)}$, Eliana Costa e Silva, and João Ferreira Santos

CIICESI - Center for Research and Innovation in Business Sciences
and Information Systems, School of Technology and Management,
Polytechnic of Porto, Porto, Portugal
{8130097,eos}@estg.ipp.pt

Abstract. The current competitive environment requires greater customer orientation and better business performance to match the clients' expectations with varied products and profitable sales prices. This has led to Supply Chain Management (SCM) as a source of competitive advantage. However, given the complexity of Supply Chain (SC) planning, optimization based quantitative models emerged as a support for decision-making. This article presents an empirical study with real data of a Portuguese company of the retail sector. In order to contribute with *"tailor-for"* quantitative models for improving decision-making in the company's purchasing management, in this paper the industrial challenge and the company are introduced, followed by a preliminary data analysis. This analysis has require for deciding the real instances that will be used on the validation of the quantitative model that will be developed in future.

Keywords: Supply Chain Management · Quantitative model
Decision-making · Purchasing management · Planning

1 Introduction

SCM can be seen as a complex network of interdependent organizations both upstream and downstream of SC, in which mutually beneficial relationships are established, and where the performance of an organization depends, in an effective and efficient manner, on cooperation between all partners [1]. SC planning is a critical business problem, since it implies an inter-organizational integration and different decision domains: procurement, warehousing, distribution and sales. Seuring [2] states that only about 12% of SCM articles apply quantitative models, therefore, there is still the need for developing these types of models.

The main objective of this paper is to contribute to the development of *"tailor-for"* quantitative models with the purpose of improving decision-making in the company's purchasing management of a Portuguese company of the retail

© Springer International Publishing AG, part of Springer Nature 2019
J. Machado et al. (Eds.): HELIX 2018, LNEE 505, pp. 786–792, 2019.
https://doi.org/10.1007/978-3-319-91334-6_107

sector. Specifically in this paper the real challenge is introduced and a preliminary data analysis of the data provided by the company is performed, with the objective of deciding on the construction of real instances that will be used for the validation of the quantitative model that will be develop in future.

The paper is organized as follows: Sect. 2 presents a literature review on SCM and addresses the evolution of the concept of SCM, the decisions and processes that involve the planning matrix of a SC and, the quantitative models already developed for decision support, contributing to effective and efficient SC planning; in Sect. 3 an empirical study will be carried out, based on real data of the retail company; Sect. 4 presents the preliminary model formulation; and finally, Sect. 5 shows the conclusions.

2   Literature Review

SCM was first introduced in 1982 [3], and since then it has received an increasing interest, both in academy and business contexts [4]. Several definitions of SCM have been proposed over the years, but none is universal [5], mainly due to the different points of view of the researchers and the multidisciplinary nature of the concept [4]. SCM and logistics terms were often confused, and it is not clear in what aspects they differ [6,7]. Logistics may be defined as *"the process of planning, implementing, and controlling the efficient, effective flow and storage of goods, services and related information from the point of origin to the point of consumption"* [6, p. 426]. However, it is focused on the coordination of activities within the organization and the movements of output of the products, not including the coordination and collaboration of functions within the company with external channels [8]. For this, and also due to increased global competition, the information technology revolution and increasing concerns of companies to increase efficiency, during the 1980s and early 1990s [5,8], it became essential to introduce the SCM concept.

SCM can be defined as planning/management of all activities involved in sourcing and acquisition, conversion, and all logistics management activities. It also includes manufacturing operations and integration of processes/activities with marketing, sales, product design, finance and information technology. SCM also comprises coordination and collaboration with channel partners (suppliers, intermediaries and customers), being responsible for business functions and processes within and between companies, in a cohesive and high performance model [9]. Special attention should be given to the *"collaborative paradigm"*, since it leads to gains through the transfer of resources, knowledge and information among all SC actors, allowing rapid adaptation to changes of the market [1,5].

The current competitive environment requires greater customer orientation and better performance, as consumers demand high availability of products and services at the best prices. In turn, retailers strive to offer varied products with cost-effective sales prices and low costs [10]. To answer this, organizations use outsourced companies, specialized in certain processes and with more knowledge [11]. Thus, with the increasing division of labor, a company's performance depends on its ability to maintain relationships and cooperate with all SC partners [12]. However, the planning of all tasks in a comprehensive and global model would not be possible in an optimal way due to the complexity of the system and interdependence between the activities [13]. For this reason, Hübner et al. [10] enunciates hierarchical planning in order to break global planning into partial planning modules, facilitating decision-making.

Fleischmann et al. [14] developed a SC planning matrix that classifies horizontally the flows of products between procurement, warehousing, distribution and sales and vertically, distinguishes the tasks in the short, medium and long term, was developed. Vertically, long-term planning prepares decisions reflected over several years, reflecting company strategies [13]. Medium-term determines regular operations for a period of 6 to 12 months [10], and is responsible for balancing demand with SC's capabilities [4]. Short-term specifies all activities for execution and control in a daily or weekly period. Horizontally, in the procurement, it is important to define outsourcing strategies that stipulate a suppliers and establish their contracting, based on factors such as price, reliability or lead time [10]. The acquisition is concerned with the supply of resources for the whole SC, having acquired a strategic role, being considered as a potential source of competitive advantage [14]. The warehouse makes decisions about the size and number of the warehouses, the organization of production processes/systems and its capacity [13]. For retailers, store orders are produced in the warehouse, which means that fixed, transport, stock and picking costs must be considered. The storage is made through: direct delivery in store; Cross-Docking (XD); Picking By Line (PBL); and Picking By Store (PBS) [10]. In addition, the size of the lots also plays a significant role [4]. Regarding distribution, the structuring of the distribution system and warehouse locations should be determined and the use of different distribution channels and vehicle routes [13]. This domain should consider the trade-off between infra-structure costs, stock maintenance and transportation and, should seek to optimize the frequency of delivery of the product. Finally, sales includes strategies regarding the type of store and planning of its location and size and increase in density of the SC network. The sales forecasting must be carried out proactively since the consumer only makes his final decision in the dealer's shop, i.e., sales domain is the basis for data entry for the remaining planning areas. It is still important to emphasize that the matrix should incorporate the relationship of cooperation with suppliers, as well as the integration of the consumer [10]. Information technology is a vital component for the successful SCM, enabling the immediate exchange of information between partners [4]. Since the uncertain increases for longer planning horizon [13], and given the complexity of SC, quantitative models have played a key role, contributing to support for decision-making [3].

A large number of mathematical models were proposed, however there is the necessity of new models (see [2,15,16]). Quantitative models for SCM can be classified as: Mathematical Programming; Simulation, Heuristics; Hybrid and Analytics [15]. Mathematical Programming is one of the most commonly used, and it can be divided into single-objective or multiobjective, and as Linear

Programming (LP), Mixed Integer Linear Programming (MILP), Nonlinear Programming (NLP). Likewise, Heuristic include Artificial Intelligence and Meta-Heuristic techniques, and solutions based on Genetic Algorithms and Simulated Annealing. Analytical can include techniques such as Multi-Criteria Decision Making, as well as Analytic Hierarchy Process and Analytic Network Process solutions. Mixed Integer Programming is the most applied technique, with 30.5% of articles propose LP and only about 7% deal with NLP [16,17]. In general, three types of decision variables are used: (i) strategic - with decisions on locations or capacities of warehouses and stores; (ii) tactical - refering to the allocation of resources or products and their planning; and (iii) operational - with decisions related to the size of the lots, quantities of stocks, among others [16]. SCM capacity and strategic purchasing and supply are less frequent [1], although of great relevance. Large number of the quantitative models are single-objective (87.6%) and only a small are multiobjective (12.4%) [16]. The most studied sectors are, Electronic (17.6%) and Energy (16.5%), while Retail (3%) and Transport (2%) are the least studied [15]. In fact, only a few studies are based on empirical research [2].

3   Empirical Study

The company under study is recognized worldwide as one of the best food distribution groups, having marked the Portuguese market with some of the most important innovations in the sector. Currently, the group is present in 16 countries and its main vision is to improve the purchasing power and quality of life of largest number of clients. In Portugal, the group has three distinct activities, namely management and construction of shopping centers, banking activity and the management of the chain of hypermarkets. The present work focus in the last activity with emphasis on purchasing. Purchase manager is responsible for the processes of acquisition and supply of products throughout the SC. Its mission is to ensure customer satisfaction with products that match their needs, purchased at the best prices, delivered to sales channels and to suppliers offering the best service. SC has several stakeholders, not only the consumer, the sales channels, the purchasing or sales direction and the suppliers, but also the transport and logistics platforms. Logistics platforms are essential for the entire process. The objective is to supply the sales channels with the lowest operating cost. It incorporates activities such as the reception, storage, preparation and dispatch, being present in different parts of the country and grouped taking into account the typology of the products. Therefore, since the purchasing direction represents the entrance in the SC and its decisions must be based on the different SC domains, the company proposed a challenge whose objective is to optimize the decision-making in the purchases.

The company provided information concerning two suppliers: I351, with 21 products and once a week deliveries, and P940, with six products and twice a week deliveries. These suppliers were selected by the company, since they present different lots dimensions and storage modes, and therefore represent a greater challenge in the decision-making of the purchasing managers. The products can be ordered in FC[1] or pallet, the number of units ranges from 6 to 36 in the case of FC and from 396 to 1980 in the case of pallets. Furthermore PBL or XD storage may be used for delivery in FC and PBS on FC or pallets. There is also information of: the unit price of the item ex-supplier (**Exw**); the value of the pallet (**Pal**), which decreases as the number of pallets to be ordered increases, and the value of a pallet is also added to the FC; the rent of the warehouse (**Rent**) that varies according to the size of the lots, with smaller value for pallet. The requests from the platform can only be performed in FC. Finally, the decision process should also consider the costs related to: picking (**Pick**), that varies according to the storage mode of the product, i.e., lower in XD and higher in PBS; transportation (**Trans**); and replacement (**Repl**). The monthly sales historical information will be used as future demand, however the developed model is easily adjusted for different sales forecast, such as time series techniques.

August to November present a larger variety of products sold, while from December to July fewer products are sold (Fig. 1 (left)). The quantity of I351 products is much higher than the ones from P940, in fact I351 supplies 21 products, while P940 only six (Fig. 1 (right)). Furthermore the number of units from P940 is approximately constant over the months, while I351 shows variation. The monthly sales suggest that there are months with similar behavior. Hierarchical cluster analysis, with standardized sales, Ward method and collinear distance was used to detect homogeneous groups in the data. The results show that the months can be grouped in three clusters: (**1**) September, October and November; (**2**) December and, (**3**) the remaining months. Based on this analysis we will start by considering instance problems extracted from the data provided by the company for one representative month of each group. The analysis was performed in R version 3.3.3. [18].

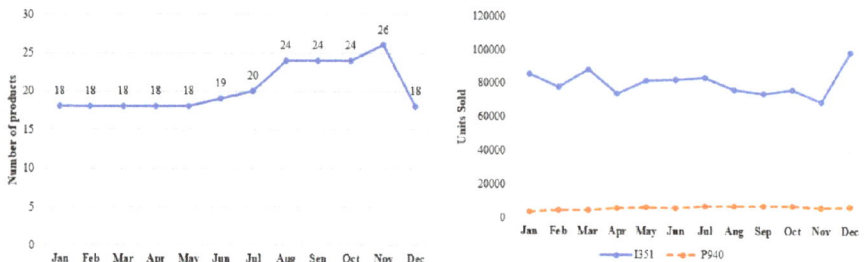

Fig. 1. Quantity of products sold (left) and units sold by supplier (right)

4 Preliminary Model Formulation

In future, a MILP model will be develop. Here we present some insights for that model. Indices and sets are denoted by: $i \in Products$ - products; i_{I351} and

---

[1] Purchase factor - quantity of products to the unit present in the order logistics unit.

i_{P940} - products of I351 and P940, respectively; $s \in \{XD, PBL, PBS\}$ - storage type. The parameters are: D_i - demand; Exw_i - unit cost of product i; Pal^P - unit cost of pallet; $Rent^{s'}$ - unit cost of warehouse rent for FC (s'=XD/PBL) or pallet (s'=PBS); $Pick^s$, $Trans$, $Repl$ - unit cost of picking, transportation and replacement, respectively; f_i and p_i - number of units per FC and per pallet, respectively; Q_i - maximum number of FC/pallet that the supplier can provide.

The company seeks to minimize the SC costs, therefore the decision should allow purchasing managers to determine the number of products to purchase and the storage mode. Decision variables are: $x_i^{s''}$, number of FC of product i with storage $s'' \in \{XD, PBL\}$; y_i, number of pallet of product i with storage PBS. Also there are some obligations of ordering for some suppliers, thus we define z_j as a binary variable which is 1 if at least one product of supplier j is order. With the above we define the objective function as:

$$Cost = \sum_i \left(C_i^{Exw} + C_i^{Pal} + C_i^{Pick} + C_i^{Trans} + C_i^{Repl} + C_i^{Rent} \right), \qquad (1)$$

where $C_i^{Exw} = Exw_i \left(f_i \sum_{s''} x_i^{s''} + p_i y_i \right)$; $C_i^{Pal} = Pal_i^{1P} f_i \sum_{s''} x_i^{s''} + Pal_i^P p_i y_i$; $C_i^{Pick} = \sum_{s''} Pick_i^{s''} x_i^{s''} + Pick_i^{PBS} p_i y_i$; $C_i^{Tans} = Trans_i \left(f_i \sum_{s''} x_i^{s''} + p_i y_i \right)$; $C_i^{Repl} = Repl_i \left(f_i \sum_{s''} x_i^{s''} + p_i y_i \right)$; $C_i^{Rent} = Rent_i^{s''} f_i \sum_{s''} x_i^{s''} + Rent_i^{PBS} p_i y_i$.

The model will include constrains, such as: customers' demand satisfaction; maximum and/or minimum number of units to be ordered; frequency products' delivery; and, product's maximum stock. The model will be implemented using the AMPL modelling language [19]. The numerical resolution of the real instances will be solved using Gurobi [20].

5   Conclusions

The constant changes in the business environment made essential to group all the company's activities into a continuous system [17]. In this context, SC management and planning has been viewed as a source of competitive advantage, gaining increasing relevance [3]. SCM is a complex network, involving planning in four domains: procurement, warehousing, distribution and sales [14]. For this reason, quantitative models have emerged as support for decision making, ensuring effective and strategically well-planned management.

This article intends to make a literature review on the subject and present an empirical study, with real data, of a Portuguese retail company. The objective of the company was the development of quantitative models that will improve decision-making in the company's purchasing management. The preliminary data analysis here presented suggests although there is variability between the monthly sales, some months present similar behavior. By cluster analysis three groups were found: (1) September, October and November; (2) December; and, (3) the remaining months.

In the future, based the data analysis, several real instance extracted from the data will be used for the validation of the MILP, for which we present here a preliminary formulation.

References

1. Gold, S., Seuring, S., Beske, P.: Sustainable supply chain management and inter-organizational resources: a literature review. Corp. Soc. Responsib. Environ. Manag. **17**(4), 230–245 (2010)
2. Seuring, S.: A review of modeling approaches for sustainable supply chain management. Decis. Support Syst. **54**(4), 1513–1520 (2013)
3. Barbosa-Póvoa, A.: As cadeias de abastecimento e a sustentabilidade. Boletim APDIO **55**, 5–9 (2016)
4. Stadtler, H.: Supply chain management and advanced planning—basics, overview and challenges. Eur. J. Oper. Res. **163**(3), 575–588 (2005)
5. Mehmeti, G.: A literature review on supply chain management evolution. In: Economic and Social Development: Book of Proceedings, p. 482 (2016)
6. Lummus, R.R., Krumwiede, D.W., Vokurka, R..I.: The relationship of logistics to supply chain management: developing a common industry definition. Ind. Manag. Data Syst. **101**(8), 426–432 (2001)
7. Stock, J.R., Boyer, S.L.: Developing a consensus definition of supply chain management: a qualitative study. Int. J. Phys. Distrib. Logist. **39**(8), 690–711 (2009)
8. Hou, H., Chaudhry, S., Chen, Y., Hu, M.: Physical distribution, logistics, supply chain management, and the material flow theory: a historical perspective. Inf. Technol. Manag. **18**(2), 107–117 (2017)
9. CSCMP. Supply chain management - terms and glossary (2013)
10. Hübner, A.H., Kuhn, H., Sternbeck, M.G.: Demand and supply chain planning in grocery retail: an operations planning framework. Int. J. Retail Distrib. Manag. **41**(7), 512–530 (2013)
11. Joyce, W.B.: Accounting, purchasing and supply chain management. Supply Chain Manag. **11**(3), 202–207 (2006)
12. Scheuermann, A., Leukel, J.: Task ontology for supply chain planning-a literature review. Int. J. Comput. Integr. Manuf. **27**(8), 719–732 (2014)
13. Fleischmann, B., Meyr, H.: Planning hierarchy, modeling and advanced planning systems. In: Handbooks in Operations Research and Management Science, vol. 11, pp. 455–523 (2003)
14. Fleischmann, B., Meyr, H., Wagner, M.: Advanced planning. In: Supply Chain Management and Advanced Planning, pp. 81–106. Springer (2008)
15. Brandenburg, M., Govindan, K., Sarkis, J., Seuring, S.: Quantitative models for sustainable supply chain management: developments and directions. Eur. J. Oper. Res. **233**(2), 299–312 (2014)
16. Govindan, K., Soleimani, H., Kannan, D.: Reverse logistics and closed-loop supply chain: a comprehensive review to explore the future. Eur. J. Oper. Res. **240**(3), 603–626 (2015)
17. Mansouri, S.A., Gallear, D., Askariazad, M.H.: Decision support for build-to-order supply chain management through multiobjective optimization. Int. J. Prod. Econ. **135**(1), 24–36 (2012)
18. R Core Team: R: A Language and Environment for Statistical Computing. R Foundation for Statistical Computing, Vienna (2016)
19. Fourer, R., Gay, D., Kernighan, B.: AMPL: A Modeling Language for Mathematical Programming, vol. 117. Boyd & Fraser, Danvers (1993)
20. Gurobi Optimization: Gurobi optimizer 5.0. Gurobi (2013). http://www.gurobi.com

One Dimensional Trim Loss Decision Support Tool

Luís Ferreirinha[1], Sara Baptista[1], Ângela Pereira[1],
Luís Pinto Ferreira[1], and Maria Teresa Pereira[2(✉)]

[1] ISEP - School of Engineering, Polytechnic of Porto, Porto, Portugal
l.ferreirinha96@gmail.com, sara.raquel.mb@gmail.com,
gimirra@hotmail.com, lpf@isep.ipp.pt
[2] Research Center of Mechanical Engineering (CIDEM)/ISEP - School
of Engineering, Polytechnic of Porto, Porto, Portugal
mtp@isep.ipp.pt

Abstract. The topic Cutting & Packing (C & P) has been published in many papers, and its importance within many industrial sectors is undeniable. For these reasons, several studies have been carried out on the optimization of the cutting process in order to improve the utilization of the useful area or the useful length of the raw material. Thus, this work aims to develop a computational tool that resolves the problem of minimizing losses in one-dimensional cuts. This problem is often encountered in industrial processes, where the beginning of the transformation and production of the raw material corresponds to the cutting process. To do so, the cutting problem will be formulated, and the Open Solver supplement will be used to obtain the solution.

Keywords: Cutting & packing · Trim loss · Linear programming
One-Dimensional cutting

1 Introduction

Cutting & Packing problems are optimization problems that consist of finding the best way to cut larger units (raw material) into smaller units (item) in order to satisfy the demand and optimize some criterion, for example, to minimize the loss generated by the cutting patterns. Cut patterns describe ways of cutting the raw material to obtain various items [1].

This type of problem appears in the most diverse types of industries, where the raw material usually corresponds to steel bars, paper and aluminum reels, metal and wood plates, printed circuit boards, glass sheets, fiberglass, leather, etc. [2, 3].

This paper presents a decision support tool that aims to respond to the problems of minimizing waste in one-dimensional cuts. Consider the following parameters: number of orders, number of raw materials in stock, width and demand order, number of cuts, and wastage. As for the decision variables, we defined a binary variable that decides whether a reel is used or not, as well as a variable that decides the quantity of an item in a certain raw material.

© Springer International Publishing AG, part of Springer Nature 2019
J. Machado et al. (Eds.): HELIX 2018, LNEE 505, pp. 793–799, 2019.
https://doi.org/10.1007/978-3-319-91334-6_108

The remaining sections of this paper are organized as follows: in Sect. 2 the Cutting & Packing problems are approached. The general formulation is presented in Sect. 3. The tool developed is presented in Sect. 4. The computational study is shown in Sect. 5. In Sect. 6 is where the paper finally presents some conclusions and provides some ideas for future works.

2 Industrial Cutting Problems

The C&P study covers a wide range of theoretical and, mainly, practical problems in many areas of human knowledge, such as Engineering, Administration, Economics, Mathematics, Operational Research, Informatics, among others [4].

Briefly, the main objective of this type of problem is to search for and obtain a geometrically efficient arrangement of figures, that is, smaller areas inserted in a larger area. Within this view, the C&P study interprets a geometrically efficient array that generates the least amount of raw material waste or greater yield of use of the original raw material [1].

The work published by Gilmore and Gomory can be considered the most important in the literature to solve the problem of cutting, and are still constantly referenced [5, 6]. In the scientific literature, the problems of C & P are denominations of problems that aim to optimize a certain combination of items. There are, however, several types of problems that contain this same logical structure among them [4, 7]:

- Trim Loss Problem;
- Bin Packing;
- Knapsack Problem;
- Vehicle Loading;
- Container and Pallet Loading;
- Layout Problems;
- Nesting and Partitioning Problems;
- Budgeting Capital Problem;
- Assembly Line Balancing;
- Memory Allocation Problems;
- Multiprocessor Scheduling Problems.

Although all these types of problems have the same basic logical structure, the addition of constraints, computational techniques, and algorithms to solve a specific problem, to the base formulation, makes each of these types of problems unique [4, 7].

In the flat product industries such as paper, cloth, leather, plastics and laminates, the problems related to raw material waste reach disturbing levels, which has led to the study of solutions to reduce the amount of wasted material in the cutting process [3, 4].

The raw material is usually supplied in rolls or sheets of different widths, this raw material has to be cut into several pieces of smaller widths and different values. The arrangement of the cuts is, most of the time, performed by practical methods that do not employ an optimization method. These procedures generate unacceptable amounts of leftover raw material and this directly influences the final cost of production. The reduction of this waste can be solved as an optimization problem and solved from the use of linear mathematical programming techniques [4, 8, 9].

Thus, the solution to this type of problem is to determine what should be the best combination and also by the determination of the cut sequencing, so that the minimization of raw material wastes is satisfied [4, 10, 11].

3 General Formulation

At this point, the general formulation incorporated in the tool is presented in order to solve problems of minimization of waste in one-dimensional cuts [2, 12, 13].

Parameters

n Number of orders;
m Number of raw material in stock;
l_j Order width j;
v_j Order search j;
L_i Raw material width i;
F_i Value that limits the number of cuts in the raw material;
P_i Part not used in the raw material i.

DV

$$x_{ij} = \begin{cases} k, \text{if the quantity k of item j is attributed to the raw material i} \\ 0, \text{otherwise} \end{cases}$$

$$y_i = \begin{cases} 1, \text{if the raw material i is used} \\ 0, \text{otherwise} \end{cases}$$

OF

$$Minz : f(x) = \sum_{i=1}^{m} P_i + \sum_{i=1}^{m} y_i \tag{1}$$

ST

$$\sum_{j=1}^{n} l_j x_{ij} \leq L_i y_i, i \in M \tag{2}$$

$$0 \leq P_i \leq \min_{j=1,\dots,n}\{l_j\} \quad \text{where } P_i = y_i L_i - \sum_{j=1}^{n} l_j x_{ij}, i \in M \tag{3}$$

$$\sum_{i=1}^{m} x_{ij} \geq v_j, j \in N \tag{4}$$

$$y_i = 0 \text{ ou } 1, i \in M \tag{5}$$

$$x_{ij} = 0 \, ou \, k, i \in M, j \in N \tag{6}$$

$$\sum_{j=1}^{n} x_{ij} \leq F_i, i \in M \tag{7}$$

$$N = \{1, 2, \ldots, n\}, M = \{1, 2, \ldots, m\} \tag{8}$$

The objective function (1) consists in minimizing the function f (x), that is, minimizing waste as well as minimizing the number of raw materials used, resulting in a set of cutting patterns where loss and number of used raw materials, is minimal. The restriction (2) ensures that the various cutting patterns created have a width less than the width of the existing raw materials in stock. The constraint (3) requires that the losses have values between zero and the lowest ordering width, this ensures that the losses never exceed any of the ordered widths, since if this happened the pattern was incomplete. Restriction (4) ensures that all demand is met. The constraint (5) requires that the use of the raw materials is a binary variable, where the value zero means that raw material i is not used, and one that the raw material i is used. The constraint (6) requires that the quantity of a given item in a particular raw material is zero if it does not exist, or k if exists, where k is a positive and integer number. The restriction (7) ensures that the number of items in a raw material is less than the value that limits the number of cuts in the raw material. The constraints (8) ensure that the number of orders and number of raw materials in stock is an integer and a positive number.

4   Tool Description

In this section of the paper will be presented the tool, developed in Excel, that solves the model considering the simulated real context.

4.1   Orders

In an initial phase, an Excel sheet was developed, entitled "Orders", where the data related to the order date, the respective quantity and measure of the order will be inserted. In order to facilitate the validation of the tool, a random order generator was developed which specifies the value of the order values, the number of orders to be generated and the date of those orders. At a later stage of this document, the resolution of 10 randomly generated instances will be applied and the results obtained will be compared.

When inserted, or generated, all the desired orders, the table is filled and goes to the sheet "Model".

4.2   Optimization Model

In this sheet, the user must select the date (s) he wants to optimize from the existing list. Once the orders to optimize have been selected, the user must enter the data related to the quantity and length of the raw material and the limit of cuts per raw material, if there is no cut-off limit, a very high value should be assumed, for example, 99999.

By entering the raw material length value, the tool automatically calculates the minimum raw material quantity to satisfy orders, so that the user must enter a quantity of raw material equal to or higher than that value. After inserting the data into the model, the tool will adjust it so that it has the rows and columns needed to solve the problem. Given the dimensions of the variables, it is recommended to use the Open Solver supplement to solve the problem.

5   Computational Study

At this point, in order to validate the operation of the application, 10 random instances will be generated through the random order generator and the results obtained in each instance will be compared.

5.1   Data

With the objective of validating the operation of the application, 10 random instants were generated, considering that the raw material has a width of 1000 mm and that the measures of the orders vary uniformly between 150 mm and 550 mm. For this example, it was decided to generate order quantities with values between 1 and 15. The data obtained are shown in Table 1.

Table 1. Randomly generated data

Group 1		Group 2		Group 3		Group 4		Group 5	
Quantity	Measure (mm)	Quantity	Measure (mm)	Quantity	Measure (mm)	Quantity	Measure (mm)	Quantity	Measure (mm)
6	438	14	519	6	295	11	193	1	250
1	396	3	293	1	475	2	489	10	272
11	289	7	225	9	514	10	521	14	448
11	515	10	284	11	197	1	232	1	445
5	341	9	349	13	170	3	300	1	207
Group 6		Group 7		Group 8		Group 9		Group 10	
Quantity	Measure (mm)	Quantity	Measure (mm)	Quantity	Measure (mm)	Quantity	Measure (mm)	Quantity	Measure (mm)
12	378	7	323	3	459	14	446	2	427
8	339	12	192	1	436	10	240	3	385
11	397	8	550	10	359	13	219	5	393
3	442	11	501	11	431	6	321	15	464
3	512	3	380	6	206	4	318	8	520

5.2   Results

After all values were entered in the model, the solutions presented in the graphs of Fig. 1 were obtained.

From the analysis of Fig. 1 it is possible to conclude that 15, 18, 12, 11, 11, 19, 19, 13, 16 and 17 units of raw material with 1000 mm each are necessary to satisfy orders, and a waste of 1427, 2074, 457, 1093, 562, 4011, 2232, 231, 348 e 1386 mm is

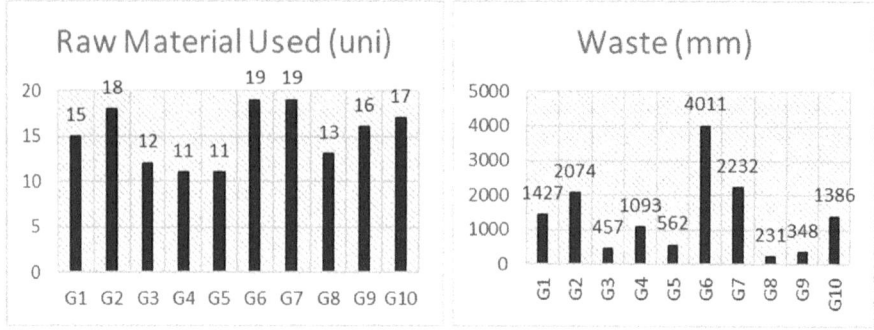

Fig. 1. Obtained results

assumed for each data group, G1, G2, G3, G4, G5, G6, G7, G8, G9, and G10, respectively. Sometimes the waste can be high since the developed model considers as waste all items with dimensions smaller than the smallest measure of the orders.

It is also stated that in G1 no stocks were generated and for G2 the stock of 1 unit of 225 mm is assumed. In G3, there is a need to assume 1 unit of 295 mm, as well as in G4 where there are 2 units of 232 mm. In G5 the stock of 2 units of 272 mm is assumed, as for G6, the stock of 1 unit of 512 mm is assumed. It is also stated that in the G7, 6 units of 192 mm of stock were generated and the G8 assumes the stock of 5 units of 206 mm and 1 unit of 359 mm. Finally, in the G9 is assumed the stock of 3 unit of 321 mm, as for the G10 is assumed the stock of 1 unit of 520 mm.

6   Conclusions

Industry and organizations have always sought better and greater profitability from their production processes. With the globalization process underway, the world market has become even more competitive, requiring cost reductions, better levels of productivity and less waste. In the flat product industries, such as paper, cloth, leather, plastic and laminate, the problems related to raw material wastes are of concern; which has led to the study of solutions to reduce the amount of waste material in the cutting process. Thus, as a reflection of the current industrial requirements, the development of decision support tools is crucial.

The solution of the problem of minimization of wastes in unidimensional cuts proposed began with a generic formulation that allows solving problems of this type. In this formulation are enumerated the parameters, the decision variables, objective function and, finally, the associated constraints. Subsequently, a simple and intuitive, user-oriented tool was developed that allows the resolution of problems that fit into the aforementioned formulation.

To validate the operation of the application, 10 instances of randomized orders were generated in certain values. The results obtained demonstrate the effectiveness of the developed tool since the solutions that minimized the waste were obtained for all data groups.

As future work, it is intended that the tool considers the stocks occurred in the cutting process, as well as its application in an industrial environment.

Acknowledgments. We acknowledge the financial support of CIDEM- Research Center of Mechanical Engineering, funded by FCT – Portuguese Foundation for the Development of Science and Technology, Ministry of Science, Technology and Higher Education, under the Project UID/EMS/0615/2016.

References

1. Martinovic, J., Scheithauer, G., de Carvalho, J.V.: A comparative study of the arcflow model and the one-cut model for one-dimensional cutting stock problems. Eur. J. Oper. Res. **266** (2), 458–471 (2018)
2. Pileggi, G.C.F.: Abordagens para otimização integrada dos problemas de geração e sequen-ciamento de padrões de corte. Doctoral dissertation, Universidade de São Paulo (2002)
3. Pinto, M.J.: Algumas contribuições à resolução do problema de corte integrado ao problema de sequenciamento dos padrões. Doctoral dissertation, São José dos Campos (2004)
4. Cunha, R.R.: Um algoritmo de minimização de sobras em corte unidimensional. Masters dissertation, Universidade Federal de Santa Catarina (1998)
5. Gilmore, P.C., Gomory, R.E.: A linear programming approach to the cutting-stock pro-blem. Oper. Res. **9**(6), 849–859 (1961)
6. Gilmore, P.C., Gomory, R.E.: A linear programming approach to the cutting stock problem —Part II. Oper. Res. **11**(6), 863–888 (1963)
7. Leao, A.A., Furlan, M.M., Toledo, F.M.: Decomposition methods for the lot-sizing and cutting-stock problems in paper industries. Appl. Math. Model. **48**, 250–268 (2017)
8. Golden, B.L.: Approaches to the cutting stock problem. AIIE Trans. **8**(2), 265–274 (1976)
9. Erjavec, J., Gradisar, M., Trkman, P.: Assessment of stock size to minimize cutting stock production costs. Int. J. Prod. Econ. **135**(1), 170–176 (2012)
10. Cugo, A.P., Furtado, J.C.: Otimização do problema de corte unidimensional na indústria usando algoritmos genéticos. Masters dissertation, Centro Universitário Fransiscano (2001)
11. Dyckhoff, H., Finke, U.: Cutting and packing in production and distribution: a typology and bibliography. Springer, Heidelberg (1992)
12. Júnior, J.A.S., Pinheiro, P.R., Thomaz, A.C.F.: Otimização das perdas em cortes guilhotinados para bobinas de aço na indústria metalmecânica. Sociedade Brasileira de Pesquisa Operacional, São João del Rei (2004)
13. Delorme, M., Iori, M., Martello, S.: Bin packing and cutting stock problems: mathematical models and exact algorithms. Eur. J. Oper. Res. **255**(1), 1–20 (2016)

Asymptotic Properties of the Spectra of a Strongly Regular Graph

Luís Vieira[1,2,3](✉)

[1] Faculty of Engineering of Porto, Section of Mathematics, Porto, Portugal
`lvieira@fe.up.pt`
[2] University of Porto, Porto, Portugal
[3] Center of Mathematics of University of Porto,
Street Dr Roberto Frias, 0351, 4200-465 Porto, Portugal
`https://sigarra.up.pt/feup/pt/func_geral.formview?p_codigo=211359`

Abstract. Let G be a strongly regular graph with three distinct eigenvalues and A his matrix of adjacency. In this work we associate a three dimensional real Euclidean Jordan algebra V with rank three to A and next we consider a Jordan frame B of idempotents of V. Next we analyse the spectra of a particular convergent Hadamard series of A^2 and establish asymptotic inequalities over the spectra and the parameters of G.

Keywords: Euclidean Jordan algebras · Graph theory
Strongly regular graphs

1 Introduction

Euclidean Jordan algebras had many applications on various branches of Mathematics, namely on the developing theory for interior-point methods [1–3], on the formalism of quantum mechanics [4] and on combinatorics and statistics [5–10].

In this work we establish asymptotic inequalities on the spectra and on the parameters of a strongly regular graph in the environment of Euclidean Jordan algebras.

This paper is organized as follows. In Sect. 2 we expose the principal results on Euclidean Jordan algebras. Next in Sect. 3 we present the principal concepts on strongly regular graphs necessary for a clear exposition of this paper. Finally in Sect. 4 we associate a three dimensional real Euclidean Jordan algebra $\mathcal{A}$ to a strongly regular graph and next we establish asymptotic inequalities on the spectra and on the parameters of a strongly regular graph, see respectively (5) and (6) of Theorems 3 and 4 respectively.

2 A Short Introduction to Euclidean Jordan Algebras

In this section relevant concepts about Euclidean Jordan algebras, which can be seen, for instance in [11,12], are shortly surveyed. Let $\mathcal{A}$ be a real vector space

© Springer International Publishing AG, part of Springer Nature 2019
J. Machado et al. (Eds.): HELIX 2018, LNEE 505, pp. 800–804, 2019.
https://doi.org/10.1007/978-3-319-91334-6_109

with finite dimension and a bilinear mapping $(u, v) \mapsto u \circ v$ from $\mathcal{A} \times \mathcal{A}$ to $\mathcal{A}$, such that for each $u \in \mathcal{A}$ the algebra spanned by u is associative. Then, $\mathcal{A}$ is called a real *power associative* algebra. If $\mathcal{A}$ contains an element, e, such that for all u in $\mathcal{A}$, $e \circ u = u \circ e = u$, then e is called the *unit* element of $\mathcal{A}$. Considering a bilinear mapping $(u, v) \mapsto u \circ v$, if for all u and v in $\mathcal{A}$ we have (J_1) $u \circ v = v \circ u$ and (J_2) $u \circ (u^2 \circ v) = u^2 \circ (u \circ v)$, with $u^2 = u \circ u$, then $\mathcal{A}$ is called a *Jordan algebra*. If $\mathcal{A}$ is a Jordan algebra with unit element, then $\mathcal{A}$ is power associative (cf. [11]). Given a Jordan algebra $\mathcal{A}$ with unit element e, if there is an inner product $< \cdot, \cdot >$ that verifies the equality $< u \circ v, w > = < v, u \circ w >$, for any u, v, w in $\mathcal{A}$, then $\mathcal{A}$ is called an *Euclidean Jordan algebra*. An element c in an Euclidean Jordan algebra $\mathcal{A}$, with unit element e, is an *idempotent* if $c^2 = c$. Two idempotents c and d are *orthogonal* if $c \circ d = 0$. We call the set $\{c_1, c_2, \ldots, c_k\}$ a *complete system of orthogonal idempotents* if (i) $c_i^2 = c_i, \forall i \in \{1, \ldots, k\}$; (ii) $c_i \circ c_j = 0, \forall i \neq j$ and (iii) $c_1 + c_2 + \cdots + c_k = e$.

Let $\mathcal{A}$ be an Euclidean Jordan algebra with unit element e. Then, for every u in $\mathcal{A}$, there are unique distinct real numbers $\lambda_1, \lambda_2, \ldots, \lambda_k$, and an unique complete system of orthogonal idempotents $\{c_1, c_2, \ldots, c_k\}$ such that

$$u = \lambda_1 c_1 + \lambda_2 c_2 + \cdots + \lambda_k c_k, \tag{1}$$

with $c_j \in \mathbb{R}[u]$, $j = 1, \ldots, k$ (see [11], Theorem III.1.1). These λ_j's are the eigenvalues of u and (1) is called the *first spectral decomposition* of u.

The *rank* of an element u in $\mathcal{A}$ is the least natural number k, such that the set $\{e, u, \ldots, u^k\}$ is linearly dependent (where $u^k = u \circ u^{k-1}$), and we write $\mathrm{rank}(u) = k$. This concept is expanded by defining the rank of the algebra $\mathcal{A}$ as the natural number $\mathrm{rank}(\mathcal{A}) = \max\{\mathrm{rank}(u) : u \in \mathcal{A}\}$. The elements of $\mathcal{A}$ with rank equal to the rank of $\mathcal{A}$ are the *regular* elements of $\mathcal{A}$. This set of the regular elements is an open and dense subset in $\mathcal{A}$. If u is a regular element of $\mathcal{A}$, with $r = \mathrm{rank}(u)$, then the set $\{e, u, u^2, \ldots, u^r\}$ is linearly dependent and the set $\{e, u, u^2, \ldots, u^{r-1}\}$ is linearly independent. Thus we may conclude that there exist unique real numbers $a_1(u), \ldots, a_r(u)$, such that $u^r - a_1(u)u^{r-1} + \cdots + (-1)^r a_r(u)e = 0$, where 0 is the null vector of $\mathcal{A}$. Making the necessary adjustments we obtain the polynomial in λ

$$p(u, \lambda) = \lambda^r - a_1(u)\lambda^{r-1} + \cdots + (-1)^r a_r(u), \tag{2}$$

that is called the *characteristic polynomial* of u, where each coefficient a_i is a homogeneous polynomial of degree i in the coordinates of u in a fixed basis of $\mathcal{A}$. Although the characteristic polynomial is defined for a regular element of $\mathcal{A}$, we can extend this definition to all the elements of $\mathcal{A}$, since each polynomial a_i is homogeneous and the set of regular elements of $\mathcal{A}$ is dense in $\mathcal{A}$. The roots of the characteristic polynomial of u, $\lambda_1, \lambda_2, \ldots, \lambda_r$, are called the eigenvalues of u. Furthermore, the coefficients $a_1(u)$ and $a_r(u)$ of the characteristic polynomial of u, are called the *trace* and the *determinant* of u, respectively.

Let $\mathcal{A}$ be a real Euclidean Jordan algebra with unit element $\mathbf{e}$. An element c in $\mathcal{A}$ is an idempotent if $c^2 = c$. Two idempotents c and d are orthogonal if $c \bullet d = 0$. The set $\{c_1, c_2, \ldots, c_l\}$ is a complete system of orthogonal idempotents if the following three conditions hold: (i) $c_i^2 = c_i$, for $i = 1, \ldots, l$, (ii) $c_i \bullet c_j = 0$, if $i \neq j$, and (iii) $\sum_{i=1}^{l} c_i = \mathbf{e}$. An idempotent c is primitive if it is a nonzero idempotent of $\mathcal{A}$ and if it can't be written as a sum of two non-zero idempotents. We say that $\{c_1, c_2, \ldots, c_k\}$ is a Jordan frame if $\{c_1, c_2, \ldots, c_k\}$ is a complete system of orthogonal idempotents such that each idempotent is primitive.

Theorem 1 *([11], p. 43). Let $\mathcal{A}$ be a real Euclidean Jordan algebra. Then for u in $\mathcal{A}$ there exist unique real numbers $\lambda_1, \lambda_2, \ldots, \lambda_k$, all distinct, and a unique complete system of orthogonal idempotents $\{c_1, c_2, \ldots, c_k\}$ such that*

$$u = \lambda_1 c_1 + \lambda_2 c_2 + \cdots + \lambda_k c_k. \tag{3}$$

The numbers λ_j's of (3) are the eigenvalues of u and the decomposition (3) is the first spectral decomposition of u.

Theorem 2 *([11], p. 44). Let $\mathcal{A}$ be a real Euclidean Jordan algebra with* $rank(\mathcal{A}) = r$. *Then for each u in $\mathcal{A}$ there exists a Jordan frame $\{c_1, c_2, \cdots, c_r\}$ and real numbers $\lambda_1, \cdots, \lambda_{r-1}$ and λ_r such that*

$$u = \lambda_1 c_1 + \lambda_2 c_2 + \cdots + \lambda_r c_r. \tag{4}$$

The decomposition (4) is called the second spectral decomposition of u.

Now let consider a n-finite dimensional real Euclidean Jordan algebra $\mathcal{A}$ such that $rank(\mathcal{A}) = n$. Let x be a regular element of $\mathcal{A}$. Then, there exists a unique Jordan frame $S = \{c_1, c_2, \cdots, c_n\}$ such that $x = \sum_{i=1}^{n} \lambda_i c_i$. But since any two elements of S are orthogonal then S is a linear independent set of $\mathcal{A}$ and therefore S is a basis of $\mathcal{A}$. The fact that an n-dimensional Euclidean Jordan $\mathcal{A}$ with $rank(\mathcal{A}) = n$ has a basis that is a Jordan frame is important for the establishment of the asymptotical inequalities (5) and (6) of Theorems 3 and 4 respectively presented in the Sect. 4.

3   Some Notions on Strongly Regular Graphs

Along this paper we consider only non-empty, simple and not complete graphs. By simple graphs we mean graphs without loops and parallel edges. Strongly regular graphs were firstly introduced by R. C. Bose in the paper [13]. Let G be a graph of order n. We denote its vertices set by $V(G)$ and its edge set by $E(G)$. An edge whose endpoints are u and v is denoted by uv and, in such case, the vertices u and v are adjacent or neighbors. The numbers of vertices of G, $|V(G)|$, is called the order of G. If all vertices of G have k neighbors, then G is called a k-regular graph. G is called a (n, k, λ, μ) strongly regular graph if is k-regular and any pair of adjacent vertices have λ common neighbors and any pair of non-adjacent vertices have μ common adjacent vertices.

Let G be a (n, k, λ, μ) strongly regular graph. The adjacency matrix of G, $A = [a_{ij}]$, is a binary matrix of order n such that $a_{ij} = 1$, if the vertex i is adjacent to j and 0 otherwise. The adjacency matrix of G satisfies the equation $A^2 = kI_n + \lambda A + \mu(J_n - A - I_n)$, where J_n is the all one matrix of order n.

It is well known (see, for instance, [14]) that the eigenvalues of G are k, θ and τ, where θ and τ are given by $\theta = (\lambda - \mu + \sqrt{(\lambda - \mu)^2 + 4(k - \mu)})/2$ and $\tau = (\lambda - \mu - \sqrt{(\lambda - \mu)^2 + 4(k - \mu)})/2$, (see [14]).

4 Asymptotical Inequalities Associated to a Strongly Regular Graph

Let G be a (n, k, λ, μ)-strongly regular graph with $0 < \mu < k < n - 1$ and A be its adjacency matrix with the distinct eigenvalues, namely k, θ and τ. Herein, k and θ are the positive eigenvalues and τ is the negative eigenvalue. Now we associate a three dimensional real Euclidean Jordan algebra with rank three. We consider the Euclidean Jordan algebra $\mathrm{Sym}(n, \mathbb{R})$ with the Jordan product $u \bullet v = \frac{uv + vu}{2}$ and with the inner product $< u, v >= \mathrm{tr}(uv)$, where uv and vu are the usual product of matrices u and v and the usual product v and u. Let $\mathcal{A}$ be the Euclidean Jordan subalgebra of $\mathrm{Sym}(n, \mathbb{R})$ spanned by I_n and the natural powers of A. Since A has three distinct eigenvalues, then $\mathcal{A}$ is a three dimensional real Euclidean Jordan algebra with $\mathrm{rank}(\mathcal{A}) = 3$. Let $\mathcal{B} = \{E_1, E_2, E_3\}$ be the unique complete system of orthogonal idempotents of $\mathcal{A}$ associated to A, with $E_1 = 1/nI_n + 1/nA + 1/n(J_n - A - I_n)$, $E_2 = (|\tau|n + \tau - k)/(n(\theta - \tau))I_n + (n + \tau - k)/(n(\theta - \tau))A + (\tau - k)/(n(\theta - \tau))(J_n - A - I_n)$, and $E_3 = (\theta n + k - \theta)/(n(\theta - \tau))I_n + (-n + k - \theta)/(n(\theta - \tau))A + (k - \theta)/(n(\theta - \tau))(J_n - A - I_n)$.

Now we present some notation. $M_n(\mathbb{R})$ we denote the set of square matrices of order n with real entries. For $B = [b_{ij}]$, $C = [c_{ij}]$ in $M_n(\mathbb{R})$, we denote by $B \circ C = [b_{ij} c_{ij}]$ the Hadamard product of matrices B and C and by $B \otimes C = [b_{ij} C]$ the Kronecker product of matrices B and C. For any nonnegative integer number k and for any matrix $B \in M_n(\mathbb{R})$ we define $B^{\circ k}$ in the following way: $B^{\circ 0} = J_n, B^{\circ 1} = B$ for $k \geq 2$ we define $B^{\circ k} = B^{\circ(k-1)} \circ B$ (see [15]).

Let x be a real positive number, and let consider the binomial Hadamard series $S_x = \sum_{j=0}^{+\infty}(-1)^j \binom{-x}{j} \left(\frac{A^{\circ 2}}{k^2 + \mu}\right)^{\circ j}$. Now $S_x = \sum_{i=1}^{3} q_x^i E_i$ is the spectral decompositions of S_x respectively to the Jordan frame $\mathcal{B}$ of $\mathcal{A}$. By an asymptotical analysis of the eigenvalues q_x^2 and q_x^3 we establish the inequalities (5) and (6) of Theorem 3 and of Theorem 4 respectively.

Theorem 3. *Let G be a (n, k, λ, μ)-strongly regular graph with $0 < \mu < k < n - 1$ and with the distinct eigenvalues k, θ and τ. If $\mu > \lambda$ then*

$$\left(\frac{k + \mu - \lambda}{k}\right)^r \leq \frac{k}{\mu}. \tag{5}$$

Theorem 4. *Let G be a (n, k, λ, μ)-strongly regular graph with $0 < \mu < k < n - 1$ and with the distinct eigenvalues k, θ and τ. If $\lambda > \mu$ then*

$$\left(\frac{k}{k + \mu - \lambda}\right)^{|s|} \leq \frac{k}{\mu}. \tag{6}$$

References

1. Cardoso, D.M., Vieira, L.A.: On the optimal parameter of a self-concordant barrier over a symmetric cone. Eur. J. Oper. Res. **169**, 1148–1157 (2006)
2. Faybusovich, L.: Linear systems in Jordan algebras and primal-dual interior-point algorithms. J. Comput. Appl. Math. **86**, 149–175 (1997)
3. Faybusovich, L.: Euclidean Jordan Algebras and Interior-point Algorithms. Positivity **1**, 331–357 (1997)
4. Jordan, P., Neuman, J.V., Wigner, E.: On an algebraic generalization of the quantum mechanical formalism. Ann. Math. **35**, 29–64 (1934)
5. Mano, V.M., Vieira, L.A.: Admissibility conditions and asymptotic behaviour of strongly regular graph. Int. J. Math. Models Methods Appl. Sci. **5**(6), 1027–1033 (2011)
6. Mano, V.M., Martins, E.A., Vieira, L.A.: On generalized binomial series and strongly regular graphs. Proyecciones J. Math. **4**, 393–408 (2013)
7. Mano, V.M., Vieira, L.A.: Alternating schur series and necessary conditions for the existence of strongly graphs. Int. J. Math. Models Methods Appl. Sci. Methods. **8**, 256–261 (2014)
8. Massam, H., Neher, E.: Estimation and testing for lattice condicional independence models on Euclidean Jordan algebras. Ann. Stat. **26**, 1051–1082 (1998)
9. Vieira, L.A.: Euclidean Jordan algebras and inequalities on the parameters of a strongly regular graph. In: AIP Conference Proceedings, vol. 1168, pp. 995–998 (2009)
10. Vieira, L.A., Mano, V.M.: Generalized Krein parameters of a strongly regular graph. Appl. Math. **6**, 37–45 (2015)
11. Faraut, J., Korányi, A.: Analysis on Symmetric Cones. Oxford Mathematical Monographs. Clarendon Press, Oxford (1994)
12. Koecher, M.: The Minnesota Notes on Jordan Algebras and Their Applications. Springer, Berlin (1999)
13. Bose, R.C.: Strongly regular graphs, partial geometries and partially balanced designs. Pac. J. Math. **13**, 389–419 (1963)
14. Godsil, C., Royle, G.: Algebraic Graph Theory. Chapman & Hall, New York (1993)
15. Horn, R.A., Johnson, C.R.: Topics in Matrix Analysis. Cambridge University Press, Cambridge (1991)

Inverse Box-Cox and the Power Normal Distribution

Rui Gonçalves[1,2]([✉])

[1] LIAAD INESC TEC, Porto, Portugal
[2] Universidade do Porto, Faculdade de Engenharia,
R. Dr. Roberto Frias s/n, 4200-465 Porto, Portugal
rjasg@fe.up.pt
https://sigarra.up.pt/feup/en

Abstract. In this paper we consider the power-normal (PN) family of distributions. This family is generated by inverting the Box-Cox [1] power transformation. If Y is a left truncated normal (TN) random variable then the variable $X = (\lambda Y + 1)^{1/\lambda}$ has a PN distribution with parameters μ and σ. We study the case where $0 < \lambda < 1$. We obtain a formula for the rth ordinary moment of the power normal distribution. We examine the bivariate power normal distribution and we calculate the marginal and conditional distributions. We give a formula for the correlation curve and we provide a numerical illustration.

Keywords: Box-Cox transformation · Power-normal distribution
Correlation curve

1 Introduction

The PN distribution is a parametric class of distributions that includes the truncated normal and the lognormal. Let X be a random variable taking only non-negative values. Using the Box-Cox transformation [1] we define Y as,

$$Y = \begin{cases} \frac{X^\lambda - 1}{\lambda}, \ \lambda > 0 \\ \ln(X), \ \lambda = 0 \end{cases}$$

This transformation is widely used, for example, in time series modelling, to stabilize variance. Although widely use, very few authors have investigated the Inverse Box-Cox scale or PN distribution. The PN family of distributions was first introduced by Goto and Inoue [2]. In that paper, the shape of the PN density is studied and an expression for the mean in terms of infinite series of gamma functions is obtained. Taylor [3] gives an estimator for the mean of Y when X proceeds from a symmetric family of distributions. More recently, Freeman and Modarres [4] calculate the mean and variance of the PN for various values of λ. They also compute the quantile functions and a quantile measure of skewness for the PN distribution.

© Springer International Publishing AG, part of Springer Nature 2019
J. Machado et al. (Eds.): HELIX 2018, LNEE 505, pp. 805–810, 2019.
https://doi.org/10.1007/978-3-319-91334-6_110

Due to its nature, the PN distribution is only defined for positive values. The consequence of this restriction on X is that the transformed variable Y has to be equal or greater than $-1/\lambda$ and therefore Y has a truncated normal (TN) instead of a normal distribution. However, the issue of truncation may be sidestepped under certain conditions. For instance, using large means (μ) and considering the case $\lambda > 0$, the $P(Y < -1/\lambda)$ can be very small so that Y is practically normal distributed. Nevertheless, in this work, we will use a valid approach for any value of μ so we use a TN distributed random variable instead of normal distributed.

So X is said to have a PN distribution, $X \approx PN(\lambda, \mu, \sigma)$, if Y has a TN distribution, $Y \approx TN(\lambda, \mu, \sigma)$. Let $0 < \lambda < 1$ and, let ϕ be the standard normal pdf and Φ the standard normal cumulative distribution function (cdf). The PN's pdf is given by,

$$f_X(x) = \frac{x^{\lambda-1}}{\sigma A(\lambda, \mu, \sigma)} \phi\left(\frac{x^{(\lambda)} - \mu}{\sigma}\right), x > 0. \tag{1}$$

where $x^{(\lambda)} = \frac{x^{\lambda}-1}{\lambda}$. Using the same notation of [5] we define,

$$A(\lambda, \mu, \sigma) = \begin{cases} \Phi(k) & , \lambda > 0 \\ 1 & , \lambda = 0 \\ \Phi(-k) & , \lambda < 0 \end{cases}$$

where k is the standardized truncation point of the TN distribution, $k = (\lambda\mu + 1)/(\lambda\sigma)$. In the next section we give a formula for the ordinary moments of order r for the PN distribution. In Sect. 3, we present the bivariate PN distribution and its marginal densities that turned out not to have a univariate PN distribution. In Sect. 4 we present the conditional moments and the correlation curve with a numerical illustration.

2   Moments of the PN Distribution

Let X be a PN distributed random variable, the ordinary moment of order r for the PN distribution, $E(X^r)$ is equal to the expected value of the function $(\lambda Y + 1)^{r/\lambda}$ and is given by,

$$
\begin{aligned}
E[(\lambda Y + 1)^{r/\lambda}] &= \frac{1}{A(\lambda, \mu, \sigma)\sigma\sqrt{2\pi}} \int_{-1/\lambda}^{+\infty} (\lambda y + 1)^{r/\lambda} \exp\left[-\frac{1}{2}\left(\frac{y-\mu}{\sigma}\right)^2\right] dy \\
&= \frac{1}{A(\lambda, \mu, \sigma)\sigma} \int_{-1/\lambda}^{+\infty} (\lambda y + 1)^{r/\lambda} \phi\left(\frac{y-\mu}{\sigma}\right) dy \\
&= \frac{1}{A(\lambda, \mu, \sigma)\sigma} \int_{-1/\lambda}^{+\infty} \sum_{n=0}^{\infty} \binom{r/\lambda}{n} (\lambda y)^n \phi\left(\frac{y-\mu}{\sigma}\right) dy \\
&= \frac{1}{A(\lambda, \mu, \sigma)} \int_{-k}^{+\infty} \sum_{n=0}^{\infty} \lambda^n \binom{r/\lambda}{n} (\sigma z + \mu)^n \phi(z) dz
\end{aligned}
$$

$$= \frac{1}{A(\lambda,\mu,\sigma)} \sum_{n=0}^{\infty} \binom{r/\lambda}{n} (\lambda\mu)^n \int_{-k}^{+\infty} \left(\frac{\sigma}{\mu}z + 1\right)^n \phi(z)\, dz$$

$$= \frac{1}{A(\lambda,\mu,\sigma)} \sum_{n=0}^{\infty} \binom{r/\lambda}{n} (\lambda\mu)^n \int_{-k}^{+\infty} \sum_{i=0}^{n} \binom{n}{i} \left(\frac{\sigma}{\mu}z\right)^i \phi(z)dz$$

$$= \frac{1}{A(\lambda,\mu,\sigma)} \sum_{n=0}^{\infty} \binom{r/\lambda}{n} (\lambda\mu)^n \left\{ \sum_{i=0}^{n} \binom{n}{i} \left(\frac{\sigma}{\mu}\right)^i \int_{-k}^{+\infty} z^i \phi(z)dz \right\}.$$

We used the substitution $z = \frac{y-\mu}{\sigma}$ so that $dz = \frac{dy}{\sigma}$. Note that $k = \frac{1+\mu\lambda}{\mu\sigma}$. We can see that this calculation has to be done numerically.

3   The Bivariate Power Normal

In order to extend the PN for two dimensions consider the random vector $X = (X_1, X_2)$ where $X_1, X_2 > 0$. We will follow the notation of [5]. Let $Y = (Y_1, Y_2)$ be the power transformed vector of X. The bivariate truncated normal (BTN) vector Y has parameters $\mu = (\mu_1, \mu_2)$ and $\sigma = (\sigma_1, \sigma_2)$ and the variance-covariance matrix Σ is given by,

$$\Sigma = \begin{pmatrix} \sigma_1^2 & \rho\sigma_1\sigma_2 \\ \rho\sigma_1\sigma_2 & \sigma_2^2 \end{pmatrix}$$

where ρ is the correlation parameter between Y_1 and Y_2. We say that the pair (X_1, X_2) has a bivariate PN distribution (BPN) if the joint pdf is,

$$g(x_1, x_2) = \frac{x_1^{\lambda_1 - 1} x_2^{\lambda_2 - 1}}{A(\lambda, \mu, \Sigma)} f(x_1^{(\lambda_1)}, x_2^{(\lambda_2)}) \tag{2}$$

for $x_1, x_2 > 0$. We will study the case where $0 < \lambda_1, \lambda_2 < 1$.

The function f is the BTN pdf,

$$f(y_1, y_2) = \frac{1}{2\pi\sigma_1\sigma_2\sqrt{1-\rho^2}} \exp\left\{ -\frac{Q(y_1, y_2)}{2} \right\}$$

where,

$$Q(y_1, y_2) = \frac{1}{1-\rho^2} \left\{ \left(\frac{y_1 - \mu_1}{\sigma_1}\right)^2 - 2\rho \left(\frac{y_1 - \mu_1}{\sigma_1}\right) \left(\frac{y_2 - \mu_2}{\sigma_2}\right) + \left(\frac{y_2 - \mu_2}{\sigma_2}\right)^2 \right\}.$$

The truncated proportional constant term $A(\lambda, \mu, \sigma)$ depends on the truncation values for X_1, $-1/\lambda_1$ and X_2, $-1/\lambda_2$. Assuming that $0 < \lambda_1 < 1$ and $0 < \lambda_2 < 1$ using Table 1 of [5],

$$A(\lambda, \mu, \sigma) = \int_{-k_2}^{+\infty} \int_{-k_1}^{+\infty} \phi_2(x_1, x_2) dx_1 dx_2$$

where ϕ_2 is the bivariate standard normal (but in this case with a not necessarily unit correlation ρ between X_1 and X_2) so,

$$\phi_2(x_1, x_2) = \frac{1}{2\pi\sqrt{1-\rho^2}} \exp\left\{-\frac{x_1^2 - 2\rho x_1 x_2 + x_2^2}{2(1-\rho^2)}\right\}$$

and the standardized truncation points k_1 and k_2 are,

$$k_j = \frac{\lambda_j \mu_j + 1}{\lambda_j \sigma_j}, \quad j = 1, 2.$$

As it occurs in the univariate case, the bivariate case presents different shapes according to different values of the parameters λ_1 and λ_2, [5].

Next we calculate both marginal and conditional densities associated with the BPN distribution. Let $g_j(x_j)$ be the univariate PN pdf for each variable X_j, $j = 1, 2$,

$$g(x_j : \lambda_j, \mu_j, \sigma_j) = \frac{x_j}{\sigma_j A(\lambda_j, \mu_j, \sigma_j)} \phi\left(\frac{x_j^{\lambda_j} - \mu_j}{\sigma_j}\right) \tag{3}$$

Integrating the BTN pdf in x_2 we obtain the marginal pdf of variable x_1,

$$g_1(x_1) = \frac{x_1^{\lambda_1 - 1}}{\sigma_1 A(\lambda, \mu, \Sigma)} \phi\left(\frac{x_1^{(\lambda_1)} - \mu_1}{\sigma_1}\right) \Phi\left[\frac{1}{\sqrt{1-\rho^2}}\left(\rho \frac{x_1^{(\lambda_1)} - \mu_1}{\sigma_1} + k_2\right)\right] \tag{4}$$

Integrating the BTN pdf in x_1 we obtain the marginal pdf of variable x_2,

$$g_2(x_2) = \frac{x_2^{\lambda_2 - 1}}{\sigma_2 A(\lambda, \mu, \Sigma)} \phi\left(\frac{x_2^{(\lambda_2)} - \mu_2}{\sigma_2}\right) \Phi\left[\frac{1}{\sqrt{1-\rho^2}}\left(\rho \frac{x_2^{(\lambda_2)} - \mu_2}{\sigma_2} + k_1\right)\right] \tag{5}$$

As we can see, the marginal pdf's are not PN univariate distributed. This is in agreement with corresponding functions of the TN distribution. The conditional pdf associated to the BPN pdf is,

$$g_{x_2|x_1}(x_2) = \frac{g(x_1, x_2)}{g_1(x_1)} = \frac{x_2^{\lambda_2 - 1} f(x_2^{(\lambda_2)} | x_1^{(\lambda_1)})}{\Phi\left[\frac{1}{\sqrt{1-\rho^2}}\left(\rho\frac{x_1^{(\lambda_1)} - \mu_1}{\sigma_1} + k_2\right)\right]} \tag{6}$$

where,

$$f(x_2^{(\lambda_2)} | x_1^{(\lambda_1)}) = \frac{1}{\sqrt{2\pi}\sigma_2\sqrt{1-\rho^2}} \exp\left[-\frac{1}{2\sigma_2^2(1-\rho^2)}\left\{x_2^{(\lambda_2)} - \mu_2 - \rho\frac{\sigma_2}{\sigma_1}(x_1^{(\lambda_1)} - \mu_1)\right\}^2\right]. \tag{7}$$

Here we note the marginal pdfs for the BPN distribution are not univariate PN. The same is valid for the corresponding marginals.

We notice that the conditional pdf associated to the BPN pdf has not univariate PN distribution.

4   Conditional Moments and Correlation Curve

The correlation curve [6] of X_2 given X_1 is the curve with coordinates $(x_1, E[X_2|X_1 = x_1])$. This curve gives us a description of the dependency of a variable on the other. The expected value X_2 given X_1 is,

$$E[X_2|X_1] = \frac{\int_0^\infty x_2^{\lambda_2} f(x_2^{(\lambda_2)}|x_1^{(\lambda_1)})dx_2}{\Phi\left[\frac{1}{\sqrt{1-\rho^2}}\left(\rho\frac{x_1^{(\lambda_1)}-\mu_1}{\sigma_1} + k_2\right)\right]} \tag{8}$$

In Fig. 1, we give a representation of the correlation curve for a pair of PN random variables with parameters, $\mu_1 = 2.5$, $\mu_2 = 1.2$, $\sigma_1 = 1$, $\sigma_2 = 0.5$ and for several values of λ from 0.4 to 0.75. The correlation parameter is $\rho = 0.95$.

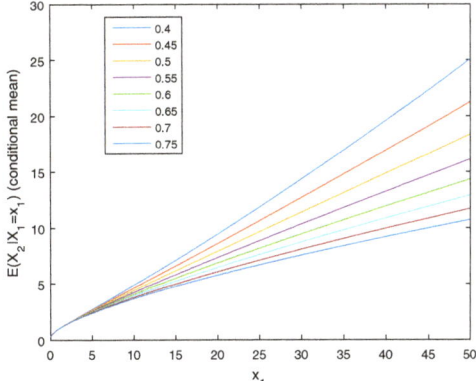

Fig. 1. Conditional mean $E(X_2|X_1 = x_1)$ for different values of λ from 0.4 to 0.75.

In Fig. 2, we give a representation of the conditional variance for x_1 for various values of λ.

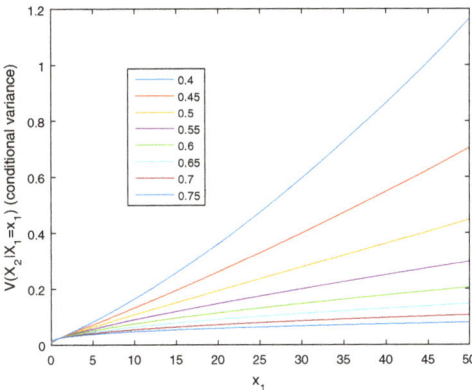

Fig. 2. Conditional variance $V(X_2|X_1 = x_1)$ for different values of λ from 0.4 to 0.75.

5 Conclusions Final Remarks and Future Work

The PN family of distributions is an important family of distributions because of the extended use of the Box-Cox transformation. In the first section we presented the PN family of distributions, the one that can be found by inverse Box-Cox transformation. We have given its pdf and we selected for analysis the case $0 < \lambda < 1$.

In Sect. 2 we gave a formula to calculate the moments of the PN distribution. Since that it is not possible to have a closed form to the primitive this calculation has to be done numerically. In Sect. 3 we presented the bivariate PN distribution and its marginal and conditional densities. We noted that both marginal and conditional distributions are not univariate PN. This is agreement with the BTN distribution i.e. the distribution of the transformed variable. In Sect. 4 we presented the conditional moments of the bivariate PN pdf. We also presented for an example the graphic representation of the correlation curve and of the conditional standard deviation.

Future work on this topic will involve a research on the effect of the Box-Cox transformation in correlated data.

References

1. Box, G.E.P., Cox, D.R.: An analysis of transformations. J. Roy. Statist. Soc. Ser. B **26**, 211–243 (1964)
2. Goto, M., Inoue, T.: Some properties of power-normal distribution. J. Biom. **1**, 28–54 (1980)
3. Taylor, J.: The retransformed mean after a fitted power transformation. J. Am. Stat. Assoc. **81**(393), 114–118 (1986)
4. Freeman, J., Modarres, R.: Inverse Box-Cox: the power normal distribution. Stat. Probab. Lett. **76**, 764–772 (2006)
5. Goto, M., Hamasaki, T.: The bivariate power-normal distribution. Bulletin Inform. Cybern. **34**(1), 29–49 (2002)
6. Murteira, B.: Probabilidade e Estatística, vol. 1, pp. 28–55. MCGraw-Hill (1990)

How a CRM Tool Can Contribute to a Better Business Performance: The Case of a Shipping Company

Joana Fróis[1], M. Teresa Pereira[2], and Fernanda A. Ferreira[3(✉)]

[1] Polytechnic of Porto, 4200-072 Porto, Portugal
joaninhafrois@gmail.com
[2] Research Centre of Mechanical Engineering (CIDEM),
School of Engineering, Polytechnic of Porto, 4200-072 Porto, Portugal
mtp@isep.ipp.pt
[3] School of Hospitality and Tourism of Polytechnic Institute of Porto,
Applied Management Research Unit (UNIAG),
4480-876 Vila do Conde, Portugal
faf@esht.ipp.pt

Abstract. Globalization is revolutionizing our way of living and the way business companies work. Technology is the reason for it, because it provides solutions that help the workers perform more effectively and efficiently and that allow for better execution. The development of relationships with the customer gives companies a better knowledge and understanding of their needs and leads to new strategies that aim at better results for the customer. Customer Relationship Management (CRM) tools are taken as important tools for this, since they help analyzing the companies' Key Performance Indicators (KPIs). In this article we use the case of a shipping company in order to understand how such systems can contribute to such improvement. By analyzing the company's CRM we suggest an adjustment to the business in order to get a better performance from it.

Keywords: CRM · Shipping · Customer service · Performance

1 Introduction

The maritime transport has grown worldwide and is, nowadays, the only way of transport capable to keep up with the growth in road transport. It is also more environmental friendly and, therefore, it is considered as an alternative to road transport [1]. In fact, most of the world's trade is carried out by sea [2]. Maritime transportation also contributes for the globalization of supply chains and to the development of peripheral areas (because it allows door-to-door deliveries) because it is relatively cheap, when compared to other modes, due to the large capacity of ships [3].

In the 1960's, when containers appeared, no one would guess the impact they would have in commercial trade worldwide. Container adoption led to a revolution in transport: not only did it allow an easier handling of the merchandise, the goods could now be moved using any kind of transport: road, rail and sea, or a combination of these

© Springer International Publishing AG, part of Springer Nature 2019
J. Machado et al. (Eds.): HELIX 2018, LNEE 505, pp. 811–817, 2019.
https://doi.org/10.1007/978-3-319-91334-6_111

with no need to reorganize the cargo [2]. In the shipping industry this change also meant an increase in the number of companies and an increase of their fleet and capacity. At the same time, because not all ports can receive these new bigger vessels, a new service arose: the feeder service, performed by smaller vessels that deliver the goods by the smaller ports, assuring, then, a door to door service.

2 Literature Review

2.1 Customer Relationship Management (CRM)

In the search of competitive advantage, companies are looking for the newest technologies, the latest advances that will give them what they need to be better. Among there there is a system that has been more and more used in companies since the 1990s: the Customer Relationship Management (CRM). With origins in both the United States and in Northern Europe, it is viewed differently by the companies: for some it is merely a technological tool, for others it is a key part of the business [4]. These different point of views result from the several meanings of the concept itself that varies from author to author. These are some of the definitions found among the literature: CRM is about acquiring customers, knowing them well, providing services and anticipating their needs [4]; *CRM is making the organisation customer centric, using new techniques and technologies, and making the customer an integral part of the organisation. (...) CRM is about getting to know your clients better through the use of technology (...)* [5]; *CRM is a synthesis of many existing principles from relationship marketing and the broader issue of customer-focused management* [6]; *CRM is a process/application that permits organisations to gather and analyse customer data rapidly while seeking to improve customer retention and profitability via targeted products and services* [7].

Despite this diversity, there is a consensus regarding the customer and its importance: they are the essence of every business and, therefore, they should be the focus of every organization. Because CRM tools allow the collection of client information, assuring that every worker has access to it and that information is not lost [6], and because knowledge of the customer means competitive advantage, such systems become valuable assets. However, technology should be seen as a tool and not as a simple solution, not only because the adoption of a new system does not necessarily mean success [8] but also because *simply throwing software or technology at an organisation for CRM implementation will cause more problems than solutions* [5].

In the literature we identify three different types of CRM: operational CRM (customer oriented), analytical CRM (collects and analyses data) and collaborative CRM (integrates all points of contact between company and customer) [4, 9, 10]. We also find these referred to as a continuum perspective [11], because such tools must be adjusted to the business and have to grow up in order to unleash all their potential. CRM tools are, then, ongoing processes that ambition to provide the customer with the best possible experience and that should incorporate all the workers of an organization, because it is that integration that will lead to success [12].

2.2   CRM, Customer Service and Business Performance

A company's goal is to provide unique value [13]. In order to achieve that, organizations look for new ways to distinguish themselves from their competitors, and they do that by improving customer service, because that is how differentiation is achieved [14]. Because *customers create sales and the most successful companies are those that win the most customers and keep them* [14], customer service gains special importance, in a time when it is fairly easy for products or services to be replaced by those of a competitor. The customer's choice is no longer based only on price and features of the product, but on the quality of the service provided by the company [14]. It is also important to ensure that the product gets to the customers, because if it is not available the product has no value. Availability can, thus, be described as the factor that makes customer service [14]. In what the shipping industry is concerned we may consider vessel frequency, transit time and equipment and space availability as key factors.

CRM systems help companies gather information that has to be analyzed to improve customer service. However, it is not enough for an organization to have a CRM strategy and a CRM tool installed. It is absolutely vital to find ways of measuring and monitoring not only the system's but also the company's performance [15].

[T]*he use of CRM applications is (...) associated with (...) improved customer satisfaction* [16] and satisfied customers affect the companies' performance. According to the 80/20 Pareto rule, the customers that create more value represent a small group of the organization's customer portfolio, companies need to identify these customers to adjust their marketing strategies that will hopefully *lead to better firm performance* [17]. Thus, performance measurement is essential to help organizations manage their resources and effectively control their goals [18].

Companies need, then, to measure business performance and to understand if the CRM tool is helping them getting better results. Companies should, therefore, *measure organizational performance through customer satisfaction, efficiency (profitability), and market effectiveness* [19]. These can be considered any company's Key Performance Indicators (KPIs). One of the most common ways to measure KPIs is the Balance Score Card (BSC), *a famous performance measurement concept (...) that considers both financial and non-financial aspects and further divides performance measurement into the four perspectives of financial, customer, internal process, and learning and growth* [18]. However, regardless of the method used, the most important is that the CRM strategy focuses on the customers and on the creation of lasting and loyal relationships with them [18].

2.3   CRM and the Shipping Industry

Any company's goal is to deliver the right product, to the right customer, on the right quantity, in the right conditions, to the right place and at the right price. That is also true for a company whose core business is the delivery of goods so, it is vital that organizations quickly meet their customers' desires in order to improve their satisfaction and loyalty [18]. However, in such an industry the human factor cannot be forgotten [20]. Although a CRM technology may be desired and helpful it should not replace the previously established relationships, based on direct contact [21]. The technology has to be understood as a way to improve relationships and to help the company achieve its goals, it should be *an assistant in creating satisfaction for the shipper but not the sole source of satisfaction* [21]. In spite of all the widely recognized advantages of CRM tools and despite the fact that the CRM strategy is to improve the company's performance [18], the reality demonstrates that *most carriers and logistic providers have yet to launch a formal CRM program* [22], being the lack of software adapted to the transportation business the main reason found.

3 Case Study

3.1 The Company

In order to carry out this research we looked at an international company of the shipping industry, that has in the maritime transportation of goods (in containers, bulk, energy, and vehicles, among others) its main business area. Around the globe, this company has more than 700 employees and more than 500 vessels able to carry daily over 40 million tons of goods. In Portugal, the company has two offices, one in Lisbon and another in Porto. The office located in Porto, where the research took place, has 17 workers who use several information systems on a daily basis. One of the most important ones is the vessel management system, through which employees communicate with several external entities, through which containers are booked and in which information regarding customers is introduced.

Although there is a CRM system in the company's offices in Portugal, installed with the goal of keeping and analysing valuable data (regarding clients and their relationship with the company), we realised that it is hardly used. The communication with the customer is still mostly made by email or by phone.

3.2 The Company's CRM and Key Performance Indicators (KPIs)

The CRM tool was installed in Portugal in 2015 with the aim of keeping and working customer data (related to the request, offer and acceptance or non-acceptance of values) that up until then was lost. By using this tool to send quotations to the customers, this system would not only be able to feed a database of customer information but would also provide the company with strategic information difficult to obtain otherwise.

However, the system was not designed or adjusted to fit the company's business. It was established by the information technology (IT) department of a logistics and transport company and resulted of the adaptation of a tool used for the road sector. Despite some adjustments made, the workers of the sales and customer service departments consider that the system is not ready to deal with the specificities of the shipping industry: it is impossible to send more than one quote at a time (even if the only change is in the place of delivery) and it does not tolerate the introduction of several places of delivery (when the contents of one single container are to be delivered to different places). This, the lack of training of the workers, the non-existence of an instructions' or users' guide, the response time of the system and the constant unavailability of the IT department led to a gradual abandonment of the tool. To this also contributed the fact that instead of facilitating the employees' work, the use of the CRM system meant more work (the same information had to be entered not only in the vessel management system and in the company's system, but also in the CRM).

One of the reasons for the installation of the CRM system came from the need to get information that could not be obtained until then, information that can be considered as the company's KPIs (best clients, most requested service, most loaded merchandise or most requested destination, among others). Nevertheless, the information provided by the system was not considered by the company as being the information needed because it resulted from information of the proposals sent to clients (customer, proposal's number and state, origin, port of origin, destination, port of destination, place of origin and of delivery, among others), instead of resulting from the actual signed contracts and performed services.

4 Analysis

Among other aspects, the usage of a CRM tool grants the company with relevant pieces of information, that allow for a better management of the relationship with the customers and better results overall. Such system is undoubtedly important for this company in order to obtain information that facilitates the relationships with the customers, thus guaranteeing their loyalty.

Unlike what is desired, this system is not accessible to all the employees of the office and it is, mainly, a system that gathers customer data. Therefore, it is considered an operational system. Its growth towards becoming an analytical system and a collaborative system would be appropriate [4, 9, 10, 11]. Another aspect that can be improved is the integration. The tool is not integrated with any other system used, but if it were, both the sales and customer service departments' work would be easier: by simply using the offer's number all the information inserted in the CRM proposal could be automatically associated to the booking created in the vessel management system. In fact, only such integration would allow the getting of key information regarding the KPIs.

Concerning the KPIs, the managers referred aspects such as total percentage of quotations accepted, percentage of quotations accepted per customer, percentage of quotations given per service, percentage of quotations given per port of origin and destiny, percentage of quotations given per commodity among others, however these aspects were not taken into account when the system was developed. The information the system provides is nothing but a summary of the proposals sent. However, if both the CRM and the vessel management system worked together such information could be easily obtained. In order to better help the company improving their business performance it is urgent that these aspects are changed and taken into account.

5 Conclusion

Technological systems are considered because they allow companies to improve their business performance [6]. As we have seen before, technology has everything in order to help companies improve their performance. However, to do so technology must be adjusted to the company's business and needs and not the other way around. Moreover, it needs to be presented, explained and accepted by all its workers, because every new system installed means a change in their way of working. Nevertheless, these changes aim to improve the company's performance instead of cutting with the past. The introduction of CRM tools does not only look for an improvement of the relationship with the customer, but also to an improvement of the company's knowledge on the customer itself and in what the customer means to it.

By analysing this company's system, we realise that it is a young system that has a lot of room to improve, because in spite of satisfying the company's initial requests it is not fitted to its specificities. In what the KPIs is concerned, the information obtained by this company is scarce and does not help it getting a better performance. However, if some changes are made (namely the proposal's summary items that can be conferred with no need of actually opening a proposal) that information can easily be achieved and, therefore, can be put up in use in order to generate more profit, that is to say; that if the company gets to know which customers load more containers, it can arrange for better prices or if the company knows what is the most requested service, it can find new marketing strategies in order to highlight the other services.

Despite the different vision each organization has regarding technology and its use, the important thing is that there is enough space for the system to improve and to better fit the company's needs and its customers' wishes. Having met our goals in analysing this system, we suggest that the company's managers meet with the IT department in order to clarify the important aspects of the business so that statistical information regarding its KPIs can be acquired through the use of such system.

It is true that companies that adopt CRM systems benefit from them [16], however the most important aspect to bear in mind is that, *[i]n the end, what matters more is not the display of technology, but how technology is put to practical use for delivering superior quality service* [21].

Acknowledgments. We acknowledge the financial support of CIDEM, R&D unit funded by the FCT – Portuguese Foundation for the Development of Science and Technology, Ministry of Science, Technology and Higher Education, under the Project UID/EMS/0615/2016 and UNIAG, R&D unit funded by the FCT – Portuguese Foundation for the Development of Science and Technology, Ministry of Science, Technology and Higher Education, under the Project UID/GES/04752/2016.

References

1. Douet, M., Cappuccilli, J.: A review of short sea shipping policy in the European Union. J. Transp. Geogr. **19**, 968–976 (2011)
2. Lee, C.Y., Song, D.P.: Ocean container transport in global supply chains: overview and research opportunities. Transp. Res. Part B **95**, 442–474 (2017)
3. Paixão, A., Marlow, P.: Strengths and weaknesses of short sea shipping. Mar. Policy **26**, 167–178 (2002)
4. Teo, T.S., Devadoss, P., Pan, S.L.: Towards a holistic perspective of customer relationship management (CRM) implementation: a case study of the housing and development board, Singapore. Decis. Support Syst. **42**(3), 1613–1627 (2006)

5. Boon, O., Corbitt, B., Parker, C.: Conceptualising the requirements of CRM from an organisational perspective: a review of the literature. In: Proceedings of the 7th Australian Workshop on Requirements Engineering, AWRE 2002, pp. 83–95. Deakin University (2002)
6. Hendricks, K.B., Singhal, V.R., Stratman, J.K.: The impact of enterprise systems on corporate performance: a study of ERP, SCM, and CRM system implementations. J. Oper. Manag. **25**(1), 65–82 (2007)
7. Payton, F.C., Zahay, D.: Understanding why marketing does not use the corporate data warehouse for CRM applications. J. Database Market. Cust. Strategy Manag. **10**(4), 315–326 (2003)
8. Ang, L., Buttle, F.: CRM software applications and business performance. J. Database Mark. Cust. Strategy Manag. **14**(1), 4–16 (2006)
9. Hayley, M.: A literature review on CRM – definitions, benefits, components, and implementation. Aust. J. Manag. Fin. Res. **1**(1), 26–34 (2016)
10. Greenberg, P.: CRM at the speed of light. Oborne/McGrawHill, Berkeley (2002)
11. Payne, A., Frow, P.: A strategic framework for customer relationship management. J. Mark. **69**(4), 167–176 (2005)
12. Chase, P. R.: Why CRM implementations fail and what to do about it. Scribe Software Corporation (2000)
13. Christopher, M.: Payne, A., Ballantyne, D.: Relationship marketing: Bringing quality customer service and marketing together (1991)
14. Christopher, M.: Logistics and Supply Chain Management, 2nd edn. Financial Times. Prentice Hall, Upper Saddle River (1998)
15. Smith, A.: CRM and customer service: strategic asset or corporate overhead? Handb. Bus. Strategy **7**(1), 87–93 (2006)
16. Mithas, S., Krishnan, M.S., Fornell, C.: Why do customer relationship management applications affect customer satisfaction? J. Mark. **69**(4), 201–209 (2005)
17. Ryals, L.: Making customer relationship management work: the measurement and profitable management of customer relationships. J. Mark. **69**(4), 252–261 (2005)
18. Wu, S.I., Lu, C.L.: The relationship between CRM, RM, and business performance: a study of the hotel industry in Taiwan. Int. J. Hosp. Manag. **31**(1), 276–285 (2012)
19. Chang, W., Park, J.E., Chary, S.: How does CRM technology transform into organizational performance? a mediating role of marketing capability. J. Bus. Res. **63**(8), 849–855 (2010)
20. Pereira, T., Fróis, J., Ferreira, F.A.: Analysis of a customer relationship management tool in a shipping company. In: Proceedings of the International Conference on Industrial Engineering and Operations Management, pp. 1–8. Bandung, Indonesia, March 6–8, 2018 (Forthcoming)
21. Durvasula, S., Lysonski, S., Mehta, S.C.: Technology and its CRM implications in the shipping industry. Int. J. Technol. Manag. **28**(1), 88–102 (2004)
22. Fakhredaei, N.: The factors affecting adoption of CRM at the organizational level in Iran's shipping industry (2007)

A Model for the Multi-depot Online Vehicle Routing Problem with Soft Deadlines

Álvaro Silva[1], Luís Pinto Ferreira[1], Maria Teresa Pereira[2(✉)], and Fábio Neves-Moreira[3]

[1] School of Engineering, Polytechnic of Porto, 4200-072 Porto, Portugal
[2] Research Center of Mechanical Engineering (CIDEM), School of Engineering, Polytechnic of Porto, 4200-072 Porto, Portugal
mtp@isep.ipp.pt
[3] Faculty of Engineering, University of Porto, 4200-465 Porto, Portugal

Abstract. In many companies in the automotive industry there are challenges in some key processes in their logistic departments, mainly in internal logistics. These challenges happen due to poorly defined rules for the transportation of goods, resulting in a great cost associated with the time lost in the process. Also, the optimization of these processes, incrementing the efficiency of internal logistics can bring competitive advantages to the companies. For that matter, this study was developed at a major tire manufacturing company and proposes a model for the optimization of in-bound logistics, viewed as an online vehicle routing problem with soft deadlines (OVRPSD), using multiple depots. The main goal of this study is the increase of efficiency in logistic, optimizing the number of vehicles to supply the machines in order to reduce the stopping time of machines due to the lack of tires to consume.

Keywords: Vehicle · Routing · Deadlines · Logistic

1 Introduction

Nowadays, one of the most important operation decisions faced by industrial companies is the transportation of goods to a set of customers, using a set of vehicles that travel routes with the minimum distance or time possible. This is the most basic definition of a capacitated vehicle routing problem (CVRP) and can be applied to real world problems in logistic systems or supply chains, with some modifications for each specific problem [1]. Therefore, and since this kind of problems require advanced planning methods to increase the efficiency in logistics, the proposed mathematical model has as an objective: the reduction of the stopping time of machines due to lack of tires to consume, satisfying the demands of the machines, without violating any restriction, for example regarding to the capacity of the vehicles. It should also be possible to optimize the number of vehicles needed to supply the machines.

This paper is divided into five sections: the first section is an introductory chapter; the second section presents a literature review on the vehicle routing problem (VRP) and its variant; the third section presents a description of the real problem for this study; The forth section present the mathematical model developed in this study for

© Springer International Publishing AG, part of Springer Nature 2019
J. Machado et al. (Eds.): HELIX 2018, LNEE 505, pp. 818–824, 2019.
https://doi.org/10.1007/978-3-319-91334-6_112

the real-world problem; the fifth section describes some considerations about future work and the approaches that will be used to solve the problem.

2 Literature Review

The vehicle routing problem can be described as the determination of a set of routes for the transportation of goods to a set of costumers, using a fleet of vehicles. Usually, the main goal is to find the routes with the minimal distance or minimal time, to reduce operational costs. The classic VRP is also a generalization of the Travelling Salesman Problem (TSP) considering the use of more than one salesman. Therefore, the VRP can also be seen as a multiple TSP (mTSP) [2].

The VRP is a very known class of combinatory optimization problems, and was introduced, for the first time, in the last year of 1950's by Dantzig and Ramser [3]. In this work, the authors proposed a mathematical formulation for their truck dispatch problem which described a real-world application for the delivery of gasoline to a set of costumers (service stations) [3, 4].

Since 1959, the VRP has been extensively studied in the optimization literature. Many papers, have been published on the VRP, like the ones by Toth and Vigo [5], Lacome et al. [6], Pearn [7], Longo et al. [8] Assad et al. [9], Chabot et al. [10], Chabot et al. [11], between others.

According to Farahni et al. [12] the most common variations of the classic VRP are: The Distance-Constrained and Capacitated VRP (DCVRP), the VRP with Time Windows (VRPTW), the VRP with Backhauls (VRPB), the VRP with Pickup and Delivery (VRPPD), the Open VRP (OVRP), the Multiple Depot VRP (MDVRP), the Mixed Fleet VRP (MFVRP), the Split Delivery VRP (SDVRP) and the Periodic VRP (PVRP) [10]. However, there is no mention of the Consistent vehicle routing problem (ConVRP) and the authors only focused on the off-line VRP, also excluding the On-line vehicle routing problem.

The Consistent VRP is an off-line variation of the classic VRP, present in companies that want the customers to be always visited by the same driver in the same route. This can be seen as an advantage, since consistency is granted. So, in this variation, the order to visit the customers (route) is static, the same vehicles travel the same routes and serve customers at the same time period each day [11].

Kovacs et al. [13] applied the consistent VRP using various objective functions to attain several objectives including: improving driver consistency or minimizing the cost of the routes. The authors solved instance with up to 199 customers and a schedule of 5 days using multi direction large neighborhood search and concluded that an increase in the cost of the travels improves the consistency by 70% and that arrival consistency and driver consistency could be improved at the same time [13].

The on-line VRP approach, according to Lipmann [14] is used when the information about the requests is not know from the start but is known over time. In this sense, the routes are calculated for a specific time interval and actions executed cannot be revoked. The status of the system is being updated constantly, in real-time, and the requests are server depending on the arrival order, meaning that the system can only complete one request at a time and has no information on how many requests will exist. The author considered the on-line routing for a TSP and for a dial-a-ride problem [14].

3   Problem Description

The challenge present in the mentioned company has two depots and the clients in this model correspond to the machines, which are grouped in lines, from line A till line T. However, from line M till line T, the number of machines per line are doubled, due to an extra line in series with each respective line. One line of machines can contain from 8 to 12 machines. Each machine has two cavities that consume tires at a known rate. The company has a total of 538 cavities, which means that there is a total of 269 machines.

In this challenge, the vehicles correspond to logistic trains, that can transport a maximum amount of 5 cars per vehicle, like the one represented in Fig. 1.

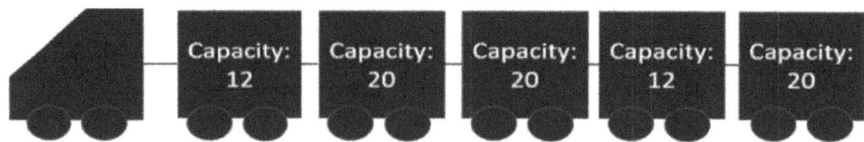

Fig. 1. Vehicle used for the transportation of tires

These cars have capacity of 12 or 20, depending on the size of the tires to transport, and with each car having only tires of the same type. Also, it is known that each vehicle must supply, approximately 30 machines, which can be translated as 3 lines of machines. In this sense, the route of a vehicle is similar to the one presented in Fig. 2.

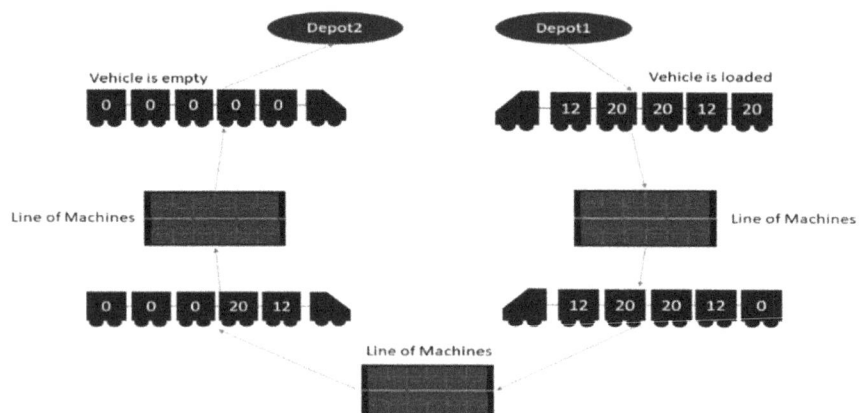

Fig. 2. Route of a vehicle

As described in Fig. 2, the vehicle departs from the depot, with all 5 cars loaded with tires, having some cars of capacity 12 and other of capacity 20. Next, the vehicle travels to the line of machines with the cavities he must supply. If the cavities are busy, the car with the type of tires that is being consumed is left near the machines and the vehicle proceeds to the next line, picking up empty cars to bring to the depot. This process is done until the vehicle is empty, which means the cars it carries are, also, empty. When this happens, the vehicle returns to the depot where it started from. The route ends there and starts again, the same way as described. This is done for all the nine vehicles used for the transportation of tires to the cavities in the machines.

4 Mathematical Formulation

The problem was modeled on a graph G = (N, A) where $N = \{1, 2, 3. . ., n\}$ is the set of nodes in the problem and $A \{(i,j)| i,j \in N \ and \ i \neq j\}$ is the set of arcs. The nodes are the cavities. Also, let there be: $K = \{1, 2, 3. . ., 9\}$ the set of vehicles, T the set of types of cavities and O the set of depots.

In turn, the parameters used in the problem are the following:

- c_{ij} - Distance, in meters, to travel the arc with origin in i and destination in j
- t_{ij} - Time to travel the arc with origin in i and destination in j
- p_i - Penalization at the cavity in node i
- f_i - Consumption rate of the cavity in node i
- s_i - Initial Stock for consumption at the cavity in node i
- d_i - Deadline to arrive at the cavity in node i
- v_i - Type of car consumed at the cavity in node i
- cap_i - Capacity of the cavity in node i
- st_i – Service time at node i
- l_h - Capacity of the car of type h
- b - Maximum number of cars allowed per vehicle
- M - Constant of high value that assures that the restrictions are respected

For the parameters presented, the indexes i and j must be different at all times, in order to exist a path between nodes. As for the decision variables in this model, they are defined as:

$$x_{ij}^k = \begin{cases} 1, & \textit{if the arc that start in i and ends in j is traveled by the vehicle k} \\ 0, & \textit{if the arc that start in i and ends in j is not traveled by the vehicle k} \end{cases}$$

$$z_i^k = \begin{cases} 1, & \textit{if node i is visited by vehicle k} \\ 0, & \textit{if node i is not visited by vehicle k} \end{cases}$$

$$r_i^k = \begin{cases} 1, & \textit{if vehicle k starts at depot i} \\ 0, & \textit{if vehicle k doesn't start at depot i} \end{cases}$$

$$y_h^k = \textit{Number of cars of type h in vehicle k}$$

$$w_{ki} = \textit{Arrival time of vehicle k to the cavity in node i}$$

$$\phi_i = \textit{Delay time in the arrival to the cavity in node i}$$

$$q_{ki} = \textit{Amount of tires to be delivered by vehicle k in the cavity of node i}$$

$$u_i = \textit{Lost production at the cavity in node i}$$

The objective of this model, which is classified as an OVRPSD, is to optimize the number of vehicles needed and to reduce the stopping of the machines caused by lack of tires to consume.

$$min \sum_{k \in K} \sum_{i \in N} \sum_{j \in N} c_{ij} x_{ij}^k + \sum_{i \in N} p_i \phi_i + \sum_{i \in N} u_i \tag{1}$$

s.t.

$$\sum_{j \in N} x_{ij}^k = z_i^k, \qquad \forall i \in N, k \in K \tag{2}$$

$$\sum_{j \in N} x_{ji}^k = z_i^k, \qquad \forall i \in N, k \in K \tag{3}$$

$$\sum_{j \in N} x_{ij}^k = \sum_{j \in N} x_{ji}^k, \qquad \forall i \in N, k \in K \tag{4}$$

$$\sum_{j \in N} x_{ij}^k \leq r_i^k, \qquad \forall i \in O, k \in K \tag{5}$$

$$\sum_{j \in O} x_{ij}^k = 1, \qquad \forall i \in N, k \in K \tag{6}$$

$$w_{ki} + t_{ij} + st_i \leq w_{kj} + M.\left(1 - x_{ij}^k\right), \qquad \forall (i,j) \in A, k \in K \tag{7}$$

$$w_{ki} + t_{ij} + st_i \geq w_{kj} - M.\left(1 - x_{ij}^k\right), \qquad \forall (i,j) \in A, k \in K \tag{8}$$

$$w_{ki} \leq d_i + \phi_i, \qquad \forall i \in N, k \in K, \tag{9}$$

$$q_{ki} = f_i.w_{ki} + u_i + cap_i - s_i, \qquad \forall i \in N, k \in K \tag{10}$$

$$q_{ki} \leq z_i^k.cap_i, \qquad \forall i \in N, k \in K \tag{11}$$

$$\sum_{i \in N:l=v_i} q_{ki} \le l_h \cdot y_k^h, \qquad \forall h \in T, k \in K \tag{12}$$

$$\sum_{h \in T} y_h^k \le b, \qquad \forall k \in K \tag{13}$$

$$x_{ij}^k, z_i^k, r_i^k \in \{0,1\}; \quad y_k^h \in \mathbb{N}^+; \quad w_{ki}, q_{ki}, \phi_i \in \mathbb{R}^+ \tag{14}$$

The objective function (1) minimizes the total travelled distance, the total penalty for delayed replenishments and the lost production, respectively. The vehicle flow conservation in each isle entrance is ensured by constraints (2), (3) and (4). Constraints (5) and (6) are used to allow the vehicles to depart and return to the depots in the start and end of the routes, respectively. Constraints (7) and (8) are used for tracking the arrival times of each vehicle and for the elimination of subtours. Constraints (9) compute the delay of each replenishment. This delay is to be penalized in the objective function. The quantities to be delivered to each cavity is defined in constraints (10). We assume that each cavity is totally fulfilled in replenishment. Additionally, the model captures the lost production incurred after a machine stops production. Constraints (11) limit the quantity delivered to each cavity. Constraints (12) define the number of cars of each type to be pulled by each vehicle. Constraints (13) limit the number of cars to the maximum of b per vehicle. Finally, constraints (14) define the domain of each variable.

5   Conclusion and Future Work

The model presented in this study is still in its early stages of development and is intended to represent the real-world problem that exists currently in the tire company mentioned before. For that matter, it is important that the objective function and the restrictions used in the model can define the main goal and limit the problem with a certain degree of reliability, allowing for the mimicking of the behavior of the system. This study contributed for the comprehension of the vehicle routing problem, proposing a new approach for the OVRPSD, showing that it is a common problem in most companies and that modeling the behavior of the system is the first step for solving it, which can improve efficiency in the transportation of goods, allowing also, for a transportation cost reduction.

The planned future work involves programming the model in Python 2.8 along with the software **IBM ILOG CPLEX** Optimization Studio 12.8, solving it using exact methods, for the clients defined. We will consider two different approaches that consist in comparing the gains obtained viewing the problem as a consistent VRP or viewing it as an Online VRP. On one hand, the expected main advantage of the on-line approach will be the definition of different routes for different times of the day, allowing for some versatility in the service of the machines, satisfying the demands in real-time. Nevertheless, this approach faces operational challenges because, in the company mentioned, there is no way of knowing in real-time the number of tires in stock. On the other hand, the consistent approach will allow for the creation of static routes and allocation of those routes to the same vehicles. This will be useful since the type of tires consumed by the machines doesn't change during a week. For that matter, the creation of routes would be defined at the start of the week and would only be done one every week. However, there would be no guarantee that the consistent routes created would be the ones with minimum cost or that the number of vehicles would be the optimum amount to satisfy machine demand. Also, this approach would not allow for much control on reducing the lost production due to machine stoppage. Further along, depending on the due date for the project, since it is being developed in the context of a master thesis, it may be possible to create an algorithm that returns the best routes for each individual available vehicle, taking into consideration which one of the two depots is the starting depot for those vehicles.

Acknowledgements. We acknowledge the financial support of CIDEM – Research Center of Mechanical Engineering. COMPETE: POCI-01-0145-FEDER 007073 and FTC – Portuguese Foundation for the Development of Science and Technology, Ministry of Science, Technology and Higher Education, under the Project UID/EMS/0615/2016.

References

1. Amberg, A., Voss, S.: A hierarchical relaxations lower bond for the capacitated arc routing problem. In: 35th Hawaii International Conference on System Sciences, Hawaii (2002)
2. Baldacci, R., Mingozzi, A., Roberti, R.: Recent exact algorithms for solving the vehicle routing problem under capacity and time window constraints. Eur. J. Oper. Res. **2018**(2012), 1–6 (2012)
3. Dantzig, G.B., Ramser, J.H.: The truck dispatching problem. Management Science **6**(1), 80–91 (1959)
4. Toth, P., Vigo, D.: The Vehicle Routing Problem, pp. 1–4. Society for Industrial and Applied Mathematics, Philadelphia (2002)
5. Toth, P., Vigo, D.: The Vehicle Routing Problem, pp. 5–8. Society for Industrial and Applied Mathematics, Philadelphia (2002)
6. Lacomme, P., Prins, C., Randame, W.: Competitive memetic algorithms for arc routing problems. Ann. Oper. Res. **131**, 159–185 (2004)
7. Pearn, W.L.: Augment-insert algorithms for the capacitated arc routing problem. Comput. Oper. Res. **18**(2), 189–198 (1991)
8. Longo, H., Poggi de Aragão, M., Uchoa, E.: Solving capacitated arc routing problems using a transformation to the CVRP. Comput. Oper. Res. **33**(2006), 1823–1837 (2006)
9. Assad, A., Golden, B.L., Pearn, W.: The capacitated Chinese postman problem: lower bounds and solvable cases. Am. J. Math. Manag. Sci. **7**(1–2), 63–88 (1987)
10. Chabot, T., Lahyani, R., Coelho, L., Renaud, J.: order picking problems under weight, fragility and category constraints. CIRRELT-2015-49, September 2015
11. Chabot, T., Coelho, L., Renaud, J., Côté, J.: Mathematical models, heuristic and exact method for order picking in 3D-Narrow aisles. CIRRELT-2015-49, June 2015
12. Farahani, R.Z., Rezapour, S., Kardan, L.: Logistics Operations and Management: Concepts and Models, 1st edn. Elsevier, Waltham (2001)
13. Kovacs, A.A., Parragh, S.N., Hartl, R.F.: The multi-objective generalized consistent vehicle routing problem. Eur. J. Oper. Res. **247**, 441–458 (2015)
14. Lipmann, M.: On-line Routing. PhD Thesis, Technical University of Eindhoven (2003)

Integration of Risk and Uncertainty on Levelized Cost of Electricity Calculation

Jorge Cunha[✉] and Paula Ferreira

ALGORITMI Research Center, University of Minho, Guimarães, Portugal
{jscunha, paulaf}@dps.uminho.pt

Abstract. The electric sector is still largely dependent on non-renewable energy sources. The importance of using renewable energies is increasingly recognized all across the world yet they are not fully ready to compete with the mature and ancient technologies that use non-renewable energies. The economic characteristics of different energy technologies can be compared by using the method of levelized cost of electricity (LCOE). LCOE represents the total cost of a power plant including investment and operation and maintenance costs over the assumed life-cycle and discounted to account for the time-value of money. In this paper, an analysis of the levelized costs is proposed for two renewable technologies in Portugal: wind power and solar photovoltaic. Firstly, a deterministic value of LCOE was computed for both technologies. Secondly, recognizing the uncertainty associated with all the assumed parameters, a probabilistic risk analysis was conducted with Monte Carlo simulation to complement the analysis. The results show the high variability of the obtained LCOE values, largely influenced by the investment values and load factors.

Keywords: Levelized cost of electricity (LCOE) · Probabilistic risk analysis Renewable energy sources

1 Introduction

Electricity plays an important role for the achievement of sustainable development of countries with major impacts on the state of the environment and the climate. However, fossil fuels and nuclear are the energy sources that still dominate electricity production in most countries [1]. Notwithstanding, the European Union is focused on changing this situation [2] and the use of renewable energy sources (RES) for electricity generation is seen as the sustainable future that will help leave a better world for future generations. However, these poses some challenges since the integration of variable renewable energy based electricity production, like wind and solar for example, is estimated to be costlier than non-variable resources, given the required flexibility of the system to respond to RES intermittency. To assess the generation costs of electricity from different technologies, the levelized cost of electricity (LCOE) financial model has been used [3]. It calculates the typical levelized costs of generating electricity over the entire operating life of the power plant for a given technology [4], taking into account key cost components, such as capital costs, fuel costs and operations and maintenance (O&M) costs [5]. Moreover, this analytical framework allows precise cost factors, such

© Springer International Publishing AG, part of Springer Nature 2019
J. Machado et al. (Eds.): HELIX 2018, LNEE 505, pp. 825–831, 2019.
https://doi.org/10.1007/978-3-319-91334-6_113

as contingency, decommissioning and carbon prices to be considered. LCOE can be interpreted as the price that must be received per unit of output as payment for producing power in order to reach a pre-defined financial return [6], or, in other words, the price that project must earn per megawatt hour in order to break-even. The LCOE calculation normalizes the units of measuring the lifecycle costs of producing electricity, thus facilitating the comparison of the cost of producing one megawatt hour by each technology [7]. Since the LCOE reveal the main cost factors of alternative generation options, and as long as many cost components vary extensively from location to location and project to project, sensitivity analyses may be performed to evaluate the impact of changes in key parameters on the costs of generating electricity. Thus, LCOE can assess the costs of producing electricity from a new power plant or for a given technology and evaluating the numerous generation options presented to investors in a given market. Therefore, LCOE can be used as a benchmarking or ranking tool to assess the cost-effectiveness of different energy generation technologies [8], resulting on an important tool for policy makers in figuring out the main cost drivers of an electricity system and in evaluating the importance of policies for generation costs. Investment projects would be better analyzed by associating LCOE with a more comprehensive risk analysis, in which multiple risks are taken into account. In fact, the uncertainty surrounding RES is still high given the possibility of technologies development and even due to the uncertainty of climate conditions which affect the underlying RES resources such as wind, solar intensity or rainfall.

Given this context, this paper aims to make a contribution to the assessment and integration of risk and uncertainty on LCOE calculation, presenting a probabilistic risk analysis of these values. Given the availability of data and the importance of both wind and solar power for the future of the Portuguese electricity system, both solar photovoltaic (PV) and wind power were selected for the study.

2 LCOE for Wind and Solar PV in Portugal

In order to determine the LCOE of wind onshore and solar PV, the following Eq. (1) was used:

$$LCOE = P_{Electricity} = \frac{\sum_{t=1}^{N}(Investment_t + O\&M_t + Decommissioning_t)(1 + r)^{-t}}{\sum_{t=1}^{N} Electricity_t (1 + r)^{-t}}$$

(1)

where Electricity$_t$: amount of electricity produced in year "t"; P$_{Electricity}$: required electricity price over the life time of the power plant to cover all costs; r: discount rate; N: life time of the plant; Investment$_t$: investment in year "t"; O&M$_t$: operations and maintenance costs in year "t"; Decommissioning$_t$: decommissioning cost in year "t".

For the calculations of the LCOE, the following assumptions were made. Firstly, installation of a 10 MW onshore wind and a 2 MW solar PV power plants. Secondly, theoretical operating time of 8,760 h in a year. Thirdly, data for investment, O&M and decommissioning costs for both wind and solar technologies were collected from [9].

Fourthly, the expected power output and the corresponding load factor for both technologies were obtained from [10, 11]. The historical values since 1998 were used allowing assessing the variability of these resources. Fifthly, for this type of investments, the literature tends to assign values of 5 or 10% to the discount rate (NEA, 2015). In this study, a rate of 5% was used. Finally, after the collection of the values of electricity generation and installed power, for both solar PV and wind onshore, from year 1998 to 2015 in Portugal, the average load factor, LF, of each technology, for each year was estimated using Eq. (2).

$$LF = \text{Electricity production}/(\text{Installed power} * 8,760) \tag{2}$$

Under a deterministic approach, the average values of investment and O&M costs, and load factor are usually considered. However, this approach fails to recognize the uncertainty associated with these parameters driven by technological and climatic changes. Therefore, a risk analysis, where multiple risk factors are taken into account, should be pursued based on the probabilistic assessment of the assumed values for costs and load factor parameters for both technologies. This analysis provides very important information for both energy policy decision-makers and investors operating in the electricity production market. In order to conduct the statistical analysis and the probabilistic risk assessment, the software @Risk was used, since it allows to perform a risk analysis using Monte Carlo simulation showing the several possible outcomes in the spreadsheet model, as well as the likelihood of their occurrence. Notwithstanding the distribution fitting of data series undertaken, a suitable probability distribution function (PDF) for the investment and O&M costs of wind onshore and for the investment and O&M costs and load factor of solar PV could not be found. For this reason and because of the existence of few data, a triangular distribution was considered. This type of distribution is determined by the maximum, the minimum and the mode. In the case of wind, a normal distribution was assumed based on the mean and the standard deviation for the load factor. Figures 1, 2 and 3 represent the data series and the assigned PDF.

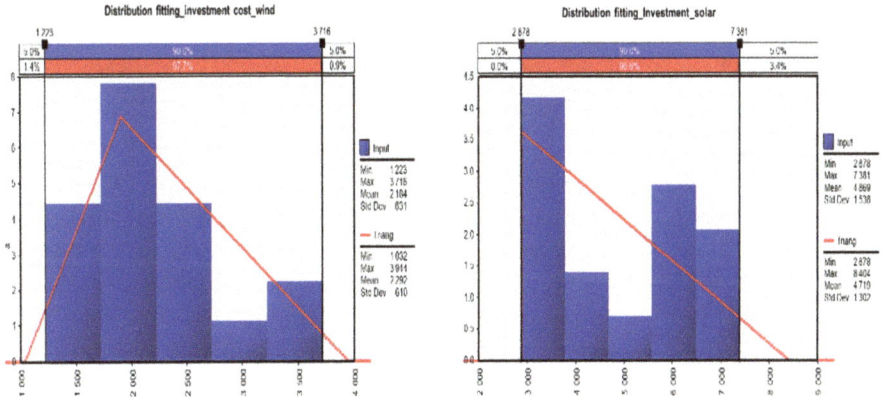

Fig. 1. Distribution fitting for the investment costs of wind and solar PV, respectively.

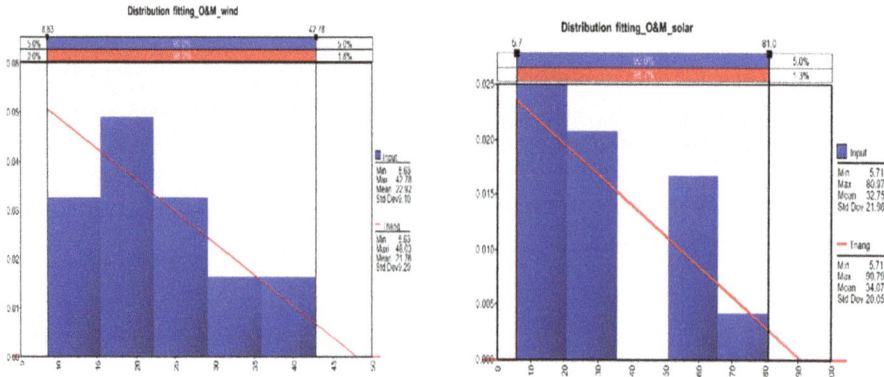

Fig. 2. Distribution fitting for the O&M of wind and solar PV, respectively.

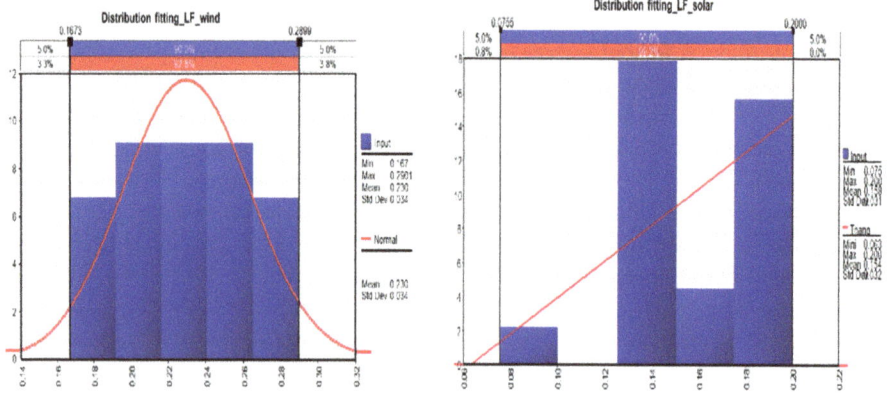

Fig. 3. Distribution fitting for the load factor of wind and solar PV, respectively.

The statistical parameters for each one of the PDF for the risk parameters for both wind and solar PV were used for the combined risk analysis of the LCOE as described in the next section.

3 Risk Analysis

The risk analysis is based on a Monte Carlo simulation combining the main sources of risk and in order to obtain the probabilistic distribution of the LCOE for both wind and solar power technologies in Portugal. Figure 4 shows the obtained PDF function for the LCOE for wind onshore, combining the risk effect on investment cost, O&M cost and load factor, as well as the tornado chart for the probabilistic sensitivity analysis.

According to the previously presented deterministic analysis, LCOE for wind power should be slightly less than 103 USD/MWh (93 €/MWh, according to average exchange

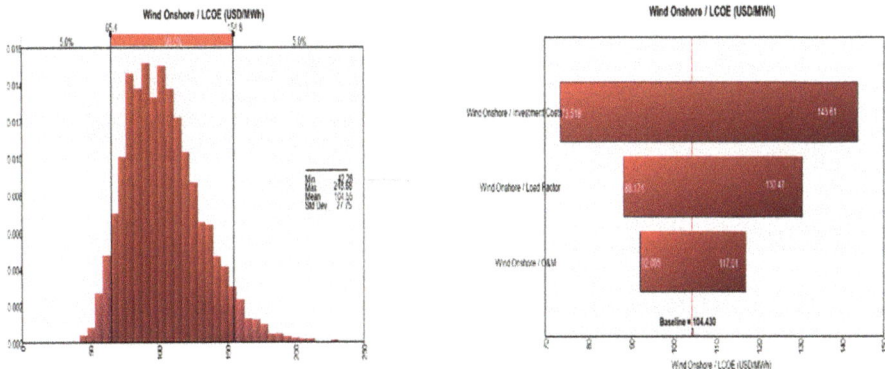

Fig. 4. Probability distribution function (left) and tornado chart (right) for LCOE wind onshore.

rate in March 2015, from www.bportugal.pt consulted in April 2016). However, Fig. 4 shows that the probability of reaching a value higher than this one is more than 50%, thus posing important challenges for investors. This is particularly relevant as average feed-in-tariffs (FIT) for wind power in Portugal were until recently approximately 75 €/MWh (data obtained from www.dgeg.pt consulted in April 2016), a value much lower than the obtained mean. In fact, the probability of reaching a LCOE value much lower than the FIT is less than 15%. The tornado chart, also shown in Fig. 4, demonstrates the relative importance of each one of the simulated parameters, showing that the investment cost has a major role on the final value of the LCOE. This demonstrates the importance of looking for technology development or scale economies as under the best conditions the LCOE would only be slightly higher than 67 €/MWh, with high impact on the investment profitability. The impact of the load factor is also highly relevant and the development of wind evaluation and forecasting models should have a high contribution on ensuring that the best spots are selected for the wind plants location, resulting in higher power output and, consequently, in lower LCOE. Although the O&M costs are the least influent ones, they should not be overlooked as they can represent more than 30% of the total cost. Wind availability is uncertain and this factor has a high influence on the LCOE production from wind generators. Probabilistic calculations for LCOE for wind power show a variation between −54% and +104% comparatively to the deterministic value. The LCOE of wind can be reduced by the upgraded performance of wind turbines, as well as their location in higher wind speed locations, that help improve the average capacity factor.

Figure 5 represents the obtained PDF function for the LCOE for solar PV, combining the risk effect on investment cost, O&M costs and load factor, as well as the tornado chart for the probabilistic sensitivity analysis.

From the deterministic analysis, LCOE for solar power should be slightly less than 282 USD/MWh (236 €/MWh, according to average exchange rate in March 2015, from www.bportugal.pt consulted in April 2016). Figure 5 shows that the probability of reaching a value higher than that is more than 30% and the average FIT for large PV power plants was until recently between 310–317 €/MWh (data obtained from www.dgeg.pt consulted in April 2016). The values show a more stable regime for solar

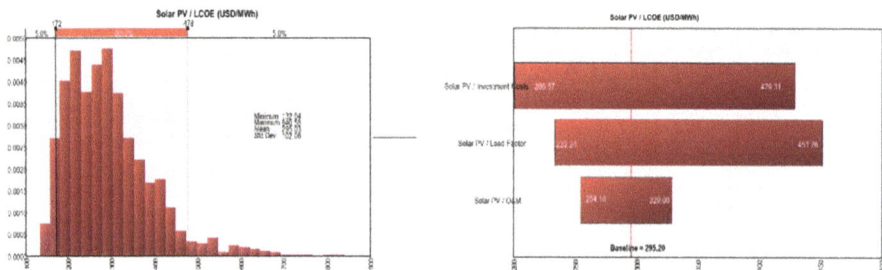

Fig. 5. Probability distribution function (left) and tornado chart (right) for LCOE solar PV.

power than for wind power with most the probabilistic LCOE values being lower than the deterministic one. The probability of reaching a LCOE value lower than the FIT is more than 60%. The tornado chart (Fig. 5) shows that in the case of solar PV the investment cost is also what influence the most the final value of the LCOE. The impact of the load factor is also highly relevant, as well as the O&M costs. System technology and design can also affect LCOE, for the similar systems in the same location may produce very different financial results. Probabilistic calculations for LCOE for PV power show a variation between −51% and +196% comparatively to the deterministic value. Although this interval is larger than the one obtained for wind power, the values seem to be more concentrated on the left side of the PDF resulting in a higher positive skew. The costs and financial returns of solar PV systems should be calculated by taking into account factors such as levels of solar insolation, government and utility incentives, and the cost of grid produced electricity, for they vary based on location. Location is not the only relevant parameter, but also the size, supplier and geographical attributes of the PV system can influence the prices of electricity. As costs of solar PV systems drop, in more and more locations, these become more cost-effective from the experience curve. To estimate if the installation of a solar PV system is profitable in a particular location, then a gathering of detailed information about the place in question, about incentives, grid produced electricity rates and levels of solar must be made [12].

4   Conclusion

The decision-making process to invest on the best projects for electricity production and for selecting the technology and resources is strongly driven by the economic factors and the use of methods relying on the LCOE can bring important information. LCOE allows to compare different power plants with different technical characteristics and cost structures. In this study, two renewable technologies have been selected and for each one the LCOE was calculated assuming the Portuguese climatic conditions. Initially, a deterministic analysis was conducted allowing computing the expected LCOE for both technologies. Afterwards, a probabilistic risk assessment was undertaken aiming to compute the PDF of the LCOE for both technologies. The results showed the high cost to implement and deploy RES, and in most cases higher than the FIT established for Portugal. This represents an important drawback for investors, as if those cannot have some guarantees of properly rewarding their investments, then they will be reluctant to participate in these projects, regardless of the environmental benefits. For the case of wind power, LCOE showed to be much behind the FIT offered in Portugal being highly affected by the high cost and also by the variability of the load factor. As for solar PV, although the range of values is higher, the LCOE tends to concentrate on the left side of the PDF showing a higher probability of FIT to compensate the LCOE. Given the low availability of data, the results obtained should be looked with caution. However, the major contribution of this paper to the literature is the proposed method for the definition of a probabilistic LCOE, especially well-suited for RES project characterized by high uncertainty of the resources and also by the still expected relevant technological development.

Acknowledgments. This work has been supported by COMPETE: POCI-01-0145-FEDER-007043 and FCT – Fundação para a Ciência e Tecnologia within the Project Scope: UID/CEC/00319/2013.

References

1. EIA: International Energy Outlook 2017, Report Number: DOE/EIA-0484 (2017)
2. European Commission: Directive of the European Parliament and of the Council on the promotion of the use of energy from renewable sources. European Commission, Brussels (2017)
3. Branker, K., Pathak, M.J.M., Pearce, J.M.: A review of solar photovoltaic levelized cost of electricity. Renew. Sustain. Energ. Rev. **15**, 4470–4482 (2011)
4. Edenhofer, O., Hirth, L., Knopf, B., Pahle, M., Schlomer, S., Schimd, E., Ueckerdt, F.: On the economics of renewable energy sources. Energ. Econ. **40**, S12–S23 (2013)
5. Matos, E., Monteiro, T., Ferreira, P., Cunha, J., García-Rubio, R.: An optimization approach to select portfolios of electricity generation projects with renewable energies. Appl. Math. Inf. Sci. **9**(2L), 1–7 (2015)
6. Larsson, S., Fantazzini, D., Davidsson, S., Kullander, S., Hook, M.. Reviewing electricity production cost assessments (2013)
7. Hernández-Moro, J., Martínez-Duart, J.M.: Analytical model for solar PV and CSP electricity costs: present LCOE values and their future evolution. Renew. Sustain. Energ. Rev. **20**, 119–132 (2012)
8. Ueckerdt, F., Hirth, L., Luderer, G., Edenhofer, O.: System LCOE: what are the costs of variable renewables. Energy **63**, 61–75 (2013)
9. NEA: Projected Costs of Generating Electricity, NEA No. 7057. OECD Publications, Paris (2015)
10. DGEG: Renovaveis - estatísticas rápidas - Dezembro 2005 (2005). www.dge.pt
11. DGEG: Renovaveis - estatísticas rápidas - Abril 2015 (2015). www.dge.pt
12. Peters, M., Schmidt, T.S., Wiederkehr, D., Schneider, M.: Shedding light on solar technologies – A techno-economic assessment and its policy implications. Energ. Policy **39**, 6422–6439 (2011)

Evaluating the Static Relative Positioning Accuracy of a GPS Equipment by Linear Models

M. Filomena Teodoro[1,2(✉)] and Fernando M. Gonçalves[3]

[1] CINAV, Portuguese Naval Academy, Portuguese Navy,
Base Naval de Lisboa, Alfeite, 2810-001 Almada, Portugal
maria.alves.teodoro@marinha.pt
[2] CEMAT - Center for Computational and Stochastic Mathematics,
Instituto Superior Técnico, Lisbon University,
Avenida Rovisco Pais, n. 1, 1048-001 Lisboa, Portugal
[3] NGI, Nottingham Geospatial Institute, University of Nottingham,
Triumph Rd., Nottingham NG7 2TU, UK

Abstract. The processing of baselines with considerable length may not be successful due several problems, for example, the ionospheric and tropospheric delays estimates are not adequate. To reduce this problem, there exists some models to minimize the biases. The first-order ionospheric biases can be reduced by 98% taking the combination of L_1 and L_2 carrier-phase. The equipment under evaluation uses this solution to the most baselines considered in our work. Still is necessary to reduce the tropospheric bias. An improved and advanced tropospheric bias mitigation strategy is used as alternative to a simpler one. The reduction of bias is verified and quantified using the rate of successful baselines processed by the GPS equipment which uses an improved strategy with a zenith tropospheric scale factor per station. We have built some models by general linear models to evaluate the performance of the equipment. We are aware that 1D and 2D present different behaviors, we analyzed both cases individually with each strategy. In this article, we present partially such analysis for 2D case.

Keywords: Baselines · Bias · General linear models · Performance GPS equipment

1 Introduction

The objective of this research is to evaluate GPS static relative positioning [18], regarding accuracy, as the equivalent of a network real time kinematic (RTK) and to address the practicality of using either a continuously operating reference station (CORS) or a passive control point for providing accurate positioning control. The precision of an observed 3D relative position between two global

© Springer International Publishing AG, part of Springer Nature 2019
J. Machado et al. (Eds.): HELIX 2018, LNEE 505, pp. 832–839, 2019.
https://doi.org/10.1007/978-3-319-91334-6_114

navigation satellite system (GNSS) antennas, and how it depends on the distance between these antennas and on the duration of the observing session, was studied.

This goal was achieved by comparing the outputs from the Leica Geo Office v5 (LGO) software and the Ordnance Survey (OS) active stations coordinates, assumed as true [10,11]. The methodology followed was using observation files from OS active stations to simulate different scenarios for the baseline length, in order to answer the question of how long should be the observing session for the LGO to process those baselines within a pre-establish threshold of accuracy.

A brief introduction of GPS, the navigation system used in the present work, can be found in [2,4,5,16] where some details about observation modeling of systematic biases and errors affecting GPS measurements are described. Other significant references can be found easily. For example, [3,7,8] are significant references where a description and discussion about the navigation system are given.

This work investigates the performance of commercial equipment when the baselines are processed in static mode. This article is a continuation of a preliminary approach presented in [13,14,16]. The parameter to be tested is the time of observation necessary to achieve a certain accuracy (1D and 2D) for each baseline lengths. In [13] were presented some numerical results organized using some descriptive statistics techniques, also some proportion tests were performed so the authors could compare the distinct scenarios. Completing this preliminary approach, in [16] was applied the analysis of variance method so one could confirm the earlier results. A statistical approach by general liner models (GLM) was proposed in [14]. This methodology was applied and the results were discussed for the 1D accuracy case, considering each one of the four different scenarios that were defined in [15]. In this manuscript we continue such approach, but we detail and discuss the results for the 2D accuracy case.

This article is comprised of an introduction and final remarks sections, Sect. 2 contains the description of statistical approach, Sect. 3 containing the description of data and methodology. The empirical application can be found in Sect. 4.

2   Generalized Linear Models

In the classical linear model, a vector X with p explanatory variables $X = (X_1, X_2, \ldots, X_p)$ can explain the variability of the variable of interest Y (response variable), where $Y = Z\beta + \epsilon$. Z is a specification matrix with size $n \times p$ (usually $Z = X$, considering an unitary vector in first column), β a parameter vector and ϵ a vector of random errors ϵ_i, independent and identical distributed to a reduced Gaussian.

The data are in the form (y_i, x_i), $i = 1, \ldots, n$, as result of observation of (Y, X) n times. The response variable Y has expected value $E[Y|Z] = \mu$.

GLM is an extension of classical model where the response variable, following an exponential family distribution [17], do not need to be Gaussian. Another extension from the classical model is that the function which relates the expected value and the explanatory variables can be any differentiable function. Y_i has expected value $E[Y_i|x_i] = \mu_i = b'(\theta_i)$, $i = 1, \ldots, n$.

It is also defined a differentiable and monotone link function g which relates the random component with the systematic component of response variable. The expected value μ_i is related with the linear predictor $\eta_i = z_i^T \beta_i$ using the relation

$$\mu_i = h(\eta_i) = h(z_i^T \beta_i), \qquad \eta_i = g(\mu_i) \tag{1}$$

where h is a differentiable function; $g = h^{-1}$ is the link function; β is a vector of parameter with size p (the same size of the number of explanatory variables); Z is a specification vector with size p.

There are different link functions in GLM. When the random component of response variable has a Poisson distribution, the link function is logarithmic and the model is log-linear. In particular, when the linear predictor $\eta_i = z_i^T \beta_i$ coincides with the canonical parameter θ_i, $\theta_i = \eta_i$, which implies $\theta_i = z_i^T \beta_i$, the link function is denominated as canonical link function. Sometimes, the link function is unknown being estimated simultaneously with the linear component of the semi-parametric model for electricity spot prices. A detailed description of GLM methodology can be found in several references such as [9,17].

3   Data and Methodology

To analyze how the time of observation and length of each baseline are related were used OS active stations. The GPS equipment whose performance is under study have processed a total of 105 baselines classified in six groups (R_i, $i = 1, \ldots, 6$) depending on their lengths in kilometers (see Table 1).

Table 1. Length interval for each range R_i, $i = 1, \ldots, 6$ and respective baselines count.

Range	Length (Km)	Number of baselines
R_1	[000−100]	5
R_2	[100−200]	14
R_3	[200−300]	27
R_4	[300−400]	29
R_5	[400−500]	14
R_6	[500−900]	16

All the stations are permanent stations of clear sky visibility and with low multipath conditions. The quality of the data is therefore expectantly high. Day 13/06/2013 of receiver independent exchange (RINEX) data of GPS week 1744 was downloaded from the data archive of the active GPS network of Ordnance Survey (OS Net) for each of the 106 stations [11]. These RINEX data include phase measurement of the carrier waves L_1 and L_2, P_1, P_2 and C/A pseudo-range code at a 30 s interval.

For this experiment, were considered all active stations considered 'healthy' on 13/06/2013, the 105 baselines formed by ABEP, the reference station. For each one of these baselines was taken into account 24 h of dual-frequency GPS carrier phase observations. Also all remaining active stations, designated as rover, from OS Net were used. The 105 baselines of ABEP have the range in length from 61 Km to 898 Km. The data of 24-h session for each baseline was split into different periods of time: 1, 2, 3, 4, 6, 8, 12 and 24 h as we can see in Table 2, where the two first digits represent the beginning of the observation period and the last two the end.

The idea of considering intervals of time with distinct lengths allowed to evaluate the performance of the software for different periods of the day.

The criteria followed to select the reference station were primarily based on location. Thus ABEP, on the west coast of England, was chosen, because of its high altitude and location, providing a well distributed range of radial vectors to all the other active stations, either in latitude and longitude. Its 3D positional coordinates were fixed to the official values adopted by OS.

Table 2. Intervals of time considered for each exposure session: length and notation.

Session	Intervals				
1 h	[0001]	[0607]	[1213]	[1819]	
2 h	[0002]	[0608]	[1214]	[1820]	
3 h	[0003]	[0609]	[1215]	[1821]	
4 h	[0004]	[0408]	[0812]	[1216]	[1620] [2024]
6 h	[0006]	[0612]	[1218]	[1824]	
8 h	[0008]	[0816]	[1624]		
12 h	[0012]	[1224]			
24 h	[0024]				

In order to evaluate at what range of baseline lengths the use of precise ephemerides become worthwhile, both results using broadcast and precise ephemerides[1] are presented as well. The corresponding SP3 files were downloaded from the data archive of [6]. The files data include precise ephemerides

---

[1] In astronomy and celestial navigation, an ephemeris (plural: ephemerides; from Latin ephemeris, "diary", from Greek: ephemeris, "diary, journal") gives the positions of naturally occurring astronomical objects as well as artificial satellites in the sky at a given time or times. Historically, positions were given as printed tables of values, given at regular intervals of date and time. Modern ephemerides are often computed electronically from mathematical models of the motion of astronomical objects and the Earth. Even though the calculation of these tables was one of the first applications of mechanical computers, printed ephemerides are still produced, as they are useful when computational devices are not available. (Cited from https://en.wikipedia.rg/wiki/Ephemeris).

at a sampling interval of 15 min and the high-rate precise satellite clocks with a sampling of 30 s.

Hence, the four different scenarios can be compared as follows:

- Direct comparison of the results obtained using the broadcast ephemerides and the precise ephemerides (BH versus PH and BC versus PC);
- Direct comparison of the results obtained using Hopfield model and computing the troposphere (BH versus BC and PH versus PC).

At starting points 1D, 2D and 3D accuracy criteria were established for each baseline, as only successful processed baselines are of interest for this research. The chosen values were set to 1D and 2D accuracies to be better than 3 cm and 3D better than 4.5 cm. These are realistic values, as the OS active stations have 1D accuracy of about 2 cm in magnitude and close to 1 cm in 2D. Therefore, assuming the 3 cm as 1D and 2D threshold seems to be reasonable due the fact that this tolerance allows for the 'absorption' of errors inherent to the coordinates of the stations. Despite how perfectly the baseline was calculated an error of up to 4 cm in height and 2 cm in plan could arise due to the uncertainty associated with the coordinates.

The published coordinates of each of these stations (in Cartesian format on the header of the corresponding RINEX file) are assumed as 'true' and used to compute the errors (1D, 2D and 3D) in the solutions processed by LGO equipment.

One of the major challenges in processing high-accurate long baselines is the presence of un-modelled ionospheric and tropospheric delays. There are effective mitigation strategies for ionospheric biases, such as the ionosphere-free linear combination of L_1 and L_2 carrier-phase, which can remove about 98% of the first-order ionospheric biases. With few exceptions this was the solution found by LGO for the 11760 baselines processed in this research. Therefore, for successful results, the appropriated approach to the mitigation of biases due to tropospheric delays is vital.

4   Empirical Application

In [16] was studied the relation for single baselines between lengths ranges and between the different ranges and the observation time required to obtain high-accurate positioning, using commercial software LGO using analysis of variance technique [12]. The results are considered valid for the applied software and under the conditions of the experiments.

In [15] the same four different strategies were established and evaluated through the processing of a total of 11760 baselines. The data processing and testing used several options concerning the best thresholds for accuracy. A brief analysis for different amplitudes of time interval of exposure, considering the four strategies is reproduced partially in such paper, where the results using GLM modeling and considering 1D accuracy were described and discussed.

4.1 Numerical Results

The results were obtained by GLM modeling. For the 2D case, in all strategies and for each class, the success increases with session length, by opposite 1D such trend is not so evident.

In 2D, PC behaves better than BC even for the shortest range, although the difference is not statistically significant.

When we consider R_1 class, sessions of 1 h are enough to produce accurate results. For R_2 to R_4 classes is necessary 4 h sessions (for the case R_4, we got 75% and 86% of success with BC and PC respectively). For baselines of R_5 class the session duration should be 12 h.

In Fig. 1 the yellow triangles represent successful baselines in 3D (1D and 2D simultaneously) and the blue triangles represent successful baselines only in 2D. The color scale bar is scaled in centimeters. Results attained processing baselines from the 2D perspective follow a linear trend, which is, higher rates of success are always associated with longer observing sessions. In the same way shortening baseline length also improves the success rate for identical periods of observation. In other words, the differential errors associated with 2D performance are temporally and spatially correlated. Figure 1 were chosen with a view to strengthen some of the conclusions already discussed.

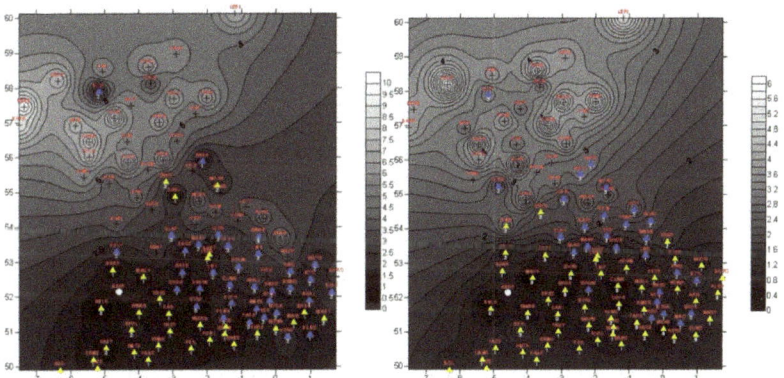

Fig. 1. Successful baselines. Yellow triangles represent successful baselines in 3D (1D and 2D simultaneously); blue triangles represent successful baselines only in 2D. Scale bar in centimeters. PHP [0004] (left); PCP [0004] (right).

An obvious conclusion would be that differences inside each strategy are more evident than between strategies, for example, PHP0004 (See Fig. 1 on left) is very much different from PHP0408 but quite similar to PCP0004 (See Fig. 1 on right). For identical sessions, solutions produced either by the Hopfield model or by computing the troposphere are statistically identical, because in both cases, for all baselines, LGO used the L_3 combination, the main difference here is due to tropospheric delays, as all the other biases and errors due to constellation' geometry, number of satellites, multipath effects and receivers' noise are common.

5   Final Remarks

A generalized linear model [17] was implemented so we could relate the exposure time, number of success and base lines distance restricted to 2D case. We have performed an analysis evaluating the contributions to the rate of baselines successfully processed by adopting an advanced tropospheric bias mitigation strategy as opposed to a sample tropospheric bias mitigation approach. For 2D case, the differences inside each strategy are more evident than between strategies.

CQ indicators provided by LGO post-processing are often overlay optimistic in all components, when compared with the correspondent 1D, 2D and 3D errors found throughout this research, and should therefore be used with caution.

The statistical analysis and remaining details of such modeling will be presented in an extended version of this manuscript, identifying and quantifying the explanatory variables with more influence in baselines success range, such as interval length of time exposure or part of day. A complementary approach using multivariate analysis [1] taking into consideration all available data was done. The analysis of such approach it is not complete but the preliminary results corroborate with the results obtained in this paper.

Acknowledgements. This work was supported by Portuguese funds through the FCT, *Center for Computational and Stochastic Mathematics* (CEMAT), University of Lisbon, Portugal, project UID/Multi/04621/2013, and *Center of Naval Research* (CINAV), Naval Academy, Portuguese Navy, Portugal.

References

1. Anderson, T.W.: An Introduction to Multivariate Analysis. Wiley, New York (2003)
2. Bingley, R.M.: GNSS Principles and Observables: Systematic Biases and Errors. Short Course, NGI, University of Nottingham, Nottingham (2013)
3. Hofmann-Wellenhof, B., Lichtenegger, H., Wasle, H.: GNSS-Global Navigation Satellite Systems GPS, GLONASS, Galileo, and More. Springer Verlag-Wien, New York (2008)
4. Hoque, M.M., Jakowski, N.: Ionospheric propagation effects on GNSS signals and new correction approaches. In: Shuanggen, J. (ed.) Global Navigation Satellite Systems: Signal, Theory and Applications, pp. 381–405. InTech, Rijeka-Croatia (2012). https://doi.org/10.5772/30090
5. Hopfield, H.S.: Tropospheric effect on electromagnetically measured range: prediction from surface weather data. Radio Sci. **6**(3), 357–367 (1971)
6. IGS, International GNSS Service. http://igscb.jpl.nasa.gov/components/prods_cb.html). Accessed 24 Feb 2018
7. Kaplan, E.D., Hegarty, C.J.: Understanding GPS: Principles and Applications. Artech House, Norwood (2006)

8. Leick, A.: GPS Satellite Surveying. Wiley, New Jersey (2004)
9. McCullagh, P., Nelder, J.A.: Generalized Linear Models. Chapman and Hall, Londres (1989)
10. OS Net Business and Government: Ordnance Survey. http://www.ordnancesurvey.co.uk/oswebsite/products/os-net/index.html. Accessed 24 Feb 2018
11. OS Net Business and Government: Ordnance Survey. http://www.ordnancesurvey.co.uk/gps/os-net-rinex-data/. Accessed 24 Feb 2018
12. Tamhane, A.C., Dunlop, D.D.: Statistics and Data Analysis: From Elementary to Intermediate. Prentice Hall, New Jersey (2000)
13. Filomena Teodoro, M., Gonçalves, F.M.: A preliminary statistical evaluation of GPS static relative positioning. In: Quintela, P., et al. (eds.) Progress in Industrial Mathematics at ECMI 2016. ECMI book series of Mathematics in Industry, Subseries The European Consortium for Mathematics in Industry. Springer, Zurich (2017). https://doi.org/10.1007/978-3-319-63082-3_110
14. Teodoro, M.F., Gonçalves, F.M., Correia, A.: Performance analysis of a GPS equipment by general linear models approach. In: Ntalianis, K. (ed.) Applied Mathematics and Computer Science, vol. 1836(1), 020084-1-020084-5. AIP, Melville, New York (2017). https://doi.org/10.1063/1.4982024
15. Teodoro, M.F., Gonçalves, F.M., Correia, A.: Analyzing the performance of a GPS device. World Sci. Eng. Acad. Soc. Trans. Environ. Develop. 13, pp. 150–157 (2017)
16. Filomena Teodoro, M., Gonçalves, F.M., Correia, A.: Performance analysis of a GPS equipment. In: Oliveira, T., et al. (eds.) Recent Studies in Risk Analysis and Statistical Modeling. Contribution to Statistics, Chap. XXl. Springer, Zurich (2018). https://doi.org/10.1007/978-3-319-76605-8_21
17. Turkman, M.A., Silva, G.: Modelos Lineares Generalizados da teoria a prática. Sociedade Portuguesa de Estatística, Lisboa (2000)
18. Wells, D.E. and et al.: Guide to GPS Positioning. Canadian GPS Associates, Fredericton (1986). http://plan.geomatics.ucalgary.ca/papers/guide_to_gps_positioning_book.pdf. Accessed 24 Feb 2018

Scheduling in an Automobile Repair Shop

Maria de Fátima Pilar$^{(\boxtimes)}$, Eliana Costa e Silva, and Ana Borges

CIICESI - Center for Research and Innovation in Business Sciences
and Information Systems, School of Technology and Management,
Polytechnic of Porto, Porto, Portugal
{8120109,eos,aib}@estg.ipp.pt

Abstract. Scheduling has been extensively studied since its correct application and development is a decisive factor in an organization's success. Its application can be found in several areas of industry. In this work a real case of an automobile sector firm, more specifically the repair shop, is presented. First, a brief literature review is presented in order to gain some sensitivity on the scheduling thematic. Following with a description of the firm, detailing its repair shop, and the statistical analysis of the database provided by the firm. This analysis is intended to understand the firm's current situation, in particular, the characterization of repairs carried out during the period under analysis. This exploratory analysis serves as the basis for a mathematical model, that will be developed in a future work, in order to support the vehicles repairment scheduling decision.

Keywords: Automobile sector · Real application · Scheduling
Job shop · Optimization

1 Introduction

Scheduling is a form of decision-making that plays an important role in many disciplines [1]. More than any other area in Operations Research (OR), scheduling is motivated by situations where scarce resources must be allocated to activities over time. In the current competitive environment, this concept has become a necessity for survival in the market, because the firms commit themselves to their customers in relation to a delivery date, and their fail to comply may represent loss of goodwill, or even the loss of the client [2]. Several types of scheduling problems have been investigated by practitioners and researchers, with the aim of producing optimal or near-optimal solution methodologies that can serve as reference to similar problems [3].

The challenge addressed in this work portrays the case of a Portuguese firm of the automobile sector, focus on its repair shop. The firm has branches throughout the country, however in the present work the repair shop located in the Porto area will be considered. The firm provided information concerning the mechanical repairs carried out over a time horizon period of one year (from August 2016

J. Machado et al. (Eds.): HELIX 2018, LNEE 505, pp. 840–846, 2019.
https://doi.org/10.1007/978-3-319-91334-6_115

to July 2017). The possibility of working on the modeling of a real industrial challenge, with real data provided by the firm, present it self as the innovative contribution of the present work. The aim of the present work is to perform a statistical analysis of the real data in order to understand the current scheduling process of the firm. More precisely, the number of Repair Orders (RO's) in the repair shop, the fulfillment of the stipulated delivery dates established with the customers, the *downtimes* of the repairs (i.e., the periods of time that the automobiles are without any intervention) and the total time actually spent in the repairs. This analysis is intended for a more in-depth approach that will be carried out in near future with the development of a tailor-made mathematical model of the firm's real challenge. Given that customer satisfaction is the firm's most valued criteria, minimizing the clients waiting times is crucial. In near future, a Mixed Integer Linear Programming (MILP) model will be developed to support the firm's decision on the best scheduling of automobile's repairment to be adopted, taking into account the environment of this repair shop.

This paper is organized as follows: Sect. 2 presents a literature review on scheduling, in order to contextualize the real challenge at hand. Next, the description of the firm is presented, followed by the some primary results of the provided data. Finally, Sect. 4 presents some conclusions and points directions for future work.

2   Literature Review

Scheduling theory has received much attention from OR practitioners, management scientists, production and operations research workers, as well as mathematicians [4]. It may be defined as the allocation of *resources* over time to perform *tasks* in order to optimize some or several performances measures [1,4]. The resources and tasks in an organization can be represented in diverse ways. In fact, *resources* may be machines in a workshop, runways at an airport, crews at a construction site, etc.; while *tasks* can be operations in a production process, take-offs and landings at an airport, stages in a construction project, etc. [2]. Grocery delivery organization, creation of work routes for hospitals or for bus or train services, organization of university examination schedules or equipment maintenance scheduling are some applications of scheduling [5]. An example of a scheduling problem that has been quite studied in literature is the *job shop* problem (JSP) [5], where *resources* are often called *machines* and *tasks* are called *jobs*. Furthermore, sometimes jobs can be made up of various tasks called *operations*. According to [4], the JSP can be described as: A set of n jobs $\{J_1, \ldots, J_n\}$ need to be executed in m available machines $\{M_1, \ldots, M_m\}$. The processing of each job J_j on a machine M_i is called an *operation*, and is represented by O_{ij}. Typically, the subscript j refers to a *job* and the subscript i refers to a *machine*. Each operation, to which a job j belongs and a machine i, has an associated *processing time*, p_{ij}. Each job has a *ready time* (or *release date*) r_j, as well as a due date d_j (at which point the job J_j must be completed). A schedule consists, therefore, of assigning jobs to machines over time, and the JSP consists of finding a schedule that optimizes some performance measure [4].

Performance measures can be based on *completion time* and *due date* [6].

Some examples of the first case are: the total flowtime $\sum_{i=1}^{n} C_i$, where C_i represents the task completion time of task i, the maximum flowtime (F_{max}), makespan (C_{max}), mean flowtime ($\bar{F}$) and mean complete time ($\bar{C}$); while the maximum lateness (L_{max}), maximum tardiness (T_{max}), mean lateness ($\bar{L}$) and mean tardiness ($\bar{T}$) are *due date* based measures. Further examples of performance measures can be found in e.g. [2,4,7,8]. JSP is NP-hard problem since efficient algorithms are not known for these problems due and they present exponential complexity [9,10]. The scheduling problems can be characterized as being static (the jobs to be executed do not change over time) or dynamic. Furthermore, the methods used to solve the scheduling problems can be deterministic, if the working conditions are represented by known constant parameters, or otherwise stochastic [7].

In highly competitive environments, optimal scheduling can provide several benefits, including insights on the full potential of a production system or inputs for decision making on expansion projects [11]. Regarding the mathematical models, the first formulations for scheduling problems were planned starting in 1960 [12,13]. Scheduling methods can be classified as combinatorial optimization, artificial intelligence, simulation-based scheduling with dispatching rules, heuristics-oriented, and multi-criteria decision making [14].

There are several works in different application areas, such as: [15] that present an MILP formulation of a flexible job shop problem in petrochemical industry, and use two heuristic methods (Tabu Search and Simulated Annealing) for solving the problem while minimize the makespan; [3] that present four MIP formulations, based on four different types of decision variables, in the context of parallel machine scheduling problems; [16] present a decomposition heuristic for problems with Linear Program (LP) formulations, for various types of JSP, both with objectives that are function of completion time of the job as well as the completion times of the intermediate operation; [17] address the case of a military supply chain, in particular in its integrated demand-responsive scheduling in its maintenance and transportation operations, and for their resolution, the authors develop novel integrated iterative approach, using mixed integer programming (MIP) for small and medium-scale instances and a combined simulation-optimization approach for large instances, in order to minimize the total completion time; [11] address the case of an ice-cream processing facility that is solved using a MILP model in a multi-week context, with the objective of minimizing the makespan; [18] present a MILP model for scheduling of repairment of aircraft engines in a context of a flexible job shop problem while minimize the total tardiness and makespan; [19] address a problem in the petroleum industry by using a MILP model, aiming to minimize the total investment and the constraints of submarine terrain and production process.

3   Problem Description

This work was motivated by a challenge proposed by a Portuguese firm in the automobile sector, with activities on vehicle sales and maintenance. The firm has shops located in several cities in Portugal. The challenge and the data provided concerns the repair shop, located in Porto area, which performs all types of repairs, including scheduled maintenance and repairs due to failure. Most of the repairs (about 75%) are pre-scheduled and only a small percentage are scheduled on the same day (about 25%). The repair shop consists of 28 workers, distributed in the areas of repair, administration and quality control, and several resources, such as, the tools of each mechanic, the repair machines, the elevators, etc. The repair area is divided in: Mechanics, Collision, Service Station, Preparation of New Vehicles (PNV), Diagnostics and Electronics. The mechanical area consists of 10 workers: six senior mechanics who can perform any task except diagnoses; two senior mechanic specialist who perform any task, including diagnostic; and two junior mechanics, who can only perform some specific tasks. The collision area consists of seven workers: 3 sheet metal workers, 3 painters and 1 chef of collision. The Service Station area consists of two washers. The PNV area as one single employee. The Administrative area is made up of three receptionists, by the manager, a responsible for quality control, a warranty operator, two officials responsible for billing. The receptionists perform the reception of vehicles, when the repair order form is filled, identifying the necessary repairs requested by the customer, and an expected delivery date is agreed. The manager is responsible for making the bridge between the reception and the repair shop. He distributes the vehicles to the available mechanics according to the type of repair required and the work already distributed. The quality control employee tests the vehicles and ensures that the repairs has been properly performed. Each mechanic introduces the repair situation of the vehicle into the computer system, for making it is possible to know whether the vehicle will be ready by the date initially agreed, or if it is necessary to postpone delivery.

The dataset provided by the firm contains information on vehicle repairs from August, 2016 to July, 2017. It includes information of Repair Orders (RO's), the respective time of opening and closing, the interventions (specifying each operation that has been made) and repairment duration. Since the objective of the firm is clients' satisfaction, minimizing the waiting times is essential, for this reason, particular attention will be given to the analysis of execution times of each RO. The database had information on 6583 RO's, however, due to the lack of information it was necessary to exclude some RO's ending with a database of 5843 RO's.

There is a large variation of the number of RO's that enter daily in to the repair shop, specially in October, November and February (Fig. 1). August and June are the months that present lower and higher number of RO's. The months with the largest number of RO's are May, June and July. Taking into account firm's information, on Monday's and at the beginning of each month, it is registered a large number of RO's entries, justifiable by weekend breakdowns, and because RO's are usually scheduled, first on Mondays or at the begging of each month. Furthermore, it was verified that 79% of the RO's complied with expected delivery dates, 13% did not and 8% of the RO's had no scheduled delivery date stipulated by the customer.

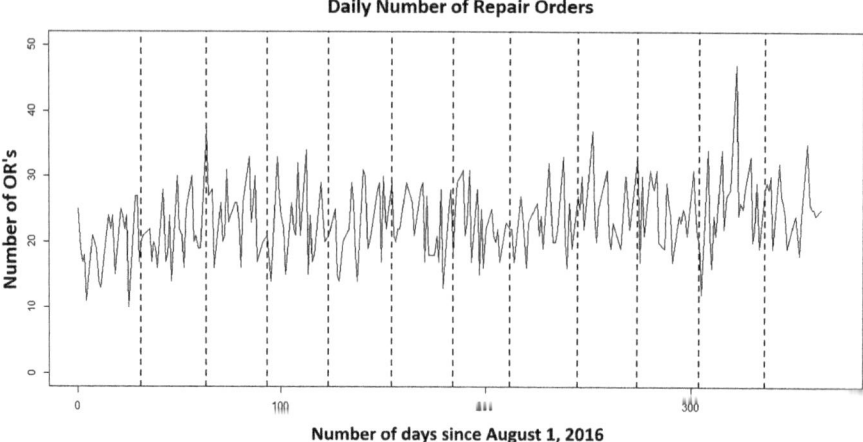

Fig. 1. Daily number of RO's from August 2016 to July 2017 (vertical dashes represent the beginning of each month).

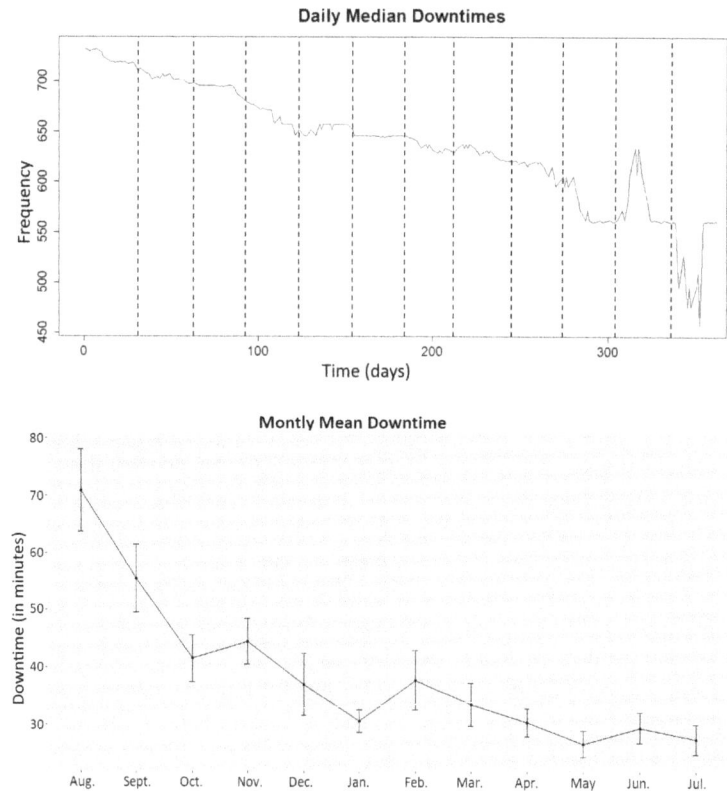

Fig. 2. Daily median *downtimes* (top) and monthly mean *downtime* with standard deviation error bars (bottom)

It is also important to determine the *downtime*, since it represents the period of time the vehicles were at the repair shop without any intervention. This is computed taking into account: the time each RO remained in the repair shop, the sum of the execution times of all the interventions performed in each RO, the night period, and the mechanics' lunch hour. The *downtime* have been decreasing since the beginning of the period under analysis (Fig. 2 (top)), due to the implementation of the times of vehicles' immobilization, and the hiring of a trainee mechanic. In the last two months of analysis (June and July), an increase is observed, followed by a fall and again an increase of these times. Regarding the first situation, the firm has no justification for the increase observed, the second corresponded of two mechanics entered on vacation, which reduced the productivity of the repair shop. Figure 2 (bottom) presents the monthly mean *downtime* variation, which elucidates the high variability of the data that, however, decreased throughout the period of analysis. The *downtime* of the vehicles varied from 1 h and 8 min to 2234 h and 13 min, with a mean of 38 h and 11 min.

4   Conclusions and Future Work

Literature on scheduling is diverse in great part due to the importance of the concept that occurs in several industrial sectors. The JSP consists on determining the schedule of jobs, with sequences of operations pre-specified in a multi-machine environment. Most scheduling problems are complex combinatorial optimization problems and therefore difficult to solve [15]. The existence of several environments related to scheduling makes these problems very complex and difficult to solve, so their correct classification, makes the correct measures be applied, thus allowing the achievement of good results, and continuous improvement.

The real challenge portrayed in this paper concerns a scheduling problem on a firm of the automobile sector whose most valued criteria is clients' satisfaction. To attain this goal and in order to get some insight on the industrial challenge, the analysis of the real data was presented in this paper. This will contribute to the understanding of the challenge, which is essential to the development of the mathematical model that will be developed in a near future, whose aim is determining the best scheduling and sequencing of automobile repairment taking into account the nature of the firm's repair shop environment. Specifically, a MILP model will developed with simultaneous two objectives: customers' satisfaction and maximization of firm's resources. Therefore, the minimization of the tardiness will be consider, so that there are no delays in the vehicles repairment, or that these delays are as small as possible. On the other hand, *downtimes* were observed, and although these times have decreased over time, it still indirectly affects customers' satisfaction, given the possibility of delays in the delivery of their vehicles, therefore it is imperative to reduce these times.

References

1. Leung, J.Y.T.: Handbook of Scheduling: Algorithms, Models, and Performance Analysis. CRC Press, Boca Raton (2004)
2. Pinedo, M.L.: Scheduling: Theory, Algorithms, and Systems. Springer, New York (2008)
3. Unlu, Y., Mason, S.J.: Evaluation of mixed integer programming formulations for non-preemptive parallel machine scheduling problems. Comput. Ind. Eng. **58**(4), 785–800 (2010)
4. Maccarthy, B.L., Liu, J.: Addressing the gap in scheduling research: a review of optimization and heuristic models in production scheduling. Int. J. Prod. Res. **31**(1), 59–79 (1993)
5. Hart, E., Ross, P., Corne, D.: Evolutionary scheduling: a review. Genet. Program. Evolvable Mach. **6**(2), 191–220 (2005)
6. French, J.: Sequencing and Scheduling: An Introduction to the Mathematics of the Job-Shop. Wiley, New York (1982)
7. Baker, K.R., Trietsch, D.: Principles of Sequencing and Scheduling. Hoboken, Wiley (2009)
8. Arenales, M., Armentano, V., Morabito, R., Yanasse. H.: Pesquisa Operacional. Elsevier, Rio de Janeiro (2011). (in Portuguese)
9. Garey, M.R., Johnson, D.S., Sethi, R.: The complexity of flowshop and jobshop scheduling. Math. Oper. Res. **1**(2), 117–129 (1976)
10. Pereira, I.A.S.: Sistema inteligente para escalonamento assistido por aprendizagem. Ph.D. Thesis in Electrical and Computer Engineering, UTAD, Vila Real (2015). (in Portuguese)
11. Wari, E., Zhu, W.: Multi-week MILP scheduling for an ice cream processing facility. Comput. Chem. Eng. **94**, 141–156 (2016)
12. Bagheri, A., Zandieh, M., Mahdavi, I., Yazdani, M.: An artificial immune algorithm for the flexible job-shop scheduling problem. Future Gener. Comput. Syst. **26**(4), 533–541 (2010)
13. Demir, Y., İşleyen, S.K.: Evaluation of mathematical models for flexible job-shop scheduling problems. Appl. Math. Model. **37**(3), 977–988 (2013)
14. Srinoi, P., Shayan, E., Ghotb, F.: Scheduling of flexible manufacturing systems using fuzzy logic. Int. J. Prod. Res. **44**(11), 1–21 (2002)
15. Fattahi, P., Mehrabad, M.S., Jolai, F.: Mathematical modeling and heuristic approaches to flexible job shop scheduling problems. J. Intell. Manuf. **18**(3), 331–342 (2007)
16. Bülbül, K., Kaminsky, P.: A linear programming-based method for job shop scheduling. J. Sched. **16**(2), 161–183 (2013)
17. Tsadikovich, D., Levner, E., Tell, H., Werner, F.: Integrated demand-responsive scheduling of maintenance and transportation operations in military supply chains. Int. J. Prod. Res. **54**(19), 5798–5810 (2016)
18. Cerdeira, J.O., Lopes, I.C., Costa e Silva, E.: Scheduling the repairment of aircrafts. In: ICCAIRO 2017, pp. 259–267. IEEE (2017)
19. Zhang, H., Liang, Y., Ma, J., Qian, C., Yan, X.: An MILP method for optimal offshore oilfield gathering system. Ocean Eng. **141**, 25–34 (2017)

Innovation and Collaborative
Arrangements

Science and Technology Studies and Engineering Innovation: The Concepts of Future and Diversity

Emilia Araújo$^{(\boxtimes)}$ (iD)

Communication and Society Research Centre,
University of Minho, Braga, Portugal
era@ics.uminho.pt

Abstract. This paper seeks to demonstrate how engineering can benefit from science and technology studies. It focuses on four key issues: responsibility, sustainability, diversity, and public participation. Science and technology studies encompass a plethora of research concerning the various ways that science and technology relate to society. The main area for discussion is the idea that policy makers and practitioners in general must consider the social and the cultural context, alongside specific science, and technology issues. In other words, society, and culture - which constitute the framework for producing, using and engaging with knowledge - are embedded within the fields of science and technology, in multiple manners. It is therefore fundamental to develop innovation in a sustainable and coherent manner.

Keywords: Science and technology studies · Diversity · Innovation Future

1 Introduction

This paper aims to define the field of science and technology studies in order to clarify how they have contributed to exploring several important sociocultural dimensions of science and technology. The text focuses more specifically on the context of engineering innovation and entrepreneurship. Science and technology are two perennial forces that shape society. Their influence has been increasingly observable since the industrial revolution. Today, scientists as well as practitioners working in the fields of technological innovation and engineering enhancement, are aware of their tremendous impact on shaping social and cultural lives, including the fields of regulation and politics. In this context, this paper has three main objectives: (i) define social studies of science and technology, highlighting their core achievements, for Science, Technology, Engineering and Mathematics (STEM) in general, and for engineering in particular; (ii) demonstrate the conceptual and practical contributions made by social studies of science, for engineering innovation and entrepreneurism, and ascertain their importance for the present and future of societies, in particular in terms of innovation engineering's capacity to embrace diversity; (iii) present a range of actions that can be enhanced in various settings and STEM areas – i.e. engineering and technology that can help

© Springer International Publishing AG, part of Springer Nature 2019
J. Machado et al. (Eds.): HELIX 2018, LNEE 505, pp. 849–855, 2019.
https://doi.org/10.1007/978-3-319-91334-6_116

practitioners develop innovation and entrepreneurism practices that are responsible, sustainable and safeguard alternative and better futures.

2   Science and Technology Studies

Science and technology studies today constitute a broad field that combines many scientific fields, interlinking different epistemological and methodological paradigms. They have been developed more systematically since the 1950s, through the concept of socio-technical systems [1]. The basic achievement of this approach has been to defend the need to take society in general in consideration when designing and implementing technological endeavours [2]. Since this period [3], several research projects have accompanied the greater complexity of science and technological developments across several areas, ranging from the health sector to the environment. The number of science and technology studies expanded dramatically in the 1990s, partly as a result of increasing investment in the internet, globalization, and the emergence of new social and cultural difficulties allied to the advent of the knowledge economy. In a seminal paper in Portugal about the relationship between sociology and engineering, Brandão Moniz pointed out that major questions, such as changes in work organization should be considered altogether as sociotechnical problems, widely considered in the field of industrial sociology [4]. The questions surrounding public understating of science and participation in scientific and technological innovation began to be highlighted within science and technology studies since the nineties onwards [5, 6]. Research also began about the place of people in society, the modes of relationship between human beings and machines, changing values, and reconfiguration of leisure and working times, educational methods, and lifestyles [7]. Social studies of science are now even more committed to tackling key challenges now facing humanity, such as natural and human catastrophes, new mental and physical disorders, global warming, the water supply, food contamination, new viruses, regulation, terror, and surveillance systems. These fields are all profoundly linked to scientific and technological development. They pose new risks and challenges that require attention from society in general and necessitate increasingly responsible research and innovation procedures [8]. In parallel, studies of science and technology have come to call attention to many important queries imbedded in science and technological development and with effects on it, such as gender, race, or social class.

2.1   Core Achievements for STEM Areas

As may be inferred from this context, several contributions may be identified in terms of STEM areas. However, most of them are currently excluded from practices developed in the context of engineering R&D units. Table 1, below, shows the most relevant theoretical issues raised by science and technology studies in relation to STEM.

Science and technology cannot be understood as deterministically driven forces per se. The social and the cultural background plays a fundamental role in designing and implementing technology and evaluating its current conditions for development. Even areas with a faster rate of technological evolution - such as transportation and mobility,

Table 1. Core theoretical issues for STEM areas

Concepts to be revised	Main arguments
Determinism	Science and technology are socially in-built elements
	Science and technology are co-produced with society
Linearization and certainty	Science and technology are dependent on society and culture to produce the most socially adequate results
	Public understanding of science is extremely relevant
	Science and technology are broadly uncertain fields. Society is always at risk of having to overcome the perverse effects of science and technology developments
Futurism	Access to science and technology is unequal across social groups, and countries
	Science and technology depend upon politically-driven funding criteria
	The future encompasses many issues that still require sectorial and national responses which are not always compatible with the imagined future. Part of these answers are dependent on the ability to avoid or end the exploitation of natural resources
Scientific knowledge *versus* common sense/lay knowledge	Science and technological effects depend on human actors and their skills to use them
	Science and technology must consider the risks associated to their use in the present and future
	Science and technology must consider lay knowledge and invest in greater information and translation of scientific knowledge

energy, clothing, and mechanics - are dependent on their respective social contexts. This means that we cannot expect innovation to be naturally adapted or accepted. Neither can we necessarily expect that it will produce its planned effects. Society and culture make a substantial contribution in this regard, since science and technology are co-produced realities, that vary throughout history. Society and culture must be studied and considered from the earliest phases of designing research and planning innovation initiatives, as co-producers of knowledge that fuel its ideals and expectations, rather than being mere consumers [9]. Scientists and innovators are normally aware of the uncertain nature of their achievements due to multiple factors, some of which are internal to the research itself - such as the limited sizes of the samples - while others are related to the external environment in which the innovation will be developed, until its final use and evaluation. Scientists and innovators also know that time-to-innovation may be very slow, and that innovation is not always incremental and accumulative. In practice, however, as outcomes of a complex mode of market-driven priorities, science and technology are widely connoted with linearity and certainty [6]. In other words, they are regarded as sources of permanent development, which will be increasingly different and better. For this reason, science and technology studies show that innovation faces complex problems of a global and structural nature. Such problems persist on the horizon as major dilemmas facing humanity, such as climate change. As a result, scientists and innovators must consider fragmentation, discontinuities, and generation of alternatives. Similar reasoning can be developed in relation to the conflicts between the short- and long-term, since some of the effects of current discoveries and inventions will only be evident over the long-term. For example, until only very recently, microwaves were considered to be a source of greater energy-based spending than had been initially expected. The association between food consumption, water pollution by chemical residues and the increased number of people suffering from cancer is a contemporary social problem. There are therefore many aspects that we need to deal with, and that are not pre-planned or may be estimated, because society and culture are unpredictable. In the wake of the aforementioned considerations on linearity and certainty, science studies call attention to the need to disentangle the artistic and imaginative dimensions from the technological and scientific worlds; from the concrete and social worlds where people live, thereby helping them to understand the complexity of the changes and improve their ability to access innovation. They also signal the need to de-linearize the future, thereby fostering the creativity of scientists and innovators, so as to design alternative futures that depend upon the action of future generations and are not completely conditioned by current options.

3   The Issue of Diversity

There are many approaches to discussing the place and role of diversity in science and technology, since diversity intersects with freedom, equality, and pluralism. If taken only from the perspective of economic logic, diversity is often regarded as a mode of capitalising on differences. This idea leads to the application of methods and strategies that are strictly orientated to reducing the cost of innovation processes and projects, by fostering and facilitating participation. However, from the point of view of Science and Technology Studies, diversity is linked to communication, and sensibility, as well as to the ability to combine different approaches and contributions which reflect different ways of living and experiencing the world. Therefore, diversity can be approached from the point of view of the subjects of study, as referred to; or from the point of view of the actors participating in the real-life process, including scientists and innovators. Pivotal interest is placed on key characteristics - such as the scientific field of study, gender, age, or ethnical differences. In this manner, diversity within the fields of science and technology is linked to responsibility, because it considers the disparate levels of development in which people are living, as well as the different and unequal modes of accessing innovation which may depend on a wide range of variables, including gender. This also means that diversity is linked to the distinct audiences that can be potential users of innovation - whether as direct consumers, or wearers of innovative solutions. In sum, the question of future implications is linked to the real practice of responsible research and innovation that should normatively to take into consideration sociological variables and their effects on the consequences brought about by its uses. All this is particularly important nowadays, in the context of the debate on sustainability.

4   A Guide for Practitioners

The exercise of responsible innovation is mainly achieved by promoting greater public understanding of science and technology. If science and technology are social and cultural realities, innovators have the undertaking to listen to social actors and ascertain what they know and what they need to know, and how they can participate in the design of sustainable innovation. Security and surveillance, prevention, diseases, or other problems are examples of the complex and deeply entrenched and socially in-built territories, where engineering innovation is confronted by social and cultural questioning. The same thing occurs in the areas of smart transportation, clothing, digital tourism, or human/machine interface - where innovation is delivering many important enquiries related to education, sociability, or family relationships. It is a fact that engineering innovation seems to be committed to highlighting, and is concerned by social elements, but the question that is raised by most authors of Science and Technology Studies is that commitment and responsibility are often exercised in accordance with market considerations and short-term laws, following a strongly top-down perspective. In fact, the questions that need to be posed are of a different nature: Who is going to use these new means of transportation? Where? Who can afford this? What adjustments are going to be demanded from drivers and families? What would the generalization of driverless cars mean to people, and to what extent are policy makers and society in general effectively prepared to deal with them, and in which manner? These are some of the questions that collaborative and multidisciplinary innovation engineering can respond to [10].

In fact, most of these questions are related to the role of public participation in shaping innovation engineering, as well as the way that engineering collaborates with other areas, such as sociology, anthropology or geography. Science and technology studies tend to recommend implementation of a participation process that fosters greater information and knowledge about science and technology in society and collects the views of stakeholders and the public in general about the nature, appropriateness, and potential of application of new innovations. At this level, Science and Technology Studies emphasise the need for greater public involvement and engagement - whether in terms of students at educational institutions, or journalists in charge of disseminating content. There are many participative strategies that can be implemented to make people more aware of the use of the latest technological innovations, as well as their effects and risks. Most of them, as already noted, imply direct contact with the population and the use of research techniques that can simultaneously reduce anxieties or dispel misunderstandings about aspects of daily life that are linked to the lack of scientific culture, and that help reproduce contra-productive practices.

Table 2 summarises the principal measures that engineering innovation can develop to respond to the main factors of responsible innovation.

Table 2. Main areas of Science and Technology Studies' contribution to responsible innovation in engineering.

Themes	Possible measures	Objectives
Sustainability, future, and ecological and social risk	Improve knowledge of innovators in matters of culture and society Actively participate in forms of approaching innovation from a pedagogical stance	Raise awareness about the futures created in the present
Implications, values, social change	Apply and use a long-term outlook	Raise awareness concerning the social responsibility of science and technology
Public understanding, diversity, and participation	Invest in promoting social participation in multiple manners, including social networks, forums, meetings, and other forms of receiving opinions from the population	Promoting diversity, making social expectations real Increase cooperation and dialogue between common sense knowledge and scientific knowledge
The relation with media and science mediatisation	Raise scientific culture for public in general, including mediators, media, and journalists Improve the modes of publicising science and innovation	Promoting critical views about science and technology that can help rethink and reframe innovation

5   Concluding Remarks

The papers calls attention to many important variables shaping science and technology at different levels, including the social [11]. Such studies shed light on the need to uphold a linear way of thinking about development, fostering the need to ensure that innovation responds to sociocultural diversity, by fostering public understating of science and technology, as well as effective and broad public participation in technology and science matters [12–14]. Issues such as sustainability, responsibility and care for the future are of utmost importance when thinking about, and generating, innovation. This implies incorporating within the process of engineering innovation a plethora of topics and questions that are closely linked to respect for diversity, in terms of social class, culture, gender, or ethnicity. It is assumed that by combining different actors, with different views, who are living and experiencing different realties, innovation can increase its potential to generate originality, and also, more significantly, respond to societal and individual needs, in a sustainable and emancipative manner.

References

1. Trist, E., Bamforth, K.: Some social and psychological consequences of the longwall method of coal getting. Hum. Relat. **4**, 3–38 (1951)
2. Trist, E., Murray, H. (eds.): The Social Engagement of Social Science, vol II. University of Pennsylvania Press, Philadelphia. http://www.moderntimesworkplace.com/archives/archives.html. Accessed 1 Jan 2018
3. Trist, E.: The evolution of sociotechnical systems (1981). http://sistemas-humano-computacionais.wdfiles.com/local–files/capitulo%3Aredes-socio-tecnicas/Evolution_of_socio_technical_systems.pdf. Accessed 18 Jan 2018
4. Moniz, B.: A contribuição da Sociologia para a formação em Engenharia [Contribution of sociology to the engineering training], Munich Personal RePEc Archive, 8103, University Library of Munich (2002)
5. Felt, U., Fouché, R., Miller, C., Smith-Doerr, L. (eds.): The Handbook of Science and Technology Studies. MIT, Cambridge (2017)
6. Jasanoff, S. (ed.): States of Knowledge: The Co-production of Science and Social Order. International Library of Sociology, London, Routledge (2010)
7. Felt, U.: The temporal choreographies of participation: thinking innovation and society from a time-sensitive perspective. Pre-print, Department of Studies of Science and technology, University of Vienna. http://sts.univie.ac.at/publications. Accessed 1 Jan 2018
8. Adam, B.: Towards a new sociology of future (s/d). http://www.cardiff.ac.uk/socsi/futures/newsociologyofthefuture.pdf. Accessed 16 Jan 2018
9. Adam, B., Groves, C.: The Future Matters. Brill, London (2010)
10. Stilgoe, J.: Machne learning, social learning and the governance of self-driving cars. Soc. Stud. Sci. **48**(1), 25–56 (2018)
11. Rajagopalan, R., Nelson, A., Fujimura, J.: Race and science in the twenty-first century. In: Felt, U., Fouché, R., Miller, C.A., Smith-Doerr, L. (eds.) The Handbook of Science and Technology Studies, pp. 349–406. MIT, Cambridge (2017)
12. Miller, C.: Engaging with societal challenges. In: Felt, U., Fouché, R., Miller, C.A., Smith-Doerr, L. (eds.) The Handbook of Science and Technology Studies, pp. 909–913. MIT, Cambridge (2017)
13. Konrad, K., Van Lente, I., Groves, C., Selin, C.: Performing and governing the future in science and technology. In: Felt, U., Fouché, R., Miller, C.A., Smith-Doerr, L. (eds.) The Handbook of Science and Technology Studies, pp. 465–493. MIT, Cambridge (2017)
14. Ottinger, G., Barandiaran, J., Kimura, A.: Environmental justice: knowledge, technology, and expertise. In: Felt, U., Fouché, R., Miller, C.A., Smith-Doerr, L. (eds.) The Handbook of Science and Technology Studies, pp. 1029–1057. MIT, Cambridge (2017)

Determinants of the Well-Succeeded Crowdfunding Projects in Brazil: A Study of the Platform Kickante

Paulo Jorge Reis Mourão[1], Marco Antonio Pinheiro da Silveira[2(✉)], and Rodrigo Santos de Melo[3]

[1] Universidade do Minho & NIPE, UMinho, Braga, Portugal
paulom@eeg.uminho.pt
[2] Universidade Municipal de São Caetano do Sul,
USCS, São Caetano do Sul, Brazil
marco.pinheiro@prof.uscs.edu.br
[3] Universidade Municipal de São Caetano do Sul,
USCS/Universidade Federal do Piauí, UFPI, São Caetano do Sul, Brazil
rodrigosantos@ufpi.edu.br

Abstract. This work analyzed the determinants for the success of crowdfunding projects in the Brazilian platform Kickante, using the model proposed by Mourão and Costa (2015). We found that the total raised value per project increases with the number of investors ('kickantes') and with the minimum value ensuring the maximum prize in each project. In relation to the percentage of achieved target, it increases with the number of investors; but it reduces in case of projects supported by an informal group, or for projects regarding books, musical editions or music concerts. The value per investor raises with the minimum value invested for the maximum prize and with certain types of promoters (favoring informal groups or firms/companies).

Keywords: Crowdfunding · Crowdsourcing · Networking

1 Introduction

Crowdfunding is a recent phenomenon, which has expanded throughout the world, including Brazil. It is a model of funding based on online plataforms, which is related to the accomplishment of collaborative works, that is, some helping others in the elaboration of their projects. Crowdfunding takes place in social networks, which allow the accomplishment of collaborative work and use of the intelligence of the crowds. The crowdfunding proposal is related to another practice, called crowdsourcing, which consists of solving problems or promoting innovations, counting on the collaboration of other people through social networks [1].

According to Mourão and Costa [2], "in crowdfunding initiatives, the investor may or may not receive something in exchange for his or her help, either in the form of tangible benefits, financial return or even receive what has lent more interest, depending on the type of the project model. There are two forms of design modalities. In the first, only if the project is fully financed, in the agreed period, is that the promoter of this one gets the money. In the other modality, the creator of the project gets everything they can get, regardless of whether or not their goal is achieved."

For those with an entrepreneurial spirit, the crowdfunding model represents an interesting possibility of leveraging their projects, as it allows them to obtain financing under very favorable conditions. For these potential entrepreneurs, it is important to know what factors determine success in obtaining funders. Some studies have already tried to identify factors that lead potential investors to support projects. Mourão and Costa [2] defined a model that was adopted to carry out this research, which considers the Amount Collected in crowdfunding projects as dependent variable, and nine independent variables.

From the point of view of someone who is willing to contribute to a crowdfunding project, one of the relevant factors for their decision is the reward they will receive for contributing. Araújo and Verschoore Filho [3] sought to understand what rewards strategies become differential in successful crowdfunding projects. Among the results of the research, they highlight that the strategies contribute to the success of the project, being: exclusivity (when the project uses limited edition rewards, which remain available only during the project campaign), pre- (when a person buys the product in advance for a value below what will be released on the market). However, the authors state that the main success factors of a crowdfunding project are not in reward strategies.

The objective of this work was to analyze the determinants of the projects financed by crowdfunding in the Brazilian platform Kickante. This platform was used in the research for being the most collected in Latin America. There were more than 72,000 campaigns between 2014 and 2017 and obtained more than R$ 57 million in contributions. The site has three main categories: cause, creative and entrepreneurship, with a wide range of subcategories [4]. One of Kickante's most successful campaigns was: Bel Pesce: Legacy The Valley Girl, who was aiming to raise funds for a lecture tour about her latest book, was able to raise $ 889,385.37, 342% above the stipulated target [5]. Other Brazilian platforms also stand out, such as Cartase, Kickstarter, Indiegogo, StartMeUp, Broota, Impulso, Idea.me, and Bicharia [4].

In other words, the paper seeks to identify the reasons for some projects to raise more funds than others because some projects are funded and others are not. The structure of this paper is as follows. In addition to this introductory section, there are 3 more sections. Section 2 reviews the literature. Section 3 presents the empirical effort to identify the variables influencing various indicators of the Kickante projects in Brazil (from the total amount financed up to, value per donor or percentage of the initially requested value). Section 4 concludes.

2 Literature Review

2.1 Crowdfunding

The crowdfunding model is essentially based on the search for grants or investments to projects. Those who donate are made to do so for different reasons. It is a question of placing funding in the hands of the crowd [2].

The Kickante platform presents the definition of crowdfunding as the act of obtaining capital for collective interest initiatives. According to its organizers, the platform allows artists, NGOs seeking donations, engaged citizens, athletes in need of sponsorship, companies seeking to raise funds, have the opportunity to receive money to develop their project.

Crowdfunding can be seen as the process leading to attracting financial contributions from investors, sponsors, or donors to fund initiatives or companies (whether or not for profit). It is an approach to raise capital for new projects or projects requesting input from a large number of stakeholders [3].

According to Zuquetto [6], crowdfunding can be seen as the process leading to the raising of financial contributions from investors, sponsors or donors to finance initiatives or companies (whether or not for profit). It is an approach to raise capital for new projects or projects soliciting input from a large number of stakeholders.

It is well known that not all projects placed on a crowdfunding platform are successful. The Catarse platform was successful in 55% of the projects, while Kickstarter had 48.1% [6]. The author considers that the fact that they validate the data and evaluate the quality of the project information previously contributed to the success of Catarse.

Considering that those who provide the resource for a project are not regular investors or consumers, participation in a crowdfundinfg project is considered a social activity that results in benefits to the community. Thus, investors/givers feel themselves to be part of a special or privileged group of investors or consumers who are providing benefits [3].

2.2 Determinants for a Well-Succeeded Crowdfunding Project

The reasons that lead a person or company to contribute to a cause or project may vary depending on the context.

In his seminal study, Mourão [7] studied the motivations that lead people to contribute to a public campaign to support the fight against cancer. The study found that while donations are positively affected by economic growth, there are other influences, such as the fact that older people tend to contribute less. This observation shows that in the different situations in which a person or company is willing to make a contribution.

Mourão and Costa [2] analyzed the determinants of crowdfunding projects in the case of Portugal's PPL Crowdfunding, using a model that considered three dependent variables (Total raised by project, Value raised by investor, Goal percentage), representing the results obtained by the projects, and ten independent variables (Year, Duration of application, Number of supporters, Type of promoters, Population density,

Number of applicants, Project validity, Minimum invested with maximum premium, with minimum premium and Valencias).

Zuquetto [6] presents a distinct set of variables that are considered to be successfully correlated, based on a bibliographical survey: Size of the social network of the entrepreneur, Campaign quality signals, Platform page highlight, Creative local population, Initial funding from family and friends, Requested value, Project duration, Number of projects supported by the entrepreneur, Seals of approval, Culture.

A study by Agrawal et al. [8] indicated that the circle of friends and family has great importance in the financial investment at the beginning of the project.

3   Method

To analyze the success factors of the crowdfunding projects that were available in the Kickante platform, the Minimum Ordinary Squares Method was used, according to Mourão and Costa [2]. We analyzed data from 788 projects that were launched between 2014 and 2017 on the platform.

4   Empirical Section and Conclusion

In this section, we will discuss the main results we got after testing the set of potential determinants of the three dependent variables. We remind these dependent variables are: "total raised value per project", "percentage of target", and "raised value per investor". The independent variables are according to the review of literature. Therefore, we will use variables focused on the project's characteristics [9], on the number of investors [10] and on the prizes available to the investors [11]. The independent variables are: the number of investors ('kickantes'), the type of projects (1: Social; 2: Books; 3: Musical editions; 4: Trips/Events; 5: Theater pieces/Music concerts; 6: Science and Technology); the type of promoters (1: individuals; 2: group of individuals; 3: NGO; 4: Firms/Companies); the existence of multiple ends attributed to the raised money, the minimum invested value with prize, and the minimum invested value with the maximum prize. Table 1 exhibits the main descriptive statistics of our variables. We observe that the average project promoted by https://www.kickante.com.br/ has the following characteristics: it received 20537 *reais*, being excessively funded 49% above the required value, and each one of the 194 investors gave around 169 reais to this representative project. Most likely it was funded in 2015, and promoted by an individual. This representative project did not involve buildings repairing. The minimum invested value that granted a prize to the investor was around 28 reais and the minimum invested valued that granted the maximum prize to the investor had the median value of 3618 reais.

The correlation matrix of our variables (revealed if requested) exhibited some coefficients with a value higher than 0.60 (which suggests the possibility of endogeneous regressors). We also run the Durbin-Hu-Hausman tests against the endogeneity of the variables more likely to introduce endogeneity troubles and we were able to reject the hypotheses of exogenous regressors; full details available under request.

Table 1. Descriptive statistics

	Number of cases	Mean	Standard deviation	Minimum	Maximum
Total raised value per project (reais)	788	20536.9	48639.1	364	889385.4
Excess over target (=Funded Value/Required Value)	788	1.498	2.246	0.360	53.562
Raised value per investor (reais)	788	168.96	529.38	5.696	10307.1
Year	788	2015.7	0.931	2013	2017
Number of investors	788	194.13	413.18	1	4856
Type of promoters[a]	788	1.793	1.078	1	4
Type of projects[b]	788	2.826	1.740	1	6
Multiple functions	788	1.302	0.707	1	5
Validity of the project (1, yes; 0, no)	788	0.876	0.328	0	1
Minimum invested value with prize (reais)	788	28.858	183.98	0	5000
Minimum invested value with the maximum prize (reais)	788	3617.23	36309.35	0	1000000
Municipality (discrete vv)	788	104.019	39.129	1	154
Project focused on construction/repair of buildings	788	0.048	0.214	0	1
Comments on the project's kickante page	582	46.577	98.644	0	893

[a]Type of promoters: 1, a single promoter; 2, an informal group of promoters; 3, a non-profit organization; 4, a private company/firm
[b]Type of projects - 1: Social; 2: Books; 3: Musical editions; 4: Trips/Events; 5: Theater pieces/Music concerts; 6: Science and Technology

Therefore, OLS is not appropriate and we have to estimate our regressions using Two Stages Least Squares [12]. Table 2 exhibits these results, estimated by Two Stages Least Squares with heteroscedasticity robust standard errors.

Instruments: i. municipality, population density of the surrounding region, number of comments

Results from Table 2 show that the Total raised value per project increases with the number of investors ('kickantes') and with the minimum value ensuring the maximum prize in each project. Books and Events are type of projects which receive lower values. It is very challenging the evidence suggesting that the type of promoters does not influence the raised value per project. The overall R^2 is highly satisfactory for the firstly exhibited regression (0.77) and the F-val from the 1st stage regression is statistically significant at a level of 5% (considering the critical values for the 2SLS

Table 2. Estimations for total funded value, excess of funding, and value per investor (site: https://www.kickante.com.br/, 2013–2017)

	Total raised value per project	Percentage of target	Raised value per investor
Constant	6574.0	1.528	208.11***
Project focused on construction/repair of buildings	3412.2	−0.180	−104.19
Number of investors	69.066***	0.0003*	−0.131***
Type of promoters	1: omitted 2: −969.3 3: −1795.9 4: −2544.4	1: omitted 2: −0.146* 3: −0.050 4: 0.137	1: omitted 2: 41.951 3: 142.6* 4: 259.01*
Type of projects	1: omitted 2: −8443.0*** 3: −15670.7 4: −9030.5* 5: −2934.7 6: 532.4	1: omitted 2: −0.307*** 3: −0.253** 4: −0.016 5: −0.257*** 6: 0.025	1: omitted 2: −91.65** 3: −79.22 4: −71.39* 5: −45.64 6: 104.25
Multiple functions/ends for the invested money	−2197.69	−0.052	−31.91
Minimum invested value with prize	−1.938	−2.7e−5	0.028
Minimum invested value with the maximum prize	3.683***	9.9e−6	0.005***
R2/Number of cases	0.773/582	0.047/582	0.036/582
Wald chi2	438.65***	30.73***	61.06***
F-val (1st stage regression)	19.866**	19.86**	19.94**

Note: Robust heteroskedasticity standard errors between parentheses. Significance levels: *10%; **5%; ***1%.

relative bias) which favor the actual option for the chosen instruments. The test on endogeneity does not allow to reject the null hypothesis that the variables are exogenous.

Regarding the other columns (i.e., related to the achieved target and to the value per investor), additional evidences arise. First, the target increases with the number of investors; but it reduces in case of projects supported by an informal group, or for projects regarding books, musical editions or music concerts. Second, the value per investor raises with the minimum value invested for the maximum prize and with certain types of promoters (favoring informal groups or firms/companies).

References

1. Araújo, M.D.M.: Crownfunding: O que as campanhas de sucesso fazem diferente? Uma análise comparativa com uso de Fuzzy Set. Tese (Mestrado em Administração) – Programa de Pós Graduação em Administração, Universidade de São Leopoldo. São Leopoldo da Universidade Vale do Rio dos Sinos (UNISINOS), p. 89 (2017)
2. Mourão, P., Costa, C.: Investors or givers? the case of a portuguese crowd-funding site. Adv. Intell. Syst. Comput. **373**, 113–120 (2015)
3. Araújo, M.D.M., Verschoore Filho, J.R.S.: O Impacto das Estratégias de Recompensas no Resultado de Projetos de Crowdfunding: uma Análise Comparativa Qualitativa com Utilização de Conjuntos Fuzzy. In: ENANPAD 2017. Inovação, Co-Produção e Criatividade, São Paulo (2017)
4. Crowdfunding no Brasil. https://www.crowdfundingnobrasil.com.br/
5. Kickante. https://www.kickante.com.br/
6. Zuquetto, R.D.: Redes ego centradas e os projetos de crowdfunding: uma relação entre as características estruturais da rede social do empreendedor e o sucesso de proejtos de financiamento coletivo no Brasil. Dissertação de Mestrado. Programa de pos Graduação em Administração, Universidade do Vale do Rio dos Sinos (2015)
7. Mourão, P.: Portuguese public collections and the economic cycle: a seminal study. Int. J. Soc. Econ. **34**(12), 961–976 (2007)
8. Agrawal, A., Catalini, C., Goldfarb, A.: Family, friends, and the flat world: the geograph of crowdfunding. SSRN Eletron. J. **10–08**, 62 (2011)
9. Poetz, M., Schreier, M.: The value of crowdsourcing: can user really compete with professionals in generating new product ideas? J. Prod. Innov. Manag. **29**(2), 245–256 (2012)
10. Schenk, E., Guittard, C.: What can be crowdsourcing to the crowd and why? <halshs-00439256v1> (2009)
11. Kappel, T.: Ex ante crowdfunding and the recording industry: a model for the US. Loyola Los Angel. Entertain. Law Rev. **29**, 375 (2008)
12. Wooldridge, M.J.: Introdução a Econometria. Thomson Learning, São Paulo (2007)

Management of Intellectual Property, Technology Transfer and Entrepreneurship: Analysis of the Experiences of Universities in Brazil and Chile

Anapatricia Morales Vilha(✉), Fabio Danilo Ferreira(✉),
Luiz Fernando Baltazar(✉), Debora Maria Rossi de Medeiros(✉),
and Alberto Suen(✉)

UFABC Innovation Agency, São Paulo, Brazil
{anapatricia.vilha, fabio.ferreira, luiz.baltazar,
debora.medeiros, alberto.suen}@ufabc.edu.br

Abstract. Nowadays, innovation, technology and entrepreneurship have been increasingly valued as generators of wealth, competitiveness and development of contemporary societies, thus composing the political agenda of developed and emerging countries. The objective that guided the development of this work is to analyze the practices of Intellectual Property (IP) management, technology transfer (TT) and entrepreneurship in teaching and research institutions in Brazil and Chile. For that, the Federal University of ABC, located in Brazil, and the University of Chile, in Chile, were selected. In July 2017, interviews were conducted with professionals from the areas of TT, IP, and entrepreneurship in both universities. The results showed that the two institutions face, in a period of maturity and consolidation, some challenges related to the extension of IP management mechanisms, consolidation of partnership practices with business actors, structuring of performance indicators that reflect the intensity of the efforts, results and performance of the actions in technology management and innovation in both universities.

Keywords: Intellectual property · Technology transfer · Entrepreneurship
Universities

1 Introduction

One of the dimensions in which the word innovation appears most often is related to the production of new products, processes and knowledge-intensive services. This dimension is related to scientific and technological development, which is why the term Science, Technology and Innovation (ST&I) is used to describe the main elements of this dynamic process that offers environmental, economic and social impacts [1].

Nevertheless, it can be observed that more systematic studies on technological-based entrepreneurship and technological and innovative development processes are accommodated in the trajectory of innovation systems whose scientific, technological and innovative dynamics are more dense and distinct in countries like Brazil.

© Springer International Publishing AG, part of Springer Nature 2019
J. Machado et al. (Eds.): HELIX 2018, LNEE 505, pp. 863–869, 2019.
https://doi.org/10.1007/978-3-319-91334-6_118

Underlying, in the nascent technological companies (commonly characterized as startups), one of the main limiters is the low capacity of management and business knowledge for the markets. In this perspective, it is the role of universities to also offer opportunities, create channels, structures and form critical knowledge (conceptual and instrumental) in the generation and consolidation of entrepreneurship. This effort is certainly moving towards fulfilling an institutional mission that includes support for economic and social advancement.

In view of the aforementioned elements, the objective that guided the development of this work was to analyze the practices of IP management, TT and entrepreneurship in teaching and research institutions in Brazil and Chile. For that, the Federal University of the ABC exam, located in Brazil, was selected; and University of Chile, located in Chile, being public institutions of education and research with notorious recognition in the scientific and technological scenario of Latin America.

In this sense, the work is based on the assumption that these teaching and research institutions have gone through a trajectory of IP development generated by their researchers; transfer gave its technological knowledge to other actors of the country's innovation system, as well as fostering the entrepreneurship of technological content. Nevertheless, they face the challenge of maturing the efficient management of their technology and innovation processes.

For this purpose, interviews were conducted with professionals from the areas of TT, IP, entrepreneurship in July 2017 and most of them were recorded, after authorization of the interviewees. To conduct the face-to-face interviews, we developed a structured research tool, with open and closed questions, which structured research axis and which include:

- History of universities and their areas dedicated to the management of technology and innovation.
- Interactions with the productive sector for technology and innovation.
- Intellectual protection processes of developed solutions.
- Support for technological content entrepreneurship.

2 Elements of Literature Review

In Brazil, public politics that support Science, Technology and Innovation (ST&I) have been acknowledged as relevant for the competitive insertion of the country in the international scenario. Because of its scope, complexity and diversity of social actors and institutional arrangements, the issues that involve interrelations inherent to ST&I activities play an important role in the national political agenda [2].

Nevertheless, the fact is that innovation matters have gain strength nationally only by the beginning of year 2000, providing the main tools to stimulate science, technology and innovation in the country. If we look at the elements belonging to the path of the industry development, as well as the actors creation, institutions and ST&I policies in Brazil, it's noticeable the establishment of a late industrialization process and guided by the model of replacement of importation, based on the acquisition of machines, equipment and technology from other countries, with low capacity of generating knowledge locally. Other aspects contribute to this scenario, such as low coordination of ST&I activities and the lack of articulation in the institutions, created to achieve goals in the area during their respective era, not being able to evolve and adapt to changes in the social, economic and technological scenarios [3].

In Chile, the scientific and technological production is concentrated in the third level education system, composed by 61 universities (16 among them being of the states responsibility) along with a council of Rectors of the public universities, 25 in total, which value most of the specialized human resources and infrastructure to develop R&D and generate knowledge to the country [4]. One must notice that, between 2006 and 2011, the universities were responsible for 91.2% of all national scientific production, although they still have low recognition because of the small environment for business in most of the country's regions [5].

Besides that, since 2008, Chile introduced changes in the regulations that encourage private investments in R&D activities, causing an impact in the participation of companies in the global financing of applied research and experimental development. The novelty of this approved initiative was the introduction of fiscal incentives for companies that perform R&D activities. Those incentives didn't exist in Chile. With Law n. 20.241, published in the official bulletin in January, 2008, modified in 2012, coming into operation with Law n. 20.570, the fiscal incentive for research is created [4].

In the context of the knowledge society, the university expands its traditional focus in teaching, research and extension, adding to its mission the direct actuation in the process of economic and social development of the society. Therein, in this context, we discuss the role of IP, TT and entrepreneurship support processes stimulated by them.

Regarding IP, its economic relevance is directly related to the fact of establishing a property right and, that way, providing its object with features of appropriateness and transferability. When the object consists of a technological innovation, the legal protection to IP becomes a way to take the profits resulting from the innovation, what can be seen as an incentive to innovative activity and expends with R&D [6].

From this discussion, arises the role of innovation as an important element to competitivity among companies and countries. However, the innovative dynamics is not limited to companies and, so, the universities are also relevant players in the knowledge production and in the development of new technologies – moreover, its interactions with the productive sector are proven to be important for the technological progress [7].

The TT can be seen as the process that allows scientific and technological knowledges to be accessible to a higher number of players that are able to develop and explore them in the form of new products, materials and services [8]. Beyond the company role – as the place for the innovative activity – this context reinforces the universities role (and research institutes) as producers and disseminators of knowledge and formation of qualified professionals to actuate in innovation activities.

The ability to undertake a business or the entrepreneurs' way of thinking stimulate the ability to perceive opportunities, be at risk and innovate [9]. Under the perspective of the role of universities to generate new entrepreneurships, activities concerning incubation of companies with technological content are one of the mechanisms of technological infrastructure to disseminate innovative activities.

3   Report of the Experiences of University of Chile and Federal University of ABC in the Management of IP, TT and Entrepreneurship in Their Institutions

3.1   History of the Universities and Their Innovation and Technology Management Areas

The University of Chile is a public third level education institution of national nature, being the oldest third level education institution in the country and one of the most esteemed ones in Latin America, as pointed out by several national and international awards. The university has an excellent team of professors, with high productivity level in scientific field and is permanently connected to consideration and action about national issues. The institutional mission of the university comprises targets such as creation, development, integration and communication of the knowledge in all areas and culture domains. Regarding the technological and innovative development, the university represents 28% of all deposited patents in Chile, turning itself in a relevant institution concerning all the efforts carried out regarding IP of the technologies developed in the country.

The University of Chile received 45 million dollars to create technological development programs and strategic innovation centers in key areas in the institution. At the same time, in 2017, the university have formed and qualified initiatives to support the use of intellectual protection, throughout the attendance of 700 participants in specialized courses including professors and researchers. In despite of the university positioning, it's noticeable the presence of structural limitations found in the country related to contact with companies abroad, financial sustainability of the institution for innovation activities, as well as attractiveness of researchers to conduct projects in innovation and technology development projects.

The Federal University of ABC (UFABC) was created in 2005 though the sanction of a Federal Law and arose from the perception about the importance of the interaction among several knowledge areas, and with the mission to conduct efforts to the concerning and serving of national and regional issues, throughout the cooperation among the public sphere, other education and research institutions and the productive sector.

In this context, in order to fill out its community demands for innovation, entrepreneurship, IP and TT, as well as to satisfy legal requirements, specifically the Law n. 10.973, from 2004, December 2nd, UFABC has created, in 2010, by means of a resolution in the University Council, its Innovation Agency, whose main purpose is to manage the institutional politics of IP, TT and entrepreneurship.

Basically, the Innovation Agency of UFABC seeks to be the main mechanism of bonding between companies and the university, promoting activities about technological research, entrepreneurship for researchers and students. However, the area faces the challenge of increasing its set of actions on interaction between the productive sector and the institutional arrangements to technology and innovation actions.

3.2 Interactions with the Productive Sector for Technology and Innovation

Regarding the TT processes, all interviewed institutions revealed a insufficiency of the productive sector in Chile, which requires additional efforts in the performance of initiatives on the interaction with companies, such as: licensing, collaborative technological development, or even technical services.

Under this perspective, most of the licensing in University of Chile are performed with national companies. With a relatively structured flow to this kind of interaction, the university informed that have established the following procedure for technological transference: (i) process assessment by a IP committee (formed by technical professors from the university); (ii) creation of a reference term; (iii) signing of an accord that goes to an area similar to juridical procuracy; (iv) sending of the antecedents to the university controller-ship; (v) publication of the transference.

In the Innovation Agency of UFABC, the TT Division is the main responsible for partnership between the productive sector and university researchers, performing tasks such as technology negotiation, invention exploration agreement, technical support in the establishment of accords, technological partnership prospections and analysis of technical and economic viability of technologies. Despite of the short age of the university, partnerships important to the technological research development were celebrated with companies such as BASF, Petrobrás, Thyssenkrupp, General Motors of Brazil, Mercedes-Benz, Braskem, among others. The partnership with Braskem resulted in a patent application deposited in the name of both parts.

Nowadays, we have 130 valid agreement with several public and private institutions, but, considering all of the established agreements since the beginning of the activities, this number reaches 232 partnerships assuming several types of collaborations. There are two licensing settlements for technology exploration established in 2018 and one IP cession term in negotiating with the development partner.

A relevant aspect relies on the fact that, with predicted exceptions, every technology contract celebration should be preceded by a public offer, from official electronic mail from the university in order to guarantee the equality of opportunities among the interested entities. This is a legal obligation which UFABC must respect due to its juridical nature (public university), which is a complicating factor for the agility in the processes and relationships with interested companies.

3.3 Intellectual Protection Processes of Developed Solutions

Regarding the intellectual protection processes of developed solutions, in 2015, University of Chile established an innovation policy which manage the patent request solicitations, whose flow comprises: (i) solicitation of a textual description of the solution state-of-art; (ii) following, a preliminary internal search is performed in the university; (iii) the search for anteriority which consists of the hiring of an external office, which is also responsible for the patent writing.

The management of IP is one of the main responsibilities of the Innovation Agency and, along with other activities, it is assigned to the IP Division. Its main responsibilities include the support in the identification of research projects with IP potential, assessment of agreement terms, partnership or celebrated contracts between UFABC and external institutions that involve rights upon an IP, promoting of events to disseminate the area, follow up of all processes related to the university and so on.

3.4 Support to Entrepreneurship of Technological Content

Concerning the nature of the efforts to support entrepreneurship, the University of Chile aids the creation of academic spinoffs, in despite of the absence of a company incubator, and being far from the private funds to perform these actions.

At UFABC, the promotion of initiatives that aim to foment and incentive the entrepreneurship culture inside the university is assigned to the Technological Entrepreneurship Division at the Innovation Agency. Among the main support activities in this area, one of the most important is the entrepreneurship challenge, which is a competition among business plans, the technological incubator for new business of technological content. On the other side, there are challenges in the promotion of integration, cooperation among technological research groups in the university, its undergraduate and graduate courses, students and professors, that can engage in the challenge of fomenting innovation actions. The arising opportunities should improve the cooperation between the private sector, public sector and education and research institutions like UFABC.

4   Conclusions

Innovation, technology, knowledge and competitivity. We are at a moment in the global economy in which these elements have been increasingly more valuable as stimulators of wealth, competitivity and development of modern societies, compounding the political agenda of developed and developing countries.

The purpose that guided the development of this work was to analyse the practices of management of IP, TT and entrepreneurship in education and research institutions in Brazil and Chile. Therefore, the Federal University of ABC and University of Chile were selected to be examined due to their notable acknowledgement in the technological and scientific scenario in Latin America.

So, this work assumes that these education and research institutions traversed a path of IP development provided by its researchers; transference has given its technological knowledge to other players in the country's innovation system, as well as the foment to entrepreneurship of technological content. Regardless, they face the challenge to mature management of their innovation and technology processes.

Still in this perspective, this work establishes the IP issues, TT and entrepreneurship in the center of the actions and strategies of universities that see in technology, innovation and entrepreneurship, important players in the performance of their mission in teaching, research and extension.

In face of the elements brought by the performed interviews, as well as the analysis of the institutional materials from both universities, we deduce that both institutions now face, during a phase of maturation and consolidation, some challenges related to:

- Amplification of the managing mechanisms for IP provided by its researchers, in local scope and, specifically aiming international market.
- Consolidation of practices for partnership with entrepreneur players, aiming the accomplishment of projects in technological development and licensing of technologies developed in the universities. Particularly, this challenge assumes a local environment, regarding Chile, and international when we examine interaction practices between UFABC and international companies.
- Structuring of performance indicators that make it possible to observe the intensity of the efforts, results and performance of the actions in the management of technology and innovation in both universities.
- Improving of a institutional policy of innovation that effectively encourage innovation in the academic environment.
- Establishing a culture of better interaction between entrepreneurship agents, technology and innovation by means of the accomplishment of interchange visits, internships and technical visits.

References

1. Fuck, M.P., Vilha, A.M.: Inovação Tecnológica: da definição à ação. Revista Contemporâneos **9**, 1–21 (2011)
2. Vilha, A.M., Maskio, S.: Trajetória das Políticas de CT&I no Brasil e o Impacto do Atual Ajuste Fiscal, XXI Encontro Nacional de Economia Política (2016)
3. Vilha, A.M., Fuck, M.P., Bonacelli, M.B.: Aspectos das trajetórias das políticas públicas de CT&I no Brasil. In: Marchetti, V. (Org.) Políticas públicas em debate. MP Ed., Santo André (2013)
4. Serralta, L.M.P.: Sistemas y estratégias de innovación em regiones de Chile. Tese de doutoramento apresentada à Universidad de Valladolid, Chile (2016)
5. Comisión Nacional de Investigación Científica y Tecnológica, CONICYT: Principales indicadores cienciométricos de la actividad científica chilena 2011. Informe 2013, Chile (2013)
6. Mello, M.T.L.: Propriedade Intelectual e Concorrência. Revista Brasileira de Inovação **8**(2), 371–402 (2009)
7. Vilha, A.M.: Relação Universidade-Empresa no Brasil: Reflexões sobre divergências e alinhamentos na formação de arranjos voltados à inovação. In: Zimerman, A. (Org.) Pesquisa na Universidade e o setor produtivo. 1 edn, vol. 1, pp. 145–166. Universidade Federal do ABC, Santo André (2013)
8. Bozeman, B.: Technology transfer and public policy: a review of research and theory. Res. Policy **29**(4), 627–655 (2000)
9. Caron, A.: Inovação tecnológica em pequenas e médias empresas. Revista FAE Bus. **8**, 25–29 (2004)

Best Practices of Corporate Governance in a Network of Tourism Firms in Brazil

Márcio Jacometti[1(✉)], Ellen C. W. Lago[1], Antônio G. Oliveira[1],
Luiz C. Oliveira[2], and Leandro R. C. Bonfim[3]

[1] Universidade Tecnológica Federal do Paraná, Curitiba, PR, Brazil
{jacometti,ellenlago,agoliveira}@utfpr.edu.br
[2] Universidade de Coimbra, Coimbra, Portugal
luizcesar3515@gmail.com
[3] Universidade Federal do Paraná, Curitiba, PR, Brazil
lrcbonfim@ufpr.br

Abstract. This article presents the results of a study held in a network of firms from the Tourism industry in the cities of Quatro Barras, and Campina Grande do Sul, located in State of Paraná, Brazil. The purpose of the research to identify the best practices of corporate governance that are emergent in this network. The method of investigation was participant observation and document analysis from secondary sources. The analysis technique was the content analysis of field notes from the participant observations and from the collected documents. The level of analysis was sectoral, and the unities of analysis were the managers from the tourism firms. The data analysis provided evidence for the identification of the basic principles of governance applied by the network according to the concepts of disclosure, fairness, accountability, and compliance. It was also noticed that the best practices of corporate governance might be applied to the model of network governance.

Keywords: Corporate governance · Best practices of governance
Network governance

1 Introduction

One of the strategies used by companies to minimize risks arising from competitiveness are the industrial agglomerations, whose main aspect to territorial proximity of economic, political, and social agents [1, 2], emerging as a possibility to increase the chances of survival and growth of organizations, and being fundamental for the existence of micro and small enterprises [1].

The Quatro Barras and Campina Grande do Sul tourism cluster, called the "Route of the Countryside Sensation", has been carrying out a social action with the entrepreneurs from the tourism industry. Thus, the cluster plays a key role in the economic and sustainable development of the region, as these municipalities are located in environmentally protected area, where the economy is restricted to those non-polluting activities, ruling out the possibility of the installation of large industries with economic potential but that do not conform to the local environmental legislation.

© Springer International Publishing AG, part of Springer Nature 2019
J. Machado et al. (Eds.): HELIX 2018, LNEE 505, pp. 870–876, 2019.
https://doi.org/10.1007/978-3-319-91334-6_119

To promote the constant development of the cluster, adopting the best governance practices needed to enable greater control over the actions of enterprises that are part of the Tourism Route, providing effective cooperation and results expected by the companies. In this study, it understood that those best practices originated from four basic principles: disclosure, fairness, accountability and corporate responsibility [3, 4]. Thus, this article presents the following research question: "to what extent the best corporate governance practices are manifested in a network of companies seeking to organize and strengthen the tourism sector in the region of the Route?".

2 Corporate Governance Practices and Network Governance

Corporate governance may be understood by different points of view, as for the company, governance is the control and transparency; while for executives, it is responsibility and commitment; for shareholders, it is democracy and justice, and for investors, governance is protection and security [5].

The Brazilian Institute of Corporate Governance (IBCG) defines it as a system by which organizations are directed, monitored and encouraged, so that the owners, council of administration, directorates, and control bodies are related [4]. Thus, governance can be understood as shared power or managed collective action, being appropriate for organizations of cooperative, democratic and associative nature, which aims to ensure transparency in the management and reduction of risks for investors, good practices of the professionalization of business management such as accountability and fairness [6, 7].

The best corporate governance practices originate from four basic principles: disclosure, fairness, accountability, and compliance [3, 4]. These practices transform the principles of corporate governance into objective recommendations so that interests aligned for the preservation and optimization of the value of organizations [4].

The first practice, disclosure, means transparency of information, which impacts business and may involve risks [3]. Transparency "refers to relevant and reliable information concerning the activities of the company, analysis of results and financial position as well as the prospects of the company" [8:68]. Fairness, the second-best practice, refers to the sense of equity in the treatment of shareholders, without providing any privileged information for anyone [3, 4, 8]. The third practice, accountability, is based on best accounting practices [3], in the sense that organizational information should be clear, concise, comprehensible and timely, and governance actors should fully assume the consequences of their actions. The fourth, compliance, means taking into account and being according to existing standards and regulations [3].

Just as in business, governance in industrial agglomerations arises for providing SMEs with greater chances of survival in the industry in which they operate [1]. Thus, studying governance in networks is important to comprehend the strategies adopted by those companies organized in clusters.

Governance in interorganizational networks is not a legal requirement as is the case of corporate governance. However, it is fundamental to the effectiveness of networks that are managed without the benefit of hierarchy or ownership [9]. The main difference between network governance and corporate governance, in the strict sense, lies in the fact that in network governance the governed actors are firms, whereas in corporate governance the governed actors are individuals [10].

Provan and Kenis [9] have presented three models of network governance: shared governance, lead-organization governance, and governance through a network administrative organization. In the shared governance, groups of organizations operate as networks, but they do not have a formal and exclusive administrative structure. In this case, governance can occur through formal meetings with representatives of companies or informally through actions of those who have an interest in the success of the network. In this case, the organizations themselves are responsible for managing the internal relationships with external actors [10].

Lead-organization governance in turn generally occurs in vertical relationships between customer-supplier, where there are a larger and stronger organization and a smaller and weaker set of firms [9]. In this model, network members share common goals, while maintaining their individual ones [10]. Finally, the model of network administrative organization arises from the inefficiency of previous models and works from the management of the network through a third-party administrative entity in order to coordinate and sustain the network [10].

3   Methods: Study Context and Research Design

The Association for Industry and Commerce of Quatro Barras and Campina Grande do Sul (QBCAMP) is a nonprofit organization formalized in 1988 by businessmen from both cities in order to gather the interests of the class and assist in the economic and social development of both municipalities [11], which are located in the metro area of Curitiba, the Capital of the State of Paraná, Southern Brazil.

It currently has 140 members from commerce, industry and service sectors, being twelve of these enterprises comprising the Center for Tourism Entrepreneurs, eight from the city of Quatro Barras and four located in the neighboring municipality of Campina Grande do Sul [11].

In order to promote the development of the collective interests of Brazilian tourism enterprises, in 2009 the Federal Government formalized a cluster of tourism in these municipalities with the purpose to implement joint actions to identify, order, promote and strengthen the relationship between family agriculture and tourism, promoting the insertion and qualification of products and services of family agriculture in the Brazilian tourism market. In general, clusters involve a productive specialization in the region in which they are embedded [1]. The enterprises belonging to the cluster of tourism from Quatro Barras and Campina Grande do Sul are characterized by Rural Tourism, Ecotourism, Adventure Tourism, Religious Tourism and Historical-Cultural Tourism [12]. The implementation of the Route of the Countryside Sensation (RCS) in the region resulted in a major change in the enterprises belonging to the cluster since they adopted new collective practices to strengthen the network and consequently the growth of the companies belonging to the Route. For investigating the extent in which best corporate governance practices as are manifested in a network of companies seeking to organize to strengthen the tourism sector in the region of Quatro Barras and Campina Grande do Sul, in the state of Paraná, participant observations were made during the regular meetings of the RCS, being held between August 2015 and September 2016, through field notes, in-depth interviews with cluster leaders, minutes of meetings and documents generated by the QBCAMP.

Thus, the researchers analyzed the mentioned data, employing descriptive-qualitative procedures through documental analysis of secondary sources and thematic content analysis [13] of field notes from the participant observations. In this sense, free observation and the content analysis method were adopted for being decisive instruments for studying the processes and products in which the researcher is interested in [14]. The level of analysis of the research is sectorial since it addressed the governance of the companies in the network and the units of analysis are the managers of the twelve associated companies.

4   Results

The first participant observation at the QBCAMP Tourism Center, in August 2015, took place in a meeting with thirteen associated enterprises and a representative of the public administration of the municipality of Quatro Barras. The purpose of the meeting was to demonstrate the group's goals and the benefits available and accessible to members as means to attract new members, according observed in others clusters in Paraná state [2]. As result of this meeting, the Center of Tourism Entrepreneurs divulged their activities to the visiting entrepreneurs, but there were no new members joining the group. It observed that the presentation of the work demonstrated whereas disclosure and accountability are the existing practice of corporate governance in the network, generating sense of security and confidence in the governance of the cluster.

In the monthly meeting of the Tourism Center held in September 2015, representatives of thirteen tourism enterprises were present. At that meeting, issues related to the execution of proposals previously decided by the network discussed, such as the formalization and delivery of claims to the respective of the municipalities, aiming at improvements in signalization and infrastructure for facilitating the access of tourists.

In meetings observed that the Tourism Center has somehow sought to comply with the norms and reiterated requests by the competent public agencies, since the installation of the signs identifying each commerce was a promise made to the entrepreneurs to facilitate the access and to improve the visibility and dissemination of their firms. As a result, the Secretary of Tourism of Quatro Barras informed the members of the Tourism Center that the signposting road project would be included in the financial budget of the Secretary for 2016, in order to meet the aspirations of the entrepreneurs.

During the observation of the monthly meeting of the Center of Tourism Entrepreneurs in October 2015, eleven tourist enterprises were present, three university professors (from UTFPR, ISAE/FGV, and UFPR), a representative of the Secretary of Tourism of Quatro Barras, the Secretary of Tourism of Campina Grande do Sul and two representatives of CACIASPAR. At the meeting, issues addressed related to the possibility of formalizing a Local Productive Arrangement (LPA) of Tourism from the network of existing tourism businesses.

As result of this meeting, the members of the Tourism Center were favorable to the attempt to implement a Tourism LPA in Quatro Barras, and Campina Grande do Sul, based on guidelines established for the formation of LPAs [14], but later, the group decided to decline the proposal. In this meeting, it was identified the adoption of the fairness as corporate governance practice in the RCS since all members present were entitled to manifest itself equally and the final decision was taken after the voting of all present member with equal weights on the decision.

In a monthly meeting held in May 2016 at the headquarters of the QBCAMP, representatives of seven enterprises were present with the objective of adopting strategies to attract new entrepreneurs to the group. It was evaluated the tactics of dissemination carried out in the year 2015 that did not have the desired effect, as well as discussed new strategies to strengthen the Tourism Center. There was a consensus on the need to create new graphic materials for the dissemination of the ventures belonging to the RCS. One of the entrepreneurs suggested the creation of a Winter Festival, where all the RCS members could exhibit their products and services, creating a local culture to give prestige to the region's enterprises.

The group adhered to the suggestion and began to develop ideas for the organization of the event, counting on a partnership of the Executive Branch of the Municipality of Quatro Barras. As a result, the responsibility for the preparation of materials for the dissemination and organization of the RCS Winter Festival defined. At the meeting, it was possible to identify the adoption of corporate governance practices regarding accountability when the goals and strategies that did not have the effect expected by the network assessed. The adoption of fairness and justice as best corporate governance practices also observed since all the members present had the right to express themselves and give their opinion on the issues addressed in an egalitarian way.

In the monthly meeting held in September 2016, on the property of one of the entrepreneurs belonging to the Route, eight representatives of tourism enterprises were present to decide on the details for the preparation of a tourist map for the dissemination of the enterprises. Given the divergence with respect to the amount and size of the material to be printed, it held a vote among the members to define what kind of material would be made. Since there was no consensus in the voting, the majority of the voting members were respected, showing the fairness among the entrepreneurs, given that each enterprise was entitled to one vote, regardless of the size of the company, which shows that all entrepreneurs are treated in the same way, with equality and justice in the decision of strategies that will be adopted for the group as a whole. The best governance practices adopted in the arrangement are summarized in Table 1, according Andrade and Rossetti [3]. The technique of participant observation provided evidence for the identification of the RCS as a shared governance model, where the companies are operating as a network. But, the network does not possess a formal and exclusive administrative structure [15]. Governance in the Tourism Center is held by holding formal meetings with representatives of companies or informally, through actions of those who have an interest in the success of the network, and the organizations are responsible for management of internal relationships with external actors.

It was also possible to notice that the main advantage of this model is the participation of all partners in decision-making [15] and the main disadvantage, as reported by Provan and Kenis [9], refers to the fact that this model is not very efficient, since it depends on the efforts of actors with diverse degrees of commitment.

Table 1. Summary of the best practices of corporate governance held in the RCS-QBCAMP (from August 2015 to September 2016)

Principles	Purposes	Observed actions
Disclosure	Transparency and fairness in disclosure of data and accounting reports	Disclosure of work done for group members and guests
Fairness	Sense of justice towards all shareholders regardless of whether they are majority or not, and also with other stakeholders	Members, irrespective of the size of the enterprise, have the right to decide and vote on the future of the Route, on an equal basis All members of the group have the right to express their opinion in an equitable way Opinions are taken into account; but decision-making follows the majority
Accountability	Responsibility to account based on the accounting and auditing practices	Presentation of results for group members and guests Assessment of previous year's goals
Compliance	Legal compliance and compliance with regulatory standards contained in internal regiments	Collective actions to promote compliance with the norms regarding the commitment of the municipal entity to install signaling boards

5 Concluding Remarks

The participant observation and analysis of field noted permitted to identify that disclosure and accountability are corporate governance best practices present in the RCS. Best corporate governance practices were identified as recurring in the meetings of the Tourism Center, so that disclosure, accountability, fairness, and compliance are constant practices in the network, demonstrating that they can be implemented by any organization, in any segment of the market. The present research made it possible to identify that although the RCS is framed as a model of shared network governance, best corporate governance practices can also be applied for strengthening the tourism sector in the region of Quatro Barras and Campina Grande do Sul, Paraná. Thus, the results of this study can generate a greater understanding of the advantages obtained by the adoption of best governance practices, regardless of company size and business sector, so that such principles are implemented effectively and efficiently, providing sustainable growth of organizations. Further research could be carried out to measure the economic and social results of good governance practices in networks.

Acknowledgements. This paper is resulting of the research "Análise da Sustentabilidade e da Governança em Arranjos Produtivos Locais: a institucionalização do APL de turismo de Quatro Barras e Campina Grande do Sul/PR", supported by Brazilian National Council for Scientific and Technological Development (CNPq) - Universal Call MCTI/CNPq 01/2016, process no. 402361/2016-4.

References

1. Lastres, H.M.M., Cassiolato, J.E.: Glossário de Arranjos e Sistemas Produtivos e Inovativos Locais. Instituto de Economia da Universidade Federal do Rio de Janeiro (2003)
2. Vale, G.M.V., Castro, J.M.: Clusters, Arranjos Produtivos Locais, Distritos Industriais: Reflexões sobre Aglomerações Produtivas. Análise Econômica 28(53), 81–97 (2010)
3. Andrade, A., Rossetti, J.P.: Governança Corporativa: Fundamentos, Desenvolvimento e Tendências, Atlas, São Paulo (2004)
4. IBGC. Instituto Brasileiro de Governança Corporativa: Código das Melhores Práticas de Governança Corporativa. IBGC, São Paulo (2004). http://www.ecgi.org/codes/documents/ibgc_may2004_pt.pdf
5. Vieira, S.P., Mendes, A.G.S.T.: Governança Corporativa: uma Análise de sua Evolução e Impactos no Mercado de Capitais Brasileiro. Revista do BNDES 11(22), 103–122 (2004)
6. Gadelha, P.: 10 Vantagens da Governança Corporativa (2013). http://bobsoftware.com.br/sitemap.xml
7. Jacometti, M.: Considerações sobre a Evolução da Governança Corporativa no Contexto Brasileiro: uma Análise a partir da Perspectiva Weberiana. Revista de Administração Pública 46(3), 753–773 (2012)
8. Jesus, S.M., Alberton, L.: O Processo de Implementação da Governança Corporativa nas Empresas de Capital Aberto: um Estudo com Ênfase na Auditoria. Revista Contemporânea de Contabilidade 4(8), 67–84 (2007)
9. Provan, K., Kenis, P.: Modes of network governance: structure, management and effectiveness. J. Public Adm. Res. Theor. 18(2), 229–252 (2008)
10. Roth, A.L., Wegner, D., Antunes Jr., J.A.V., Padula, J.A.: Diferenças e Inter-relações dos Conceitos de Governança e Gestão de Redes Horizontais de Empresas. Revista de Administração 47(1), 112–123 (2012)
11. QBCAMP. Associação Industrial e Comercial de Quatro Barras e Campina Grande do Sul (2016). http://www.qbcamp.com.br/
12. Quatro Barras: Secretaria Municipal de Meio Ambiente, Agricultura e Turismo (2017)
13. Bardin, L.: Análise de Conteúdo. Lisboa: Edições 70 (2010)
14. Triviños, A.N.S.: Introdução à Pesquisa em Ciências Sociais: a Pesquisa Qualitativa em Educação, Atlas, São Paulo (1987)
15. Rodrigues, A.L., Malo, M.C.: Estruturas de Governança e Empreendedorismo Coletivo: o Caso dos Doutores da Alegria. Rev. Administração Contemporânea 10(3), 29–50 (2006)

From Handshakes to Hugs: How Brazilian Business Fairs Enhance Relationships Between Exhibitors

Débora Regina Schneider Locatelli[1],
Marco Antonio Pinheiro da Silveira[2](✉) ⓘ,
and Paulo Jorge Reis Mourão[3] ⓘ

[1] Universidade Federal da Fronteira Sul, UFFS, Chapecó, Brazil
deboraschneider@gmail.com
[2] Universidade Municipal de São Caetano do Sul, USCS,
São Caetano do Sul, Brazil
marco.pinheiro@prof.uscs.edu.br
[3] Universidade do Minho & NIPE, Braga, Portugal
paulom@eeg.uminho.pt

Abstract. Business fairs are strategic tools for companies. However, when studying the Brazilian trade-fair literature, we observed a gap related to the discussion of inter-organizational relationships. This study investigated how business fairs promote the development of relationships and joint actions among exhibitors. Using an exploratory and qualitative approach, this paper focused on Brazilian metal-working companies from two Brazilian states (Santa Catarina and Rio Grande do Sul), which have been exhibitors at trade fairs. The study enabled the identification of a complexity of joint activities and different levels of integration resulting from companies' participation in these fairs.

Keywords: Business fairs · Interorganizational relationships · Joint actions

1 Introduction

The literature on interorganizational relationships has emphasized how a company's participation in events such as trade fairs can be beneficial. Relationships between companies can range from a simple exchange of information to the constitution of solid forms of collaborative work that allow dynamic competitive advantages such as cluster generation, strategic alliances, and joint ventures, among others. At the same time that companies participate in competitive systems, they seek to form connections with their peers [1].

These networks encompass a set of strategic relationships, both horizontal and vertical, with other organizations, whether suppliers, customers, competitors, or other entities, including relationships with other sectors and countries.

Inserting itself into the thematic of organizational networks, this study investigates relations between trade fairs exhibitors and their potential for the development of relationships and joint activities among the exhibitors and their ability or lack thereof to

© Springer International Publishing AG, part of Springer Nature 2019
J. Machado et al. (Eds.): HELIX 2018, LNEE 505, pp. 877–883, 2019.
https://doi.org/10.1007/978-3-319-91334-6_120

form organizational networks. As shown in Fig. 1, a varied set of actors participates in a business fair. As shown in Fig. 1, a varied set of actors participate in a business fair.

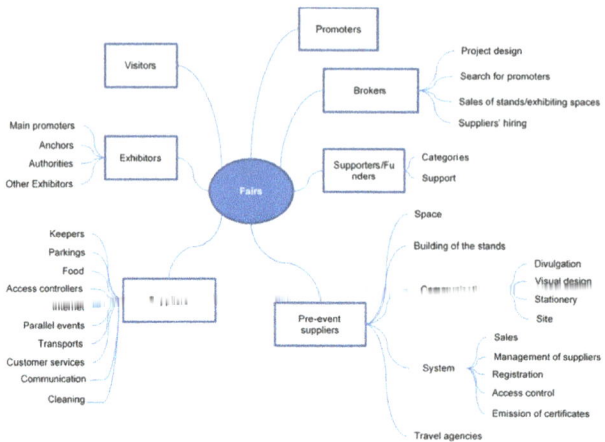

Fig. 1. Agents in a business fair [2].

As also presented [3], trade fairs are used as a strategic tool to win commercial, technological, and even collaborative partners. The remaining structure of this work comprises Sect. 2, which reviews the literature on interorganizational networks and joint activities. Section 3 presents the exploratory qualitative methodology used in the auscultation that we conducted among metalworking companies that participate in trade fairs in the states of Rio Grande do Sul and Santa Catarina. Section 4 analyzes the results obtained, while Sect. 5 presents the conclusion.

2   Literature Review

2.1   Interorganizational Networks

A relationship can be considered successful or not according to several factors. If it is considered satisfactory and provides benefits, partners will usually engage in efforts to make the relationship long lasting [3]. Vilmányi and Hetesi [4] report that the most important factors found in interorganizational relationships are the perceived quality of the products or services, satisfaction, trust, commitment, fairness, and loyalty.

For Oliver [5], organizational networks are formed by a set of companies that are interrelated through ties represented by continuing transactions carried out in their environments. These ties and transactions must be formalized between a limited number of interrelated firms by an administrative structure [5] or established with informal contracts [6].

Networking creates additional features for businesses, such as information and knowledge. Network partners are able to take advantage of these features depending on their positions in the network, their access to these resources, their reputations, and the strength of their links with other network partners [7].

Camarinha-Matos and Afsarmanesh [8] propose the concept of collaborative networks (RCs) that can be composed of different organizations or individuals. Most of the time, these are autonomous; they can be distributed geographically; and they are heterogeneous; that is, they have different forms of operationalization and different cultures, social capital, and objectives. However, they unite in order to improve their ability to reach common targets, and their interactions are supported by computer networks.

There are a variety of types and forms of RCs, ranging from simple negotiations between companies (B2B) to internationalization initiatives to dynamic collaboration networks supported by virtual organization-creation environments [8].

2.2 Joint Actions and Levels of Integration

Camarinha-Matos and Afsarmanesh [9] suggest that joint work can exist at different levels, although this is not widespread. The levels of integration are part of a gradual coalition process, presented in Fig. 2, which shows how each type of coalition constitutes a block that supports the next definition. Coordination extends networking; cooperation extends coordination; and collaboration extends cooperation. Thus, an integration process can develop and move from one stage to another [9].

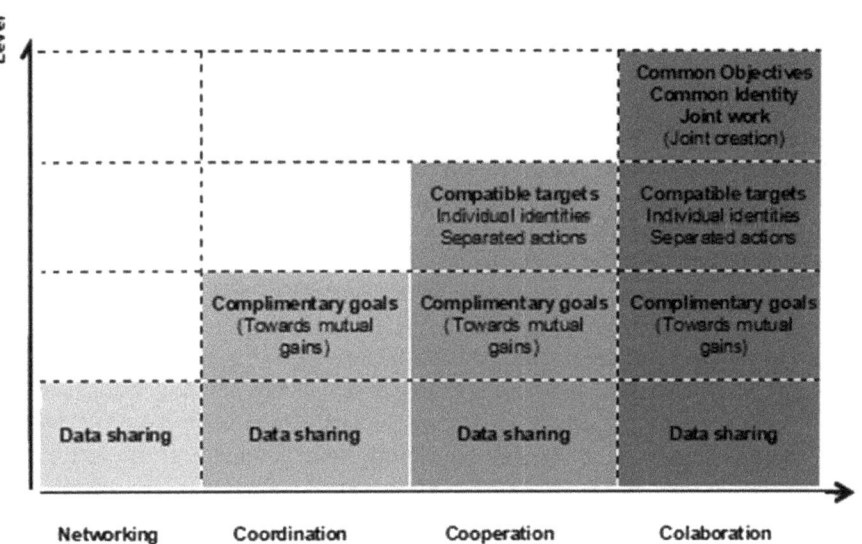

Fig. 2. Levels of integration of the Joint Actions [9].

Networking involves communication among those involved, with the exchange of information for mutual benefit. An example would be a group of companies that share information about their experience using a specific tool, like a type of software, but there is necessarily some common goal or structure that influences their work individually. In this type of coalition, there is no common generation of value [9].

In network coordination, besides the exchange of information, there is some kind of alignment of the activities of the companies involved, so that those involved gain more efficient results. It involves a harmonious way of working between partners. For example, coordinated activities occur when a number of heterogeneous organizations share information and together establish an activity of common interest that, in some way, maximizes a factor that will have an impact on them [9]. Coordination develops an alignment of activities so that better and more efficient results can be obtained; it works as a lever for efficiency but does not guarantee efficiency [10].

Cooperation involves not only exchanging information and adjusting activities but also sharing resources to achieve compatible objectives. There is a common project, which in most cases is not defined in collaboration but may have been planned by a single organization and requires some level of co-working, at least at points where one partner's results are delivered to another [9, 11].

Finally, network collaboration is the process by which organizations share information, resources, and responsibilities so that they can jointly plan, execute, and evaluate activities to achieve a common goal [9].

It is important to clarify the main differences between cooperation and collaboration. The first involves, in addition to the exchange of information and the adjustment of activities, the sharing of resources in order to achieve joint objectives. Collaboration is more comprehensive and refers to the process by which participants share information, resources, and responsibilities to achieve a common goal, sharing and rewards [12].

3   Method

To conduct the present study, an exploratory and qualitative approach was adopted. We used the perspective of the research participants and their knowledge about the research object [13]. We interviewed 20 managers or owners of metalworking companies that have been exhibitors at business fairs located in the west of Santa Catarina and north of Rio Grande do Sul, Brazil. The definition of the number of respondents followed the saturation tendency, which predicts that, at some point in the study, new interviews will add little to the results of the research due to the repetition of content. This is because even if the experiences seem appropriate to an individual, they are the consequence of a social process [14]. Further details on script design, structure of the protocols, records of interviews, and categorization from the records were omitted here due to the page limit, but these details will be provided upon request.

4   Analysis and Discussion

The 20 surveyed companies participate in different types of business fairs. The largest types are agribusiness and meat processing (with six representatives in each segment of the fair). Another six companies had participated as exhibitors at international fairs, and an additional five have visited trade fairs outside the country in their own areas of activity.

All of the surveyed companies had already undertaken several joint activities with other exhibiting companies. The following joint activities were conducted in the numbers shown in parentheses: the nomination of other exhibitors (9), the development of a product together (4), prospects for resellers and/or representatives (4), prospects for suppliers (4), potential mergers or acquisitions (1), the exchange of experience (1), and the potential creation of a new company (1).

Our research findings were also analyzed starting from the levels of integration established [9], as shown in Table 1. It should be noted that, according to these authors, an integration process can develop and change from one stage to another.

Table 1. Level of integration provided by joint activities carried out based on participation in a trade fair (source: prepared by the authors).

Joint actions	Networking	Coordination	Cooperation	Collaboration
Identification of partners	X			
Development of a product	X	X	X	X
Prospection of delegates	X	X	X	
Prospection of suppliers	X			
Marry sale	X	X	X	X
Joint export	X	X	X	
Prospection of a customer	X			
Potential fusion/acquisition	X	X	X	X
Share of experiences	X			
New company	X	X	X	X

The first level is networking. All of the joint activities developed go through this level of integration, which comprises the following activities: indications of other exhibitors, prospecting for suppliers, and prospecting for potential clients. These activities are fixed at this level of integration.

The joint activities identified at the coordination level are: joint participation in trade fairs, product development, prospecting for resellers and/or representatives, marry sales, joint exporting, potential mergers or acquisitions, and the potential formation of new companies.

Cooperation, which already involves sharing resources to achieve compatible objectives, is represented by the activities of prospecting for resellers and/or representatives and joint exporting.

The highest level of integration, collaboration, is one in which companies aim to achieve a common goal. They can achieve this level of integration by developing a unique image in the market involving all partners. Three of the joint activities developed by the exhibitors are included in this framework: (i) product development, (ii) married sales, potential mergers or acquisitions, and (iii) the potential formation of a new company. Figure 3 illustrates the levels of integration and the activities related to each level.

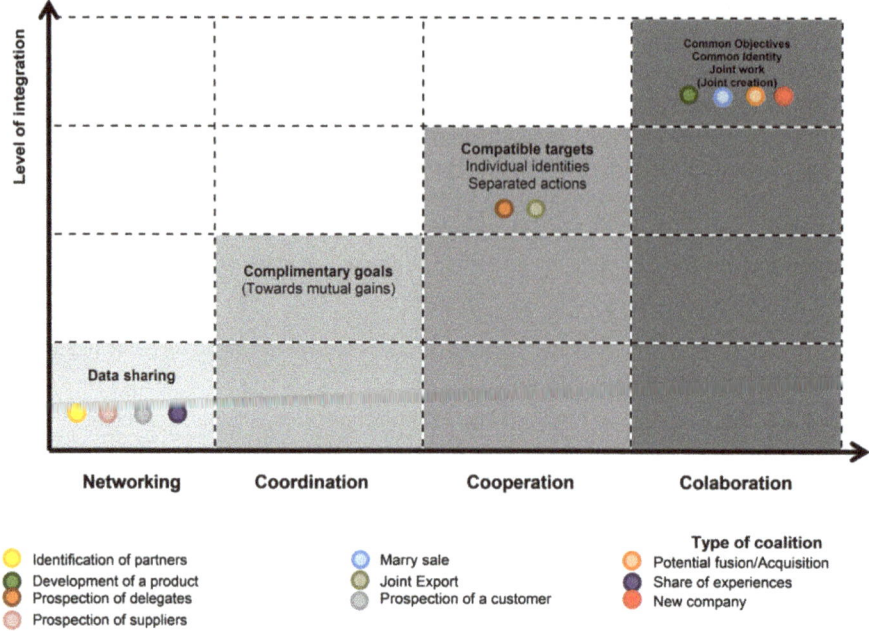

Fig. 3. A figure caption is always placed below the illustration. Short captions are centered, while long ones are justified. The macro button chooses the correct format automatically.

5 Conclusion

This work discussed the interorganizational relationships observed at business fairs in Brazil. It was observed that business fairs are complex events in which agents from various economic sectors participate. These agents work to establish different relationships, which can vary from occasional networking relationships to complex systems of collaboration. The results of the present study were obtained using qualitative analysis. The selection process was based on saturation selection. Our results demonstrate that business fairs are complex events where networking is most often established but where activities involving coordination and cooperation are also observed.

Thus, while the relationships established between companies at trade shows may seem ephemeral, the present study has shown that, in many cases, such relationships actually develop into long-term joint activities with a high level of integration, thereby forming collaborative coalitions.

A suggestion for new studies related to the topic developed in this article is the deepening through the use of quantitative analysis and exploring other regions and productive chains.

References

1. Barabasi, A.-L.: Linked: The New Science of Networks. Perseus BooksGroup, Cambridge (2003)
2. Locatelli, D.R., Silveira, M.P., Barbacovi, N.E.: As feiras de negócios como palco para a construção de parcerias entre empresas: o caso das empresas de produção de eventos. Revista Eletrônica de Administração e Turismo – REAT, pp. 30–42 (2017)
3. Giacaglia, M.C.: Organização de eventos: teoria e prática. Cengage Learning, São Paulo (2012)
4. Vilmányi, M., Hetesi, E.: The effect of dynamic relationship capabilities on B2B lolyalty. Theory Methodol. Pract. **12**, 79–88 (2016)
5. Oliver, C.: Determinants of interorganizational relationships: integration and future directions. Acad. Manag. Rev. **15**, 241–265 (1990)
6. Grandori, A., Soda, G.: Inter-firm network: antecedents, mechanisms and forms. Organ. Stud. **16**, 183–214 (1995)
7. Geigenmüller, A.: The role of virtual trade fairs in relationship value creation. J. Bus. Ind. Mark. **25**, 284–292 (2010)
8. Camarinha-Matos, L.M., Afsarmanesh, H.: Brief historical perspective for virtual organizations. In: Camarinha-Matos, L.M., Afsarmanesh, H., Ollus, M. (eds.) Virtual Organizations: Systems and Practices, pp. 3–10. Springer, Heidelberg (2005)
9. Camarinha-Matos, L., Afsarmanesh, H.: Collaborative networks: value creation in a knowledge society. Prolamat, pp. 26–40 (2006)
10. Denise, L.: Collaboration vs. C-Three (cooperation, coordination and communication). Innovating Reprint. vol. 7, no. 3 (1999)
11. Balestrin, A.: A dinâmica da complementaridade de conhecimentos no contexto das redes interorganizacionais, Porto Alegre (2005)
12. Loss, L.: Um arcabouço para o aprendizado de redes colaborativas de organizações: uma abordagem baseada em aprendizagem organizacional e gestão do conhecimento. Florianópolis (2007)
13. Flick, U.: Desenho da pesquisa qualitativa. Artmed, Porto Alegre (2009)
14. Gaskell, G.: Entrevistas individuais e grupais. In: Bauer, M., Gaskell, G. (eds.) Pesquisa qualitativa com texto, imagem e som: um manual prático. Vozes, Petrópolis (2008)

Elements of Entrepreneurship Under the Cultural Aspect: A Comparison of Culture of Portuguese Speaking Countries with Spanish Speaking Countries

Silveli Cristo-Andrade[1] and Ricardo Gouveia Rodrigues[2(✉)]

[1] Universidade da Beira Interior, Covilhã, Portugal
[2] NECE-UBI, Universidade da Beira Interior, Covilhã, Portugal
rgrodrigues@ubi.pt

Abstract. The objective of this study is to show aspects related to the entrepreneurship of cultures of Portuguese speaking countries, when compared to cultures of Spanish speaking countries. We investigate the literature on national culture and entrepreneurship and survey the theoretical connections between the national culture and opportunity perception, previous competencies, risk tolerance, society's perceptions, entrepreneurial intent and job satisfaction. The database on entrepreneurship is based on the Global Entrepreneurship Monitor. Based on regression analysis, results showed differences between residents in culture of Portuguese and Spanish speaking countries for all variables measured. There is evidence of distinct perceptions between cultures of Portuguese and Spanish speaking countries, which may affect how to manage entrepreneurship in such countries. Finding differences in perceptions among distinct cultures can facilitate the development of assertive public policies for the promotion of entrepreneurship.

Keywords: National culture · Entrepreneurship · Opportunity perception
Risk tolerance · Societal perceptions · Entrepreneurial intent · Job satisfaction

1 Introduction

Among the elements that influence the entrepreneurial development of a nation, we can highlight national culture. It influences greatly values and social behaviors, influencing all the activities of the respective society [1]. National culture affects the perceptions of individuals on the various aspects related to entrepreneurship, such as opportunity perception, risk tolerance, developed knowledge, society's perceptions regarding undertaking a business venture, the motivations of doing so and even the satisfaction of individuals towards their life and their job [2].

Countries which were colonized, tended to have acquired the culture of the colonizing country. It is reasonable to assume that an ex-colony adopts the culture of the country that gave rise to its colonization [3]. Countries with a history of colonization, such as Portugal and Spain, spread their culture throughout their former colonies, leading such countries to have a national culture like that of the colonizing country.

© Springer International Publishing AG, part of Springer Nature 2019
J. Machado et al. (Eds.): HELIX 2018, LNEE 505, pp. 884–890, 2019.
https://doi.org/10.1007/978-3-319-91334-6_121

This study aims to show aspects related to the entrepreneurship of cultures of Portuguese speaking countries, when compared to cultures of Spanish speaking countries. We addressed the aspects of opportunity perceptions, previous competencies, low risk tolerance, society's entrepreneurial perceptions, motivations to entrepreneurial intent and job satisfaction [1, 2].

2   Theoretical Framework

2.1   National Culture and Entrepreneurship

We can identify culture in four groups [1]: occupational culture, organizational culture, industrial culture and national culture. Because of the boundaries that often divide groups, regions or societies, we can also classify culture by aspects of civilizations, and cultural differences can originate in the ancestors of each civilization, thus, it generates what we consider the national culture, which we approach in this study.

In this study, we define culture as a set of values, beliefs, and behaviors that are divided and shared, whether socially or politically [2]. We can also consider it as a complex junction of knowledge, laws, customs, skills and abilities that people living in a society acquire, share, modify and transmit to others in their group [4]. We can say then that people are predisposed to act and think in the same way when they share this complex roll of values, beliefs, customs, being possible that in face of business opportunity or risks their behaviors are similar or close [4].

We define entrepreneurship as a dynamic process of creation, where the individual instils a great part of themselves, of passion and energy, solving problems, generating new ideas, having a vision of the best opportunities and risks [5].

2.2   Aspects of Entrepreneurship and Its Relation to National Cultures

When we focus on the individual so study entrepreneurship [1], motivations that drive a person to start-up a business are paramount. If viewed from the market perspective, we can observe laws, organizational environment and entrepreneurial activity. Finally, when we think of entrepreneurship for countries, culture comes across as an illuminating phenomenon of the motivations of the people under a given culture.

Authors such as [6] emphasize that entrepreneurship can be observed from two different perspectives, one of discovery and another of creation. Using one of these two approaches, the individual can discover a profitable opportunity and act on it to generate value, or else, they can create the opportunity themselves and consequently a value for their business. Regardless of whether we discover or create the business opportunity, the situation is the result of an action or reaction of an individual. Studies [2] have pointed to the national culture as one of the variables responsible for the level of entrepreneurship in certain countries.

Different people will identify the opportunities in different ways; they will act differently in the presence of a profitable opportunity and may take different modes of action to exploit them. Another fact we shall consider are the previous competencies that everyone possesses. As highlighted by [7], entrepreneurial individuals have some characteristics that distinguish them, for example, energy, passion and perseverance are elements that are part of their behavior, yet the previous competencies a person acquires bring a load of confidence that predisposes them to accept challenges more easily [8].

Entrepreneurs are more likely to accept risk than non-entrepreneurs are [1]. They will not undertake such risk without prior judgment of their benefit. Perceptions of risk are influenced by national culture, because certain cultures perceive risk much more tolerantly than other cultures, being much more risk-averse [9].

Relating corruption to entrepreneurial intent, [9] observed that society's perceptions on the act of starting-up a business are strongly influenced by the national culture, through the levels of corruption in each country. The cultural element influences society's perceptions with respect to entrepreneurship, regarded by some cultures as positive and by others as a form of domination of the poor by the rich [3].

Companies that have effective professionals have a high success rate. It is due to this positioning that the motivations to entrepreneurial intent is also an important aspect to consider. Among the factors that are pointed out as drivers for an individual to undertake a business venture, competitiveness has been one of the main cited motivations, with the perceptions of individuals shaped according to their national culture [1].

Another element that studies address regarding culture and entrepreneurship, such as in [2, 4], is the meaning of work for individuals who are under a particular national culture. Work can be perceived as an obligation or as a means of career advancement [4]. The satisfaction of individuals with their jobs suffers cultural influence, which may or may not favor entrepreneurship. If the individual is not satisfied with the job he/she has, this may simply be because the individual would like not having the need to work, or this could mean the will of wanting to grow and thrive in their career. Both depend on how the national culture perceives work [2]. In the second case, entrepreneurship can clearly be fostered as a way of overcoming individuals' dissatisfaction with their job and career [4]. In the first situation, the stimuli to entrepreneurship will probably have little effect, and the act of undertaking a business venture tends to be due to necessity [4]. Therefore, job satisfaction, combined with other elements that evidence entrepreneurial intentions, may reveal the role of national culture as an influencer of entrepreneurship.

3   Methodology

The original Global Entrepreneurship Monitor 2013 database has 66,114 observations. We excluded all non-Portuguese nor Spanish speaking countries, resulting in 3 Portuguese speaking countries and 11 Spanish speaking countries. Observations with missing data or duplicate data were also excluded, resulting in 2,639 observations.

Seven variables were selected, with the national culture being the dependent variable and opportunity perception, previous competencies, low risk tolerance, society's perception, entrepreneurial intent and job satisfaction, the independent variables. We also used four control variables: age, gender, education and the income of respondents. The following are all study variables with their respective descriptions.

National Culture (NAC): refers to the country of origin of the respondent, with 0 taken for the culture of Spanish speaking countries and 1 for the culture of Portuguese speaking countries. Opportunity Perception (OPP): refers to the individual perceiving business opportunities in the next 6 months in the region where they live, with 0 taken when negative, and 1 when otherwise. Previous Competencies (PRC): refers to the individual if they have the necessary skills to start a new business, with 0 taken when negative, and 1 when otherwise. Low risk tolerance (LRT): refers to the individual's fear of starting a new business, with 0 taken when negative, and 1 when otherwise. Society's Perception (SOP): refers to the individual's perceptions of how society perceives the entrepreneur; this variable was composed of 4 questions, whose answers were 0 when negative, and 1 when otherwise. It also addressed aspects related to the perceptions of the local society on status, career, standard of living and the dissemination of successful businesses; the value of this variable was the mean, by observation, of the answers of the 4 questions. Entrepreneurial Intent (ENI): refers to the motivations of the individual to undertake a business venture and measure the duality of opportunity versus need; the answers were on a scale of 3 points, with 1 being purely opportunity and 3 purely necessity. Job Satisfaction (JOS): refers to the individual's satisfaction with their current job; this variable was composed of 5 questions accompanied by a scale of agreement, with 1 being totally disagree and 5 totally agree. It also addresses aspects such as income, personal satisfaction, stress, autonomy and personal meaning of the work carried out; the value of this variable was the mean, by observation, of the answers of the 5 questions.

Control Variables: gender (0 for male, 1 for female), age, income (scale ranging from 0 if the respondent were in the 33% who earn less to 2 if the respondent were in the 33% who earn more) and level of schooling (scale ranging from 0 if the respondent had only pre-primary education to 6 if the respondent had masters or more).

To answer the objective of this study, we adopted the multiple linear regression statistical method with the ordinary least squares method, with the national culture being the dependent variable and the remaining ones being the independent variables. We initially generated the descriptive statistics of the variables, allowing initial analysis of the data. Subsequently, we compared the means of independent samples, to observe differences between the cultures Portuguese and Spanish speaking cultures. The model we adopted to answer the objective of this study was:

$$NAC = k + \beta * OPP + \beta * PRC + \beta * LTR + \beta * SOP + \beta * ENI + \beta * JOS + Controls + \varepsilon$$

4   Analysis and Discussion of Results

Comparing the means between the samples from cultures of Portuguese and Spanish speaking countries, we found that all mean differences were significant.

Results of regression analysis are shown in Table 1.

Although all items we measured have significant differences, the effects are small, which may indicate some cultural similarity between the two cultures analyzed, as

Table 1. Multiple linear regression (N = 2639)

Dependent variable: independent variable	NAC model without controls (Beta)	Model with controls (Beta)
OPP	0.07**	0.06**
PRC	−0.07**	−0.06**
LRT	0.06**	0.06**
SOP	0.12**	0.11**
ENI	−0.08**	−0.09**
JOS	−0.06**	−0.05**
Control variables Age		−0.05*
Gender		0.03
Education		−0.04*
Income		0.07
R^2	0.04	0.04

Caption: NAC = National Culture/OPP = Opportunity Perception/PRC = Previous Competencies/LRT = Low Risk Tolerance/SOP = Society's entrepreneurial perceptions/ENI = Entrepreneurial intent/JOS = Job satisfaction.
**The correlation is significant at 0.01 level. *The correlation is significant at 0.05 level.

expected, since they share Latin roots; possibly the differences that have emerged may be historically explained by the Arab influence, much stronger in Spain than in Portugal.

Respondents of cultures of Portuguese speaking countries assume a better perception of the opportunities, which may result in a greater entrepreneurial tendency among countries whose national culture is of Portuguese origin, highlighting cultural differences as pointed out earlier by [3]. On the other hand, those under the culture of Portuguese speaking countries consider themselves less prepared in terms of previous competencies. They perceive the opportunities better, but are less prepared, indicating a need for improvement in developing competences aimed at entrepreneurship [7].

Individuals living in culture of Portuguese speaking countries tend to be less risk tolerant, reinforcing findings by [9]. We see here a paradox, since individuals of Portuguese speaking countries, assume to perceive opportunities, but demonstrate more fear to take risks. It may be because they feel less prepared to start-up businesses.

Regarding society's perceptions, which includes respondents' opinion on status, career, standard of living and the dissemination of successful businesses; we find that individuals from cultures of Portuguese speaking countries perceive what society thinks about entrepreneurship better than individuals who find themselves in cultures of Spanish speaking countries. This aspect can be reinforced through public policies favorable to the valorization of the entrepreneur [9], since in the view of the respondents, Portuguese speaking countries tend to perceive the entrepreneur more positively.

Another result connected to the perception of opportunities was the variable that observed entrepreneurial intent. On a scale where 1 represented the opportunities and 3 the need to undertake a business venture, the result achieved indicates that residents in cultures of Portuguese speaking countries potentially start-up businesses due to opportunity more than due to necessity when compared to residents from cultures of Spanish speaking countries. This result indicated once again that those who are in cultures of Portuguese speaking countries tend to observe opportunities more than those who are in cultures of Spanish speaking countries, revealing another cultural difference between the two, something previously highlighted [2].

Job satisfaction, on the other hand, showed a tendency of respondents of cultures of Portuguese speaking countries to be less satisfied with their own work when compared to respondents of culture of Spanish speaking countries. The reasons for this lower satisfaction refer mainly to autonomy at work, the meaning of the actual job for the individual and income, since the means of these variables that made up the variable job satisfaction were significantly different between respondents from cultures of Portuguese and Spanish speaking countries with the average results of cultures of Spanish speaking countries being always higher. We can explain this by the results of the other variables measured. Individuals of cultures of Portuguese speaking countries, when compared to others of Spanish speaking countries, perceive opportunities better, but tend to be less risk-tolerant and consider themselves less prepared to undertake business ventures. Therefore, lower satisfaction may be because people coming from cultures of Portuguese speaking countries observe opportunities, but fear to take advantage of them because they are less prepared, which may result in lower job satisfaction, since they would like to take advantage of the opportunities that arise to them, similar to the one pointed by [2, 4]. Finally, we should note the low R^2 of the tested model. This value was naturally expected, since the national culture tends to be explained by several variables in addition to the variables tested here, because the culture appears to be a complex element and of difficult measurement.

When comparing the perceptions of individuals residing in cultures of Portuguese speaking countries with individuals residing in cultures of Spanish speaking countries, perceiving opportunities seems to be more common among people of Portuguese speaking countries. However, tolerating risk and assuming oneself as competent to undertake business ventures tends to be present more among people of cultures of Spanish speaking countries, even if entrepreneurship is due to necessity. This potentially causes lower job satisfaction of individuals whose origin is from a culture of Portuguese speaking countries. On the other hand, society's perceptions on entrepreneurship have been more positive in the vision of residents of cultures of Portuguese speaking countries, which may favor the development of entrepreneurship in these countries, when compared to countries whose culture is Spanish.

5 Conclusions

The objective of this study was to highlight aspects related to the entrepreneurship of cultures of Portuguese speaking countries, when compared to cultures of Spanish speaking countries. After analyzing six distinct aspects, we can conclude that individuals from cultures of Portuguese speaking countries, when compared to individuals residing in cultures of Spanish speaking countries, they assume to perceive opportunities and that society values entrepreneurial activity. On the other hand, we find that they tend to be less risk-tolerant and feel less prepared to undertake business ventures. This may be reflected in their job satisfaction, since respondents from Portuguese speaking countries declared themselves less satisfied with their job compared to respondents from Spanish speaking countries. There was evidence of distinct perceptions between Portuguese and Spanish cultures, which may affect how to manage entrepreneurship in countries with such cultures.

The main limitation of this study refers to the database used. Firstly, the base is from 2013, somewhat old, but it is the most recent basis made available by GEM. It is also worth noting the low explanatory power of the model, but this can be justified by the complexity of the dependent variable, the national culture, which requires several variables that need explaining and GEM's database provides few variables that can explain the national culture. In summary, such limitations reduce the generalization of results, but they provide evidence that national culture tends to play a relevant role in the development of entrepreneurship in a country.

Furthermore, we suggest adding more years to the database, performing a longitudinal study, with panel data, observing differences over time. Another suggestion is to perform the analysis of other cultures, such as Asian or Anglo-Saxon culture, since the cultures analyzed here are naturally close because they are of Latin roots. Finally, we suggest expanding the number of analyzed variables, using the actual GEM or other databases that also address aspects of entrepreneurship, such as the World Bank or the Doing Business Database.

References

1. Hofstede, G., Noordehaven, N.G., Thurik, A.R., Uhlaner, L.M., Wennekers, A.R.M., Wildeman, R.E.: Cuture's role in entrepreneurship: self-employment out of dissatisfaction. In: Innovation, Entrepreneurship and Culture: The Interaction Between Technology, Progress and Economic Growth, pp. 162–203. Edward Elgar, Cheltenham (2002)
2. Hayton, J.C., George, G., Zahra, S.A.: National culture and entrepreneurship: a review of behavioral research. Entrep. Theory Pract. **26**(4), 33–52 (2002)
3. Taras, V., Kirkman, B.L., Steel, P.: Examining the impact of culture's consequences: a three-decade, multilevel, meta-analytic review of Hofstede's cultural value dimensions. J. Appl. Psychol. **95**, 405 (2010)
4. Huggins, R., Thompson, P.: Culture, entrepreneurship and uneven development: a spatial analysis. Entrep. Reg. Dev. Int. J. **26**(9–10), 726–752 (2014)
5. Kuratko, D., Audretsch, D.: Strategic entrepreneurship: exploring different perspectives of an emerging concept. Entrep. Theory Pract. **44**(812), 611–634 (2009)
6. Alvarez, S.A., Barney, J.B.: Discovery and creation: alternative theories of entrepreneurial action. Strateg. Entrep. J. **1**(1–2), 11–26 (2007)
7. Thomas, A.S.: A case for comparative entrepreneurship: assessing the relevance of culture. J. Int. Bus. Stud. **2**, 287–301 (2000)
8. Acs, Z.J., Audretsch, D.B., Braunerhjelm, P., Carlsson, B.: Growth and entrepreneurship. Small Bus. Econ. **39**(2), 289–300 (2012)
9. Costa, L.D.A., Mainardes, E.W.: The role of corruption and risk aversion in entrepreneurial intentions. Appl. Econ. Lett. **23**(4), 290–293 (2016)

Collaboration Between University and Enterprise for Innovation in the Greater ABC Region, Brazil

Sandro Renato Maskio[1(✉)] and Anapatricia Morales Vilha[2]

[1] UFABC, Economic Observatory of UMESP, São Paulo, Brazil
sandro.maskio@ufabc.edu.br
[2] UFABC Innovation Agency, São Paulo, Brazil
anapatricia.vilha@ufabc.edu.br

Abstract. With the rise of flexible manufacturing and discussions about economic development models with the recognition that they depend on innovation capacity, this paper assesses the specific relationship between universities and enterprises in the innovation system of the industrial region of Greater ABC, Brazil. What is observed is that, despite the region's prominence in the country's industrialization process, relations between universities and companies focused on the transformation of knowledge into technology and innovation have not reached the desired maturity.

Keywords: Regional innovation · University-Enterprise Interaction
Technology and development

1 Introduction

In recent decades the industrial production model has promoted the rise of flexible manufacturing with the introduction of new technologies. According to Lahorgue [1] in this new pattern countries and their regions had to train themselves in producing and spreading knowledge, and in forming skilled labor capable of mastering new technologies.

These changes have brought a new horizon for discussion about economic development models with the recognition that they depend on a sustainable capacity for innovation.

Thus, the economic development associated with innovation has steadily influenced economic, social and technological debates, emphasizing relations between economic agents and their role in the political and economic scenario, whether at the regional, national or global level.

The objective of this article is to discuss the connections of the local innovation system in the regional scope, especially the relations between University and Company in the industrial region of the Great ABC, Brazil.

J. Machado et al. (Eds.): HELIX 2018, LNEE 505, pp. 891–897, 2019.
https://doi.org/10.1007/978-3-319-91334-6_122

2 The Relationship Between University and Enterprise in the Innovation System

According to Cassiolato and Lastres [2], the ineffectiveness of the neoliberal policies adopted in the 1990s in Brazil to increase the competitiveness of the national productive sector has broadened the need to structure public policy actions geared towards economic development. This context, according to the authors, has brought greater insight into the importance of innovation and knowledge as key determinants in development processes.

In this sense, CEPAL's structuralist approach, according to Ferrer [3], already pointed out that the change in the economic systems of developing countries requires a profound change in the productive structure in order to incorporate, in economic activity, knowledge as a fundamental instrument for development. This vision finds in the neo-Schumpeterian discussions a contribution on how changes in the technical-economic paradigms alter the technological frontier and create new sets of patterns, practices and productive processes, with the purpose of stimulating the development.

In the interpretation of Freeman and Soete [4] such changes will only be achieved by obtaining the capacity to create and improve, as opposed to the simple use of imported technologies. The development of this approach, as opposed to the argument of technoglobalism, recognized in the localized, national and regional environment the potential for the generation, assimilation and diffusion of innovation.

In this context, faced with the challenge presented, according to Nelson [5] the interest in innovation systems and their influence on the performance of economies intensified. These systems evolve a set of actors, including companies, suppliers, customers, political actors, universities, technology incubators and research centers.

According to Tatsch [6], access to knowledge, particularly those of science and technology, as well as the ability to apprehend, accumulate and use them, defines the degree of competitiveness and development of nations, regions, and sectors. This will be based on the organization, the characteristics and potentialities of the actors and the local productive process, and the effectiveness of the interrelation established between companies, universities and research and training institutions (Fig. 1).

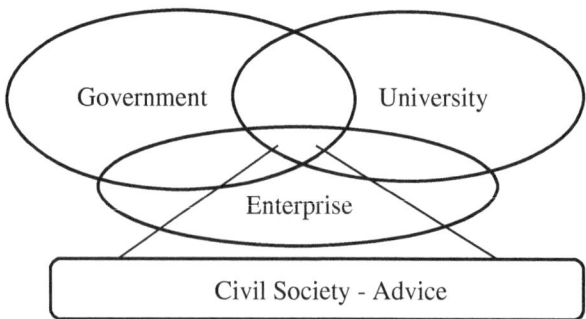

Fig. 1. The Triple Helix social structure (Source: Etzkowitz [7])

In the Triple Helix (TH) model developed by Etzkowitz and Leydesdorff [8], universities appear as key pieces in the interaction with company and government to improve environmental conditions that favor innovation, differing from the model proposed by the Sábato Triangle, according to which government takes the leading position because of the lack of availability of resources and the ability to coordinate to create a knowledge-based industry, according to Etzkowitz [7].

The research conducted by Lester [9] of the Massachusetts Institute of Technology (MIT) relates to the Local Innovation System identified fourth "University-Enterprise Interaction Channels", namely:

– Education and Training
– Codified Knowledge
– Local capacity to solve scientific and technological problems
– Discussion Area

Currently, according to Lester [9], universities' main focus on their interactions with companies has been technology transfer, licensing and startup development, influenced by success stories such as California's Silicon Valley and Route 128 in Boston, among others.

At the same time, Ferreira [10] points out that it is not all the regions that can become centers of technological entrepreneurship, since universality alone is not a factor that guarantees a successful innovation environment.

In spite of Ferreira's observation [10] that universities are helping to put knowledge into use, in favor of innovation and development, the following items will bring an x-ray of the interaction between companies and universities in Brazil, especially in the region of Great ABC, cradle of the industrialization of the country in the XX century.

3   The Relationship Between University-Enterprise in Brazil

According to data from the Innovation Survey (PINTEC), conducted by the Brazilian Institute of Geography and Statistics (IBGE) since 2000, the index of innovation in the country is low. According to the latest issue of the Global Innovation Index, Brazil ranked 69th in a total of 140 countries.

In the last 15 years, encompassed by the six editions of PINTEC, only 35% of the processing industries have made innovations in products or processes. Of these, only about 11% established some kind of cooperation for innovation, but only 4% have cooperated with universities or research centers. Despite the assertion of Ferreira [10] on the expansion of the participation of Brazilian universities in the generation and application of knowledge in Brazil, PINTEC data reveal that there is still a great challenge regarding the relationship between universities and companies.

Even in the state of São Paulo, where there is the highest level of innovation activity, more than 75% of the 1.25% of GDP spent on R & D in the country is allocated, the situation is no different. In the same period of analysis, according to PINTEC information, only 34% of companies innovate in products or processes, of which only 12% performed some kind of cooperation, and only 4.4% cooperated with Universities or Research Centers with the aim of promoting innovation.

In addition to the low level of interaction between companies and universities, most managers of companies that have made innovation evaluated the degree of importance of information received from universities and research centers as low or even insignificant, as shown in Fig. 2.

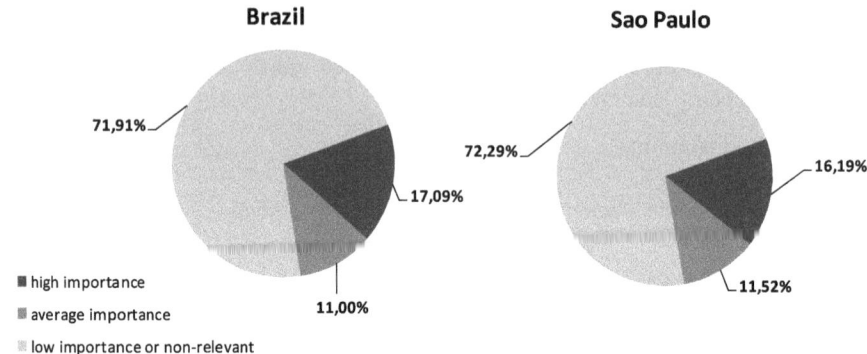

Fig. 2. Degree of importance of information received from universities and research centers (Source: IBGE/PINTEC - 2000, 2003, 2005, 2008, 2011, 2014 [11])

In addition to the different objectives and values between universities and companies, as well as the financial and temporal limitations that tend to hinder the relationship between both, in Brazil also weigh the late creation of universities and research institutions in the country and the late character of the industrialization of the economy Brazilian, depends on multinational capital, according to Suzigan and Albuquerque [12].

Even in the region of the Great ABC, the cradle of Brazilian industry in the twentieth century, the relationship between universities and companies did not evolve to a mature stage.

4   Deficiencies in the Interaction Between University and Enterprise in the Greater ABC Region

The region of Grande ABC, located in the State of São Paulo, was the cradle of Brazil's industrialization, carried out through the Import Substitution System, and presented its most advanced stage between the 1950s and the mid-1970s.

The main regional impulse was given by the installation and expansion of the automotive sector, which had high foreign direct investment by multinational companies in the sector, as well as strong public sector stimulus.

As a result, the region of Greater ABC became the largest producer of automobiles in the country, developing the vocation for metallurgy, and later for the petrochemical, rubber and plastic segments in particular.

However, despite the emergence of technological institutes and engineering courses, with emphasis on the Mauá Institute of Technology and the Faculty of Industrial Engineering, there was no establishment of solid links for adaptation and/or technology generation. Another indicator of this deficiency is that only in the decade of 2000 was the installation of a federal public university, with a more intense performance in leading research and the training of doctors.

In this context, interactions with the objective of promoting innovation in the Greater ABC region were small, and showed immatures along the industrial trajectory of the region, based at a great deal on passive learning and low propensity to transform knowledge into innovation.

In spite of this characteristic, PINTEC data referring to the last decades for a special sample of companies of the industrial sector of the region show a degree of interaction between universities and companies greater than at national level.

Considering the period between 1998 and 2014, about 63% of the region's industry companies claimed to have made product or process innovation. Approximately 40% of them performed some kind of cooperation with the aim of innovating. However, just fewer than 60% of them have established cooperation with universities and/or research centers (Table 1).

Table 1. Companies that innovate in products and processes in the ABC region and the cooperative relations with Universities and Research Centers

Edition	Period	Total of companies	Made innovations	A	B
2000	1998–2000	230	62.17%	33.57%	16.78%
2003	2001–2003	222	62.16%	44.20%	29.71%
2005	2004–2005	252	75.00%	43.39%	27.51%
2008	2006–2008	305	60.98%	25.27%	15.59%
2011	2009–2011	247	56.68%	45.00%	27.14%
2014	2012–2014	208	61.54%	53.13%	25.00%

Source: IBGE/PINTEC - 2000, 2003, 2005, 2008, 2011, 2014 - [11]
A - Percentage of innovative companies that have cooperated
B - Percentage of innovative companies that have cooperated with Universities and/or Research Centers.

Nevertheless, PINTEC also reveals that the main sources of information for innovation in the region, according to its managers, are customers and consumers, other areas of the company, suppliers, information networks, fairs and exhibitions, internal R & D departments, conferences, competitors and other group companies. Only after these appear, in the scale of importance, the information obtained from research institutions, technological centers and universities. Situation characteristic of the behavior of companies focused on technological incorporation and adaptation as innovation strategy, according to PINTEC construction methodology, based on the Oslo Manual.

For about 54% of the companies that have established innovation in the Greater ABC the information received from universities and research centers is of little importance or even of no relevance.

Even presenting a higher level of interaction between universities and companies for innovation, partly explained by the industrial vocation of the regional trajectory, it was not able to foster the formation of a mature regional innovation system in the strict sense of transformation of knowledge in innovation.

Nevertheless, the region reveals some capacity for generating incremental or adaptive technologies, largely leveraged by the multi-national companies themselves, such as the automobile sector according to Consoni and Quadros [13] and Amatucci and Bernardes [14].

5 Final Considerations

In the last decades the nuclei of the high technology have become an alternative to the development of the economies, at national and regional level. In order to do so, local characteristics and availability have been seen as a key factor for innovative capacity and economic development, as well as the effectiveness of local vocations and potentialities to create environments for innovation and technological diffusion, with repercussions on productive development and local economic progress.

For that, the deepening of the relationship between companies, universities and research centers is essential. However, in Brazil the difficulties faced by this relationship involve bureaucracy, slowness, restraint of resources, lack of entrepreneurial vision and a technology policy in universities; as well as the lack of knowledge of opportunities for interaction by a large part of the business sector, ignorance of processes and interaction procedures. Recently the Law of Innovation (Law 10.973/2004) that regulates the interaction between universities, companies and research centers has brought advances in order to broaden the interactions.

Specifically in the Greater ABC, in addition to the points mentioned above, it is worth mentioning the fact that the companies with the greatest capacity for innovation are multinational, and it is not among their priorities to develop endogenous technologies in the region, especially those of a disruptive nature. This also reduces the opportunities for interaction with local universities, even with the smaller companies that make up the production chains, which often meets pre-established and determined requirements of larger firms.

References

1. Lahorgue, M.A.: Pólos tecnológicos no Brasil: espontaneidade ou inovação social?. In: I Congreso Iberoamericano de Ciência, Tecnología, Sociedad e Innovación CTS+I, México. Organización de Estudios Iberoamericanos (2006)
2. Cassiolato, J.E., Lastres, H.M.M.: Sistemas de inovação e desenvolvimento: as implicações de política. São Paulo em perspectiva 19(1), 34–45 (2005)
3. Ferer, A.: Raúl Prebisch y el delema del desarollo en el mundo global. Revista CEPAL. Santiago – Chile, no. 101 (2010)
4. Freeman, C., Soete, L.: A Economia da Inovação Industrial. Tradutores: André Luiz Sica de Campos e Janaina Oliveira Pamplona da Costa. Editora da Unicamp. Campinas (2008)
5. Nelson, R.R.: As fontes de crescimento econômico. Tradutora: Adriana Gomes de Freitas. Editora da Unicamp. Campinas (2006)

6. Tatsch, A.L.: A dimensão local e os arranjos produtivos locais: conceituações e implicações em termos de políticas de desenvolvimento industrial e tecnológico. Ensaios FEE, Porto Alegre, vol. 27, no. 2, pp. 279–300 (2006)
7. Etzkowitz, H.: The Triple Helix: University-Industry-Government Innovation in Action. Routledge, New York (2008)
8. Etzkowitz, H., Mello, J.M.C., Almeida, M.: Towards, "Meta-Innovation" in Brazil: the evolution of the incubator and the emergence of a Triple Helix. Res. Policy **34**(4), 411–424 (2005)
9. Lester, R.K.: Universities, innovation, and the competitiveness of local economies: a summary report from the Local Innovation Systems Project. MIT – Industrial Performance Center Cambridge, Working Paper 05-010, December 2005
10. Ferreira, A.: Desenvolvimento regional: limites e possibilidades institucionais: um estudo de caso da região do Vale do Paraíba. Tese (doutorado) – Universidade Federal do Rio de Janeiro, Instituto de Economia, Programa de Pós-Graduação em Políticas Públicas, Estratégias e Desenvolvimento, UFRJ, Rio de Janeiro (2012)
11. IBGE. Technological Innovation Research - PINTEC. Editions 2000, 2003, 2005, 2008, 2011 and 2014. http://www.pintec.ibge.gov.br/
12. Suzigan, W., Albuquerque, E.M.: A interação entre universidades e empresas em perspectiva histórica no Brasil (Texto para discussão). Cedeplar-UFMG, Belo Horizonte (2008)
13. Consonni, F.L., Quadros, R.: Desenvolvimento de Produtos na Indústria Automobilística Brasileira: Perspectivas e Obstáculos para a Capacitação Local. Revista de Administração Contemporânea, Rio de Janeiro – RJ, vol. 6, no. 1, pp. 39–61 (2002)
14. Amatucci, M., Bernardes, R.C.: Impacto do Desenvolvimento de Produtos sobre a estratégia da Subsidiária: o caso do setor automotivo brasileiro. XXXII Encontro da ANPAD, Rio de Janeiro – RJ (2008)

Coworking and the Generation of Employment and Income

Milton Carlos Farina[(⊠)], Alvaro Francisco Fernandes Neto,
Antonio Aparecido de Carvalho, and Davi de França Berne

Universidade de São Caetano do Sul – USCS, São Caetano do Sul, Brazil
{milton.farina,alvaro.fernandes,antonio.carvalho,
davi.berne}@uscs.edu.br

Abstract. Coworking spaces have been growing in Brazil, due to the search for cost reduction, collaborative work, job and income generation. The common space utilizes by companies allow the shared use of resources, ideas and services. The aim of this research was to identify the elements for a Coworking space to create an innovative and entrepreneurial environment conducive to the generation of jobs and income. The research was qualitative phenomenological, with the proposal to describe the experience lived by coworkers in what concerns the identification of the fundamental elements in a space of services that allow innovation and business development. Interviews were conducted with representatives of five coworking spaces and the analyses of research data identified that the spaces of coworking provide the innovation when they offer spaces that allow the collaborative work, thus contributing to the generation of employment and income. The incentive to the coworkers to participate of the events that happen in the coworking space deepens the relationship between them and bring possibilities of solutions to their business.

Keywords: Coworking spaces · Employment · Income

1 Introduction

Munhoz et al. [1] state that in the coworking spaces there is a sharing of ideas, resources, and services among the participants, factors that in addition to the reduction of costs bring the collaborative character, allowing a stimulus for the formation of partnerships. Petch [2] describes that proximity has emerged as a key factor in coworking since it allows sharing of resources and also, through the physical arrangement of space, that facilitates contact, which in turn increases collaboration. This contributes to the development of links of interactions that can lead to innovation. The present research focuses on the following question: What are the fundamental elements for a coworking space to create an innovative and entrepreneurial environment that can lead to the generation of jobs and income based on the premises of Capdevila [3]? The objective was to identify the fundamental elements that a coworking space has to encourage innovation and the entrepreneurial character among the participants contributing to the generation of employment and income.

© Springer International Publishing AG, part of Springer Nature 2019
J. Machado et al. (Eds.): HELIX 2018, LNEE 505, pp. 898–904, 2019.
https://doi.org/10.1007/978-3-319-91334-6_123

In order to achieve this goal, the research made use of the Capdevila proposal [3], which brings the existence of four main elements for the coworking space: (1) Place: the chosen place must possess the appropriate resources and conditions that make possible the collaborative work; (2) Space: it is the element that is linked to the group's cognitive approach, which allows the individual interests to become complementary; (3) Events: it is the variable that allows the members of a shared space to participate in congress, internal meetings, or specific activities; and (4) Projects: these are resources that arise from the development and integration of the participants and the relationship with the external community.

The study is justified because, according to the census of Coworking Brazil [4], released in March 2018, coworking spaces showed growth of 114%. Compared to in relation to the 2016, were 378 spaces and in 2017 the number increased to 810 spaces. The research show that the spaces offer 56 thousand workstations, were moved in 2016 the value of R$82 million and generated 2.326 direct jobs and 1.174 indirect jobs. The reduction of jobs from 2015 can find in spaces of coworking contribution for generating employment and income. The research can provide theoretical content that allows other spaces of coworking to be used to increase the generation of employment and income as well as the adoption of publics policies that can accelerate the generation of employment and income.

2 Theoretical Background

2.1 Coworking

According to Koevering [5], the five core values that reflect the philosophy of a coworking are: (1) Collaboration: willingness to cooperate with others to create shared value; (2) Openness: free sharing of ideas, information and people; (3) Community: a group with a common purpose and shared thoughts; (4) Accessibility: coworking spaces arc accessible, socially and economically, for each type of worker and (5) Sustainability: resources are used together, which brings economic and ecological benefits.

In Santos' view [6], coworking consists of spaces offered by entrepreneurs to other entrepreneurs, professionals and people from diverse origins. This new element is differentiated from the traditional context, since space is used collectively, without the presence of bosses, surveillance or hierarches, ruled by values laded by a collaborative lifestyle.

According to Capdevila [3], coworking spaces (CWS) act as intermediaries between creative individuals and innovative companies, and there are several types of coworking spaces studied according to their characteristics of innovation: (a) First type of coworking space: no innovative community can be identified; (b) Second type of coworking space: there is a certain degree of collaboration among members who develop innovative collective projects that lead to the emergence of a sense of community; and (c) Third type of coworking space: it is the most innovative due to the emergence of a highly innovative community.

The creative economy involves the creation of products, effective production and distribution. Wünsch ([7] reports that knowledge, technology and innovation form the tripod that drives the generation of employment and income, as it promotes the development of new products or innovative services.

3   Methodology

The primary data were obtained from interviews with partners, owners of the coworking spaces located in the cities of São Paulo and São Caetano do Sul. This research is qualitative with a phenomenological methodology, which according to Ribeiro [8], p. 68, is "the doctrine that intellect bring immediately and absolutely a certainty about the essence of things" that means phenomenology brings the knowledge that people have about things".

The focus was to identify the fundamental elements for the creation of innovative environment offered by the coworking spaces, following the premises of Capdevila [3], regarding the degree of innovation. The type of research was phenomenological and according to Gil [9] can be used in the administrative area to research work satisfaction, or membership in an organization.

Five coworking spaces were searched, four of them did not authorize the disclosure of their names, henceforth they will be called coworkings A, B, C and D. The Cube space authorized the disclosure. The research question was: "What are the fundamental elements for a coworking space to create an innovative and entrepreneurial environment conducive to the generation of jobs and income, based on the premisses of Capdevila [3]?"

The script of the interviews was based on questions that sought to understand the predominant motives that cause people and companies to seek shared spaces. The research also sought to understand how the relationship between residents of shared spaces is, how shared spaces contribute to the innovation process and as a consequence if there is a contribution to the generation of employment and income.

4   Presentation and Analysis of Results

Based on the interviews the results are presented and analyzed for each coworking space.

Coworking A
Located in a commercial building in the city center of São Caetano in the State of São Paulo. The interview with the owner partner showed that the main focus of the space is to contribute to small businesses and microentrepreneurs participating in a space that allows cost reduction and mainly provides networking among its participants. The space offers training and events where it is possible to information exchange. Average length of stay for coworkers is up to six months.

According to the premises of Capdevila [3], this space belongs to the second type of coworking space, where there is a certain degree of collaboration between participants that can lead to creativity and innovation.

The strengths raised in this space of shared offices are the strengthening of networks of relationship and cooperation in projects that bring new products and mainly bring generation of employment and income.

Coworking B

Located in the city of São Caetano in a commercial building, which offers auditorium and space for coffee. The interviewee was the manager of the space who reported that the demand for shared space increased, mainly due to the need that professionals and microentrepreneurs have to reduce costs and to increase their relationship network also the relationship network. The space also focuses on the events that unite people from different areas of activity, generating the possibility of merging business and ideas.

The integration between coworkers occurs in the events that the direction of space promotes. In relation to the creation of an environment that promotes creativity and innovation, the interviewee affirms that many ideas arise mainly in the events and that there is a collaborative work among the participants.

It was possible to identify the concern that space has like with micro entrepreneurs, since resident coworkers or visitors are mostly micro entrepreneurs who are in the process of starting a business, are looking for reduced costs and need to strengthen relationships that can add value to the business. This coworking space is also in the second type, which according to Capdevila [3], offers a certain degree of collaboration that allows the generation of creativity and innovation.

In this shared space the focus is on offering participation in events with the intention of uniting people from different areas of activity, providing the possibility of merging business and ideas.

Coworking C

Located in Vila Olimpia, in the city of São Paulo, it has a direct relationship with Cubo Coworking, considering that it is the manager of this. He was the pioneer in Coworkings spaces in Brazil, with the inauguration in 2012.

The interviewee reported that the focus of the coworkers is the sharing of ideas, that the space offers from various events including the presence of coworkers from other spaces of coworking. He also informed that the area of coexistence that the space offers facilitates the exchange of information. In the interview were reported cases of coworkers who launched new products and services in the market. The space hosts hundreds of companies and the focus is to collaborate on creating new business. There is also the offer of events, especially those of technology and entrepreneurial culture, is one of the main hubs that foster entrepreneurship.

The mission of space is to connect people, leverage ideas and businesses. The type of coworking space in this case, according to Capdevila [3], is type 3, innovative type due to the participation of highly creative and innovative people, with an environment conducive to dissemination of information, mainly due to the various events proposed.

Here the emphasis is on the sharing of ideas between coworkers and in the area of coexistence that the space offers that favors the exchange of information.

Coworking D

Located in the center of the city of São Paulo, Coworking D is a space that hosts startups, is open to the public that through accreditation can use the common areas with the offer of free Wi-Fi. It was made available the possibility of visiting the space and interviewing the users, people who make use of services for free. On the occasion, an interview was made with a visitor who said she used the services of the coworking space because the work at home office did not have the desired effect, mainly due to the lack of contact with other people. The interviewee said that (attending) the space allows interaction with others and favors the creativity of their projects, as well as collaborating with projects of people from different areas. Coworking D focuses on innovative designs.

When participating in the event Coworking Brazil 2017, which brought together the coworkings of Brazil, it was possible to talk informally with one of the representatives of the space of Coworking D, who informed that the space has a focus on the creative and innovative process, a fact that corroborates the data provided by visitor interviewed.

From the report of the visitor and the informal conversation with the representative of the space it was possible to verify that the Coworking D is in stage 3, highly innovative, according to Capdevila [3].

Coworking Cube

The Coworking Cube is located in the region of Vila Olímpia, also in the city of São Paulo. The interview was made with the director of Cubo Coworking, who was entrusted with the implementation of said project created by Banco Itaú. The space has been in operation for two years, and has approximately 300 coworkers, who look for the space to start their startups. Space is a hub for entrepreneurship.

Applicants that intend to use the Coworking Cube space go through an evaluation process that considers: business models, acceptance by B2B and B2C companies, are companies whose products or services can bring solutions to the final consumer or to other undertakings; synergy with other startups, the focus is to make the partner companies in the development of new products or services and in the conquest of new clients; startup differentials such as brand and value invested initially.

The Coworking Cube offers events that count on the participation of other Coworkings, also receives universities interested in knowing the premises of the coworking.

According to the premises of Capdevila [3], the Coworking Cube is in stage 3 type highly innovative, because the said space exactly seeks that the collaborative and innovative process occurs among its participants. The focus of the Coworking Cube is to foster partnerships between companies for the development of new products and to attract new customers.

The research showed that all the interviewees come to collaboration, interaction, creativity and innovation, but D and Cubo spaces are conducive to collaborative work and that seek technological innovations and generation of new products and services.

5 Conclusion

The interviews revealed that the fundamental elements that propagate the generation of innovation and employment and income are the interrelation between people, participation in events and networks of relationship resulting from the use of the same space for work.

A factor considered limiting to the research was in relation to the fact that the subject is considered relatively new in Brazil, because there is little national literature about it, the data were obtained mainly from international articles. Much more than saving administrative costs, the spaces of coworking bring the possibility of partnerships, exchanges of knowledge, exchanges of information and the maintenance of a network of relationships. Shared offices present themselves as business platforms, with openness to improve and disseminate knowledge and stimulate creativity. The precepts established by Capdevila [3], are present in all coworking spaces surveyed and indicated different development stages. The innovations offered by the spaces promote the generation of employment and income, which meet the requirements established by Wünsch [7].

Coworking support for coworkers is not limited to tables and chairs, or availability of desks, telephones and mailing addresses. The greatest support is in the events whether they are in the Coworking space or other spaces. Some Coworkings play the role of business accelerators, such as the Cube Coworking, others choose to startups that can generate business with scale in the near future. Whatever the focus of Coworking, research shows that relationships between coworkers is essential.

Based on the interviews, it was possible to verify that each of the shared offices tends to provide improvements in order to have greater interaction among the participants, among them the physical expansion of the spaces, allowing the entry of a larger number of participants, fact which increases the possibilities of contacts and interactions resulting in co-creation and innovation. Another item that became evident is the encouragement and dissemination of events from the leaders of shared office spaces.

As a proposal to increase the generation of employment and income, it is suggested that public policies be created for the implementation and creation of public spaces regionally.

For future research it is suggested that the research be applied to other business centers in other states, so that it is possible to identify the different reality of Brazil.

References

1. Munhoz, A.C.C., Sengia, B.O., Fazzio, B.J., Oliveira, G.P.S.: Coworking and crowdsourcing: how innovative business models influence start-ups development. In: XVI Semeade, São Paulo, pp. 1–15 (2013)
2. The Urban Planner's Guide To Coworking: A Case Study Of Toronto, Ontario
3. Capdevila, I.: Typologies of localized spaces of collaborative innovation. SSRN Electron. J. 1–28 (2013)
4. Censo Coworking Brasil (2017). https://coworkingbrasil.org/censo/2017

5. The preferred characteristics of coworking spaces the relation between user characteristics and preferred coworking space characteristics: an attribute based stated choice experiments
6. Santos, R.M., Barbosa, F.C., Morais, H.M., Vieira, R.S.G.: Coworking and creative environment: organizational model for creative communities. In: XXXIX EnANPAD, São Paulo, pp. 1–12 (2015)
7. Innovation and job creation, as a phenomenon of the knowledge society. http://emporiododireito.com.br/backup/inovacao-e-geracao-de-emprego-como-um-fenomeno-da-sociedade-do-conhecimento. Accessed 19 Nov 2017
8. Ribeiro Jr., J.: Phenomenology. Pancast Editorial, São Paulo (1991)
9. Gil, A.C.: Como elaborar projetos de pesquisa. Editora Atlas, São Paulo (2017)

Barriers that Difficult the Entrepreneurial Action in the Process of Internationalization of Cosmetic Micro and Small Companies of the State of São Paulo

Milton Carlos Farina, Alvaro Francisco Fernandes Neto,
Antonio Aparecido de Carvalho[(✉)], and Davi de França Berne

Universidade de São Caetano do Sul – USCS, São Caetano do Sul, Brazil
{milton.farina,alvaro.fernandes,antonio.carvalho,
davi.berne}@uscs.edu.br

Abstract. Micro and small companies are essential for the economy, as they generate employment and income. The cosmetics industry in Brazil has been emphasizing international trade and micro and small companies are looking for other markets to conquer new clients and new partners, but the process for internationalization finds barriers. The focus of this paper is to identify and verify if the barriers are similar for both micro and small entrepreneurial companies. To meet the objective, a research was carried out with 41 micro and 115 small cosmetics companies. The data collection was carried out through a questionnaire, whose results are shown to be the same for micro and small enterprises, as they are: the difficulty to make changes in the products destined to external customers' demands, exchange rate fluctuations, delay in receiving the payments from commercialization and difficulties in obtaining capital.

Keywords: Micro and small companies · Cosmetics · Barriers

1 Introduction

According to the Brazilian Support Service for Micro and Small Companies - Sebrae [1] from the 1980s the Micro and Small Companies were gaining ground in Brazil and over time were gaining representativeness. On December 14, 2006, Law 123 was created establishing the Statute of Micro and Small Companies, which advocates the establishment of a differentiated tax burden for such companies.

Once companies operate regionally, many of them seek new markets and strive to win international clients. Rasmussan and Madsen [2] emphasize the need for prior planning, knowledge of market, legislation requirements and of tariff and non-tariff barriers.

Cosmetics companies operating in Brazil, have shown growth in exports. Data from the Brazilian Association of the Personal, Perfumery and Cosmetics industry - ABHI-PEC show that, from 2006 to 2016, exports grew more than 293% in the period [3].

This article seeks to answer the following key question: What are the barriers that difficulty the process of internationalization of micro and small cosmetics companies in

© Springer International Publishing AG, part of Springer Nature 2019
J. Machado et al. (Eds.): HELIX 2018, LNEE 505, pp. 905–911, 2019.
https://doi.org/10.1007/978-3-319-91334-6_124

the state of São Paulo? The objectives are to identify the barriers that end up making it difficult for micro and small companies in the cosmetics sector of the state of São Paulo to internationalize and identify if there are differences in the barriers when comparing micro and small companies.

The choice of theme is justified by the importance that micro and small companies bring to the economic scenario, generating employment and in-come to Brazilian GDP and to the growth of cosmetic Brazilian industry.

The research method was quantitative, descriptive and non-probabilistic and data were collected by questionnaire answered by 41 micro and 115 small cosmetic producers in the state of São Paulo.

2   Theoretical Framework

The term entrepreneur is of French origin, for Vries [4], entrepreneur is the one who makes decisions, assumes responsibilities and is subject to risks. Aguiar [5] describes the entrepreneur as being the person who accomplishes things always in different ways, a fact that brings satisfaction in the execution of tasks. Schumpeter [6] makes the association of entrepreneurship and innovation, both terms go together, in the author's view the entrepreneur is always in search of new opportunities. Ribeiro and Teixeira [7] bring the idea that the entrepreneur has the ability to combine all productive resources in the making of products or in the provision of services that can bring satisfaction to users. Micro and small enterprise companies present themselves to the society in the face of the capacity to create jobs, as well as the generation of foreign exchange through exports made to the international market.

Sebrae [8] establishes the difference between micro and small companies based on the number of employees that the companies have. For the industrial branch, it is considered micro-enterprise that has up to 19 employees and a small company with 20 to 99 employees.

Costa and Leandro [9] report that micro and small businesses currently number in Brazil is 14.812.460, in the State of São Paulo from January to November 2016 in absolute values was R$ 532.2 billion. The Micro and Small Companies in the state of São Paulo represent 98% of companies, account for 49% of jobs and 37% of payroll, compared to 27% of the state's GDP.

According to Teixeira and Pincchai [10] the approach of international entrepreneurship brings the combination of innovative behavior in the micro and small companies, seeking to cross borders with the minimization of risks, however there are barriers that can make the internationalization process unfeasible.

Machado [11] lists three stages of the internationalization process: in the first stage the company analyzes what has already been explored in the domestic market; in the second stage the company seeks the expansion of foreign business and in the third stage, company is already part of the international market.

The author points out that in the first stage the company has four determining factors for the search of the international market: 1 - better search for supplies abroad; 2 - government incentives and exchange rate fluctuations; 3 - reduction of costs due to transport and communication innovations and 4 - follow the tendency of competitors as a defense.

According to Garcia [12], the study of barriers to internationalization identifies a number of factors, such as: lack of technical capacity to develop and produce products that meet the needs and/or desires of external markets. It is noticed the insufficient existence of laboratories and centers of research and development. These include barriers to the lack of a well-known, respected, desired brand or not yet have distribution channels that guarantee the delivery of the products within the correct term and conditions.

Cordeiro [13] lists other elements that hinder the internationalization of a micro or small enterprise as the lack of an export culture, as well as the failure to develop a technological standard consistent with the market to which it intends to export.

Caetano and Paiva [14] describe barriers to the internationalization of micro and small companies; they are the endogenous and exogenous barriers.

The endogenous barriers are those related to the internal market such as: access to information, country cost, obtaining financing/government investment, organizational management, economy of scale and technical knowledge/labor.

Exogenous barriers include those related to entry into destination countries, such as cultural barriers, national protectionism, partnerships, legislation, import quotas, packaging, exchange rate fluctuations, economic instability, sanitary and phytosanitary barriers, trade agreements, distribution channels, price competition, technological level, antidumping laws, global economic instability, logistics, suppliers and sanctions.

Faced with such barriers, it can be understood that micro and small companies require greater effort to adapt to the insertion in the international market.

The representativeness of the Brazilian cosmetics, perfumes and personal hygiene products industry is demonstrated by the data extracted from the Brazilian Association of the Personal, Perfumery and Cosmetics Industry - ABIHPEC [3] from 1996 to 2014. There was a significant evolution in net sales, from R $ 4.9 billion to R $ 43.3 billion.

Data provided by ABIHPEC [3] show that Brazil's 2016 representativeness in HPPC (Sector Development Program for Personal Hygiene, Perfumery and Cosmetics) is 49,1% of the market, while other Latin American countries represent 50, 9%. In comparison with the rest of the world, Brazil reaches 6.6% of participation.

Research carried out by ABHIPEC [3] with the support of Euromonitor, shows that Brazil occupies the 4th position in the ranking among the top 10 global consumers of HPPC (personal hygiene, perfumery and cosmetics), amounting to US $ 29.3 billion (twenty-nine billion and three hundred million dollars). In relation to the contribution of the sector to the movement of the economy, Brazil stands out as the 1st industrial sector that invests more in advertising and it is also the second industrial sector that invests more in innovation.

In relation to exports, the research shows that in the year 2016, the values exported reached US $ 278.8 million, but compared with the year 2015, exports dropped of 13.7%. Sebrae (2017) reported that exports in the last ten years showed a growth of 293.5%. Brazil occupies the 24th position in the world ranking So that the sector can improve its position with the support of ABHIPEC, APEX, SEBRAE and other support entities.

In this sense, ABHIPEC [3] reported that cosmetic manufacturing MPEs show growth in the international market, exports in the first quarter of 2017 compared to the same period of 2016 increased by 0.7%.

3   Research Method

In order to identify the barriers that end up making the process of internationalization of small and micro cosmetic companies in the state of São Paulo impossible or difficult, a quantitative, descriptive study with a non-probabilistic sample was made. The data collection was based on the application of a questionnaire divided into two parts, the first one aims to identify the characterization of the companies, the second part focuses on the barriers that prevent or hinder the internationalization process, was applied to micro and small companies of cosmetics of state of São Paulo, answered by partners, owners and managers. Eleven variables were utilized, based on the surveys of Garcia [12], Rezende and Serpa [15] and Cordeiro [13]. The variables are: delayed payment of sales, tariff barriers, bureaucracy in processes, logistical costs, cultural differences, difficulty in obtaining capital, difficulty to provide after-sales service, lack of government incentive, need to make changes to the product to meet requirements foreign clients, exchange rate fluctuations and restrictions imposed by international law. Based on the statements elaborated on the variables listed above, the respondent gave his degree of agreement, ranging from 0 to 10, with 0 being "totally disagree" and 10, "totally agree".

The questionnaire was inserted into the Google Forms software and based on an electronic search, it was identified 17.000 companies in the cosmetic segment of the state of São Paulo. A message with the research link was sent to the electronic mail of these companies.

A total of 156 questionnaires were received and data were entered into the SPSS 23 software to carry out the statistical analyses. Micro companies that answered the research totaled 41 and small companies totaled 115.

4   Discussion of Results

To achieve the objectives outlined, the research had the application of statistical analyzes of frequency to identify the profile of the companies and the barriers that were difficulties in the process of internationalization of the MPEs in the cosmetics field.

4.1   Company Profile

The time of operation in the national market shows that 15 (36.6%) of the surveyed companies have a time of activity between 5 and 7 years and 13 (31.7%) are in the market in a period of 3 to 5 years. The results show that 13 (31.7%) are exporting their products in the period between 1 and 3 years, 11 (26.8%) from 3 to 5 years, 9 (22%) less than 1 year, 5 (12.2%) from 5 to 7 years and 3 (7.3%) from 7 to 10 years. The main export market are the United States, with 19.5%, Canada with 17.1%, United Arab Emirates with 14.6% and Italy with 9.8%.

4.2   Barriers that Can Jeopardize the Internationalization of Micro and Small Companies

Frequency statistics based on data analysis showed that the factors that cause the most difficulties, cited by the micro companies: difficulty in proceeding with changes in the product due to external customer requirements (22%); exchange rate fluctuations (17.1%), difficulties in after-sales service (14.6%), delay in receipt of the amounts coming from products sold (12.2%). The least important barriers are: logistic costs (4.9%), bureaucracy in processes, cultural differences, government incentives and international legal restrictions (2.4% each one).

The most important barriers indicated by the small companies are in order of relevance: exchange rate fluctuations (15.7%), delay in receiving amounts related to exports (13.9%); tariff barriers and difficulties in obtaining capital presented a percentage of 13% each one and the need to carry out changes in the products due to external customer requirements and logistic costs are presented with 7.8% each. Legal restrictions of an international character were mentioned by 7% of respondents. The less significant barriers were cultural differences (4.3%), process bureaucracy (3.5%), and government incentives (1.7%).

The difficulty in proceeding with changes in the product due to external customer requirements, was indicated by 94% of companies that had the percentage of revenues coming from exportation, below than 50%. With the same reasoning, 58% of the companies that indicated the difficulty "delay in receipt of the amounts coming from products sold" had less than 50% of revenues coming from exportation. The difficulty "tariff barriers" was indicated by companies of which 74% had less than 50% of revenues coming from exportation and finally the difficulty "exchange rate fluctuations" was declared by 65% of that companies. To the other difficulties indicated, there were not differences based on the percentage of revenues coming from exportation.

The research showed that both the micro and small companies in the sector have three common barriers that make the internationalization process unfeasible: the exchange rate fluctuations that affect the value of exports, the delay in receiving the values derived from the sales of the products, and the necessity to do changes in products according to customer requirements. The barriers that have the least impact on the internationalization process common to both sizes of firms are cultural differences and government incentives.

Based on the next steps that companies intend to adopt it is important to mention that the objective "to expand the company" was mentioned by 30 companies of which 80% had less than 50% of revenues coming from exportation. The objective "to create new product lines" was mentioned by 31 companies and of which 77% hat that origin revenues. The objective of "to search for new partnerships" was indicated by 35 companies from what 67% had more than 50% of revenues coming from exportation. The crossing data analysis enables a better understanding of the profiles of the companies that export.

5   Final Considerations

Based on bibliographical research, it was possible to verify that micro and small enterprises are indispensable for the Brazilian economy, since they generate employment and income and increase the values of GDP.

The objective of the research was to identify the barriers that difficult or harm micro and small cosmetic companies in the internationalization process, making a comparison between both kinds of companies.

The results demonstrate that according to the statistical calculations the variables researched were relevant to reach the objective. The sample of the micro companies surveyed point out that the barriers that are most unfeasible or detrimental in the internationalization process are when there is a need to make changes in the product by customer demand, the exchange rate swings that are susceptible, after sales service and delay in receipt of values of the products marketed.

The small companies surveyed have the result that, like micro-enterprises, the exchange rate swings are considered barriers, the delay in receiving the values from the commercialization, the tariff barriers, after-sales service and difficulty in obtaining capital.

It is therefore considered that the variables used by Garcia [12], Rezende and Serpa [15] and Cordeiro [13] about the barriers to internationalization, indicate the difficult or the impossibility for both micro and small companies, that is, the variables are present in the two types of companies surveyed.

The limitation of the research is related to the number of respondents, since it is a non-probabilistic research. For future research it is recommended that the questionnaire can be applied to companies considered of medium size and other segments.

References

1. Panorama dos pequenos negócios. https://m.sebrae.com.br/Sebrae/PortalSebrae/UFs/SP/Pesquisas/PanoramadosPequenosNegocios2017.pdf
2. Rasmussan, E., Madsen, T.: The born global concept. In: Proceedings of the 28th EIBA Annual Conference. European International Business Academy, Athens (2002)
3. Panorama do setor 2015. https://abihpec.org.br/publicacao/panorama-do-setor-2015/. Accessed 25 Aug 2017
4. Vries, M.F.R.K.: The entrepreneurial personality: a person at the crossroads. J. Manag. Stud. **14**(1), 34–57 (1977)
5. Aguiar, M.A.: Psicologia aplicada à administração: uma introdução à Psicologia organizacional. Atlas, São Paulo (1981)
6. Schumpeter, J.A.: Teoria do desenvolvimento econômico: uma investigação sobre lucros, capital, crédito, juro e o ciclo econômico. Abril Cultural, São Paulo (1982)
7. Ribeiro, R.A.M., Teixeira, M.R.C.: From small farmer to successful entrepreneur: "The history of Aguinaldo's pamonharia". In: 1st National Congress of Entrepreneurship - CONENPRE, p. 10. CONENPRE, Florianópolis (2003)
8. Os donos do negócio no Brasil, por regiões e unidades da federação. https://m.sebrae.com.br/Sebrae/PortalSebrae/Anexos/DN_regiao_unidades_federa%C3%A7%C3%A3o.pdf. Accessed 09 Dec 2017

9. The current scenario of micro and small companies in Brazil (2016)
10. Teixeira, M.J., Picchiai, D.: Analysis of the internationalization process of micro and small companies in Campinas/SP, in the light of behavioral theories. Euronet Mag. **1**,1–16 (2015)
11. Machado, C.: Factores de Internacionalização das Empresas. Textos de Apoio à matéria de Estratégias de Internacionalização de Empresas da EEG da Universidade do Minho (2004)
12. Garcia, R.: Internacionalização comercial e produtiva na indústria de cosméticos: desafios competitivos para empresas brasileiras. Rev. Prod. **15**(2), 158–171 (2005)
13. O processo de internacionalização de pequenas e médias empresas de perfumaria, cosméticos e higiene pessoal do estado do Paraná (2011)
14. Caetano, L.M., Paiva, D.L.: Barriers to the internationalization of Brazilian medium, small and micro companies. Rev. de Iniciação Científica Tecnol. e Artística **5**(5), 33–46 (2016)
15. Rezende, O., Serpa, C.A.: Analysis of the performance of a cosmetic exporter group from a business perspective: a case study of the Minas Beauty Group. Internext – Electron. J. Int. Bus. **4**(1), 79–99 (2009)

Destaylorization: Thinking an Organizational Context to Combating Poverty and Social Exclusion

Paula Cristina Salgado Pereira Rodrigues Vieira(✉)

Instituto Superior de Serviço Social do Porto, Av. Dr. Manuel Teixeira Ruela, 370, 4460-362 Senhora da Hora, Portugal
paula.vieira@isssp.pt

Abstract. In this short reflection are discussed some of the potential effects of destaylorization in the organization and production of social services, taking advantage of the knowledge produced about this process in certain high-end sectors of industrial production. This reflection will be made by analyzing the probable benefits of the adoption in these organizations of non-taylorist logics of conception and organization of work that can substantially improve the quality of the services provided, materializing the conditions of solidarity that enable the effective autonomization of individuals who they turn to. This presupposes a rupture with the logics of organizational functioning that, instead of contradicting the inversion of the trajectories of social marginalization and self-destruction that mark so many times the lives of social service users, reinforce their processes of social segregation. What modalities of conception and work organization can be mobilized by organizations that provide social services that allow the experimentation and validation of socio-therapeutic intervention models capable of preventing and/or counteracting states of poverty, disqualification, and social desfiliation?

Keywords: Destaylorization · Work organization · Social services

1 Introduction

Producing more and better at the lowest possible cost, ensuring the maximum profit, has been a determining and structuring axis of all economic activity, such is the centrality that the productive activity plays in western societies.

How to ensure the realization of this nuclear purpose and/or also how to reduce the social and human costs that its implementation has produced over recent history are two of the main issues that have occupied scientific and technological research.

To a large extent subsidiary to economic activity, scientific research has always invested significant efforts in determining the technical and organizational conditions that make it possible to make the most of the productive process. This happens decisively in the field of exact sciences, with special emphasis on the technological sciences, but also in the social and human sciences. The Scientific Work Organization of F. Taylor is a paradigmatic example, and perhaps one of the most notable, of the

© Springer International Publishing AG, part of Springer Nature 2019
J. Machado et al. (Eds.): HELIX 2018, LNEE 505, pp. 912–918, 2019.
https://doi.org/10.1007/978-3-319-91334-6_125

practical application of a supposed scientific study to the productive process, with the aim of giving it a rationality that makes possible the optimization of production.

In the industrial era, scientific research searched to provide enterprises with models of work organization that ensure the production of goods on a large scale at the lowest cost. In the post-industrial era (information societies), the central issue now is how to make the productive process capable of integrating and producing new information, a fundamental condition to ensure the economic survival of companies.

In fact, in this new stage of the development of the capitalist mode of production, post-capitalism as Drucker [3] call it, the knowledge, or rather, the business ability to systematically incorporate new knowledge plays a key role in an increasingly competitive environment.

With new rules regulating the functioning of the company [5, 13][1] and the market, changes in the productive organizations tend to occur both at the level of the conception of work, appealing increasingly to their cognitive dimensions and to cooperation [1] not compatible with decomposition/recomposition operations typical of the taylorization of the productive process, as at the organizational level.

In this second plan the tendency is to question the strictly hierarchical forms of management/coordination and the rigid separation between design work and work of achievement. In fact both appear to be misadjusted faced by the need to produce, not in mass, but to produce better, with more quality and with great flexibility as to the quantity of goods and services to produce and, even, as to the type/nature of these goods.

We share the idea that the complex mutations that are affecting capitalist enterprises are more a sign of the improvement of the capitalist mode of production than a sign of the emergence of a new economic order. Yet, we know for sure that these mutations are introducing significant changes in the organization of systems production systems.

In other words, although the conditions under which the company can maximize profit in order to strengthen its power in the economic field, on the contrary, it seems that we are witnessing "a gradual and orderly change in the work process", without it implying, however, a "substantial transformation of working structures" [1]. In this regard, several authors of the sociology of work [2, 6, 14] speak of a destaylorization movement.

Are there lessons to be learned from the restructurings with which capitalist production units are confronted that can help rethink the organization of work in services?

---

[1] The authors Petit and Dubois [13] refer to other important mutations that are occurring at company level in the face of the uncertainty of the current economic context: "Intense competition between organizations is no longer only in quantity, but in quality of goods and services. The possibility of rapidly changing the nature of products and their characteristics becomes a must. It is necessary to be able to produce simultaneously or successively different product ranges, while maintaining the benefit of the mass production, and find a way to ensure a floating production, combining speed and diversity (…). If specific dimensions of work are introduced, reflections on polyvalence, polycompetence and polyfunctionality, are mainly to satisfy the requirements of optimization of production and the demand for total quality".

2 An Organization Able to Reflect on Itself and to Stimulate the Discovery of Solutions Adapted to the Social Problems

The uncertainties that mark the current context of socio-technical and economic transformations, highlighting the unpredictability of the markets, obliges companies to develop their adaptive capacity and to become flexible.

The need to regularly associate new technologies and new information with production strategies to ensure satisfactory competitiveness standards, challenges the ability of enterprises to innovate, at least in two areas: in terms of re-qualification of jobs, mainly translated into "capacity to define and choose the procedures and operative knowledge to be mobilized, in a situation of uncertain work" [1]; in organizational terms, adopting practices that call for the initiative and the capacity of conception of the operators [1].

In the foreground, the substitution of Taylorism rationality points to the exhaustion of one of its fundamental principles, "the right man for the right place", with identified knowledges adapted to a type of task. The use of low-skilled workers, even though they have a manual know-how with meaning and usefulness on an assembly line, seems to be in complete extinction.

Although it is not the purpose of this reflection to analyze the social effects of the "new" productive logics, one cannot fail to mention, with perplexity, one of these most harmful effects, the unemployment[2].

Also not problematizing these effects, Alter [1] gives some contributions to define the profile of the new worker, or the "new professional"[3] in the current context. According to this author, since knowledge becomes the central economic resource for production, itself uncertain and always changeable, the competence of the new professionals will have to be defined in terms of ability to draw theoretical and practical lessons from their intervention in the work process and to systematically learn.

Thus, professionals required by the movement of destaylorization are defined less by the school certificate or professional they hold or by the function they assume within the company (hierarchical position or by statute) than by the evidence of knowledge demonstrated in the concrete circumstances of their activity.

If we focus on organizations providing social services, the question we can ask is whether the professional profile that is drawn in the framework of productive processes touched by innovations in procedures or products will not be the only one compatible with the challenges faced by social intervention professionals.

Is it possible to provide material and psychosocial support to people on the basis of stereotyped responses that are provided independently of the analytical elements crucial

---

[2] This is a very serious social problem that does not seem to have a solution, since without firm regulation of the labor market by government entities, companies tend to put workers' training and/or retraining on the last line of their priorities, preferring to discard them in a purely economic logic. The revival of the liberal principle of self-regulation of the market and the economy it is used, among other things, to legitimize the possibility of capitalist enterprise to use labor on a contingent basis in the light of economic fluctuations. What to do, then, of the huge "reserve army" that every day increases with further massive lay-off?

[3] In the terminology of Dubar [4].

for understanding the impact of material, relational, and symbolic deprivations that mark their lives? Or that they apply without a previous reflection on the experience? A reflection that allows interpreting the specificities that concrete problems presents and that often escape to the explanation that the theoretical laws give about phenomena[4]?

As pointed out by Germinet [7] regarding the training of engineers, who are also challenged to construct solutions to concrete problems, the social worker who is confronted in his daily work with processes of social vulnerability must develop the "real intelligence". This means that the professional will have to exercise (I) the ability to uncover which leads to a complex set of events or facts, (II) the ability to clarify the essential components of reality that can be modified and (III) the ability to discover the combinations that the action must produce to achieve the expected results. Only the renewed and controlled application of the "observation-experimentation-conceptualization-observation" circle will allow the construction and validation of more effective solutions to complex social problems.

The new professional is therefore at the crossroads of two fundamental axes of work organization: design and execution, whether a technician with operational functions or supervisory functions, not directly involved in the field work. In turn, his position in the hierarchy will result much more from his knowledge and competence than the status resulting from the rarity of their training or specialization. Knowledge and competence measured by its capacity to invest knowledge in action, to reflect on experience and to mobilize it to better solve problem-situations.

Assuming that professional ways of thinking and acting are to a large extent conditioned by the objective conditions that the organization creates, what organizational structure should be implemented in order to stimulate reflective and empathetic competence of the professionals? And also the spirit of mission that necessarily should be developed to recover individuals whose human dignity is seriously compromised?

According to the *Green Book of Innovation* [9], the innovation of organizational models in the productive sector is to a large extent due to the need to "associate workers with technological changes and their consequences in the organization of production".

With very different degrees of rupture with the old taylorist-fordist model, the new organizational models results from modifications introduced into the organizational devices that make productive activity possible as a structured process. Briefly, the most important change trends in each of these devices are presented:

– content of the work: weakening the separation between design work and execution work, calling for the initiative and capacity of design of the operators, enrichment of work contents and bet on a culture of learning;

---

[4] As Malglaive [10] points out, practice requires the invention and mastery of other knowledge that does not automatically and spontaneously arise from existing theories, however deep and solid they may be. More concretely, in order to transform a good theory into a guide for action, it is necessary to find out which are the different domain of action to privilege, their respective connections (procedural knowledge), as well as to discover the know how to do and the specific techniques that must be mobilized in each one of these domains of action (know how to do).

- motivation for work: increased motivation through a greater involvement of the worker in the decisions related with the productive process. Recognition and encouragement of the efforts of individuals and work teams, investing in the attribution of material and symbolic benefits;
- company culture: production of a socio-cultural context of strong identification with the company and its objectives;
- circulation and distribution of information: faster and more transparent;
- decision-making and work coordination mechanisms: more flexible and decentralized;
- operational processes: rationalized, investing in information and communication technologies.

Analyzing the changes introduced in each of these devices, Kovács [8] characterizes four models of organizational innovation: "lean production", "reflective or anthropocentric model", "reengineering" and "networks and virtual organizations".

Without losing sight of the objective of the reflection proposed here - learning from the new models of work organization that are being formulated for the productive sector that may serve the improvement of the organizations that provide social services - we elect for analysis the anthropocentric model. From among the enunciated models, this one seems to be the one that presents greater virtualities to help rethink the traditional pattern on which the work organization in this type of organizations is based.

Of the many characteristics that this model includes, three stand out.

Firstly, the subordination of the work organization to the workers' capacity for understanding and reflection on the entire production process. By requiring systematic improvements in the knowledge base from the training of reflective capacity over experience, "reflexive production"[5] appeals to the worker's knowledge and intelligence.

It is clear that this model is not exempt from internal contradictions[6] having serious social costs. Particularly when it tries to establish a core of skilled and excellent workers, and, at the same time, to discard the less productive because they are less prepared to take ownership of the new contents of the tasks. If this principle is applied to the social services production, it is clear that the structuring and coordination of work in this way will limit the tendency of institutional actors to reproduce bureaucratic attitudes. That means attitudes neglecting the experimentation and validation of differentiated solutions according to the specificity of concrete problems.

Secondly, consonant with the centrality that gains knowledge in the organization of productive work, the organic and flexible character of the organizational structure. This type of structure is favorable to the work organization in small teams, enhancers of circulation and sharing of information and knowledge about operative modes and their

---

[5] Designation used by Freyssenet [6] for the anthropocentric model.

[6] The "paradigm of flexibility" in the field of labor management, as Freire [5] calls it, is based on contradictory principles, for example in the need to manage the ambivalence between, on the one hand, the importance attached to human capital and its qualification to achieve better productive performances and, on the other hand, the weight of wage costs in a very open and ruthless competitive framework.

chaining. Also the decision processes, much more horizontalized and decentralized as Freire [5] point out, are significantly closer to the operational centers. That is to say, there is still a command function, but this one is not disconnected from the production of new results or innovative solutions.

Mintzberg [12] characterizes power in this new organizational context as being the place or position of someone whose authority is based on competence. Someone, individual or team, that assumes the conflicts resolution, the mobilization around the productive project and the reflection about the complex and imponderable situations.

Finally, there is another feature that the anthropocentric organizational model contains that, when applied to the production of social services, it can greatly help institutional actors to question their routinized ways of acting and the production of external solutions to the problems. We refer to learning and the prominent role that it is given to it in the productive process.

Indeed, in this organizational context, the creation of devices that select the most pertinent solutions and place them at the disposal of the organizational collective is especially important. Micelli [11] explains that it's the operators themselves that determines the variety of knowledge and solutions that arises around the problems concerning the productive process. Before being formalized in procedures or artifacts they must be filtered through group discussions according to technological and economic criteria. Actors and organizations participate jointly in the selection and retention of knowledge mechanisms.

In the case of the organizations providing social services, to link the professionals to the conception and implementation dynamic instruments of action monitoring will surely be a way for them to evaluate the coherence and effectiveness of their operative modes. Only like this it will be possible to discover, select and retain the technical and organizational solutions that effectively contribute to promote the reestablishment of the social bond of large sectors of the portuguese population. And, also, only like this it will be possible to produce knowledge and philosophy of action about the implementation of these solutions.

References

1. Alter, N.: Innovation et organisations: deux légitimités en concurrence. Rev. Fr. Sociol. **34** (2), 175–197 (1993)
2. Demailly, L.: Simplifier ou complexifier? Les processus de racionalisation du travail dans l'administration publique. Sociol. di«u Travail **34**(4), 429–450 (1992)
3. Druker, P.: Sociedade Pós-Capitalista. Difusão Cultural, Lisboa (1993)
4. Dubar, C.: La Socialisations. Armand Colin, Paris (1991)
5. Freire, J.: Sociologia do trabalho, uma introdução. Ed. Afrontamento, Porto (2001)
6. Freyssenet, M.: La production reflexive, une alternative à la production de masse et à la production au plus juste? Sociol. Travail **37**(3), 365–388 (1995)
7. Germinet, R.: Aprendizagem pela acção. Instituto Piaget, Lisboa (1999)
8. Kovács, I.: Inovação e organização. In: Caraça, J., Amaral, J.F. do (coord.) Sociedade, Tecnologia e Inovação Empresarial (debates Presidência da República), Lisboa (2000)
9. Livro Verde sobre a Inovação. CE, Bruxelas (1995)
10. Malglaive, G.: Ensinar adultos. Porto Editora, Porto (2003)

11. Micelli, S.: Système productifs: les modèles en question. Sociol. Travail **37**(3), 345–409 (1995)
12. Mintzberg, H.: Le Management, Voyage au Centre des Organisations. Les Editions d' Organisations, Paris (1990)
13. Petit, F., Dubois, M.: Introdução à psicossociologia das organizações. Instituto Piaget, Lisboa (2000)
14. Thévenot, L.: L'action en plan. Sociol. Travail **37**(3), 411–434 (1995)

Regional Distribution and Economic Activity of the Brazilian Software Industry: A Study for the Period 2007/2016

Fernando Semenzato$^{(\boxtimes)}$ ⓘ and Luis Paulo Bresciani ⓘ

Universidade Municipal de São Caetano do Sul, São Caetano do Sul, Brazil
{fernando.semenzato, luis.bresciani}@prof.uscs.edu.br

Abstract. Software development represents a growing activity in several emergent or developed economy countries. This study presents the economic trajectory of software development companies and the changes in the geographical distribution of these companies, which occurred in the period from 2007 to 2016 in Brazil. With a documentary research that compared information obtained from governmental and non-governmental organizations, it was possible to accomplish a sectorial analysis of the Brazilian software industry within the period. Starting from the development of this industry in the 1990s, it was observed the continuous growth until 2014, followed by a decline in the growth rate between 2015 and 2016. It is observed a concentration of activity mainly in the Southeast region states, the industry growth in all regions of Brazil, and the increase of the economic share of the South and Northeast regions.

Keywords: Software industry · Economic activity · Regional distribution

1 Introduction

There is an increasing importance of Information Technology (IT) activity in Brazil, achieving a revenue level of US$ 39.6 billion in 2016 (including hardware, software, services and exports). This figure represented 1.9% of total IT investments worldwide, equivalent to 2.1% of the Brazilian Gross Domestic Product (GDP) in 2016 [1]. It is a percentage similar to countries like China and India, which like Brazil belong to the BRICS group (acronym for Brazil, Russia, India, China and South Africa) [1].

According to the Brazilian Association of Software Companies (ABES), the Brazilian exports of IT business in 2016 represented US$ 1.1 billion, less than 2.8% of the total moved by activity, characterizing this focus on the Brazilian domestic market [1]. Among the activities encompassed by IT stands out the software development, with knowledge development as an important competitive factor, and opportunities for large, small and medium-sized companies [2]. In Brazil, the software development market also represents an emerging activity, with a revenue level of US$ 8.7 billion in 2016 and annual growth of 1.2% in relation to 2015, in the same period that IT investments retracted for 3.6% [1, 3].

In order to know more deeply this activity, the objective of this article is to describe the changes in the economic activity and the geographic distribution of software

© Springer International Publishing AG, part of Springer Nature 2019
J. Machado et al. (Eds.): HELIX 2018, LNEE 505, pp. 919–924, 2019.
https://doi.org/10.1007/978-3-319-91334-6_126

development companies in Brazil, during the period between 2007 and 2016, seeking to verify if they characterize a sectorial decentralization and a greater economic relevance of this activity in the country.

The study was based on literature research related to the software development sector, as well as in the analysis considering governmental and non-governmental databases, including ABES sectoral reports. We highlight the data obtained in the Annual Report of Social Information (RAIS) for 2007, 2014 and 2016, searching three classes of software development activities. The data were analyzed considering a few specific classifications in the National Registry of Economic Activities (CNAE): 6201-5 (Computer development programs on demand), 6202-3 (Development and licensing of customizable computer programs) e 6203-1 (Development and licensing of non-customizable computer programs).

2 The Software Industry Since the 1990s

Brazil developed a broad software sector focused on the internal demand during the 1990s [4]. Accompanying an economic opening of this period, this activity began to be executed on a large scale, and Stefanuto [5] uses the term "software industry" to represent the private sector companies.

Sabóia [6] reminds us that industrial development did not occur in an equitable manner in all Brazilian regions, during that decade. Using the same database (RAIS), it was observed that the number of jobs decreased in most of the Southeast, North and Northeast regions, while the Midwest and South regions received new industries and observed growth in the jobs number. Sabóia also affirmed that this process generated an industrial decentralization at that time in Brazil, but the state of São Paulo remained the main industrial center of the country. But what happened to the software industry regional distribution, belatedly born in Brazil? Was it concentrated in the major industrial centers, or was it established in a more distributed way in the Brazilian territory, following the decentralization observed by the referred author? [6].

The IT economic share in Brazilian GDP can be compared to other countries from BRICS in 2016, such as China and India, but the numbers of Brazilian growth are very far from those observed in the Chinese and Indian software business, as shown by reports from the Chamber of Deputies [7], and studies by Schware [8], Arora and Gambardella [4] on the distinct trajectories of Brazil and India.

Also are visible the reasons that explain the different growth rates when observing the development policy of the activity in China, with the creation of 11 software parks since 1978. Its main park in Dalian grew more than 50% from 1998 to 2005, with sales increasing from US$ 25 million to US$ 1.28 billion. Among the factors of success are public incentives, workers training, investments in infrastructure, incentives for business investment and innovation [9].

3   The Software Development Activity in Brazil Between 2007 and 2016

Considering the absence of a historical series with a regular pattern when observing the evolution of this activity in Brazil, it was necessary to use some basic data as well as the interpretation of some institutional reports published in the country. Then we sought to compare different information sources, together with official indicators of economic activity. Analyzing the RAIS database, it is observed in Fig. 1 a solid and constant growth in the number of companies up to 2014, starting from 3,865 companies in 2007, and reaching a peak of 9,213 firms, followed by a decrease in subsequent years.

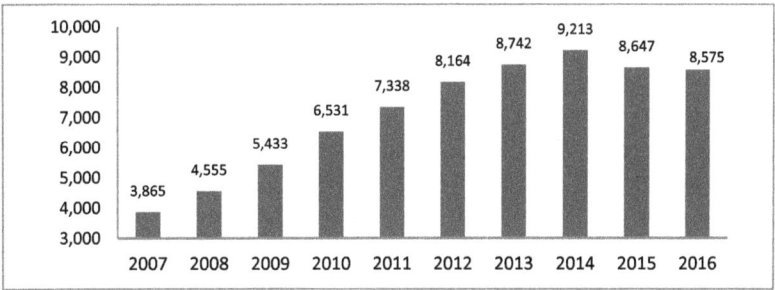

Fig. 1. Companies in the sector year by year (source: elaborated based on RAIS [10]).

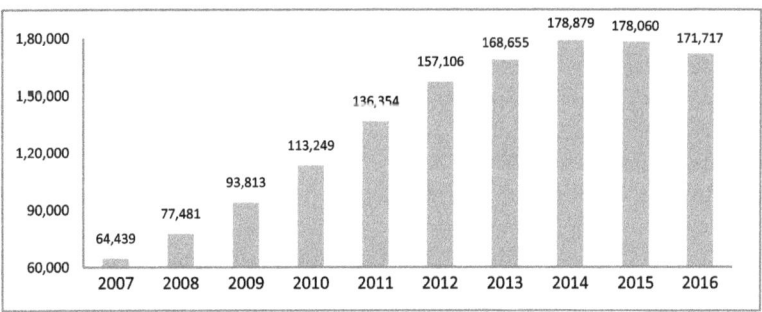

Fig. 2. Number of formal employment in the sector, year by year (source: elaborated based on RAIS [10]).

Figure 2 shows the formal employment of this industry. The figures show a similar trajectory to Fig. 1 (number of companies), with the number of employees growing between 2007 and 2014 (with a peak of 178,879 employees in 2014), decreasing in the following years. The data in Figs. 1 and 2 indicate a rupture in the growth trajectory of software development activity in Brazil, a move that accompanied a decline in the Brazilian GDP in 2015 and 2016, as observed in Table 1.

Table 1. Brazilian GDP between 2007 and 2016 (sources: world bank [11] and BACEN [12]).

	2007	2008	2009	2010	2011	2012	2013	2014	2015	2016
GDP world bank US $ billion	1,397	1,696	1,667	2,209	2,616	2,465	2,473	2,456	1,804	1,796
GDP central bank of Brazil R$ billion	5,456	5,734	5,726	6,158	6,402	6,525	6,721	6,755	6,501	6,267

4   The Regional Distribution of the Software Industry in Brazil

According to data from RAIS 2016 [10], the six Brazilian states with the largest number of software companies are São Paulo, Minas Gerais, Santa Catarina, Rio Grande do Sul, Paraná and Rio de Janeiro. In a total of 8,575 companies in Brazil in 2016, the state of São Paulo has 3,233 companies, almost 38% of the national total. Regarding the main aspects that can explain the regional distribution of these companies, it is observed that geographic proximity is very important to improve the cooperation linkages between companies and research institutes [13]. Studies about software clusters in very different cities, like Sunderland [14] and Dalian [9], show that a large number of companies will be located next to the provision of specialized professionals. Etzkowitz and Leydesdorff also pointed out that innovative companies consider the universities as relevant sources of technology in the software industry, gaining competitive advantages, and intensifying the regional economic development [15].

In fact, data from the INEP 2016 census [16] shows that the state of São Paulo has the highest number of professionals coming from university regular courses focused on the software business. The state of Minas Gerais occupies the second place, the same position observed in the states ranking of software companies.

In a regional perspective, Table 2 shows the software developers figure by region in 2007 and 2016, and also in 2014 (when the peak of this indicator was registered in Brazil). In 2007, the regions with the largest number of companies were: Southeast, South, Northeast, Midwest and North, and these placements were maintained in within 2014 and 2016. But we observed in Table 3 that between 2007 and 2014, the region with the highest growth figure in companies' number was the Northeast (255%), followed by the North (218%), Midwest (166%), South (156%) and Southeast (115%).

Table 2. Number of software companies by Brazilian region in 2007, 2014 and 2016 (source: based on RAIS [10]).

Region	2007	2014	2016
South	889	2,280	2,131
Southeast	2,416	5,190	4,814
Midwest	253	672	629
Northeast	256	909	858
North	51	162	143

Table 3. Growth of the software companies number by region between 2007 and 2014 (source: based on RAIS [10]).

Region	2007	2014	%
Brazil	3,865	9,213	138%
Southeast	2,416	5,190	115%
South	889	2,280	156%
Northeast	256	909	255%
Midwest	253	672	166%
North	51	162	218%

Although the ranking positions were maintained, the growth rate was higher in the three less developed regions of Brazil according to these criteria (Northeast, North and Center-West) in relation to the two more developed regions (Southeast and South). There is a variation in the representativeness on four of the five regions, since in 2007 the Southeast represented 62% of the software companies in Brazil in total number, decreasing in 2016 to 56%, still bigger than the other regions. In fact, the South region moved from 23% to 25%, the Northeast from 7% to 10% and the Northern region from 1% to 2%. The Midwest maintained its representation in 7%.

The general growth rate in the number of companies at this industry in Brazil was 138% within 2007 to 2014, and between 2014 and 2016 almost all the regions showed a decrease around 6% and 7%, except for the North region, with an expressive decrease of 12%.

5   Conclusion

The article confirms the increasing economic relevance of software development activity in Brazil between 2007 and 2016, with a strong evolution in the number of companies, coming from 3,865 firms in 2007 to 8,575 in 2016, with a growth rate of 222% in the companies' figures, and 266% in the formal employment level. The growth of this economic activity was continuous, even in the periods of GDP retraction in Brazil. The international representativeness of the Brazilian software development business shows an important difference compared to India and China, because these countries are more oriented to the international market and software exportation, while Brazil mainly focuses its internal demand.

Regarding the geographical distribution, it is observed the predominance of companies' presence in the Southeast and South regions, but with higher growth in the others Brazilian regions, altering the national share of each region, and accompanying the conclusions of Sabóia [6] about the uneven industrial development between regions, as well as following the trend of industrial decentralization in Brazil.

It is concluded that in the period between 2014 and 2016 there was a rupture in the growth trajectory of the software development industry and market in Brazil, after almost ten years of continuous growth [10], a movement that followed the pattern observed for the retraction of Brazilian GDP in the same period [11, 12]. But it's relevant to emphasize the significant growth in the number of companies and formal employment in this sector, leading to the great importance of public policies supporting its development in the country.

Based on these conclusions, this study will be followed by a new research comprising the main features of software development clusters in the Brazilian territory, the roles of micro and small software development companies in local and regional development, as well as the cooperation linkages to public organizations and universities in their recent trajectories.

References

1. Brazilian Association of Software Companies (ABES): Brazilian Software Market: Scenarios and Trends 2017. ABES, São Paulo (2017)
2. Britto, J., Stallivieri, F.: Innovation, cooperation and learning in the Brazilian software industry: exploratory analysis based on the concept of local productive systems (LPSs). Economia e Soc. **19**(2), 315–358 (2010)
3. Brazilian Association of Software Companies (ABES): Brazilian Software Market: Scenarios and Trends 2016. ABES, São Paulo (2016)
4. Arora, A., Gambardella, A.: The globalization of the software industry: perspectives and opportunities for developed and developing countries. Innov. Policy Econ. **5**, 1–32 (2005)
5. Stefanuto, G.N.: The Softex Program and the Brazilian Software Industry. Universidade Estadual de Campinas, Campinas (2004)
6. Saboia, J.: Descentralização industrial no Brasil na década de noventa: um processo dinâmico e diferenciado regionalmente. Nova Economia **11**(2), 85–122 (2001)
7. Brazilian Chamber of Deputies. http://bd.camara.gov.br/bd/
8. Schware, R.: Software industry entry strategies for developing countries: a walking on two legs proposition. World Dev. **20**(2), 143–164 (1992)
9. Jan, C.-G., Chan, C.-C., Teng, C.-H.: The effect of clusters on the development of the software industry in Dalian, China. Technol. Soc. **34**(2), 163–173 (2012)
10. RAIS database. http://bi.mte.gov.br/bgcaged/login.php
11. The World Bank: World Development Indicators (2016). https://data.worldbank.org/data-catalog/gdp-ranking-table
12. Banco Central do Brasil (BACEN): consolidated economic indicators database. http://www.bcb.gov.br/pec/Indeco/Port/indeco.asp
13. Garcia, R., Araujo, V., Mascarini, S.: The role of geographic proximity for university-industry linkages in Brazil: an empirical analysis. Australas. J. Reg. Stud. **19**(3), 433–456 (2013)
14. Dunn, D., et al.: Sunderland software city: the impact of a collaborative project to develop the software industry within the north east of England. GSTF Int. J. Comput. (JoC) **3**(2), 96 (2013)
15. Etzkowitz, H., Leydesdorff, L.: The endless transition: a 'Triple Helix' of university industry government relations. Minerva **36**(3), 203–208 (1998)
16. Instituto Nacional de Estudos e Pesquisas Educacionais Anísio Teixeira (INEP): Statistical Synopsis of Higher Education (2016). http://portal.inep.gov.br/web/guest/sinopses-estatisticas-da-educacao-superior

Circular and Collaborative Economies as a Propulsion of Environmental Sustainability in the New Fashion Business Models

Solange Fernandes[1][✉] , José Lucas[1][✉], Maria José Madeira[2][✉],
Alexandra Cruchinho[3][✉], and Isabela Dias Honório[4][✉]

[1] FibEnTech R&D Unit, University of Beira Interior, Covilhã, Portugal
embaixadorafrubi@gmail.com, jlucas@ubi.pt
[2] NECE, University of Beira Interior, Covilhã, Portugal
maria.jose.madeira@ubi.pt
[3] Polytechnic Institute of Castelo Branco, Castelo Branco, Portugal
alexacruchinho@gmail.com
[4] State University of Londrina, Londrina, PR, Brazil
isabela.honorio@hotmail.com

Abstract. The objective of this work is to explore convergence and divergence between business models of circular and collaborative economies in fashion industry and the benefits to sustainable development. The methodological approach is exploratory through bibliographic review and case study of two business models. The first, Rent the Runway, makes use of the Services and Products System (PSS), being a circular model using eco-efficient services with the potential to replicate and to compete with the 'fast-fashion' industry. The case study was done using data from Rent the Runway website. The second, Wardrobe, a P2P platform whose study was conducted with personal information with the company co-founder Germanno Teles. Wardrobe business models have a high potential for scalability, as they can be replicated anywhere, have no physical inventory, as users own their clothes. Thus, circular economy is considered to be holistic and adaptive, representing an evolution of the linear economy through the perception and need of a new sustainable strategy in response to the environmental degradation caused by the traditional linear economy. Collaborative economy can be considered an environmentally sustainable socio-economic system, built through the use of digital networks to connect and share goods and products, as well as human, financial and physical capital.

Keywords: Circular economy · Collaborative economy
Sustainable business models

1 Introduction

The concept of sustainable development ensued from the ECO-92 conference. By 2015, all countries have committed to reduce emissions in defence of climate and salvation of the planet [1]. The promise was to limit the global average temperature

© Springer International Publishing AG, part of Springer Nature 2019
J. Machado et al. (Eds.): HELIX 2018, LNEE 505, pp. 925–932, 2019.
https://doi.org/10.1007/978-3-319-91334-6_127

increase up to 2 °C (Paris Climate Change Agreement). The main objectives of COP22 2016, and COP23 2017, were to follow the application of the Paris Agreement - the period of this agreement will go until 2030 [2]. All countries will have to act to achieve the objectives, being vital to transform the current model of production and consumption: Make it sustainable.

Thus, this paper aims to analyse the points of convergence and divergence between business models of circular and collaborative economies in the fashion industry and the benefits to sustainable development. In this context, textile and apparel industry is a major sector in the global economy, providing jobs to more than 300 million people worldwide [3], but it is considered the second most polluting sector in the world following oil.

The textile and clothing production system not only depletes and degrades natural resources, but also uses dangerous substances affecting workers and consumers health. The textile industry relies mainly on non-renewable resources, such as oil to produce synthetic fibres, fertilizers and chemical products, like insecticides, herbicides and fungicides in the cultivation of cotton, for example.

The actual production and distribution model operates in a linear fashion, accumulating raw materials and end products that could be reused; according to [3], p. 3, "large amounts of non-renewable resources are extracted to produce clothes that are often used for a short time, after which they are mostly sent to landfill or incinerated. More than USD 500 billion of value is lost every year due to underutilisation clothing and the lack of recycling. Furthermore, this take-make-dispose model has numerous negative environmental and societal impacts".

As Ellen MacArthur Foundation (EMF) reported in 2017 [3], the gas emitted by the textile production system worldwide is around 1.2 billion tons per year. These and other problems were further aggravated by the 2000s, when fast fashion chains came up, with a fast pace of production and consumption, allowing manufacturing fashion products with shorter life cycles.

The fast fashion industrial production is governed by financial results: increase production and generate higher profits. Consequently, in addition to ecological harms, social problems due to exploitation of cheap labour and poor working conditions in underdeveloped countries also rise. The fast fashion chain has generated easy access to low price fashion products [4]. Thus, clothes became disposable, increasing textile waste. According to [5], "discard economy contributes significantly to climate change." This model has negative impacts, which inevitably leads to potential environmental catastrophe increase.

According to the aforesaid and identified info gaps in the scope of sustainable growth, the present exploratory work contributes to the improvement of knowledge on convergence and divergence between circular and collaborative economy models, and to foster the analysis and benefits identification for sustainable development. Following [6], there is a lack of studies that analyse the wide range of activities and business models in the collaborative economy.

1.1 Circular Economy

The Circular Economy is pragmatic, grounded and contextualized in the planet current reality. Its purpose is to preserve natural resources, optimize the ones available to us and guarantee those essential for the future.

In 2012, the report "Towards circular economy: economic and business rationality for an accelerated transition" published by the EMF [7], and also, the European Commission published the "Manifesto for a Resource-Efficient Europe", which addresses the need for Europe to move to a circular, regenerative economy. In 2015, the European Commission launched the "circular economy package". According to the Circular Europe Network, the concept of circular economy refers not only to recycling, but is part of the basic tripod of the 3Rs: Reduce, Reuse, Recycle, and enabling to reach the hierarchy of multi-R: Rethink, Redesign, Repair, Remaking, and Redistribute, Recover [8].

The circular economy goes beyond the waste problem: assumes a closed cycle management of all natural and energy resources, where inputs are reduced in production, use and reuse. It proposes environment, nature, society and economy integration.

EMF presented in 2017 [3], the report "A New Textiles Economy: Redesigning Fashion's Future" showing a vision for a new long term integrated system, based on the principles of circular economy. The goal is to transform the textile and apparel economy into an opportunity that integrates better economic, social and ecological outcomes. This is a great ambition. It requires a broad change:

(1) Raw material obtaining, with reduction of natural resources, as water, use of renewable inputs, avoids toxic substances.
(2) Textile production, disposal of synthetic fibres derived from plastic and toxic detergents, dyes and bleaches; fabrics and fibres which retain their quality during use and which may be reusable.
(3) Creation of fashion design with an Eco-design approach, to avoid environmental impacts associated with clothing during its life cycle, zero waste, multifunction, modular, and customization design [9, 10]. Upcycling allows to create new clothes with those to be discarded and waste raw materials, such as patchwork, fabric leftovers. Upcycling is also a market opportunity to increase the circularity of clothing. Restrict the use of blended fabrics to an ease end of the line recycling.
(4) Clothing manufacturing sector should make use of innovative technologies.
(5) Distribution by means of bioethanol, biodiesel, electric, hybrid low environmental impact transportations. Use of less environmental impact packaging. New point-of-sale models such as "fitting showrooms" store tester or the pop up brand can be presented in partnership with other physical outlets, the products are exposed for a short period and follow another itinerary to make the brand known and the consumer purchasing products through e-commerce [9].
(6) Firstly, when wearing clothes, consumers should be aware of the production system, use of natural resources and labour exploitation: "Ask the companies who does my clothes? (Fashion Revolution, 2018 [11])," rethinking their consumption habits and asking themselves why they have so much clothing in the closet and at

the same time have nothing to wear, why always the need to buy something new? Investing in the style consulting service may be a good option, since many clothes are discarded due to lack of knowledge of the body biotype. The staff helps to choose clothes that better dress each body type, make repairs and transformations of clothes, choose collaborative consumption, redistribute through online platforms to sell, exchange, lend and rent, do it yourself, and buy quality products with durability.

(7) For the disposal, the solution still is the redistribution through new business models, since 95% of discarded clothes have good wear conditions. Keep the products collected in the country, as when sending them to underdeveloped ones will cause difficulty to improve local economies. Recycling is an end-of-pipe option, turning these products into by-products. Currently, with excess production and consumption of fashion products and wear reduction, often the garments are still discarded with the label.

The circular economy encompasses the collaborative economy also called shared economy, through the access of goods and services that allow the increase of wearing clothes by the collaborative consumption.

1.2 Collaborative Economy

The collaborative economy has exponential growing "the relevance of the phenomenon is linked with sustainable consumption that encourages resources efficiency, waste reduction, better development and surplus reuse of products due to overproduction and overconsumption" [12].

According to [8], p. 22, the collaborative economy covers several areas: "collaborative production (easing the design and production of goods through bricolage, fablabs and marketspaces), collaborative consumption (networks for exchange, rental, loans, donation or interchange of goods and services) and collaborative development". In 2001, Rifkin published the "The Age of Access" book, in which he presented future transitions from economic activities based on access to goods and services through shared use or collaborative production [*apud* 13]. The term collaborative consumption was consecrated in [13]. Collaborative consumption involves trading, lending, negotiating, leasing, donating and exchanging from person to person (P2P). In [13] the collaborative consumption is divided into three systems: Product Services System (SSP), Redistribution Markets, and Collaborative Lifestyles. Gansky, is also driving the collaborative economy with her work called "Mesh" [5]. This is a term that refers to the interconnectivity of people through the digital technology used to provide access to goods and services. For [14], collaborative economy is a way of doing business; it is an exchange of social value that allows access rather than possession. The collaborative economy is formed by decentralized P2P networks that make use of idle assets and create new markets. One move that is driving the collaborative economy is OuiShare, which began in 2011 and has expanded. "OuiShare is a global community empowering citizens, public institutions and companies to build society based on openness, collaboration and sharing" [15].

In 2016 the European Commission (EU) launched "The future of the EU collaborative economy - Using scenarios to explore future effects for employment." It recognizes the potential of collaborative economy to create new sources of income and benefits for users through new business models for the temporary use of goods and services, such as housing and sharing [16]. Also, in 2011, Time Magazine was pointed the Collaborative Economy as one of 10 ideas that will change the world. According to [13], collaborative economy has become more popular since the global financial crisis of 2008; the recession and unemployment drove consumers to use platforms for sharing goods and services. With increased use of smartphones with mobile internet access, this resulted in an explosion of software applications, (apps). The main users, drivers and influencers of the collaborative economy are the "Millennials", born from 1982, as they dominate digital tools. The motivations refer to the economic advantages and interpersonal benefits of social interaction, trust relationship through the reputation generated by user assessment.

The distinction of collaborative economy is that it involves the use of internet technologies to connect groups of people in order to make better use of goods, services, skills, create collaborative companies and projects.

2 Case Study: Circular and Collaborative Business Models – Rent the Runway (SSP) and Wardrobe (P2P)

According to [17], "Companies may have different reasons to review their business models. Some can exploit value-added occasions by switching to circular design, providing more durable and efficient products, retrieving end-of-life ones, etc. For others, the business model may be the start of new service opportunities that foster greater circularity of products". This was the case of Rent the Runway, founded in 2008, a traditional like business model of rental of party dresses and luxury accessories that, by market perception, started to offer a new service: clothes for the day to day, named Rent the Runway Unlimited, charging a monthly membership fee. This type of offer is characteristic of collaborative economy business models. The unlimited membership automatically renews and is charged on the same date every month, enabling the member to cancel at any time. Members have more than 350 high-end items for US$159 a month, can rent three dresses, blouses, skirts or accessories at a time, keeping them as long as they like, and have personalized style consulting. The RTR College Rep Program is specific for students.

Clothes are dry-cleaned with non-toxic products. Shipping packages are used by the brand as reused ecological clothing bags. The return of the clothes is free. Members have insurance covering most of the setbacks. Customers also can download the application on the phone as the brand makes use of social networks Instagram, Facebook, Twitter, and Pinterest. There are currently five Chicago, NYC Flagship, San Francisco, Topanga and Washington DC stores [18].

Wardrobe is an app designed to meet the party sector whereby users can make quickly, securely, easily, intuitively and wisely low cost business without initial infrastructure investment. Those interested should photograph the clothes that will be shared, register in the mobile application, post the clothes, giving a brief description and value of them. Wardrobe provides for the possibility of partnerships with out-sourced services (delivery, laundry, repairs, etc.) and can also be used by fashion design students, serving as a showcase for the work of future stylists.

The Wardrobe makes use of "Mach" tool, which helps with clothes' sizes. To know if the clothes "fit well" height, bust, waist and hip measures are enough. The Wardrobe Technological Differential is an application version with cognitive intelligence and ability to understand and know the behaviour of users, so that the Wardrobe can suggest clothes in accordance to individual tastes [19]. Culminates with the creation of a Big Data capable of collecting, analysing, interpreting the current data of the platform by the users. Thus, using Artificial Intelligence, one can develop an individual personal style, guiding users in their choices, according to their habits and needs.

Finally, results are analysed, presenting more prominent intended information so that, interpretations, conclusions, limitations and expectations regarding the subject matter of study are developed.

3   Discussion and Final Considerations

Rent the Runway is a circular business model with the potential to compete with the fast-fashion business model. It makes use of the Service and Products System (PSS) through the dematerialization of products: it offers a wide variety of quality and design clothing models and eco-efficient services. If it is to be expanded to other countries, it will bring environmental benefits, increase products life cycle, extend the life of materials and minimize the use of natural resources. The financial gain is centralized and strictly follows existing regulations, as taxes, insurance, licensing, etc.

Wardrobe is a P2P platform to reallocate party clothing, keeping in circulation products previously idle or stocked. It operates in a collaborative way, allowing users to monetize clothes without being discarded. It is recommended for special occasions needing durable products that remain with quality for long time. It offers the possibility of partnerships with outsourced services (delivery, laundry, repairs, etc.).

As the Wardrobe does not have physical inventory, users being their clothing owners, gives it a high potential for scalability, since the business model can be replicated anywhere on the planet. To increase the use of fast-fashion products and to intercept early disposal is the collaborative consumption of exchange platforms and loans.

Circular economy is thus holistic and adaptive, representing a sustainable solution to the environmental degradation caused by the traditional linear economy. Although the circular economy encompasses the collaborative one, it is perceived that both are independent lines of research, with many points of convergence that complement each other. They accelerate the transition to sustainable development through new business models that increase the use of fashion products. But, at a long period, rethinking the entire value chain is needed: clothes will have to be designed to last.

Collaborative economy can be considered as a socio-economical sustainable system built on a digital network connecting and sharing goods and products, and human, financial and physical capital. It originated naturally and consumers became prosumers, creating their own businesses, often for the financial need, opportunity to optimize goods and services, share knowledge, always through an online platform. In the collaborative economy, the economic and financial result is distributed among those involved. In 2016, the EU recognizes the collaborative economy potential [16]. However, collaborative business models, while having all the positive aspects discussed above, are still far from being in line with current regulations.

As study limitations, it should be noted that cases associated to circular and collaborative economies were analysed and that the observations in the adopted methodology are always partial and are not exhaustive in a single study. Other variables can be considered in case analysis, and can be improved and broadened to allow for more detailed information. As proposal for future research, a case-study with more cases and its comparison with the present study is suggested. This can also be complemented by descriptive statistical analysis on the impact on sustainable development.

Acknowledgements. The authors thanks to Santander-Totta for the IDB/ICI-FE/Santander-Totta-UBI/2017 doctorate grant, to FibEnTech R&D Unit, Covilhã, Portugal, and to Design, Sustainability and Innovation Research Group (Design), UEL Design Department.

References

1. COP21 2015, Homepage. https://nacoesunidas.org/cop21/. Accessed 29 June 2016
2. COP23 2017, Homepage. https://cop23.com.fj/. Accessed 29 June 2017
3. Ellen MacArthur Foundation: A New Textile Economy Full Report-Updated, p. 3 (2017). Homepage. https://www.ellenmacarthurfoundation.org/assetsdownloads/publications/ANewTextileEconomyFullReport-Updated_1-12-17.pdf. Accessed 25 Jan 2018
4. Fletcher, K., Grose L.: Moda & Sustentabilidade; Design para mudança. Senac, São Paulo (2011)
5. Gansky, L.: Mesh: Por que o Futuro dos Negócios é Compartilhar, p. 75. Atlas Books, Rio de Janeiro (2011)
6. Plewnia, F., Guenther, E.: Mapping the sharing economy for sustainability research. Manag. Decis. **56**, 570–583 (2018)
7. Ellen MacArthur Foundation: Towards circular economy: economic and business rationality for an accelerated transition. Homepage. https://www.ellenmacarthurfoundation.org/assets/downloads/towardscircular economy_Updated_08-12-15.pdf. Accessed 10 Dec 2017
8. Circular Europe Network: Orientação gerais para a implementação de Estratégias Integradas de Economia Circular Nível Regional (SD)
9. Salcedo, E.: Moda ética para um futuro sustentável. Ed. Gustavo Gili, São Paulo (2014)
10. Gwilt, A.: Moda Sustentável-um guia prático. Ed. Gustavo Gili, São Paulo (2014)
11. Fashion Revolution Homepage. http://fashionrevolution.org/. Accessed 4 Mar 2018
12. Toni, M., et al.: Understanding the link between collaborative economy and sustainable behavior. J. Clean. Prod. **172**, 4467–4477 (2018)
13. Botsman, R., Rogers, R.: O que é meu é seu: como o consumo colaborativo vai mudar o nosso mundo. Bookman, Porto Alegre (2011)
14. Stokes, K., et al.: Making Sense of the UK Collaborative Economy, (2014). Homepage. http://www.nesta.org.uk/sites/default/files/making_sense_of_the_uk_collaborative_economy_14.pdf. Accessed 20 Nov 2017
15. Ouishare Homepage. http://academy.ouishare.net/. Accessed 20 Jan 2018

16. Bock, A., et al.: The future of EU collaborative economy-using scenarios to explore future implications for employment. JCR Science for Policy Report (2016)
17. Weetman, C.: A Circular Economy Handbook for Business and Supply Chains: Repair, Remake, Redesign, Rethink, p. 70. Kogan Page, New York (2017)
18. Rent the runway Homepage. https://www.renttherunway.com/unlimited. Accessed 23 Jan 2018
19. Wardrobe Homepage. http://www.wardrobe.com.br/promover-roupa. Accessed Jan 2018

Technology Transfer by Transnational Corporations: A Discussion of the Importance of Cooperative Arrangements in Foreign Direct Investment

Luiz César Fernandes da Silva and Paulo Reis Mourão[✉]

University of Minho, Braga, Portugal
luizcesarfs@gmail.com, paulom@eeg.uminho.pt

Abstract. Foreign direct investment (FDI) is recognized by many economic actors, and particularly by the State as a relevant factor of economic growth. One aspect that supports this statement is that, for these investments, there is a technology transfer between countries that promotes increased productivity at the microeconomic level and generates a "competitiveness shock" for local businesses. From the internationalization process of capitalist economies since the 80s and 90s, new opportunities for exchange (and markets for achievements) of the operations of Transnational Corporations (TNC) through FDI have been intensified. The technological transfers were carried out by horizontal and vertical FDI, with both positive and negative aspects in this process. The major problem in this research will be the identification of these significant positive and negative aspects of FDI under the flows operated by TNC. To this end, the literature points to the transfer of technology to the recipient economies and to the role of government policy. It is done through a literature review and descriptive research.

Keywords: Technology transfer · Foreign Direct Investment
Cooperative arrangements

1 Introduction

Foreign Direct Investment (FDI) is recognized by many "actors," and particularly by the State as a possible factor in the receiving economies. One aspect that supports this statement is that, for these investments there is the technology transfer between countries, promoting increased productivity at the microeconomic level and generating a "competitiveness shock" of local businesses. The possibilities of transfer of technology are made by: Indirect form, by intra-sectoral competitive relations (horizontal overflow); The North-South model, where the leading economy that generates innovation and that invests abroad (IDE), and the South, the follower economy that imitates the innovations created in the North and; Directly, through the acquisition of a local firm and the transfer of technology and management methods to the new branches (inter-sector relations – vertical effect).

© Springer International Publishing AG, part of Springer Nature 2019
J. Machado et al. (Eds.): HELIX 2018, LNEE 505, pp. 933–938, 2019.
https://doi.org/10.1007/978-3-319-91334-6_128

With the depending of the internationalization process of capitalist economies, opportunities began to emerge for Transnational Corporations (TNCs) to conduct technology transfer through FDI operations. In this new, more integrated and competitive business scenario, these companies started to establish new collaborative strategies with respect to such capital contributions. In this context, this work, we will especially focus on the Government policies in order to enhance the technological transfer into the economy through Foreign Direct Investment. This is a descriptive research through a bibliographic review. The structure is as follows. Section 2 reviews the literature about the strategic action through FDI, Sect. 3 details FDI and the transfer of new technologies. Section 4 focuses on the role of appropriate Government policies to optimize FDI's externalities. Section 5 concludes.

2 Literature Revision - The Strategic Action ThroughFDI

The strategies for a greater integration in foreign markets through FDI can be categorized into three types regarding objectives, strategies and production structure: vertical, horizontal and differentiated goods [1]. The vertical type will occur in places where one gains advantages in the prices of the factors used in each step of the production process. It is characterized by a centralized management that controls the allocation of resources to subsidiaries with significant trade between the parent company and subsidiary firms. The benefit of vertical FDI to TNCs, compared to the national firm, is the best cost-benefits through allocation at its branches. It is linked to the idea of complementarily of trade, taking advantage of the comparative advantages of the related countries. These productive linkages, in particular, can provide a direct channel for a diffusion of knowledge, which in turn can help local companies to carry out a technological and capacity upgrade with transshipment effects for an entire economy. This diffusion of knowledge is of particular importance for companies still seeking to achieve international competitiveness.

The horizontal IDE is characterized by the production of different finished goods at various locations. It occurs when firms decide to achieve economies of scale in the receiver plant by investing there in various activities such as R&D. An array of technology, financial, administrative, engineering and R&D services are centered in one place, unlike other services and activities of the company. They can be transferred to other production units at no cost. The horizontal FDI consists of companies that produce homogeneous goods and have several production units, and one of them is at the company's headquarters [1]. Another aspect of the horizontal transference (related to the endogenous growth models) is induced by the progress of technological knowledge, considering the existence of two countries - the North, the leading economy that generates innovation and that invests abroad (FDI), and the South, the follow-up economy that imitates the innovations created in the North and that receives foreign investment [2].

The third type of FDI, called differentiated goods, occurs when there is acquisition by TNCs in foreign factories (A&F) whose final goods are differentiated and intended for both the local market and for exports. This type of strategy is encouraged when the fixed costs of the plant are high. Regardless of the way in which TNCs operate, the end goal of this contribution is related to the benefits of economies of scale, market integration and differences in appropriations of productive factors [3].

In studies that seek to identify technology transfer through the IDE, Pack and Kamal [4] focusing on vertical technology transfer from a multinational to its suppliers, have shown that technology diffusion among suppliers can benefit foreign firms sourcing components. Goh [5] finds, however, that diffusion of knowledge to other potential suppliers can either encourage or discourage technology transfer depending on the incumbent suppliers cost of technological effort. Using firm-level data from Lithuania, Javorcik [6] finds evidence of spillovers from foreign affiliates to their local suppliers in upstream sectors, but only for projects with shared domestic and foreign ownership (not for fully foreign owned investments).

Airtken et al. [7] explore the idea that technology spillovers ought to increase the marginal product of labor and this increased productivity should show up as higher wages. Employs data from manufacturing firms in Venezuela, Mexico, and the United States. The study finds no positive impact of FDI on the wages of workers employed by domestic firms. The authors report a small negative effect for domestic firms, whereas the overall effect for the entire industry is positive.

In a critical discussion of the plant-level studies of horizontal spillovers from FDI, Moran [8] argues that there is a difference in operating characteristics between subsidiaries that are integrated into the international sourcing networks of the parent multinationals, and those that serve protected domestic markets and are prevented by policy restrictions (such as mandatory joint venture and domestic content requirements) from being so integrated. These different operating characteristics include size of plant, proximity of technology and quality control procedures to industry best practices, speed with which production processes are brought to the frontier, efficiency of operations, and cost of output. Argues that while the former have a positive impact on the host country, often accompanied by vertical backward linkages and externalities, the latter may have a negative impact.

Blomström and Fredrik [9] find that the degree of foreign ownership did not affect the productivity of local partners or spillovers to domestic firms in Indonesia for 1991. Yet having any foreign participation at all did matter: plants with no foreign participation were less productive. FDI is attracted to more productive plants. Balasubramanyam et al. [10] find the growth stimulating effects of FDI are stronger for countries that pursue export promotion rather than import substitution policies. So trade policy seems to affect the benefits of FDI, although trade orientation could proxy for other unmeasured differences across countries. For export promoting countries, FDI stimulated growth more than domestic investment. Borensztein et al. [11] find that FDI contributes more to economic growth than domestic investment for countries that have a sufficient stock of human capital.

3   FDI and Transfer of New Technologies

Since the strategic possibilities of performance of TNCs in foreign markets, technology transfer, via contribution of FDI, is also part of this context. This can occur through TNCs investing in their branches/subsidiaries (the transfer of imported machinery and equipment — technology embodied in capital goods); through manpower training in the subsidiaries (tacit knowledge transfer). Thus, between different conjunctures, possible effects related to the impacts on the transfer of the technology to the receiving countries, via FDI, become diverse, as presented in the Table 1 (below).

Table 1. FDI and technology transfer; main arguments

Positive effects	Negative effects
Cause increased productivity in local companies [12] Performance improvements enterprises. New transferred technologies reduce costs of research and development in the recipient companies, making them more competitive Reducing the cost of products and production processes, and a decrease in the technological gap between countries Increased exports and employment, the possibility of acquiring a technology at a marginal cost, which is especially advantageous because it avoids fixed costs that would have been incurred if the technology had been developed locally Innovations that spread to the rest of the economy, generating positive externalities and encouraging private, domestic activity	Local businesses are dependent on technologies introduced by multinationals and lose interest in their own production. Multinational with attitudes that discourage technological development to keep its advantages Extensive manipulation of profits by the use and abuse of transfer pricing; Use of inadequate technology (or very sophisticated or very obsolete technology) according to objectives and resources; real economic development is impossible in the host country due to difficult industrial relationships with the local sector [3]

Source: Self elaboration upon several authors.

Some features are necessary for the benefits to be realized, resulting in increased economic efficiency in host countries: profile of adequate labor; ability of domestic firms to absorb technological knowledge (efforts in R&D); stable institutional environment, possibility of adaptation of technology, protection of intellectual property, tax incentives, among others. The State exercises a primacy of function and importance of policies that enhance these factors, because these characteristics are not derived by the free market. Thus, the types of implemented policies and institutional differences result in different results for the allocation of these transfers. The government through political action becomes important, as we are going to analyze at next section.

4   The Government Policy

The lower the level of technological capacity of the country, the greater the possibility of TNCs imposing their bargaining power aimed only at their own benefits. Directly or indirectly, governments have an important influence on technology transfer, with the possibility of introducing policies in various fields to achieve specific results. In general, the effectiveness of any policy depends on the state's power over the supplier of technology and the technological capacity of the host.

In addition to those mentioned in the table, other authors list measures that may be a tool to stimulate the interaction between the productive sector and research institutions to promoter technological innovation. The main ones are: legalization of activities undertaken by foundations linked to federal institutions of higher education; the sharing of infra structure of federal institutions' R&D with the productive sector; stimulation of business incubation; creation of mechanisms that generate greater interaction between the private sector and knowledge-generating sector in the absorption of public institution-generated research, encouragement of the transfer of technology; stimulation of the culture of innovation through a new treatment of intellectual property in the context of educational institutions and public research, in particular through the implementation of technological innovation centers; provision of financial resources to the productive sector in the form of subsidies and tax incentives for the development of innovative products or processes; establishment of public-private partnerships for the development of scientific and technological projects for the commercialization of new technologies.

Beyond, it is necessary to encourage cooperative efforts between universities and companies, and between the companies themselves, in the form of cooperative research networks, shared centers, common infrastructure, etc. This reinforces positive externalities by creating a favorable environment for R&D and innovation activities. Even so, one must invest in overcoming technical barriers to trade and infrastructure, creating programs that support activities related to the development of technological innovations, such as; external acquisition of R&D products; acquisition of other external knowledge; and the acquisition of machinery and other equipment necessary for the development of technological innovations.

There are numerous differences in policies that can be, in each case, used by governments to encourage improvement in the results of technology transfer to the recipient countries. Notably, each country has its specific institutional features, with many different variables in both the macro and micro level, and will act within its means to take best advantage of its strengths and weaknesses. To this end, it should be noted that, often, major reforms will be needed.

5   Conclusion

When running through market strategies and rationalization of production, TNCs create intra-firm and intra-industry trade that allows them to transfer technology via FDI through these channels and in various forms. These companies tend to invest in economies where the economic factors may be more likely and thus, most of the developing countries are considered a strategic market and a relevant source of raw materials. There are positive and negative possibilities posed by these transfers to recipient countries. For benefits to be achieved, some characteristics are required, such as: a skilled profile of the workforce; ability of domestic firms to absorb technological knowledge (R&D efforts); good physical and S&T infrastructures; stable institutional environment, adaptability of technology, norms and laws that fit the local characteristics to encourage and enable the recipient companies to absorb the technology, etc.

The states will play a key role in inducing this route via political mechanisms. Through these mechanisms, there are numerous ways to improve results. It must be remembered that each country has its own economic characteristics and is at different stages in terms of technological advances (one distinct range of microeconomic, macroeconomic and natural resources). There are institutional differences and differences in the framework of legislation, but, in general, government action, a priori, establishes the possibility of successful technological transfers. Thus, for that to happen, there is a need for an adequate governmental structure. It has been shown that government can act in many spheres and in diverse ways to get positive results that may best benefit its domestic market via positive externalities posed by spillover from technology transfer. To suggest future research, case studies focused on the impacts of technology transfer through FDI in response to government policies with a quantitative approach are relevant. It will eventually respond to the specificities of the various local variables.

References

1. Jorge, C.A.: The impact of trade integration on foreign direct investment. Masters dissertation, Department of Economics, Federal Fluminense University, UFF, Rio de Janeiro (2007)
2. Barbosa Teixeira, V.A: O impacto do investimento direto estrangeirono I&D, produtividade e crescimento economico. Master thesis, Economic e Management, University of Porto (2017)
3. Dunning, J.H.: Explaining changing patterns of international and productions: in defense of the eclectic theory. Oxford Bull. Econ. Stat. **41**, 269–295 (1979)
4. Pack, H., Kamal, S.: Vertical technology transfer via international outsourcing. J. Dev. Econ. **65**, 389–415 (2001)
5. Goh, A.-T.: Knowledge diffusion, input supplier's technological effort and technology transfer via vertical relationships. J. Int. Econ. **66**, 527–540 (2005)
6. Smarzynska Javorcik, B.: Does foreign direct investment increase the productivity of domestic firms? In search of spillovers through backward linkages. Am. Econ. Rev. **94**, 605–627 (2004)
7. Aitken, B., Harrison, A., Lipsey, R.E.: Wages and foreign ownership: a comparative study of Mexico, Venezuela, and the United States. J. Int. Econ. **50**, 345–371 (1996)
8. Moran, T. How does foreign direct investment affect host country development: do we already know the answer? Using industry case studies to make reliable generalizations. In: Magnus, B., Edwardm, G., Theodore, M. (eds.), The Impact of Foreign Direct Investment on Development: New Measures, New Outcomes, New Policy Approaches. Institute for International Economics, Washington DC (2004)
9. Blomström, M., Fredrik, S.: Technology transfer and spillovers: does local participation with multinationals matter? Eur. Econ. Rev. **43**, 915–923 (1999)
10. Balasubramanyam, V.N., Salisu, M., Sapsford, D.: Foreign direct investment and growth in EP and IS Countries. Econ. J. **106**, 92–105 (1996)
11. Borensztein, E., De Gregorio, J., Lee, J.-W.: How does foreign direct investment affect economic growth? J. Int. Econ. **45**, 115–135 (1998)
12. Saggi, K.: On technology transfer from trade and foreign direct investment. World Bank Res. Obs. **17**, 191–236 (2002)

Social Design - A Path to Regional Innovation and Entrepreneurship in the Development of Depressed Regions

Inês Cerqueira Mendes de Oliveira
and Maria da Graça Pinto Ribeiro Guedes(⊠)

Universidade do Minho, Guimarães, Portugal
id5549@alunos.uminho.pt, mgg@det.uminho.pt

Abstract. Everything that is produced through the natural and social resources of a given region or place comes to represent its material culture and population behaviours. The present work intends to highlight the importance of a regional sustainable development based on local natural resources and skills, exposing, in a bibliographical research, the point of view of several authors on regional innovation or entrepreneurship initiatives, with applicability in regions of low growth, highlighting the important role of cooperativism and inclusion or even a time that shows culture. At the end of it, it is intended to understand if social design and its benefits can be synonymous of innovation and regional value, greater visibility, identity, economic growth and a way to create new products with a view to the global market.

Keywords: Regional innovation · Entrepreneurship · Social design

1 Introduction

A region must be seen as a civic place for cooperation and inclusion. This should promote the development of infrastructure, connectivity and create conditions for different lifestyles where one can grow intellectually, where there are purchasing power, jobs and leisure opportunities.

It is in this context that concepts such as regional innovation arise, combine planning and urban interventions, the projection of infrastructures and promote cultural events. These are key elements that, under a sustainable and long - term vision, have the capacity to retain inhabitants and capture new ones, boost tourism growth and, consequently, lead to the increase of investment and business establishment [1].

Understood as a set of structured actions to Plonski [2], a process of regional innovation must be seen as a multidisciplinary field applied to knowledge and to practise of administration, economics, and engineering among others: "Innovation generates new realities" [2].

A few decades, the word innovation was associated to an almost exclusive way of launching new industrial products, processes of making or technology.

Nowadays, innovation allied to design is considered the main goal by several economical sectors, regional development and others social segments, highlighting the

© Springer International Publishing AG, part of Springer Nature 2019
J. Machado et al. (Eds.): HELIX 2018, LNEE 505, pp. 939–945, 2019.
https://doi.org/10.1007/978-3-319-91334-6_129

search for a innovation in society in a context of exploration of new resources and new opportunities which emphasize positive results in the process of creating news realities.

The contribution of social design to the regional development is still feeble. However, present economies face a strong recession, which inevitably strikes all Countries (low income taxes), present a fundamental characteristic and might work as leverage: huge regional asymmetries. Even in the most developed Countries it is possible to find less developed regions and despite the efforts to support their sustainable development the asymmetries still maintain.

Facing the economical, social and environmental crises a new path was opened, new guidelines to the development and new paradigms, which change the application of resources, social systems related to work and social security and offer and global distribution of goods and services (with the undeveloped countries fighting against the overdeveloped for innovation and design skills), Since the beginning of this century other focuses arose in economics centred on technology, information, communication, knowledge and innovation. Society related to these factors along with creativeness and the concept of "Blue Economy" influence the appearance of new systems of values and it is clear that future society will be much more diverse than the one in which we grew up.

According to Canclini [3], the culture of a region can act as an aid in the transformation of the social system, that is, it can function as a process capable of improving ways of living, ideas and values. All these initiatives promote the success of a region, and the population (individually or collectively) becomes satisfied as an integral part of a decision, become more conscious and participating, are proud of the achievements of their region, willing to live or invest in it. However, cultural issues are not restricted only to the values and way of living of a society, they are also related to material, economic and technological conditions available.

To Krabbendam [4], the need to stimulate regional development and a creation of an innovating social system, capable of generating social and economical well-being to population in-need is more and more a smart and disturbing question for multiple reasons (unavailable income, elderly population and absence of foreign investments). Once allied to design and focused on those sustainable systems as well as in orienteering and maintaining the increase of people's well being will result in a new structural social system based on human, social and natural regional resources: "The social factor is actually the key-factor which leads design to assume a more ethical dimension" [4].

Thus, through the view of several authors, it will be pertinent to see if social design can contribute as a tool for a sustainable regional development based on local natural resources and skills and if the development of new products will be enough to promote a regional cultural identity.

2 Social-Economic Growth and the Sustainable Development of a Region

When talking about sustainable development, this is an issue that affects and influences living standards of a society.

According to Barbosa [5], sustainable development turns into the combination of three components (economic, social and environmental quality) and a concept of sustainability present in the economy and in the social sphere which enables the development and choice of sustainable strategies that can contribute to improve a population's quality of life (Fig. 1). The goal will be to develop gradually the value of goods and natural resources that can be transformed and promote their insertion into planning and into daily economy. There is no economy if natural resources disappear, so the ideal would be the linking of these two areas [6].

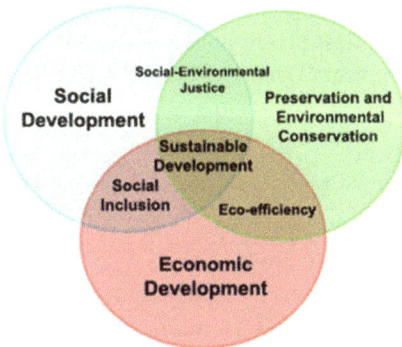

Fig. 1. Parameters to achieve sustainable development (adaped from [5])

A process of regional development, characterized by its autonomy and resources, should be a process of socio-structural change by searching a bio-cultural-economic development site and that is configured as something local-regional-global sustainable. A development based on sustainability, besides revealing a concept of regional development and entrepreneurship, shows the performance of a society [7].

However, when talking about regional innovation linked to sustainability, it may also indicate the creation of innovative and sustainable solutions, which generate positive economic, social and environmental results.

Bringing, for example, the handcrafted tradition into modern times has promoted the development of new products. Maintaining its cultural essence, together with designers' thinking, many traditional techniques and materials can be combined into new production processes as an answer to market needs [8]. Handicrafts reveal the identity and the culture of a particular region, and an interdisciplinary approach can enhance the reinvention of products. Design can be positioned as a communication interface between an inherited past (handicrafts) and a desired future (innovation) [9].

The innovation factor has the ability to generate differentiation and power in the market; therefore, sustainable innovations should not be seen only in their ecological sense, but one can also invest in regional innovation and influence social, cultural and economic aspects.

3 Entrepreneurship, Quality of Life and Attractiveness in a Region

Since the establishment of the European Union, and particularly in recent years, a lot of considerations have emerged around the convergence and cohesion processes in relation to regional development. There has been a huge economic growth in the most developed countries however regional asymmetries still remain [10].

The persistence of regional imbalances is a factor to be taken into account when discussing issues such as growth and regional development, whenever possible, sustainable and social. The idea and importance of human capital is getting more and more visibility [11] and from an infrastructure policy, we should move to a service policy, with human capital reinforcement (Table 1). "From a regret economy, we need to move to a business economy" [1?]

Table 1. Capacity to generate entrepreneurship (adapted from [19])

	Factors of social entrepreneurship
1	Cooperation (community work)
2	Develop a strategic vision on current local resources and value them
3	Establish a favorable network to local development

Due to the past economic development, there are now new challenges to be taken into account in the creation of a new regional development policy: reinforcing cohesion, tackling globalization, restructuring the economy, strengthening production quality and establishing new partnerships. However, one must take into account territorial differences, population or local economy, without compromising the quality of life of its citizens.

Quality of life is understood as the situation, the cultural context and the values in which a certain region lives in relation to its living standards, expectations and concerns.

As the principles of sustainable regional development must be based on an economic production using all natural and local resources that a certain region possesses, when it comes to regional development from the point of view of sustainability, it is expected an increase in the quality of life and attractiveness to the region.

The search for sustainable regional development should be based on getting profits for the benefit of the population in the region, establishing a social-economic cultural process. However, to reach this effect, it is necessary to set goals and the local community can contribute to a social control of the economic activities involved and redirect the profits obtained to an investment in the region, increasing the quality of life in it.

"Regional development must be composed of three interconnected elements: the cultural ability to think about oneself and to innovate; the political-administrative ability to make decisions and organize their implementation; the ability of production according to the social objectives" [6].

In conclusion, sustainable economic management can also contribute to an evolution of thinking from the social point of view that is a process involving human, material and cultural resources, transforming the regional area into the ethical development that rapidly reaches public opinion, increasing the attractiveness of the region.

4 Social Design as a Factor of Regional Innovation and Entrepreneurship

Integrated in the contemporary world, design becomes the center of the relationship between economy and culture [13]. This should be used to value local identities and create competitive and symbolic differences in the market, in order to maintain a traditional culture alive [14].

The development of entrepreneurship strategies and the creation of regional products for the development and economic growth of some regions have been a concern in recent years. The consumer begins to develop ideas of goods and services within their place and the development of sustainable solutions, as synonymous of entrepreneurship and regional innovation, begin to generate a new approach of consumption. Fagianni [15] argues that, through the effects of globalization, the quality factor is no longer a differential factor, product innovation must appeal to the emotional.

In order to promote this type of solutions, we need the contribution of several skills and social design can reinforce this dynamics and interactivity with a logical collaborative and inclusive innovation, opening the way for the intervention of society. More than an art, design is a cultural phenomenon [16].

Design has the function of making "communicable" and symbolic products in relation to their function [17] and according to this new perspective, in association with a culture and region of origin, a product can be placed competitively in the market and become an over- valued product. For Pichler [18], design should be responsible for the creation and innovation of products which represent the material culture of a given location and the behaviours of the population where it is inserted.

"Innovation towards sustainability requires social participation" [19]. This means, the responsibility of the designer is to instil in a society new consumption criteria based on quality rather than quantity.

It is important to know how to communicate the quality and the social-environmental content of a product. The consumer can participate in the idea and creation of it but he is also influenced in the process of choice and appreciation. Presenting information that really shows the sustainability of a product is fundamental to stimulate the development of the producer - consumer - region relationship.

When design contributes to the development of a product for the benefit of a region, it should contribute both to the benefit of the producers and consumers of the region. Social design emerges here as an ally of regional innovation promoting cooperativism and inclusion, enhancing territorial capital, making resources in the region profitable and favorable to development, creating entrepreneurship and increasing the supply and demand of the global market [19].

5 Conclusion

Nowadays there is a growing interest in sustainability and entrepreneurship policies, that is, more and more companies compete for high economic performance and require a high level of creativity and innovation, betting on this dynamic of entrepreneurship and regional innovation as a competitive advantage in the global market.

Having finished the research carried out on the development of strategies in terms of entrepreneurship and regional innovation, creation of a local identity and the influence of new design concepts, it was possible to conclude that we are enabling economic and social development, cooperation and inclusion, respecting the culture and tradition of each region to be explored.

Although it is a brief analysis, and a generic way of approaching the topic in question, it does not make this research meaningless. The present study, besides opening the way to new researches on the subject, represents a contribution in the analysis of the relationship between entrepreneurship, regional innovation and the creation of competitive advantage in the depressed territorial areas based on social knowledge.

It is notorious throughout the research that the various authors argue that consumption and sustainable production contribute to a strong regional renewal while promoting their products.

The communication of historical elements, associated to a product, allows us to know the "history of the product", and the influence of the design can be the answer for its valorization. Social design can be essential when we talk about the development of a sustainable value chain.

So we can concluded that design relies on several levels and, in order to have regional innovation, create a competitive and creative advantage in the global market and the opportunity for entrepreneurship, there must be sustainable development, both economically and socially. It is important to understand what the favorable elements are in a certain region to be able to exploit it in an adequate way, to favour and stimulate (economic and social) relationships, to make a sustainable economy grow.

Acknowledgments. This work is financed by FEDER funds through the Competitivity Factors Operational Programme - COMPETE and by national funds through FCT – Foundation for Science and Technology within the scope of the project POCI-01-0145-FEDER-007136.

References

1. Abreu, J., et al.: *Success Full* - Casos de Sucesso nos Municípios Portugueses, 1º edição. *Idioteque* Edição, Portugal (2017)
2. Plonski, G.: Inovação em transformação, Estudos Avançados nº 31, Faculdade de Economia Administração e Contabilidade Universidade de São Paulo, São Paulo, Brasil (2017)

3. Canclini, N.: Culturas Híbridas, 4ª edição. Edição Universidade de São Paulo, Brasil (2008)
4. Krabbendam, D., et al.: Sustainist Design Guide, 2nd edn. BIS Publishers, Amsterdam (2013)
5. Barbosa, M.: O desafio do desenvolvimento sustentável, Revista Visões, 4ª edição, nº 4, vol. 1, Brasil (2008)
6. Alves, J.: O Desenvolvimento regional sob a óptica da sustentabilidade: uma reflexão sobre a economia e o meio ambiente, Ágora - Revista de Divulgação Cientifica, vol. 17, nº 2, Portugal (2010)
7. Sachs, I.: Estratégias de Transição para do século XXI: Desenvolvimento e Meio Ambiente. In: *Studio Nobel* – Fundação para o desenvolvimento administrativo, Brasil (1993)
8. Delgado, M., et al.: The contribution of regional costume in fashion. In: 6th International Conference on Applied Human Factors and Ergonomics and the Affiliated Conferences, Portugal (2015)
9. Albino, C., et al.: Reinterpretar os valores da tradição do território do Minho: o artesanato como chave de leitura (Reinterpretation of tradition values in Minho territory: Handcraft – a reading key). Strateg. Des. Res. J. (2011)
10. Costa, et al.: Convergência económica e coesão social e territorial da península ibérica na união europeia, 10º Colóquio Ibérico de Geografia - A Geografia Ibérica, Portugal (2005)
11. Fonseca, M.: A política regional da União Europeia: uma utopia viável? Universidade do Porto, Portugal (2004)
12. Gonsalez, A.: (citação em discurso politico), Membro representante do Governo da Galiza para o Planeamento e Fundos Comunitários. In: International Conference on the Role of Government in Regional Economic Development, Espanha (2005)
13. Gomez, B.: La Cultura del Diseno, estrategia para la generacion de valor e innovacion en la PyMe del Area Metropolitana del Centro Occidente, nº 34, Cent. Estud. Diseno Comum., Colômbia (2010)
14. Hall, S.: Identidade cultural na pós-modernidade, 11ª Ed, Brasil (2006)
15. Fagianni, K.: O poder do design. Brasília: *Thesaurus*, Brasil (2006)
16. Schneider, B.: Design – uma introdução: O design no contexto social, cultural e económico, São Paulo: *Blucher*, Brasil (2010)
17. Ono, M.: Design, cultura e identidade, no contexto da globalização, Revista Design em Foco, vol. I, Nº 01, Salvador   Universidade do Estado da Bahia, Brasil (2004)
18. Pichler, R.: O Design e a Valorização da Identidade Local, Universidade Federal de Santa Maria, Brasil (2012)
19. Krucken, L.: Design e território – valorização de identidade e produtos locais, Editora *Studio Noble* (2009)

Nonprofit Organizations at the Crossroads of Offline Fundraising and Social Innovation: The Influence of Promoter Behavior on the Success of Donation-Based Crowdfunding Through Digital Platforms

Noelia Salido-Andres[1(✉)], Marta Rey-Garcia[1],
Luis Ignacio Alvarez-Gonzalez[2], and Rodolfo Vazquez-Casielles[2]

[1] University of A Coruña, A Coruña, Spain
{noelia.sandres,martarey}@udc.es
[2] University of Oviedo, Oviedo, Spain
{alvarezg,rvazquez}@uniovi.es

Abstract. The purpose of this research consists of exploring the extent to which the behavior of promoters or fundraisers - explaining the success of offline fundraising campaigns for charitable causes - may influence as well the success of donation-based crowdfunding (DCF) campaigns promoted through digital platforms. First, the literature on the influence of the promoter behavior on the success of offline fundraising campaigns for charitable causes is reviewed. Second, hypotheses are proposed linking the determinants identified in the literature to DCF campaigns. Thirdly, their explanatory capacity is measured through quantitative analysis based upon a database of 360 campaigns fostered by nonprofits via Microdonaciones, a donation-based crowdfunding digital platform, for the period 2012–2017. Logistic regression analysis is used to test the hypotheses proposed. Results confirm in the light of the promoters behavior, that fundraising campaigns fostered by nonprofits organizations for charitable causes behave similarly offline and online.

Keywords: Donation-based crowdfunding · Charitable causes
Digital platforms · Promoter behavior · Determinants of success

1 Introduction

Crowdfunding (CF) emerges in the new digital sphere as a tool for the online funding of resources, goods and services. Belleflamme et al. [1] define CF as an open call via the Internet for the provision of financial resources to support the realization of initiatives for specific purposes. CF is based on *donation* model when funders donate to causes just for the sake of supporting them, with no expected (material) compensations [2].

The usage of donation-based crowdfunding (DCF) through digital platforms - on-line portals to connect fundraisers and funders - is growing rapidly for nonprofit organizations (NPO). In a context of reduced access to traditional income sources and economic strains, this online tool complements, or substitutes, the usage of traditional

© Springer International Publishing AG, part of Springer Nature 2019
J. Machado et al. (Eds.): HELIX 2018, LNEE 505, pp. 946–952, 2019.
https://doi.org/10.1007/978-3-319-91334-6_130

fundraising instruments to develop campaigns for charitable causes [3]. Furthermore, DCF campaigns through digital platforms are fully situated in the intersection between new social marketing practices by NPO and the emerging forms of civic participation facilitated by Information and Communication Technologies (ICT). They should be thus considered as social innovation in themselves, social innovation being defined as a combination of both, the development and implementation of new ideas (products, services and models) to meet social needs, and the simultaneous creation of new social relationships or collaborations [4]. Far from being once-only events, DCF campaigns become actual content within digital platforms since the use of technological devices guarantees immediate and permanent access to a wide variety of opportunities for contribution and participation. Potential online donors will therefore discriminate social causes according to their particular interests, unlimited by space and time, neutralizing any possibility of campaign fatigue characteristic of offline contexts [5–7]. However, and despite the increasing importance of DCF for NPO to promote prosocial values and successfully advance social causes, specific emerging literature is scarce and predominantly focuses on aspects related to funding behaviors, the effects of giving on reputation, donor satisfaction and self-esteem, or tax incentives effects on donation [8].

In light of the above, the identification of the determinants explaining the success of DCF campaigns remains largely unexplored. In this context, the main objective of this analysis consists on exploring the extent to which factors related to the behavior of promoters explaining the success of offline fundraising campaigns for charitable causes, may also explain the success of DCF campaigns fostered by NPO through digital platforms.

2 Literature Review and Hypothesis Formulation

2.1 Crowdfunding Platforms and DCF Campaigns Promoted by NPO

CF campaigns consist of open calls by promoters or fundraisers to contribute to a wide variety of causes with different objectives (i.e. technological, scientific, creative, business, cultural, artistic or social objectives, among many others). These can follow an *all or nothing* modality - the monetary contributions are effective for contributors or funders (charged to bank accounts) and promoters (effective payment) only if the total amount requested is achieved in the due date -, or a *keep it all* modality - when monetary contributions are effective to contributors and promoters, regardless of the amount raised relative to target -.

As previously noted, CF campaigns specifically based on donation are a frequently model used by NPO for financing charitable causes. However, in addition to a funding channel, they serve as well as tools to increase their social support, disseminate nonprofit initiatives, and generate the optimal conditions to create stable fundraising communities linked to NPO beyond the funding of occasional projects [9].

Resulting from the development of Web 2.0-based technologies, CF campaigns are frequently channeled through new electronic spaces – *e-marketplaces* – via digital platforms. A digital platform is an online portal where users' authentication is required and commercial or noncommercial transactions between the parties involved are handled. It often provides other kind of services such as media hosting or social networking, increasing the online visibility of the operations and the variety of potential contacts between users and contributors [10]. Crowdfunding platforms (CFP) can be *own* platforms – launched by the same promoters of the campaigns (i.e. individuals, entities or businesses) – or *external* – when third parties act as intermediaries between promoters and funders -. The latter type is especially recommended when promoters have no prior experience launching and managing CF calls. Depending on the variety of campaign categories, CFP can be *specialists* when they host campaigns from a same category (i.e. charity), or *general*, when they host campaigns from a wide variety of categories (i.e. cultural, creative, social, technology). In terms of the geographical scope of the owner, CFP can be either *national* or *franchises* of global or international platforms.

In the context of the scarce literature specific on DCF campaigns, we argue that it is reasonable to assume that some of the factors that influence the success of offline fundraising campaigns for charitable causes, as those related to the behavior of the promoter or fundraiser – *How?* -, may also explain to some extent the outcomes of online campaigns.

2.2   The Behavior of the Promoting NPO as Determinant of Success of Offline Fundraising Campaigns in the Context of DCF for Charitable Causes

Information asymmetry characterizes the relationship between donors and NPO, since the former are usually deprived of full - and updated - information on how charities use their contributions [11]. The information provided by NPO and their effective and transparent behavior, results crucial to minimize the effects of *toxic charity* in the intention of giving, especially when NPO provides charitable services abroad [6, 12–14]. In response to this information gap, potential donors may collect information on potential beneficiaries, as well as on the governance and previous performances of promoters, whether they intend to contribute offline or online. In this sense, quantity, quality and accessibility of information is crucial [12].

Donors seem to contribute less when organizations are run inefficiently or the distribution of aid to victims is irresponsible and unfair, and consequently their contributions will not make a big real change [12, 15]. NPO should therefore stimulate giving by behaving transparently and accountably, since private donors look for guarantees that their contributions will reach the target beneficiaries as efficiently and effectively as possible [7], especially when the monetary target requested aims multiple ends [16]. Provision of relevant information should not only be limited during the event, but also extend to the post-event stage, when donors will need to reassure that their money has been effectively spent. In line with the aforementioned, we expect that

Hypothesis 1a (H1a). DCF campaigns for charitable causes through digital platforms where the promoting NPO provides information on the advances of the campaign are more likely to succeed

and,

Hypothesis 1b (H1b). DCF campaigns for charitable causes through digital platforms where the promoting NPO provides information on the funding uses are more likely to succeed.

Conceptual model is depicted as follows (Fig. 1).

Fig. 1. Determinants of successful offline fundraising campaigns related to the behavior of the promoting NPO, driving the success of DCF campaigns for charitable causes via digital platforms.

3   Methodology

3.1   Selection of the DCF Platform for Charitable Causes

The focus of this research is on the *external*, *specialist*, *national* and *all-or-nothing* DCF platform Microdonaciones.net, a digital platform launched by Fundacion Hazloposible in 2012. Microdonaciones promoted giving of small amounts of money to charitable campaigns fostered by mostly Spanish NPO. Donors could contribute either to a specific charitable campaign, or through regular and monthly contributions to a portfolio of campaigns. There was neither a minimum nor a maximum amount for giving. All the campaigns hosted were active on the platform for a fixed period of five weeks, distributed by categories according to their final purpose (e.g. education, social exclusion, childhood, environment, health). In addition to centralize the resulting monetary transactions, the platform also intermediated in the communication processes among NPO – normally small-medium size with limited ICT skills - and (potential) donors.

3.2   Data Collection and Sample Description

In order to test our hypothesis, data on independent and dependent variables were gathered from Microdonaciones official website. In particular, details of charitable campaigns were collected from the Microdonaciones online datasheets. A database was built storing a total of 360 successful and unsuccessful charitable campaigns fostered by NPO since March 19, 2012, when the first charitable campaign hosted started, until March 22, 2017, when the last campaign finished. In this 5-year time frame period, campaigns attracted over 9.300 online donations via the platform, resulting in 262 successful charitable campaigns, 73% of the total promoted (Table 1).

NPO updated information on the campaign advances in 60% of the cases, and in 19% they provided details on the final funding uses raised during the campaign (Table 2).

Table 1. Microdonaciones activity effects during the period analyzed (2012–2017).

Total campaigns	Total successful campaigns	Total unsuccessful campaigns	Total requested (€)	Total raised (€)	Total donors	Total beneficiaries
360	262	98	681.733	516.448	8.413	149.477

Source: Authors' own elaboration from Microdonaciones website [17]

Table 2. Sample description.

Descriptors	Volume of campaigns in Microdonaciones (N = 360)
How: The promoter behavior	
Information on advances	59.7
Information on funding uses	18.9

3.3 Measuring the Model Variables

Our dependent variable - success of DCF campaigns - was operationalized as the *attainment of the monetary goal requested* in due time within the platform. As a dichotomous dependent variable, two possible values can be adopted: 1, when the monetary goal was achieved (successful campaign) and 0 on the contrary case, if funding raised was not enough to reach the target goal (unsuccessful campaign).

For each of the 360 charitable campaigns data for the predictor variable were recorded, attending to the additional information voluntarily provided by NPO in each respective datasheets. In the particular case of the advances, additional disclosures on the *advances for each campaign* in order to encourage the potential donors' commitment (i.e. ongoing thank-you messages, motivational messages encouraging keep giving, etc.) were recorded. In the case of additional information on the *uses of funding* finally raised, data were gathered attending to the narrative and campaign imagery added once the campaigns were over (e.g. listing of initial purchases vs. final purchases, audiovisual reports on preparations and/or results of activities scheduled, audiovisual reports on symbolic laying of "first stones" or final look of infrastructures, buildings or technologies, imagery from actual beneficiaries thanking/receiving/celebrating the target reached, etc.).

4   Results

Results of a logit model using STATA 13.0 MP for Windows are detailed in Table 3.

Results reveal the strong link among the inclusion of additional information voluntarily provided by the promoting NPO and the success of DCF campaigns. The voluntary inclusion of additional and updated information on the advances (p = 0.000) and the end-uses of funds raised (p = 0.001) in the context of the DCF campaigns, fully determines (p < .05) their success. Therefore, and as expected, H1a and H1b are supported.

Table 3. Effects of determinants of success of offline fundraising campaigns for charitable causes related to promoter behavior, on success of DCF campaigns via digital platforms.

Variable	Coef.	Std. Err.	P > |z|
How: The promoter behavior			
Information on advances	1.719	0.27	0.000
Information on funding uses	3.286	1.02	0.001
Constant	−0.15	0.174	0.389
Log likelihood	−168.84603		
N	360		
LR chi2	85.15		
Prob > chi2	0.0000		
Pseudo R2	0.2014		

5 Implications and Conclusions

Successful DCF campaigns include details and information on the advances and the end-uses of the volume of contributions finally raised. In this regard, relevant implications for the promoting NPO emerge from a managerial perspective. Digital accountability and transparency seem key not only in order to minimize the characteristic information asymmetry in the relationship between NPO and their donor communities, but also for providing and sharing contents that stimulate the creation of stable, long-term relations, and bring donations. In particular, the information that is voluntarily disclosed in the campaign (e.g. narrative claims and description and picture and video imagery) may be useful to effectively impact potential donors and help them empathize with the charitable cause and its beneficiaries. In view of the results we can conclude that, taking the behavior of promoters or fundraisers as reference, fundraising campaigns promoted by NPO for charitable causes behave similarly offline and online.

References

1. Belleflamme, P., Lambert, T., Schwienbacher, A.: Crowdfunding. Tapping the right crowd (2012). http://ssrn.com/abstract=1578175
2. Massolution. The Crowdfunding Industry report (2012). http://www.crowdsourcing.org/document/crowdfunding-industry-report-market-trends-composition-and-crowdfunding-platforms/14277
3. Rey-Garcia, M., Alvarez, L.I., Valls, R.: The evolution of national fundraising campaigns in spain: nonprofit organizations between the state and emerging civil society. Nonprofit Volunt. Sect. Q. **42**, 300–321 (2013)
4. Sanzo, M.J., Alvarez-Gonzalez, L.I., Rey-Garcia, M., García, N.: Business-nonprofit partnerships: a new form of collaboration in a corporate responsibility and social innovation context. Serv. Bus. Int. J. **9**, 611–636 (2015)
5. Kinnick, K.N., Krugman, D.M., Cameron, G.T.: Compassion fatigue: communication and burnout toward social problems. Journal. Mass Commun. Q. **73**(3), 687–707 (1996)

6. van Leeuwen, M.H.D., Wiepking, P.: National campaigns for charitable causes: a literature review. Nonprofit Volunt. Sect. Q. **42**(2), 219–240 (2013)
7. Wiepking, P., van Leeuwen, M.H.D.: Picturing generosity: explaining the success of national campaigns in The Netherlands. Nonprofit Volunt. Sect. Q. **42**(2), 262–284 (2013)
8. Gleasure, R., Feller, J.: Emerging technologies and the democratization of financial services: a metatriangulation of crowdfunding research. Inf. Organ. **26**, 101–115 (2016)
9. Salvetti Llombart. El perfil del donante en España (2013). https://www.afundacion.org/docs/socialia/informe_perfil_donante_espa%c3%b1a.pdf
10. Danmayr, F.: Archetypes of Crowdfunding Platforms: A Multidimensional Comparison. Springer Fachmedien Wiesbaden, Wiesbaden (2014). https://doi.org/10.1007/978-3-658-04559-3
11. Beldad, A., Gosselt, J., Hegner, S., Leushuis, R.: Generous but not morally obliged? Determinants of Dutch and american donors' repeat donation intention (REPDON). Voluntas **26**, 442–465 (2015)
12. Tremblay-Boire, J., Prakash, A.: Will you trust me?: how individual american donors respond to informational signals regarding local and global humanitarian charities. Voluntas **28**, 621–647 (2017)
13. Hou, J., Zhang, C., Allen, R.: Understanding the dynamics of the individual donor's trust damage in the philanthropic sector. Voluntas **28**, 648–671 (2017)
14. Bekkers, R., Wiepking, P.: A literature review of empirical studies of philanthropy: eight mechanisms that drive charitable giving. Nonprofit Volunt. Sect. Q. **40**(5), 924–973 (2011)
15. Einolf, C., Philbrick, D., Slay, K.: National giving campaigns in the United States: entertainment, empathy, and the national peer group. Nonprofit Volunt. Sect. Q. **42**, 241–261 (2013)
16. Mourao, P., Costa, C.: Investors or givers? The case of a Portuguese crowdfunding site. In: Omatu, Si, et al. (eds.) Distributed Computing and Artificial Intelligence, 12th International Conference, pp. 113–120. Springer, Cham (2015). https://doi.org/10.1007/978-3-319-19638-1_13
17. Microdonaciones. http://microdonaciones.hazloposible.org/

Exploring Circular Economy in the Hospitality Industry

Jorge Julião[1]([✉]), Marcelo Gaspar[2], Benny Tjahjono[3],
and Sara Rocha[1]

[1] Católica Porto Business School,
Universidade Católica Portuguesa, Porto, Portugal
jjuliao@porto.ucp.pt
[2] Escola Superior de Tecnologia, Instituto Politécnico, Castelo Branco, Portugal
[3] Centre for Business in Society, Coventry University, Coventry, UK

Abstract. This paper explores the role of Circular Economy (CE) in the hospitality industry, namely in hotels and restaurants. Today's hospitality consumers have become more ecologically conscious than ever before, and the demand for eco-friendly products and services has grown. This has imposed many hospitality companies into the adoption of Green Practices (GP) and Circular Economy (CE) principles. In particular, CE has been gaining popularity among governments and academia. However, current research shows that CE principles are being mainly discussed and applied in production and manufacturing. The application of CE principles into services, particularly in the hospitality industry, seem to receive little attention. The paper presents the CE concept and discusses it from the perspectives of both companies and consumers and debates the adoption of green practices by the hospitality industry. Our findings indicated that the consumer awareness of sustainable issues has a direct impact on companies' adoption of CE and GP practices. The paper also identifies that green practices are gaining increased attention in the hospitality industry, and that a wide variety of green practices are already being adopted. However, the role of CE in hospitality industry appears to be uncovered by current research and obscured in green practices. The paper, therefore, identifies a research gap and calls for further investigation in the application of CE in the hospitality industry.

Keywords: Circular economy · Green practices · Hospitality industry
Sustainability · Tourism

1 Introduction

Depletion of the earth's natural resources and sustainability issues have imposed governments and business to adopt policies that aim to achieve resource-efficient economies. In this context, Circular Economy (CE) and Green Practices (GP), are gaining attention worldwide as a way for the society to increase prosperity, while reducing dependence on natural resources and energy [1], and to leave the current production and consumption model based on continuous growth and increasing resource throughput [2]. CE is a sustainable development strategy, or economic model,

© Springer International Publishing AG, part of Springer Nature 2019
J. Machado et al. (Eds.): HELIX 2018, LNEE 505, pp. 953–960, 2019.
https://doi.org/10.1007/978-3-319-91334-6_131

that attempts to conceptualize the integration and interaction of economic activity and environmental issues in a sustainable way, balancing economic, environmental, technological and social aspects [3].

In a Multiple Helix (MH) system a common ground is set between ecology, knowledge, and innovation, creating the synergies between economy, society, and democracy [4]. This approach argues that the development of green products and services are influenced by the interrelationship between four main actors, namely, university, industry, government and consumers [5]. Within this paper, the interactions between companies and consumers are explored, addressing their needs, expectations, and attitudes, towards CE and Green Practices (GP).

Triggered by the shifts in consumer behavior towards GP, hospitality industry is increasingly motivated and willing to take steps to greater environmental responsibility [6]. Therefore, some of these companies are adopting proactive environmental management and practices to improve their competitiveness [7]. However, the applications of CE principles in hospitality industry are particularly lacking in existing research. Considering the motivation of these industries towards the adoption of GP, the growing of the service sector and the current research in CE, the paper calls for further research in the applications of CE principles in the hospitality industry.

2 Circular Economy Concept

The CE is an industrial and social evolutionary concept that pursues holistic sustainability goals through a culture of no waste, that proposes a closed-loop of material flows in the economy, in opposition to the linear economy [8]. From the environmental economic perspective, CE is based on the balance principle, which infers that all material flows need to be accounted for, although it will be the economic values, not the physical flows, that guide their management [9].

CE attempts to harmonize the ambitions of both environmental conservation and economic growth, considering multiple perspectives, *i.e.*, levels of analyses and life cycle phases. CE can be applied to an economic sector, or individual process, and comprising the environmental impact of the entire activity system, *e.g.*, production, design, transportation, distribution, consumption, recycling, and disposal. Although the term of CE is clearly conceptualized, there is no CE definition commonly accepted. For example, there are definitions of CE that focus on the economic aspects [10], the 3R principles [11], and the industrial ecology [12]. Based on an extensive review and case studies analysis, the Ellen MacArthur Foundation proposes a more comprehensive definition that considers both the environmental and economic advantages simultaneously. CE is hereby defined as "an industrial system that is restorative or regenerative by intention and design" [8].

3 Companies and Their Lead-Role in the Circular Economy Model

Companies play a key role in CE because they are responsible for the creation of products and services. Although the linear economic model, which is based on throughput optimization and cost efficiency, appears to prevail in most companies, some of them are starting to implement strategies that integrate sustainability issues [8]. The challenge is to identify drivers and arguments that impel companies towards CE and circular innovations, particularly, adopting cleaner production and eco-design.

Mainly driven by public awareness and regulations [13], companies are increasing their efforts to integrate environmental sustainability issues into their products and services. A way of enabling this integration is through product design. One of the principles CE is based on, is that manufactured durable products (and services), need to be designed from the start for reuse, *i.e.*, designed in a way that optimizes disassembly and recirculation. An argument that may drive companies towards circular products, raised by MacArthur [8], is the resources price and volatility, which have been increasing in recent years, as populations grow and raw materials become scarce, hence more expensive than ever before. The underlying argument is that, if companies manage to manufacture products that require fewer raw materials and have a high percentage of reuse, the company will be less exposed to resources price and volatility.

The CE-initiatives are often seen as constraints to industrial activities rather an opportunity for sustainable business and growth. Nonetheless, it has been confirmed by several researchers that there is a direct correlation between the integration of environmentally sustainable solutions and the gain of competitive advantage [14]. Moreover, according to Porter and van der Linde [15] reducing environmental impact at lower costs could be perceived as an opportunity by companies, mainly by redesigning products, processes, and/or operation methods. As such, the relationship between industry and environment is critical for industrial business performance [16]. This means that although CE may present different challenges and opportunities to companies, it can offer positive impacts on their business. The main challenge for companies is how to adopt the business models that aim to reap the benefits from existing resources and reduce dependency on new, virgin resources.

The above arguments imply that by adopting CE strategies and principles, companies can expect to have a direct gain in costs reduction and competitive advantage, and ultimately to contribute to the environment protection and social wellbeing. These could be important drivers for the growth of CE initiatives. Moreover, current research seems to focus on CE applied to production and manufacturing sectors, and little has been devoted to the roles of CE in services, particularly in the hospitality industry.

4 Consumers as the Target of Circular Economy

The commercial success of products and services that integrate CE principles is crucial in driving companies and society towards environmental sustainability [17]. However, most consumers may still relatively be unaware of the CE principles despite the increased awareness of sustainability issues due to regulations, scientific publications and public discussions. This unawareness may be due to the fact that many conceptualizations of the CE seem to exclude a large part of the social dimension [18], in particular consumers. Despite the fact that consumers may have a great deal of contribution to the implementation of CE principles, since they are the ones who purchase products/services and could influence the governments.

Consumers tend to select products based on the perceived value for money according to its price and quality ratio [19]. As discussed by Witjes and Lozano [20] quality criteria may include other non-economic criteria, e.g. environmental criteria. Thus, the increased consumer awareness for sustainability may increase the demand for circular products, driving the shift in business models from convention to circular products. This perceived value, however, needs to be communicated to consumers in an efficient way.

Although consumers have information about environmental advantages of CE and gains associated with well-being and health, have little knowledge and experience with products/services that integrate CE principles. This lack of experience may increase consumers doubts regarding product/service specifications, claims and added value, affecting their purchase intention. Environmental labeling is an effective way of communicating to customers the specific benefits and characteristics of the product and the claim, which can be displayed by using symbols or messages [21]. Thus, the acceptance of circular products and services by consumers may be enhanced if sustainability messages are communicated effectively, which in turn will promote the integration of CE in the manufacturing of products.

5   Green Practices in Hotel Industry

The tourism industry growth has a direct impact on natural resources consumption, and environment [22]. Hotels are one of the most energy-intensive building types due to their multi-usage functions and around the clock operations [23]. According to the Pacific Gas and Electric's Food Service Technology Center (FSTC), hotels use almost 5 times more energy per square foot compared to any other commercial buildings. Consequently, the adoption of green practices is gaining attention in the hotel industry [24]. Many hotels are committed to a variety of environmental initiatives such as reducing energy and water consumption, and decreasing waste outputs [25], which has led to the emergence of 'green hotels'. This market niche is expanding, and being embraced not only by leading hotel brands, but also by small and medium-sized businesses [26]. Some of these companies acknoweldged that green management can improve cost saving, employee loyalty, and customer retention, and help to meet short-term operational targets [27].

Although the demand for green hotels is still relatively low, it is gradually increasing [28]. Customers are becoming more aware of pollution and waste issues, and they look for hotels that adopt green practices [29]. To meet this emerging demand for new more sophisticated customers, hotels are incorporating products and services that are more environmentally friendly [30]. Guests are nowadays willing to pay a premium for environmentally friendly products and services [31], and green hotel consumers show a willingness to sacrificing their convenience, comfort and some luxury standards through the process [25]. This consumer behavior is positively influencing the adoption of green practices by the hotel industry. Some surveys indicate that the majority of hotels have a medium to high experience in green practices, and understand the need and the value of adopting green practices [32].

The above arguments show that the adoption of green practices is gaining attention in the hotel industry, and that a wide variety of these sustainable practices are being adopted. However, current research does not demonstrate if the hotel industry is aware of CE practices, nor applying its principles.

6 Green Practices in Restaurant Industry

Like for hotels, green practices are becoming a trend in restaurants. Restaurants consume a vast amount of disposable products, water and energy, so they need to adopt a new way of thinking that will allow the business to benefit by improving their environmental performance [33]. The business activity in restaurants is dependent on a strong image and brand, which makes the adoption of green practices crucial [34]. Research indicates that these practices have a positive effect on brand image, which promotes financial benefits and contributes to the economic sustainability of the local community [34].

Thus, similarly to hotels, the concept of green restaurant was created. Green restaurants promote recycling and composting, water and energy efficiency, and waste management, offer the option of locally grown or organic food on the menu [35], devote effort to the three Rs (reduce, reuse, and recycle) and the two Es (energy and efficiency) [36]. Restaurants employ a large variety of green practices, for example, implementing equipment and rules that save energy and water, reduce and recycling waste, forbidding disposable containers, using locally-grown and organic raw materials, and training employees to adopt green practices [34]. These green restaurant practices are implemented according to a strategic perspective, which can be health, environmental or social concerns [37].

A survey conducted by the National Restaurant Associations in 2011, showed that consumers are more likely to spend their money at a restaurant if they know it is green. Moreover, green food and green practices are important factors that influence patronage decisions [38]. Consumers that are more health-conscious value food-focused practices, such as organic or locally-grown food [38], which have a positive and significant impact on consumers' preferences [39]. Administration-focused practices, such as green certifications and CSR (cooperate social responsibility), also positively influences consumer decision, since it reduces consumers' perceived uncertainty and risks with food products.

7 Summary and Conclusions

CE is a concept that becomes popular within governments and businesses as an approach to promote sustainability, based on restorative and regenerative production and consumption systems. In particular, North America and European countries, China and Japan, are strongly investing in and promoting the CE initiatives. The number of research papers being published in the body of literature demonstrated an increased attention to CE. Most of this research is devoted to manufacturing and focuses on the development of CE solutions for products and their production processes. However, the service sector, directly and indirectly, is responsible for a significant impact on the environment. Nevertheless, attention to the application of CE principles into the service sector seems to be lacking.

The paper identifies that, nowadays, consumers are becoming more aware of environmental issues, which is reflected in their behaviour when buying products and services. Moreover, it shows that consumers have a positive impact on companies' adoption of CE and GP practices. Thus, studies towards the identification of drivers and arguments that impose hotels and restaurants towards CE and circular innovations, would benefit from surveying customers to better understand their motivations when purchasing products and services in these industries. The paper also identifies that green practices are gaining increased attention in the hotel and restaurant industries, and that a wide variety of green practices are already being adopted. However, the role of CE in these industries, appears to receive little attention by current research and could be obscured in green practices. The paper, therefore, identifies a research gap in the application of CE principles in the tourism industry and calls for further research in this field.

References

1. MacArthur, E., Zumwinkel, K., Stuchtey, M.R.: Growth within: a circular economy vision for a competitive Europe. Ellen MacArthur Foundation (2015)
2. Ghisellini, P., Cialani, C., Ulgiati, S.: A review on circular economy: the expected transition to a balanced interplay of environmental and economic systems. J. Clean. Prod. **114**, 11–32 (2016)
3. Murray, A., Skene, K., Haynes, K.: The circular economy: an interdisciplinary exploration of the concept and application in a global context. J. Bus. Ethics **140**, 369–380 (2015)
4. Carayannis, E.G., Sindakis, S., Walter, C.: Business model innovation as lever of organizational sustainability. J. Technol. Transf. **40**, 85–104 (2014)
5. Julião, J., Gaspar, M., Tjahjono, B.: Key factors on green product development: influence of multiple elements. In: Peris-Ortiz, M., Ferreira, J.J., Farinha, L., Fernandes, N.O. (eds.) Multiple Helix Ecosystems for Sustainable Competitiveness, pp. 75–90. Springer (2016)
6. Bohdanowicz, P.: A study of environmental impacts, environmental awareness and pro-ecological initiatives in the hotel industry (2003)
7. Claver-Cortés, E., Molina-Azorín, J.F., Pereira-Moliner, J.: The impact of strategic behaviours on hotel performance. Int. J. Contemp. Hosp. Manag. **19**, 6–20 (2007)
8. MacArthur, E.: Towards the circular economy: opportunities for the consumer goods sector (2013)

9. Andersen, M.S.: An introductory note on the environmental economics of the circular economy. Sustain. Sci. **2**, 133–140 (2007)
10. Stahel, W.R., Reday-Mulvey, G.: Jobs for Tomorrow: The Potential for Substituting Manpower for Energy. Vantage Press, Burlington (1981)
11. Yuan, Z., Bi, J., Moriguichi, Y.: The circular economy: a new development strategy in China. J. Ind. Ecol. **10**, 4–8 (2006)
12. Geng, Y., Doberstein, B.: Developing the circular economy in China: challenges and opportunities for achieving' leapfrog development'. Int. J. Sustain. Dev. World Ecol. **15**, 231–239 (2008)
13. Dangelico, R.M., Pujari, D.: Mainstreaming green product innovation: why and how companies integrate environmental sustainability. J. Bus. Ethics **95**, 471–486 (2010)
14. Wong, S.K.-S.: The influence of green product competitiveness on the success of green product innovation. Eur. J. Innov. Manag. **15**, 468–490 (2012)
15. Porter, M., van der Linde, C.: Green and competitive: ending the stalemate. Harv. Bus. Rev. **73** (1995)
16. Lieder, M., Rashid, A.: Towards circular economy implementation: a comprehensive review in context of manufacturing industry. J. Clean. Prod. **115**, 36–51 (2016)
17. Hall, J., Clark, W.W.: Environmental innovation. J. Clean. Prod. **11**, 343–346 (2003)
18. Geissdoerfer, M., Savaget, P., Bocken, N.M.P., Hultink, E.J.: The circular economy – a new sustainability paradigm? J. Clean. Prod. **143**, 757–768 (2017)
19. Mandese, J.: New study finds green confusion. Advert. Age. **62**, 1–56 (1991)
20. Witjes, S., Lozano, R.: Towards a more circular economy: proposing a framework linking sustainable public procurement and sustainable business models. Resour. Conserv. Recycl. **112**, 37–44 (2016)
21. D'Souza, C., Taghian, M., Lamb, P., Peretiatkos, R.: Green products and corporate strategy: an empirical investigation. Soc. Bus. Rev. **1**, 144–157 (2006)
22. Wu, K.S., Teng, Y.M.: Applying the extended theory of planned behavior to predict the intention of visiting a green hotel. Afr. J. Bus. Manag. **7**, 7579–7587 (2011)
23. Huang, Y., Song, H., Huang, G.Q., Lou, J.: A comparative study of tourism supply chains with quantity competition. J. Travel Res. **51**, 717–729 (2012)
24. Han, H., Hsu, L.-T.J., Sheu, C.: Application of the theory of planned behavior to green hotel choice: testing the effect of environmental friendly activities. Tour. Manag. **31**, 325–334 (2010)
25. Rahman, I., Reynolds, D.: Predicting green hotel behavioral intentions using a theory of environmental commitment and sacrifice for the environment. Int. J. Hosp. Manag. **52**, 107–116 (2016)
26. Rahman, I., Park, J., Chi, C.G.: Consequences of "greenwashing": consumers' reactions to hotels' green initiatives. Int. J. Contemp. Hosp. Manag. **27**(6), 1054–1081 (2015)
27. Chen, Y.S.: The driver of green innovation and green image - green core competence. J. Bus. Ethics **81**, 531–543 (2008)
28. Chan, E.S.W., Hsu, C.H.: Environmental management research in hospitality. Int. J. Contemp. Hosp. Manag. **28**(5), 886–923 (2016)
29. Manaktola, K., Jauhari, V.: Exploring consumer attitude and behaviour towards green practices in the lodging industry in India. Int. J. Contemp. Hosp. Manag. **19**(5), 364–377 (2007)
30. Dief, M.E., Font, X.: The determinants of hotels' marketing managers' green marketing behaviour. J. od Sustain. Tour. **18**, 157–174 (2010)
31. Kang, K.H., Lee, S., Huh, C.: Impacts of positive and negative corporate social responsibility activities on company performance in the hospitality industry. Int. J. Hosp. Manag. **29**, 72–82 (2010)

32. Al-Aomar, R., Hussain, M.: An assessment of green practices in a hotel supply chain: a study of UAE hotels. J. Hosp. Tour. Manag. **32**, 71–81 (2017). https://doi.org/10.1016/j. jhtm.2017.04.002

33. Revell, A., Blackburn, R.: The business case for sustainability? An examination of small firms in the UK's construction and restaurant sectors. Bus. Strateg. Environ. **16**, 404–420 (2007)

34. Schubert, F., Kandampully, J., Solnet, D., Kralj, A.: Exploring consumer perceptions of green restaurants in the US. Tour. Hosp. Res. **10**, 286–300 (2010)

35. Jang, Y.J., Kim, W.G., Bonn, M.A.: Generation Y consumers' selection attributes and behavioral intentions concerning green restaurants. Int. J. Hosp. Manag. **30**, 803–811 (2011)

36. Gilg, A., Barr, S., Ford, N.: Green consumption or sustainable lifestyles? Identif. sustain. consum. Futures. **37**, 481–504 (2005)

37. Choi, G., Parsa, H.G.: Green practices II: measuring restaurant managers'psychological attributes and their willingness to charge for the "green practices". J. Foodserv. Bus. Res. **9**, 41–63 (2006)

38. Hu, H.-H., Parsa, H.G., Self, J.: The dynamics of green restaurant patronage. Cornell Hosp. Q. **51**, 344–362 (2010)

39. Vieregge, M., Scanlon, N., Huss, J.: Marketing locally grown food products in globally branded restaurants. J. Foodserv. Bus. Res. **10**, 67–82 (2007)

Entrepreneurship and Internationalization

Intangible Resources, Absorptive Capabilities, Innovation and Export Performance: Exploring the Linkage

Orlando Lima Rua[1]([⊠]), Rubén Fernández Ortiz[2], Alexandra França[3], and Mónica Clavel San Emeterio[2]

[1] Polytechnic of Porto, ISCAP, CEOS.PP/UNIAG, Porto, Portugal
orua@iscap.ipp.pt
[2] University of La Rioja, Logroño, Spain
{ruben.fernandez,monica.clavel-san}@unirioja.es
[3] University of Vigo, Vigo, Spain
franca.alexandra@gmail.com

Abstract. The main goal of this study is to analyse the influence of intangible resources and absorptive capabilities on export performance, considering the mediating effect of innovation. Based on survey data from 247 Portuguese small and medium-sized enterprises (SMEs) findings suggest that: (1) Intangible resources has a positive influence on absorptive capabilities, innovation and on export performance; (2) Absorptive capabilities has a positive influence on innovation and on export performance; (3) Innovation has a positive influence on export performance; and (4) Innovation has a mediating effect on the relationship between intangible resources and export performance, the same does happens on the relationship between absorptive capabilities and export performance. This study provides novel insights into strategic management literature by trying to understand the SMEs business growth. Moreover, this paper presents further evidences of the strategies that small firm managers should pursue and policy makers should promote.

Keywords: Intangible resources · Absorptive capabilities · Innovation
Export performance · SMEs · PLS-SEM

1 Theoretical Framework and Hypotheses Derivation

The literature suggests that resources and capabilities are related to each other [1, 2]. Integration, reconfiguration and learning resources only become significant when resources are abundant, thus improving firms' dynamic capabilities. Moreover, innovation is an ability that can attract the necessary resources to exploit opportunities. These resources can thus promote, support and facilitate innovation, allowing firms to innovate and prosper, contributing to the construction of healthy and enduring business [3]. The benefits of innovation may result in the development of products and processes that occurs in multiple stages (multi-stage process), requiring a complete set of resources for an innovative firm [4]. [5] highlights resources as assets, capabilities, organizational processes, firm attributes, information and knowledge, which, according

© Springer International Publishing AG, part of Springer Nature 2019
J. Machado et al. (Eds.): HELIX 2018, LNEE 505, pp. 963–970, 2019.
https://doi.org/10.1007/978-3-319-91334-6_132

to this author, are valuable, rare, imperfectly imitable and non-substitutable (VRIN). The resource-based perspective conceptualizes innovation as a complex and dynamic process through which firms consistently develop innovation capabilities by exploring new resources or new combinations of resources [4]. These ideas serve as the basis to suggest our next hypotheses: H1: *Intangible resources influence positively absorptive capabilities*; H2: *Intangible resources influence positively innovation*; and H3: *Absorptive capabilities influence positively innovation*.

[6] believe that innovation is important for organizational success both to local and foreign markets. According to these authors, success in the global market requires creativity and risk-taking. Literature suggests that innovation has a positive influence in business performance [7], since it increases firms' engagement to, for example, create new products and services, seek new opportunities and new markets [3, 8]. In this sense, innovative firms have an extraordinary performance and can even be seen as a country's engine of economic growth [9]. Thus, these firms can control markets by mastering distribution channels and building brand recognition. Hence, we intended to confirm this relationship and test the following working hypothesis: H4. *Innovation influences positively export performance*.

RBV scholars argue that variations in firms' performance result from the possession of heterogeneous resources. This heterogeneity leads to performance imbalances and affects firms' ability to design and implement competitive strategies [5, 10]. Thus, and according to this theory, the possession of heterogeneous resources and capabilities directly affects firms' performance [4, 11]. In the same sense, dynamic capabilities enable firms to achieve superior long-term performance [12]. [4, p. 516] defined dynamic capabilities as the "firm's ability to integrate, build, and reconfigure internal and external competences to address rapidly changing environments". Dynamic capabilities thus reflect firms' ability to achieve new and innovative forms of competitive advantage. Absorptive capabilities (ACAP) is a dynamic capability found in organizational processes that enables firms to reconfigure their core resources, react to environmental dynamics and build competitive advantage [13]. The RBV posits that variations in firms' performance result from the possession of heterogeneous resources. This heterogeneity of resources and capabilities leads to performance imbalances and affects firms' ability to design and implement competitive strategies [5, 10]. Thus, this theory suggests that heterogeneous resources and capabilities have a direct effect on firms' performance [4, 11]. Thus, we tested the following hypotheses: H5: *Intangible resources influences positively export performance*; and H6: *Absorptive capabilities influences positively with export performance*.

Innovation requires different strategic orientations, technological resources and processes. Innovation provides a mechanism to effectively manage change by repeatedly pursue and achieve both disruptive and incremental innovation [14]. This test is used to determine whether the indirect effect of the independent variable on the dependent variable via the mediator is significantly different from zero. Thus, based on the explained above, the following hypotheses can be posed: H6. *Innovation mediates the relationship between intangible resources and export performance; and H7. Innovation mediates the relationship between absorptive capabilities and export performance.*

2 Method

The population of this empirical study has been drawn from Portuguese textile industry firms. Questionnaires were used as primary data sources and were carried out over the period of February 16 to April 30, 2016. The identification of companies was done through the Portugal's Textile Association (ATP) database. So, in this study we use a non-probabilistic and convenient sampling of 247 companies.

Well-validated scales from previous studies were used to operationalize the key constructs and adapted them to the particular context of our empirical setting: (1) *Independent variables* – To assess intangible resources we adopted [15]'s and to measure absorptive capabilities' construct we use [14]'s scale; (2) *Mediator* – To assess innovation we adopted [16]'s measurements; and (3) *Dependent variable* – Export performance's construct was assessed using [17]'s measurement instrument. All constructs were assessed on a five-point Likert scale.

3 Results

3.1 Evaluation of Measurement Model

Results from Table 1 show that the measurement model meets all general requirements. First, all reflective items have a load higher than 0.707, which means that the reliability of individual indicators (loading^2) are higher than 0.5. Second, all composite reability values and Cronbach's alpha values are higher than 0.70, suggesting acceptable model reliability. Third, the average variance extracted (AVE) values of all constructs are higher than 0.50, indicating an adequate convergent validity and implying that our set of indicators represent the same underlying construct [18].

Table 1. Measurement model

First-order constructs	Items	Factor loading	Item loading^2	Alpha de Cronbach	Composite reliability	AVE
Intangible resources						
Reputational resources	REP1	0.928	0.861			
	REP2	0.915	0.837	0.905	0.934	0.779
	REP3	0.847	0.717			
	REP4	0.835	0.697			
Access to financial resources	FIN1	0.940	0.884			
	FIN1	0.940	0.884			
	FIN2	0.962	0.925	0.964	0.974	0.902
	FIN3	0.942	0.887			
	FIN4	0.956	0.914			

(continued)

Table 1. (*continued*)

First-order constructs	Items	Factor loading	Item loading[2]	Alpha de Cronbach	Composite reliability	AVE
Human resources	HUM1	0.875	0.766			
	HUM2	0.889	0.790	0.932	0.952	0.832
	HUM3	0.943	0.889			
	HUM4	0.939	0.882			
Cultural resources	CULT1	0.922	0.850			
	CULT2	0.914	0.835	0.891	0.932	0.821
	CULT3	0.880	0.774			
Relational resources	REL1	0.963	0.927			
	REL2	0.916	0.839	0.951	0.965	0.872
	REL3	0.934	0.872			
	REL4	0.922	0.850			
Informational resources	INF1	0.875	0.766			
	INF2	0.822	0.676	0.881	0.917	0.734
	INF3	0.864	0.746			
	INF4	0.866	0.750			
Absorptive capabilities						
Acquisition*	ACAQ1	0.729	0.531			
	ACAQ2	0.694	0.482			
	ACAQ3	0.830	0.689	0.782	0.850	0.532
	ACAQ4	0.687	0.472			
	ACAQ6	0.698	0.487			
Assimilation	ACAS1	0.819	0.671			
	ACAS2	0.932	0.869	0.847	0.907	0.766
	ACAS3	0.871	0.759			
Transformation*	ACTR2	0.827	0.684			
	ACTR3	0.873	0.762			
	ACTR4	0.795	0.632	0.874	0.908	0.665
	ACTR5	0.854	0.729			
	ACTR6	0.721	0.520			
Exploitation	ACEX1	0.791	0.626			
	ACEX2	0.765	0.585			
	ACEX3	0.866	0.750	0.897	0.922	0.663
	ACEX4	0.836	0.699			
	ACEX5	0.703	0.494			
	ACEX6	0.909	0.826			
Innovation	INNOV1	0.813	0.661			
	INNOV2	0.892	0.796	0.827	0.896	0.742
	INNOV3	0.876	0.767			

(*continued*)

Table 1. (*continued*)

First-order constructs	Items	Factor loading	Item loading^2	Alpha de Cronbach	Composite reliability	AVE
Export performance	EP1	0.873	0.762			
	EP2	0.889	0.790			
	EP3	0.837	0.701	0.927	0.945	0.775
	EP4	0.915	0.837			
	EP5	0.887	0.787			

* The variables ACAQ5 and ACTR1 corresponding to Acquisition and Transformation factor's were excluded from the measurement model due to low values. Accordingly, values lower than 0.7 generate a low correlation and threaten the reliability of the scale.

The results shown in Table 2 confirm the existence of discriminant validity in our study.

Table 2. Latent constructs correlation (Fornell-Larcker criterion)

Dimensions	1.	2.	3.	4.	5.	6.	7.	8.	9.	10.	11.	12.
1. Acquisition	**0.730**											
2. Assimilation	0.307	**0.875**										
3. Cultural resources	0.510	0.271	**0.906**									
4. Exploitation	0.357	0.656	0.524	**0.815**								
5. Export performance	0.390	0.466	0.337	0.584	**0.880**							
6. Financial resources	0.481	0.444	0.511	0.345	0.435	**0.950**						
7. Human resources	0.557	0.243	0.810	0.534	0.406	0.636	**0.912**					
8. Informational resources	0.613	0.351	0.557	0.370	0.220	0.496	0.666	**0.857**				
9. Innovativeness	0.280	0.290	0.387	0.488	0.513	0.212	0.334	0.195	**0.861**			
10. Relational resources	0.593	0.236	0.698	0.412	0.409	0.622	0.715	0.606	0.296	**0.934**		
11. Reputational resources	0.514	0.268	0.639	0.484	0.468	0.589	0.660	0.459	0.526	0.656	**0.882**	
12. Transformation	0.412	0.620	0.516	0.848	0.640	0.331	0.550	0.383	0.520	0.437	0.443	**0.816**

In the following Tables 3 and 4, we present the results of reliability, convergent validity and discriminant validity corresponding to the second order model. All data confirm the strength of our model.

Table 3. Convergence validity and reliability indexes of the second-order model

Constructs	Cronbach's alpha	Composite reliability	AVE
Absorptive capabilities	0.849	0.900	0.694
Export performance	1.000	1.000	1.000
Innovation	0.832	0.875	0.742
Intangible resources	0.908	0.929	0.687

Table 4. Discriminant validity index of the second-order model

Constructs	Absorptive capabilities	Export performance	Innovation	Intangible resources
Absorptive capabilities	0.662			
Export performance	0.650	0.880		
Innovation	0.515	0.511	0.861	
Intangible resources	0.621	0.476	0.407	0.750

To study our dependent variable, the value that we have to maximize is R^2. This coefficient measures the amount of construct variance that is explained by the model, where values of 0.5 are considered to be moderate and 0.25 weak. In our model, the R^2 coefficient is 0.385 for AC, 0.277 for INNOV and 0.469 for EP, so we can assert that these values are more than satisfactory.

Finally, and applying the non-parametric bootstrapping test, we evaluated the significance of mediation effects. The results show significance of coefficients shown (Table 5).

Table 5. Significant testing results of the structural model path coefficients

Hypotheses	Original Sample (O)	Sample Mean (M)	Standard Deviation (STDEV)	T Statistics (| O/STDEV|)	P Values	2.5%	97.5%
H1. **IR → AC**	0.621	0.622	0.056	11.108	0.000*	0.509	0.719
H2. **IR → INNOV**	0.143	0.140	0.064	2.224	0.027**	0.010	0.262
H3. **AC → INNOV**	0.426	0.432	0.059	7.193	0.000*	0.310	0.543
H4. **AC → EP**	0.479	0.487	0.039	12.306	0.000*	0.413	0.567
H5. **INNOV → EP**	0.230	0.226	0.042	5.464	0.000*	0.137	0.305
H6. **IR → EP**	0.085	0.079	0.044	1.920	0.055***	-0.014	0.157
H7. **IR → INNOV → EP**	0.391	0.395	0.040	9.697	0.000*	0.319	0.481
H8. **AC → INNOV → EP**	0.098	0.097	0.021	4.593	0.000*	0.057	0.138

*p < 0.001; **p < 0.05; ***p < 0.1.

4   Discussion and Conclusions

Our findings confirm that indeed resources and capabilities are related to each other, corroborating the studies of [1, 2], supporting H1 ($\beta = 0.621$; $p < 0.001$). The literature suggests that innovation has a positive and direct influence in business performance. Conversely, results do not support [4], since these authors claim that innovation requires not only the property of intangible resources, but also its exploration (H2 supported; $\beta = 0.143$; $p < 0.05$). The DCV conceptualizes innovation as a complex and dynamic process through which firms consistently develop innovation capabilities by exploring new resources or new combinations of resources [4]. Thus, this research is in line with this perspective (H3 supported; $\beta = 0.426$; $p < 0.001$). Additionally, innovation has a positive and significant impact on export performance (H4 supported; $\beta = 0.479$; $p < 0.001$), confirming [7] beliefs. Moreover, this confirms the commitment to innovation, supported by [3, 8], regarding the creation of new products and services, search for new opportunities and opening of new markets; and with proactiveness, since only proactives firms will be able to achieve superior performance compared to competition.

The possession of heterogeneous resources and capabilities directly affects firms' results [4, 11], leading to performance imbalances and affecting the ability to design and implement competitive strategies [5, 10], as previously mentioned. In this study intangible resources have a positive influence on export performance (H5 supported; $\beta = 0.230$; $p < 0.001$). This study also demonstrated that the firms' absorptive capabilities has also a positive influence on export performance (H6 supported; $\beta = 0.085$; $p < 0.1$). Thus, the analyzed firms are able to acquire, assimilate, transform and dynamic capabilities. This perspective is also consistent with [19] regarding the uniqueness of paths. The results of this study does confirm that absorptive capabilities enable firms to achieve superior long-term performance [12].

Success in the global market requires creativity, ingenuity and risk taking, both in domestic markets and in foreign ventures [6]. Indeed, innovative firms have an extraordinary performance and can even be seen as a country's engine of economic growth [9]. Our findings support this statement partially, that is, in the present study the mediating role of innovation in the relationship between intangible resources and export performance was confirmed (H7 supported; $\beta = 0.391$; $p < 0.001$) and also in the relationship between absorptive capabilities and export performance (H8 supported; $\beta = 0.098$; $p < 0.001$).

Our study is responsive to the call of [21] which suggests that, in international market context, firms' survival and expansion, and consequent economic growth of many countries, is strongly dependent on a better understanding of the strategic determinants that influence export performance. Moreover, our study confirms the important complementarity of intangible resources and dynamic capabilities, thus not diverging from RBV and DCV [19]. We also highlight the contribution of this study to the theory of strategic management. Additionally, our findings provide guidance to business practitioners, since they indicate that intangible resources, absorptive capabilities and innovation are predictors of performance.

The fact that the sampling is non probabilistic and convenience is a limitation. Therefore we advise prudence in the generalization of results.

Further studies with a longitudinal perspective would be of added value to research why these differences persist. In other words, to find how and why some small exporters become highly successful while others, in the same industry, struggle to raise their export strengths.

References

1. Dhanaraj, C., Beamish, P.W.: A resource-based approach to the study of export performance. J. Small Bus. Manag. **41**(3), 242–261 (2003)
2. Morgan, N., Kaleka, A., Katsikeas, C.S.: Antecedents of export venture performance: a theoretical model. J. Mark. **68**, 90–108 (2004)
3. Miller, D.: The correlates of entrepreneurship in three types of firms. Manag. Sci. **29**(7), 770–791 (1983)
4. Teece, D.J., Pisano, G., Shuen, A.: Dynamic capabilities and strategic management. Strateg. Manag. J. **18**(7), 509–533 (1997)
5. Barney, J.: Firm resources and sustained competitive advantage. J. Manag. **17**(1), 99–120 (1991)
6. Zahra, S., Garvis, D.: International corporate entrepreneurship and firm performance: the moderating effect of international environmental hostility. J. Bus. Ventur. **15**, 469–492 (2000)
7. Wiklund, J., Shepherd, D.: Entrepreneurial orientation and small business performance: a configurational approach. J. Bus. Ventur. **20**(1), 71–91 (2005)
8. Lumpkin, G., Dess, G.: Clarifying the entrepreneurial orientation construct and linking it to performance. Acad. Manag. Rev. **21**(1), 135–172 (1996)
9. Schumpeter, J.A.: The Theory of Economic Development. New Brunswick (U.S.A) and London (U.K.): Transaction Publishers (1934)
10. Peteraf, M.: The cornerstones of competitive advantage: a resource-based view. Strateg. Manag. J. **14**(3), 179–191 (1993)
11. Makadok, R.: Toward a synthesis of the resource-based and dynamic-capability views of rent creation. Strateg. Manag. J. **22**(5), 387–401 (2001)
12. Teece, D.J.: Explicating dynamic capabilities: the nature and microfoundations of (sustainable) enterprise performance. Strateg. Manag. J. **1350**, 1319–1350 (2007)
13. Zahra, S., George, G.: Absorptive capacity: a review, reconceptualization, and extension. Acad. Manag. Rev. **27**(2), 185–203 (2002)
14. Jansen, J.J.P., Van Den Bosch, F.A.J., Volberda, H.W.: Managing potential and realized absorptive capacity: how do organizational antecedents matter? Acad. Manag. J. **48**(6), 999–1015 (2005)
15. Morgan, N., Vorhies, D.W., Schlegelmilch, B.B.: Resource–performance relationships in industrial export ventures: the role of resource inimitability and substitutability. Ind. Mark. Manag. **35**(5), 621–633 (2006)
16. Covin, J., Slevin, D.: Strategic management of small firms in hostile and benign environments. Strateg. Manag. J. **10**(1), 75–87 (1989)
17. Okpara, J.: Entrepreneurial orientation and export performance: evidence from an emerging economy. Int. Rev. Bus. Res. Pap. **5**(6), 195–211 (2009)
18. Hair, J., Hult, G.T.M., Ringle, C.M., Sarstedt, M.: A Primer on Partial Least Squares Structural Equation Modeling (PLS-SEM). SAGE Publications, Inc., Thousand Oaks (2013)
19. Eisenhardt, K.M., Martin, J.A.: Dynamic capabilities: what are they? Strateg. Manag. J. **21**(10–11), 1105–1112 (2000)
20. Ambrosini, V., Bowman, C.: What are dynamic capabilities and are they a useful construct in strategic management? Int. J. Manag. Rev. **11**(1), 29–49 (2009)
21. Sousa, C.M.P., Martínez-López, F.J., Coelho, F.: The determinants of export performance: a review of the research in the literature between 1998 and 2005. Int. J. Manag. Rev. **10**(4), 343–374 (2008)

Innovative Strategies on External Media: Dialogues Between Architecture, Graphic Design and Visual Arts

Angela Maria dos Santos$^{(\boxtimes)}$ and Gisela Belluzzo de Campos

Design, Anhembi Morumbi University, São Paulo, Brazil
angesan@gmail.com, giselabelluzzo@uol.com.br

Abstract. The paper analyzes the innovative strategies on the graphic design developed for external media in São Paulo city which promote dialogues between architecture, graphic design and the installation language, a concept proper to the visual arts field. For the analysis, two examples of installation in public spaces are shown: the artistic installation "Chromatic Program" developed by the artist Amelia Toledo for the passageway of Arcoverde Underground Station at Rio de Janeiro in 1998, and an external media installation designed for the Secretariat of Tourism, Sports and Entertainment of Pernambuco to improve this northeast province tourism, that was located at the Angelica Avenue Underground Station passageway in December 2017. Using a comparison methodology based on artistic installation concepts, we verify a close relationship between the artistic and the external media installation strategies which enhance the understanding of contemporary graphic design acting in urban landscapes.

Keywords: Graphic design · Visual arts · Installation

1 Introduction

This paper is part of a doctoral thesis where the research focuses in the dialogue between architecture, graphic design and artistic installation, a language proper to the visual arts field. The examples of graphic design studied fall into external media category and are basically located in the city of São Paulo west region. Living in a contemporary metropolis means being in a constant movement from one place to another, an intensity of sounds, a confluence of different ways of transportation, a rushed time, distinct architecture memories, international urban furniture standard and extensive information network. The inclusion of São Paulo in the map of global cities is, according to Canclini [1, p. 157], "due to the fact that it has become the decisive focus of global economic and communication networks". The city itself has become the place where several layers of visual communication overlap and coexist.

Graphic design plays a prominent role in the contemporary cities visual communication, and concerning the designer's work, Frascara [2, pp. 19–20] states that "as an activity, it is the action of conceiving, programming, designing and performing visual communications, generally produced through industrial means and destined to convey

© Springer International Publishing AG, part of Springer Nature 2019
J. Machado et al. (Eds.): HELIX 2018, LNEE 505, pp. 971–977, 2019.
https://doi.org/10.1007/978-3-319-91334-6_133

specific messages to specific groups [...] the graphic designer works the interpretation, ordering and the visual representation of messages".

Among the great diversity of function carried out by visual communication messages in urban spaces, we highlight the external media, a term originated from the English expression 'outdoor advertising' and used to promote and stimulate the sale of new products. To emphasize the external media differences from other communication vehicles, Mendes [3, p. 51] states that it is "a way of communication visible in the public space, used to advertise information, ideas and products" [...] street signs, signposts, traffic signs, panels and artistic drawings, graffiti, for example, are not advertising information vehicles and therefore do not fit into the definition of external media".

In a metropolis like São Paulo, the external media is one of the main visual stimuli offered that involuntarily imposes itself on peoples' attention. Mendes [3] says that external media can be classified according to its functions and characteristics, and following these classifications, our example falls into the *Disclosure* category which uses a public space to disseminate an idea in partnership with its client and is also known as "out-of-home" advertising. In the example analyzed here, the client is the Secretariat of Tourism, Sports and Entertainment of Pernambuco state, and the installation was shown at Angelica Ave. Station hallway in São Paulo Line 4 Yellow Underground in December 2017.

The second example to be analyzed is the *Chromatic Program*, an artistic installation conceived by the artist Amelia Toledo (1926–2017) for the Arcoverde Underground Station hallway in Copacabana, Rio de Janeiro, in 1998. Through the analyses of both we highlight the strategic conceptual dialogue proposed by the thesis, which is the use of artistic languages by graphic designers to potentiate the external media message.

The São Paulo Line 4 Yellow Underground is the newest in the city, managed by a public/private partnership and transport an average of 700 hundred passengers daily. It presents high technology in the development and control of the trains and a visual communication which emphasizes the moving image displayed in flat screens. The advertisements have different sizes and compositions and are strategically positioned for better visibility; the line also has the "in tunnel" system, a Canadian technology where advertisements are presented in a tunnel between two stations, and simulate films shown when the train passes by. All the orientation and traveling guidance is mostly displayed in screen. Large format photos are used mainly in advertisements alongside the moving walkway and there, they usually have a more traditional design in the text and images use; in exceptional occasions, they present advertising campaigns as installations in some stations' hallways, like the example analyzed here; the train wagons are also completely covered by adhesive photos, performing quite immersive installations as in the Pernambuco Tourism campaign (see Fig. 1).

Fig. 1. View of wagon from Fernando de Noronha Island images. Pernambuco Tourism Campaign. Source: https://diariodotransporte.com.br/2017/11/19/linha-4-amarela-do-metro-de-sp-divulga-turismo-de-pernambuco/, last accessed 2017/11/19

2 Methodology

The chosen methodology is the analyses of the artist Amelia Toledo's *Chromatic Program* installation (see Figs. 2 and 3), and the external media (see Fig. 4), using the following aspects of contemporary artistic installation concept: classification, references related to art history such as the Situationist movement, and the description of contemporary art installation specificities.

3 Installation Language

The term installation appears in the visual arts in the 1960s, designating an environment specially built for art works in galleries and museum spaces. At the same period, the artistic movement known as Land-Art produced art in natural landscapes outside of the traditional exhibition art spaces, changing the art conception and affecting the art market. By that time, the concepts of site (art works done in a particular location anywhere in the world) and non-site (documents resulting from these sites, such as photos, videos, artist books or any resulted product that were later exhibited in the museums and art galleries) were conceived. Historically, installations were already

Fig. 2. *Chromatic Program* installation. Source: http://www2.uol.com.br/ameliatoledo/08.htm? imagem=8&total=11&pagina=1, last accessed 2018/01/04

Fig. 3. *Chromatic Program* installation. Source: http://www2.uol.com.br/ameliatoledo/08.htm? imagem=8&total=11&pagina=1, last accessed 2018/01/04

found in the early twentieth century in Kurt Schwitters (1887–1948) and Marcel Duchamp's (1887–1948) artistic production; however, it was only in the 1980s and 1990s that the installation definitely established itself as an artistic language, occasion where a large number of distinct installations began to emerge together with the term definition.

Installation can be defined as a procedure where the art works are produced for and in accordance with the architectural space where it will be seen, thus creating an environment or scene that proposes a differentiated perception to the public.

Oliveira et al. [4, p. 8], present the installation concept as follow: "[…] a kind of art making which rejects concentration on one object in favor of a consideration on the relationship between a number of elements or on the interaction between things and contexts". To the authors "the history of installation perception is born from the narratives presented in architecture, painting, sculpture, theatre and performance and they

Fig. 4. View of Fernando de Noronha Island with the floor simulating the sand beach. Angelica Ave. Underground Station. Source: Angela Santos archive, December 2017

also affirm [4, p. 7] that […] the practice is therefore defined by its 'hybrid' quality, concentrating diverse, even contradictory notions within its influence".

In the 1980s Krauss [5] focus her writings in the relation between public spaces and sculpture and thus expanded its concept in regards to art in open spaces.

In Brazil, the artist Amelia Toledo created many projects in close dialogue with public spaces and architecture, using a wide variety of materials that included stones, crystals and others. In 1998 she developed art works for Arcoverde Underground Station external and internal areas, and together with the architect, made the decision on the walls finishing. *Chromatic Program* (see Fig. 2) is an installation with a sequence of colored panels that occupies 200 linear meters of an intern passageway. Her intention was to promote visual rhythms as we move through the colored hallway, and thus, mix them and transform their combination into new ones. The space amplitude, the panels' size and the color shades determined by the artist, motivates the passersby to experiment a new visual perception in the urban space and alter the everyday condition of subway users.

The images (see Figs. 4 and 5) show the external media installation that exemplifies the dialogues between art and graphic design discussed here. The advertisement aims to improve the tourism of Pernambuco northeast regions and presents six different places followed by a calendar of events: the city of Olinda, Porto de Galinhas and Carneiros beaches, Fernando de Noronha Island, street Carnival and the São João festivities. The external media installation occupied 40 linear meters of walls and floor and was part of a campaign that also had six wagons with walls and floor completely covered with the touristic places adhesive photographs. Logos and texts accompany the images and reinforce the messages, calling the clients attention to the varied activities and places to visit in the northeast state.

Fig. 5. Hallway installation view of a Carnival scene at Recife streets, Pernambuco. Angelica Ave. Underground Station. Source: Angela Santos archive, December 2017

4   Final Considerations

The artistic installation characteristics seem to follow the art history, starting at the 20th century till the 21st century. We mentioned the -1960s - as an important moment where the installation performed major changes in its proposal of site and non-site experiences and of taking the art out of its comfortable established places. However we would like to point to a previous moment in the -1950s - when the Situationist International movement had its apex through its idealizer Guy Debord [6]. Together with artists and writers he published a journal where the first issue in 1958 had a list of definitions where "some of the terms list bear significantly on the development of an aesthetic that sees art in the context of everyday life than as operating in the realm distinct from it". Oliveira et al. [4, p. 26].

This art concept matches with the *Chromatic Program* installation by Amelia Toledo, an art scape specially designed for a public space that operates in the people's everyday life. Classified as *site specif,* it is an installation proper to a certain environment and it cannot be built in a different place unless the same conditions are applied. A similar classification can be applied to the external media installation in the São Paulo Underground hallway, as only in this architectural condition and user's activity can be fully experienced. In addition to creating an immersive installation by the space covered by the adhesive photographs.

Regarding the contemporary art installation specificities, we point the user's participation increase, especially in the works conceived for digital technologies, virtual or immersive ambiences, moving image, and shown in art institutions were the viewer can see the works for a while, leave and come back later, just pass by or experience a film in broken or multiple narratives [7]. Although the installations here do not use the digital technology or the film language as mentioned above, they do play with the users' participation, for both depend on people's attention and walk to accomplish the messages. Still in the participatory subject, we point the users' freedom to enter the external media installation space from different ways, in a discontinuous time and/or fragmented reading, reflecting the contemporary way we perceive the urban landscapes and the city. That is also similar to the way the public experience artistic installations within museums and art galleries.

By the external media analysis in the light of artistic installation concepts, we realize that it loses its previous specificities and exposes a hybrid nature with elements of both languages. These new strategies that stablish dialogues between graphic design and art involve the users, expand peoples' perception of art and advertising in public spaces and bring a wider understanding on contemporary external media acting.

References

1. Canclini, N.G.: A globalização imaginada. Iluminuras, São Paulo (2003)
2. Frascara, J.: Diseño gráfico y comunicación, 7th edn. Ediciones Infinito, Buenos Aires (2000)
3. Mendes, C.: Paisagem urbana: uma mídia redescoberta. Editora SENAC, São Paulo (2006)
4. Oliveira, N., Oxley, N., Petry, M.: Installation Art. Thames & Hudson Ltd., London (1997)
5. Krauss, R.: Sculpture in the expanded field. In: The Anti-aesthetics: Essays on Postmodern Culture, org. Hal Foster. Bay Press, Seattle/Washington (1983)
6. Debord, G.: A Sociedade do Espetáculo. 9ª reimpressão, Contraponto Editora Ltda., Rio de Janeiro (2007)
7. Dubois, P.: Sobre o efeito cinema nas instalações contemporâneas de fotografia e vídeo. In: Transcinemas, org. Katia Maciel, Contra Capa Livraria, Rio de Janeiro (2009)

Internationalization Strategies for Tourism in Portugal

Anabela Oliveira$^{(\boxtimes)}$, Vítor Braga, and Eliana Costa e Silva

CIICESI, ESTG/P.PORTO - Center for Innovation and Research in Business
Sciences and Systems Information, School of Technology and Management,
Polytechnic of Porto, Porto, Portugal
{8130100,vbraga,eos}@estg.ipp.pt

Abstract. Tourism is one of the activities of great relevance on a global scale;
with an increasing and fierce competition, needing to attract national and
international tourists to maintain its distinctive position. Being considered as the
Portuguese economy's booster, and as a strong strategic activity for the future
development of the country it is required to verify if: the agglomeration of
tourism companies, the continuous investment on the qualification of human
resources and the inherited and complementary resources influence the decision
of the internationalization of the Portuguese hotels. The multivariate linear
regression (MLR) analysis carried out for the Portuguese context, using a
database of 2015, concludes that the agglomeration is statistically significant,
whereas only some endogenous and knowledge resources are statistically significant for the internationalization of the sector.

Keywords: Tourism · Internationalization · Knowledge · MLR analysis

1 Introduction

In recent years, due to the gradual increase in the interest to study the internationalization of tourism, some models which explain the main factors that encourage the
internationalization and globalization of tourism companies have been presented [1, 2].
However, the globalization of tourism market, required the sector to implement
strategies to improve its competitive position through cooperation for innovating and
differentiating the product and/or service offered [3]. Despite the importance of the
internationalization and survival of the sector in the global market, tourism is also
considered an engine for regional development, which improves the endogenous
characteristics of each region and promote the natural, cultural and historical heritage.

Considering these facts studied by [1, 4], the present study intends to contribute to
the identification of factors that may influence the decision of companies in the Portuguese tourism sector to expand abroad. To achieve the central objective, it is intended
to respond the following specific objectives: (a) Analyze the relationship between
variables validated by literature, as well as the role of the concentration of companies of
the sector in the Territorial Units (TU) of Portugal Continental (NUTS III) in the
decision making for the internationalization of hotel chains; (b) Analyze the extent to
which the agglomeration of companies and the training offer in the tourism sector in
NUTS III generates knowledge and promotes employment, qualification, enhancement

© Springer International Publishing AG, part of Springer Nature 2019
J. Machado et al. (Eds.): HELIX 2018, LNEE 505, pp. 978–984, 2019.
https://doi.org/10.1007/978-3-319-91334-6_134

of human resources and influences internationalization; (c) Identify the factors that motivate tourism companies to expand internationally, in terms of natural, historical-cultural and capital resources.

With a business structure mainly constituted by small and medium enterprises (SMEs), the Portuguese tourism sector reveals a huge capacity in the development, qualification of resources and territorial requalification, which allows the expansion of the Portuguese tourism sector. The decision to focus the study on TU and Portuguese tourism is due to the fact, that this is a segment driven by the external market and the first European country to be distinguished as the best touristic destination in 2017, by World Travel Awards. Every year, Portugal attracts millions of tourists, due to complementary product companies, but also to inherited and acquired resources [5]. If the previously proposed objectives are achieved, the study will be able to elucidate the capacity that the agglomeration and the tourism resources exert in the process of internationalization of the segment.

The rest of this document is structured as followed: In Sect. 2, a theoretical approach based on several literary studies, and specifically in the study of [1] with the aim of understanding the motivations for the internationalization and globalization of Portuguese hotel chains. Moreover, the hypotheses for further investigation will also be proposed. In the third and fourth section, the methodology used and the main results of the study will be described. The final section presents the conclusions.

2 Literature Review

For companies, internationalization is the main strategy adopted to increase competitiveness, employment and to provide an environment conducive to economic growth and development of innovations. In general, this process represents their survival and prosperity in the market, by increasing long-term profitability and sustainability of competitiveness [6]. In order to explain the initial decision of internationalization, resulting from several internal and external factors, there are different theories in the literature, namely, Uppsala model, based on the acquisition of knowledge, learning and experience resources resulting from the involvement with other external markets [7]; the theory of networks, which enhances the competitive capacity, learning and specialization through the establishment of collaboration networks [8]; and the eclectic paradigm that considers as main advantages for the company to act in the international market, property, location and internalization [9]. In this sense, the success of the internationalization strategy is directly related to the capacities and resources that a company holds [10].

Considered by organizations as a relevant factor in the decision-making process, the agglomeration represents externalities, i.e. access to resources, better suppliers of goods and services, expansion of the network of relationships and knowledge, and improvements in the demand [11]. In this way, hotels tend to be located in urban areas or tourist destinations, where tourism is one of the main economic sectors, and where geographic characteristics [5] are favourable, forming an economy of location.

With the stagnation of the domestic market and small opportunities provided by it, the companies that belong to these agglomerations have competitive advantages at the national level and tend to be influenced to penetrate external markets [12], to strengthen its competitive capacity and to exploit its advantages at an international level, through its own resources and the ones provided by the country in which it intends to operate. Considering this, the first hypothesis proposed for research will be: *H1. the number of internationalized hotels will be higher in the TUs with a greater degree of tourism companies agglomeration.*

Seen as one of the most important intangible resource for organizations, knowledge can be generated through internal learning processes or acquired through cooperation with more experienced organizations that belong to the sector of activity, and it is also considered as one of the most important factors in internationalization theories [13], such as the Uppsala model, in which it is considered as an instrument to reduce uncertainty in the commitment to the market and to identify opportunities at the international level. The increasing competitive advantages resulting from the knowledge and qualified human and intellectual resources in the area, influence the company to expand to new markets, which leads to the formulation of the following hypothesis: *H2. the number of internationalized hotels will be higher in the TUs that generate knowledge related to the sector,* measured by the number of universities (*H2a*), professional training institutions (*H2b*) and research and development (R&D) centres (*H2c*), which generate specific knowledge for the sector.

Although the knowledge resources are important for the sustainability of tourism companies, the characteristic resources of a region contribute to attract tourists, since they present themselves as more attractive and more successful destinations. The services and hotels, seek to locate in territories close to the markets and potential customers, as well as in central locations where there are several services and tourist attractions, access to inherited resources [5] and infrastructure with complementary products [14]. This motivated the following research hypothesis: *H3. the number of internationalized hotels will be higher in the territorial units that have more resources,* specifically: more natural resources, i.e. beaches and protected areas (*H3a*), greater number of historical-cultural resources, i.e. monuments and museums (*H3b*) and greater number of capital resources, i.e. activities and thematic parks (*H3c*).

3   Research Methodology

Previously applied by several studies in tourism [15, 16], MLR analyses the relationship between a dependent variable and a set of independent variables. The statistical software SPSS (version 23) was used. The purpose was explaining to what extent the sources of technical and specific knowledge for the tourism sector and tourism resources in Portugal are significant or not for the internationalization of Portuguese hotel chains in 2015. To test the hypotheses of the previous section, a MLR model is proposed, focused on the internationalization of Portuguese hotel, for *H1* to *H2*:

$$HOT\_Int = \beta_1 Agg + \beta_2 UI + \beta_3 MLI + \beta_4 R\&D + \varepsilon, \tag{1}$$

and for **H3**:

$$HOT\_Int = \beta_5 NR + \beta_6 HCR + \beta_7 CR + \varepsilon, \tag{2}$$

where β_1 .... β_7 are the coefficients of the explanatory variables, ε is the residual variable with normal distribution, HOT_Int the number of internationalized hotels; Agg the agglomeration, UI the university institutions; MLI the middle-level institutions; $R\&D$ the centres of Research and Development; NR the natural resources; HCR the historical-cultural resources and CR the capital resources.

The population comprises all level III TUs from Portugal. With support in the main Portuguese databases, namely the *Instituto Nacional de Estatística* (INE), 25 TUs were identified, two of which were excluded from the study[1]. Thus, the study sample comprises 23 TUs, 10 located in the coastal and the remaining 13 in interior areas.

Dependent Variable. To identify the hotels, that took the decision to internationalize, belonging to each TU in study, the database of *Turismo de Portugal* (TP) was used, where was collected the list of hotels existing in Portugal Continental. This list and the respective website of each hotel, allowed recognizing the TU of origin of each internationalized hotel. Only 59 internationalized hotels for a total of 1024 were identified, and 49 are concentrated in the great centres and on the coastal zone.

Independent Variables. As proposed by [1], the agglomeration of each TU was calculated as follows[2]:

$$Agg_i = \frac{\text{Employment sector tourism in TU i}}{\text{Total employment in TU i}} : \frac{\text{Employment sector tourism in Portugal}}{\text{Total employment in Portugal}}$$

To measure knowledge resources, the number of higher education and professional training institutions, as well as R&D centres related to the tourism sector, was considered in each TU, and all the information, was collected directly from the website of *Direção Geral de Estatística e Educação para a Ciência* and *Direção Geral dos Estabelecimentos Escolares*. Knowledge resources were measured as: Universities and Polytechnics - number of institutions in each TU that confer academic degree in tourism, relativized by 100,000 inhabitants; Middle-level schools - number of institutions in each TU with vocational training in tourism, per 100,000 inhabitants; R&D centres - number of public and/or private research centres in each TU in the tourism segment.

It is also intended to study the resources that may influence the decision to internationalize. The data collection was made from the TP website and from the statistical portal INE. The resources were measured as: Natural resources - number of protected areas, gardens, parks and beaches; Historical-cultural resources - number of monuments and museums; Capital resources - number of tourist activities and theme parks of each TU.

---

[1] The autonomous regions were excluded due to lack or confidentiality of information.

[2] Data on employment and unemployment in the country, of each TU in study, corresponding to economic activity codes 55 and 56 - hotels, catering and tourism.

4   Analysis and Discussion of the Results

As the study sample is small and presents a non-normal distribution, the non-parametric Spearman correlation test was used to study the correlation between the dependent variable – *HOT_Int* – and each of the independent variables. Table 1 shows the existence of a possible positive association between all the independent variables and the response variable, which is statistically significant for most of the variables. These values suggest the validation of the hypotheses previously stated.

Table 1. Spearman Correlation (non-parametric test) with the dependent variable *HOT_Int*.

Variables	Coefficients	Variables	Coefficients
Agglomeration	0.407*	PhD tourism	0.403*
University inst.	0.068	Professional courses	0.383*
Prof. training inst.	0.192	Natural resources	0.208
R&D centres	0.345	Historical-cultural resources	0.431**
Graduation tour.	0.433**		
Masters tourism	0.634***	Capital resources	0.436**

*** $p < 0.01$; ** $p < 0.05$; * $p < 0.1$

The results for the MLR models are depicted in Table 2. Model 1 indicates that the agglomeration explains 56.3% of the variability of the internationalization of hotels. Including in this model the variables related to knowledge resources (model 2), 68.4% of hotels internationalization is explained by independent variables, while in model 3, the explained percentage is 57.4%, suggesting that knowledge generated by the companies presented in a cluster is an important factor for the internationalization. Finally, model 4 indicates that 48.2% of hotels internationalization is explained by natural, historical-cultural and capital resources. Using stepwise method for eliminating the non-statistically significant variables obtained for the models (1 to 3), the agglomeration is the only variable with a positive coefficient and statistically

Table 2. MLR model results.

Independ. Var.	Model 1	Model 2	Model 3	Independ. Var.	Model 4
Agglomeration	0.639***	1.215***	1.897***	Natural resources	0.025**
University Inst		−0.171			
Prof. Inst.		−0.317**		Historical-cultural resources	0.001
R&D Centres		0.139	0.542*		
Grad + Master			−0.017		
PhD Tourism			0.077	Capital resources	−0.001
Prof. Courses			−0.013		
F	30.675***	13.473***	7.186***	F	8.126***
R_a^2	0.563	0.684	0.574	R_a^2	0.482
DW	1.406	1.591	1.897	DW	1.488

*** $p < 0.,01$; ** $p < 0.05$; * $p < 0.1$

significant, i.e. the hotel chains that tend to internationalize, in TU with higher agglomeration, while contrary to the expected the professional training institutions have a negative impact.

As for variables that measure knowledge generated through universities (*H2a*) and R&D centres (**H2c**), studied by [1] for Spain, do not have a representation in the Portuguese context, which indicates the non-verification of *H2*, i.e., there is no statistical significance that the Portuguese hotels take advantage of the knowledge to internationalize. Given that the variables under study for this topic did not allow to verify the hypothesis, four proxy variables were introduced, namely the number of courses (graduation, masters, PhD and professional courses) that each institution offers in tourism areas. Model 3 shows that there is a positive relationship between the courses offered in the area (PhD) and R&D centres, in the internationalization of hotels, however, only R&D centres are statistically significant, which leads to the verification of *H2c*. Finally, natural and historical-cultural resources have a positive impact on the internationalization of Portuguese hotels, while capital resources have a negative impact, only the natural resources variable is statistically significant, with *H3a* being verified. For all models, the Durbin-Watson (DW) indicator shows that the residuals are not correlated (DW < 2). By means of the Kolmogorov-Smirnov test, it was verified that for a level of significance of 10%, the residual variable follows a normal distribution. It was also observed the existence of one multivariate outlier in model 2 (Leverage value >0.5), three in model 3 and in model 4, two multivariate outliers. These outliers can contain relevant information about population, reason why they were not excluded. In all models, the VIF values are less than 10 and the Condition Index is small, thus verifying the absence of multicollinearity.

5   Conclusions

In general, industries and hotel sector, tend to locate in territories in which exist, or are generated, resources favourable to the activity i.e. in areas where the concentration of similar or complementary activities are agglomerated, allowing the creation of competitive advantages internally or through interaction with companies with specific knowledge, belonging to the cluster. In this sense, the results presented confirm, from the practical point of view and for the Portuguese hotel context, the hypothesis that hotels tend to internationalize when they are in TU with more agglomeration.

Despite the fundamental role attributed to knowledge in the literature, in the process of internationalization, and in [1] for the Spanish territory, it was not possible to obtain similar evidences for Portugal. We believe that there are disparities in terms of geography, population, employment and infrastructures in each TU, which may contribute to the skew of the study. Thus, according to the results obtained for this hypothesis, the specific knowledge in the activity sector does not have a vital role in the internationalization decision of the hotel sector, although it has a certain importance given the need for qualified resources and research for the business development. Also considering the introduction of new variables in the study, as natural, historical-cultural and capital resources, may have an influence on the location of tourism enterprises, since it contributes to attracting tourists and the improvement of the supply, but do not have in totality a significant role in the decision to internationalize. In the future, we intend to refine the model by reclassifying the resources present in each category of study, as well as expanding the sample under study.

In summary, for the Portuguese hotel sector, only the agglomeration translates into a strategic advantage for internationalization, while knowledge and endogenous resources are neglected, evidencing the existence of factors with greater importance in the internationalization decision. In theoretical terms it can translate into survival to the country's economic crisis, saturation of opportunities in the domestic market, or just a means to mitigate uncertainty or increase the profitability of the business.

References

1. Marco-Lajara, B., Zaragoza-Sáez, P., Claver-Cortés, E., Úbeda-García, M., García-Lillo, F.: Tourist districts and internationalization of hotel firms. Tour. Manag. **61**, 451–464 (2017)
2. Chen, M.-H.: A quantile regression analysis of tourism market growth effect on the hotel industry. Int. J. Hosp. Manag. **52**, 117–120 (2016)
3. Smeral, E.: The impact of globalization on small and medium enterprises: new challenges for tourism policies in European countries. Tour. Manag. **19**, 371–380 (1998)
4. Marco-Lajara, B., Claver-Cortés, E., Úbeda-García, M., Zaragoza-Sáez, P.C.: Hotel performance and agglomeration of tourist districts. Reg. Stud. **50**, 1016–1035 (2014)
5. Mira, M.R., Moura, A., Breda, Z.: Destination competitiveness and competitiveness indicators. TÉKHNE – Rev. Appl. Manag. Stud. **14**, 90–103 (2016)
6. Kubíčková, L., Votoupalavá, M., Toulová, M.: Key motives for internationalization process of small and medium-sized enterprises. Procedia Econ. Financ. **12**, 319–328 (2014)
7. Johanson, J., Vahlne, J.E.: The internationalization process of the firm: a model of knowledge development and increasing foreign market commitments. J. Int. Bus. Stud. **8**, 23–32 (1997)
8. Coviello, N., Munro, H.: Network relationships and the internationalisation process of small software firms. Int. Bus. Rev. **6**, 361–386 (1997)
9. Dunning, J.H.: The eclectic paradigm of international production: a restatement and some possible extensions. J. Int. Bus. Stud. **19**, 1–31 (1998)
10. Wernerfelt, B.: A resource-based view of the firm. Strateg. Manag. J. **5**, 171–180 (1984)
11. Canina, L., Enz, C.A., Harrison, J.: S.: agglomeration effects and strategic orientations: evidence from the U.S. lodging industry. Acad. Manag. J. **48**, 565–581 (2005)
12. Wang, S.Q., Dulaimi, M.F., Aguria, M.Y.: Risk management framework for construction projects in developing countries. Constr. Manag. Econ. **22**, 237–252 (2004)
13. Prashantham, S.: Toward a: knowledge-based conceptualization of internationalization. Int. Entrep. Manag. J. **3**, 37–52 (2005)
14. Papatheodorou, A.: Why people travel to different places. Ann. Tour. Res. **28**, 164–179 (2001)
15. Chi, C.G.-Q., Qu, H.: Examining the structural relationships of destination image, tourist satisfaction and destination loyalty. Tour. Manag. **29**, 624–636 (2008)
16. Gross, J., Brown, G.: An empirical structural model of tourists and places: progressing involvement and place attachment into tourism. Tour. Manag. **31**, 1141–1151 (2008)

Linking Sustainable Tourism and Electric Mobility – Moveletur

George Ramos[(⊠)], Rogério Dionísio, and Paula Pereira

Instituto Politécnico de Castelo Branco,
Av. Pedro Álvares Cabral, n.º 12, 6000-084 Castelo Branco, Portugal
gramos@ipcb.pt

Abstract. This paper approaches the permanent struggle that less favoured regions must deal with regarding economic opportunities, job creation, income and regional production increase. Since an increased demand for nature and protected areas is taking place in a more and more urban society, some innovation potential is emerging. The study we have developed is focused on sustainable tourism practices in a specific natural area (Malcata Mountain Reserve), using electric mobility, which is known for its zero emission, no polluting and noise-free travelling. The broader study is carried out under the Interreg Funding Program in the Moveletur Project. Our aims are to promote a model of sustainable and clean tourism for visitors of natural areas, to create a network of green tourism itineraries connecting sites of natural and/or cultural value using electric vehicles and to empower tourism sector entrepreneurs with a new added-value service for their activity. Joint work with other natural areas is required to increase results. After the project is finished (by the end of 2018) there will be an improved knowledge about natural and cultural values that natural areas hold and that can be used for visitors' enjoyment. There will be a more respectful way of 'doing tourism' in natural areas and hopefully it will address employment creation and improved territorial competitiveness. Finally, tourism experiences will have more quality and the project will promote smart villages' further development by using technological components.

Keywords: Sustainable tourism · Electric mobility · Natural areas

1 Introduction

Sustainable mobility and, in particular, the electric one is gaining importance in our days. Natural spaces are configured as ideal spaces to nurture this type of mobility, where no emissions, contaminants, or noise are produced. Moreover, electric vehicles make possible for people with reduced mobility to access these parks, into places they could not reach before by their own means.

The MOVELETUR project (see logo in Fig. 1) is committed to the promotion of innovation in emerging sectors in the rural economy. The emergence of leisure and entertainment activities in natural spaces, calls into action new forms of occupation of the territory, which have and will have a strong territorial, economic and social impact. In face of these changes, the use of land for tourism purposes can experience some new

© Springer International Publishing AG, part of Springer Nature 2019
J. Machado et al. (Eds.): HELIX 2018, LNEE 505, pp. 985–991, 2019.
https://doi.org/10.1007/978-3-319-91334-6_135

arrangements. The challenge comes from knowing how to integrate it with traditional uses in a way that creates wealth without destroying values or resources. In short, the rural world is undergoing a series of important changes and must seek new directions, new forms of development and new activities to bring together the different stakeholders.

Fig. 1. Logo of the project

The project also pays attention to protected natural spaces as an area of economic potential. All the protected natural areas included in the scope of the project belong to the Natura 2000 ecological network. This network was created in European Union in order to guarantee the long-term conservation of the most threatened species and habitats in Europe, contributing to prevent threats to biodiversity [1, 2]. It is the main trans-European instrument for nature conservation.

The project pays attention to the development of sustainable tourism by creating tours that bring together natural and cultural values through electric mobility networks. Although focusing a bordering territory, there is mutual unawareness of the natural and cultural resources on both sides of the region.

1.1 Intervention Area

The geographical scope of the project covers seven protected natural areas of the Portuguese-Spanish border region, with the area, number of municipalities and population detailed in Table 1.

It is a rural cross-border territory, since the great majority of the population lives in population centres of less than 5,000 inhabitants, that is, in rural centres. In short, this region forms an extensive rural territory to renew, characterized by:

- Low population density. In the case of Spanish protected areas, it is approximately 17 inhab/km^2, compared to 91 inhab/km^2 in Spain. In the case of Portuguese protected areas it is 12 inhab/km^2, compared to 112 inhab/km^2 in Portugal.
- Depopulation. The loss of population in Alto Tâmega has gone at a faster rate than the loss of population in the rest of the Portuguese country. In the case of the Serra da Malcata Natural Reserve, the situation is even more pronounced, in which no inhabitant lives. The municipalities of the Spanish protected areas within the scope of the project have also suffered a population decline, a trend that has been much more pronounced in recent decades.

- Aging population. The municipalities integrating the Spanish protected areas hold a 29% aging rate compared to the Spanish aging rate which stands at 16%. In the case of municipalities of Portuguese protected areas, they also have aging rates that almost double the national average. The consequences of this massive emigration due to the rural exodus have been and still are devastating the municipal censuses.
- High significance of agricultural activity, over the services sector. In the case of Spanish municipalities, 18% of the population affiliated with social security is in the agricultural sector, compared to 1.14% at the state level. In the Portuguese case, agriculture has also a high weight.
- Activity rates lower than national averages. In the case of the municipalities in the border region, they stand at 37.1% and 28.4%, respectively, compared to 55.6% in Spain and 48.4% in Portugal.

Table 1. Targeted areas.

Natural park/reserve (Region and country)	Protected area (Ha)	Municipalities (N.er)	Population (Inhabitants)
Sanabria Lake (Zamora, Spain)	22,365	4	2,720
Arribes del Duero (Zam./ Salamanca, Spain)	106,105	37	16,514
Batuecas-Sierra de Francia (Salam., Spain)	32,300	15	5,578
Sierra de Gredos (Ávila, Spain)	86,236	28	22,229
Peneda-Gerês (Alto Támega, Portugal)	70,290	5	9,099
Montesinho (Alto Támega, Portugal)	75,000	2	9,000
Malcata (Sabugal/Penamacor, Portugal)	16,348	2	2

2   Challenges

The MOVELETUR project has a markedly transboundary nature, meaning that the actions that are to be carried out find meaning in a coordinated way and managed jointly. The cross-border elements of the project are:

1. The protected territories on both sides of the border are characterized by their peripheral nature. They are therefore geographical areas that are characterized by depopulation dynamics, aging, weak territorial articulation, economic atony, etc.
2. The project partners on both sides of La Raya manage equipment for public use regarding attention and information to visitors located in protected areas, which eases up communication with potential users of electric vehicles and the management of recharging points.
3. The implementation of green itineraries requires dialogue and coordination with sustainable hospitality entrepreneurs (or other tourism services) located in protected areas on both sides of the border.

4. The Portuguese and Spanish partners need to develop all the activities together in order to obtain a successful connectivity result through the offer of truly trans-border tourist itineraries. This will provide electric connectivity among natural spaces.
5. Border territories are also spaces of opportunity: they have a natural and cultural heritage of enormous value, on which cross-border initiatives have been carried out; they are spaces where accessibility and territorial articulation have improved (road, rail network); they are areas of economic cooperation, due to successive cross-border cooperation programs (namely INTERREG).

3   Objectives

The main challenges identified in the previous section encompass the establishment of work methodologies that contribute to the development of a concept of transboundary natural spaces. This involves the appreciation of natural, cultural and landscape resources together with the development of nature tourism. In parallel, it also aims creating new employment opportunities and to allow the use of technologies and innovation for the management and enhancement of natural spaces.

These challenges structure a set of opportunities that can enhance the importance of having MOVELETUR or similar projects researching and operating in natural spaces. It is well known that natural spaces have great potential (specifically in bordering areas) to develop nature tourism/sustainable tourism activities and services. Tourism destinations that safeguard their cultural and/or natural heritage have been noticed as tourism attractors and hold a better-quality tourism infrastructure. The natural elements are valued in the market and may increase the economic capacity of territories [3–5]. In Portugal, protected areas show interesting number of visitors, as Fig. 2 reveals.

Tourism activities can contribute to territorial development, but they are not harmless [6, 7]. Tourism activities must be inclusive with local development and local economic structure to prevent impacts and risks. Conversely, local territories must allow renewal of traditional sectors and the emergence of new activities, as long as they respect the specific characteristics of natural areas.

Fig. 2.   Number of visitors in Protected Areas, in Portugal, per year (1996–2016). Source: ICNF.

For this purpose, new technologies offer the possibility of developing new tourist services which in turn can contribute to new economic activities and jobs. Based on these challenges and opportunities, the project MOVELETUR aims at the following goals:

- to create a tourism destination image of excellence and environmental sustainability through electric mobility, which allows reducing the environmental footprint of tourism (no noise or CO_2 emissions),
- to create a new tourism product through the definition of electric mobility tours and itineraries, inside specific natural areas (bicycles or similar) or amongst cross-border natural areas (automobiles),
- to use innovation grounded on the use of electric vehicles for rural areas and natural spaces as an emerging technology, especially suitable for natural environments,
- to use new technologies (ICT) to manage the new sustainable mobility/electric mobility tourism service, through the creation of smart mobility App and software for the electric mobility system,
- to train both technicians and students in the process of adapting to a low carbon economy, to become electric mobility equipment managers and maintenance services operators, and
- to create direct employment by contracting maintenance services and management of electric mobility equipment.

4   Case Study: Malcata Natural Reserve

The work carried out so far allowed to identify several good practices for natural and cultural resources valorization (mainly in EU) through integrated itineraries and electric vehicles use. These are: La Metropoli Verde, Spain; Werfenweng, Austria; Gorensjka, Slovenia; Krka National Park, Croatia; Luberon Natural Regional Park, France; Sintra, Portugal; and also, National Parks Initiative in the USA [8]. From these good practices we can state that, although in the political agenda, efforts to promote electric mobility in natural areas are disappointing. Nevertheless, it is expectable due to land morphology, difficulties of access, areas with low number of residents/visitors, problems arising from charging facilities installation in protected areas, among other problems. It is also possible to conclude that it is not normal (for now, at least) that public funding programs regarding electric mobility address environmental institutions or organizations. Usually public funding is used hoping that demonstration effect takes place in communities and incentive people to adopt electric mobility.

Serra da Malcata Natural Reserve (SMNR) is located between the small villages of Penamacor and Sabugal in Portugal centre area, neighbouring Spain. Its main symbol is the Iberian lynx, the most endangered feline in Europe. The development of the project focusses the creation of electric vehicles' tourism itineraries, considering the connection between the most important natural and cultural elements, accessibility for different types of electric vehicles and battery charging stations location. For this reason, an App and a management system is being developed, although this paper does not focus these issues. The territory had already been subject of intervention regarding walking and cycling itineraries, turning easier the work to be carried out: instead of creating new itineraries, it was decided (even for nature safekeeping sake) to start from the existing itineraries and work from that point on (Fig. 3).

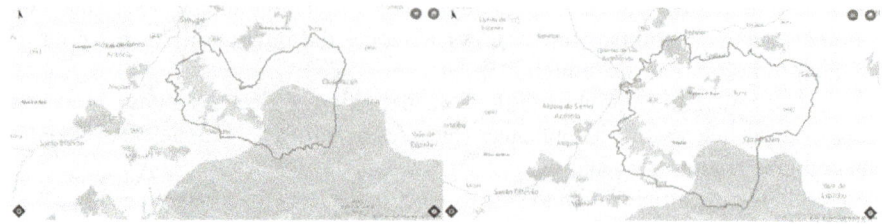

Fig. 3. Example of planned electric itineraries (north area of SMNR).

Although previously planned to comprehend a small set of both bicycles and automobiles charging stations, near public services or hospitality facilities, it was decided to use only automobiles charging stations (see Fig. 4). The reason that underlies this change is the fact that todays' systems for bicycle charging are (almost) plug and play systems and the distances within the Reserve are affordable for the bicycles' existing autonomy. In this sense, the project was oriented for connectivity between the project's natural areas. This proposal matches some other projects that are undergoing throughout the territory regarding slow/soft mobility.

Fig. 4. Electric car charging station and map with the location of the charging stations.

5 Concluding Remarks

Halfway gone after starting the project's implementation, and deriving from the partners' exchange of experiences, it is possible to say that there's a difference regarding the way it is institutionally managed, in Portugal and Spain. This difference affects common decisions and led to different approaches regarding charging stations' location. The need for this kind of facilities it is also different once that there are more scattered charging stations in Portugal than in Spain in the areas of intervention.

The MOVELETUR project proposes an opportunity for sustainable and innovative tourism development in protected cross-border spaces. It considers an alternative transport model without emissions of polluting gases and allowing to discover and enjoy the natural and cultural values through a tourism product. This product is based on principles of cross-border cooperation, and as an opportunity and formula to help fixing population on both sides of the border.

The MOVELETUR project aims to protect and enhance the natural and cultural heritage as a support for the economic base of the transboundary region, focusing on sustainable mobility in natural areas. MOVELETUR aims to promote a model of sustainable public use for visitors from cross-border natural areas for which a network of green tourist itineraries is being developed. This network links sites of natural and cultural value of this region through electric vehicles. Complementarily, the project intends to train entrepreneurs in natural areas so that they can offer a tourism offer related to electric mobility. To achieve so, a plan of professional training is being developed, including e-learning and classroom teaching. Further on, marketing and communication actions will be provided for the implementation of the aforementioned project, leading up to an Electric Tour linking all participant natural areas.

Acknowledgments. The authors would like to acknowledge EP INTERREG V A España-Portugal Program (POCTEP) for financing project MOVELETUR.

References

1. Evans, D.: Building the European Union's Natura 2000 network. Nat. Conserv. **1**, 11–26 (2012). https://doi.org/10.3897/natureconservation.1.1808
2. Sundseth, K., Creed, P.: Natura 2000: Protecting Europe's Biodiversity. Office for Official Publications of the European Communities, Oxford (2008)
3. Boley, B., Green, G.: Ecotourism and natural resource conservation: the 'potential' for a sustainable symbiotic relationship. J. Ecotour. **15**(1), 36–50 (2016)
4. Mckercher, B.: Sustainable tourism development - guiding principles for planning and management. National Seminar on Sustainable Tourism Development, Bishkek, Kyrgyzstan, 5–9 November 2003 (2003). https://www.researchgate.net. Accessed 08 Feb 2018
5. Saner, R., Yiu, L., Filadoro, M.: Tourism development in least developed countries: challenges and opportunities. In: Camillo, A.A. (ed.) Handbook of Research on Global Hospitality and Tourism Management, pp. 234–261. IGI Global Publ., Hershey (2015)
6. Prats, L.: Concepto y gestión del patrimonio local. Cuadernos de Antropologia Social **21**, 17–35 (2005)
7. Lambas, M.E.L., Ricci, S.: Planning and management of mobility in natural protected areas. Procedia – Soc. Behav. Sci. **162**, 320–329 (2014)
8. Fundación Patrimonio Natural de Castilla y León: Benchmarking report on electric mobility for sustainable use in natural areas (Spanish version) - Moveletur Project (2017). http://patrimonionatural.org/proyectos/turismo-sostenible-y-movilidad-electrica-en-espacios-naturales-moveletur. Accessed 02 Mar 2018

Opportunities for Cooperation Between Brazil and China in the Brazilian Energy Sector

Humberto Medrado Gomes Ferreira[1](✉) (iD),
Isabel Cristina dos Santos[2] (iD),
and Caroline de Melo Campos Neves[3] (iD)

[1] Faculdade Arthur Sá Earp Neto, Petrópolis, Brazil
hmedrado@gmail.com
[2] Universidade Municipal de São Caetano do Sul, São Caetano do Sul, Brazil
isa.santos.sjc@gmail.com
[3] Universidade do Porto, Porto, Portugal
carolcampos26@hotmail.com

Abstract. With the increase of global competitiveness, the countries that reach a better economic situation tends to attract a network of international cooperation resulting from foreign direct investment. In this regard, China has achieved high level of attractiveness. As a partner of Brazil in a strategic sector such as energy, China has been acting in a way that has a relevant flow of exchange among countries. Energy is an important productive input and an item of great importance in household consumption. To guarantee adequate amount of energy for consumption requires strategic partnerships to ensure the energy security. Through a bibliographic analysis and based on secondary data, one can infer the range of opportunities offered by this partnership, from the process of merger and acquisition of companies to the perspective of job and income generation.

Keywords: Opportunities for cooperation · Brazil and china · Energy sector

1 Introduction

Brazil has one of the most diversified matrix of energy on the planet. Ensuring the sustainability of the energy supply, as well as the permanence of reserves, is a challenge that must be addressed at the national and international levels.

From this point of view, and with the minimization of US dependence, China has been increasing its participation in Latin America and especially in Brazil, through investments not only in the form of credit but also in the form of investments in natural resources [1] and has become an important international player, with its companies State Grid and China Three Gorges bought in Brazil, The Paulista Company of Power and Light - CPFL - and Duke Energy, with investments that involved values about US$ 10 billion.

© Springer International Publishing AG, part of Springer Nature 2019
J. Machado et al. (Eds.): HELIX 2018, LNEE 505, pp. 992–997, 2019.
https://doi.org/10.1007/978-3-319-91334-6_136

1.1 Theoretical Context

Chinese investments on a global scale in the energy sector are justified not only because it is one of the largest oil producers in the world, but also because it is among the largest consumers on the planet.

Contextualizing this new and more aggressive phase of internationalization of Chinese companies is fundamental. Thus, we believe that the slowdown in the economy, leading to low growth in the industry and real estate, coupled with idle capacity, reduced investment opportunities in the real estate sector and the growing number of protectionist measures against acquisitions [2].

The emerging of Chinese economy presents new opportunities for cooperation in the energy sector for Brazil, where "China's high economic dynamism has repositioned this country, making it one of the protagonists of the world economy" [1, 3] reinforces that, from the moment the Chinese industrial park demands energy for production, one can expect an increase in the opportunities created by the Chinese productive intensity. These investments find justification in two main elements: the need for China to sustain its growth and, the institutional crisis that still haunts the Brazilian economy.

The Brazilian crisis cannot be seen as a matter for cooperation, according to [3]. On the contraire, "The Brazilian crisis opened up opportunities for investments in Brazilian infrastructure, which was once dominated by large contractors in Brazil. China is nowadays the only country in the world that has the financial availability and willingness to invest in the current Brazil risk. In addition to bringing capital to Brazil, these investments create jobs and generate wealth, helping to maintain the economic activity that currently requires reinforcement" [3].

The most promising Chinese futures investments include the following actions presented [3]:

1. Belo Monte transmission lines for the southeastern region of Brazil;
2. Purchase of new hydroelectric plant for approximately US $ 10 billion;
3. Purchase of the concession of a transmission lot of Eletrosul;
4. Auctions of energy derived from waste;
5. Opportunity for new companies to settle in Brazil;
6. Thermoelectric project fueled by natural gas;
7. Investment in the Brazilian wind farm to generate 700 MW of energy;
8. Manufacturing of substations in Sorocaba/SP;
9. Manufacture of wind turbines;
10. Manufacture of solar panels;
11. Investment in factory of wires and electric cables;
12. Factory assembly of solar energy converters and inverters;

It resides in the evaluation of Chinese investments in Brazil and the opportunities for cooperation between countries in the national energy sector, the foundation of the present work, supported by a theoretical contextualization based on secondary data that allow to infer the status quo of this relationship of cooperation and future projects.

The way in which China has guided its international relations is remembered as one of the foundations for the development of the Chinese global expansion project, with a business policy based on the possibility of exporting equipment and services fruit of

Chinese productive capacity. The exchange between of equipment and services of Chinese productive capacity. The exchange between exports and the investments made in the countries with which it has a commercial relationship forms the basis of the win-win philosophy of the Chinese development project [3].

In five years, between 2005 and 2010, China gained prominence with the increase of exports and imports, rivaling the USA in the position of main commercial artery in the global productive fabric, especially with the countries and development in the form of investment and financing, especially in the oil sector [4].

In Brazil, the point of change of this level came from 2014, with Chinese investments in the Brazilian electric sector and in the renewable energy sector, albeit in a subtle way [2]

> The sectoral characteristics of Chinese FDI abroad have shown that the relative scarcity of natural resources in the country has made investments in these activities, as well as in energy, appear as a necessary and priority option. In this sense, the government has developed an aggressive policy of external investments of the Resource Seeking type - oriented towards natural resources [5].

Data from the Secretary for International Affairs - SAI - [6] of the Brazilian Ministry of Planning, Development and Management, already present an indicative of Chinese investments in the period between August and September 2017, which are:

1. São Simão hydroelectric plant (brownfield) in the state of Minas Gerais and Goiás: State Power Investment Corporation Overseas - Pacific Hydro (state-owned company) acquired the right to operate the hydroelectric plant for US $ 2.25 billion.
2. Two areas for oil exploration: At the 14th National Petroleum Agency (ANP) Exploratory Block Bidding Round, China's state-owned CNOOC Limited and private company Tek Oil and Gas closed off two greenfield areas (in the Recôncavo Baiano and Espírito Santo, respectively), with a total contribution of US $ 7.72 million.
3. Acquisitions by State Grid of shares of Companhia Paulista de Força e Luz (CPFL) raising its shareholding in the company from the current 54.64% to 94.75%.
4. Operating agreement between Shanghai Electric and Eletrosul companies for new investments in 2,100 km of electric expansion lines

In this scenario, several states were included with Chinese investments, such as São Paulo, Rio de Janeiro, Alagoas, Rio Grande do Norte, Minas Gerais, and Mato Grosso, each responding to a strategic segment of the Chinese investment portfolio.

Other relevant data, extracted from the Secretariat of International Affairs of the Brazilian government, represents the percentage by segment invested and the total number of projects by investment area, as shown in Table 1.

The data in Table 1 make it possible to glimpse the profile of Chinese investments in Brazil: 47% of the total investments are made in the energy sector, ratifying the importance of this segment for strategic purposes and enabling, in the medium and long term, Brazil market its productive surplus to China itself. It can also be seen that 43 projects, out of a total of 215 planned for Brazil, are focused on the energy sector, equivalent to 20% of the total projects. The telecommunications sector (11.63%),

Table 1. Chinese direct investment. Source: Adapted [6].

Areas	Percent age by sector	Number of projects
Energy (electricity, gas, generation, transmission and distribution)	23	27
Extraction of petroleum and natural gas	24	16

capital goods (10.23) and financial services (10.23%) complete the list of the most relevant, with half of the projects foreseen for the energy sector.

The energy sector is also relevant in the context of cooperation between countries when it promotes technological diffusion, as manifested by [7] for whom "Brazil can contribute to China's technological advancement in the field of petroleum, energy, minerals and food". According to [7], "China is the largest exporter of technology-intensive products to the world, as well as the largest producer of equipment for wind energy production".

The SAI data reflects China's strategic need vis-à-vis its trading partners and described in the newsletter of the Secretariat for International Affairs: 85% of confirmed investments are concentrated in the energy sector.

This form of foreign investment appropriation, compared to the amount invested by Brazil in that country, shows the imbalance between them, as indicated in Table 2.

Table 2. Comparison of investments China X Brazil. Source: [10].

Investments by country	2009 (U$ billions)	2010 (U$ billions)
China	5,1	18,9
Brazil	3,0	9,0

They contribute to the asymmetry of investments presented in Table 2, according [8], the following factors:

1. The methodological structure for the appropriation of investments by the Central Bank of Brazil;
2. Scarcity of information on Brazilian investments in China and;
3. The slow process of internationalization of Brazilian companies with China, but with potential for growth.

Investment in clean energy is also part of the Chinese project to mitigate fossil dependence, investing considerable sums and seeking the efficiency of the country's traditional energy sources [9].

Between 2004 and 2014, global investment in renewable energy increased by 286.11% (from US $ 36 billion to US $ 139 billion) for developed countries, while for developing countries the increase was in the order of 1,355.56% (from US $ 9 billion to US $ 131 billion). The high volumes justify the global concern with the perennial of the energy supply [10].

By investing in self-interest areas and boosting the local economy, China creates conditions for trade between countries, increasing productive efforts in sectors that demand the majority of Chinese exports.

> The expansion of trade, financial and productive relations between China and Brazil has been taking shape at a time of significant international transformations - changes in the division of labor, trade and financial flows, and political arenas - that alter the status of certain national states in the hierarchy of the world system. Return and emergence of actors in global political and economic dispute spaces, such as Brazil, Russia, India, and notably China [7].

Brazil, in this scenario, has strategic participation as an indispensable channel for supplying the energy needs of its industrial policy as a result of the expansion of Chinese industry and guaranteeing effective support for energy supply [5].

2 Final Considerations

The perspective of partnership in the Brazilian energy sector involving China translates into cooperative elements of a relationship that can elevate Brazil to the category of global supplier - caused by the pre-salt reserves and their derivations.

The collective interest in the partnership is given by the guarantee of supply that covers the Chinese industrial policy and offers the country conditions of productive planning for the energy supply coming from the Brazilian reserves.

From this proposal emerges elements that configure the potentialities arising from the established relations between China and Brazil that encompass the economic challenges between the commercial partners, in particular the legitimacy of the institutionally weakened Brazilian government; the ability to mobilize resources among agents within the current economic environment; innovative projects that reduce the cost of production and maximize energy efficiency; increasing relations with research centers and universities, promoting the diffusion of technology and knowledge; productive investments in areas that demand governmental attention, promoting local development, reducing inequalities, and, finally, establishing a model of governance for the national energy sector that contemplates propositive actions for the development of partnerships and effective projects planned for the sector.

Some points emerge for cooperation:

1. The devaluation of the Brazilian currency makes foreign investments attractive;
2. The relatively low cost of acquiring domestic companies (a privatization process that should be conducted by the Brazilian government in a transparent way, to guarantee the transfer of technology and knowledge);
3. Perspectives of research funding to solve the urgent demands of both countries;
4. Possibilities to generate employment and income from the construction of factories for the supply of materials and equipment for the energy production chain.
5. The growth of the Chinese industrial park will imply in the consequent reduction of the ozone emissions, creating possibilities for the Brazilian companies to develop equipment for such.
6. The prospect of reducing Chinese fossil dependence may open up research fronts on new energy sources or even increase existing ones with the expectation of cheapening production costs.

References

1. Pinto, E.C., Cintra, M.A.M.: Latin America and China: economic and political limits to development. Texts for Discussion, no. 12 (2015)
2. CEBC: An analysis of Chinese investments in Brazil: 2007–2012, Rio de Janeiro (2013)
3. Tang, C.: A bulletin of the conjuncture of the energy sector - Chinese investments in the Brazilian energy sector: opportunities for Brazil. FGV Energia, August 2017
4. De Medeiros, C.A., Cintra, M.R.V.P.: Impact of the Chinese rise on the Latin American countries. Rev. Econ. Polit. 35(1), 28–42 (2015)
5. IPEA: Institute of applied economic research internationalization of Chinese enterprises: the priorities of Chinese direct investment in the world. Communiqués of IPEA, no. 84 (2011)
6. Brazil: Secretary for International Affairs of the Brazilian Ministry of Planning, Development and Management. Chinese Investments in Brazil. Brasília (DF), August/September 2017
7. Acioly, L., Pinto, E.C., Cintra, M.A.M., Calixtre, A.B.: Brazil-China bilateral relations: the rise of China in the world system and the challenges for Brazil (2011)
8. Hiratuka, C., Sarti, F.: Economic relations between Brazil and China: analysis of foreign direct trade and investment flows. World Time, p. 83 (2016)
9. Pedrozo, G.E., Da Silva, M.: The Chinese cooperation for the development of green technologies in front of the "new normal". Braz. J. Int. Relat. 6(3), 491–521 (2018)
10. Bhattacharya, M., Paramati, S.R., Ozturk, I., Bhattacharya, S.: The effect of renewable energy consumption on economic growth: evidence from top 38 countries. Appl. Energy 162, 733–741 (2016)

Impacting Factors in Decision Making Internationalization of Micro and Small Entrepreneurship Companies of the Machines Sector Agricultural Equipment

Milton Carlos Farina, Maria do Carmo Romeiro,
Alvaro Francisco Fernandes Neto, Antonio Aparecido de
Carvalho[✉], and Davi de França Berne

Universidade de São Caetano do Sul – USCS, São Caetano do Sul, Brazil
{milton.farina, mromeiro, alvaro.fernandes,
antonio.carvalho, davi.berne}@uscs.edu.br

Abstract. This article aimed to compare the micro and small companies in the agricultural equipment sector with regard to the factors that impact on the decision making of the internationalization of the companies in the segment. The objective was to identify the different reasons that led companies to undertake the process of internationalization of the agricultural machinery and equipment sector. A non-probabilistic descriptive and quantitative survey was used. The work was based on the Uppsala model, the reasons and the critical factors that lead to the decision to export from the authors Moreira [1] and Lirani [2]. From a questionnaire applied to 159 companies in the industry sector, it was possible to identify the reasons that most impact the decision-making process of internationalization. For micro-enterprises, the reasons were knowledge of the market, the need for business expansion, conditions for internationalization and support for internationalization. Already for the small companies the reasons that led to the internationalization were different: company strategies, support for the internationalization process and conditions of the internal market.

Keywords: Entrepreneurship · Micro and small enterprises
Internationalization

1 Introduction

Entrepreneurship according to Dornelas [3] is not a new administrative theory, but a behavior that encompasses organizational processes directing the company in search of new opportunities. Many companies can seek the international market according to some factors, such as those described by Moreira [1], such as saturation of the national market, participation in new markets, diversification of products or services among others. In view of the above, we try to answer the following question: What are the impact factors for the internationalization of micro and small entrepreneurial companies in the field of agricultural machinery and equipment? The objective of this article is to identify the impact factors that lead micro and small entrepreneurs in the segment of

© Springer International Publishing AG, part of Springer Nature 2019
J. Machado et al. (Eds.): HELIX 2018, LNEE 505, pp. 998–1004, 2019.
https://doi.org/10.1007/978-3-319-91334-6_137

agricultural machinery and equipment in addition to acting in the national market, seek the internationalization of their businesses and verify if the factors prevail for both micro and small companies. According to data extracted from the Brazilian Association of Machinery and Equipment - ABIMAQ [4], exports increased by 57.6% in relation to the previous year.

2   Theoretical Framework

According to Farah et al. [5], in Brazil, taxonomy of organizations takes into account, number of employees, brand value, assets and liabilities, among other items, being the denomination of organizations based on their size can be: micro, small, medium or big enterprise. Duarte [6] reports that the entrepreneurial spirit is responsible for the growth of the number of companies in Brazil.

Microenterprise is understood to be that registered in the competent bodies with annual gross revenue of less than or equal to three hundred and sixty thousand reais. If the revenue is higher than this value, the company will have the denomination of Small Business. Albuquerque [7] points out that micro and small companies from the 1980s began to gain the attention of governments, which created public policies to encourage society to embark on new business lines. According to Sebrae [8] micro and small companies in 2001 accounted for 23.2% of GDP and ten years later the share was 27.0%. Between 2007 and 2016 the number of micro and small companies increased 283.78%.

The search for internationalization according to Johanson and Vahlne [9] is explained by the Uppsala model, which emphasizes that the entrepreneur makes the decision to invest in other markets and gradually acquires knowledge of the markets, a fact that would minimize the occurrence of risks. The model starts from three assumptions: lack of knowledge, which is the biggest obstacle; the knowledge is acquired from operations with the intended market and the company invests gradually in the process of internationalization. Carneiro and Dib [10] included to the Uppsala model 5 questions that impact on the internationalization decision: why, what, when, where and how. Lirani [2] cites critical success factors of micro and small companies, highlighting: the knowledge that the manager of the company has about his company and the external market, and the quality of the product offered. Moreira [1] lists the reasons why micro and small companies seek foreign markets, such as growth need, financial strategy, prior knowledge of the international market, experiences in other cultures, technological advantage, competitive pressures, domestic market, ease of information, communication technologies and internal financial crisis.

The evolution of the segment of agricultural machinery and equipment is based on data from the Confederation of Agriculture and Livestock of Brazil of 2005, which indicated that from 1995 to 2005, agribusiness was a significant highlight in the national economy, mainly due to the expansion of the machinery and equipment industry agricultural activities. Data from ABIMAQ [11] showed that from 2000 to 2004 there was an increase in the turnover of the segment's industries by 100%.

Toledo and Simões [12] describe the characteristics of companies in the segment: family companies, or that are still undergoing a professional management process; the focus is on product diversification and innovation; the volume of production by type of product is low; there is seasonality by demand; need to create specific solutions due to the existing types of farming, soil, climate and labor.

Research by ABIMAQ [13] shows that exports of agricultural machinery and equipment had a 40.7% growth between January and February 2017. The data show that the foreign market has excellent receptivity to the products of the segment.

3 Research Method

With the objective of comparing the factors that impact the micro and small companies of the segment of agricultural machinery and equipment in the decision-making process to internationalize their businesses, the research was quantitative, descriptive, non probabilistic and the procedure used was a field survey. To collect the data was used, a questionnaire with 40 open and semi-open questions, multiple choice and questions with a level of agreement varying from 0 (totally disagree) to 10 (totally agree). The issues were based on the works of Moreira [1] and Lirani [2]. The questionnaire was divided into three blocks following the Uppsala model with the objective of answering the 5 questions elaborated by Carneiro and Dib [10]: Why? What? When? At where? As?

The questionnaire was applied in electronic form with the use of the Google Forms tool, aiming at reaching the largest number of respondents possible, based on the set of 9.700 companies in the agricultural equipment segment of the State of São Paulo.

In order to carry out the comparative analysis of the reasons that lead to internationalization, the data were analyzed in two categories: Micro and Small Companies. For the 159 questionnaires received, the software Statistic Package for the Social Sciences - SPSS version 23 was used. The descriptive statistical analyzes were used and the factorial analysis was used to examine the relations between the variables.

Grades varied on a scale of 0 to 10, with a value of 0 meaning "totally disagree" and 10, "I totally agree".

The variables used in the research were based on the works of Moreira [1] and Lirani [2] and adapted to indicate the reasons that led to internationalization. (1) business expansion, (2) growth strategy, (3) internal competition, (4) Brazilian economic crisis, (5) suppliers suggestion, (6) partners suggestion, (7) support of associations, (9) consultancy advice, (10) professional experience abroad, (11) academic experience abroad, (12) foreign exchange factors, (13) innovative product and (14) support from development agents.

4 Discussion of Results

4.1 Statistical Data

Block 1 Characterization of Respondent Companies: Regarding micro and small enterprises, of the total of 159 companies, 74 (47.2%) are micro enterprises and 85 (52.8%) are small companies. The research showed that the products marketed by the researched micro companies are accessories and equipment, the predominant export type is direct export. In relation to small enterprises the predominant type of export is also direct export, the type of product exported are machinery and equipment. The destination of exports to both micro and small enterprises is centered in China.

Block 2 Respondent's knowledge about the international market: The applied variables are related to international experiences, be they professional or academic. For both micro and medium-sized companies, the results presented similar averages, ranging from 5.80 to 6.0 with very close standard deviations, which indicates that previous market knowledge was important for the sample surveyed, considering that the agreement level is from 0 to 10.

Block 3 Reasons for the internationalization of micro-enterprises: This block is based on the study of Moreira [1], which presents eight reasons for MPEs to seek internationalization: the need to expand the business, technological advantages that the company possesses, competitive pressure, saturation of the domestic market, internal financial crisis, strategy for financial gain and differentiated product. Factorial analysis was applied both to the responses given by microenterprises and to small firms. The correlation matrix, the KMO test and the Bartlett sphericity test, the anti-image matrix and commonalities, and the rotated matrix of the components were analyzed. A total of 14 variables were inserted with the following description: the internationalization decision was due to: business expansion; strategic decision making; competition in the national market; Brazilian economic crisis; suppliers suggestion; suggestion of the partners; customer suggestion; support of some association; suggestion of advice; exchange rate factors; academic experience abroad; professional experience abroad; innovative product and support from some development agent.

Results of the micro enterprises: the KMO was 0.772 (value indicating the adequacy of the factorial analysis). The results of the 14 variables according to the anti-image matrix presented values ranging from 0.725 to 0.845, adequate for the analysis to be continued. The commonality of the variables presented a result superior to 0.5 (from 0.610 to 0.828). The matrix of the components after the rotation or rotated loads represents according to Fávero et al. [14] that the variables can be grouped in 4 factors namely:

Factor 1: Expansion of business:
    Need for expansion;
    Growth strategy;
    Internal competition and
    Member's will
Factor 2: Support Internationalization:
    Support from development agents;
    Support from associations and
    Suppliers suggestion
Factor 3: Market knowledge:
    Exchange rate factors;
    Professional experience abroad;
    Academic experience abroad;
    Consulting suggestion and
    Brazilian economic crisis.

Factor 4: Conditions for internationalization:
    Innovative product and
    Customer suggestion

Results of the small companies: the analysis of the 14 variables presented the KMO value of 0.807, the anti-image matrix presented results ranging from 0.695 to 0.909. The commonality presented results as follows: 12 variables present values higher than 0.5 ranging from 0.537 to 0.768, two variables presented values lower than 0.5, they are: internationalization was motivated by suggestion of consulting (0.439) and motive which led to the internationalization was to companies having an innovative product (0.422).

The factorial analysis demonstrates that the variables can be grouped into three distinct factors:

Factor 1: Strategy
    Growth strategy;
    Expansion of business;
    Suggestion of members and
    Suppliers suggestion.
Factor 2: Knowledge of the international market
    Association support;
    Customer suggestion;
    Professional experience abroad;
    Exchange role factors;
    Academic experience abroad and
    Support from development agents.
Factor 3: Internal Market
    Internal competition and
    Brazilian economic crisis.

The research made it possible to outline the different reasons for internationalization and their differences when comparing the motives of micro and small companies that lead to the decision-making process for internationalization. The variables are grouped into four factors to micro companies, the first factor being expansion of business, the second factor is support internationalization, the third factor is represented by the market knowledge, and finally the fourth factor is the conditions of internationalization.

The reasons for small companies to export are expressed in three factors, the first one is based on the strategy, or the need for the company to grow in its field of activity; the second factor is support for internationalization, considering support from associations, suggestion of suppliers and clients, experience gained in the international market and the exchange band; the third factor is represented by domestic market conditions, mainly as a result of the Brazilian economic crisis.

In view of the results, it can be seen that the common factor between micro and small companies is the need to expand the business to the domestic market, so it is a strategy for both micro and small businesses for their survival.

5 Final Considerations

The result of this research, which sought to identify the different reasons for internationalization in relation to micro and small enterprises and their differences from the manufacturers of agricultural machinery and equipment was successful. It was possible to identify that, for micro companies, the variables can be grouped into four factors and the main reasons for the decision of the internationalization process are the need for expansion, support for internationalization, knowledge of the market and conditions for internationalization. For small companies, the variables were grouped into three factors: internationalization occurs through the strategic decision of the companies, support to the internationalization process and internal market conditions, such as an internal economic crisis.

The innovative product factors and consulting suggestion do not represent importance in the process of decision making in the internationalization process of small companies.

The factors that stand out for both are: membership suggestion, expansion need, growth strategy, market saturation due to internal competition, innovative product, customer suggestion, consulting suggestion, economic crisis and supplier suggestions.

Carneiro and Dib [10] included the Uppsala model 5 issues that affect the internationalization decision. The survey answered the 5 questions: 1. Why? The factorial analysis of the data grouped the components. The so-called strategy component answers the question: For business expansion; 2. What? Statistical data indicate that companies market accessories and equipment; 3. When? The answer lies in the so-called strategic component, which indicates that the internationalization decision was due to internal market conditions; 4. Where? The result of the survey showed that the highest percentage of exports goes to China; 5. How? Direct exports are evidenced by the greater number of respondents.

The limitation of the research was because it is a non-probabilistic sample, for future research it is suggested to replicate it in other segments of the market and in medium-sized companies.

References

1. Difficulties encountered in the internationalization process: study in the MPE of furniture in the municipality of Paços de Ferreira and Paredes (2014)
2. Critical success factors of the small company: a survey in companies of the retail sector of São Carlos/SP (2014)
3. Dornelas, J.C.A.: Corporate Entrepreneurship: How to Be Entrepreneur, Innovate and Differentiate in Established Organizations. Elsevier, Rio de Janeiro (2003)
4. Indústria brasileira de bens de capital mecânicos - Indicadores conjunturais (2016). http://www.abimaq.org.br/Arquivos/Download/Upload/1561.pdf
5. Farah, O.E., Cavalcanti, M., Marcondes, L.P.: Entrepreneurship: Survival Strategy for Small Companies. Saraiva, São Paulo (2012)
6. Entrepreneurship in micro and small enterprises: a study applied to the city of Pará de Minas - MG (2013)

7. Albuquerque, E.M.N., Cardoso, L.A., Carmo Filho, M.M., Cavalcante, T.S.B., Souza, W.A. R.: O uso de indicadores e relatórios contábeis para tomada de decisão nas micro e pequenas empresas do Estado do Amazonas. Revista de Administração da FATEA **6**(6), 6–21 (2013)
8. Panorama dos pequenos negócios. https://m.sebrae.com.br/Sebrae/PortalSebrae/UFs/SP/Pesquisas/PanoramadosPequenosNegocios2017.pdf
9. Johanson, J., Vahlne, J.: The internationalization process of the firm: a model of knowledge development and increasing market commitment. J. Int. Bus. Stud. **8**, 23–32 (1977)
10. Carneiro, J., Dib, L.A.: Avaliação comparativa do escopo descritivo e explanatório dos principais modelos de internacionalização de empresas. Rev. Int. Bus. **2**(1), 1–25 (2007)
11. Apresentação Dos Indicadores Conjunturais - Apresentação (2005). http://www.abimaq.org.br/site.aspx/Apresentacao-Conjuntural2. Accessed 25 Aug 2017
12. Toledo, J.C., Simões, J.C.S.: Product development process management in small and medium-sized Brazilian companies in the agricultural machinery and implement industry sector. Revista Gestão e Produção **17**(2), 257–269 (2010)
13. Indústria Brasileira de Bens de Capital - Indicadores conjunturais (2017). http://www.abimaq.org.br/Arquivos/Download/Upload/1539.pdf
14. Fávero, L.P., Belfiore, P., Silva, F.L., Chan, B.L.: Análise de dados: modelagem multivariada para tomada de decisões. Elsevier, Rio de Janeiro (2009)

How Smart Specialisation Strategies in Geotourism Contribute to Local Economic Development? The Case of Atlantic Geoparks

António Duarte[1], Vitor Braga[2(✉)], and Carla Susana Marques[3]

[1] Centre for Transdisciplinary Development Studies (CETRAD),
School of Technology and Management (IPP),
Associação Geoparque Arouca (AGA),
University of Trás-os-Montes e Alto Douro (UTAD), Arouca, Portugal
antonio.duarte@aroucageopark.pt
[2] School of Technology and Management (IPP),
Center for Research and Innovation in Business Sciences
and Information Systems (CIICESI), Porto, Portugal
vbraga@estg.ipp.pt
[3] UTAD & CETRAD, Vila Real, Portugal
smarques@utad.pt

Abstract. This study aims to explore how can a Smart Specialisation Strategy contribute to the local economic growth. Through 10 case studies: the Atlantic UNESCO Global Geoparks, with primary and secondary data collected from this Geoparks, it is possible to suggest the existence of positive impacts caused by the smart specialisation on this concept and approach, suggesting the interest in their promotion. The results of geotourism activities impact at different levels, but they increase, in particular, the local entrepreneurial ecosystem.

Keywords: Geotourism · Smart specialisation strategy
Local economic growth · Geoparks

1 Introduction

Currently, within the framework of European public policies, smart specialization strategies are planned and developed, whose regional governance must be based on the quadruple helix model that is conducive to the development of entrepreneurial discovery processes, which allows [1, 2] to verify the collective interaction between the different stakeholders (Government, enterprises, University and Society) and the exchange of knowledge. We have recently witnessed a specialization in the field of Geotourism as a strategic tourism product and an approach to emerging territorial development. According to [3], Geotourism should be developed in the context of the sustainable development of local, national and international tourism. [4] states that the partnerships inherent in geotourism facilitates an environment that is conducive to economic development and can, mutually, benefit all partners. At the entrepreneurship and innovation ecosystems level, in the area of geotourism, [5] refers that the emergence of new artisan products (called geoproducts) promote the local economy and enrich the geotourism offer. However, it has been verified, through the review a

© Springer International Publishing AG, part of Springer Nature 2019
J. Machado et al. (Eds.): HELIX 2018, LNEE 505, pp. 1005–1011, 2019.
https://doi.org/10.1007/978-3-319-91334-6_138

literature, that there is a space for research due to the lack of scientific studies relating Smart specialization, geotourism and growth. Based on this gap, this study aims to explore the extent to which ecosystems of innovation, entrepreneurship and public investments, inherent in a geotourism specialization, stimulate demand and contribute to smart, sustainable and inclusive growth. In order to pursue such analysis, we used to take as a case study - the UNESCO Global Geoparks located in the European Atlantic Area. The rest of this paper is structured as follows: the next section presents the review of the literature. Section three describes the methodology used. In the fourth section, the results are analysed, and the last section presents the study's conclusions, implications and limitations, as well as suggestions for future lines of research.

2 Literature Review

The concept of smart specialization was developed as an academic concept in the mid-2000s [6], and later developed by the "Knowledge for Growth Expert Group" (chosen by the European Commission), resulting in the concept currently used ([7–9], Foray et al. 2011). With the European growth strategy, by 2020, the European Commission wants the EU to become "a smart, sustainable and inclusive economy". These three mutually reinforcing priorities should assist the EU and the Member States to provide high levels of employment, productivity and social cohesion" [10]. [11] states that the success of a Smart strategy depends on its ability to transform innovation and knowledge into the implementation of regional strategies that exploit its own unexplored resources and, at the same time, pursue a policy of cohesion and competitiveness. The literature on economic geography, entrepreneurship and innovation suggests that entrepreneurship tends to be lower in regions with lower population densities; lower in regions dominated by a small number of large firms; lower in regions with international firms; and lower in regions with low market potential. On the other hand, innovative urban policy paradigms related to intelligent cities interrelate issues of green cities, connected life, intelligent communities, innovation ecosystems, and environmental and social sustainability with urban growth [12]. Therefore, innovation ecosystems are supported by actions that foster the relations of agents in innovation processes and diversify the forms of creation, diffusion and production of knowledge, stimulating coevolution and co-specialization of certain territories [13].

According to [4], geotourism partnerships between government, local residents, private sectors, local businesses, outside business, travel agencies, restaurants, tourist accommodation, promote the development of the economy and can benefit all partners. In terms of entrepreneurship and innovation, within geotourism, [5] refer to the emergence of new crafted products (called geoproducts) as a way to promote the local economy and enrich the geotourism offer. This study has as its theoretical basis, in three theories that have inspired the literature review:

- the theory of endogenous growth [14], where progress is the driving force of economic growth and it is modelled endogenously;
- the theory of regional Smart specialization advocated by [15] based on the potential of regions, through their resources and assets (with characteristics of inimitability

and non-transference), able to create tradable and constructed products and services and competitive advantages; and

– the Tourism-Led Growth (TLG) theory which, according to [16], argues that tourism is an important determinant of long-term global economic growth.

Figure 1 displays the relationship between the theories adopted in this study and the purpose of the study:

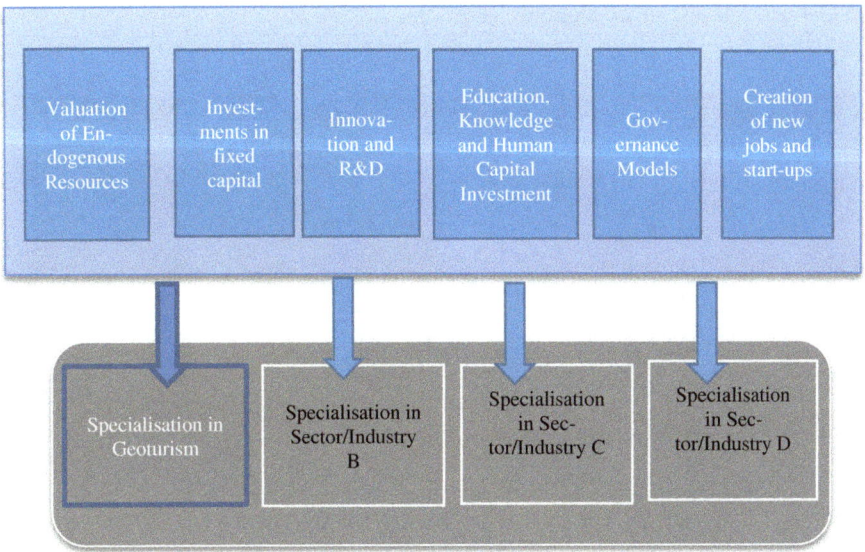

Fig. 1. Determinants of Specialisation. Source: Authors

3   Methodology

The concept of Smart Specialization Strategy is based on the distinctiveness of the local resources and the potential to generate competitive advantages in the production and commercialization of differentiating products and services. This distinctiveness applies, intrinsically, to the natural and cultural heritage, which includes the geological heritage, and constitutes an opportunity for valorisation of endogenous resources, through a wide range of economic activities. Therefore, and as a result of the literature review, there is a research opportunity for deepening the knowledge about the relationship between regional specialization in geotourism and growth. In line with the such perspective our research question is: can a regional strategy of Smart specialization, based on geo-tourism effectively promote and contribute to the smart, sustainable and inclusive growth of low-density regions in the European Atlantic Area? [3] shows that, in line with the geotourism's growth trends, the potential of this segment in several territories, the emergent partnerships of this new reality, there are some examples of excellence in its interpretation, and highlights the main tool for geodevelopment: the Global Geoparks Network. According to [17], UNESCO Global Geoparks are unique and contiguous geographic areas where sites and landscapes are of international geological importance and are integrated into a holistic territorial strategy that includes a logic of protection, education and sustainable development. Geoparks are an ascending territorial approach that consists of combining conservation with sustainable development and involving local communities. There are, currently, 127 UNESCO Global Geoparks in 35 countries and five continents. Our study includes primary and secondary data from 10 Geoparks classified by UNESCO, located in the European Atlantic Area. The data was analysed through qualitative methods, and shows the reported impacts of these Geoparks in the local economy (Fig. 2).

Fig. 2. UNESCO Global Geoparks in Europe. Source: [18]

4   Analysis of Results

The results of this study allow to point out that the territories where the geoparks are located are organized and composed of local partnerships, mostly of a bottom-up type, in which public and private organizations as well as local communities are called to participate in the definition and implementation of their sustainable territorial development strategies and their action plans. This model of sharing information, knowledge, heritage and/or resources has been shown to be an intelligent territorial development approach, as it allows the preservation of the natural and cultural heritage, the promotion of education for sustainability and growth through geotourism.

Our results clearly indicate that the geotourism, as a Smart Specialisation Strategy, fits into the quintuple helix paradigm, as one can see in the Fig. 3 below:

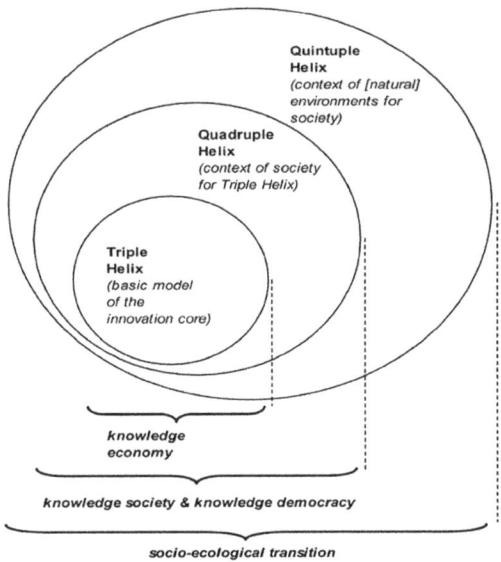

Fig. 3. Quintuple Helix. Source: [19]

There is a relationship with the geotourism offer, through a tourism product that is based on the valorisation of the geological heritage and that offers directly and indirectly related products and services, tradable through the provision of visitation infrastructures, interpretation centres, tourism visits, lodging, restaurants, handicrafts, among others, and with the demand based on the flows of geotourists who, for various reasons, move to geotouristic destinations.

5 Conclusions

Despite the predominance of Global Geoparks, these have been neglected by the academia in terms of exploring their contribution to the local economic growth, and how they may be seen as a smart specialisation strategy.

In light with such finding, this paper has explored how a Smart specialisation of geotourism in Geoparks impacts on the local development, at different levels, but where the impact can be felt at the entrepreneurial system, job creation, wealth accumulation, public funding attraction and, ultimately, the quality of life of people.

This study has used a case study approach, analysing the impact caused by 10 geoparks in Europe, which may not be fully representative of the total Geoparks in the world. However, such methodological approach has allowed to investigate, further, and to develop into new research. These limitations could be the starting point for future research, which would then contribute further to the existing literature on this topic. The present study needs to be extended to the 127 UNESCO Global Geoparks, where more generalizable results may be achieved.

Funding. The authors wish to thank Programa Operacional do Norte (N2020) for providing financial support of Project U.Norte Inova, n.º NORTE-01-0246-FEDER-000005, which was co-financed by the Fundo Europeu de Desenvolvimento Regional (FEDER) through Norte 2020 (Programa Operacional Regional do Norte 2014/2020). This work was also supported by European Structural and Investment Funds via FEDER through the Operational Competitiveness and Internationalisation Programme (COMPETE 2020 [Project No. 006971 (UID/SOC/04011)]) and by national funds through the FCT – Portuguese Foundation for Science and Technology (Project UID/SOC/04011/2013).

References

1. Carayannis, E., Rakhmatullin, R.: The quadruple/quintuple innovation helixes and smart specialisation strategies for sustainable and inclusive growth in Europe and beyond. J. Knowl. Econ. **5**(2), 212–239 (2014)
2. Carayannis, E., Grigoroudis, E.: Quadruple innovation helix and smart specialization: knowledge production and national competitiveness. Foresight STI Gov. **10**(1), 31–42 (2016)
3. Dowling, R.: Geotourism's global growth. Geoheritage **3**, 1–13 (2011)
4. Dowling, R.: Geotourism's contribution to local and regional development. In: Neto de Carvalho, C., Rodrigues, J.C. (eds.) Geotourism and Local Development, pp. 15–37. Câmara Municipal, Idanha-a-Nova (2009)
5. Farsani, N.T., Coelho, C., Costa, C.: Geotourism and geoparks as novel strategies for socio-economic development in rural areas. Int. Tour. Res. **13**(1), 68–81 (2011)
6. McCann, P., Ortega-Argilés, R.: Smart specialization, regional growth and applications to European union cohesion policy. Reg. Stud. **49**(8), 1291–1302 (2015)
7. David, P., Foray, D., Hall, B.: Measuring smart specialisation: the concept and the need for indicators. Knowl. Growth Expert Group (2009)
8. Foray, D., David, P. A., Hall, B.: Smart specialisation the concept. In: Knowledge Economists Policy Brief, no. 9 (2009)
9. Foray, D., Goddard, J., Beldarrain, X., Landabaso, M., McCann, P., Morgan, K., Nauwelaers, C., Ortega-Arguilés, R.: Guide to Research and Innovation Strategies for Smart Specialisation (RIS 3), S3 Platform, Sevilha (2012)
10. European Commission: Fact Sheet National Regional Innovation Strategies for Smart Specialisation (RIS3) - Cohesion Policy 2014–2020 (2014)
11. Capello, R., Kroll, H.: From theory to practice in smart specialization strategy: emerging limits and possible future trajectories. Eur. Plann. Stud. **24**(8), 1393–1406 (2016)
12. Zygiaris, S.: Smart city reference model: assisting planners to conceptualize the building of smart city innovation ecosystems. J. Knowl. Econ. **4**, 217–231 (2013)
13. Marques, T., Santos, H.: Lugares e redes de inovação na área metropolitana do Porto. Revista da Faculdade de Letras – Geografia – Universidade do Porto, III série, v. 2, pp. 203–225 (2013)
14. Romer, P.: Endogenous technological change. J. Polit. Econ. **98**(5), 71–102 (1990)
15. Foray, D., David, P.A., Hall, B.H.: Smart specialisation – from academic idea to political instrument, the surprising career of a concept and the difficulties involved in its implementation. Mtei-Working_Paper-2011-001, pp. 1–16 (2011)
16. Balaguer, J., Cantavella-Jorda, M.: Tourism as a long-run economic growth factor: the Spanish case. Appl. Econ. **34**, 877–884 (2002)

17. UNESCO. UNESCO Global Geoparks (2017). http://www.unesco.org/new/en/natural-sciences/environment/earth-sciences/unesco-global-geoparks/. Accessed 21 Dec 2017
18. European Geoparks Network (2015). http://www.europeangeoparks.org/?page_id=168. Accessed 01 Nov 2015
19. Carayannis, E., Barth, T., Campbell, D.: The Quintuple Helix innovation model: global warming as a challenge and driver for innovation. J. Innov. Entrepreneurship **1**(2), 1–12 (2012). A Systems View Across Time and Space

Universities in Entrepreneurial Ecosystems – The Recent Experience of IT UFABC in Brazil

Alberto Sanyuan Suen$^{(\boxtimes)}$, Debora Medeiros, and Anapatricia Vilha

UFABC Innovation Agency, Santo André, SP, Brazil
{alberto.suen, debora.medeiros,
anapatricia.vilha}@ufabc.edu.br

Abstract. The purpose of this paper is to describe the emerging concept of entrepreneurial ecosystems, reviewing definitions of ecosystems found within the academic literature. In addition, the concept of technological incubators is also examined. Finally, the paper shows the case of IT (Technological Incubator) UFABC (Federal University of ABC), an initiative developed at Federal University of ABC in Brazil and concludes discussing the challenges and perspective for the region of ABC, an important industrial neighborhood, which is part of the metropolitan region of Sao Paulo, the main city of the country.

Keywords: Entrepreneurial ecosystems · Entrepreneurship · Innovation

1 Introduction

Universities around the world have played an important role in promoting entrepreneurship, encouraging the transformation of the knowledge developed in their research laboratories into innovations that can reach the market, generating economic and social development. In this effort, a trend that has been observed is the development of innovation habitats, which offer entrepreneurs and their startups support and collaboration, so that new ideas and new business proposals effectively can be transformed into successful enterprises.

Such innovation habitats are called in the academic literature of business incubators or technology parks or business accelerators, or other creative names. Authors such as [1] consider these habitats as important tools to support innovation and entrepreneurship, thus contributing to their connected universities becoming an important member of the entrepreneurial ecosystem.

The purpose of this paper is to investigate the emerging concept of entrepreneurial ecosystems, reviewing definitions of ecosystems found within the academic literature. In addition, the concept of technological incubators is also examined. Finally, the paper shows the case of IT (Technological Incubator) UFABC (Federal University of ABC), an initiative developed at Federal University of ABC in Brazil and concludes discussing the challenges and perspective for the region of ABC, an important industrial area which is part of the metropolitan region of Sao Paulo, the main city of the country.

© Springer International Publishing AG, part of Springer Nature 2019
J. Machado et al. (Eds.): HELIX 2018, LNEE 505, pp. 1012–1017, 2019.
https://doi.org/10.1007/978-3-319-91334-6_139

2   The Role of Universities in the Entrepreneurial Ecosystems

Ecosystem has been one of most enduring approaches adopted by social sciences from biology, where this fundamental unit is a heuristic tool used to describe the interaction between living and non-living components of a given habitat [2].

From the biology perspective an ecosystem is a community of living and non-living organisms and means the system where one lives, that is, the set of physical characteristics, chemical and biological factors that influence the existence of an animal or plant species. It is therefore a natural and stable system, consisting of a non-living part such as water, atmospheric gases, mineral salts and solar radiation, and another living part composed of plants and animals, including micro-organisms that interact or relate to each other [2].

This concept can be transported by analogy to the social sciences field, and more specifically to economics, where entrepreneurial ecosystems are understood as the environments formed by the most diverse stakeholders of entrepreneurship that act in network, where there are interconnections and dynamism.

The entrepreneurial ecosystem consists of companies, government, research and teaching institutions, incubators, accelerators, class associations and service providers, and entrepreneurs [3]. As in biology, the various participants in the entrepreneurship ecosystem also need each other to survive and knowledge in this field is accumulated through learning and interaction among the various stakeholders. Thus, the importance of the diffusion of ideas, experiences and information is emphasized.

The importance of a system or ecosystem that encourages the entrepreneurial environment is recognized from the origin of modern entrepreneurship, when the Austrian economist and political scientist Joseph Schumpeter published his works "The Theory of Economic Development" in 1934. Currently the leading academic reference on entrepreneurship ecosystems is Professor Daniel Isenberg of Babson College. Then, fostering entrepreneurship has become a core component of economic development in countries and cities around the world. The predominant metaphor for fostering entrepreneurship as an economic development strategy is the "entrepreneurial ecosystem."

Mason and Brown [4] define an entrepreneurial ecosystem, based on the synthesis of definitions found in the literature, it as follows:

> "a set of interconnected entrepreneurial actors (both potential and existing), entrepreneurial organizations (e.g. firms, venture capitalists, business angels, banks), institutions (universities, public sector agencies, financial bodies) and entrepreneurial processes (e.g. the business birth rate, numbers of high growth firms, number of serial entrepreneurs) which formally and informally coalesce to connect, mediate and govern the performance within the local."

Nowadays, there is a great number of models of entrepreneurial ecosystems. In recent years a particularly influential approach has been developed by Professor Daniel Isenberg at Babson College. The approach constitutes a novel and cost-effective strategy for stimulating economic prosperity and it potentially becomes a "pre-condition" for the successful deployment of cluster strategies, innovation systems, knowledge economy or national competitiveness policies [5].

Isenberg identifies six domains within the entrepreneurial system: a conducive culture, enabling policies and leadership, availability of appropriate finance, quality human capital, venture markets for products, and institutional supports. Each ecosystem emerges under a unique set of conditions and circumstances.

The information and communication technology revolution, the emergence of the knowledge economy, the turbulence of the economy and consequent funding conditions, all have thrown new light and new demands on higher education systems across the world.

Research universities around the world play a key role in encouraging entrepreneurship, promoting the transformation of the knowledge developed in their laboratories into innovations that reach the market, generating economic development.

Academic literature is plenty of examples in which universities have played an important role in successful entrepreneurial ecosystems.

The role of Stanford University in the Silicon Valley entrepreneurial ecosystem and the University of Texas at Austin contribution for the development of the Great Austin entrepreneurial ecosystem are very well known.

In Europe, we could mention the contribution of the University of Oxford for the London entrepreneurial ecosystem. One significant European response is seen in the development, in concept and in practice, of the Entrepreneurial University, characterized by innovation throughout its research, knowledge exchange, teaching and learning, governance and external relations [6].

In Hong Kong, we can give as an example the EYE program, that was conceived to help build connections between the investors, well established industry members, the media, physical hubs like business incubators and accelerators. A program based at the City University of Hong Kong became the platform where local entrepreneurs and their startups come together [7].

3   Technological Incubators as Part of Entrepreneurial Ecosystems

Business incubators can be defined as an organization designed to provide business incubation programs that would contribute for the growth and success of new enterprises.

The programs are compounded by a variety of business support resources and services that could include physical space, capital, coaching, common services, and networking connections, and often are sponsored by private or public entities, such as colleges and universities.

Their goal is to help create and grow young businesses by providing them with necessary support and financial and technical services.

Business incubators figure as one of the main mechanisms for the dissemination of innovative activities in organizations [8].

The business incubators provide support to for the development of new firms providing managerial, legal, technological support, so that they are strengthened to develop innovations and manage their business to act in the market [9].

In this sense, the environment of an incubator is a relevant habitat for the nascent companies, considering the synergy created by the concentration of entrepreneurs who have the goal of business success.

Among the advantages offered by business incubators are: (i) physical space for installation of the company; (ii) human resources and specialized services to assist in the management of business, technological innovation, marketing, marketing, fundraising, intellectual property, among others; (iii) training and training of entrepreneurs; and (iv) access to university laboratories and libraries [9].

Specifically on the role of technology-based incubators for regional development, the study produced by Anprotec [10] indicates that incubators have the potential to contribute to the generation of product, process and service technologies in an integrated way with the actors scientific, technological and local innovations, as well as the possibility of contributing to the increase of employment and income generation, creation of new businesses with technological content and articulation with the public sector and other institutions of a given region.

4   The Experience of UFABC (Federal University of ABC) in Brazil: The Technological Incubator Established in Early 2017

The Federal University of ABC (UFABC) was created in 2005 from the perception of the importance of the interaction of the various areas of knowledge, focusing on the reflection and attention to regional and national issues, through cooperation with the public sphere, with other teaching and research institutions and with the productive sector. Today, just eight years later, UFABC has been standing out as an institution of excellence in teaching and research activities, and is among the Brazilian institutions best evaluated in national and international university rankings.

In order to foster technological and innovative development, since 2010, UFABC has a core of technological innovation (NIT), whose mission is to stimulate innovative activity. In 2013, UFABC added to NIT's mission to stimulate technological entrepreneurship, thus transforming NIT into an Innovation Agency - more robust and diversified in its missions related to intellectual property, technology transfer and technological entrepreneurship.

With the commitment to stimulate the generation of innovations and build for the business, technological and economic development of the region of Greater ABC, the Innovation Agency of Federal University of ABC created the UFABC-IT UFABC Technological Incubator.

Launched in early 2017, the initiative provides support and accompaniment to entrepreneurs linked to research and development (R&D) activities and innovative new business ideas developed by the UFABC community and society in general.

The first incubation cycle of IT UFABC, which was offered in 2017, includes a non-resident incubation modality, in which the entrepreneur does not physically settle in the university, but uses the services offered. The term of stay of the agent in the incubation in this modality is of 6 months, renewable for less or equal period, limited to one year.

To the best of authors´ knowledge, it is an incubation model that has not been explored in literature and the results still need to be better analyzed and validated.

At the beginning of the process, incubators are trained and oriented to develop some monthly scheduled deliveries, such as (i) canvas of the business model, (ii) market study and elaboration of access strategies, (iii) operational plan; (iv) marketing mix; (v) projection of cash flows, and (vi) consolidation of the business plan and elaboration of the pitch elevators.

At the beginning of 2018, the Innovation Agency launched the cycle of a resident incubation in a coworking environment, which process allows the entrepreneur or company incorporated physically in the dedicated coworking space for the incubator, with a maximum duration of 18 (eighteen) months.

5   Challenges for IT UFABC and Conclusions

The region of the Great ABC, in the metropolitan area of Sao Paulo, the major city in the country passes since the mid-1990s, by an intense process of changes in its industrial park.

To contribute for a jump in the generation of technological content in its industries, it is fundamental that actions such as the creation of IT UFABC are contained in a broader set of public policies aimed at transforming the current industrial base to a more technologically and innovative density.

The challenges in creating entrepreneurship and innovation habitats, such as IT UFABC, range from fostering the process of innovation and articulation with actors in regional systems, alignment with public policy and financial resources, and consolidating the management of the incubator.

Thus, IT UFABC can be characterized as an important development vector in the Greater ABC region to:

- Stimulate startups in the early stages of their lives.
- Develop the entrepreneurial spirit in UFABC and society in general;
- Enable the entrepreneur to use the services of the incubator;
- Empower entrepreneurs in the generation of technological and managerial innovations;
- To stimulate the integration between the entrepreneurs, be it between them or between the partners that support the incubator, seeking the development and the exchange of technology;
- Support and train entrepreneurs by providing mentoring with entrepreneurs, consultants, professors and researchers;
- Spreading the culture of entrepreneurship in society.

Although the local society of the ABC region has received IT UFABC as a solution for the development of new business, many are still the challenges to be faced. We can appoint some of them:

- Developing a closer relationship with the venture capital and private equity community;
- Improve the relationship with the local public power;
- Integrate the incubator with the university's research laboratories.

  As conclusions of the paper, we can mention:

- The concept of entrepreneurial ecosystems, so relevant in the academic literature, has inspired UFABC, local business leaders and the public sector to work together building an entrepreneurial ecosystem in the ABC region in Sao Paulo, Brazil.
- The Innovation Agency of UFABC would continue to research best practices all over the world making its contribution for the development of the region.

References

1. Mian, S., Lamine, W., Fayolle, A.: Technology business incubation: an overview of the state knowledge. Technovation **50–51**, 1–12 (2016)
2. Moran, E.F.: Environmental Social Science. Wiley, Hoboken (2010)
3. Bontillier, S., Carré, D., Leuratto, N.: Entrepreneurial Ecosystems. Wiley, Hoboken (2016)
4. Mason, C., Brown, R.: Entrepreneurial Ecosystems and Growth Oriented Entrepreneurship, The Hague (2014)
5. Isenberg, D.: The entrepreneurship ecosystem strategy as a new paradigm for economy policy: principles for cultivating entrepreneurship, Babson Entrepreneurship Ecosystem Project, Babson College, Babson Park, MA (2011)
6. OCDE: Higher Education Management and Policy. Special Issue: Entrepreneurship, vol. 17, no. 3. OCDE Publishing, Paris (2005)
7. Dowejko, M., Au, K., Shen, N.: Entrepreneurship Ecosystem of Hong Kong (2014)
8. Vedovello, C., Figueiredo, P.N.: Incubadora de inovação: que nova espécie é essa? RAE-eletrônica, **4**(1), 1–18 (2005). Article ID: 10
9. Ceia, A.M.: Avaliação da implantação do modelo de gestão nas incubadoras de base tecnológica fluminenses. Dissertação de mestrado apresentada à Universidade Federal do Rio de Janeiro (2006)
10. Anprotec, Associação Nacional de Entidades Promotoras de Empreendimentos Inovadores: Estudo, Análise e Proposições sobre as Incubadoras de Empresas no Brasil, MCTI, Brasília (2012)

Factors That are Highlighted in the Decision-Making in the Process of Internationalization of Micro and Small Companies of the Cosmetics Segment of the State of São Paulo

Milton Carlos Farina, Maria do Carmo Romeiro,
Alvaro Francisco Fernandes Neto,
Antonio Aparecido de Carvalho[(⊠)], and Davi de França Berne

Universidade de São Caetano do Sul – USCS, São Caetano do Sul, Brazil
{milton.farina, mromeiro, alvaro.fernandes,
antonio.carvalho, davi.berne}@uscs.edu.br

Abstract. This article focuses on the study of micro and small cosmetic companies in the state of São Paulo, due to the representativeness that these companies have in the exports of their products. The objective is to identify the elements that stand out in the decision-making process for the internationalization of its business from the export of its products. Behavioral Theory was used as the internationalization model of McDougall [1]. Based on the research, it was identified that the elements that stand out most in decision making for micro companies are the need for business expansion due to the saturation of the domestic market and the competition, whereas for small companies the highlight is in strategic factors for business maintenance and suggestions from suppliers, customers and consultants.

Keywords: Micro and small enterprises · Internationalization
Cosmetics sector

1 Introduction

According to Brazilian Service of Support to Micro and Small Companies - Sebrae [2] from the 1980s, Micro and Small Companies were gaining ground in Brazil and over time were gaining representativeness. Once companies have established themselves, many seek new markets, they strive to win international clients, but Rasmussem and Madson [3] emphasize the need for prior planning, knowledge of the market in which to participate, as well as legislation requirements as well as possible tariff and non-tariff barriers.

Research from Sebrae [4] show that micro and small companies in the state of São Paulo represent 98% of companies, account for 49% of jobs and 37% of payroll, compared to 27% of the state's GDP. The survival rate is 76.3%, only 1 in 5 companies

© Springer International Publishing AG, part of Springer Nature 2019
J. Machado et al. (Eds.): HELIX 2018, LNEE 505, pp. 1018–1023, 2019.
https://doi.org/10.1007/978-3-319-91334-6_140

closes their activities before 2 years. The highest survival rate per sector is 81.4%, followed by construction (80.5%), commerce (76.3%) and services (74.1%).

Cosmetics companies operating in Brazil, have shown growth in exports, research from Brazilian Association of Personal Hygiene, Perfumery and Cosmetics – ABHIPEC [5] show that from 2006 to 2016 exports grew more than 293% in the period. In this sense ABHIPEC [5] reported that cosmetic manufacturing of micro and small companies show growth in the international market, exports in the first quarter of 2017 compared to the same period of 2016 increased by 0.7%. Note that growth rates were high, despite the decline due to the recession the sector continues to grow.

This article aims to answer the following question: What are the factors that stand out in the decision-making process for micro and small companies in the state of São Paulo in the cosmetics sector to seek internationalization? The objective is to identify the factors that stand out so that the micro and small companies of the cosmetics sector of the state of São Paulo make the decision to seek internationalization and also to determine if the factors are the same for micro and small companies.

The choice of theme is justified because of the relevance that micro and small companies bring to the economic scenario, either through the generation of employment and income or in the participation in the Brazilian GDP and by the growth that the cosmetics sector has presented and also by the representativeness of the exports of micro and small companies in the cosmetics sector.

2   Theoretical Framework

According to Vale, Wilkinson and Amâncio [6] when a entrepreneur aims to create a business, taking advantage of a gap in the market. Dornelas [7] argues that it is the one who makes things happen, anticipates the facts and has a future vision of the organization. For Vries [8], it is the one who makes decisions, assumes responsibilities and is subject to risks. Aguiar [9] describes the undertaking as the person who accomplishes things always in different ways, a fact that brings satisfaction in the execution of the tasks. Schumpeter [10] makes the association of entrepreneurship and innovation and both terms go together. Ribeiro and Teixeira [11] bring the idea that the entrepreneur has the ability to combine all productive resources in the making of products that can bring satisfaction to users.

Vieira (2004) states that micro and small enterprises are a source of jobs, especially for people with low qualifications, contributing to the generation of jobs and income. Complementary Law 123/2006, known as the General Law of Micro and Small Companies, divides small businesses into four categories and prioritizes tax benefits. The categories are as follows: Individual microentrepreneur (MEI) with annual gross revenues up to R$ 60,000.00; Microenterprise (ME) with annual gross revenues up to R$ 360,000.00; Small Business Company (EPP) with annual gross sales between R$ 360,000.00 and R$ 3,600,000.00 and Small Rural Producer (PPR) with gross annual sales up to R $ 3,600,000.00. Sebrae [4] reported that small businesses currently number 14.812.460 companies, the participation of micro and enterprises of the State of São Paulo in 2014 was: services 41%, commerce 37%, industry 12%, construction 7% and agriculture 3%.

According to Andrade et al. [12], internationalization consists in the insertion of the organization or the product in the international market. The company needs to know the market in which it intends to act and the strategies must be aligned with all those involved in the process, especially the market entry strategy. The present article opted for the study of Behavioral Theories, adopting the McDougall International Entrepreneurship model of 1989 that brings the combination of innovative behavior, seeking to cross borders with risk minimization. In this approach the entrepreneur must be extremely willing and able to learn new concepts and ideas, be able to adapt to new cultural realities and be able to seek in the international market processes that bring transformation to his company.

Pinho and Martins [13] cite barriers that hinder the process, such as: lack of information and knowledge of the external market, few financial resources, poor qualification of employees, product quality, need to adapt to international requirements, difficulties of direct and indirect export, competition and industrial structure. Cordeiro [14] lists elements that hinder the internationalization of micro or small companies as the lack of an export culture and the failure to develop a technological standard consistent with the market to which it intends to export. Teixeira and Picchai [15] argue that the process of internationalization for small and micro companies faces five factors: the country does not have an export culture, has no knowledge about the external market, too much process bureaucracy, low productive capacity and imposition of international barriers.

Research by ABHIPEC [5] shows that Brazil is the 4th among the top 10 consumers worldwide personal hygiene, perfumery and cosmetics. Regarding the contribution of the sector to the economy, Brazil stands out as the 1st industrial sector that invests more in advertising, it is also the second industrial sector that invests more in innovation. Regarding exports, the survey shows that in 2016, the export value reached US\$ 278.8 million. Sebrae [4] reported that exports in the last ten years showed a growth of 293,5% and that national and international markets are giving preference to ecological cosmetics, whose composition takes natural ingredients and rely on rigorous production processes, with the cosmetics market growing on average at 20% per year.

3   Research Method

With the purpose of identifying the elements that stand out in the process of decision making in the process of internationalization of small and micro companies of the cosmetics sector of the state of São Paulo, a descriptive quantitative research was done. The sample is non-probabilistic, with the use of a questionnaire. The model was based on behavioral theory and adopted McDougall's [1] approach to international entrepreneurship. Ruzzier et al. [16] describe this approach as a combination of innovative behavior, which seeks risk and new markets. Kiss and Danis [17] argue that in environments characterized by change and turmoil, entrepreneurs adopt a survival stance.

Prior to the application of the questionnaire, a pre-test was applied. The questionnaire was sent to 17.000 companies in the cosmetics segment of the State of São Paulo, 410 questionnaires were received, of which 156 are from the industrial branch,

41 are micro enterprises and 115 are small companies. The others 254 are from the commerce and services sector. The survey focused on micro and small companies in the industry. For the data analysis SPSS 23 software was used for the factorial analysis with the use of the 20 variables related to the internationalization process.

The variables selected to identify the most important factors for internationalization are: business expansion needs, growth strategy, high internal competition, Brazilian economic crisis, suppliers suggestion, partners suggestion, customer suggestion, support of some type of association, participation in seminars and fairs, work experience abroad, academic training abroad, foreign market knowledge, technological innovation, information and communication technologies, exchange factors, product quality, product price, innovative product and support agents, based on Behavioral Theory and the McDougall model [1].

4   Discussion of Results

The separate data for the 41 micro and 115 small companies analyzed are presented below.

4.1   Micro Companies Factor Analysis

The KMO test that indicates the correlation between the variables resulted in 0.808, according to Fávero et al. [18] indicates a good correlation between the 20 variables. The anti-image matrix showed numbers between 0.472 and 0.930, so according to Hair (2005) all variables are valid. For the definition of the factors that group the variables was used the Varimax method, which according to Corrar et al. [19], seeks to minimize the number of variables, it was verified that the variables can be explained by only four factors. The variables were grouped into factor 1: Expansion - composed of need for expansion, technological innovation and internal competition; Factor 2 - Strategy, consisting of support from associations, training abroad, participation in fairs and events, growth strategy, suppliers suggestion and work experience abroad; Factor 3 - Third parties, consisting of national economic crisis, suggestion of the members, suggestion of the clients, knowledge of the external market, suggestion of advice and price; Factor 4 - Product by product quality, support from development agents, exchange factors, innovative product and communication and information technology.

4.2   Small Business Factor Analysis

The correlation between the variables presented KMO of 0.941, a fact that indicates high correlation; the anti-image matrix presented results ranging from 0.911 to 0.962, it is concluded that all the variables used are valid. In order to group the variables into factors, the Varimax method was used, the result obtained presents the variables distributed in four distinct factors, denominated: Factor 1 - Strategy, composed of support of associations, will of the partners, expansion of business, growth strategy and competition internal. Factor 2 - Suggestion of Third Parties, consisting of suggestion of advice, suggestion of suppliers, price, customer suggestion and economic crisis; Factor

3 - Experience in the external market, composed of technological innovation, training abroad, work experience abroad and innovative product; Factor 4 - Product, consisting of product quality, information and communication technology, support of development agents, participation in fairs and events, knowledge of the foreign market and exchange factors.

5 Final Considerations

The individualized surveys between micro and small companies showed that the variables according to the respondents of the micro enterprises were grouped into four factors, and the factors considered most relevant were the need to expand the businesses linked to the need for innovation and intense internal competition that makes that micro companies seek other markets, following the strategic question, followed by the suggestions of third parties, the variables of lesser relevance are those linked to the quality of the product to the support of the development agents, the exchange factors, the company presenting an innovative product and information and communication technologies.

Differently for the respondents of the small companies stands out the strategic factor composed of support of the associations, partners' will, business expansion, strategy for growth and high domestic competition, followed by the suggestion of third parties, external market knowledge and variables which are presented with less relevance, composed of product quality, information and communication technology, participation in international fairs and events, knowledge of the foreign market and exchange factors, the variables of lesser relevance are very similar to those of micro enterprises, they are: product quality, communication and information technologies, support from development agents and foreign exchange factors, in addition to participation in international fairs and events and knowledge of the foreign market.

It is understood that both companies considered micro and small companies in the cosmetics industry seek solidification in the domestic market, but need to seek new markets to continue their business, according to Behavioral Theories, which considers the International Entrepreneurship model of McDougall, whose approach is that entrepreneurs need to be extremely willing and able to learn new concepts and ideas, adapt to new cultural realities, and be able to seek processes in the international market that bring transformation to their company.

It is considered that the research reached the objectives outlined, it is suggested that new research be done with other branches of activity and in other States and even if the study be extended to medium-sized companies. It is also suggested that the research seeks to understand the reasons that make the process of internationalization of micro and small enterprises unfeasible.

The limitation found in the research is relative to the size of the sample because it was made non-probabilistically.

References

1. McDougall, P.P.: International versus domestic entrepreneurship: new venture strategic behavior and industry structure. J. Bus. Ventur. **4**(2), 387–400 (1989)
2. Yearbook of Work in Micro and Small Business. http://www.sebrae.com.br/Sebrae/PortalSebrae/Anexos/Anuario-dowork-namicro-e-pequenaempresa-2014.pdf. Accessed 09 Dec 2017
3. Rasmussan, E., Madsen, T.: The born global concept. In: Proceedings of the 28th EIBA Annual Conference, European International Business Academy, Athens (2002)
4. Panorama dos pequenos negócios. https://m.sebrae.com.br/Sebrae/PortalSebrae/UFs/SP/Pesquisas/PanoramadosPequenosNegocios2017.pdf. Accessed 2017
5. Associação das Industrias de Higiene Pessoal, Perfumaria e Cosméticos - Panorama do setor 2017. https://abihpec.org.br/publicacao/panorama-do-setor-2017/. Accessed 25 Aug 2017
6. Vale, G.V., Wilkison, J., Amancio, R.: Entrepreneurship, innovation and networks: a new approach. RAE Electron. **7**(1), 1–16 (2008)
7. Dornelas, J.C.A.: Empreendedorismo: transformando idéias em negócios. Campus, Rio de Janeiro (2001)
8. Vries, M.F.R.K.: The entrepreneural personality: a person at the crossroads. J. Manag. Stud. **14**(1), 34–57 (1977)
9. Aguiar, M.A.: Psicologia aplicada à administração: uma introdução à Psicologia organizacional. Atlas, São Paulo (1981)
10. Schumpeter, J.A.: Teoria do desenvolvimento econômico: uma investigação sobre lucros, capital, crédito, juro e o ciclo econômico. Abril Cultural, São Paulo (1982)
11. Ribeiro, R.A.M., Teixeira, M.R.C.: From small farmer to successful entrepreneur: "The history of Aguinaldo's pamonharia". In: I National Congress of Entrepreneurship - CONENPRE, p. 10. CONENPRE, Florianópolis (2003)
12. Andrade, M.A.R., de Almeida, P.F., Freitas, L.F.C.: Internationalization as a competitive strategy for small and medium enterprises in Brazil: a bibliographic review. In: SEGeT XI Symposium on Excellence in Management and Technology, pp. 1–12 (2014)
13. Pinho, J.C., Martins, L.: Exporting barriers: insights from Portuguese small - and medium-sized exporters and non-exporters. J. Int. Entrepreneurship **8**, 254–272 (2010)
14. O processo de internacionalização de pequenas e médias empresas de perfumaria, cosméticos e higiene pessoal do estado do Paraná. last accessed
15. Teixeira, M.J., Picchiai, D.: Analysis of the internationalization process of micro and small companies in Campinas/SP, in the light of behavioral theories. Euronet Mag. **1**, 1–16 (2015)
16. Ruzzier, M., Hisrich, R.D., Antoncic, B.: SME internationalization research: past, present, and future. J. Small Bus. Enterpr. Dev. **13**(4), 476–497 (2006)
17. Kiss, A.N., Danis, W.M.: Social networks and speed of new venture internationalization during institutional transition: a conceptual model. J. Int. Entrepreneurship **8**, 273–287 (2010)
18. Fávero, L.P., Belfiore, P., Silva, F.L., Chan, B.L.: Análise de dados: modelagem multivariada para tomada de decisões. Elsevier, Rio de Janeiro (2009)
19. Corrar, L.J., Paulo, E., Filho, J.M.D.: Análise multivariada: para os cursos de administração, ciências contábeis e economia. Atlas, São Paulo (2007)

The Speed of Internationalization of SME's in Peripheral Regions of Europe: A Cognitive Approach

Carina Silva[1,2,3(✉)], Miguel González-Loureiro[2,3], and Vítor Braga[1,2]

[1] School of Technology and Management (ESTG),
Polytechnic Institute of Porto (IPP), Felgueiras, Porto, Portugal
{ccs,vbraga}@estg.ipp.pt
[2] Center for Research and Innovation in Business Sciences
and Information Systems (CIICESI), Felgueiras, Portugal
mloureiro@uvigo.es
[3] University of Vigo, Vigo, Spain

Abstract. Currently, the pressure to internationalize is big. Selling outside the domestic market is a relevant goal for many Small and Medium Enterprises (SMEs). Being the decision makers' attitudes and preferences the center of the internationalization activities of SMEs, understanding how the management team's cognitive system and organizational ambidexterity determine the speed of internationalization, is particularly important. We propose, therefore, to contribute to a better understanding of the factors that influence the speed of internationalization of SMEs.

Keywords: Speed, internationalization, cognitive system
Ambidexterity organizational · Small and medium-sized enterprises

1 Introduction

The speed of internationalization has been the most important dimension in the research on the internationalization of companies [1] gaining, in the last decade, a central position [2, 3]. Given that rapid international growth can compromise the scarce resources available to SMEs and consequently their performance, the speed of internationalization is the key to internationalization strategies [4].

The cognitive system is a factor with a strong influence in international activities [5], since information processing and decision-making depend on two cognitive systems. In addition, the literature indicates that the company should be able to compete in the short term (resource exploitation activities and current knowledge) while designs how it will compete in the future (exploration of new markets, resources and knowledge). This ambidexterity was not frequently considered in the study of the speed of internationalization.

The literature that relates the cognitive system and the international behavior of companies is scarce [5, 6] and nonexistent when it is intended to evaluate the influence of the cognitive system on the speed of internationalization of companies.

© Springer International Publishing AG, part of Springer Nature 2019
J. Machado et al. (Eds.): HELIX 2018, LNEE 505, pp. 1024–1030, 2019.
https://doi.org/10.1007/978-3-319-91334-6_141

The aim of this study is to analyze whether the manager's cognitive system and the organizational ambidexterity affect the speed of internationalization of SMEs. Managers with highly developed cognitive systems are expected to be able to learn more quickly in the process of accumulating international experience, which will accelerate the internationalization process. In addition, these managers will also influence the company's ability to have a high organizational ambidexterity, thus being faster and more successful in the internationalization process (exploiting what the company already has and exploring new opportunities in the international market). We will use a survey of a sample of SMEs from the Euro-region Galicia and Northern Portugal.

The rest of this paper is structured as follows: the next section presents a review of the literature. Section three describes the methodology used. In the fourth section, the results are analyzed, and the last section presents the study's conclusions, implications and limitations, as well as suggestions for future lines of research.

2   Literature Review

2.1   Speed of Internationalization: Concept and Determinants

In spite of its increasing research, the speed of internationalization shows a lack of conceptual clarity and of measures [7], being approached according to several perspectives [8]. Regarding the measurement, it became quite common to differentiate between the initial input speed (earliness) and the post-input speed (post-internationalization speed) [1, 9]. In addition to the fact that in most studies age and time of internationalization were measured in the same way [7], after we analyzing the empirical studies that intend to explain the speed of internationalization, we verified that speed was operationalized and measured differently by several authors, considering several approaches.

Some of the most relevant studies in this area of research measure the speed of internationalization like (1) the difference between the year of constitution the company and the year of the first export [10]; (2) number of days between the launch of new country-specific sites [11]; (3) elapsed time between entries in new markets [12]; (4) the average speed at which the company has expanded its sales or points of sale internationally [7, 13]; (5) ratio of international sales and total sales [14]; (6) relation to international sales and total sales, divided by the time [9]; (7) age of entry into the international market [3]; (8) number of foreign direct investments per year [15]; (9) time needed for the company to reach a percentage of 20% of total sales in the international market [16]. Moreover, it was possible to verify that most of the empirical studies use one-dimensional measures to measure the speed of internationalization. However, there are authors who have combined some of the above measures [17, 18]. In fact, both conceptual studies [2], and empirical studies [18] have suggested that the speed of internationalization is a complex and multi-dimensional concept.

2.2   Relation Between Cognitive System and Speed of Internationalization

According to Zhou and Wu [19] the speed of internationalization results essentially from the entrepreneurial orientation of the manager. Knowing that individuals are the central element in any strategic decision making in companies [6], there is an increasing interest in the study of the different competences that these should have for the creation of new businesses [20].

Several authors recognize that the cognitive traits of individuals promote the recognition of opportunities and the selection of the way through the way they process the information [3]. This processing involves the interaction of two cognitive systems - System -X or experiential and System-C or rational [21].

The literature shows that experiential managers are characterized by greater creativity and capacity to solve problems [6]. As such, they will be able to detect opportunities more readily than rational managers, because they needing detailed analysis, which delays decision-making. However, the literature shows that there is a third type of management style - integrated, which uses the analytical and experiential decision according to the context [22]. Therefore, the use of cognitive systems will be related to the speed of internationalization.

2.3   Relation Between Organizational Ambidexterity and Speed of Internationalization

'Exploration' and 'Exploitation' are two major organizational orientations that affect learning processes at different levels [23].

It is common to associate 'exploitation' with the appreciation of existing potentials and the 'exploration' to the discovery of new possibilities [24]. Thus, we can conclude that 'exploitation' activities seek to increase the efficiency and utilization of current resources and capacities, and 'exploration' activities allow the discovery of new opportunities outside the organization, in order to be successful in the long term [25], including especially opportunities in international markets [26]. An 'exploration' orientation leads to the development of new products and/or the search for new markets and an 'exploration' orientation is intended to respond to current environmental conditions, adapting existing technologies and focusing on current customers [25, 26].

There are several authors who argue that the long-term success of companies requires an 'ambidexterity' orientation [27, 28]. Ambidextrous companies simply consider 'exploration' and 'exploitation' activities [27], that presenting as two distinct but complementary perspectives [28]. Therefore, the more ambitious the organization, the more capacity it will have to exploiting the current resources and exploring new international markets successfully, thus helping to accelerate the internationalization process.

3   Methodology

3.1   Data and Sample

The research analysis unit are the international SMEs from the Northern Euro-region of Portugal/Galicia. In order to select the companies, we will start from the SABI (Iberian

Balance Analysis System) database that contains credible and current quantitative information from Portuguese and Spanish companies. As an exploratory study, we focus on traditional manufacturing sectors, footwear and textile, because they are heavily internationalized sectors (NACE códigos 14 e 15).

In addition to the quantitative data available in a secondary source statistical database, we will send a questionnaire to the selected SME management team.

3.2 Measures

Target Variable: Speed of Internationalization. According to several contributions from the literature [18, 29] that consider the speed of internationalization as the time required for a company to reach a certain degree of internationalization, we propose to adopt this multidimensional metric. Thus, we measure the degree of internationalization of companies following the suggestion of Kowalik *et al.* [17] which considers the internationalization speed according to 3 dimensions (speed, scale and scope).

Antecedents: Use of Cognitive Systems (STSS Scale) and International Organizational Ambidexterity (IOA). To measure the use of the *cognitive system* we adopted the scale *Situation-Specific Thinking Style* (SSTS) of Novak and Hoffman [30], where respondents are invited to assess how the team responsible for internationalization activities tend to making decisions. The STSS scale measures the use of cognitive systems with 10 items for the experiential system (System X) - and 10 for the rational system (System C). It is a 5-point Likert scale ranging from 1 (I disagree at all) to 5 (I agree entirely). To measure the *International Organizational Ambidexterity (IOA) we adopted the Lubatkin et al.* scale [25], where respondents are invited to evaluate the company's OA for the past 3 years. It considers 6 items for 'explorative' orientation and 6 for 'exploitative' orientation through a 5-point Likert scale ranging from 1 (disagree at all) to 5 (fully agree). This scale is explicitly referred to the orientation of companies in the combination of company's markets (domestic and international).

Control Variables. The present study controls the following factors: age, qualifications, professional experience, international experience, institutional support and networking.

4 Analysis of Results

Based on the literature review, we constructed a model (cf. Fig. 1) to test whether the Cognitive System of the management team and OA affect the speed of internationalization of SMEs in peripheral regions.

Depending on the acceptance of hypotheses, the model can be purely or partially mediated. In the first case, the impact of cognitive systems on the speed of internationalization will be mediated by the OA (no direct impact of cognitive systems on speed). The second case would entail a direct effect of cognitive systems on speed plus an indirect effect through OA.

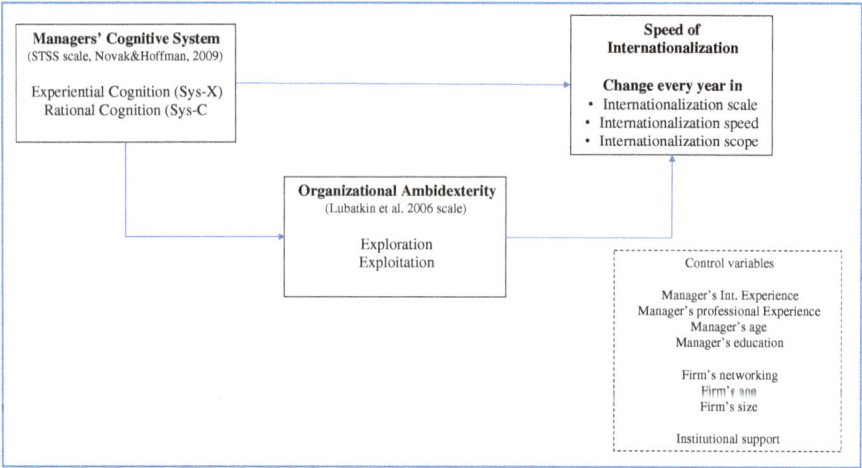

Fig. 1. Research model of the study

The results highlight the positive impact that an ambidextrous organization may obtain from this capability when it comes to speed up the process of internationalization. In addition, experiential knowledge stemming from system-X will also entail the possibility of accelerating this process since expertise-based intuition implies a higher ability to detect new opportunities in the international market. Yet this would be useless unless the manager be able to use that combined with the analytical cognitive system in order to make better-informed decisions.

5   Conclusion

This paper has sought to shed light on the micro-foundations of the speed of internationalization by including the manager's cognitive system in the equation. In addition, we dig deeper in whether an ambidextrous organization can speed up its process of internationalization after the first entry.

This study is limited in scope. First, we only have tested this among a sample of manufacturing firms from a traditional industry. Further analysis should be conducted in other industries to validate whether this impact is similar or the extent to which cognitive systems and organizational ambidexterity may be an industry-specific feature. Second, we have checked these hypotheses among SMEs located in a peripheral region of the EU. Firms located more proximal to the centre of this common market may need a different combination of cognitive systems and explorative-exploitative activities. Accordingly, this research will benefit from enlarging the geographical scope of the sample.

The theoretical implications imply that the firm's speed of internationalization is influenced by both managerial cognition and the organizational capability to compete for today (exploitation) while preparing to compete for the future (exploration). Accordingly, a portion of the explanation of the emergence of accelerated patterns of internationalization are rooted in both managerial and organizational dimensions. Or to put it differently, the manager's ability in terms of entrepreneurship and the organization's ability to perform exploitative and explorative activities.

Practical implications relate to the idea of emergent and deliberate strategies. While some firms may go international to seize opportunities in an unplanned manner, they do must devote time and resources to plan the exploitative and explorative activities required for competing internationally. Furthermore, they must learn to cope with the trade-off between allocating resources to exploit extant resources and capabilities and to explore new opportunities outside the firm and domestic markets. In addition, it seems that going internationally at a quick pace may require a certain type of managerial cognition: a combination of experiential and analytic cognition. In summary, to get a quick speed of internationalization firms seem to need certain combination of individual and organizational capabilities.

References

1. Prashantham, S., Young, S.: Post-entry speed of international new ventures. Entrep. Theory Pract. **35**(2), 275–292 (2011)
2. Casillas, J.C., Acedo, F.J.: Speed in the internationalization process of the firm. Int. J. Manag. Rev. **15**(1), 15–29 (2013)
3. Acedo, F.J., Jones, M.V.: Speed of internationalization and entrepreneurial cognition: Insights and a comparison between international new ventures, exporters and domestic firms. J. World Bus. **42**(3), 236–252 (2007)
4. Chetty, S., Johanson, M., Martín, O.M.: Speed of internationalization: conceptualization, measurement and validation. J. World Bus. **49**(4), 633–650 (2014)
5. Maitland, E., Sammartino, A.: Managerial cognition and internationalization. J. Int. Bus. Stud. **46**(7), 733–760 (2015)
6. González-Loureiro, M., Vlačić, B.: International business decisions and manager's cognitive style: opening up research avenues from cognitive behavioural strategy. Gestão e Soc. **10** (27), 1522 (2016)
7. Hilmersson, M., Johanson, M., Lundberg, H., Papaioannou, S.: Time, temporality, and internationalization: the relationship among point in time of, time to, and speed of international expansion. J. Int. Mark. **25**(1), 22–45 (2017)
8. Jørgensen, E.J.B.: Internationalisation patterns of border firms: speed and embeddedness perspectives. Int. Mark. Rev. **31**(4), 438–458 (2014)
9. Autio, E., Sapienza, H.J., Almeida, J.G.: Effects of age at entry, knowledge, intensity, and immutability on international growth. Acad. Manag. J. **43**(5), 909 (2000)
10. Musteen, M., Francis, J., Datta, D.K.: The influence of international networks on internationalization speed and performance: a study of czech SMEs. J. World Bus. **45**(3), 197–205 (2010)
11. Schu, M., Morschett, D., Swoboda, B.: Internationalization speed of online retailers: a resource-based perspective on the influence factors. Manag. Int. Rev. **56**(5), 733–757 (2016)
12. Hutzschenreuter, T., Kleindienst, I., Guenther, C., Hammes, M.: Speed of internationalization of new business units: the impact of direct and indirect learning. Manag. Int. Rev. **56**(6), 849–878 (2016)
13. Mohr, A., Batsakis, G.: Intangible assets, international experience and the internationalisation speed of retailers. Int. Mark. Rev. **31**(6), 601–620 (2014)

14. Hauser, C., Moog, P., Werner, A.: Internationalisation in new ventures – what role do team dynamics play? Int. J. Entrep. Small Bus. **15**(1), 23–38 (2012)
15. Nadolska, A., Barkema, H.G.: Learning to internationalise: the pace and success of foreign acquisitions. J. Int. Bus. Stud. **38**(7), 1170–1186 (2007)
16. Zhou, L.: The effects of entrepreneurial proclivity and foreign market knowledge on early internationalization. J. World Bus. **42**(3), 281–293 (2007)
17. Kowalik, I., Danik, L., Král, P., Řezanková, H.: Antecedents of accelerated international-isation of polish and czech small and medium-sized enterprises. Entrep. Bus. Econ. Rev. **5**(3), 31–48 (2017)
18. Pla-Barber, J., Escribá-Esteve, A.: Accelerated internationalisation: evidence from a late investor country. Int. Mark. Rev. **23**(3), 255–278 (2006)
19. Zhou, L., Wu, A.: Earliness of internationalization and performance outcomes: exploring the moderating effects of venture age and international commitment. J. World Bus. **49**(1), 132–142 (2014)
20. Kickul, J., Gundry, L.K., Whitcanack, L.: Intuition versus analysis? testing differential models of cognitive style on entrepreneurial self-efficacy and the new venture creation process. Entrep. Theory Pract. **212**, 439–453 (2009)
21. Kahneman, D., Frederick, S.: Representativeness revisited: attribute substitution in intuitive judgment. In: Gilovich, T., Griffin, D., Kahneman, D. (eds.) Heuristics and Biases, pp. 49–81. Cambridge University Press, Cambridge (2002)
22. Agor, W.H.: Using intuition to manage organizations in the future. Bus. Horiz. **27**(4), 49–54 (1984)
23. Levitt, B., March, J.G.: Organizational Learning. Ann. Rev. Soc. **14**(1), 319–338 (1988)
24. Stephan, M., Kerber, W.: "Ambidextrie": Der unternehmerische Drahtseilakt zwischen Ressourcenexploration und -exploitation ('Ambidexterity': Keeping the Balance between Resource Exploration and Exploitation), 1st edn (2010)
25. Lubatkin, M.H., Simsek, Z., Ling, Y., Veiga, J.F.: Ambidexterity and performance in small-to medium-sized firms: the pivotal role of top management team behavioral integration. J. Manag. **32**(5), 646–672 (2006)
26. Prange, C., Verdier, S.: Dynamic capabilities, internationalization processes and perfor-mance. J. World Bus. **46**(1), 126–133 (2011)
27. Raisch, S., Birkinshaw, J.: Organizational ambidexterity: antecedents, outcomes, and moderators. J. Manag. **34**(3), 375–409 (2008)
28. Voss, G.B., Voss, Z.G.: Strategic ambidexterity in small and medium-sized enterprises: implementing exploration and exploitation in product and market domains. Organ. Sci. **24**(5), 1459–1477 (2013)
29. Hilmersson, M., Johanson, M.: Speed of SME internationalization and performance. Manag. Int. Rev. **56**(1), 67–94 (2016)
30. Novak, T.P., Hoffman, D.L.: The fit of thinking style and situation: new measures of situation-specific experiential and rational cognition. J. Consum. Res. **36**(1), 56–72 (2009)

The Role of a Business Incubator in Supporting the Needs of Innovative Business

Paulo Roberto Silveira Machado[(⊠)], Igor Ceratti Treptow,
Marta de Oliveira Roveder, Roberto Schoproni Bichueti,
and Nilza Zampieri

Federal University of Santa Maria, Santa Maria, Brazil
paulorosm@gmail.com, iceratti@hotmail.com

Abstract. Business incubators present themselves as an alternative to small and medium-sized enterprises, as they provide basic support for the assistance of these enterprises during their initial period of activities. This study has as its main objective to analyze the role of a technological incubator in the formation process of incubated companies. In order to do that, an exploratory case study of qualitative approach was carried out through semi-structured interviews with four incubated companies formed with the coordination of the analyzed incubator. The results show that the incubator has a central role in the development of enterprises and small businesses in their initial phase, besides it contributing to the regional development.

Keywords: Triple Helix · Incubators · Innovation

1 Introduction

Incubators are intended to support entrepreneurs, such as infrastructure and managerial support [1]. According to Fonseca [2], they play an important role in promoting interaction among public and private entities, stimulating cooperation amongst universities, private companies, and public authorities. The popularity of incubators suggests that they are effective tools to support new enterprises [3].

According to ANPROTEC (Associação Nacional de Entidades Promotoras de Empreendimentos Inovadores) [1], Brazil has 369 incubators, 2,310 incubated enterprises, and 2,815 formed companies, which generates 53,280 jobs and revenues that exceed BRL 15 billion. The technological bias is present in 67% of enterprises [2].

In this context, this study aims to analyze the role of an incubator in the formation process of enterprises. An exploratory case study of a qualitative approach was carried out through documentary research and semi-structured interviews with formed companies and with the coordinator of a technological incubator.

2 Theoretical Framework

According to Lima and Santos [4], universities play a crucial role in the incubation process, since they are the main inducers of this innovation mechanism, as they are part of the innovation ecosystem and key actors in the creation/processing of knowledge explored and traded in the market [5].

© Springer International Publishing AG, part of Springer Nature 2019
J. Machado et al. (Eds.): HELIX 2018, LNEE 505, pp. 1031–1035, 2019.
https://doi.org/10.1007/978-3-319-91334-6_142

This understanding is in line with the Triple Helix theory, which considers innovation to be the result of a complex and continuous process of relations among science, technology, research and development in universities, industries, and government [6].

Enterprises approved to enter incubators usually have an idea that is transformed into a project. However, they have few resources and low degree of information about administrative processes [7].

In this context, Weele et al. [3] list tangible and intangible resources that incubators should offer, such resources can be seen in Table 1.

Table 1. Startup resource needs and incubator support.

Resource needs	Incubator support to fulfill resource needs
Physical capital	Office space
	University equipment and library
Financial capital	Seed capital in Exchange for equity
	Access to investors
Knowledge	Provide technological knowledge through the proximity to university groups and laboratories
	Provide business knowledge through coaching and training
Social capital	Facilitate the creation of external networks by organizing events, creating partnerships and making introductions
	Facilitate the creation of a community through co-location, social events and introductions
Legitimacy	Association with an established incubator

3   Method

For the purposes of this study, we chose to use the strategy of a case study [8]. This research has a qualitative approach [9] and an exploratory nature [10].

Data were collected through semi-structured interviews [11], which were adapted from Weele et al. [3]. The interviews were carried out with the entrepreneurs and the coordinator of the studied incubator. In order to complement the data collection, a documentary search of secondary sources was done through websites, social networks and scientific papers [10].

4   Analysis and Discussion of Results

The Santa Maria Technological Incubator (ITSM) is connected to the Federal University of Santa Maria, located in the city of Santa Maria in the state of Rio Grande do Sul - Brazil. It has been operating since 1999 with the purpose of supporting new entrepreneurs and transforming ideas into businesses, seeking to contribute to the formation of an innovative entrepreneurial mindset.

Enterprise A was founded in 2010, its main product consists of a collar that remotely monitors herds of cattle. In 2017, it was the first enterprise to receive funding from the largest investment fund in Latin America. With the beginning of its activities in 2007, enterprise B has as its product virtual reality games, being a reference in the field and partnering with major internationally renowned companies such as Samsung. Enterprise C remained at ITSM from 2003 to 2007 and it develops personalized solutions for information systems in a variety of market segments. Enterprise D began its activities in October 2011, and it offers an online food delivery service. Still incubated, it began the process of franchising for entrepreneurs in other cities.

4.1 Entrance in the Incubator

This paper suggests that the enterprises that participated in this study sought the technological incubator for different reasons, although all of them pointed to the physical structure and networking as being fundamental for their entry.

None of the interviewees mentioned difficulties in participating in the selection process for entering the incubator besides the protocol to be followed by the entrance notice. Moreover, the enterprises were in the pre-incubation period when they sought for the entrance to ITSM.

All entrepreneurs had a connection with the university to which the incubator is connected so that enterprise A and C were results of re-search developed within the scope of The Federal University of Santa Maria.

4.2 Incubator Support to Incubated Companies

The incubator provides the physical support necessary for the enterprises to carry out their activities, except the internet access, which should be improved, according to an interviewee. Providing adequate infrastructure leads to savings and lower costs for investment in core activities [12]. Additionally, an environment for carrying out the enterprises' activities was pointed out by the interviewees as a motivational factor.

In the regard of financial capital, results show that the incubator only provides guidelines for the participation of public notices. Additionally, results also show that the ITSM operation can improve in this aspect, since financial constraints are the most critical obstacle for small enterprises in early stages, as many of them close due to lack of capital [13].

Within its possibilities, the incubator offers the training described by Peters et al. [14] such as training, educational workshops, seminars and/or programs, and others. One of the interviewees suggested the implementation of a mentoring process, this suggestion can be made without cost with voluntary participation entrepreneurs formed by the incubator itself.

The environment provided important networking for all the enterprises that participated in this study. Bruneel et al. [15] point out that the networking offered by this type of environment facilitates access to technology, and to professionals and financial agents that otherwise would not be within the reach of new enterprises.

According to the interviewees, to be associated with the incubator brings legitimacy at the beginning of the enterprises due to the recognition of the importance of the

Federal University of Santa Maria. However, when the enterprises acquire maturity, it becomes a negative aspect, as they are seen as still immature enterprises. According to Lai and Lin [16], most incubated enterprises believe that the services and offered resources only give assistance to overcome initial barriers, similarly, legitimacy as an intangible resource, presents itself as a positive factor. However, in the final stage of incubation, it brings discredit to the business.

5   Final Considerations

In view of the results obtained, it is considered that the Santa Maria Technological Incubator (ITSM) fulfills its objective of providing assistance to entrepreneurs and small enterprises in their initial stage. Additionally, it was possible to verify that the incubator contributes to the regional development by fomenting the generation of jobs in local enterprises, retaining intellectual capital. In the regard to the resources offered, considering the financial and knowledge constraint experienced by the incubator, it is considered that the administrative body seeks to meet the demands of the enterprises.

The limitations of the study are the limited number of enterprises. As a suggestion for future studies, the total number of enterprises assisted by the incubator should be investigated.

References

1. Anprotec (2017). http://anprotec.org.br/site/menu/incubadoras-e-parques/
2. Fonseca, M.L.M.: Análise das incubadoras de empresas de base tecnológica como promotora do desenvolvimento regional brasileiro: uma abordagem teórica. In: XXIV Seminário nacional de Parques Tecnológicas e Incubadoras de Empresas. 22–26 set (2014)
3. Weele, M., Rijnsoever, F.J., Nauta, F.: You can't always get what you want: how entrepreneur's perceived resource needs affect the incubator's assertiveness. Technovation. **59**, 18–33 (2017)
4. Lima, F.V.R., Santos, J.A.B.: Mapeamento dos Bens de Propriedade Intelectual em Empresas de Base Tecnológica Vinculadas a Incubadoras. Revista FSA. **25**(4), 3–31 (2017)
5. Redondo, M., Camarero, C.: Dominant logics and the manager's role in university business incubators. J. Bus. Ind. Mark. **32**(2), 282–294 (2017)
6. Valente, L.: Hélice tríplice: metáfora dos anos 90 descreve bem o mais sustentável modelo de sistema de inovação. Conhecimento & Inovação, Campinas, vol. 6, no. 1 (2010)
7. Vivaldini, M., Soriano, J.E.: Processos de negócios na cadeia de suprimentos: um estudo de caso em incubadoras de empresas. Raimed – Revista de Administração IMED, vol. 4, no. 3, ago/dez, pp. 286–299 (2014)
8. Yin, R.K.: Estudo de caso: planejamento e métodos, 3rd edn. Bookman, Porto Alegre (2010)
9. Minayo, D.S.: Pesquisa social: Teoria, método e criatividade, 28th edn. Rj: Vozes, Petropolis (2010)
10. Gil, A.C.: Como Elaborar Projetos de Pesquisa, 5ª edn. Atlas, São Paulo (2010)
11. Prodanov, C.C.E., Freitas, C.F.: Metodologia do trabalho científico: métodos e técnicas da pesquisa e do trabalho acadêmico, 2nd edn. Feevale, Novo Hamburgo (2013)
12. Mrkajic, B.: Business incubation models and institutionally void environments. Technovation **68**, 44–55 (2017)

13. Hong, J., Jinfeng, L.: Assessing the effectiveness of business incubators in fostering SMEs: evidence from China. Int. J. Entrepreneurship Innov. Manag. **20**(1/2), 45–60 (2016)
14. Peters, L., Rice, M., Sundarajan, M.: The role of incubators in the entrepreneurial process. J. Technol. Transfer **29**, 83–91 (2014)
15. Bruneel, J., Ratinho, T., Clarysse, B., Groen, A.: The evolution of business incubators: Comparing demand and supply of business incubation services across different incubator generation. Technovation **32**, 110–121 (2012)
16. Lai, W.H., Lin, C.C.: Constructing business incubation service capabilities for tenants at post-entrepreneurial phase. J. Bus. Res. **68**, 2285–2289 (2015)

Challenges of Venture Capital Managers to Invest in Biotech Companies in Brazil

Reinaldo Igarashi, João Henrique de Souza Pereira[(⊠)],
Sérgio Kannebley Júnior, and Geciane Silveira Porto

School of Economics, Business Administration and Accounting,
University of São Paulo, Ribeirão Preto, São Paulo, Brazil
{joaohs,skj,geciane}@usp.br

Abstract. Venture capital is an important source of fundraising for technology based-firms by providing temporary financial and management support in exchanges for equity. In Brazil, the venture capital market is under explored compared to mature ecosystems, resulting in restriction of risk investments. The contribution of this research is the better understanding of the difficulties of the relation between venture capital managers and entrepreneurs of biotech companies in Brazil. This can help to improve the relationship between these ones and, consequently, to benefit the future development of the biotech sector in habitats like the one studied. The method includes a collection of primary data through interviews with venture capital managers. It was concluded that the difficulties are present in the venture capital activities of fundraising and investments, that are influenced by the limited culture of venture capital investments in Brazil and the economic variables, as the interest rate, by its impact in the investors cost of capital. Also, was identified that, among several criteria investigated, the team characteristics was the one of greater value for most of the venture capital fund managers analyzed.

Keywords: Venture capital · Biotechnology · Technology based firms

1 Introduction

Private equity (PE) and venture capital (VC) refer to the capital contribution made by investment funds in companies with high growth potential in exchange for participation in the corporate society. The PE is related to investments in companies in advanced stages, while the VC in the early stages. Fundraising in the early stages of development is a critical factor for the growth and survival of start-up companies, especially in technology based enterprises (TBE) [1].

The companies invested by VC funds has faster growth, especially when they are active in competitive markets. In the short and long term, the results in profit and growth rate are higher compared to non-invested companies [2]. The invested ones have a higher level of efficiency by the support received from the VC funds [3]. The small Brazilian PE and VC market has a direct impact on these companies, as the financing is important for the technology innovations, with long-term return and great uncertainty [4]. However, in Brazil and other similar environments, there are little

© Springer International Publishing AG, part of Springer Nature 2019
J. Machado et al. (Eds.): HELIX 2018, LNEE 505, pp. 1036–1042, 2019.
https://doi.org/10.1007/978-3-319-91334-6_143

amount of risk investment available and this increases the need for works, such as this one, to try to understand and approximate risk investors and entrepreneurs [5–7].

Despite the importance of VC for TBE, the topic is still few explored in research in Brazil, especially when considering investments in the initial stage of development. At the global level, many studies have ignored the differences in investment decision criteria in this study area [8].

In this way, the aim of this research is to investigate why venture capital investments in biotech companies in Brazil are restricted, with the identification of the challenges in the fundraising and investment activities.

This paper is organized in 4 sections, the first one is this introduction. The Sect. 2 presents the private equity and venture capital characteristics. Section 3 shows the analysis of results in the activities of fundraising and investments. Section 4 presents the final considerations, limitations and suggestions for future work.

2   The Private Equity and Venture Capital

PE and VC investments are made in companies with high potential for growth and profitability, through corporate participation with the objective of obtaining significant capital gains in the medium and long term [9]. It is a temporary capital contribution, made by a private equity or venture capital fund, through the participation in the firm [4].

The resource provided by the PE and VC aims at the growth and maturation of the business in a period, so that, the investor obtains his return on leaving the company. The role of the PE and VC is not restricted to capital resources, it also includes guidance in the management of the enterprise, assisting the founders in the management aspects that in most cases have a technical background without the same managerial and commercial skills of the investor. Management support becomes as important to companies as it is to financial resources [10, 11], consuming approximately 60% of the time of the investment fund managers in portfolio management activities [12].

Despite its importance, the PE and VC market still has a small share of investments in Brazil. It started in the 80s, and only after the stabilization of inflation and the Real Plan (series of measures that introduced the new currency - Real - in Brazil), in 1994, the PE and VC market grown up, in Brazil.

2.1   The VC Market Cycle

After the formation of the investment fund, the management organization of the fund has the activity of raising capital from independent investors whose may be public or private. At this stage, the investor contracts are made with the funds, to define the risks and returns of each player in the investment.

One of the reasons for the low volume of VC investments, in Brazil, may be the high interest rate in the recent years, which discouraged investors from applying their financial resources in high risk projects [4].

For [13], the cultural factor also has influence in the VC market. The lack of expansion of the VC market reflects in the absence of an investment history, making many investors afraid of investing their resources in the VC market, having the traditional investments as the main option.

In the investments activity, the managers of the VC funds carry out the selection process, whose structure uses some criteria such as: characteristics of the team; characteristics of the product or service; characteristics of the target market; financial potential of the company; and geographic distance from the fund managers.

There is a great importance of the characteristics of the founders for the success of the invested companies, such as experience, personality, ability to react to risk, and familiarity with the target market [14]. In cases where the founders do not have such skills, they must be able to assemble a team with these abilities and be able to lead it. Also, experienced entrepreneurs perform better and the ones who have already had companies with VC resources have a greater chance of success than the initial entrepreneurs and the ones who failed initially [15, 16].

Generally, one company does not need a product with great improvement over others in the market, it is enough to have an adequate competitive advantage. However, for biotech companies, the financial criterion seems to be one of the most importance in decision making [8].

For [17], the funds prefer to invest in geographically near firms as a way of minimizing information asymmetry, transaction costs and have gains with local market characteristics such as the ability of researchers, knowledge overflow and network of contacts.

3   Analysis of Results

This research is a multiple case studies in VC funds with investments in biotech companies. Intentional sample was used, in which the participants choice was based on their characteristics, experiences, attitudes and perceptions. For the interviews, was used the semi-structured technique, with specific questions and openness for the exposition of the participant's thinking [18]. Was used the content analysis technique to explore the data [19].

According to [9], there are 27 VC fund managers in Brazil, covering the early, seed and start-up stages. When considering only VC fund managers with interest in the biotech sector, this number reduces to 8. This research analyzed 5 from these 8 funds. The summary of these 5 is described below.

Performa Investments. Founded in 2005, manages two funds for selection of projects. The first, FMIEE-I Performa Investments, has a maximum equity of R$ 26 million and a range of investments of up to R$ 5 million. The second is Peforma-Key Environmental Innovation Fund, with investment range of R$ 5 to R$ 20 million, with a 10-year disinvestment period.

SP Ventures. Founded in 2007, constituted the Paulista Innovation Fund in 2013, with an expected duration of 8 years and investments up to R$ 6 million in companies of different sectors, including biotech. His investments in the biotech sector is made by Criatec Fund.

Inseed Investments. Founded in 2009 by Instituto Inovação S.A., invested up to R$ 5 million in 36 companies, with the average expectation of divestment for 10 years. For the future, this Fund will invest up to R$ 10 million in each company with a six-year disinvestment period.

Pitanga Fund. Created in 2011, by a group of entrepreneurs with the objective of investing in innovative companies with high growth potential, without restrictions regarding the company's sector of activity or its geographic location. This fund uses only private capital and does not have limits or deadlines for investments and disinvestment.

Trivèlla Investments. Manages 4 VC funds: Trivella M3 FIP, Trivella Guarani, Cypress M3 and FIEC (Investment Fund for Participations in the Creative Economy). It has investment range with average of R$ 1 to R$ 6 million, and duration of 4 to 8 years.

3.1 Fundraising

For the fundraising, the most relevant topics identified for the funds composition were economic variables and investment culture.

Economic Variables. The interviewees affirmed the existence of influence of the macroeconomic variables in the activities of raising funds for the constitution of the investment funds. The variation of interest rate, has a direct influence on the impact of the cost of capital, encouraging or discouraging investors from bringing resources into VC funds. A high interest rate, or the expectation of raising, causes a natural preference of investors in rate-indexed applications, making VC funds less attractive. The fall, or downward trend, in the interest rate leads to a reduction in the return on interest rate-indexed investments, making VC funds more attractive to investors.

  The VC fund managers compete with public securities at the time of fundraising. By this, the low liquidity of the VC market and the need to apply the funds to long terms (approximately 8 years) obliges VC funds guarantee a much greater financial return than public bonds, which already have a high remuneration rate in Brazil due to the high interest rate. This corroborates [4], when they affirm that a reduction of the interest rate practiced in Brazil can encourage a greater volume of resources in the VC market. For most VC funds, the interest rate is the most influential variable for fundraising.

Investment Culture. There was a unanimous perception, of the difficulty in raising funds due to the lack of culture and investment history. Brazilian VC fund managers do not have great experiences in the VC market and there are a small number of successful cases in the market. This corroborates with [13]. Due to the recent expansion of the VC market in Brazil, successful investment experiences are a minority, causing investors to fear to provide resources to VC funds. Many investors are unaware of the VC market or are unfamiliar with regulation, which makes investors afraid to invest in the VC market. The search for managers with successful experiences is a way to minimize their risks, however, there are few managers with experience and cases of success of companies invested in the Brazilian market.

3.2   Investments

The main characteristics considered by the VC funds to invest in companies are:

Characteristics of the Team. Is considered by the interviewees as one of the main criterion for selecting companies. A team with unsatisfactory team causes the company to be discarded at the beginning of the selection process, even if it has a technology with great marketing and financial potential. This corroborates with [14, 20, 21].

Characteristics of the Product or Service. The characteristics considered by the managers interviewed associated with the product or service are the competitive differential business scalability, regulatory framework and risk of development. In the vision of the Inseed Investments and Pitanga Fund it is relevant that the companies have an innovative idea that presents a different solution from those already existing in the market.

For most of the interviewees, scalability of the development of a product or service is the variable with main importance to increase the company's profitability.

Characteristics of the Market. The wider the market, the larger the consumer potential and lower entry barriers, the greater is the chance of the company to receive the investment.

SP Venture and Trivèlla Investments consider the existing business chain in the market to determine the impact that a given technology will bring to it and which players will benefit and be harmed in the business chain. This analysis allows the identification of possible entry barriers and the acceptance of the technology in the market, verifying players that can create competitive difficulties. For Trivèlla Investments, the existence of these barriers does not represent that the company will not succeed, however, for the VC fund that aims at an accelerated growth of the invested company, it becomes an additional risk.

Company's Financial Potential. This criterion is important, but was identified that it does not has the highest relevance, and this refutes [8].

According to Performa Investments it is needed to do a case-by-case analysis, not having defined criteria specifically for the biotech sector. The analysis needs to compare other biotech companies, considering the revenues, the survey of the company's expenses, the expectation of working capital for the next year, the impact of the sales variation to elaborate several scenarios in which the manager can perform a sensitivity analysis based on expectations and projections, instead of considering financial indicators commonly used in the financial market and capital markets.

On the other hand, SP Venture use financial 3 criteria: profitability; capital requirement; and size of the market.

Geographic Distance. The geographic distance between the manager's headquarters and the company is important only for Performa Investments and SP Venture. The Performa analyses the travel costs and the demand for time. The SP Venture prioritizes companies that are closer to the VC fund manager head office.

4 Final Considerations

This research identified that the managers of Brazilian VC funds, in the biotech area, faces challenges in the activities of fundraising and investment. These challenges are related to the missing of a culture of VC investments in Brazil, in addition to the influence of economic variables that discourage investors from operating in this market. The interest rate has the greatest influence for the VC fundraising, by the impact in the investors cost of capital. The lack of a history of investments with success market cases also has influences on fundraising. Many investors are still unaware of the VC market and do not have a culture of high risk investments.

Regarding the criteria of investments was found that the team characteristics is the one with greater weight for most of the VC funds analyzed. In accordance with the particularities of the company, there is the possibility of inclusion or exclusion of criteria, becoming a flexible process, without strong rules and rigid steps. It was also verified that there are no specific selection criteria for the biotech companies and the fund's managers carry out the analysis on a case-by-case basis.

Was identified that the Brazilian VC funds do not have the capacity to contribute with large capital volumes in a single company. One of the alternatives to increase the capital volume capacity and minimize the VC funds risk is the realization of co-investments. However, when seeking for co-investments, there is the problem of the small number of VC funds available in the Brazilian biotech market.

The geographical distance between the headquarters of the VC fund manager and the company has little influence on the selection of firms and it is not a restrictive reason for the resource contribution, for most of the VC funds manager in the sample of this research, being considered by only 2 of them.

One of the limitations of this research is the small number of VC funds manager in the biotech area, in Brazil. The present study analyzed 5 VC funds from one universe with 8 funds. Also, there are few national studies conducted before about this theme.

For future work, it is suggested to analyze the particularities of the national VC funds in the areas beyond the biotech, compare the differences of the Brazilian VC market with other countries, structure a model with quantitative and qualitative criteria used by VC managers in the selection of companies and verify the influence of the agent-principal problem and asymmetry of information in VC fund managers for biotech and other areas.

References

1. Ozmel, U., Robinson, D.T., Stuart, T.E.: Strategic alliances, venture capital, and exit decisions in early stage high-tech firms. J. Financ. Econ. **107**(3), 655–670 (2013)
2. Inderst, R., Mueller, H.M.: Early-stage financing and firm growth in new industries. J. Financ. Econ. **93**, 276–291 (2009)
3. Chemmanur, T.J., Krishnan, K., Nandy, D.K.: How does venture capital financing improve efficiency in private firms? A look beneath the surface. Rev. Financ. Stud. **24**, 4037–4090 (2011)

4. Meirelles, J.L.F., Pimenta Júnior, T., Rebelatto, D.A.N.: Venture Capital e Private Equity no Brasil: alternativa de financiamento para empresas de base tecnológica. Gestão e Produção **15**, 11–21 (2008)
5. Florida, R., Mellander, C.: Rise of the startup city: the changing geography of the venture capital financed innovation. CMR – Calif. Manag. Rev. **59**(1), 14–38 (2017)
6. Florida, R., King, K.M.: Rise of the global startup city: the geography of venture capital investment in cities and metros across the globe. Martin Prosperity Institute, Rotman School of Management, University of Toronto, Toronto (2016)
7. Florida, R., Mellander, C.: Rise of the startup city: the changing geography of the venture capital financed innovation. Martin Prosperity Institute, Rotman School of Management, University of Toronto, Toronto (2014)
8. Jung, S., Rauch, M.G., Koch, G.: Decoding VCs decision making behavior in biotechnology: the relative importance of established decision criteria and the role of start-ups' knowledge networks. Social Science Research Network Working Paper Series (2011)
9. ABVCAP: Brazilian Private Equity and Venture Capital Association. Industria PEVC, sobre o setor (2010)
10. Sapenza, H.J.: When do venture capitalists add value. J. Bus. Ventur. **7**, 9–27 (1992)
11. Gompers, P.: Optimal investment, monitoring and the staging of venture capital. J. Financ. **50**(5), 1461–1489 (1995)
12. Gorman, M., Sahlman, W.: What do venture capitalists do. J. Bus. Ventur. **4**(4), 231–248 (1989)
13. Judice, V.M.M., Baêta, A.M.C.: Modelo Empresarial, Gestão de Inovação e Investimentos de Venture Capital em Empresas de Biotecnologia no Brasil. Revista de Administração Contemporânea **9**(1), 171–191 (2005)
14. Colombo, M.G., Grilli, L.: On growth drivers of high-tech start-ups: exploring the role of founders' human capital and venture capital. J. Bus. Ventur. **25**, 610–626 (2010)
15. Gompers, P., Kovner, A., Lerner, J., Scharfstein, D.: Performance persistence in entrepreneurship. J. Financ. Econ. **96**(1), 18–32 (2010)
16. Gompers, P., Lerner, J.: The venture capital revolution. J. Econ. Perspect. **15**(2), 145–168 (2001)
17. Sorenson, O., Stuart, T.E.: Syndication networks and the spatial distribution of venture capital investments. Am. J. Sociol. **106**, 1546–1588 (2001)
18. Cooper, D.R., Schindler, P.S.: Métodos de Pesquisa em Administração, 10th edn. Bookman, Porto Alegre (2011). 784 p.
19. Bardin, L: Análise de Conteúdo, 5th edn (2008)
20. Macmillan, I.C., Zemann, L., Narasimha, P.N.S.: Criteria distinguishing successful from unsuccessful ventures in the venture screening process. J. Bus. Ventur. **2**, 123–138 (1987)
21. Macmillan, I.C., Siegel, R., Narasimha, P.N.S.: Criteria used by venture capitalists to evaluate new venture proposals. J. Bus. Ventur. **1**, 119–128 (1985)

Cash Holdings Determinants from Brazilian Listed Firms

Flávio Morais[1], Ana Nave[2], and Ricardo Gouveia Rodrigues[3(✉)]

[1] CEFAGE and NECE, University of Beira Interior, Covilhã, Portugal
[2] University of Beira Interior, Covilhã, Portugal
[3] NECE Research Centre, University of Beira Interior, Covilhã, Portugal
rjagr@ubi.pt

Abstract. Recognizing the newness of the research on topics related to the level of cash holdings held by firms in emerging economies, this study has as main aim to present an empirical evidence on the determinants/factors, which explain the cash ratio presented by Brazilian listed non-financial firms in 2016. With the support of a theoretical framework based on the arguments of the Trade-off Theory and Pecking Order Theory, the determinants to be considered in the study were established. Using a sample of 164 firms, the results from the multiple linear regression show that the liquid assets substituting for cash holdings and the level of leverage lead to lower cash ratios. Contrary to expectations, larger firms have larger cash holdings, emphasizing a positive relationship between the variables. Both Pecking Order Theory and Trade-off Theory are relevant for explaining the decisions regarding cash holdings, having the first major relevance. The study does not present evidence in favor of the precautionary motive, which it has strong implications for the literature on cash holdings.

Keywords: Cash holdings · Pecking order · Trade-off · Emerging markets

1 Introduction

The last decades have proved to be particularly useful for researches that focus on the topics related to the level of cash holdings as a percentage of the asset [1]. According to Gill and Shah [2], cash holdings can be defined as money in cash or readily available for the day-to-day activities of the firm, encompassing also the equivalents that can easily be converted into cash. Although a wide range of studies is available on the subject of cash holdings, the majority are concentrated in firms from developed countries, especially in the USA [3], in United Kingdom [4] and other European countries [5]. The determinants of cash holdings is a fertile area for new researches, since little attention has been given to the emerging markets, in fact lacking to research the reality in firms of those countries [6]. In an attempt to fill the identified gap, the main aim of this study is to research the determinants of cash ratio of non-financial listed firms from Brazil in 2016. Specifically, supported by the main theories (Trade-off Theory and Pecking Order Theory), the purpose is to provide empirical evidence on the variables that determine the cash of Brazilian firms. The year of 2016 is marked by the confirmation of the country's recession, which may have a strong influence on the cash

© Springer International Publishing AG, part of Springer Nature 2019
J. Machado et al. (Eds.): HELIX 2018, LNEE 505, pp. 1043–1050, 2019.
https://doi.org/10.1007/978-3-319-91334-6_144

holdings of Brazilian firms, leading to the accumulation of cash for precautionary reasons [7]. Simultaneously, the pertinence of studying the phenomenon in Brazilian listed firms is due to the fact that it is one of the key countries in the international business and in the global economy [8]. Using a sample of 164 firms, it is evident that both liquid assets substituting cash holdings and leverage have a negative impact on the cash ratio. Contrary to expectations, the size of the firm has a positive and significant effect on cash holdings. This study has direct contributions to the growing and recent literature that researches the determinants of cash holdings in emerging markets.

2   Literature Review

2.1   Theoretical Framework

In the distant work of Keynes [9] are mentioned two main reasons for firms to keep cash: (i) transaction motive and (ii) precautionary motive. The first concerns the need of money for current business transactions as a result of inflows-outflows, seeking to minimize the transaction costs of having to resort to external financing. The precautionary motive derives from the desire for security in the face of uncertain events. Two main theoretical models discuss the preference of firms for liquidity. The Trade-off Theory [10] argues that firms are looking for an optimal level of cash, which balances their costs and benefits. Otherwise, Pecking Order Theory [11] argues that in order to reduce financing costs, the firm must first finance itself with internal liquidity, when the former are exhausted, it must resort to debt and only in the latter case should it resort to issuing shares. The study of the determinants of the cash ratio in emerging markets was driven by the research of Al-Najjar [8], where the author promoted a comparison of listed firms from Brazil, Russia, China and India with a control sample of USA and UK firms. Currently, the studies that examine the determinants of cash ratio exclusively on firms from emerging markets resorts to listed firms from Saudi Arabia [6]; Turkish listed firms [12]; Brazilian non-listed firms of sugarcane industry [13].

2.2   Variables and Hypotheses

Cash Ratio: The dependent variable of the study is the cash ratio (CASH). As most studies [14], the CASH variable is obtained as the cash and cash equivalents to total assets ratio. As a measure of robustness is used the variable CASH2, initially proposed by Opler et al. [15], which divides the value of cash and cash equivalents by the total assets net of cash and cash equivalents.

Size: SIZE is calculated as the natural logarithm of the total asset. According to the trade-off model, larger firms have less need for cash to avoid financial distress [16].

H1: Size has a negative effect on cash ratio.

Growth Opportunities: GROWTHOP variable is a proxy for growth opportunities and it is calculated through the market-to-book ratio. According to Pecking Order Theory, firms with greater growth opportunities should have higher cash holdings.

H2: Growth opportunities have a positive effect on cash ratio.

Leverage: LEV variable attends as a proxy for leverage and is calculated as the ratio of total debt to total assets. Pecking Order Theory points for a negative relationship between the variables.

H3: Leverage has a negative effect on cash ratio.

Debt Structure: STDEBT variable was computed as the ratio between short-term debt and total debt. According to Trade-off Theory and for precautionary motives, firms with a predominance of short-term debt will maintain higher levels of cash holdings.

H4: Short-term debt has a positive effect on cash ratio.

Cash Flow: CFLOW variable represents the cash flow of the period and is obtained by the item representing the operating cash flows generated in the year, available in the financial statements accessed through the BOVESPA. According to Pecking Order Theory, firms with higher cash flows reinforce their liquidity levels.

H5: Cash flow generated by the firm has a positive effect on cash ratio.

Internal Liquidity: Net Working Capital (NWC) variable is computed as the ratio between net working capital and total assets and serves as a proxy for internal liquidity other than cash. According to Trade-off Theory, we hypothesize that:

H6: Net working capital has a negative effect on cash ratio.

Capital Expenditures: CAPEX variable attends as a proxy for capital expenditures and was obtained as the annual expenditures in fixed assets plus expenditures in intangible assets divided by the total asset. According to Pecking Order Theory, a negative relationship between capital expenditure and cash holdings should be found.

H7: Capital expenditure has a negative effect on cash ratio.

Asset Tangibility: TANG variable was measured as tangible assets to total assets ratio. According to Trade-off Theory, the following hypothesis is established:

H8: Tangibility has a negative effect on cash ratio.

Dividends: To proxy for dividend payment the Div_dummy assumes the value of 1 if the firm pays dividends in 2016 and 0 otherwise. According to Trade-off Theory, firms that pays dividends can obtain funds to finance itself at reduced costs [15].

H9: Dividends have a negative effect on cash ratio.

3   Method and Data

To test the hypotheses empirically was used the accounting and financial information available for the year of 2016 for Brazilian firms listed on BOVESPA - the São Paulo Stock Exchange. Public data were collected through the BOVESPA Financial Statements repository. The data were treated in SPSS 25, where the statistical analysis was performed. Only firms with positive assets and sales during the year are considered and it was required that complete information about the variables of the study was available [14]. Finally, financial firms were removed from the sample because they are subject of specific rules and regulations that influence their cash holdings policy [14]. After applying the above criteria, a sample of 164 firms was obtained, representing the same number of observations. In order to study the determinants of cash ratio for only one year, the multiple linear regression method was adopted. As Uyanık and Güler [17], the base regression model used is as follows:

$$y_i = \beta_0 + \beta_1 x_{i1} + \cdots + \beta_k x_{ik} + \cdots + \beta_K x_{iK} + \varepsilon_i \tag{1}$$

Where y is the dependent variable, the x's are independent variables, and ε is the error in equation. β_1 through β_K are parameters that indicate the effect of a given x on y; β_0 is the intercept. By placing the variables of the study in the above equation, the following is obtained:

$$CASH_i = \beta_0 + \beta_1 LEV_{i1} + \beta_2 NWC_{i2} + \beta_3 SIZE_{i3} + \beta_4 GROWTHOP_{i4} + \beta_5$$
$$STDEBT_{i5} + \beta_6 CAPEX_{i6} + \beta_7 TANG_{i7} + \beta_8 CFLOW_{i8} + \beta_9 DIV\_dummy_{i9} + \varepsilon_i \tag{2}$$

4   Results

4.1   Univariate Analysis

The descriptive analysis of main variables is presented in Table 1. It is possible to see that the mean of CASH is 0.0794, i.e., the sample of firms has a cash ratio around 7.94% in 2016. This value is lower than what the literature mentions for developed countries [14, 18], being also lower than the values reported for emerging markets [6, 12].

Table 1. Descriptive statistic

Variables	N	Minimum	Maximum	Mean	Median	Std. deviation
CASH	164	0	0.2624	0.0794	0.0610	0.0673
SIZE	164	10.7417	20.5063	15.0296	15.0067	1.6473
GROWTHOP	164	0.2470	6.3664	1.4934	1.1249	0.9873
CFLOW	164	−0.3407	0.4810	0.0929	0.0885	0.1061
LEV	164	0.1020	0.9940	0.5778	0.5701	0.2032
STDEBT	164	0.0170	0.9992	0.4813	0.4448	0.2061
NWC	164	−0.5841	0.6752	0.0580	0.0354	0.1966
CAPEX	164	0	0.1768	0.0344	0.0268	0.0316
TANG	164	0.0001	0.7897	0.2410	0.1931	0.2160

4.2   Multivariate Analysis

The developed econometric tests confirm the assumptions for the use of linear regression [17]. Specifically, according to Osborne [19], in order to guarantee the normality of the residual term, the dependent variable was submitted to a positive square root, which allowed for greater symmetry and a more standardized distribution of the errors. The regression results are presented in Table 2. In the regression model 1, the base model of the study shows that the NWC is the main determinant of the Brazilian firms' cash ratio for the year 2016. Other factors that are able to explain the levels of cash holdings of the firms under analysis are the size and level of leverage. Specifically, the NWC variable, which serves as a proxy for liquid assets substituting for cash holdings, has a significant negative coefficient (significant at 0.01 level), which translates into a negative impact on CASH variable. The results show that the increase of, for example, 0.10 in the NWC variable, ceteris paribus, results in a decrease of approximately 0.39% points (CASH variable was transformed, so to interpret the coefficient their value were raised square keeping the negative signal when it was present) in cash ratio. This result confirms the expected negative relation between liquid assets substitute and the cash ratio established in hypothesis H6. As advocated by the Trade-off Theory, Brazilian firms with more liquid assets have less need to accumulate cash, because in case of shortage of money they can more easily convert these assets into cash. The results corroborate the study of Guizani [6], on emerging markets. Comparing to evidence collected in Brazil, the negative relationship runs counter to the non-significant result presented by Al-Najjar [8], but is in line with the results of Manoel et al. [13].

Table 2. Regression Results

Independent Variable	1 - Entered Model	2 - Reduced Model	3 - CASH$_2$ Model
(Constant)	0.122	0.152	0.113
SIZE	0.013*	0.012**	0.013*
GROWTHOP	0.001	–	0.002
CFLOW	0.105	–	0.110
LEV	−0.105*	−0.114**	−0.117*
STDEBT	0.013	–	0.014
NWC	−0.197***	−0.214***	−0.224***
CAPEX	−0.141	-	−0.227
TANG	0.019	-	0.022
DIV_DUMMY	−0.008	–	−0.008
R^2	0.124	0.114	0.117
N	164	164	164

***Significant at the 0.01 level. **Significant at the 0.05 level. *Significant at the 0.1 level.

The positive and statistically significant coefficient at a level of 0.10 for the variable SIZE presents some support for a positive impact of the size of the firm on the levels of cash holdings. However, its impact on CASH variable is low, for instance, the increase of 0.10 in the SIZE variable, ceteris paribus, determines an increase of approximately 0.0017% points in cash ratio. Indeed, the positive relation is in line with the arguments of the Pecking Order Theory that large Brazilian firms, as they are apparently more profitable, can accumulate more cash than smaller firms. This result contradicts the majority of empirical studies [15] and is contrary to the established in the H1 hypothesis. In the study of Manoel et al. [13] evidences of a non-significant relation between size and cash ratio were obtained. The positive relationship evidenced corroborates the result presented by Guizani [6].

The LEV variable presents a negative coefficient, statistically significant at a level of 0.10, which allows obtaining some evidence of a negative relation between the level of leverage of Brazilian firms and their level of cash holdings, confirming the hypothesis H3. The impact of LEV on CASH is not very high, namely an increase of 0.10 in the LEV variable, ceteris paribus, leads to a decrease of approximately 0.11% points in the cash ratio. Pecking Order Theory, explains that the negative relation between the variables is a sign that the internal sources of liquidity have been exhausted and that the firm needs to resort to debt to finance itself [20]. In contrast to Al-Najjar [8] and Manoel et al. [13], this study shows a significant relationship between the variables and the negative sign presented is the most usual result in literature, even in emerging markets [6]. The remaining variables presented in model 1, although showing the expected signal, are not significant in explaining cash ratio. As a measure of robustness, alternative models were proposed. Model 2 represents a reduced model, where only the significant variables (NWC, SIZE and LEV) are used. Model 3 uses the CASH2 variable (being also this variable submitted to a positive square root) as an alternative measure to the dependent variable used in model 1. Overall, the results presented in models 2 and 3 are coincident with the evidence extracted from model 1. Overall, the results point to a greater preponderance of the arguments of the Pecking Order Theory in the explanation of the decisions of Brazilian listed firms regard their cash holdings policy. In particular, the negative effect of leverage and the positive one of size on cash ratio provide support for Pecking Order Theory. The presence of the arguments of the Trade-off Theory are visible in the substitution effect between net working capital and cash holdings. The collected evidence allows us to reject the precautionary motive as being fundamental for the accumulation of cash holdings during the year of 2016. The non-significant impact of the short-term debt, the cash flow generated, and the growth opportunities in the cash ratio show that the cash holdings policies do not reflect: (i) the desire of firms to maintain currency in order to avoid default when short-term debt is high; (ii) the need to increase liquidity reserves from cash flows, in order to avoid future uncertainty; (iii) nor the concern to maintain high levels of cash holdings that allow investing in projects with positive NPV that may arise. Therefore, precautionary motive that has been presented in the literature as the most cited argument for high cash holdings [7, 14] do not find evidences in its favor in this study. So, transaction motive is highlighted here.

5 Conclusion

The study analyses the determinants of the Brazilian firms' cash ratio for the year 2016, a period marked by the recession that affects the country and which stems largely from political uncertainty. The sample is composed by 164 Brazilian non-financial firms listed on BOVESPA. The level of cash holdings shown by the firms in the sample is on average 7.94% of total assets, a figure lower than that reported by most of the studies that have focused on firms from developed countries and those in emerging markets.

Using a multiple linear regression method, there is evidence that the main determinants of the Brazilian firms' cash ratio are the liquid assets substitutes for cash holdings (Net Working Capital is the most significant variable in the study), the size of the firm and the level of leverage. Objectively, Brazilian firms with higher levels of net working capital and leverage have lower levels of cash holdings, while unexpectedly larger firms present higher cash ratios, a result less common in the literature. Robustness tests allows confirming the previous findings and concluding that both, Pecking Order Theory and Trade-Off Theory are important in explaining the decisions adopted by studied firms, with particular preponderance to the arguments presented by the first. Regarding the motives for holding currency, the study finds no evidence in favor of the motive that has been indicated as highly determinant for the explanation of cash holdings: the precautionary motive. In this study, the transaction motive enjoys greater primacy. This study contributes directly to the scientific community, especially to the recent literature that researches the determinants of cash ratio in emerging markets. The finding that cash holdings are not accumulated for precautionary motives raises important implications for the literature and this finding should be subject of future researches to confirm if the precautionary motive has been losing preponderance over the last decade, in both developing and developed countries. The study limitation is the use of only one year to observe the determinants of cash ratio, leaving open the possibility for future researches to use time series that allows the construction of a panel data.

References

1. Kariuki, S.N., Namusonge, G.S., Orwa, G.O.: Firm characteristics and corporate cash holdings: a managerial perspective from Kenyan private manufacturing firms. Int. J. Adv. Res. Manag. Soc. Sci. **4**(4), 51–70 (2015)
2. Gill, A., Shah, C.: Determinants of corporate cash holdings: Evidence from Canada. Int. J. Econ. Financ. **4**(1), 70–79 (2012)
3. Bao, D., Chan, K.C., Zhang, W.: Asymmetric cash flow sensitivity of cash holdings. J. Corp. Financ. **18**(4), 690–700 (2012)
4. Al-Najjar, B., Belghitar, Y.: Corporate cash holdings and dividend payments: evidence from simultaneous analysis. Manag. Decis. Econ. **32**, 231–241 (2011)
5. Bigelli, M., Sánchez-Vidal, J.: Cash holdings in private firms. J. Bank. Financ. **36**(1), 26–35 (2012)
6. Guizani, M.: The financial determinants of corporate cash holdings in an oil rich country: evidence from Kingdom of Saudi Arabia. Borsa Istanb. Rev. **17**(3), 133–143 (2017)
7. Lian, Y., Sepehri, M., Foley, M.: Corporate cash holdings and financial crisis: an empirical study of Chinese companies. Eurasian Bus. Rev. **1**(2), 112–124 (2011)
8. Al-Najjar, B.: The financial determinants of corporate cash holdings: evidence from some emerging markets. Int. Bus. Rev. **22**(1), 77–88 (2013)
9. Keynes, J.M.: The General Theory of Employment, Interest and Money. Harcourt Brace, London (1936)
10. Tobin, J.: The interest-elasticity vs transactions demand for cash. Rev. Econ. Stat. **38**(3), 241–247 (1956)

11. Myers, S.C., Majluf, N.S.: Corporate financing and investment decisions when firms have information that investors do not have. J. Financ. Econ. **13**(2), 187–221 (1984)
12. Uyar, A., Kuzey, C.: Determinants of corporate cash holdings: evidence from the emerging market of Turkey. Appl. Econ. **46**(9), 1035–1048 (2014)
13. Manoel, A.A.S., Moraes, M.B., Santos, D.F.L., Neves, M.F.: Determinants of corporate cash holdings in times of crisis: insights from Brazilian sugarcane industry private firms. Int. Food Agribus. Manag. Rev. **21**, 1–17 (2017)
14. Bates, T.W., Kahle, K.M., Stulz, R.M.: Why do U.S. firms hold so much more cash than they used to? J. Financ. **64**(5), 1985–2021 (2009)
15. Opler, T., Pinkowitz, L., Stulz, H., Williamson, R.: The determinants and implications of corporate cash holdings. J. Financ. Econ. **52**, 3–46 (1999)
16. Titman, S., Wessels, R.: The determinants of capital structure choice. J. Financ. **43**(1), 1–19 (1988)
17. Uyanık, G.K., Güler, N.: A study on multiple linear regression analysis. Procedia – Soc. Behav. Sci. **106**, 234–240 (2013)
18. Gao, H., Harford, J., Li, K.: Determinants of corporate cash policy: insights from private firms. J. Financ. Econ. **109**(3), 623–639 (2013)
19. Osborne, J.W.: Improving your data transformations: applying the Box-Cox transformation. Pract. Assess. Res. Eval. **15**(12), 1–9 (2010)
20. Ferreira, M.A., Vilela, A.S.: Why do firms hold cash? Evidence from EMU countries. Eur. Financ. Manag. **10**(2), 295–319 (2004)

Capital Structure on Portuguese SMEs: The Role of Innovation and Internationalization During Financial Crisis

Luís Soares$^{(\boxtimes)}$, Fábio Duarte, and Ana Borges

CIICESI, ESTG/P.PORTO, Porto, Portugal
luisfhsoares@gmail.com

Abstract. This study explicitly examines the relation between financing decisions, innovation and internationalization in the case of Portuguese small and medium sized enterprises (SMEs) from the manufacturing sector in a period of financial crisis (between 2010 and 2016). Using a panel data analysis (Fixed effects model) with a sample of 1768 firms, this study tests the determinants of the Pecking Order Theory (POT) in comparison with Agency Cost Theory (ACT) and the Trade-Off Theory (TOT) This study provides first-hand results regarding the relation between both innovation and internationalization with SMEs debt financing. Results broadly support the POT but results show no significance of Internationalization to the econometric model. Innovative practices are positively related with debt.

Keywords: Panel data · Capital structure · Internationalization
Innovation

1 Introduction

Since [1, 2] capital structure has been a broadly studied topic. There is however a lack of firm-level analysis linking both innovation and internationalization. The objective is to contribute to the empirical literature as our goal is to understand their part on SMEs financial decisions. In one hand, innovation and internationalization plays a crucial role for SMEs to reduce their performance dependence related to the domestic demand and to maintain competitive advantages. In the other hand, innovative and international practices may contribute to diversify sources of finance with impact on firms' capital structure. Thus, we should expect differences in terms of capital structure between innovative/internationalized and non-innovative/non-internationalized. This study aims to explore the relation between firm level determinants (with a special focus on innovation and internationalization) and Total Debt (TD) while testing support for Pecking Order Theory (POT), Agency Costs Theory (ACT) and Trade-Off Theory (TOT). Moreover, we aim to answer the following research question: Are there significant differences between innovative Vs. non-innovative firms and internationalized Vs. non-internationalized firms. Regarding the structure of the remainder of this study: Sect. 2 provides a brief literature review about the mainstream capital structure theories, Sect. 3 presents the hypothesis, Sect. 4 provides the methodology of the study, Sect. 5 reports the results and discussion and in Sect. 6 we present the conclusions and future research.

© Springer International Publishing AG, part of Springer Nature 2019
J. Machado et al. (Eds.): HELIX 2018, LNEE 505, pp. 1051–1057, 2019.
https://doi.org/10.1007/978-3-319-91334-6_145

2 Literature Review

2.1 Capital Structure Theories: TOT, ACT and POT

Decisions about the capital structure (or financing decisions) concerns the proportion of equity and debt that any company use to finance itself. Several theories grew up and each of them provides approaches that contribute to shed more light on the capital structures subject (e.g., TOT, ACT and POT). However, given the ambiguous results provided by the literature, there is no consensus regarding an optimal capital structure, being the POT from Myers [3] a puzzle still actual. The TOT claims the existence of an optimal capital structure that firms aim to reach with the objective of maximize their value based on the assumption that there are benefits and costs of debt. Once the interests of debt financing are deducted from the net income, the value of a levered firm is higher than the value of an unlevered firm [2]. There is however a trade-off between the tax benefits of debt and potential costs of financial distress. The trade-off balances essentially the tax deductibility of interest paid [2] and the excessive amount of debt with consequent potential bankruptcy costs [4]. Firms have an optimal debt ratio which they attempt to sustain that is maximized when the benefits from tax shields are equal to the marginal cost of debt. The ACT states that the optimal capital structure of each firm depends on the value of debt that mitigates the conflicts between the principal and the agent [5]. The stockholder-manager agency costs of free cash-flow push firms towards more debt in order to reduce the "free cash at managers" disposal while the stockholder–debtholder agency costs of underinvestment and asset substitution push firms towards less leverage, since large debt levels may be an incentive for rejecting value-increasing projects [6] and pursuing risky projects [7]. In relation to POT, [3] argue that firms do not hold an optimal capital structure. Nonetheless, financing decisions are not irrelevant for their value. Due to information asymmetries between firms' managers, stockholders and potential outside financiers, firms tend to adopt a perfect hierarchical order of financing: first, they use internal funds; second, if they need external financing, firms issue low-risk debt and as a last resort, new shares are issued. In the absence of investment opportunities, firms retain earnings and build up financial slack to avoid having to raise external finance in the future. The preference for internal finance is usually explained by the existence of supply-side constraints in debt markets due to information asymmetries [8].

2.2 Capital Structure, Innovation and Internationalization

A firm wanting to compete through innovation may not have the internal resources to cover the entire cost of such an investment. However, in an "imperfect" world dominated by asymmetric information, bankruptcy risks and agency conflicts, external financing may be highly costly, specially to riskier/innovative projects, thus a firm's investment behavior might be constrained [9]. Aiming to validate the POT, only few studies approached the innovation as a determinant of the capital structures (e.g., [9–11]) and results provide an ambiguous result about this causal relation. A robust export performance is increasingly a critical factor for a firm general performance and survival. Internationalization involves high-risk decisions (e.g., sunk costs, exposition to exchange rate, changes on external markets, cultural barriers) [12]. However, also allows firms to create new growth opportunities which might reduce cash-flow volatility and the dependence on domestic demand, decreasing the expected costs of bankruptcy. Furthermore, international firms might have access to alternative and cheaper sources of finance. [13] argue that multinational corporations face higher agency costs of debt than domestic firms.

3   Hypotheses

3.1   Firm Age (AGE) and Size (SZ)

Younger firms tend to have higher loan rates and this might be due to lack of financial information, unknow types of investment opportunities or uncertainty on management behavior implying adverse selection and moral hazard costs combined with a lack of trading history and collateralizable assets that restricts access to external debt [14]. Young firms turn to debt more often, contrary to older firms, who can rely on retained earnings accumulated over time [15]. The hypothesis is formulated: **H1** – AGE and TD are negatively related. Bigger firms are inclined to have more diversified business activities which means they will have a lower probability of default, less agency costs [3] and a better reputation [16] increasing the attractiveness of debt. The hypothesis is formulated: **H2** – SZ and TD are positively related.

3.2   Firm Growth (GRO), Tangibility (TANG) and Profitability (PROF)

POT suggests a positive relation between growth and debt, since firms in a growing stage may have depleted retained earnings and need to borrow to fund undergoing investment projects [17]. Several studies provided evidences regarding the positive relationship between growth and debt ratios (e.g., [5]). The hypothesis is formulated: **H3** – GRO and TD are positively related. If a large portion of assets are tangible then it increases liquidity of a firm as they provide collateral in case of bankruptcy [18] and therefore less agency problems will arise consequently increasing leverage [19] and reducing risk [14]. It should be expected a negative relationship between tangibility and debt ratios as predicted by TOT. POT suggests a positive relation since the use of collaterals lessens informational asymmetry. The hypothesis is formulated: **H4** – TANG and TD are positively related. The higher the profits the more the available internal funds because firms tend to accumulate profits over time [20]. POT suggests the preferred method of finance is internal funds so a firm with retained earnings would use less debt [3] and consequently debt and profitability are negatively related. However, TOT predicts that debt ratios and profitability should be positively related. With the objective to offset corporate tax profitable firms are likely to raise their level of leverage thus shielding income from taxation. The hypothesis is formulated: **H5** – PROF and TD are negatively related.

3.3   Firm Innovation (INNOV) and Internationalization (INT)

Previous studies reported that investment in R&D and leverage are negatively related (e.g., [21]) as these types of investments create primarily intangible assets, not serving as collateral. Financial slack should then be a strategic requirement for firms who compete predicated in innovation [11]. Hence, funds are available when needed to finance continuous streams of investment in R&D. [22] discussed that innovative strategies were associated with the lowest debt level. However, [11] argued that a firm's expenditures on R&D may not be as important in defining capital structure as the strategic importance of innovation to the firm[1]. The hypothesis is formulated: **H6** – INNOV and TD are negatively related. The relationship between internationalization and debt have proven to be a conflicting issue. [23] reported lower cost of capital for firms more internationally diverse resulting in a positive relation between debt ratios and the level of internationalization. However, [24] among others, found a negative relation. It stems from the fact that more internationalized firms face higher agency costs [24]. For the Portuguese SMEs industry, [12] found a negative relation and concluded that export-oriented firms have lower levels of debt. The hypothesis is formulated: **H7** – INT and TD are negatively related.

4   Methodology

This study relies on a dataset between 2010–2016, collected from SABI[2] database regarding Portuguese SMEs (EU recommendations 2003) operating in the manufacturing sector (CAE[3] 10 to 33). Inactive and negative equity firms were withdrawn and after accounting for missing values, the final sample is composed by 1768 firms. To test the empirical hypotheses this study uses a panel data methodology. Based on the Hausman test we apply for the Fixed effects estimator. Capital Structure will be measured through the dependent variable TD ratio, in line with [17] (*TotalLiabilities/TotalAssets*) calculated from book value. The regression model can be generalized as follows:

$$Y_{it} = \alpha_i + \beta_1 AGE_{it} + \beta_2 SZ_{it} + \beta_3 GRO_{it} + \beta_4 TANG_{it} + \beta_5 PROF_{it} + \beta_6 INNOV_{it} + \beta_7 INT_{it} + \varepsilon_{it} \quad (1)$$

where Y = {TD}, α_i represents the variability between SMEs and ε_{it} are N i.i.d. realizations of N $(0; \tau^2)$, representing the measurement error (variability non- specified). As measures for independent variables: AGE = Log(year data-year foundation); SZ = Log (total assets in euros); GRO = [(total assets $_n$ - total assets $_{n-1}$)/total assets $_{n-1}$]; TANG = Tangible Assets/Total Assets; PROF = EBIT/Total Assets; INNOV = Log

---

[1] Innovation is considered implemented when brought to use in firm's operations. NCRF6 is the Portuguese adaptation of the IAS 38 norm that provides the guidelines for accounting treatment of Intangible Assets. It requires the recognition of the asset only if there is possibility to generate economic benefits in the future (if it can be sold or used internally). Therefore, in this study, the measure for Innovation will be Expenditures in Development Projects.

[2] SABI ("Sistema de Análise de Balanços Ibéricos" – Iberic Balance Sheet analysis system).

[3] CAE ("Classificação de Atividade Económica" – Economic Activity Classification).

(development projects in euros); INT = Exports/Total Sales. A t-test is performed to identify mean differences in terms of debt between innovative Vs. non-innovative firms and internationalized Vs. non-internationalized firms.

5   Results and Discussion

Table 1 reports the results of the panel data model. The coefficients of the independent variables are statistically significant except for Internationalization. Furthermore, the explaining power of the dependent variable TD ($R^2 = 21.6\%$), as suggested by [17] the weak explanatory power of the models should be postponed, applying dynamic panel data models[4].

Table 1. Estimate regression results (fixed effects model)

	AGE	SZ	GRO	TANG	PROF	INNOV	INT	R^2	F
TD	−0.010***	0.080***	0.087***	0.070***	−0.507***	0.010***	−0.008	0.216	416.13
Sd	(0.000)	(0.007)	(0.004)	(0.011)	(0.014)	(0.002)	(0.007)		
p-value	0.000	0.000	0.000	0.000	0.000	0.000	0.249		

Note: *p < 0,10; **p < 0,05; ***p < 0,01

AGE and TD are negatively related as expected supporting POT and validating H1. Firm maturity allows to save funds avoiding resorting to debt and thus, younger firms cannot retain earnings [25] forcing them to borrow more. SZ and TD are positively related as larger firms, who have more diversity of activities, find it easier to obtain debt due to less information asymmetries, lower bankruptcy risks and lower financing costs [12, 17] thus having incentive to raise debt levels supporting both POT and TOT and validating H2. From the results, GRO and TD are positively related therefore H3 is validated. POT is supported as firms with higher investments to fund their growth accumulate more debt [15]. This is more severe in the case of smaller and younger firms which have not yet generated retained earnings enough [17] and may have depleted retained earnings with a need to borrow more. Results show a positive relation between TANG and TD in line with [15, 17] thus, we validate H4 and found support to both the POT and TOT. Firms with higher levels of tangible assets are capable to provide creditors with collateral, reducing the problems of adverse selection [17] and information asymmetries as well as these assets can act as collateral in case the firm enters in bankruptcy process [12, 19]. The relation between PROF and TD is negative validating H5 and with support to POT. Higher level of profitability offsets the need for external financing as Portuguese SMEs prefer to finance through internal funds [17] and only when these are insufficient firms resort to debt [12]. INNOV is positively and significantly related to TD, hence we reject H6. [26] debated that credit suppliers may restrict debt if internal development of R&D attempts to preserve critical information.

---

[4] The correlation matrix was analyzed but not reported by space limitation. The results do not suggest multicollinearity problems.

As we use a distinct measure for innovation it may cause an alteration in debt financing decisions as development projects are already recognized as assets thus credit suppliers will not restrict loans allowing innovative firms to borrow more. Finally, we do not find any empirical evidence to support H7 regarding the relation between INT and TD. The t-tests results lead us to conclude that there are significant differences on both mean values of TD between internationalized/non-internationalized firms (t = −4.502; p-value < 0.001) and innovative/non-innovative (t = −6.1949; p-value < 0.001) where internationalized and innovative firms have significant higher levels of debt when compared to their counterparts. Financial constraints should affect R&D investments more severely because of the high degree of uncertainty characterizing innovation output [9]. However, development projects are assets already recognized thus, the uncertainty is lowered and innovative firms can increase debt easier. As for internationalized firms [23] reported that the degree of international diversification results in lower overall cost of capital which results in higher levels of debt.

6 Conclusion and Future Research

The results of this study show that firm age, size, growth profitability and innovation explain the debt financing whereas internationalization has no significance to the model. Innovative firms have significant higher levels of Debt when compared with non-innovative firms. The impact of financial crisis and the new norms introduced by BASEL III accords have changed the credit environment therefore, an extension of this study would be to test and explicitly control the effect of the international financial crisis on funding decisions. Moreover, to the extent that innovation affect strategy of the firm it is necessary to explore new dimensions for measuring innovation as well as its interaction with internationalization.

References

1. Modigliani, F., Miller, M.H.: The cost of capital, corporation finance and the theory of investment. Am. Econ. Rev. **48**(3), 261–297 (1958)
2. Modigliani, F., Miller, M.H.: Corporate income taxes and the cost of capital: a correction. Am. Econ. Rev. **53**(3), 433–443 (1963)
3. Myers, S.: The capital structure puzzle. J. Financ. **39**(3), 575–592 (1984)
4. Kraus, A., Litzenberger, R.H.: A state-preference model of financial leverage. J. Financ. **28**, 911–922 (1973)
5. Ramalho, J., Silva, J.: A two-part fractional regression model for the financial leverage decisions of micro, small, medium and large firms. Quant. Financ. **9**(5), 621–636 (2009)
6. Myers, S.: Determinants of corporate borrowing. J. Financ. Econ. **5**, 147–175 (1977)
7. Jensen, M.C., Meckling, W.H.: Theory of the firm: managerial behaviour, agency costs and ownership structure. J. Financ. Econ. **3**, 305–360 (1976)
8. Chittenden, F., Hall, G., Hutchinson, P.: Small firm growth, access to capital markets and financial structure: review of issues and an empirical investigation. Small Bus. Econ. **8**(1), 59–67 (1996)

9. Bartoloni, E.: Capital structure and innovation: causality and determinants. Empirica **40**, 111–151 (2013)
10. Prędkiewicz, K., Prędkiewicz, P.: Pecking order theory and innovativeness of companies. In: Procházka, E.D. (ed.) New Trends in Finance and Accounting, pp. 631–642. Spinger, Prague (2017)
11. O'Brien, J.P.: The capital structure implications of pursuing a strategy of innovation. Strateg. Manag. J. **24**, 415–431 (2003)
12. Pacheco, L.: Capital structure and internationalization: the case of Portuguese industrial SMEs. Res. Int. Bus. Financ. **38**, 531–545 (2016)
13. Singh, M., Davidson, W.N., Suchard, J.: Corporate diversification strategies and capital structure. Q. Rev. Econ. Financ. **43**, 147–167 (2003)
14. Duarte, F., Gama, M., Paula, A., Esperança, J.P.: The role of collateral in the credit acquisition process: evidence from SME lending. J. Bus. Financ. Acc. **43**(5–6), 693–728 (2016)
15. Hall, G., Hutchinson, P., Michaelas, N.: Determinants of the capital structures of European SMEs. J. Bus. Financ. Acc. **31**(5 & 6), 711–728 (2004)
16. Degryse, H., de Goeij, P., Kappert, P.: The impact of firm and industry characteristics on small firms' capital structure. Small Bus. Econ. **38**, 431–447 (2012)
17. Matias, F., Serrasqueiro, Z.: Are there reliable determinant factors of capital structure decisions? Empirical study of SMEs in different regions of Portugal. Res. Int. Bus. Financ. **40**, 19–33 (2017)
18. Harris, M., Raviv, A.: The theory of capital structure. J. Financ. **46**, 297–355 (1991)
19. Rajan, R.G., Zingales, L.: What do we know about capital structure? Some evidence from international data. J. Financ. **50**(5), 1421–1460 (1995)
20. Kayhan, A., Titman, S.: Firms' histories and their capital structures. J. Financ. Econ. **83**(1), 1–32 (2007)
21. Simerly, R.L., Li, M.: Environmental dynamism, capital structure and performance: a theoretical integration and an empirical test. Strateg. Manag. J. **21**(1), 31–49 (2000)
22. Jordan, J., Lowe, J., Taylor, P.: Strategy and financial policy in UK small firms. J. Bus. Financ. Acc. **25**, 1–27 (1998)
23. Singh, M., Nejadmalayeri, A.: Internationalization, capital structure, and cost of capital: evidence from French corporations. J. Multinatl. Financ. Manag. **14**, 153–169 (2004)
24. Chkir, I.E., Cosset, J.-C.: Diversification strategy and capital structure of multinational corporations. J. Multinatl. Financ. Manag. **11**, 17–37 (2001)
25. López-Gracia, J., Sogorb-Mira, F.: Testing trade-off and pecking order theories financing SMEs. Small Bus. Econ. **31**, 117–136 (2008)
26. Vicent-Lorente, J.D.: Specificity and opacity as resource-based determinants of capital structure: evidence from Spanish manufacturing firms. Strateg. Manag. J. **22**, 157–177 (2001)

Market Orientation and Hotel Industry: Developing a Measurement Model

Carlos Alberto Fernandes Sampaio[1(✉)],
José Manuel Hernández-Mogollón[2], and Ricardo Gouveia Rodrigues[3]

[1] Instituto Politécnico de Castelo Branco, Castelo Branco, Portugal
cfsampaio@ipcb.pt
[2] Universidad de Extremadura, Cáceres, Spain
jmherdez@unex.es
[3] Universidade da Beira Interior, Covilhã, Portugal
rgrodrigues@ubi.pt

Abstract. This study seeks to fill a gap in market orientation literature about the hotel industry and deals with the construction of a market orientation scale tailored based on a sample obtained from the hotel industry in a multicultural context. The proposed model has three dimensions: intelligence generation, intelligence dissemination and coordinated response to the client, competition and market structure domains. Research results indicate that the scale has good psychometric indicators. Content validity was assessed by questioning a group of marketing experts across Western Europe. Working data indicates that the proposed model holds convergent and discriminant validity as well reliability.

Keywords: Hotel industry · Market orientation · Scale construction

1 Introduction

Narver and Slater [1] and Kohli and Jaworski [2] established the foundations for the following research on market orientation. During the 1990s, several contributions were made that shaped the present state of market orientation study [1–6].

Regardless of the extensive literature developed during the 1990s and early 2000s, market orientation research on services sector, particularly empirical research about service provider companies as hotels and other leisure providers, were scarce.

On the other hand, empirical studies on market orientation in hotel industry uses the MARKOR scale [1] or the MKTOR scale [4], or some kind of adaptation. Despite the high acceptance of these models, several criticisms to its psychometrics characteristics, namely to validity and reliability issues were highlighted.

Accordingly, this work seeks to contribute to the development of market orientation literature in hotel industry companies. The authors develop a market orientation measurement scale, tailored using a sample obtained from hospitality sector firms. Working data were obtained from an international context. A sample of hotels from Western Europe was used (France, Ireland, Italy, Portugal, Spain and the United Kingdom).

© Springer International Publishing AG, part of Springer Nature 2019
J. Machado et al. (Eds.): HELIX 2018, LNEE 505, pp. 1058–1065, 2019.
https://doi.org/10.1007/978-3-319-91334-6_146

2 Literature Review

2.1 Measuring Market Orientation

Two main proposals were made in developing market orientation definition. Market orientation as the "organizational culture that most effectively and efficiently creates the necessary behaviors for creating superior value for buyers and, thus, continuous superior performance" [1], and the "organizationwide generation of market intelligence pertaining to current and future customer needs, dissemination of intelligence across departments, and a organizationwide responsiveness to it" [2].

Despite several criticisms made to the MKTOR and MARKOR scales, the main question about measuring market orientations is related with the measurement model accuracy. Currently, empirical research on market orientation uses former market orientation measurement models, mainly MARKOR and MKTOR models, or adaptations based on them.

The hospitality sector has been overlooked in the development of existing market orientation measurement models. In general, the scales used to measure this matter in the hotel industry are also based in the MARKOR scale [7–9] and in the MKTOR scale [10–14], or adaptations. Only recently some proposals were settled to improve and adapt these models to the hotel industry [9, 14].

On the other hand, empirical research on the relationship between market orientation and business performance, notwithstanding some inconsistent results [15, 16], states a positive relation between market orientation and business performance [7, 9–11, 14, 17, 18], reflecting a certain unanimity about it.

3 Methodology

3.1 Item Generation and Content Validity

Churchill [19], Webb [20] and Nunnaly and Bernstein [21] recommendations were followed in order to develop the proposed scale.

First, a qualitative step was conducted. An analysis to the existent market orientation literature was conducted.

Literature analysis enabled to specify the construct domain [19]. Several market orientation measurement models were analyzed, original scales, or adaptations of existing scales.

This procedure enabled to identify the proposed scale dimensions and its measurement indicators.

Consequently, based on the literature analysis, it was proposed market orientation as a multidimensional construct with three dimensions: intelligence generation, intelligence dissemination and coordinated response to market information. These dimensions should measure three domains: client domain, competition domain and market structure domain.

Next, content validity was tested. Content validity analysis is essentially a subjective process and largely based upon opinions of various users [21]. Therefore, the proposed scale was sent to a group of experts in marketing. Experts were PhD holders from universities from Western Europe countries, and were asked to evaluate the model items.

Based on the experts' feedback, a minor change was made in one item. The final proposed model was based on twenty-one items and three dimensions: intelligence generation; intelligence dissemination; and coordinated response, each one with seven items.

3.2 Questionnaire and Data

Data used in this research were collected using an online survey sent to the managers of 32377 hotels around Western European countries.

The statistical population was composed by the set of hotel companies from, France, Ireland, Italy, Portugal, Spain and the United Kingdom.

Hotel contacts were obtained in government tourism departments and yellow pages' services around the study context.

Data collection was performed between October 2013 and January 2014.

All valid questionnaires were answered by the hotels' directors, marketing department directors or direction assistants.

4   Results

4.1   Measurement Model Purification

Market orientation construct was defined as a second order construct with three dimensions, intelligence generation, intelligence dissemination and coordinated response. Therefore, to conduct a confirmatory analysis, data were computed in IBM SPSS Amos 24.

Conducting a factor analysis during the early stages of developing measurement models could produce many more dimensions than those conceptually identified [19]. Therefore, prior to the CFA analysis, items correlations to underlying constructs were evaluated.

An item to construct correlations below 0.55 indicates that the variable shares little in common with the other measurements, that is of questionable value in defining the component [22]. Therefore, variables with correlation less than 0.60 were removed from the model.

Computed data were evaluated and five variables were removed from the model. MODDI1 (0.54), MODGI3 (0.56) and MODGI7 (0.47), and MODRpC1 (0.55) and MODRpC3 (0.18).

4.2   Confirmatory Factor Analysis

To assess the confirmatory factor analysis, the second order construct (Fig. 1) was evaluated and a multivariate normality test was conducted (Fig. 2).

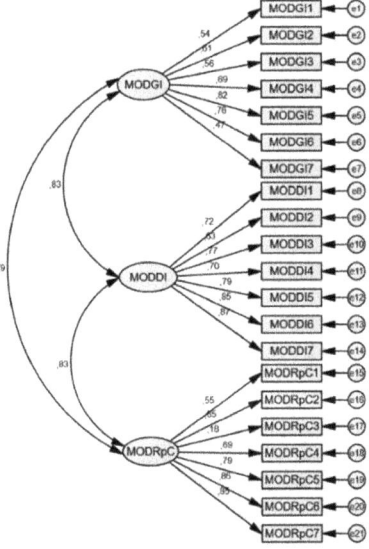

Fig. 1. Second tier measurement model

Variable	skew	c.r.	kurtosis	c.r.
MODGI2	-0.643	-5.551	-0.802	-3.46
MODGI4	-1.206	-10.411	0.794	3.428
MODGI5	-0.925	-7.98	0.149	0.643
MODGI6	-0.901	-7.78	0.005	0.02
MODDI1	-0.716	-6.181	-0.701	-3.025
MODDI2	-0.267	-2.304	-1.36	-5.868
MODDI3	-0.872	-7.53	0.132	-1.862
MODDI4	-0.039	-0.338	-1.269	-5.474
MODDI5	-0.705	-6.087	-0.584	-2.521
MODDI6	-0.569	-4.907	-0.775	-3.347
MODDI7	-0.668	-5.768	-0.671	-2.896
MODRpC2	-1.488	-12.841	1.704	7.356
MODRpC4	-0.346	-2.987	-0.91	-3.926
MODRpC5	-0.346	-2.985	-0.61	-2.631
MODRpC6	-0.452	-3.9	-0.635	-2.741
MODRpC7	-0.539	-4.653	-0.496	-2.142
Multivariate			96.813	42.643

Fig. 2. Multivariate normality test

According to the multivariate normality test conducted, data did not follow a normal multivariate distribution. In order to work around this issue, a 2000 resampling bootstrap procedure was computed [23].

Data from confirmatory factor analysis were evaluated (Fig. 3 – model 1). Results were modest, thus, once the model was previously purified, modification indices were analysed.

	x^2	df	p	x^2/df	CFI	GFI	NFI	PGFI	PNFI	PCFI	RMSEA
Model 1	491.345	101	0	4.865	0.92	0.86	0.90	0.64	0.75	0.77	0.09
Model 2	403.012	99	0	4.071	0,93	0.89	0.91	0.65	0.75	0.77	0,08

Fig. 3. Model fit indicators

Modification indices revealed an obvious relation between variables MODGI4 and MODGI6, and variables MODRpC4 and MODRpC5. Thus, a second model (model 2) was constructed, and errors from variables MODGI4 and MODGI6, and from variables MODRpC4 and MODRpC5 were correlated, and the second fit values were evaluated (Fig. 3).

Global fit indices of both models (Fig. 3) confirm that the second model presents a better fit.

4.3   Convergent Validity, Discriminate Validity and Reliability

Average variance extracted (AVE) values were evaluated to assess convergent validity. Convergent validity is achieved if AVE values are above 0.50 [24]. Extracted AVE values indicates convergent validity, MODGI = 0.543, MODDI = 0.583, and MODRpC = 0.583.

As for reliability, extracted data presented values above the 0.70 recommended value. Composite reliability were MODGI = 0.824, MODDI = 0.906, and MODRpC = 0.873.

Finally, in order to assess discriminant validity, a chi-square difference test was computed. According Segars [25], if the chi-square difference is significant, it confirms discriminant validity. According the extracted data from the chi-square difference, discriminant validity was confirmed.

Figure 4 describes the market orientation scale after purification and model fit.

Intelligence generation (AVE = 0.543; Composite reliability = 0.824)	
MODGI2	We track our competitors activities and offers at least two times a year.
MODGI4	Our company is fully committed, when dealing with customers, in gathering information about their desires and needs, even if they didn't recognize them yet.
MODGI5	Our company seeks to obtain quickly market information that could change clients' perceptions about our products and services.
MODGI6	Our company seeks to know quickly our competitor's new products and services, each time they bring them to the market.
Intelligence dissemination (AVE = 0.583; Composite reliability = 0.906)	
MODDI1	Customer information is quickly disseminated to all firm's departments.
MODDI2	In our company we have a formal information dissemination procedure, among all firm's departments, about our clients.
MODDI3	If a firm's department gets key intelligence about our clients, it spreads the information quickly to all the other departments.
MODDI4	Our company usually organizes formal meetings targeted to discuss our competitors activities and offers.
MODDI5	Information about our competitors advantages known by one of this firm departments is quickly spread to all the other departments.
MODDI6	This firm's departments are fully committed in sharing information about market information and trends affecting our business.
MODDI7	Information about market structure changes (trends, regulation, etc.), obtained by one of our firm's departments, spreads quickly to all the other departments.
Coordinated response (AVE = 0.583; Composite reliability = 0.873)	
MODRpC2	When it is need to act targeted to clients needs, all departments in our company participate.
MODRpC4	We always respond to competitive activities from our competitors.
MODRpC5	Competitive actions from our competitors have a quick coordinated response from our company.
MODRpC6	All this company departments respond quickly to changes in market structure (trends, regulations, etc.)
MODRpC7	Response to market changes is done in a coordinated way by all this firm's departments.

Fig. 4. Market orientation scale after purification process

5 Conclusions, Implications and Limitations

The developed market orientation scale was based on three dimensions, intelligence generation, intelligence dissemination and coordinated response, related with three domains: client, competition and market structure.

Furthermore, an empirical analysis was conducted using a sample obtained from hotels from Western Europe. Firstly, the proposed market orientation scale suffered a purification process, during which five items were removed from de model. Next, a confirmatory factor analysis was carried out.

Moreover, market orientation scale validity and reliability were tested. Results confirm that the proposed scale presents convergent and discriminant validity. Composite reliability results, above the threshold of 0.70 [26], indicate that the scale is reliable.

The performed research found a scale able to obtain good psychometric indicators in a multicultural context. Particularly in a context that accounts in a large amount to the international tourism.

The proposed scale implies two main developments. Firstly, the scale was built based solely on taking a sample from the hotel sector's companies. Moreover, the scale was developed based in the international context, culturally somehow far from the

MAKROR and MKTOR developing context. On the other hand, a substantial part of hotels in Europe are small companies with less than 10 employees [27] and increasing and maintaining a certain degree of market orientation is a complex process that requires considerable expenditure of money and time [28].

This work deals with several important questions related to market orientation study in the hotel industry. Despite the enlightenment it carries, further research is needed. One line of study is to evaluate the relationship between market orientation and business performance using the proposed measurement model to assess market orientation.

References

1. Narver, J.C., Slater, S.F.: The effect of a market orientation on business profitability. J. Mark. **54**, 20–35 (1990)
2. Kohli, A.K., Jaworski, B.J.: Market orientation: the construct, research propositions and managerial implications. J. Mark. **54**, 1–18 (1990)
3. Jaworski, B.J., Kohli, A.K.: Market orientation: antecedents and consequences. J. Mark. **57**, 53–70 (1993). https://doi.org/10.2307/1251854
4. Kohli, A.K., Jaworski, B.J., Kumar, A.: MARKOR: a measure of market orientation. J. Mark. Res. **30**, 467–477 (1993)
5. Ruekert, R.W.: Developing a market orientation: an organizational strategy perspective. Int. J. Res. Mark. **9**, 225–245 (1992). https://doi.org/10.1016/0167-8116(92)90019-H
6. Deshpandé, R., Farley, J.U., Webster Jr., F.E.: Corporate culture, customer orientation, and innovativeness in Japanese firms: a quadrad analysis. J. Mark. **57**, 23–37 (1993)
7. Qu, R., Ennew, C.T.: An examination of the consequences of market orientation in China. J. Strateg. Mark. **11**, 201–214 (2003). https://doi.org/10.1080/0965254032000133449
8. Quintana-Déniz, A., Beerli-Palacio, A., Martín-Santana, J.: Human resource systems as antecedents of hotel industry market orientation: an empirical study in the Canary Islands, Spain. Int. J. Hosp. Manag. **26**, 854–870 (2007). https://doi.org/10.1016/j.ijhm.2006.07.007
9. Polo-Peña, A.I., Frías-Jamilena, D.M., Rodrigues-Molina, M.Á.: Validation of a market orientation adoption scale in rural tourism enterprises. Relationship between the characteristics of the enterprise and extent of market orientation adoption. Int. J. Hosp. Manag. **31**, 139–151 (2012). https://doi.org/10.1016/j.ijhm.2011.06.005
10. Agarwal, S., Erramilli, M.K., Dev, C.S.: Market orientation and performance in service firms: role of innovation. J. Serv. Mark. **17**, 68–82 (2003). https://doi.org/10.1108/08876040310461282
11. Sin, L.Y.M., Tse, A.C.B., Heung, V.C.S., Yim, F.H.K.: An analysis of the relationship between market orientation and business performance in the hotel industry. Int. J. Hosp. Manag. **24**, 555–577 (2005). https://doi.org/10.1016/j.ijhm.2004.11.002
12. Haugland, S.A., Myrtveit, I., Nygaard, A.: Market orientation and performance in the service industry: a data envelopment analysis. J. Bus. Res. **60**, 1191–1197 (2007). https://doi.org/10.1016/j.jbusres.2007.03.005
13. Zhou, K.Z., Brown, J.R., Dev, C.S.: Market orientation, competitive advantage, and performance: a demand-based perspective. J. Bus. Res. **62**, 1063–1070 (2009). https://doi.org/10.1016/j.jbusres.2008.10.001

14. Campo, S., Díaz, A.M., Yagüe, M.J.: Market orientation in mid-range service, urban hotels: how to apply the MKTOR instrument. Int. J. Hosp. Manag. **43**, 76–86 (2014). https://doi.org/10.1016/j.ijhm.2014.08.006

15. Sargeant, A., Mohamad, M.: Business performance in the UK hotel sector - does it pay to be market oriented? Serv. Ind. J. **19**, 42–59 (1999). https://doi.org/10.1080/0264206990 0000029

16. Au, A.K.M., Tse, A.C.B.: The effect of marketing orientation on company performance in the service sector: a comparative study of hotel industry in Hong Kong and New Zealand. J. Int. Consum. Mark. **8**, 77–87 (1995)

17. Sandvik, I.L., Sandvik, K.: The impact of market orientation on product innovativeness and business performance. Int. J. Res. Mark. **20**, 355–376 (2003). https://doi.org/10.1016/j.ijresmar.2003.02.002

18. Gray, B.J., Matear, S.M., Matheson, P.K.: Improving the performance of hospitality firms. Int. J. Contemp. Hosp. Manag. **12**, 149–155 (2000). https://doi.org/10.1108/095961100103 20643

19. Churchill, G.A.: A paradigm for developing better measures of marketing constructs. J. Mark. Res. **16**, 64 (1979). https://doi.org/10.2307/3150876

20. Webb, J.R.: Understanding and Designing Marketing Research, 2nd edn. Thomson Learning, Stamford (2002)

21. Nunnaly, J.C., Bernstein, I.H.: Psychometric Theory, 3rd edn. McGrawHill, New York (1994)

22. Falk, R.F., Miller, N.B.: A Primer for Soft Modeling. University of Akron Press, Akron (1992)

23. Bollen, K.A., Stine, R.A.: Bootstrapping goodness-of-fit measures in structural equation models. Sociol. Methods Res. **21**, 205–229 (1992). https://doi.org/10.1177/00491241920 21002004

24. Fornell, C., Larker, D.F.: Evaluating structural equation models with unobservable variables and measurement error. J. Mark. Res. **18**, 39–50 (1981). https://doi.org/10.2307/3151312

25. Segars, A.H.: Assessing the unidimensionality of measurement: a paradigm and illustration within the context of information systems research. Omega **25**, 107–121 (1997). https://doi.org/10.1016/S0305-0483(96)00051-5

26. Hair, J.F., Ringle, C.M., Sarstedt, M.. PLS-SEM: indeed a silver bullet. J. Mark. Theory Pract. **19**, 139–152 (2011). https://doi.org/10.2753/MTP1069-6679190202

27. Eurostat: Annual enterprise statistics by size class for special aggregates of activities (NACE Rev. 2) (2017). http://ec.europa.eu/eurostat/web/tourism/data/database

28. Slater, S.F., Narver, J.C.: Does competitive environment moderate the market orientation-performance relationship? J. Mark. **58**, 46–55 (1994). https://doi.org/10.2307/1252250

A Multivariate Approach
to Entrepreneurial Intentions

Oscarina Conceição[1,2,4(✉)] (iD), Teresa Dieguez[1,3] (iD),
and Márcia Duarte[1,4] (iD)

[1] Polytechnic Institute of Cávado and Ave, Barcelos, Portugal
oconceicao@ipca.pt
[2] Instituto Universitário de Lisboa (ISCTE-IUL),
DINÂMIA'CET-IUL, Lisbon, Portugal
[3] Polytechnic Institute of Porto, Porto, Portugal
[4] UNIAG, APNOR, Lisbon, Portugal

Abstract. This study intends to explain the entrepreneurial intention using a
multivariate model. The purpose is to generate a more complete explanatory
model that can describe the entrepreneurial intentions. It was applied a questionnaire to the 40 master students who attended the curricular unit
Entrepreneurship in the 2016/2017 school year. The results show that personal
background (gender and attendance of an entrepreneurship course), business
knowledge (involvement in patenting activities and protection of intellectual
property, possess analytical skills and possess the ability to think critically),
entrepreneurial motivations (satisfy a market need and create something for
oneself), and the institutional environment (knowledge of IPCA structures
support to entrepreneurship) contribute for entrepreneurial intentions of master
students. These results are then discussed in terms of theoretical and practical
implications for entrepreneurship.

Keywords: Business knowledge · Entrepreneurial intention
Entrepreneurial motivations

1 Introduction

Entrepreneurship is a research's field with multiple concepts and theories, but without
theoretical consensus [1]. However, entrepreneurship contributes to social and economic development. Higher Education can have an important role in this process, by
providing a culture of positivity that allows leverage the entrepreneurial activity
through the development of new and creative ideas and business opportunities. Institutions like Academy has an unquestionable role in this kind of metamorphosis [2].

O. Conceição and M. Duarte—UNIAG, R&D unity supported by FCT – Foundation for Science
and Technology (FCT), Ministry of Science, Technology and Higher Education by the project
UID/GES/04752/2016.

This paper aims to understand which variables explain the entrepreneurial intentions. Most studies have focused on identifying a single variable or a limited set of variables. The objective is to apply a multivariate model to the explanation of the entrepreneurial intentions of the IPCA students and identify the variables that have a greater explanatory weight.

2 Theorical Background

The most used theoretical framework in the study of entrepreneurial intentions is the Theory of Planned Behavior [3] which defends the strength of intention as an immediate antecedent of behavior.

Entrepreneurial intention can be evaluated regarding five main dimensions, namely: (1) personal background, (2) business knowledge, (3) entrepreneurial motivations, (4) entrepreneurial self-efficacy and (5) institutional environment [4, 5].

Concerning the individual characteristics of entrepreneurs, some examples are, the economic literature on entrepreneurship highlights the demographic characteristics, the family and professional antecedents, the formation and academic qualification, the attitudes, the values and the motivations [5]. Some studies concluded that female students have lower propensity to entrepreneurial career intentions [6]. For Davidsson, attitudes, values and personal achievement explained differences between males and females [7].

H1:   Personal background is related to entrepreneurial intentions in case of IPCA students.
      Regarding business knowledge, the needed valued skills for creation and development of a business are the identification and use of a business opportunity; the ability to relate it to relevant business people, through the communicational and leadership skills, for example; conceptual skills, which involve decision-making and problem-solving; the formulation of business strategy and objectives [4]. Several studies [8, 9] support the thesis's that entrepreneurs are more minded to personal achievement than the general population.

H2:   Business knowledge is related to entrepreneurial intentions in case of IPCA students.
      Carsrud and Brannback argued that entrepreneurial motivations remain largely under research despite its critical importance in predicting entrepreneurial behaviors [10]. People with high need for achievement prefer tasks that involve skill and effort, provide clear performance feedback, and moderate risk or challenge [9]. In a meta-analysis, Collins, Hanges, and Locke found support for achievement's need in predicting entrepreneurial activity and performance in an entrepreneurial role [11].

H3:   Entrepreneurial motivations are related to entrepreneurial intentions in case of IPCA students.
      Most studies on entrepreneurial intentions included, as an explanatory variable, self-efficacy, that represents the belief in one's capabilities to organize and execute the courses of action required to manage prospective situations [12, 13].

H4:   Self-efficacy is related to entrepreneurship intentions in case of IPCA students. The institutional environment, in particular the supports, the initiatives and the units in the Higher Education Institutions, can create an environment that can lead to a development of student's entrepreneurial spirit [14, 15].

H5:   Institutional environment is related to entrepreneurship intentions in case of IPCA students.

3 Methodology and Measures

The main purpose of this work is to evaluate the entrepreneurial intentions of master students who attended the curricular unit of Entrepreneurship. Previous empirical studies used only one class of explanatory variable [16] but our purpose is to generate a more complete explanatory model that describes the entrepreneurial intention of the students who attended the curricular unit of Entrepreneurship, inspired in the original model of Davidsson [7].

Following the model of Carvalho and González [4], five dimensions to measure the entrepreneurial intentions have been analyzed: the personal background; the business knowledge; the entrepreneurial motivations; the entrepreneurial self-efficacy and the institutional environment.

The final sample is composed by 40 master students which attended, in the 2016/2017 school year, the Master Degree in Tourism Management and Master Degree in Business Management in the Polytechnic Institute of Cávado and Ave (IPCA). The students were mainly female (55%) and mainly with ages between 30 and 40 years (63%). The Master's Degree in Tourism Management was represented by 43% of the respondents and the Master's Degree in Business Management by 57%. Concerning familiar background, 65% of the students had relatives with previous experiences on entrepreneurship. From all the 40 students, 26 (65%) had previous professional experience of less than five years. Relating previous attendance of an entrepreneurship curricular unit, 5% of the respondent had it on Technical Education and 70% had it at university or Polytechnic.

The data obtained from the questionnaire enabled us to build several variables that are used as multidimensional measures of the personal background (scientific area, year, student status, age, gender, professional experience, entrepreneurial family background, parents' academic qualifications); the business knowledge (opportunity, strategy, relational skills, conceptual skills); the entrepreneurial motivations (need for independence, need for personal development, perception of wealth instrumentality, need for approval); the entrepreneurial self-efficacy (expectations of future success) and the institutional environment (encouragement to put in practice entrepreneurial ideas, knowledge of the existence and use of IPCA entrepreneurship support units).

Considering that our goal was to analyze the entrepreneurial intention we used, as a dependent variable, a measure of the intention to "start your own business or work for yourself". The variable Entrepreneurial Intention is a categorical variable that distinguishes between students who have stated an interest in starting their own business/self-employment and the students that did not. As independent variables we used the five dimensions measures, however, are indented.

4 Results

Please note Non-parametric tests were performed to test the individual influence of the explanatory variables on the entrepreneurial intention (dependent variable). Regarding the Personal Background dimension, the non-parametric test indicates that two categories individually, influence on the entrepreneurial intention: "gender" and "have attended an entrepreneurship discipline in higher education" (*p-value* of 0.032 and 0.079, respectively). Concerning the Business Knowledge, dimension the non-parametric test indicate that three categories individually influence the entrepreneurial intention: "Already been involved in the patenting of a technology or in the protection of intellectual property", "possess analytical skills" and "possess the ability to think critically" (*p-value* of 0.086; 0.036 and 0.022, respectively). Relating to the Entrepreneurial Motivations dimension the non-parametric test indicates that none of the categories that integrate this dimension have individually influence on entrepreneurial. Regarding the Entrepreneurial Self-Efficacy dimension the non-parametric test indicate that the category "have the skills to rigorously estimate the costs of implementing a new project" has, individually, influence on entrepreneurial intentions (*p-value* of 0.061). Finally, about the Institutional Environment dimension the non-parametric test indicate that none of the categories that integrate this dimension have individually influence on entrepreneurial intentions.

In order to analyze the simultaneous behaviour of the five dimensions as predictors of entrepreneurial intention, a multivariate analysis was performed. Considering the dichotomous nature of the dependent variable entrepreneurial intention (1 = Intends to start your own business or work for yourself) we run the model using a stepwise logistic regression. Given the Omnibus Tests of Model Coefficients the models provide a good fit to the data; the chi-squared goodness-of-fit test for the change in the −2Loglikelihood value revealed to be statistically significant (*chi-square = 26.414, p-value = 0.009*). The significance value of less than 0.05 provides support for acceptance of the model.

Regarding the Pseudo-R2 Nagelkerke, the model explains 78% of the variance (see Table 1). Concerning the Personal Background dimension, the results show that "being" female decreases the odds of intends to start your own business or work for yourself, as the proportionate change of odds (Exp b) is below 1. Concerning the Entrepreneurial Motivations dimension the results show that the personal motivation to satisfy a market need and create something for oneself increases the odds of intends starting their own business/self-employment. About the Institutional environment dimension the results show that know the IPCA structures for support entrepreneurship increases the odds of intends to start or work own business (Table 1).

Table 1. Results of logistic regression.

Dimensions	Variables	Exp (B)
Personal background	Age	,956
	Gender	,022*
	Familiar entrepreneurial context	1,017
	Previous professional experience	,593
	Previous education on entrepreneurship	,000
Business knowledge	Training/active learning	1107,913
	Experimentation/experimental learning	1,046
Entrepreneurial motivations	Perceived benefits	15309491541,766*
	Perceived difficulties	,000
Entrepreneurial self efficacy	Design thinking, innovation and management	7,049
Institutional environment	Managerial and technological support	1364,944*
	Technological transfer, incubation, internationalization	,000
	Constant	,021
Pseudo-R² Nagelkerke		,778
Valid N		40

*Sig ≤ 0.05

5 Discussion

The results of non-parametric tests show that some categories of the five dimensions are related to entrepreneurial intentions in this sample. These are important results and deserve attention and discussion.

Regarding the personal background dimension some studies have suggested that women are less inclined to men to create her own businesses [6]. The lack of confidence on their abilities, the perceived lack of support and the fear of failure are important antecedents that explain the hesitation of women in decisions to find their own ventures. In fact, the non-parametric tests show that the attendance of an Entrepreneurship course is a predictor of entrepreneurial intentions. This study seems to reinforce the relevance of the Entrepreneurship Education in formation of entrepreneurial intentions supporting H1 [15, 17].

Concerning the business knowledge dimension, the variable level "Already been involved in the patenting of a technology or in the protection of intellectual property" has, individually, influence on entrepreneurial intentions. This result shows that the students who already start to put in practice their business plans have more intentions to start a business. Regarding the variables "possess analytical skills" and "possess the ability to think critically" they have, individually, influence on entrepreneurial intentions. Analytical skills and critical thinking refer to conceptual and abstract skills required to a good entrepreneur. Most of these students possess solid management knowledges acquired in higher education courses and perceive the importance of analytic thinking, complex decision-making, environmental analysis, and critical and long term thinking for the success of a business [4]. These business knowledges are strengthened by the attendance of entrepreneurship courses. These results globally show that perceiving to have previous business knowledges, individually, are a significant predictor of entrepreneurial intentions of master students (support H2).

Self-efficacy is identified in most studies about entrepreneurial intentions. In non-parametric tests the variable "have the skills to rigorously estimate the costs of implementing a new project" has, individually, influence on entrepreneurial intentions (support H4).

Regarding the entrepreneurial motivations dimension, the results of logistic regression show that the personal motivation to satisfy a market need and create something for oneself are predictors of the entrepreneurial intentions (support H3). The need for autonomy and independence are entrepreneurship motivations well identified in literature [5, 10, 11] as well the need for achievement. Satisfying a market need can mean that the individual has a strong need for achievement and personal development.

Finally, the regression analysis also shows that institutional environment, namely the knowledge of IPCA structures that support entrepreneurship, explains entrepreneurial intentions of master students (support H5). IPCA has structures that support entrepreneurship, namely, PRAXIS XXI e G3E. These structures help students in the business idea development, in the creation of spin-offs, give support in trademark registration and offer training in entrepreneurship, for example.

6 Conclusions

The main contribution of this study is the use of a multivariate model to explain entrepreneurial intentions of IPCA master students. Our model shows that personal background (namely gender and attendance of an entrepreneurship course), business knowledge (involvement in patenting activities and protection of intellectual property, possess analytical skills and possess the ability to think critically), entrepreneurial motivations (satisfying a market need and create something for oneself), and the institutional Environment (knowledge of IPCA structures support to entrepreneurship) explains the entrepreneurial intentions of master students. By the point of view of Entrepreneurship Education, the Higher Education Institutions should invest in the development of soft skills, particularly in female students.

In terms of future research, it would be interesting to analyse the impact of the course of Entrepreneurship from the perspective of a longitudinal study. Future research might examine the entrepreneurial commitment as the missing link between intentions and behaviours [2].

References

1. Palma, P.J., Cunha, M.P.: New challenges in entrepreneurship: introduction to the special issue. Comport. Organ. Gest. **12**(1), 3–6 (2006)
2. Fayolle, A., Liñán, F.: The future of research on entrepreneurial intentions. J. Bus. Res. **67**, 663–666 (2014)

3. Ajzen, I.: From intentions to actions: a theory of planned behavior. In: Kuhl, J., Beckmann, J. (eds.) Action-Control: From Cognition to Behavior, pp. 11–39. Springer, Heidelberg (1985)

4. Carvalho, P.M., González, L.: Modelo explicativo sobre a intenção empreendedora. Comport. Organ. Gest. **12**(1), 43–65 (2006)

5. Dinis, A., Ussman, A.M.: Empresarialidade e empresário: revisão da literatura. Comport. Organ. Gest. **12**(1), 95–114 (2006)

6. Shinnar, R.S., Giacomin, O., Janssen, F.: Entrepreneurial perceptions and intentions: the role of gender and culture. Entrep. Theory Pract. **36**, 465–493 (2012)

7. Davidsson, P.: Determinants of entrepreneurial intentions. Paper prepared for the RENT IX Workshop, Piacenza, Italy, pp.23–24, November 1995

8. Jaafar, M., Abdul-Aziz, A., Maideen, S., Mohd, S.: Entrepreneurship in the tourism industry: issues in developing countries. Int. J. Hosp. Manag. **30**(4), 827–835 (2011)

9. McClelland, D.C.: The Achieving Society. Van Nostrand Reinold Princeton (1961)

10. Carsrud, A., Brannback, M.: Entrepreneurial motivations: what do we still need to know? J. Small Bus. Manag. **49**(1), 9–26 (2011)

11. Collins, C.J., Hanges, P.J., Locke, E.A.: The relationship of achievement motivation to entrepreneurial behavior: a meta-analysis. Hum. Perform. **17**(1), 95–117 (2004)

12. Bandura, A.: Self-Efficacy mechanism in human agency. Am. Psychol. **37**(2), 122–147 (1982)

13. Wilson, F., Kickul, J., Marlino, D.: Gender, entrepreneurial self-efficacy, and entrepreneurial career intentions: implications for entrepreneurship education. Entrep. Theory Pract. **31**, 465–493 (2007)

14. Bae, T.J., Qian, S., Miao, C., Fiet, J.: The relationship between entrepreneurship education and entrepreneurial intentions: a meta-analytic review. Entrep. Theory Pract. **38**, 217–254 (2014)

15. Martin, B.C., McNally, J.J., Kay, M.: Examining the formation of human capital in entrepreneurship: a meta-analysis of entrepreneurship education outcomes. J. Bus. Ventur. **28**, 211–224 (2013)

16. Autio, E., Keeley, R.H., Klofsten, M., Ulfstedt, T.: Entrepreneurial intent among students. Testing an intent model in Asia, Scandinavia and USA. In: Frontiers of Entrepreneurship Research, Proceedings of the 17th Annual Babson College Entrepreneurship Research Conference (1997)

17. Heuer, A., Kolvereid, L.: Education in entrepreneurship and the theory of planned behaviour. Eur. J. Train. Dev. **38**(6), 506–523 (2013)

The Attractiveness of Retail Clusters

Denis Donaire$^{(\boxtimes)}$, Milton Carlos Farina,
and Edson Keyso de Miranda Kubo$^{(\boxtimes)}$

Universidade Municipal de São Caetano do Sul,
Av. Goias, 3400, São Caetano do Sul 09550051, Brazil
edsonkubo@uscs.edu.br

Abstract. The objective was to evaluate the perception of consumers in the purchase process on the attributes considered responsible for the attractiveness of retail cluster. In this sense, a quantitative research was done in which 240 consumers were surveyed in two different retail clusters. The proposed model used the modeling of structural equations based on the partial least squares and the results indicate that the customer service was the most important variable. On the other hand, there was low value for the location, which is usually considered one of the most important factors of the retail mix. Thus, it can be concluded that if there is a retail cluster in which we have good service, prices, conditions of purchase and reliable stores, its location can be made in any region that will result in a strong attractiveness for consumers.

Keywords: Retail cluster · Attractiveness · Perception of consumer

1 Introduction

The location decision is crucial for retailers. It is a fundamental strategic decision because it is the first consideration of consumers in the purchase process [1]. There is a quote from Baron Sieff that emphasizes this: "There are three important things in retail: location, location, and location" [2].

It is known that the location decision is crucial for the retailer. In this sense, it is a fundamental strategic decision, because this is the first consideration that the consumer realizes when deciding on the purchase process [1].

In the same line of reasoning, the location choice is considered one of the most important decisions in the retail sector, which is why a great number of studies have been directed at this issue, seeking the development of techniques and models that help in determining the best retail location [3]. Although a good location is not enough to compensate for mediocre strategies, the disadvantages of a poor location are extremely difficult to overcome.

However, although the retail point of sale is very important, when choosing where to make purchases, the consumer also considers other aspects, such as the images they have of stores, purchase conditions, prices, product attributes (which may be unique to a single store), services, parking, etc. This multiplicity of factors is responsible for the perceived attractiveness of a retail company.

© Springer International Publishing AG, part of Springer Nature 2019
J. Machado et al. (Eds.): HELIX 2018, LNEE 505, pp. 1073–1079, 2019.
https://doi.org/10.1007/978-3-319-91334-6_148

Considering all these aspects that must be evaluated in the behavior of the consumer when being attracted to choose a retail cluster, the question that is intended to elucidate in this research is the following:

"Under a retail cluster, which presents characteristics different from those observed at the level of stores, how can we assess the multi-attributes concerning the attractiveness of the retail over in consumer's shopping experience?"

The research objective can be defined as follows: to evaluate at the time of purchase the perceptions of consumers about the attributes considered responsible for the attractiveness exerted by the retail cluster.

2 Theoretical Framework

2.1 Retail

The retail sector is a set of organizations involved in the process of making products or services available for use or consumption. In addition to reducing distribution costs, retailers increase the alignment and value of the offer to consumers, as they act as important links in the distribution channels by providing products and services from manufacturers to end consumers [4]. Consumers often evaluate products and stores using certain criteria, known as 'attributes' [5]. Although the set of attributes varies according to each type of retail store, one can classify them in accordance with the 6Ps of the retail mix: products, prices, presentation, promotions, personnel, and point (location), as shown in Table 1.

Table 1. Example attributes for evaluation of retailers

6P's of retail mix	Attributes
P – Product mix	-variety -quality -exclusivity
P – Presentation	-lay out - decoration and atmosphere
	- visual communication -comfort
P – Price	-price to benefit/costs-time frame and method of payment
P – Promotion	-advertising - promotion at the point of sale
	- loyalty programs
P – Personnel	-service-interest and courtesy - qualification
P – Point and location	-proximity – accessibility
	-complementarity with other stores

As we are interested in identifying the relevance of such attributes to retail clusters, we will consider the six variables such as products, prices, stores, service, purchasing conditions and location. This way it replaces presentation by the stores, personnel by service and promotion by conditions of purchase.

Location. Location is a key factor in consumers' process of selecting where to buy goods. Consumers, in general, exclude from their selection of alternatives those shops for which there is a need to travel tens of kilometers [5]. Thus, distance is important but is not definitive, as other factors (such as quality, assortment and price) can significantly influence the choice of place of purchase. In general, there are three possibilities for retail locations: isolated stores, shopping centers (including clusters) and shopping centers (regional, community or neighborhood) [6]. Depending on the business type, isolated stores may be the best option in many cases, as they are not competitive and have cheaper rental costs, but they may have high costs in attracting buyers to their points of sale. Malls serve large areas of influence and are characterized by a central administration and a distribution of stores containing one or more anchor stores, as well as a number of convenience stores. Commercial areas represent traditional commerce based on the city's central region and have a large flow of people, which, over time, gives rise to specific conglomerates selling related products – retail clusters, which are the focus of this work.

Attractiveness and the Consumer Behavior. The consumer goes through several stages in the buying decision process [2]. These stages have been cemented in the literature on consumer behavior by authors [7–11] we can say that consumer behavior goes through five stages in their purchasing decision:

1. Recognition of Need
2. Information Search
3. Evaluation of Alternatives
4. Purchase Decision
5. Behavior after Purchase.

When evaluating consumer behavior, it should be considered that although the stages of the purchasing decision process can be presented in a compartmentalized way, in reality, they can occur concomitantly and may not be unidirectional: that is, stages may be resumed after they have been passed [4].

After the evaluation of the alternatives, the consumer feels able to make their purchase decision. At that moment, prior to the actual purchase, the intention to buy is located, which results from the convergence between the behavior of the consumer and the behavior of the retailer (the six variables of the retail mix), which in this article we are defining as attractiveness. Attractiveness is equal to the convergence between consumer behavior and retailer behavior.

This intention of purchase, which reflects the attractiveness of the product, leads the consumer to choose the retailer company that best reflects the needs and desires of his customers: shopping malls, isolated stores or retail clusters.

In this research, we are interested in establishing exactly the purchase intention point that should coincide with the attractiveness resulting from the convergence between consumer behavior and retailer behavior, seeking to identify what retailer attributes it considers important when addressing a retail cluster. Figure 1 below shows this theoretical configuration.

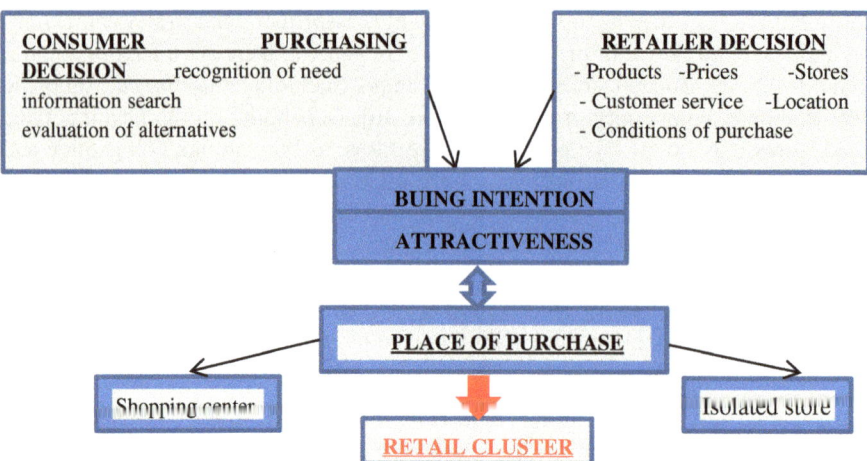

Fig. 1. The attractiveness of retail clusters

3   Research Method

This intention of purchase, which reflects the attractiveness of the product, leads the consumer to choose the retailer company that best reflects the needs and desires of his customers: shopping malls, isolated stores or retail clusters.

In order to evaluate the attributes responsible for the attractiveness of a retail cluster, a quantitative descriptive study was carried out with consumers who were physically present in two important retail clusters in the city of S. Paulo. The influence of e-commerce was not considered in the assessment of the attractiveness of a retail cluster.

The results were export to the SPSS version 21 (Statistical Package for the Social Science) and PLS-PM 3.0 (Partial Least Squares Path Modeling) applications. The treatment technique for the data analysis was the Modeling of Structural Equations, in order to enable the understanding of the relations between the formative latent variables (product, price, store, location, purchase and service conditions) and the reflexive latent variable named attractiveness of the cluster. In order to identify the formative latent variables, a certain model [12] was initially used, which identifies the existing foundations in a cluster that has increased its capacity to compete. Given that consumers would have difficulties interpreting these fundamentals and their unfolding, we chose to identify the resulting observable effects, which were evaluated in an empirical research carried out with the consumers that were located within a certain cluster (Fig. 2).

For the application of this model, a total of 240 consumers were surveyed, of which 120 were found in the bridal retail cluster located at Rua São Caetano (preferably attended by women) and another 120 were found in the electronics retail cluster located at Rua Santa Ifigênia (attended mostly by men). Both clusters were located in São Paulo.

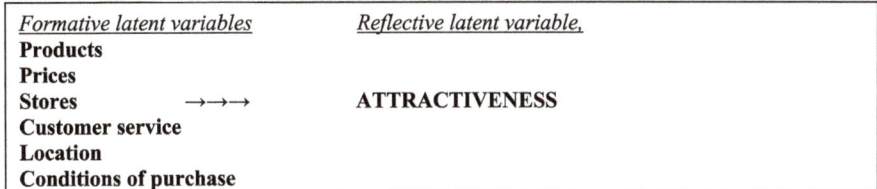

Formative latent variables	Reflective latent variable,
Products	
Prices	
Stores →→→	**ATTRACTIVENESS**
Customer service	
Location	
Conditions of purchase	

Fig. 2. Retail cluster latent variables

4 Presentation and Analysis of the Results

Considering the logic of the PLS-PM application, the treatment and validation of formative latent variables (which refer to the variables of the retail mix, in our case: products, prices, stores, service, location and conditions of purchase), the results were developed in the measurement model, which establishes the Beta coefficient values of the factorial load of the observable effects on each of the six variables of the retail mix, as well as the degree of influence of each variable on the attractiveness of consumers in the retail clusters.

Student's t values were checked for the variables under study, indicating acceptance for the coefficients found both for the observable effects and for the chosen variables.

4.1 Measurement Model - Influence of Latent Formative Variables

The six dimensions of the variables of the retail mix were evaluated in their influence on the degree of attractiveness of the retail clusters surveyed (Fig. 3).

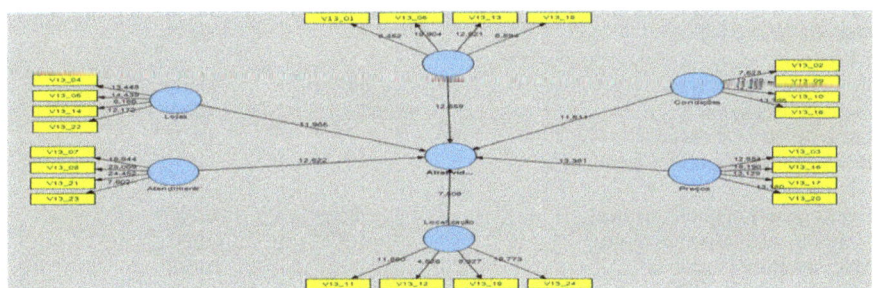

Fig. 3. PLS-PM model: beta coefficients

The dimensions found in the PLS-PM model indicate:

- Variable CUSTOMER SERVICE (Atendimento) with factorial load 0.233
- Variable PRICES (Preços) with factorial load 0.214
- Variable CONDITIONS OF PURCHASE (Condições) with factorial load 0.209
- Variable STORES (Lojas) with factorial load 0.201
- Variable PRODUCTS (Produto) with factorial load 0.189
- Variable LOCATION (Localização) with factorial load 0.159.

It is possible to observe that, in the views of the consumers, the variables of the customer service, prices, conditions of purchase and stores were more significant than products and location. It is noteworthy the fact that customer service was the variable considered most important for attracting consumers to the retail cluster. On the other hand, it is worth noting the low value found for location, which is usually considered one of the most important factors of the retail mix. Taking into account the opinions of the consumers of the two clusters surveyed, it can be considered that if there is a retail cluster in which we have good customer service, low prices, and good stores, its location can be made in any region that will result in strong attractiveness for consumers (Fig. 4).

1. **Customer Service**:		
Good quality of service	charge	0.843
More Specialty Sellers		0.836
Better technical service		0.827
Reliability in the delivery deadline		0.660
2.**Prices**:		
Most honest prices	charge	0.781
Cheapest prices		0.764
Product price range		0.731
Better Payment Terms		0,730
3.**Conditions of Purchase**:		
Search Conditions for Purchases	charge	0.787
Quicker purchasing conditions		0.740
It is more pleasant to shop		0.740
Further information on prices and products		0,669
4.**Stores**:		
Largest diversity of stores of each type	charge	0,774
Increased competition among stores		0.770
Appropriate working hours		0.723
More Reliable Stores		0.678
5.**Products**:		
Best quality products	charge	0.813
More product news		0,768
Large variety of products		0,674
Best technical support (of products)		0.591
6.**Location**:		
Higher quality of public services	charge	0.797
More ease of transport to reach		0.760
More safety for purchases		0.721
Largest parking facilities		0.530

Fig. 4. Influence of observables effects on each variable in the retail mix

5 Final Considerations

Knowledge of how the consumer is drawn to shopping in a retail cluster is a very important aspect to research. Although a very large number of studies on this subject have already been conducted in stores, in this article, the objective was to verify how this occurs when the purchase intention is directed not at a specific store but at a cluster of them, which have related products, as is the case with retail clusters.

The field research was able to verify the significant influence of customer service as the most important attraction factor for a consumer to shop in a retail cluster. This fact underscores the importance of the workforce in a store located in a retail cluster, which should be the target of continuous improvement and training. In addition, it was evident in the research that the factor of location is greatly minimized in these clusters, with the evidence that wherever a retail cluster is formed, there will always be sufficient attraction through other variables, such as customer service, prices and purchase conditions, which presented more-significant values.

References

1. Levy, M., Weitz, B.A.: Retailing Management, 10h edn. Atlas, New York (2000)
2. Mattar, F.N.: Administração de varejo, 4th edn. Elsevier, Rio de Janeiro (2011)
3. Botelho, R.V.: Os modelos de localização e os *shopping centers*. In: 3th SEMEAD – Seminários de Estudo em Administração, pp. 1–16. FEA-USP, São Paulo (2007)
4. Oliveira, B.: Gestão de Marketing, 2nd edn. Pearson Prentice Hall, São Paulo (2011)
5. Parente, J.: Varejo no Brasil: gestão e estratégia, 2nd edn. Atlas, São Paulo (2000)
6. Las Casas, A.L.: Marketing de varejo, 3rd edn. Atlas, São Paulo (2004)
7. Howard, J.A., Sheth, J.N.: A theory of buyer behavior in changing marketing systems. In: Consumer, Corporate And Government Interfaces: Proceedings of the Winter Conference of Marketing Association. AMA, Chicago (1967)
8. Rivas, J.A.: Comportamiento del consumidor, 8th edn. Esic, Madrid (1997)
9. Schiffman, L., Kanuk, L.L.: Consumer Behavior, 8th edn. Prentice Hall, Upper Saddle River (2000)
10. Solomon, M.R.: Consumer Behavior: Buying, Having and Being, 4h edn. Prentice-Hall, Upper Saddle River (2002)
11. Engel, J., Blackwell, R., Miniard, P.: Consumer Behavior, 9th edn. Thomson Learning, Southwestern (2005)
12. Zaccarelli, S.B., Telles, R., Siqueira, J.P.L., Boaventura, J.M.G., Donaire, D.: Clusters e Redes de Negócios: uma nova visao para a gestão dos negócios, 1st edn. Atlas, São Paulo (2008)

Innovative Strategic Cluster for the Smart Field of Mechatronics and Cyber - MixMecatronics in Romania - MECHATREC

Gh. Ion Gheorghe(✉)

MECHATREC Cluster, 6-8 Pantelimon Road, 2nd District, Bucharest, Romania
geocefin@yahoo.com

Abstract. MECHATREC cluster has been designed, built and developed in Romania since 2008 as a value-added growth vector in the Bucharest-Ilfov Region and as an integrating vector of strategic innovation in Romania, based on effective collaboration protocols with over 15 clusters, 6 regional development agencies (ADR) and 7 chambers of commerce and industry in Romania and by extension to the European Union based on strategic collaboration protocols with over 12 clusters from EU countries.

Keywords: MECHATREC cluster · Cyber-Mixmechatronics
Innovation in clusters

1 Preamble

Accelerating structural change and innovation in industry is at the forefront of Europe's growth agenda. Public and private investments need to be more strategic, smarter, and better targeted.

To achieve this, Europe's regions must offer entrepreneurs a favorable business environment and adequate support to their specific needs. This is why cluster strategies and smart specialization strategies are at the heart of the EU's growth strategy, as they contribute to boosting employment, growth of SME's and investment.

Smart specialization strategies are key elements of the new, result-oriented, innovation-driven results, which is currently being implemented at the level of the European Union. They help regions at different stages of development to focus on their strengths, position themselves in global value chains, and build strategic partnerships in Europe with other regions that have complementary strength in similar priority areas.

Clusters and cluster policies can facilitate this process and help maximize the impact of European structural and investment funds dedicated to investment in research and innovation, accompanied by other smart specialization strategies. The implementation of these strategies promises to trigger industrial upgrading processes and create more productive, diversified and visible economies. It should, in particular, help to provide better support and investment opportunities to small and medium-sized enterprises - the backbone of the EU economy.

In the vast majority of cases, clusters are not "created", but appear in the context of regional development needs, rather different locations offer different types of

© Springer International Publishing AG, part of Springer Nature 2019
J. Machado et al. (Eds.): HELIX 2018, LNEE 505, pp. 1080–1086, 2019.
https://doi.org/10.1007/978-3-319-91334-6_149

opportunities for certain companies to invest, succeed and grow. Clusters are the result of cumulative processes in which the success of a company or cluster opens the way for other companies or other clusters.

Such processes take longer and are inherently unpredictable. Cluster development is a natural process, in most cases, success depends on creating specific business environment that offers a unique and lasting location and advantage.

2   The Principle and the Model for MECHATREC Cluster

MECHATREC cluster is built on the principles and concepts of the four-leaf clover model, adapted and transformed from the European triple helix. The "four-leaf clover" model combines research, industry, local government and catalyst organizations - consultancy firms specialized in technology transfer and innovation, technology transfer centers, and so on.

At present, there are 144 entities (2018) in the cluster: 13 research and development entities, 9 public authorities, 22 catalyst institutions and 100 industrial enterprises (see Fig. 1).

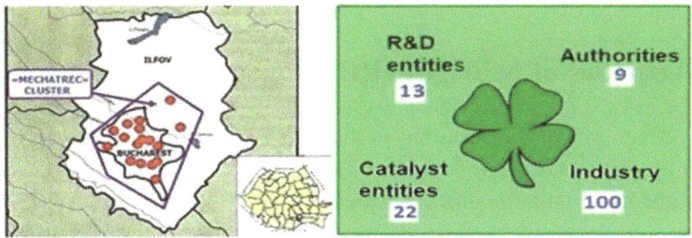

Fig. 1. MECHATREC cluster overview

MECHATREC has been accredited at European level, obtaining the Bronze label. In order to obtain the Bronze label, MECHATREC has been evaluated on multiple levels for the types of member organizations and their degree of involvement, organization, main activities, services offered, management team profile, funding available, attracted projects. The MECHATREC Cluster is currently audited for the Silver Label.

2.1   General Objectives

- developing the institutional capacity of the MECHATREC Cluster, improving competitiveness both for the Bucharest-Ilfov Region and the national and international SME members, and improving the internationally competitive position of both the cluster and its members;
- attracting public-private investments in the Bucharest-Ilfov region and other regions of Romania and increasing the number of employees in the innovative research, industry and business sector in the region, from Romania and Europe.

- attracting international members (and European members) to define international partnership (and European) of Mechatrec Cluster (wish you to become a member in this Cluster).

2.2   Specific Objectives

- Focusing the resources on key priorities and creating opportunities for developing the institutional capacity of the MECHATREC Cluster, as well as improving the regional and national competitiveness, as well as internationally, of the member SMEs;
- Increasing the volume of investments made by cluster companies;
- Increased exports by high-tech manufacturing/SMEs based on the intensive use of knowledge of smart cluster domains;
- Increased professional intelligence in advanced areas;
- Increasing labor productivity and cluster jobs;
- Increasing trade relations between cluster members and development regions in Romania;
- Development of cross-sectoral, national and international networks for the Mechatrec Cluster and other clusters in Romania and the EU.

3   Main Activities of MECHATREC Cluster

- research - development - innovation for the creation and development of innovative high-tech products and technologies;
- smart fabrication, collaboration between entities and support of commercial activities;
- information, communication, promotion, branding and public relations;
- comprehensive technological transfer and implementation by capitalizing and valorizing the results of RDI activities;
- training, qualification and skills development of human resources, both in the cluster and outside of it;
- domestic and international cooperation of SME's and the competent units of the cluster and other clusters in Romania and the EU;
- networking and internationalization, exchange of good practices.

4   Innovation in MECHATREC Cluster

Over the last decade, Europe has become a global leader in the use of cluster-based economic development tools. There have been numerous group initiatives supported by a wide range of government programs. In Romania, most of the strategies included clustering as a measure to stimulate innovation and strengthen competitiveness and regional development. Cluster initiatives were supported in Romania by the Sectoral Operational Program "Increasing Economic Competitiveness 2007–2013", co-financed by the European Regional Development Fund, Priority Axis 1 "An innovative and eco-efficient production system", Major field of intervention D1. 3 "Sustainable Development of Entrepreneurship" Operation "Support for enterprise integration in supplier or cluster chains".

The result of such a complex process is the MECHATREC Innovative Strategic Cluster in response to the business environment requirements of the Bucharest-Ilfov Region and Romania.

The MECHATREC cluster is positioned at the level of small and medium-sized enterprises within the reference sector, in particular Mechatronics and Cyber-MixMechatronics.

The focus of Bucharest, the capital of Romania, the most developed center of industrialization, with the highest participation in the country's budget, and the extension to Ilfov County of the border area - where a number of SMEs originally moved to Bucharest, has led to the creation of premises for a sectoral intervention targeting local SMEs like SMEs. These firms have know-how and experience that are becoming more interesting for multinationals.

This created the premises of a very valuable collaboration on joint projects with a high degree of technical complexity and at the same time they put together specialized human resources and high-tech research infrastructures that successfully contribute to the access and implementation of large projects complex in the field of Mechatronics and Cyber-MixMecatronics at both national and international level.

A concrete example is that for 6 RDAs of Romania, MECHATREC Cluster has become a Strategic Partner for important contributions to the elaboration of Smart Specialized Development Strategies (ADR Bucharest-Ilfov, ADR South-Muntenia, AR Centre, RDA North East Galati, etc.).

Another concrete and good practice example is the participation of the MECHA-TREC Cluster, with project proposals, at national and international competitions (e.g. the "Practice, a step towards employment" project and the project "Towards employment through practice" in the POCU Program the project "Increasing the Capacity of NGOs and Social Partners" - in the POCA Program 111/1/1, the project "Danube Transnational Cluster in Mechatronics" - in the INTERREG Danube Program, etc.)

The industrial sector covered by the MECHATREC cluster is in the category of emerging industries, defined at EU level either as new industrial sectors or as existing industrial sectors that are developing or reuniting in new industries. These are defined as "the creation of a completely new industrial value chain, or the radical reconfiguration of an existing one, focused by an integrative idea (or convergence of ideas), leading to the transformation of these ideas/opportunities into new value-added smart products (e.g. smart mechatronic systems and cyber-mix-mechatronic systems for the automotive industry, high-performance smart equipment, hydronic and pneumatic equipment, state-of-the-art optical and electrical equipment, general equipment and devices, etc.).

MECHATREC cluster has intervened in this market with the experience of multiple approaches (the creation of a network of subcontractors before launching the cluster, and collaborations and partnerships with similar entities active at European level such as: the Pimurmanget cluster, Hungary, the Futuralia International cluster, MECHA-TREC Cluster, Belgium, Mechatronic Cluster, Austria, International Mechatronics Centre, Linz, Austria).

The impact of MECHATREC cluster actions on innovation capacity is revealed by:

Increasing the innovative capacity of the research institutes materialized by capitalizing the own research in the industry towards the high-tech producers in the Bucharest Ilfov region and other development regions of Romania.

Establishing partnerships between research and development institutions universities - SMEs for the joint development of innovative applications for the smart specialized field.

Increasing the interaction between producers, research and education centers, investors and consultants, non-governmental organizations and public authorities so that the research topics have on one hand a closer connection with the concrete requirements of the moment and on the other hand a shorter time implementing products and technologies implemented and implemented.

Recognizing the results of R & D & I activities achieved by cluster members in the national and international scientific community.

5 MECHATREC Cluster Mission

Developing the cooperation of science and research with the business environment, creating conditions conducive to innovation and advanced technological transfer, increasing the competitiveness of the members, supporting innovation to achieve a profitable position within the value chains of regional, national and international research in the high-tech field of Smart Mechatronics and Cyber-MixMecatronics Integrating and internationalizing businesses under a strong common brand and promoting an appropriate research and production environment, as well as attracting young people looking for a career in research as well as optimizing resources and fostering the attraction of skilled labor [1–4].

6 MECHATREC Cluster Vision

The vision of the MECHATREC cluster is to become one of the most credible suppliers of high-tech innovative products and services in Romania, Europe and the world. The motto of the "Mechatronics & Cyber-MixMechatronics for the Future!" Cluster demonstrates the importance of smart high-tech, the formation of human capital in excellence and innovation research, as well as the actual R & D activity, producing knowledge, innovations and smart products and advanced technologies coupled with smart economic growth, but above all the importance of a cluster-type concentration with a national and international vocation to develop this innovative field.

The impact of MECHATREC cluster actions on international competitiveness is revealed through:

- creating a viable SME network with high-tech activities in the region, linked to existing technologies at international level and major EU targets for the 2020 and 2030 horizons;
- increasing exports of cluster members;

- increasing the degree of interaction of foreign investments in the high-tech field in the Ilfov region with the local business environment and cluster member companies;
- developing partnerships with leading universities that cooperate with R & D institutes and other EU Member States in RDI activities for new and advanced technologies.

7 MECHATREC Cluster Offer

Smart Products and Cyber-Mix-Mechatronic Systems: smart mechatronic/integrated/adaptive and cyber-mix-mechatronic smart devices, systems, devices and other mechatronic/integrated/smart instruments for applications in industrial and automotive processes (auto, pneutronics, hydronics, agrotronics, energotronics, medtronics, etc.); smart micro-nano-mechatronic and smart micro-nano-mechatronic micro-cyber applications with the use of lasers in civil, aerospace, military, etc.; smart mechatronic/integrated/adaptive and cyber-mix-mechatronic devices for micro and nano medical, biomedical, environmental applications, etc.; smart industrial robots based on mechatronic/integrated/adaptive and/or cyber-mix-mechatronic principles, for general and specialized applications; medical, biomedical equipment, devices and appliances; systems, equipment and components for hydronics and pneutronics; systems, equipment and components for hot technologies; equipment for environmental control and monitoring; components and software specialized for industrial applications.

Advanced Technologies: smart and innovative mechatronic/integrated/adaptive and cyber-mix-mechatronic technologies for industrial and laboratory applications; smart and innovative mechatronic/integrated/adaptive and cyber-mix-mechatronic specialized technologies for measuring, control, calibration, adjustment and testing in industrial processes, smart manufacturing lines and laboratories; unconventional smart and innovative technologies: rapid prototyping, for alternative energy sources, specialized for hot and cold processes (deposition of biocompatible materials, laser processing, etc.); transfer smart and innovative technology to RDI activities.

High-Tech Services for: commissioning, testing, calibration, maintenance, mechatronic/integrated/adaptive and/or cyber-mix-mechatronic smart equipment and systems, industrial and specialized laboratories. characterization of micro-nanostructured, smart and targeted properties, obtained and exploited by innovative high-tech mix-mechatronic and cyber-mix-mechatronic technologies; professional training/training/evaluation/attestation of specialized personnel with medium and high qualifications, in the fields of mechatronics/integronics/adapronics, cyber-mix-mechatronics, etc.; environmental monitoring and smart control; laboratory and industrial testing, trials and calibration; technical consultancy, analysis, synthesis and strategies, etc.; publishing and promoting technical, scientific and commercial materials; enrolment in the Regiconia platform for international promotion.

Collaborations for: valorization of RDI results, dissemination and transfer of knowledge, technologies and smart products and specialized technical assistance; developing projects in smart high-tech areas, strategies, market studies, development and feasibility, etc. in the framework of the National RDI Program (PNCDI III), the European Horizon 2020 Program, the Program with Romanian and Foreign Economy Agents, etc.; creation of consortia for the elaboration of RDI project proposals and their implementation in the national and international programs; creating national and international investment partnerships and consortia.

8 Conclusions

MECHATREC Cluster in its progressive development is a complex structure with real chances of success. The transition to modern cluster policies that interact with smart specialization strategies adds another level of development and complexity. This is a very important way to focus activities and develop strategic capabilities to stimulate cluster potential within regions of national and international value chains to accelerate the development of emerging industries in Romania and Europe.

The MECHATREC cluster is strategic and innovative, focuses on the development of new industrial sectors by encouraging appropriate cross-sectoral collaboration and transformation and development of existing industrial sectors.

Within the MECHATREC cluster, trust was considered to be essential for success in achieving these goals, as it is essential to motivate different stakeholders to cooperate. We focused on the strengths identified by the SWOT analysis of the cluster development strategy, we added value and connected the right people.

In conclusion, we are looking forward to you becoming members of Mechatrec Cluster!

References

1. Gheorghe, Gh.: Inovarea în clusterul strategic inovativ Mechatrec – Vector de creştere a valorii adăugate. In: 9th Forum for Innovation, Bucharest (2017)
2. Gheorghe, Gh.: Noi ecosisteme Inteligente Mecatronice si Cyber-MixMecatronice pentru transferul rezultatelor catre mediul industrial, economic si societal prin IC Mecatron. In: 9th Forum for Innovation, Bucharest (2017)
3. Gheorghe, Gh.: From mechatronics to cyber-mechatronics and from mechatronics systems to cyber-mechatronics systems. In: Proceedings of the XIth Edition of the Annual Conference the Academic Days of Technical Sciences Academy of Romania, pp. 344–356 (2016). ISSN 2066-6586
4. Constantin, A., Gheorghe, Gh.: CMOS transducer with linear response using negative capacitance for the force measurement in human walking analysis with applications in MEMS and NEMS technologies. In: Proceedings of CONTROLO 2016, Guimaraes, Portugal (2016)

Oriented Education for Innovation, Engineering and/or Entrepreneurship

Mechanical Educational System for Automatic Area Observation and Firing Control Techniques

Zdeněk Úředníček[1]([✉]) [iD], Roman Vítek[2], and Jiří Zátopek[1] [iD]

[1] Tomas Bata University in Zlín, Nad Stráněmi, 4511, 760 05 Zlín
Czech Republic
zatopek@fai.utb.cz
[2] University of Defence in Brno, Kounicova, 65, 662 10 Brno, Czech Republic

Abstract. The article deals with the description and utilization of a biaxial mechanical educational system for familiarization of observation and weapons systems control techniques. Its application in education is divided into three sections. The first part deals with the model design and kinematic/dynamic analysis of the whole mechanical structure. A mathematical model is simplified into a form that still reflects the dynamics of the real system. The next part solves an influence of mechanical and regulation part, using a physical model for its simulation. Each degree of freedom can be separated and has its own simulation model. The last part applies measurement results from the real educational system which, besides adjusting a feedback control, also includes a possibility of mechanical system parameterization. These data are compared with simulation results. The similarity between the real system and the physical model is demonstrated in the final comparison.

Keywords: Simulation · Observation · Firing control · Motion control
Biaxial system

1 Introduction

Firing control systems are one of the most significant areas for creating a students' knowledge base and for preparing technical specialists not only for the Czech army, but also for other security forces such as advanced specialists in a particular field [1, 2].

These include not only knowledge about the technical means enabling effective application of firing and, for example, identifying all the parameters needed in order to solve the collision task, but also all the processes connected with firing problems and their control [3, 4]. However, previous research has held back and only unprofessional inaccurate models/systems available in the area of education [5]. Although many articles have been published [6–9] on the topic of image processing and its subsequent use for motion control, students have no chance to compare the simulated data with a real system response.

In our article, we emphasize weapon systems and their control at the University of Defense in Brno, Department of Weapons and Ammunition. For this purpose, it is

© Springer International Publishing AG, part of Springer Nature 2019
J. Machado et al. (Eds.): HELIX 2018, LNEE 505, pp. 1089–1096, 2019.
https://doi.org/10.1007/978-3-319-91334-6_150

necessary to analyze the above-mentioned firing processes control systems. Accordingly, the aim is to find and introduce a constructional solution of an observation system or weapon mounting, with possible parasitic influences of the mechanical construction. Its control should be feasible without a deep theoretical analysis; nevertheless, with the opportunity to familiarize students with the system behavior by simulations and also by practical measurements, e.g., by optical means [2]. Because of that, the article deals with the description and utilization of a biaxial mechanical educational system for familiarization of observation and weapon control system techniques.

2   Mechanical Arrangement of the Educational System

The required device design is based on several essential requirements:

1. The mechanical part of the device respects the conventional biaxial arrangement of a weapon or observation system; i.e. a system with two degrees of freedom around two orthogonal axes of rotation, commonly referred to as azimuth-elevation setting.
2. Supplemental properties are deliberately added into the mechanical system to ensure commonly occurring parasitic effects:
   a. lever mechanism enabling to present the influence of these elements on the dynamic behavior,
   b. a consciously inappropriate constructional solution, consisting in the incorrect position of inertia elements - directly on the rotor of the actuator,
   c. adjustable backslash and stiffness in torque transmission.

A device presented in Fig. 1a is the result of 3D CAD design. Each link has an associated local coordinate system in accordance with the Denavit-Hartenberg (DH) convention [4, 10]. The azimuthal system is created by a structure similar to a ball bearing with random friction, varying with the angle of azimuth rotation. It is driven by a positional actuator with a permanent magnet synchronous motor (PMSM), including gearbox and cascade regulation with vector torque control [4, 11, 12].

The elevation system is created by a similarly controlled PMSM servomotor; however, its output includes a relatively large inertial mass consisting of a variable stiffness and mechanical backslash. A ball screw that actuates the carriage of a translational system coupled with a lever (see Fig. 2a) and moves the platform of the effector (local coordinate system O_{x6}, $_{y6}$, $_{z6}$) is behind the clutch.

The DH table is derived from the illustrated elevation parameters in Fig. 2.a and from the table, the individual transformation matrices are determined [10]. All parameters of the transformation matrices can be derived; furthermore, the resulting transformation matrix from the local coordinate system of the effector to the global coordinate system can be also ascertained [10].

Equations (1) and (2) summarize the motion equations of this biaxial system, showing the mutual dynamic relationships between the azimuth and elevation masses (see Fig. 2b). Note that the carriage position is determined by the rotation of the ball screw input, where the clutch's inertial mass acts on the elevation of the servomotor.

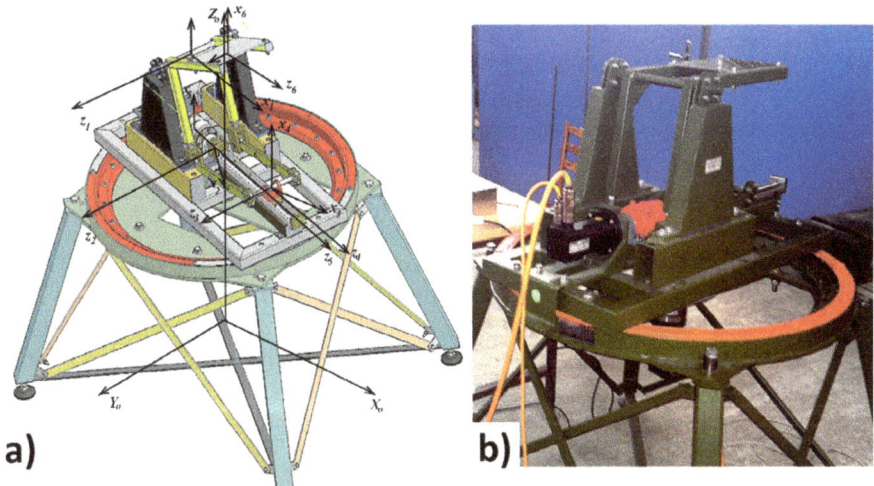

Fig. 1. Educational system structure – 3D CAD model (a) and the real system configuration (b)

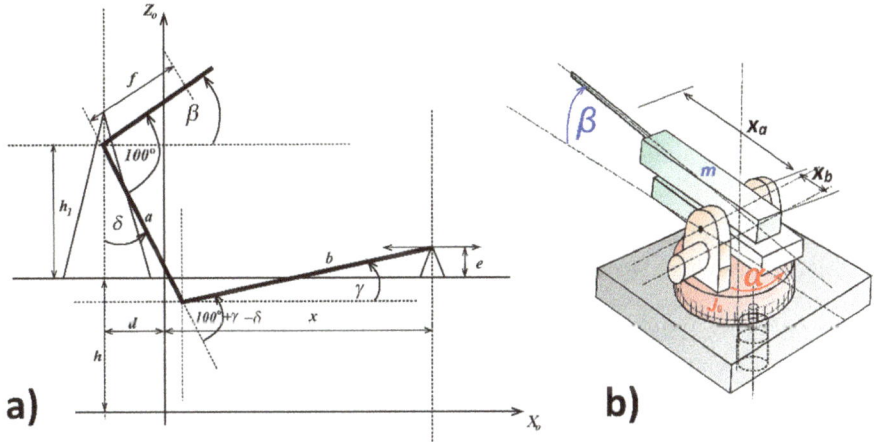

Fig. 2. Elevation part parameters (a) and simplified mechanical system (b)

$$m_\alpha = \left\{ J_0 + \frac{1}{3} m \cdot \left[(x_a - x_b)^2 + x_a x_b \right] \cdot \cos^2 \beta \right\} \cdot \frac{d^2\alpha}{dt^2} - \frac{1}{3} m \cdot \left[(a - b)^2 + ab \right] \cdot \sin 2\beta$$
$$\cdot \frac{d\alpha}{dt} \cdot \frac{d\beta}{dt}$$

$$(1)$$

$$m_\beta := \frac{1}{3}m \cdot \left[(x_a - x_b)^2 + x_a x_b\right] \cdot \frac{d^2\beta}{dt^2} - \frac{1}{6}m \cdot \left[(x_a - x_b)^2 + x_a x_b\right] \cdot sin2\beta \cdot \left(\frac{d\alpha}{dt}\right)^2$$
$$+ m \cdot g \cdot \frac{x_a - x_b}{2} \cdot cos\beta$$

$$(2)$$

3   Simulation Model of the Educational System

The principles of physical modeling based on physical schemas were used for the educational system simulation analysis, which uses the sub models of the individual parts. Their connections were performed by simulation software. Some simulation software uses these principles, and the Dynast[1] simulation environment was used for the described purposes [1].

The system simulation model was divided into two parts because analysis of the dynamic equations revealed that interactions between moving masses were negligible and would have been reflected at very high azimuth movement speeds. The azimuth and the elevation simulation environments were created separately.

As a result, motion control design was implemented as two autonomous servo-drive systems, wherein the axes' mutual influences were considered as disturbances.

3.1   Azimuth Simulation Model

Software simulation environment enabling a student to perform simulation experiments with a model of the educational device is in Fig. 3.

A student can use a mouse to modify the model parameters, especially the parameters of the cascade control structure, which corresponds with the real environment of the device control. The aims of using a simulation model are simulation experiments allowing users to:

- understand the principles of device operation
- determine the impact of the individual parameters
- set regulator constants, which will be further used in a real device.

An example of the simulation experiment result for the azimuth at the desired $\alpha = 15°$ angle jump is illustrated in Fig. 4.

The azimuth behavior with the appropriate setting of the control parameter can be seen from the simulation results.

---

[1] Dynast is simulation software environment for modeling and simulation experiments with models of mechatronic systems, where the task is based on multipolar models of individual parts of the system. More on https://sites.google.com/site/dynasthelp/

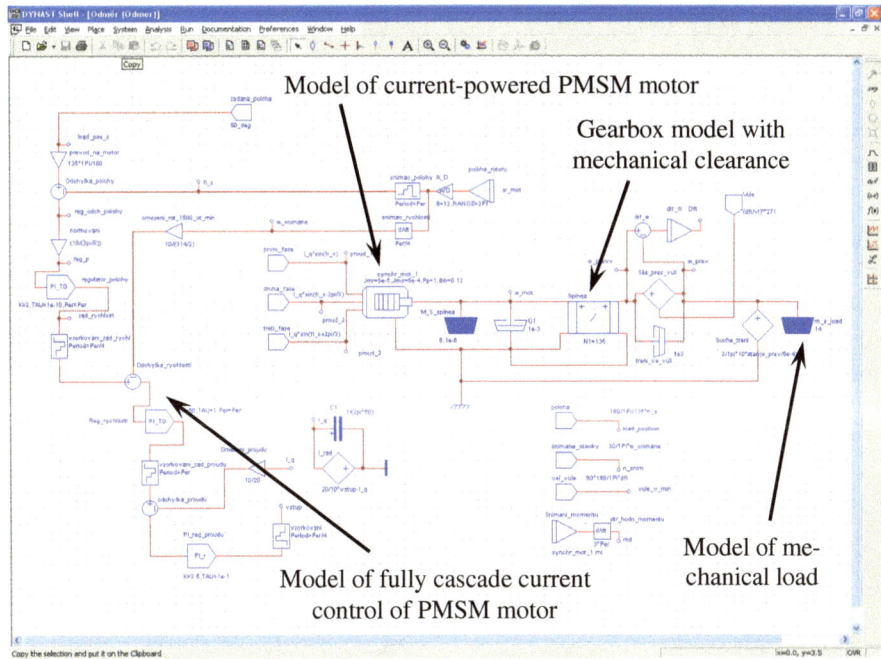

Fig. 3. The azimuth simulation environment

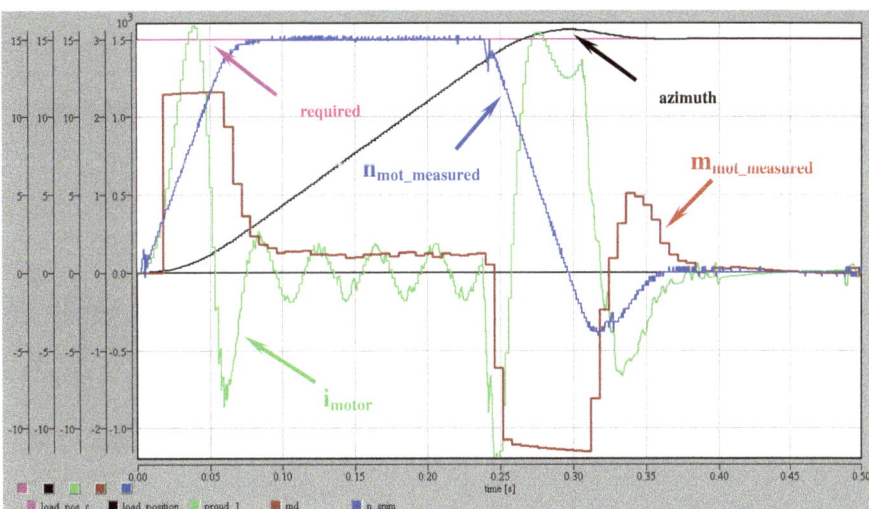

Fig. 4. The results of the azimuth simulation experiment

3.2   Elevation Simulation Model

The simulation environment with an elevation model is very similar to the azimuth model. The simulation experiment results in the required jump in the elevation (ramp) to $\beta = 50°$ are shown in Fig. 5. At such a significant demand, the carriage reached the mechanical stop (the required movement was outside of the device working area).

4   Measurement on the Real System

The integral part of the current educational system is the simulation experiment result verification with a measured variable. The mentioned comparison (see Fig. 6) can be performed by comparing the simulated values with the captured values of the individual control subsystems. Figure 6 shows simulated and measured values of position

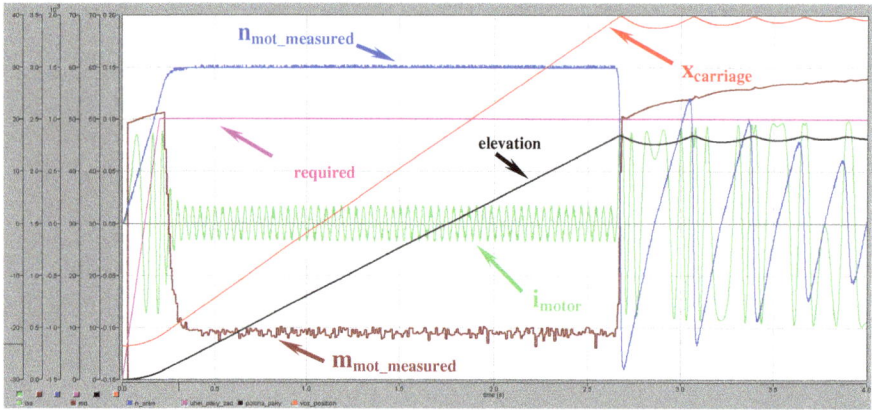

Fig. 5. The results of the elevation simulation experiment

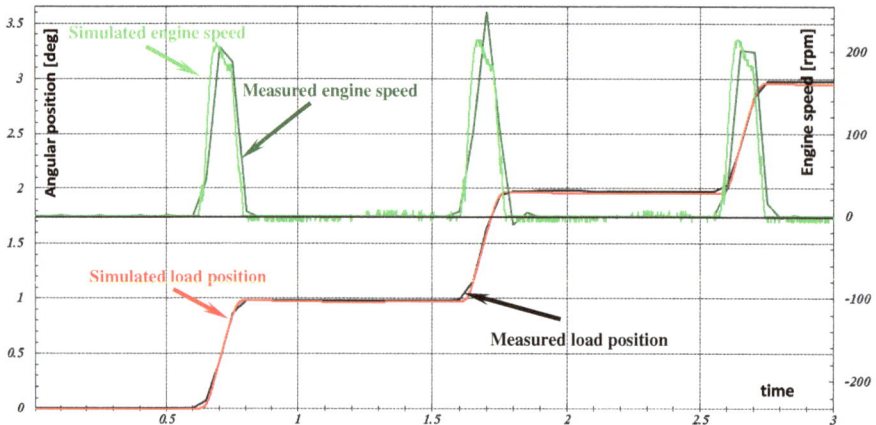

Fig. 6. The comparison of the simulation experiment and the real azimuth measurement

and velocity in azimuth for a series of jump values of about 1°. The real educational system is displayed in Fig. 1b.

5 Conclusion

Experience acquired in the process of building engineering skills, partly based on basic knowledge of natural and technical disciplines, but mainly on qualified engineering skills to solve problems of current industries, shows that in the area of process control supported by technical instruments with a high degree of automation, it is necessary to provide sufficient knowledge about this instrument behavior and its technical possibilities.

If the task of any specialists is an effective acquisition of human and technical means cooperation, it is not always objectively possible to assume profound knowledge of technical instrument theory, and it is even unnecessary to consider this information as essential. What is obviously needed is sufficient ability to use existing technical capabilities based on understanding how to use them and on practical experience with its dynamic behavior. For this reason, the UNOB Brno educational system was created, consisting of a specific hardware arrangement simulating a biaxial guidance system. This is applied in a semi-automatic or fully automatic mode of space mapping and in areas of automatically controlled weapon systems.

The second part of this system is a computer-oriented simulation environment. This enables its users to simulate a system by dynamic behavior of a physical simulation of the proposed hardware. The aforementioned fact allows students to understand not only the system behavior in the context of digital control parameters, but also in the manner of the influence of common mechanical system properties (mechanical backslash, stiffness torque transmission, lever transmission non-linearity and other). They have a major impact on the system dynamic behavior and its resulting accuracy.

References

1. Mann, H.: A versatile modeling and simulation tool for mechatronics control system development. In: Proceedings of the IEEE International Symposium on Computer-Aided Control System Design, pp. 524–529 (1996)
2. Liu, P., Zheng, S., Wei, B.: Study on the simulation and training system state transition method of the complex weapon equipment on operation-oriented. In: Lecture Notes in Electrical Engineering, vol. 423, pp. 949–957 (2018)
3. Safonov, M.G.: Stability and Robustness of Multivariable Feedback Systems. MIT Press, Cambridge (1980)
4. Mostyn, V., Skarupa, J.: Improving mechanical model accuracy for simulation purposes. Mechatronics 14(7), 777–787 (2004)
5. Mohamed, R.M., Muthuramalingam, T.: Tracking and locking system for shooter with sensory noise cancellation. IEEE Sens. J. 18(2), 732–735 (2018)
6. Cai, C., Weng, X., Zhu, Q.: Sea-skyline-based image stabilization of a buoy-mounted catadioptric omnidirectional vision system. EURASIP J. Image Video Process. 2018, 1 (2018)

7. Ristic, B., Arulampalam, S., Wang, X.: Measurement variance ignorant target motion analysis. Inf. Fusion **43**, 27–32 (2018)
8. Liang, F., Liu, Y., Yao, G.: Recognition of blurred license plate of vehicle based on natural image matting. In: Proceedings of SPIE - The International Society for Optical Engineering, vol. 7495(749527) (2009)
9. Dorri, M.K., Roshchin, A.A.: Multicomputer research desks for simulation and development of control systems. In: IFAC Proceedings Volumes, vol. 41(2), pp. 15244–15249 (2008)
10. Úředníček, Z.: Robotics, 1st edn. Tomas Bata University in Zlín, Zlín (2012)
11. VanAntwerp, J.G., Braatz, R.D., Sahinidis, N.V.: Globally optimal robust control for systems with time-varying nonlinear perturbations. Comput. Chem. Eng. **21**(Suppl. 1), S130 (1997)
12. Mohan, N., Robbins, W.P., Undeland, T.M., Nilssen, R., Mo, O.: Simulation of power electronic and motion control systems-an overview. In: Proceedings of the IEEE, vol. 82(8), pp. 1287–1302 (1994)

Using Real-Time Laboratory Models in the Process of Control Education

Frantisek Gazdos[✉]

Faculty of Applied Informatics, Tomas Bata University in Zlin,
Nam. T.G. Masaryka 5555, 760 01 Zlin, Czech Republic
gazdos@utb.cz

Abstract. The paper describes usage of real-time scale-models in the process of education in the field of Automatic Control at Tomas Bata University in Zlin, Czech Republic. It presents both hardware and software solutions utilized in the presented laboratory together with some typical tasks for students related to a selected process. Teacher's experiences and students' feedback from a regular course in this lab are also given, as well as some future plans. Consequently, this work can serve as an inspiration for other similarly-oriented departments.

Keywords: Process control · Control Engineering · Education
Experimental laboratories · Real-time systems

1 Introduction

The positive role of experiments in the process of education is indisputable [1–3]. When students learn in an interactive way, either by trial and error or more sophisti-cally, it definitely leaves a deeper trace in their memories than simple memorization of theoretical facts [1, 2]. Nowadays, due to the explosion in the field of information and communication technologies we are facing growing interest in the virtual and remote laboratories, e.g. [2, 4–7]. In virtual laboratories it is possible to perform experiments safely, cheaply and quickly using modelling and simulation tools and some computing power. Remote laboratories connected to the Internet enable to make experiments simply on the other side of the globe. However, direct experimentation, if possible, brings best practical skills – when students are forced to interact with real environment and processes, facing real disturbances and challenges, they are consequently more confident under real conditions in practice. Therefore universities try their best to offer prospective students top-quality laboratories reflecting state of the art in a given field of study. The process of education in the area of Automatic Control/Process Control/Control Engineering is not an exception. Besides learning basics from vari-ous subjects, such as Systems theory, Control theory, Mathematical modelling, Sim-ulation, System identification and many others, at the same time, students are taught to test and use their knowledge practically and laboratories are the best places for this. Here, students are forced to work with real hardware, identify it, implement suggested control algorithms and face real challenges, such as stochastic disturbances, signal saturations, nonlinearities, etc. Real-time laboratory scale models have proven to be

© Springer International Publishing AG, part of Springer Nature 2019
J. Machado et al. (Eds.): HELIX 2018, LNEE 505, pp. 1097–1103, 2019.
https://doi.org/10.1007/978-3-319-91334-6_151

beneficial for this purposes. As the mathematical apparatus of Control and Systems theory is general and abstract enough, the laboratories usually contain some typical processes to prepare prospective control engineers for various types of systems and conditions [8, 9].

This contribution presents author's experiences with one such laboratory at the Department of Process Control, Faculty of Applied Informatics, Tomas Bata University in Zlin, Czech Republic [10]. Here, students of the Automatic Control and Informatics Master's degree programme gain practical skills related to process control. The paper briefly introduces motivation and purpose of the laboratory, presents actual equipment for both hardware and software and offers some typical students' assignment for the real process models presented. Some feedback from students and teachers is also given together with a brief outlook for the near future of the lab. As a result, this contribution can serve as an inspiration for other similarly oriented departments when planning similar laboratory, or just to compare ways and means of education in the field of Automatic Control/Process Control/Control Engineering throughout our global education space.

The paper is structured as follows: after this introductory section it goes on describing motivation and purpose of the presented laboratory. Further, its equipment is described in detail, including both hardware and software, then some typical students' assignment for a selected process is presented, together with more detailed description of the process. Finally some feedback from both teachers and students is also given and the contribution ends with a brief outlook for the near future of the laboratory.

2 Laboratory of Real Models

2.1 Motivation and Purpose of the Laboratory

The laboratory entitled "Laboratory of Real Models" was established at our department as an effort to offer students a place to test and train their practical skills in the field of Automatic Control. A soon as our institution had accredited studies in the field of Process Control, the department started building systematically a practically-oriented laboratory containing some typical processes. As the space provided was limited, scale models of various systems were gradually acquired instead of building real industrial lines. The scale models, although smaller and simpler than their real industrial counterparts, offer enough possibilities to train practical skills related to control of various types of systems [8].

In our faculty [10] the lab is used by all levels of students – while Bachelor's and Master's degree students often use these scale models for their final works (Bachelor's & Diploma theses), Ph.D. students use the laboratory for research purposes related to all aspects of Automatic Control, as well as academic staff of the faculty. Regular lessons here run for Master's degree students in their second (last) year of studies. These students already have strong theoretical foundations from their previous studies, including also some practical skills. They should have completed e.g. various courses on programming (Programming, Object-oriented Programming, Programmable Logic Computers, Microcomputer Programming, MATLAB & Simulink), some control &

systems oriented courses (such as Automation, Optimisation, System Theory, Discrete Control, State-space and Algebraic Control Theory) and also courses related to hardware, sensors and actuators (e.g. Electrotechnics & Industrial Electronics, Microelectronics, Sensors, Technical Means of Automation), and others. Here the students complete the course entitled "Real Process Control", where they go through the whole procedure of designing reliable and effective control systems – from getting to know the process, its input and output signals and their limits, identification of static and dynamic properties of the system, verification of the resultant model, control system design, simulation tuning and testing and finally real-time implementation and fine-tuning. This complex task tests thoroughly skills of the prospective control engineers and teaches them (with the help of a qualified instructor) best practices usable in their future career.

2.2   Laboratory Equipment

At this moment the laboratory is equipped with several scale models from different companies offering control education models worldwide. These include TecQuipment, AMIRA, Feedback, Leybold and Armfield models. TecQuipment [11] represents the richest collection of scale models in our laboratory:

- CE107 Engine Speed Control Apparatus
- CE108 Coupled Drives Apparatus
- CE120 Controller
- CE150 Helicopter Model
- CE151 Ball and Plate Apparatus
- CE152 Magnetic Levitation Model.

The AMIRA [12] models include:

- DTS200 Three-Tank-System
- DR300 Speed Control with Variable Load
- PS600 Inverted Pendulum.

Then there is one each model from Feedback [13], Leybold [14] and Armfield [15]:

- 33-007-PCI Twin Rotor MIMO System
- T 8.2.1.4 Gas Flow Control
- PCT40 Multifunction Process Control Teaching System.

This collection represents various types of systems to offer students different control challenges they can further encounter in practice.

2.3   Software Equipment and Interface

All the models are connected to a common PC with a universal multifunction I/O card. For real-time control, simulations and computations related to control system design and tuning the MathWorks' MATLAB & Simulink environment, e.g. [16–18], is fruitfully exploited. More specifically, Simulink real-time application libraries are used for the real-time control and testing.

3   A Typical Students' Assignment

In this section, one of the models is introduced in more details together with some typical tasks intended for the students.

The 'CE108 Coupled Drives Apparatus' scale-model presented in Fig. 1 represents electric systems related to industrial material transport problems as they occur in magnetic tape drives, textile machines, paper mills, strip metal production plants, etc. where the material is processed in continuous lengths, it is transported through work stations by drive systems and the material speed and tension have to be controlled within defined limits at all times.

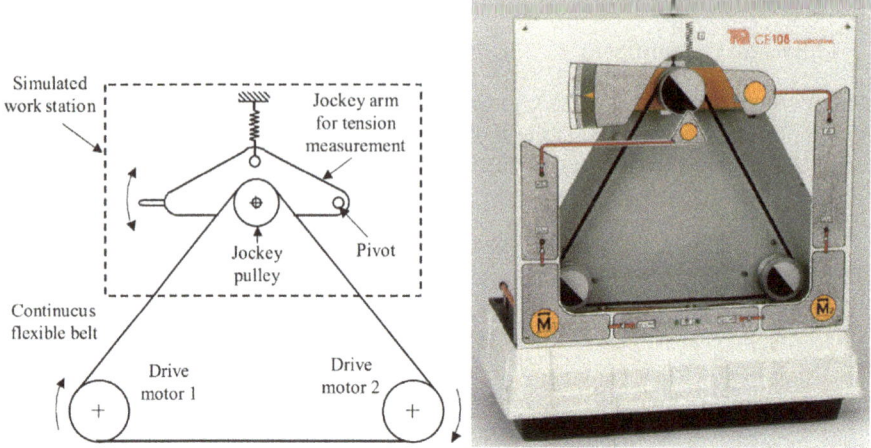

Fig. 1. Scheme of the coupled drives system (left) and the model by TecQuipment (right)

The system has two drive motors operating together to control the speed of the continuous flexible belt that goes round pulleys on the drive motor shafts and so called *jockey pulley*. The jockey pulley is mounted on a swinging arm that is supported by a spring. The deflection of the arm is a measure of the tension in the drive belt. The pulley and arm assembly represents a *work station* where material that the belt represents can be processed. The main control problem here is to regulate the belt speed and tension by varying the motor torques. The CE108 coupled drive apparatus is the product of TecQuipment Inc. [11] and it is designed to have characteristics seen in industrial drives, but it is not any particular industrial application – it is a prototype for all industrial coupled drive applications. From the control theory point of view, the system is a multivariable plant with two inputs (power supply into the two drive motors) and two outputs (pulley speed and belt tension). There is a strong coupling between the inputs and outputs given by the fact that both motors change both outputs (due to the drive belt). A students' typical task here can be like this:

- *Identify the system for controlling speed of the flexible belt using the provided scheme for decoupling.*

This task usually includes measurements of the system's static characteristics, choice of a suitable operating point for both identification and control purposes, choice of a model structure, identification experiment(s), evaluation of the results, and validation of the identified model (if not successful, choice of a new model and next steps above follow again).

- *Design a controller for speed control of the belt.*

Here, based on the identified model, students have to design a suitable controller – both structure and its parameters, test it by simulation means first and then implement it under real condition with appropriate fine-tuning.

- *Identify the system for controlling tension of the belt (use the same scheme for decoupling, interchange the inputs only); for the belt tension measurements use a suitable filter.*

In this task, again, students are expected to choose a suitable operating point for identification and control of the belt tension and follow a similar procedure as in the first task. In addition, students have to deal with a suitable filter design.

- *Design a controller for control of the belt tension.*

Again, based on the identified model, students follow a controller design procedure as in the second task, including both, simulation and real-time fine-tuning.

- *Design a set-up for simultaneous control of both belt speed and tension.*

This last task represents a complex problem of designing a control system for a multivariable (MIMO) plant with two inputs and two outputs with strong coupling. Here, based on the previous steps, including SISO identification and control design, students are expected to suggest a set-up and compensators for simultaneous control of both speed and tension of the belt simultaneously.

All the assignments are general enough to provide students sufficient space to test all their knowledge gained during studies and, at the same time, do not restrict them to follow only one 'right' solution.

4   Teachers' Experiences and Students' Feedback

The course "Real Process Control" is donated by 5 credits and Master's students spend 4 h/week here. They usually have 2 weeks for one model and have to go through 6 typical processes during the semester. The first week is typically devoted to getting familiar with the model, its inputs/outputs and identification. The second week is usually spent by controller design and testing. A variable option offered to all students is to choose only one model and "play" with it the whole semester, including more advanced forms of identification and control. This choice is not frequently utilized, however, some skillful students choose this option occasionally. Although the tasks in this laboratory are not simple and represent complex problems from the broad area of process control, most of the students quite quickly adapt to the conditions and work independently on the given assignments, consulting only when necessary. From each of the assignments a written report is needed together with all the measured data and simulation & real-time schemes. This material is checked by a lecturer and revised by a student, if necessary, which may take more than one iteration.

The process of education and quality of teaching at our university is a subject of several regular evaluation procedures. One of these represents students' evaluation (anonymous feedback) given at the end of each semester. If we look at the survey results of the course "Real Process Control" in the introduced laboratory, it is possible to find such feedback comments from the students: "*...course that was probably most beneficial...time demands are higher, but it makes sense...why not more such courses?*", or similar. This confirms usefulness of this course and justifies costs, effort and time invested into development of the laboratory and course concept. It turns out that students can not only test their knowledge practically here, but they can also learn best practices in the field usable further in their control engineer's practice.

5 A Brief Outlook for the Near Future

As some of the equipment in the lab is gradually getting obsolete (computers, I/O cards, SW and some models) we are planning to upgrade it soon. This will include:

- new HW & SW for computers,
- new multifunction I/O cards,
- update of the MATLAB license,
- several typical industrial process controllers,
- and 4 new scale models.

The process controllers will enable students to get familiar with the latest products in the field, with some typical areas of usage prepared as new students' assignments. Four new models will probably be *Active Suspension & 3DOF Gyroscope* from Quanser [19] and *3D Crane & Two-Wheeled Unstable Transporter* from INTECO [20]. These models will represent new challenges for prospective students. The reconstruction is funded from a project and it is planned in the summer time, so that it does not interfere with the lectures in the laboratory.

Another plan is to include a similar course in the laboratory also in Bachelor's studies with corresponding simplification of the students' tasks and assignments relevant to their knowledge with the aim to motivate students for further studies of Automatic Control in the Master's degree programme, and prepare them for practice well.

6 Conclusion

In this paper, the concept of an experimental laboratory for advanced training in the field of Automatic Control is introduced. It describes both hardware and software solutions utilized together with some teacher's experiences and students' feedback from a regular course in this laboratory. These are generally positive which justifies costs, effort and time invested into the laboratory development and course preparation.

Some typical students' tasks for a selected process are also given together with more detailed description of the model. Future plans are also discussed briefly. As a result, this contribution can serve as an inspiration for other departments educating in the same field or as means of comparison between them.

References

1. Dormido, S., Dormido-Canto, S., Dormido, R., Sánchez, J., Duro, N.: The role of interactivity in control learning. Int. J. Eng. Educ. **21**(6), 1122–1133 (2005)
2. Bencomo, S.D.: Control learning: present and future. Annu. Rev. Control **28**, 115–136 (2004)
3. Gany, A.: The role of laboratory experiments in engineering education. In: Proceedings of the 9th Biennial Conference on Engineering Systems Design and Analysis, vol. 3, pp. 645–647. American Society of Mechanical Engineers, New York (2009)
4. Hough, M., Marlin, T.: Web-based interactive learning modules for process control. Comput. Chem. Eng. **24**, 1485–1490 (2000)
5. Sánchez, J., Morilla, F., Dormido, S., Aranda, J., Ruipérez, P.: Virtual control lab using Java and Matlab: a qualitative approach. IEEE Control Syst. Mag. **22**(2), 8–20 (2002)
6. Shin, D., Yoon, E.S., Lee, K.Y., Lee, E.S.: A web-based, interactive virtual laboratory system for unit operations and process systems engineering education: issues, design and implementation. Comput. Chem. Eng. **26**, 319–330 (2002)
7. Hu, W., Lei, Z., Zhou, H., Liu, G.P., Deng, Q., Yhou, D., Liu, Z.W.: Plug-in free web-based 3-D interactive laboratory for control education. IEEE Trans. Ind. Electron. **64**(5), 3808–3818 (2017)
8. Horáček, P.: Laboratory experiments for control theory courses: a survey. Annu. Rev. Control **24**, 151–162 (2000)
9. Leva, A.: A simple and flexible experimental laboratory for automatic control courses. Control Eng. Pract. **14**, 167–176 (2006)
10. Faculty of Applied Informatics, Tomas Bata University in Zlin. http://www.utb.cz/fai-en. Accessed 11 Jan 2018
11. TecQuipment Homepage. http://www.tecquipment.com. Accessed 11 Jan 2018
12. Amira Homepage. http://www.ict.com.tw/AI/Amira/amira/home_e.htm. Accessed 11 Jan 2018
13. Feedback Homepage. http://www.feedback-instruments.com. Accessed 11 Jan 2018
14. LD Didactic Homepage. http://www.ld-didactic.de/en. Accessed 11 Jan 2018
15. Armfield Homepage. http://discoverarmfield.com. Accessed 11 Jan 2018
16. Tyagi, A.K.: MATLAB and SIMULINK for Engineers. Oxford Higher Education, Oxford (2012)
17. Xue, D., Chen, Y.: Modeling, Analysis and Design of Control Systems in MATLAB and Simulink. World Scientific Publishing Co., Singapore (2014)
18. Magda, Y.: MATLAB and Simulink: a practical introduction to real-time system design (2012)
19. Quanser Homepage. http://www.quanser.com. Accessed 17 Jan 2018
20. INTECO Homepage. http://www.inteco.com.pl. Accessed 17 Jan 2018

Perspectives of Entrepreneurship in Engineering Education: An Exploratory Study

Celina P. Leão[1(✉)], Carina Andrade[2], Ana Trigo[3],
and Filomena Soares[4]

[1] IEEE WIE-PT, Department of Production and Systems, Algoritmi Centre,
University of Minho, Guimarães, Portugal
celinapleao@ieee.org
[2] IEEE WIE-PT, Department of Information Systems, Algoritmi Centre,
University of Minho, Guimarães, Portugal
[3] IEEE-PT ExCom, Algoritmi Centre, University of Minho, Guimarães, Portugal
[4] IEEE WIE-PT, Department of Industrial Electronics, Algoritmi Centre,
University of Minho, Guimarães, Portugal

Abstract. Entrepreneurship relevance is recognized and the governments investments are visible through the startup programs volume that have been arising all over the world. Universities can promote the entrepreneurial spirit and help to develop entrepreneurship skills in their engineering students. This could increase engineering students intentions to create their own business or to be intrapreneurs in the employee companies, helping the development of their future careers. This work is the result of the first intervention of the authors on engineering schools. A survey was distributed to engineerings students and young professionals in order to analyse their entrepreneurship perspectives. Early findings show that students should learn more about entrepreneurship, however, opposing results were obtained related with how these issues should be addressed in an engineering course. This on going work covers important topics that need a more detailed analysis.

Keywords: Entrepreneurship · Entrepreneurship education
Entrepreneurship in engineering education · Survey entrepreneurship education

1 Introduction and Motivation

Entrepreneurship is accepted as a promoter to innovation and new technologies development. In 2003 in Portugal, it was made an effort to have entrepreneurship courses in the Portuguese universities [1]

Entrepreneurship is a trending topic, with an important role in Portugal through the economic crisis in 2008, which encouraged the search for self-employment [2]. Despite of the efforts, in 2015, it is still identified in OECD diagnosis report about skills strategy, the need of promoting entrepreneurship skills in Portugal for developing an environment of innovation, economic growth and job creation [3]. In the paper

© Springer International Publishing AG, part of Springer Nature 2019
J. Machado et al. (Eds.): HELIX 2018, LNEE 505, pp. 1104–1110, 2019.
https://doi.org/10.1007/978-3-319-91334-6_152

"Does self-employment reduce unemployment?", the authors identified a relation between a decrease in self-employment with an increase in unemployment [4].

In engineering school of University of Minho, courses of the 2nd cycle have, in their majority, at least one class of entrepreneurship or business models development. Furthermore, there are some extra curricular programs that students can enroll. But, in a time where Portugal has been followed with interest by international companies, such as Google, Amazon, Fujitsu, among others [5], could lead engineering students to be less interested in having their own business. This study intends to diagnose the intentions of engineering students and recent engineering graduated professionals on following an entrepreneur career.

Considering the topics mentioned above the authors decided to develop an entrepreneurship diagnosis among the engineering students and young professionals. In order to make a diagnosis, it was applied an existing and validated survey to engineering students and young professionals. At this moment, the number of answers of the survey was not considered enough to do an extended analysis of the entrepreneurship in engineering education, but it allows to do a first analysis of the entrepreneurial environment and interests among engineering students.

2 Study Design

This section aims to explain the process that was followed to accomplish this study. It will be presented the main objectives of this work, the data collection process that was performed and the followed research methodology.

2.1 Objectives of the Work

In the related work search performed, an interesting journal paper was found as being of maximum interest for the goal of this work.

The work of Duval-Couetil et al. (2011) aims to investigate "how entrepreneurship education impacts a broad range of attitudes and outcomes of entrepreneurship education for engineering students" [6]. In other words, with this study, the authors purpose was to understand the influence of entrepreneurship in education on the further behavior and attitudes of the engineering students. To understand this, the authors considered necessary to have information about the entrepreneurial activities in their academic program creating the questions listed below:

- "To what extent do engineering students participate in entrepreneurship education and related activities?"
- "To what extent is entrepreneurship addressed in their engineering programs?"
- "What are engineering student attitudes toward entrepreneurship as a career?"
- "Why are students interested or not interested in entrepreneurship?"
- "How familiar are engineering students with entrepreneurship terms and concepts?"
- "What are engineering student perceptions of their entrepreneurship-related abilities?"
- "What are the characteristics of engineering students participating in entrepreneurship education?"

In this sense, some information about students demographic and academic program is also needed to distinguish the answers within several groups, such as students that participated or not in entrepreneurship courses; age; familial entrepreneurial background; or, engineering curricular units.

Taking into account this related work, the authors intend to deeply understand the relationship between entrepreneurship education and the engineering courses and schools, in Portugal.

Thereby, to achieve the work goal it is explored the existence of curricular units in engineering courses or, the existence of other individual courses, outside of the engineering courses. On the other hand, it is also explored the degree of students encouragement to participate in varied entrepreneurship activities such as give an elevator pitch, write a business plan, take an entrepreneurship course or workshops. Furthermore, it is also analyzed the students conscious or unconscious involvement with the entrepreneurship topic exploring a reflection about their engineering courses in an entrepreneurship view.

Another analyzed perspective on the relationship between the engineering students and the entrepreneurship is directly related to their post-graduation options, understanding if they want or not to start their own business and, why they want it or not.

Finally, the last analysis perspective is about the knowledge presented by students/younger engineers about related entrepreneurship areas and their perception of entrepreneurship abilities or skills.

In this study, and to enable the analyze, several research questions should be answered, including:

- Do the engineering courses include entrepreneurship curricular units?
- Why students/younger engineers do not start a business?
- What are the major areas related to entrepreneurship?
- How students/younger engineers rate their entrepreneurial ability?
- What are the major skills related to entrepreneurship ability?

2.2   Design of the Instrument for Data Collection and Methodology

A survey in this area, published in [6] was used as a guide for this study. To proceed with the data collection, a survey was created in the Google Forms [7] with the selected questions from the followed survey.

The decision about the exclusion or adaptation of some question was weighted and indispensable for the creation of the survey to the Portuguese reality.

The following list shows the questions that were kept in this survey, for the Portuguese students/younger engineers, even with small adaptations in the English writing, for a better understanding by the contributors because some words are not frequently used in Portuguese environment (e.g., "major" in the original second question).

A. Characterization
1. University (where you are studying or studied);
2. Country of Origin;
3. What is your area of study? (Select one or more choices);
4. Age;

5. Sex;
6. Are any of your parents entrepreneurs?
7. Is anyone in your family (not including your parents) an entrepreneur?
B. Entrepreneurship Program
1. How many entrepreneurship courses have you taken in your academic career, outside of engineering course?
2. If you have taken any entrepreneurship courses outside of engineering (see previous question), please list the course name(s).
3. Have you participated in any of the following types of academic entrepreneurship programs? If you have taken any entrepreneurship courses outside of engineering, please list the course name(s).
4. Rate your level of agreement (from 1- Strongly Disagree to 5- Strongly Agree) with the following: In general, in my engineering course... (a list of 9 sentences).
C. Student Entrepreneurship Perspective
1. Consider your post-graduation options and please rate your level of agreement (from 1- Strongly Disagree to 5- Strongly Agree) with the following: I plan to... (a list of 7 sentences);
2. Please rate your level of agreement (from 1- Strongly Disagree to 5- Strongly Agree) with the following (a list of 7 sentences);
3. Please check the answer that best fits your current situation (a list of 6 sentences);
4. If you are interested in being an entrepreneur, what type of business are you interested in starting? Please describe the product or service and industry or market; (open question)
5. Please rate your level of agreement (from 1- Strongly Disagree to 5- Strongly Agree) with the following: I would start a business in order to... (a list of 12 sentences)
6. If there are any reasons unlisted above for why you would start a business, please list them here; (open question).
7. Please rate your level of agreement (from 1- Strongly Disagree to 5- Strongly Agree) with the following: I would NOT start a business due to... (a list of 14 sentences)
8. If there are any reasons unlisted above for why you would start a business, please list them here; (open question)
D. Entrepreneurship Knowledge and Entrepreneurial Ability
1. Please rate your level of knowledge or skill (from 1- Poor to 5- Excellent) in the following areas related to entrepreneurship; (a list of 37 sentences)
2. For each statement select a number from 0 (0% Not at all confident) to 10 (100% Completely confident) to indicate how confident you are that you could perform that skill or ability now; (a list of 15 sentences)
3. Overall, how would you rate your entrepreneurial ability? (from 1- Poor to 5- Excellent)
4. How would you rate your ability to start a business now? (from 1- Poor to 5- Excellent)
5. Rate your skill levels (from 1- Poor to 5- Excellent) in the following areas... (a list of 6 sentences);

In addition to this list, other three questions were included in the survey, because the authors considered these questions as being of great interest to the survey context. These questions are:

1. Academic Degree (achieved or to achieve);
2. Year of Graduation (finished or expected to finish);
3. In your course, there is/was some curricular unit that addressed entrepreneurship?

Moreover, the race question in the followed survey ("What is your ethnic/racial background? Select all that apply.") was excluded from this study because could cause embarrassment to the contributors.

Regarding the followed methodology, the survey was published online and was available for three weeks. During these weeks, this survey was released by institutional e-mail in some Universities. In addition, the survey was also released on the IEEE contacts and by personal networks, enforcing the idea "word of mouth" to answer the survey.

With these methods, the authors are aware of the convenience sample and snowball sampling. The first one allowed the selection of an accessible sample in a faster way. On the other hand, the snowball sampling allowed an increase in the quantity of responses by the "word of mouth". As disadvantages, the difficulty to determine the sampling error.

3 Early Results and On-Going Work

So far few students and younger engineers answered to the challenge and 51 valid questionnaires were received. It was authors' believe to achieve a higher number of fulfilled questionnaires within a month, allowing to infer on the perspectives of Entrepreneurship in Engineering, however it was not the case. Nevertheless, the data collected would be used to reflect the trend in these issues.

The majority of the respondents are from the School of Engineering of the University of Minho (76%) followed by Faculty of Engineering of University of Porto (14%) and the remaining 10% divided between others Higher Education Institutions namely, University of Aveiro, Faculty of Science of University of Lisbon, School of Engineering of Polytechnic of Porto; in the areas of Electrical/Electronic Engineering (38%), Computer Engineering (32%), Industrial/Systems Engineering (12%), Biomedical Engineering (6%), Mechanics Engineering (6%), Information Systems (4%) and Civil Engineering (2%); with an average age of 26.7 (range 19–42); only 31.4% of whose respondents are women; 56.8% of the respondents already have an academic degree; 90.2% are from Portugal, the others came from five different countries (Canada, Brazil, Venezuela, Cape Verde and Romania). Only 17.6% of the respondents have entrepreneurs parents. This value increase to 35.3% when extended to others family members. The majority of the respondents (64.7%) mentioned to attend some curricular units that addressed entrepreneurship. 33.3% are not interested in starting a business, only 13.7% have their own business at the present moment and 47.1% do not have any plans to start a business.

As illustrated in Fig. 1, students and younger engineers have opposed views regarding if and how an engineering course should address issues related to entrepreneurship, although the similarity on the maximum and minimum obtained in all the nine items in analysis (5, strongly agree, and 1, strongly disagree, respectively). An exception is found in the opinion that students should learn more about entrepreneurship with 92.2% agreement (with the highest agreement average obtained of 4.19, see Table 1). The information on Table 1 complements the statistics data regarding the nine items in analysis. Even no significant differences were found between male and female, in each of these items (all the p-values obtained, after the Mann-Whitney U test for the agreement in the male and female students and young engineers, were > 0.05), female students and young engineers scores higher than their male counterparts in some items, especially in those related to motivational aspects (i.e. "Students should learn more about entrepreneurship", female with an average of 4.47 and 4.06 for male).

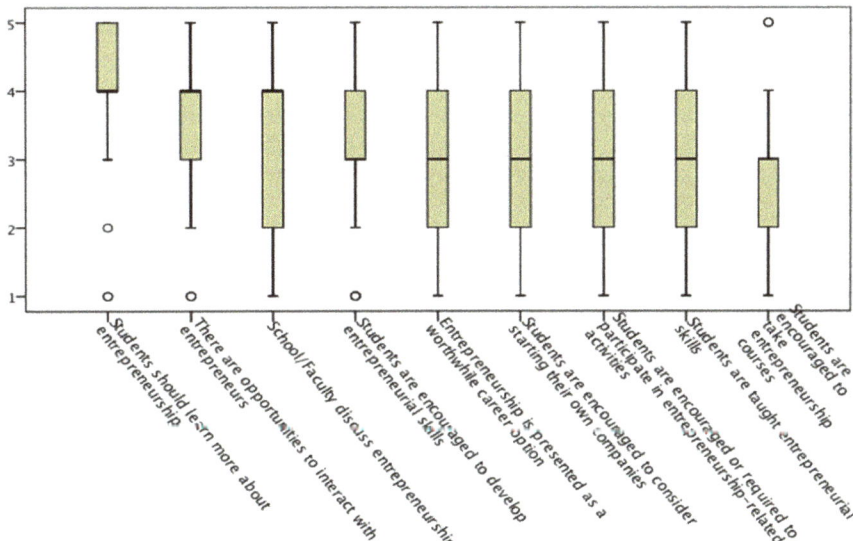

Fig. 1. Agreement rate distribution: "In general, in my engineering course..."

Level of risk tolerance was identified as the skill, related to entrepreneurship ability, where the respondents present a poor or below average rate (33%), in opposition to analytical skills with 47.9% of excellent and above average rate, followed by communication and presentation skills with 47.7% each.

This is an ongoing work and the results presented corresponds to the first insig, awaiting new and more results allowing to draw more consolidated and deep conclusions.

Table 1. Main statistical values for total data

Item	Mean	Std. deviation	1st quartil	3rd quartil	Median
Students are: … taught entrepreneurial skills	2.85	.95	2	4	3
… encouraged to develop entrepreneurial skills	3.25	1.02	2.5	4	3
… encouraged to take entrepreneurship courses	2.79	.95	2	3	3
… encouraged or required to participate in entrepreneurship-related activities	2.89	1.01	2	4	3
… encouraged to considered starting their own companies	3.06	1.06	2	4	3
School/Faculty discuss entrepreneurship	3.30	1.10	2	4	4
There are opportunities to interact with entrepreneurs	3.32	1.03	2.5	4	4
Students should learn more about entrepreneurship	4.19	.92	4	5	4

Acknowledgements. The authors would like to express their acknowledgments to all the engineering students and professional who accepted to collaborate in this study and to national funds by FCT Strategic Project PEst2015–2020, reference UID/CEC/00319/2013 and the doctoral scholarships (PD/BDE/135101/2017 and PD/BDE/135106/2017), for all the support provided.

References

1. Redford, D.T.: Educação para o Empreendedorismo no Ensino Superior em Portugal. Entrepreneurship Education in Higher Education in Portugal. https://www.dges.gov.pt/pt/pagina/educacao-para-o-empreendedorismo-no-ensino-superior-em-portugal-entrepreneurship-education
2. Daniel, A.D., Cerqueira, C., Ferreira, J.J.P., Preto, M.T., Afonso, P., Quaresma, R.: Ensino do Empreendedorismo – Teoria & Prática - Reflexão das I Jornadas do Ensino do Empreendedorismo em Portugal (2015)
3. OECD skills strategy diagnostic report executive summary Portugal (2015)
4. Thurik, A.R., Carree, M.A., van Stel, A., Audretsch, D.B.: Does self-employment reduce unemployment? J. Bus. Ventur. **23**, 673–686 (2008). https://doi.org/10.1016/j.jbusvent.2008.01.007
5. Ministry of Foreign Affairs: Portugal, an investment destination. https://www.riade.embaixadaportugal.mne.pt/en/the-embassy/news/portugal,-an-investment-destination
6. Duval-Couetil, N., Reed-Rhoads, T., Haghighi, S.: The engineering entrepreneurship survey: an assessment instrument to examine engineering student involvement in entrepreneurship education. J. Eng. Entrepreneurship, **2**(2), 40–63 (2011)
7. IEEE WiE-PT: Entrepreneurship in Engineering Survey (2018). https://goo.gl/forms/Ud8ONXWK1rJe8gzF3

Low Cost CNC Equipment Supporting Teaching/Learning Activities

João Peixoto and Caetano Monteiro[(⊠)]

University of Minho, Guimarães, Portugal
{D7426, cmonteiro}@dem.uminho.pt

Abstract. The present paper addresses the use of numerical control (CNC) control equipment to support the training of professionals for industry in Portugal. The typical architecture of these equipment is approached, possible strategies of resources optimization is presented considering the design of low cost equipment, as didactic resource for training. Also the importance of this approach for learning operation, maintenance and design of industrial CNC machine tools is emphasized, and the work being done at the University of Minho is around this subject is depicted.

Keywords: CNC · Low cost machines · Training

1 Introduction

Technological modernization in nowadays industry largely depends and follows CNC equipment integration, which assumes a preponderant role in companies' development. Acquiring this kind of equipment constitutes a rapid way to obtain, and incorporate, the new the technology they provide for the prosecution of their normal activity. However most manufacturing companies in Portugal do not master the machine design technology, not only because most of the equipment is imported from a short list of developed countries, but also because they are mainly concerned with part production issues, in order to make the machine acquisition profitable. In general manufacturing companies are then dependent form machine designers/builders, both in technical support and spare components supplies to maintain the new equipment updated and profitable. This dependency is worst because spare parts and technical support often can only be found in foreign countries, then making maintenance activities more expensive. It also involves a permanent technological and economic dependence from the countries of machines origin.

When looking to machine tools obsolescence, one can notice it mainly derives from electronic components and software obsolescence. Then, in perspective of resource optimization, machine retrofitting of existing equipment is a feasible alternative to decrease the need to continuous renewal of the machines becoming obsolete, due to the significant investment implied in the acquisition of new equipment. To be able to keep up with advanced technological improvements, qualified and skilled personnel are needed, mastering equipment design, operation and parameterization. Thereof is then

© Springer International Publishing AG, part of Springer Nature 2019
J. Machado et al. (Eds.): HELIX 2018, LNEE 505, pp. 1111–1117, 2019.
https://doi.org/10.1007/978-3-319-91334-6_153

mandatory to develop strategies to motivate young people to the importance of the knowledge about machine design technology.

2   General Considerations About CNC Equipment

By CNC equipment is meant any computer controlled system independently capable of executing manufacturing or forming operations by means of a set of instructions (part program), previously introduced, that can be repeated cyclically and easily replaced to perform a new set operations. Industrial CNC manufacturing machine tools enable high productivity of accurate parts production, also ensuring real-time process control [5].

Figure 1 presents the operating principle of a CNC machine tool.

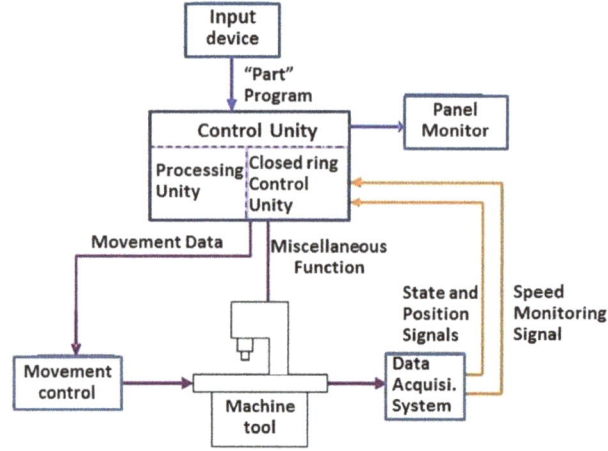

Fig. 1. Operating principle of a CNC machine tool (adapted from [9])

There is a control unit responsible for reading (or introduction) the part programs, which processes (interprets) the instructions contained therein and which sends manufacturing data (distance to run, speed, etc.) to the control drive system of the machine. This system is responsible for managing the motorized axes drivers, for driving the solenoids and other existing organs. At the same time there is a data acquisition system responsible for, in real time, supplying the control unit with updated data regarding the speed and position of each axis and the state of each actuator. The control unit is also responsible for the dialog with the operator through the man-machine interface (panel/monitor).

Originally the man-machine interface computerized system consisted in builder proprietary equipment but nowadays builders tend to use of-the-shelf computers, as a strategy to lower equipment cost. This trend will probably continue, with the machine designers/builders concerned with the performance of their peripheral electronics and its control software to enhance the machine system characteristics.

3 The Pedagogical Importance of CNC Equipment

Vocational training centres or vocational secondary education ensuring the training of qualified workforce through professional courses (Level 4), provide the trainees with the technical skills envisaging integration in the labour market, mainly as machine operators or part programmers. These courses also provide competences for pursuing to the higher education level, naturally by following specialized studies oriented for this area (engineers, researchers, etc.), and at this higher education level some attention is also being given to CNC [6].

However, there are several obstacles to the use of these resources in teaching/learning activities: first there is the issue of this equipment high cost, then coupled with the shortage of available skilled human resources in this area. Educational institutions at all levels face budgetary constraints, which condition the acquisition of this kind of necessary equipment for their training activity either in quantity and diversity.

Facing this scenario, companies associate themselves to create and finance training institutions (called vocational training centres), in order to gather the necessary resources for skilled operators training. On the other hand, the equipment manufacturers or their representatives or subsidiaries, also undertake formative activities directed towards the use of their particular equipment. These actions intend to give an answer to the persistent lack of skilled labourers. However, these are sparse periodic actions, with a limited number of training places, with associated costs non accessible to everyone. Furthermore, being carried out by specific equipment manufacturers, they take place in specific training centres (namely in Portugal), or in equipment installed in companies using it, and so ensuring the operation of the specific equipment they commercialize and their programming. It lacks universality of accessibility.

Given the growing proliferation of autonomous equipment and the consequent need for labour force, the resources involved in training are quite insufficient! On the other hand, it is important to attract youngsters interest in autonomous equipment, not only as passive users (as they often are), but also to involve them in the development of their own equipment, in order they acquire sensibility and gain interest to develop competences in this field.

Schools can also use this strategy to motivate and train their students, while bridging potential resource gaps. Robotics clubs are popular and may constitute an example, but the manufacture of low-cost machines can be also used in the training of technical disciplines, linked to the programming of CNC equipment, to the operation of CAD/CAM systems, to machine maintenance knowledge development, etc.

Although the design construction of equipment similar to the existing in the industry (accuracy and fastness of production), may be not the main objective in teaching activities, it must be intended that knowledge on this subject be acquired, to prepare a skilled population, mastering a series of concepts enabling the country for continuous technological evolution in the autonomous equipment field. By acquiring knowledge it is possible to gather competences not only for designing new equipment but also for repairing faults and maintaining the existing ones in operational conditions.

Many broken components (namely for electronic devices) are no longer marketed or supported, or even their manufacturers no longer exist. As an alternative of the usage of the equipment being stopped, there is the possibility of its actualization, including remodelling of the entire system, which generally implies great investments.

Another attractive alternative approach includes the possibility of characterizing the faulty component, to repair it or to design another component to perform the desired equipment function. It is therefore important to stimulate this possibility among the training entities and graduates launching the "seed" that in the future may germinate in manufacturing skills. The role of CNC machine tools equipment is in this context constitutes a learning field of utmost importance [11, 12].

4   Designing Low-Cost CNC Equipment

The main block elements integrating a CNC system and the respective data flow between them are shown in Fig. 2.

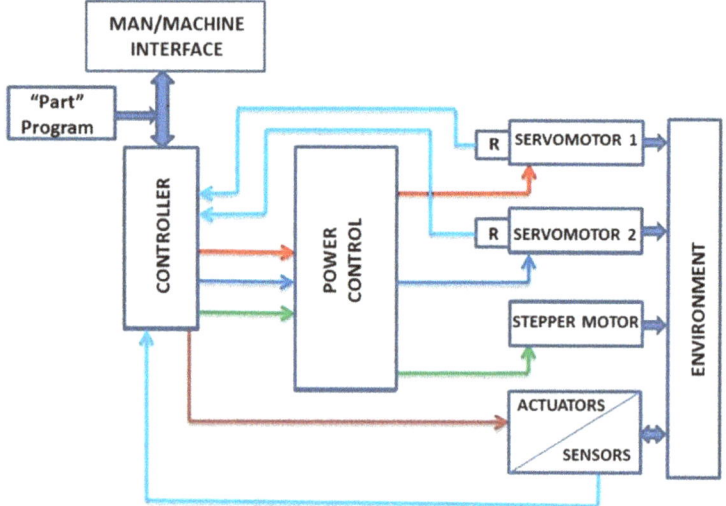

Fig. 2. CNC constitution of the CNC machine tool [9]

The schematic includes a man-machine interface through which the operator interacts with the system. On most systems the interface is integrated into the con-troller. In other cases the interface is installed on an off-the-shelf computer and the controller is another separate device with lower capacity (processing speed and storage) than the computer, more specifically a microcontroller, which interacts with the computer. In this case the computer supports the interface and can support the G-code interpreter required to interpret the instructions contained in the part program in ISO language [9].

Personal computer software simulating CNC environment underwent a widespread development in the last few years.

There are several approaches to the actual functions of the controller, since it can take distinct tasks depending on the system architecture. It is necessary to give instructions to the actuators, either directly (solenoid valves, solenoids, relays, etc.) or through the power controller (in the case of driving step motors or servo motors) and to monitor the behaviour of the system by reading sensors (encoders, resolvers), and the presence or state sensors, like temperature, pressure, etc. The power control block is responsible for driving the axes according to the instructions (speed and displacement) it receives from the controller.

In the present work the microcontroller devices and power controllers (drivers) are be approached in the design perspective of low cost equipment. Industrial equipment whose access is not available through free-standing platforms is excluded.

4.1 Available Resources

The development and design of low-cost equipment is facilitated by utilizing components out-of-service equipment (like computers, printers and other peripherals) as well as components available on the market at affordable cost, pooling enough resources to foster creativity. This is an excellent way to provide learning through experimentation, with the added satisfaction of intervening in creation. The issue is to be addressed from the point of view of the design concepts, rather than from the industrial effectiveness, and so for the driver components (stepper motors, servomotors, etc.), usually presenting considerable costs, the above cheaper solutions may be useful to develop didactic low cost equipment [1, 8, 9].

Low Cost Solutions for Drives
This objective can be achieved by using end-of-life equipment, such as computer equipment, printing or reproduction. It should be kept in mind that the simple typology of some drives is not the most favourable for stopping control in a certain position, and so it is necessary to study and use control gain laws over the power supply of the motor in order to compensate the eventual lack of torque at the low feed requires for the axis speed.

There are already drives on the market prepared for this type of need, with compact and economical solutions.

Drive and Control Boards
Printed circuit boards are available for various applications for the drive control and performance. These boards include microcontrollers that can be programmed through specific software. The potential of the microcontrollers comes through development boards, where they are integrated, giving the physical structure and robustness (pins of connection power and command voltage, etc.), facilitating their integration in the various applications. Considering the vast diversity available, the choice of free programming hardware and software are is advisable. Examples of such equipment include: **Raspberry PI** (www.raspberrypi.org), Arduino platform (www.arduino.cc); and for the drive control, the boards: Easy Driver (www.schmalzhaus.com/Easy-Driver), L298N controller (**MotoMama** - www.botnroll.com/pt), etc. may be useful [9, 10].

Control and Interface Development Software

There is control software available for free download from the internet with the two blocks (interface and interpreter) included. LinuxCNC (*EMC2*) is one of them, developed to work under the Linux system (*ubuntu*). It allows control up to 3 axes (www.linuxcnc.org). Although not free, Mach4-Hobdy software (www.machsupport. com/shop/mach4-hobby), in a non-industrial version, is available for the approximate value of €200, but many others can be found for starting a low cost CNC developing lab. These software and some other interface models developed, as the result of research or students work that, provide excellent basis to start a hands-on learning platform, to catch students attention for the difficulties for developing, operating and maintaining CNC systems. The main aspects of machine control development can then be addressed at an affordable cost [2, 9, 10].

Work in Progress

Machine structures have been constructed in the UMinho Labs, for pedagogical and learning purposes. They are being used for improving students' skills and their scientific potential. A number of MSc dissertations has been produced as the outcome of the work developed [1, 2, 9, 10, 13].

Also alternative control systems have been under study and development, for the retrofitting of obsolete or damaged CNC equipment using the learning activities to seek the final objective of restoring the normal functioning of the equipment [4, 8].

One can also include as a result of the continuous teaching effort being carried out, the CNC equipment that also being developed at the local industry [3, 7].

By now low cost CNC equipment is being developed covering CNC machining, Rapid Prototyping, Coordinate Measurement, Laser Cutting, and also the retrofitting of old machines in the UMinho Labs is under study, including software improvement, electronic devices updating and development and mechanical repair and updating.

5   Conclusion

Low cost CNC equipment is an excellent resource increasingly used by individuals, educational institutions and other entities in the development of their activities. Whether as a resource of didactic support or complement in the accomplishment of tasks with relative autonomy, the CNC equipment have gained special importance nowadays. Its construction involves reduced costs, since in its design normally recycled components of equipment "End of life" can be used. Electronic devices are based on embedded, high-performance, low-cost systems and control programs are developmentally free, with almost no associated costs.

On the other hand, easy accessibility to the components associated with the creativity facilitates the proliferation of this kind equipment and the acquisition of knowledge and experience in the field of design and operation of CNC systems.

The knowledge acquired is also important and very useful for the development, retrofitting and maintenance of industrial equipment. This work developed at university of Minho by giving the opportunity to actually make, operate and maintain machines is improving the professionals' competence joining the industry, for it increased the interest for machine development and manufacturing issues among the students from Mechanics, Mechatronics and Electronics.

References

1. Antunes, J., et al.: Development of a laboratorial diamond wire stone cutting machine prototype. In: 23rd International Congress of Mechanical Engineering, 23rd ABCM. COBEM 2015 (2015)
2. Fernandes, R.J.: Máquina de medir por coordenadas de baixo custo. MSc Dissertation, University of Minho, Guimarães (2012)
3. Ferrão, J.: Projeto e conceção de um equipamento para gravação de chapa metálica com recurso a laser. MSc Dissertation, University of Minho, Guimarães (2017)
4. Freitas, L.: Requalificação de uma Máquina-ferramenta. MSc Dissertation, University of Minho, Guimarães (2015)
5. InovCreative: Computer Numeric Control (CNC). Hong Kong: Hong Kong Polytechnic University (2009)
6. Machado, J., et al.: The Role of Superior Education Institutions on Post-secondary (Non Superior) Education (2012)
7. Moreira, A.: Máquina-Ferramenta CNC para Gravação – Desenvolvimento de Protótipo de Baixo-Custo. MSc Dissertation, University of Minho, Guimarães (2015)
8. Noversa, J.T.: Desenvolvimento e actualização do controlo de um equipamento de prototipagem rápida. MSc Dissertation, University of Minho, Guimarães (2015)
9. Peixoto, J.M.: Análise e Controlo de Máquinas CNC de Baixo Custo. MSc Dissertation, University of Minho, Guimarães, Portugal (2017)
10. Pereira, J.D.: Máquinas CNC de Baixo-Custo - Desenvolvimento do Controlo Triaxial Linear. MSc Dissertation, University of Minho, Guimarães (2012)
11. Pruvot, F.: Conception et calcul des machines-outils. Presses polytechniques et universitaires romandes, Lausane (1993)
12. Pruvot, F.: Concepção e Cálculo das Máquinas-Ferramenta (1993). Tradução: Português
13. Silva, E.J.: Contribuição para Equipamento Didático de Máquina de Medição por Coordenadas. MSc Dissertation, University of Minho, Guimarães (2015)

Effectiveness of SCRUM in Project Based Learning: Students View

José Dinis-Carvalho$^{(\boxtimes)}$, Ana Ferreira, Catarina Barbosa,
Cláudia Lopes, Helena Macedo, and Paulo Tereso

Production and Systems Department, University of Minho, Braga, Portugal
dinis@dps.uminho.pt, anaf3ferreira@gmail.com,
catarina.cacote@gmail.com,
claudia.lopes822@gmail.com,
helena_macedo@outlook.com, paulo151295@gmail.com

Abstract. Project and team management play a major role in the student team's project performance. This is more evident when projects last a long time and the teams are large. In this work, a student team accepted to use SCRUM as their project management methodology during their Project Based Learning (PjBL) experience. This PjBL experience took place on the 7th semester of the Integrated Master in Industrial Engineering and Management degree. The team had a short period of time to train the technique and apply it throughout the entire semester. Although not very enthusiastic in the beginning of the project, the team gradually became aware of the advantages of SCRUM features, recognizing the feeling of having the project under control and gaining management effectiveness throughout the semester. In the end, the team performed well above the average, being one of the two teams with the highest score of the class.

Keywords: Project Based Learning · SCRUM · Project management

1 Introduction

Project Based Learning (PjBL) [1] can be a very worthy experience for students, especially when the projects take place in real context. One of the complex issues of those experiences is the way that student teams manage themselves and their projects. A very typical problem is that student teams end up having most of the work taking place just a few days before the due dates for deliverables. The quality of the deliverables could be better with student anxiety and stress reduced if they could manage their work in more effective ways. The nature of these projects held in real context, so complex and open, with unpredictable results, are virtually impossible to plan in reasonable detail, so traditional project management methodologies are not applicable.

The objective of this study is to evaluate the effectiveness of SCRUM when used with the project and team management tool in a PjBL environment.

A randomly selected team involved in the PjBL semester in industry context went through a quick training in SCRUM and then the team members where interviewed about their expectations regarding the methodology effectiveness in their project and the same type of interview took place at the end of the semester.

J. Machado et al. (Eds.): HELIX 2018, LNEE 505, pp. 1118–1124, 2019.
https://doi.org/10.1007/978-3-319-91334-6_154

2   Framework

The engineering practices require the application of transversal competences or soft skills such as autonomy, leadership or interaction with others in interdisciplinary teams, negotiation or solving conflicts, communication in effective ways, as well as project management skills [2]. It is also recognised that the development of engineering competences requires the application of technical knowledge in specific contexts linked to the professional practice. Some examples of learning processes closer to the professional experience in higher education have resulted primarily from practical training in final years of formation or even after the end of initial formation. However, it can be argued by following this approach, students may lose the possibility to develop these competences integrating academic technical knowledge with professional knowledge.

2.1   Project Based Learning

Project-based learning (PjBL) is a student-centred learning method focused on the development of a project in a real professional context, in which teams of students solve an interdisciplinary problem by articulating theories and practicing the development of a project [3, 4]. Other characteristics of Project-based learning can also be referred as being focused on the integration of several areas of knowledge (including technical and transversal skills) for solving a problem linked to real situations, during a long period of time (e.g. a semester), resulting in a specific final result [5].

The students develop several competencies: Knowledge, Skills and Attitudes when they go through PjBL in real context [6]. PjBL in real world problems and context is believed to motivate students to identify and apply research concepts and information.

2.2   SCRUM

Agile methodologies were used to handle the challenges of managing complex projects during the development phase. In 1993, Jeff Sutherland had used SCRUM for software development projects for the first time. In fact, 94% of all organizations practice some form of Agile and SCRUM is one of the most popular method of Agile because 86,9% of Agile SCRUM users observe increased profits [7]. However, a study of Agile software development shows that only 3% of the existing scientific evidence on Agile software development focuses on SCRUM [8].

The objective of SCRUM is achieved by optimizing the development process by identifying the tasks, managing time more effectively, and setting-up teams.

The SCRUM team consists of a Product Owner, a SCRUM Master, and Development Team Members. The Product owner represents the business, customers or users, being responsible for managing the Product Backlog, the list of requirements of the product. He/She has also to ensure that the team understands the goals of the project and performs at a high level. The SCRUM Master instructs the Development Team on creating clear Product Backlog items and communicates with the team to ensure that the team understands the long-term plans of the project. The Development Team is responsible for implementing and delivering the releasable product at the end of each "Sprint," which is a period of time to create a usable increment of the product.

Some items of the Product Backlog created by the SCRUM Master are selected for the specific sprint, and it is designated the Sprint Backlog. The goals of each sprint are planned in the Sprint Planning Meeting, which occurs once a month prior to each sprint. Moreover, there is a daily SCRUM meeting where team members update one another on their progress, difficulties they have experienced and their future goals. Therefore, Sprints contain and consists of the Sprint Planning, Daily SCRUMs, the Sprint Review and the Sprint Retrospective. Hence, the Development team delivers products iteratively and incrementally, maximizing the feedback they receive. As the Development Team members, it has been found that the ideal size is seven members [9].

The use of SCRUM principles and features to improve the learning effectiveness is also materialized by eduSCRUM, developed by Willy Wijnands, a professor of Physics and Chemistry at Ashram College located in the Netherlands [10, 11].

3   Description of the PjBL Model

The Project-bases Learning (PjBL) model adopted in this case is a methodology inspired by the PLE model [4] which aims to solve interdisciplinary problems based on a real situation where learning is student centered [12]. Throughout the project students are able to develop technical and soft skills with great learning effectiveness.

The project is carried out in teams, supported by a teaching staff team responsible for the curricular units and for the tutoring of each team. The role of the teaching staff is to provide technical support to the project allowing its development based on the respective course content. The tutor's role is in monitoring the project's progress leading the students in the desired direction and helping them in project and team management.

At the University of Minho, this methodology is adopted in the seventh semester of the Integrated MSc in Industrial Engineering and Management. The projects have been developed in student teams of 9–11 members, in cooperation with a company where the main objective is to deal with a real situation to improve the core competencies for professional practice [13]. In this project, students must analyze the respective company production system and present well founded suggestions for improvements.

The assessment of the project (PIEGI II) is based on 2 reports (70%) and 3 presentations (30%). Throughout the semester student teams also perform self-assessment and peer assessment with impact in the final individual grading. To improve the reliability of the project work a set of milestones were defined to monitor all project teams.

The main difficulties encountered by students during the course of this project are mostly related to interpersonal relations and project management. As for the interpersonal relationships, the main difficulties come for managing conflicts caused by differences of opinions, individual goals and the lack of communication within the group. The project management level, the challenges are focused on reconciliation schedules, deadlines greetings and planning organization of project tasks. Thus, it is necessary to apply project management methodologies in order to overcome these difficulties.

4   SCRUM Applied in a PjBL Team

In order to carry out the project it was felt the need to introduce a methodology able to plan and manage all the tasks to be performed in a quick and simple way. The need for a methodology of this kind stems from the fact that it is a very complex project, which in turn entails several tasks that need to be carried in time so that there is no accumulation of work that could result in future problems.

The initial training and reading material about SCRUM as well as the help from the tutor it was possible to understand the basics of the SCRUM methodology, until then unknown to all team members. After this initial phase dedicated to the understanding of SCRUM the student team built their own version of a SCRUM framework divided into four columns, Plans/Problems/Ideas, Portfolio Tasks, Sprint and Done, as can be seen in Fig. 1. The use of team board placed in the team room space dedicated to the project as well as other visual management tools are encouraged by the teaching staff.

Fig. 1. SCRUM board adopted by the team

In addition to these four columns, the SCRUM framework also includes a section devoted to waiting tasks, for assignments that were not completed on time or waiting from data from external entities. The team also adopted the burndown charts to monitor the progression of the completion of the planned sprint tasks with the average theoretical sprint pace. Figure 2 shows a team's burndown chart example for one sprint of one week. As can be observed the theoretical pace is represented by a red straight line and the real progression is represented by the other set of blue line segments. The pattern seen in the Fig. 2 was a typical pattern for all sprints because the team had less classes to attend to on Wednesdays so more work could be done on that day and more tasks were completed.

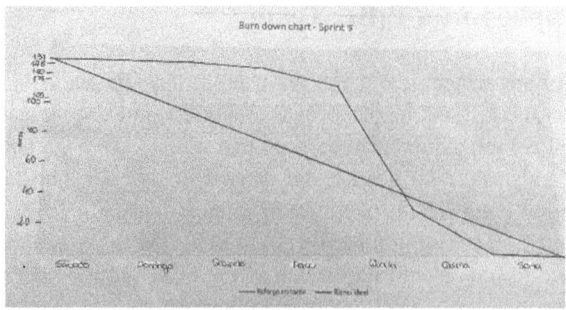

Fig. 2. Example of a burndown chart

The team decided to have the sprint review and sprint planning at the end of every Friday because no other time was possible to agenda restrictions.

The first point of the meeting was the retrospective of the previous Sprint, having more emphasis to the column of completed and waiting tasks. Each element related what went worse and better throughout the week as well as some proposals improvement so the next week could be better. In this first point it is also finished the burndown chart, which allows the group know the number of hours worked and if it met the expectations. The second meeting point was focuses on planning the next sprint for next week, an element is chosen every week, which will play the role of SCRUM Master. They are then planned all the necessary tasks to be performed, and are assigned to different elements as well as the number of hours corresponding to their achievement. With help of SCRUM Poker application with the Fibonacci sequence, each group member chooses a number that corresponds to the total of hours required to perform a given task. Subsequently ideas are discussed and long-term problems that will be placed in the first column above the SCRUM.

This meeting was one of the key points of this methodology when applied to PjBL as it allows the group to have a perception of the tasks performed, and thus it is possible to perceive the development state of the project and set tasks that require priority.

5   Results

The first contact with the SCRUM tool caused reluctance to all group members, since this approach never been displayed or used before. The lack of practice in the application of this method has a very long planning stage, due to the existence of different perceptions of the operation and concepts associated with SCRUM. However, the interest in the practice of new methods motivated the group to understand and improve their planning skills, organization and interaction by sharing ideas, research and group tutor support.

During the project, the group faced several difficulties in adapting to this tool. The long time spent in the sprints, the lack of understanding of the assignment of the number of hours devoted to each task and the lack of differentiation between "not started" tasks and task "sprint" were the main obstacles to overcome.

In order to fill these gaps, the team received additional support from the tutor and a workshop with a Product Owner (from a computer company) who clarified doubts and gave feedback on how SCRUM should be applied by the working team. After answering questions, the group honed his SCRUM techniques and the great advantages of this method began to show good results. Although it was necessary to spend a large amount of time in the weekly sprints, this very visual practice, proved fruitful in meeting deadlines, in performing each task as well as the assignment of responsibilities in the implementation of the same.

The use of the burndown chart graphic, to complement the way all this practice, proved to be an asset, since this tool is very visual and allows for easy perception of the amount of work performed by the group throughout the week. The results taken from these weekly charts show a large discrepancy compared with graphics related to the use of SCRUM methodology because is not applied in the field of education, since the work team only had the Wednesday and Friday to focus on achievement from the project. In the remaining days of the week the amount of work performed by group members was very low, reflecting only small individual work done outside of school.

The SCRUM methodology when applied to PjBL provides the basis for a careful organization that enables easy management of all tasks, which in turn, results in an a more productive use of time. This can be said even when the SCRUM methodology was not being applied fully because of the environment condition of this study. Team member did not have proper training on the methodology neither the teaching staff had the necessary experience to assist them. The SCRUM master role was improvised by a team member without proper training and experience to perform that role in an effective way. The daily meetings to plan the day was not applied because the team had no agenda to do it. Finally, the availability of team members to work was limited most of the days and varying frequently due to other academic duties.

6   Conclusion

The challenge that was faced in this work was to adopt the SCRUM methodology in a team of students under a Project Based Learning environment based on industrial problems to be solved. The project is the most important part of the semester in a way that failing the project represents a most certain failure in most other curricular units that support the project. The expectations and motivation are high among the students but the tension is high as well since every team wants to perform well and ideally better than other teams. It is difficult to put more variables in the game so student are normally against trying different methodologies to manage the project and the team. It happens that the randomly selected team, although without enthusiasm and with a lot of doubts, accepted to experiment the use of SCRUM.

The team started with many difficulties in accepting the amount of time spent in planning the sprints without seeing apparent benefit resulting from that effort. Gradually, the team learned how to do it faster and more effectively and felt the comfort of having everything under control and having the ability to increase their speed from sprint to sprint. After the first four or five weeks, the team was unable to manage themselves anymore without the SCRUM and its events and artifacts.

Even with the some limitations regarding the proper use of the SCRUM methodology, in the end, the team ended up doing a lot of valuable work being one of the two teams with better score on the project and probably the adoption of the SCRUM methodology played a role on that success.

Acknowledgements. This work has been partially supported by projects COMPETE-POCI-01-0145-FEDER-007043 and FCT-UID-CEC-00319-2013, from Portugal.

References

1. Dym, C.L., Agogino, A.M., Eris, O., Frey, D.D., Leifer, L.J.: Engineering design thinking, teaching, and learning. J. Eng. Des. **94**(1), 103–120 (2005)
2. UNESCO: UNESCO Engineering Report: Issues Challenges and Opportunities for Development. UNESCO Publishing, Paris (2010)
3. de Graaf, E., Kolmos, A.: Management of Change: Implementation of Problem-Based and Project-Based Learning in Engineering. Sense Publishers, Roterdam (2007)
4. Powell, P., Weenk, W.: Project-Led Engineering Education. Lemma, Utrecht (2003)
5. Helle, L., Tynjälä, P., Olkinuora, E.: Project-based learning in post-secondary education - theory, practice and rubber sling shots. High. Educ. (2006). https://doi.org/10.1007/s10734-004-6386-5
6. Lima, R.M., Dinis-Carvalho, J., Sousa, R.M., Arezes, P., Mesquita, D.: Development of competences while solving real industrial interdisciplinary problems: a successful cooperation with industry. Producao **27**(Specialissue) (2017). https://doi.org/10.1590/0103-6513.230016
7. Grieser, B.: PMO advisory introduces SCRUM training courses in the metro New York (2016). https://www.scrumalliance.org/why-scrum. Accessed 1 Jan 2018
8. Dybå, T., Dingsøyr, T.: Empirical studies of agile software development: a systematic review. Inf. Softw. Technol. **50**(9–10), 833–859 (2008). https://doi.org/10.1016/J.INFSOF.2008.01.006
9. Sutherland, J., Schwaber, K.: The Scrum Guide, the Definitive Guide to scrum: The Rules of the Game (2017)
10. de Jager, T.: Using eduScrum to introduce project-like features in Dutch secondary Computer Science Education. Utrecht University (2015)
11. Delhij, A., van Solingen, R., Wijnands, W.: The eduScrum Guide. EduScrum Team (2015)
12. Powell, P.C.: Assessment of team-based projects in project-led education. Eur. J. Eng. Educ. **29**(2), 221–230 (2004). https://doi.org/10.1080/03043790310001633205
13. Lima, R.M., Mesquita, D., Rocha, C.: Professionals' demands for production engineering: analysing areas of professional practice and transversal competences. In: 22nd International Conference on Production Research, ICPR 2013 (2013)

Instructional Design in Electrotechnical Engineering: A Case Study on Integrated Online Approaches

Mónica Régio$^{(\boxtimes)}$, Marcelo Gaspar, and Margarida Morgado

Instituto Politécnico de Castelo Branco, Castelo Branco, Portugal
{monicaregio, calvete, marg.morgado}@ipcb.pt

Abstract. Higher Education Institutions are now preparing students for a globalized working area where they must be prepared for communicating effectively in a foreign language. Considering CLIL is an approach that integrates the teaching and learning of a content subject and a language at the same time and that engineering students respond well to this methodology, a task sequence in a Moodle platform is presented in this paper. The perspectives of a language teacher and a content lecturer are offered. A task sequence, being the result of a three year integrated collaborative approach between a language teacher and a content teacher, was tested in an Electro technical Engineering course. The preferred methodology was blended learning, as face-to-face classes and online digital Moodle platform were used.

Keywords: Higher Education · CLIL · Instructional design
Online learning environments · Blended learning

1 Introduction

1.1 Background to Using the CLIL Approach in Engineering

The CLIL approach in Higher Education.
Higher Education Institutions are now preparing their students for a globalized working area where their students will need to be prepared to communicate effectively in a foreign language in future working environments. English presents itself as the language for global communication and for students to become pluriliterate users of academic disciplines [1]. This is the reason why non-native English-speaking students must be prepared and HE institutions need to prepare them to confidently use English to think, work and communicate. This is commonly recommended through exposing HE students to rich content-based input, as well as allowing them sufficient time for interaction and production in the Foreign Language (FL).

The CLIL approach, being an innovative methodology in which students combine the learning of a certain subject content with the necessary language skills to do so, could be a key solution that outstands those of extra Foreign Language (FL) classes and supplements the use of English as Medium of Instruction (EMI), which is becoming wider in HEIs [2]. Engineering students, used to function with task-based methods, and

© Springer International Publishing AG, part of Springer Nature 2019
J. Machado et al. (Eds.): HELIX 2018, LNEE 505, pp. 1125–1130, 2019.
https://doi.org/10.1007/978-3-319-91334-6_155

engineering teachers, willing to prepare their students to work as engineers, understand CLIL as one viable answer for preparing students for the job market requirements.

This paper will present a perspective of a three-year adjunct CLIL experiment between a language and a content teacher with engineering students at the Superior School of Technology in the Polytechnic Institute of Castelo Branco. Adjunct CLIL has been described as simultaneous language support coordinated with subject content studies, which also implies that there is joint class or course planning by the language and the subject content teachers and that the learning outcomes apply to both content and language [3, 4].

The guiding question for applying CLIL to HE engineering courses is the foreseeable impact on the cognitive skill development of students, increased student motivation, and student preparing for plurilingual and multicultural work environments by learning to learn, think and work in a second — or foreign — language.

This paper will showcase an effective tested task sequence in adjunct CLIL, which results from joint class planning by an English ESP B1 (CEFR) teacher and an electrotechnical engineering teacher. The perspective offered is that of teachers who have tested this task sequence as a consequence of a three-year long engagement in CLIL.

1.2   Adjunct CLIL: Working with a Partner Teacher

The first stage (first year) of this three-year experience [5–8] was to have engineering teachers sit together with language (English) teachers through a CLIL training course to design learning sequences and strategies they would implement and monitor as to their own and student satisfaction. As familiar users of web-based tools for learning, such as a Moodle e-platform, engineering teachers showed no resistance to designing online instructional sequences for blended learning online and in the classroom. However, the training focused, firstly, on how to collaborate with the English teacher and, secondly, on how to engage students in effective communicative uses of English (as a Lingua Franca) in meaningful scenarios while learning specific engineering contents and developing linguistic communicative skills in English both for academic and professional purposes.

Since needs analyses surveys had made clear that engineering undergraduate and postgraduate students were not proficient in English (ranging from A1 to B2 levels CEFR), considered learning English at HEI a waste of time and had little to no motivation to follow the English for Specific Purposes (ESP) course on offer, and were usually unable to show good progress in language skills through the two-semester course in ESP they took, the challenges for both language and content teachers, in the framework of blended e-instruction, became those of designing and implementing e-tasks and task sequences that would raise student motivation and engage students in using English to think about concrete authentic engineering activities and problems, while using English to solve them.

The experiment thus required initially the training of both engineering and English teachers, who were expected to collaborate. The 15-h staff training sessions, which aimed at constituting a Community of Practice (CoP) (Wenger) uncovered the teachers' conceptions about education, methods, attitudes and skills, while they invited them to revisit their own and students' needs and linguistic skills, as well as motivation to learn in and through English. The design of instructional activities included classroom-based activities and e-activities. The sequences of instructions and tasks given to students were trialled and implemented during six semesters and each time evaluated through data collected (students' assignments and feedback, notes taken by teachers).

In the second year of CLIL implementation the partner teachers redesigned e-strategies to further capture students' attention, interest and motivation. Online questionnaires, simulated online interviews or video discussions were examples of the co-developed activities. In addition, the language and content teachers decided to create a common project across their subjects in which students would be evaluated in both English and the content subject simultaneously.

This paper will mainly focus on the third year of this experiment. Considering students' answers to questionnaires and feedback, as well as the teachers' notes and observations, both teachers decided to create varied online activities. In doing so, the lecturers expected the sequence of tasks to increase student participation and motivation.

2   Task Sequence on a Moodle Platform

To develop students' content and language skills simultaneously the teachers co-designed a task sequence on the school's Moodle platform. Lecturers' main concerns were related to students' motivation to work on a Moodle platform and to learners' subject acquisition and language need awareness.

2.1   Content and Language Integrated Tasks

With the goal of improving students' skills in both content and language at the same time while sustaining interest and motivation, diversity of the tasks was given pride of place, so that students expected something new and different every class.

The first task was a collaborative report in which students were supposed to write about the content topic discussed in the face-to-face class. The report would show how students had learned the content topic, the vocabulary and linguistic structures associated, as they had to express their thoughts in the foreign language.

In the next class students were proposed a show and tell speaking exercise based on a given topic; they were afterwards invited to write a brief summary online, as it would help them to learn the new terminology and how to express their ideas in English.

The third assignment consisted on choosing a video, about the face-to-face class taught topic, and uploading it in a forum. All the students were asked to comment other students' posts, so they were able to develop their communication skills and synthesize the learned contents.

Afterwards, an online chat challenge issued. Every student had to contribute to the discussion with at least ten entries. The purpose of this task was to create the opportunity for students to improve their communication skills on a given subject.

The fourth activity was an online questionnaire for students to search online for a specific subject. After this they had to write a small report online about their findings. In the following task students were asked to offer two different ideas to solve a given problem in a forum. After doing that all the students had to choose their colleagues' best idea.

An online glossary was proposed for students to collect the most relevant vocabulary of the semester. This was an open task since students should insert the relevant terms every week.

At the end of the semester, and to prove they had learned the most important subjects, students had to create a video pitch representing their project. That was the starting point of the English speaking exercise, which aim was to evaluate if the students had learned the contents and, at the same time, acquired the communication skills associated to them.

2.2 Student Work

CLIL and Task-Based Language Teaching (TBLT) approaches consider that language and meaning are inseparable. While in CLIL meaning is connected to content learning, in TBLT it is associated with experimental learning. They both have in common the fact that the teachers and the students interact and collaborate to create situations in which the learning of language is maximized to promote the creation of meaning [9].

Considering the importance of TBLT for engineering students' learning process, the co-teachers designed a task sequence for the blended course supported by the Moodle platform. Students were asked to answer a set of tasks in face-to-face classes and online scenarios. The in person assignments were mainly speaking exercises in which the language teacher helped the students to gain linguistic competence while leaning a new content topic. The online works were thought to be different from one another in order to promote different learning scenarios. Teachers normally introduced students to the tasks helping them to understand what was really asked. Most of the times students didn't need additional support, nevertheless in some occasions teachers had to change the face-to-face class plan and support students while they were responding to the Moodle tasks. Lecturers concluded that this was essential to move the course on and maintain students motivated and willing to do the next assignments.

3 Discussion and Reflection

A task sequence, being the result of a three year integrated collaborative approach between a language teacher and a content teacher, was tested in an Electro technical Engineering course. The preferred methodology was blended learning, as face-to-face classes and online digital Moodle platform were used. During the first semester of 2017–2018 school year an engineering teacher and an English lecturer co-worked and co-designed a blended CLIL task sequence in which students were supposed to learn specific content related topics while they simultaneously developed communicative English skills.

Data was collected to assess student motivation toward blended learning courses and for further improvement in future CLIL courses combining face-to-face methodologies with the use of online digital platforms.

4   Conclusions

The designed blended task-sequence with ICT in adjunct CLIL catered for three important aspects: introducing online learning environments, such as Moodle, as it seemed to be more motivating for students than the alternative face-to-face classes; supporting students' learning through variety of tasks and scaffolded tasks (both in linguistic as in content); and, thirdly, supporting student involvement and participation through face-to-face coaching, using face-to-face classroom type to explain task sequences and either initiate, develop or complete tasks when needed.

So, task design with ICT in adjunct CLIL follows the same rules as for standard CLIL practices, although engineering students seem to become more motivated when they can use digital platforms and online learning support, while not giving up completely on face-to-face guidance. This may be due to their own preferred academic cultures of learning, to which the EFL or ESP teacher has to adapt. Although not implemented or piloted, computer-based instruction seems to open up new pedagogic avenues for customization of student learning and for independent practice of the FL at the student's own pace.

What emerged clearly from the practice described is that students claimed they could follow all teaching better as the computer-based instruction helped them keep up with the face-to-face classes they had missed. Also, the Moodle platform used as part of the blended task sequence not only supported adaptive learning, but also shortened student waiting time for task assessments results, which increased motivation.

As the learning sequence required 'active explaining' by students, they were more engaged and motivated than they would have been if they had been taught in a teacher-centered way; having to explain online and to colleagues and teachers further developed their communicative skills and provided a connection between the conceptual continuum and the communicative continuum [11], which is characteristic of the CLIL approach.

Guidelines for adjunct CLIL to be deduced from this three year long experiment would probably encompass adjunct-CLIL teachers finding the right balance to address the content and language needs of students, their motivation and their learning. Adjunct-CLIL teachers have to experiment widely with the resources that will work for a CLIL task sequence, as the CLIL approach is not just about the topic content, but also about how that content-based instruction can be articulated with a focus on language and how the language will be used in context. Resources need to support student cognitive development and language acquisition and use. Thus, in this particular case it was important to expose students to rich content-based input, while keeping the communicative skills relatively simple (given the average B1 level of students) while giving explicit instructions on how to communicate their knowledge effectively in English and supporting their learning in face-to-face contexts.

References

1. Lister, R.: Preface. In: Valcke, J., Wilkinson, R. (eds.) Integratting Content and Language in Higher Education. Perspectives on Professional Practice, pp. 7–14. Peter Lang, Frankfurt am Main (2017)
2. Valcke, J., Wilkinson, R.: Introduction - ICLHE, professional practice, disruption, and quality. In: Valcke, J., Wilkinson, R. (eds.) Integrating Content and Language in Higher Education. Perspectives on Professional Practice. Peter Lang, Frankfurt am Main (2017)
3. Content and Language Integrated Learning (CLIL). LanQua
4. Ni, N., Jauni, H.: Teacher perceptions of teaching CLIL courses. In: Valcke, J., Wilkinson, R. (eds.) Integrating Content and Language in Higher Education. Perspectives on Professional Practice. Peter Lang, Frankfurt am Main (2017)
5. Gaspar, M.C., Régio, M., Morgado, M.: Learning English through 3D-printing : a case study with engineering and design students in higher education, pp. 5132–5138 (2016)
6. Gaspar, M.C., Régio, M., Morgado, M.: The co-construction of a bilingual glossary of terms for engineering and design technologies as part of learning in engineering through a collaborative CLIL approach, pp. 3214–3218 (2017)
7. Gaspar, M.C., Régio, M., Morgado, M.: Lean-green manufacturing: collaborative content and language integrated learning in higher education and engineering courses. J. Educ. Cult. Soc. 208–217 (2017). https://doi.org/10.15503/jecs20172.208.217
8. Morgado, M., Régio, M., Gaspar, M.: Content, language and intercultural challenges in engineering education: (e-)strategies to improve instructional design. In: New Trends and Issues Proceedings on Humanities and Social Sciences, pp. 153–161 (2017)
9. Ortega, L.: Researching CLIL and TBLT interfaces. System **54**, 103–109 (2015). https://doi.org/10.1016/j.system.2015.09.002
10. Lyster, R.: Learning and Teaching Languages through Content: A Counterbalanced Approach. John Benjamins Publishing Company, Amsterdam/Philadelphia (2007)
11. Meyer, O., Coyle, D., Halbach, A., et al.: A pluriliteracies approach to content and language integrated learning – mapping learner progressions in knowledge construction and meaning making. Lang. Cult. Curric. **28**, 41–57 (2015)

Proposal of Curricular Program to Introduce BIM in a Civil Engineering School

Alcínia Z. Sampaio[✉]

Department of Civil Engineering, University of Lisbon, Lisbon, Portugal
zita@civil.ist.utl.pt

Abstract. The attention of Civil Engineering is currently oriented to the methodology Building Information Modelling (BIM), a new concept that is based on recent scientific advances concerning the ability of computer systems to store, manage and manipulate large amounts of information, and established as the big breakthrough in the construction industry. Being the school the main actor in the formation of new engineers, it has the mission to offer disciplines or training workshops to prepare students for the professional activity, giving the knowledge concerning the current most advanced computer technology. In the context of implementing BIM, the future engineer should acquire the ability to apply basic BIM tools in different domains and recognize the advantages in developing collaborative projects provided by BIM platforms. In Portugal, the involvement in terms of design offices and construction companies, have been growing and as so this interest the Construction industry is an incentive that justifies the inclusion of BIM within the curriculum of the Master Course of Civil Engineering.

Keywords: Advanced technology · BIM · Curricular program
Pedagogic methodology

1 Motivation

The Building Information Modelling (BIM) concept focuses on the development of engineering projects, on the basis of collaborative work involving a single digital model that archives in an organized way, the information coming from the different linked subjects. The information, required in each design phase, is transposed from the central BIM model and handled in such a way, supporting the development of follow design stages. The processes of transfer, manipulation and generation of data, are carried out using basic BIM tools resources, and require an effective team work coordination and information management. The amount and variety of information produced during the life cycle of a project, makes its management, without the proper tools, a complex process [1].

BIM implementation covers various sectors of the construction industry. The engineer should know what basic BIM tools are available in the market that can be used to support the development of his expertise. BIM methodology interferes with all aspects involved in the project in engineering: at the initial stage when generating the geometric form of the building (architecture); at different phases of structural analysis

© Springer International Publishing AG, part of Springer Nature 2019
J. Machado et al. (Eds.): HELIX 2018, LNEE 505, pp. 1131–1137, 2019.
https://doi.org/10.1007/978-3-319-91334-6_156

(structural solution design and detailed reinforced drawings production); at the budget estimation (material take-off of materials); at the construction planning stage (linking the Gantt map with milestone components of the model); or, later, controlling the building occupation (supporting the management and maintenance activity and the establishment of repair or rehabilitation projects).

In order to add the BIM new topic to the course offered to students, at the University of Lisbon, it was proposed the introduction of the "BIM Methodology" discipline. A correct insertion of this curricular unit, taking the form of a discipline and involving the different specialties, was at the level of last year of the Civil Engineering course.

2 Contextualization

The proposed curricular program was established to encompass: the introduction to the topic and to the parametric modeling process, pointing out the benefits and limitations; the reference to the sectors in which the BIM has had greater visibility and success; the use of BIM, which base their potential and specificities; the development of each expertise in a BIM environment; overlapping components and conflicts analysis between disciplines; the verification of the correctness of data transferred between components of the project; the analysis of the consequences and benefits introduced in project planning; the establishment of methodologies of collaborative work required in the process of implementation.

It turns out that the student, in the final stage of his formation, is more aware of the subject involved and recognizes that this competence will be useful for them in the job market. These aspects justify and reinforce the incorporation of the courses in the last school year. The academic discipline is elective and is offered to all branches of the Master curriculum. Besides the fields of building and structures, the sectors of architecture, management, and hydraulic systems may also be involved. The training of collaborative working teams, involving this diversity of domains is, naturally, enrichment for students, for the Department of Civil Engineering and for the school.

The author is the responsible of the discipline. The teaching is given in two semesters for teachers of different areas with a greater involvement of the teachers of the Department of Civil Engineering, and the students are finalists, attending the last year of the course of Civil Engineering. In the pilot year just a class is supposed to be created.

3 Teaching Methodology

The BIM issue requires the enlightenment of concepts and its applicability and level of internalization to students. However, the practical component is what makes the student able to act in different contexts and phases of the project. So, the practical use of teaching basic BIM tools is essential in the course. The training provided should enable the student to: carry out modelling of different specialties, namely data transfer between phases, reuse and create new information; be critical in every step regarding the correction of data transferred and know how to perform in accordance; carry out the addition of components on the centralized BIM model, meet its organization and learn how to establish effective management of information; be able to extract the model documentation inherent to a project in the way usually required in the construction sector.

As an approach of training, and learn how the BIM involves various aspects of the project, namely, the initial architectural component, it was proposed the implementation of a complete project over a real case study. The student is prompted to initiate the contact with the BIM methodology with the generation of the architectural model, using the first parametric objects. This model serves as the basis for the development of the following components: Generation of structural models (analysis, design and detailed drawings); Definition of MEP (mechanical, electrical and plumbing) models concerning service networks of water, sewerage, electrical and air conditioning installations, and the corresponding analysis clash detection between disciplines; Construction process planning (associating several model components to the steps of the Gantt map of, creating a 4D model); Extraction information from the model in order to obtain tables of quantities of components and material. The student must, sequentially, perform the listed steps. In the last years, different stages of the model process were performed concerning architecture, structures, MEP and construction (Fig. 1):

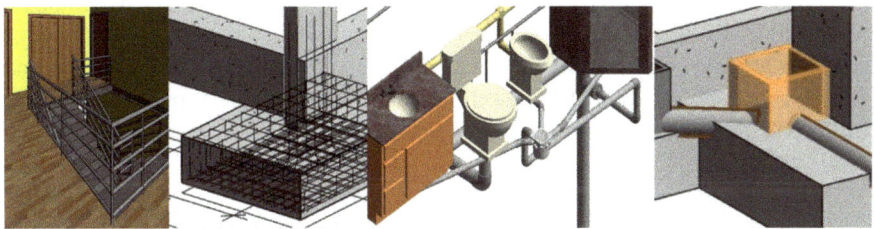

Fig. 1. Examples of didactic exercises.

The student carries out, at the same time, the same exercises and after he is invited to develop its real case. The workshops and the individual exercise have been held in the computer lab, equipped with the updated software required in the course.

3.1 Programmatic Content

The curricular program of the syllabus consists of different modules trying to address all topics that will be useful to the engineer in his future profession. The sequence of items starts from the generic concepts reaching a comprehensive use of all the information added to the centralizer BIM model (Fig. 2), throughout the all course. The BIM model is composed of all disciplines and must contains all the information that later is needed to define drawings and maps of quantities or to support construction planning and building maintenance. Below are listed the considered items:

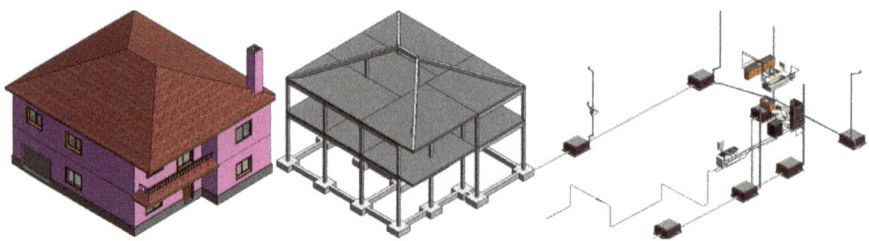

Fig. 2. A real case developed in several modelling steps (architecture, structure, and MEP).

- *Introduction to BIM concept*: State-of-the-art, application, benefits and limitations; Parametric modeling process; Model detail levels and project phases; BIM implementation strategies in enterprises.
- *Interoperability capacity and standardization*: Information transfer formats; Collaborative platforms and systems managers of BIM projects; Limitations of BIM interoperability in design management.
- *Model of architecture:* BIM base tools frequently used; General aspects of BIM base software (interface, parametric objects and unit definition); Initial settings of the modelling process (orthogonal grids and levels of floors), Selection and edition of parametric objects (walls, windows, doors, floors, and roofs); Display views of the model (drawing plants, elevations, horizontal and vertical sections, axonometric perspective); Getting information query (type of objects or materials proprieties).
- *Model of structures:* Re-use of information of the architectural model; Selection and edition of structural parametric objects (columns, slabs, foundations); Association of material and physical properties to objects; Transposition of geometric model to analytical structural model.
- *Structural analyses:* Transposition of structural model to a structural analysis software; Verification of the correctness of the transferred information; Structural analyses; definition of reinforcement details; Transfer of structural results information to the initial BIM model; Elaboration of reinforcement details drawings.
- *MEP model*: Networks and services equipment; MEP software-mechanical, electrical, and plumbing engineering; Modelling of water and sewerage networks; Air conditioning model generation; Analysis of conflicts between components.
- *Construction planning:* Establishing milestones in Gantt map; Generation of the 4D/BIM model; Use of BIM viewers;
- *Extraction of information from the model*: Maps of quantities of materials; Definition of technical drawings; Budgets estimation.
- *Coordination and management of information:* collaborative project (team and BIM work methodology); Analysis of advantages and limitations in BIM processes.

3.2   Evaluation and Expected Result

The knowledge assessment is carried out on a final draft. This work should include: Elaboration of architectural, structural and MEP models; Overlapping components and corresponding analysis of conflict; Structural analyses including representation of detailed reinforcement drawings; Planning the construction process by defining the correspondent 4D model; Getting maps of quantities take-off and elaboration of budgets concerning several design stages.

For the preparation of individual BIM design real cases, a set of initial information is provided, concerning drawings and characteristics of the building components for a posterior comparison, between traditional way and BIM environment. The last procedure is the ideal teaching methodology BIM, because it makes students more likely to work later in the context of collaborative project, the genesis of BIM. The biggest drawback in this process is that each element does not develop in a balanced way their capabilities in relation to different specialties. However the collaboration is needed and, bearing, transmission of knowledge is somehow made between the members of the working team. Performing a complex job and obtaining success is of course challenging and a cause for satisfaction for the student, making him more confident in his ability with BIM projects in his future activity.

As a result the student adds to his skills in Civil Engineering the subject BIM. Along the course he acquires knowledge of the applicability of BIM and its benefits in various areas of the project, as well as the working methodology inherent to BIM concept. The students are able to set important guidelines in the process of implementation of BIM in an engineering office; he is able to indicate which are the changes and the adjustments that must be made on the team and on usual the work methodology, in order to establish collaborative platforms. The knowledge and training gained enables him to perform with success the projects as may be requested by means of an innovative methodology, for which only a few engineers are currently trained. In Portugal and in the current early stage of the BIM implementation, the student, as a future professional, constitutes an important difference in a world of scarcity of job opportunities, motivated by the current reduced activity of the construction.

4   Previous Experience in Introduction BIM at School

The Computer-Aided Design syllabus offered to 1st year Civil Engineering students was recently improved including this advanced technology, in order to adjust its issue to the current social and scientific requirements. From 2014–2015 curricular year, the program of the discipline incorporates the BIM component, for a trial period of 1 week (4,5 h), and it was extend for four weeks the following year.

However, BIM cannot eliminate completely the practice in computer graphic systems, like *AutoCAD*, because this is still a frequent tool used in design offices and as so the students' training must continue to contain this strand. As this discipline is taught at the level of the first academic year of the Master course, just the modeling BIM perspective, of the architectural [2] and structural components [3], is taught in tutorial lessons and only this part can well be understood by the young students. As mentioned, the complexity of BIM methodology and its whole understanding is only perceptible when developing a more comprehensive project.

In the last years, the search for the BIM theme, as a Master thesis topic, has been great. The thesis' supervision has been one of the teachers' activity most involved in the BIM subject. All the completed thesis reports were inserted in the digital platform of IST allowing its consult. Below are listed some of the most recent research works (Fig. 3):

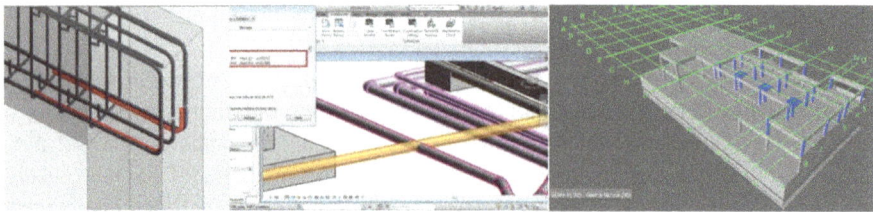

Fig. 3. Applications nD/BIM – Structural analyses [3], management [6] and planning [5].

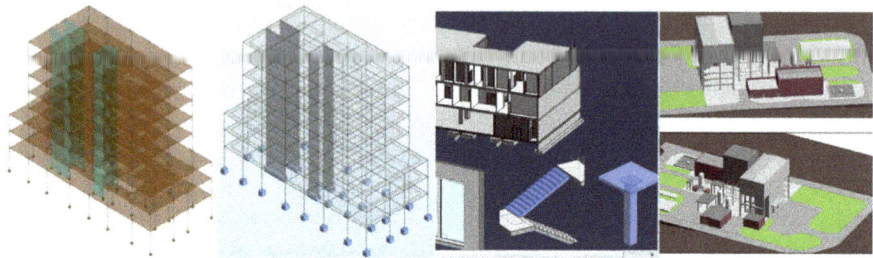

Fig. 4. Imagens of the short course: structural modeling, architectural modeling, 4D model construction planning.

- Oliveira, "BIM model management within the structural design", 2016 [3].
- Silva, "Quantities take-off supported in BIM methodology", 2016 [4].
- Mota, "4D/BIM model construction planning based on BIM", 2016 [5].
- Silva, "Construction management supported in BIM: central energy", 2015 [6].
- Berdeja, "Conflict analysis based in the BIM methodology", 2014 [7].
- Simões, "Maintenance of buildings supported on a model BIM", 2013 [8].

Additionally, in the school some workshops, short courses and seminars have been organized within the FUNDEC DECivil activity. The FUNDEC course presented several uses of BIM supporting distinct activities that are usually operated over a project (Fig. 4).

Other activity, concerning the dissemination of BIM within the Department of Civil Engineering, behind the coordination of the author, is the realization of workshops. Two events were already made in March and November 2017. The forum Civil, a student association, has been asking the author to offer a BIM workshop to Civil Engineering finalist students. The session "BIM: Concept and practice" of one day initializes with a brief introduction followed by an applied lesson: Building Information Modelling (Concept of BIM methodology, implementation and application); Practice of basic BIM tool handling. The course had a great acceptance from the students, a clear sign that the issue is attractive and that it is important to them to acquire BIM skills as it will be useful on their future activity in the labor market.

This follows other experiences mainly in the University of Oporto with the initiative "Session of introduction to BIM", Dec, 2013 [9] and in the ESTG of Leiria "BIM 2018 – Course", Apr–Jun, 2018 [10].

5 Conclusions

The school of Civil Engineering follows the newest advances in technology that can be applied in the Construction activity. The BIM methodology, its concept and based tools are important issues to be taught to students. The present report describes the curricular program approach of an academic discipline proposed in order to introduce this emergent topic in the scholar activity. The motivation and contextualization are referred in detail as well as the teaching methodology. Some examples that were worked out in classes by students follow the detailed programmatic content; some images presented along the text illustrate the positive result achieved with the course. Other activities supporting the course are also described, namely, workshops and FUNDEC short courses. As referred initially BIM implementation is a new process that requires study, practice, research and dissemination. The described teaching work is a positive contribution that goes in the right way.

References

1. Barlish, K., Sullivan, K.: How to measure the benefits of BIM: a case study approach. Automation in Construction **24**, 129–159 (2012). https://doi.org/10.1016/j.autcon.2012.02.008
2. Sampaio, A.Z.: BIM model: generation of architectural model (Civil). Didactic text of Computer-Aided Design discipline. IST, Lisbon (2016)
3. Sampaio, A.Z.: BIM model: generation of structural model (Civil). Didactic text of Computer-Aided Design discipline. IST, Lisbon (2016)
4. Silva, B.: Quantities take-off supported in BIM methodology, MSc thesis. University of Lisbon, IST, Lisbon (2016)
5. Mota, C.: 4D/BIM model construction planning based on BIM technology, MSc thesis. University of Lisbon, IST, Lisbon (2015)
6. Silva, D.: Construction management supported in BIM: central of energy, MSc thesis. University of Lisbon, IST, Lisbon (2015)
7. Berdeja, E.: Conflict analysis based in the BIM methodology, MSc thesis. University of Lisbon, IST, Lisbon (2014)
8. Simões, D.G.: Maintenance of buildings supported on a model BIM, MSc thesis. University of Lisbon, IST, Lisbon (2013)
9. Session of introduction to BIM, Oporto, Portugal (2013). https://sigarra.up.pt/feup/pt/noticias_geral.ver_noticia?p_nr=28267
10. BIM course 2018, Leiria, Portugal (2018). https://www.cursobim.com/#curso

Characteristics and Attitude Entrepreneurs: Development of Entrepreneurship Education in Graduation Students in a Brazilian University

Italo F. Minello[1], Cristiane Krüger[1(✉)] (iD), Denise A. Johann[1] (iD), and Rafaela E. Bürger[2] (iD)

[1] Federal University of Santa Maria, Santa Maria, RS, Brazil
`cris.kruger@hotmail.com`
[2] Federal University of Santa Catarina, Florianópolis, SC, Brazil

Abstract. This article aims to present the Project Education and Entrepreneurial Attitude of a Brazilian University, the focus of the project is to stimulate education and entrepreneurial attitude, initially, of the students of the Federal University of Santa Maria (UFSM), aiming to develop a foundation for the creation of a University Entrepreneur (UE). The first stage of the research, called informal research, identified teachers who, in the students' opinion, develop singular activities in the classroom. For the quantitative approach two validated questionnaires were applied. The results show actions carried out through the project. The project has already included more than 8 mil students, distributed in 116 UFSM undergraduate courses. To date, 19 Entrepreneurial Attitude courses have been created in courses not geared to the business world.

Keywords: Entrepreneurial education · Characteristics
Attitude and entrepreneurial intent · Undergraduate students

1 Introduction

In a society characterized by constant transformations there is an increasing dependence on the capacity of intellectual assets to generate ideas that aim at economic and social development [1]. Even in other times, rescuing Schumpeter [2], entrepreneurs were already playing a relevant role in these fronts, since they were characterized as possessing skills capable of driving such a development process.

These individuals, in McClelland's [3] view, have a motivational structure differentiated by the marked presence of the need for achievement, which encourages them to pursue goals that involve challenging yet calculated-risk activities. Given the relevance of entrepreneurs to society and, since attributes of entrepreneurial behavior can be improved through learning processes [4, 5], universities are identified as main influencers for the development of entrepreneurial individuals.

In this sense, teaching in undergraduate courses should provide knowledge that broadens the basic higher education. They must also be in tune with the demands of society, as well as with the generation of knowledge in the area of entrepreneurship,

© Springer International Publishing AG, part of Springer Nature 2019
J. Machado et al. (Eds.): HELIX 2018, LNEE 505, pp. 1138–1145, 2019.
https://doi.org/10.1007/978-3-319-91334-6_157

through scientific research. Because of this, the importance of universities is emphasized. Nothing more conducive than teaching and research institutions to act on this front as fomenters of the entrepreneurial attitude in the students, which increases the economic and social development, supported by the constitutional principle that aims to guarantee the quality of teaching, in its different levels and modalities, established by Art. 206 of the Federal Constitution [6], regulated by the National Education Guidelines and Bases Law [7]. In order to contribute to the commitment of those involved in such actions and projects, to stimulate and develop entrepreneurship within the university [8].

The university emerges as a propeller of the scientific development of knowledge and stimulating its application in an experiential way, through its teaching, research and extension triad, as well as the fourth element for the transformation and/or creation of an entrepreneurial university, which is the quest for economic and social development [9].

From the foregoing, it can be seen that the actions promoting entrepreneurship, within the universities, involve intense participation of the agents that interact in the process, contributing to the development of the entrepreneurial mindset at all levels of teaching, research and extension. In this way, [9] states that universities are currently undergoing a "second revolution", in which social and economic development is incorporated as part of the university's mission. Thus, the university integrating economic and social development as an additional function has been called "Entrepreneurial University" [10]. For [11] universities should create internal structures to promote and coordinate such actions, establishing strategies to articulate teaching, research and extension with society.

The need for the university to play its role and to evolve in this field is strongly emphasized. Thus, this article presents the reality of entrepreneurship in Brazilian higher education institutions, addressing the UFSM context. For this, a project was developed to stimulate entrepreneurial attitudes in UFSM undergraduate students through activities and integrated actions among students, teachers and managers of the institution, aiming to develop a foundation for the creation of an Entrepreneurial University (UE). Faced with a context in which society yearns for professionals increasingly able to modify the reality in which they find themselves, the objective of this article is to present the Education and Entrepreneurial Attitude Project.

2 Entrepreneurial Education

Entrepreneurs are not born but developed, it becomes necessary to include in the curricula the teaching of entrepreneurship. Thus, according to [5, 12], entrepreneurial characteristics can be improved through entrepreneurial learning processes. In this process of learning, the importance of entrepreneurial education in the curricula of several courses of Higher Education Institutions [13] is observed, which, should play a role of promoter, disseminating an entrepreneurial culture at all levels. Thus, entrepreneurial education can allow the student to see and evaluate a given situation, taking a proactive position in front of it, enabling him to elaborate and plan ways and strategies to interact with what he came to realize.

The [14] presents a framework for entrepreneurship education in higher education divided into three objectives: (a) developing entrepreneurship among students, (b) training students to start a business and manage it, (c) develop entrepreneurial skills necessary to identify and exploit business opportunities. This relationship is demonstrated in Fig. 1.

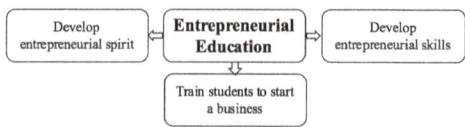

Fig. 1. Objectives of Teaching Entrepreneurship (Source: European Commission Enterprise and Industry Directorate-General 2007).

3 Entrepreneurship

3.1 Entrepreneurial Behavioral Characteristics

The entrepreneurial individual manifests a series of characteristics that identifies him, distinguishing him according to the skill set closest to his form, [15] affirms that the entrepreneurial characteristics make with which the entrepreneur is a dynamic social actor in behavioral issues. Given this, entrepreneurial characteristics are traits of personality that distinguish the detaining people and make them more susceptible to adopting behavior and entrepreneurial attitudes.

For [3], whose premise is the confrontation of challenges and the perception of opportunities of individuals, classified the society into two groups, those that are predisposed to undertake a minority of the population, and those who would not be willing to take such risks, which represent the majority of the population. [5] point out that this profile is marked by entrepreneurial actions, and such actions reflect the entrepreneurial behavior [3] segmented the entrepreneurial characteristics into three sets of actions: achievement, planning and power. These sets point to a series of competencies, characterized by the entrepreneurial behavior faced with the challenges experienced in their daily lives. Such sets and characteristics are identified in Fig. 2.

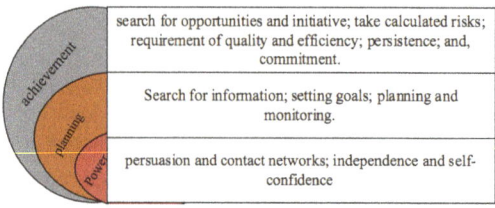

Fig. 2. Entrepreneurial Behavioral Characteristics. Source: elaborated based on McClelland [16].

These characteristics can help individuals to face the challenges of undertaking [3] as well as the lack of them, can make the formation of a business unfeasible. In view of the above, by identifying the entrepreneurial characteristics, the students are awakened to actions that lead to entrepreneurial behaviors and attitudes. This is corroborated by [17] when they affirm that when developing methodologies that help entrepreneurs in the identification of entrepreneurial characteristics, initial conditions are created for interventions that promote the development of the entrepreneurial potential. This encourages the creation of new businesses, generating employment, income and promoting economic and social development.

Therefore, the challenge and desire of an entrepreneur is to transform entrepreneurial characteristics into attitudes, so that this happens, needs to awaken to the importance of their role, develop an entrepreneurial mindset, and acquire knowledge that makes them capable of carrying out their activities. Entrepreneurial characteristics can be developed, and learning proves to be of great importance in achieving this process. The teaching of entrepreneurship becomes, at the same time, the role and goal of the entrepreneurial university.

3.2   Entrepreneurial Attitude

The entrepreneurial attitude explained through Ajzen's Theory of Planned Behavior [18], which defines this as an evaluation of an object of stimulus, influenced by beliefs. According to [18] people's attitudes spontaneously and consistently follow beliefs accessible in memory and then guide the corresponding behavior.

Entrepreneurial attitude, according to [19], is defined as the "predisposition learned, or not, to act in an innovative, autonomous, planned and creative way, establishing social networks". The entrepreneur is anchored in the ability of individuals to create their own businesses, taking risks, capitalizing on results, and taking advantage of the opportunities that arise.

The Entrepreneurial Action Measurement Instrument (IMAE) [19], was based on four characteristics: Planning, Innovation, Realization and Power. With the exception of the "Innovation" feature, the others were based on McClelland's Entrepreneurial Behavioral Characteristics [16], as explained above. In this perspective, the IMAE grouped the entrepreneurial behavioral characteristics in two dimensions: Prospecting and Innovation, and Management and Persistence, as can be seen in Table 1.

Table 1. Behavioral characteristics and dimensions.

Behavioral Characteristics		Dimensions
Setting goals; Information search, Planning, monitoring	Planning	Prospecting and Innovation
Creativity; Innovation	Innovation	
Search for opportunities, Initiative, Persistence, Acceptance of risks; Commitment	Realization	Management and Persistence
Persuasion; Establishment of contact networks; Leadership, Independence; Self confidence	Power	

4   Methodology

The methodology presented refers to the Entrepreneurship Education and Attitude Project. The project presents a qualitative and quantitative approach, of the descriptive and exploratory type, based on theoretical-empirical research. The population of this research is made up of 28.966 students, distributed in 267 UFSM undergraduate courses, 1.942 professors, 2.764 administrative and education technicians.

Considering a sample error of 1%, the minimum sample is 6.477 students. Figure 3 shows the minimum samples, proportionally distributed among the UFSM teaching units.

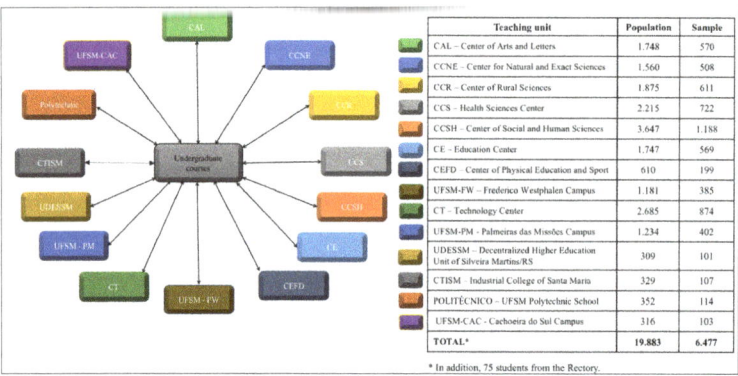

Teaching unit	Population	Sample
CAL – Center of Arts and Letters	1.748	570
CCNE – Center for Natural and Exact Sciences	1.560	508
CCR – Center of Rural Sciences	1.875	611
CCS – Health Sciences Center	2.215	722
CCSH – Center of Social and Human Sciences	3.647	1.188
CE – Education Center	1.747	569
CEFD – Center of Physical Education and Sport	610	199
UFSM-FW – Frederico Westphalen Campus	1.181	385
CT – Technology Center	2.685	874
UFSM-PM – Palmeiras das Missões Campus	1.234	402
UDESSM – Decentralized Higher Education Unit of Silveira Martins/RS	309	101
CTISM – Industrial College of Santa Maria	329	107
POLITÉCNICO – UFSM Polytechnic School	352	114
UFSM-CAC - Cachoeira do Sul Campus	316	103
TOTAL*	19.883	6.477

* In addition, 75 students from the Rectory.

Fig. 3. Minimum population samples per UFSM teaching unit.

From Fig. 3, it can be observed that the university in question is divided into 14 (fourteen) centers, distributed evenly between the cities of Santa Maria, Silveira Martins, Frederico Westphalen and Palmeira das Missões and Cachoeira do Sul, located in southern of Brazil.

The first step in data collection is to conduct an informal interview between researchers and students of each UFSM undergraduate course. In the informal survey, information about students' perceptions about the activities carried out by teachers, suggestions and expectations about entrepreneurial activities is presented. This information is analyzed qualitatively and are compared with the results of the instruments applied in the formal research. The demands pointed out in the informal survey together with the results of the formal research, form the basis for the creation of the Entrepreneurial Attitude disciplines. In the formal collection, validated instruments are applied [16, 19]. Qualitative data are analyzed by means of content analysis and statistical tests are performed for the quantitative data.

5   Analysis and Discussion of Results

To date, a series of activities have been carried out in order to develop the proposed actions. From this perspective, the precursor action of the process was the presentation of the project to the teaching community of each research unit. At this stage, meetings were held to demonstrate the importance of developing students' entrepreneurial attitude, as well as a proposal to achieve this goal.

In addition, informal and formal data were considered among the undergraduate students. To date, 4.681 students were contemplated in the informal phase and 4.126 students were contemplated in the formal stage. After analyzing the informal research, the professors appointed as entrepreneurs in their respective teaching centers, were gathered in order to discuss the creation of the college subjects of Entrepreneurial Attitude and its didactic-pedagogic menus.

In the sequence, entrepreneurship disciplines were created. These disciplines were structured based on three perspectives: (i) the behavioral needs evidenced in the students from the data analysis of the formal collection, composed by the instruments [16, 19]; (ii) the demands perceived by the students about the activities carried out in the course of their academic life; and (iii) the future professional practices of the respective course.

In this sense, the discipline aims to stimulate, in a practical way, the future professionals to undertake in any environment that they choose. In addition, differing from the conventional model of teaching, disciplines are taught by at least two teachers from different areas. The results show actions carried out through the project. The project has already included more than 8 mil students, distributed in 116 UFSM undergraduate courses. To date, 19 Entrepreneurial Attitude courses have been created in courses not geared to the business world.

In the discipline the student should develop an innovative project, based on opportunities and gaps perceived by the students in their professional environment and that could bring some benefit to the educational institution. At the end of the school year, a meeting was held with the teachers participating in the disciplines and the project, in order to exchange experiences and reports from the disciplines taught.

In summary, action was taken to encourage entrepreneurship, among them the use in the classroom as didactic-pedagogical tools, the study blog, the creation of educational videos made by the students themselves for the community, the formation of groups of research, exchange experiences with successful entrepreneurs, mini courses adapted to the needs of each course/center, junior company, poker tournament, business plan and integration events between the community, university and companies.

These actions break the traditional classroom, stimulating the development of entrepreneurial characteristics, as [17], considering entrepreneurship education as a driving force for creative thinking, for the generation of innovations and for the growth of a sense of self-esteem and responsibility, which is seen as still more necessary in higher education institutions.

6 Final Considerations

The objective of this article was contemplated, the education project and entrepreneurial attitude was presented. It was possible to verify a series of actions with the objective of promoting and stimulating the education and entrepreneurial attitude. Among these actions can mention: Disciplines of entrepreneurial attitude, teaching capacities, entrepreneurial actions of undergraduate students and exchange of ideas among the teachers.

As to the contribution of this study, it should be pointed out that the results already obtained serve as the basis for a longitudinal study, goal of the Entrepreneurial Attitude Project, which allows monitoring the evolution of the students who study the subjects during the semesters, types of courses of the research institution and comparisons with others. For further studies it is suggested the replication of the present research in institutions of higher education and institutions of basic education in other cultures.

References

1. Quandt, C.O., Silva, H.F.N., Ferraresi, A.A., Frega, J.R.: Programas de gestão de ideias e inovação: as práticas das grandes empresas na região sul do Brasil. Revista de Administração e Inovação, São Paulo **11**(3), 176–199 (2014)
2. Schumpeter, J.A.: O fenômeno fundamental do desenvolvimento econômico. In: A teoria do desenvolvimento econômico. Nova Cultura, Rio de Janeiro (1985)
3. McClelland, D.C.: A sociedade competitiva: realização & progresso social. Expressão e Cultura, RJ (1972)
4. Dolabela, F., Filion, L.J.: Fazendo revolução no Brasil: a introdução da pedagogia empreendedora nos estágios iniciais da educação. REGEPE **3**(2), 134–181 (2013)
5. Zampier, M.A., Takahashi, A.R.W.: Competências e aprendizagem empreendedora em MPE's educacionais. Revista Pensamento Contemporâneo em Administração **8**(3), 1–22 (2014)
6. Brasil: Constituição da República Federativa do Brasil. Senado Fed, Brasília (1988)
7. Brasil: Lei de Diretrizes e Bases da Educação: Lei nº 9.394/96 – 24 de dez (1966)
8. Clark, B.: The character of the entrepreneurial university. Int. Higher Educ., n. 38 (2015)
9. Etzkowitz, H.: Anatomy of the entrepreneurial university. Soc. Sci. Inf. **52**(3), 486–511 (2013)
10. Casado, F.L., Siluk, J.C.M., Zampieri, N.L.V.: Universidade empreendedora e desenvolvimento regional sustentável: proposta de um modelo. Rea UFSM, pp. 633–650. Dez (2012)
11. Ferreira, G.C., Soria, A.F., Closs, L.: Gestão da interação Universidade – Empresa: o caso PUCRS. Revista. Sociedade e Estado **27**(1), 79–94 (2012)
12. Politis, D.: The process of entrepreneurial learning: a conceptual framework. Entrepreneurship Theory Pract. **29**, 399–424 (2005)
13. De Oliveira, A.G.M., Melo, M.C.O.L., Muylder, C.F.: Educação Empreendedora: O Desenvolvimento do Empreendedorismo e Inovação Social em Instituições de Ensino Superior. Revista Administração em Diálogo RAD **18**(1), 29–56 (2016)
14. Comissao Europeia: Repensando educação: investindo em habilidades para melhores resultados socioeconômicos (2012)

15. Mansfield, R.S., McClelland, D.C., Spencer, J.L.M., Santiago, J.: The identification and assessment of competencies and other personal characteristics of entrepreneurs in developing countries. Final report. McBer and Company, Massachusetts (1987)
16. Marinho, E.S., Minello, I.F.: Incubação De Empresas: Desenvolvendo O Comportamento Empreendedor. Novas Edições Acadêmicas, vol. 1, 164p. (2017)
17. Lima, E., Lopes, R.M.A., Nassif, V.M.J., Silva, D.: Ser seu Próprio Patrão? Aperfeiçando-se a educação superior em empreendedorismo. RAC, Rio de Janeiro **19**(4), 419–439 (2015). art. 1
18. De Souza, E.C.L., Lopez Jr., G.S.: Atitude empreendedora em proprietários-gerentes de pequenas empresas: construção de um instrumento de medida. REA **11**(6), 1–21 (2005). ed. 48
19. Ajzen, I.: From intentions to actions: a theory of planned behavior. In: Kuhl, J., Beckman, J. (eds.) Action-Control: From Cognition to Behavior, pp. 11–39. Springer, Heidelberg (1985)

To Improve Education in Order to Break the Vicious Cycle of Inequality: An Urgent Investment

Elsa Montenegro Marques[(⊠)] and Paula Cristina Salgado Vieira

Instituto Superior de Serviço Social do Porto, Av. Dr. Manuel Teixeira Ruela,
370, 4460-362 Senhora da Hora, Portugal
{elsa.montenegro, paula.vieira}@isssp.pt

Abstract. The present article reflects on the role of the educational system in the (re) production of social inequalities, a phenomenon that has been the object of a large scientific production by the social sciences, in particular by the sociology of education. Our observation, as social professionals, of the interactions between teachers and students in a school context, leads us to recognize *in situ* the phenomenon of school elimination that Bourdieu [1] called "the excluded from the inside". This observation allows us to verify how and how much the school institution can contribute to generate a rise of completely unrealistic expectations, in the precise measure in which it refuses to invest seriously in the learnings of the students that it certifies. We refer in particular to those students who are less equipped to appropriate the curricular knowledge that it conveys. This short reflection highlights some pedagogical practices dominant in the school that tend to reproduce the (unequal) order that marks the access to school certification. At the same time, it also points out ways of improving the modalities of teaching, according the theories about teaching through problems and competences [2–5, 8].

Keywords: Education · Inequality · Competences · Learning

1 Introductory Notes

At a time when social cleavages are increasingly becoming more pronounced and there is a collapse of the links that give rise to the social fabric, it seems appropriate to reflect on the role of the education system at this level. In other words, it seems pertinent to see if the public school has succeeded in being a social system capable of promoting the social bond between the younger generations and society, providing them with the resources to make a transition to life adult education, or whether, on the contrary, the school has contributed to the dissociation of many young people - especially those of the most disadvantaged social origin - from within the school and, consequently, from society as a whole.

Among the metamorphoses that occurred in the school, we highlight the democratization and massification policies of education that, despite having made possible access to secondary education by social sectors previously excluded, can not guarantee successful experiences for all students. In fact, the extension of school trajectories has

© Springer International Publishing AG, part of Springer Nature 2019
J. Machado et al. (Eds.): HELIX 2018, LNEE 505, pp. 1146–1152, 2019.
https://doi.org/10.1007/978-3-319-91334-6_158

not been synonymous for all pupils of the internalization of the knowledge necessary for a dignify progression within academic careers and, later, for insertion in the labor market [1].

On the contrary, the dominant model of the educational system has been submitting individuals from families deprived of cultural capital to successive experiences of school retention and failure. The high rates of failure, absenteeism and early school leaving[1] are all the more worrying as we know that the absence of school certification today imposes serious constraints on a dignified and qualified transition into adult life.

Perceiving the reason for these problems presupposes be attentive to a multitude of factors, among which we emphasize the ways and processes used by the school institution to communicate with the socio-cultural diversity of its public. It is interesting to reflect, in particular, on the pedagogical practices employed by the school that difficult (or even impede) the success of students whose cultural configurations are different from the official culture transmitted by it. The question is extremely complex and it is not within the scope of this reflection to deepen it exhaustively, but only to highlight it in order to emphasize the close relationship between educational practices and the reproduction of social inequality, thus seeking to break with the established prejudice that "access is enough to succeed" in school [6].

2   The Debate on the Relationship Between Knowledge and Action: How to Improve Education?

The school supports the education of the new generations above all in the transmission of knowledge, with very little concern for establishing connections between transmitted knowledge and the cultural universes of students, and also between them and the real contexts in which they may be applicable and useful. Acting in this way the school will continue to generate disinterest and failure among many children and youngsters, especially those whose pattern of reception and sense is less familiar with abstract knowledge. Precisely because their culture of origin has not allowed them to acquire the dispositions favorable to the valorization of knowledge, from knowledge to knowledge, their passage through the school will tend to be marked by malaise and suffering. Malaise and suffering that will be greater, the less qualified the students feel to decode the meaning of the disciplines and the contents. A failure subjectively felt but also reflected by teachers and peers to each new difficulty, each new failure. Without combating all forms, even the most subtle ones, of hierarchizing practices and knowledges, namely those that distinguish between "pure" and "applied", between "theoretical" and "practical" or "technical", the school institution will never be a place where all students can have the opportunity to experience success [7].

Philippe Perrenoud [2, 8] is one of the authors who has most reflected on the impact on learning of the absence of relationship between knowledge and its practical application.

---

[1] The Eurostat data for 2016 on Portugal reaffirm the severity of the problem of early school leaving, ie before compulsory schooling was completed: 14% of pupils dropped out of school before completing their 12th year in Portugal, while that the rate recorded in the EU28 is 10,7%.

According to Perrenoud [2, 8], the school should serve to learn things directly related to life, otherwise it would compromise the development of competences to understand and act on the concrete problems. This author points out that the opposition between the perspective that attributes to the school the function of teaching knowledge and the perspective that, on the contrary, attributes the development of competences conceals a misunderstanding that consists in believing that teaching skills presupposes giving up transmitting knowledge. Knowledge and skills are closely related and are closely complementary. This means that there can be no conflict of priority between them, particularly with regard to the division of working time in the classroom. Equating the question in these terms refers to two equally reducionists views of the school curriculum. One that privileges the widest possible domain of knowledge, without worrying about its mobilization in a given situation. The other that accepts to limit the amount of knowledge taught and required, to intensively exercise their mobilization in a complex situation. The definition of competence proposed by Perrenoud [8] refers to the ability to act effectively in a particular type of situation, which is supported by knowledge but not limited to it. It means that skills are not in themselves knowledge, they use them, integrate them, or mobilize them. But if a competence does not correspond to the pure and simple "rational" implementation of knowledge, its formation /acquisition can not underestimate its assimilation. By the same token, the appropriation of numerous knowledge does not, by itself, allow its mobilization in situations of action [8].

We consider it fundamental that the educational system evolves towards the development of competences so that all students can give meaning to the school, which, in turn, requires important changes in programs, didactics, assessment, classroom functioning, of the teacher and of the student office. Transformations that provoke the resistance, passive or active, of many agents, more interested in the continuity of the practices or in the preservation of the acquired advantages than in the effectiveness of the formation. First of all, this transformation of the educational system would require a rupture with the interpretation that conceives the approach by the competences as an obstacle to the acquisition of a general culture.

The idea that general culture can be reduced to an accumulation of knowledge devalues the role it has in understanding and transforming the world in which we live. At the same time, it also brings out an arbitrary hierarchy of knowledge, rendering a poor service to literary works and their authors. Restrict culture to familiarization with classical works or the assimilation of basic scientific knowledge, without extracting the lessons that allow us to face, with dignity, with critical sense, with intelligence, with autonomy and with respect for others, the various situations of life, is the same as conceiving art, culture and science as contemplative and uprooted activities. Bourdieu [7] emphasizes the importance of this perspective on knowledge, in order to legitimize the school elimination precisely of those who were not trained to enter into the rhetorical game of words, dissociated from the reality lived.

Returning to the competences, if these are formed by the practice, this necessarily implies the use of concrete situations, with contents, contexts and identified risks. It is, however, necessary to emphasize that the fact that the language of competences is invading the programs does not mean that this corresponds to taking all the demands and implications of this type of learning seriously. To designate a competence, it is not enough to refer to an action in which a given knowledge can be used. Let us take the case of the mother tongue, for example, by highlighting the various specific skills that need to be mobilized in order to respond to a job posting or apply for a given job: choosing the right information, organizing the text (date, header, greetings, etc.), ensure consistency between sentences (verb tense, linking, etc.), use appropriate vocabulary based on content, construct correct sentences, respect the spelling conventions, and organize the text satisfactorily on the page. Forming competencies implies a considerable transformation of teachers' relationship with knowledge, their way of teaching and, after all, their identity and their own professional skills.

The competency approach requires a focus on the student, differentiated pedagogies and active methods; requires knowledge to be addressed as resources to be mobilized for action; requires working on the basis of problems, negotiating and conducting projects with students; requires the teacher to adopt flexible lesson planning and be able to improvise; who practices a formative assessment in a work situation; in short, requires the assumption of less disciplinary compartmentalisation. Such a competence approach determines the place of knowledge in action, which is to say, consider them resources to identify and solve problems, as well as to prepare and make decisions.

It should be clarified, however, that some teachers who accept the notion of competence continue to think that students should be given basic knowledge before mobilizing them for a given situation. In this understanding, the knowledge must be acquired independently of the real situations that cause its use. According to Perrenoud's analysis [2, 8], this interpretation is due to the fact that teachers are very accustomed to a disciplinary approach and to face real difficulties in "transmitting their subject in relation to a problem". The pedagogical tradition leads them to autonomize the exposition of knowledge and to conceive the situations of its application of implementation as simple exercises of understanding or memorization. However, the formation of competencies implies a rupture with the perspective that distinguishes and opposes, in dichotomy, a logic of teaching and a logic of training.

The small "cultural revolution" inherent in learning by competences consists in understanding a relatively simple postulate: that competences are built through their exercise in complex situations, which is possible "learn by doing, what we do not know how to do ". Such an approach involves important teacher identity changes, so that the teacher no longer considers the pragmatic relation to knowledge as a lesser relation. To let the teacher fail to devalue applied knowledge and rebuild a relationship with knowledge and action does not rest on the idea that erudite knowledge, that is, decontextualized, is superior to the knowledge derived from experience, and understands that knowledge is always somehow anchored in action.

The relation between knowledge and action is also at the heart of Jean Piaget's theory [3][2]. Without neglecting the impact of understanding on action, Piaget [3] points

---

[2] In his work "Reussir et Comprendre" [3], he makes a distinction between two types of knowledge: `knowledge as action' (reussir) and `knowledge as conceptualization' (comprendre). The first consists in `understanding in action a given situation' while the second means `to dominate in thought the same situation in order to know how to solve problems with regard to the' why `and the' how `of the connections found in the action'. To understand `consists in extracting reason from things whereas being successful amounts to only using them successfully' [3].

out that it is "knowing to do" the "motor" and "nourishment" of understanding, which means that in order to master certain situations in the explain and solve problems associated with them, one must be able to "grasp in action" those same situations.

However, the problem of many pedagogical practices that are developed in the school lies precisely in the fact that students are often asked to understand a given action or situation in the first place, instead of leading them to do the same and, only then, in an inductive way, lead them to understand why they succeeded in accomplishing it. Proof of this is the predominance of pedagogical and didactic routines that neglect "knowledge as action". Through pedagogies whose sole purpose seems to be to memorize knowledge to guarantee passage in the tests, many are the students who are faced with forms of conjugation, historical facts, grammatical rules, processes, physical laws, geographic phenomena without knowing in what circumstances and in time to apply them. This is the case for us to ask ourselves: what is the use of passing tests, without the knowledge being really apprehended? Without students being prepared to solve problems, make decisions, achieve goals? [2, 8].

Our experience in the accompaniment of children and adolescents with school difficulties, as social interveners, led us to observe students who spend their lives in school without stimuli to ask questions, to ask questions about the subjects, to discover and find out for themselves the rules, the facts, the phenomena that are taught in class. To exacerbate this problem, we also find some interpretations of teachers explaining school failure as the result of lack of aptitude in students or lack of interest in learning, without ever questioning their practices and without invest with conviction in the results of the students who present more learning's difficulties.

It is, in this way, that the educational system masks the social factors that are the origin of the cultural inequalities brought by the students when they reach school, as well as their role of reproduction of these same inequalities. Instead of constructing ways of decoding its reality, in order to make it intelligible to all the cultures it hosts, the school institution makes impossible the communication with the children whose experiences, values, social relations, orders of signification, language codes, finally, whose cultures are different from yours. Without instruments to decode/decipher the new information that the school imposes on them, there is no way for these students to incorporate it and thus to redo their knowledge structures. How can they be oriented in the school reality, if the map they have is considered inappropriate and if they are not given the means to reconstruct a new one?

3   Final Considerations

According to the theorisations about teaching by problems and competences [2–5], working for the formation of competences implies that the teacher leaves the traditional concept of class, providing the information that is necessary for the student to begin to think. In this case, the teacher's skill is not to expose knowledge in a discursive way, but to suggest the links between knowledge and concrete situations. The teacher cannot constantly replace the student, otherwise he will not be able to learn. He must renounce the expository class, as well as the word ex cathedra, which, however, does not remove the role of "driver of the process". However, it requires a capacity for renewal and variation, since problem situations must be stimulating and surprising. Problem situations in which the student is faced with the need to make decisions to achieve a goal that he has chosen or proposed. Problem situations organized around overcoming an obstacle previously identified by the class and requiring students to invest their knowledge and previous representations. Problem situations that require the teacher to help the student identify the obstacles, the axial point of the pedagogical action, being able to provide all the resources that avoid the feeling of impotence and discouragement. When Perrenoud [8] speaks of obstacles he is referring to strongly structured erroneous convictions, which have a status of truth in the student's mind and temporarily block their learning process. Only through a work of strong imbalance can the student accommodate himself to an inverse conviction.

According to this perspective, being a teacher would not consist in *teaching, but rather in making learn* [8], that is, in creating favorable situations that increase the probability of thinking.

This is a mode of work that requires a greater mastery of the discipline itself, of the founding questions that constitute it and organize it as such. Deliberately structuring obstacles or anticipating them in a task that is part of a given process requires a great capacity for analysing the student's situations, tasks and mental processes, an ability to forget his own skill to put himself in the place of the student and understand what blocks it. It also implies a strong ability to communicate with the student, to help verbalize what bothers or blocks, and to manage the classroom in a complex and heterogeneous environment. It also implies the production of school materials, the design of spaces, the use of equipment that allows not only to make the student autonomous and active, but also to confront him with obstacles that impose new learning.

In short, the modes of functioning of the classes in which the teacher, instead of communicating, would rather "make communications" and "deposits"[3] would be abandoned. In other words, it is urgent to break educational practices in which the educational agents themselves familiarize students with a game of "make-believe" that is a fraudulent game that will lead to its elimination from the segment of the qualified and valued labor market.

References

1. Bourdieu, P., Champagne, P.: Les exclus de l'intérieur. Actes de la Recherche en Sciences Sociales **91**(92), 71–75 (1992)
2. Perrenoud, P.: L'approche pas compétences, une réponse à l'échec scolaire? Réussir au Collégial **38**(2), 6–11 (2000)
3. Piaget, J.: Réussir et comprendre. Presses Universitaires de France, Paris (1974)
4. Charlot, B.: Da relação com o saber. Artmed, Porto Alegre (2000)

---

[3] The expression is of Paulo Freire [9] when he refers to "banking education" as an oppressive and dominating pedagogy, in which teaching consists of depositing, transmitting, transferring values and knowledge.

5. Novak, J.: Aprender, criar e utilizar o conhecimento: mapas conceptuais como ferramentas de facilitação nas escolas e nas empresas. Coleção Plátano Universitária, Lisboa (2000)
6. Bourdieu, P.: A Reprodução. Elementos para uma teoria do sistema de ensino. Editora Vega, Lisboa (1982)
7. Bourdieu, P.: Propostas para o Ensino do Futuro. Cadernos de Ciências Sociais **5**, 101–119 (1987)
8. Perrenoud, P.: Construir as competências desde a escola. Tred Editora, Porto Alegre (2007, 1999)
9. Freire, P.: Pedagogia do Oprimido. Edições Afrontamento, Porto (1975)

Turning Knowledge into Business Ideas: Insights from the IdeaLab Business Accelerator

Helena Moura[1], Marta Catarino[1], Paulo Afonso[2],
and Manuel Nunes[2(✉)]

[1] TecMinho, Campus of Azurém, 4800-058 Guimarães, Portugal
{hmoura,mcatarino}@tecminho.uminho.pt
[2] Department of Production and Systems, University of Minho,
Campus of Azurém, 4800-058 Guimarães, Portugal
{psafonso,lnunes}@dps.uminho.pt

Abstract. The IdeaLab is a laboratory for accelerating business ideas. It results from a pioneering initiative in Portuguese universities, which started in 2009. So far, there have been 17 editions, in which 255 ideas from 570 entrepreneurs have been developed. Of these ideas, more than forty have materialized into companies. The IdeaLab enables participants to (1) enhance the business potential of their ideas and (2) develop their entrepreneurial skills. A team of trainers/consultants ensures thematic workshops, coaching, networking and pre-incubation. The promoters are offered methodologies and tools to improve the design of innovative products, services and businesses. In this article, the IdeaLab is analyzed and discussed, as a catalyst of the process of turning knowledge (technology-based and other) into a product/service and/or sustainable businesses. Based on the analysis of the 17 editions, the results presented highlight the relationship between the promoters' background and profiles and the type of ideas. Their potential to materialize into effective companies is also analyzed.

Keywords: Entrepreneurship · Business accelerator · Laboratory of ideas

1 Introduction

IdeaLab is a TecMinho business acceleration laboratory. Since 2009, and throughout 17 editions, it has supported the development of 255 business ideas and has offered entrepreneurial training to 570 entrepreneurs. About forty projects resulted in companies, some of them with the spin-off status of the University of Minho. For example, Earboxwear, which integrates headphones into urban garments, Geojustiça, which provides services related to geographic information in support of dispute resolution, SilicoLife, which creates computational biology solutions for Life Sciences, and Fermentum, which produces craft beer.

IdeaLab's project ideas and participants have come from a diversity of areas of knowledge, namely: Architecture, Biotechnology, Education, Management, International

© Springer International Publishing AG, part of Springer Nature 2019
J. Machado et al. (Eds.): HELIX 2018, LNEE 505, pp. 1153–1159, 2019.
https://doi.org/10.1007/978-3-319-91334-6_159

Business, Psychology and Sociology, Sciences and Engineering. It is noteworthy that there has been an increase in projects from the field of social sciences.

At the IdeaLab, participants developed their behaviour and entrepreneurial skills and tested and improved the business potential of their ideas. They acquired knowledge, methodologies and tools that enable them to design innovative products, services and businesses. Technical and market conditions are also important to be considered. Participants are monitored by a team of trainers, who provide thematic workshops, coaching, networking and pre-incubation. Over a period of five months, the promoters of business ideas are accompanied by a team of consultants who provide them with a set of customized tools that enable them, on the one hand, to test vocations and entrepreneurial skills through mechanisms based on creative processes and strategies and, on the other hand, allow the validation of ideas through the elaboration of a business plan. The activities of IdeaLab take place on the Campus of Azurém, at the University of Minho.

In this study, the diversity of entrepreneurs' competencies and their business ideas was carried out, as well as the results achieved in terms of company creation. This interconnection enhances the process of maturing business ideas and allows the process of validation, maturation, evolution and change of ideas throughout each edition of IdeaLab.

2   Context

Idealab can be considered a laboratory of ideas, entrepreneurship training and acceleration of ideas in order to transform ideas into potential business opportunities.

Entrepreneurship education has traditionally focused on individualized education, but these activities are lately being geared towards a supported approach in action, emphasizing project learning. The cases analyzed by several researchers have shown that entrepreneurship education is increasingly less focused on teaching in a classroom and more geared to the project-based and networked teaching process [1–3]. There are different initiatives with different objectives, such as the acquisition of skills by entrepreneurs, the definition of new business and the commercialization of research carried out at the university. There are several relevant aspects in the definition of an action-based entrepreneurship training program, taking into account the different objectives previously mentioned.

Furthermore, it is interesting to note that entrepreneurship training programs aimed at accelerating ideas and incubating business can also be done in the private domain and not only in the context of public or teaching institutions as is the obvious and paradigmatic case of universities. In fact, there is already an industry specialized in accelerating and incubating business based on different business models. This industry is a catalyst of such a process and introduces economic criteria for project selection and risk management of these investments, establishing an effective bridge between the promoters of ideas, entrepreneurs who will create new businesses, financiers and investors and the market itself. On the other hand, they tend to develop new business development processes making it not only more effective but more efficient, reducing time-to-market and business development costs. The "business development process" can be based on steps or gates as it is commonly presented in the process of developing new products and may be developed as the latter in several steps, typically 4 to 8 steps [4].

Acceleration programs are characterized as programs of limited duration, i.e., less than three months. These helped the promoters of start-ups. Acceleration programs also provided a diversity of networking opportunities, i.e. with other entrepreneurs and mentors, who can be successful entrepreneurs, venture capitalists, business angels, or even business managers. Finally, most of the programs end with an event, a demo day where promoters pitch a broad audience of qualified investors [5].

The limited duration of these programs implies the participation of entrepreneurs in groups. The group selection process focuses on throttle marketing and occurs on relevant dates. In addition, the use of open innovation attracts a large number of entrepreneurs. Entrepreneurs sometimes change to other programs in order to participate in top-level programs. Top acceleration programs accept only a very small percentage of applicants.

Guidance and education are cornerstones of acceleration programs and often the main reason for entrepreneurs to participate. The educational component includes seminars on a wide range of entrepreneurship topics. These seminars are usually provided by program directors or guest speakers, who often provide individualized guidance. Mentoring is also often cited as a relevant aspect of acceleration programs, but it varies substantially across programs. Generally, the development of a network is referred to as an important aspect of accelerator participation. Finally, program managers and tutors provide guidance throughout the program, helping entrepreneurs to acquire and apply the knowledge gained through follow-up meetings, seminars, and other means and actions [6].

3   IdeaLab Business Accelerator

IdeaLab was a pioneering initiative in Portuguese universities, which emerged in 2009, as the first program to accelerate and develop ideas of long-term business. It is a laboratory of experimentation and validation, but also of generation and development of innovative business ideas based on technology or on intensive knowledge. In addition to testing the commercial potential of their ideas, participants can also assess their vocation and entrepreneurial skills.

Promoters of ideas, individually or in groups, up to a maximum of five participants per group, use the Laboratory for a maximum period of five months in order to mature their business idea. Participants are accompanied by a team of consultants/trainers, which ensure a set of thematic workshops related to the study and validation of business ideas, tutoring, networking and coaching. The team of consultants provides a set of tools and methodologies tested and validated by IdeaLab, that validates the original idea and contribute to the elaboration of a pre-business plan.

The Start-Up Workshops component integrates five thematic workshops: the voice of the customer, market and industry analysis, strategic analysis and development of the business model and finance. At the end of these thematic workshops, participants will estimate investment needs and analyze the project's economic and financial viability.

In the elevator pitch workshop, the promoters acquire the skills to present a coherent and persuasive argumentation of their business idea.

Throughout the IdeaLab process, the promoters receive individualized follow-up by the team of tutors and mentors, which allows the development of a coherent business plan.

4   Results and Discussion

Throughout the 17 editions of IdeaLab, there has been a significant majority of entrepreneurs that come from the field of Engineering, with greater focus in the years 2011, 2012 and 2014. In addition, the area of Health Sciences appears in second place of prominence, followed by the areas of communication and human sciences.

The relationship between the field of knowledge of the business idea and the academic field shows that there is a direct connection between them. For example, entrepreneurs with academic knowledge in the field of health sciences tend to develop ideas in the same domain, namely in the areas of Bioengineering, Bioinformatics, Biochemistry, Physics, Nanotechnology, Neurology and Optometry. The same happens in the areas of Engineering and related fields. Therefore, it can be affirmed that academic knowledge presupposes important technical skills for the operational development of the idea.

Furthermore, the relationship between the ideas that result in products versus services and the background of the promoters was also analyzed. Interesting results were found (Fig. 1).

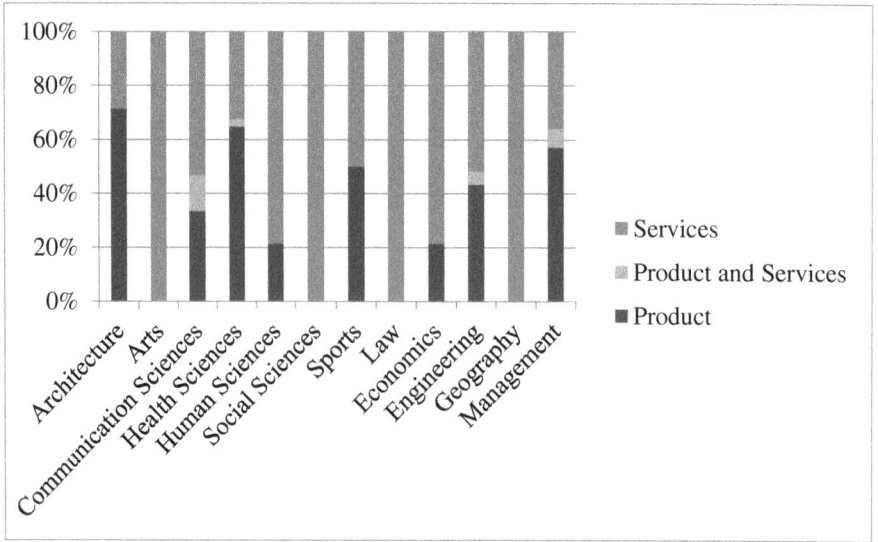

Fig. 1. Relationship between the academic knowledge and the typology of the idea.

On the one hand, as can be seen in Fig. 1, the typology of ideas and the development of products are more related to the areas of health sciences, architecture and management. On the other hand, services are linked to the areas of arts, law, geography and social sciences. Regarding the cases where product and services are combined, the prevalent areas are engineering, management and communication sciences.

Furthermore, considering gender and age, the promoters that have been developing their projects in the IdeaLab appear to be a good sample of the people that participate in these programs of business acceleration. Throughout the various editions of the Idea-Lab there has been a significant majority of male participants (67%) and in every three entrepreneurs, two are 20 to 40 years old.

Taking into consideration the number and type of companies created it is important to highlight some interesting results (Fig. 2).

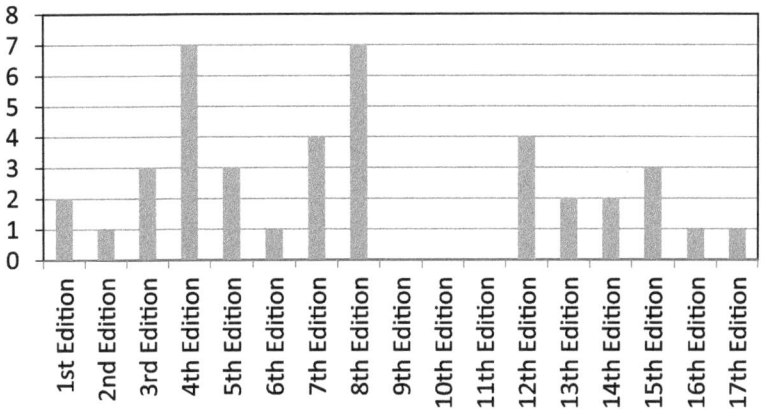

Fig. 2. Number of companies created per edition of IdeaLab.

The areas of activity of the companies are based on: Biological Engineering; Biology; Biotechnology; Catering & Restaurant; Digital Marketing; Education; Electronics; Food Engineering; Geography; Informatics; Mechanical Engineering; Molecular Biology; Nanotechnology; Neuroscience; Optometry; Physiotherapy; Polymers; Product design; Psychology; Law; Social Communication; Sociology; Textiles; Transportation and Logistics and Web Design.

Of these areas, those that create more companies are associated to Biological Engineering, Biotechnology and Informatics, this means, technological and intensive knowledge companies. The number of promoters per company tends to be on average two per team and the knowledge profile of the promoters of these entrepreneurial projects is more related to their academic background, as well as to the results and research processes developed within the University of Minho.

Compared with other editions of IdeaLab, the 4th and 8th editions resulted in a larger number of companies. This means that of the 15 business ideas presented in each of the editions, seven companies were created. However, the average time span between participation in the IdeaLab and the creation of the company is about one to two years.

In addition, and although in the 9th, 10th and 11th editions no companies were formally constituted, at least one team per edition continues to work on their business idea, essentially on technical aspects and prototyping.

The ideas that result in new companies are related to the type of idea, the state of maturity of the idea, the ideas' market potential and, most importantly, the willingness of promoters to start a business.

5 Conclusions

Business accelerators play an important role in economic development. They are relevant in the promotion of entrepreneurship, by providing education and training in entrepreneurship and promoting activities that develop entrepreneurial skills in the promotors of the ideas.

TecMinho's Laboratory of Business Ideas is an initiative that aims to define an integrated strategy to stimulate entrepreneurship at the University of Minho. This initiative is capable of developing entrepreneurial skills in its students and promoting entrepreneurship as an attractive career alternative. This initiative combines training and mentoring, theoretical and practical components (Start-up workshops and Pre-Incubation). It also allows the testing and development of innovative and differentiated business models and acquisition of entrepreneurial skills with the support of a team of specialized consultants.

If, on the one hand, technology-based companies generate products with high added value that facilitate their sustainability in an increasingly global economy. On the other hand, the significant changes that are occurring in the world of work seem to appeal to the development of the individuals' attitudes, values and skills favourable to the practical and proactive search for solutions to technical, economic or social problems, as well as the identification, analysis, planning, development and implementation of marketable products or services.

Acknowledgements. This work has been supported by COMPETE: POCI-01-0145-FEDER-007043 and FCT – Fundação para a Ciência e Tecnologia within the Project Scope: UID/CEC/00319/2013.

References

1. Katz, J.A.: The chronology and intellectual trajectory of American entrepreneurship education 1876–1999. J. Bus. Ventur. **18**(2), 283–300 (2003)
2. Peterman, N., Kennedy, J.: Enterprise education: influencing students' perceptions of entrepreneurship. Entrep. – Theor. Pract. **28**(2), 129–144 (2003)
3. Etzkowitz, H.: Research groups as 'quasi-firms': the invention of the entrepreneurial university. Res. Policy **32**(1), 109–121 (2003)
4. Stevens, G., Burley, J.: Piloting the rocket of radical innovation. Res. Technol. Manag. **46**, 16–25 (2003)

5. Cohen, S.: What do accelerators do? insights from incubators and angels. Innovations **8**(3/4), 19–25 (2013)
6. Burgers, J., Van Den Bosch, F.A.J., Henk, W., Volberda, H.W.: Why new business development projects fail: coping with the differences of technological versus market knowledge. Long Range Plan. **41**(1), 55–73 (2008)

Scratch 3 – Beginners Programming Course in 3ʳᵈ Year of Primary School

António Marques[1], Carina Guimarães[1(✉)], and Ana Salgado[2]

[1] Universidade da Beira Interior, Estrada do Sineiro, 6200-209 Covilhã, Portugal
{antonio.marques, scmg}@ubi.pt
[2] Escola Superior de Saúde do Porto – P.Porto,
Rua Dr. António Bernardino de Almeida, 400, 4200-072 Porto, Portugal
ais@ess.ipp.pt

Abstract. In the 2015/2016 academic year, Portugal's General Directorate of Education launched the pilot project IP1 – Beginners Programming in Primary School, addressing the goal that students learn to program and with this experience develop cognitive and social skills.

Programming is becoming popular in different levels of school in many countries due to its numerous advantages such as the development of competences for problem solving, divergent ways of thinking, creativity, communication and group work.

The work presented here is an initial work, implemented within the frame of a doctoral degree project that describes a program of 21 fifty-minute sessions', using Scratch tools, designed to develop the computational thinking of students in the third year of primary school. This paper also describes the instruments that will be built to evaluate the program's effectiveness.

This program can be a pedagogical reference for implementation of project IP1 in other primary schools. Final considerations will be raised about future steps.

Keywords: Computational thinking · Primary school · Intervention programs

1 Introduction

Nowadays, any service or product that we consume interrelates with, whether directly or indirectly, some process or decision determined by a computer chip. Almost everyone, irrespective of their age, needs some basic technological knowledge. Some authors [1, 2] maintain that computational thinking is literacy for the 21ˢᵗ century. With this technology so very present, a new learning profile is appearing, the so called "digital native", a person who is accustomed to working on various tasks simultaneously, with easy and fast access to information, interconnected via networks and based on immediate and frequent gratifications [3]. Since a very high percentage of professions require digital competences [4] people should not act only as consumers, they also need to be creators. Countless education reports point in this direction [5]. The contribution of technology to educational context is not only necessary but also an unmistakeable reality for our schools.

© Springer International Publishing AG, part of Springer Nature 2019
J. Machado et al. (Eds.): HELIX 2018, LNEE 505, pp. 1160–1166, 2019.
https://doi.org/10.1007/978-3-319-91334-6_160

Despite Papert [6] having introduced a programming language for educational purposes and transferable to other spheres of life in 1980, it was only in 2006 that Jeannette Wing popularize the term *"computational thinking"* and set out the necessary theoretical framework. This concept "involves solving problems, designing systems, and understanding human behaviour, by drawing on the concepts fundamental to computer science" [7].

Multiple definitions and updates were to follow in search of a consensus that is still lacking in the field [8]. One the most consensual definition of CT – computational thinking, include the "creative, critical and strategic human capacity to know how to use the fundamentals of computing in the most diverse areas of knowledge, with the purpose of identifying and resolving problems, whether individually or collaboratively, through clear steps, so that either a person or a machine may execute them efficiently" [9].

For the majority of people, programming and coding are perceived as complex actions based on what are considered to be advanced levels of technological knowledge and thus accessible only those with certain abilities. However, a number of authors defend an opposing idea. Papert [6], for example, advocates that programming languages should have different levels, from the simple to the complex, and thus be accessible to all.

The increasing number of programming languages, such as *Scratch* [10] or *Alice* [11], free and user friendly, easier to grasp and begin using, raises the interest of programming among pre university level. In recent years, primary and secondary schools, mainly all in Anglo-Saxon countries, have embarked on initiatives striving to introduce CT. Worldwide, school curricula have been updated in this field to include students at younger ages via an enormous variety of educational interventions [12, 13]. Valente [14] systematized these as follows: (a) activities without resource of technologies; (b) programming in Scratch; (c) pedagogical robotics; (d) producing digital narratives; (e) designing games and (f) using of simulators. However, even though these technologies are diverse and susceptible to multiple applications, these interventions frequently emerge as sporadic or they only target an elite group or they lack overall intentionality.

In Portugal the most recent initiatives launched by the Ministry of Education and Science in this field include the pilot project *"Beginners Programming in Primary School* and the inventory and support for *Programming* and *Robotics Clubs* existing nationwide" [15]. However, the literature states that there are only occasional studies on such interventions in K12 school contexts (pre-university) [8] and that these studies are especially rare in primary school [16].

In Brazil, Campos and his colleagues [17] carried out one of the first attempts to measure improvements to CT, through pre- and post-testing. The results did not return any statistically significant improvement but the activities were described as gratifying. In Spain, Roman [18] developed a CT test with 28 items with four response choices for a 45 min test for school children aged between 12 and 13. Faber and his colleagues [19] developed a six-session program in 26 primary school establishments in the Netherlands and received positive feedback from teachers and students but there was no metric for participants' levels of satisfaction or statistical data on changes in CT performance standards.

Specifically in Portugal, there have been diverse studies carried out to evaluate the IP1 program in order to boost the effectiveness of future proposals [15]. The conclusions convey a very positive general evaluation by school headmasters, teachers and school initiative coordinators with regard to the level of attainment of the project goals and also highlighted the importance of the guidelines drawn up, which helped reinforce the continuity of the program. However, Ramos and Espadeiro [15] remind us of the need to evaluate the impact of these programs in terms of the changes in student competences and knowledge.

In accordance with this framework, this study seeks to describe a program fostering CT and to analyse its effectiveness.

2   Program_Scratch 3

The program presented here stems from the pilot project "Beginners Programming in Primary School" [20], run by the GDE - General Directorate of Education, as an intervention proposal for a field that is integrated into the education program of a number of schools through recourse to the Scratch programming application (3rd and 4th years).

The literature details various ways of applying Scratch to different levels of teaching [21] and, in terms of the challenges set by the pilot project, each teacher and school established their own interventions. What occurred in practice was that in many schools, the program featured some random activities, detached from the theoretical framework that would have endowed them with meaning or coherence. Another issue was that there were are also many teachers engaged who lacked any sort of guide for orienting their work, and consequently were unable to evaluate the effectiveness of their interventions.

In order to avoid these problems, a group of five computer teachers (one is author of this paper), who are developing the pilot project in 3rd year of primary school, designed originally this program which intends to guide the work of teaching programming in partnership with teachers from other subject areas (see Table 1). The program to foster computational thinking among participants students follow literature review, guidelines handed down by the Ministry of Education (ME) for the program and the curricular targets to be attained by third year students. Specific goals (e.g. activities planning, symbolic representation) and soft skills (e.g. group work, communication) proposed by ME are also integrated. These 21 sessions, 50 min each, through use of the Scratch tools (e.g. characters, stages, scenarios, movements, sound, sensors, etc.) allow students to develop fundamental concepts in the field of programming such as sequences, parallel operations, events, decision conditions/structures, cycles, operators, data/variables, and simple functions, among others.

According to Callear [22] when teaching programming is relevant to: (a) slow pace teaching to facilitate the concepts integration; (b) teach one by one topic to avoid confusions; (c) review previous topics and (d) sort the subjects by degree of difficulty. These principles guided this specific program.

The training is planned around exercises with increasing levels of difficulty (most of the times with the introduction of new blocks of code, which are explored at the

Table 1. Program general plan

Session	Activity	Sequences	User passive interaction	Repetition cycles	Conditional statements	Communication and synchronized	Boolean logical	Random numbers	Variables	User direct interaction
					Programming concepts					
1	Scratch Presentation	X	X							
2	Customize actors	X	X	X						
3	Insert stages	X	X							
4	Conditional statements	X	X		X					
5	Messages and sounds	X	X	X	X	X				
6	Effects	X	X	X						
7	Sensors	X	X	X	X	X				
8	Sensors II	X	X	X	X	X				
9	Customize stages	X	X	X	X	X				
10	Visibility (yes/no)	X	X				X			
11	Customize stages and actors	X	X							
12	Synchronized communication	X	X			X				
13	Actions overlap	X	X	X		X				
14	Change of stages	X	X	X	X	X				
15	Synchronized change of stages and actors	X	X			X				
16	Random numbers	X	X	X	X	X	X	X		
17	Variables	X	X	X	X	X	X	X	X	
18	Actors and stages interaction	X	X	X	X	X	X	X	X	
19	Customizing the cursor and using multiple variables	X	X	X		X		X	X	
20	Direct interaction with the user	X	X	X	X	X	X	X	X	X
21	Assessment	X	X	X	X	X	X	X	X	X

beginning of each session) and in their majority, ended with options that students have to analyse, often by breaking down the problem into smaller sections in order to arrive their best decision. This allows students to consolidate the knowledge that they have thus far acquired and to develop problem-solving strategies.

3 Evaluation of Program Effectiveness

Several instruments will be use to evaluate the effectiveness of the program.

The *Computational Thinking Test,* originally developed by Roman [18] for application in Spain involves measuring the level of CT development with a focus on the fundamental concepts of programming such as sequences and repetitions. The adaptation and validation of the Portuguese version will be done within the scope of this work (with both experts and prior pilot studies). In Brazil, there have also been psychometric studies applying this test [23, 24]. The test contains 28 exercises: four questions about simple sequences; four questions about the cycle of repetition, by the number of times; four questions about the cycle and repetition through to a determined condition; four questions about simple conditions, (condition – action); four questions about conditions composed (condition   action = exception); four questions about the cycle of repetition as a condition for verification; and four questions about simple functions so as to divide a problem by grouping tasks. The level of complexity and difficulty of the exercises rises throughout the test and always presents four response options (A, B, C and D), of which only one is correct. The test lasts for 45 min and begins with three practice exercises. Carrying out this test also explores the capacity to formulate and resolve problems either directly or in phased approaches.

Self-efficacy in Computational Thinking is measured through a specific scale developed by the authors in accordance with the literature [e.g. 25]. The scale evaluates participant perceptions in regard to their competences in the computer science field (e.g. "*I know how to symbolically represent sequences of actions for activities*") with responses made according to a five-point Likert scale (1 – totally disagree to 5 – totally agree). Validation for the portuguese population shall comply with all the requirements for its appropriate application.

In addition, *sociodemographic data* (e.g. *academic qualifications of parents*) as well as a *technology user information* (e.g. "*Do you use a computer daily?*") were also collected.

The data gathering process also extends to the participants' academic records in three subjects: Portuguese, Mathematics and Environmental Studies.

We would further add that evaluating the program's efficiency spans two phases: pre (M1) and post (M2) application into two groups of students.

4 Final Considerations

Computational thinking in education is, to some extent, an inevitable trend. There are various studies demonstrating how students with training and education in programming are better prepared for future daily professional tasks and challenges [10]. Based on their study findings, Sáez-López and Sevillano-García [21] recommend that those responsible for education policies integrate them into primary school teaching programs and into subject areas of art, music and technology.

However, such practices are not always reflective and intentional, and very often the teaching of programming languages teaching only a very reduced number of students or only happens on an occasional basis. The IP1, proposed by the ME, is an opportunity for schools to develop structured CT programs. However, according to literature [e.g. 16], is also important to assess these programs, in areas like computational thinking skills and their impact on learning. important evaluate this programs. Therefore, this paper is only one first step, within the frame of a doctoral degree project doctoral, presenting one program and their process of evaluation. In the near future authors will be obtained the authorizations from the respective School Board and parents of the children participated (3rd and 4th years of primary school) in two studies: evaluation of program and psychometric study of instruments. After gathering all data, statistical analysis of the instruments and the program will be carried out according to the goals defined. The results and discussion will continue to be just a small contribution towards building knowledge in this area. More research is clearly required to establish how to teach children use programming tools and how they influence their development (e.g. problem-solving and computational thinking abilities).

References

1. Heintz, F., Mannila, L., Farnqvist, T.: A review of models for introducing computational thinking, computer science and computing in K–12 education. In: Frontiers in Education Conference (FIE), pp. 1–9, Erie, US (2016)
2. Rose, S., Habgood, J., Jay, T.: An exploration of the role of visual programming tools in the development of young children's computational thinking. Electron. J. e-Learning **15**, 297–309 (2017)
3. Prensky, M.: Digital natives, digital immigrants. Horizon **9**(5), 1–6 (2001)
4. European Commission. E-Skills for Jobs in Europe: measuring progress and moving ahead. http://eskills-monitor2013.eu/results/. Accessed 12 Dec 2017
5. Johnson, L., Becker, S., Estrada, V., Freeman, A.: NMC Horizon Report: 2014 K-12 Edition. Austin, Texas: The New Media Consortium. https://www.learntechlib.org/p/147472/. Accessed 7 Feb 2018
6. Papert, S.: Mindstorms: Children, Computers, and Powerful Ideas. Basic Books, New York (1980)
7. Wing, J.: Computational thinking. Commun. ACM **49**(3), 33–36 (2006)
8. Grover, S., Pea, R.: Computational thinking in K-12: a review of the state of the field. Educ. Res. **42**(1), 38–43 (2013)
9. Kurshan, B.: Thawing from a long winter in computer science education. Forbes, pp. 1–2 (2016)
10. Resnick, M., Maloney, J., Monroy-Hernandez, A., Rusk, N., Eastmond, E., Brennan, K., Millner, A., Rosenbaum, E., Silver, J., Silverman, B., Kafai, Y.: Scratch: programming for all. Commun. ACM **52**(11), 60–67 (2009)
11. Graczyńska, E.: ALICE as a tool for programming at schools. Natural Sci. **2**(2), 124–129 (2010)
12. Lye, S.Y., Koh, J.H.L.: Review on teaching and learning of computational thinking through programming: What is next for K-12? Comput. Hum. Behav. **41**, 51–61 (2014)
13. Bers, M.I., Flannery, L., Kazakoff, E.R., Sullivan, A.: Computational thinking and tinkering: exploration of an early childhood robotics curriculum. Comput. Educ. **72**, 145–157 (2014)
14. Valente, J.: A. Integração do Pensamento Computacional no Currículo da Educação Básica: Diferentes Estratégias Usadas e Questões de Formação de Professores e Avaliação do Aluno. Revista e-Curriculum, vol. 14(3). https://revistas.pucsp.br/index.php/curriculum/article/view/29051. Accessed 12 Dec 2017

15. Ramos, J.L., Espadeiro, R.G.: Iniciação à Programação no 1.º Ciclo do Ensino Básico. Estudos de Avaliação do projeto-piloto. DGE – Direção Geral de Educação, Portugal (2016)
16. Maya, I., Pearson, J.N., Tapia, T., Wherfel, Q.M., Reese, G.: Supporting all learners in school-wide computational thinking: a cross-case qualitative analysis. Comput. Educ. **82**, 263–279 (2015)
17. Campos, G., Cavalheiro, S., Foss, L., Pernas, A., Piana, C.F., Aguiar, M., Du Bois, A., Reiser, R.: Organização de informações via pensamento computacional: relato de atividade aplicada no ensino fundamental. 3º Congresso Brasileiro de Informática na Educação (CBIE). 20.º Workshop de informática na escola (WIE), UFGD, Brasil (2014)
18. Román, M.: Test de Pensamiento Computacional: principios de diseño, validación de contenido y análisis de ítems. In: Murga Menoyo, M.A. (eds.) Perspectivas y Avances de la investigacion, pp. 279–302. UNED, Madrid (2015)
19. Faber, H., Wierdsma, M., Doornbos, R., van der Ven, J.S., de Vette, K.: Teaching computational thinking to primary school students via unplugged programming lessons. J. Europ. Teach. Educ. Netw. **12**, 13–24 (2017)
20. Educação, D.G.: Iniciação à programação no 1. ciclo do ensino básico. Linhas orientadoras gerais. http://www.erte.dge.mec.pt/sites/default/files/Projetos/Programacao/IP1CEB/linhas_orientadoras.pdf. Accessed 24 Jan 2017
21. Sáez-López, J.M., Sevillano-García, M.L.: Sensors, programming and devices in Art Education sessions. One case in the context of primary education. Cultura y Educación **29** (2), 350–384 (2017)
22. Callear, D.: Teaching programming: some lessons from prolog. In: Proceedings of the LTSN-ICS 1st Annual Conference, Heriot Watt, Edinburg, UK (2000)
23. Boucinha, R.M.: Aprendizagem do pensamento computacional e desenvolvimento do raciocínio. Tese de doutoramento, Universidade Federal do Rio Grande do Sul (2017)
24. Brackmann, C.P.: Desenvolvimento do pensamento computacional através de atividades desplugadas na educação básica. Tese de doutoramento, Universidade Federal do Rio Grande do Sul (2017)
25. Joet, G., Usher, E., Bressoux, P.: Sources of self-efficacy: an investigation of elementary school students in France. J. Educ. Psychol. **103**(3), 649–663 (2011)

Author Index

© Springer International Publishing AG, part of Springer Nature 2019
J. Machado et al. (Eds.): HELIX 2018, LNEE 505, pp. 1167–1171, 2019.
https://doi.org/10.1007/978-3-319-91334-6

Lightning Source UK Ltd.
Milton Keynes UK
UKHW02n1232190618
324466UK00001B/14/P